ALGEBRA FOR COLLEGE STUDENTS

IGNACIO BELLO

Hillsborough Community College
Tampa, Florida

W. B. SAUNDERS COMPANY / *Philadelphia* / *London* / *Toronto*

W. B. Saunders Company: West Washington Square
Philadelphia, PA 19105

1 St. Anne's Road
Eastbourne, East Sussex BN21 3UN, England

1 Goldthorne Avenue
Toronto, Ontario M8Z 5T9, Canada

Records cited on pages 49, 59, 60, 61, 99, 104, 105, 106, 107, 108, 113, 115, 118, 240, 241, 348, 435, and 519 from *Guinness Book of World Records,* copyrighted by Sterling Publishing Company, Inc., New York.

Front cover illustration of a quipu, courtesy of the Musée de L'Homme, Paris, France

Albegra for College Students ISBN 0-7216-1685-2

Last digit is the print number: 9 8 7 6 5 4 3

PREFACE

This book is intended for a one-term course in algebra for the college or junior college student with little or no background in the subject. The principal objective of the book is to develop the algebraic skills needed for subsequent or concurrent courses.

Although the topics covered are essentially the traditional ones of intermediate algebra, the treatment is quite different from that in most current books. Each section starts with an actual application of the topic to be discussed, or with a related cartoon, to arouse the student's interest. Most students are not enthusiastic about studying mathematics per se, and they constantly ask, "Why do we have to learn this?" We have tried to answer this question by giving the natural motivation, real problems taken from newspapers, magazines, and other sources. The last section of most chapters is concerned with applications pertinent to the topics covered in the chapter.

We have presented axioms and definitions to the student only as they are needed and have not thrown a whole mass of this material at the reader at one time. The comments, notes, and remarks in the margins of the pages will be found quite helpful in the student's learning process. The historical notes that precede each chapter are concerned with the material of the chapter, and will be of interest to anyone with a curiosity about the origins of the material to be studied.

Each section is introduced by a set of objectives giving a brief verbal description of the topics to be covered, and the section ends with a short Progress Test. This test is keyed as closely as possible to the objectives of the section. A large number of examples are worked out in complete detail in each section. The exercises at the end of the section provide the necessary practice in handling the types of problems and the basic principles being studied. Besides the Progress Test at the end of each section, there is a Self-Test at the end of the chapter. Answers to all the questions in the test, as well as answers for the odd-numbered problems, are given in the book. A solutions manual, containing the answers to the even-numbered problems and additional tests for each of the chapters, is available to instructors. Finally, a semiprogrammed workbook containing additional problems, fill-in statements, chapter reviews, and tests for each chapter is also available to accompany the text.

ACKNOWLEDGMENTS

I would like to express my appreciation to the following people:

The reviewers provided by W. B. Saunders—Professors Carole Bauer, Calvin Lathan, and Ara Sullenberger.

Professors George Kosan and Donald Rose II, who offered many suggestions to improve the manuscript, Professor P. Shamm, who provided constant encouragement during the development of the book and James Gard, who solved the problems in the exercises.

Gladys Penaranda, who typed the manuscript, and our assistants, Vincent Cardoso, Irene Rodeiro, and Cynthia A. Scheff.

Mr. John Snyder, the patient editor of W. B. Saunders, and his assistant, Ms. Juliana Kremer, who handled the details of securing permissions, cartoons, and illustrations and who coordinated the production of the book.

The staff of W. B. Saunders Company, especially Mr. Tom O'Connor and Mr. Frank Polizzano of the Production Department and Ms. Lorraine Battista of the Design Department.

Finally, I gratefully acknowledge the careful editing of Jack Britton, who endured—and improved—the writing.

TO THE STUDENT

How to Use the Book

Students with minimal or no algebra background might use the text as follows:

1. Read the objectives provided at the beginning of each Section to determine what is expected.
2. Study the Section, paying particular attention to definitions and examples.
3. Do the Progress Test at the end of the Section.
4. If the Progress Test results are satisfactory (the answers are provided), do the odd-numbered problems in the last part of the exercise set. Then go to the next Section.
5. If the Progress Test results are not satisfactory, restudy the Section and then do the odd-numbered problems in the exercises.
6. Take the Self-Test at the end of the Chapter.

CONTENTS

ALGEBRA FOR COLLEGE STUDENTS

ALGEBRA FOR COLLEGE STUDENTS

CHAPTER 1

HISTORICAL NOTE

"Number" is one of the earliest concepts developed by the human race. No primitive tribe has ever been found to lack the ability to distinguish between different numbers of objects, although many tribes have had no well-defined system of counting. The first known concepts of number date back to as long ago as the Old Stone Age, and some time in the dim dawn of civilization, numerical terms came slowly into use to offer a distinction between one, two, and many. The development of crafts and commerce, for which numerical records were first kept by simply tallying, eventually forced the creation of symbols to denote the counting numbers one, two, three, and so on. A good example of tallying is found among the Incas of Peru, who tied knots in a series of strings (a quipu) as a means of census-taking. With the continued progress of civilization, the needs of daily life began to require the measurement of various quantities such as length, volume, weight, and time. Because the counting, or natural, numbers did not fully satisfy these simple needs, people invented the numbers that we call fractions. The ancient Egyptians avoided the computational difficulties encountered with fractions by representing all fractions, except the fraction $\frac{2}{3}$, as a sum of unit fractions—that is, fractions with a numerator of 1. They accomplished this representation by the use of tables such as the one that appears in the Rhind Papyrus, a document found in the ruins of an ancient building at Thebes.

Although the Egyptians—and the Babylonians as well—made considerable progress in the development of a numeration system, the negative numbers were completely unknown to them. These numbers seem to have occurred first in the work of the Hindus in about 600 A.D. The Hindus indicated a negative number by putting a dot or a circle above or around a numeral: $\dot{2}$, $\mathring{2}$, or ②. In the nineteenth century, the work of Richard Dedekind, Georg Cantor, and Giuseppe Peano showed how the number system which we are studying in Chapter 1 can be derived from a postulate set for the natural numbers. Some of these postulates or axioms are discussed in this chapter.

A quipu used by the Incas of Peru as a method of census taking. The color of each string indicates the sex of the person being counted. (Musée de L'Homme, Paris.)

THE ALGEBRA OF REAL NUMBERS

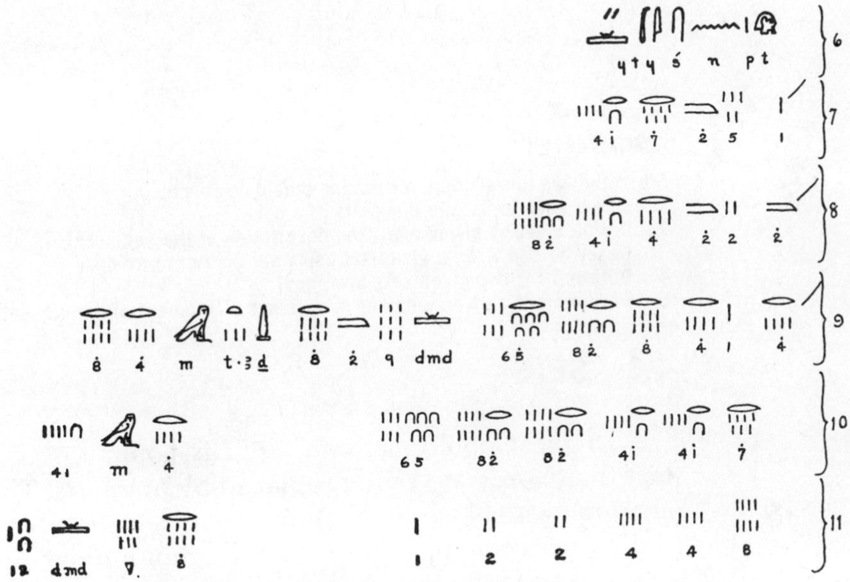

A single page of the great papyrus Rhind. (From A. B. Chace et al., eds., *The Rhind Mathematical Papyrus.* Vol. II, p. 56.)

Conchy by James Childress, courtesy of Field Newspaper Syndicate.

OBJECTIVES 1.1

After studying this section, the reader should be able to:
1. **Use braces to list the elements of a set.**
2. **Define a set by giving a written description of the set.**
3. **Describe and write a given set using set builder notation.**
4. **Determine if two given sets are equal.**
5. **Determine if a given element is or is not an element of a given set.**

1.1 SETS

In the Conchy cartoon, Duff is complaining about the fact that words are being replaced by numbers. The study of algebra, which we are beginning, is a natural extension of ordinary arithmetic where we will often find it convenient to replace words by numbers and numbers by letters, such as Duff did when he replaced the number "100" by the letter "A."

In arithmetic, the numbers which we use to count objects (one, two, three, and so on) are written using the numerals* 1, 2, 3, and so on. These counting numbers are an example of a *set* called the *set of natural numbers*. The concept of a *set* is one of the basic ideas in mathematics. You see or hear the word "set" almost every day. We see an ad for a *set* of dishes or hear that somebody bought a new *set* of golf clubs. More generally, a set is a collection (group, assemblage, aggregate) of objects called the *elements* or *members* of the set.

A set is a collection of objects called the elements or members of the set.

We use capital letters, such as A, B, C, X, Y, and Z, to denote sets, and lowercase letters, such as a, b, c, x, y, and z, to denote the elements of the set. It is customary, when

*A numeral is a symbol that is the name of a number. However, from here on we shall refer to the *number* 1 (or 5, or 37) rather than to "the number represented by the numeral 1 (or 5, or 37)."

practicable, to list the elements of a set in braces and to separate these elements by commas. Thus, A = {1, 2, 3} means that "A is the set consisting of the elements 1, 2, and 3." To indicate the fact that "2 is an element of the set A," we write 2 ∈ A (read "2 is an element of the set A"). To indicate that 7 is not an element of A, we write 7 ∉ A.

2 ∈ A means that 2 is an element of the set A

7 ∉ A means that 7 is not an element of the set A.

In defining the set A, we simply listed the elements of A and separated these elements by commas. It is also possible to give a *description* for the set A. For example, we can say that "A is the set of natural numbers less than 4." A third way to describe a set is to use a "defining property" for the set and write the set in *set-builder* notation. For example, the set A can be described in set-builder notation as follows:

{x | x is a natural number less than 4}

Read "the set of all x such that x is a natural number less than 4."

Note that the vertical bar after the first x is translated as "such that." From this discussion you can see that there are three ways in which we can define a set:

(1) We can *list* the elements of the set. For example:
   (a) {1, 2, 3, 4, 5}
   (b) {1, 2, 3, . . .} (The three dots indicate that the enumeration continues without end.)
   (c) {1, 2, 3, . . . , 99} (The three dots here mean that the enumeration continues until 99 is reached.)

The set {1, 2, 3, ...} is an infinite set. The set {1, 2, 3, . . . 99} is a finite set.

(2) We can give a *description* of the set. Those named in (1) can be respectively described as:
   (a) The set of natural numbers less than 6
   (b) The set of natural numbers
   (c) The set of natural numbers less than 100

(3) We can write the set using "*set-builder*" notation. The notation for each of the sets in (1) is as follows:
   (a) {x | x is a natural number less than 6}
   (b) {x | x is a natural number}
   (c) {x | x is a natural number less than 100}

Example 1

Write a description of the following sets:
(a) {a, b, c, . . . , z}
(b) {2, 4, 6, . . .}

Solution

(a) The set of letters in the English alphabet
(b) The set of even natural numbers

Example 2

Use braces to list the elements in the following sets:
(a) {x | x is a letter in the word "Tallahassee"}
(b) {x | x is a counting number between 1 and 2}

Solution

(a) {T, a, l, h, s, e}. Note that although the word contains letters that are repeated, we list each letter only once.
(b) { }. Note that the set { } has no elements, since there is no counting number between 1 and 2. The set { } will be denoted by the symbol ∅ and will be called the *empty*, or *null*, set.

∅ is the empty, or null, set.

We can reason that the order in which the elements of a set are listed is of no significance. For example, if you ask a friend to list the set of digits of the year in which Columbus discovered America, your friend might write {1, 4, 9, 2,}. On the other hand, another friend, when asked the same question, might write {1, 2, 4, 9}. Obviously, both answers are correct! Hence we agree that {1, 4, 9, 2} = {1, 2, 4, 9}. In general, two sets are equal if they have the same members. Thus, {1, 2, 3} = {3, 2, 1}, and {b, a, t} = {t, a, b}.

{1, 4, 9, 2} = {1, 2, 4, 9}

Example 3

Let A = {2}, B = {t, o, w}, and C be the set of letters in the word "too." Which of these sets
(a) contains two elements?
(b) equals {t, w, o}?
(c) equals the set of letters in the word "toot"?

Solution

(a) Since C = {t, o}, C contains two elements.
(b) Since the order in which the elements are listed is of no significance, {t, w, o} = {t, o, w} = B.
(c) Since the set of letters in the word "toot" is {t, o}, the set C = {t, o} equals this set.

We shall now mention certain sets of numbers which will occur throughout algebra and shall be discussed in some detail later in the chapter. These sets are:

1. The set N of natural numbers, which we have already mentioned.
2. The set W of whole numbers, whose elements consist of the natural numbers and 0, that is, W= {0, 1, 2, 3, . . .}.
3. The set I of integers, whose elements consist of the natural numbers, their negatives, and 0, that is, I = {. . . ,−1, 0, 1, . . .}.
4. The set Q of rational numbers, whose elements consist of all the numbers which can be written in the form $\frac{a}{b}$, where a and b represent integers and b is not 0, that is, $Q = \{r \mid r = \frac{a}{b}, a, b \in I, b \neq 0\}$.
5. The set H of irrational numbers, whose elements consist of those numbers which can not be expressed in the form $\frac{a}{b}$, where a and b are integers. For example, $\sqrt{2}$, π, and $\sqrt{7}$ are irrational.
6. The set R of real numbers, which is the set of all rational and irrational numbers.

N = {1, 2, 3, . . .}

W = {0, 1, 2, 3, . . .}

I = {. . ., −1, 0, 1, . . .}

$Q = \left\{q \,\middle|\, q = \frac{a}{b}, a, b \in I \text{ and } b \neq 0\right\}$

PROGRESS TEST*

1. Use braces to list the elements in the set of whole numbers less than 6: ______________________.
2. A description for the set A= {0, 1, 2} is: ______________________.
3. When written in set-builder notation, the set of positive integers is written as ______________________.
4. The set {r, a, t} ________ (is, is not) equal to the set {t, a, r}.
5. −12 ________ (∈, ∉) I, the set of integers.

EXERCISE 1.1

In Problems 1 through 4, let A= {1, 3, 5, 7}. Which are correct statements?

Illustration: (a) 5 ∈ A
(b) 7 ∉ A

Solution: (a) 5 ∈ A is a correct statement
(b) 7 ∉ is not a correct statement

1. 3 ∈ A

PROGRESS TEST ANSWERS

*Answers to the Progress Tests will always appear at the end of the page.

1. {0, 1, 2, 3, 4, 5}
2. The set of whole numbers less than 3
3. {x | x is a positive integer}
4. is
5. ∈

2. $8 \notin A$

3. $5 \notin A$

4. $A \notin A$

In Problems 5 through 10, use braces to list the elements in the given sets.

Illustration: The set of natural numbers less than 3
Solution: $\{1, 2\}$

5. The set of natural numbers less than 8

6. The set of whole numbers less than 8

7. The set of integers less than 8

8. The set of digits of the year in which the Declaration of Independence was signed

9. The set of letters in the word "Mississippi"

10. The set of women who have been President of the U.S.

In Problems 11 through 18, give a description of the set.

Illustration: $\{1, 2, 3\}$
Solution: The set of natural numbers less than 4

11. $\{0, 1\}$

12. $\{1, 2\}$

13. $\{-1, -2, -3\}$

14. {Christopher Columbus}

15. $\{a, z\}$

16. $\{m, a, n\}$

17. $\{4, 2, 5, 1, 3\}$

18. $\{1, 3, 5, \ldots\}$

In Problems 19 through 24, write the set using set-builder notation.

Illustration: $\{1, 2, 3\}$
Solution: $\{x \mid x \text{ is a natural number less than } 4\}$

19. $\{1, 3, 5\}$

20. $\{2, 4, 6\}$

21. $\{1, 2, 3, 4, 5\}$

22. $\{5, 10, 15, \ldots\}$

23. $\{w, o, m, a, n\}$

24. $\{b, o, y\}$

In Problems 25 through 30, fill in the blank with $\in$ or $\notin$ so that the resulting statement is true.

Illustration: 3 ______ $\{1, 3, 5\}$
Solution: 3 __$\in$__ $\{1, 3, 5\}$

25. 1 ______ $\{1, 3, 5\}$

26. 8 ______ $\{1, 3, 5\}$

27. 5 ______ $\{1, 3, 5\}$

28. 9 ______ $\{1, 3, 5, \ldots, 21\}$

29. 8 ______ $\{1, 2, 3, \ldots, 21\}$

30. 8 ______ $\{1, 2, 3, \ldots\}$

In Problems 31 through 34, fill in the blank with $=$ or $\neq$ so that the resulting statement is true.

Illustration: $\{1, 3, 5\}$ ______ $\{5, 3, 1\}$
Solution: $\{1, 3, 5\}$ __$=$__ $\{5, 3, 1\}$

31. $\{a, b, c\}$ ______ $\{c, a, b\}$

32. $\{1, 3, 1\}$ ______ $\{3, 1\}$

33. $\{2, 5\}$ ______ $\{25, 5\}$

34. $\{15, 5, 5\}$ ______ $\{5, 15\}$

35. Let P be the set of odd counting numbers less than 7 and $Q = \{1, 3, 5, 7\}$. Fill in the blanks with $=$ or $\neq$ so that the resulting statement is true.
(a) P ______ Q
(b) $\{1, 3, 5\}$ ______ P

In Problems 36 through 44, let A= $\{-4, \sqrt{2}, \frac{1}{3}, 0, \pi\}$ and follow the instructions given.

Illustration: Use braces to list the real numbers less than 1 in A.

Solution: $\{\frac{1}{3}, 0, -4\}$

36. Use braces to list the whole numbers in A.

37. Use braces to list the natural numbers in A.

38. Use braces to list the integers in A.

39. Use braces to list the irrational numbers in A.

40. Use braces to list the negative real numbers in A.

41. Use braces to list the real numbers between 0 and 1 in A.

42. Use braces to list the common fractions in A.

43. Use braces to list the negative numbers in A.

44. Use braces to list the non-negative numbers in A.

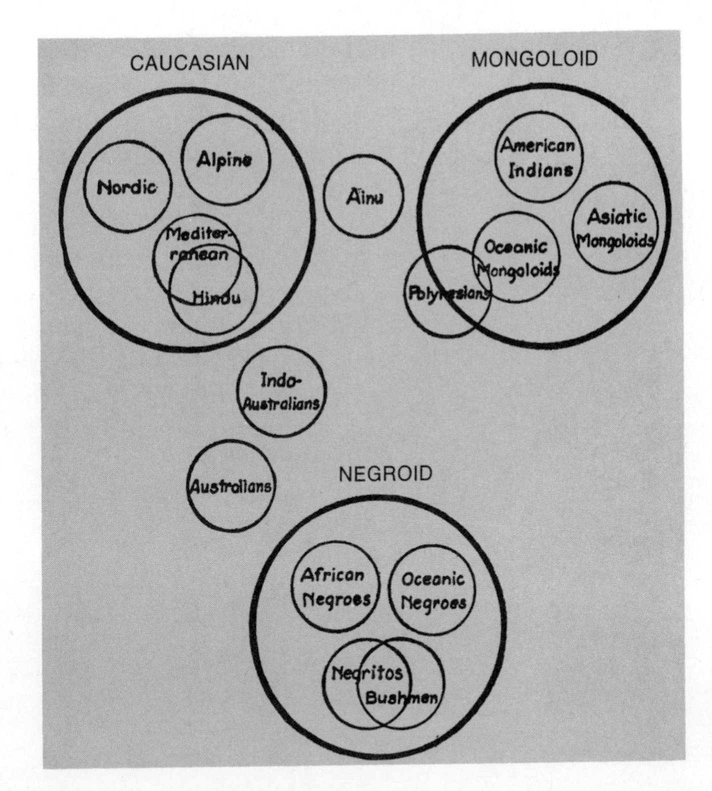

From A. L. Kroeber, Anthropology. New York: Harcourt Brace Jovanovich, Inc., 1923, p. 47.

OBJECTIVES 1.2

After studying this section, the reader should be able to:

1. **List all the subsets of a given set.**
2. **Illustrate by means of a Venn diagram the concept of subset.**
3. **Find the union and intersection of two or more given sets.**

1.2 OPERATIONS WITH SETS

The illustration at the beginning of this section shows the races of mankind. The three major racial stocks are Caucasian, Mongoloid, and Negroid. Notice that these three major racial stocks, represented by the three large circles, have no elements or members in common.

Inside the three circles representing the members of the three major racial stocks, there are other, smaller circles. These are *subraces* of the major racial stocks. For example, the Mediterraneans and the Hindus are *subraces* of the Caucasian race. On the other hand, the Australians are *not* a subrace of any of the three major racial stocks. You can also see that some races, like the Mediterraneans and the Hindus, are closely related. These races may even have some mixed members that belong to both races.

In mathematics, we classify sets by a process similar

to the one used in the illustration. For example, if C is the set of students in your class, and S is the set of students in your school, every element of C is also in S (because every student in your class is a student in your school). In such cases we say that C is a *subset* of S. We denote this by writing $C \subseteq S$ (read "C is a subset of S"). Thus, the set of natural numbers N is a subset of the set of integers, since every natural number is also an integer. Similarly, the Mediterranean people in the illustration at the beginning of this section are a subset of the Caucasians, and the American Indians are a subset of the Mongoloid race.

We make the idea of a subset precise in the following definition.

DEFINITION 1.1

The set A is a subset of the set B, denoted by $A \subseteq B$, if every element of A is also in B. If, in addition, B contains at least one element not in A, then A is a proper subset of B, denoted by $A \subsetneq B$.

$A = B$ means that $A \subseteq B$ and $B \subseteq A$.

We can see from this definition that $A = B$ when $A \subseteq B$ and $B \subseteq A$. Furthermore, since every element of A is contained in A, $A \subseteq A$. Thus, if $A = \{1, 2, 3\}$ and $B = \{3, 2, 1\}$, then by Definition 1.1, $\{1, 2, 3\} \subseteq \{3, 2, 1\}$ and $\{3, 2, 1\} \subseteq \{1, 2, 3\}$. In symbols $A \subseteq B$ and $B \subseteq A$, so $A = B$. Also, $\{1, 2, 3\} \subseteq \{1, 2, 3\}$, so clearly $A \subseteq A$.

$\emptyset \subseteq A$

Definition 1.1 is often stated in the form: $A \subseteq B$ if there is no element of A that is not in B. It follows from this that $\emptyset \subseteq A$, because $\emptyset$ is empty and there is no element of $\emptyset$ that is not in A.

The set of natural numbers is sometimes called the set of counting numbers.

In order to discuss all of the subsets of a given set, we need to refer to a *universal set*. A universal set $\mathcal{U}$ is simply the set of all elements under discussion. Thus, if we agree to discuss the letters in the English alphabet, $\mathcal{U} = \{a, b, c, \ldots z\}$. On the other hand, if we wish to talk about the natural numbers, $\mathcal{U} = N = \{1, 2, 3, \ldots\}$. The following examples illustrate this idea.

Example 1 Find all the subsets of $\mathcal{U} = \{1, 2, 3\}$.

Solution We have to form subsets of $\mathcal{U}$ by assigning some, none, or all

of the elements of $\mathscr{U}$ to the subsets. We organize the work as follows:
Subsets of 0 elements: $\emptyset$ (remember, $\emptyset$ is a subset of any set)
Subsets of 1 element: $\{1\}$, $\{2\}$, $\{3\}$
Subsets of 2 elements: $\{1, 2\}$, $\{1, 3\}$, $\{2, 3\}$
Subsets of 3 elements: $\{1, 2, 3\}$

Note that the set $\mathscr{U}$ of Example 1 has 3 elements and $2^3 = 8$ subsets. In general, a set with n elements has 2^n subsets (see Problem 31).

The idea of a subset can be illustrated graphically by the use of drawings called *Venn diagrams* (after John Venn, an English mathematician and logician). In these diagrams, the universal set $\mathscr{U}$ is usually represented by a rectangle, and the subsets being considered by simple closed curves (usually circles) drawn inside this rectangle. For example, if A is a subset of a universal set $\mathscr{U}$, we can represent this universal set by the set of points in the interior of the rectangle shown in Figure 1.1(a). The interior of the circle represents the set of points in A. The situation can also be shown as in Figure 1.1(b).

As we mentioned in the illustration given at the beginning of this section, certain subsets have some elements in common. For example, the circles representing the set of Mediterraneans and Hindus overlap, indicating that there may be some members or elements that are common to both races. In mathematics, the set consisting of the elements that two given sets A and B have in common is called the *intersection* of A and B and is denoted by writing $A \cap B$ (read "A intersection B.") Thus, if $A = \{1, 3, 5, 7\}$ and $B = \{2, 3, 5, 6\}$, $A \cap B = \{3, 5\}$ since 3 and 5 are the only elements common to both A and B. On the other hand, if we wish to find the elements which are in A or in B (or possibly in both A and B) we would form a different set, called the *union* of A and B, denoted by $A \cup B$ (read "A union B"). Thus, the union of A and B—that is, the elements that are in A or B (or both)—is the set $\{1, 2, 3, 5, 6, 7\}$. Here are the definitions of intersection and union.

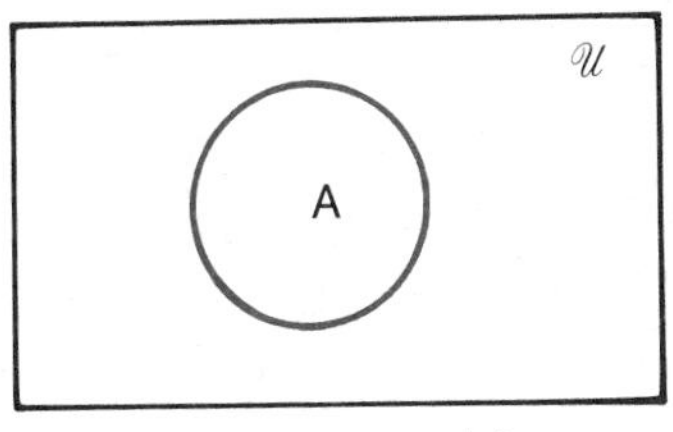

A is a subset of $\mathscr{U}$
FIGURE 1.1(a)

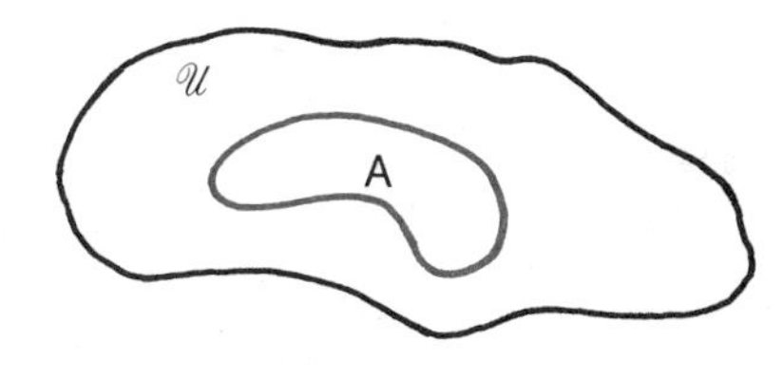

A is a subset of $\mathscr{U}$
FIGURE 1.1(b)

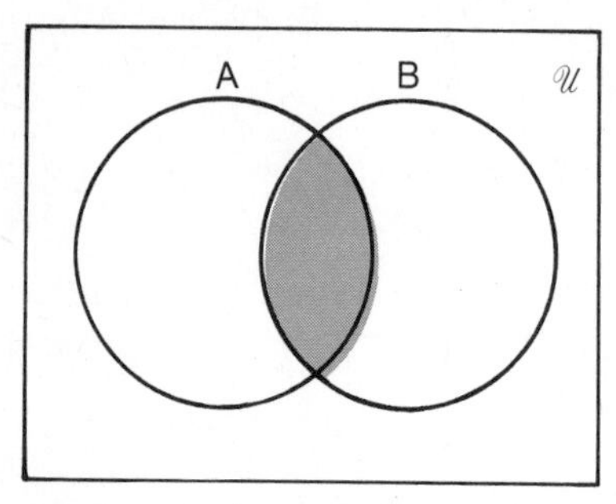

$A \cap B = \{x \mid x \in A \text{ and } x \in B\}$ is represented by the shaded region

FIGURE 1.2(a)

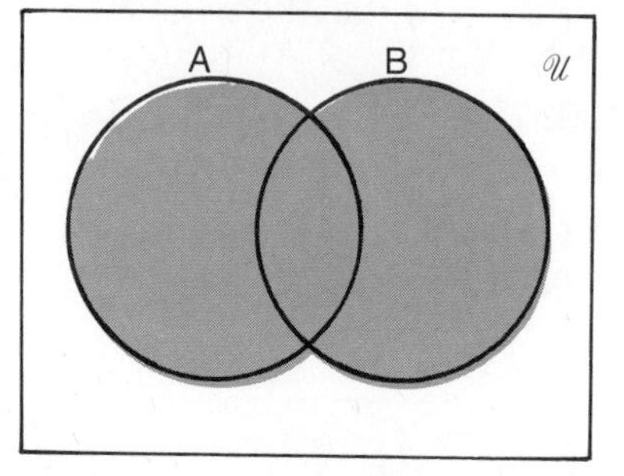

$A \cup B = \{x \mid x \in A \text{ and/or } x \in B\}$ is represented by the shaded region

FIGURE 1.2(b)

DEFINITION 1.2

Definition of $A \cap B$

If A and B are sets, the intersection of A and B, denoted by $A \cap B$, is the set of all elements common to both A and B, that is, $A \cap B = \{x \mid x \in A \text{ and } x \in B\}$.

DEFINITION 1.3

Definition of $A \cup B$

If A and B are sets, the union of A and B, denoted by $A \cup B$, is the set of all elements that are either in A or in B or in both, that is, $A \cup B = \{x \mid x \in A \text{ and/or } x \in B\}$.

The intersection and union of two sets, A and B, are illustrated by means of Venn diagrams in Figure 1.2(a) and 1.2(b), respectively.

Example 2

Let $A = \{1, 3, 5, 7\}$, $B = \{2, 4, 6, 8\}$, and $C = \{1, 4, 5, 6\}$ Find:
(a) $A \cap C$
(b) $A \cup C$
(c) $A \cap B$
(d) $A \cup B$
(e) $(A \cap C) \cup B$
(f) $(A \cup C) \cap B$

Solution

(a) $A \cap C$ is the set of all elements common to A and C, that is, $\{1, 5\}$. Thus, $A \cap C = \{1, 5\}$.
(b) $A \cup C$ is the set of all elements in A or in C or in both, that is, $\{1, 3, 4, 5, 6, 7\}$. Thus, $A \cup C = \{1, 3, 4, 5, 6, 7\}$.
(c) A and B have no elements in common, so $A \cap B = \emptyset$. If

two sets have no elements in common, we say that the sets are *disjoint.*

$A \cap B = \emptyset$ means that A and B are disjoint.

(d) $A \cup B = \{1, 2, 3, 4, 5, 6, 7, 8\}$
(e) From part (a), $A \cap C = \{1, 5\}$, so $(A \cap C) \cup B = \{1, 5\} \cup \{2, 4, 6, 8\} = \{1, 2, 4, 5, 6, 8\}$
(f) From part (b), $A \cup C = \{1, 3, 4, 5, 6, 7\}$, so $(A \cup C) \cap B = \{1, 3, 4, 5, 6, 7\} \cap \{2, 4, 6, 8\} = \{4, 6\}$

PROGRESS TEST

1. If $\mathscr{U} = \{a, b, c\}$, list all the subsets of $\mathscr{U}$: ________.
2. Draw A as a proper subset of B in the accompanying Venn Diagram.

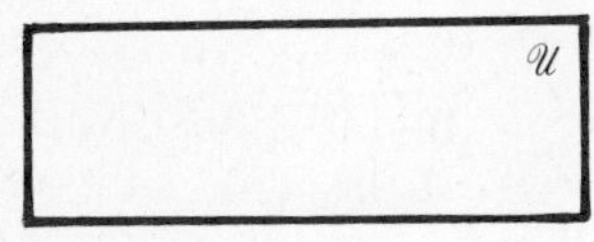

3. If $A = \{1, 6, 7, 9\}$ and $B = \{6, 9, 10\}$, $A \cap B$ equals ________.
4. If A and B are as in the preceding Problem, $A \cup B$ equals ________.
5. If $A = \{a, b, c, d, e\}$, $B = \{a, d\}$ and $C = \{b, c\}$, $(A \cap B) \cup C$ equals ________.

EXERCISE 1.2

In Problems 1 through 6, let $A = \{1, 2, 3\}$, $B = \{2, 3, 4\}$, and $C = \{1, 4\}$. Which statements are correct?

Illustration: $\{3\} \subseteq \{1, 2, 3\}$
Solution: $\{3\} \subseteq \{1, 2, 3\}$ is a correct statement.

1. $\{1\} \subseteq \{1, 2, 3\}$

2. $\{2, 3\} \subset B$

3. $\{1, 2, 3\} \subset A$

4. $\{1, 2, 3\} \subset \{2, 3, 4\}$

5. $C \subseteq B$

6. $\emptyset \subseteq A$

PROGRESS TEST ANSWERS

1. $\emptyset$, $\{a\}$, $\{b\}$, $\{c\}$, $\{a, b\}$, $\{a, c\}$, $\{b, c\}$, $\{a, b, c\}$
2.

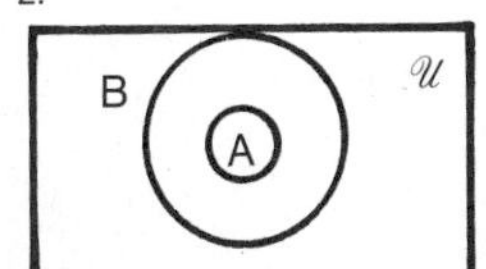

3. $\{6, 9\}$
4. $\{1, 6, 7, 9, 10\}$
5. $\{a, b, c, d\}$

In Problems 7 through 12, let A, B, and C be as in Problems 1 through 6, and find:

Illustration: $A \cap C$

Solution: $A \cap C = \{1\}$, since 1 is the only element common to A and C.

7. $A \cap B$

8. $A \cup B$

9. $C \cap A$

10. $A \cup C$

11. $(A \cap B) \cup C$

12. $(A \cup B) \cap C$

In Problems 13 through 16, let $A = \{\{1, 2\}, 3\}$, $B = \{1, 2, 3\}$, $C = \{1, 2\}$, and find:

Illustration: $A \cup C$

Solution: $A \cup C = \{\{1, 2\}, 1, 2, 3\}$

13. $A \cap B$

14. $(A \cup B) \cup C$

15. $A \cap (B \cup C)$

16. $(A \cap B) \cup (A \cap C)$

17. Let $U = \{1, 2, 3, 4\}$. Find all the subsets of U containing two elements.

18. If $U = \{1, 2, 3\}$, find all the non-empty proper subsets of U.

19. Let A be the set of numbers that are divisible by 2 and B the set of numbers that are divisible by 4.
(a) Is $A \subseteq B$?
(b) Is $B \subseteq A$?

20. Let A be the set of numbers that are divisible by 2, and E be the set of even numbers.
(a) Is $A \subseteq E$?
(b) Is $E \subseteq A$?
(c) In view of your answers to (a) and (b), what can you say about A and E?

21. If A = {1, 2, 3}, B = {2, 3, 4}, and C = {1, 3, 5}, find the smallest set that will serve as a universal set for A, B and C.

22. If A, B, and C are as in Problem 21 and, in addition, D = {1, 2, 3, 4, 5}, find the smallest set that will serve as a universal set for A, B, C, and D.

In Problems 23 through 25, let:
U be the set of employees of a company
M be the set of males who are employees
F be the set of females who are employees
D be the set of employees who work in the data-processing department
T be the set of employees who are under 21
S be the set of employees who are over 65
Ø be the empty set

23. Find a single letter to represent each of the following sets:
(a) $M \cup F$
(b) $M \cap F$
(c) $T \cap S$

24. Describe in words each of the following sets:
(a) $F \cap T$
(b) $M \cap T$
(c) $F \cap D$
(d) $M \cap D$

25. Find a set representation for the set of:
(a) Employees in data processing who are over 65.
(b) Female employees who are under 21.
(c) Male employees who work in data processing.

26. A zoology book lists the following characteristics of giraffes and okapis:

Giraffes	Okapis
tall	short
long neck	short neck
long tongue	long tongue
skin-covered horns	skin-covered horns
native to Africa	native to Africa

Let G be the set of characteristics of giraffes, and O be the set of characteristics of okapis.
(a) Find $G \cap O$

(b) What characteristics are common to okapis and giraffes?

27. Draw a Venn diagram and shade $A \cup B$ when:
(a) $A = B$
(b) $A \subset B$

28. Draw a Venn diagram and shade $A \cap B$ when:
(a) $A = B$
(b) $A \subset B$

29. Forecasters in the National Weather Service use the ideas of Venn diagrams. In the accompanying map, for example, rain, showers, snow, and flurries are indicated with different types of shading. Find the states in which:
(a) It is warm and there are showers.
(b) It is warm and raining.

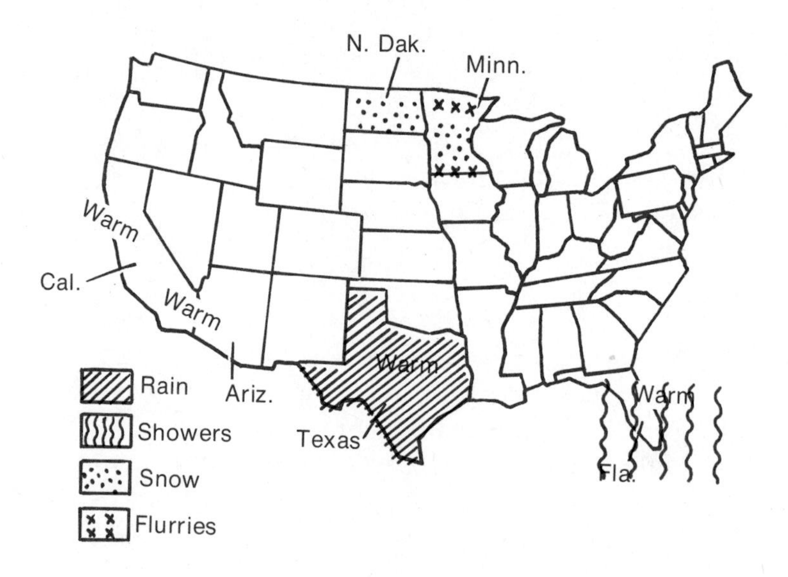

30. Referring to Problem 29, find the states in which:
(a) It is raining or warm.
(b) It is warm only.
(c) It is snowing or there are showers.

31. If a set has no elements (it is empty), it has only one subset, $\emptyset$. If a set has one element, it has two subsets, $\emptyset$ and itself. If a set has three elements, it has $8 = 2^3$ subsets (see Example 1). Here are these relationships:

Number of elements	Number of subsets
0	1
1	$2 = 2^1$
2	
3	$8 = 2^3$

(a) How many subsets does a set with two elements have?
(b) How many subsets does a set with four elements have?
(c) How many subsets do you think a set with n elements has?

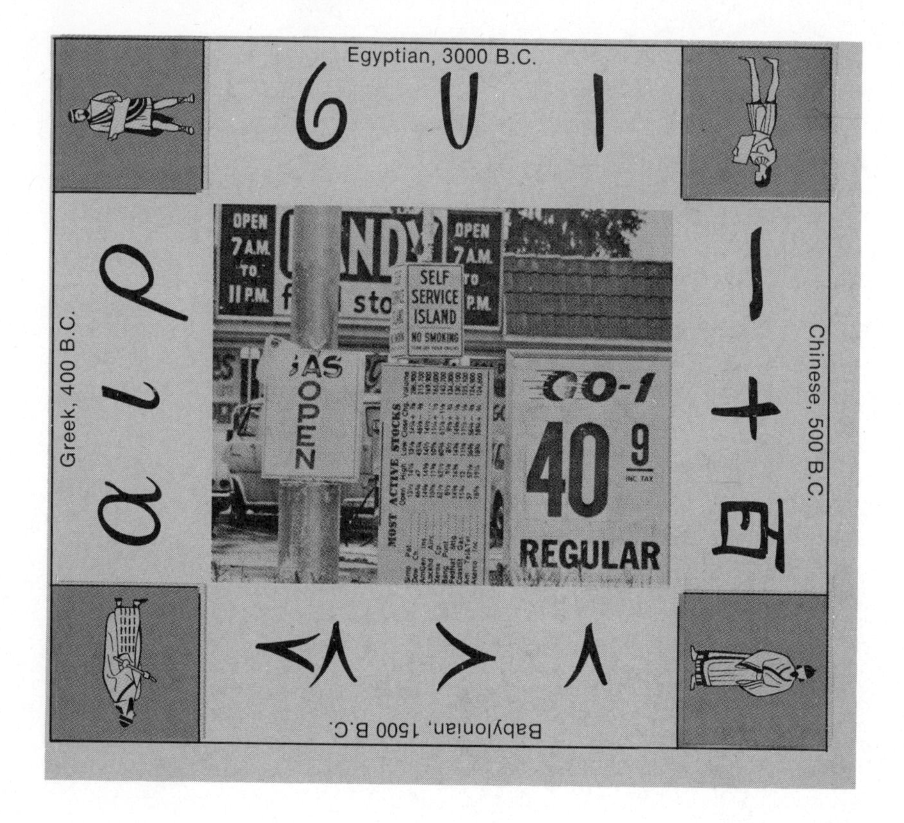

OBJECTIVES 1.3

After studying this section, the reader should be able to:

1. **Graph a given point on the number line.**
2. **Find the coordinates of a given point graphed on the number line.**
3. **Write a rational number of the form $\frac{a}{b}$ in decimal form.**
4. **Write a terminating or a repeating decimal in the form $\frac{a}{b}$.**
5. **Classify a given number as rational or irrational.**

1.3 THE REAL NUMBERS

One of the earliest concepts developed by the human race is the concept of number. As you can see in the illustration, this concept is used by both primitive and modern societies. In this section we shall use the language of sets to discuss some of the sets of numbers used in algebra.

N = {1, 2, 3, . . .} is the set of natural numbers.

The first set of numbers we become acquainted with is the set of natural numbers N = {1, 2, 3, . . .}. This set is also called the set of *positive integers;* it is the set of numbers used for ordinary counting purposes. As civilization developed, these numbers became inadequate to satisfy everyday needs. For example, the invention of the thermometer occasioned the use of numbers other than positive integers to measure low temperatures. The centigrade scale has 100 divisions between the freezing point of water at sea level (0°C) and the boiling point of water at sea level (100°C). Temperatures below 0° are preceded by a − (read "negative") sign. Thus, there is a set of *negatives* of the positive integers, the set of *negative integers* {−1, −2, −3, . . .}. We have now added an important concept to the idea of a number – the concept of direction. To help us visualize this concept, we first draw a horizontal straight line, choose a point on this line and label it 0, the *origin*. We then measure successive equal intervals to the right of 0 and label them with the positive integers in their order, 1, 2, 3, and so on. Those to the left of 0 are labeled −1, −2, −3 and so on (see Figure 1.3).

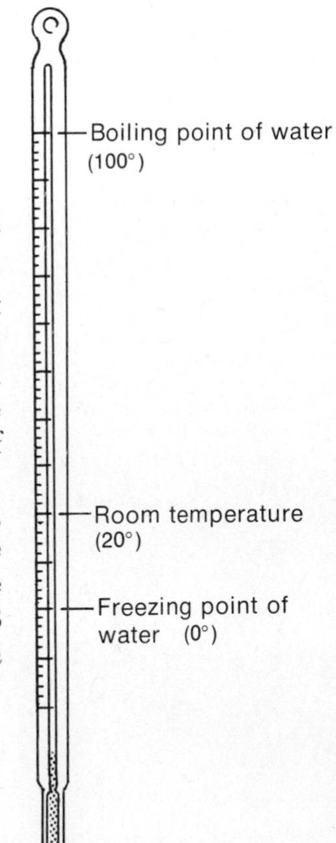

This geometric representation is called a *number line*, or *line graph*. The number corresponding to a point on the number line is called the *coordinate* of the point, and the point is called the *graph* of the number. Thus, in Figure 1.3 the coordinate of point A is 2, and that of B is −1. The *graph* of these points is shown in Figure 1.3.

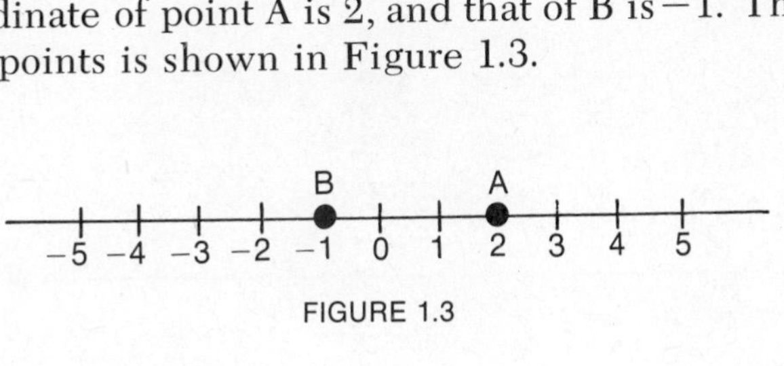

FIGURE 1.3

Example 1

Graph the following points on the number line:
(a) −4 (b) 0 (c) 3

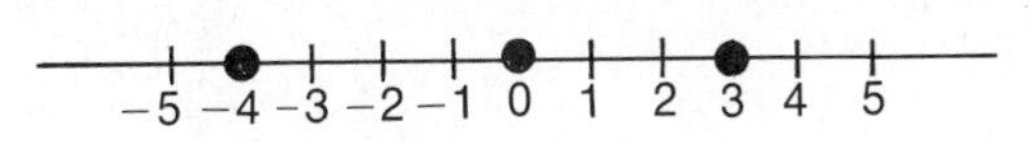

Solution

The graph of the three numbers is shown.

Example 2

Give the coordinates of the points A, B, and C:

A B C
−5 −4 −3 −2 −1 0 1 2 3 4 5

Solution

The coordinates of A, B and C are −5, −3, and 1 respectively.

The number line illustrates the fact that the set of integers I = {. . . ,−1, 0, 1, . . .} can be separated into three subsets: the set of positive integers, the set of negative integers, and the set consisting of 0—that is,

The set of integers is {. . ., −1, 0, 1, . . .}.

$$I = \{1, 2, 3, \ldots\} \cup \{0\} \cup \{-1, -2, -3, \ldots\}$$

Rational numbers—numbers that can be expressed as the *ratio* of two integers—can also be associated with points on the number line. For example, to locate $\frac{1}{3}$ on the number line, we divide the segment from 0 to 1 into 3 equal parts. Each of these parts will be $\frac{1}{3}$ of a unit. The point of division closest to 0 corresponds to the number $\frac{1}{3}$ (see Figure 1.4).

Rational numbers are numbers which can be written as the ratio of two integers. $\frac{1}{5}$, $\frac{7}{4}$, $\frac{3}{1}$, $\frac{-8}{9}$, and $\frac{-3}{4}$ are rational numbers.

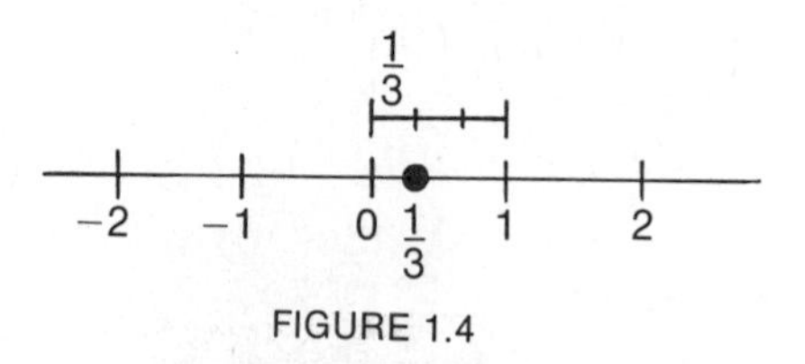

FIGURE 1.4

Example 3 Graph the points $\frac{1}{2}$ and $\frac{-3}{4}$ on the number line.

Solution To graph the point $\frac{1}{2}$, we divide the line segment from 0 to 1 into 2 parts. The point at the end of the first part corresponds to the point $\frac{1}{2}$, and its graph is shown below.

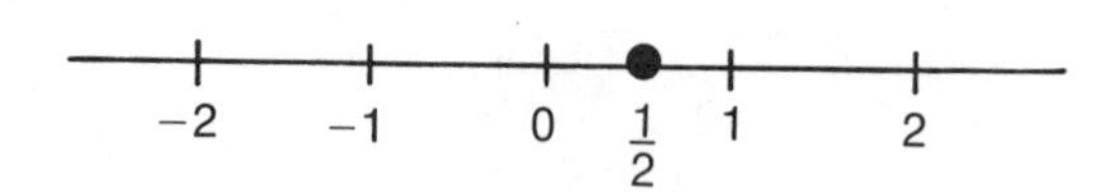

To graph $\frac{-3}{4}$, we divide the line segment from −1 to 0 into 4 equal parts. The graph of the point $\frac{-3}{4}$ is shown below.

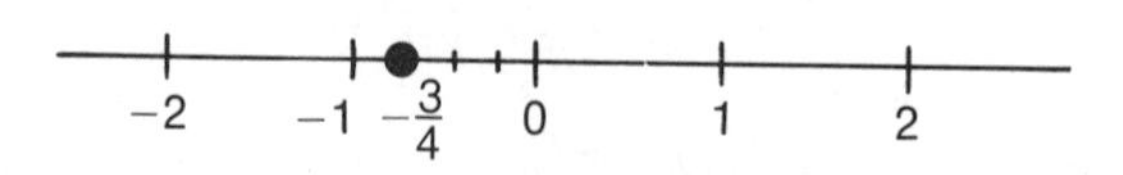

The number $\frac{1}{2}$ can also be written in decimal form, as 0.5.

Since all rational numbers can be written in the form $\frac{a}{b}$, where a and b are integers, it is always possible to change a rational number $\frac{a}{b}$ to its decimal form by simply dividing a by b. When this division is performed, you will get either a *terminating decimal* or a *repeating decimal.** Here is why: When you divide a by b, there are only two possibilities.

(1) You eventually obtain a zero remainder. In this case you have a *terminating decimal.*

(2) You keep on dividing, but you never get a zero remainder. In this case you have a *repeating decimal.* The reason for this is that in a division problem the number of remainders is limited. (In particular, when dividing a by b, the possible remainders are $0, 1, 2, \ldots, b-1$.) If you do not obtain a zero remainder, but you continue the division process, then sooner or later a remainder will be repeated! At this point, the digits in the quotient will start repeating.

The preceding discussion shows the following:

Every rational number can be written either as a terminating decimal or as a repeating decimal.

0.5 is a <u>terminating</u> decimal, but 0.555... is a <u>repeating</u> decimal.

Rational numbers can be written in the form $\frac{a}{b}$, with a and b integers and $b \neq 0$.

$\frac{1}{4} = 0.25$ is a <u>terminating</u> decimal. $\frac{2}{3} = 0.666\ldots$ is a <u>repeating</u> decimal.

Let us illustrate the second case. Suppose you wish to write $\frac{1}{7}$ as a decimal. We start by dividing 1 by 7. The division looks like this:

$$7 \overline{)1.000000} = 0.142857$$

	Division	
	7	
The remainder is 3	30	The possible new remainders now are 1, 2, 4, 5 and 6.
	28	
The remainder is 2	20	The possible new remainders now are 1, 4, 5 and 6.
	14	
The remainder is 6	60	The possible new remainders now are 1, 4 and 5.
	56	
The remainder is 4	40	The possible new remainders now are 1 and 5.
	35	
The remainder is 5	50	The possible new remainder now is 1.
	49	
The remainder is 1	1	All remainders have occurred. The numbers will start repeating.

*When we refer to a repeating decimal, we assume that the decimal is nonterminating. Thus, 0.333 . . . is a repeating (nonterminating) decimal, but 0.33 is *not* a repeating decimal.

Thus, $\frac{1}{7} = 0.142857142857\ldots$ For convenience, we write $\frac{1}{7} = 0.\overline{142857}$, with the bar over the 142857 to indicate that the number 142857 repeats indefinitely.

Example 4 Write the following rational numbers as decimals and state if they are repeating or terminating.

(a) $\frac{1}{6}$

(b) $\frac{2}{3}$

(c) $\frac{3}{8}$

Solution (a) Dividing 1 by 6, we obtain

$$\begin{array}{r} 0.1666\ldots \\ 6\overline{)1.0000} \\ \underline{6} \\ 40 \\ \underline{36} \\ 40 \end{array}$$

Same remainder (arrows pointing to the two 40s)

Thus, $\frac{1}{6} = 0.1666\ldots = 0.1\overline{6}$, a repeating decimal.

(b) Dividing 2 by 3, we obtain

$$\begin{array}{r} 0.666\ldots \\ 3\overline{)2.000} \\ \underline{1\,8} \\ 20 \end{array}$$

Same remainder (arrows pointing to the 2 and the 20)

Thus, $\frac{2}{3} = 0.666\ldots = 0.\overline{6}$, a repeating decimal.

(c) Dividing 1 by 8, we obtain

$$\begin{array}{r} 0.375 \\ 8\overline{)3.0} \\ \underline{2\,4} \\ 60 \\ \underline{56} \\ 40 \\ \underline{40} \\ 0 \end{array}$$

Thus, $\frac{1}{8} = 0.375$, a terminating decimal.

From our examples we can conclude that every rational number can be written either as a terminating decimal or as a repeating decimal. Is the converse of this statement true—that is, is a decimal that terminates or repeats a rational number? If the decimal is terminating, the number can be written as a rational number with a multiple of 10 as a denominator. For example, $0.41 = \frac{41}{100}$, and $\frac{7}{10} = 0.7$. If the decimal is repeating, then we can proceed as follows: Suppose we wish to write 0.555 . . . as a ratio of two integers. The idea is to get rid of the part that repeats. If $x = 0.555\ldots$, then $10x = 5.555\ldots$.

Thus,
$$\begin{aligned} 10x &= 5.555\ldots \\ x &= 0.555\ldots \end{aligned}$$

Subtracting, we get
$$9x = 5$$

so
$$x = \frac{5}{9}$$

Example 5

Write 5.2313131 . . . as a ratio of two integers.

Solution

If $x = 5.2313131\ldots$, then $10x = 52.313131\ldots$, and $1{,}000x = 5231.313131\ldots$.

Thus,
$$\begin{aligned} 1{,}000x &= 5231.313131\ldots \\ 10x &= 52.313131\ldots \end{aligned}$$

Subtracting, we get
$$990x = 5179$$

so
$$x = \frac{5179}{990}$$

Notice that in this example, *we multiplied x by multiples of 10 in order to obtain numbers with identical decimal parts.*

We have shown that terminating decimals such as 0.41 and repeating decimals such as 5.2313131 . . . can be written as ratios of two integers. The same procedure can be used to write *any* other terminating or repeating decimal as a ratio of two integers, that is, as a rational number. Thus, we can say the following:

Every terminating or repeating decimal is a rational number.

Combining this with the statement we made earlier (every

rational number can be written as a terminating or repeating decimal), we can reach the following conclusion:

Note that b cannot be 0, since division by 0 is not allowed.

The set of rational numbers is the same as the set of terminating or repeating decimals.

This means that rational numbers can be discussed from two viewpoints: as decimals, or as numbers of the set $Q = \{r \mid r = \frac{a}{b}, a \in I, b \in I, b \neq 0\}$.

Is the number 0.101001000 . . . formed by writing successive multiples of 10 (that is, 10, 100, 1000, etc.), a rational number? The answer is no, because it will never repeat. This type of number is called an *irrational number.* Another example of an irrational number is the number $\sqrt{2}$, which also has a non-repeating decimal representation (1.4142135 . . .). The set of irrational numbers is denoted by the letter H. The union of the set of rational numbers Q with the set of irrational numbers H is called the set of real numbers R; $R = Q \cup H$. Thus, real numbers may be classified as rational or irrational. The rational numbers are the numbers that can be expressed as ratios of two integers and have terminating or repeating decimal representations. The irrational numbers, however, cannot be expressed as ratios of two integers and have nonterminating, nonrepeating decimal representations.

H represents the set of irrational numbers.

Example 6

Classify the following numbers as rational or irrational.

(a) $\frac{-9}{4}$ (b) $\sqrt{2}$ (c) 0.12

(d) $0.\overline{145}$ (e) 0.12120120012000 . . . (f) $\sqrt{5}$

Solution

(a) $\frac{-9}{4}$ is the ratio of two integers, so it is rational.
(b) $\sqrt{2}$ is irrational.
(c) 0.12 is a terminating decimal. Thus, 0.12 is rational.
(d) $0.\overline{145}$ is a repeating decimal, so it is rational.
(e) 0.12120120012000 . . . is a nonterminating, nonrepeating decimal, and, as such, is irrational.
(f) $\sqrt{5}$ is irrational.

As a final point, we should mention that irrational numbers such as $\sqrt{2}$ can also be graphed on the number line, using the ideas employed by the Greeks in discovering irrational numbers over 2,000 years ago. The Greeks found that the diagonal of a square of unit length is not a rational number (see Figure 1.5). The symbol used to represent the length of this diagonal is $\sqrt{2}$. We can then superimpose this square on the number line and obtain the point corresponding to $\sqrt{2}$, as shown in Figure 1.6.

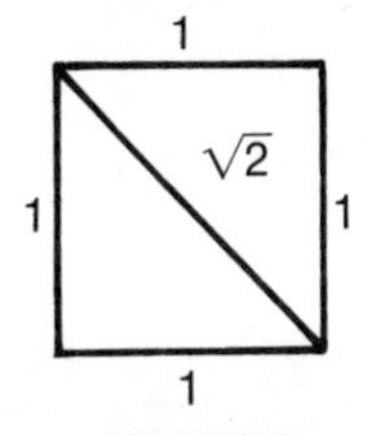

FIGURE 1.5

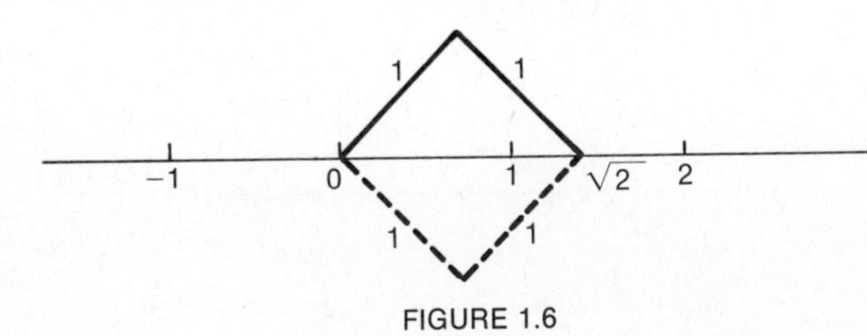

FIGURE 1.6

PROGRESS TEST

1. Graph the points −2, 2, and 5 on the number line:

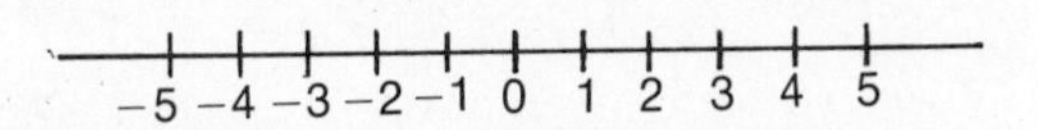

2. The coordinates of the points A, B, and C are, respectively, ________.

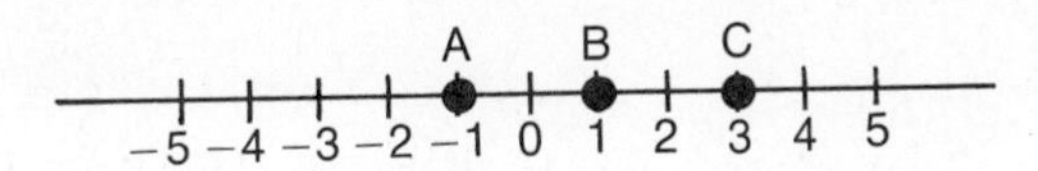

3. When written as decimals, $\frac{2}{7}$ and $\frac{3}{5}$ equal ________ and ________.

4. Classify as rational or irrational.
   (a) $\sqrt{7}$
   (b) $\frac{3}{9}$
   (c) 0.414414441 . . .
   (d) 0.123123123 . . .

5. When written as the ratio of two integers, 2.191919 . . . equals ________.

PROGRESS TEST ANSWERS

1.

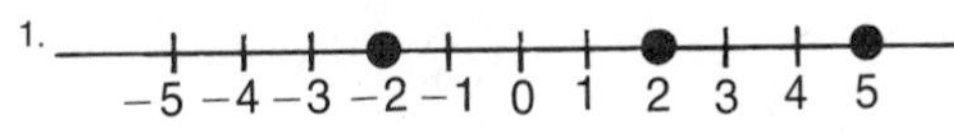

2. −1, 1, and 3
3. $0.\overline{285714}$, 0.6
4. (a) irrational; (b) rational; (c) irrational; (d) rational
5. $\frac{217}{99}$

EXERCISE 1.3

In Problems 1 through 8, graph the given number on the number line.

Illustration: (a) $\frac{-3}{2}$

(b) $\frac{5}{2}$

Solution:

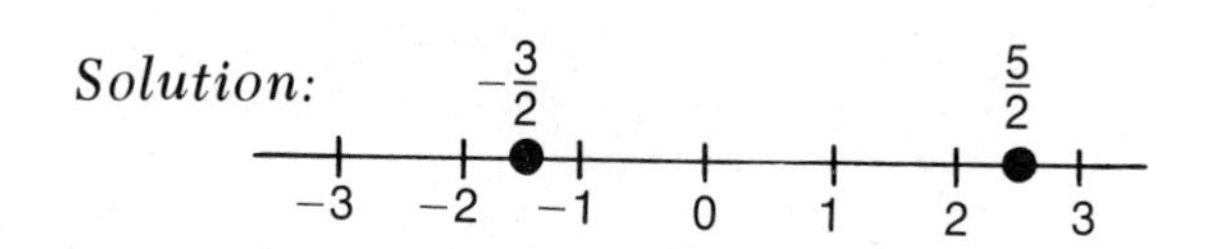

1. -7
2. $\frac{-1}{3}$
3. $\frac{3}{4}$
4. $3\frac{1}{2}$
5. $-3\frac{1}{4}$
6. $\frac{1}{5}$
7. $4\frac{1}{3}$
8. $5\frac{1}{2}$

In Problems 9 through 16, find the coordinate of the specified point.

Illustration: A

A B C D x E F G H
−5 −4 −3 −2 −1 0 1 2 3 4 5

Solution: The coordinate of A is $-4\frac{1}{2}$

9. B
10. C
11. D
12. x
13. E
14. F
15. G
16. H

In problems 17 through 22, use division to write the given rationals as decimals.

Illustration: (a) $\frac{2}{3}$

(b) $\frac{3}{5}$

Solution: (a) $3\overline{)2.00}$ with quotient 0.66, thus $\frac{2}{3} = 0.666\ldots = 0.\overline{6}$

$$\begin{array}{r} 0.66 \\ 3\overline{)2.00} \\ \underline{1\,8} \\ 20 \\ \underline{18} \\ 20 \end{array} \leftarrow \text{The division repeats}$$

(b) $5\overline{)3.0}$ with quotient 0.6, thus $\frac{3}{5} = 0.6$

$$\begin{array}{r} 0.6 \\ 5\overline{)3.0} \\ \underline{3\,0} \\ 0 \end{array}$$

17. $\frac{1}{3}$

18. $\frac{2}{5}$

19. $\frac{4}{7}$

20. $\frac{4}{5}$

21. $\frac{5}{6}$

22. $\frac{5}{8}$

In Problems 23 through 30, write the given decimals as the ratio of two integers.

Illustration: (a) 0.2

(b) $0.\overline{2}$

Solution: (a) $0.2 = \frac{2}{10} = \frac{1}{5}$

(b) If $x = 0.\overline{2}$, then $10x = 2.\overline{2}$.

Thus, $10x = 2.\overline{2}$

$x = 0.\overline{2}$

Subtracting, $9x = 2$

or $x = \frac{2}{9}$

23. 0.11

24. $0.\overline{7}$

25. $0.\overline{73}$

26. 0.43

27. $2.\overline{345}$

28. $4.\overline{12}$

29. (a) $1.\overline{83}$ (b) $2.\overline{142857}$ (c) $3.\overline{285714}$

30. (a) $4.\overline{3}$ (b) $2.\overline{6}$ (c) $3.\overline{16}$

In Problems 31 through 42, classify the given number as rational or irrational.

Illustration: (a) $\sqrt{11}$
(b) 0.131313 . . .
(c) 0.121201200 . . .

Solution: (a) $\sqrt{11}$ is irrational.
(b) 0.131313 . . . is a repeating decimal and thus it is a rational number.
(c) 0.121201200 . . . is nonrepeating and nonterminating and thus it is irrational.

31. $\frac{1}{7}$

32. 0

33. $\sqrt{13}$

34. $\sqrt{8}$

35. $\sqrt{9}$

36. $\frac{-5}{16}$

37. 0.121212 . . .

38. 0.303003000 . . .

39. $0.\overline{68}$

40. 83.242402400

41. 83.242424 . . .

42. $731.\overline{81}$

43. In a certain factory, the ratio of men to women is $3.\overline{16}$. Write this number as the ratio of two integers.

44. In 1982, the ratio of solar to lunar eclipses will be $1.\overline{3}$. Write this number as the ratio of two integers and determine how many solar and how many lunar eclipses will occur in 1982, if it is known that the total number of eclipses in 1982 will be 7.

45. The price to earnings ratio of a certain stock is $1.1\overline{6}$. Write this number as the ratio of two integers.

B. C. by permission of John Hart and Field Enterprises, Inc.

OBJECTIVES 1.4

After studying this section, the reader should be able to indicate which axiom or theorem is being used to justify a given statement.

1.4 PROPERTIES OF THE REAL NUMBERS

In the preceding section, we discussed how to graph elements of the set of real numbers on the number line. In this section, we shall concentrate on some of the operations on real numbers and the relations that exist between real numbers.

The set of real numbers, together with the four fundamental operations and certain relations that exist between these numbers, is an example of a *mathematical system.* In general, a mathematical system is composed of:

Definition of a mathematical system.

(1) A set of elements (in our case R, the set of real numbers);
(2) One or more operations (for the time being, addition and multiplication);
(3) One or more relationships between the elements (for example, equality).

In order to undertake the study of a mathematical system, we first must make some formal assumptions about the elements of the set under consideration. These assumptions are called *axioms* or *postulates.* The words *property, law,* and *principle* are used sometimes to indicate assumptions. However, these terms are usually reserved to indicate certain consequences which can be deduced from the axioms. The first assumption we shall make about the real numbers has to do with *equality.*

Definition of equality.

An *equality* is simply a mathematical statement indicating that two or more symbols are names for the same thing. Thus, if a and b are real numbers, the statement a = b (read "a is equal to b") means that a and b represent the same number. Similarly, $5 + 2 = 3 + 4$ is a mathematical statement indicating that $5 + 2$ and $3 + 4$ represent the same number. In an equality, the symbolic expression on the left side of the equals sign is called the *left-hand* member of the equality, and the symbolic expression on the right side of the equals sign is called the *right-hand member.* Thus, in the equality $5 - 2 = 1 + 2$, $5 - 2$ is the left-hand member and $1 + 2$ the right-hand member. We shall now assume that the = relationship has the following properties.

a	=	b
Left-hand member		Right-hand member

AXIOMS FOR EQUALITY

If a, b, c ∈ R:

E-1	a = a	Reflexive Law
E-2	If a = b, then b = a.	Symmetric Law
E-3	If a = b and b = c, then a = c.	Transitive Law
E-4	If a = b, then a may be replaced by b (or b by a) in any statement without changing the truth or falsity of the statement.	Substitution Law

The Symmetric Law enables us to say that if 1,000 meters equal one kilometer, then one kilometer equals 1,000 meters. Similarly, if $x - 7 = y$, then $y = x - 7$.

In a different application, if it is known that 2.54 centimeters equal one inch, and that one inch equals $\frac{1}{12}$ of a foot, we can use the Transitive Law to conclude that 2.54 centimeters equal $\frac{1}{12}$ of a foot. (More concisely, if 2.54 cm. = 1 inch, and 1 inch = $\frac{1}{12}$ ft., then 2.54 cm. = $\frac{1}{12}$ ft.) Similarly, if $32 = a$ and $a = b$, then $32 = b$.

As a final illustration of these axioms, suppose that $x = y - 1$ and that we know that $y - 1$ is an even number. Then, as a consequence of the Substitution Law, the statement "x is an even number" is true.

Do we need any other axioms? To answer this question, let us try to solve a very simple puzzle.

In the cartoon at the beginning of this section, a subject is asked to think of a number between 1 and 10. You have probably seen a magician (or friend) ask a person to think of a number and to make several calculations involving it. Then, without knowing the original number, the magician will reveal the number with which his subject has ended up! Here is one of these puzzles, as well as the operations being performed at each step.

Think of a number	n
Add 1 to it	$n+1$
Multiply the result by 5	$5 \bullet (n+1)$
Subtract 5 times the number you started with	$5 \bullet (n+1) - 5 \bullet n$
The result is 5	

Note that we use parentheses and the dot • to indicate that 5 is to be multiplied by $(n+1)$.

Do you see why this works? To prove that the result we have obtained, $5 \bullet (n+1) - 5 \bullet n$, is equal to 5, we have to use axioms pertaining to the operations of addition and multiplication. For example, we first must be able to multiply 5 by the quantity $n+1$, that is, to find $5 \bullet (n+1)$. The axiom we use to do this is the Distributive Law, which states that if a, b, and c are real numbers, $a \bullet (b+c) = (a \bullet b) + (a \bullet c)$. We now have: $5 \bullet (n+1) - 5 \bullet n = (5 \bullet n + 5) - 5 \bullet n$. Our next step is to try to combine the n's. To do this, we need them within the same parentheses. So we first change the order of $5 \bullet n$ and 5 within the first parentheses. Here we are using the Commutative Law, $a+b=b+a$. Our problem now looks like this:

$$\begin{aligned} 5 \bullet (n+1) - 5n &= (5 \bullet n + 5) - 5 \bullet n \\ &= (5 + 5 \bullet n) - 5 \bullet n \end{aligned}$$

We then associate the 5n and the $-5n$ by using the Associative Law of Addition, $(a+b)+c = a+(b+c)$. We obtain $(5+5 \bullet n) - 5 \bullet n = 5 + (5 \bullet n - 5 \bullet n)$. We finally note that if a number is subtracted from itself, the answer is 0. This follows from the fact that $a - b = a + (-b)$ and from the additive inverse axiom, which states that $a + (-a) = 0$. So, $5n - 5n = 0$. But zero *added* to any number yields the same number, because 0 is the additive identity. Thus, $5 + (5n - 5n) = 5$. Here are all the steps and the reasons for each:

$5 \bullet (n+1) - 5 \bullet n = (5 \bullet n + 5) - 5 \bullet n$	Distributive Law $a \bullet (b+c) = (a \bullet b) + (a \bullet c)$
$= (5 + 5 \bullet n) - 5 \bullet n$	Commutative Law $a+b=b+a$
$= 5 + (5 \bullet n - 5 \bullet n)$	Associative Law $(a+b)+c = a+(b+c)$
$= 5 + 0$	Additive inverse axiom: $a + (-a) = 0$, and the

fact that $a - b = a + (-b)$

$= 5$ Identity element for +; $a + 0 = a$

Of course, we assumed that if we started with a real number and performed certain operations involving this number, the result would be a real number. This means that we assume that the set of real numbers is *closed* under certain operations.

a • b is usually written as ab.

We now list as axioms the properties of the sum $a + b$ and the product $a \bullet b$ (usually written as ab). Mathematical systems satisfying these properties are called *fields*. In these axioms, a, b, and c represent real numbers.

FIELD AXIOMS FOR R

	ADDITION AXIOMS		MULTIPLICATION AXIOMS	NAME
F-1	$a + b$ is a unique real number	F-2	$a \bullet b$ is a unique real number	Closure
F-3	$a + b = b + a$	F-4	$a \bullet b = b \bullet a$	Commutative Laws
F-5	$(a + b) + c = a + (b + c)$	F-6	$(a \bullet b) \bullet c = a \bullet (b \bullet c)$	Associative Laws
F-7	There exists a unique number 0 such that $a + 0 = a = 0 + a$	F-8	There exists a unique number 1 such that $a \bullet 1 = a = 1 \bullet a$	Identity element
F-9	For each real number a, there exists a unique real number $-a$ (the negative or opposite of a) such that $a + (-a) = 0 = (-a) + a$	F-10	For each real number a except $a = 0$, there exists a unique real number $\frac{1}{a}$ (the reciprocal of a) such that $a \bullet \frac{1}{a} = 1 = \frac{1}{a} \bullet a$	Inverses

A final axiom relates the operations of addition and multiplication.

F-11 $a(b + c) = (ab) + (ac)$ Distributive Law

Example 1 The given statements are consequences of the axioms F-1 through F-11. In each case, indicate which axiom applies. Assume all letters (variables) are real numbers.

(a) $5 \cdot x$ is a real number
(b) $x + 5 = 5 + x$
(c) $(x + 9) + 0 = (x + 9)$
(d) $(x + y) \cdot \frac{1}{(x + y)} = 1$, where $x + y \neq 0$
(e) $2(x + y) = 2x + 2y$

Solution

(a) $5 \cdot x$ is a real number because the set of real numbers is closed under multiplication; axiom F-2.
(b) $x + 5 = 5 + x$ because the addition of real numbers is commutative; axiom F-3.
(c) $(x + 9) + 0 = (x + 9)$ because 0 is the identity element for addition; axiom F-7.
(d) $(x + y) \cdot \frac{1}{(x + y)} = 1$ because $(x + y)$ and $\frac{1}{(x + y)}$ are reciprocals; axiom F-10.
(e) $2(x + y) = 2x + 2y$ because of the Distributive Law; axiom F-11.

The axioms of equality (E-1 through E-4), together with F-1 through F-11, can be used to deduce other properties of the real numbers which will be used later. Properties deduced from the axioms are usually called *theorems*. Theorems consist of two parts:

A theorem is a property derived from an axiom.

1. The "if" part, or *hypothesis*
2. The "then" part, or *conclusion*

The argument used to verify a theorem is called a *proof*. Thus, to prove a theorem means to use some of the axioms (and perhaps some previously proved theorems) to verify that if the hypothesis is true then the conclusion must also be true.

A proof is an argument that veri-ies a theorem.

We now consider a theorem which will be used later. In the remainder of this section the letters (variables) used represent real numbers.

THEOREM 1.1

ADDITION LAW OF EQUALITY

If $a = b$, then $a + c = b + c$ and $c + a = c + b$

PROOF:

	Statement	Reason
1.	$a = b$	Hypothesis
2.	$a + c$ is a real number	Since the real numbers are closed under addition, Axiom F-1
3.	$b + c$ is a real number	Same reason as in Step 2
4.	$a + c = a + c$	Reflexive Law, Axiom E-1
5.	$a + c = b + c$	Substituting b for a in Step 4, Axiom E-4
6.	$c + a = c + b$	Using the Commutative Law, Axiom F-3

Thus, if the hypothesis $a = b$ is true, the conclusion $a + c = b + c$ and $c + a = c + b$ is also true.

Theorem 1.1 can be used to find the value of x in equations such as $x + 5 = 0$. If x is a real number, then $x + 5$ is a real number, by the closure axiom, F-1. Thus,

$(x + 5) + (-5) = 0 + (-5)$ by Theorem 1.1 with $a = x + 5$, $b = 0$ and $c = -5$

$x + [5 + (-5)] = 0 + (-5)$ by the Associative Law, F-5

$x + 0 = 0 + (-5)$ since 5 is the negative or opposite of -5; F-9

$x = -5$ since 0 is the identity; F-7

We also have a Multiplicative Law of Equality whose proof is left as an exercise.

THEOREM 1.2

MULTIPLICATION LAW OF EQUALITY

If $a = b$, then $ac = bc$ and $ca = cb$

Next, we prove the Cancellation Law for addition.

THEOREM 1.3

CANCELLATION LAW FOR ADDITION

If $a + c = b + c$ then $a = b$

PROOF:

	Statement	Reason
1.	$a + c = b + c$	Hypothesis
2.	$(a + c) + (-c) = (b + c) + (-c)$	Theorem 1.1
3.	$a + [c + (-c)] = b + [c + (-c)]$	Associative Law, F-5
4.	$a + 0 = b + 0$	Since $c + (-c) = 0$ by the Inverse Law (F-9), we substitute 0 for $c + (-c)$ in Step 3
5.	$a = b$	Since $a + 0 = a$ and $b + 0 = b$; F-7

Theorem 1.3 can be used to find the value of x in equations such as $x + 9 = 3 + 9$. Since $x + 9$ and $3 + 9$ are real numbers by the Closure Law, $x = 3$ by Theorem 1.3.

There is also a Cancellation Law for Multiplication whose proof is similar to that of Theorem 1.3. This theorem is stated next and the proof is left as an exercise.

THEOREM 1.4

CANCELLATION LAW FOR MULTIPLICATION

If $ac = bc$ and $c \neq 0$, then $a = b$

We now give two more laws that will be used frequently in our later work. The first one depends on Axiom F-9. If we let $a = -5$ in this axiom, we have

$$(-5) + [-(-5)] = 0$$

$$\text{But } (-5) + 5 = 0 \text{ also;}$$

so, since each number has a unique inverse, $-(-5) = 5$. In general, we have the rule given below.

THEOREM 1.5

$-(-a) = a$

Thus, $-(-19) = 19$, $-\left(\frac{-1}{5}\right) = \frac{1}{5}$, and $-(-0.\overline{319}) = 0.\overline{319}$.

Finally we list a theorem (whose proof is left as an exercise) which will be used in Section 2.1.

THEOREM 1.6

$(-1) \cdot a = -a$

Thus, $(-1) \cdot 4 = -4$, $(-1) \cdot (5) = -5$, and $(-1) \cdot (-5) = -(-5) = 5$.

PROGRESS TEST

1. If we know that 2 pints = 1 quart and then conclude that 1 quart = 2 pints, we have used ____________.
2. If 1 mile = 1.6 kilometers and 1.6 kilometers = 1600 meters, we can conclude that 1 mile = 1600 meters by using ____________.
3. 1. $3(n+1) - 3n = (3n+3) - 3n$
   2. $= (3+3n) - 3n$
   3. $= 3 + (3n - 3n)$
   4. $= 3 + 0$
   5. $= 3$

   The axioms used to justify each of the steps are, respectively, ____________.
4. $(-1) \cdot (-6)$ equals ____________.

EXERCISE 1.4

In Problems 1 through 25, each given statement is a consequence of one of the axioms (E-1 through E-4 or F-1 through F-11). In each case indicate which axiom has been used. As usual, the letters used represent real numbers.

Illustration: (a) $x + 0 = x$
(b) $2(x+1) = 2x + 2 \cdot 1$

Solution: (a) $x + 0 = x$ by Axiom F-7
(b) $2(x + 1) = 2x + 2 \cdot 1$ by the Distributive Law, Axiom F-11

1. $7 + x = x + 7$

2. $\frac{-1}{2} + y = y + \left(\frac{-1}{2}\right)$

PROGRESS TEST ANSWERS

1. the Symmetric Law, E-2
2. the Transitive Law, E-3
3. F-11, F-3, F-5, F-9, and the fact that $a - b = a + (-b)$, F-7
4. 6

3. $7 + 9 \in R$

4. $\frac{-1}{5} + 0.\overline{33} \in R$

5. $3 \cdot 0.888 \ldots = 0.888 \ldots \cdot 3$

6. $\frac{-1}{2} \cdot \frac{4}{5} = \frac{4}{5} \cdot \left(\frac{-1}{2}\right)$

7. If $A = 2 \cdot \pi \cdot r$ and $r = 3$, then $A = 2 \cdot \pi \cdot 3$

8. If $P = 4s$ and $s = 2$, then $P = 4 \cdot 2$

9. If $x = y$ and $x + 5 = 7$, then $y + 5 = 7$

10. $3x + y = y + 3x$

11. $(3x)y = 3(xy)$

12. $a + (2 + y) = (a + 2) + y$

13. $(2x + y) + z = 2x + (y + z)$

14. $(5x + y) + 2 = (y + 5x) + 2$

15. $(2x + b) + c = c + (2x + b)$

16. $3 \cdot 0.666 \ldots \in R$

17. If $\frac{3}{4} = 0.75$ and $0.75 = 0.75\overline{0}$ then $\frac{3}{4} = 0.75\overline{0}$

18. $(8x) + (-8x) = 0$

19. $0 + \frac{1}{7} = \frac{1}{7}$

20. $x \cdot \frac{1}{x} = 1$

21. $1 \cdot (3 + x) = 3 + x$

22. $1 \cdot (5 + y) = 5 + y$

23. If $-9 = y$, then $y = -9$

24. If $a + b = z$ and $z = 0$, then $a + b = 0$

25. If $a = 3$ and $7b = 5a$, then $7b = 5 \cdot 3$

In Problems 26 through 30, the given statement is a consequence of one of the theorems (1.1 through 1.6). In each case indicate which theorem has been used.

Illustration: $-(x+1) = (-1)(x+1)$
Solution: Theorem 1.6, with $a = x+1$

26. If $x = 7$, then $x + (-7) = 7 + (-7)$

27. If $x + 1 = y + 1$, then $x = y$

28. If $3x = 3(2y)$, then $x = 2y$

29. If $\frac{2}{3} = 0.666\ldots$, then $\frac{3}{2} \cdot \frac{2}{3} = \frac{3}{2} \cdot 0.666\ldots$

30. $-(-0.31) = 0.31$

31. Prove Theorem 1.2.

32. Prove Theorem 1.4.

33. Prove Theorem 1.6. (Hint: Use the fact that for any real number a, $a \cdot 0 = 0$.)

Courtesy of International Business Machines Corporation.

OBJECTIVES 1.5

After studying this section, the reader should be able to:

1. **Find the sum or difference of two or more real numbers.**
2. **Find the absolute value of any real number.**
3. **Verify, by using the number line, the answer obtained when adding or subtracting two integers.**

Since in addition we combine two numbers to obtain a third number, the sum, addition is said to be a binary operation.

1.5 ADDITION AND SUBTRACTION OF REAL NUMBERS

The illustration at the beginning of this section shows a photograph of the first adding machine. This machine was invented by the French mathematician Blaise Pascal at the age of 19 to add long columns of figures. In this section we shall concentrate on the addition and subtraction of numbers.

Addition is a process by which we can combine two real numbers a and b and arrive at a number a + b called the *sum* of a and b. When writing the sum of a and b, the + sign is used to indicate the operation of addition. In algebra, the + sign has a further meaning. For example, in a Centigrade thermometer, which has a scale giving readings *above* and *below* zero, a reading of 15° *above* zero is indicated by writing +15°, and a reading of 5° *below* zero is written as −5°. Here the plus (+) and minus (−) signs are *not* used as signs of operation but indicate the *sense* or *direction* in which the numbers are to be taken. These ideas can be made clear by using the number line.

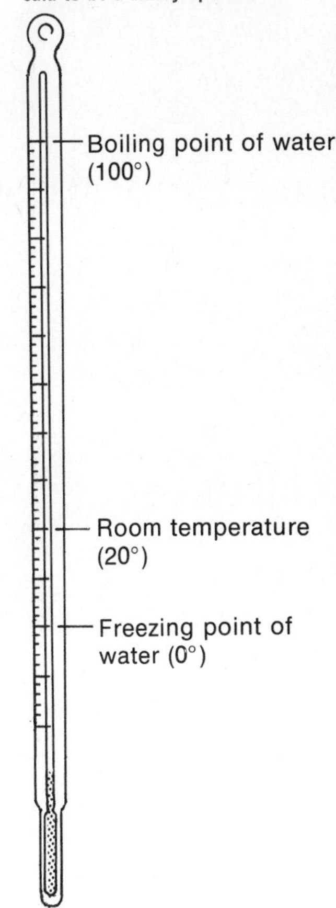

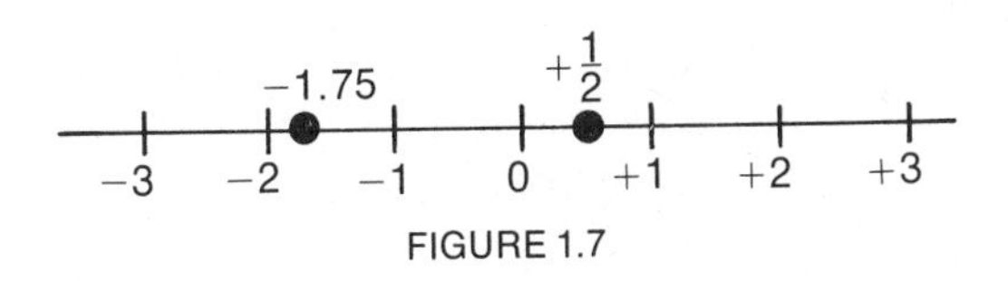

FIGURE 1.7

Usually, if a number is positive no sign is placed in front of it. Thus +8 is written as 8, $+\frac{1}{7}$ as $\frac{1}{7}$.

As you can see, the point midway between 0 and 1 represents the number $\frac{1}{2}$. To emphasize that this number is *positive*, we have labeled it $+\frac{1}{2}$. The point three-fourths of the way from -1 to -2 represents the number -1.75. The fact that this number is negative is indicated by placing a $-$ sign in front of the number. Any number to the *right* of 0 represents a *positive* number; any number to the *left* of 0 represents a *negative* number.

The addition of two positive numbers in algebra is done just as in arithmetic. The addition $2 + 3 = 5$, for example, can be represented on the number line by starting at 0, moving 2 units in the positive direction (to the right), followed by moving 3 more units to the right, ending at 5 (see Figure 1.8). Thus, $2 + 3 = 5$.

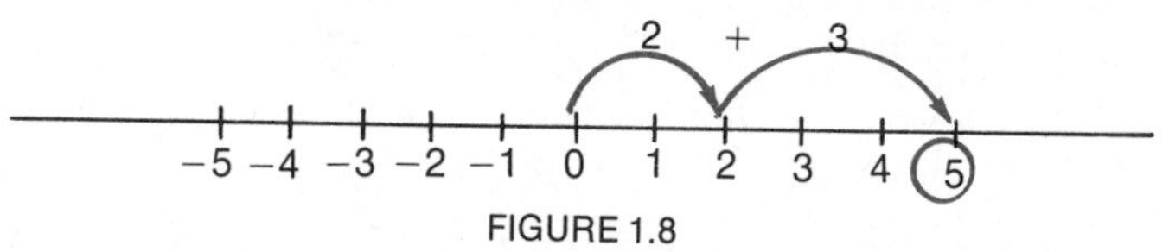

FIGURE 1.8

To add -2 and -3, we follow a similar pattern. We start at 0, but this time move 2 units in the *negative* (left) direction, and follow by moving 3 more units to the left (see Figure 1.9). The result is -5. Thus, $(-2) + (-3) = -5$.

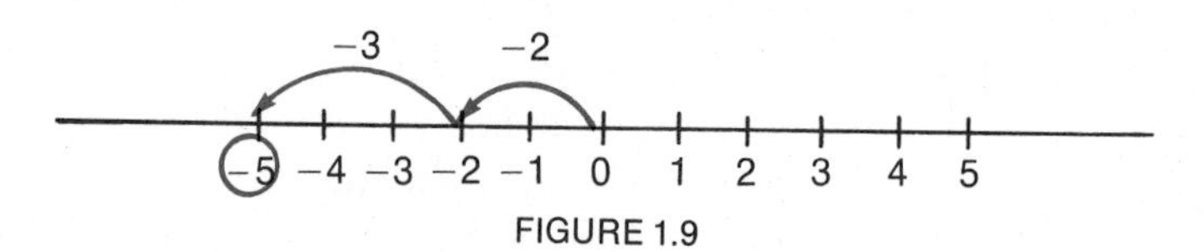

FIGURE 1.9

The absolute value of a number x, denoted by $|x|$, is the number with the sign disregarded. Thus, $|3| = 3$, and $|-5| = 5$.

From these and similar examples, we can easily deduce a procedure for adding real numbers with the same sign, but first we need a simple way to describe the distance of a number from the origin on the number line. This distance is called the *absolute value* of the number. In simple terms, the absolute value of a signed number is the number with the sign disregarded. For example, the absolute value of $+5$ is 5, and that of -3 is 3.

Mathematically, the absolute value of x is defined as follows:

DEFINITION 1.4

$$|x| = \begin{cases} x, & \text{if } x \text{ is positive} \\ -x, & \text{if } x \text{ is negative} \\ 0, & \text{if } x \text{ is } 0 \end{cases}$$

We now give the rule for adding real numbers with the same sign.

RULE 1.1

To add real numbers with the same sign, add their absolute values and give the sum the common sign.

Thus, $3 + 9 = +(|3| + |9|) = 12$ and $(-5) + (-4) = -(|-5| + |-4|) = -9$.

Example 1

Add the following numbers.

(a) $7 + 11$
(b) $(-3) + (-7)$
(c) $0.3 + 0.8$
(d) $(-0.7) + (-0.8)$

Solution

(a) $7 + 11 = +(|7| + |11|) = 18$
(b) $(-3) + (-7) = -(|-3| + |-7|) = -10$
(c) $0.3 + 0.8 = +(|0.3| + |0.8|) = 1.1$
(d) $(-0.7) + (-0.8) = -(|-0.7| + |-0.8|) = -1.5$

How do we add a positive and a negative number? To do this, we again consider the number line. For example, to add $5 + (-3)$, we begin at 0, move 5 units to the right, and then move 3 units to the left, obtaining 2. Thus, $5 + (-3) = 2$, as shown in Figure 1.10.

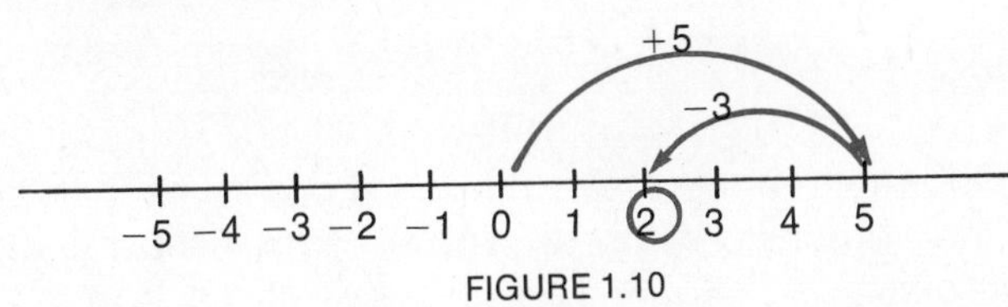

FIGURE 1.10

Now consider the sum $(-5) + 3$. We begin at 0, move 5 units to the left, and then move 3 units to the right. The result is -2, as shown in Figure 1.11.

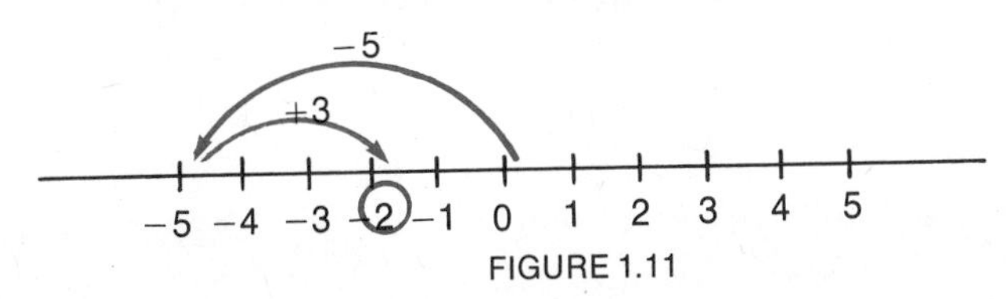

FIGURE 1.11

Note that $5 + (-3) = 5 - 3 = 2$, and that $(-5) + 3 = -(|-5| - |3|) = -2$. These two examples lead to the following rule.

RULE 1.2

To add a positive and a negative real number,
1. find the absolute value of the numbers;
2. subtract the number with the smaller absolute value from the one with the greater absolute value;
3. use the sign of the number with the greater absolute value for the result obtained in Step 2.

Example 2

Add:
(a) $6 + (-3)$
(b) $(-8) + 4$
(c) $(-3) + 9$
(d) $5 + (-11)$

Solution

(a) $6 + (-3) = |6| - |-3| = 6 - 3 = 3$
(b) $(-8) + 4 = -(|-8| - |4|) = -4$
(c) $(-3) + 9 = (|9| - |-3|) = 9 - 3 = 6$
(d) $5 + (-11) = -(|-11| - |5|) = -(11 - 5) = -6$

Of course, in actual practice most of these operations are carried out mentally.

We now define the *difference* of two real numbers.

DEFINITION 1.5

If a and b are real numbers, the difference of a and b, denoted a − b, is the number c with the property that a = b + c, that is, a − b = c means that a = b + c.

Thus, $8 - 3 = 5$ because $8 = 3 + 5$, and $11 - 4 = 7$ because $11 = 4 + 7$.

Since we have developed a procedure to add real numbers, it will be convenient to do subtraction in terms of addition. We do this in Theorem 1.7.

THEOREM 1.7

If a and b are real numbers, a − b = a + (−b).

PROOF:

	STATEMENT	REASON
1.	$a - b = c$ means that $a = b + c$	Definition 1.5
2.	$a + (-b) = (b + c) + (-b)$	Addition Law of Equality, Theorem 1.1
3.	$a + (-b) = (-b) + (b + c)$	By the Commutative Law, F-3
4.	$a + (-b) = [(-b) + b] + c$	By the Associative Law, F-5
5.	$a + (-b) = 0 + c$	Since $(-b) + b = 0$; F-9
6.	$a + (-b) = c$	F-7
7.	$a - b = a + (-b)$	Since in Step 1 $a - b = c$ and in Step 6, $a + (-b) = c$, we use the Transitive Law to conclude that $a - b = a + (-b)$

Thus, $6 - (-3) = 6 + 3 = 9$, $(-7) - (2) = (-7) + (-2) = -9$, and $(-9) - (-5) = (-9) + 5 = -4$.

Example 3

Use Theorem 1.7 to write each difference as a sum.
(a) $7 - 3$
(b) $x - y$
(c) $-x - 3y$

Solution

(a) $7 - 3 = 7 + (-3)$
(b) $x - y = x + (-y)$
(c) $-x - 3y = -x + (-3y)$

Example 4

As you can see from the chart in page 46, in 1971 the U.S. balance of payments showed a *deficit* of 30 billion dollars. In 1947, there was a *surplus* of 5 billion dollars. If a 30-billion deficit is denoted by -30 and a 5-billion surplus by $+5$, find:
(a) The difference (in billions) in the U.S. balance of payments between the years 1971 and 1947.
(b) The difference (in billions) in the U.S. balance of payments between the years 1971 and 1950, if it is known that in 1950 there was a 5-billion-dollar deficit.

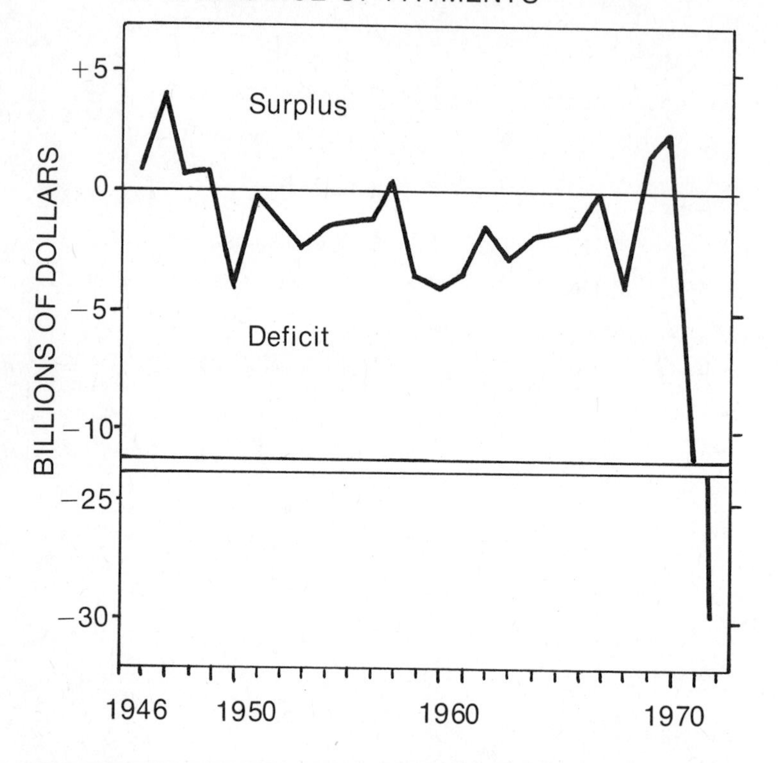

From *Economics*, 9th Ed., by Paul A. Samuelson. Copyright 1973. Used with permission of McGraw-Hill Book Company.

Solution

(a) The difference is $-30-5=-35$ (billions)

(b) The difference is $-30-(-5)=-30+5=-25$ (billions)

PROGRESS TEST

1. The sum $8+19$ equals ______.
2. The sum $(-7)+(-11)$ equals ______.
3. The absolute value of 17 is ______.
4. The absolute value of -13 is ______.
5. $9-11$ equals ______.
6. $(-3)-(7)$ equals ______.
7. $(-4)-(-9)$ equals ______.
8. $5-(-9)$ equals ______.

PROGRESS TEST ANSWERS

1. 27
2. −18
3. 17
4. 13
5. −2
6. −10
7. 5
8. 14

EXERCISE 1.5

In Problems 1 through 10, find the sum. Verify your answer using the number line.

Illustration: (a) $4 + (-5)$
(b) $0.1 + (0.2)$

Solution: (a) $4 + (-5) = -1$

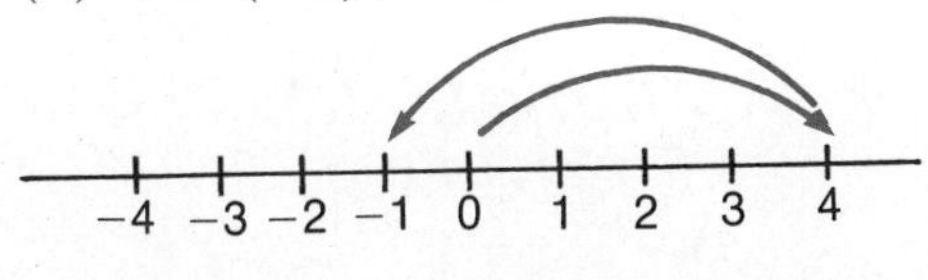

(b) $0.1 + 0.2 = 0.3$

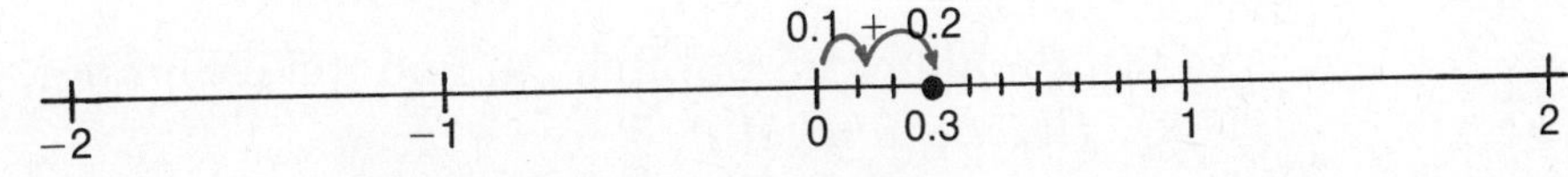

1. $9 + 11$
2. $0.4 + 0.9$
3. $(-0.3) + 0.2$
4. $(-8) + 5$
5. $(-4) + 6$
6. $(-0.2) + 0.3$
7. $(-0.5) + (-0.3)$
8. $(-7) + (-11)$
9. $\left(\frac{-1}{5}\right) + \frac{2}{5}$
10. $\left(\frac{-4}{7}\right) + \frac{2}{7}$

In Problems 11 through 16, evaluate:

Illustration: (a) $|5|$
(b) $\left|\frac{-1}{5}\right|$

Solution: (a) $|5| = 5$
(b) $\left|\frac{-1}{5}\right| = \frac{1}{5}$

11. $|9|$
12. $\left|\frac{3}{4}\right|$
13. $|-0.\overline{33}|$
14. $|-9|$
15. $|-(-5)|$
16. $\left|-\left(\frac{-1}{7}\right)\right|$

In Problems 17 through 24, rewrite each difference as a sum.

Illustration: (a) $4 - 9$
(b) $2x - y$

Solution: (a) $4 - 9 = 4 + (-9)$
(b) $2x - y = 2x + (-y)$

17. $3 - 7$

18. $4 - (-3)$

19. $2x - 3z$

20. $-5y - 4x$

21. $-3a - 2b$

22. $-4x - 5y$

23. $-x - (-7y)$

24. $-y - (-7)$

In Problems 25 through 34, find the difference.

Illustration: (a) $5-7$
(b) $0.3-(-0.4)$

Solution: (a) $5-7=5+(-7)=-2$
(b) $0.3-(-0.4)=0.3+0.4=0.7$

25. $14 - 6$

26. $0.24 - 0.10$

27. $8 - 11$

28. $0.5 - 0.9$

29. $-0.6 - 0.3$

30. $-3 - 9$

31. $-4 - (-9)$

32. $-0.5 - (-0.3)$

33. $\frac{1}{3} - \left(\frac{-1}{3}\right)$

34. $\frac{1}{7} - \left(\frac{-3}{7}\right)$

In Problems 35 through 50, perform the indicated operations.

Illustration: (a) $3 - 4 + 5$
(b) $-0.5 + 0.9 + 0.4$

Solution: (a) $3 - 4 + 5 = (3 - 4) + 5$
$= -1 + 5$
$= 4$
(b) $-0.5 + 0.9 + 0.4 = (-0.5 + 0.9) + 0.4$
$= 0.4 + 0.4$
$= 0.8$

35. $4 - 7 + 3$

36. $0.5 - 0.8 + 0.3$

37. $0.3 - 0.4 - 0.6$

38. $2 - 5 - 3$

39. $-0.5 + (-0.4) - (-0.6)$

40. $10 - (-4) + (-5)$

41. $0.7 - (0.5 - 0.3)$

42. $6 - (8 - 1)$

43. $(0.7 - 0.4) - 0.9$

44. $(9 - 3) - 7$

45. $(5 - 3) + (-4 + 2)$

46. $(8 - 9) + (5 - 2)$

47. $(5 - 7 + 3) - (7 - 9)$

48. $(18 - 7) + (9 - 13 - 1)$

49. $(7 + 7 - 11) + (6 - 9 + 3)$

50. $(-9 + 5 - 1) - (8 - 3 + 7)$

51. Mount Pico in the Azores rises 7615 feet above sea level. If the distance from the ocean floor to the crest of Mt. Pico is 23,615 feet, how many feet below sea level is the base of this mountain?

52. The deepest point of the Dead Sea is 2,600 feet below the Mediterranean (Guinness). If the shore of the Dead Sea is 1,291 feet below sea level, how many feet deep is the Dead Sea at its deepest point?

The following information will be used in Problems 53 and 54.

Important historical dates

323 B.C.	Alexander the Great dies
216 B.C.	Hannibal defeats the Romans
476 A.D.	Fall of the Roman Empire
1492 A.D.	Columbus discovers America
1776 A.D.	The Declaration of Independence is signed
1939 A.D.	World War II starts

53. If negative integers represent years B.C., find the number of years between:
(a) The fall of the Roman Empire and the defeat of the Romans by Hannibal
(b) The start of World War II and the death of Alexander the Great

54. Referring to the Table of Important Historical dates, find the number of years between:
(a) The signing of the Declaration of Independence and Columbus' discovery of America
(b) The death of Alexander the Great and Columbus' discovery of America

55. The highest point on earth is Mt. Everest, rising about 9 kilometers above sea level. The lowest point on earth is the Marianas Trench in the Pacific, 11 kilometers below sea level. Find the difference between these two extremes.

"Say, I think I see where we went off. Isn't eight times seven fifty-six?"

OBJECTIVES 1.6

After studying this section, the reader should be able to:

1. **Find the product or quotient of two or more real numbers.**
2. **Find the indicated product when a number is written using exponents.**
3. **Write a number of the form $\frac{a}{b}$ as a product of an integer and the reciprocal of another integer.**

1.6 MULTIPLICATION AND DIVISION OF REAL NUMBERS

In the cartoon at the beginning of this section, the scientists have forgotten that the *product* of 8 and 7 is 56, that is, $8 \cdot 7 = 56$. As you can see, multiplication is a binary operation in which two real numbers a and b are paired to form a third number denoted by ab (or $a \cdot b$), called the *product* of a and b. The numbers a and b are called *factors.* Thus, in the

Since in multiplication we combine two numbers to obtain a third number, the product, multiplication is a binary operation.

In the problem $3 \cdot 5 = 15$, 3 and 5 are factors, 15 is the product.

problem in the cartoon, 8 and 7 are *factors* and 56 is the *product*. From this example, and from your previous experience in arithmetic, it is evident that *the product of two positive numbers is positive*. Thus, $5 \cdot 4 = 20$, $8 \cdot 9 = 72$ and $21 \cdot 4 = 84$.

What can be said about the product of a positive and a negative number? For example, what is $3 \cdot (-2)$?

Multiplication can be thought of as repeated additions.

Since the multiplication of integers can be thought of as a process of repeated addition, $3 \cdot (-2)$ can be interpreted as the addition of three -2's, that is,

$$3 \cdot (-2) = \underbrace{(-2) + (-2) + (-2)}_{\text{three } -2\text{'s}} = -6$$

Thus, $3 \cdot (-2) = -6$, that is, *the product of a positive and a negative number is a negative number.* Of course, since multiplication is commutative, $(-2) \cdot (3) = -6$, that is, *the product of a negative and a positive number is a negative number.*

Now, what about the product of two negative numbers? In particular, how shall we define the product of (-2) and (-3)? We shall now show that if the axioms and properties of real numbers are assumed to hold, the product of $(-2)(-3)$ should be 6. Here is our argument:

Statement	Reason
1. $3 + (-3) = 0$	Axiom F-9, $a + (-a) = 0$
2. $-2[3 + (-3)] = -2 \cdot 0$	Multiplication Law of Equality
3. $(-2)(3) + (-2)(-3) = 0$	Distributive Law and the fact that $a \cdot 0 = 0$ for any a
4. $(-2)(-3)$ is the inverse (opposite) of $(-2)(3) = -6$	Since $(-2)(3) + (-2)(-3) = 0$
5. 6 is the inverse (opposite) of -6	Since $6 + (-6) = 0$
6. $(-2)(-3) = 6$	Since $(-2)(-3)$ and 6 are both inverses (opposites) of $(-2)(3)$, they must be equal (Axiom F-9)

This procedure can be generalized to show that if a and b are any positive numbers, $(-a)(-b) = ab$ – i.e., that the product of two negative numbers is a positive number. This discussion is summarized in the following theorem.

THEOREM 1.8

The product of two nonzero numbers with like signs is positive. The product of two nonzero numbers with unlike signs is negative.

Thus, according to Theorem 1.8, we have:

(a) $(4)(5) = 20$
(b) $(-5)(-4) = 20$
(c) $(-5)(4) = -20$
(d) $(5)(-4) = -20$

Example 1

Find the following products:

(a) $(3)(-5)$
(b) $(-6)(4)$
(c) $(-8)(-4)$

Solution

(a) $(3)(-5) = -15$
(b) $(-6)(4) = -24$
(c) $(-8)(-4) = 32$

If the *factors* in a product are identical, we may use *exponents* to write the product. An *exponent* is a number which tells us how many times another number, called the *base*, is to be used as a factor in an expression. Thus,

An exponent is a number which tells you how many times the base is to be used as a factor.

$5 \cdot 5 \cdot 5 = 5^3 \leftarrow$ exponent
$\uparrow$
base

5^3 is read "5 to the third power" or "5 cubed."

$6 \cdot 6 \cdot 6 = 6^3$ The 3 is the exponent, the 6 is the base
$x \cdot x \cdot x = x^3$ The 3 is the exponent, the x is the base
$7 \cdot 7 = 7^2$ The 2 is the exponent, the 7 is the base
$x \cdot x = x^2$. The 2 is the exponent, the x is the base

7^2 is read "7 to the second power" or "7 squared."

Example 2

Find:

(a) 2^4
(b) $(-2)^4$
(c) -2^4

Solution

(a) $2^4 = 2 \cdot 2 \cdot 2 \cdot 2 = 16$
(b) $(-2)^4 = (-2) \cdot (-2) \cdot (-2) \cdot (-2) = 16$
(c) $-2^4 = -2 \cdot 2 \cdot 2 \cdot 2 = -16$

Note that $(-2)^4$ and -2^4 have different meanings, so the parentheses are very important in the expression $(-2)^4$.

We are now in a position to determine the procedure for finding the quotient of two nonzero numbers.

As you recall, to check the division problem $2\overline{)6}$ (with quotient 3), we multiply $2 \cdot 3$ to obtain 6. That is, instead of saying "Divide 6 by 3," we ask, "What number must be multiplied by 2 to obtain 6?" The answer to both problems is 3. Thus, the operation of division is defined as follows:

DEFINITION 1.6

Division is a binary operation in which two real numbers a and b ($b \neq 0$) are paired with a unique third number q, called the *quotient* of a and b and denoted by $a \div b$ $\left(\text{or } \frac{a}{b}\right)$, with the property that $a = bq$; that is,

$$\frac{a}{b} = q \text{ means that } a = bq$$

The number a is called the *dividend* and b the *divisor.*

Thus,

$$\frac{8}{4} = 2 \text{ because } 8 = 4 \cdot 2$$

$$\frac{8}{-4} = -2 \text{ because } 8 = (-4)(-2)$$

$$\frac{-8}{4} = -2 \text{ because } -8 = (4)(-2)$$

$$\frac{-8}{-4} = 2 \text{ because } -8 = (-4)(2)$$

Note that in Definition 1.6 the divisor b is restricted to nonzero real numbers, for if $b = 0$ and $a \neq 0$, then there is no q such that $a = 0 \cdot q$. On the other hand, if $b = 0$ and $a = 0$, then for any q, $0 = 0 \bullet q$ and the quotient is not unique. Thus, *division by zero is not defined.* Moreover, since division is defined in terms of multiplication, we have the following Theorem:

Division by zero is not defined.

THEOREM 1.9

The quotient of two nonzero numbers with like signs is positive. The quotient of two nonzero numbers with unlike signs is negative.

Thus,

(a) $\frac{4}{2} = 2$

(b) $\frac{-4}{-2} = 2$

(c) $\frac{-4}{2} = -2$

(d) $\frac{4}{-2} = -2$

Example 3

Find the following quotients:

(a) $\frac{15}{3}$

(b) $\frac{-16}{-8}$

(c) $\frac{-36}{4}$

(d) $\frac{9}{-3}$

Solution

(a) $\frac{15}{3} = 5$

(b) $\frac{-16}{-8} = 2$

(c) $\frac{-36}{4} = -9$

(d) $\frac{9}{-3} = -3$

The quotient of two numbers a and b can also be expressed as the product of a and the reciprocal of b:

$$\frac{a}{b} = a \cdot \left(\frac{1}{b}\right)$$

This fact will be used later to divide one rational number by another by multiplying the dividend by the reciprocal of the divisor. In the meantime we see that

$$\frac{3}{4} = 3 \cdot \left(\frac{1}{4}\right)$$

$$\frac{2}{3} = 2 \cdot \left(\frac{1}{3}\right)$$

$$\text{and } \frac{5}{7} = 5 \cdot \left(\frac{1}{7}\right)$$

Example 4 Write the following quotients as the product of an integer and the reciprocal of another integer.

(a) $\frac{-2}{5}$

(b) $\frac{3}{-7}$

Solution

(a) $\frac{-2}{5} = -2 \cdot \left(\frac{1}{5}\right)$

(b) $\frac{3}{-7} = 3 \cdot \left(\frac{1}{-7}\right)$

PROGRESS TEST

1. (7) (−3) equals ______.
2. (−8) (9) equals ______.
3. (−5) (−6) equals ______.
4. $\frac{10}{-5}$ equals ______.

5. $\frac{-9}{3}$ equals ______.

6. When written as a product of an integer and the reciprocal of another integer, $\frac{-5}{6}$ equals ______.

EXERCISE 1.6

In Problems 1 through 14, find the product.

Illustration: (a) $(-2)(3)$
(b) $(-3)(-4)$

Solution: (a) $(-2)(3) = -6$
(b) $(-3)(-4) = 12$

1. $(-5)(8)$

2. $(-9)(6)$

3. $(4)(-3)$

4. $(6)(-8)$

5. $(-10)(-5)$

6. $(-6)(-9)$

7. $(-3)(4)(-5)$

8. $(-5)(2)(3)$

9. $(-4)(-2)(5)$

10. $(-2)(-5)(9)$

11. $(-3)(5)(-2)$

12. $(-3)(10)(-2)$

13. $(4)(-5)(2)$

14. $(10)(-3)(6)$

In Problems 15 through 24, first write each product without using exponents and then multiply out.

Illustration: (a) $(-3)^2$
(b) -3^2

Solution: (a) $(-3)^2 = (-3) \cdot (-3) = 9$
(b) $-3^2 = -3 \cdot 3 = -9$

15. $(-4)^2$

16. -4^2

17. -5^3

18. $(-5)^3$

19. -6^4

20. $(-6)^4$

21. $2^3 \cdot (-2)^2$

22. $2^2 \cdot (-2)^3$

23. $3^3 \cdot 3^2$

24. $-3^3 \cdot 3^2 \cdot 3$

PROGRESS TEST ANSWERS

1. -21
2. -72
3. 30
4. -2
5. -3
6. $-5 \cdot \left(\frac{1}{6}\right)$

In Problems 25 through 40, find the quotient.

Illustration: (a) $\frac{-8}{4}$

(b) $\frac{-16}{-4}$

Solution: (a) $\frac{-8}{4} = -2$

(b) $\frac{-16}{-4} = 4$

25. $\frac{-18}{9}$

26. $\frac{-32}{16}$

27. $\frac{20}{-5}$

28. $\frac{36}{-3}$

29. $\frac{-14}{-7}$

30. $\frac{-24}{-8}$

31. $\frac{0}{-3}$

32. $\frac{0}{-9}$

33. $\frac{4}{0}$

34. $\frac{-7}{0}$

35. $-\left(\frac{-4}{-2}\right)$

36. $-\left(\frac{-10}{-5}\right)$

37. $-\left(\frac{-27}{3}\right)$

38. $-\left(\frac{-9}{3}\right)$

39. $-\left(\frac{15}{-5}\right)$

40. $-\left(\frac{18}{-6}\right)$

In Problems 41 through 46, use definition 1.6 to rewrite the given quotients in the form $a = bq$.

Illustration: $\frac{-9}{3} = -3$

Solution: $\frac{-9}{3} = -3$ means that $-9 = (3)(-3)$

41. $\frac{-3}{-3} = 1$

42. $\frac{-18}{-9} = 2$

43. $\frac{-16}{4} = -4$

44. $\frac{-48}{6} = -8$

45. $\frac{-56}{8} = -7$

46. $\frac{-54}{6} = -9$

In Problems 47 through 60, rewrite each quotient as the product of an integer and the reciprocal of another integer.

Illustration: $\frac{8}{9}$

Solution: $\frac{8}{9} = 8 \cdot \left(\frac{1}{9}\right)$

47. $\frac{5}{7}$

48. $\frac{6}{11}$

49. $\frac{3}{8}$

50. $\frac{7}{10}$

51. $\frac{41}{43}$

52. $\frac{11}{100}$

53. $\frac{-7}{11}$

54. $\frac{-8}{19}$

55. $\frac{-2}{7}$

56. $\frac{-7}{31}$

57. $\frac{4}{-3}$

58. $\frac{7}{-9}$

59. $\frac{5}{-6}$

60. $\frac{2}{-9}$

Problems 61 through 65 require multiplication for their solution.

61. The largest vehicle in the world was the U.S. Overland Train Mk. II (Guinness). This vehicle had 9 times as many wheels as an ordinary 6-wheel truck. How many wheels did the U.S. Overland Train Mk. II have?

62. The vehicle of Problem 61 was 26 times as long as a regular 22-foot-long truck. How long was the vehicle?

63. Sir John Robertson, who was five times Prime Minister of New South Wales, Australia, bought three pints of rum every morning for 35 years. Assuming that a year has 365 days, how many pints of rum did he buy? By the way, he did not drink it all himself! He drank one pint on the spot, gave one pint to his horse, and poured the third pint into his riding boot as a preventive against rheumatism.

64. Can you pull 90 times your own weight? The *Passalid* beetle can (Guinness). If the beetle's weight were to reach 2 ounces and it could maintain its phenomenal strength, how much (in ounces) could it pull?

65. When driving on a highway, you should stay about 18 feet from the car ahead of you for each 10 miles per hour of your speed. Find the distance (in feet) that you should keep from the car ahead when traveling:
(a) 30 miles per hour
(b) 40 miles per hour
(c) 55 miles per hour

Problems 66 through 70 require division for their solution.

66. The Price/Earning (P/E) ratio of a stock is defined as the market price of the stock divided by its actual or indicated annual earning per share. If the market value of a share of American Construction is $54 and the stock earned $9 per share, find the P/E ratio for the stock.

67. Do you know what your I.Q. (intelligence quotient) is? The highest I.Q. ever recorded is that of Kim Ung-Yong of South Korea. At the age of 4 years and 8 months, he had an I.Q. of 210 (Guinness). The I.Q. of a child is computed by dividing the mental age (M.A.) of the child by the chronological age (C.A.) and then multiplying by 100. If the mental age of a child is 8 years and his chronological age is 5 years, what is his I.Q.?

68. The highest recorded shorthand writing speed under championship conditions is 1,500 words in five minutes (Guinness). How many words per minute is that?

69. The highest rate ever offered to a writer was \$30,000 to Ernest Hemingway for a 2,000-word article on bullfighting by Sports Illustrated in January, 1960 (Guinness). What was the rate per word?

70. The highest price ever paid for a bottle of wine of any size was 55,000 francs (\$13,200) in Paris, for a jeroboam of Chateau Mouton Rothchild 1870 (Guinness). The bottle contained approximately 30 glasses of wine; in dollars, what was the price of each glass of wine?

OBJECTIVES 1.7

After studying this section, the reader should be able to:

1. **Evaluate a numerical expression involving two or more operations and one or more sets of grouping symbols.**
2. **Determine the correct order in which operations should be performed in an expression containing two or more of the fundamental operations.**

1.7 ORDER OF OPERATIONS

Imagine that you were given this clue to solving a problem: "The answer can be found by taking four plus seven multiplied by two." Do you think this would be a helpful hint? Actually, it isn't, because the sentence is ambiguous. It could mean four and seven should be added together and the result multiplied by two; it could just as easily mean that four should be added to the product of seven and two. Written in numerical form, our ambiguous sentence would look like this: $4 + 7 \cdot 2$. How can we clarify in what order we should perform the indicated operations? In written English, we would modify the wording of our sentence—for example, it could be changed to read "The answer can be found by first taking four plus seven and then multiplying this sum by two." In mathematics, we indicate the order of operations with parentheses. The expression $4 + 7 \cdot 2$, written to match our changed sentence, is arranged as follows:

$$(4 + 7) \cdot 2$$

In the expression $(4 + 7) \cdot 2$, we multiply 4 by 7 first.

The parentheses in the expression $(4 + 7) \cdot 2$ indicate that addition should be performed *first*, followed by multiplication. If we wished to multiply first and then add, we would use parentheses as shown:

$$4 + (7 \cdot 2)$$

In the expression $4 + (7 \cdot 2)$, we multiply 7 by 2 first.

As you can see, the parentheses *group* together certain terms in our original expression. We have grouped the terms in two different ways, and because the parentheses indicate order of operation, we have two different solutions, as shown below:

$(4 + 7) \cdot 2$	$4 + (7 \cdot 2)$
$11 \cdot 2$	$4 + 14$
22	18

From this discussion we can conclude that the *parentheses are grouping symbols* that are used to indicate which operations are to be performed first. The square brackets [ ] and the braces { } are also grouping symbols which can be used in the same manner as the parentheses. Thus,

The parentheses are grouping symbols that tell you which operations should be performed first.

$$4 \cdot (3+2),\ 4 \cdot [3+2],\ \text{and } 4 \cdot \{3+2\}$$

all mean that we must first add 3 and 2, and then multiply this sum by 4.

Example 1

Evaluate:
(a) $(4 \cdot 5) + 6$
(b) $4 \cdot (5+6)$

Solution

(a) $(4 \cdot 5) + 6$
$20 + 6$
26
(b) $4 \cdot (5+6)$
$4 \cdot 11$
44

The placement of parentheses is equally important in evaluating numerical expressions involving multiplication and division. For example, the expressions $(32 \div 4) \cdot 2$ and $32 \div (4 \cdot 2)$ have different evaluations. Thus,

$(32 \div 4) \cdot 2$	$32 \div (4 \cdot 2)$
$8 \cdot 2$	$32 \div 8$
16	4

Example 2

Evaluate:
(a) $48 \div (4 \cdot 3)$
(b) $(48 \div 4) \cdot 3$

Solution

(a) $48 \div (4 \cdot 3)$
$48 \div 12$
4
(b) $(48 \div 4) \cdot 3$
$12 \cdot 3$
36

The numerical expressions discussed so far include only three numbers. There are other expressions which may contain more than three numbers. For example, in the accompanying ad, a mattress and boxspring and a matching bedspread cost \$88 + \$14. If Ms. Perez decides to buy two of each and, in addition, a lamp, her purchase (in dollars) will amount to $[2 \cdot (88 + 14)] + 12$. Since only two numbers may be added at one time, the first step in the evaluation of $[2 \cdot (88 + 14)] + 12$ is to find the sum of 88 and 14. The evaluation is shown below.

$$\begin{aligned} [2 \cdot (88 + 14)] + 12 \\ [2 \cdot (102)] + 12 \\ 204 + 12 \\ 216 \end{aligned}$$

Thus, her purchase will amount to \$216.

Example 3 Evaluate $[5 \cdot (6 + 4)] + 9$

Solution

$$\begin{aligned} [5 \cdot (6 + 4)] + 9 \\ [5 \cdot 10] + 9 \\ 50 + 9 \\ 59 \end{aligned}$$

Expressions involving subtraction can be evaluated similarly. Thus, $[5 \cdot (9 - 2)] - 5$ is evaluated as follows:

$$\begin{aligned} [5 \cdot (9 - 2)] - 5 \\ [5 \cdot 7] - 5 \\ 35 - 5 \\ 30 \end{aligned}$$

Example 4

Evaluate $[10 \cdot (8-3)] - 9$

Solution

$[10 \cdot (8-3)] - 9$
$[10 \cdot 5] - 9$
$50 - 9$
41

In some cases we have to evaluate expressions in which a bar is used to indicate division. For instance, if we wish to convert a temperature which is given in degrees Fahrenheit to degrees Celsius, we have to evaluate the expression

$\frac{a}{b}$ means "a divided by b."

$$\frac{5 \cdot (F-32)}{9}$$

where F represents the temperature in degrees Fahrenheit. Thus, if the temperature is 77° F, the corresponding Celsius temperature is

Degrees Celsius are also called degrees Centigrade.

$$\frac{5 \cdot (77-32)}{9}$$

$$\frac{5 \cdot (45)}{9}$$

$$\frac{225}{9}$$

$$25°$$

You can also do this by dividing 45 by 9 first, and then multiplying the result, 5, by 5, obtaining 25.

Example 5

In September, 1933, a freak heat flash struck the city of Coimbra, in Portugal. On this day, the temperature rose to 158° F for 120 seconds. How many degrees Celsius is that?

Solution

In this case F = 158, thus the Celsius temperature is:

$$\frac{5 \cdot (158-32)}{9}$$

$$\frac{5 \cdot (126)}{9}$$

$$\frac{630}{9}$$

70° Celsius

You can also do this by dividing 126 by 9 first, and then multiplying the result, 14, by 5, obtaining 70.

Finally, if an expression does not contain parentheses or brackets, we must establish an order in performing mathematical operations. For example, the expression $6 + 9 \div 3$ can be evaluated in two ways. If we add 6 and 9 first and divide the result by 3, the answer is 5. On the other hand, if we divide 9 by 3 first, and then add 6 to the result, the answer is 9. In order to avoid this ambiguity, we agree to perform any sequence of operations in the following order.

1. Simplify expressions by performing first the operations inside any grouping symbols, starting with the innermost grouping symbols and the operations above and below division bars.
2. Perform multiplications and divisions as they occur from left to right.
3. Perform additions and subtractions as they occur from left to right.

With this convention, the expression $6 + 9 \div 3$ is evaluated as follows:

$$6 + 9 \div 3$$
$$6 + 3$$
$$9$$

Example 6

Evaluate the expression $6 \cdot 2 + \frac{(4-8)}{2} + 10 \div 5$.

Solution

$$6 \cdot 2 + \frac{(4-8)}{2} + 10 \div 5$$

$$6 \cdot 2 + \frac{-4}{2} + 10 \div 5$$ Performing the operation inside parentheses

$$12 + (-2) + 2$$ Performing multiplications and divisions as they occur from left to right

$$10 + 2$$ Since $12 + (-2) = 10$

$$12$$ Since $10 + 2 = 12$

PROGRESS TEST

1. When evaluated, $(5 + 2) \bullet 7$ equals ______.
2. When evaluated, $5 + (2 \bullet 7)$ equals ______.
3. When evaluated, $(54 \div 6) \bullet 3$ equals ______.
4. When evaluated, $54 \div (6 \bullet 3)$ equals ______.
5. When evaluated, $8 \bullet (17 + 3) + 2$ equals ______.
6. When evaluated, $2 \bullet (14 - 4) - 2$ equals ______.
7. When evaluated, $7 - 5 \bullet 8 \div 5 + 2$ equals ______.
8. When evaluated, $6 + \frac{(8-4)}{-2} + 3 \bullet 2$ equals ______.

EXERCISE 1.7

In Problems 1 through 12, evaluate the given expression.

Illustration: $4(3 - 1) - 2$

Solution:
$$\begin{aligned} &4(3-1)-2\\ &4(2)-2\\ &8-2\\ &6 \end{aligned}$$

1. (a) $(10 \bullet 3) + 4$
(b) $10 \bullet (3 + 4)$

2. (a) $(6 \bullet 4) + 6$
(b) $6 \bullet (4 + 6)$

3. (a) $(36 \div 4) \bullet 3$
(b) $36 \div (4 \bullet 3)$

4. (a) $(28 \div 7) \bullet 2$
(b) $28 \div (7 \bullet 2)$

5. $[5 \bullet (8 + 2)] + 3$

6. $[7 \bullet (4 + 3)] + 1$

7. $7 + [3 \bullet (4 + 5)]$

8. $8 + [3 \bullet (4 + 1)]$

9. $[6 \bullet (4 - 2)] - 3$

10. $[2(7 - 5)] - 8$

PROGRESS TEST ANSWERS

1. 49
2. 19
3. 27
4. 3
5. 162
6. 18
7. 1
8. 10

11. $3 - [8 \cdot (5 - 3)]$

12. $7 - [3(4 - 5)]$

In Problems 13 through 30, evaluate the given expression.

Illustration: $\dfrac{3(5-1)}{2} + 4 \div 2$

Solution: $\dfrac{3(5-1)}{2} + 4 \div 2$

$$\frac{3 \cdot 4}{2} + 4 \div 2$$

$$\frac{12}{2} + 4 \div 2$$

$$6 + 2$$

$$8$$

13. $5 \cdot 6 - 6$

14. $5 \cdot 2 - 2$

15. $7 \cdot 3 \div 3 - 3$

16. $36 \cdot 2 \div 18 - 4$

17. $(20 - 5 + 9 \div 3) \div 6$

18. $(10 - 2 + 10 \div 5) \cdot 4$

19. $\dfrac{8 + (-3)}{5} - 1$

20. $\dfrac{7 + (-3)}{2} - 4$

21. $\dfrac{4 \cdot (6 - 2)}{-8} - \dfrac{6}{-2}$

22. $\dfrac{5 \cdot (6 - 2)}{-4} - \dfrac{16}{-4}$

23. $8[3 - 2(4 + 1)] + 1$

24. $6[7 - 2(5 - 7)] - 2$

25. $48 \div [4(8 - 2[3 - 1])]$

26. $96 \div [4(8 - 2[1 - 3])]$

27. $\left[\dfrac{9-(-3)}{8-6}\right]\left[\dfrac{3+(-8)}{7-2}\right]$

28. $\left[\dfrac{6+(-2)}{3+(-7)}\right]\left[\dfrac{8+(-12)}{2-4}\right]$

29. $\dfrac{3-5\left(\dfrac{4+2}{2+1}\right)-2}{-4+3\left(\dfrac{4-2}{4-6}\right)-2}$

30. $\dfrac{8+2\left(\dfrac{9-15}{3-1}\right)-2}{-4+8\left(\dfrac{6-3}{1-4}\right)+12}$

In Problems 31 through 40, evaluate each expression for the given values of the letters.

Illustration: $\frac{9}{5}C+32;\ C=15$

Solution: $\frac{9}{5}\cdot 15+32$

$27+32$

59

31. $\dfrac{5(F-32)}{9};\ F=50$

32. $RT;\ R=50,\ T=2$

33. $\dfrac{R+r}{r};\ R=10,\ r=5$

34. $IR;\ I=10,\ R=5$

35. $2W+2L;\ W=4,\ L=5$

36. $LW;\ L=3,\ W=7$

37. $\dfrac{ab}{a+b};\ a=10,\ b=15$

38. $\dfrac{I-O}{I};\ I=15,\ O=3$

39. $5H-190;\ H=60$

40. $\dfrac{1bh}{3};\ b=9,\ h=2$

SELF-TEST—CHAPTER 1

1. Use braces to list the elements of the given sets.

______ (a) The natural numbers between 4 and 10.

______ (b) The odd natural numbers less than 10.

______ (c) The set of letters in the word "Tallahassee."

---

2. Write the given set using set builder notation.

______ (a) $\{1, 2, 3, 4\}$

______ (b) $\{a, e, i, o, u\}$

---

3. Let $A = \{2, 4, 6, 8, 10\}$, $B = \{1, 2, 3, 4, 5\}$, $C = \{1, 3, 5, 7, 9\}$. Find:

______ (a) $A \cap C$

______ (b) $A \cup B$

______ (c) $(A \cap B) \cup C$

---

4. Classify each number as rational (Ra) or irrational (Ir). If the number is rational, write it in the form $\frac{a}{b}$.

______ (a) $\sqrt{2}$

______ (b) $0.\overline{31}$

______ (c) $0.0101001000\ldots$

______ (d) 3.14

______ (e) 0.41

---

5. Each of the given statements is a consequence of one of the axioms (or laws) we have studied. In each case, state the name of the axiom (or law).

______ (a) $a + (b + c) = (a + b) + c$

______ (b) $3 + (-3) = 0$

______ (c) $3(4 + 5) = 3 \cdot 4 + 3 \cdot 5$

---

6. Find:

______ (a) $|-4|$

______ (b) $|8|$

______ (c) $-|-2|$

7. Find the indicated sum or difference.

_______________ (a) $4-(-2)$

_______________ (b) $3-(4+2-9)$

_______________ (c) $-3+(-3)-(-5)$

---

8. Find the indicated product or quotient.

_______________ (a) $(-8)(-3)$

_______________ (b) $(8)(-7)$

_______________ (c) $\frac{-64}{-4}$

_______________ (d) $\frac{81}{-3}$

_______________ (e) $\frac{2+(-8)}{3}+\frac{18}{-9}$

---

9. Carry out the indicated operations:

_______________ (a) $\frac{4(3-2)}{-4}-\frac{27}{-3}$

_______________ (b) $\left[\frac{7+(-2)}{-2-3}\right]\left[\frac{8+(-12)}{2-4}\right]$

---

10. Carry out the indicated operations:

_______________ (a) $8-3\cdot 4\div 3+2$

_______________ (b) $(8+3\cdot 2\div 2-4)\div 7$

CHAPTER 2

HISTORICAL NOTE

In this chapter we shall learn how to solve equations by using certain techniques discussed in the text. Actually, the solution of algebraic equations is not a new idea. The Rhind Papyrus, mentioned in the historical note accompanying Chapter 1, contains one of the first algebra problems known to have been solved by man. The problem, which appears in the hieroglyphics of the papyrus, can be translated as follows:

Aha, its whole, its seventh, it makes 19

The significance of the document is its exhibition of the first approach to algebraic symbolism; a standard character is used for "aha," meaning "quantity," representing the unknown to be found. Many of the problems in the Rhind Papyrus indicate that as early as 1700 B.C. men and women were concerned with the kinds of problems that we solve with algebra today. These problems deal with questions regarding the strength of ingredients in bread and beer, with feed mixtures for cattle, and with the storage of grain.

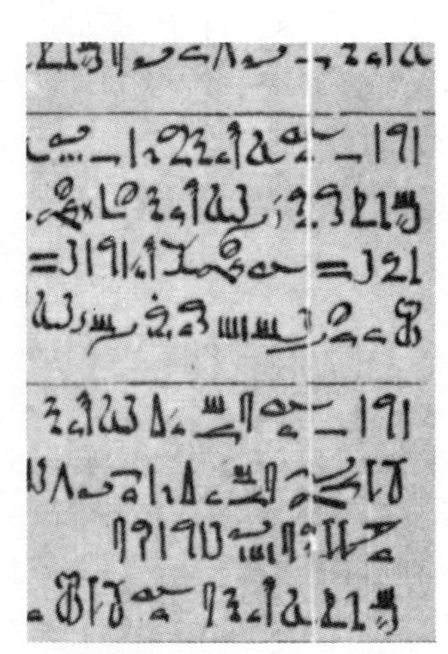

The Rhind Papyrus (detail) is one of the earliest works on algebra. It was prepared by Ahmes about 1650 B.C. (Courtesy of the Trustees of the British Museum.)

The Rhind Papyrus also contains symbols to indicate *plus* and *minus.* The first of these symbols is represented by a pair of legs walking from right to left, the normal direction of Egyptian writing, while the minus sign is shown by a pair of legs walking from left to right, opposite to the direction of Egyptian writing.

The Arabs also made many contributions to the study of algebra. They used positive and negative symbols and developed fractions such as we use today. Much of their work was collected and improved by al-Khowârizmî, a teacher in the mathematical school of Baghdad. The arabic word *algebra* was actually derived from the title of his work, Hisab al-jabr w'al-muqabalah, which can be translated as "the science of the reunion and the opposition."

FIRST-DEGREE EQUATIONS IN ONE VARIABLE

Tomb of Omar Khayyam in Nishapur. Khayyam, a Persian poet and astronomer, wrote one of the first algebra books.

TIGER

OBJECTIVES 2.1

After studying this section the reader should be able to:
1. **Classify a given sentence as an open (algebraic) sentence or as a statement.**
2. **Find the number of terms in a given algebraic expression.**
3. **Identify the terms and the coefficients in a given algebraic expression.**
4. **Simplify an algebraic expression.**

2.1 ALGEBRAIC EXPRESSIONS

x, y, and z are called variables.

It has been said before that mathematics is the language of science. As such, it has many things in common with other languages. A language has pronouns, such as *he, she,* and *it,* to be used in place of nouns. In algebra, we use *variables,* such as x, y, and z, in place of numbers. A language has verbs, which indicate actions. In algebra, actions are expressed by using operations. In a language, we put words together to form sentences. In algebra, we use verbs and numbers to form word sentences, such as "Two plus two equals four" shown in the cartoon at the beginning of this section. Here are some other word sentences:

(1) Two plus four equals six.
(2) The difference of five and two is three.
(3) Three multiplied by zero is three.

These word sentences can be classified as *true* or *false.* Thus, (1) and (2) are true, but (3) is false.

A statement is a sentence which can be classified as true or false.

In algebra, a sentence which can be classified as true or false (but not both) is called a *statement.* Thus, the word sentences (1), (2), and (3) are *statements.* On the other hand, if x represents a real number, expressions (4), (5), (6), and (7) are *not* statements, because they are not even sentences.

(4) $x + 9$
(5) $9x^2 - 2x - 5$
(6) $(x - 9) \div 4$
(7) $3 + 5$

These expressions are called *algebraic*, or *open*, expressions and are defined as follows:

DEFINITION 2.1

An algebraic (or open) expression is a meaningful collection of numbers, variables representing real numbers, and signs of operation.

When an algebraic expression is written in the form $A + B + C + \ldots$, A, B, C, etc. are called the *terms* of the expression. Thus, the terms of the expression $x + 9$ are x and 9. Moreover, the numerical *factors* of a term are called the *coefficients* of the term. Thus, the coefficients in $9x^2 - 2$ are 9 and -2, respectively. If no other coefficient appears in a term, the coefficient is assumed to be 1. For example, in the expression $x + y^2 + z$, the coefficient of x is 1, the coefficient of y^2 is 1, and the coefficient of z is also 1.

In the expression $A + B + C \ldots$, A, B, and C are called terms. In the expression ab, a and b are factors. The coefficient of $9x^2$ is 9. The coefficient of -2 is -2. In general, the coefficient of a number c is c.

The coefficient of x is understood to be 1.

Example 1

Determine the number of terms in each of the following expressions and give the coefficient of each term.
(a) $x - x^2 + 5$
(b) $(x + y)^5$

Solution

(a) $x - x^2 + 5$ has three terms. The coefficient of x is 1; the coefficient of x^2 is -1, since $x - x^2 + 5$ can be written as $x + (-x^2) + 5$; the coefficient of 5 is 5.
(b) $(x + y)^5$ is an indicated product, so it has only one term. Thus, the coefficient of $(x + y)^5$ is 1.

Since some algebraic expressions are sums of terms representing real numbers, the axioms, rules, and theorems discussed in Chapter 1 can be used to *simplify* such expressions. For example, by using the Symmetry Law of Equality, we can write the Distributive Law, $a(b + c) = ab + ac$, as $ab + ac = a(b + c)$, which, by the Commutative Law, can be expressed as $ba + ca = (b + c)a$. Thus:

$$\begin{aligned} 5x + 8x &= (5 + 8)x \\ &= 13x \end{aligned}$$

Similarly,

$$\begin{aligned} 9y - 3y - 2y &= 9y + (-3y) + (-2y) \\ &= [9 + (-3) + (-2)]y \\ &= 4y \end{aligned}$$

Terms that differ only in their numerical coefficients are like terms. For example, $3x$ and $-2x$ are like terms.

As you can see from these two illustrations, terms that differ only in their numerical coefficients can easily be combined. These terms are called *like terms,* and the simplification shown here is referred to as *combining like terms.* Of course, the expression obtained after combining like terms will have the same value as the original expression for all real numbers replacing the variable or variables involved.

Example 2

Simplify:
(a) $5x + 3x - 2x$
(b) $9y^2 - 2y^2 + 3y^2$

Solution

(a) $$\begin{aligned} 5x + 3x - 2x &= (5 + 3 - 2)x \\ &= 6x \end{aligned}$$
(b) $$\begin{aligned} 9y^2 - 2y^2 + 3y^2 &= (9 - 2 + 3)y^2 \\ &= 10y^2 \end{aligned}$$

Recall that by Theorem 1.6, $-1 \cdot a = -a$.

If there is a minus sign in front of parentheses, the parentheses can be removed by changing the signs of all the terms inside the parentheses. This is equivalent to multiplying each term inside the parentheses by -1.

Expressions of the form $a - (b + c)$, in which a set of parentheses is preceded by a negative sign, can first be written as

$$a + [-(b + c)] \quad \text{since } x - y = x + (-y)$$

Then, since $-(b + c) = -1 \cdot (b + c) = -b - c$, we have

$$a - (b + c) = a - b - c$$

Thus,

$$\begin{aligned} 9x - (x + 2) &= 9x - x - 2 \\ &= 8x - 2 \end{aligned}$$

Similarly, it can be shown that $a - (b - c) = a - b + c$. Thus,

$$\begin{aligned} 8x - (2x - 4) &= 8x - 2x + 4 \\ &= 6x + 4 \end{aligned}$$

Example 3

Simplify:
(a) $7x - (9x + 1)$
(b) $3x - (2x - 4)$

Solution

(a) $7x - (9x + 1) = 7x - 9x - 1$
$= -2x - 1$
(b) $3x - (2x - 4) = 3x - 2x + 4$
$= x + 4$

We are now ready to simplify expressions such as $2x + 5(3x + 1) + 3$. In these expressions, the first step is to remove the parentheses by using the Distributive Law. Then, similar terms are added. The procedure is as follows:

$2x + 5(3x + 1) + 3$	
$= 2x + 5 \cdot 3x + 5 \cdot 1 + 3$	By the Distributive Law
$= 2x + 15x \quad + 5 \quad + 3$	Simplifying
$= 17x + 8$	Combining like terms

Example 4

Simplify $3x - 2(x - 1) - 7$.

Solution

$3x - 2(x - 1) - 7$
$= 3x - 2x - 2 \cdot (-1) - 7$
$= 3x - 2x + 2 - 7$
$= x - 5$

Example 5

A chemist concentrated a solution by boiling x milliliters and thus evaporating off two milliters.
(a) Write an algebraic expression representing the volume of solution left.
(b) After boiling the solution, the chemist doubles it by adding an equal volume of the original solution. Write an algebraic expression for the volume of solution she now has, and then simplify this expression.

Solution

(a) The amount left is $x - 2$ milliliters.
(b) If she doubles the amount of solution, she has $2(x-2) = 2x-4$ milliliters of solution.

PROGRESS TEST

1. The number of terms in the expression $x^2 + x - 2$ is ______, and the coefficient of x^2 is ______.
2. When simplified, $7x + 3x - 12x$ equals ______.
3. When simplified, $3x - (2x + 1)$ equals ______.
4. When simplified, $5x - 3(x - 2) + 4$ equals ______.

EXERCISE 2.1

In Problems 1 through 10, determine the number of terms in the given expression and give the coefficient of each term.

Illustration: (a) $x^2 + 3x$
(b) $4(x + 1)$

Solution: (a) $x^2 + 3x$ has two terms. The coefficient of x^2 is 1, and the coefficient of $3x$ is 3.
(b) $4(x + 1)$ has one term. The coefficient of $(x + 1)$ is 4.

1. $x + 2$

2. $x^2 - 7x + 1$

3. $2x - x^2$

4. $3x + 3$

5. $3(x + 1)$

6. $2(x + 1)^2$

7. $(x + 2)^2 + 2$

8. $2(x + 1) - (x + 1)^2$

9. $a(b + c)$

10. $ab + ac$

PROGRESS TEST ANSWERS

1. 3, 1
2. $-2x$
3. $x - 1$
4. $2x + 10$

In Problems 11 through 30, simplify the given expression.

Illustration: $4(x - 1) + 3x$

Solution:

$4(x - 1) + 3x$

$= 4x - 4 + 3x$ By the Distributive Law

$= 7x - 4$ Combining like terms

11. $9x + 7x$

12. $8x - 3x$

13. $3y - 7y$

14. $3y - 2y + 5y$

15. $5x - 7x + 2x$

16. $8x - 10x + 11x$

17. $3xy - 2xy + 5xy$

18. $7xy - xy + 3xy$

19. $3x - (2x + 1)$

20. $5x - (3x + 4)$

21. $8x - (5x - 2)$

22. $9x - (6x - 5)$

23. $3x + 3(x + 1)$

24. $5x + 2(x + 2)$

25. $5(x - 1) - 3$

26. $6(x - 3) - 10$

27. $5x - 2(x - 1) + 3x$

28. $9x - 4(x - 1) - 3x$

29. $x - [3x + (1 - x)]$

30. $5x - [2x - 2(x + 1)]$

31. During a child's period of growth, the size of the head doubles. If h is the initial size, write an algebraic expression representing the new size.

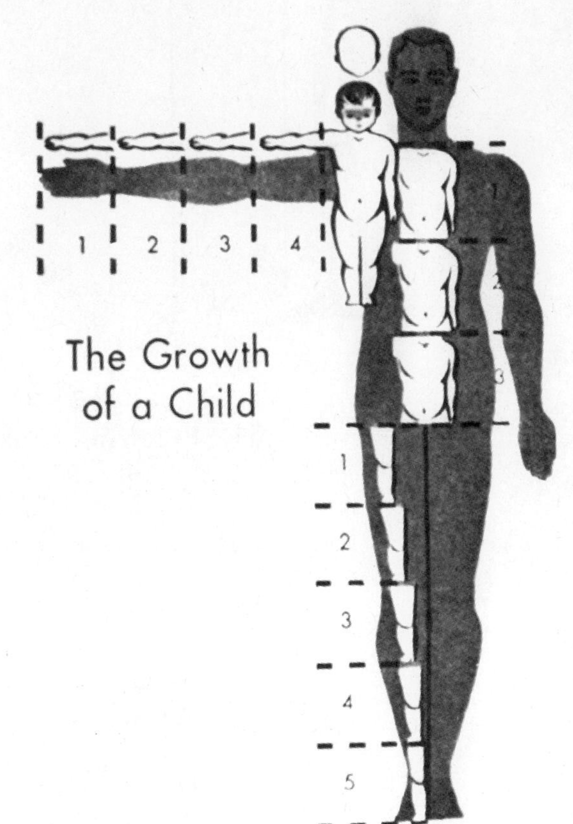

Kahn, F.: Man in *Structure and Function.* Volume 1. New York: Alfred A. Knopf, Inc.

32. During a child's period of growth, the legs become five times as long as they were originally. If L is the original length, write an algebraic expression representing the new length.

33. Anthropologists can determine a person's height in life by using his or her unearthed bones as a clue. For example, the height of a man (in centimeters) with a humerus bone of length h can be obtained by multiplying 2.89 by the length h and adding 70.64 to the result. Write an algebraic expression representing the height of this person.

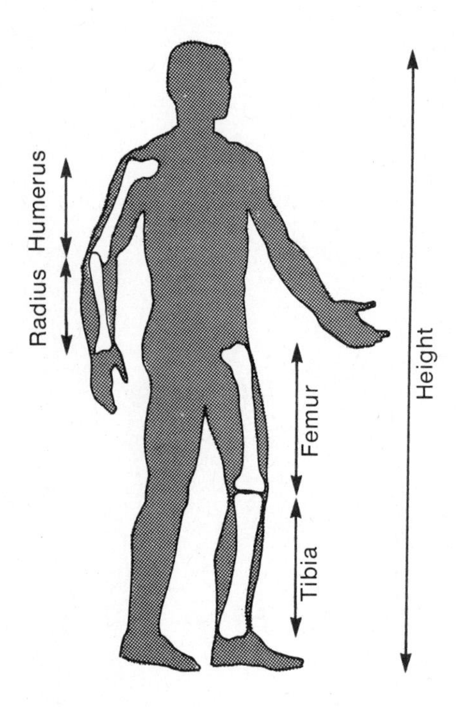

34. Referring to Problem 33, anthropologists multiply the length h of a humerus by 2.75 and add 71.48 to find the height of a female (in centimeters). Write an algebraic expression representing the height of the female.

35. The number of passengers in a 747 plane is 83 more than in a 707 plane carrying p passengers. Write an algebraic expression giving the number of passengers in the 747.

36. The total number of persons on board a plane is obtained by adding the number of passengers p and the

13 crew members. Write an algebraic expression representing this number.

37. The cost of an article is found by deducting the tax t from the total amount a paid for the article. Write an algebraic expression representing the cost of the article.

38. The numbers of hours a growing child should sleep is determined by finding the difference between a constant, 17, and one half the age a of the child. Write an algebraic expression to represent this quantity.

39. The weight of a person can be estimated by finding the difference between 5 times the height h of the person (in inches) and 190. Write an algebraic expression to represent the weight.

40. The cephalic index is obtained by dividing the product of 100 and the width w of a head by the length L of the head. Write an algebraic expression representing this index.

41. The intelligence quotient is obtained by dividing 100 times the mental age M by the chronological age C. Write an algebraic expression representing the intelligence quotient.

OBJECTIVES 2.2

After studying this section, the reader should be able to:
1. **Determine the replacement set for a variable being used in an equation.**
2. **Classify a given equation as a conditional equation or an identity.**
3. **Determine the solution set of a given conditional equation.**
4. **Determine if a given number is a solution of a given equation.**

2.2 EQUATIONS IN ONE VARIABLE

In the cartoon at the beginning of this section, Sally is about to begin the study of equations. In Chapter 1, we learned that the equality $a = b$, where a and b are real numbers, indicates that a and b represent the same number. In this section, we shall combine algebraic and numerical expressions with verb phrases to form sentences such as:

(1) $8 + x = 7 - 2$
(2) $x + 9 = 2x$
(3) $x^2 = 1$

x^2 means $x \cdot x$.

These sentences are not statements, since they can not be classified as true or false. However, because of the close relationship between algebraic (open) expressions and the sentences given in (1), (2), and (3), these sentences are called *open sentences.* An open sentence which uses the verb "is equal to" is called an *equation.* Thus, (1), (2), and (3) are examples of equations. We make this idea precise in the following definition.

$8 + x = 7 - 2$ and $x + 9 = 2x$ are open sentences. An equation is an open sentence which uses the verb "is equal to."

DEFINITION 2.2

An equation in the variable x is a statement of the form

$$E = F$$

where E and F are algebraic or numerical expressions at least one of which contains x.

Thus the following are equations:

(4) $x + 9 = 10$

(5) $x + 8 = 8 + x$
(6) $x + 1 = x + 2$
(7) $\frac{x}{x} = 1$

The replacement set of a variable, that is, the numbers which can replace the variable in the equation, is assumed to be the set of real numbers for which the algebraic expressions in the equation are defined. Thus, in Equations (4) through (6), the replacement set is the set of real numbers, since each of the algebraic expressions involved is defined for any real number. On the other hand, the left member of Equation (7) is not defined for $x = 0$; hence the replacement set of Equation (7) is the set of all real numbers *except* 0.

The replacement set of a variable is the set of numbers for which the algebraic expressions in the equation are defined.

Example 1

Find the replacement set for each of the following equations:

(a) $\frac{3}{x-1} = 5$

(b) $5 + \frac{1}{x} = \frac{2}{x+2}$

Solution

(a) The algebraic expression $\frac{3}{x-1}$ is defined for all real numbers except 1. Thus, the replacement set of $\frac{3}{x-1} = 5$ is the set of all real numbers except 1.

(b) The algebraic expressions $\frac{1}{x}$ and $\frac{2}{x+2}$ are not defined for $x = 0$ and $x = -2$ respectively. Thus, the replacement set of $5 + \frac{1}{x} = \frac{2}{x+2}$ is the set of all real numbers except 0 and -2.

We have now established that there are certain numbers that can be used as the replacement set for a variable. It may be that certain elements in this replacement set make the equation a true statement when replaced for the variable. For example, if x is replaced by 1 in the equation $x + 9 = 10$, the resulting statement $1 + 9 = 10$ is true. For this reason, we say that the *solution* (or root) of the equation $x + 9 = 10$ is 1. A number (such as 1) that is a solution of an equation is

The solution of $x + 9 = 10$ is 1.

said to *satisfy* the equation. Moreover, the set of all solutions of an equation is called the *solution set* of the equation. Thus, the number 1 satisfies the equation $x + 9 = 10$, and consequently the solution set of $x + 9 = 10$ is $\{1\}$.

1 satisfies the equation $x + 9 = 10$, since $1 + 9 = 10$.

In Equation (5), x can be replaced by any real number and the result will be a true statement. Thus, the solution set of $x + 8 = 8 + x$ is the set of all real numbers. On the other hand, there is no real number which can be substituted for x in Equation (6) to obtain a true statement, so the solution set of the equation $x + 1 = x + 2$ is the empty set, $\emptyset$.

In Equation (7), any number *except* 0 will satisfy the equation. Therefore, the solution set of $\frac{x}{x} = 1$ is the set of all real numbers except 0.

We can summarize this discussion by pointing out that the equations $x + 9 = 10$ and $x + 1 = x + 2$ are false statements for at least one element in the replacement set. Such equations are called *conditional equations.* On the other hand, the equations $x + 8 = 8 + x$ and $\frac{x}{x} = 1$ are true statements for every element in the replacement set. Such equations are called *identities.* To reiterate, $x + 9 = 10$ and $x + 1 = x + 2$ are *conditional* equations, and $x + 8 = 8 + x$ and $\frac{x}{x} = 1$ are *identities.*

$x + 9 = 10$ and $x + 1 = x + 2$ are conditional equations. $x + 8 = 8 + x$ and $\frac{x}{x} = 1$ are identities.

Example 2

Determine the solution of each conditional equation. If the equation is an identity, so state.

(a) $2x + 1 = 3$

(b) $2(x + 1) = 2x + 2$

Solution

(a) Since $2x + 1 = 3$, $2x$ must be 2, and the solution must be 1.

(b) Since, by the Distributive Law, $2(x + 1) = 2x + 2$, all real numbers satisfy the equation. Thus, $2(x + 1) = 2x + 2$ is an identity.

Example 3

Anthropologists can determine if a humerus bone could have come from a certain male by finding whether the length of the bone satisfies the equation $H = 2.89h + 70.64$, where H is the height of the man (in centimeters) and h is the length of the humerus (in centimeters). The body of a male measuring 175 centimeters and a 30-centimeter humerus are found. Can the humerus belong to this body?

Solution

If the bone belongs to the body, $h = 30$ will satisfy the equation $175 = 2.89h + 70.64$. But $(2.89)(30) + 70.64 = 157.34$, and thus the bone cannot belong to this body.

PROGRESS TEST

1. The replacement set for the equation $x + \frac{x}{5 - x} = 3$ is ______________.
2. The solution set of the equation $5x + 1 = 11$ is ______.
3. The solution set of the equation $x + 1 = 1 + x$ is ______________.
4. The solution set of the equation $x - 1 = x - 2$ is ______.
5. The equation $x + 1 = 1 + x$ is called an ______________.
6. The equation $5x + 1 = 11$ is called a ______________ equation.
7. If the humerus in Example 3 is found to be 36.11 centimeters, could the bone come from the 175 cm. body found with it?

EXERCISE 2.2

In Problems 1 through 6, find the replacement set of the given equation.

Illustration: $\frac{7}{x + 9} = 2$

Solution: The replacement set of $\frac{7}{x + 9} = 2$ is the set of all real numbers except -9.

1. $\frac{4}{x + 8} = 7$

2. $\frac{3(x + 1)}{x + 2} = 7x$

3. $\frac{5}{x - 3} = 9$

4. $\frac{x}{x - 1} = \frac{2}{x + 4}$

PROGRESS TEST ANSWERS

1. The set of all real numbers except 5.
2. $\{2\}$
3. The set of real numbers.
4. $\emptyset$
5. identity
6. conditional
7. Yes. $(2.89)(36.11) + 70.64 = 174.9979$

5. $\dfrac{3}{(x-1)(x+1)} = \dfrac{5}{x-1}$

6. $\dfrac{8x}{x+4} = \dfrac{9x}{x+5}$

In Problems 7 through 16, determine if the given number is a solution of the given equation.

Illustration: (a) $-2;\ x+2=0$
(b) $-3;\ -x+3=0$

Solution: (a) If -2 is substituted for x in the equation $x+2=0$, we obtain $-2+2=0$, a true statement. Thus, -2 is a solution of $x+2=0$.
(b) Since $-(-3)+3=6$, -3 is *not* a solution of $-x+3=0$.

7. $3;\ 2x+8=14$

8. $2;\ 5x+5=10$

9. $-1;\ -2x+1=3$

10. $-2;\ -3x+4=10$

11. $3;\ 2y-5=y-2$

12. $4;\ 3y-7=y+1$

13. $\dfrac{-1}{5};\ \dfrac{4}{5}t-1=5t$

14. $\dfrac{-1}{3};\ 6t+1=t+\dfrac{2}{3}$

15. $3;\ \dfrac{1}{2}x+5=5-\dfrac{1}{3}x$

16. $3;\ \dfrac{-1}{3}x+1=x-3$

In Problems 17 through 26, find the solution of each conditional equation. If the equation is an identity, so state.

Illustration: (a) $x+2=4$
(b) $-(x+1)=-x-1$

Solution: (a) $x + 2 = 4$ is a conditional equation whose solution is 2.
(b) $-(x + 1) = -x - 1$ is an identity.

17. $2x + 2 = 4$

18. $3x + 6 = 9$

19. $5x - 1 = 9$

20. $6x - 2 = 10$

21. $4(x - 1) = 4x - 4$

22. $8 - (2 - 3x) = 3(x + 2)$

23. $5(x - 1) = 4x - 4$

24. $\frac{x + 1}{x + 1} = 1$

25. $x + 1 = x + 2$

26. $x - 3 = x - 2$

27. The speed V_a of an automobile involved in an accident can be determined by performing a simple test consisting of driving the car at a predetermined speed, skidding to a stop, and measuring the length of the skid. If the length of the skid at the scene of an accident is known, the equation for the speed of the car at the time of the accident is

$$V_a^2 = \frac{L_a V_t^2}{L_t}$$

Assume:

The length of the skid mark on the test, L_t, is 36 feet.

The length of the skid mark at the time of the accident, L_a, is 144 feet.

The speed on the test, V_t, is 30 miles per hour.

The driver claims that his speed at the time of the accident was 50 miles per hour. Does this number satisfy the given equation? (Hint: 30 miles per hour is equivalent to 44 feet per second, and 50 miles per hour is equivalent to $73\frac{1}{3}$ feet per second.)

28. Using the information given in Problem 27, find the speed of the car at the time of the accident.

29. The temperature (in degrees Fahrenheit) and the number of chirps per minute of a cricket satisfy the equation

$$C = 4(F - 40)$$

where F is the temperature in degrees Fahrenheit and C is the number of chirps per minute. A farmer claimed that a cricket chirped 150 times in a minute when the temperature was 80 degrees. Is this possible?

30. Referring to Problem 29, if the temperature was 80 degrees, how many times per minute should the cricket chirp?

OBJECTIVES 2.3

After studying this section, the reader should be able to:
1. Determine if two given equations are equivalent.
2. Use the five-step procedure given in the text to solve a linear equation.

2.3 SOLVING EQUATIONS IN ONE VARIABLE

Reprinted by permission of the *Tampa Tribune*, September 11, 1975.

The illustration at the beginning of this section shows that the Dow Jones average went down 5 points, to 812.66. If we wish to know what the average x was the day before (Sept. 10), we could solve the equation

$$x - 5 = 812.66$$

The average today is 5 points less than the average yesterday (arrow to $x - 5$)

The average today is 812.66 (arrow to 812.66)

$ax + b = c$, $a \neq 0$ is a first-degree linear equation.

In this section we shall develop a procedure which can be used to find the solution set of a first-degree linear equation, that is, an equation which can be written in the form $ax + b = c$, where $a \neq 0$. This procedure is usually referred to as "solving the equation" and consists of finding an equation whose solution set is obvious and identical to the solution set of the given equation. For example, each of the equations $x + 5 = 10$ and $x = 5$ has $\{5\}$ as its solution set, but the solution set of $x = 5$ is obvious. Equations such as these are called *equivalent equations*, and are defined as follows.

$x + 5 = 10$ and $x = 5$ are equivalent equations.

DEFINITION 2.3

Two equations are equivalent if they have identical solution sets.

Thus, $x + 2 = 4$ and $x = 2$ are equivalent equations, since the solution set of each equation is $\{2\}$. Similarly, $3x = 9$ and $x = 3$ are equivalent equations, since the solution set of each equation is $\{3\}$.

The procedure used to solve linear equations depends on the fact that in such equations the left and right members are expressions for the same number. Consequently, the Addition and Multiplication Laws of Equality, Theorems 1.1 and 1.2 respectively, imply the following properties.

If $P(x)$, $Q(x)$, and $R(x)$ represent algebraic expressions, then, for all values of x for which these expressions are defined, the statement

$$P(x) = Q(x)$$

is equivalent to

$$P(x) + R(x) = Q(x) + R(x) \quad \text{Property 1}$$
$$P(x) \cdot R(x) = Q(x) \cdot R(x),\ R(x) \neq 0 \quad \text{Property 2}$$

Property 1 means that you can add or subtract the same expression to or from each member of an equation and obtain an equivalent equation.

Property 2 means that you can multiply or divide both sides of an equation by a nonzero expression and obtain an equivalent equation.

Thus, to solve the equation $x - 5 = 812.66$, where x represents the Dow Jones average for September 10, we proceed as follows:

$$x - 5 = 812.66$$

1. $(x - 5) + 5 = 812.66 + 5$ — Using Property 1 to add 5 to each member of the equation
2. $x + (-5 + 5) = 817.66$ — By the Associative Law of Addition
3. $x + 0 = 817.66$ — Since -5 and 5 are additive inverses
4. $x = 817.66$ — Since 0 is the identity and $x + 0 = x$

Hence the solution set of the equation $x - 5 = 812.66$ is $\{817.66\}$, which means that the Dow Jones average on September 10 was 817.66.

Example 1 Solve the equation $x + 5 = 7$.

Solution

$$x + 5 = 7$$

1. $(x + 5) + (-5) = 7 + (-5)$ — Using Property 1 to add -5 to both sides of the equation
2. $x + [5 + (-5)] = 2$ — By the Associative Law of Addition
3. $x + 0 = 2$ — Since 5 and -5 are additive inverses
4. $x = 2$ — Since $x + 0 = x$

Thus, the solution set of $x + 5 = 7$ is $\{2\}$.

To solve the equation $7x = 2$, we use Property 2 and multiply both sides of the equation by $\frac{1}{7}$, the reciprocal of 7, obtaining

1.	$\frac{1}{7}(7x) = \frac{1}{7}(2)$	Using Property 2 to multiply both members of the equation by $\frac{1}{7}$
2.	$\left(\frac{1}{7} \cdot 7\right)x = \frac{2}{7}$	By the Associative Law for multiplication
3.	$1 \cdot x = \frac{2}{7}$	Since $\frac{1}{7}$ and 7 are reciprocals
4.	$x = \frac{2}{7}$	Since 1 is the multiplicative identity, $1 \cdot x = x$

Thus, the solution set of $7x = 2$ is $\left\{\frac{2}{7}\right\}$. In actual practice, the procedure shown here can be considerably shortened by omitting Steps 2 and 3, and writing Step 4 immediately after Step 1.

To solve the equation $4x - 7 = x + 4$, we have to use Properties 1 and 2. To do this, we first add 7 to both members of the equation, and then add the additive inverse (opposite) of x to both members so that all the variables appear on the left side of the equation. These two steps can be combined as shown:

1.	$(4x - 7) + (7 - x) = (x + 4) + (7 - x)$	Using Property 1 to add $7 - x$ to both members of the equation
2.	$(4x - x) + (-7 + 7) = (x - x) + (4 + 7)$	Using the Commutative and Associative Laws
3.	$3x + 0 = 0 + 11$	Since $-7 + 7 = 0$, $x - x = 0$, and $4x - x = 3x$
4.	$3x = 11$	Since $3x + 0 = 3x$
5.	$\frac{1}{3}(3x) = \frac{1}{3}(11)$	Using Property 2 to multiply both members of the equation by $\frac{1}{3}$

6. $\left(\frac{1}{3} \cdot 3\right)x = \frac{11}{3}$ — By the Associative Law of Multiplication

7. $1 \cdot x = \frac{11}{3}$ — Since $\frac{1}{3}$ and 3 are reciprocals

8. $x = \frac{11}{3}$ — Since $1 \cdot x = x$

Example 2 Find the solution set of $2(2x - 3) = x + 6$.

Solution To solve this equation, we first simplify the left member as follows:

1. $4x - 6 = x + 6$ — Since $2(2x - 3) = 2 \cdot 2x - 2 \cdot 3 = 4x - 6$ by the Distributive Law

2. $(4x - 6) + (6 - x) = (x + 6) + (6 - x)$ — Using Property 1 to add $6 - x$ to both members

3. $3x = 12$ — Since $(4x - \not{6}) + (\not{6} - x) = 3x$ and $(\not{x} + 6) + (6 - \not{x}) = 12$

4. $\frac{1}{3}(3x) = \frac{1}{3}(12)$ — Using Property 2 to multiply both members by $\frac{1}{3}$

5. $x = 4$

Thus, the solution set of $2(2x - 3) = x + 6$ is $\{4\}$. This fact can be easily checked by substituting the value 4 for x in the original equation: $2(2 \cdot 4 - 3) = 4 + 6$, a true statement, since $2(2 \cdot 4 - 3) = 10$ and $4 + 6$ is also 10.

$2(2 \cdot 4 - 3) = 2(8 - 3)$
$= 2 \cdot 5$
$= 10$

When fractions are present in an equation, we first multiply each term by the smallest number which is a multiple of each denominator—that is, by the lowest common denominator (LCD) of the fractions involved. For example, the LCD of $\frac{1}{3}$ and $\frac{1}{4}$ is 12, the LCD of $\frac{1}{7}$ and $\frac{1}{5}$ is 35, and the LCD of $\frac{1}{x}$ and $\frac{3}{8}$ is 8x. This procedure is illustrated in solving the equation

$$\frac{x}{4} + 5 = \frac{7x+8}{6}$$

1.	$\overset{3}{\cancel{12}} \cdot \frac{x}{4} + 12 \cdot 5 = \overset{2}{\cancel{12}}\left(\frac{7x+8}{\cancel{6}}\right)$	Multiplying each term by 12 (the LCD of 4 and 6) and simplifying
2.	$3x + 60 = 14x + 16$	By the Distributive Law
3.	$(3x + 60) + [-14x + (-60)]$ $= (14x + 16) + [-14x + (-60)]$	Adding $-14x + (-60)$ to both members
4.	$-11x = -44$	Collecting like terms
5.	$\frac{-1}{11} \cdot (-11x) = \frac{-1}{11}(-44)$	Using Property 2 to multiply both members by $\frac{-1}{11}$
6.	$x = 4$	

Thus, the solution set of $\frac{x}{4} + 5 = \frac{7x+8}{6}$ is {4}. This can be easily shown: $\frac{4}{4} + 5 = 6$, and $\frac{7 \cdot 4 + 8}{6} = 6$, so $\frac{4}{4} + 5 = \frac{7 \cdot 4 + 8}{6}$; the solution checks.

The procedure we have used can be generalized as follows to solve any equation of the form $ax + b = c$.

PROCEDURE FOR SOLVING LINEAR EQUATIONS

1. If there are fractions in the equation, multiply each term by the lowest common denominator of the fractions.
2. Simplify both members of the equation if necessary.
3. Add an algebraic expression to both members of the equation so that the right hand member contains numbers only and the left hand member contains variables only.

4. If the coefficient of the variable is not 1, multiply each member by the reciprocal of this coefficient.
5. Check the solution by substituting in the original equation the number obtained in Step 4.

We illustrate the use of this procedure in solving the next example.

Example 3 Find the solution of $\frac{x}{x-2} = 2 + \frac{2}{x-2}$.

Solution We proceed in five steps as suggested.

$$\frac{x}{x-2} = 2 + \frac{2}{x-2}$$

1. $(x-2)\left(\frac{x}{x-2}\right) = (x-2)2 + (x-2)\left(\frac{2}{x-2}\right)$ — Multiplying each member by $x-2$, the LCD of the expressions

2. $x = 2x - 4 + 2$ or $x = 2x - 2$ — Simplifying by using the Distributive Law

You can see from this example the importance of checking your solution. Even though it appears that a solution has been obtained, the solution set may be empty!

3. $x + (-2x) = 2x - 2 + (-2x)$ or $-x = -2$ — Adding $-2x$ to both members

4. $x = 2$ — Multiplying both members by $\frac{1}{-1}$, the reciprocal of -1

5. Since the solution seems to be 2, we substitute this value in the original equation, obtaining $\frac{2}{0} = 2 + \frac{2}{0}$, which is not defined. Thus, the solution set of $\frac{x}{x-2} = 2 + \frac{2}{x-2}$ is $\emptyset$.

PROGRESS TEST

1. The solution set of $x - 5 = 8$ is ______.
2. The solution set of $x + 3 = 9$ is ______.
3. The solution set of $3x - 1 = x + 2$ is ______.
4. The solution set of $2(2x + 3) = 3 + x$ is ______.
5. The solution set of $\frac{x}{3} + 2 = \frac{5x - 3}{4}$ is ______.
6. The solution set of $\frac{x}{x - 3} = 2 + \frac{3}{x - 3}$ is ______.

EXERCISE 2.3

In Problems 1 through 24, find the solution set of the given equation. (Check your answers.)

Illustration: $\frac{3}{4} + \frac{x}{6} = x - \frac{7}{4}$

Solution:

1. $12 \cdot \frac{3}{4} + 12 \cdot \frac{x}{6} = 12 \cdot x - 12 \cdot \frac{7}{4}$ — Multiplying each term by 12, the LCD of $\frac{3}{4}$, $\frac{x}{6}$, and $\frac{7}{4}$

2. $9 + 2x = 12x - 21$ — Simplifying

3. $(9 + 2x) + (-12x - 9) = (12x - 21) + (-12x - 9)$ or $-10x = -30$ — Adding $-12x - 9$ to both members

4. $\left(\frac{-1}{10}\right) \cdot -10x = \frac{-1}{10} \cdot -30$ or $x = 3$ — Multiplying each member by $\frac{-1}{10}$

5. Check: $\frac{3}{4} + \frac{3}{6} = \frac{5}{4}$, and $3 - \frac{7}{4} = \frac{12}{4} - \frac{7}{4} = \frac{5}{4}$; thus $\{3\}$ is the solution set of $\frac{3}{4} + \frac{x}{6} = x - \frac{7}{4}$.

1. $x + 4 = 12$

2. $x + 3 = -4$

PROGRESS TEST ANSWERS

1. $\{13\}$
2. $\{6\}$
3. $\left\{\frac{3}{2}\right\}$
4. $\{-1\}$
5. $\{3\}$
6. $\emptyset$

3. $5x - 4 = 6$

4. $3x - 1 = -7$

5. $2x + 5 = x + 6$

6. $3x + 2 = 2x + 8$

7. $2x - 1 = 3x - 7$

8. $3x - 2 = 6x - 8$

9. $-3x + 1 = -9$

10. $-5x + 1 = -13$

11. $2x = 8$

12. $3x = -2$

13. $2x + \frac{1}{5} = 3x - \frac{4}{5}$

14. $3x + \frac{7}{2} = 4x + \frac{1}{2}$

15. $-2x + \frac{1}{4} = 2x + \frac{4}{5}$

16. $6x + \frac{1}{7} = 2x - \frac{2}{7}$

17. $-4x + \frac{1}{2} = 6\left(x - \frac{1}{8}\right)$

18. $-6x + \frac{2}{3} = 4\left(x - \frac{1}{5}\right)$

19. $-6 = 6(x - 1) + 8x$

20. $\frac{-1}{3} = 8x - 3$

21. $\frac{-x}{2 - x} = -2 - \frac{2}{2 - x}$

22. $\frac{-x}{3 - x} - 2 = \frac{-3}{3 - x}$

23. $\dfrac{x}{x+1} = 1 + \dfrac{2}{x+1}$

24. $\dfrac{5}{x+1} = 1 + \dfrac{3}{x+1}$

In Problems 25 through 40, solve for the specified variable in the given equation.

Illustration: t in I = Prt

Solution: Dividing both members of the equation I = Prt by Pr, we obtain

$$\frac{I}{Pr} = \frac{Prt}{Pr}$$

Thus,

$$t = \frac{I}{Pr}$$

25. I in $E = IR$

26. g in $v = gt$

27. h in $V = \pi r^2 h$

28. L in $A = LW$

29. L in $P = 2L + 2W$

30. s in $A = \pi r^2 + \pi rs$

31. x in $y = mx + b$

32. a in $L = a + (n-1)d$

33. W in $F = \dfrac{Wv^2}{gr}$

34. h in $A = \dfrac{bh}{2}$

35. V_2 in $\dfrac{V_2}{V_1} = \dfrac{P_1}{P_2}$

36. b in $\dfrac{a}{b} = \dfrac{c}{d}$

37. r in $s = \dfrac{a}{1-r}$

38. R in $I = \dfrac{E}{R + nr}$

39. a in $f = \dfrac{ab}{a+b}$

40. H in $S = \dfrac{f}{H-h}$

Photo credit : Sterling Publishing Company, Inc., New York.

OBJECTIVES 2.4

After studying this section, the reader should be able to use the five-step procedure given in the text to solve (a) a given word problem of a general nature or (b) a "mixture" problem.

2.4 INTRODUCTION TO WORD PROBLEMS

In this section we shall be concerned with solving problems that are stated in words. This will require us to represent verbal sentences symbolically, that is, to translate word sentences into equations that correspond to the stated problems and then solve these equations using the methods learned in Section 2.3.

The photograph at the beginning of this section shows Don Koehler, the world's tallest man, meeting Mihaly Mesyaros (known as "Mishu") during the television special "The Second David Frost Presents the Guinness Book of World Records" in 1974. As you can see from the photo, Mr. Koehler is almost three times as tall as Mishu! To put it exactly, if one inch is subtracted from three times Mishu's height, the result will be Mr. Koehler's height, 98 inches. Can you find Mishu's height?

To solve this problem, we can proceed as follows:

1. Read the problem carefully and decide what numbers we are asked for. In our problem we are asked for Mishu's height.
2. Let h represent Mishu's height.
3. According to the problem:

If one inch is subtracted from three times Mishu's height, the result will be Mr. Koehler's height, 98 inches.

$$3h \quad - \quad 1 \quad = \quad 98$$

4. We solve the equation $3h - 1 = 98$.

$3h - 1 + 1 = 98 + 1$ Adding 1 to both members

or $3h = 99$

$\frac{1}{3} \cdot 3h = \frac{1}{3} \cdot 99$ Multiplying both members by $\frac{1}{3}$

$h = 33$

Thus, Mishu's height is 33 inches.

5. We check these results with the requirements of the problem. Is it true that if one inch is subtracted from three times Mishu's height the result is 98? That is,

$$3 \cdot 33 - 1 \stackrel{?}{=} 98$$

$$99 - 1 \stackrel{\checkmark}{=} 98$$

Since the result is a true statement, our answer is correct.

The procedure used to solve the preceding problem is now restated so that it can be used in the solution of other word problems.

PROCEDURE TO SOLVE WORD PROBLEMS

1. Read the problem carefully and decide what number is asked for.
2. Select a variable (letter) to represent the number asked for in the problem.
3. Write an equation using two expressions representing the same number.
4. Solve the equation obtained in Step 3.
5. Check the result by making sure that your answer satisfies the requirements of the problem.

Example 1

The first stage of the Saturn V vehicle shown in the picture is such that if six feet is added to four times its diameter, the result will equal its length, 138 feet. What is the diameter of the first stage of the Saturn V?

NASA photo no. 66–H–182.

Solution

We proceed by steps.

1. We are asked for the diameter of the first stage.
2. Let this number be d.
3. According to the problem,

If six feet is added to four times its diameter, the result is 138 feet.

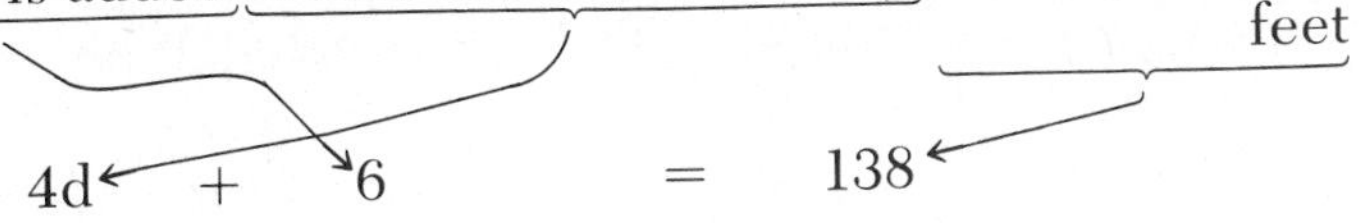

4. We solve the equation:

$$4d + 6 = 138$$

$$4d + 6 + (-6) = 138 + (-6) \quad \text{Adding } -6 \text{ to both members}$$

$$\text{or } 4d = 132$$

$$\frac{1}{4} \cdot 4d = \frac{1}{4} \cdot 132 \quad \text{Multiplying both members by } \frac{1}{4}$$

$$d = 33$$

5. Since six feet added to four times the diameter is 138 feet, that is, $4(33) + 6 = 138$, our result is correct.

We are now ready to discuss another type of problem—mixture problems. In this type of problem, two or more substances are blended to make a mixture. For example, the man shown in the photo may mix two types of coffee of different prices to obtain a blend with a predetermined price. In solving these problems, we must keep in mind that the sum of the weights (or values) of the original ingredients must equal the weight (or value) of the final mixture.

Photo by Jim Goetz.

Example 2

The man in the photo has coffee worth \$1.30 per pound and coffee worth \$1.50 per pound. How many pounds of each should he mix in order to obtain 50 pounds of a blend worth \$1.45 per pound?

Solution

1. We are asked for the number of pounds of each coffee to be used.
2. Let P be the number of pounds of the \$1.30 coffee. Then, since we have to make 50 pounds, 50 − P will be the number of pounds of \$1.50 coffee used.
3. We arrange our information in a chart.

	Price per pound	× Pounds =	Total Price
\$1.30 coffee	\$1.30	P	1.30 P
\$1.50 coffee	\$1.50	50 − P	1.50 (50 − P)
Mixture	\$1.45	50	\$72.50

The total value of the mixture can be expressed by 1.30 P + 1.50 (50 − P) or by \$72.50. Thus the value of the \$1.45 coffee, expressed as 1.30 P + 1.50 (50 − P), equals \$72.50:

$$1.30\,P + 1.50\,(50 - P) = 72.50$$

4. Solving this equation, we have

$$\begin{aligned} 1.30P + (1.50)(50) - 1.50P &= 72.50 \\ 1.30P - 1.50P &= 72.50 - 75 \\ -0.20P &= -2.50 \\ P &= \frac{-2.50}{-0.20} = 12.50 \end{aligned}$$

Thus, he must use 12.50 pounds of the \$1.30 coffee and 50 − 12.50 = 37.50 pounds of the \$1.50 coffee.

5. Since (1.30)(12.50) + (1.50)(37.50) = 16.25 + 56.25 = \$72.50, our result is correct.

Example 3

A dietician in a health spa wishes to prepare a meal containing 60% protein. How many grams of a certain fiber containing 10% protein should be added to 300 grams of a certain food containing 70% protein to obtain the desired 60% protein mixture?

Solution

1. We are asked for the number of grams of fiber to be added.
2. Let this amount be g.

3. We arrange our information in a chart:

	% of Protein ×	Amount of Substance =	Amount of Protein
Fiber	0.10	g	0.10g
70% Mixture	0.70	300	210
60% Mixture	0.60	g + 300	0.60(g + 300)

4. Since the amount of protein is given by 0.10g + 210 or

$$0.60(g + 300) = 0.10g + 210$$
$$0.60g + 180 = 0.10g + 210$$
$$0.50g = 30$$
$$g = \frac{30}{0.50} = 60$$

Thus, the amount of fiber containing 10% protein to be added should be 60 grams.

5. The verification of the answer is left to the reader.

PROGRESS TEST

1. The length of the Saturn V is about 280 feet. This length is 40 feet more than 3 times the length of the second stage. How long is the second stage?
2. The most expensive tea in the world is Oolong tea, retailing for $13 per pound in 1974. How many pounds of Oolong tea should be added to 50 pounds of regular tea selling at $4 per pound to obtain a mixture selling for $7 per pound?
3. How many ounces of regular vodka (40% alcohol) should be added to 30 ounces of Poland White Spirit Vodka (80% alcohol) to obtain vodka containing 70% alcohol?

PROGRESS TEST ANSWERS

1. Let x be the length of the second stage:

$3x + 40 = 280$ $3x = 240$ $x = 80$ feet

2. Let x be the amount of Oolong tea to be added

	Price per pound ×	Amount =	Total Price
Oolong	$13	x	13x
Regular	$ 4	50	200
Mixture	$ 7	x + 50	7(x + 50)

$13x + 200 = 7(x + 50)$ $13x + 200 = 7x + 350$ $6x = 150$ $x = 25$

Thus, 25 pounds of Oolong tea should be added.

3. Let x be the amount of regular Vodka to be added

	% Alcohol ×	Amount =	Total Alcohol
Polish	0.80	30	24
Regular	0.40	x	0.40x
Mixture	0.70	30 + x	0.70(30 + x)

$0.70(30 + x) = 24 + 0.40x$ $21 + 0.70x = 24 + 0.40x$ $0.30x = 3$ $x = 10$

Thus, 10 ounces of regular vodka should be added.

EXERCISE 2.4

In Problems 1 through 20, follow the five-step procedure given in the text to solve the problem.

Illustration: The sum of three consecutive integers is 14 less than four times the smallest of the three integers. Find the integers.

Solution:

1. We are asked for three integers.
2. Let x be the first integer. Since successive integers differ by 1, the three successive integers are x, $x+1$, and $x+2$.
3. According to the problem, the sum of three integers is 14 less than four times the smallest of the three integers:

$$x+(x+1)+(x+2)=4x-14$$

4. We solve the equation:

$$x+(x+1)+(x+2)=4x-14$$

1. $3x+3 = 4x-14$
2. $3x+3+[-3+(-4x)] = 4x-14+[-3+(-4x)]$
3. $-x=-17$
4. $x=17$

$$x+1=18$$
$$x+2=19$$

5. Since $17+18+19=4 \cdot 17-14$, our solution is correct.

1. The sum of three consecutive integers is 33. Find the integers.

2. The sum of three consecutive integers is 17 less than four times the smallest of the three integers. Find the integers.

3. The sum of three consecutive integers is 156. Find the integers.

4. The larger of two numbers is six times the smaller. Their sum is 147. Find the numbers.

5. Elvis Presley is the artist for whom the most gold records are claimed (Guinness). By 1975 he had 120 gold records. If he has eight fewer 45-rpm gold records than seven times the number of his L.P. gold records, how many gold singles and how many L.P. gold records does he have?

6. The greatest weight difference ever recorded in a major boxing bout was 140 pounds, in a match between John Fitzsimmons and Ed Punkhorst (Guinness). If the combined weight of the contestants was 484 pounds, find Fitzsimmons' weight (he was the lighter of the two).

7. The estimated length of prehistoric sharks, whose jaws could enclose a fully upright man, is 80 feet (Guinness). This length is about five times as big as the largest fish ever caught with a rod. Find the approximate length of the largest fish ever caught with a rod.

8. The crew of Apollo 17 spent 31 more hours in their lunar exploration module, the Challenger, than in "extra-vehicular activity." If their total stay on the lunar surface was 75 hours, how many hours did they spend in "extra-vehicular activity"?

9. The astronauts of Apollo 17 brought back 40 more pounds of lunar rocks and soil than did the astronauts of Apollo 16. If the total weight of the rocks and soil brought back by both crews is 466 pounds, how many pounds did the astronauts of Apollo 17 bring back?

10. The combined cost of the U.S. and Soviet manned space programs has been estimated at 71 billion dollars. If the U.S. program cost 19 billion less than the Soviet program, find the cost of each.

11. The weight of an object on the moon is one-sixth of the weight of that object on earth. The crew of Apollo 16 collected lunar rocks and soil weighing 35.5 pounds on the moon. What is the weight of the rocks and soil on earth?

12. If an object is thrown vertically upward with an initial velocity of 192 feet per second, its velocity at time t is given by $v = 192 - 32t$. In how many seconds will the object reach its highest point? (Hint: At its highest point, $v = 0$.)

13. A car rental agency charges $15 per day and 10¢ per mile. Thus, the cost C in dollars per day is given by $C = 0.10m + 15$ (m is the number of miles). Find the number of miles traveled by a person who paid $23 on a certain day.

14. A businessperson makes a $10,000 initial investment in a manufacturing plant. Production costs are $3 per item. If the items are to be sold for $7, how many should be sold before the person breaks even?

15. In 1960, the cost-of-living index was seven more than twice what it was in 1940. If it was 103 in 1960, find the cost-of-living index in 1940.

16. A student's report card showed 51 quality points. Each credit hour of A work is worth 4 quality points, and each credit hour of B work is worth 3 quality points. If the student was enrolled for 15 credit hours and received A's and B's only, how many credit hours of A work was the student earning?

17. The largest painting in the world used to be the Panorama of the Mississippi, by John Banvard (Guinness). If the length of this painting was 4,988 feet more than its width and its perimeter was 10,024 feet, find the dimensions of this rectangular painting. Hint: The perimeter of a rectangle is $2L + 2W$, where L is the length and W is the width of the rectangle.

18. The largest painting now in existence is probably the Battle of Gettysburg (Guinness). The length of this painting exceeds its width by 340 feet. If the perimeter of this rectangular painting is 960 feet, find its dimensions.

19. The scientific building with the greatest capacity is the Vehicle Assembly Building (VAB) at the John F. Kennedy Space Center (Guinness). The width of this building is 198 feet less than its length. If the perimeter of the rectangular building is 2,468 feet, find its dimensions.

20. The largest Fair Hall is in Hanover, Germany (Guinness). The length of this hall exceeds its width by 295 feet. If the perimeter of this rectangular hall is 4,130 feet, find its dimensions.

In Problems 21 through 35, use the five-step Procedure in the text to solve the given problem.

Illustration: A car radiator contains 32 quarts of a 60% antifreeze solution. How many quarts of this solution should be drained and replaced with water so that the new solution is 40% antifreeze?

Solution:

1. We are asked for the number of quarts of water to be added.
2. Let x be this number. Then, $32 - x$ is the number of quarts of solution remaining.
3. We chart our information:

	% Anti-freeze ×	Quarts of Solution =	Quarts of Antifreeze
Original	0.60	32 − x	0.60(32 − x)
Water	0	x	0
New Solution	0.40	32	12.8

Since the number of quarts of antifreeze can be represented by $0.60(32 - x)$ or by 12.8, we have $0.60(32 - x) = 12.8$.

4. Solving this equation, we obtain

$$19.2 - 0.60x = 12.8$$
$$-0.60x = -6.4$$
$$x = \frac{-6.4}{-0.60} = 10\frac{2}{3}$$

Thus, $10\frac{2}{3}$ quarts should be replaced by water.

5. The verification of this result is left to the reader.

21. A car radiator contains 30 quarts of a 50% antifreeze solution. How many quarts of this solution should be drained and replaced with water so that the new solution is 30% antifreeze?

22. Repeat Problem 21 for a radiator containing 25 quarts of a 40% antifreeze solution.

23. According to the Guinness Book of Records, the costliest perfume in the world is "Adoration," manufactured by Nina Omar of Puerto Real, Cadiz, Spain, and costing $185 per half ounce. How many ounces of "Adoration" should be mixed with another perfume selling at $160 per ounce to produce a blend selling for $300 per ounce? (Hint: The cost of one ounce of "Adoration" is $370.)

24. A candy-maker has peanuts worth 50 cents per pound and chocolate worth 80 cents per pound. How many pounds of each should he mix to make 90 pounds of chocolate-and-peanut candy selling for 67 cents per pound?

25. In September, 1975, the price of copper was about 65 cents per pound, and the price of zinc was approximately 40 cents per pound. How many pounds of copper and zinc should be mixed to make 50 pounds of brass selling for 50.75 cents per pound?

26. In 1974, Oolong tea retailed for $13 per pound. How many pounds of Oolong tea should be mixed with regular tea selling for $3 per pound to produce 50 pounds of tea selling for $9 per pound?

27. How many gallons of a 10% salt solution should be added to 15 gallons of a 20% salt solution to obtain a 16% solution?

28. How many gallons of a 30% salt solution must be added to 12 gallons of a 15% salt solution to obtain a 20% salt solution?

29. A dietician wishes to prepare a special meal containing 50% carbohydrates. How many grams of a certain food containing 20% carbohydrates should be added to 100 grams of another food containing 60% carbohydrates to obtain the desired 50% mixture?

30. How many ounces of vermouth containing 10% alcohol should be added to 10 ounces of gin containing 40% alcohol so that the resulting pitcher of martinis will contain 30% alcohol?

31. Manhattans are made by mixing bourbon and sweet vermouth. How many ounces of Manhattans containing 40% vermouth should a bartender mix with Manhattans containing 20% vermouth to obtain one quart (32 ounces) of Manhattans containing 30% vermouth?

32. How much brass containing 30% zinc should be melted with brass containing 35% zinc to obtain 50 pounds of brass containing 32% zinc?

33. How many gallons of rocket fuel containing 90% hydrogen peroxide should be mixed with rocket fuel containing 98% hydrogen peroxide to obtain 3000 gallons of fuel containing 96% hydrogen peroxide?

34. How many quarts of pure acid must be added to 30 quarts of a 15% acid solution to obtain a 20% solution?

35. How much distilled water should be added to an 8% acid solution to obtain 100 cubic centimeters of a solution that is 6% acid?

Photo by Alan Koch, U. S. Air Force.

OBJECTIVES 2.5

After studying this section, the reader should be able to use the five-step procedure given in the text to solve (a) a "uniform motion" problem or (b) an "investment" problem.

2.5 UNIFORM MOTION AND INVESTMENT PROBLEMS

In some instances, it is necessary to supplement the information given in a word problem by a formula that will enable us to write an equation using the facts provided in the problem. For example, if an object is moving in a straight line at a constant speed (uniform motion), the relationship between the distance D that the object moves, the rate R at which the object is traveling, and the time T it takes to travel this distance is given in the formula

(1) $\quad D = RT$

Problems involving the use of Formula (1) are called *uniform motion problems*. Of course, in using Formula (1), the units of measurement must be consistent. Thus, if the rate of travel is measured in miles per hour, then the time must be measured in hours.

The procedure used in solving uniform motion problems follows the five steps given in the preceding section and is illustrated in the following examples.

Example 1

An unidentified bomber flying at 600 miles per hour flies over an observation post. One-half hour later, a Lockheed SR-71 (see photograph at the beginning of this section) scrambles in pursuit of the bomber. How long will it take the Lockheed, flying at 1,800 miles per hour, to intercept the bomber?

Solution

We proceed by these steps:

1. We are asked for the time it will take the Lockheed to catch the bomber.
2. Let this time be T.
3. We arrange the given information in a chart:

	RATE $\times$	TIME $=$	DISTANCE
Lockheed	1,800	T	1,800T
Bomber	600	$T+\frac{1}{2}$	$600\left(T+\frac{1}{2}\right)$

We wish to find the time T at which both planes have traveled the same distance. From the entries in the last column, the distances traveled are equal when $1800T = 600\left(T+\frac{1}{2}\right)$

4. We solve the equation $1,800T = 600\left(T+\frac{1}{2}\right)$.

$$1,800T = 600T + 300$$
$$1,200T = 300$$
$$T = \frac{1}{4}$$

Thus, the plane intercepts the bomber after $\frac{1}{4}$ of an hour (15 minutes).

5. After $\frac{1}{2}+\frac{1}{4}=\frac{3}{4}$ hours, the bomber has traveled $\frac{3}{4}\bullet 600 =$ 450 miles, the same distance traveled by the Lockheed in $\frac{1}{4}$ of an hour $\left(\frac{1}{4}\bullet 1800 = 450\right)$; thus our result is correct.

Example 2

Tommie C. Smith, a world track record holder, completed a race in 20 seconds. Another runner took 22 seconds to finish the same race. If Smith's speed is 1 yard per second more than the other runner's, find their speeds and the distance run.

Solution

1. We are asked for their rates.
2. Let R be the rate of the slow runner.
3. We arrange the given information in a chart:

	RATE $\times$	TIME	= DISTANCE
Slower Runner	R	22 sec.	$22R$
Smith	$R+1$	20 sec.	$20(R+1)$

Since both runners covered the same distance, $22R = 20(R+1)$.

4. We now solve $22R = 20(R+1)$.

$$\begin{aligned} 22R &= 20R + 20 \\ 2R &= 20 \\ R &= 10 \end{aligned}$$

Thus, the slower runner had a rate of 10 yards per second, and Smith had one of $10 + 1 = 11$ yards per second. Since the slower runner ran at 10 yards per second for 22 seconds, the distance covered was $10 \cdot 22 = 220$ yards.

We are now ready to discuss investment problems. In the ad on page 113, a certain bank claims to give the highest interest, $7\frac{1}{2}$ per cent, on the lowest deposits in town. If we assume the interest is paid annually, this means that the depositor receives as interest $7\frac{1}{2}$ cents for each dollar (100 cents) left in the bank for a period of one year. The amount of annual interest I that is earned when P dollars are invested at a rate of R per cent per year is given in the formula

$$(2) \qquad I = PR$$

Problems involving the use of Formula (2) are called *investment* or *interest* problems. As before, in solving these problems we shall follow the five-step procedure developed in the preceding sections.

Example 3

According to the Guiness Book of Records, in 1972 the highest bank interest rate was that of Brazil at 20 per cent and the lowest that of Morocco at 3.5 per cent. A man deposited part of $10,000 in a Brazilian bank and the remaining part in another bank in Morocco. If his annual income from these investments is $1,670, how much does he have deposited in each country?

Solution

1. We are asked for the amount deposited in each country.
2. Let P be the amount deposited in Brazil at 20 per cent. Then the amount deposited in Morocco will be 10,000 − P.
3. We arrange the given information in a chart:

	Amount Deposited	× Rate	= Interest
Brazil	P	20 per cent = 0.20	0.20P
Morocco	10,000 − P	3.5 per cent = 0.035	0.035(10,000 − P)

Since the annual income from both deposits is $1670, we have

$$0.20P + 0.035(10{,}000 - P) = 1670$$

Recall that 0.035 • 10,000 = 350 (moving the decimal point 4 places to the right). Also,

$$\begin{array}{r} 0.200P \\ -0.035P \\ \hline 0.165P \end{array}$$

$$\begin{array}{r} 8{,}000 \\ .165.\overline{)1320{,}000} \\ 1320 \\ \hline 0 \end{array}$$

4. Solving this equation, we get

$$\begin{aligned} 0.20P + (0.035 \cdot 10{,}000) - 0.035P &= 1670 \\ 0.20P - 0.035P + 350 &= 1670 \\ 0.165P + 350 &= 1670 \\ 0.165P &= 1320 \\ P &= \frac{1320}{0.165} = \$8000 \end{aligned}$$

NOW EARN THE HIGHEST INTEREST ON THE LOWEST DEPOSITS IN TOWN.

7½ PERCENT.

Reprinted, by permission, from *The Tampa Tribune*, November 16, 1973, p. 5E.

Since \$8000 was deposited in Brazil, the rest, \$10,000 − \$8000, or \$2000, was deposited in Morocco.

5. \$8000 deposited at 20 per cent gives $I = (0.20)(8000) =$ \$1600 interest. \$2000 deposited at 3.5 per cent gives $I = (0.035)(2000) =$ \$70 interest. Thus, the annual income from both deposits is \$1600 + \$70, or \$1670.

Example 4 A women invested \$5000 at 6 per cent. What additional amount should she invest at 9 per cent so that her total annual yield is 8 per cent?

Solution

1. We are asked for the additional amount to be invested.
2. Let this amount be P.
3. We write our information in a chart:

	Amount Invested ×	Rate =	Interest
Original Invest.	5000	0.06	300
New Invest.	P	0.09	0.09P
Total Invest.	5000 + P	0.08	0.08(5000 + P)

Since the total interest in the last column can be represented by $300 + 0.09P$ or by $0.08(5000 + P)$, we have

$$300 + 0.09P = 0.08(5000 + P)$$

4. Solving this equation, we get

$$\begin{aligned} 300 + 0.09P &= (0.08)(5000) + 0.08P \\ 300 + 0.09P &= 400 + 0.08P \\ 0.09P - 0.08P &= 400 - 300 \\ 0.01P &= 100 \\ P &= \frac{100}{0.01} = 10{,}000 \end{aligned}$$

Thus, \$10,000 more dollars should be invested at 9 per cent.

5. The verification of this result is left to the reader.

PROGRESS TEST

1. A Highway Patrol officer is in pursuit of a car traveling 60 miles per hour on a straight road. If the car has a 15-minute $\left(\frac{1}{4}\text{ hour}\right)$ lead on the officer, who is traveling at 80 miles per hour, how long will it take the officer to intercept the other car?
2. It takes 30 minutes ($\frac{1}{2}$ hour) to fly from Tampa to Miami in a jet. A private plane makes the trip in 4 hours. If the speed of the jet exceeds that of the plane by 420 miles per hour, find the speed of each plane and the distance from Tampa to Miami.
3. A bus traveling at 50 miles per hour leaves Boston for New Haven, a distance of 260 miles. An hour later, a car traveling 55 miles per hour leaves New Haven for Boston. How long does it take before the vehicles meet?
4. A man had $8000 to invest. He invested part at 6 per cent and the rest at 8 per cent. If his annual income from these investments is $600, how much money did he have invested at each rate?
5. A man invested $6000 at 8 per cent. What additional amount should he invest at 10 per cent so that his total annual interest is $880?
6. A woman had $10,000. She invested $5000 at 5 per cent, and $2500 at 6 per cent. At what rate of interest should she invest the rest of her money so that her annual interest is $600?

See following page for answers.

EXERCISE 2.5

In Problems 1 through 10, follow the five-step procedure in the text to solve the given problem.

Illustration: The world's longest and most expensive ocean liner is the France (Guinness). Assume it leaves Le Havre for New York traveling at 31 knots (nautical miles per hour). Ten hours later, the United States, the flagship of the United States Lines Company, leaves New York for Le Havre traveling at 35 knots. If the distance from New York to Le Havre is 2,950 nautical miles, how many hours will the France have been at sea when it meets the United States?

PROGRESS TEST ANSWERS

1. We follow these steps:

	R	×	T	=	D
Officer	80		$T - \frac{1}{4}$		$80(T - \frac{1}{4})$
Car	60		T		60T

$80\left(T - \frac{1}{4}\right) = 60T \qquad 80T - 20 = 60T \qquad 20T = 20 \qquad T = 1$

It takes the officer $\frac{3}{4}$ of an hour to intercept the car.

2. We follow these steps:

	R	×	T	=	D
Plane	R		4		4R
Jet	R + 420		$\frac{1}{2}$		$\frac{1}{2}\left(R + 420\right)$

$4R = \frac{1}{2}\left(R + 420\right) \qquad 4R = \frac{1}{2}R + 210 \qquad \frac{7}{2}R = 210 \qquad R = 60$

The speed of the plane is 60 miles per hour and that of the jet is $60 + 420 = 480$. The distance from Tampa to Miami is $\frac{1}{2} \cdot 480 = 240$ miles.

3. We follow these steps:

	R	×	T	=	D
Bus	50		T		50T
Car	55		T − 1		55(T − 1)

$55(T - 1) + 50T = 260 \qquad 55T - 55 + 50T = 260 \qquad 105T - 55 = 260$

$105T = 315 \qquad T = 3$

It takes 2 hours after the car leaves before the vehicles meet.

4. Let P be the amount invested at 6 per cent.

	Amount	×	Rate	=	Interest
6% Inv.	P		0.06		0.06P
8% Inv.	8000 − P		0.08		0.08(8000 − P)

$0.06P + 640 - 0.08P = 600 \qquad -0.02P = -40 \qquad P = 2000$

$2000 was invested at 6%. $6000 was invested at 8%.

5. Let P be the additional amount invested.

	Amount	×	Rate	=	Interest
8%	$6000		0.08		480
10%	P		0.10		0.10P

$480 + 0.10P = 880 \qquad 0.10P = 400 \qquad P = \frac{400}{0.10} = 4000$

An additional $4000 should be invested at 10%.

6. Let R be the rate at which the last $2500 is invested.

	Amount	×	Rate	=	Interest
1st Inv.	5000		0.05		250
2nd Inv.	2500		0.06		150
Last Inv.	2500		R		2500R

$250 + 150 + 2500R = 600 \qquad 400 + 2500R = 600 \qquad 2500R = 200$

$R = \frac{200}{2500} \qquad R = \frac{8}{100} = 8\%$

Thus, the remaining $2500 should be invested at 8%.

Solution:

1. We are asked for the number of hours spent at sea by the France before the ships meet.

2. Let T be the number of hours traveled by the France. Since the United States left 10 hours later, it traveled T − 10 hours.

3. We arrange the given information in a chart:

	RATE	× TIME =	DISTANCE
France	31 knots	T	31T
U.S.	35 knots	T − 10	35(T − 10)

When the ships meet, the number of miles travelled must be 2,950. This distance is also given by 31T + 35(T − 10), so 31T + 35(T − 10) = 2950.

4. Solving,

$$31T + 35T - 350 = 2950$$
$$66T - 350 = 2950$$
$$66T = 3300$$
$$T = 50$$

Thus, the France will have been at sea 50 hours when the ships meet.

5. Since the France traveled for 50 hours, it traveled 50 • 31 = 1550 nautical miles, while the United States traveled 40 • 35 = 1400 nautical miles. The sum of these two distances is 2950 nautical miles, so our result is correct.

1. Two hours after a car leaves a certain town traveling at an average speed of 60 kilometers per hour, a Highway

Patrolman leaves from the same starting point to overtake the car. If the average speed of the Patrolman is 90 kilometers per hour, how long will it be before he overtakes the car?

2. A group of smugglers crosses the border traveling in a car at 96 kilometers per hour. An hour later, the Border Patrol starts after them in a light plane flying 144 kilometers per hour. (a) How long will it be before the Border Patrol reaches the smugglers? (b) At what distance from the border will the smugglers be reached?

3. A jet plane flies from Boston to Philadelphia in $\frac{1}{2}$ hour. A small plane takes two hours for the same trip. If the speed of the jet is 150 miles per hour less than 5 times the speed of the small plane, find the speed of each plane.

4. Beau Madison, a speed record holder for horses, completed a race in 45 seconds (Guinness). Another horse took 50 seconds to complete the same race. If Madison ran 4 miles per hour faster than the other horse, find their speeds and the distance run.

5. At 9 A.M., a plane traveling 480 miles per hour leaves San Antonio for San Francisco, a distance of 1632 miles. At 10 A.M., another plane traveling at the same speed leaves San Francisco for San Antonio. At what time do the two planes pass each other?

6. Two airplanes leave their respective airports at Houston and Boston at the same time and travel toward each other. The Houston-bound plane is 24 miles per hour faster than the other. If the planes pass each other after 4 hours, find the distance from Houston to Boston.

7. A plane has gasoline for 7 hours of flight. If the plane uses up this gasoline by flying to a target at 240 miles per hour and returning at 320 miles per hour, how far is the target from the starting point?

8. Two cars leave the same town at the same time traveling in opposite directions. The first car travels at 56 miles per hour and the second one at 64 miles per hour. How long will it be before they are 480 miles apart?

9. A hitchhiker started walking toward a town 15 miles away at a rate of 4 miles per hour. She was then picked up by a car traveling 40 miles per hour. If the whole trip took 1.5 hours, how long did she ride?

10. A train leaves from A for B traveling at 64 miles per hour and returns from B to A traveling at 96 miles per hour. if the round trip takes 8 hours, what is the distance from A to B?

In Problems 11 through 20, follow the five-step procedure given in the text to solve the given problem.

Illustration: A woman had $8000. She invested $4000 at 6 per cent and $2500 at 5 per cent. At what rate of interest should she invest the rest of her money so that she can receive an annual interest of $470?

Solution:

1. We are asked for a rate of interest.
2. Let R be this rate.
3. We write our information in a chart:

Amount Invested ×	Rate	= Interest
$4000	0.06	$240
$2500	0.05	$125
$1500	R	1500R

4. Since the total interest must be $470, we have

$$\begin{aligned} 240 + 125 + 1500R &= 470 \\ 365 + 1500R &= 470 \\ 1500R &= 105 \\ R &= \frac{105}{1500} = \frac{7}{100} \\ &= 7 \text{ per cent} \end{aligned}$$

5. Since $(0.06)(4000) + (0.05)(2500) + (0.07)(1500) = 240 + 125 + 105 = 470$, our result is correct.

11. Two sums of money totaling $15,000 earn, respectively, $2\frac{1}{2}\%$ and $3\frac{1}{2}\%$ annual interest. If together they earn $435 in interest, find the two amounts.

12. A woman invested $2000, part at 6% and the rest at 8%.

Find the amount invested at each rate if the annual income from the two investments is $130.

13. A woman invested $8000 in a stock yielding 5% annually. How much more money should she invest in another stock yielding 10% annually so that her total annual yield rate is 6%.

14. John Q. Public desired to have an overall return of 11% on his investments. He has already invested $2400 at 10%. What additional amount must he invest at 15% to obtain his desired 11% return?

15. A jeweler paid $6000 for an order consisting of two kinds of rings. He made a 15% profit on the first kind of ring and 10% on the second. If his annual profit rate on the entire order was 11%, how much did he pay for each kind of ring?

16. Two sums of money totaling $15,000 earn 5% and 7% interest, respectively. If the annual yield rate for the total investment is 5.8%, how much is invested at each rate?

17. $20,000 is split and invested, part at 5% and the rest at 6%. If the return on the 5% investment exceeds that on the 6% investment by $10 per year, how much is invested at each rate?

18. An investor receives $1200 annually from two investments. He has $1000 more invested at 8% than at 6%. Find the amount he has invested at each rate.

19. One third of a certain sum is invested at 12% and the rest at 8%. The annual interest is $784. How much money is invested?

20. A certain sum is split and invested at 15% and 10%. The amount of interest from the 10% investment is 4 times as much as the amount from the 15% investment. If the total annual interest is $11,000, how much is invested at each rate?

Sears **SAVE \$10.11 to \$20.07 on Great Performing 10-Speed Racers**

OBJECTIVES 2.6

After studying this section, the reader should be able to:

1. **Determine the relationship between two given integers by using the Trichotomy Law.**
2. **Classify a given inequality as an absolute (unconditional) inequality or a conditional inequality.**
3. **Find the solution set of a given linear inequality in one variable.**
4. **Use the two given theorems (2.1 and 2.2) to solve a linear inequality in one variable involving absolute values.**
5. **Graph the solution set of the inequalities discussed in objectives 3 and 4.**

2.6 ORDER RELATIONS

The Sears ad at the beginning of this Section says that you can save \$10.11 to \$20.07 on 10-speed racers. If s is the amount of money saved when buying a racer, then the sentence in the ad can be written as $10.11 \leq s \leq 20.07$, read "10.11 is less than or equal to 20.07." The notation $10.11 \leq s$ means that 10.11 is *less than or equal to* a certain quantity s. This fact can also be expressed as $s \geq 10.11$, which means that a quantity s is *greater than or equal to* 10.11.

$s \geqq 10.11$ means that s is greater than or equal to 10.11.

$a > b$ means that a is greater than b. $a < b$ means that a is less than b.

The order relations, "greater than" and "less than" ($>$ and $<$), were already mentioned in Chapter 1. We can now illustrate different properties of these relations on the number line shown in Figure 2.1.

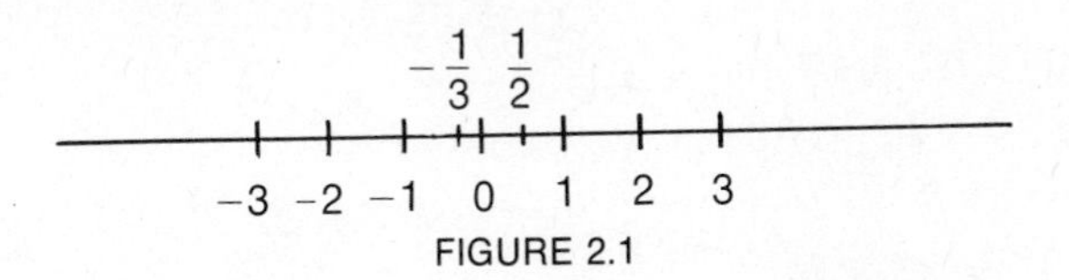

FIGURE 2.1

We *order* the numbers on this line by considering their position: Any number whose corresponding point on the number line is to the right of a point corresponding to a second number is said to *greater than* the second number; we also say that the second number is *less than* the first number. Thus:

If a is to the right of b on the number line, $a > b$ or, equivalently, $b < a$.

$$3 > 1 \text{ or } 1 < 3$$

and

$$\frac{1}{2} > \frac{-1}{3} \text{ or } \frac{-1}{3} < \frac{1}{2}$$

since the point corresponding to the number 3 is to the *right* of the point corresponding to the number 1, and the point corresponding to the number $\frac{1}{2}$ is to the right of the point corresponding to the number $\frac{-1}{3}$ (see Figure 2.1). Of course, since $3 > 1$, the difference $3 - 1$ is positive. Similarly, the difference $\frac{1}{2} - \frac{(-1)}{3}$ is also positive. This discussion can be summarized in the following definition.

DEFINITION 2.4

If a and b are real numbers, a is less than b ($a < b$), or, equivalently, b is greater than a ($b > a$), if $b - a$ is a positive number.

$3 < 5$ and $-5 < -1$ are inequalities of the same sense.

$3 < 5$ and $-1 > -5$ are inequalities of the opposite sense.

Thus, $3 < 5$ (or $5 > 3$), since $5 - 3 = 2$, a positive number. Similarly, $-5 < -1$ (or $-1 > -5$), since $-1 - (-5) = 4$, a positive number. The inequalities $3 < 5$ and $-5 < -1$ are said to be *of the same sense,* because in each case the left-hand member is less than the right-hand member. In contrast, the inequalities $3 < 5$ and $-1 > -5$ are said to be *of opposite sense.*

We now adopt the following axioms of order for the real numbers.

ORDER AXIOMS FOR R

If a, b, and c are real numbers:

0–1	Exactly one of the following relations is true: $a < b$, $a = b$, or $a > b$	Trichotomy Law
0–2	If $a < b$ and $b < c$, then $a < c$	Transitive Law

The solution set of an inequality is the set of numbers that make the inequality a true statement when the variable is replaced by a number in the replacement set.

As in the case of first-degree (linear) equations, the *solution set* of an inequality is the set of numbers that makes the inequality a true statement when the variable is replaced by a number in the replacement set. Thus, the *solution set* of the inequality $x + 1 > 4$ is the set of all numbers greater than 3. This set is sometimes written as $\{x \mid x > 3\}$. The graph of this solution set is shown in Figure 2.2. Note that the number 3 is excluded from the graph. This fact is shown by drawing a small open circle around the number 3.

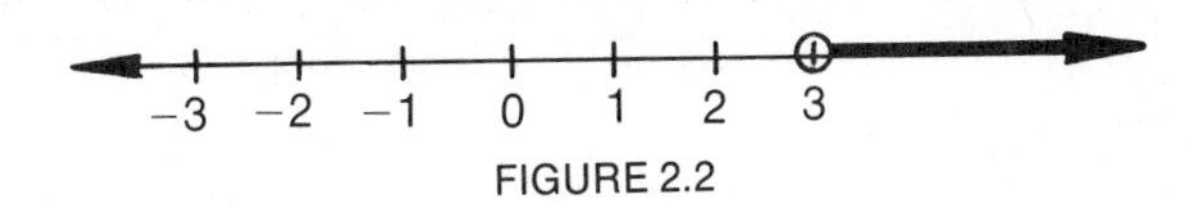

FIGURE 2.2

By convention, and unless otherwise specified, the replacement set of an inequality is the set of real numbers for which the members of the inequality are defined. Inequalities that are *true* for every number in the replacement set—for example, $x^2 \geq 0$—are called *unconditional* or *absolute* inequalities. On the other hand, inequalities that are *not* true for every element in the replacement set, such as $x + 1 > 4$, are called *conditional* inequalities.

The replacement set of an inequality is the set of real numbers for which the inequality is defined.

$x^2 \geqq 0$ is an unconditional, or absolute, inequality.

$x + 1 > 4$ is a conditional inequality.

Most of the properties discussed in connection with the solution of equations also apply to inequalities. For example,

(1) Equal expressions may be added to (or subtracted from) each member of an inequality, producing an inequality of the same sense.

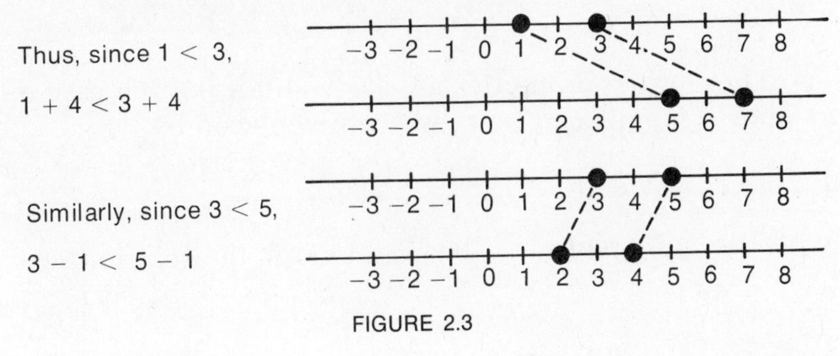

FIGURE 2.3

The procedure used to solve conditional inequalities is similar to that used to solve linear equations, and consists of finding an inequality that is equivalent to the given one and whose solution set is obvious. For instance, as you can see from Figure 2.3, the result of adding or subtracting an expression representing a real number from both members of an inequality is a translation (a move) of both points in the same direction and the same distance along the number line, and thus the result is an equivalent inequality.

Two inequalities are equivalent if their solution sets are identical.

When we multiply (or divide) both members of an inequality by an expression representing a real number, the result depends on the sign of the expression.

(2) Multiplying (or dividing) both sides of an inequality by an expression representing a positive number produces an equivalent inequality in the same sense.

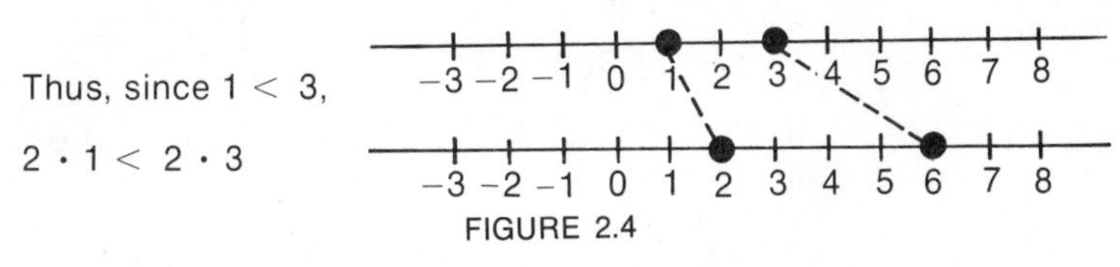

FIGURE 2.4

(3) Multiplying (or dividing) both sides of an inequality by an expression representing a negative number produces an equivalent inequality in the opposite sense.

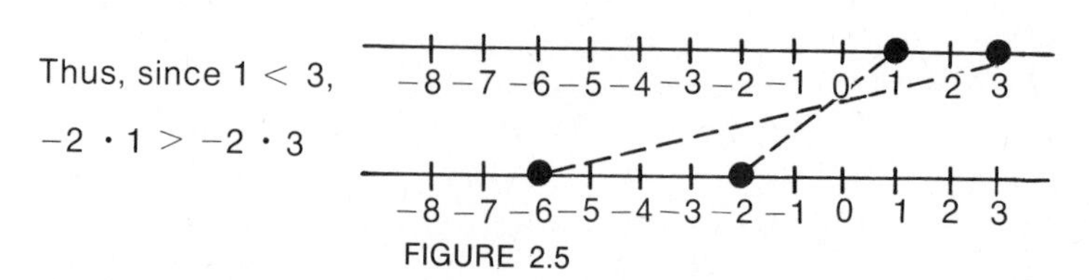

FIGURE 2.5

This discussion can be summarized as follows:

> If P(x), Q(x), and R(x) represent algebraic expressions, then, for all values of x for which these expressions are defined, the inequality $P(x) < Q(x)$ is equivalent to:
> 1. $P(x) + R(x) < Q(x) + R(x)$
> 2. $P(x) \cdot R(x) < Q(x) \cdot R(x)$, for $R(x) > 0$
> 3. $P(x) \cdot R(x) > Q(x) \cdot R(x)$, for $R(x) < 0$

These three properties are also valid if in each case the symbol $<$ is replaced by $\leq$ and $>$ is replaced by $\geq$.

Example 1 Solve the inequality $\frac{x+1}{3} \leq \frac{3}{2}$ and graph the solution set.

Solution

1. $6\left(\frac{x+1}{3}\right) \leq 6\left(\frac{3}{2}\right)$ Using Property 2 to multiply both members by 6, the LCD of the fractions

or $2(x+1) \leq 9$

2. $2x + 2 \leq 9$ Using the Distributive Law
3. $2x \leq 7$ Adding −2 to both members (Property 1)
4. $x \leq \frac{7}{2}$ Multiplying both members by $\frac{1}{2}$ (Property 2)
5. The solution set is $\left\{x \mid x \leq \frac{7}{2}\right\}$; the graph of this solution set is shown in Figure 2.6. Note that the number $\frac{7}{2} = 3.5$

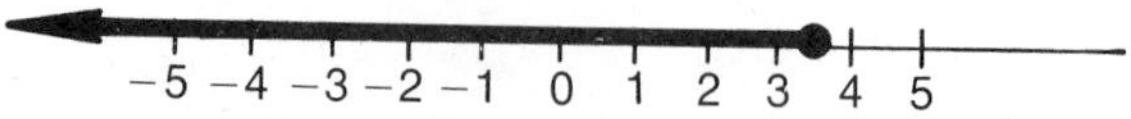

−5 −4 −3 −2 −1 0 1 2 3 4 5

FIGURE 2.6

is included in the graph of the solution set, as indicated by the solid dot.

Example 2

Find and graph the solution set of $\frac{x+2}{3} > x + 4$.

Solution

1. $3\left(\frac{x+2}{3}\right) > 3(x+4)$ Using Property 2 to multiply both members by 3
2. $x + 2 > 3x + 12$ By the Distributive Law
3. $x + 2 + (-3x - 2) > 3x + 12 + (-3x - 2)$ or $-2x > 10$ Adding $(-3x - 2)$ to both members (Property 1)
4. $x < -5$ Multiplying both members by $\frac{-1}{2}$ and reversing the sense of the inequality
5. The solution set is $\{x \mid x < -5\}$; the graph of this set is shown in Figure 2.7.

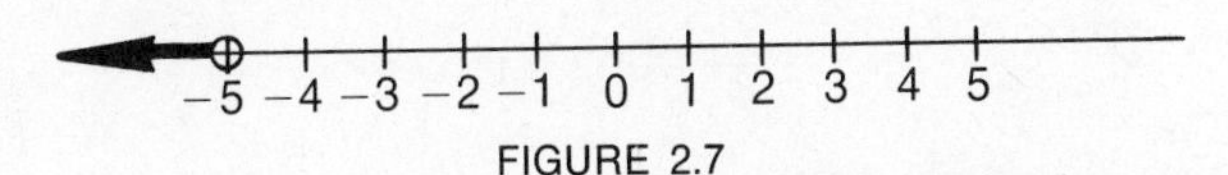

FIGURE 2.7

Sometimes inequalities appear in the form $-8 \leq 2x \leq 10$. The solution set of such inequalities is obtained in a manner similar to the one used above. In our case, each expression can be multiplied by $\frac{1}{2}$, obtaining $-4 \leq x \leq 5$. Thus, the solution set of $-8 \leq 2x \leq 10$ is $\{x \mid -4 \leq x \leq 5\}$; the graph of this solution set is shown in Figure 2.8.

FIGURE 2.8

Example 3

Find and graph the solution set of $-15 \leq 5x < 20$.

Solution

Dividing each member by 5, we have $-3 \leq x < 4$. The solution set is $\{x \mid -3 \leq x < 4\}$; the graph for the solution set is shown in Figure 2.9.

FIGURE 2.9

We turn now to the solution of equations and inequalities involving absolute values. As you recall, the absolute value of a number x, denoted by $|x|$, is the number with the sign disregarded. Thus, $|2| = 2$, $|-5| = 5$, and $|0| = 0$. More precisely, we have

$$|x| = \begin{cases} x, & \text{if } x \geq 0 \\ -x, & \text{if } x < 0 \end{cases}$$

$|x| = 2$ is an absolute value equation.

With this definition, we can see that the absolute value equation $|x| = 2$ has $\{2, -2\}$ as its solution set, since

$$|2| = 2$$
$$|-2| = 2$$

The graph of the solution set for the equation $|x| = 2$ is given in Figure 2.10.

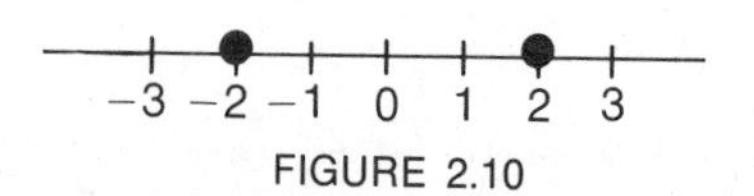

FIGURE 2.10

Note that if $|x| < 2$ then $-2 < x < 2$.

As you can see from this figure, the solution set of $|x| = 2$ consists of all points which are exactly 2 units away from 0. For this reason, the solution set of $|x| < 2$ will consist of all points which are *less than* 2 units away from 0. The graph of the solution set for $|x| < 2$ is shown in Figure 2.11, and it consists of the numbers between -2 and 2, that is,

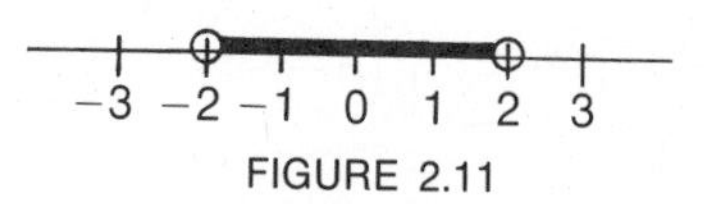

FIGURE 2.11

Recall that if $x \geqq 0$, $|x| = x$.

$$-2 < x < 2$$

Recall that if $x \leqq 0$, $|x| = -x$.

This fact can be verified by using the following argument:

If $x \geq 0$, $|x| = x$, but x must be less than 2. Thus $0 \leq x < 2$.

If $x < 0$, $|x| = -x$; since $|x| < 2$, then $-x < 2$. Thus $x > -2$, so $-2 < x < 0$.

Consequently, if $|x| < 2$, then $-2 < x < 0$ or $0 \leq x < 2$. That is, $-2 < x < 2$. In general, we have:

THEOREM 2.1

If $|x| < a$, then $-a < x < a$, for $a > 0$.

Example 4

Use Theorem 2.1 to solve $|x-1| < 3$, then graph the solution set.

Solution

If $|x| < a$, then $-a < x < a$. If $|x-1| < 3$, then $-3 < x-1 < 3$. Adding 1 to each term, we have $-2 < x < 4$. The graph of the solution set appears in Figure 2.12.

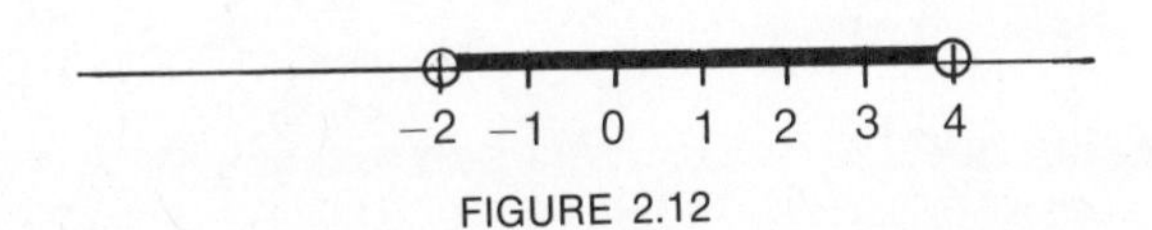

FIGURE 2.12

To find the solution set of $|x| > 3$, we recall that the solution set of $|x| = 3$ consists of all points exactly 3 units from 0. Thus, the solution set of $|x| > 3$ must consist of all points that are *more than* 3 units away from 0. The graph of this solution set appears in Figure 2.13.

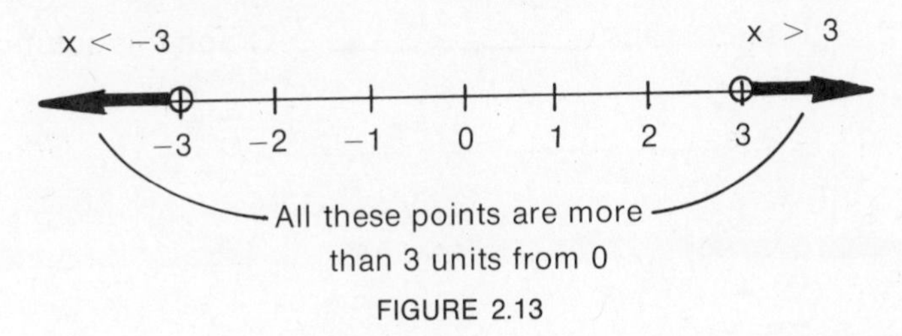

FIGURE 2.13

Since the solution set consists of all numbers greater than 3 or smaller than -3, the solution can be written as $\{x \mid x < -3\} \cup \{x \mid x > 3\}$.

From this discussion we can deduce the following theorem.

THEOREM 2.2

If $|x| > a$, then $x < -a$ or $x > a$.

Example 5

Find and graph the solution set of $|2x+1| \geq 3$.

Solution

By Theorem 2.2, if $|x| > a$, then $x < -a$ or $x > a$. Thus, if $|2x+1| \geq 3$, then $2x+1 \leq -3$ or $2x+1 \geq 3$. Solving $2x+1 \leq -3$ or $2x+1 \geq 3$, we have

$$\begin{aligned} 2x &\leq -4 \quad \text{or } 2x \geq 2 \\ x &\leq -2 \quad \text{or } \ x \geq 1 \end{aligned}$$

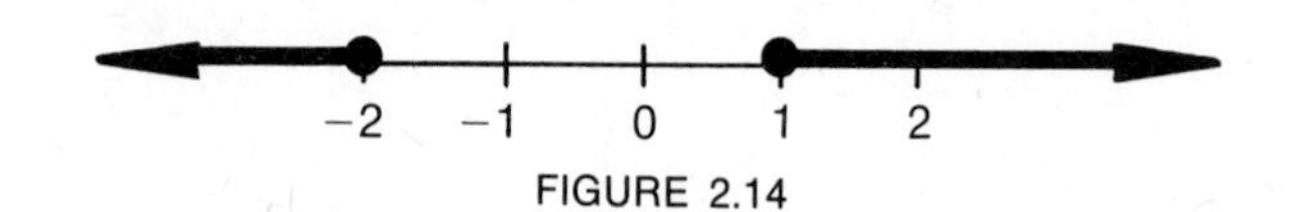

FIGURE 2.14

The solution set, then, is $\{x \mid x \leq -2\} \cup \{x \mid x \geq 1\}$. The graph of this solution set appears in Figure 2.14.

PROGRESS TEST

1. The solution set of $\frac{x+2}{2} < \frac{3}{4}$ is ______________. Graph this solution set.
2. The solution set of $\frac{x-1}{2} > x+3$ is ______________. Graph this solution set.
3. The solution set of $-14 \leq 7x < 21$ is ______________. Graph this solution set.
4. The solution set of $|x-2| < 1$ is ______________. Graph this solution set.
5. The solution set of $|x+1| \geq 2$ is ______________. Graph this solution set.

EXERCISE 2.6

In Problems 1 through 20, solve the given inequality and graph the solution set.

1. $x + 1 \leq 3$
2. $2x + 5 \leq 3$
3. $5x - 3 > 7$
4. $7x - 2 > 5$
5. $\frac{x}{3} + 2 < 3$
6. $\frac{x}{4} + 1 < \frac{-1}{4}$
7. $\frac{9x-2}{4} \geq 4$

PROGRESS TEST ANSWERS

1. $\{x \mid x < -\frac{1}{2}\}$
2. $\{x \mid x < -7\}$
3. $\{x \mid -2 \leq x \leq 3\}$
4. $\{x \mid 1 < x < 3\}$
5. $\{x \mid x \geq 1\} \cup \{x \mid x \leq -3\}$

8. $\frac{10x + 5}{3} \geq 5$

9. $\frac{3x - 3}{2} < 3$

10. $\frac{5x - 7}{3} < 1$

11. $6 - \frac{x}{2} > \frac{5x}{3} - 7$

12. $\frac{x}{2} - 2 > \frac{x}{4} - 1$

13. $\frac{5x}{4} - 3 < \frac{3x}{4} - 1$

14. $\frac{3x}{7} - 2 < x - \frac{2}{7}$

15. $\frac{x}{3} + 2 > x - 1$

16. $\frac{x}{4} + 1 > 2x - \frac{5}{2}$

17. $-3 \leq 6x \leq 12$

18. $-22 < 11x < 33$

19. $-1 \leq \frac{x}{2} \leq 2$

20. $-2 < \frac{x + 2}{3} < 1$

In Problems 21 through 30, use Theorems 2.1 and 2.2 to solve the given inequalities.

Theorem 2.1 If $|x| < a$, then $-a < x < a$ $(a > 0)$

Theorem 2.2 If $|x| > a$, then $x < -a$ or $x > a$ $(a > 0)$

Illustration: Solve:

(a) $|x + 2| < 3$

(b) $|x - 1| \geq 4$

Solution: (a) $|x| < a$ is equivalent to $-a < x < a$.

So $|x + 2| < 3$ is equivalent to $-3 < x + 2 < 3$. Adding -2 to each term, we have $-5 < x < 1$.

(b) $|x| > a$ is equivalent to $x < -a$ or $x > a$. So $|x - 1| \geq 4$ is equivalent to $x - 1 \leq -4$ or $x - 1 \geq 4$. Solving these inequalities, we have: $x \leq -3$ or $x \geq 5$.

21. $|x + 5| < 2$

22. $|x + 2| < 1$

23. $|2x + 1| \leq 3$

24. $|2x + 3| \leq 7$

25. $|x - 1| \geq 2$

26. $|x - 3| \geq 4$

27. $|2x - 1| \geq 3$

28. $|3x - 1| \geq 2$

29. $|3 - 4x| > 7$

30. $|5 - 3x| > 11$

In Problems 31 through 35, use the five-step procedure given in the text to solve the given problem.

Illustration: A student in a math class needs at least 240 points to obtain a grade of B. What grade on her third test would give her a B for the course if her scores on the other two tests were 72 and 86 respectively?

Solution:

1. We are asked for the possible scores on her third test.
2. Let this score be s.
3. The sum of her 3 scores is $72 + 86 + s$. This sum must be at least 240. So,

$$72 + 86 + s \geq 240$$

4. Solving this inequality, we have:

$$\begin{aligned} 72 + 86 + s &\geq 240 \\ 158 + s &\geq 240 \\ s &\geq 240 - 158 \\ s &\geq 82 \end{aligned}$$

Thus her score must be at least 82.

5. Since $72 + 86 + 82 = 240$, our answer is correct.

31. A student in an English class needs at least 360 points to obtain a grade of A. What can the score on his fourth test be so that he can make an A if the scores on the other three tests were 92, 90, and 84, respectively?

32. A man wishes to invest $10,000 in two companies paying 8% and 9% annual return on his money. If he wishes to obtain more than $840 annual interest, what is the maximum amount he could invest in the company paying 8% return? (Hint: $I = PR$.)

33. A man has $2,500 invested at 8%. What is the least amount of money he can invest at 10% in order to receive at least 9% on his total investment?

34. A man wishes to invest $20,000, part at 6% and part at 8%. What is the least amount he can invest at 8% if he wishes an annual return of at least $1,320?

35. A woman has $2500 invested at 8%. What is the least amount she can invest at 12% in order to receive at least a 10% return on her investment?

SELF-TEST–CHAPTER 2

1. Simplify:

    ____________ (a) $7x + 5 - 3x - 9$

    ____________ (b) $3x - 3(x - 2)$

---

2. Find the replacement set for the given equation.

    ____________ (a) $x + \frac{x}{2 + x} = 1$

    ____________ (b) $\frac{-3}{x + 2} = 7$

---

3. Solve:

    ____________ (a) $x - 2 = 6$

    ____________ (b) $3x - 2 = x + 1$

    ____________ (c) $\frac{x}{5} - \frac{x}{2} = 9$

---

4. Find the solution set of the given equation.

    ____________ (a) $\frac{x}{x - 4} = 4 + \frac{4}{x - 4}$

    ____________ (b) $\frac{x}{x - 2} = 2 - \frac{2}{x - 2}$

---

5. In an election between two candidates, 680 votes were cast. If the winner received 150 more votes than the loser, find:

    ____________ (a) The number of votes received by the winner.

    ____________ (b) The number of votes received by the loser.

---

6. How many gallons of a 30% salt solution must be mixed with 40 gallons of a 12% salt solution to obtain a 20% solution? ________

---

7. A woman invested \$10,000 at 8%. What additional amount must she invest at 6% so that her annual interest is \$830? ________

---

8. A train traveling at 50 miles per hour leaves for a certain town. One hour later, a car traveling at 60 miles per hour leaves for the same town and arrives at the same time as the train. If both the car and the train traveled in a straight line, how far is it to the town?

---

9. Solve and graph the solution set of:

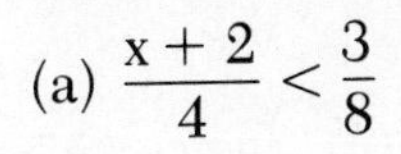

(a) $\frac{x+2}{4} < \frac{3}{8}$

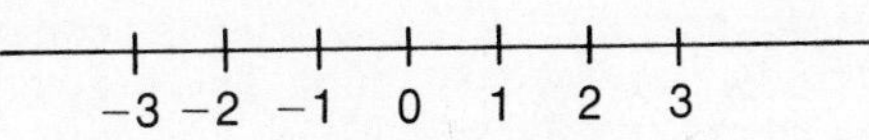

(b) $-6 \leq 2x \leq 4$

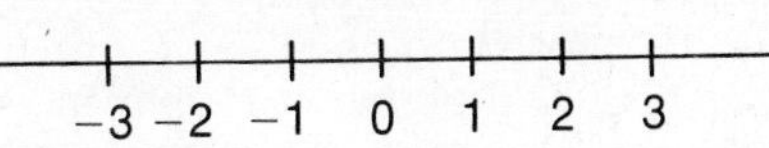

---

10. Graph the solution set of:

(a) $|x - 1| < 2$

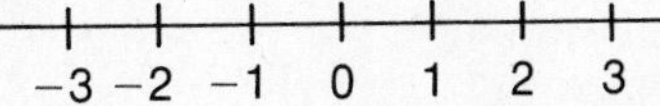

(b) $|x - 1| > 1$

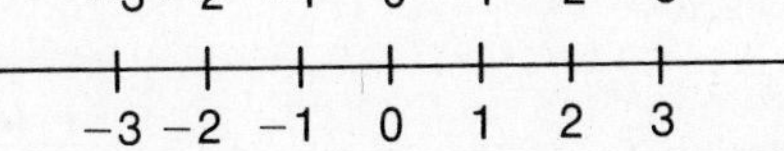

CHAPTER 3

HISTORICAL NOTE

A polynomial equation is an equation like the following:

$$2x^4 + 6x^3 - 7x^2 + 3x - 4 = 0$$

The left side is a sum of powers of the unknown, each multiplied by a number (called a coefficient), and the right side is zero. This kind of equation is named by the highest power of the unknown that occurs. Thus, the preceding equation is called a quartic (fourth-degree) equation. We have already learned how to solve linear (first-degree) equations, and we shall learn in Chapter 6 how to solve quadratic (second-degree) equations.

A problem that was of considerable interest to the ancient Greeks was that of duplicating the cube, that is, constructing a cube with its volume double that of a given cube. This problem leads to the cubic (third degree) equation $x^3 - 2a^3 = 0$. There is a story that the inhabitants of the Greek island of Delos were once told by their oracle that to rid themselves of a certain plague they must double the size of Apollo's cubical altar. The problem was given to the Greek mathematicians, and during the years 400 to 200 B.C., a number of them were able to devise geometrical methods for obtaining a solution.

One of the most outstanding mathematical accomplishments of the sixteenth century was the discovery, by Italian mathematicians, of algebraic solutions of the cubic and the quartic equations. This feat spurred many famous mathematicians to try to solve the general equation of the fifth and higher degrees, and it was not until the early nineteenth century that an Italian, P. Ruffini, and a Norwegian, N. H. Abel, proved that the general equation of any degree higher than four cannot be solved by *algebraic* means. However, equations with given specific numbers for the coefficients can always be solved by methods of approximation.

N. H. Abel, a Norwegian mathematician who proved that the general equation of degree higher than four cannot be solved algebraically. (From Eves, H.: Introduction to the History of Mathematics, 3rd ed. New York, Holt, Rinehart and Winston, 1969, p. 219.)

THE ALGEBRA OF POLYNOMIALS

Delos is remembered for its altar. The inhabitants of the city, after consulting the oracle, were ordered to double the size of the altar of Apollo, which was a cube. (Reproduced by permission from Scully, V.: *The Earth, The Temple, The Gods: Greek Sacred Architecture.* New Haven: Yale University Press, 1962.)

Photo credit : Sterling Publishing Company, Inc., New York.

OBJECTIVES 3.1

After studying this section, the reader should be able to:
1. **Classify a given polynomial as a monomial, binomial, or trinomial.**
2. **Find the degree of a given polynomial.**
3. **Rewrite a given polynomial using exponents.**
4. **Evaluate a given polynomial at a point.**

3.1 MONOMIALS, BINOMIALS, AND TRINOMIALS

In the photograph at the beginning of this section, the man is jumping from an altitude of 118 feet. Do you know how high above sea level he will be after falling downward for t seconds? His height above sea level after t seconds is given in the expression

P(t) is read as "P of t."

$-16t^2 + 118$ is a polynomial. $\frac{5}{x}$, $\sqrt{x}$, and $\frac{x+y}{z}$ are not polynomials.

$$P(t) = -16t^2 + 118$$

The expression $-16t^2 + 118$ is an algebraic expression called a *polynomial.* In general, a polynomial is an expression in which the operations involved consist exclusively of addition, subtraction, and multiplication (except that division by a non-zero number is allowed) and in which all variables have non-negative integers as exponents. Those parts of the polynomial separated by + or − signs are called the *terms* of the polynomial. Thus, the polynomial $-16t^2 + 118$ has *two* terms, $-16t^2$ and 118. A polynomial consisting of *two* terms is called a *binomial.*

A polynomial of two terms is a binomial.

As you recall from Section 1.6, an *exponent* is a number which tells us how many times another number, called the *base,* is used as a factor in an expression. Thus,

$$4 \cdot 4 \cdot 4 = 4^3 \leftarrow \text{exponent}$$
↑
base

In the expression 4^3, 3 is the exponent, and 4 is the base.

$$6 \cdot 6 \cdot 6 \cdot 6 = 6^4 \leftarrow \text{exponent}$$
↑
base

In the expression 6^4, 4 is the exponent, and 6 is the base.

$$x \cdot x = x^2 \leftarrow \text{exponent}$$
↑
base

In the expression x^2, 2 is the exponent, and x is the base.

In general, if x is a real number and n is a natural number,

In the expression x^n, n is the exponent, and x is the base.

$$x^n = \underbrace{x \cdot x \cdot x \ldots \cdot x}_{n \text{ factors}}$$

In the expression x, 1 is understood to be the exponent, and x is the base.

If no exponent appears, the exponent 1 is understood, that is, $x = x^1$, $2 = 2^1$, and $8 = 8^1$. The expressions x^n and x are polynomials with *one* term. A polynomial consisting of only *one* term is called a *monomial.* Thus, x, x^n, and $-3y^7$ are monomials. Finally, a polynomial with *three* terms is called a *trinomial.* Hence, $x + y + z$, $x^2 + 3xy + yz^2$, and $y^2 - 2z + 9$ are examples of trinomials.

A polynomial of one term is a monomial.

A polynomial of three terms is a trinomial.

Example 1

Classify the following polynomials as monomials, binomials, or trinomials.

(a) $x + x^2$
(b) $-9x$
(c) $3x^2 - y + 3xyz$

Solution

(a) Since $x + x^2$ has two terms, $x + x^2$ is a binomial.
(b) Since $-9x$ has one term, $-9x$ is a monomial.
(c) Since $3x^2 - y + 3xyz$ has three terms, $3x^2 - y + 3xyz$ is a trinomial.

In Example 1, the polynomials $x + x^2$ and $-9x$ contain only one variable, x. Polynomials in one variable can be classified according to the highest exponent of the variable that occurs. This number is called the *degree* of the polynomial. Thus, $-8x^5$ is of *fifth* degree, $-3x^3 + x^2 + x$ is of *third* degree, and $-5x$ is of *first* degree. By convention, a non-zero number such as -3 or 9 is called a polynomial of degree 0. The number 0 is called the *zero polynomial* and is not assigned a degree.

The degree of a polynomial in one variable is the highest exponent of the variable it contains.

-3 is a polynomial of degree 0. 0 is called the zero polynomial.

The notion of a polynomial can be extended to include algebraic expressions in more than one variable. For example, the expression $3x^2 - y$ is a polynomial in *two* variables, x and y. The polynomial $3x^2 - y + 3xyz$ is a polynomial in three variables, x, y, and z. The *degree* of a polynomial in several variables is *the largest sum of the exponents in any one term of the polynomial.* For example, the degree of $3x^2 - y$ is 2, the degree of the first term; the degree of $3x^2 - y + 3xyz$ is $1 + 1 + 1 = 3$, the degree of the third term; and the degree of $-x^3y^2z + x^4yz^2$ is $4 + 1 + 2 = 7$, the degree of the second term.

Recall that $3xyz = 3x^1y^1z^1$.

Example 2

Find the degree of the given polynomials.
(a) $-5x^2 + 3x^5 + 9$
(b) $-x^2 + xy^2z^3 - x^5$
(c) 3
(d) 0

Solution

(a) The degree of $-5x^2 + 3x^5 + 9$ is 5.
(b) The degree of $-x^2 + xy^2z^3 - x^5$ is $1 + 2 + 3 = 6$
(c) The degree of 3 is 0.
(d) 0 has no degree.

Example 3

Rewrite each of the given polynomials using exponents and state the degree of the polynomial.
(a) $(x \bullet x) + (x \bullet y \bullet y) + (x \bullet x \bullet y \bullet y)$
(b) $8xxxyy - 2xxyyzzz$

Solution

(a) $(x \cdot x) + (x \cdot y \cdot y) + (x \cdot x \cdot y \cdot y)$ can be written as $x^2 + xy^2 + x^2y^2$. The degree of this polynomial is $2 + 2 = 4$.
(b) $8xxxyy - 2xxyyzzz$ can be written as $8x^3y^2 - 2x^2y^2z^3$. The degree of this polynomial is $2 + 2 + 3 = 7$.

In mathematics, polynomials in one variable are sometimes represented by using symbols such as $P(t)$, $Q(x)$, and $D(y)$, where the symbol in parentheses indicates the variable being used. For example, we may have:

$$P(t) = -16t^2 + 118$$
$$Q(x) = x^2 - 3x$$
$$D(y) = -3y - 9$$

With this notation, it is easy to indicate the value of a polynomial for specific values of the variable. Thus, $P(1)$ represents the value of the polynomial $P(t)$ for $t = 1$. Similarly, $Q(3)$ represents the value of $Q(x)$ for $x = 3$. Thus, since

P(1) represents the value of the polynomial P(t) when t = 1. We also say that we are evaluating P(t) at t = 1.

$$P(t) = -16t^2 + 118,$$
$$P(1) = -16(1)^2 + 118 = 102.$$

Similarly, since

$$Q(x) = x^2 - 3x,$$
$$Q(3) = 3^2 - 3(3) = 0.$$

Using these ideas, we can find the height above sea level of the man mentioned at the beginning of this section. As you recall, the altitude of the man after t seconds was given as $P(t) = -16t^2 + 118$. Thus, after 1 second his altitude will be

$$P(1) = -16(1)^2 + 118 = 102 \text{ feet}$$

Here we are evaluating P(t) when t = 1.

After 2 seconds it will be

$$P(2) = -16(2)^2 + 118 = -64 + 118 = 54 \text{ feet}$$

Here we are evaluating P(t) when t = 2.

Example 4

Find the altitude of the man mentioned at the beginning of this section after 3 seconds.

Solution

After 3 seconds the altitude of the man will be $P(3) = -16(3)^2 + 118 = -26$ feet, that is, 26 feet *below* sea level. (Since the water is only 12 feet deep at this point, divers cannot continue to free fall after they hit the surface.)

PROGRESS TEST

1. A polynomial of one term is called a ____________.
2. A polynomial of two terms is called a ____________.
3. A polynomial of three terms is called a ____________.
4. $y + 3xy$ is a ______, but $y^2 + 3x + 3y$ is a ____________.
5. The degree of $x^2 + x^5 - 6x$ is ______.
6. The degree of $5xyz + x^2z^3 + y^7$ is ______.
7. If $P(x) = x^2 - 2x + 3$, $P(0)$ equals ______________.

EXERCISE 3.1

In Problems 1 through 10, classify the given polynomial as a monomial, binomial, or trinomial and determine its degree.

Illustration: (a) $x^2 + 2$
(b) $(y + z)^2$
(c) $xy^2z^2 + x^3 + z^4$

Solution: (a) $x^2 + 2$ is a binomial. The degree is 2.
(b) Just as x^2 is a monomial of degree 2, $(y + z)^2$ is also a monomial of degree 2.
(c) $xy^2z^2 + x^3 + z^4$ is a trinomial. Its degree is the degree of the first term, $1 + 2 + 2 = 5$.

1. xyz^2

2. u^2vw^3

3. $x^2 + yz^2$

4. $x^2y + z^3 - x^6$

5. $(x + y + z)^3$

6. $(x + y)^2 + y^3$

7. $x^2yz - xy^3 - u^2v^3$

8. 8

9. 0

10. $3xyz - uv^2 + v^7$

In Problems 11 through 16, rewrite each of the polynomials using exponents.

PROGRESS TEST ANSWERS

1. monomial
2. binomial
3. trinomial
4. binomial, trinomial
5. 5
6. 7
7. $0^2 - 2(0) + 3 = 3$

Illustration: (a) $x \cdot x + y \cdot y \cdot y$
(b) $(x - y)(x - y)$

Solution: (a) $x \cdot x + y \cdot y \cdot y = x^2 + y^3$
(b) $(x - y)(x - y) = (x - y)^2$

11. $x \cdot x \cdot x + y \cdot y \cdot y \cdot y$

12. $(a + b)(a + b) - (x - y)(x - y)$

13. $xxxx + yyyy + zzzz$

14. $2aabb - 4(xy)(xy)$

15. $3xx - 4yyy + 5(x + y)(x + y)$

16. $9x - 6(x + y + z)(x + y + z)$

In Problems 17 through 20, evaluate the given polynomial for $x = 2$, $y = -1$, $z = -2$.

Illustration: (a) x^2
(b) xy^2
(c) $(x + y + z)^2$

Solution: (a) $x^2 = (2)^2 = 4$
(b) $xy^2 = (2)(-1)^2 = (2)(1) = 2$
(c) $(x + y + z)^2 = (2 - 1 - 2)^2 = (-1)^2 = 1$

17. z^3

18. $(xy)^3$

19. $(x - 2y + z)^2$

20. $(x - y)(x - z)$

In Problems 21 through 25, evaluate the given polynomial for the specified value of the variable.

Illustration: If $P(t) = 3t^2 - 4t$, find $P(1)$.

Solution: $P(1) = 3(1)^2 - 4(1) = 3 - 4 = -1$

21. If $P(x) = 4x^2 + 4x - 1$, find $P(-1)$

22. If $P(x) = -3x^2 + 3x + 2$, find $P(0)$.

23. If $Q(y) = y^2 - 7y - 2$, find $Q(-2)$.

24. If $P(x) = 2x^2 + 3x$ and $Q(y) = -3y^2 - 7y + 1$, find $P(0) + Q(1)$.

25. If $R(u) = -u^2$ and $S(v) = 2v^3$, find $R(-1) - S(1)$.

B.C. by Johnny Hart, courtesy of Field Enterprises, Inc.

OBJECTIVES 3.2

After studying this section, the reader should be able to find the sum or difference of two or more polynomials.

3.2 ADDITION AND SUBTRACTION OF POLYNOMIALS

The cartoon at the beginning of this section illustrates the fact that

$$2c + 3c = 5c$$
$$\text{and } 5c - 5c = 0$$

Like terms are terms that differ only in their numerical coefficients.

These examples show that to add or subtract polynomials, only similar (or like) terms—that is, terms differing only in their numerical coefficients—can be combined. For example, suppose we wish to add $(5x^2 + 3x + 9) + (7x^2 + 2x + 1)$. For our solution, we use the commutative, associative, and distributive properties studied in Section 1.4 and the fact that (by the Symmetric and Commutative Laws) the Distributive Law can also be written as

$ab + ac = a(b + c)$; thus, $ba + ca = (b + c)a$

$$ba + ca = (b + c)a$$

With these facts, the terms in the expression $(5x^2 + 3x + 9) + (7x^2 + 2x + 1)$ can be added as follows:

Here we grouped like terms.

Here we added like terms.

$$\begin{aligned}(5x^2 + 3x + 9) + (7x^2 + 2x + 1) &= (5x^2 + 7x^2) + (3x + 2x) + (9 + 1)\\ &= (5 + 7)x^2 + (3 + 2)x + (9 + 1)\\ &= 12x^2 + 5x + 10\end{aligned}$$

This addition is sometimes done by writing the terms of the polynomials in order of descending (or ascending) degree, and then placing like terms in a column, as shown.

$$\begin{array}{r} 5x^2 + 3x + 9 \\ + \quad 7x^2 + 2x + 1 \\ \hline 12x^2 + 5x + 10 \end{array}$$

As we have already seen, $a - b = a + (-b)$. Thus, the signs in a polynomial are always taken to indicate positive or negative coefficients, and the operation involved is assumed to be addition. Thus,

$$\begin{aligned}(6x^2 - 9x - 8) + (x^2 + 3x - 1) &= [6x^2 + (-9x) + (-8)] \\ &\quad + [x^2 + 3x + (-1)] \\ &= (6x^2 + x^2) + (-9x + 3x) \\ &\quad + [-8 + (-1)] \\ &= (6 + 1)x^2 + (-9 + 3)x \\ &\quad + [-8 + (-1)] \\ &= 7x^2 + (-6)x + (-9) \\ &= 7x^2 - 6x - 9\end{aligned}$$

Recall that $6x^2 - 9x - 8 = 6x^2 + (-9x) + (-8)$, and that $x^2 + 3x - 1 = x^2 + 3x + (-1)$.

Recall that $x^2 = 1 \cdot x^2$.

Using the column method, this problem can be shortened to

$$\begin{array}{r} 6x^2 - 9x - 8 \\ + \quad x^2 + 3x - 1 \\ \hline 7x^2 - 6x - 9 \end{array}$$

It is important to remark that the scheme used to add polynomials is dependent on the fact that the same laws used in the addition of numbers also apply to polynomials. We list these laws here for your convenience.

If $P(x)$, $Q(x)$, and $R(x)$ are polynomials,

1. $P(x) + Q(x) = Q(x) + P(x)$	Commutative Law
2. $[P(x) + Q(x)] + R(x) = P(x) + [Q(x) + R(x)]$	Associative Law
3. $P(x)[Q(x) + R(x)] = P(x)Q(x) + P(x)R(x)$ $[Q(x) + R(x)]P(x) = Q(x)P(x) + R(x)P(x)$	Distributive Laws

Example 1

Add $10x^3 + 8x^2 - 7x - 3$ and $9 - 4x + x^2 - 5x^3$.

Solution

We write $9 - 4x + x^2 - 5x^3$ in descending order: $-5x^3 + x^2 - 4x + 9$. We then place like terms in a column and add as shown next.

$$\begin{array}{r} 10x^3 + 8x^2 - 7x - 3 \\ + \quad -5x^3 + x^2 - 4x + 9 \\ \hline 5x^3 + 9x^2 - 11x + 6 \end{array}$$

Example 2 Add $-8x^2 + 7x - 8$ and $3x^2 + 5$.

Solution The polynomials are already in descending order, so we place like terms in a column and add as follows:

$$\begin{array}{rr} & -8x^2 + 7x - 8 \\ + & 3x^2 \qquad + 5 \\ \hline & -5x^2 + 7x - 3 \end{array}$$

To subtract polynomials, we first recall that

$$a - (b + c) = a - b - c$$

Thus,

$$\begin{aligned}(3x^2 + 4x - 5) - (5x^2 + 2x + 3) &= 3x^2 + 4x - 5 - 5x^2 - 2x - 3 \\ &= -2x^2 + 2x - 8\end{aligned}$$

Note that to subtract $(5x^2 + 2x + 3)$ from $(3x^2 + 4x - 5)$, we "changed the sign" of each term in $(5x^2 + 2x + 3)$ and then added. This procedure can also be done in columns as shown:

$$\begin{array}{rr} & 3x^2 + 4x - 5 \\ - & 5x^2 + 2x + 3 \\ \hline \end{array} \quad \text{or} \quad \begin{array}{rr} & 3x^2 + 4x - 5 \\ + & -5x^2 - 2x - 3 \\ \hline & -2x^2 + 2x - 8 \end{array}$$

Example 3 Subtract $9x^3 - 7x^2 + 3x - 5$ from $6x^3 + 2x^2 + 5$.

Solution

$$\begin{array}{rr} & 6x^3 + 2x^2 \qquad + 5 \\ - & 9x^3 - 7x^2 + 3x - 5 \\ \hline \end{array} \quad \text{or} \quad \begin{array}{rr} & 6x^3 + 2x^2 \qquad + 5 \\ + & -9x^3 + 7x^2 - 3x + 5 \\ \hline & -3x^3 + 9x^2 - 3x + 10 \end{array}$$

Of course, the same result can be obtained by using the usual rules of signs. Thus,

$$\begin{aligned}&(6x^3 + 2x^2 + 5) - (9x^3 - 7x^2 + 3x - 5) \\ &= (6 - 9)x^3 + [2 - (-7)]x^2 - 3x + [5 - (-5)] \\ &= -3x^3 + 9x^2 - 3x + 10\end{aligned}$$

PROGRESS TEST

1. The addition of $(7x^2 - 5x + 2)$ and $(2x^2 + 3x - 8)$ gives ________.
2. When $3x^2 - 2x - 5$ is subtracted from $8x^2 + 5x + 2$, the result is ________.

EXERCISE 3.2

In Problems 1 through 25, perform the indicated operations.

1. $(x^2 + 4x - 8) + (5x^2 - 4x + 3)$
2. $(3x^2 + 2x - 1) + (8x^2 - 7x + 5)$
3. $(5x^2 + 3x + 4) + (-4x^2 - 5x - 8)$
4. $(-5x^2 + 4x - 3) + (6x^2 + 4x - 7)$
5. $(4x^2 + 7x - 5) - (x^2 + 3x + 4)$
6. $(8x^2 - 6x + 3) - (2x^2 + 4x - 6)$
7. $(3y^2 - 6y - 5) - (8y^2 + 7y - 2)$
8. $(4y^2 - 5y - 2) - (5y^2 - 3y + 6)$
9. $(x^3 - 6x^2 + 4x - 2) + (3x^3 - 6x^2 + 5x - 4)$
10. $(-6x^3 + 2x^2 - 3x + 2) + (2x^3 - 6x^2 + 8x - 4)$
11. $(-8y^3 + 7y^2 + 5y - 5) + (8y^3 + 7y - 6)$
12. $(5y^3 + 3y - 8) + (-9y^3 + 2y^2 - 6y + 3)$
13. $(6v^3 - 3v^2 + 2v - 5) - (3v^3 - v^2 + v + 2)$
14. $(3v^3 - 7v^2 + 3v - 1) - (5v^3 + 3v^2 - 6v + 4)$
15. $(4u^3 - 5u^2 - u + 3) - (9u^3 + 2u - 7)$
16. $(x^3 + y^3 - 8xy + 3) + (2x^3 - y^3 + 10xy - 6)$
17. $(x^3 + y^3 - 6xy + 7) + (3x^3 - y^3 + 8xy - 8)$

PROGRESS TEST ANSWERS

1. $9x^2 - 2x - 6$
2. $5x^2 + 7x + 7$

18. $(2x^3 - y^3 + 3xy - 5) - (x^3 + y^3 - 3xy + 9)$

19. $(x^3 - y^3 + 5xy - 2) - (x^2 - y^3 + 5xy + 2)$

20. $(4x^2 + y^2 - 3x^2y^2) - (x^3 + 3y^2 - 3x^2y^2)$

21. $(a + a^2) + (9a - 4a^2) + (a^2 - 5a)$

22. $(2x + 5x^2) + (7x - x^2) - (x + x^2)$

23. $2y + (x + 3y) - (x + y)$

24. $8y - (y + 3x) + 7y$

25. $(3x^2 + y) - (x^2 - 3y) + (3y + x^2)$

In Problems 26 through 35, justify each of the given equalities by using one of the three laws given in the text.

Illustration: (a) $x^2 + 7x = 7x + x^2$
(b) $(3x + 5)2 = 6x + 10$

Solution: (a) $x^2 + 7x = 7x + x^2$, by the Commutative Law
(b) $(3x + 5)2 = 6x + 10$, by the Distributive Law

26. $3x^2 + 9x = 9x + 3x^2$

27. $-y^3 + 7y = 7y + (-y)^3$

28. $(8x + 9)4 = 32x + 36$

29. $(7 - 2x)(-3) = -21 + 6x$

30. $x^2 + (x + 5) = (x^2 + x) + 5$

31. $(y^3 + y^2) + 4 = y^3 + (y^2 + 4)$

32. $x^7 + (3 + x) = x^7 + (x + 3)$

33. $(y + 7) + y^3 = y^3 + (y + 7)$

34. $3(x^2 + 5) = 3x^2 + 15$

35. $8(x^3 - 5x) = 8x^3 - 40x$

In Problems 36 through 40, let $P(x) = x^2 - 2x + 3$ and $Q(x) = 2x^2 + 3x - 1$. Work out the given expression.

Illustration: (a) $P(x) + Q(x)$
(b) $P(2) + Q(1)$

Solution: (a) $P(x) + Q(x) = (x^2 - 2x + 3) + (2x^2 + 3x - 1)$
$= 3x^2 + x + 2$

(b) $P(2) = (2)^2 - 2(2) + 3$
$= 4 - 4 + 3 = 3$
$Q(1) = 2(1)^2 + 3(1) - 1$
$= 2 + 3 - 1 = 4$
Thus, $P(2) + Q(1) = 3 + 4 = 7$

36. $P(x) - Q(x)$

37. $P(0) + Q(0)$

38. $P(1) - Q(-1)$

39. $P(x) - P(x)$

40. $[P(x) + Q(x)] + P(x)$

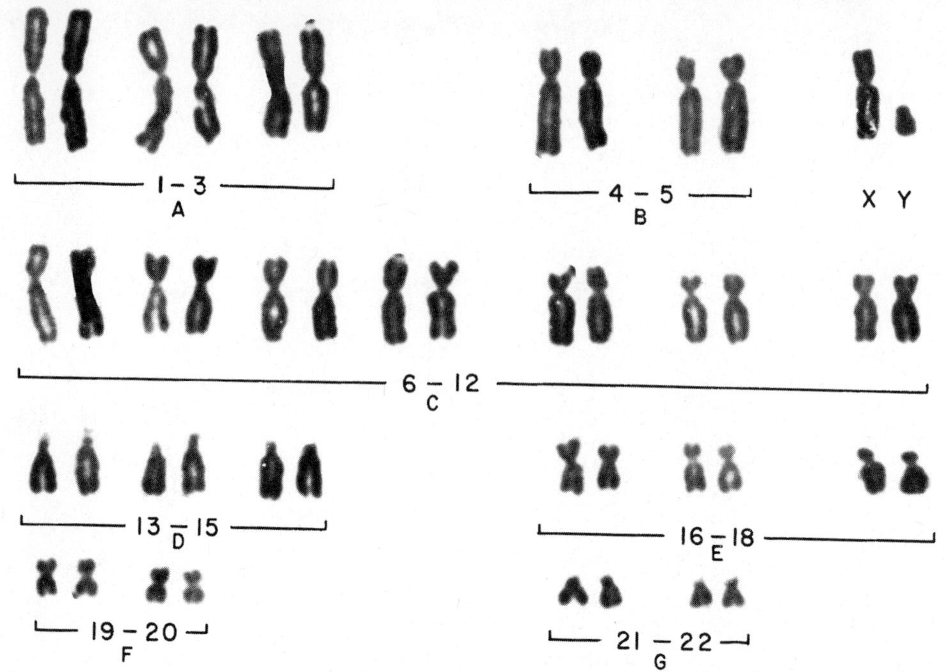

From D. Federman: *Abnormal Sexual Development* (Philadelphia: W. B. Saunders Company, 1967).

OBJECTIVES 3.3

After studying this section, the reader should be able to:

1. **Find the product or quotient of two monomials by correctly using the Laws of Exponents.**
2. **Raise a given monomial to a power.**

3.3 MULTIPLICATION AND DIVISION OF MONOMIALS

The photograph at the beginning of this section shows the 23 pairs of chromosomes present in a human cell (the X and Y chromosomes in the upper right are pair 23). As you are probably aware, chromosomes are the carriers of certain hereditary characteristics such as the color of your hair, the color of your eyes, and the height of your body. Do you know how many different characteristics *each* of your parents can provide? Since there are 23 pairs of chromosomes, the answer is 2^{23}, read "2 to the twenty-third power."

You can obtain 2 characteristics from the first chromosome pair and 2 from the second pair. Thus there are $2 \cdot 2 = 2^2 = 4$ possible characteristics. Then you can obtain 2 more characteristics from the third pair, making the total $2 \cdot 2 \cdot 2 = 2^3$, etc.

As for the total number of characteristics provided by parents, we can see that if each parent can provide 2^{23} different characteristics, the number of possible combinations for a child is

$$2^{23} \times 2^{23} = \underbrace{2 \times 2 \times 2 \ldots \times 2}_{23 \text{ times}} \times \underbrace{2 \times 2 \times 2 \ldots \times 2}_{23 \text{ times}} = 2^{46}$$

Similarly, $2^4 \cdot 2^6 = 2^{4+6} = 2^{10}$. In general, if

$$a^m = \underbrace{a \cdot a \cdot a \ldots a}_{m \text{ factors}}$$

$$\text{and } a^n = \underbrace{a \cdot a \cdot a \ldots a}_{n \text{ factors}}$$

$$a^m \cdot a^n = \underbrace{a \cdot a \cdot a \ldots a}_{m+n \text{ factors}}$$

Thus, we have the following theorem.

THEOREM 3.1

If m and n are natural numbers and a is a real number, then

$$a^m \cdot a^n = a^{m+n} \qquad \textbf{(1)}$$

First Law of Exponents

Equation (1) is sometimes called the *First Law of Exponents,* and it is used to simplify a product involving powers of the same base by simply adding the exponents and using this sum as an exponent for the base. For example,

$$x^6x^9 = x^{6+9} = x^{15}$$
$$y^3yy^5 = y^{3+1+5} = y^9$$

These ideas can be used in the multiplication of two monomials. For example, to multiply $4x^2y^3$ by $2xy^2$, we proceed as follows:

$$(4x^2y^3)(2xy^2) = 4 \cdot 2 \cdot x^2 \cdot x \cdot y^3 \cdot y^2$$
$$= 8x^3y^5$$

by using the Commutative and Associa- Laws

Example 1

Simplify each of the following products.

(a) $(5xy^3)(3x^2y^7)$

(b) $(-2xy^2z^4)(3x^2yz^3)$

(c) $(x^4)^2$

Solution

(a) $(5xy^3)(3x^2y^7) = 15x^{1+2}y^{3+7} = 15x^3y^{10}$
(b) $(-2xy^2z^4)(3x^2yz^3) = -6x^{1+2}y^{2+1}z^{4+3} = -6x^3y^3z^7$
(c) $(x^4)^2 = (x^4)(x^4) = x^{4+4} = x^8$

In part (c) of the preceding example, we notice that the second power of x^4 is x^8. As you can see, the exponent 8 is the product of 2 and 4. Similarly, $(y^4)^3 = y^4 \bullet y^4 \bullet y^4 = y^{4+4+4} = y^{3 \cdot 4} = y^{12}$
In general, we have the following theorem.

THEOREM 3.2

If m and n are natural numbers and a is a real number, then

$$(a^m)^n = a^{m \cdot n} \qquad \textbf{(2)}$$

Thus, $(3^2)^3 = 3^{2 \cdot 3} = 3^6$, $(x^4)^5 = x^{4 \cdot 5} = x^{20}$, and $(y^2)^7 = y^{2 \cdot 7} = y^{14}$.

Example 2

Simplify each of the following products.
(a) $(3x^3y^5)(x^2)^3$
(b) $(y^3)^2(-4xy^3)$

Solution

(a) $(3x^3y^5)(x^2)^3 = (3x^3y^5)(x^{2 \cdot 3})$ Using Equation (2)
$= (3x^3y^5)(x^6)$
$= 3x^9y^5$ Using Equation (1)

(b) $(y^3)^2(-4xy^3) = (y^{3 \cdot 2})(-4xy^3)$ Using Equation (2)
$= (y^6)(-4xy^3)$
$= -4xy^9$ Using Equation (1)

If we use the definition of an exponent, we can see that

$$(ab)^3 = (ab)(ab)(ab) = (a \bullet a \bullet a)(b \bullet b \bullet b) = a^3 \bullet b^3$$

This idea can be generalized so that it applies to any natural number n. The result is shown in Theorem 3.3.

THEOREM 3.3

If a and b are real numbers and n is a natural number, then

$$(ab)^n = a^n b^n \tag{3}$$

Thus, $(xy)^7 = x^7y^7$, $(uv)^4 = u^4v^4$, and $(pq)^9 = p^9q^9$

Example 3

Simplify each of the following products.
(a) $(4xy^3)(xy)^7$
(b) $(x^3)^5(xy)^4$
(c) $(x^3y^2)^4(xy)^3$

Solution

(a) $(4xy^3)(xy)^7 = (4xy^3)(x^7y^7)$ Using Equation (3)
$= 4x^8y^{10}$ Using Equation (1)

(b) $(x^3)^5(xy)^4 = (x^{15})(x^4y^4)$ Using Equations (2) and (3)
$= x^{19}y^4$ Using Equation (1)

(c) If we let $a = x^3$, $b = y^2$, and $n = 4$ in Theorem 3.3, we have

$$(x^3y^2)^4 = (x^3)^4(y^2)^4 = x^{12}y^8$$

Thus, $(x^3y^2)^4(xy)^3 = (x^{12}y^8)(xy)^3$
$= (x^{12}y^8)(x^3y^3)$ Using Equation (3)
$= x^{15}y^{11}$ Using Equation (1)

The ideas contained in Theorem 3.3 can be extended to cover any number of factors. For example,

$$(x^2y^3z^5)^4 = (x^2)^4(y^3)^4(z^5)^4 = x^8y^{12}z^{20}$$

Similarly,

$$(a^2b^3c^4d^5)^6 = (a^2)^6(b^3)^6(c^4)^6(d^5)^6$$
$$= a^{12}b^{18}c^{24}d^{30}$$

Example 4 Simplify each of the following products.

(a) $(x^2yz^3)^4(-3xy^2z)^3$
(b) $(-2x^2)^3(-3y)^4$
(c) $(-x^3y^2)^3(-xy)^2(x^4y^3)^4$

Solution

(a)
$$\begin{aligned}(x^2yz^3)^4(-3xy^2z)^3 &= [(x^2)^4(y)^4(z^3)^4][(-3)^3(x)^3(y^2)^3(z)^3]\\ &= [x^8y^4z^{12}][-27x^3y^6z^3]\\ &= -27x^{11}y^{10}z^{15}\end{aligned}$$

(b)
$$\begin{aligned}(-2x^2)^3(-3y)^4 &= [(-2)^3(x^2)^3][(-3)^4(y)^4]\\ &= [-8x^6][81y^4]\\ &= -648x^6y^4\end{aligned}$$

(c)
$$\begin{aligned}&(-x^3y^2)^3(-xy)^2(x^4y^3)^4\\ &= [(-x^3)^3(y^2)^3][(-x)^2(y^2)][(x^4)^4(y^3)^4]\\ &= [-x^9y^6][x^2y^2][x^{16}y^{12}]\\ &= -x^{27}y^{20}\end{aligned}$$

We are now ready to consider the division of monomials. As you recall, any number of the form $\frac{a}{b}$ $(b \neq 0)$ can be written as $a \cdot \left(\frac{1}{b}\right)$.

Thus, we can write

$$\frac{a^5}{a^2} = a^5 \cdot \left(\frac{1}{a^2}\right)$$

Recall that $a^5 = a^3 \cdot a^2$.

$$= a^3 \cdot a^2 \cdot \left(\frac{1}{a^2}\right) \quad \text{By using Equation (1)}$$

$$= a^3 \cdot \left(a^2 \cdot \frac{1}{a^2}\right) \quad \text{By using the Associative Law}$$

$$= a^3 \cdot 1 \quad \text{Since the product of a number and its reciprocal is 1}$$

$$= a^3$$

Note that $\frac{a^5}{a^2} = a^3 = a^{5-2}$.

The same argument will hold for the quotient

$$\frac{a^m}{a^n}, \quad a \neq 0$$

where m and n are natural numbers, m > n, and a is a real number. This result is stated formally in Theorem 3.4.

THEOREM 3.4

If m and n are natural numbers, m > n, and a is a nonzero real number, then

$$\frac{a^m}{a^n} = a^{m-n} \tag{4}$$

Thus, $\frac{3^5}{3^2} = 3^{5-2} = 3^3$, $\frac{x^7}{x^5} = x^{7-5} = x^2$, and $\frac{y^4}{y^3} = y^{4-3} = y$.

Example 5

Simplify each of the following quotients:

(a) $\frac{21x^7y^3z^4}{7xy^2z^3}$

(b) $\frac{(-10^3y^4)(-4x^3y^2)}{8x^2y^3}$

(c) $\frac{(4ab^2c^3)^3}{-16abc}$

Solution

(a) $\frac{21x^7y^3z^4}{7xy^2z^3} = \frac{21}{7}x^{7-1}y^{3-2}z^{4-3} = 3x^6yz$

(b) $\frac{(-10x^3y^4)(-4x^3y^2)}{8x^2y^3} = \frac{(-10)(-4)x^{3+3}y^{4+2}}{8x^2y^3} = \frac{40x^6y^6}{8x^2y^3}$
$= 5x^4y^3$

(c) $\frac{(4ab^2c^3)^3}{-16abc} = \frac{4^3a^3b^6c^9}{-16abc} = \frac{64a^3b^6c^9}{-16abc} = \left(\frac{64}{-16}\right)a^{3-1}b^{6-1}c^{9-1}$
$= -4a^2b^5c^8$

PROGRESS TEST

1. When simplified, the product $(-2x^2y^3z)(4xy^2z^5)$ equals ________.
2. When simplified, the product $(4x^3y^2)(x^2)^3$ equals ________.
3. When simplified, the product $(-5xy^2)(x^2y)^4$ equals ________.
4. When simplified, the product $(-3xy)^2(4x^2y)^3$ equals ________.
5. When simplified, the quotient $\frac{24x^2y^5z^3}{8xy^2z}$ equals ________.
6. When simplified, the quotient $\frac{(3x^2y)^2(-xy^2z^2)}{9x^2y^3z}$ equals ________.

EXERCISE 3.3

In Problems 1 through 20, find the indicated products. Assume that variables in the exponents represent natural numbers.

Illustration: (a) $(3xy)^2(x^{n-2})$
(b) $(a^{n+1}b^n)(a^2b)^n$

Solution: (a) $(3xy)^2(x^{n-2}) = (3^2x^2y^2)(x^{n-2})$
$= 9x^ny^2$

(b) $(a^{n+1}b^n)(a^2b)^n = (a^{n+1}b^n)(a^{2n}b^n)$
$= a^{(n+1)+2n}b^{n+n}$
$= a^{3n+1}b^{2n}$

1. $(5a)(4a)$

2. $(8a)(-2a^2)$

3. $(4xy)(3x^3y)$

4. $(-2x^2y)(7xy^2)$

5. $(8u^2vw^3)(-3uv^3w)$

6. $(u^n)(u^{n+1})$

7. $(x^{n-1})(x^{n+2})$

8. $x^{2n-2} \bullet x^{n+3}$

PROGRESS TEST ANSWERS

1. $-8x^3y^5z^6$
2. $4x^9y^2$
3. $-5x^9y^6$
4. $576x^8y^5$
5. $3xy^3z^2$
6. $-x^3yz$

9. $(x^2)^7$

10. $(y^3)^2$

11. $(a^2b)^3$

12. $(-3ab^2)^3$

13. $(-2ab^2)^4$

14. $(-3a^2bc^3)^3$

15. $(4x^2)(3y^2)^3$

16. $(-2x)^2(3y)^4$

17. $(3xy)^2(-4x^2y)^3$

18. $(2x^2yz^2)^3(-xyz)^4$

19. $(2xyz)^{2n}(x^2yz)^n$

20. $(nx^n)^2(nx^n)^4$

In Problems 21 through 32, find the indicated quotients. Assume that variables in the exponents represent natural numbers and that variables in the denominator are nonzero real numbers.

Illustration: (a) $\dfrac{a^{2n+1}}{a^{n-1}}$

(b) $\dfrac{(a^2b^3)^n(4ab)^2}{8ab^n}$

Solution: (a) $\dfrac{a^{2n+1}}{a^{n-1}} = a^{2n+1-(n-1)} = a^{n+2}$

(b) $$\begin{aligned}\frac{(a^2b^3)^n(4ab)^2}{8ab^n} &= \frac{(a^{2n}b^{3n})(16a^2b^2)}{8ab^n}\\ &= \frac{16a^{2n+2}b^{3n+2}}{8ab^n}\\ &= 2a^{2n+1}b^{2n+2}\end{aligned}$$

21. $\dfrac{a^8}{a^3}$

22. $\dfrac{-a^5}{a^3}$

23. $\dfrac{6x^2y^3}{3xy^2}$

24. $\dfrac{36x^5y^2}{6x^2y}$

25. $\dfrac{9x^2y^5z^3}{3xy^2z}$

26. $\dfrac{54x^9y^2z^7}{-9x^2yz^3}$

27. $\dfrac{(2x^2y^3)(-3x^5y)}{6xy^3}$

28. $\dfrac{(-3x^3y^2z)(4xy^3z)}{6xy^2z}$

29. $\dfrac{(xy^2)^3(x^2y^7)}{xy^8}$

30. $\dfrac{(-x^2y)(x^3y^2)}{x^3y}$

31. $\dfrac{(a^2bc^2)^3(-3ab)^2}{3a^3b}$

32. $\dfrac{(-4a^2b^2c^3)^2(3abc)^4}{8ab^3c^5}$

OBJECTIVES 3.4

After studying this section, the reader should be able to:

1. **Use the commutative, associative, and distributive laws to find the product of a monomial and a polynomial.**
2. **Use the four special products mentioned in the text to find the product of two binomials.**

3.4 MULTIPLICATION OF POLYNOMIALS

In the cartoon at the beginning of this section, the boy wishes to know what the result is when two polynomials are multiplied together. The result, of course, is another polynomial.

In multiplying polynomials such as $6x^2$ and $2x + 3xy$, we use the Commutative, Associative, and Distributive Laws of Multiplication for polynomials. These laws are generalizations of the Commutative, Associative, and Distributive Laws discussed in previous chapters and are stated as follows:

If P(x), Q(x), and R(x) are polynomials,	
1. $P(x)Q(x) = Q(x)P(x)$	Commutative Law
2. $P(x)[Q(x)R(x)] = [P(x)Q(x)]R(x)$	Associative Law
3. $P(x)[Q(x) + R(x)] = P(x)Q(x) + P(x)R(x)$ $[P(x) + Q(x)]R(x) = P(x)R(x) + Q(x)R(x)$	Distributive Laws

To multiply $6x^2y$ by $2x + 3xy$, we proceed as follows:

$6x^2y(2x + 3xy) = 6x^2y(2x) + 6x^2y(3xy)$	Using the Distributive Law
$= 12x^3y + 18x^3y^2$	Multiplying

We can multiply $4x^2$ by $3x^3 + 7x^2 - 4x + 8$ in a similar way. Thus,

$$\begin{aligned} 4x^2(3x^3 + 7x^2 - 4x + 8) &= 4x^2(3x^3) + 4x^2(7x^2) + 4x^2(-4x) + 4x^2(8) \\ &= 12x^5 + 28x^4 - 16x^3 + 32x^2 \end{aligned}$$

Example 1

Multiply:
(a) $5x^2(3x^3 + 3x^2 - 2x - 3)$
(b) $-2x^3y(x^2 + 7xy - 2y^2)$

Solution

(a) $5x^2(3x^3 + 3x^2 - 2x - 3) = 15x^5 + 15x^4 - 10x^3 - 15x^2$
(b) $-2x^3y(x^2 + 7xy - 2y^2) = -2x^5y - 14x^4y^2 + 4x^3y^3$

To multiply $(3x + 2)(x + 3)$, we use the fact that

$$a(b + c) = ab + ac$$

and multiply each term of $x + 3$ by $3x + 2$, obtaining

$$\begin{aligned} (3x + 2)(x + 3) &= (3x + 2)(x) + (3x + 2)(3) \\ &= 3x^2 + 2x + 9x + 6 \\ &= 3x^2 + 11x + 6 \end{aligned}$$

This multiplication can also be done by arranging the work as in ordinary multiplication and placing like terms in the same column. The procedure looks like this:

$$\begin{array}{rl} x + 3 & \\ \underline{3x + 2} & \\ 3x^2 + 9x \quad & \text{Multiplying } 3x \text{ by } x + 3 \\ \underline{\quad 2x + 6} & \text{Multiplying 2 by } x + 3 \\ 3x^2 + 11x + 6 & \text{Adding like terms} \end{array}$$

Since there are certain types of factors whose multiplications is very common in algebra, it is convenient to list the general results—the "special products"—obtained when the multiplication is carried out. The verification for these products appears in the margin.

$$\begin{array}{r} x + b \\ \underline{x + a} \\ x^2 + bx \quad \\ \underline{ax + ab} \\ x^2 + (a + b)x + ab \end{array}$$

SPECIAL PRODUCT 1

$(x + a)(x + b) = x^2 + (a + b)x + ab$

Thus, $(x+5)(x+3) = x^2 + (5+3)x + 5 \cdot 3 = x^2 + 8x + 15$ and $(x+8)(x+4) = x^2 + (8+4)x + 8 \cdot 4 = x^2 + 12x + 32$.

If in Special Product 1 we let $a = b$, we obtain the following:

SPECIAL PRODUCT 2

$$(x+a)(x+a) = (x+a)^2 = x^2 + 2ax + a^2$$

$$\begin{array}{r} x + a \\ x + a \\ \hline x^2 + ax \quad\quad \\ ax + a^2 \\ \hline x^2 + 2ax + a^2 \end{array}$$

Using Special Product 2, we can see that

$$\begin{aligned}(x+5)^2 &= x^2 + 2 \cdot 5x + 5^2 \\ &= x^2 + 10x + 25\end{aligned}$$

and that

$$\begin{aligned}(x+3)^2 &= x^2 + 2 \cdot 3x + 3^2 \\ &= x^2 + 6x + 9\end{aligned}$$

If in Special Product 2 a is replaced by $-a$, we obtain the following:

SPECIAL PRODUCT 3

$$(x-a)^2 = x^2 - 2ax + a^2$$

Note that $(x-a)^2 \neq x^2 - a^2$.

Thus,

$$\begin{aligned}(x-1)^2 &= x^2 - 2 \cdot 1x + 1^2 \\ &= x^2 - 2x + 1\end{aligned}$$

and

$$\begin{aligned}(x-2)^2 &= x^2 - 2 \cdot 2x + 2^2 \\ &= x^2 - 4x + 4\end{aligned}$$

Example 2

Find:
(a) $(2x+y)(2x+5y)$
(b) $(3x+2y)^2$
(c) $(2x-3y)^2$

Solution

(a) Using $2x$ instead of x, y instead of a, and $5y$ instead of b in Special Product 1, we have:

$$\begin{aligned}(2x+y)(2x+5y) &= (2x)^2 + (y+5y)2x + y \cdot 5y \\ &= 4x^2 \quad + 12xy \quad\quad + 5y^2\end{aligned}$$

(b) Using 3x instead of x and 2y instead of a in Special Product 2, we have:

$$(3x + 2y)^2 = (3x)^2 + 2(2y)(3x) + (2y)^2$$
$$= 9x^2 + 12xy + 4y^2$$

(c) Using 2x instead of x and 3y instead of a in Special Product 3, we have:

$$(2x - 3y)^2 = (2x)^2 - 2(3y)(2x) + (3y)^2$$
$$= 4x^2 - 12xy + 9y^2$$

$$\begin{array}{l} x - y \\ \underline{x + y} \\ x^2 - xy \\ \underline{\quad + xy - y^2} \\ x^2 \qquad - y^2 \end{array}$$

SPECIAL PRODUCT 4

$(x + y)(x - y) = x^2 - y^2$

If we use Special Product 4 with $x = 2a$ and $y = 3b$, we have:

$$(2a + 3b)(2a - 3b) = (2a)^2 - (3b)^2$$
$$= 4a^2 - 9b^2$$

Example 3 Find:

(a) $(3a + b)(3a - b)$

(b) $(4a^2 + 3b^3)(4a^2 - 3b^3)$

Solution (a) Using Special Product 4 with $x = 3a$ and $y = b$, we have:

$$(3a + b)(3a - b) = (3a)^2 - b^2$$
$$= 9a^2 - b^2$$

b) Using Special Product 4 with $x = 4a^2$ and $y = 3b^3$, we obtain:

$$(4a^2 + 3b^3)(4a^2 - 3b^3) = (4a^2)^2 - (3b^3)^2$$
$$= 16a^4 - 9b^6$$

PROGRESS TEST

1. $5xy^2(3x - 2xy)$ equals ______.
2. $-3x^2y(x^2 - 2xy + 3y)$ equals ______.
3. $(2x + 1)(3x - 1)$ equals ______.
4. $(x + 2)(x + 7)$ equals ______.
5. $(2x + 4y)^2$ equals ______.
6. $(3x - 5y)^2$ equals ______.
7. $(3x + y)(3x - y)$ equals ______.

EXERCISE 3.4

In Problems 1 through 20, do the indicated multiplication.

Illustration: (a) $x^{3n}(x^{n-1} + x^2 - x)$
(b) $(x + a)(x^2 - ax + a^2)$

Solution: (a) $x^{3n}(x^{n-1} + x^2 - x)$

$$= x^{3n+(n-1)} + x^{3n+2} - x^{3n+1}$$
$$= x^{4n-1} + x^{3n+2} - x^{3n+1}$$

(b) $(x + a)(x^2 - ax + a^2)$

$$= (x + a)x^2 + (x + a)(-ax) + (x + a)(a^2)$$
$$= x^3 + \cancel{ax^2} - \cancel{ax^2} - \cancel{a^2x} + \cancel{a^2x} + a^3$$
$$= x^3 + a^3$$

This problem can also be done by using the column method. Thus,

$$\begin{array}{r} x^2 - ax + a^2 \\ x + a \\ \hline x^3 - ax^2 + a^2x \qquad \\ + ax^2 - a^2x + a^3 \\ \hline x^3 \qquad\qquad\quad + a^3 \end{array}$$

1. $3x(4x - 2)$

2. $4x(x - 6)$

3. $-3x^2(x - 3)$

4. $-5x^3(x^2 - 8)$

PROGRESS TEST ANSWERS

1. $15x^2y^2 - 10x^2y^3$
2. $-3x^4y + 6x^3y^2 - 9x^2y^2$
3. $6x^2 + x - 1$
4. $x^2 + 9x + 14$
5. $4x^2 + 16xy + 16y^2$
6. $9x^2 - 30xy + 25y^2$
7. $9x^2 - y^2$

5. $-8x(3x^2 - 2x + 1)$

6. $-4x^2(3x^2 - 5x - 1)$

7. $(3x + 3)(3x + 1)$

8. $(x + 5)(2x + 7)$

9. $(5x - 4)(x + 3)$

10. $(2x - 1)(x + 5)$

11. $(3a - 1)(a + 5)$

12. $(3a - 2)(a + 7)$

13. $(y + 5)(2y - 3)$

14. $(y + 1)(5y - 1)$

15. $(x - 3)(x - 5)$

16. $(x - 6)(x - 1)$

17. $(2x - 1)(3x - 2)$

18. $(3x - 5)(x - 1)$

19. $(2x - 3a)(2x + 5a)$

20. $(5x - 2a)(x + 5a)$

In Problems 21 through 50, use special products 1, 2, 3, and 4 to multiply.

Illustration: (a) $-3x(x + 3)(x + 5)$
(b) $x(x + 5)^2$
(c) $(x - 12)^2$
(d) $(x + y)(x - y)x^3$

Solution: (a) $-3x(x + 3)(x + 5)$
$= -3x(x^2 + 8x + 15)$
$= -3x^3 - 24x^2 - 45x$
(b) $x(x + 5)^2 = x[x^2 + 10x + 25]$
$= x^3 + 10x^2 + 25x$
(c) $(x - 12)^2 = x^2 - 24x + 144$
(d) $(x + y)(x - y)x^3 = (x^2 - y^2)x^3$
$= x^5 - x^3y^2$

21. $(x+7)(x+8)$

22. $(x+1)(x+9)$

23. $(2a+b)(2a+4b)$

24. $(3a+2b)(3a+5b)$

25. $(4u+v)^2$

26. $(3u+2v)^2$

27. $(2y+z)^2$

28. $(4y+3z)^2$

29. $(3a-b)^2$

30. $(4a-3b)^2$

31. $(a+b)(a-b)$

32. $(a+4)(a-4)$

33. $(5x-2y)(5x+2y)$

34. $(2x-7y)(2x+7y)$

35. $-(3a-b)(3a+b)$

36. $-(2a-5b)(2a+5b)$

37. $3x(x+1)(x+2)$

38. $3x(x+2)(x+3)$

39. $-3x(x-1)(x-3)$

40. $-2x(x-5)(x-1)$

41. $x(x+3)^2$

42. $3x(x+7)^2$

43. $-2x(x-1)^2$

44. $-5x(x-3)^2$

45. $(2x + y)(2x - y)y^2$

46. $(3x + y)(3x - y)x^2$

47. $x[x - (3x + 2) - (x + 1)]$

48. $-x[2x + (x - 2) - (3x + 1)]$

49. $-[a - 2(a + 1) - (3a + 1)]$

50. $-[(a + 1) - 3(2a - 1) + 1]$

OBJECTIVES 3.5

After studying this section, the reader should be able to:

1. **Write a composite number as a product of primes.**
2. **Find a monomial factor which is common to the terms of a given polynomial and then write the polynomial as a product of this monomial and another polynomial.**

3.5 INTRODUCTION TO FACTORING

In the cartoon at the beginning of this section, Franklin is asking Patty what we will get if we multiply x times y and a times b. The correct answers are, of course, xy and ab. As you recall from Section 1.6, x and y are called *factors* of the product xy. Similarly, because $80 = 2 \cdot 40$, 2 and 40 are factors of 80. Before discussing the factorization of algebraic expressions, we need to introduce the idea of a "prime factor." As you may be aware, if a natural number greater than 1 has as its only factors itself and 1, the number is said to be *prime*. Thus, 2, 3, 5, 7, 11, and 13 are prime numbers, since their only factors are themselves and 1. Any natural number greater than 1 that is not prime is said to be *composite*. Thus, 4, 6, 8, 9, and 10 are *composite*. The number 1 is neither a prime nor a composite number. A composite number is said to be in completely factored form when it is written as a product of prime numbers only. The prime factorization of a composite number can be performed by "dividing out" the prime factors of the number, starting with the smallest factor. Thus,

x and y are factors of xy.

A number is prime if its only factors are itself and 1.

A number greater than 1 that is not prime is composite.

A composite number is completely factored if it is written as a product of primes.

$$18 = 2 \cdot 9 = 2 \cdot 3 \cdot 3$$
$$\text{and } 40 = 2 \cdot 20 = 2 \cdot 2 \cdot 10 = 2 \cdot 2 \cdot 2 \cdot 5$$

$18 = 2 \cdot 3^2$

$40 = 2^3 \cdot 5$

Some people prefer to write these computations as follows:

18	2 ← divide by 2
9	3 ← divide by 3
3	3 ← divide by 3
1	

40	2 ← divide by 2
20	2 ← divide by 2
10	2 ← divide by 2
5	5 ← divide by 5
1	

In every case, "dividing out" a composite number yields a unique series of prime factors. This is stated formally in the next Theorem.

THEOREM 3.5

FUNDAMENTAL THEOREM OF ARITHMETIC

Every composite number can be expressed as a unique product of primes (disregarding the order of the factors).

Example 1 Express the given number as a product of primes.
(a) 126
(b) 210

Solution (a)

126	2
63	3
21	3
7	7
1	

Thus, $126 = 2 \cdot 3 \cdot 3 \cdot 7 = 2 \cdot 3^2 \cdot 7$

(b)

210	2
105	3
35	5
7	7
1	

Thus, $210 = 2 \cdot 3 \cdot 5 \cdot 7$

As in the case of numbers, an algebraic expression can be written as the product of other algebraic expressions, called *factors* of the original. For example, Special Product 1 states that

$$x^2 + (a + b)x + ab = (x + a)(x + b)$$

In this case, $(x + a)$ and $(x + a)$ are the *factors* of $x^2 + (a + b)x + ab$. Similarly, the factors of the expression $x^2 - y^2$ are $(x + y)$ and $(x - y)$, since $x^2 - y^2 = (x + y)(x - y)$. In the remainder of this section we shall study different methods of finding the factors of a polynomial with integer coefficients. This process is called *factoring* the polynomial. If a polynomial does not have any factors with integer coefficients except itself and 1, the polynomial is said to be *prime*.

The process of writing a polynomial as a product of its factors is called factoring the polynomial.

The easiest type of factorization involves using the Distributive Law,

$$ax + ay + az = a(x + y + z)$$

Thus, to factor the polynomial $4x^2 + 6x + 10$, we observe that the number 2 is a factor of each term. We first write

$$4x^2 + 6x + 10 = 2(\qquad\qquad)$$

and insert in the parentheses the polynomial that will maintain the equality. To do this, we have to find those monomials that when multiplied by 2 will yield $4x^2$, $6x$, and 10 respectively. The final result will be

We shall show in Section 3.6 that $2x^2 + 3x + 5$ is not factorable.

$$4x^2 + 6x + 10 = 2(2x^2 + 3x + 5)$$

Similarly, to factor $6x^3 + 8x^2 - 4x$, we must first find the factor that is common to $6x^3$, $8x^2$, and $4x$. This factor is $2x$, so we write

$$6x^3 + 8x^2 - 4x = 2x(3x^2 + 4x - 2)$$

Example 2

Factor
(a) $18x^3y^4 + 21xy^2$
(b) $3a^3b^2c^4 - 9a^2bc^3 - 6abc^2$

Solution

(a) Since the common factor of $18x^3y^4$ and $21xy^2$ is $3xy^2$, we have:

$$18x^3y^4 + 21xy^2 = 3xy^2(6x^2y^2 + 7)$$

(b) The common factor of $3a^3b^2c^4$, $-9a^2bc^3$, and $-6abc^2$ is $3abc^2$; thus,

$$3a^3b^2c^4 - 9a^2bc^3 - 6abc^2 = 3abc^2(a^2bc^2 - 3ac - 2)$$

In some cases, it is impossible to find a common factor for every term of a polynomial, but the terms can be grouped in such a way that each group has a common factor. For example, the polynomial $ax + ay + bx + by$ does not have a factor common to all terms. However,

$$(ax + ay) + (bx + by) = a(x + y) + b(x + y)$$

and now the right member has a common factor: $(x + y)$. Thus,

$$\begin{aligned} ax + ay + bx + by &= a(x + y) + b(x + y) \\ &= (x + y)(a + b) \end{aligned}$$

Similarly,

We shall show in Section 3.6 that $x^2 + y^2$ is not factorable.

$$\begin{aligned} 4x^2 + 4y^2 + ax^2 + ay^2 &= 4(x^2 + y^2) + a(x^2 + y^2) \\ &= (x^2 + y^2)(4 + a) \end{aligned}$$

Example 3

Factor:

(a) $3a^2 - 5a + 3ab - 5b$
(b) $5xy - 5yz + x - z$
(c) $18 - 3x^2 - 6x + x^3$

Solution

(a) $3a^2 - 5a + 3ab - 5b = a(3a - 5) + b(3a - 5)$
$= (3a - 5)(a + b)$

(b) $5xy - 5yz + x - z = 5y(x - z) + (x - z)$
$= (x - z)(5y + 1)$

(c) $18 - 3x^2 - 6x + x^3 = 3(6 - x^2) - x(6 - x^2)$
$= (6 - x^2)(3 - x)$

PROGRESS TEST

1. When written as a product of primes, 480 equals __________.
2. When factored, the expression $9x^3 - 6x^2 + 3x$ equals __________.
3. When factored, $3x + 3y + 6x^2 + 6xy$ equals __________.
4. When factored, $8x^3 - 6x^2 + 20x - 15$ equals __________.

PROGRESS TEST ANSWERS

1. $2^5 \cdot 3 \cdot 5$
2. $3x(3x^2 - 2x + 1)$
3. $3x + 3y + 6x^2 + 6xy$
$= 3(x + y + 2x^2 + 2xy)$
$= 3[(x + y) + 2x(x + y)]$
$= 3[(x + y)(1 + 2x)]$
4. $8x^3 - 6x^2 + 20x - 15$
$= 2x^2(4x - 3) + 5(4x - 3)$
$= (4x - 3)(2x^2 + 5)$

EXERCISE 3.5

In Problems 1 through 20, factor the given expression.

Illustration: (a) $-5x + 25$
(b) $x^n - x^{n+1}$
(c) $16x^2y - 24xy^2$

Solution: (a) $-5x+25=-5(x-5)$. Note that we can also factor $-5x+25$ as $5(-x+5)$. However, $-5(x-5)$ is the preferred answer, because the first term of the binomial $x+5$ has a positive sign.

(b) $x^n - x^{n+1} = x^n(1-x)$

(c) $16x^2y - 24xy^2 = 8xy(2x-3y)$

1. $4x + 16$

2. $6x + 18$

3. $8a - 16$

4. $7a - 28$

5. $-6x + 42$

6. $-9y + 90$

7. $-3a - 15$

8. $-5x - 10$

9. $6x^2y + 12x^2$

10. $8xy^2 + 24y^2$

11. $6x^2y - 9xy^2 + 6xy$

12. $4x^3y + 6xy^3 - 4x^2y^2$

13. $8a^4b^4 - 6a^2b^2 + 4a^3b^3$

14. $5a^3b^2 - 10a^2b^3 + 15a^4b^4$

15. $18a^6y^4 - 24a^3b^3 + 12a^4b$

16. $4a^3b + 8a^2b^2 - 16ab^3$

17. $-10u^3r^2s - 25u^2r^2s^2 - 15ur^3s$

18. $8u^2r^3 - 16u^3r + 4u^3r^3$

19. $3x^2y - 6xy^2 + 12xy$

20. $3x^2y - 6xy^2 + 4xy$

In Problems 21 through 35, factor the given expression.

Illustration: (a) $18x^3 + 12x^2 - 15x - 10$
(b) $ax^2 + bx^2 + ay^2 + by^2$

Solution: (a) $18x^3 + 12x^2 - 15x - 10$
$= 6x^2(3x + 2) - 5(3x + 2)$
$= (3x + 2)(6x^2 - 5)$

(b) $ax^2 + bx^2 + ay^2 + by^2$
$= x^2(a + b) + y^2(a + b)$
$= (a + b)(x^2 + y^2)$

21. $x(x + 1) + y(x + 1)$

22. $x(x + 5) + y(x + 5)$

23. $y(2x - 1) + z(2x - 1)$

24. $a(7x - 4) + b(7x - 4)$

25. $5x(2x - y) - 3(2x - y)$

26. $a^2(3x - 2y) + b(3x - 2y)$

27. $6ax - 2ay - 18bx + 6by$

28. $5xy - 10xz - 6ay + 12az$

29. $5a^3 - 5a^2 + 5a - 5$

30. $5ab - 10ac - 5bd + 10cd$

31. $x^{n+1} + xy^n + x^n + y^n$

32. $x^{n+1} + xy^n - x^n - y^n$

33. $ax^{n+1} + bx^{n+1} + cx^{n+1}$

34. $ax^{n-1} - bx^{n-1} + cx^n$

35. $x^{n+3} - x^{2n}$

OBJECTIVES 3.6

After studying this section, the reader should be able to:

1. **Factor a second-degree polynomial using the four special products given in the text.**
2. **Determine if a given second-degree polynomial is factorable.**
3. **Factor a second-degree polynomial in which the coefficient of the second-degree term is not 1.**

3.6 FACTORING SECOND-DEGREE POLYNOMIALS

In the cartoon, the school building has learned how to factor second-degree polynomials. In this section we shall try to do the same. The simplest type of factorization occurs when the polynomial to be factored is one of the special products studied in Section 3.5. For example, since the polynomial $x^2 + (a + b)x + ab$ is Special Product 1, it can be factored as shown in Formula F-1.

$$x^2 + (a + b)x + ab = (x + a)(x + b) \qquad \text{(F-1)}$$

As you can see, any polynomial $x^2 + px + q$ whose last term q is the product of two expressions a and b and the coefficient of whose middle term is the sum of these two expressions can be factored using Formula F-1. Thus, $x^2 + 7x + 12$ is factorable, since in this case $a = 3$, $b = 4$, and $x^2 + 7x + 12 = x^2 + (3 + 4)x + 3 \cdot 4 = (x + 3)(x + 4)$. In essence, to factor a polynomial of the form shown in F-1, we have to find two expressions whose product is the last term and whose sum is the coefficient of the middle term.

Example 1

Factor:
(a) $x^2 + 3x - 10$
(b) $x^2 - 2x - 15$

Solution

(a) To factor $x^2 + 3x - 10$, we need to factor -10 as a product of factors whose sum is 3. These factors are 5 and -2, since $5(-2) = -10$ and $5 + (-2) = 3$. Thus, $x^2 + 3x - 10 = (x + 5)(x - 2)$.

(b) To factor $x^2 - 2x - 15$, we need two factors whose product is 15 and whose sum is -2. These factors are -5 and 3. Thus, $x^2 - 2x - 15 = (x - 5)(x + 3)$.

A trinomial whose first and last terms are perfect squares can be factored provided the middle term is the product of twice the two expressions whose squares appear as first and last terms. For example, the expression $x^2 + 6x + 9$ has two perfect square terms, x^2 and $(3)^2 = 9$. Since the middle term is $2(3)x = 6x$, the expression is factorable. We know that the factors must have a product of 9 and must add up to 6. These factors are 3 and 3. Thus,

$9x^2$ is a perfect square, since $9x^2 = (3x)^2$.

$4y^2$ is a perfect square, since $4y^2 = (2y)^2$.

$$x^2 + 6x + 9 = (x + 3)(x + 3)$$

Note that this is a form of Special Product 2, from which we take the following formula:

$$x^2 + 2ax + a^2 = (x + a)^2 \qquad \text{(F-2)}$$

Thus, since $x^2 + 8x + 16$ can be written as $x^2 + 2 \cdot 4x + 4^2$, we have $x^2 + 8x + 16 = (x + 4)^2$.

Example 2

Factor:

(a) $x^2 + 10x + 25$

(b) $9x^2 + 12xy + 4y^2$

Solution

(a) $x^2 + 10x + 25$ can be written as $x^2 + 2 \cdot 5x + 5^2$. Thus, $x^2 + 10x + 25 = (x + 5)^2$.

(b) $9x^2 + 12xy + 4y^2$ can be written as $(3x)^2 + 2(3x)(2y) + (2y)^2$
Thus, $9x^2 + 12xy + 4y^2 = (3x + 2y)^2$.

A special case of Formula F-2 occurs when a is replaced by $-a$. In this case, we obtain

$$x^2 - 2ax + a^2 = (x - a)^2 \qquad \text{(F-3)}$$

For example, $x^2 - 6x + 9 = (x - 3)^2$ and $4a^2 - 20ab + 25b^2 = (2a - 5b)^2$.

Example 3

Factor:
(a) $x^2 - 12x + 36$
(b) $25x^2 - 20xy + 4y^2$

Solution

(a) Since $x^2 - 12x + 36$ can be written as $x^2 - 2 \cdot 6x + 6^2$, we have $x^2 - 12x + 36 = (x - 6)^2$.
(b) Since $25x^2 - 20xy + 4y^2$ can be written as $(5x)^2 - 2(5x)(2y) + (2y)^2$, we have $25x^2 - 20xy + 4y^2 = (5x - 2y)^2$.

To factor $x^2 - 16$, we recall that by Special Product 4, $(x + 4)(x - 4) = x^2 - 16$. The difference of two squares can always be factored as follows:

$$x^2 - a^2 = (x + a)(x - a) \qquad \text{(F-4)}$$

Thus, $u^2 - 36 = (u + 6)(u - 6)$
and $100x^2 - 49y^2 = (10x + 7y)(10x - 7y)$

Example 4

Factor:
(a) $9x^2 - 1$
(b) $4a^2 - 49b^2$

Solution

(a) Since $9x^2 - 1$ can be written as $(3x)^2 - 1^2$, we have $9x^2 - 1 = (3x + 1)(3x - 1)$.
(b) Since $4a^2 - 49b^2$ can be written as $(2a)^2 - (7b)^2$, $4a^2 - 49b^2 = (2a + 7b)(2a - 7b)$.

The polynomial $6x^2 - 7x + 2$ is *not* of the same type as the ones studied previously. How do we know if this polynomial is even factorable? It can be shown that a polynomial of the form $ax^2 + bx + c$ is factorable if the product ac has factors whose sum is b. Thus, $6x^2 - 7x + 2$ *is* factorable, because $6 \cdot 2 = 12$ has the factors -3 and -4, and their sum is -7. To complete the factorization, we proceed as follows:

Use the factors -3 and -4 to rewrite

$6x^2 - 7x + 2$

Note that $-4x + 2 = -(4x - 2)$.

as $6x^2 - 3x - 4x + 2$

$= (6x^2 - 3x) - (4x - 2)$ Grouping the terms

$= 3x(2x - 1) - 2(2x - 1)$ Factoring the groups

$= (2x - 1)(3x - 2)$ Factoring $(2x - 1)$ from both expressions

Note that if you write $2x^2 + x + 6x + 3$ the answer is equivalent, since $2x^2 + x + 6x + 3 = x(2x + 1) + 3(2x + 1) = (2x + 1)(x + 3)$ and $(2x + 1)(x - 3) = (x + 3)(2x + 1)$.

To factor $2x^2 + 7x + 3$, we note that $2 \cdot 3 = 6$ has factors 1 and 6, whose sum is 7. Thus,

$2x^2 + 7x + 3$

$= 2x^2 + 6x + x + 3$ Rewriting $7x$ as $6x + x$

$= (2x^2 + 6x) + (x + 3)$ Grouping

$= 2x(x + 3) + 1 \cdot (x + 3)$ Factoring

$= (x + 3)(2x + 1)$ Factoring $(2x + 1)$ from both expressions

---

Example 5

(a) $6x^2 - 7x - 3$
(b) $3x^2 - x + 1$

Solution

(a) Since $6 \cdot (-3) = -18$, and -18 has factors 2 and -9, whose sum is -7, we have:

$$\begin{aligned} &6x^2 - 7x - 3 \\ &= 6x^2 + 2x - 9x - 3 \\ &= 2x(3x + 1) - 3(3x + 1) \\ &= (3x + 1)(2x - 3) \end{aligned}$$

(b) $3 \cdot 1 = 3$ has *no* factors whose sum is -1. Thus, $3x^2 - x + 1$ is *not* factorable using integer coefficients.

PROGRESS TEST

1. When factored, $x^2 + 4x - 21$ equals ______________.
2. When factored, $x^2 + 4x + 4$ equals ______________.
3. When factored, $16x^2 + 24xy + 9y^2$ equals ______________.

4. When factored, $x^2 - 16x + 64$ equals ______________.
5. When factored, $x^2 - 1$ equals ______________.
6. When factored, $81x^2 - 49y^2$ equals ______________.
7. When factored, $10x^2 - x - 2$ equals ______________.
8. When factored, $7x^2 - 23x + 6$ equals ______________.

EXERCISE 3.6

In Problems 1 through 56, factor completely.

Illustration: (a) $x^2 - 13x + 30$
(b) $81x^2 + 36x + 4$
(c) $49x^2 - 42x + 9$
(d) $100x^2 - 121y^2$
(e) $8x^2 - 2x - 1$

Solution: (a) We have to find factors whose product is 30, and whose sum is -13. These factors are -10 and -3. Thus, $x^2 - 13x + 30 = (x - 10)(x - 3)$.

(b) $81x^2 + 36x + 4 = (9x)^2 + 2 \cdot 2 \cdot 9x + 2^2$
$= (9x + 2)^2$

(c) $49x^2 - 42x + 9 = (7x)^2 - 2 \cdot 3 \cdot 7x + 3^2$
$= (7x - 3)^2$

(d) $100x^2 - 121y^2$
$= (10x)^2 - (11y)^2$
$= (10x + 11y)(10x - 11y)$

(e) $8x^2 - 2x - 1 = 8x^2 - 4x + 2x - 1$
$= 4x(2x - 1) + 1 \cdot (2x - 1)$
$= (2x - 1)(4x + 1)$

1. $x^2 + 5x + 6$

2. $x^2 + 15x + 56$

3. $a^2 + 7a + 10$

4. $a^2 + 10a + 24$

5. $x^2 + x - 12$

6. $x^2 + 5x - 6$

7. $x^2 + x - 2$

8. $x^2 + 7x - 18$

9. $x^2 - x - 2$

10. $x^2 - 5x - 14$

11. $x^2 - 3x - 10$

12. $x^2 - 4x - 21$

13. $a^2 - 16a + 63$

14. $a^2 - 4a + 3$

15. $y^2 - 13y + 22$

16. $y^2 - 12y + 11$

PROGRESS TEST ANSWERS

1. $(x - 3)(x + 7)$
2. $(x + 2)^2$
3. $(4x + 3y)^2$
4. $(x - 8)^2$
5. $(x + 1)(x - 1)$
6. $(9x + 7y)(9x - 7y)$
7. $(2x - 1)(5x + 2)$
8. $(x - 3)(7x - 2)$

17. $x^2 + 2x + 1$

18. $x^2 + 20x + 100$

19. $y^2 + 22y + 121$

20. $y^2 + 14y + 49$

21. $4x^2 + 4x + 1$

22. $9x^2 + 6x + 1$

23. $9x^2 + 30xy + 25y^2$

24. $25x^2 + 30xy + 9y^2$

25. $36a^2 + 48a + 16$

26. $9a^2 + 60a + 100$

27. $y^2 - 2y + 1$

28. $y^2 - 10y + 25$

29. $x^2 - 14x + 49$

30. $x^2 - 100x + 2500$

31. $49a^2 - 28ax + 4x^2$

32. $4a^2 - 12ax + 9x^2$

33. $16x^2 - 24xy + 9y^2$

34. $9x^2 - 42xy + 49y^2$

35. $y^2 - 64$

36. $y^2 - 121$

37. $a^2 - c^2$

38. $x^2 - d^2$

39. $64 - b^2$

40. $81 - b^2$

41. $36a^2 - 49b^2$

42. $36a^2 - 25b^2$

43. $x^2 - y^2$

44. $x^2 - 9y^2$

45. $9x^2 + 37x + 4$

46. $2x^2 + 5x + 2$

47. $3a^2 - 5a - 2$

48. $8a^2 - 2a - 21$

49. $2y^2 - 3y - 20$

50. $6y^2 - 13y - 5$

51. $4x^2 - 11x + 6$

52. $16x^2 - 16x + 3$

53. $6x^2 + x - 12$

54. $20y^2 + y - 1$

55. $21a^2 + 11a - 2$

56. $18x^2 - 3x - 10$

B.C. by Johnny Hart, courtesy of Field Enterprises, Inc.

OBJECTIVES 3.7

After studying this section, the reader should be able to:

1. **Factor the sum of two cubes.**
2. **Factor the difference of two cubes.**
3. **Factor a polynomial of four or six terms by finding a binomial factor and writing the polynomial as a product of this binomial and another polynomial, or by first writing the terms of the polynomial as the difference of two squared binomials.**

3.7 FACTORING OTHER POLYNOMIALS

In the cartoon at the beginning of this section, the man has used a bouillon cube. In this section we shall learn how to factor binomials that are sums or differences of two cubes.

In Illustration (a) at the beginning of Exercise 3.4, we showed that $(x + a)(x^2 - ax + a^2) = x^3 + a^3$. Consequently, we may write

$$x^3 + a^3 = (x + a)(x^2 - ax + a^2) \qquad \text{(F-5)}$$

Thus, since $x^3 + 27$ can be written as $x^3 + (3)^3$, we have

$$\begin{aligned} x^3 + 27 &= (x + 3)(x^2 - 3x + 3^2) \\ &= (x + 3)(x^2 - 3x + 9) \end{aligned}$$

Similarly,

$$\begin{aligned} x^3 + 8 &= (x + 2)(x^2 - 2x + 2^2) \\ &= (x + 2)(x^2 - 2x + 4) \end{aligned}$$

Example 1

Factor:
(a) $8x^3 + 27$
(b) $27x^3 + 64y^3$

Solution

(a) Since $8x^3 + 27 = (2x)^3 + 3^3$,
$$8x^3 + 27 = (2x + 3)(4x^2 - 6x + 9).$$

(b) Since $27x^3 + 64y^3 = (3x)^3 + (4y)^3$,
$$27x^3 + 64y^3 = (3x + 4y)(9x^2 - 12xy + 16y^2).$$

Since the product $(x - a)(x^2 + ax + a^2) = x^3 - a^3$, we see that the difference of two cubes can be factored as follows:

$$x^3 - a^3 = (x - a)(x^2 + ax + a^2) \qquad \text{(F-6)}$$

Thus, since $x^3 - 8 = x^3 - 2^3$, we have

$$\begin{aligned} x^3 - 8 &= (x - 2)(x^2 + 2x + 2^2) \\ &= (x - 2)(x^2 + 2x + 4) \end{aligned}$$

Similarly, $x^3 - 125$ can be written as $x^3 - 5^3$, so

$$\begin{aligned} x^3 - 125 &= (x - 5)(x^2 + 5x + 5^2) \\ &= (x - 5)(x^2 + 5x + 25) \end{aligned}$$

Example 2

Factor:
(a) $64x^3 - 27$
(b) $125x^3 - 64y^3$

Solution

(a) $64x^3 - 27$ can be written as $(4x)^3 - 3^3$, thus $64x^3 - 27 = (4x - 3)(16x^2 + 12x + 9)$.
(b) $125x^3 - 64y^3$ can be written as $(5x)^3 - (4y)^3$, so $125x^3 - 64y^3 = (5x - 4y)(25x^2 + 20xy + 16y^2)$.

In Section 3.5, we discussed a method of factoring polynomials in which terms having a common factor are grouped together. A similar procedure can be used to factor some polynomials which are not written in one of the six forms we have studied. For example, the polynomials $x^2 + 2xy + y^2 - 1$ can be factored as follows:

$$\begin{aligned} x^2 + 2xy + y^2 - 1 &= (x^2 + 2xy + y^2) - 1 \\ &= (x + y)^2 - 1^2 \qquad \text{Factoring } x^2 + 2xy + y^2 \\ &= (x + y + 1)(x + y - 1) \end{aligned}$$

Similarly,

$$\begin{aligned} x^2 + 2xy + y^2 - z^2 &= (x^2 + 2xy + y^2) - z^2 \\ &= (x + y)^2 - z^2 \\ &= (x + y + z)(x + y - z) \end{aligned}$$

Example 3

Factor:
(a) $x^2 - 6x + 9 - y^2$
(b) $x^2 + 6xy + 9y^2 - 4$

Solution

(a) $$\begin{aligned} x^2 - 6x + 9 - y^2 &= (x^2 - 6x + 9) - y^2 \\ &= (x - 3)^2 - y^2 \\ &= (x - 3 + y)(x - 3 - y) \\ &= (x + y - 3)(x - y - 3) \end{aligned}$$

(b) $$\begin{aligned} x^2 + 6xy + 9y^2 - 4 &= (x^2 + 6xy + 9y^2) - 4 \\ &= (x + 3y)^2 - 4 \\ &= (x + 3y + 2)(x + 3y - 2) \end{aligned}$$

Example 4

Factor $4x^2 - 4xy + y^2 - a^2 + 2ab - b^2$.

Solution

$$\begin{aligned} &4x^2 - 4xy + y^2 - a^2 + 2ab - b^2 \\ &= (4x^2 - 4xy + y^2) - (a^2 - 2ab + b^2) \\ &= (2x - y)^2 - (a - b)^2 \\ &= [(2x - y) + (a - b)][(2x - y) - (a - b)] \\ &= (2x - y + a - b)(2x - y - a + b) \end{aligned}$$

In certain cases, some of the terms in a polynomial can be grouped and factored as per one of our formulas (F-1 through F-6), so that a common factor can then be obtained. For example, the polynomial $x^2 - y^2 + x - y$ can be factored as follows:

$$\begin{aligned} x^2 - y^2 + x - y &= (x^2 - y^2) + (x - y) \\ &= (x + y)(x - y) + (x - y) \\ &= (x - y)(x + y + 1) \end{aligned}$$

Note that $(x - y)$ is common to $(x + y)(x - y)$ and $(x - y)$.

Similarly,

$$\begin{aligned} 4a^2 - 9b^2 + 2a - 3b &= (4a^2 - 9b^2) + (2a - 3b) \\ &= (2a + 3b)(2a - 3b) + (2a - 3b) \\ &= (2a - 3b)[(2a + 3b) + 1] \\ &= (2a - 3b)(2a + 3b + 1) \end{aligned}$$

As a final point, we must mention that in some instances polynomials may factor into more than two factors. For example,

$$\begin{aligned} x^8 - y^8 &= (x^4 + y^4)(x^4 - y^4) \\ &= (x^4 + y^4)(x^2 + y^2)(x^2 - y^2) \\ &= (x^4 + y^4)(x^2 + y^2)(x + y)(x - y) \end{aligned}$$

Example 5

Factor completely:
(a) $x^6 - 64$
(b) $x^6 - 2x^3 + 1$

Solution

(a)
$$\begin{aligned} x^6 - 64 &= x^6 - 2^6 \\ &= [x^3]^2 - [2^3]^2 \\ &= (x^3 + 2^3)(x^3 - 2^3) \\ &= [(x + 2)(x^2 - 2x + 4)][(x - 2)(x^2 + 2x + 4)] \end{aligned}$$

(b)
$$\begin{aligned} x^6 - 2x^3 + 1 &= (x^3)^2 - 2x^3 + 1^2 \\ &= (x^3 - 1)^2 \\ &= [(x - 1)(x^2 + x + 1)]^2 \\ &= (x - 1)^2(x^2 + x + 1)^2 \end{aligned}$$

PROGRESS TEST

1. When factored, $x^3 + 216$ equals ________________.
2. When factored, $27x^3 + 8y^3$ equals ________________.
3. When factored, $x^3 - 64$ equals ________________.
4. When factored, $a^2 + 2ab + b^2 - c^2$ equals ________________.
5. When factored, $9x^2 - 6xy + y^2 - a^2 + 2ab - b^2$ equals ________________.
6. When factored, $x^4 - y^4$ equals ________________.
7. When factored, $x^6 - y^6$ equals ________________.

PROGRESS TEST ANSWERS

1. $(x + 6)(x^2 - 6x + 36)$
2. $(3x + 2y)(9x^2 - 6xy + 4y^2)$
3. $(x - 4)(x^2 + 4x + 16)$
4. $(a + b + c)(a + b - c)$
5. $(3x - y + a - b)(3x - y - a + b)$
6. $(x^2 + y^2)(x + y)(x - y)$
7. $[(x + y)(x^2 - xy + y^2)][(x - y)(x^2 + xy + y^2)]$

EXERCISE 3.7

In Problems 1 through 10, factor the given polynomial.

Illustration: (a) $27a^3 + 125b^3$
(b) $125x^3 - 27y^3$

Solution: (a)
$$\begin{aligned} 27a^3 + 125b^3 &= (3a)^3 + (5b)^3 \\ &= (3a + 5b)(9a^2 - 15ab + 25b^2) \end{aligned}$$

(b) $125x^3 - 27y^3$
$= (5x)^3 - (3y)^3$
$= (5x - 3y)(25x^2 + 15xy + 9y^2)$

1. $x^3 + 125$
2. $x^3 + 64$
3. $a^3 + 1$
4. $a^3 + 343$
5. $8x^3 + y^3$
6. $125x^3 + 8y^3$
7. $x^3 - 1$
8. $x^3 - 216$
9. $125a^3 - 8b^3$
10. $216a^3 - 125b^3$

In Problems 11 through 20, factor the given polynomial.

Illustration: (a) $9a^2 + 12ab + 4b^2 - c^2$
(b) $z^2 - 4x^2 + 4xy - y^2$

Solution: (a) $9a^2 + 12ab + 4b^2 - c^2$
$= (3a + 2b)^2 - c^2$
$= (3a + 2b + c)(3a + 2b - c)$
(b) $z^2 - 4x^2 + 4xy - y^2$
$= z^2 - (4x^2 - 4xy + y^2)$
$= z^2 - (2x - y)^2$
$= (z + 2x - y)(z - 2x + y)$

11. $a^2 + 4ab + 4b^2 - c^2$
12. $9a^2 + 6ab + b^2 - 1$
13. $4x^2 - 4xy + y^2 - 1$
14. $9x^2 - 30xy + 25y^2 - 9$
15. $9y^2 - 12xy + 4x^2 - 25$
16. $16y^2 - 40xy + 25x^2 - 36$
17. $16a^2 - (x^2 + 6xy + 9y^2)$
18. $25a^2 - (4x^2 - 4xy + y^2)$
19. $y^2 - a^2 + 2ab - b^2$
20. $9y^2 - 9x^2 + 6xz - z^2$

In Problems 21 through 30, factor the given polynomial.

Illustration: (a) $a^2 - b^2 + a + b$

(b) $16x^4 - 81y^4$

Solution: (a) $a^2 - b^2 + a + b$

$= (a^2 - b^2) + (a + b)$

$= (a + b)(a - b) + 1 \cdot (a + b)$

$= (a + b)(a - b + 1)$

(b) $16x^4 - 81y^4$

$= (4x^2 + 9y^2)(4x^2 - 9y^2)$

$= (4x^2 + 9y^2)(2x + 3y)(2x - 3y)$

21. $x^2 - 4y^2 + x + 2y$

22. $x^2 - 9y^2 + x + 3y$

23. $4x^2 - 9y^2 + 2x - 3y$

24. $9x^2 - 4y^2 + 3x - 2y$

25. $x^4 - 16$

26. $x^4 - 1$

27. $16x^4 - 81$

28. $81x^4 - 16y^4$

29. $x^6 + 2x^3 + 1$

30. $x^6 - 4x^3 + 4$

SELF-TEST–CHAPTER 3

1. Identify each polynomial as a monomial, binomial, or trinomial. Give the degree of each polynomial.

_____________ a. $3x^3 - 7$

_____________ Degree

_____________ b. $3x^4 + x^5 + 7x$

_____________ Degree

_____________ c. x^2y^5z

_____________ Degree

---

2. If $P(x) = x^2 - 2x + 3$, find:

_______ a. $P(0)$

_______ b. $P(-1)$

_______ c. $P(0) + P(-1)$

---

3. Simplify:

_____________________ a. $(4x^2 + 2) - (5x + 2) + (x^2 + 6x)$

_____________________ b. $2xy + x^2y + 6xy$

---

4. Find the product or quotient.

_____________ a. $3x(x - 2y)$

_____________ b. $(5x + 1)(2x + 3)$

_____________ c. $(x + y)(x - y)$

_____________ d. $\dfrac{a^2b^3c^3}{-abc}$

_____________ e. $\dfrac{x^{5n}}{x^{3n}}$

---

5. Simplify each of the following expressions.

_______ a. $(2xy^2)^2(xy)^5$

_______ b. $\frac{(4abc^2)^3}{-16abc^2}$

---

6. Factor completely.

_______ a. $2x + 6$

_______ b. $x^2 - 16$

_______ c. $4x^2 - 5x + 1$

_______ d. $x^3 + y^3$

_______ e. $x^3 - y^3$

---

7. Factor completely.

_______ a. $ax^2 + x + ax + 1$

_______ b. $y^3 - (3x)^3$

_______ c. $x^4 - 16$

_______ d. $-5x + 10$

_______ e. $4x^2 + 8x$

---

8. Evaluate the given expression.

_______ a. $2x + y^2$, for $x = 3$, $y = -1$

_______ b. $(2x - y)^2$, for $x = 1$, $y = -1$

_______ c. $(2x - y)^2$, for $x = 1$, $y = 0$

---

9. a. In an expression of the form $A + B + C + \ldots$, A, B, C, etc. are called the _______ of the expression.

b. The number of terms in the expression $x + y + z - 7$ is _______.

---

10. If $P(x) = x^2 + x - 4$ and $Q(x) = x + 2$, find

_____________ a. $P(x) + Q(x)$

_____________ b. $P(x) - Q(x)$

CHAPTER 4

HISTORICAL NOTE

As we have mentioned before, the progress of civilization and the need to measure quantities such as length, volume, weight, and time precipitated the invention of the numbers that we call fractions. One of the most remarkable aspects of Egyptian arithmetic was its handling of operations with fractions. In this system, all fractions were reduced to sums of the so-called unit fractions—that is, fractions with a numerator of 1. The only exception to this rule was the fraction $\frac{2}{3}$, which was written as $1 - \frac{1}{3}$, and for which there was a special symbol. The computations made by the Egyptians were applied to a variety of everyday problems such as determining the amounts of grain needed for making beer or bread and the finding of areas and volumes. The solution of such problems required nothing more than simple equations, and these were generally solved by a method that was later known as the *rule of false position.* For example, to solve $x + \frac{x}{7} = 16$, we assume that x equals a convenient value, say 7. Then $x + \frac{x}{7} = 8$, instead of 16. Because we must multiply 8 by 2 to get 16, the correct value of x must be $2 \cdot 7 = 14$.

The Babylonians, who used a sexagesimal (base 60) system consisting of two symbols, the wedge 𒁹, representing the number 1, and the symbol 𒌋, representing 10, also used fractions. Their fractions were written in descending powers of 60, in the same manner in which we write decimal fractions in descending powers of 10. Thus, while in our system the number 2.34 means to us $2 + 3\left(\frac{1}{10}\right) + 4\left(\frac{1}{100}\right)$, the number 𒁹𒁹 𒌋𒁹 𒁹 may have meant $2 + 11\left(\frac{1}{60}\right) + 1\left(\frac{1}{3600}\right)$. The advantage of this notation is that computations with fractions are done in the same manner as computations with whole numbers, which is the same advantage we have when working with decimals. In this chapter we shall study fractions, how to operate with them, and the methods we can use to solve equations similar to $x + \frac{x}{7} = 16$.

THE ALGEBRA OF FRACTIONS

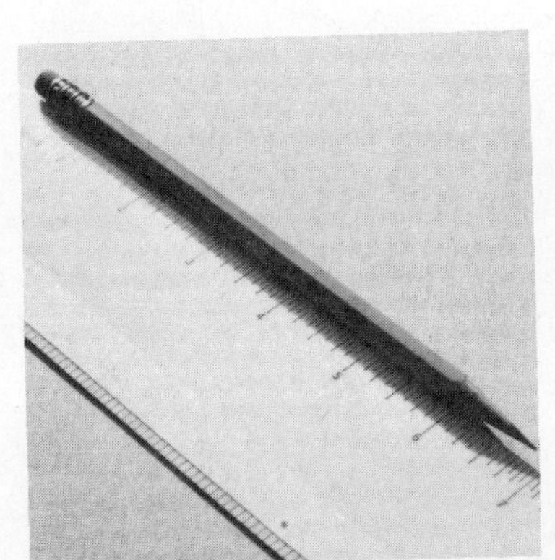

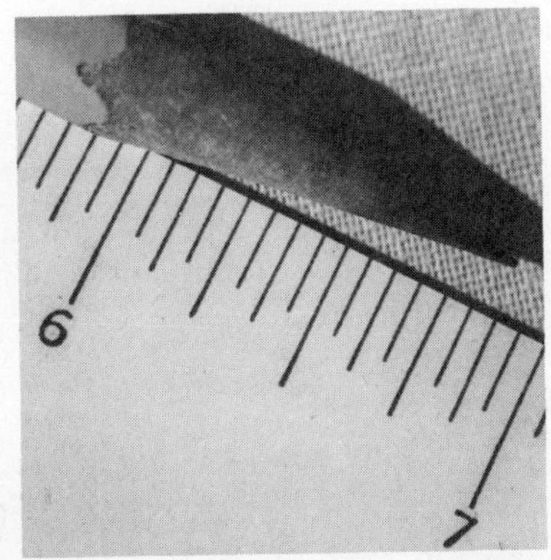

Using fractions of a unit makes it possible to get an accurate measurement. In the photograph at the left, for example, the pencil measures between 6 and 7 inches long. Because it is closer to 7 inches on the ruler, we might say it is 7 inches long. But in the photo at the right, we see that the pencil measures somewhat short of 7 inches. A more accurate measurement of its length is $6\frac{7}{8}$ inches.

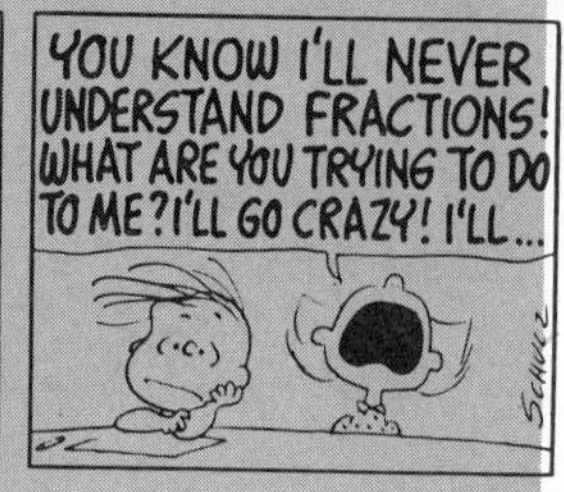

OBJECTIVES 4.1

After studying this section, the reader should be able to:
1. **Determine if two given fractions are equivalent.**
2. **Write a given fraction as an equivalent one with the indicated denominator.**
3. **Write a given fraction in standard form.**
4. **Reduce a given fraction to lowest terms.**
5. **Use long division to obtain the quotient of two given polynomials.**

4.1 THE CONCEPT OF A FRACTION

In the Peanuts cartoon, Sally is very upset with the idea of learning about fractions. The word "fraction" is derived from the Latin word "fractio," which means "to break" or "to divide." The latter meaning tells us that a fraction is an expression denoting a quotient. We know that any fraction of the form $\frac{a}{b}$ (with a and b as integers and $b \neq 0$) is a *rational number.* We can extend this idea to polynomial expressions: if the numerator and denominator of a fraction are polynomials, then the fraction will be called a *rational expression.* For example,

In the fraction $\frac{a}{b}$, a is the numerator and b is the denominator.

A rational expression is a fraction in which the numerator and denominator are polynomials.

$$\frac{x+y}{z},\ \frac{x^2-2x+1}{x+2},\ \frac{1}{y^2-1},\ \text{and}\ \frac{x}{x-2}$$

are rational expressions. Of course, since any polynomial can be considered as the quotient of itself and 1, all polynomials are rational expressions.

$x^2+x+1=\frac{x^2+x+1}{1}$, $5x=\frac{5x}{1}$, and $9=\frac{9}{1}$; indeed, every polynomial is a rational expression.

The rational expressions $\frac{x+y}{z}$, $\frac{x^2-2x+1}{x+2}$, $\frac{1}{y^2-1}$, and $\frac{x}{x-2}$ represent real numbers for those replacements of the variable for which the denominator is not zero. If the denominator *is* 0, the expression is said to be *undefined.* In

If $z=0$, $\frac{x+y}{z}$ is undefined, since $\frac{x+y}{z}=\frac{x+y}{0}$.

regard to undefined expressions, please note the statement below:

> We shall assume from now on that the variables in an expression may not be replaced by values that will cause the denominator to be zero.

As you can see from the accompanying photograph, the fractions $\frac{2}{4}$ and $\frac{1}{2}$ represent the same number. Similarly, since 4 divided by 2 is 2, and 6 divided by 3 is also 2, we know that $\frac{4}{2}$ and $\frac{6}{3}$ represent the same number, 2. Thus, we say that these fractions are equivalent. We have

$$\frac{2}{4}=\frac{1}{2} \quad \text{and} \quad \frac{4}{2}=\frac{6}{3}$$

Here we are "cross multiplying": $\frac{2}{4} \times \frac{1}{2}$ $\quad \frac{4}{2} \times \frac{6}{3}$

In general, two fractions are equivalent if they represent the same number. Note that if $\frac{2}{4}=\frac{1}{2}$, then multiplying the numerator of the left-hand fraction by the denominator of the right-hand fraction will give us a product equal to that obtained by multiplying the denominator of the left-hand fraction by the numerator of the right-hand fraction: $2 \cdot 2 = 4 \cdot 1$. Likewise, in $\frac{4}{2}=\frac{6}{3}$, $4 \cdot 3 = 2 \cdot 6$. The following Theorem tells us how "cross multiplying" can indicate whether two fractions are equivalent.

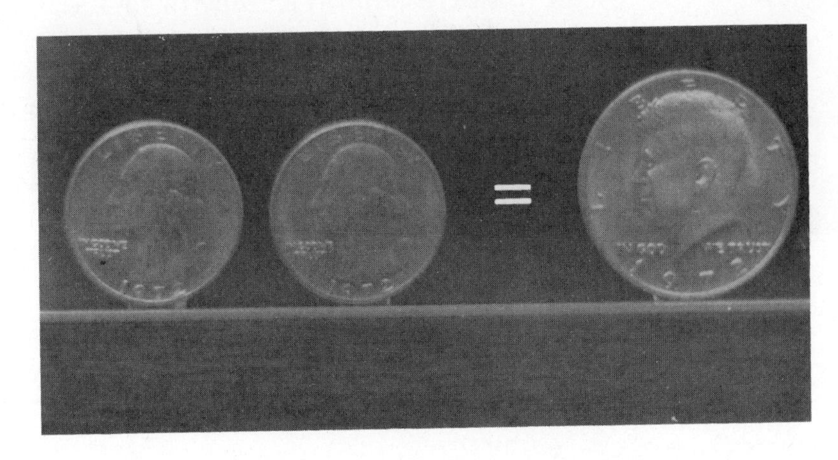

Photo by DeAngelis.

THEOREM 4.1

The fraction $\frac{a}{b}$ is equivalent to the fraction $\frac{c}{d}$ if and only if $ad = bc$. (In symbols, $\frac{a}{b} = \frac{c}{d} \leftrightarrow ad = bc$.)

The statement of this theorem has an "if and only if" part, so we must prove two things:

1. If $\frac{a}{b} = \frac{c}{d}$, then $ad = bc$

and 2. If $ad = bc$, then $\frac{a}{b} = \frac{c}{d}$

PROOF OF PART 1

Statement	Reason
1. $\frac{a}{b} = \frac{c}{d}$	Given
2. $bd \cdot \left(\frac{a}{b}\right) = bd \cdot \left(\frac{c}{d}\right)$	Multiplying both members by bd
3. $bda \cdot \left(\frac{1}{b}\right) = bdc \cdot \left(\frac{1}{d}\right)$	Since $\frac{a}{b} = a \cdot \left(\frac{1}{b}\right)$ and $\frac{c}{d} = c \cdot \left(\frac{1}{d}\right)$
4. $ad \cdot b\left(\frac{1}{b}\right) = bc \cdot d\left(\frac{1}{d}\right)$	Using the Associative and Commutative Laws
5. $ad \cdot 1 = bc \cdot 1$	Since $b \cdot \left(\frac{1}{b}\right) = 1$ and $d \cdot \left(\frac{1}{d}\right) = 1$
6. $ad = bc$	Since 1 is the multiplicative identity

Since each of the steps in this argument is reversible, we also conclude that if $ad = bc$, then $\frac{a}{b} = \frac{c}{d}$.

Theorem 4.1 provides a means of identifying equivalent fractions. Thus,

$$\frac{3}{9} = \frac{1}{3} \quad \text{because} \quad 3 \cdot 3 = 9 \cdot 1$$

$$\frac{-4}{16} = \frac{-1}{4} \quad \text{because} \quad -4 \cdot 4 = 16 \cdot (-1)$$

and $\frac{8}{9} \neq \frac{15}{16}$ because $8 \cdot 16 \neq 9 \cdot 15$

Example 1

Use Theorem 4.1 to determine if the following fractions are equal.

(a) $\frac{3}{7}$ and $\frac{9}{21}$

(b) $\frac{13}{52}$ and $\frac{1}{3}$

(c) $\frac{1}{2}$ and $\frac{1 \cdot 3}{2 \cdot 3}$

Solution

(a) $\frac{3}{7} = \frac{9}{21}$ because $3 \cdot 21 = 7 \cdot 9$

(b) $\frac{13}{52} \neq \frac{1}{3}$ because $13 \cdot 3 \neq 52 \cdot 1$

(c) $\frac{1}{2} = \frac{1 \cdot 3}{2 \cdot 3}$ because $1 \cdot (2 \cdot 3) = 2 \cdot (1 \cdot 3)$

In part (c) of Example 1, we can see that $\frac{1}{2} = \frac{1 \cdot 3}{2 \cdot 3}$. This fact is an application of a basic theorem which states that *the numerator and denominator of a fraction may be multiplied (or divided) by the same nonzero number to obtain an equivalent fraction.* This theorem is stated as follows:

FUNDAMENTAL THEOREM OF FRACTIONS

$\frac{a}{b} = \frac{a \cdot k}{b \cdot k}$; a, b, and k are real numbers and $b, k \neq 0$.

The proof of this theorem follows from Theorem 4.1:

$$\frac{a}{b} = \frac{a \cdot k}{b \cdot k} \quad \text{because } a \cdot (b \cdot k) = b \cdot (a \cdot k)$$

For example, the fraction $\frac{4}{5}$ can be written as an equivalent fraction with a denominator of 10 by multiplying the numerator and denominator by 2. Hence $\frac{4}{5} = \frac{4 \cdot 2}{5 \cdot 2} = \frac{8}{10}$. Similarly, $\frac{6}{9}=\frac{2 \cdot \cancel{3}}{3 \cdot \cancel{3}}=\frac{2}{3}$ when the numerator and denominator are divided by 3. We usually write $\frac{\overset{2}{\cancel{6}}}{\underset{3}{\cancel{9}}}=\frac{2}{3}$.

Example 2

Write the given fraction as an equivalent one with the indicated denominator.

(a) $\frac{3}{4}$, denominator 16

(b) $\frac{-a}{b}$, denominator b^2

Solution

(a) To write the fraction $\frac{3}{4}$ with 16 as a denominator, we have to multiply the denominator, and hence the numerator, by 4, obtaining

$$\frac{3}{4}=\frac{3 \cdot 4}{4 \cdot 4}=\frac{12}{16}$$

(b) $\frac{-a}{b}=\frac{-a \cdot b}{b \cdot b}=\frac{-ab}{b^2}$, multiplying the numerator and denominator by b.

The sign before the fraction $-\frac{3}{5}$ is negative. The sign of the numerator is positive and the sign of the denominator is positive.

It is important to know that there are three signs associated with a fraction:

1. The sign before the fraction
2. The sign of the numerator
3. The sign of the denominator

Using our definition of a quotient and the Fundamental Theorem of Fractions, we can conclude that:

$$\frac{-a}{b}=\frac{a}{-b}=-\frac{a}{b}=-\frac{-a}{-b} \text{ and}$$

$$\frac{a}{b}=\frac{-a}{-b}=-\frac{a}{-b}=-\frac{-a}{b}$$

$\frac{-a}{b}$ and $\frac{a}{b}$ are in standard form.

The forms $\frac{-a}{b}$ and $\frac{a}{b}$, in which the sign of the fraction and that of the denominator are positive, are called the *standard forms* of the fractions. Thus, $\frac{-2}{9}$ and $\frac{4}{7}$ are in standard form, but $\frac{2}{-9}$ and $\frac{-4}{-7}$ are not. Of course, in expressions with more than one term in the numerator or denominator, there are alternative standard forms. For example,

$$\frac{-1}{x-y} = \frac{-1}{-(y-x)} = \frac{1}{y-x}$$

Recall that $-(y-x) = -y + x = x - y$.

Thus, either $\frac{-1}{x-y}$ or $\frac{1}{y-x}$ can be used as the standard form.

Example 3

Write the following fractions in standard form.

(a) $\frac{x}{-2}$

(b) $-\frac{-3}{y}$

(c) $-\frac{x-y}{5}$

Solution

(a) $\frac{x}{-2} = \frac{-x}{2}$

(b) $-\frac{-3}{y} = \frac{3}{y}$

(c) $-\frac{x-y}{5} = \frac{-(x-y)}{5}$ or $\frac{y-x}{5}$

The Fundamental Theorem of Fractions can also be used to reduce fractions—that is, to write fractions as equivalent ones in which no integers other than 1 can be divided into both the numerator and denominator. For example, the fraction $\frac{14}{21}$ can be reduced by proceeding as follows:

A fraction is in lowest terms if the numerator and denominator have no common factors other than 1.

$$\frac{14}{21} = \frac{2 \cdot \overset{1}{\cancel{7}}}{3 \cdot \underset{1}{\cancel{7}}} = \frac{2}{3}$$

Here we are dividing the numerator and denominator by 7.

We usually write $\frac{\overset{2}{\cancel{14}}}{\underset{3}{\cancel{21}}}$.

Example 4 Reduce each fraction to lowest terms:

(a) $\dfrac{x^3y^4}{xy^6}$

(b) $\dfrac{x-y}{x^2-y^2}$

(c) $\dfrac{2x+xy}{x}$

Solution (a) $\dfrac{x^3y^4}{xy^6} = \dfrac{x^2 \cdot \overset{1}{\cancel{xy^4}}}{y^2 \cdot \underset{1}{\cancel{xy^4}}} = \dfrac{x^2}{y^2}$

(b) $\dfrac{x-y}{x^2-y^2} = \dfrac{1 \cdot (x-y)}{(x+y)(x-y)} = \dfrac{1}{x+y}$. Sometimes this reduction is written as $\dfrac{x-y}{x^2-y^2} = \dfrac{\overset{1}{\cancel{(x-y)}}}{(x+y)\underset{1}{\cancel{(x-y)}}} = \dfrac{1}{x+y}$.

(c) $\dfrac{2x+xy}{x} = \dfrac{\overset{1}{\cancel{x}}(2+y)}{\underset{1}{\cancel{x}}} = 2+y$

The division of one polynomial by another often furnishes an example of reducing a fraction to lowest terms. Thus, to reduce the fraction $\dfrac{3x^2-5x-2}{x-2}$ we proceed as follows:

$$\frac{3x^2-5x-2}{x-2} = \frac{(3x+1)\overset{1}{\cancel{(x-2)}}}{\underset{1}{\cancel{x-2}}} = 3x+1$$

This result can also be obtained by a method similar to the long division used in arithmetic. First we write:

Note that both polynomials are in descending order.

1. $x-2\,\overline{)\,3x^2-5x-2}$

2. $x-2\,\overset{3x}{\overline{)\,3x^2-5x-2}}$ Dividing the first term of the divisor (x) into the first term of the dividend ($3x^2$).

3.
$$\begin{array}{r}
3x\phantom{{}-5x-2}\\
x-2\,\overline{)\,3x^2-5x-2}\\
\underline{3x^2-6x\phantom{{}-2}}\\
x-2
\end{array}$$

Multiplying $x-2$ by $3x$, lining up like terms, subtracting, and bringing down the -2.

$a-b=a+(-b)$; thus, to subtract $3x^2-6x$ from $3x^2-5x$, you can "change the sign" of $3x^2-6x$ and add, obtaining:

$$\begin{array}{r}
3x^2-5x\\
\underline{-3x^2+6x}\\
x
\end{array}$$

4.
$$\begin{array}{r}
3x+1\phantom{{}-2}\\
x-2\,\overline{)\,3x^2-5x-2}\\
\underline{3x^2-6x\phantom{{}-2}}\\
x-2\\
\underline{x-2}\\
0
\end{array}$$

Since x divides x one time, we enter $+1$ in the quotient, multiply $x-2$ by 1, and subtract this product from $x-2$, obtaining a zero remainder.

Example 5

Divide:
(a) $3x^2+7x-6$ by $x+3$
(b) $5x^2+7x-5$ by $x+2$

Solution

(a)
$$\begin{array}{r}
3x-2\phantom{{}-6}\\
x+3\,\overline{)\,3x^2+7x-6}\\
\underline{3x^2+9x\phantom{{}-6}}\\
-2x-6\\
\underline{-2x-6}\\
0
\end{array}$$

The quotient is $3x-2$, with zero remainder.

(b)
$$\begin{array}{r}
5x-3\phantom{{}-5}\\
x+2\,\overline{)\,5x^2+7x-5}\\
\underline{5x^2+10x\phantom{{}-5}}\\
-3x-5\\
\underline{-3x-6}\\
1
\end{array}$$

The quotient is $5x-3$, and the remainder is 1. We write this remainder as in arithmetic; for instance, when we divide 16 by 3

$$\begin{array}{r}
5\\
3\,\overline{)\,16}\\
\underline{15}\\
1
\end{array}$$

We write the answer $\frac{16}{3}=5+\frac{1}{3}$. In this case, we write

$$\frac{5x^2+7x-5}{x+2}=5x-3+\frac{1}{x+2}$$

PROGRESS TEST

1. The fraction $\frac{3}{7}$ ______ (is, is not) equal to $\frac{21}{49}$.

2. When written with a denominator of 35, the fraction $\frac{3}{7}$ equals ______.

3. When written with a denominator of x^2, the fraction $\frac{y}{-x}$ equals ______.

4. When written in standard form, $\frac{a}{-3}$ equals ______.

5. When written in standard form, $-\frac{b-a}{7}$ equals ______.

6. When reduced, $\frac{x^4y^3}{xy^7}$ equals ______.

7. When reduced, $\frac{4a-b}{16a^2-b^2}$ equals ______.

8. When reduced, $\frac{ab+a^2}{a}$ equals ______.

9. $\frac{x^2+3x+5}{x+1}$ equals ______.

EXERCISE 4.1

In Problems 1 through 10, write an equivalent fraction with the indicated denominator.

Illustration: (a) $\frac{x}{y}$; denominator of $2y^2$

(b) $\frac{2-x}{1-x}$; denominator of $x-1$

Solution:

(a) $\frac{x}{y}=\frac{x \cdot 2y}{y \cdot 2y}$ Multiplying the numerator and denominator by 2y

$=\frac{2xy}{2y^2}$

PROGRESS TEST ANSWERS

1. is
2. $\frac{15}{35}$
3. $\frac{-xy}{x^2}$
4. $\frac{-a}{3}$
5. $\frac{a-b}{7}$ or $\frac{-(b-a)}{7}$
6. $\frac{x^3}{y^4}$
7. $\frac{1}{4a+b}$
8. $b+a$
9. $x+2+\frac{3}{x+1}$

(b) $\frac{2-x}{1-x} = \frac{(-1)(2-x)}{(-1)(1-x)}$ Multiplying the numerator and denominator by (-1)

$= \frac{x-2}{x-1}$ Since $(-1)(2-x) = -2+x = x-2$ and $(-1)(1-x) = -1+x = x-1$

1. $\frac{2x}{3y}$; denominator $6y^3$

2. $\frac{-3y}{2x}$; denominator $8x^2$

3. $\frac{a}{(x-y)}$; denominator $y-x$

4. $\frac{-b}{y-x}$; denominator $x-y$

5. $\frac{x}{x+y}$; denominator x^2-y^2

6. $\frac{-y}{x-y}$; denominator x^2-y^2

7. $\frac{-x}{y-x}$; denominator x^2-y^2

8. $\frac{-4x}{y-x}$; denominator x^2-y^2

9. $\frac{-x}{2x-3y}$; denominator $3y-2x$

10. $\frac{-x}{-2x-y}$; denominator $y+2x$

In Problems 11 through 20, write each fraction in standard form.

11. $-\frac{y}{-2}$

12. $-\frac{x-3}{y}$

13. $-\frac{x}{x-5}$

14. $\frac{2x-y}{-x}$

15. $-\dfrac{-2x}{-5y}$

16. $\dfrac{-x}{-y}$

17. $\dfrac{-(x+y)}{-(x-y)}$

18. $-\dfrac{-(3x+y)}{-(x-5y)}$

19. $\dfrac{-1}{-(x-2)}$

20. $\dfrac{-y}{-(x+1)}$

In Problems 21 through 34, reduce each fraction to lowest terms.

Illustration: (a) $\dfrac{x^3+y^3}{x^2-y^2}$

(b) $\dfrac{y^2+5y+6}{y+3}$

(c) $\dfrac{3x^3-6x^2+9x}{3x}$

Solution:

(a) $\dfrac{x^3+y^3}{x^2-y^2} = \dfrac{\overset{1}{\cancel{(x+y)}}(x^2-xy+y^2)}{\underset{1}{\cancel{(x+y)}}(x-y)} = \dfrac{x^2-xy+y^2}{x-y}$

(b) $\dfrac{y^2+5y+6}{y+3} = \dfrac{(y+2)\overset{1}{\cancel{(y+3)}}}{\underset{1}{\cancel{(y+3)}}} = y+2$

(c) $\dfrac{3x^3-6x^2+9x}{3x} = \dfrac{\overset{1}{\cancel{3x}}(x^2-2x+3)}{\underset{1}{\cancel{3x}}} = x^2-2x+3$

21. $\dfrac{x^4y^2}{xy^5}$

22. $\dfrac{x^5y^3c^2}{x^2y^6c^4}$

23. $\dfrac{3x-3y}{x-y}$

24. $\dfrac{4x^2}{4x-4y}$

25. $\dfrac{3x-2y}{9x^2-4y^2}$

26. $\dfrac{4x^2-9y^2}{2x+3y}$

27. $\dfrac{(x-y)^3}{x^2-y^2}$

28. $\dfrac{x^2-y^2}{(x+y)^3}$

29. $\dfrac{ay^2 - ay}{ay}$

30. $\dfrac{a^3 + 2a^2 + a}{a}$

31. $\dfrac{x^2 + 2xy + y^2}{x^2 - y^2}$

32. $\dfrac{x^2 + 3x + 2}{x^2 + 2x + 1}$

33. $\dfrac{y^2 - 8y + 15}{y^2 + 3y - 18}$

34. $\dfrac{y^2 + 7y - 18}{y^2 - 3y + 2}$

In Problems 35 through 40, reduce each fraction to lowest terms.

Illustration: $\dfrac{(3 - y)x}{y - 3}$

Solution:

$$\frac{(3-y)x}{y-3} = \frac{(-1)(3-y)x}{(-1)(y-3)}$$ Multiplying numerator and denominator by (−1)

$$= \frac{(y-3)x}{(-1)(y-3)}$$ Since $(-1)(3-y) = y-3$

$$= \frac{x}{(-1)} = -x$$

35. $\dfrac{2 - y}{y - 2}$

36. $\dfrac{3(x - y)}{4(y - x)}$

37. $\dfrac{9 - x^2}{x - 3}$

38. $\dfrac{25 - 9x^2}{3x - 5}$

39. $\dfrac{y^3 - 8}{2 - y}$

40. $\dfrac{2 + x}{x^3 + 8}$

In Problems 41 through 50, perform the indicated division.

Illustration: $(x^3 + 3x^2y + 3xy^2 + 2y^3) \div (x^2 + 2xy + y^2)$

Solution:

$$\begin{array}{r|l} & x + y \\ \hline x^2 + 2xy + y^2 & x^3 + 3x^2y + 3xy^2 + 2y^3 \\ & x^3 + 2x^2y + xy^2 \\ \hline & x^2y + 2xy^2 + 2y^3 \\ & x^2y + 2xy^2 + y^3 \\ \hline & y^3 \end{array}$$

$$\text{Thus, } \frac{x^3 + 3x^2y + 3xy^2 + 2y^3}{x^2 + 2xy + y^2} = x + y + \frac{y^3}{x^2 + 2xy + y^2}$$

41. $(x^2 - 5x + 6) \div (x - 3)$

42. $(x^2 + 3x + 2) \div (x + 1)$

43. $(y^2 + 5y + 5) \div (y + 1)$

44. $(y^2 + 8y + 11) \div (y + 2)$

45. $(x^2 - 2x - 15) \div (x - 5)$

46. $(x^2 - 3x - 40) \div (x - 8)$

47. $(y^2 - 13y + 29) \div (y - 3)$

48. $(y^2 - 2y - 14) \div (y + 3)$

49. $(x^3 + 8x^2 + 9x + 2) \div (x^2 + 7x + 2)$

50. $(x^3 - x^2 - 10x - 8) \div (x - 4)$

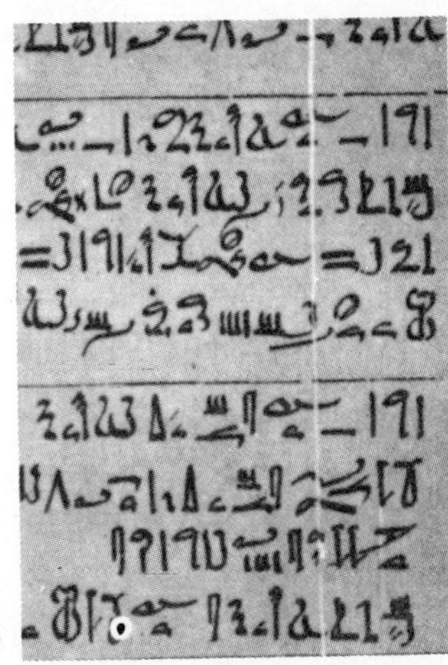

Courtesy of the Trustees of the British Museum.

OBJECTIVES 4.2

After studying this section, the reader should be able to:

1. **Find the sum and difference of two rational expressions.**
2. **Find the LCM of two numbers and use it to find the sum or difference of two fractions having the given numbers as denominators.**

4.2 ADDITION AND SUBTRACTION OF RATIONAL EXPRESSIONS

The photograph at the beginning of this section shows part of the Rhind Papyrus, a document found in the ruins of an ancient building at Thebes. In this document, which contains problems written with Egyptian numerals, fractions with unit numerators are represented by a rounded symbol above the denominator. Thus, $\frac{1}{3}$ appears as $\overset{\circ}{3}$, and $\frac{1}{2}$ as $\overset{\circ}{2}$. It is interesting to note that the Rhind Papyrus includes no fraction other than $\frac{2}{3}$ with a numerator different from one.

The addition of fractions with the same denominator can be accomplished by simply adding the numerators and retaining the same denominator. Thus, if you have one quarter and then you earn two more quarters, you now have three quarters. In symbols,

$$\frac{1}{4}+\frac{2}{4}=\frac{1+2}{4}=\frac{3}{4}.$$

In general, to add $\frac{a}{c}+\frac{b}{c}$ we proceed as follows:

$$\frac{a}{c}+\frac{b}{c}=a\cdot\left(\frac{1}{c}\right)+b\cdot\left(\frac{1}{c}\right)$$

Since $\frac{a}{c}=a\cdot\left(\frac{1}{c}\right)$ and $\frac{b}{c}=b\cdot\left(\frac{1}{c}\right)$

$$=(a+b)\cdot\left(\frac{1}{c}\right)$$

By the Distributive Law

$$=\frac{a+b}{c}$$

Hence, if a, b, and c are real numbers, and $c \neq 0$, then

$$\frac{a}{c}+\frac{b}{c}=\frac{a+b}{c}$$

Thus, $\frac{1}{5}+\frac{2}{5}=\frac{3}{5}$, $\frac{x}{7}+\frac{y}{7}=\frac{x+y}{7}$, and $\frac{a}{x+y}+\frac{b}{x+y}=\frac{a+b}{x+y}$.

Of course, the final answer should be in reduced form, as in the next example.

Example 1

Add:

(a) $\dfrac{3}{2x} + \dfrac{5}{2x}$

(b) $\dfrac{7}{3(x+1)} + \dfrac{2}{3(x+1)}$

Solution

(a) $\dfrac{3}{2x} + \dfrac{5}{2x} = \dfrac{3+5}{2x} = \dfrac{\overset{4}{\cancel{8}}}{\underset{1}{\cancel{2}x}} = \dfrac{4}{x}$

(b) $\dfrac{7}{3(x+1)} + \dfrac{2}{3(x+1)} = \dfrac{7+2}{3(x+1)} = \dfrac{\overset{3}{\cancel{9}}}{\underset{1}{\cancel{3}}(x+1)} = \dfrac{3}{x+1}$

The LCD is the smallest positive integer that is exactly divisible by each denominator. The LCD is sometimes called the least common multiple (LCM) of the denominators.

When adding fractions that do not have the same denominator, we must write equivalent fractions with a common denominator. To keep all fractions as simple as possible, we use the *least common multiple* (LCM) of the denominators (also called the least common denominator, or LCD). The LCD is the smallest positive integer that is exactly divisible by each denominator present. Thus, the LCD of $\dfrac{1}{12}$ and $\dfrac{1}{18}$ is 36, the smallest integer that is exactly divisible by 12 and 18.

To find the LCM of a set of natural numbers, we proceed as follows:

1. Write each number as a product of primes.
2. Select those prime factors raised to the highest power in the products of step 1.
3. The product of the factors selected in Step 2 is the LCM.

We use this procedure to find the LCM of 12, 18, and 24.

Note that the powers of 2 are written in the same column.

Step 1. $12 = 2^2 \cdot 3$
$18 = 2 \cdot 3^2$
$24 = 2^3 \cdot 3$

Step 2. From the first column, the 2's, we select 2^3 (the highest power of the prime 2 in this factoring). From the second column, we pick 3^2.

Step 3. The LCM is $2^3 \cdot 3^2 = 8 \cdot 9 = 72$. Thus, the LCM of 12, 18, and 24 is 72.

To find the sum of $\frac{1}{12} + \frac{1}{18} + \frac{1}{24}$, we can rewrite each of these fractions as an equivalent one, using 72, the LCM of 12, 18, and 24, as the denominator. Hence,

$$\frac{1}{12} = \frac{1 \cdot 6}{12 \cdot 6} = \frac{6}{72}$$

$$\frac{1}{18} = \frac{1 \cdot 4}{18 \cdot 4} = \frac{4}{72}$$

$$\frac{1}{24} = \frac{1 \cdot 3}{24 \cdot 3} = \frac{3}{72}$$

Thus,

$$\frac{1}{12} + \frac{1}{18} + \frac{1}{24} = \frac{6}{72} + \frac{4}{72} + \frac{3}{72} = \frac{6+4+3}{72} = \frac{13}{72}$$

The LCM of a set of polynomials can be found by using a method similar to that used to find the LCM of a set of natural numbers. For example, to find the LCM of x^2, $x^2 - 9$, and $x^2 - 6x + 9$, we proceed as follows:

1. Factor each expression, writing factors common to two expressions in the same column.

	1	2	3
$x^2 =$	x^2		
$x^2 - 9 =$		$(x+3)$	$(x-3)$
$x^2 - 6x + 9 =$			$(x-3)^2$

2. Select the factors with the greatest exponent in each column, that is, x^2, $x + 3$, and $(x - 3)^2$.
3. The LCM is $x^2(x + 3)(x - 3)^2$. (This fact will be used in Example 3.)

Example 2

Find the LCM of $x^2 + 4x + 4$ and $x^2 - 4$.

Solution

	1	2
$x^2 + 4x + 4 =$	$(x+2)^2$	
$x^2 - 4 =$	$(x+2)$	$(x-2)$

The LCM is $(x+2)^2(x-2)$

To add the fractions $\frac{1}{x^2+4x+4}$ and $\frac{1}{x^2-4}$, we first write each fraction as an equivalent one, using the LCM $(x+2)^2(x-2)$ as the denominator. Thus,

Here we factored $x^2 + 4x + 4$ first, then multiplied the numerator and denominator by $(x-2)$.

$$\frac{1}{x^2+4x+4} = \frac{1}{(x+2)^2} = \frac{(x-2)}{(x+2)^2(x-2)}$$

We first factored $x^2 - 4$, then multiplied the numerator and denominator by $(x+2)$.

and $$\frac{1}{x^2-4} = \frac{1}{(x+2)(x-2)} = \frac{(x+2)}{(x+2)^2(x-2)}$$

Hence,

$$\frac{1}{x^2+4x+4} + \frac{1}{x^2-4} = \frac{(x-2)}{(x+2)^2(x-2)} + \frac{(x+2)}{(x+2)^2(x-2)}$$

$$= \frac{x-2+x+2}{(x+2)^2(x-2)}$$

$$= \frac{2x}{(x+2)^2(x-2)}$$

Example 3

Add $\frac{1}{x^2} + \frac{1}{x^2-9} + \frac{1}{x^2-6x+9}$.

Solution

To add $\frac{1}{x^2} + \frac{1}{x^2-9} + \frac{1}{x^2-6x+9}$, we first write all the fractions as equivalent ones, with $x^2(x+3)(x-3)^2$ as the denominator. (Recall that the LCM of x^2, x^2-9, and x^2-6x+9 is $x^2(x+3)(x-3)^2$.)

Here we multiplied the numerator and denominator of $\frac{1}{x^2}$ by $(x+3)(x-3)^2$.

$$\frac{1}{x^2} = \frac{(x+3)(x-3)^2}{x^2(x+3)(x-3)^2}$$

We first factored $x^2 - 9$, then multiplied the numerator and denominator by $x^2(x-3)$.

$$\frac{1}{x^2-9} = \frac{1}{(x+3)(x-3)} = \frac{x^2(x-3)}{x^2(x+3)(x-3)^2}$$

We first factored $x^2 - 6x + 9$, then multiplied the numerator and denominator by $x^2(x+3)$.

$$\frac{1}{x^2-6x+9} = \frac{1}{(x-3)^2} = \frac{x^2(x+3)}{x^2(x+3)(x-3)^2}$$

Therefore,

$$\frac{1}{x^2}+\frac{1}{x^2-9}+\frac{1}{x^2-6x+9}$$

$$=\frac{(x+3)(x-3)^2}{x^2(x+3)(x-3)^2}+\frac{x^2(x-3)}{x^2(x+3)(x-3)^2}+\frac{x^2(x+3)}{x^2(x+3)(x-3)^2}$$

$$=\frac{(x+3)(x^2-6x+9)+x^3-3x^2+x^3+3x^2}{x^2(x+3)(x-3)^2}$$

$$=\frac{(x^3-6x^2+9x)+(3x^2-18x+27)+x^3-3x^2+x^3+3x^2}{x^2(x+3)(x-3)^2}$$

$$=\frac{3x^3-3x^2-9x+27}{x^2(x+3)(x-3)^2}$$

$$=\frac{3(x^3-x^2-3x+9)}{x^2(x+3)(x-3)^2}$$

To subtract fractions with the same denominators, we can use the fact that for any real numbers x and y,

$$x-y=x+(-y)$$

Thus,

$$\frac{a}{c}-\frac{b}{c}=\frac{a}{c}+\frac{-b}{c}$$

$$=\frac{a-b}{c}$$

Hence,

$$\frac{a}{c}-\frac{b}{c}=\frac{a-b}{c}$$

The procedure used to find the difference of two fractions is similar to that used for finding the sum. For example, to find $\frac{3x+1}{5x}-\frac{2x-2}{5x}$, we write

$$\frac{3x+1}{5x}-\frac{2x-2}{5x}=\frac{(3x+1)-(2x-2)}{5x}$$

$$=\frac{3x+1-2x+2}{5x}=\frac{x+3}{5x}$$

Recall that $-(2x-2)=-2x+2$.

To find $\frac{2x+5}{x+5}-\frac{x-1}{x^2-25}$, we proceed as follows:

1. Find the LCD of $x + 5$ and $x^2 - 25$. Since $x + 5 = x + 5$, and $x^2 - 25 = (x + 5)(x - 5)$, the LCD is $(x + 5)(x - 5)$.

2. $$\frac{2x+5}{x+5} - \frac{x-1}{x^2-25} = \frac{(2x+5)(x-5)}{(x+5)(x-5)} - \frac{(x-1)}{(x+5)(x-5)}$$
$$= \frac{(2x^2-5x-25)-(x-1)}{(x+5)(x-5)}$$
$$= \frac{2x^2-5x-25-x+1}{(x+5)(x-5)}$$
$$= \frac{2x^2-6x-24}{(x+5)(x-5)}$$
$$= \frac{2(x^2-3x-12)}{(x+5)(x-5)}$$

Example 4

Find:

(a) $$\frac{2x-5y}{3(x+y)} - \frac{x-3y}{3(x+y)}$$

(b) $$\frac{1}{x^2-y^2} - \frac{1}{x^2+2xy+y^2}$$

Solution

(a) Since both fractions have the same denominator, we have:

$$\frac{2x-5y}{3(x+y)} - \frac{x-3y}{3(x+y)} = \frac{(2x-5y)-(x-3y)}{3(x+y)}$$
$$= \frac{2x-5y-x+3y}{3(x+y)}$$
$$= \frac{x-2y}{3(x+y)}$$

(b) We first find the LCM of $x^2 - y^2$ and $x^2 + 2xy + y^2$.

$$x^2 - y^2 = (x+y)(x-y)$$
$$x^2 + 2xy + y^2 = (x+y)^2$$

The LCM is $(x+y)^2(x-y)$. Now we can subtract:

$$\frac{1}{x^2-y^2} - \frac{1}{x^2+2xy+y^2} = \frac{1}{(x+y)(x-y)} - \frac{1}{(x+y)^2}$$

$$= \frac{(x+y)}{(x+y)^2(x-y)} - \frac{(x-y)}{(x+y)^2(x-y)}$$

$$= \frac{(x+y)-(x-y)}{(x+y)^2(x-y)}$$

$$= \frac{x+y-x+y}{(x+y)^2(x-y)}$$

$$= \frac{2y}{(x+y)^2(x-y)}$$

From these examples, the procedure for the addition (or subtraction) of fractions may be stated as follows:

1. Change each fraction to an equivalent fraction with the LCD as denominator.
2. Add (or subtract) the numerators of the equivalent fractions.
3. Write the expression obtained in step 2, divided by the LCD, in reduced form.

PROGRESS TEST

1. When added, $\frac{3}{2(x-1)} + \frac{9}{2(x-1)}$ equals __________.
2. When added, $\frac{3x+2}{2x} + \frac{x+4}{2x}$ equals __________.
3. The LCM of 12, 16, and 18 is __________.
4. $\frac{1}{12} + \frac{1}{16} + \frac{1}{18}$ equals __________.
5. $\frac{a}{b+a} - \frac{b}{b+a}$ equals __________.
6. $\frac{a}{a^2-b^2} + \frac{b}{a+b}$ equals ____________________.
7. $\frac{a-b}{(a-b)^2} - \frac{a+b}{a^2-b^2}$ equals __________.

PROGRESS TEST ANSWERS

1. $\frac{6}{(x-1)}$
2. $\frac{2x+3}{x}$
3. $2^4 \cdot 3^2 = 144$
4. $\frac{29}{144}$
5. $\frac{a-b}{b+a}$
6. $\frac{a+ab-b^2}{(a+b)(a-b)}$
7. 0

EXERCISE 4.2

In Problems 1 through 10, write each sum or difference as a single fraction in lowest terms.

Illustration: (a) $\frac{2(x+1)}{3x} + \frac{4(x+1)}{3x}$

(b) $\dfrac{a}{a-b} - \dfrac{b}{b-a}$

Solution:

(a) $$\frac{2(x+1)}{3x} + \frac{4(x+1)}{3x} = \frac{2(x+1)+4(x+1)}{3x}$$

$$= \frac{\overset{2}{\cancel{6}}(x+1)}{\cancel{3}x}$$

$$= \frac{2(x+1)}{x}$$

(b) $$\frac{a}{a-b} - \frac{b}{b-a} = \frac{a}{a-b} + \frac{-b}{b-a}$$ Since $x - y = x + (-y)$

$$= \frac{a}{a-b} + \frac{b}{a-b}$$ Since $\dfrac{-b}{b-a} = \dfrac{(-1)\cdot(-b)}{(-1)\cdot(b-a)} = \dfrac{b}{a-b}$

$$= \frac{a+b}{a-b}$$

1. $\dfrac{x}{5} + \dfrac{2x}{5}$

2. $\dfrac{x+1}{3x} + \dfrac{2x+7}{3x}$

3. $\dfrac{7x}{3} - \dfrac{2x}{3}$

4. $\dfrac{2x-1}{5x} - \dfrac{x+1}{5x}$

5. $\dfrac{3}{5x+10} + \dfrac{2x}{5(x+2)}$

6. $\dfrac{2x+1}{3(x+2)} + \dfrac{3x+1}{3x+6}$

7. $\dfrac{2x+1}{2(x+1)} - \dfrac{x-1}{2x+2}$

8. $\dfrac{3x-1}{4(x-1)} - \dfrac{4x-1}{4x-4}$

9. $\dfrac{2x+1}{3(x-1)} + \dfrac{x+3}{3x-3} - \dfrac{x-1}{3(x-1)}$

10. $\dfrac{3x-1}{5(x+1)} - \dfrac{x+1}{5x+5} + \dfrac{2x-5}{5(x+1)}$

In Problems 11 through 20, find the LCM.

Illustration: (a) 8, 14, 20
(b) x^2-1, $(x+1)^2$, $(x-1)^2$

Solution: (a)

	1	2	3
8 =	2^3		
14 =	2		7
20 =	2^2	5	

The LCM is $2^3 \cdot 5 \cdot 7 = 280$

(b)

	1	2
$x^2-1=$	$(x+1)$	$(x-1)$
$(x+1)^2=$	$(x+1)^2$	
$(x-1)^2=$		$(x-1)^2$

The LCM is $(x+1)^2(x-1)^2$

11. 14 and 210

12. 80 and 92

13. 12, 18, and 30

14. 12, 15, and 20

15. $2x$, $4y$, and $6x^2y$

16. $6xy$, $8x^2y$, and $12x^2y^2$

17. $a-b$ and a^2-b^2

18. $a^2-2ab+b^2$ and a^2-b^2

19. x^2, $(x+y)^2$, and x^2+xy

20. $2x^2-8$, $2x^2$, and x^2+4x+4.

In Problems 21 through 50, write each sum or difference as a fraction in lowest terms.

Illustration: (a) $\frac{x+y}{x^2+2xy+y^2}+\frac{x-y}{x^2-2xy+y^2}$

(b) $\frac{x}{(x+2)(x-2)}-\frac{2}{(2-x)(x+2)}$

Solution:

(a) $\frac{x+y}{x^2+2xy+y^2}=\frac{x+y}{(x+y)^2}=\frac{1}{x+y}$ Reducing the fraction

$\frac{x-y}{x^2-2xy+y^2}=\frac{x-y}{(x-y)^2}=\frac{1}{x-y}$ Reducing the fraction

Thus,

$$\frac{x+y}{x^2-2xy+y^2}+\frac{x-y}{x^2-2xy+y^2}=\frac{1}{x+y}+\frac{1}{x-y}$$

$$=\frac{x-y}{(x+y)(x-y)}+\frac{x+y}{(x+y)(x-y)}$$

$$=\frac{2x}{(x+y)(x-y)}$$

(b) $\frac{x}{(x+2)(x-2)}-\frac{2}{(2-x)(x+2)}$

$$=\frac{x}{(x+2)(x-2)}+\frac{-2}{(2-x)(x+2)}$$

$$=\frac{x}{(x+2)(x-2)}+\frac{2}{(x-2)(x+2)}$$

$$=\frac{x+2}{(x+2)(x-2)}$$

$$=\frac{1}{x-2}$$

21. $\frac{x+2}{x^2-4}+\frac{x+3}{x^2-9}$

22. $\frac{x-3}{x^2-9}+\frac{x+3}{x^2+6x+9}$

23. $\frac{a-4}{a^2-16}+\frac{a+3}{a^2+5a+6}$

24. $\frac{a+3}{a^2+5a+6}+\frac{a+2}{a^2+6a+8}$

25. $\frac{a+3}{a^2+5a+6}-\frac{a-4}{a^2-16}$

26. $\frac{3a+3}{a^2+5a+4}-\frac{a-3}{a^2+a-12}$

27. $\frac{2a}{5a-7b}-\frac{5a+7b}{25a^2-49b^2}$

28. $\frac{5a-15}{a^2+2a-15}-\frac{a^2+5a}{a^2+8a+15}$

29. $\frac{3}{y^2-9}+\frac{2y}{y-3}$

30. $\frac{y}{y^2-1}+\frac{y}{y-1}$

31. $\frac{3y}{y^2-4}-\frac{y}{y+2}$

32. $\frac{3y+1}{y^2-16}-\frac{2y-1}{y-4}$

33. $\frac{3x-5y}{2x-3y}+\frac{2x-3y}{2x+3y}$

34. $\frac{5x+2y}{5x-2y}+\frac{5x-2y}{5x+2y}$

35. $\frac{x+3y}{x-5y}-\frac{x+5y}{x-3y}$

36. $\frac{3x-y}{2x-y}-\frac{2x+y}{3x+y}$

37. $\frac{a+3}{a^2+a-6}+\frac{a-2}{a^2+3a-10}$

38. $\frac{x+3}{x^2-x-2}+\frac{x-1}{x^2+2x+1}$

39. $\frac{8x}{x^2-4y^2}-\frac{2x}{x^2-5xy+6y^2}$

40. $\frac{x+1}{x^2-x-2}-\frac{x}{x^2-5x+4}$

41. $\frac{3}{x^2-4}+\frac{1}{2-x}-\frac{1}{2+x}$

42. $\frac{2}{5+x}+\frac{5x}{x^2-25}+\frac{7}{5-x}$

43. $\frac{1}{x^2+x-12}+\frac{2}{x^2+2x-15}+\frac{3}{x^2+9x+20}$

44. $\frac{x}{(x-y)(2-x)}+\frac{y}{(y-x)(x-2)}-\frac{y}{(x-y)(2-x)}$

45. $\frac{a}{(b-a)(c-a)}+\frac{b}{(c-b)(a-b)}+\frac{c}{(a-c)(b-c)}$

46. $\frac{4a^2-9b^2}{4a^2-12ab+9b^2}+\frac{12a+18b}{4a^2+12ab+9b^2}-\frac{15a-17b}{2a+3b}$

47. $\frac{x+2y}{x^3+8y^3}+\frac{5}{x+2y}+\frac{2x-3y}{x^2-2xy+4y^2}$

48. $\frac{x+5}{x^3+125}+\frac{x-5}{x^2-25}-\frac{1}{x+5}$

49. $\frac{a}{a+3}+\frac{a-2}{a^2-3a+9}+\frac{5a-7a}{a^3+27}$

50. $\frac{5a-7a}{a^3+27}-\frac{a-2}{a^2-3a+9}-\frac{a}{a+3}$

OBJECTIVES 4.3

After studying this section, the reader should be able to find the product and quotient of two rational expressions and express the answer in lowest terms.

4.3 MULTIPLICATION AND DIVISION OF RATIONAL EXPRESSIONS

In the ad at the beginning of this section, we can see that each package of peas weighs $1\frac{1}{2} = \frac{3}{2}$ pounds. If we have a recipe that calls for $2\frac{1}{2}$ packages of peas, how many pounds is this? Since $2\frac{1}{2} = \frac{5}{2}$, we need to multiply $\frac{5}{2}$ by $\frac{3}{2}$. In elementary arithmetic, we learned that the product of $\frac{5}{2}$ and $\frac{3}{2}$ is written as $\frac{5}{2} \cdot \frac{3}{2} = \frac{5 \cdot 3}{2 \cdot 2} = \frac{15}{4}$.

In general, if a, b, c, and d are real numbers, we have:

$$\frac{a}{b} \cdot \frac{c}{d} = \frac{a \cdot c}{b \cdot d}$$

Thus, $\frac{3}{5} \cdot \frac{2}{7} = \frac{3 \cdot 2}{5 \cdot 7} = \frac{6}{35}$ and $\frac{8x}{y} \cdot \frac{xy^3}{2} = \frac{\overset{4}{\cancel{8}}x^2\overset{y^2}{\cancel{y^3}}}{\underset{1}{\cancel{2y}}} = 4x^2y^2$.

Example 1 Write the given products in lowest terms.

(a) $\frac{7}{16} \cdot \frac{4}{21}$

(b) $\frac{x-1}{2x-3y} \cdot \frac{4x^2-9y^2}{2x^2-x-1}$

(c) $\frac{x^2-4}{4x^2-9y^2} \cdot \frac{2x^2-3xy}{2x+4}$

Solution

(a) $\frac{7}{16} \cdot \frac{4}{21} = \frac{\overset{1}{\cancel{7}} \cdot \overset{1}{\cancel{4}}}{\underset{4}{\cancel{16}} \cdot \underset{3}{\cancel{21}}} = \frac{1}{12}$

(b) $\frac{x-1}{2x-3y} \cdot \frac{4x^2-9y^2}{2x^2-x-1}$

$= \frac{x-1}{2x-3y} \cdot \frac{(2x+3y)(2x-3y)}{(x-1)(2x+1)}$ Factoring

$= \frac{(2x+3y)}{(2x+1)} \cdot \frac{\overset{1}{[\cancel{(2x-3y)(x-1)}]}}{\underset{1}{[\cancel{(2x-3y)(x-1)}]}}$ By the Commutative and Associative Laws

$= \frac{2x+3y}{2x+1}$

(c) $\frac{x^2-4}{4x^2-9y^2} \cdot \frac{2x^2-3xy}{2x+4}$

$= \frac{(x+2)(x-2)}{(2x-3y)(2x+3y)} \cdot \frac{x(2x-3y)}{2(x+2)}$ Factoring

$= \frac{x(x-2)}{2(2x+3y)} \cdot \frac{\overset{1}{[\cancel{(x+2)(2x-3y)}]}}{\underset{1}{[\cancel{(x+2)(2x-3y)}]}}$ By the Commutative Law

$= \frac{x(x-2)}{2(2x+3y)}$

As you recall from Chapter 1, $x \div y = z$ means that $x = y \cdot z$. If x, y, and z are fractions,

(1) $\frac{a}{b} \div \frac{c}{d} = q$ means $\frac{a}{b} = \left(\frac{c}{d}\right) \cdot q$

(2) $\left(\frac{d}{c}\right)\left(\frac{a}{b}\right) = \left(\frac{d}{c}\right)\left(\frac{c}{d}\right) \bullet q$ — Multiplying both members by $\frac{d}{c}$

(3) $\left(\frac{d}{c}\right)\left(\frac{a}{b}\right) = q$ — Since $\left(\frac{d}{c}\right)\left(\frac{c}{d}\right) = 1$

(4) Thus, $q = \left(\frac{d}{c}\right)\left(\frac{a}{b}\right) = \frac{a \bullet d}{b \bullet c}$ — Using the Symmetric and Commutative Laws

(5) Substituting the value of q from Equation (4) into Equation (1), we have

$$\frac{a}{b} \div \frac{c}{d} = \frac{a \bullet d}{b \bullet c}$$

Briefly, this rule says to invert the divisor and multiply. For example,

$$\frac{2}{5} \div \frac{3}{8} = \left(\frac{2}{5}\right)\left(\frac{8}{3}\right) = \frac{16}{15}$$

and
$$\frac{3x^2}{2y} \div \frac{6x}{4y^2} = \left(\frac{3x^2}{2y}\right)\left(\frac{4y^2}{6x}\right)$$

$$= \frac{\overset{1x\ \ y}{\cancel{12x^2y^2}}}{\underset{1}{\cancel{12xy}}}$$

$$= xy$$

Note that answers should be given in reduced form.

Example 2

Write the given quotients in lowest terms.

(a) $\frac{x^3y}{z} \div \frac{x^2y}{z^4}$

(b) $\frac{5x^2 - 5}{3x + 6} \div \frac{x + 1}{3}$

(c) $\frac{x + 3}{x - 3} \div (x^2 + 6x + 9)$

Solution

(a) $\frac{x^3y}{z} \div \frac{x^2y}{z^4} = \left(\frac{x^3y}{z}\right)\left(\frac{z^4}{x^2y}\right)$

$$= \frac{\overset{x\ \ z^3}{\cancel{x^3yz^4}}}{\cancel{x^2yz}}$$

$$= xz^3$$

(b) $\dfrac{5x^2-5}{3x+6} \div \dfrac{x+1}{3} = \dfrac{5(x+1)(x-1)}{3(x+2)} \bullet \dfrac{3}{x+1}$

$= \dfrac{5(x-1) \bullet [\overset{1}{\cancel{3(x+1)}}]}{(x+2) \bullet [\underset{1}{\cancel{3(x+1)}}]}$

$= \dfrac{5(x-1)}{x+2}$

(c) $\dfrac{x+3}{x-3} \div (x^2+6x+9) = \dfrac{x+3}{x-3} \div \dfrac{x^2+6x+9}{1}$

$= \left(\dfrac{x+3}{x-3}\right)\left(\dfrac{1}{(x+3)(x+3)}\right)$

$= \dfrac{1 \bullet [\overset{1}{\cancel{x+3}}]}{(x-3)(x+3) \bullet [\underset{1}{\cancel{x+3}}]}$

$= \dfrac{1}{(x-3)(x+3)}$

PROGRESS TEST

1. $\dfrac{-4}{5} \bullet \dfrac{3}{7}$ equals __________.

2. $\dfrac{9x}{y^2} \bullet \dfrac{x^2y}{-3}$ equals __________.

3. $\dfrac{-(x+3)}{3x-2y} \bullet \dfrac{9x^2-4y^2}{2x^2+5x-3}$ equals __________.

4. $\dfrac{x^2-25}{16x^2-9} \bullet \dfrac{4x+3}{2x+10}$ equals __________.

5. $\dfrac{8}{17} \div \dfrac{-17}{2}$ equals __________.

6. $\dfrac{-x^2y^3}{z^2} \div \dfrac{-xy^6}{z^5}$ equals __________.

7. $\dfrac{x^2-1}{4x+8} \div \dfrac{x-1}{x+2}$ equals __________.

PROGRESS TEST ANSWERS

1. $\dfrac{-12}{35}$
2. $\dfrac{-3x^3}{y}$
3. $\dfrac{-(3x+2y)}{(2x-1)}$
4. $\dfrac{x-5}{2(4x-3)}$
5. $\dfrac{-16}{289}$
6. $\dfrac{xz^3}{y^3}$
7. $\dfrac{x+1}{4}$

EXERCISE 4.3

In Problems 1 through 20, write the given product in lowest terms.

Illustration: (a) $\dfrac{2-x}{x+1} \bullet \dfrac{x^2+3x+2}{x^2-4}$

(b) $\dfrac{x^2+3x+2}{x^2+5x+4} \bullet \dfrac{x^2+2x-3}{x^2+x-2}$

Solution: (a) $\dfrac{2-x}{x+1} \bullet \dfrac{x^2+3x+2}{x^2-4}$

$$= \frac{-(x-2)}{x+1} \bullet \frac{(x+1)(x+2)}{(x+2)(x-2)}$$

$$= \frac{\overset{1}{-\cancel{(x-2)}\cancel{(x+1)}\cancel{(x+2)}}}{\underset{1}{\cancel{(x-2)}\cancel{(x+1)}\cancel{(x+2)}}}$$

$$= -1$$

(b) $\dfrac{x^2+3x+2}{x^2+5x+4} \bullet \dfrac{x^2+2x-3}{x^2+x-2}$

$$= \frac{\overset{1}{\cancel{(x+2)}}\overset{1}{\cancel{(x+1)}}\overset{1}{\cancel{(x-1)}}(x+3)}{(x+4)\underset{1}{\cancel{(x+1)}}\underset{1}{\cancel{(x-1)}}\underset{1}{\cancel{(x+2)}}}$$

$$= \frac{x+3}{x+4}$$

1. $\dfrac{3}{4} \bullet \dfrac{2}{5}$

2. $\dfrac{-9}{10} \bullet \dfrac{2}{3}$

3. $\dfrac{14x^2}{15} \bullet \dfrac{5}{7x}$

4. $\dfrac{-5x^3}{7y} \bullet \dfrac{4y^3}{9x^6}$

5. $\dfrac{-2xy^4}{9z^5} \bullet \dfrac{-3z}{7x^3y^3}$

6. $\dfrac{-35x^5z}{24x^3y^9} \bullet \dfrac{84x^3y^8}{15x^4y^7z}$

7. $\dfrac{10x + 50}{6x + 6} \bullet \dfrac{12}{5x + 25}$

8. $\dfrac{x + y}{xy - y^2} \bullet \dfrac{y^2}{x^2 - y^2}$

9. $\dfrac{6y + 3}{2y^2 - 3y - 2} \bullet \dfrac{y^2 - 4}{3y + 6}$

10. $\dfrac{y^2 + 9y + 18}{y - 2} \bullet \dfrac{2y - 1}{5y + 15}$

11. $\dfrac{y - x}{x^2 + 2xy} \bullet \dfrac{5x + 10y}{x^2 - y^2}$

12. $\dfrac{2 - 2x}{9x^2 - 25} \bullet \dfrac{6x - 10}{x^2 - 1}$

13. $\dfrac{3y^2 - 17y + 10}{y^2 - 4y - 5} \bullet \dfrac{y^2 + 3y + 2}{y^2 + y - 2}$

14. $\dfrac{y^2 + 2y - 3}{y^2 - 4y - 5} \bullet \dfrac{y^2 - 3y - 10}{y^2 + 5y - 6}$

15. $\dfrac{y^2 + 2y - 8}{y^2 + 7y + 12} \bullet \dfrac{y^2 + 2y - 3}{y^2 - 3y + 2}$

16. $\dfrac{y^2 + 2y - 15}{y^2 - 7y + 10} \bullet \dfrac{y^2 - 6y + 8}{y^2 - y - 12}$

17. $\dfrac{x^3 - 8}{4 - x^2} \bullet \dfrac{x^2 + x - 2}{x^2 + 2x + 4}$

18. $\dfrac{x^3 + y^3}{y^2 - x^2} \bullet \dfrac{x - y}{x^2 - xy + y^2}$

19. $\dfrac{a^3 + b^3}{a^3 - b^3} \bullet \dfrac{a^2 + ab + b^2}{a^2 - ab + b^2}$

20. $\dfrac{a^3 - 8}{a^2 + 2a + 4} \bullet \dfrac{a^2 + 3a + 9}{a^3 - 27}$

In Problems 21 through 40, write the given quotient in lowest terms.

Illustration: (a) $\dfrac{xy^2}{z} \div \dfrac{-x^2y}{z^3}$

(b) $\dfrac{4x^2 - 4}{3x - 3} \div \dfrac{2x + 2}{6}$

Solution: (a) $\dfrac{xy^2}{z} \div \dfrac{-x^2y}{z^3} = \dfrac{xy^2}{z} \bullet \dfrac{z^3}{-x^2y}$

$$= \frac{-xy^2z^3}{x^2yz}$$

$$= \frac{-yz^2}{x}$$

(b) $\dfrac{4x^2 - 4}{3x - 3} \div \dfrac{2x + 2}{6}$

$$= \frac{4(x + 1)(x - 1)}{3(x - 1)} \bullet \frac{6}{2(x + 1)}$$

$$= \frac{4[\overset{1}{\cancel{6(x + 1)(x - 1)}}]}{[\underset{1}{\cancel{6(x + 1)(x - 1)}}]}$$

$$= 4$$

21. $\dfrac{3}{5} \div \dfrac{10}{9}$

22. $\dfrac{3}{7} \div \dfrac{-9}{14}$

23. $\dfrac{4}{5x^2} \div \dfrac{12}{25x^3}$

24. $\dfrac{6x^2}{7} \div \dfrac{30x}{28}$

25. $\dfrac{24a^2b}{7c^2d} \div \dfrac{8ab}{21cd^2}$

26. $\dfrac{16a^3b}{15a} \div \dfrac{12ab^2}{20b^4}$

27. $\dfrac{3x-3}{x} \div \dfrac{x^2-1}{x^2}$

28. $\dfrac{5x^2-45}{x^3} \div \dfrac{x+3}{x}$

29. $\dfrac{y^2-25}{y^2-4} \div \dfrac{3y-15}{4y-8}$

30. $\dfrac{y^2+y-12}{y^2-1} \div \dfrac{3y+12}{4y^2+4y}$

31. $\dfrac{a^3+b^3}{a^3-b^3} \div \dfrac{a^2-ab+b^2}{a^2+ab+b^2}$

32. $\dfrac{a^2+ab+b^2}{a^3+b^3} \div \dfrac{a^2+ab+b^2}{a^2-ab+b^2}$

33. $\dfrac{8a^3-1}{6u^4w^3} \div \dfrac{1-2a}{3u^2w}$

34. $\dfrac{-b^2c}{27a^3-1} \div \dfrac{b^3c^2}{3a-1}$

35. $\dfrac{x-x^3}{2x^2+6x} \div \dfrac{5x^2-5x}{2x+6}$

36. $\dfrac{121y-y^3}{y^2-49} \div \dfrac{y^2-11y}{y+7}$

37. $\dfrac{y^2+y-12}{y^2-8y+15} \div \dfrac{3y^2+7y-20}{2y^2-7y-15}$

38. $\dfrac{3y^2+11y+6}{4y^2+16y+7} \div \dfrac{3y^2-y-2}{y^2-y-28}$

39. $\dfrac{4x^2-12x+9}{25-4x^2} \div \dfrac{6x^2-5x-6}{6x^2+19x+10}$

40. $\dfrac{4x^2+12x+9}{9-4x^2} \div \dfrac{10x^2+27x+18}{8x^2-2x-15}$

Photo of Dr. Demento from Gordon/Casady Inc.

OBJECTIVES 4.4

After studying this section, the reader should be able to write a given complex fraction as a simple fraction in lowest terms.

4.4 COMPLEX FRACTIONS

The picture at the beginning of this section shows the disc jockey of station KMET in Los Angeles. Each hour he devotes $7\frac{1}{2}$ minutes to commercials, leaving $60 - 7\frac{1}{2}$ minutes for music. If the records he plays last an average of $3\frac{1}{4}$ minutes and it takes him about $\frac{1}{2}$ minute to get a record going, how many records can he play each hour? The answer is

$$\begin{array}{r} \text{Time allowed for music } \rightarrow \\ \text{Time it takes for playing } \rightarrow \\ \text{each record} \end{array} \frac{60 - 7\frac{1}{2}}{3\frac{1}{4} + \frac{1}{2}}$$

The fraction $\dfrac{60 - 7\frac{1}{2}}{3\frac{1}{4} + \frac{1}{2}}$ is a *complex fraction*—that is, a fraction whose numerator, denominator, or both contain other fractions. A fraction that is *not* complex is called a *simple fraction.*

A fraction is complex if its numerator, its denominator, or both contain other fractions.

If a fraction is not complex, it is simple.

Thus, the fractions $\dfrac{\frac{1}{2}}{\frac{3}{4}+\frac{1}{5}}$, $\dfrac{\frac{3x}{5}-\frac{1}{8}}{\frac{1x}{7}}$, $\dfrac{-\frac{1}{3}}{\frac{1}{9}}$, and $\dfrac{x}{\frac{7}{8}}$ are complex fractions, but $\frac{1}{7}$, $\frac{3}{5}$, and $\frac{x}{9}$ are simple fractions.

To simplify a complex fraction, it is necessary to recall that the main fraction bar indicates that the numerator of the fraction is to be divided by the denominator of the fraction.

Thus, the complex fraction $\dfrac{60-7\frac{1}{2}}{3\frac{1}{4}+\frac{1}{2}}$ means $\left(60-7\frac{1}{2}\right)\div\left(3\frac{1}{4}+\frac{1}{2}\right)$. This fraction can be simplified in either of two ways:

1. by multiplying the numerator and denominator of the complex fraction by the LCD of the simple fractions appearing; or
2. by performing the operations indicated in the numerator and denominator of the given complex fraction, and then dividing the numerator by the denominator.

We now simplify $\dfrac{60-7\frac{1}{2}}{3\frac{1}{4}+\frac{1}{2}}=\dfrac{60-\frac{15}{2}}{\frac{13}{4}+\frac{1}{2}}$ using each of these methods.

Method 1. The LCD of $\frac{15}{2}$, $\frac{13}{4}$, and $\frac{1}{2}$ is 4, so we have:

$$\frac{60-\frac{15}{2}}{\frac{13}{4}+\frac{1}{2}}=\frac{4\bullet\left(60-\frac{15}{2}\right)}{4\bullet\left(\frac{13}{4}+\frac{1}{2}\right)}$$

Multiplying numerator and denominator by 4, the LCD of $\frac{15}{2}$, $\frac{13}{4}$, and $\frac{1}{2}$.

$$=\frac{240-30}{13+2}$$

$$=\frac{210}{15}$$

$$=14$$

Method 2. $$\frac{60-\frac{15}{2}}{\frac{13}{4}+\frac{1}{2}}=\frac{\frac{120}{2}-\frac{15}{2}}{\frac{13}{4}+\frac{2}{4}}$$

$$= \frac{\frac{105}{2}}{\frac{15}{4}}$$

$$= \frac{105}{2} \div \frac{15}{4}$$

$$= \frac{\overset{7}{\cancel{105}}}{\underset{1}{\cancel{2}}} \bullet \frac{\overset{2}{\cancel{4}}}{\underset{1}{\cancel{15}}}$$

$$= 14$$

Example 1

Write $\dfrac{\frac{3}{a} - \frac{4}{b}}{\frac{1}{2a} + \frac{2}{3b}}$ as a simple fraction in lowest terms.

Solution

We first multiply the numerator and denominator of the given fraction by 6ab, the LCD of $\frac{3}{a}, \frac{4}{b}, \frac{1}{2a}$, and $\frac{2}{3b}$, obtaining

$$\frac{6ab \bullet \left(\frac{3}{a} - \frac{4}{b}\right)}{6ab \bullet \left(\frac{1}{2a} + \frac{2}{3b}\right)} = \frac{6ab \bullet \frac{3}{a} - 6ab \bullet \frac{4}{b}}{6ab \bullet \frac{1}{2a} + 6ab \bullet \frac{2}{3b}}$$

$$= \frac{18b - 24a}{3b + 4a}$$

Note that in solving Example 1, we multiplied the numerator and denominator by the LCD of the simple fractions involved. This was done because the fractions present were not too complicated, and the LCD was obvious. In Example 2, we use the second method of simplifying complex fractions.

Example 2

Write $\dfrac{\frac{x}{x-2} + x}{2 + \frac{1}{x^2 - 4}}$ as a simple fraction in lowest terms.

Solution

$$\frac{\frac{x}{x-2}+x}{2+\frac{1}{x^2-4}} = \frac{\frac{x}{x-2}+\frac{x(x-2)}{x-2}}{\frac{2(x^2-4)}{x^2-4}+\frac{1}{x^2-4}}$$

$$= \frac{\frac{x+x(x-2)}{x-2}}{\frac{2(x^2-4)+1}{x^2-4}}$$

$$= \frac{\frac{x+x^2-2x}{x-2}}{\frac{2x^2-8+1}{x^2-4}}$$

$$= \frac{x^2-x}{x-2} \div \frac{2x^2-7}{x^2-4}$$

$$= \frac{x(x-1)}{x-2} \bullet \frac{(x+2)(x-2)}{2x^2-7}$$

$$= \frac{x(x-1)(x+2) \bullet \overset{1}{\cancel{[x-2]}}}{(2x^2-7) \bullet \underset{1}{\cancel{[x-2]}}}$$

$$= \frac{x(x-1)(x+2)}{2x^2-7}$$

PROGRESS TEST

1. When written as a simple fraction in lowest terms, $\dfrac{5+\frac{1}{4}}{\frac{1}{2}-\frac{3}{8}}$ equals ________.

2. When written as a simple fraction in lowest terms, $\dfrac{\frac{a}{b}-\frac{b}{a}}{\frac{a}{b}+\frac{b}{a}}$ equals ________.

3. When written as a simple fraction in lowest terms, $\dfrac{\frac{1}{(1+x)}+2}{\frac{x}{3}-2}$ equals ________.

PROGRESS TEST ANSWERS

1. 42
2. $\dfrac{a^2-b^2}{a^2+b^2}$
3. $\dfrac{3(2x+3)}{(x+1)(x-6)}$

EXERCISE 4.4

In Problems 1 through 25, write the given fraction as a simple fraction in lowest terms.

Illustration: (a) $$\frac{\dfrac{1}{x-y}+\dfrac{1}{x+y}}{\dfrac{1}{x-y}-\dfrac{1}{x+y}}$$

(b) $$1+\frac{a}{1+\dfrac{1}{1+a}}$$

Solution:

(a) The LCD of the fractions involved is $(x-y)(x+y)$ Multiplying numerator and denominator by the LCD, we have:

$$\frac{(x-y)(x+y)\bullet\left(\dfrac{1}{x-y}+\dfrac{1}{x+y}\right)}{(x-y)(x+y)\bullet\left(\dfrac{1}{x-y}-\dfrac{1}{x+y}\right)}$$

$$=\frac{(x-y)(x+y)\bullet\dfrac{1}{x-y}+(x-y)(x+y)\bullet\dfrac{1}{x+y}}{(x-y)(x+y)\bullet\dfrac{1}{x-y}-(x-y)(x+y)\bullet\dfrac{1}{x+y}}$$

$$=\frac{(x+y)+(x-y)}{(x+y)-(x-y)}$$

$$=\frac{2x}{2y}$$

$$=\frac{x}{y}$$

(b) $$1+\frac{a}{1+\dfrac{1}{1+a}}=1+\frac{a}{\dfrac{1+a}{1+a}+\dfrac{1}{1+a}}$$ Since $1+\dfrac{1}{1+a}=\dfrac{1+a}{1+a}+\dfrac{1}{1+a}$

$$=1+\frac{a}{\dfrac{2+a}{1+a}}$$ Adding

$$=1+\frac{a(1+a)}{2+a}$$ Since $\dfrac{a}{\dfrac{2+a}{1+a}}=a\bullet\dfrac{1+a}{2+a}=\dfrac{a(1+a)}{2+a}$

$$= \frac{2 + a}{2 + a} + \frac{a(1 + a)}{2 + a}$$

$$= \frac{2 + 2a + a^2}{2 + a}$$

1. $\dfrac{\frac{3}{5}}{\frac{4}{5}}$

2. $\dfrac{\frac{-1}{7}}{\frac{3}{7}}$

3. $\dfrac{\frac{a}{b}}{\frac{c}{b}}$

4. $\dfrac{\frac{-a^2}{c}}{\frac{-b^2}{c}}$

5. $\dfrac{\frac{x}{y}}{\frac{x^2}{z}}$

6. $\dfrac{\frac{x^2}{y^2}}{\frac{x}{z}}$

7. $\dfrac{\frac{3x}{5y}}{\frac{3x}{2z}}$

8. $\dfrac{\frac{7x}{3y}}{\frac{14x}{5y}}$

9. $\dfrac{\frac{1}{2}}{2 - \frac{1}{2}}$

10. $\dfrac{\frac{1}{4}}{3 - \frac{1}{4}}$

11. $\dfrac{a - \frac{a}{b}}{1 + \frac{a}{b}}$

12. $\dfrac{1 - \frac{1}{a}}{1 + \frac{1}{a}}$

13. $\dfrac{y + \frac{2}{x}}{y^2 - \frac{4}{x^2}}$

14. $\dfrac{y - \frac{3}{x}}{y^2 + \frac{9}{x^2}}$

15. $\dfrac{\frac{x}{y^2} - \frac{y}{x^2}}{x^2 + xy + y^2}$

16. $\dfrac{\frac{x}{y^2} + \frac{y}{x^2}}{x^2 - xy + y^2}$

17. $3 - \dfrac{3}{3 - \frac{1}{2}}$

18. $2 - \dfrac{2}{2 - \frac{1}{2}}$

19. $a - \dfrac{a}{a + \dfrac{1}{2}}$

20. $a + \dfrac{a}{a + \dfrac{1}{2}}$

21. $x - \dfrac{x}{1 - \dfrac{x}{1 - x}}$

22. $2x - \dfrac{x}{2 - \dfrac{x}{2 - x}}$

23. $\dfrac{1}{1 + \dfrac{1}{2 + \dfrac{1}{3 + \dfrac{1}{4}}}}$

24. $\dfrac{1}{1 - \dfrac{1}{2 - \dfrac{1}{3 - \dfrac{1}{4}}}}$

4 out of 5 people say
Big John's Beans taste better.
We bet 20¢ you'll agree.

Big John's is a registered trademark of Hunt-Wesson Foods, Inc.

OBJECTIVES 4.5

After studying this section, the reader should be able to find the solution set of a linear equation containing rational expressions.

4.5 EQUATIONS INVOLVING FRACTIONS

In the ad at the beginning of this section, it is claimed that 4 out of 5 people say Big John's Beans taste better. If this statement was actually made by 300 people in a survey, how many people were surveyed? If x is the number of people surveyed, and 4 out of 5 people $\left(\frac{4}{5}\right)$ say that Big John's Beans taste better, then $\frac{4x}{5}$ and 300 represent the number of people making the statement. Thus,

$$\frac{4x}{5} = 300$$

To solve the equation $\frac{4x}{5} = 300$, we first multiply both sides of this equation by 5, the LCD of the fractions present, obtaining

$$5 \cdot \frac{4x}{5} = 5 \cdot 300$$

$$\text{or } 4x = 1500$$

$$x = \frac{1500}{4} = 375 \qquad \text{dividing by 4}$$

Thus, 375 people were surveyed.

To solve the equation $\frac{x}{2} + \frac{x}{3} = 10$, we use a similar procedure and first multiply each number of the equation by 6, the LCD of $\frac{x}{2}$ and $\frac{x}{3}$. Hence

$$6 \cdot \left(\frac{x}{2} + \frac{x}{3}\right) = 6 \cdot 10$$

$$\text{or } 6 \cdot \frac{x}{2} + 6 \cdot \frac{x}{3} = 6 \cdot 10$$

$$\text{or} \qquad 3x + 2x = 60$$

Simplifying, we get

$$5x = 60$$

Dividing by 5, we have $x = 12$.

As we have shown in Section 2.3, when variables occur in the denominator it is possible to multiply both members by the LCD of the fractions involved and obtain a solution of the resulting equation which does *not* satisfy the original equation. For example, if we assume that there is a solution for the equation

$$3 + \frac{1}{x-3} = \frac{1}{x-3}$$

we first multiply both members by $x - 3$, obtaining

$$(x-3) \cdot 3 + (x-3) \cdot \frac{1}{x-3} = (x-3) \cdot \frac{1}{x-3}$$

$$3x - 9 + 1 = 1$$

$$3x = 9$$

$$x = 3$$

If we replace x by 3 in the equation $3 + \frac{1}{x-3} = \frac{1}{x-3}$, we obtain $3 + \frac{1}{0} = \frac{1}{0}$. Since division by 0 is not defined, the equation $3 + \frac{1}{x-3} = \frac{1}{x-3}$ has no solution, and the solution set is $\emptyset$. This example points out the necessity of checking, by direct substitution in the original equation, any prospective solutions obtained after multiplying both members of an equation by factors containing the unknown.

Example 1

Solve:

(a) $\frac{4}{x} = \frac{6}{x+2}$

(b) $\frac{1}{x+1} = \frac{2}{x+2}$

Solution

(a) The LCD of $\frac{4}{x}$ and $\frac{6}{x+2}$ is $x(x + 2)$. Multiplying both members by this LCD, we get

$$x(x+2) \cdot \frac{4}{x} = x(x+2) \cdot \frac{6}{x+2}$$

$$\begin{aligned} (x+2) \cdot 4 &= x \cdot 6 \\ 4x + 8 &= 6x \\ -2x &= -8 \\ x &= 4 \end{aligned}$$

To check the answer, we substitute 4 for x in the original equation, obtaining $\frac{4}{4} = \frac{6}{4+2}$, or $1 = 1$. Therefore, the solution $x = 4$ is correct.

(b) The LCD of $\frac{1}{x+1}$ and $\frac{2}{x+2}$ is $(x+1)(x+2)$. Multiplying both members by $(x+1)(x+2)$, we obtain

$$(x+1)(x+2) \cdot \frac{1}{x+1} = (x+1)(x+2) \cdot \frac{2}{x+2}$$

$$\begin{aligned} x + 2 &= (x+1) \cdot 2 \\ x + 2 &= 2x + 2 \\ -x &= 0 \\ x &= 0 \end{aligned}$$

The verification that $x = 0$ is the correct solution is left to the student.

Example 2

Find the solution set of:

(a) $\frac{1}{x-4} - \frac{1}{x-2} = \frac{2x}{x^2 - 6x + 8}$

(b) $\frac{x}{x-3} - \frac{x-4}{x+2} = \frac{4x+3}{x^2 - x - 6}$

Solution

(a) We first factor the denominator of the right member, obtaining

$$\frac{1}{x-4}-\frac{1}{x-2}=\frac{2x}{(x-4)(x-2)}$$

Since the LCD of the fractions involved is $(x-4)(x-2)$, we multiply each member by this LCD and get

$$(x-4)(x-2)\bullet\frac{1}{x-4}-(x-4)(x-2)\bullet\frac{1}{x-2}$$

$$=(x-4)(x-2)\bullet\frac{2x}{(x-4)(x-2)}$$

$$\begin{aligned}(x-2)-(x-4)&=2x\\ 2&=2x\\ x&=1\end{aligned}$$

Here we are substituting $x=1$ in each member of the equation $\frac{1}{x-4}-\frac{1}{x-2}=\frac{2x}{x^2-6x+8}$.

Since $\frac{1}{1-4}-\frac{1}{1-2}=\frac{-1}{3}+1=\frac{2}{3}$, and $\frac{2\bullet 1}{1-6\bullet 1+8}=\frac{2}{3}$, our result is correct, and the solution set of $\frac{1}{x-4}-\frac{1}{x-2}=\frac{2x}{x^2-6x+8}$ is $\{1\}$.

(b) We first write $\frac{x}{x-3}-\frac{x-4}{x+2}=\frac{4x+3}{(x-3)(x+2)}$.

The LCD is $(x-3)(x+2)$, and we multiply throughout:

$$(x-3)(x+2)\bullet\frac{x}{x-3}-(x-3)(x+2)\bullet\frac{x-4}{x+2}$$

$$=(x-3)(x+2)\bullet\frac{4x+3}{(x-3)(x+2)}$$

$$\begin{aligned}(x+2)\bullet x-(x-3)(x-4)&=4x+3\\ x^2+2x-(x^2-7x+12)&=4x+3\\ 9x-12&=4x+3\\ 5x&=15\\ x&=3\end{aligned}$$

The check is important not only to catch errors but also to rule out extraneous solutions.

If x is replaced by 3 in the original equation, the term $\frac{x}{x-3}$ yields $\frac{3}{0}$, which is meaningless. Consequently, the solution set of $\frac{x}{x-3}-\frac{x-4}{x+2}=\frac{4x+3}{x^2-x-6}$ is $\emptyset$.

PROGRESS TEST

1. The solution set of $\frac{3}{x+5} = \frac{-2}{x}$ is ______.
2. The solution set of $\frac{1}{x-1} = \frac{2}{x-2}$ is ______.
3. The solution set of $\frac{4}{x-4} - \frac{1}{x-2} = \frac{x}{x^2-6x+8}$ is ______.
4. The solution set of $\frac{x}{x-1} - \frac{x-2}{x+2} = \frac{x}{x^2+x-2}$ is ______.

EXERCISE 4.5

In Problems 1 through 30, find the solution set of the given equation.

Illustration: Find the solution set of

$$\text{(a)}\ \frac{2}{3x^2-5x-2} + \frac{4}{3x+1} = \frac{1}{x-2}$$

$$\text{(b)}\ \frac{1}{x-2} = \frac{-1}{3-x}$$

Solution:

(a) We first factor the denominator of the first term and write $\frac{2}{(3x+1)(x-2)} + \frac{4}{3x+1} = \frac{1}{x-2}$. Multiplying both members by $(3x+1)(x-2)$, we get

$$(3x+1)(x-2) \cdot \frac{2}{(3x+1)(x-2)}$$

$$+ (3x+1)(x-2) \cdot \frac{4}{3x+1} = (3x+1)(x-2) \cdot \frac{1}{x-2}$$

$$2 + (x-2) \cdot 4 = 3x + 1$$
$$2 + 4x - 8 = 3x + 1$$
$$x = 7$$

Thus, the solution set is $\{7\}$.

The verification of this answer is left to the student.

(b) We multiply by the LCD, $(x-2)(3-x)$, and write

$$(x-2)(3-x) \cdot \frac{1}{x-2} = (x-2)(3-x) \cdot \frac{-1}{3-x}$$

PROGRESS TEST ANSWERS

1. $\{-2\}$
2. $\{0\}$
3. $\emptyset$
4. $\left\{\frac{1}{2}\right\}$

$$3 - x = (x - 2) \cdot (-1)$$
$$3 - x = -x + 2$$

Adding x to both members, we have $3 = 2$. Since this statement is obviously false, the solution set of $\frac{1}{x-2} = \frac{-1}{3-x}$ is $\emptyset$.

1. $\frac{x}{3} + \frac{x}{6} = 3$

2. $\frac{x}{2} + \frac{x}{4} = \frac{3}{8}$

3. $\frac{x}{5} - \frac{3x}{10} = \frac{1}{2}$

4. $\frac{x}{6} - \frac{x}{5} = \frac{1}{15}$

5. $\frac{1}{y} + \frac{4}{3y} = 7$

6. $\frac{10}{3y} - \frac{9}{2y} = \frac{7}{30}$

7. $\frac{2}{y-8} = \frac{1}{y-2}$

8. $\frac{2}{y-4} = \frac{3}{y-2}$

9. $\frac{3}{3z+4} = \frac{2}{5z-6}$

10. $\frac{2}{4z-1} = \frac{3}{2z+1}$

11. $\frac{-2}{2x+1} = \frac{3}{3x-1}$

12. $\frac{-5}{2x+3} = \frac{2}{3x-1}$

13. $\frac{-1}{x+1} = \frac{-2}{2x-1}$

14. $\frac{-5}{5x-2}=\frac{-3}{3x+1}$

15. $\frac{2}{3x+1}=\frac{4}{6x+2}$

16. $\frac{3}{2x-1}=\frac{6}{4x-5}$

17. $\frac{2}{x^2-4}+\frac{5}{x+2}=\frac{7}{x-2}$

18. $\frac{3}{x^2-9}+\frac{5}{x+3}=\frac{8}{x-3}$

19. $\frac{t+2}{t^2-3t+2}+\frac{1}{t-2}=\frac{3}{t-1}$

20. $\frac{t+3}{t^2+4t+3}+\frac{1}{t+1}=\frac{4}{t+3}$

21. $\frac{x^2}{x^2-1}=1+\frac{1}{x+1}$

22. $\frac{x^2}{x^2-9}=1+\frac{1}{x-3}$

23. $\frac{1}{x^2-4x+3}+\frac{1}{x^2-2x-3}=\frac{1}{x^2-1}$

24. $\frac{1}{x^2+3x+2}+\frac{1}{x^2+x-2}=\frac{1}{x^2-1}$

25. $\frac{x+2}{3x^2+4x+1}=\frac{x+1}{3x^2+7x+2}$

26. $\frac{x+2}{2x^2+x-1}=\frac{x-2}{2x^2+x-1}$

27. $\frac{2z+13}{2z^2+5z-3}+\frac{3}{z+3}=\frac{4}{2z-1}$

28. $\frac{z-14}{2z^2-3z-2}+\frac{3}{z-2}=\frac{4}{2z+1}$

29. $\frac{3-x}{5x^2-4x-1}+\frac{2}{5x+1}=\frac{1}{x-1}$

30. $\frac{16-x}{4x^2-11x-3}+\frac{5}{4x+1}=\frac{2}{x-3}$

Reproduction credit: Eric Shaal for Time, Inc., Courtesy of the Vatican.

OBJECTIVES 4.6

After studying this section, the reader should be able to solve word problems involving ratios by using the five-step procedure given earlier in the text.

4.6 APPLICATIONS: WORD PROBLEMS

The unfinished canvas shown at the beginning of this section was painted by Leonardo da Vinci and is entitled St. Jerome. A Golden Rectangle (black overlay) fits so neatly around St. Jerome that experts conjecture that da Vinci painted the figure to conform to those proportions. For many years it has been said that the Golden Rectangle is one of the most visually satisfying of all geometric forms. Do you know how to construct a Golden Rectangle? Such a rectangle has a special ratio of length to width, about 8 to 5. The situation can be described by writing

A ratio is an indicated quotient.

$$\frac{\text{Length of rectangle}}{\text{Width of rectangle}} = \frac{8}{5}$$

Another way to state this relationship is to say that the length and width of a Golden Rectangle are *in the ratio of 8 to 5*. In symbols, 8:5. Thus, the ratio of a number a to another number b can be written as $\frac{a}{b}$ or a : b. For example, if it takes you 5 hours to complete a certain job, you would complete $\frac{2}{5}$ of the job in 2 hours and $\frac{x}{5}$ of the job in x hours. The part of a job done by a person (or machine) in a given period of time can be represented by a *ratio* or fraction in the following manner.

$$\text{Part of job done by A} = \frac{\text{Time during which A worked}}{\text{Time needed to do entire job}}$$

If two or more persons work together to complete a job, each does a fraction of the work. In such cases the sum of the fractional parts is 1 job.

Example 1

A computer can do a job in 6 hours. Computer sharing is arranged with a second computer that can do the job in 10 hours. How many hours will it take the two computers working together to complete the job?

Solution

To do this problem, we follow the procedure given on page 100.

1. Read the problem and decide what is asked for. We wish to know the number of hours it takes the computers working together to do the job.
2. Let x be this number. Then $\frac{x}{6}$ is the part done by the first computer, and $\frac{x}{10}$ is the part done by the second computer.
3. Since they work together, the sum of the fractional parts of the job must be 1. That is,

$$\frac{x}{6} + \frac{x}{10} = 1$$

4. We now solve the equation in 3 by first multiplying both members by 30, the LCD, obtaining

$$30\left(\frac{x}{6}\right) + 30\left(\frac{x}{10}\right) = 30 \bullet 1$$

$$5x + 3x = 30$$
$$8x = 30$$
$$x = 3\frac{3}{4}$$

Thus, it takes $3\frac{3}{4}$ hours to complete the job.

5. Check: In $3\frac{3}{4}$ hours, the first computer does $\frac{3\frac{3}{4}}{6}$, or $\frac{5}{8}$, of the job; the second does $\frac{3\frac{3}{4}}{10}$, or $\frac{3}{8}$, of the job. Together they do $\frac{5}{8} + \frac{3}{8} = \frac{8}{8}$ of the job.

The ideas used in the preceding problem can be applied to uniform motion problems. As you recall, the relationship between the distance D traveled at a rate R in a time T is given by D = RT. We shall use this relationship in the next example.

Example 2

A small plane travels 220 miles with a tail wind in the same time that it takes it to travel 180 miles with a head wind. If the wind velocity is 15 miles per hour, what is the plane's speed in still air?

Solution

1. We are asked for the speed of the plane in still air.
2. Let this speed be R miles per hour. Then R + 15 is the rate with the tail wind and R − 15 the rate against the wind.
3. Since D = RT, $T = \frac{D}{R}$. The time it takes to travel with the tail wind is $\frac{220}{R + 15}$, and the time it takes to travel against the wind is $\frac{180}{R - 15}$. Since these two times are equal, we have

$$\frac{220}{R + 15} = \frac{180}{R - 15}$$

4. We solve this equation as follows:

$$(R+15)(R-15) \cdot \frac{220}{R+15} = (R+15)(R-15) \cdot \frac{180}{R-15}$$

$$\begin{aligned} (R-15) \cdot 220 &= (R+15) \cdot 180 \\ 220R - 15 \cdot 220 &= 180R + 15 \cdot 180 \\ 40R &= 15 \cdot 180 + 15 \cdot 220 \\ 40R &= 6000 \\ R &= 150 \end{aligned}$$

Thus, the speed of the plane in still air is 150 miles per hour.

5. The verification of this result is left to the student.

Example 3

The rocket used by Apollo XV consumes 15 tons of propellants per second for 150 seconds. The auxiliary engine by itself can consume the same amount of fuel in 200 seconds. If both engines were fired together, how long would the 15 tons of fuel last?

Solution

1. We are asked how long the fuel would last.
2. Let the time the fuel would last be T seconds.
3. The first engine burns $\frac{T}{150}$ of the fuel and the second one burns $\frac{T}{200}$. Together they burn $\frac{T}{150} + \frac{T}{200} = 1$.
4. Solving, we have

$$600 \cdot \frac{T}{150} + 600 \cdot \frac{T}{200} = 600 \cdot 1$$

$$\begin{aligned} 4T + 3T &= 600 \\ 7T &= 600 \\ T &= \frac{600}{7} = 85\tfrac{5}{7} \text{ seconds.} \end{aligned}$$

5. The verification is left to the student.

PROGRESS TEST

1. A bricklayer can finish a job in 4 hours and another can do the same job in 6 hours. If they work together, the time it will take them to finish the job is ______ hours.
2. A barge travels 24 miles downstream in the Mississippi in the same time it takes it to travel 12 miles upstream. If the current in the Mississippi flows at 2 miles per hour, the barge rate in still water is how many miles per hour?
3. A water tank can be filled by an intake pipe in 4 hours and can be emptied by a drain pipe in 8 hours. The time it takes to fill the tank with both pipes open is ______ hours.

EXERCISE 4.6

Use the five-step procedure given on page 100 to solve Problems 1 through 6.

Illustration: The world record for bricklaying was established in 1937 by Joseph Raglon, who placed 3,472 bricks in 1 hour (Guinness). If another bricklayer can lay 3,472 bricks in 4 hours, how long will it take both together to lay the 3,472 bricks?

Solution:

1. We are asked for the time needed to finish the job.
2. Let this time be T hours.
3. Mr. Raglon does T of the job in this time, and the other man does $\frac{T}{4}$. Together, they do $T + \frac{T}{4}$.
4. We solve the equation

$$T + \frac{T}{4} = 1$$

$$4T + T = 4$$

$$5T = 4$$

$$T = \frac{4}{5}$$

Both men can finish the job in $\frac{4}{5}$ of an hour, that is, $\frac{4}{5} \cdot 60 = 48$ minutes.

PROGRESS TEST ANSWERS

1. $2\frac{2}{5}$
2. The time upstream is $\frac{12}{R-2}$ and the time downstream is $\frac{24}{R+2}$. These times are equal: $\frac{12}{R-2} = \frac{24}{R+2}$. Solving for R, we find the rate R is 6 miles per hour.
3. $\frac{T}{4} - \frac{T}{8} = 1$

   $2T - T = 8$

   $T = 8$

5. The verification of this result is left to the reader.

1. If one typist can finish a job in 3 hours, while another typist can finish in 5 hours, how long will it take both of them working together to finish the job?

2. A carpenter can finish a job in 8 hours, while another one can do it in 10 hours. How long will it take them to finish the job working together?

3. The world record for riveting is 11,209 rivets in 9 hours, by J. Mair of Ireland (Guinness). If another man can rivet 11,209 rivets in 10 hours, how long will it take both of them working together to rivet the 11,209 rivets?

4. Mr. Gerry Harley of England shaved 130 men in 60 minutes (Guinness). If another barber can shave all these men in 5 hours, how long will it take both of them working together to shave the 130 men?

5. A printing press can print the evening paper in $\frac{1}{2}$ the time another press takes to print it. Together, they can print the paper in 2 hours. How long will it take each of them to print the paper?

6. A computer can do a job in 4 hours. With the help of a newer computer, the job is completed in 1 hour. How long will it take the newer computer to complete the job alone?

Use the five-step procedure given on page 100 to solve Problems 7 through 14.

Illustration: The world's strongest current is the Saltstraumen in Norway, reaching 18 miles per hour (Guinness). A motor boat can go 48 miles downstream in the same time it takes to go 12 miles upstream. What is the speed of the boat in still water?

Solution:
1. We are asked for the speed of the boat.
2. Let the speed be R miles per hour.
3. The rate downstream is R + 18 and upstream is R − 18. Since the time is the same on both trips, $\frac{48}{R+18} = \frac{12}{R-18}$.

4. Solving $\frac{48}{R+18} = \frac{12}{R-18}$, we have

$$\begin{aligned} 48(R-18) &= 12(R+18) \\ 48R - 48 \cdot 18 &= 12R + 12 \cdot 18 \\ 36R &= 48 \cdot 18 + 12 \cdot 18 \\ 36R &= 60 \cdot 18 \\ R &= \frac{60 \cdot 18}{36} \\ R &= 30 \end{aligned}$$

5. The verification of this fact is left to the reader.

7. A water skier travels 30 miles downstream in the same time it takes him to go 20 miles upstream. If the river current flows at 5 miles per hour, what is the skier's speed in still water?

8. A small plane goes 240 miles against the wind in the same time it takes it to go 360 miles with a tail wind. If the wind velocity is 30 miles per hour, find the plane's speed in still air.

9. A jet plane goes 700 miles against the wind in the same time it takes it to go 900 miles with a tail wind. If the wind velocity is 50 miles per hour, what is the plane's speed in still air?

10. A small plane can cruise at 120 miles per hour in still air. It takes this plane the same time to move 270 miles against the wind as it does to go 450 miles with a tail wind. What is the wind velocity?

11. A small plane can travel 200 miles against the wind in the same time it takes it to travel 260 miles with a tail wind. If the plane is cruising at 115 miles per hour, find the wind velocity.

12. An automobile travels 200 miles in the same time in which a small plane travels 1,000 miles. Find their rates of speed if the airplane is 100 miles per hour faster than the automobile.

13. A runner ran 1,000 meters in the same time that another runner ran 950 meters. If the speed of the slower runner was $\frac{1}{4}$ of a meter per second less than that of the faster one, what was the speed of the faster runner?

14. A runner covered 500 meters in the same time that another one covered 450 meters. If the speed of the faster runner was $\frac{1}{3}$ of a meter per second more than that of the slower one, what was the speed of the slower runner?

Use the five-step procedure given on page 100 to solve Problems 15 through 21.

Illustration: A pool can be filled by an intake pipe in 4 hours and can be emptied by a drain pipe in 5 hours. How long would it take to fill the tank with both pipes open?

Solution:

1. We are asked for the time it takes to fill the pool.
2. Let this time be T hours.
3. The intake pipe fills $\frac{T}{4}$ of the pool and the drain pipe drains $\frac{T}{5}$. Since both pipes are open, we must solve $\frac{T}{4} - \frac{T}{5} = 1$.
4. $$20 \cdot \frac{T}{4} - 20 \cdot \frac{T}{5} = 20 \cdot 1$$
$$5T - 4T = 20$$
$$T = 20$$
It takes 20 hours to fill the pool.
5. Since $\frac{20}{4} - \frac{20}{5} = 5 - 4 = 1$, our result is correct.

15. A tank can be filled by an intake pipe in 9 hours and drained by another pipe in 21 hours. If both pipes are open, how long would it take to fill the tank?

16. A faucet can fill a tank in 12 hours and the drain pipe can empty it in 18 hours. If the faucet and the drain pipe are open, how long would it take to fill the tank?

17. A pipe can fill a pool in 7 hours while another one can fill it in 21 hours. How long would it take to fill the tank using both pipes?

18. One pipe can fill a tank in 6 hours, and another can fill it in 4 hours. How long will it take both pipes together to fill the tank?

19. The main engine of a rocket can burn for 60 seconds on the fuel in the rocket's tank, while the auxiliary engine can burn for 90 seconds on the same amount of fuel. How long can both engines burn if they are operated together on the rocket's tank of fuel?

20. An in-flow pipe can fill a pool in 12 hours, while another pipe can drain it in 4 hours. How long will it take to empty the pool if both pipes are open simultaneously? Assume that the pool is full at the start.

21. A pipe can fill a tank in 9 hours, while the drain can empty it in 6 hours. How long will it take to empty the tank if both pipes are open simultaneously? Assume that the tank is full at the start.

SELF-TEST–CHAPTER 4

1. Write in standard form:

_______ a. $\frac{-8}{-4}$

_______ b. $\frac{-a}{a-7}$

_______ c. $-\frac{-a}{-b}$

_______ d. $\frac{-x^2}{-y}$

_______ e. $\frac{-3}{-x}$

---

2. Write each fraction on the left as an equal fraction with the denominator shown on the right.

_______ a. $\frac{-3}{b-a} = \frac{\quad}{a-b}$

_______ b. $\frac{-a}{a-b} = \frac{\quad}{b-a}$

_______ c. $-\frac{1}{a-b} = \frac{\quad}{b-a}$

---

3. Reduce to lowest terms.

_________ a. $\frac{2x + xy}{3x}$

_________ b. $\frac{x - y}{x^2 - y^2}$

_________ c. $\frac{12ab^2}{6a^3b^3c}$

---

4. Divide and indicate the quotient and the remainder.

_______________ a. $\dfrac{x^4 + x^2 + 2x - 1}{x + 3}$

Remainder ________

_______________ b. $\dfrac{3x^3 - 4x - 1}{x - 2}$

Remainder ________

_______________ c. $\dfrac{x^3 + 8}{x + 2}$

Remainder ________

---

5. Find the LCM of:

_______________ a. 6, 8, 15

_______________ b. $6xy$, $8x^2$

_______________ c. $a^2 - b^2$, $a - b$

_______________ d. $2a - 2$, $a - 1$

_______________ e. $x^2 - 1$, $2(x - 1)^2$

---

6. Find the sum or difference and write as a single fraction *in lowest terms.*

_______________ a. $\dfrac{a}{3} - \dfrac{2}{3}$

_______________ b. $\dfrac{a + 1}{b} - \dfrac{a - 1}{b}$

_______________ c. $\dfrac{3}{a^2 - 9} - \dfrac{1}{a - 3}$

_______________ d. $\dfrac{1}{b - 2} + \dfrac{1}{b^2 - 4}$

_______________ e. $x + \dfrac{1}{x - 1}$

---

7. Write each product as a single fraction in lowest terms.

_______________ a. $\dfrac{a^2}{xy} \cdot \dfrac{3x^3y}{4a}$

__________ b. $15x^2y \cdot \dfrac{3}{45xy^2}$

__________ c. $\dfrac{4x^2 + 8x + 3}{2x^2 - 5x + 3} \cdot \dfrac{6x^2 - 9x}{1 - 4x^2}$

---

8. Write each quotient as a single fraction in lowest terms.

__________ a. $\dfrac{a^2b}{c} \div \dfrac{ab^3}{c^2}$

__________ b. $24a^3b \div \dfrac{3a^2b}{7x}$

__________ c. $\dfrac{a^2 + 2a - 15}{a^2 + 3a - 10} \div \dfrac{a^2 - 9}{a^2 - 9a + 14}$

---

9. Write each complex fraction as a single fraction in *lowest terms*.

__________ a. $\dfrac{\frac{b}{c}}{\frac{b^2}{c^2}}$

__________ b. $\dfrac{1 - \frac{2}{3}}{3 + \frac{1}{3}}$

__________ c. $a - \dfrac{a}{a + \frac{1}{2}}$

---

10. Find the solution set of:

__________ a. $\dfrac{6}{x + 5} = \dfrac{-4}{x}$

__________ b. $\dfrac{x}{x + 1} - \dfrac{x + 1}{x - 2} = \dfrac{x}{x^2 - x - 2}$

CHAPTER 5

HISTORICAL NOTE

During the Middle Ages, there were several scholars who contributed to the development of mathematics. One of these was Nicole Oresme, who was born in Normandy about 1323. Oresme wrote five mathematical works and translated some of the works of Aristotle. In working on these translations, Oresme (and Thomas Brawardine of Oxford) criticized the ideas of Aristotle on motion and later extended Euclid's idea of proportions to include such notions as y being proportional to x^n.

One of the tracts written by Oresme contains the first known use of fractional exponents (although not in our present notation). This tract is said to have been reprinted several times and may have influenced Renaissance mathematicians and perhaps even Descartes.

At a later date, the Frenchman Nicolas Chuquet wrote *Three Parts of the Science of Numbers,* a book in which he used the symbol $\bar{p}$ for *plus* (the French word *plus* means *more*) and the symbol $\tilde{m}$ for *less* (*moins* means *less*). Chuquet represented $3x^2$ by $.3^{.2}$ and he wrote negative exponents such as $5x^{-2}$ as $.5^{2.\tilde{m}}$. Chuquet was one of the first mathematicians to use the law of exponents that we shall be discussing in this chapter.

The square root sign is thought by some historians to have been originated by Leonardo da Pisa (Fibonacci), while others attribute this invention to the German mathematician Christoff Rudolff. The sign used by Fibonacci (see Section 5.3) comes from the Latin word *radix*, meaning *root.* In any case, the symbol was later used to denote the square root. It was a lower-case r, an abbreviation for radix. Sometimes a capital R was used in front of the number, as R5, to indicate a square root, but by using a small letter the number can be written underneath, as *r*5, which later evolved into $\sqrt{5}$.

THE ALGEBRA OF EXPONENTS

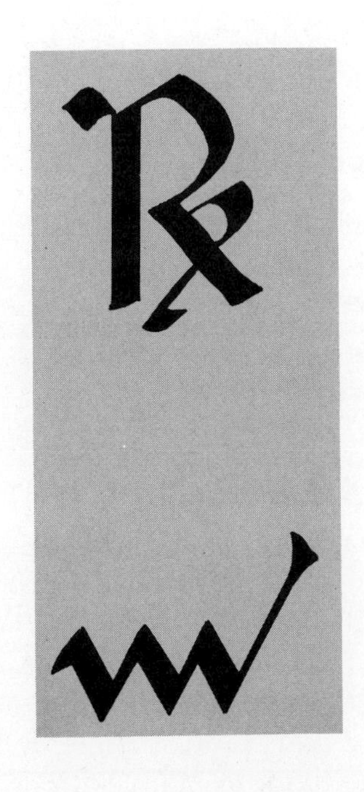

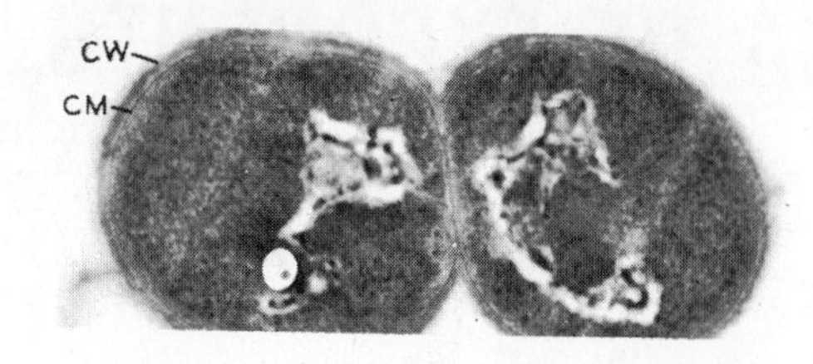

From Carpenter, P. L.: *Fundamentals of Microbiology* (Philadelphia: W. B. Saunders Company, 1972).

OBJECTIVES 5.1

After studying this section, the reader should be able to:

1. **Find the product or quotient of two monomials by correctly using the laws of exponents.**
2. **Evaluate an expression containing positive and/or negative exponents.**
3. **Write a quotient of two monomials which is raised to a given power as a product using positive exponents.**

5.1 INTEGRAL EXPONENTS

$2^4 \leftarrow$ exponent
$\uparrow$
base

The photograph at the beginning of this section shows the splitting of a single cell. When such a cell splits, there are then 2 cells. Later, when each of these two cells splits, there will be 2×2 cells. The next splitting will then bring on $2 \times 2 \times 2$ cells, and so on. Each time the entire colony splits, there are twice as many cells as before. How many cells would there be after 4 splits? The answer is, of course, $2 \times 2 \times 2 \times 2$. The number $2 \times 2 \times 2 \times 2$ can be written as 2^4. In the expression 2^4, the 2 is called the *base* and the 4 is called the *exponent*. The *exponent* 4 tells us how many times the *base* 2 is used as a factor. In general, if a is a real number and n is a positive integer,

$$a^n = \underbrace{a \bullet a \bullet a \bullet \ldots \bullet a}_{n\ a\text{'s}}$$

In Chapter Three, we developed the following two theorems, or laws, of exponents from this definition.

$$\text{I} \quad a^m \bullet a^n = a^{m+n}$$

$$\text{II} \quad \frac{a^m}{a^n} = a^{m-n} \quad (m > n,\ a \neq 0)$$

In a fraction such as $\frac{a^4}{a^7}$, Law II does not apply, because $n > m$. However, we can develop a method for handling expressions of the form $\frac{a^m}{a^n}$ where $n > m$. By the definition of a^n, we have

$$\frac{a^4}{a^7} = \frac{a \cdot a \cdot a \cdot a}{a \cdot a \cdot a \cdot a \cdot a \cdot a \cdot a}$$

$$= \frac{(1) \cdot \overset{1}{\cancel{[a \cdot a \cdot a \cdot a]}}}{\underset{1}{\cancel{[a \cdot a \cdot a \cdot a]}} \cdot [a \cdot a \cdot a]}$$

$$= \frac{1}{a \cdot a \cdot a}$$

$$= \frac{1}{a^3}$$

Thus,

$$\frac{a^4}{a^7} = \frac{1}{a^{7-4}} = \frac{1}{a^3}, \; a \neq 0$$

Similarly,

$$\frac{x^8}{x^{13}} = \frac{1}{x^{13-8}} = \frac{1}{x^5}, \; x \neq 0$$

$$\text{and } \frac{y^6}{y^{10}} = \frac{1}{y^{10-6}} = \frac{1}{y^4}, \; y \neq 0$$

In general, we have the following law, where a is a non-zero real number and n and m are positive integers.

$$\text{II(a)} \quad \frac{a^m}{a^n} = \frac{1}{a^{n-m}} \quad (n > m)$$

In Chapter Three, we also discussed the following laws of exponents.

$$\text{III} \quad (a^m)^n = a^{m \cdot n}$$
$$\text{IV} \quad (ab)^n = a^n b^n$$

$(a^m)^n = \underbrace{a^m \cdot a^m \cdot a^m \ldots a^m}_{\text{n times}}$
$= a^{m+m+m+\ldots+m}$
$= a^{m \cdot n}$

By Law III,

$$(a^3)^5 = a^{3 \cdot 5} = a^{15}$$
$$(a^2)^4 = a^{2 \cdot 4} = a^8$$
$$\text{and } (a^5)^2 = a^{5 \cdot 2} = a^{10}$$

$(ab)^n = \underbrace{(ab) \cdot (ah) \cdot (ab) \ldots (ab)}_{\text{n times}}$
$= \underbrace{(a \cdot a \ldots a)}_{\text{n times}} \underbrace{(b \cdot b \ldots b)}_{\text{n times}}$
$= a^n b^n$

By Law IV,

$$(ab)^3 = a^3b^3$$
$$(xy^2)^3 = x^3(y^2)^3 = x^3y^6$$
$$\text{and } (3x^2y^3)^4 = 3^4(x^2)^4(y^3)^4 = 81x^8y^{12}$$

We now state another law of exponents which will be used to simplify certain expressions in the examples ahead.

$$\text{V} \quad \left(\frac{a}{b}\right)^n = \frac{a^n}{b^n}$$

$$\left(\frac{a}{b}\right)^n = \frac{a}{b} \cdot \frac{a}{b} \cdot \frac{a}{b} \cdots \frac{a}{b} = \frac{a \cdot a \cdot a \ldots a}{b \cdot b \cdot b \ldots b} = \frac{a^n}{b^n}$$

For example,

$$\left(\frac{2}{3}\right)^3 = \frac{2^3}{3^3}$$
$$\left(\frac{x}{y}\right)^4 = \frac{x^4}{y^4}$$
$$\text{and } \left(\frac{-2x}{y}\right)^5 = \frac{(-2x)^5}{y^5}$$

We are now ready to simplify expressions of the form $\left(\dfrac{2x^2y}{xy^3}\right)^4$. The simplification is as follows:

$$\left(\frac{2x^2y}{xy^3}\right)^4 = \left(\frac{2x^{2-1}}{y^{3-1}}\right)^4 \qquad \text{By Laws II and II(a)}$$
$$= \left(\frac{2x^1}{y^2}\right)^4$$
$$= \frac{(2x^1)^4}{(y^2)^4} \qquad \text{By Law V}$$
$$= \frac{2^4(x^1)^4}{(y^2)^4} \qquad \text{By Law IV}$$
$$= \frac{2^4x^{1\cdot 4}}{y^{2\cdot 4}} \qquad \text{By Law III}$$
$$= \frac{16x^4}{y^8}$$

Example 1

Write each expression as a quotient with positive exponents.

(a) $\dfrac{x^4y^3}{x^6y}$

(b) $\dfrac{(-xy)^2}{(-xy^2)^3}$

Solution

(a) $\dfrac{x^4y^3}{x^6y} = \dfrac{y^{3-1}}{x^{6-4}}$ By Laws II and II(a)

$= \dfrac{y^2}{x^2}$

(b) $\dfrac{(-xy)^2}{(-xy^2)^3} = \dfrac{(-1 \bullet x \bullet y)^2}{(-1 \bullet x \bullet y^2)^3}$ Since $-a = -1 \bullet a$

$= \dfrac{(-1)^2x^2y^2}{(-1)^3x^3y^6}$ By Laws II and IV

$= \dfrac{x^2y^2}{-x^3y^6}$

$= \dfrac{-1}{x^{3-2}y^{6-2}}$ By Law II(a)

$= \dfrac{-1}{xy^4}$

In the preceding discussion, all exponents represented positive integers. From here on, though, we will need definitions for zero and for negative integer exponents as well. In order for these definitions to be most useful, we require that the meaning of a zero or a negative integer exponent be such that Laws I through V apply to it just as they do to positive exponents. In particular, if Law I is to apply when $n = 0$ and $a \neq 0$, we can see that $a^0 \bullet a^m = a^{0+m} = a^m$; that is, $a^0 \bullet a^m = a^m$. Since $1 \bullet a^m = a^m$, we must define a^0 as 1:

$$a^0 = 1$$

This definition also follows if Law II is to hold for $n = m$. In this case, we have $\dfrac{a^m}{a^n} = \dfrac{a^m}{a^m} = a^{m-m} = a^0$; that is, $\dfrac{a^m}{a^m} = a^0$. By the definition of a quotient, $\frac{a^m}{a^m} = 1$, so again $a^0 = 1$, provided, of course, $a \neq 0$. Thus, $2^0 = 1$, $\left(\frac{3}{4}\right)^0 = 1$, $(-7)^0 = 1$, and $x^0 = 1$, if $x \neq 0$. From our discussion, we can state the following definition.

DEFINITION 5.1

If a is any nonzero real number, then $a^0 = 1$.

If the First Law of Exponents is to apply to negative integers, then for $n > 0$, $a^n \cdot a^{-n} = a^{n-n} = a^0 = 1$. This shows that a^{-n} must be the reciprocal (or multiplicative inverse) of a^n, if Law I applies. Thus, we make the following definition.

DEFINITION 5.2

$$a^{-n} = \frac{1}{a^n} \quad (a \neq 0)$$

From this definition, it follows that $2^{-3} = \frac{1}{2^3} = \frac{1}{8}$, $(-3)^{-2} = \frac{1}{(-3)^2} = \frac{1}{9}$, and $x^{-3} = \frac{1}{x^3}$. Note that $\frac{1}{a^{-n}} = \frac{1}{\frac{1}{a^n}} = 1 \div \frac{1}{a^n} = \frac{1}{1} \cdot a^n = a^n$. Thus, $\frac{1}{a^{n-m}} = \frac{1}{a^{-(m-n)}} = a^{m-n}$, and Laws II and II(a) can be generalized as:

$$\text{II} \quad \frac{a^m}{a^n} = a^{m-n}, \quad a \neq 0$$

From this point on, we shall assume that all the laws of exponents apply to integer exponents.

Example 2

Evaluate, writing your answer in simplest form without exponents:

(a) $2 \cdot 3^{-4}$
(b) $2^{-2} + 2^2$

(c) $\frac{4}{2^{-4}}$

Solution

(a) $2 \cdot 3^{-4} = 2 \cdot \frac{1}{3^4}$

$= 2 \cdot \frac{1}{81}$

$= \frac{2}{81}$

(b) $2^{-2} + 2^2 = \frac{1}{2^2} + 2^2$

$= \frac{1}{4} + 4$

$= \frac{1}{4} + \frac{16}{4}$

$= \frac{17}{4}$

(c) $\frac{4}{2^{-4}} = \frac{4}{\frac{1}{2^4}}$

$= 4 \cdot 2^4$
$= 4 \cdot 16$
$= 64$

Example 3

Write the given expression as a product or quotient with positive exponents.

(a) $(x^{-4} \cdot x^6)^{-2}$

(b) $\left(\frac{x^{-2}y}{x^3y}\right)^{-3}$

Solution

(a) $(x^{-4} \cdot x^6)^{-2} = (x^{-4+6})^{-2}$ By Law I

$= (x^2)^{-2}$

$= x^{-4}$ By Law III

$= \frac{1}{x^4}$ By Definition 5.2

It makes no difference in what order the laws are used as long as they are used correctly.

(b) $\left(\frac{x^{-2}y}{x^3y}\right)^{-3} = (x^{-2-3}y^{1-1})^{-3}$ By Law II

$= (x^{-5}y^0)^{-3}$

$= (x^{-5} \cdot 1)^{-3}$ Since $y^0 = 1$

$= (x^{-5})^{-3}$ Since $x^{-5} \cdot 1 = x^{-5}$

$= x^{(-5)\cdot(-3)}$ By Law III

$= x^{15}$

PROGRESS TEST

1. When written as a quotient with positive exponents, $\frac{x^3y^4}{x^6y^2}$ equals ______.
2. $3 \cdot 2^{-3}$ equals ______.
3. $2 \cdot 3^{-2} + 3^2$ equals ______.
4. When written as a simplified quotient with positive exponents, $x^{-5} \cdot x^3$ equals ______.
5. When written as a product with positive exponents, $\left(\frac{x^{-3} \cdot y^5}{x^5 \cdot y^5}\right)^{-2}$ equals ______.

EXERCISE 5.1

In Problems 1 through 40, write the given expression as a product or quotient with positive exponents.

Illustration: (a) $\frac{(x^2 \cdot x^{-3} \cdot y^3)^3}{(x^4y^3)^2}$

(b) $\left(\frac{-x^{-3}y^2z^3}{x^{-4}y^3z^0}\right)^{-2}$

Solution:

(a) $\frac{(x^2 \cdot x^{-3} \cdot y^3)^3}{(x^4y^3)^2} = \frac{(x^{2-3} \cdot y^3)^3}{(x^4y^3)^2}$ Law I

$= \frac{(x^{-1}y^3)^3}{(x^4y^3)^2}$

$= \frac{(x^{-1})^3(y^3)^3}{(x^4)^2(y^3)^2}$ Law IV

PROGRESS TEST ANSWERS

1. $\frac{y^2}{x^3}$
2. $\frac{3}{8}$
3. $\frac{2}{9} + 9 = \frac{83}{9}$
4. $\frac{1}{x^2}$
5. x^{16}

$$= \frac{x^{-3}y^9}{x^8y^6} \qquad \text{Law III}$$

$$= x^{-3-8}y^{9-6} \qquad \text{Law II}$$

$$= x^{-11}y^3$$

$$= \frac{y^3}{x^{11}} \qquad \text{Definition 5.2}$$

(b) $\left(\frac{-x^{-3}y^2z^3}{x^{-4}y^3z^0}\right)^{-2} = (-x^{-3-(-4)}y^{2-3}z^{3-0})^{-2}$ Law II

$$= (-x^1y^{-1}z^3)^{-2}$$

$$= x^{-2}y^2z^{-6} \qquad \text{Law IV}$$

$$= \frac{y^2}{x^2z^6}$$

Since $x^{-2} = \frac{1}{x^2}$ and $z^{-6} = \frac{1}{z^6}$

1. $x^3 \bullet x^4$
2. $x^6 \bullet x^7$
3. $x^{-3} \bullet x^4$
4. $x^{-5} \bullet x^7$
5. $x^{-5} \bullet x^2$
6. $x^{-7} \bullet x^3$
7. $x^{-3} \bullet x^{-8}$
8. $x^{-2} \bullet x^{-9}$
9. $\frac{a^8}{a^2}$
10. $\frac{a^{15}}{a^7}$
11. $\frac{a}{a^7}$
12. $\frac{a^3}{a^9}$
13. $(y^3)^4$
14. $(y^5)^2$
15. $(y^{-2})^3$
16. $(y^{-3})^4$
17. $(y^3)^{-4}$
18. $(y^4)^{-3}$
19. $(x^{-4})^{-5}$
20. $(x^{-3})^{-5}$
21. $(-2xy^2)^2$
22. $(-3x^2y)^4$

23. $(2x^{-1}y^2)^{-2}$

24. $(3x^{-2}y^3)^{-2}$

25. $\left(\frac{a}{b^3}\right)^2$

26. $\left(\frac{a^2}{b}\right)^3$

27. $\left(\frac{-3a}{2b^2}\right)^3$

28. $\left(\frac{-2a^2}{3b^0}\right)^2$

29. $\left(\frac{a^{-4}}{b^2}\right)^{-2}$

30. $\left(\frac{a^{-2}}{b^3}\right)^{-3}$

31. $\frac{(x^{-4}y^{-1})^{-2}}{(x^2y^{-3})^{-3}}$

32. $\frac{(x^{-3}y^{-3})^{-3}}{(x^3y^{-1})^{-2}}$

33. $\left(\frac{x^{-4}y^3}{x^5y^5}\right)^{-3}$

34. $\left(\frac{x^{-2}y^0}{x^7y^2}\right)^{-2}$

35. $\frac{(2x)^3}{(3x)^2}$

36. $\frac{(x^2y)^3}{(x^3y)^2}$

37. $\frac{(-2x)^2(-x^2)^2}{(x^3)^2}$

38. $\frac{(-x^2)^3(2x)^2}{x^2 \cdot x^5}$

39. $\left(\frac{x^3y}{3}\right)^2\left(\frac{-3}{x^2y}\right)^3$

40. $\left(\frac{x}{y}\right)^3\left(-\frac{2}{3x}\right)^2$

In Problems 41 through 50, simplify and write the given expression as a product or a quotient.

Illustration: (a) $\frac{x^{n+2}}{x^n \cdot x^{n+1}}$

(b) $\frac{(x^{n+1})^2}{x^{n+2}}$

(c) $(x^{n+1} \cdot x^{2n-1})^2$

Solution:

(a) $\frac{x^{n+2}}{x^n \cdot x^{n+1}} = \frac{x^{n+2}}{x^{n+(n+1)}}$ Law I

$= \frac{x^{n+2}}{x^{2n+1}}$

$= x^{(n+2)-(2n+1)}$ Law II

$= x^{-n+1}$

(b) $\dfrac{(x^{n+1})^2}{x^{n+2}} = \dfrac{x^{(n+1)2}}{x^{n+2}}$ Law III

$= \dfrac{x^{2n+2}}{x^{n+2}}$ Since $(n+1)\,2 = 2n+2$

$= x^{(2n+2)-(n+2)}$ Law II

$= x^n$

(c) $(x^{n+1} \cdot x^{2n-1})^2 = (x^{(n+1)+(2n-1)})^2$ Law I

$= (x^{3n})^2$

$= x^{6n}$ Law III

41. $x^n \cdot x^{-n}$

42. $x^n \cdot x^{-2n}$

43. $\dfrac{x^n \cdot x^{3n}}{x^{n+1}}$

44. $\dfrac{x^{n+2}}{x^{2n} \cdot x^n}$

45. $\left(\dfrac{y^4 y}{y^2}\right)^n$

46. $\left(\dfrac{y^5 y^2}{y^4 y^3}\right)^{3n}$

47. $\left(\dfrac{x^n \cdot x^{2n}}{x^{3n}}\right)^2$

48. $\dfrac{x^{4n}}{(x^n x^{2n})^3}$

49. $\dfrac{(x^{-n} x^{2n})^{-2}}{x^n x^{4n}}$

50. $\dfrac{x^n \cdot x^{3n}}{(x^{-n} \cdot x^{-2n})^{-3}}$

In Problems 51 through 55, evaluate the given expression.

Illustration: (a) $(-5)^{-2}$

(b) $\left(\dfrac{3}{5}\right)^{-2}$

(c) $\dfrac{2^0 + 2^{-1}}{4^{-2} + 2^{-2}}$

Solution:

(a) $(-5)^{-2} = \dfrac{1}{(-5)^2} = \dfrac{1}{25}$

(b) $\left(\frac{3}{5}\right)^{-2} = \frac{1}{\left(\frac{3}{5}\right)^2} = \frac{1}{\frac{9}{25}} = \frac{25}{9}$

(c) $\frac{2^0 + 2^{-1}}{4^{-2} + 2^{-2}} = \frac{1 + \frac{1}{2}}{\frac{1}{16} + \frac{1}{4}} = \frac{\frac{3}{2}}{\frac{1}{16} + \frac{4}{16}}$

$$= \frac{\frac{3}{2}}{\frac{5}{16}} = \frac{3}{2} \bullet \frac{16}{5} = \frac{24}{5}$$

51. $(-4)^{-4}$

52. $\left(\frac{4}{5}\right)^{-2}$

53. $(2^{-2} + 3^{-2})^{-1}$

54. $\frac{3^{-1} + 2^0}{2^{-1} + 2^2}$

55. $\frac{3^0 + 2^0}{2^{-3}}$

In Problems 56 through 60, perform the indicated operations and simplify.

Illustration: $\frac{x^{-1} + y^{-1}}{x^{-1} - y^{-1}}$

Solution: $\frac{x^{-1} + y^{-1}}{x^{-1} - y^{-1}} = \frac{\frac{1}{x} + \frac{1}{y}}{\frac{1}{x} - \frac{1}{y}}$

$$= \frac{xy\left(\frac{1}{x} + \frac{1}{y}\right)}{xy\left(\frac{1}{x} - \frac{1}{y}\right)}$$ Multiplying numerator and denominator by xy, the LCD

$$= \frac{y + x}{y - x}$$

56. $\dfrac{x^{-1} - y^{-1}}{x^{-1} + y^{-1}}$

57. $\dfrac{x^{-1} + y^{-1}}{x^{-1}}$

58. $a^{-1}b - ab^{-1}$

59. $a^{-2}b + ab^{-2}$

60. $\dfrac{(a + b)^{-1}}{a^{-1} + b^{-1}}$

Reprinted by permission of Irwin Caplan.

OBJECTIVES 5.2

After studying this section, the reader should be able to:

1. **Find the real nth root of a given number, if it exists.**
2. **Evaluate an expression containing rational exponents.**

5.2 RATIONAL EXPONENTS AND RADICALS

In the cartoon at the beginning of this section, the robot says it hurts when he does square roots. In this section we shall consider an operation for which we frequently need *irrational numbers*, that is, numbers that are not rational. This operation is the *extraction of roots*, the inverse of raising a number to a power.

DEFINITION 5.3

If a and x are real numbers, and n is a positive integer such that $x^n = a$, then x is called an nth root of a.

For example,

(1) A square (second) root of 4 is 2, because $2^2 = 4$.
(2) Another square root of 4 is -2, because $(-2)^2 = 4$.
(3) A cube (third) root of 27 is 3, because $(3)^3 = 27$.
(4) A cube (third) root of -64 is -4, because $(-4)^3 = -64$.
(5) A fourth root of $\frac{16}{81}$ is $\frac{2}{3}$, because $\left(\frac{2}{3}\right)^4 = \frac{16}{81}$.

As you can see from (1) and (2), there are two square roots of 4: 2 and -2. Similarly, the fourth roots of $\frac{16}{81}$ are $\frac{2}{3}$ and $\frac{-2}{3}$, because $\left(\frac{-2}{3}\right)^4 = \left(\frac{2}{3}\right)^4 = \frac{16}{81}$. To distinguish between the two roots, we introduce the notion of "principal nth root" in the following definition.

DEFINITION 5.4

If n is a positive integer greater than 1, then $\sqrt[n]{a}$ denotes the principal nth root of a, and:

(i) If $a > 0$, $\sqrt[n]{a}$ is the *positive* nth root of a.

(ii) If $a < 0$, and *n is odd,* $\sqrt[n]{a}$ is the *negative* nth root of a.

(iii) $\sqrt[n]{0} = 0$.

Note that if n is even and $a > 0$, a has both a positive and a negative nth root. However, part (i) specifies that only the *positive* root be used. If we wish to refer to the *negative* nth root, we shall write $-\sqrt[n]{a}$. In Definition 5.4, $\sqrt[n]{a}$ is called a *radical expression*; $\sqrt{\ }$ is a *radical sign*, n is the *index*, and a is the *radicand.* In the case of a square root, the index 2 is understood but is not written; that is, $\sqrt{4}$ means $\sqrt[2]{4}$, and $\sqrt{25}$ means $\sqrt[2]{25}$. From Definition 5.4, we can see that:

$\sqrt{9} = 3$, and $-\sqrt{9} = -3$.

(1) $\sqrt[4]{16} = 2$ (read "the principal 4th root of 16 equals 2"), since $2^4 = 16$.
(2) $\sqrt[3]{-27} = -3$, since $(-3)^3 = -27$.
(3) $\sqrt{x^2} = |x|$, the positive square root of x^2.

In the event that $a < 0$, and n is a positive even integer, $\sqrt[n]{a}$ is not a real number. For example, $\sqrt[2]{-4}$ and $\sqrt[4]{-81}$ are not real numbers, since there is no real number whose square is -4, and there is no real number whose fourth power is -81.

Example 1

Find, if possible:

(a) $\sqrt[3]{-64}$

(b) $\sqrt{-64}$

(c) $\sqrt[3]{\left(\frac{-1}{8}\right)}$

Solution

(a) $\sqrt[3]{-64} = -4$, since $(-4)^3 = -64$.

(b) $\sqrt{-64}$ is not a real number. Note that $\sqrt{-64} \neq -8$, since $(-8)(-8) \neq -64$.

(c) $\sqrt[3]{\left(\frac{-1}{8}\right)} = \frac{-1}{2}$, since $\left(\frac{-1}{2}\right)^3 = \left(\frac{-1}{2}\right)\left(\frac{-1}{2}\right)\left(\frac{-1}{2}\right) = \frac{-1}{8}$.

In order to arrive at a reasonable meaning for rational exponents, let us use Law III to multiply $a^{1/2}$ by itself. We get

$$a^{1/2} \bullet a^{1/2} = (a^{1/2})^2 = a^{1/2 \cdot 2} = a$$

This means that if $a^{1/2}$ is to have a meaning consistent with the laws for positive integral exponents, it should be taken as one of the square roots of a. We shall take it as the *principal* square root, that is, $a^{1/2} = \sqrt{a}$, provided a is positive. In general, $(a^{1/n})^n = a^{(1/n) \cdot n} = a$. This means that $a^{1/n}$ should be an nth root of a. Accordingly, we define $a^{1/n}$ in the following manner.

$\sqrt[n]{a}$ is a radical expression. $\sqrt{\ }$ is a radical, n is the index, and a is the radicand.

DEFINITION 5.5

$a^{1/n} = \sqrt[n]{a}$ whenever n is a positive integer greater than 1 and $\sqrt[n]{a}$ is a real number.

Thus,

$$16^{1/2} = \sqrt{16} = 4$$

$$(-8)^{1/3} = \sqrt[3]{-8} = -2$$

$$\left(\frac{1}{81}\right)^{1/4} = \sqrt[4]{\frac{1}{81}} = \frac{1}{3}$$

Example 2

Find:
(a) $9^{1/2}$
(b) $(-125)^{1/3}$
(c) $\left(\frac{1}{16}\right)^{1/4}$

Solution

(a) $9^{1/2} = \sqrt{9} = 3$

(b) $(-125)^{1/3} = \sqrt[3]{-125} = -5$

(c) $\left(\frac{1}{16}\right)^{1/4} = \sqrt[4]{\left(\frac{1}{16}\right)} = \frac{1}{2}$

How shall we define $a^{m/n}$, where m and n are positive integers with $n > 1$ and $\sqrt[n]{a}$ a real number? If we assume that Law III applies, then

$$a^{m/n} = (a^{1/n})^m = (a^m)^{1/n} = (\sqrt[n]{a})^m = \sqrt[n]{a^m}$$

From this we arrive at the following definition.

DEFINITION 5.6

$a^{m/n} = (\sqrt[n]{a})^m = \sqrt[n]{a^m}$, provided m and n are positive integers with no common factors and $\sqrt[n]{a}$ is a real number.

Observe that in Definition 5.6 the numerator of the exponent is the exponent of the radical expression, and the denominator is the index of the radical. For example, $a^{1/5} = \sqrt[5]{a}$ and $a^{2/5} = (\sqrt[5]{a})^2$.

Example 3

Evaluate $8^{2/3}$.

Solution We can do the evaluation by either of two methods.

Method 1		Method 2
$8^{2/3} = (\sqrt[3]{8})^2$	By Definition 5.6	$8^{2/3} = \sqrt[3]{8^2}$
$= (2)^2$		$= \sqrt[3]{64}$
$= 4$		$= 4$

Note that for $a < 0$, and m and n positive even integers, $(a^m)^{1/n} \neq (a^{1/n})^m$. For example, $[(-4)^2]^{1/2} = 4$, but $[(-4)^{1/2}]^2$ is not defined, since $(-4)^{1/2}$ is not a real number.

The five laws of exponents (I through V, Section 5.1) are satisfied by rational exponents with one exception. If, in Law III, $a < 0$, m is even, and n is even, $(a^m)^{1/n} \neq a^{m(1/n)}$. For example, $[(-2)^2]^{1/2} = 4^{1/2} = 2$. But $(-2)^{2\cdot(1/2)} = (-2)^1 = -2$. Clearly, $[(-2)^2]^{1/2} \neq (-2)^{2\cdot 1/2}$. To avoid this, we provide the following definition.

DEFINITION 5.7

$(a^m)^{1/n} = |a|^{m/n}$ when m and n are positive even integers.

Using Definition 5.7, we have

$$\begin{aligned}[(-16)^2]^{1/4} &= |-16|^{2/4}\\ &= |-16|^{1/2}\\ &= 16^{1/2}\\ &= 4\end{aligned}$$

and

$$\begin{aligned}[(-4)^2]^{1/4} &= |-4|^{2/4}\\ &= |-4|^{1/2}\\ &= 4^{1/2}\\ &= 2\end{aligned}$$

Finally, to define negative rational exponents, we first note that if m and n are positive integers with no common factors,

$$-\frac{m}{n} = \frac{-m}{n}$$

Thus, if Law III is to apply,

$$a^{-m/n} = (a^{1/n})^{-m}$$

$$= \frac{1}{(a^{1/n})^{m}} \qquad \text{By the definition of negative exponent}$$

Hence, we have:

DEFINITION 5.8

$$a^{-m/n} = \frac{1}{(a^{m/n})}$$ (m and n positive integers, $a^{1/n}$ a real number, $a \neq 0$)

Using this definition, we have

(1) $a^{-1/2} = \dfrac{1}{\sqrt{a}}$

(2) $32^{-3/5} = \dfrac{1}{32^{3/5}} = \dfrac{1}{(\sqrt[5]{32})^{3}} = \dfrac{1}{2^3} = \dfrac{1}{8}$

(3) $1000^{-2/3} = \dfrac{1}{1000^{2/3}} = \dfrac{1}{(\sqrt[3]{1000})^{2}} = \dfrac{1}{10^2} = \dfrac{1}{100}$

Example 4

Evaluate:

(a) $16^{-3/4}$
(b) $(-8)^{-4/3}$
(c) $125^{-2/3}$

Solution

(a) $16^{-3/4} = \dfrac{1}{16^{3/4}} = \dfrac{1}{(\sqrt[4]{16})^{3}} = \dfrac{1}{2^3} = \dfrac{1}{8}$

(b) $(-8)^{-4/3} = \dfrac{1}{(-8)^{4/3}} = \dfrac{1}{(\sqrt[3]{-8})^{4}} = \dfrac{1}{(-2)^4} = \dfrac{1}{16}$

(c) $125^{-2/3} = \dfrac{1}{125^{2/3}} = \dfrac{1}{(\sqrt[3]{125})^{2}} = \dfrac{1}{5^2} = \dfrac{1}{25}$

PROGRESS TEST

1. $\sqrt[3]{-216}$ equals ______.
2. $\sqrt[3]{\frac{-1}{125}}$ equals ______.
3. $36^{1/2}$ equals ______.
4. $\left(\frac{1}{625}\right)^{1/4}$ equals ______.
5. $81^{3/4}$ equals ______.
6. $[(-4)^3]^{1/3}$ equals ______.
7. $(-32)^{-3/5}$ equals ______.

EXERCISE 5.2

In Problems 1 through 24, evaluate, if possible.

Illustration: (a) $\sqrt{49}$

(b) $\sqrt[3]{\frac{-1}{216}}$

(c) $\left(\frac{-1}{16}\right)^{1/4}$

Solution: (a) $\sqrt{49} = 7$ because $7 \cdot 7 = 49$

(b) $\sqrt[3]{\frac{-1}{216}} = \frac{-1}{6}$ because $\left(\frac{-1}{6}\right)\left(\frac{-1}{6}\right)\left(\frac{-1}{6}\right) = \frac{-1}{216}$

(c) $\left(\frac{-1}{16}\right)^{1/4} = \sqrt[4]{\frac{-1}{16}}$, not a real number.

PROGRESS TEST ANSWERS

1. -6
2. $\frac{-1}{5}$
3. 6
4. $\frac{1}{5}$
5. 27
6. -4
7. $\frac{-1}{8}$

1. $\sqrt{4}$

2. $\sqrt{25}$

3. $\sqrt[3]{8}$

4. $\sqrt[3]{125}$

5. $\sqrt[3]{-8}$

6. $\sqrt[3]{-125}$

7. $\sqrt[3]{\frac{-1}{64}}$

8. $\sqrt[3]{\frac{-1}{27}}$

9. $\sqrt[4]{16}$

10. $\sqrt[4]{625}$

11. $\sqrt[5]{32}$

12. $\sqrt[5]{\frac{-1}{243}}$

13. $9^{1/2}$

14. $16^{1/2}$

15. $(-4)^{1/2}$

16. $-4^{1/2}$

17. $27^{1/3}$

18. $125^{1/3}$

19. $81^{1/4}$

20. $16^{1/4}$

21. $\left(\frac{-1}{8}\right)^{1/3}$

22. $\left(\frac{-1}{27}\right)^{1/3}$

23. $\left(\frac{-1}{256}\right)^{1/4}$

24. $\left(\frac{1}{256}\right)^{1/4}$

In Problems 25 through 40, evaluate, if possible.

Illustration: (a) $(-8)^{4/3}$
(b) $81^{-3/4}$
(c) $[(-3)^4]^{1/4}$

Solution: (a) $(-8)^{4/3} = (\sqrt[3]{-8})^4 = (-2)^4 = 16$

(b) $81^{-3/4} = \dfrac{1}{81^{3/4}} = \dfrac{1}{(\sqrt[4]{81})^3} = \dfrac{1}{(3)^3} = \dfrac{1}{27}$

(c) $[(-3)^4]^{1/4} = |-3| = 3$

25. $27^{2/3}$

26. $(-27)^{2/3}$

27. $125^{2/3}$

28. $216^{2/3}$

29. $\left(\frac{1}{8}\right)^{2/3}$

30. $\left(\frac{1}{81}\right)^{3/4}$

31. $(-8)^{4/3}$

32. $(-27)^{4/3}$

33. $(32)^{4/5}$

34. $(-32)^{4/5}$

35. $-32^{4/5}$

36. $(-64)^{5/3}$

37. $64^{-2/3}$

38. $27^{-2/3}$

39. $[(-7)^4]^{1/4}$

40. $[(-11)^6]^{1/6}$

In Problems 41 through 44, simplify the given expression and write the answer using positive exponents only. Assume that x and y represent positive numbers.

Illustration: (a) $x^{1/5}(x + x^{4/5})$
(b) $x^{-1/2}(x^{1/2} + x)$

Solution: (a) $x^{1/5}(x + x^{4/5}) = x^{1/5} \cdot x + x^{1/5} \cdot x^{4/5}$
$= x^{6/5} + x$

(b) $x^{-1/2}(x^{1/2} + x) = x^{-1/2+1/2} + x^{1-1/2}$
$= x^0 + x^{1/2}$
$= 1 + x^{1/2}$

41. $x^{1/3}(x^{2/3} + x)$

42. $x^{4/5}(x + x^{1/5})$

43. $y^{3/4}(y - y^{1/2})$

44. $y^{2/3}(y^{1/2} - y)$

In Problems 45 through 50, simplify and write the given expression as a product or quotient with positive exponents. Assume that all the letters represent positive numbers.

Illustration: (a) $\dfrac{x^{1/2} \cdot x^{-1/3}}{x^{1/6}}$

(b) $(x^{1/2}y^{3/2})^4$

(c) $\left(\dfrac{x^{1/5}}{x^{3/5}}\right)^2$

Solution: (a) $\dfrac{x^{1/2} \cdot x^{-1/3}}{x^{1/6}} = \dfrac{x^{1/2-1/3}}{x^{1/6}}$

$= \dfrac{x^{1/6}}{x^{1/6}} = 1$

(b) $(x^{1/2}y^{3/2})^4 = (x^{1/2})^4 \cdot (y^{3/2})^4$
$= x^2y^6$

(c) $\left(\dfrac{x^{1/5}}{x^{3/5}}\right)^2 = (x^{1/5-3/5})^2$

$= (x^{-2/5})^2$

$$= x^{(-2/5)(2)}$$
$$= x^{-4/5}$$
$$= \frac{1}{x^{4/5}}$$

45. $\dfrac{x^{1/6} \cdot x^{-5/6}}{x^{1/3}}$

46. $\dfrac{(x^{1/3} \cdot x^{1/2})^2}{x^{1/2}}$

47. $\dfrac{(x^{1/3} \cdot y^{-1/2})^6}{(y^{1/2})^{-4}}$

48. $\left(\dfrac{x^{4/3} \cdot y^{1/2}}{x^{1/3}}\right)^{-1/2}$

49. $\dfrac{(x^{1/4} \cdot y^2)^4}{(x^{2/3} \cdot y)^{-3}}$

50. $\left(\dfrac{-8a^{-3}b^{12}}{c^{15}}\right)^{-1/3}$

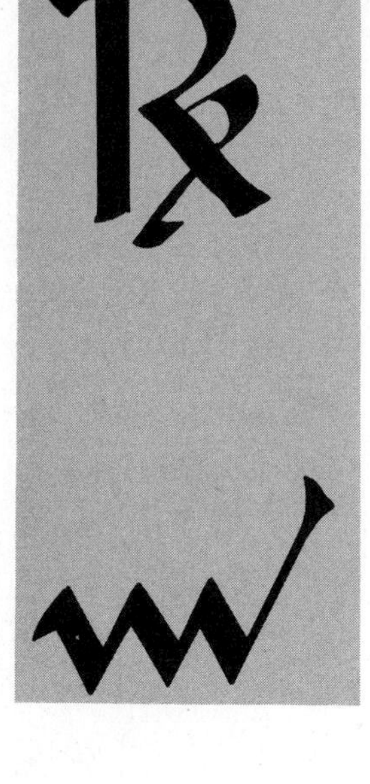

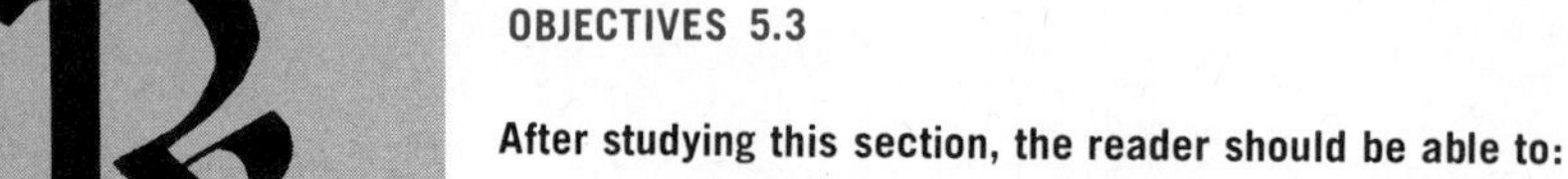

OBJECTIVES 5.3

After studying this section, the reader should be able to:

1. **Use the three relationships involving radicals given in the text to simplify expressions containing radicals.**
2. **Rationalize the denominator of a given expression.**
3. **Reduce the order of a given radical.**

5.3 PROPERTIES OF RADICALS

The illustrations at the beginning of this section show a square root sign, derived from the word *radix* (Latin for "root") and first used by Fibonacci, and a cube root sign created in 1525 by Christoff Rudolff, a German mathematician.

These two symbols, which are now written as $\sqrt{}$ and $\sqrt[3]{}$ respectively, are alternative symbols for the square root and the cube root of a number, discussed in the preceding section. In general, the nth root of a number a (see Definition 5.4) is defined so that

Recall that the index of a radical must be greater than or equal to 2.

$$a^{1/n} = \sqrt[n]{a} \quad (a > 0)$$

From this definition, we can derive three important relationships involving radicals that can be proved using the properties of exponents discussed previously. *In the discussion that follows, we shall assume that when the index of a radical is even, the radicand is non-negative.*

Law (1) $\sqrt[n]{a^n} = a \quad (a > 0)$

Law (2) $\sqrt[n]{ab} = \sqrt[n]{a}\,\sqrt[n]{b}$

Law (3) $\sqrt[n]{\dfrac{a}{b}} = \dfrac{\sqrt[n]{a}}{\sqrt[n]{b}}$

The first of these laws is equivalent to the definition of the principal nth root of a. Thus, $\sqrt[n]{a^n} = [a^n]^{1/n} = a^{n \cdot 1/n} = a$. The other two laws are obtained as follows:

$$(2) \quad \sqrt[n]{ab} = (ab)^{1/n} = a^{1/n} \bullet b^{1/n} = \sqrt[n]{a} \bullet \sqrt[n]{b}$$

$$(3) \quad \sqrt[n]{\frac{a}{b}} = \left(\frac{a}{b}\right)^{1/n} = \frac{a^{1/n}}{b^{1/n}} = \frac{\sqrt[n]{a}}{\sqrt[n]{b}}$$

We have already mentioned in Definition 5.7 that when n and m are even, $(a^m)^{1/n} = |a|^{m/n}$. Thus, if n is even and $m = n$, $(a^n)^{1/n} = |a|$. For example, $\sqrt{2^2} = |2| = 2$, $\sqrt{(-2)^2} = |-2| = 2$.

From Law (1), $\sqrt[3]{8} = \sqrt[3]{2^3} = 2$, and $\sqrt[3]{-8} = \sqrt[3]{-2^3} = -2$. Laws (1) and (2) may be used together to simplify certain expressions. For example, to simplify $\sqrt[4]{48}$ we proceed as follows:

$$\begin{aligned}\sqrt[4]{48} &= \sqrt[4]{2^4 \cdot 3}\\ &= \sqrt[4]{2^4} \cdot \sqrt[4]{3}\\ &= |2| \cdot \sqrt[4]{3} = 2\sqrt[4]{3}\end{aligned}$$

Similarly, $\sqrt[4]{324a^6b^9}$ can be simplified as follows:

$$\begin{aligned}\sqrt[4]{324a^6b^9} &= \sqrt[4]{2^2 \cdot 3^4 \cdot a^2 \cdot a^4 \cdot (b^2)^4 \cdot b}\\ &= \sqrt[4]{2^2} \cdot \sqrt[4]{3^4} \cdot \sqrt[4]{a^2} \cdot \sqrt[4]{a^4} \cdot \sqrt[4]{(b^2)^4} \cdot \sqrt[4]{b}\\ &= \sqrt[4]{2^2} \cdot 3 \cdot \sqrt[4]{a^2} \cdot |a| \cdot b^2 \cdot \sqrt[4]{b}\\ &= 3 \cdot |a| \cdot b^2 \cdot \sqrt[4]{2^2 \cdot a^2 \cdot b}\\ &= 3|a|b^2\sqrt[4]{4a^2b}\end{aligned}$$

Note that $324 = 2^2 \cdot 3^4$ when written as a product of primes.

Example 1

Simplify:

(a) $\sqrt[3]{54}$

(b) $\sqrt[3]{64a^4b^6}$

Solution

(a)
$$\begin{aligned}\sqrt[3]{54} &= \sqrt[3]{3^3 \cdot 2}\\ &= \sqrt[3]{3^3} \cdot \sqrt[3]{2}\\ &= 3 \cdot \sqrt[3]{2}\end{aligned}$$

(b)
$$\begin{aligned}\sqrt[3]{64a^4b^6} &= \sqrt[3]{4^3 \cdot a^3 \cdot a \cdot (b^2)^3}\\ &= \sqrt[3]{4^3} \cdot \sqrt[3]{a^3} \cdot \sqrt[3]{a} \cdot \sqrt[3]{(b^2)^3}\\ &= 4 \cdot a \cdot \sqrt[3]{a} \cdot b^2 = 4ab^2\sqrt[3]{a}\end{aligned}$$

The third law mentioned in this section can be used to change a radical into a form in which the radicand contains no fractions. For example,

$$\sqrt{\frac{3}{16}} = \frac{\sqrt{3}}{\sqrt{16}} = \frac{\sqrt{3}}{4}$$

In the expression $\sqrt{\frac{3}{16}}$, $\sqrt{\ }$ is the radical and $\frac{3}{16}$ is the radicand.

$$\text{and } \sqrt[3]{\frac{7}{8}} = \frac{\sqrt[3]{7}}{\sqrt[3]{8}} = \frac{\sqrt[3]{7}}{2}$$

Example 2

Simplify:

(a) $\sqrt{\frac{7}{32}}$

(b) $\sqrt[3]{\frac{9}{x^3}}$

Solution

(a) $\sqrt{\frac{7}{32}} = \sqrt{\frac{7 \cdot 2}{32 \cdot 2}} = \sqrt{\frac{14}{64}} = \frac{\sqrt{14}}{\sqrt{64}} = \frac{\sqrt{14}}{8}$

(b) $\sqrt[3]{\frac{9}{x^3}} = \frac{\sqrt[3]{9}}{\sqrt[3]{x^3}} = \frac{\sqrt[3]{9}}{x}$

Note that in (a) we multiplied the numerator and denominator of the radicand by 2 to make the denominator a perfect square.

In some cases, the denominator of the radicand is not a perfect power. For example, if we wish to simplify $\sqrt{\frac{3}{5}}$, we can use Law (3) to obtain $\sqrt{\frac{3}{5}} = \frac{\sqrt{3}}{\sqrt{5}}$. Then, by the Fundamental Principle of Fractions,

$$\frac{\sqrt{3}}{\sqrt{5}} = \frac{\sqrt{3} \cdot \sqrt{5}}{\sqrt{5} \cdot \sqrt{5}}$$

$$= \frac{\sqrt{15}}{\sqrt{5^2}} \qquad \text{By Law (2)}$$

$$= \frac{\sqrt{15}}{5}$$

This process is called rationalizing the denominator. To rationalize the denominator in the expression $\frac{\sqrt{7}}{\sqrt{3}}$, we proceed as follows:

$$\frac{\sqrt{7}}{\sqrt{3}} = \frac{\sqrt{7} \cdot \sqrt{3}}{\sqrt{3} \cdot \sqrt{3}} \qquad \text{By the Fundamental Principle of Fractions}$$

$$= \frac{\sqrt{21}}{\sqrt{3^2}} \qquad \text{By Law (2)}$$

$$= \frac{\sqrt{21}}{3}$$

Example 3

Rationalize the denominator of the given expressions.

(a) $\frac{\sqrt{11}}{\sqrt{6}}$

(b) $\frac{\sqrt{3}}{\sqrt{5x}}$, $x > 0$

Solution

(a) $\frac{\sqrt{11}}{\sqrt{6}} = \frac{\sqrt{11} \cdot \sqrt{6}}{\sqrt{6} \cdot \sqrt{6}} = \frac{\sqrt{66}}{6}$

(b) $\frac{\sqrt{3}}{\sqrt{5x}} = \frac{\sqrt{3} \cdot \sqrt{5x}}{\sqrt{5x} \cdot \sqrt{5x}} = \frac{\sqrt{15x}}{5x}$

When the radical in the denominator is of order n, we must make the denominator an exact nth power. For example, to rationalize $\sqrt[3]{\frac{5}{3x}}$, we multiply numerator and denominator by $\sqrt[3]{3^2x^2}$, obtaining

$$\sqrt[3]{\frac{5}{3x}} = \frac{\sqrt[3]{5} \cdot \sqrt[3]{3^2 \cdot x^2}}{\sqrt[3]{3x} \cdot \sqrt[3]{3^2 \cdot x^2}}$$

$$= \frac{\sqrt[3]{5 \cdot 9 \cdot x^2}}{\sqrt[3]{3^3x^3}}$$

$$= \frac{\sqrt[3]{45x^2}}{3x}$$

Example 4

Rationalize the denominator in the following expressions.

(a) $\frac{1}{\sqrt[3]{5x}}$

(b) $\sqrt[5]{\dfrac{5}{8x^4}}$

Solution

(a) $\dfrac{1}{\sqrt[3]{5x}} = \dfrac{1 \cdot \sqrt[3]{5^2x^2}}{\sqrt[3]{5x} \cdot \sqrt[3]{5^2x^2}}$

$= \dfrac{\sqrt[3]{25x^2}}{5x}$

(b) $\sqrt[5]{\dfrac{5}{8x^4}} = \sqrt[5]{\dfrac{5}{2^3x^4}}$

$= \sqrt[5]{\dfrac{5 \cdot 2^2 \cdot x}{2^3 \cdot x^4 \cdot 2^2 \cdot x}}$

$= \sqrt[5]{\dfrac{20x}{2^5 \cdot x^5}}$

$= \dfrac{\sqrt[5]{20x}}{2x}$

The *order* of a radical can sometimes be reduced by writing the radical as a power with a rational exponent, and then reducing the exponent. For example, if $x \geq 0$,

Note that $\sqrt[6]{x^3} = x^{3/6} = x^{1/2}$.

$$\begin{aligned}\sqrt[6]{x^3} &= (x^3)^{1/6}\\ &= x^{3 \cdot 1/6}\\ &= x^{1/2}\\ &= \sqrt{x}\end{aligned}$$

Note that $\sqrt[4]{64x^2y^2} = \sqrt[4]{(8xy)^2}$
$= (8xy)^{2/4}$
$= (8xy)^{1/2}$

Similarly, for $x \geq 0$ and $y \geq 0$,

$$\begin{aligned}\sqrt[4]{64x^2y^2} &= \sqrt[4]{(8xy)^2}\\ &= [(8xy)^2]^{1/4}\\ &= [8xy]^{2 \cdot 1/4}\\ &= (8xy)^{1/2} = \sqrt{8xy}\end{aligned}$$

Example 5

Reduce the order of the given expression.

(a) $\sqrt[4]{\dfrac{16}{81}}$

(b) $\sqrt[6]{27c^3d^3}$, $c \geq 0$, $d \geq 0$

Solution

(a) $\sqrt[4]{\dfrac{16}{81}} = \left[\left(\dfrac{2}{3}\right)^4\right]^{1/4} = \dfrac{2}{3}$

(b) $\sqrt[6]{27c^3d^3} = \sqrt[6]{(3cd)^3}$
$= [(3cd)^3]^{1/6}$
$= [3cd]^{1/2}$
$= \sqrt{3cd}$

PROGRESS TEST

1. $\sqrt{3^2} =$ ______.
2. $\sqrt{(-3)^2} =$ ______.
3. $\sqrt[4]{80} =$ ______.
4. $\sqrt[4]{243a^7b^{10}} =$ ______________________.
5. When simplified, $\sqrt{\dfrac{11}{36}}$ equals ______.
6. When the denominator is rationalized, $\dfrac{\sqrt{3}}{\sqrt{7x}}$ equals ______ $(x > 0)$.
7. When the denominator is rationalized, $\dfrac{\sqrt[3]{5}}{\sqrt[3]{4x^2}}$ equals ______.
8. When the order is reduced, $\sqrt[4]{36a^2}$ equals ______ $(a \geq 0)$.

EXERCISE 5.3

In Problems 1 through 24, simplify. (Assume all variables in the radicands denote positive real numbers.)

Illustration: (a) $\sqrt{(-4)^2}$

(b) $\sqrt[4]{243a}$

(c) $\sqrt{\dfrac{11}{49x^2}}$

Solution: (a) $\sqrt{(-4)^2} = |-4| = 4$

PROGRESS TEST ANSWERS

1. 3
2. 3
3. $\sqrt[4]{2^4 \cdot 5} = 2 \cdot \sqrt[4]{5}$
4. $\sqrt[4]{3^4 \cdot 3a^4 \cdot a^3 \cdot (b^2)^4 \cdot b^2}$
$= 3 \cdot |a| \cdot b^2 \cdot \sqrt[4]{3a^3b^2}$
5. $\dfrac{\sqrt{11}}{6}$
6. $\dfrac{\sqrt{21x}}{7x}$
7. $\dfrac{\sqrt[3]{10x}}{2x}$
8. $\sqrt{6a}$

(b) $\sqrt[4]{243a} = \sqrt[4]{3^4 \cdot 3 \cdot a} = 3\sqrt[4]{3a}$

(c) $\sqrt{\dfrac{11}{49x^2}} = \dfrac{\sqrt{11}}{\sqrt{49x^2}} = \dfrac{\sqrt{11}}{7x}$

1. $\sqrt{(-5)^2}$

2. $\sqrt{(5)^2}$

3. $\sqrt[3]{-64}$

4. $\sqrt[3]{-125}$

5. $\sqrt{8}$

6. $\sqrt{75}$

7. $\sqrt{12a}$

8. $\sqrt{27a}$

9. $\sqrt{18a^3b^3}$

10. $\sqrt{72a^5b^3}$

11. $\sqrt[3]{16x^3y^3}$

12. $\sqrt[3]{81x^3y^4}$

13. $\sqrt[3]{40x^4y}$

14. $\sqrt[3]{81x^3y^6}$

15. $\sqrt[4]{x^5y^7}$

16. $\sqrt[4]{162x^4y^7}$

17. $\sqrt[5]{-243a^{10}b^{17}}$

18. $\sqrt[5]{-32a^{15}b^{20}}$

19. $\sqrt{\dfrac{13}{49}}$

20. $\sqrt{\dfrac{17}{64}}$

21. $\sqrt{\dfrac{17}{4x^2}}$

22. $\sqrt{\dfrac{19}{64x^4}}$

23. $\sqrt[3]{\dfrac{3}{64x^3}}$

24. $\sqrt[3]{\dfrac{-7}{27x^6}}$

In Problems 25 through 40, rationalize the denominator. (Assume all variables in the radicands represent positive real numbers.)

Illustration: (a) $-\sqrt{\dfrac{1}{2a}}$

(b) $\sqrt[3]{\dfrac{11}{5x^2}}$

(c) $\dfrac{\sqrt{a}\,\sqrt{ab^3}}{\sqrt{b}}$

Solution: (a) $-\sqrt{\frac{1}{2a}} = -\sqrt{\frac{1 \cdot 2a}{2a \cdot 2a}}$

$$= -\sqrt{\frac{2a}{(2a)^2}} = -\frac{\sqrt{2a}}{2a}$$

(b) $\sqrt[3]{\frac{11}{5x^2}} = \sqrt[3]{\frac{11 \cdot 25x}{5x^2 \cdot 25x}} = \frac{\sqrt[3]{275x}}{5x}$

(c) $\frac{\sqrt{a}\sqrt{ab^3}}{\sqrt{b}} = \sqrt{\frac{a^2b^3}{b}}$

$$= \sqrt{a^2b^2}$$
$$= ab$$

25. $\sqrt{\frac{2}{3}}$

26. $\sqrt{\frac{4}{5}}$

27. $\frac{-\sqrt{2}}{\sqrt{7}}$

28. $\frac{-\sqrt{3}}{\sqrt{11}}$

29. $\sqrt{\frac{5}{2a}}$

30. $\sqrt{\frac{7}{36}}$

31. $\sqrt{\frac{5}{32ab}}$

32. $\sqrt{\frac{5}{8ab}}$

33. $-\sqrt{\frac{3}{2a^3b^3}}$

34. $-\sqrt{\frac{3}{8ab^3}}$

35. $\frac{\sqrt{x}\sqrt{xy^3}}{\sqrt{y}}$

36. $\frac{\sqrt{xy}\sqrt{xy^4}}{\sqrt{y}}$

37. $-\sqrt[3]{\frac{7}{9}}$

38. $-\sqrt[3]{\frac{3}{32}}$

39. $\sqrt[3]{\frac{3}{16x^2}}$

40. $\sqrt[3]{\frac{5}{16x}}$

In Problems 41 through 50, reduce the order of the given radical and simplify if possible. (Assume the variables in the radicands represent positive real numbers.)

Illustration: (a) $\sqrt{16a^4b^4}$

(b) $\sqrt[4]{81a^6b^{10}}$

Solution:

(a) $\sqrt{16a^4b^4} = \sqrt{2^4a^4b^4}$
$= (2^4a^4b^4)^{1/2}$
$= 2^2(a^4)^{1/2}(b^4)^{1/2}$
$= 4a^2b^2$

(b) $\sqrt[4]{81a^6b^{10}} = \sqrt[4]{3^4a^6b^{10}}$
$= (3^4a^6b^{10})^{1/4}$
$= 3^1 \bullet a^{6/4} \bullet b^{10/4}$
$= 3a^{3/2}b^{5/2}$
$= 3\sqrt{a^3b^5}$
$= 3\sqrt{a^2 \bullet a \bullet (b^2)^2 \bullet b}$
$= 3ab^2\sqrt{ab}$

41. $\sqrt[4]{9}$

42. $\sqrt[6]{8}$

43. $\sqrt[4]{4a^2}$

44. $\sqrt[4]{9a^2}$

45. $\sqrt[4]{25x^6y^2}$

46. $\sqrt[4]{36x^2y^6}$

47. $\sqrt[4]{49x^{10}y^6}$

48. $\sqrt[4]{100x^{10}y^{10}}$

49. $\sqrt[6]{8a^3b^3}$

50. $\sqrt[6]{27a^3b^9}$

OBJECTIVES 5.4

After studying this section, the reader should be able to:

1. **Add, subtract, multiply, and/or divide numerical expressions containing radicals.**
2. **Rationalize the denominator of a given fraction containing sums or differences involving radicals.**

5.4 OPERATIONS WITH RADICALS

In the cartoon at the beginning of this section, the man is selling square roots. In this section we shall learn to add, subtract, multiply, and divide expressions containing square roots and other radicals.

As you recall from Chapter 1, for any real numbers a, b, and c,

(1) $a(b + c) = ab + ac$ — Distributive Law

(2) $ba + ca = (b + c)a$ — By the Symmetric and Distributive Laws

Equation (2) can be used to add (or subtract) radicals that are similar in the same manner in which we added (or subtracted) like terms. For example, $3\sqrt{2} + 4\sqrt{2} = (3 + 4)\sqrt{2} = 7\sqrt{2}$, and $5\sqrt[3]{3} + 7\sqrt[3]{3} = (5 + 7)\sqrt[3]{3} = 12\sqrt[3]{3}$. Of course, if the radicals are *not* similar, we must simplify them before performing the addition. Thus, to add $\sqrt{75} + \sqrt{27}$, which are not similar, we proceed as follows:

Two radicals are similar if they can be written with identical radical factors and the same index.

Recall that $\sqrt{ab} = \sqrt{a}\sqrt{b}$.

$$\sqrt{75} + \sqrt{27} = \sqrt{25 \cdot 3} + \sqrt{9 \cdot 3}$$

$$= \sqrt{25} \cdot \sqrt{3} + \sqrt{9} \cdot \sqrt{3}$$

$$= 5\sqrt{3} + 3\sqrt{3}$$

$$= (5 + 3)\sqrt{3} = 8\sqrt{3}$$

The subtraction of similar radicals is done along the same lines. Thus,

$$\sqrt{80} - \sqrt{20} = \sqrt{16 \cdot 5} - \sqrt{4 \cdot 5}$$

$$= \sqrt{16} \cdot \sqrt{5} - \sqrt{4} \cdot \sqrt{5}$$

$$= 4\sqrt{5} - 2\sqrt{5}$$

$$= (4 - 2) \cdot \sqrt{5} = 2\sqrt{5}$$

Example 1

Perform the indicated operations.

(a) $\sqrt{175} + \sqrt{28}$

(b) $\sqrt{98} - \sqrt{32}$

Solution

(a) $\sqrt{175} + \sqrt{28} = \sqrt{25 \cdot 7} + \sqrt{4 \cdot 7}$

$$= \sqrt{25} \cdot \sqrt{7} + \sqrt{4} \cdot \sqrt{7}$$

$$= 5\sqrt{7} + 2\sqrt{7} = 7\sqrt{7}$$

(b) $\sqrt{98} - \sqrt{32} = \sqrt{49 \cdot 2} - \sqrt{16 \cdot 2}$

$$= \sqrt{49} \cdot \sqrt{2} - \sqrt{16} \cdot \sqrt{2}$$

$$= 7\sqrt{2} - 4\sqrt{2} = 3\sqrt{2}$$

The Distributive Law, used in conjunction with the fact that $\sqrt{a} \cdot \sqrt{b} = \sqrt{ab}$, can be used to simplify expressions containing parentheses. For example,

Since $a(b + c) = ab + ac$, $\sqrt{2}\,(\sqrt{3} + \sqrt{5}) = \sqrt{2}\,\sqrt{3} + \sqrt{2}\,\sqrt{5}$.

$$\sqrt{2} \cdot (\sqrt{3} + \sqrt{5}) = \sqrt{2} \cdot \sqrt{3} + \sqrt{2} \cdot \sqrt{5}$$

$$= \sqrt{6} + \sqrt{10}$$

Similarly, if $x \geq 0$, then

$$\sqrt{2x} \cdot (\sqrt{x} + \sqrt{3x}) = \sqrt{2x} \cdot \sqrt{x} + \sqrt{2x} \cdot \sqrt{3x}$$

$= \sqrt{2x^2} + \sqrt{6x^2}$

Since $\sqrt{ab} = \sqrt{a} \cdot \sqrt{b}$, $\sqrt{2x^2} = \sqrt{2} \cdot \sqrt{x^2}$.

$= \sqrt{2}\sqrt{x^2} + \sqrt{6}\sqrt{x^2}$

$\sqrt{x^2} = x$ if $x \geq 0$.

$= x\sqrt{2} + x\sqrt{6}$

$= x(\sqrt{2} + \sqrt{6})$

Since $ab + ac = a(b + c)$, $x\sqrt{2} + x\sqrt{6} = x(\sqrt{2} + \sqrt{6})$.

Example 2

Perform the indicated operations:

(a) $\sqrt{3}(\sqrt{5} + \sqrt{12})$

(b) $\sqrt{3x}(\sqrt{x} - \sqrt{5})$, $x \geq 0$

Solution

(a) $\sqrt{3}(\sqrt{5} + \sqrt{12}) = \sqrt{3} \cdot \sqrt{5} + \sqrt{3} \cdot \sqrt{12}$

$= \sqrt{15} + \sqrt{36}$

$= \sqrt{15} + 6$

(b) $\sqrt{3x}(\sqrt{x} - \sqrt{5}) = \sqrt{3x}\sqrt{x} - \sqrt{3x}\sqrt{5}$

$= \sqrt{3x^2} - \sqrt{15x}$

$= x\sqrt{3} - \sqrt{15x}$

If we wish to obtain the product of two binomials containing radicals, we first simplify the radicals involved, if possible. For example, to find the product $(\sqrt{98} + \sqrt{27})(\sqrt{72} + \sqrt{75})$, we proceed as follows:

$(\sqrt{98} + \sqrt{27})(\sqrt{72} + \sqrt{75})$

$= (\sqrt{49 \cdot 2} + \sqrt{9 \cdot 3})(\sqrt{36 \cdot 2} + \sqrt{25 \cdot 3})$

$= (7\sqrt{2} + 3\sqrt{3})(6\sqrt{2} + 5\sqrt{3})$

$= 42\sqrt{2^2} + 6 \cdot 3\sqrt{3} \cdot \sqrt{2} + 7 \cdot 5\sqrt{2}\sqrt{3} + 15\sqrt{3^2}$

$\sqrt{98} = \sqrt{49 \cdot 2}$

$\sqrt{27} = \sqrt{9 \cdot 3}$

$\sqrt{72} = \sqrt{36 \cdot 2}$

$\sqrt{75} = \sqrt{25 \cdot 3}$

$$= 42 \cdot 2 + 18\sqrt{6} + 35\sqrt{6} + 15 \cdot 3$$

$$= 84 + 53\sqrt{6} + 45$$

$$= 129 + 53\sqrt{6}$$

Example 3 Find the product $(\sqrt{63} + \sqrt{75})(\sqrt{28} - \sqrt{27})$.

Solution We first simplify the radicals:

$$(\sqrt{63} + \sqrt{75})(\sqrt{28} - \sqrt{27})$$

$$= (\sqrt{9 \cdot 7} + \sqrt{25 \cdot 3})(\sqrt{4 \cdot 7} - \sqrt{9 \cdot 3})$$

$$= (3\sqrt{7} + 5\sqrt{3})(2\sqrt{7} - 3\sqrt{3})$$

$$= 6\sqrt{7^2} + 10\sqrt{21} - 9\sqrt{21} - 15\sqrt{3^2}$$

$$= 6 \cdot 7 + \sqrt{21} - 15 \cdot 3$$

$$= 42 + \sqrt{21} - 45$$

$$= -3 + \sqrt{21}$$

We are now in a position to rationalize denominators of fractions containing sums or differences involving radicals. To do this, we first recall that $(a + b)(a - b) = a^2 - b^2$. Each of the factors $(a + b)$ and $(a - b)$ is called the *conjugate* of the other factor. Now consider the fraction $\frac{3}{3 + \sqrt{3}}$. If we use the Fundamental Principle of Fractions and multiply the numerator and denominator of this fraction by the conjugate of $3 + \sqrt{3}$, that is, $3 - \sqrt{3}$, we have

The conjugate of $3 + \sqrt{3}$ is $3 - \sqrt{3}$, and the conjugate of $\sqrt{a} - \sqrt{b}$ is $\sqrt{a} + \sqrt{b}$.

$$\frac{3}{3 + \sqrt{3}} = \frac{3 \cdot (3 - \sqrt{3})}{(3 + \sqrt{3})(3 - \sqrt{3})}$$

$$= \frac{3 \cdot (3 - \sqrt{3})}{3^2 - (\sqrt{3})^2}$$

Since $(a + b)(a - b) = a^2 - b^2$, $(3 + \sqrt{3})(3 - \sqrt{3}) = 3^2 - (\sqrt{3})^2$.

$$= \frac{3 \cdot (3 - \sqrt{3})}{9 - 3}$$

Note that $(\sqrt{3})^2 = \sqrt{3} \cdot \sqrt{3} = \sqrt{9} = 3$

$$= \frac{3 \cdot (3 - \sqrt{3})}{6}$$

$$= \frac{3 - \sqrt{3}}{2}$$

$\frac{\overset{1}{\cancel{3}} \cdot (3 - \sqrt{3})}{\underset{2}{\cancel{6}}} = \frac{3 - \sqrt{3}}{2}$

Example 4

Rationalize the denominator in the expression $\frac{\sqrt{x}}{\sqrt{x} - \sqrt{y}}$, where x and y represent positive numbers.

Solution

We first multiply numerator and denominator by $\sqrt{x} + \sqrt{y}$, the conjugate of $\sqrt{x} - \sqrt{y}$.

$$\frac{\sqrt{x}}{\sqrt{x} - \sqrt{y}} = \frac{\sqrt{x}\,(\sqrt{x} + \sqrt{y})}{(\sqrt{x} - \sqrt{y})(\sqrt{x} + \sqrt{y})}$$

$$= \frac{\sqrt{x}\,(\sqrt{x} + \sqrt{y})}{(\sqrt{x})^2 - (\sqrt{y})^2}$$

$$= \frac{\sqrt{x}\,(\sqrt{x} + \sqrt{y})}{x - y}$$

$$= \frac{\sqrt{x^2} + \sqrt{xy}}{x - y}$$

$$= \frac{x + \sqrt{xy}}{x - y}$$

PROGRESS TEST

1. When simplified, $\sqrt{44} + \sqrt{99}$ equals ________.
2. When simplified, $\sqrt{50} - \sqrt{98}$ equals ________.
3. If $x \geq 0$, the product $\sqrt{7x}\,(\sqrt{2x} + \sqrt{7x})$ equals ________.
4. $(\sqrt{27} + \sqrt{28})(\sqrt{75} - \sqrt{112})$ equals ________.
5. When the denominator of $\dfrac{-2}{3 + \sqrt{5}}$ is rationalized, this fraction equals ________.

EXERCISE 5.4

In Problems 1 through 44, perform the indicated operations.

Illustration: (a) $\sqrt[3]{54} - (\sqrt[3]{16} - \sqrt[3]{128})$

(b) $\sqrt[3]{3x}\,(\sqrt[3]{9x^2} - \sqrt[3]{18x})$

Solution: (a) $\sqrt[3]{54} - (\sqrt[3]{16} - \sqrt[3]{128})$

$$= \sqrt[3]{3^3 \cdot 2} - (\sqrt[3]{2^3 \cdot 2} - \sqrt[3]{4^3 \cdot 2})$$

$$= \sqrt[3]{3^3}\,\sqrt[3]{2} - \sqrt[3]{2^3}\,\sqrt[3]{2} + \sqrt[3]{4^3}\,\sqrt[3]{2}$$

$$= 3\sqrt[3]{2} - 2\sqrt[3]{2} + 4\sqrt[3]{2}$$

$$= 5\sqrt[3]{2}$$

(b) $\sqrt[3]{3x}\,(\sqrt[3]{9x^2} - \sqrt[3]{18x})$

$$= \sqrt[3]{3x}\,\sqrt[3]{9x^2} - \sqrt[3]{3x}\,\sqrt[3]{18x}$$

$$= \sqrt[3]{27x^3} - \sqrt[3]{54x^2}$$

$$= \sqrt[3]{3^3 \cdot x^3} - \sqrt[3]{3^3 \cdot 2 \cdot x^2}$$

$$= 3x - 3\sqrt[3]{2x^2}$$

PROGRESS TEST ANSWERS

1. $5\sqrt{11}$
2. $-2\sqrt{2}$
3. $x\sqrt{14} + 7x$
4. $-11 - 2\sqrt{21}$
5. $\dfrac{-(3 - \sqrt{5})}{2}$

1. $12\sqrt{2} + 3\sqrt{2}$

2. $15\sqrt{3} + 2\sqrt{3}$

3. $\sqrt{80a} + \sqrt{125a}$

4. $\sqrt{98a} + \sqrt{32a}$

5. $\sqrt{50} - \sqrt{32}$

6. $\sqrt{75} + \sqrt{12}$

7. $\sqrt{50a^2} - \sqrt{200a^2}$

8. $\sqrt{48a^2} - \sqrt{363a^2}$

9. $2\sqrt{300} - 9\sqrt{12} - 7\sqrt{48}$

10. $\sqrt{175} + \sqrt{567} - \sqrt{63}$

11. $\sqrt[3]{40} + \sqrt[3]{625}$

12. $\sqrt[3]{54} + \sqrt[3]{16}$

13. $\sqrt[3]{81} - 3\sqrt[3]{375}$

14. $\sqrt[3]{24} - \sqrt[3]{81}$

15. $2\sqrt[3]{-24} - 4\sqrt[3]{-81} - \sqrt[3]{375}$

16. $10\sqrt[3]{-40} - 2\sqrt[3]{-135} + 4\sqrt[3]{-320}$

17. $\sqrt[3]{3a} - \sqrt[3]{24a} + \sqrt[3]{375a}$

18. $\sqrt[3]{r^5} - \sqrt[3]{8r^5} - r\sqrt[3]{64r^2}$

19. $\frac{3\sqrt[3]{2}}{2} - \frac{\sqrt[3]{3}}{3}$

20. $\frac{4}{5} - \frac{\sqrt[3]{2}}{2}$

21. $3(5 - \sqrt{2})$

22. $-2(\sqrt{2} - 3)$

23. $\sqrt{2}\,(\sqrt{2} + 3)$

24. $\sqrt{3}\,(\sqrt{5} + 2)$

25. $2\sqrt{3}\,(7\sqrt{5} + 5\sqrt{3})$

26. $2\sqrt{5}\,(5\sqrt{2} + 3\sqrt{5})$

27. $3\sqrt{5}\,(2\sqrt{3} - \sqrt{5})$

28. $4\sqrt{2}\,(3\sqrt{5} - 3\sqrt{2})$

29. $-4\sqrt{7}\,(2\sqrt{3} - 5\sqrt{2})$

30. $-3\sqrt{2}\,(5\sqrt{7} - 2\sqrt{3})$

31. $(5\sqrt{3} + \sqrt{5})(3\sqrt{3} + 2\sqrt{5})$

32. $(2\sqrt{2} + 5\sqrt{3})(3\sqrt{2} + \sqrt{3})$

33. $(3\sqrt{6} - 2\sqrt{3})(4\sqrt{6} + 5\sqrt{3})$

34. $(3\sqrt{5} - 2\sqrt{3})(2\sqrt{5} + 3\sqrt{3})$

35. $(7\sqrt{5} - 11\sqrt{7})(5\sqrt{5} + 8\sqrt{7})$

36. $(2\sqrt{3}-5\sqrt{2})(3\sqrt{3}+2\sqrt{2})$

37. $(1+\sqrt{2})(1-\sqrt{2})$

38. $(2+\sqrt{3})(2-\sqrt{3})$

39. $(2+3\sqrt{3})(2-3\sqrt{3})$

40. $(5+5\sqrt{2})(5-5\sqrt{2})$

41. $(\sqrt{3}-\sqrt{2})^2$

42. $(\sqrt{2}-\sqrt{3})^2$

43. $(a-\sqrt{b})^2$

44. $(\sqrt{a}-b)^2$

In Problems 45 through 55, rationalize the denominator of the given fraction. (Assume that the variables represent positive numbers.)

Illustration:

(a) $\dfrac{\sqrt{x}-\sqrt{y}}{\sqrt{2x}}$

(b) $\dfrac{\sqrt{x}-\sqrt{y}}{\sqrt{x}+\sqrt{y}}$

Solution:

(a) $$\frac{\sqrt{x}-\sqrt{y}}{\sqrt{2x}}=\frac{(\sqrt{x}-\sqrt{y})\bullet\sqrt{2x}}{\sqrt{2x}\bullet\sqrt{2x}}$$

$$=\frac{\sqrt{2x^2}-\sqrt{2xy}}{\sqrt{4x^2}}$$

$$=\frac{x\sqrt{2}-\sqrt{2xy}}{2x}$$

(b) $$\frac{\sqrt{x}-\sqrt{y}}{\sqrt{x}+\sqrt{y}} = \frac{(\sqrt{x}-\sqrt{y})(\sqrt{x}-\sqrt{y})}{(\sqrt{x}+\sqrt{y})(\sqrt{x}-\sqrt{y})}$$ Multiplying numerator and denominator by the conjugate of the denominator

$$= \frac{(\sqrt{x})^2 - 2\sqrt{x}\sqrt{y} + (\sqrt{y})^2}{(\sqrt{x})^2 - (\sqrt{y})^2}$$

$$= \frac{x - 2\sqrt{xy} + y}{x - y}$$

45. $\frac{3+\sqrt{3}}{\sqrt{2}}$

46. $\frac{2+\sqrt{5}}{\sqrt{3}}$

47. $\frac{2}{3-\sqrt{2}}$

48. $\frac{6}{2-\sqrt{2}}$

49. $\frac{4a}{3-\sqrt{5}}$

50. $\frac{3a}{4-\sqrt{3}}$

51. $\frac{3a+2b}{3+\sqrt{2}}$

52. $\frac{5a+b}{2+\sqrt{3}}$

53. $\frac{\sqrt{a}+b}{\sqrt{a}-b}$

54. $\frac{a+\sqrt{b}}{a-\sqrt{b}}$

55. $\frac{\sqrt{a}+\sqrt{2b}}{\sqrt{a}-\sqrt{2b}}$

Courtesy of Library of Congress.

OBJECTIVES 5.5

After studying this section, the reader should be able to:

1. **Find the sum, difference, product, or quotient of two or more complex numbers.**
2. **Rationalize the denominator of an expression containing complex numbers.**

5.5 COMPLEX NUMBERS

The illustration at the beginning of this section shows Carl Friedrich Gauss, the first mathematician to give a sound treatment of *complex numbers.*

As we stated after Definition 5.4, if $a < 0$ and n is even, $\sqrt[n]{a}$ is not a real number. For example, $\sqrt{-2}$, $\sqrt{-16}$, and $\sqrt{-12}$ are *not* real numbers because the square of a real number is always a non-negative real number. Since negative numbers do not have real square roots, we say that the set of real numbers is *not* closed with respect to the operation of taking square roots. To avoid this difficulty, we can define, as Gauss did, a new set of numbers which contains elements that are square roots of negative numbers. We start by first defining a number i so that the square of i is -1; that is, i is such that $i^2 = -1$. We also agree to write $i = \sqrt{-1}$.

Now, for $b > 0$, $\sqrt{-b} = \sqrt{-1 \cdot b} = \sqrt{-1} \cdot \sqrt{b} = i\sqrt{b}$. Thus, the square root of any negative real number can be written as the product of a real number and the number i. For example, $\sqrt{-4} = \sqrt{-1} \cdot \sqrt{4} = i\sqrt{4} = 2i$, and $\sqrt{-2} = \sqrt{-1} \cdot \sqrt{2} = i\sqrt{2} = \sqrt{2}\,i$. We agree always to write the square root of a negative number as i times the square root of a positive number, as in these illustrations. Incorrect results may be obtained if this agreement is not followed.

Note that:
$i = \sqrt{-1}$
$i^2 = -1$
$i^3 = i^2 \cdot i = -1 \cdot i = -i$
$i^4 = i^2 \cdot i^2 = (-1)(-1) = 1$
$i^5 = i^4 \cdot i = 1 \cdot i = i$
$i^6 = -1$
$i^7 = -i$
$i^8 = 1$

$i\sqrt{2}$ is sometimes the preferred term because $\sqrt{2}\,i$ is easily confused with $\sqrt{2i}$.

Example 1 Write the given expression in the form bi ($b \in R$).

(a) $\sqrt{-9}$

(b) $\sqrt{-18}$

Solution (a) $\sqrt{-9} = \sqrt{-1 \cdot 9} = \sqrt{-1}\sqrt{9} = 3i$

Recall that $\sqrt{18} = \sqrt{9 \cdot 2} = \sqrt{9}\sqrt{2} = 3\sqrt{2}$.

(b) $\sqrt{-18} = \sqrt{-1 \cdot 18} = \sqrt{-1}\sqrt{18} = i\sqrt{18} = 3\sqrt{2}i$

If b is a positive real number, then $\sqrt{-b}$ is a pure imaginary number.

The number represented by $\sqrt{-b} = \sqrt{b} \cdot i$, where $b \in R$ and $b > 0$, is called a *pure imaginary number.* We now define the set C of *complex numbers* as the set of all numbers of the form $a + bi$, where $a, b \in R$. In set notation:

$$C = \{a + bi \mid a, b \in R, i = \sqrt{-1}\}$$

In the complex number $a + bi$, a is called the *real* part and bi the *imaginary part.* Thus the number $-3 + 4i$ is a complex number whose *real* part is -3 and whose *imaginary* part is 4i. Similarly, $2 - 3i$ is a complex number with 2 as its real part and $-3i$ as its imaginary part. The sum and the difference of complex numbers are defined next.

DEFINITION 5.9

For a, b, c, and d as real numbers,

$$(a + bi) + (c + di) = (a + c) + (b + d)i$$
$$(a + bi) - (c + di) = (a - c) + (b - d)i$$

For example, $(3 + 4i) + (8 + 2i) = (3 + 8) + (4 + 2)i = 11 + 6i$, and $(9 + 2i) - (2 + 4i) = (9 - 2) + (2 - 4)i = 7 - 2i$.

Example 2 Find:

(a) $(5 + 4i) + (7 - 2i)$

(b) $(6 + 5i) - (7 - 3i)$

Solution

(a) $(5 + 4i) + (7 - 2i) = (5 + 7) + [4 + (-2)]i = 12 + 2i$

(b) $(6 + 5i) - (7 - 3i) = (6 - 7) + [5 - (-3)]i = -1 + 8i$

To define the product of two complex numbers, we recall that $(a+b)(c+d) = ac+ad+bc+bd$. Thus, a reasonable way in which to form the product of two complex numbers is as follows:

$$\begin{aligned}(a+bi)(c+di) &= a(c+di) + bi(c+di) \\ &= ac+adi+bci+bdi^2 \\ &= ac+adi+bci-bd \\ &= (ac-bd)+(ad+bc)i\end{aligned}$$

Recall that $i^2 = -1$, so $bdi^2 = -bd$.

We summarize this discussion formally in the next definition.

DEFINITION 5.10

For a, b, c, and d as real numbers,
$(a+bi)(c+di) = (ac-bd)+(ad+bc)i$

In practice, we do not need to memorize Definition 5.10 to multiply complex numbers—we can merely follow the rule used to multiply binomials and replace i^2 by -1. Thus,

$$\begin{aligned}(3+4i)(2+3i) &= 6+9i+8i+12i^2 \\ &= 6+9i+8i-12 \\ &= -6+17i\end{aligned}$$

Since $i^2 = -1$, $12i^2 = -12$.

Example 3

Find the product:
(a) $(2-5i)(3+7i)$
(b) $-3(4-7i)$

Solution

(a) $$\begin{aligned}(2-5i)(3+7i) &= 6+14i-15i-35i^2 \\ &= 6-i+35 \\ &= 41-i\end{aligned}$$
(b) $-3(4-7i) = -12+21i$

Finally, to define the *quotient* of two complex numbers, we use the rationalizing process developed in Section 5.5 and the assumption that $\frac{a+bi}{c} = \frac{a}{c} + \frac{bi}{c}$. For example, to find $\frac{2+3i}{4-i}$, we proceed as follows:

The conjugate of $a+bi$ is $a-bi$.

$$\frac{2+3i}{4-i} = \frac{(2+3i)(4+i)}{(4-i)(4+i)}$$

Multiplying the numerator and denominator by the conjugate of $4-i$, that is, $4+i$.

$$= \frac{8 + 2i + 12i + 3i^2}{16 + 4i - 4i - i^2}$$

$$= \frac{8 + 2i + 12i - 3}{16 + 4i - 4i + 1}$$

$$= \frac{5 + 14i}{17}$$

$$= \frac{5}{17} + \frac{14i}{17}$$

In general,

$$\frac{a + bi}{c + di} = \frac{(a + bi)(c - di)}{(c + di)(c - di)}$$

$$= \frac{ac - adi + bci - bdi^2}{c^2 - cdi + cdi - d^2i^2}$$

$$= \frac{ac - adi + bci + bd}{c^2 + d^2}$$

$$= \frac{(ac + bd) + (bc - ad)i}{c^2 + d^2}$$

$$= \frac{ac + bd}{c^2 + d^2} + \frac{(bc - ad)i}{c^2 + d^2}$$

This fact is stated in the next definition.

DEFINITION 5.11

For a, b, c, and d real numbers (c and d not both 0),

$$\frac{a + bi}{c + di} = \frac{ac + bd}{c^2 + d^2} + \frac{(bc - ad)i}{c^2 + d^2}$$

Definition 5.11 gives the quotient of two complex numbers as another complex number. However, it is not necessary to memorize the formula in this definition; it is better to use the procedure of multiplying numerator and denominator by the conjugate of the denominator as we did in leading up to the definition. The important fact to know is that $(c + di)(c - di) = c^2 + d^2$; that is, the product of a complex number and its conjugate is a real number, the sum of the squares of the two real-number components. We use this fact in the next example.

Recall that the conjugate of $a + bi$ is $a - bi$.

Example 4

Find:

(a) $\dfrac{5+4i}{3+2i}$

(b) $\dfrac{2-4i}{5-3i}$

Solution

(a) $\dfrac{5+4i}{3+2i}=\dfrac{(5+4i)(3-2i)}{(3+2i)(3-2i)}$ Rationalizing the denominator

$=\dfrac{15-10i+12i-8i^2}{3^2+2^2}$ Since by Definition 5.11 the denominator is $c^2+d^2=3^2+2^2$.

$=\dfrac{15-10i+12i+8}{13}$

$=\dfrac{23}{13}+\dfrac{2i}{13}$

(b) $\dfrac{2-4i}{5-3i}=\dfrac{(2-4i)(5+3i)}{(5-3i)(5+3i)}$ Rationalizing the denominator

$=\dfrac{10+6i-20i-12i^2}{5^2+3^2}$

$=\dfrac{10+6i-20i+12}{34}$

$=\dfrac{22}{34}-\dfrac{14i}{34}$

$=\dfrac{11}{17}-\dfrac{7i}{17}$

PROGRESS TEST

1. When written in the form bi, $\sqrt{-16}$ equals ______.
2. $(-2+3i)+(5-7i)$ equals ______.
3. $(5+2i)-(-2-3i)$ equals ______.
4. $(2-3i)(4+5i)$ equals ______.
5. $\dfrac{4+i}{-2+3i}$ equals ______.

PROGRESS TEST ANSWERS

1. $4i$
2. $3-4i$
3. $7+5i$
4. $23-2i$
5. $\dfrac{-5}{13}-\dfrac{14i}{13}$

EXERCISE 5.5

Write each expression in the form a + bi, when a and b are real numbers.

Illustration: (a) $i(-2-3i)$
(b) $(3-i)(2-3i)$
(c) $\sqrt{-2}(2-\sqrt{-2})$

Solution: (a)
$$i(-2-3i) = -2i - 3i^2 = -2i + 3 = 3 - 2i$$

(b)
$$(3-i)(2-3i) = 6 - 9i - 2i + 3i^2 = 6 - 9i - 2i - 3 = 3 - 11i$$

(c) Since $\sqrt{-2} = \sqrt{-1}\,\sqrt{2} = \sqrt{2}i$,

$$\sqrt{-2}\,(2-\sqrt{-2}) = \sqrt{2}i(2-\sqrt{2}i)$$
$$= 2\sqrt{2}i - \sqrt{2} \bullet \sqrt{2} \bullet i^2$$
$$= 2\sqrt{2}i - 2i^2$$
$$= 2\sqrt{2}i + 2$$
$$= 2 + 2\sqrt{2}i$$

Note that *before* multiplying, $\sqrt{-2}$ was written as $\sqrt{2}i$; otherwise, we would have obtained an *incorrect* result.

1. $\sqrt{-25}$

2. $\sqrt{-81}$

3. $\sqrt{-50}$

4. $\sqrt{-98}$

5. $4\sqrt{-72}$

6. $3\sqrt{-200}$

7. $-3\sqrt{-32}$

8. $-5\sqrt{-64}$

9. $4\sqrt{-28} + 3$

10. $7\sqrt{-18} + 5$

11. $(4+i) + (2+3i)$

12. $(7+3i) + (2+i)$

13. $(3-2i) - (5+4i)$

14. $(4-5i) - (2+3i)$

15. $(-3 - 5i) + (-2 - i)$

16. $(-7 - 3i) + (-2 - i)$

17. $(3 + \sqrt{-4}) - (5 - \sqrt{-9})$

18. $(-2 - \sqrt{-16}) - (3 - \sqrt{-25})$

19. $(-5 + \sqrt{-1}) + (-2 + 3\sqrt{-1})$

20. $(-3 + 2\sqrt{-1}) + (-4 + 5\sqrt{-1})$

21. $(3 - 4i) + (5 + 3i)$

22. $(3 - 7i) + (3 + 4i)$

23. $(4 + \sqrt{-9}) + (6 + \sqrt{-4})$

24. $(-3 - \sqrt{-25}) + (5 - \sqrt{-16})$

25. $(2 - \sqrt{-2}) - (5 + \sqrt{-2})$

26. $(3 + \sqrt{-50}) - (7 + \sqrt{-2})$

27. $(-5 - \sqrt{-2}) - (-4 - \sqrt{-18})$

28. $(-8 - \sqrt{-125}) - (-2 - \sqrt{-5})$

29. $(-4 + \sqrt{-20}) + (-3 + \sqrt{-5})$

30. $(-7 + \sqrt{-24}) + (-3 + \sqrt{-6})$

31. $3(4 + 2i)$

32. $5(4 + 3i)$

33. $-4(3 - 5i)$

34. $-3(7 - 4i)$

35. $\sqrt{-4}(3 + 2i)$

36. $\sqrt{-9}(2 + 5i)$

37. $\sqrt{-3}(3+\sqrt{-3})$

38. $\sqrt{-5}(2-\sqrt{-5})$

39. $3i(3+2i)$

40. $7i(4+3i)$

41. $4i(3-7i)$

42. $-5i(2-3i)$

43. $-\sqrt{-16}(-5-\sqrt{-25})$

44. $-\sqrt{-25}(-3-\sqrt{-9})$

45. $(3+i)(2+3i)$

46. $(2+3i)(4+5i)$

47. $(3-2i)(3+2i)$

48. $(4-3i)(4+3i)$

49. $(3+2\sqrt{-4})(4-\sqrt{-9})$

50. $(-3+3\sqrt{-9})(-2+5\sqrt{-4})$

51. $(2+3\sqrt{-3})(2-3\sqrt{-3})$

52. $(4+2\sqrt{-5})(4-2\sqrt{-5})$

53. $\frac{3}{i}$

54. $\frac{5}{i}$

55. $\frac{6}{-i}$

56. $\frac{3}{-2i}$

57. $\frac{i}{1+2i}$

58. $\frac{2i}{1+3i}$

59. $\frac{3i}{1-2i}$

60. $\frac{4i}{2-3i}$

61. $\frac{3+4i}{1-2i}$

62. $\frac{3+5i}{1-3i}$

63. $\frac{4+3i}{2+3i}$

64. $\frac{5+4i}{3+2i}$

65. $\dfrac{3}{\sqrt{-4}}$

66. $\dfrac{-4}{\sqrt{-9}}$

67. $\dfrac{3+\sqrt{-5}}{4+\sqrt{-2}}$

68. $\dfrac{2+\sqrt{-2}}{1+\sqrt{-3}}$

69. $\dfrac{-1-\sqrt{-2}}{-3-\sqrt{-3}}$

70. $\dfrac{-1-\sqrt{-3}}{-2-\sqrt{-2}}$

SELF-TEST–CHAPTER 5

1. Write using positive exponents, with each variable occurring once.

    ______ a. $a^4 \bullet (b^3a^5)$

    ______ b. $x^3 \bullet x^{-5} \bullet x^1$

---

2. Write in simplest form using positive exponents.

    ______ a. $(x^3y^2)^4$

    ______ b. $\left(\frac{y^2}{x^3}\right)^5$

---

3. Write in simplest form using positive exponents.

    ______ a. $\frac{(-x)^3(-x^4)^2}{(x^2)^4}$

    ______ b. $\frac{y^{-2}}{z^{-3}}$

---

4. Evaluate (if possible).

    ______ a. $\sqrt[3]{-8}$

    ______ b. $(16)^{-3/4}$

---

5. Simplify:

    ______ a. $\sqrt[4]{x^6y^8}$

    ______ b. $\sqrt{81a^4b^6}$

---

6. Rationalize the denominator.

______ a. $\dfrac{2}{\sqrt{x} - \sqrt{y}}$

______ b. $\dfrac{1 + 2i}{1 - i}$

---

7. Write in simplest form using positive exponents.

______ a. $\left(\dfrac{a^9}{b^6}\right)^{2/3}$

______ b. $x^{2/5}(x^{1/5} + x^{3/5})$

---

8. Simplify:

______ a. $8^{1/3} + 4(\sqrt{2} + 1)$

______ b. $-\sqrt{4} + 7\sqrt{3} - 4\sqrt{3}$

---

9. Simplify each product.

______ a. $x^{1/2}(x^{3/4})$

______ b. $x^{-1/2}(x^{-1/3})$

---

10. Write each expression in the form a + bi

______ a. $(-4 + 3i) + (5 - 2i)$

______ b. $(5 - 2i) - (2 - 5i)$

______ c. $i(3 - 2i)$

______ d. $(1 + i)(2 - i)$

______ e. $\dfrac{3 + i}{-2 + i}$

CHAPTER 6

HISTORICAL NOTE

Problems leading to quadratic equations have interested men for almost 4000 years. Babylonian tablets that have been deciphered only in recent years exhibit problems like this: "Two squares have a total area of 1000. The side of one square is 10 less than $\frac{2}{3}$ of the side of the other square. What are the sides of the squares?"

There seems to be definite evidence that the Babylonians were already familiar with the quadratic formula that you will study in this chapter. They also knew the relationship $a^2 + b^2 = c^2$ for a right triangle with legs a and b and hypotenuse c. In fact, a remarkable tablet (Item 322 in the G. A. Plimpton collection at Columbia University), which was deciphered only in 1945, indicates that the Babylonians knew how to find integer solutions of the equation $x^2 + y^2 = z^2$. This can be done by choosing two whole numbers, u and v, and then taking $a = 2uv$, $b = u^2 - v^2$, and $c = u^2 + v^2$. For instance, $u = 2$, $v = 1$ gives $a = 4$, $b = 3$, $c = 5$, the famous 3, 4, 5 right triangle that is shown on the facing page.

Although the right triangle relationship was known to the Babylonians and perhaps to the ancient Egyptians, the first general proof of the relationship seems to have been given by the Greek mathematician Pythagoras, who lived during the sixth century B.C. This is the reason for the name Pythagorean Theorem, a theorem that gives rise to many problems involving quadratic equations, the simplest such problem being to find the third side of a right triangle when the other two sides are known. You will study the equation for this type of problem in the exercises for Section 6.6.

QUADRATIC EQUATIONS AND INEQUALITIES

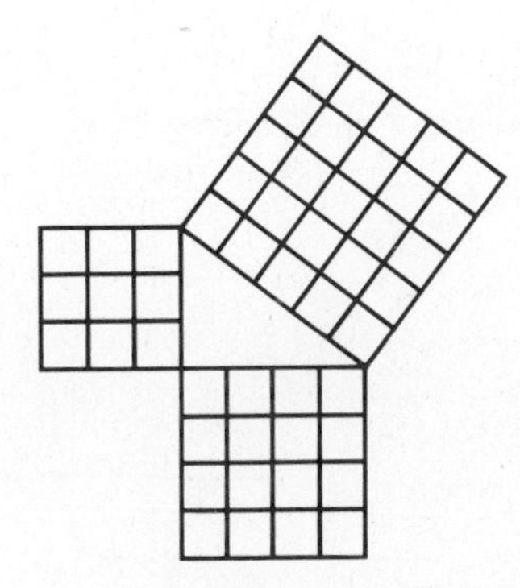

The picture shows an illustration of Pythagoras' Theorem. If a right triangle has sides that are 3 and 4 units long, the longest side, the hypotenuse, is 5 units long. Reprinted by permission of Harper & Row, Publishers, Inc.

Pythagoras, mystic, mathematician, investigator of nature to the best of his self-hobbled ability, "one tenth of him genius, nine-tenths sheer fudge." (E. T. Bell) Courtesy of the New York Public Library.

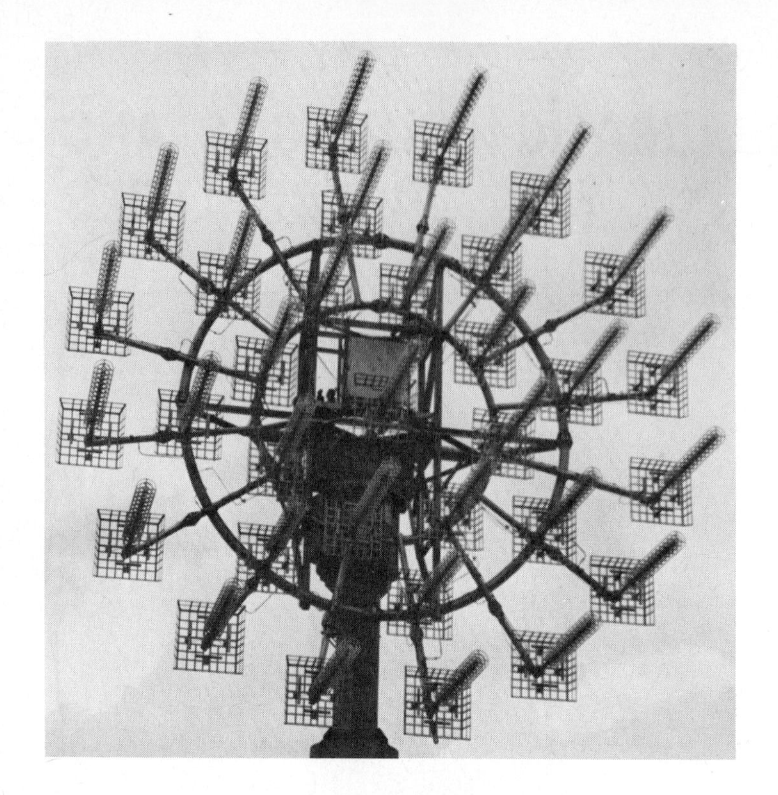

NASA Photo 62-Tiros 5-4.

OBJECTIVES 6.1

After studying this section, the reader should be able to:

1. **Solve quadratic equations of the form $x^2 = a$.**
2. **Solve quadratic equations of the form $ax^2 + b = 0$.**
3. **Solve quadratic equations which can be written in the form $(ax + b)^2 = c$, $c > 0$.**

6.1 SOLUTION OF QUADRATIC EQUATIONS OF THE FORM $(x + a)^2 = b$

In Chapter 2, we studied methods of solving *linear equations*, that is, equations in which the variable involved has an exponent of 1 (the first power). We are now ready to discuss equations containing the *second* (but no higher) power of the unknown.

$ax^2 + bx + c = 0$ is a quadratic equation written in standard form.

Such equations are called *second-degree*, or *quadratic*, equations, and are written in *standard form* as

$$ax^2 + bx + c = 0 \quad (a, b, c \text{ real numbers}, a \neq 0)$$

The procedure used to solve these equations is similar to that employed in Chapter 2, and consists of applying certain transformations to the given equations to obtain equivalent equations whose solution set is evident. For example, the antenna shown in the photo at the beginning of this section is designed to receive weather pictures from orbiting satellites. The area of each of the square "grids" on the antenna must be very precise so that the transmission can be received. If it is known that the area of one of these squares is 36 square feet, can we find the dimensions of each grid?

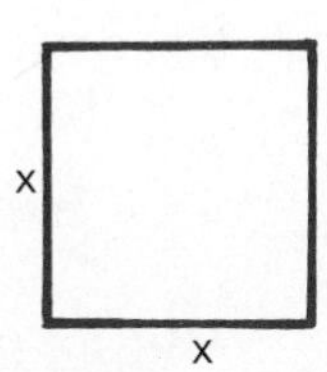

We know the area is 36 square feet. We also know that the area of a square (as in the diagram) is $x \cdot x = x^2$. In this case, we have the equation $x^2 = 36$. The equation $x^2 = 36$ can be solved by *extracting the square roots* of 36. Since a nonzero number has two square roots, we get two solutions, $x = \sqrt{36} = 6$ and $x = -\sqrt{36} = -6$; the solution set of $x^2 = 36$ is $\{6, -6\}$. The procedure for solving the equation $x^2 = 36$ is usually shortened to

Of course, if x represents a dimension of the antenna grid, the answer -6 is not acceptable, because -6 is negative.

$$x^2 = 36$$
$$x = \pm\sqrt{36} = \pm 6$$

$x = \pm 6$ means that $x = 6$ or $x = -6$.

Similarly, the solution of the equation

$$x^2 = 25$$

is $x = \pm\sqrt{25} = \pm 5$

Check: $(5)^2 = 25$
$(-5)^2 = 25$
Thus, 5 and -5 are solutions of $x^2 = 25$.

and the solution set is $\{5, -5\}$. In general, the solution set of the equation $x^2 = a$ is $\{\sqrt{a}, -\sqrt{a}\}$.

To solve the equation $x^2 - 4 = 0$, we must first write the equation in the form $x^2 = a$. This can be done by adding 4 to both members, obtaining the equivalent equation $x^2 = 4$. The procedure is as follows:

$x^2 - 4 = 0$

$x^2 = 4$ Adding 4 to both members

$x = \pm\sqrt{4} = \pm 2$ Extracting roots

Thus, the solution set of $x^2 - 4 = 0$ is $\{2, -2\}$.

Check: $(\pm 2)^2 - 4 = 4 - 4 = 0$
Thus, 2 and -2 are solutions of $x^2 - 4 = 0$.

Example 1 Find the solution set of:

(a) $x^2 = 16$
(b) $x^2 - 9 = 0$

Solution (a) $x^2 = 16$

$x = \pm\sqrt{16} = \pm 4$

The solution set is $\{4, -4\}$, which can be easily verified since $(4)^2 = 16$ and $(-4)^2 = 16$.

(b) $x^2 - 9 = 0$

$x^2 = 9$ Adding 9 to both members.

$x = \pm\sqrt{9} = \pm 3$

The solution set is $\{3, -3\}$, which can be easily verified since $(3)^2 - 9 = 0$ and $(-3)^2 - 9 = 0$.

In the preceding examples, only one transformation was needed to obtain an equivalent equation of the form $x^2 = a$. In some cases, more than one transformation may be needed. For example, to solve the equation $9x^2 - 16 = 0$, we proceed as follows.

$9x^2 - 16 = 0$

$9x^2 = 16$ Adding 16 to both members

$x^2 = \frac{16}{9}$ Dividing both members by 9

$x = \pm\sqrt{\frac{16}{9}} = \pm\frac{4}{3}$

Check this. Thus, the solution set of $9x^2 - 16 = 0$ is $\left\{\frac{4}{3}, \frac{-4}{3}\right\}$.

Example 2 Find the solution set of

(a) $4x^2 + 9 = 0$
(b) $3x^2 - 24 = 0$
(c) $5x^2 - 4 = 0$

Solution

(a) $4x^2 + 9 = 0$

$4x^2 = -9$ Adding -9 to both members

$x^2 = \frac{-9}{4}$ Dividing both members by 4

$x = \pm\sqrt{\frac{-9}{4}} = \pm\frac{3i}{2}$ Extracting roots

The solution set is $\left\{\frac{3i}{2}, \frac{-3i}{2}\right\}$.

(b) $3x^2 - 24 = 0$

$3x^2 = 24$ Adding 24 to both members

$x^2 = 8$ Dividing both members by 3

$x = \pm\sqrt{8}$

$x = \pm 2\sqrt{2}$ Since $\sqrt{8} = \sqrt{4}\sqrt{2} = 2\sqrt{2}$

The solution set is $\{2\sqrt{2}, -2\sqrt{2}\}$. Note that the answer is given in the *simplified* form $\pm 2\sqrt{2}$ instead of $\pm\sqrt{8}$.

(c) $5x^2 - 4 = 0$

$5x^2 = 4$ Adding 4 to both members

$x^2 = \frac{4}{5}$ Dividing both members by 5

$x = \pm\frac{4}{5} = \pm\frac{2}{\sqrt{5}}$

$x = \pm\frac{2\sqrt{5}}{5}$ Since $\frac{2}{\sqrt{5}} = \frac{2}{\sqrt{5}} \cdot \frac{\sqrt{5}}{\sqrt{5}} = \frac{2\sqrt{5}}{5}$

The solution set is $\left\{\frac{2\sqrt{5}}{5}, \frac{-2\sqrt{5}}{5}\right\}$. Note that the answer is given with a *rationalized* denominator.

The method used to solve equations of the form $ax^2 + c = 0$ can also be used in solving equations of the form $(x + a)^2 = b$. Thus, to solve the equation $(x + 3)^2 = 16$, we proceed as follows:

$(x + 3)^2 = 16$

$x + 3 = \pm\sqrt{16} = \pm 4$ Extracting roots

$x + 3 = 4$ or $x + 3 = -4$

$x = 1$ or $x = -7$ Solving $x + 3 = 4$ and $x + 3 = -4$

Thus, the solution set of $(x + 3)^2 = 16$ is $\{1, -7\}$.

To solve the equation $(x - 2)^2 - 9 = 0$, we add 9 to both members and then extract roots, obtaining

$$\begin{aligned}(x - 2)^2 - 9 &= 0\\ (x - 2)^2 &= 9\\ x - 2 &= \pm\sqrt{9} = \pm 3\\ x - 2 &= 3 \text{ or } x - 2 = -3\\ x &= 5 \text{ or } x = -1\end{aligned}$$

Thus, the solution set of $(x - 2)^2 - 9 = 0$ is $\{5, -1\}$.

Example 3

Find the solution set of:
(a) $(x - 1)^2 = 27$
(b) $3(x + 2)^2 + 25 = 0$

Solution

(a) $(x - 1)^2 = 27$

$$\begin{aligned}x - 1 &= \pm\sqrt{27} = \pm 3\sqrt{3}\\ x - 1 &= 3\sqrt{3} \quad \text{or} \quad x - 1 = -3\sqrt{3}\\ x &= 1 + 3\sqrt{3} \quad \text{or} \quad x = 1 - 3\sqrt{3}\end{aligned}$$

The solution set is $\{1 + 3\sqrt{3}, 1 - 3\sqrt{3}\}$. This can be verified by substituting $1 + 3\sqrt{3}$ and $1 - 3\sqrt{3}$ in the equation $(x - 1)^2 = 27$. Since $[(1 + 3\sqrt{3}) - 1]^2 = (3\sqrt{3})^2 = 3^2 \bullet 3 = 27$ and $[(1 - 3\sqrt{3}) - 1]^2 = (-3\sqrt{3})^2 = 3^2 \bullet 3 = 27$, the result is correct.

(b) $3(x + 2)^2 + 25 = 0$

$3(x + 2)^2 = -25$ — Subtracting 25 from both members

$(x - 2)^2 = \dfrac{-25}{3}$ — Dividing both members by 3

$x - 2 = \pm\sqrt{\dfrac{-25}{3}}$ — Extracting roots and rationalizing the denominator

$= \pm\dfrac{5i}{\sqrt{3}}$

$= \pm\dfrac{5\sqrt{3}i}{3}$

$x - 2 = \dfrac{5\sqrt{3}i}{3}$ or

$$x - 2 = \frac{-5\sqrt{3}i}{3}$$

$$x = 2 + \frac{5\sqrt{3}i}{3} \quad \text{or}$$

$$x = 2 - \frac{5\sqrt{3}i}{3}$$

Solving the equations

$$x - 2 = \frac{5\sqrt{3}i}{3} \text{ and}$$

$$x - 2 = \frac{-5\sqrt{3}i}{3}.$$

The solution set of $3(x + 2)^2 - 25 = 0$ is $\left\{2 + \frac{5\sqrt{3}i}{3}, 2 - \frac{5\sqrt{3}i}{3}\right\}$. The verification is left to the student.

PROGRESS TEST

1. The solution set of $x^2 = 49$ is ______________.
2. The solution set of $9x^2 + 4 = 0$ is ______________.
3. The solution set of $6x^2 - 48 = 0$ is ______________.
4. The solution set of $(x + 1)^2 = 9$ is ______________.
5. The solution set of $(x - 2)^2 = 18$ is ______________.
6. The solution set of $5(x - 1)^2 - 36 = 0$ is ______________.

EXERCISE 6.1

In Problems 1 through 20, find the solution set of the given equation.

Illustration: (a) $x^2 - 49 = 0$
(b) $7x^2 + 16 = 0$

Solution: (a) $x^2 - 49 = 0$

$x^2 = 49$ Adding 49 to both members

$x = \pm\sqrt{49}$ Extracting roots

$x = \pm 7$

The solution set is $\{7, -7\}$.

PROGRESS TEST ANSWERS

1. $\{7, -7\}$
2. $\left\{\frac{2i}{3}, \frac{-2i}{3}\right\}$
3. $\{2\sqrt{2}, -2\sqrt{2}\}$
4. $\{2, -4\}$
5. $\{2 + 3\sqrt{2}, 2 - 3\sqrt{2}\}$
6. $\left\{1 + \frac{6\sqrt{5}}{5}, 1 - \frac{6\sqrt{5}}{5}\right\}$

(b) $7x^2 + 16 = 0$

$7x^2 = -16$ Subtracting 16 from both members

$x^2 = \frac{-16}{7}$ Dividing both members by 7

$x = \pm\sqrt{\frac{-16}{7}}$

$= \pm\frac{4i}{\sqrt{7}}$ Since $\sqrt{-16} = 4i$

$= \pm\frac{4\sqrt{7}i}{7}$ Rationalizing the denominator

The solution set is $\left\{\frac{4\sqrt{7}i}{7}, \frac{-4\sqrt{7}i}{7}\right\}$.

1. $x^2 = 64$

2. $x^2 = 81$

3. $x^2 = -121$

4. $x^2 = -144$

5. $x^2 - 169 = 0$

6. $x^2 - 100 = 0$

7. $x^2 + 4 = 0$

8. $x^2 + 25 = 0$

9. $36x^2 - 49 = 0$

10. $36x^2 - 81 = 0$

11. $4x^2 + 81 = 0$

12. $9x^2 + 64 = 0$

13. $3x^2 - 25 = 0$

14. $5x^2 - 16 = 0$

15. $5x^2 + 36 = 0$

16. $11x^2 + 49 = 0$

17. $3x^2 - 100 = 0$

18. $4x^2 - 13 = 0$

19. $13x^2 + 81 = 0$

20. $11x^2 + 4 = 0$

In Problems 21 through 40, find the solution set of the given equation.

Illustration: (a) $(x + 3)^2 = -4$
(b) $(x + 8)^2 + 25 = 0$

Solution:

(a) $(x + 3)^2 = -4$

$(x + 3) = \pm\sqrt{-4} = \pm 2i$

$x + 3 = 2i$ or $x + 3 = -2i$
$x = -3 + 2i$ or $x = -3 - 2i$

The solution set is $\{-3 + 2i, -3 - 2i\}$.

(b) $(x + 8)^2 + 25 = 0$
$(x + 8)^2 = -25$ Subtracting 25 from both members

$x + 8 = \pm\sqrt{-25} = \pm 5i$
$x + 8 = 5i$ or $x + 8 = -5i$
$x = -8 + 5i$ or $x = -8 - 5i$

The solution set is $\{-8 + 5i, -8 - 5i\}$.

21. $(x + 5)^2 = 4$

22. $(x + 3)^2 = 9$

23. $(x + 2)^2 = -25$

24. $(x + 1)^2 = -16$

25. $(x - 6)^2 = 18$

26. $(x - 2)^2 = 50$

27. $(x - 1)^2 = -28$

28. $(x - 3)^2 = -32$

29. $(x - 1)^2 - 50 = 0$

30. $(x - 2)^2 - 18 = 0$

31. $(x - 5)^2 - 32 = 0$

32. $(x - 2)^2 - 4 = 0$

33. $(x - 9)^2 + 64 = 0$

34. $(x - 2)^2 + 25 = 0$

35. $3(x + 1)^2 - 96 = 0$

36. $3(x + 5)^2 - 72 = 0$

37. $7(x - 2)^2 - 350 = 0$

38. $3(x - 1)^2 - 54 = 0$

39. $7(x - 5)^2 + 189 = 0$

40. $5(x - 3)^2 + 250 = 0$

Tiger cartoon © 1974 by King Features Syndicate, Inc.

OBJECTIVES 6.2

After studying this section, the reader should be able to solve quadratic equations by the method of factoring.

6.2 SOLVING QUADRATIC EQUATIONS BY FACTORING

In the preceding section we solved the equation $x^2 - 4 = 0$ by adding 4 to both members and extracting roots. The solution of the equation $x^2 - 4 = 0$ can also be obtained by the method of *factoring.*

As you can see from the Tiger cartoon, 0 multiplied by anything remains 0. It can also be shown that if the product of two numbers is 0, at least one of the numbers is 0. For suppose $a \cdot b = 0$ and $a \neq 0$. Then, dividing both members by a, we get $b = \frac{0}{a} = 0$. Similarly, if $b \neq 0$, then $a = 0$. Thus, at least one of the numbers a or b must be 0. This is reiterated in the following theorem.

THEOREM 6.1

If $a \cdot b = 0$, then $a = 0$ or $b = 0$.

Recall that $x^2 - 2^2 = (x+2)(x-2)$. Check: $(-2)^2 - 4 = 0$ and $(2)^2 - 4 = 0$.

We can use Theorem 6.1 to solve the equation $x^2 - 4 = 0$ by first factoring the left member to get $(x + 2)(x - 2) = 0$. Next, we solve $x + 2 = 0$ or $x - 2 = 0$ (by Theorem 6.1, with $a = x + 2$, and $b = x - 2$), and $x = -2$ or 2. Similarly, if $x(x + 3) = 0$, $x = 0$ or $x + 3 = 0$—i.e., $x = 0$ or $x = -3$.

Check: $0 \cdot (0 + 3) = 0$
$(-3)(-3 + 3) = 0$

Example 1

Solve by factoring.
(a) $x^2 - 9 = 0$
(b) $x^2 + 8x = 0$

Solution

(a) $x^2 - 9 = 0$

$(x + 3)(x - 3) = 0$ Factoring

$x + 3 = 0$ or $x - 3 = 0$ By Theorem 6.1, with $a = x + 3$, and $b = x - 3$

$x = -3$ or $x = 3$ Solving the equations $x + 3 = 0$ and $x - 3 = 0$

The verification of these answers is left to the reader.

(b) $x^2 + 8x = 0$

$x(x + 8) = 0$ Factoring

$x = 0$ or $x + 8 = 0$ By Theorem 6.1, with $a = x$, and $b = x + 8$

$x = 0$ or $x = -8$ Solving the equation $x + 8 = 0$

The reader should verify that 0 and -8 are solutions of $x^2 + 8x = 0$.

The equation $x^2 + x - 6 = 0$ can also be solved by factoring. As you recall, to factor $x^2 + x - 6$, we need to find numbers whose sum is 1 and whose product is -6. These numbers are 3 and -2. Thus, we have

Note that $3 + (-2) = 1$ and that $3(-2) = -6$.

$x^2 + x - 6 = 0$

$(x + 3)(x - 2) = 0$ Factoring

$x + 3 = 0$ or $x - 2 = 0$ By Theorem 6.1

$x = -3$ or $x = 2$ Solving $x + 3 = 0$ and $x - 2 = 0$

Check:
$(-3)^2 - 3 - 6 = 9 - 3 - 6 = 0$
$(2)^2 + 2 - 6 = 4 + 2 - 6 = 0$

Example 2

Solve by factoring.
(a) $x^2 - 6x + 8 = 0$
(b) $x^2 + x = 2$

Solution

(a) We must first find numbers whose sum is -6 and whose product is 8. These numbers are -4 and -2. Thus,

$x^2 - 6x + 8 = 0$

$(x - 4)(x - 2) = 0$ Factoring

$x - 4 = 0$ or $x - 2 = 0$ By Theorem 6.1

$x = 4$ or $x = 2$ Solving $x - 4 = 0$ and $x - 2 = 0$

(b) The equation $x^2 + x = 2$ is *not* in standard form. To solve an equation by factoring, we must first write the equation in the standard form $ax^2 + bx + c = 0$. Thus, subtracting 2 from both members of $x^2 + x = 2$, we have

$x^2 + x - 2 = 0$

$(x + 2)(x - 1) = 0$ Factoring

$x + 2 = 0$ or $x - 1 = 0$ By Theorem 6.1

$x = -2$ or $x = 1$ Solving $x + 2 = 0$ and $x - 1 = 0$

In the preceding examples, the coefficient of x^2 has been 1. To solve the equation $4x^2 + 8x = 0$, we use a procedure similar to the one already discussed. Thus,

$4x^2 + 8x = 0$

$4x(x + 2) = 0$ Factoring

$4x = 0$ or $x + 2 = 0$ By Theorem 6.1

$x = 0$ or $x = -2$

Similarly, to solve the equation $6x^2 - x - 2 = 0$, we first factor $6x^2 - x - 2$ as shown:

We have to find numbers whose product is -12 and whose sum is -1, that is, -4 and 3.

$6x^2 - x - 2 = 0$

$6x^2 - 4x + 3x - 2 = 0$ Writing the middle term $-x$ as $-4x + 3x$

$2x(3x - 2) + (3x - 2) = 0$ Factoring the first and last pairs of terms

$(3x - 2)(2x + 1) = 0$ Factoring out the common factor, $3x - 2$

$3x - 2 = 0$ or $2x + 1 = 0$ By Theorem 6.1

$3x = 2$ or $2x = -1$

$x = \frac{2}{3}$ or $x = \frac{-1}{2}$

Example 3

Solve by factoring.
(a) $12x^2 + 5x - 3 = 0$
(b) $6x^2 - x = 1$

Solution

(a) $12x^2 + 5x - 3 = 0$

$12x^2 + 9x - 4x - 3 = 0$ Writing 5x using coefficients whose sum is 5 and whose product is -36, that is, $5x = 9x - 4x$

$3x(4x + 3) - (4x + 3) = 0$ Factoring the first and last pair of terms

$(4x + 3)(3x - 1) = 0$ Factoring out the common factor $4x + 3$

$4x + 3 = 0$ or $3x - 1 = 0$ By Theorem 6.1

$4x = -3$ or $3x = 1$

$x = \frac{-3}{4}$ or $x = \frac{1}{3}$

(b) $6x^2 - x = 1$ is *not* in standard form. Subtracting 1 from both members, we have

$$6x^2 - x - 1 = 0$$
$$6x^2 - 3x + 2x - 1 = 0$$
$$3x(2x - 1) + (2x - 1) = 0$$
$$(2x - 1)(3x + 1) = 0$$
$$2x - 1 = 0 \quad \text{or} \quad 3x + 1 = 0$$
$$2x = 1 \quad \text{or} \quad 3x = -1$$
$$x = \frac{1}{2} \quad \text{or} \quad x = \frac{-1}{3}$$

PROGRESS TEST

1. The solution set of $x^2 - 121 = 0$ is ______.
2. The solution set of $x(x - 1) = 0$ is ______.
3. The solution set of $x^2 - 6x + 5 = 0$ is ______.
4. The solution set of $x^2 - 3x = 10$ is ______.
5. The solution set of $10x^2 - 3x - 1 = 0$ is ______.
6. The solution set of $12x^2 + x - 1 = 0$ is ______.

EXERCISE 6.2

In Problems 1 through 34, find the solution set of the given equation.

Illustration: (a) $(x + 1)(x + 3) = 0$

(b) $\frac{x^2}{3} - \frac{x}{6} = 1$

Solution:

(a) $(x + 1)(x + 3) = 0$

$x + 1 = 0$ or $x + 3 = 0$ By Theorem 6.1

$x = -1$ or $x = -3$

The solution set is $\{-1, -3\}$.

(b) We first have to write the equation in the form $ax^2 + bx + c = 0$. Because it is usually easier to work with integer coefficients, proceed as follows:

$\frac{x^2}{3} - \frac{x}{6} = 1$

$6 \cdot \frac{x^2}{3} - 6 \cdot \frac{x}{6} = 6 \cdot 1$ Multiplying each member by 6, the LCD of $\frac{x^2}{3}$ and $\frac{x}{6}$

$2x^2 - x = 6$ Simplifying

$2x^2 - x - 6 = 0$ Subtracting 6 from both members to put the equation into standard form

$2x^2 - 4x + 3x - 6 = 0$ Writing $-x$ as $-4x + 3x$

PROGRESS TEST ANSWERS

1. $\{11, -11\}$
2. $\{0, 1\}$
3. $\{1, 5\}$
4. $\{5, -2\}$
5. $\left\{\frac{1}{2}, \frac{-1}{5}\right\}$
6. $\left\{\frac{1}{4}, \frac{-1}{3}\right\}$

$2x(x-2)+3(x-2)=0$ Factoring

$(x-2)(2x+3)=0$ Factoring $x-2$ from each term

$x-2=0$ or $2x+3=0$ By Theorem 6.1

$x=2$ or $2x=-3$

$x=2$ or $x=\frac{-3}{2}$

The solution set is $\left\{2, \frac{-3}{2}\right\}$.

1. $(x+1)(x+2)=0$
2. $(x+3)(x+4)=0$
3. $(x-1)(x+4)=0$
4. $(x+5)(x-3)=0$
5. $\left(x-\frac{1}{2}\right)\left(x-\frac{1}{3}\right)=0$
6. $\left(x-\frac{1}{4}\right)\left(x-\frac{1}{7}\right)=0$
7. $y(y-3)=0$
8. $y(y-4)=0$
9. $y^2-64=0$
10. $y^2-1=0$
11. $y^2-81=0$
12. $y^2-100=0$
13. $x^2+6x=0$
14. $x^2+2x=0$
15. $x^2-3x=0$
16. $x^2-8x=0$
17. $y^2-12x+27=0$
18. $y^2-10y+21=0$
19. $y^2+6x+5=0$
20. $y^2+3y+2=0$
21. $x^2-2x-15=0$
22. $x^2-4x-12=0$
23. $3y^2+5y+2=0$
24. $3y^2+7y+2=0$
25. $2y^2-3y+1=0$
26. $2y^2-3y-20=0$
27. $2y^2-y-1=0$
28. $2y^2-y-15=0$
29. $\frac{x^2}{12}+\frac{x}{3}-1=0$
30. $\frac{x^2}{2}-\frac{x}{12}-1=0$
31. $\frac{x^2}{3}-\frac{x}{2}=\frac{-1}{6}$
32. $\frac{x^2}{6}+\frac{x}{3}=\frac{1}{2}$
33. $\frac{x^2}{12}+\frac{x}{2}=\frac{-2}{3}$
34. $\frac{x^2}{3}+\frac{x}{3}=\frac{1}{4}$

OBJECTIVES 6.3

After studying this section, the reader should be able to:

1. **Solve quadratic equations by the method of completing the square.**
2. **Solve word problems involving quadratic equations.**

6.3 SOLUTION OF QUADRATIC EQUATIONS BY COMPLETING THE SQUARE

In the cartoon at the beginning of this section, the professor is going to have a big surprise when the square is completed. In algebra, there is a technique used to solve quadratic equations called *completing the square.*

As you recall from Chapter 2,

The coefficient of x is 2a. One half of the coefficient of x is $\frac{1}{2}$ of 2a, which is a.

$$(1) \qquad (x + a)^2 = x^2 + 2ax + a^2$$

Note that one half of the coefficient of x is a, whose square appears in the last term on the right side of Equation (1). Thus, if we wish to find a value of a so that $x^2 + 8x + a^2 = (x + a)^2$, we let $a = 4$, one half of 8, the coefficient of x. Hence, $x^2 + 8x + 4^2 = (x + 4)^2$. Similarly, to find a value of a so that $x^2 + 10x + a^2 = (x + a)^2$, we let $a = 5$, one half of 10, the coefficient of x, obtaining $x^2 + 10x + 5^2 = (x + 5)^2$.

Thus, to complete the square in the expression $x^2 + 2ax$, we must add a^2, the square of one half of the coefficient of x. This procedure can be used to write any quadratic equation in the form $(x+a)^2 = b$. The equation can then be solved by extraction of roots. For example, to solve the equation $x^2 + 6x - 16 = 0$, we proceed as follows:

$x^2 + 6x = 16$	Adding 16 to both members
$x^2 + 6x + 3^2 = 16 + 3^2$	Adding 3^2, the square of one half of the coefficient of x, to both members
$(x+3)^2 = 25$	Factoring
$x + 3 = \pm\sqrt{25} = \pm 5$	Extracting roots
$x + 3 = 5$ or $x + 3 = -5$	
$x = 2$ or $x = -8$	Solving $x + 3 = 5$ and $x + 3 = -5$

Check this.

Example 1

Solve the equation $x^2 + 10x - 24 = 0$ by completing the square.

Solution

$x^2 + 10x - 24 = 0$	
$x^2 + 10x = 24$	Adding 24 to both members
$x^2 + 10x + 5^2 = 24 + 5^2$	Adding 5^2, the square of one half of the coefficient of x, to both members
$(x+5)^2 = 49$	Factoring
$x + 5 = \pm\sqrt{49} = \pm 7$	Extracting roots
$x + 5 = 7$ or $x + 5 = -7$	
$x = 2$ or $x = -12$	Solving $x + 5 = 7$ and $x + 5 = -7$

Check this.

To complete the square in the equation $x^2 - 6x - 7 = 0$ we recall that

$$(x-a)^2 = x^2 - 2ax + a^2 \qquad (2)$$

Thus,

$x^2 - 6x - 7 = 0$	
$x^2 - 6x = 7$	Adding 7 to both members
$x^2 - 6x + 3^2 = 7 + 3^2$	Adding 3^2, the square of one half of the coefficient of x

$(x-3)^2 = 16$ Factoring

$x - 3 = \pm\sqrt{16} = \pm 4$ Extracting roots

Check this.

$x - 3 = 4$ or $x - 3 = -4$

$x = 7$ or $x = -1$ Solving $x - 3 = 4$ and $x - 3 = -4$

If the coefficient of x^2 is not 1, divide first by this coefficient.

In the preceding examples, the coefficient of x^2 has been 1. If this is not the case, we must first divide each term of the equation by the coefficient of x^2. For example, to solve the equation $3x^2 + 2x - 16 = 0$, we first divide each term by 3, obtaining:

$$x^2 + \frac{2x}{3} - \frac{16}{3} = 0$$

Then,

$x^2 + \frac{2x}{3} = \frac{16}{3}$ Adding $\frac{16}{3}$ to each member

Note that $\frac{1}{2}$ of $\frac{2}{3}$ is $\frac{1}{2} \cdot \frac{2}{3} = \frac{1}{3}$

$x^2 + \frac{2x}{3} + \left(\frac{1}{3}\right)^2 = \frac{16}{3} + \left(\frac{1}{3}\right)^2$ Adding the square of one half of $\frac{2}{3}$ to each member

$\frac{16}{3} + \left(\frac{1}{3}\right)^2 = \frac{48}{9} + \frac{1}{9} = \frac{49}{9}$

$\left(x + \frac{1}{3}\right)^2 = \frac{49}{9}$ Factoring

$x + \frac{1}{3} = \pm\sqrt{\frac{49}{9}} = \pm\frac{7}{3}$ Extracting roots

$x + \frac{1}{3} = \frac{7}{3}$ or $x + \frac{1}{3} = \frac{-7}{3}$

Check this.

$x = \frac{6}{3} = 2$ or $x = \frac{-8}{3}$ Solving $x + \frac{1}{3} = \frac{7}{3}$ and $x + \frac{1}{3} = \frac{-7}{3}$

Example 2 Solve $3x^2 + 3x - 1 = 0$ by completing the square.

Solution

$3x^2 + 3x - 1 = 0$

$x^2 + x - \frac{1}{3} = 0$ Dividing each term by 3

$x^2 + x = \frac{1}{3}$ Adding $\frac{1}{3}$ to each member

$x^2 + x + \left(\frac{1}{2}\right)^2 = \frac{1}{3} + \left(\frac{1}{2}\right)^2$ Adding the square of one half of 1 to each member

$\left(x + \frac{1}{2}\right)^2 = \frac{7}{12}$ Factoring

$x + \frac{1}{2} = \pm\sqrt{\frac{7}{12}} = \pm\frac{\sqrt{7}}{\sqrt{12}}$ Extracting roots and rationalizing the denominator

$$= \pm\frac{\sqrt{7}}{2\sqrt{3}}$$

$$= \pm\frac{\sqrt{21}}{6}$$

$$x + \frac{1}{2} = \frac{\sqrt{21}}{6} \quad \text{or} \quad x + \frac{1}{2} = -\frac{\sqrt{21}}{6}$$

$$x = \frac{-1}{2} + \frac{\sqrt{21}}{6} \quad \text{or} \quad x = \frac{-1}{2} - \frac{\sqrt{21}}{6}$$

Check this.

The method of completing the square can be used to solve problems involving projectile motion. For example, the photo on p. 322 shows Dr. Robert Goddard with the first liquid-fueled rocket. The height h (in feet) of a rocket can be expressed in the equation $h = -16t^2 + v_0t$, where v_0 is the initial velocity in feet per second and t is the time in seconds after launching.

A rocket is launched and reaches a height of 1,000 feet before the fuel is exhausted. At this time the rocket has a velocity of 800 feet per second. How many more seconds will it take for the rocket to rise another 9,600 feet? To solve this problem, we set h = 9,600 feet, $v_0 = 800$, and solve for t as follows:

$$-16t^2 + 800t = 9600$$

$t^2 - 50t = -600$ Dividing by -16

$t^2 - 50t + (25)^2 = -600 + (25)^2$ Adding $(25)^2$ to both members

Courtesy of Clark University.

$(t-25)^2 = -600 + 625 = 25$	Factoring and simplifying
$t - 25 = \pm 5$	Extracting roots
$t = 25 \pm 5$	Adding 25 to both sides
$t = 25 - 5$ or $t = 25 + 5$	
$t = 20$ or $t = 30$	

Thus, it takes the rocket 20 seconds to reach an altitude of 9600 feet. The second answer, $t = 30$, is the number of seconds for the rocket to reach the same point on its way down.

Example 3 A ball is thrown straight up in the air with an initial velocity $v_0 = 64$ feet per second. How many seconds later will the ball be at a height of 48 feet?

Solution Since $h = -16t^2 + v_0t$, we let $h = 48$ and $v_0 = 64$, obtaining

$-16t^2 + 64t = 48$	
$t^2 - 4t = -3$	Dividing each term by -16

$t^2 - 4t + 2^2 = -3 + 2^2$

$(t - 2)^2 = 1$

$t - 2 = \pm 1$

$t = 2 + 1$ or $t = 2 - 1$

$t = 3$ or $t = 1$

Thus, the ball attains a height of 48 feet after 1 second (on its way up) and then after 3 seconds (on its way down).

PROGRESS TEST

1. When solved by completing the square, the solution set of $x^2 + 2x - 3 = 0$ is __________.
2. When solved by completing the square, the solution set of $x^2 - 4x - 5 = 0$ is __________.
3. When solved by completing the square, the solution set of $2x^2 + 3x - 2 = 0$ is __________.
4. When solved by completing the square, the solution set of $2x^2 - x - 15 = 0$ is __________.
5. A ball is thrown straight up with an initial velocity of 128 feet per second. The ball will be at a height of 192 feet in ______ and in ______ seconds. (Hint: Use $h = -16t^2 + v_0t$ for the height of the ball after t seconds.)

EXERCISE 6.3

In Problems 1 through 20, find the solution set by completing the square.

Illustration: (a) $x^2 + 4x - 5 = 0$

(b) $2x^2 + 3x + \frac{5}{4} = 0$

Solution: (a) $x^2 + 4x - 5 = 0$

$x^2 + 4x = 5$

$x^2 + 4x + 2^2 = 5 + 2^2$

PROGRESS TEST ANSWERS

1. $\{1, -3\}$
2. $\{5, -1\}$
3. $\left\{\frac{1}{2}, -2\right\}$
4. $\left\{3, \frac{-5}{2}\right\}$
5. 2, 6

$$(x+2)^2 = 9$$
$$x+2 = \pm\sqrt{9} = \pm 3$$
$$x+2 = 3 \text{ or } x+2 = -3$$

The solution set is {1, −5}.

(b) $2x^2 + 3x + \frac{5}{4} = 0$

$$x^2 + \frac{3}{2}x + \frac{5}{8} = 0$$

$$x^2 + \frac{3x}{2} = \frac{-5}{8}$$

$$x^2 + \frac{3x}{2} + \left(\frac{3}{4}\right)^2 = \frac{-5}{8} + \left(\frac{3}{4}\right)^2$$

$$\left(x + \frac{3}{4}\right)^2 = \frac{-1}{16}$$

$$x + \frac{3}{4} = \pm\sqrt{\frac{-1}{16}} = \pm\frac{i}{4}$$

$$x + \frac{3}{4} = \frac{i}{4} \text{ or } x + \frac{3}{4} = \frac{-i}{4}$$

$$x = \frac{-3}{4} + \frac{i}{4} \text{ or } x = \frac{-3}{4} - \frac{i}{4}$$

The solution set is

$$\left\{\frac{-3}{4} + \frac{i}{4}, \frac{-3}{4} - \frac{i}{4}\right\}.$$

1. $x^2 + 6x + 5 = 0$

2. $x^2 + 4x + 3 = 0$

3. $x^2 + 8x + 15 = 0$

4. $x^2 + 8x + 7 = 0$

5. $x^2 + 6x + 10 = 0$

6. $x^2 + 12x + 37 = 0$

7. $x^2 - 10x + 24 = 0$

8. $x^2 + 12x - 28 = 0$

9. $x^2 - 10x + 21 = 0$

10. $x^2 - 2x - 143 = 0$

11. $x^2 - 8x + 17 = 0$

12. $x^2 - 14x + 58 = 0$

13. $2x^2 + 4x + 3 = 0$

14. $2x^2 + 7x + 6 = 0$

15. $3x^2 + 6x + 78 = 0$

16. $9x^2 + 6x + 2 = 0$

17. $25y^2 - 25y + 6 = 0$

18. $4y^2 - 16y + 15 = 0$

19. $4y^2 - 4y + 5 = 0$

20. $9x^2 - 12x + 13 = 0$

In Problems 21 through 25, use the formula $h = -16t^2 + v_0t$, where h is the height in feet, v_0 is the initial velocity in feet per second, and t is the time in seconds.

Illustration: A ball is batted up with an initial velocity of 160 feet per second. Find the maximum height of the ball.

Solution: If we find the time it takes the ball to return to the ground ($h = 0$) and divide this time in half, we can find how long it took the ball to reach its highest point. We then substitute this value of t in the equation $h = -16t^2 + v_0t$ and find the maximum height.

For $h = 0$, $v_0 = 160$

$-16t^2 + 160t = 0$

$t^2 - 10t = 0$ Dividing each term by -16

$t^2 - 10t + 5^2 = 5^2$ Adding 5^2 to both members

$(t - 5)^2 = 25$

$t - 5 = \pm 5$

$t = 5 + 5 = 10$

or $t = 5 - 5 = 0$

Thus, the ball returns to the ground after 10 seconds. When $t = \frac{10}{2} = 5$, the ball is at its highest point (see diagram). At this point, the height of the ball is

$$\begin{aligned} h &= -16(5)^2 + 160(5) \\ &= -16 \cdot 25 + 800 \\ &= -400 + 800 = 400 \text{ feet} \end{aligned}$$

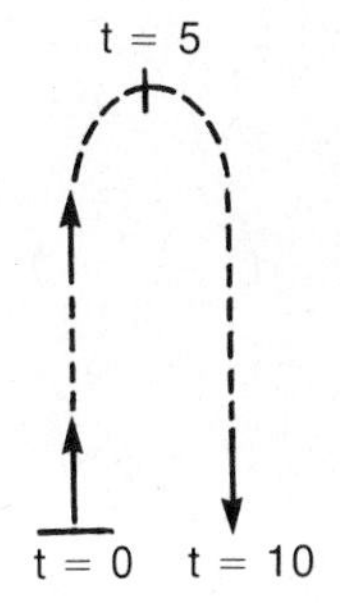

21. A ball is thrown straight up with an initial velocity of 96 ft./sec. In how many seconds will the ball be 80 feet above the ground?

22. What is the maximum height the ball in Problem 21 will attain?

23. A ball is thrown straight up with an initial velocity of 64 ft./sec. How long will it take the ball to return to the ground?

24. How long will it take the ball in Problem 23 to attain a height of 48 feet on its way *down*?

25. Can a ball thrown up with an initial velocity of 64 ft./sec. attain a height of 96 feet? Explain.

OBJECTIVES 6.4

After studying this section, the reader should be able to:

1. **Solve quadratic equations by using the quadratic formula.**
2. **Find the discriminant of a quadratic equation and determine the character of the roots of the equation.**

6.4 THE QUADRATIC FORMULA

In the cartoon at the beginning of this section, Nancy is waiting for her interest to *compound.* If Nancy had invested P dollars at a compound rate of interest r, her patience would be rewarded, say at the end of the second year, by receiving I dollars of interest. The amount of interest for the second year is expressed in the formula

$$I = Pr(1 + r)$$

Now let us assume that at the end of the two years she received interest of \$1.05 on an investment of \$20. Can we find the rate r at which her money was invested? If we substitute I = \$1.05 and P = \$20 in the formula $I = Pr(1 + r)$, we get

$$1.05 = 20r(1 + r) = 20r + 20r^2$$

In the standard form,

$$20r^2 + 20r - 1.05 = 0, \text{ or}$$

$$(1) \qquad 4r^2 + 4r - 0.21 = 0 \qquad \text{Dividing each term by 5}$$

If we wanted to solve Equation (1) by completing the square, we first would have to divide each term by 4, obtaining $\frac{0.21}{4}$ as the new last term. Instead of doing this, we shall solve the *general quadratic equation*, $ax^2 + bx + c = 0$, $a \neq 0$, by completing the square, and will then substitute the values $a = 4$, $b = 4$, and $c = 0.21$ in the solution set. Of course, the solution obtained for the equation $ax^2 + bx + c = 0$ will give us the roots or *solutions* of any quadratic equation of the form $ax^2 + bx + c = 0$ in terms of the coefficients a, b, and c. The steps needed to solve the general quadratic equation by completing the square are as follows:

$$ax^2 + bx + c = 0, \quad a \neq 0$$

$$x^2 + \frac{bx}{a} + \frac{c}{a} = 0$$ Dividing each term by a

$$x^2 + \frac{bx}{a} = \frac{-c}{a}$$ Subtracting $\frac{c}{a}$ from both members

Note that we have added the square of one-half the coefficient of x to both members. (The coefficient of x is $\frac{b}{a}$, so one-half of this is $\frac{\frac{b}{a}}{2} = \frac{b}{2a}$.)

$$x^2 + \frac{bx}{a} + \left(\frac{b}{2a}\right)^2 = \frac{-c}{a} + \left(\frac{b}{2a}\right)^2$$ Adding $\left(\frac{b}{2a}\right)^2$ to both members

$$\left(x + \frac{b}{2a}\right)^2 = \frac{b^2}{4a^2} - \frac{c}{a}$$

$$= \frac{b^2}{4a^2} - \frac{4ac}{4a^2}$$

$$= \frac{b^2 - 4ac}{4a^2}$$

$$x + \frac{b}{2a} = \pm\sqrt{\frac{b^2 - 4ac}{4a^2}}$$

$$= \pm\frac{\sqrt{b^2 - 4ac}}{2a}$$

$$x = \frac{-b}{2a} \pm \frac{\sqrt{b^2 - 4ac}}{2a}$$

$$= \frac{-b \pm \sqrt{b^2 - 4ac}}{2a}$$

Thus, the solution for *any* quadratic equation of the form $ax^2 + bx + c = 0$, $a \neq 0$, is given in the famous *quadratic formula*:

$$x = \frac{-b \pm \sqrt{b^2 - 4ac}}{2a}$$

Returning to Equation (1), $4r^2 + 4r - 0.21 = 0$, we have $a = 4$, $b = 4$, and $c = -0.21$.

$$r = \frac{-4 \pm \sqrt{4^2 - 4 \cdot 4 \cdot (-0.21)}}{2 \cdot 4}$$

$$= \frac{-4 \pm \sqrt{16 - 16 \cdot (-0.21)}}{8}$$

$$= \frac{-4 \pm \sqrt{16 + 3.36}}{8}$$

$$= \frac{-4 \pm \sqrt{19.36}}{8}$$

$\sqrt{19.36} = 4.4$, since $(4.4)^2 = 19.36$.

$$= \frac{-4 \pm 4.4}{8} = \frac{-1 \pm 1.1}{2}$$

Thus,

$$r = \frac{-1 + 1.1}{2} \text{ or } r = \frac{-1 - 1.1}{2}$$

That is,

$$r = \frac{0.1}{2} = 0.05 = 5\% \text{ or } r = \frac{-2.1}{2} = -1.05 = -105\%$$

Since r, the rate, is positive, Nancy's interest rate is 5%.

We cannot use a negative interest rate.

Example 1

Use the quadratic formula to find the solution set of
(a) $2x^2 - x - 6 = 0$
(b) $5x^2 + 4x + 1 = 0$

Solution

(a) The equation $2x^2 - x - 6 = 0$ is of the form $ax^2 + bx + c = 0$ where $a = 2$, $b = -1$, and $c = -6$. Substituting these values in the formula $x = \frac{-b \pm \sqrt{b^2 - 4ac}}{2a}$, we get

$$x = \frac{-(-1) \pm \sqrt{(-1)^2 - 4(2)(-6)}}{2 \cdot 2}$$

$$= \frac{1 \pm \sqrt{1 + 48}}{4} = \frac{1 \pm 7}{4}$$

Thus, $x = \frac{1 + 7}{4}$ or $x = \frac{1 - 7}{4}$. That is, $x = \frac{8}{4} = 2$ or $x = \frac{-6}{4} = \frac{-3}{2}$. The solution set is $\left\{2, \frac{-3}{2}\right\}$. The verification of this fact is left to the student.

(b) In the equation $5x^2 + 4x + 1 = 0$, $a = 5$, $b = 4$, $c = 1$, so

$$x = \frac{-4 \pm \sqrt{4^2 - 4(5)(1)}}{2 \cdot 5}$$

$$= \frac{-4 \pm \sqrt{16 - 20}}{10}$$

$$= \frac{-4 \pm \sqrt{-4}}{10}$$

$$= \frac{-4 \pm 2i}{10}$$

Note that $\frac{-4+2i}{10} = \frac{2(-2+i)}{10} = \frac{-2+i}{5}$.

Thus, $x = \frac{-4+2i}{10} = \frac{-2+i}{5}$ or $x = \frac{-4-2i}{10} = \frac{-2-i}{5}$. The solution set is $\left\{\frac{-2+i}{5}, \frac{-2-i}{5}\right\}$. Check:

$$5\left(\frac{-2+i}{5}\right)^2 + 4\left(\frac{-2+i}{5}\right) + 1$$

$$= 5\left(\frac{4-4i-1}{25}\right) + 4\left(\frac{-2+i}{5}\right) + 1$$

$$= \frac{3-4i}{5} + \frac{-8+4i}{5} + 1 = \left(\frac{3}{5} - \frac{8}{5} + 1\right) + \left(-\frac{4}{5} + \frac{4}{5}\right)i$$

$$= 0 + 0i = 0.$$

The verification for the other solution is left to the reader.

In part (a) of Example 1, $b^2 - 4ac$ is a *positive* number, so the solutions of the equation $2x^2 - x - 6 = 0$ are *real* numbers. In part (b), however, $b^2 - 4ac$ is negative, so the solutions of $5x^2 + 4x + 1 = 0$ are imaginary numbers. From these examples, you can see that in the quadratic formula

$$x = \frac{-b \pm \sqrt{b^2 - 4ac}}{2a},$$

1. If $b^2 - 4ac > 0$, there are two unequal *real* solutions.
2. If $b^2 - 4ac < 0$, there are two unequal *imaginary* solutions.

3. If $b^2 - 4ac = 0$, the solutions are $x = \frac{-b \pm \sqrt{0}}{2a}$, which coincide to give the one solution $x = \frac{-b}{2a}$. Such a solution is often called a *double root* or a solution of *multiplicity two.*

The number $b^2 - 4ac$ is usually called the *discriminant* of the equation $ax^2 + bx + c = 0$.

Example 2

Find the discriminant of $x^2 - \frac{x}{3} = \frac{2}{9}$ and

(a) determine the character of the roots of this equation;
(b) find these roots.

Solution

(a) Here $a = 1$, $b = \frac{-1}{3}$, and $c = -\frac{2}{9}$; the discriminant is $b^2 - 4ac = \left(\frac{-1}{3}\right)^2 - 4(1)\left(\frac{-2}{9}\right) = \frac{1}{9} + \frac{8}{9} = 1$. Thus since $b^2 - 4ac > 0$, the equation will have two unequal real roots.

(b) The solution of $x^2 - \frac{x}{3} - \frac{2}{9} = 0$ is

$$x = \frac{-b \pm \sqrt{b^2 - 4ac}}{2a} = \frac{-\left(\frac{-1}{3}\right) \pm \sqrt{1}}{2}$$

Remember that $b^2 - 4ac = 1$.

That is,

$$x = \frac{\frac{1}{3} + 1}{2} \quad \text{or} \quad x = \frac{\frac{1}{3} - 1}{2}$$

Thus,

$$x = \frac{4}{6} = \frac{2}{3} \quad \text{or} \quad x = \frac{-2}{6} = \frac{-1}{3}$$

These results are correct, since $\left(\frac{2}{3}\right)^2 - \frac{\frac{2}{3}}{3} = \frac{4}{9} - \frac{2}{9} = \frac{2}{9}$,

and $\left(\frac{-1}{3}\right)^2 - \frac{\frac{-1}{3}}{3} = \frac{1}{9} + \frac{1}{9} = \frac{2}{9}$.

PROGRESS TEST

1. If a person receives \$3.30 interest in the second year after investing \$30 at a compounded rate r, then in the formula $I = Pr(1+r)$, r would be ______ %.
2. The solution set of the equation $x^2 - 4x + 4 = 0$ is ______.
3. The solution set of the equation $5x^2 - 6x - 8 = 0$ is ______.
4. The solution set of the equation $x^2 + 6x + 13 = 0$ is ______.
5. The discriminant of $x^2 - \frac{4x}{3} = \frac{-1}{9}$ is ______.

EXERCISE 6.4

In Problems 1 through 16, use the quadratic formula to find the solution set of the given equation.

Illustration: $\frac{x^2}{12} - x = \frac{-3}{4}$

Solution: We first multiply each term by 12, the LCD of the fractions involved, and then write the equation in the form $ax^2 + bx + c = 0$, obtaining:

$$12 \cdot \frac{x^2}{12} - 12 \cdot x = 12 \cdot \frac{-3}{4}$$

$$x^2 - 12x = -9$$

$$x^2 - 12x + 9 = 0$$

Now, $a = 1$, $b = -12$, and $c = 9$, giving us

$$x = \frac{-(-12) \pm \sqrt{(-12)^2 - 4(1)(9)}}{2 \cdot 1}$$

$$= \frac{12 \pm \sqrt{108}}{2}$$

$$= \frac{12 \pm 6\sqrt{3}}{2}$$

Thus, the solution set is $\{6 + 3\sqrt{3}, 6 - 3\sqrt{3}\}$.

PROGRESS TEST ANSWERS

1. 10
2. $\{2\}$
3. $\left\{\frac{-4}{5}, 2\right\}$
4. $\{-3 + 2i, -3 - 2i\}$
5. $\frac{4}{3}$

1. $x^2 + x - 2 = 0$

2. $x^2 + 4x - 1 = 0$

3. $x^2 + 4x + 1 = 0$

4. $x^2 + 6x + 5 = 0$

5. $x^2 - 3x - 2 = 0$

6. $x^2 - 4x - 12 = 0$

7. $7y^2 - 12y + 5 = 0$

8. $7x^2 - 6x + 1 = 0$

9. $5y^2 + 8y + 5 = 0$

10. $5y^2 + 6y + 5 = 0$

11. $2y^2 + 7y + 6 = 0$

12. $2y^2 + 7y + 3 = 0$

13. $\frac{x^2}{5} - \frac{x}{2} = \frac{-3}{10}$

14. $\frac{x^2}{4} - \frac{x}{2} = -\frac{1}{8}$

15. $\frac{x^2}{7} + \frac{x}{2} = \frac{-3}{14}$

16. $\frac{x^2}{8} + \frac{x}{2} = -\frac{1}{8}$

In Problems 17 through 20, simplify the given equation, find the discriminant, and then determine the character of the roots.

Illustration: $\frac{x^2}{8} - \frac{x}{4} = \frac{1}{8}$

Solution: We first write the equation in the form

$$\frac{x^2}{8} - \frac{x}{4} - \frac{1}{8} = 0$$

Here $a = \frac{1}{8}$, $b = \frac{-1}{4}$, $c = \frac{-1}{8}$; therefore

$$b^2 - 4ac = \left(\frac{-1}{4}\right)^2 - 4\left(\frac{1}{8}\right)\left(\frac{-1}{8}\right)$$
$$= \frac{1}{16} + \frac{4}{64} = \frac{1}{8} > 0$$

Thus, there are two real and unequal roots.

17. $\frac{x^2}{2} - \frac{3x}{4} = \frac{-1}{8}$

18. $\frac{x^2}{10} - \frac{x}{5} = \frac{3}{2}$

19. $\frac{x^2}{8} + \frac{x}{4} = \frac{-1}{8}$

20. $\frac{x^2}{12} + \frac{x}{4} = \frac{-1}{3}$

In Problems 21 through 25, solve for the rate r when the interest $I = Pr(1 + r)$.

Illustration: $I = \$11$, $P = \$100$

Solution:

$$11 = 100r(1 + r)$$

$$100r^2 + 100r - 11 = 0$$

$$r = \frac{-100 \pm \sqrt{100^2 - 4(100)(-11)}}{2 \cdot 100}$$

$$= \frac{-100 \pm \sqrt{14400}}{200}$$

$$= \frac{-100 \pm 120}{200}$$

Thus, since r has to be positive,

$$r = \frac{-100 + 120}{200} = \frac{20}{200} = 10\%$$

21. $P = \$2{,}000$, $I = \$105$

22. $P = \$100$, $I = \$13.44$

23. $P = \$100$, $I = \$6.36$

24. $P = \$200$, $I = \$14.98$

25. $P = \$100$, $I = \$9.81$

OBJECTIVES 6.5

After studying this section, the reader should be able to:

1. **Solve equations involving radicals.**
2. **Solve equations involving rational expressions by finding the LCD of the expressions, multiplying by this LCD, and obtaining an equivalent quadratic equation.**

6.5 EQUATIONS THAT LEAD TO QUADRATIC EQUATIONS

The preceding cartoon illustrates the "boom" that happens when matter of mass m is converted into energy E according to Einstein's famous formula which gives the relationship between m, E, and the speed of light, c. The formula $c = \sqrt{\frac{E}{m}}$ is usually written in the form $E = mc^2$. To transform the equation $c = \sqrt{\frac{E}{m}}$ to the form $E = mc^2$, we could square both members, obtaining $c^2 = \frac{E}{m}$ or $E = mc^2$.

In algebra, the equation $\sqrt{x} = 2$ can also be solved by squaring both members and obtaining $x = 2^2 = 4$. In some cases, however, squaring both members of an equation does not result in an equivalent equation. For example, if we square both sides of the equation

(1) $\quad x = 2,$

we obtain

(2) $\quad x^2 = 4,$

which has 2 and -2 as solutions. The number -2 is a solution of Equation (2) but is *not* a solution of Equation (1). Such prospective solutions are called *extraneous solutions;* they point out the necessity of checking in the original equation. However, note that 2 is a solution of *both* equations (1) and (2). This discussion leads to the following theorem.

THEOREM 6.2

All the solutions of the equation $E = F$ (E and F algebraic expressions in x) are solutions of $E^n = F^n$ (n a positive integer).

The reader should note, however, that not all solutions of $E^n = F^n$ need be solutions of $E = F$, as we have shown in the preceding illustration.

By Theorem 6.2, all solutions of $\sqrt{x-1} = x - 7$ are solutions of $(\sqrt{x-1})^2 = (x-7)^2$. To solve $\sqrt{x-1} = x - 7$, then, we first square both members, obtaining

$$(\sqrt{x-1})^2 = (x-7)^2$$
$$x - 1 = x^2 - 14x + 49$$
$$x^2 - 15x + 50 = 0$$
$$(x-10)(x-5) = 0$$
$$x = 10 \text{ or } x = 5$$

and then check to see if we have introduced any *extraneous* roots.

Does $\sqrt{10-1} = 10 - 7$? Yes, since $\sqrt{9} = 3 = 10 - 7$.

Does $\sqrt{5-1} = 5 - 7$? No, since $\sqrt{4} = 2 \neq -2$.

Thus, the solution set of $\sqrt{x-1} = x - 7$ is $\{10\}$.

Example 1

Solve:

(a) $\sqrt{x-1} - x = -1$

(b) $\sqrt{x+1} = x - 1$.

Solution

(a) Since squaring both sides will not remove the radical, we first add x to both members in order to isolate the radical on one side. Thus, we get

$$\sqrt{x-1} = x - 1$$

$x - 1 = (x - 1)^2$	Squaring both members
$x - 1 = x^2 - 2x + 1$	Since $(x - 1)^2 = x^2 - 2x + 1$
$x^2 - 3x + 2 = 0$	Adding $-x + 1$ to both members
$(x - 2)(x - 1) = 0$	Factoring
$x = 2$ or $x = 1$	Theorem 6.1

We now check the answers in the original equation:

Does $\sqrt{2-1} - 2 = -1$? Yes, since $\sqrt{1} - 2 = -1$.

Does $\sqrt{1-1} - 1 = -1$? Yes, since $\sqrt{0} - 1 = -1$.

Thus, the solution set of $\sqrt{x-1} - x = -1$ is $\{2, 1\}$.

(b) We square both members of $\sqrt{x+1} = x - 1$, obtaining

$x + 1 = (x - 1)^2$	
$x + 1 = x^2 - 2x + 1$	
$x^2 - 3x = 0$	Adding $-x - 1$ to both members
$x(x - 3) = 0$	Factoring
$x = 0$ or $x = 3$	Theorem 6.1

Check: Does $\sqrt{0+1} = 0 - 1$? No, so 0 is not a solution.

Does $\sqrt{3+1} = 3 - 1$? Yes, since $\sqrt{4} = 2 = 3 - 1$.

Thus, the solution set of $\sqrt{x+1} = x - 1$ is $\{3\}$.

Equations involving rational fractions frequently lead to quadratic equations. However, before studying such equations, the student must understand the methods used for simplifying equations involving fractions, as well as the procedures used to obtain the LCD of two or more fractions. With this in mind, we can solve the equation

$$\frac{x-2}{x^2 - x - 6} - \frac{x}{x^2 - 4} = \frac{3}{2(x+2)}$$

$x^2 - x - 6 = (x + 2)(x - 3)$
$x^2 - 4 = (x + 2)(x - 2)$
$2(x + 2) = 2(x + 2)$

by first writing it as

$$\frac{x-2}{(x+2)(x-3)} - \frac{x}{(x+2)(x-2)} = \frac{3}{2(x+2)}$$

The LCD is $2(x-2)(x+2)(x-3)$.

and then multiplying by the LCD, $2(x-2)(x+2)(x-3)$, obtaining:

$$\begin{aligned}
2(x-2)(x-2) - 2x(x-3) &= 3(x-2)(x-3) \\
2(x^2-4x+4) - 2x^2 + 6x &= 3[x^2-5x+6] \\
\cancel{2x^2} - 8x + 8 - \cancel{2x^2} + 6x &= 3x^2 - 15x + 18 \\
3x^2 - 13x + 10 &= 0 \\
(3x-10)(x-1) &= 0 \\
x = \frac{10}{3} &\text{ or } x = 1
\end{aligned}$$

The check is left for the student to carry out.

Example 2

Solve $\dfrac{x+1}{x^2+x-2} - \dfrac{x}{x^2-1} = \dfrac{1}{2(x+2)}$.

Solution

Since $x^2+x-2 = (x+2)(x-1)$,
$x^2-1 = (x+1)(x-1)$,
and $2(x+2) = 2(x+2)$,

The LCD is $2(x+1)(x+2)(x-1)$. Multiplying each term by the LCD, we have

$$\begin{aligned}
2(x+1)(x+1) - 2x(x+2) &= (x+1)(x-1) \\
2(x^2+2x+1) - 2x^2 - 4x &= x^2 - 1 \\
\cancel{2x^2} + \cancel{4x} + 2 - \cancel{2x^2} - \cancel{4x} &= x^2 - 1 \\
x^2 - 3 &= 0 \\
x &= \pm\sqrt{3}
\end{aligned}$$

Thus, the solution set is $\{\sqrt{3}, -\sqrt{3}\}$. The check is left for the student.

PROGRESS TEST

1. The solution set of $\sqrt{x+1} = x - 5$ is ________.
2. The solution set of $\sqrt{x+1} - x = 1$ is ________.
3. The solution set of $\dfrac{6x^2 - 32}{x^2 - 4} - \dfrac{2x}{x+2} = \dfrac{3x - 4}{x - 2}$ is ________.

EXERCISE 6.5

In Problems 1 through 20, find the solution set of the given equation.

Illustration: $\sqrt{x-5} - \sqrt{x} = -1$

Solution: We first add $\sqrt{x}$ to both members, obtaining

$$\sqrt{x-5} = \sqrt{x} - 1$$

$x - 5 = (\sqrt{x} - 1)^2$ Squaring both members

$x - 5 = x - 2\sqrt{x} + 1$ Since $(\sqrt{x} - 1)^2 = x - 2\sqrt{x} + 1$

$2\sqrt{x} = 6$ Writing $2\sqrt{x}$ as left member

$\sqrt{x} = 3$ Dividing by 3

$x = 9$ Squaring both members.

Since $\sqrt{9-5} - \sqrt{9} = \sqrt{4} - \sqrt{9} = 2 - 3 = -1$, our result is correct, and the solution set is $\{9\}$.

1. $\sqrt{x+4} = x + 2$

2. $\sqrt{x+3} = x + 1$

3. $\sqrt{x+3} = x - 3$

4. $\sqrt{x+9} = x - 3$

PROGRESS TEST ANSWERS

1. $\{8\}$
2. $\{0, -1\}$
3. $\{4, -6\}$

5. $\sqrt{y+8} = 4$

6. $\sqrt{y+4} = 5$

7. $\sqrt{x+5} - x = -7$

8. $\sqrt{x+5} - x = -1$

9. $\sqrt{x-5} - x = -7$

10. $\sqrt{x-1} - x = -3$

11. $\sqrt{y+1} - y = 1$

12. $\sqrt{y-4} = 2 + \sqrt{y}$

13. $\sqrt{y+8} - \sqrt{y} = 2$

14. $\sqrt{y+5} - \sqrt{y} = 1$

15. $\sqrt{x+3} = \sqrt{x} + \sqrt{3}$

16. $\sqrt{x+5} = \sqrt{x} + \sqrt{5}$

17. $\sqrt{5x-1} + \sqrt{x+3} = 4$

18. $\sqrt{2x-1} + \sqrt{x+3} = 3$

19. $\sqrt{x-3} + \sqrt{2x+1} = 2\sqrt{x}$

20. $\sqrt{x+4} + \sqrt{3x+9} = \sqrt{x+25}$

In Problems 21 through 24, find the solution set of the given equation.

Illustration: (a) $\dfrac{4}{x^2-4} - \dfrac{1}{x-2} = 1$

(b) $\dfrac{-12}{x^2-9} + \dfrac{1}{x-3} = 1$

Solution:

(a) Multiplying each term by the LCD, $x^2 - 4 = (x+2)(x-2)$, we have

$$(x^2-4)\cdot\frac{4}{(x^2-4)} - (x^2-4)\cdot\frac{1}{x-2} = x^2-4$$

$$4-(x+2) = x^2-4 \qquad \text{Simplifying}$$

$$x^2+x-6=0 \qquad \text{Writing in standard form}$$

$$(x+3)(x-2)=0 \qquad \text{Factoring}$$

$$x=-3 \text{ or } x=2$$

However, $x=2$ is *not* a solution, since $\frac{1}{x-2}$ is not defined for $x=2$. Thus the solution set is $\{-3\}$.

(b) The LCD is x^2-9. Multiplying each term by this LCD, we have

$$(x^2-9)\cdot\frac{-12}{x^2-9} + (x^2-9)\cdot\frac{1}{x-3} = x^2-9$$

$$-12+x+3 = x^2-9 \qquad \text{Simplifying}$$

$$x^2-x=0$$

$$x(x-1)=0$$

$$x=0 \text{ or } x=1$$

The solution set is $\{0, 1\}$, as can be shown by checking in the given equation.

21. $\frac{7}{x-3} + \frac{3}{3-x} = 4$

22. $\frac{3}{x+2} - \frac{1}{x} = 0$

23. $\frac{3}{y+2} - \frac{1}{y-2} = \frac{5}{3} - \frac{1}{y^2-4}$

24. $\frac{2}{y-3} - \frac{1}{y-1} = 0$

In Problems 25 through 35, find the solution set of the given equation.

Illustration: $\dfrac{x}{x+3} + \dfrac{x}{x+4} = 0$

Solution:
We first have to write the equation in standard form. Multiplying by $(x+3)(x+4)$, the LCD, we have

$$(x+3)(x+4) \cdot \frac{x}{x+3} + (x+3)(x+4) \cdot \frac{x}{x+4} = (x+3)(x+4) \cdot 0$$

$$x(x+4) + (x+3)x = 0 \qquad \text{Simplifying}$$

$$x^2 + 4x + x^2 + 3x = 0 \qquad \text{Multiplying}$$

$$2x^2 + 7x = 0 \qquad \text{Adding like terms}$$

$$x(2x+7) = 0 \qquad \text{Factoring}$$

$$x = 0 \text{ or } 2x + 7 = 0$$

$$x = 0 \text{ or } x = \frac{-7}{2}$$

The solution set is $\left\{0, \dfrac{-7}{2}\right\}$.

25. $\dfrac{x}{x+4} + \dfrac{x}{x+1} = 0$

26. $\dfrac{x}{x+2} + \dfrac{x}{x+3} = 0$

27. $\dfrac{x-1}{x+11} - \dfrac{2}{x-1} = 0$

28. $\dfrac{x+1}{x-2} - \dfrac{8}{x-1} = 0$

29. $\dfrac{x}{x-1} - \dfrac{x}{x+1} = 0$

30. $\dfrac{x}{x+4} + \dfrac{x}{x-2} = \dfrac{-1}{2}$

31. $\dfrac{x}{x+2} - \dfrac{x}{x+1} = \dfrac{-1}{6}$

32. $\dfrac{x}{x+1} - \dfrac{x}{x-1} = \dfrac{-3}{4}$

33. $\dfrac{x}{x+4} + \dfrac{x}{x+2} = \dfrac{-4}{3}$

34. $\dfrac{2x}{x-2} + \dfrac{x}{x-1} = \dfrac{7}{6}$

35. $\dfrac{2x}{x+1} - \dfrac{4x}{x-1} = \dfrac{14}{3}$

B. C. by permission of Johnny Hart and Field Enterprises, Inc.

OBJECTIVES 6.6

After studying this section, the reader should be able to solve word problems involving quadratic equations.

6.6 APPLICATIONS: WORD PROBLEMS

In the cartoon at the beginning of this section, the man wants to know how deep his well is. As we stated in Section 6.3, the distance h (in feet) traveled by an object thrown up with an initial velocity v_0 (in feet per second) in t seconds is expressed as

(1) $h = -16t^2 + v_0t.$

Since in our case the object is dropped, the initial velocity v_0 is 0. Thus, the depth of the well is $h = -16\ (16)^2 = -4096$ ft. As you can see from Equation 1, if an object is dropped, the distance h (in feet) traveled by the object after t seconds can be expressed as

Note that −4,096 is negative—that is, 4,096 below ground level.

(2) $h = 16t^2$ (the negative sign is dropped here since we know the direction is downward)

We use Equation 2 in solving the following problem.

Example 1

The photograph on page 344 shows Henry La Motte diving into a $12\frac{1}{2}$-inch deep pool from a height of 40 feet. How long does it take him to hit the water?

Photo credit: Sterling Publishing Company, Inc., New York.

Solution Letting $h = 40$ in Equation (2), we have

$$16t^2 = 40$$

$$\text{or } t^2 = \frac{5}{2}$$

Thus,

$$t = \pm\sqrt{\frac{5}{2}}$$

Since the time t must be positive, it takes La Motte $\sqrt{\frac{5}{2}}$ seconds to hit the water. (This is approximately 1.58 seconds.)

The accompanying ad shows the amount of interest paid (on the spot) by depositing the amount indicated for a period of two years. For example, if you agreed to deposit \$5,000 for 2 years, the interest paid to you would be \$572.75. If this amount were paid *at the end* of the two years, do you know what compound interest rate that would be? The total amount of money A received at the end of two years when

EXAMPLES OF THE INTEREST YOU CAN COLLECT ON THE SPOT.

PRINCIPAL DEPOSITED	INTEREST ON-THE-SPOT
$1,200	$ 137.46
$2,500	$ 286.38
$5,000	$ 572.75
$7,500	$ 859.13
$12,000	$1,374.60
$30,000	$3,436.50
$50,000	$5,727.50

The law imposes a substantial penalty for any time deposit withdrawal prior to the maturity date you agree to

P dollars are invested at interest compounded annually for two years at a rate r is given in the formula

(3) $\quad A = P(1 + r)^2$

In our case, $A = \$5{,}000 + 572.75$, and $P = 5{,}000$. Hence

$$5{,}572.75 = 5{,}000(1 + r)^2$$

Recall that $\approx$ means "is approximately equal to."

$$\text{or } (1 + r)^2 = \frac{5{,}572.75}{5{,}000} \approx 1.115$$

$\sqrt{1.115} \approx 1.056$.

Thus,

$$1 + r \approx \pm\sqrt{1.115} \approx 1.056$$
$$\text{and } r \approx -1 + 1.056 = 0.056$$

The negative answer $r \approx -1 - 1.056$ is discarded.

The rate of interest is therefore approximately 5.6 per cent.

Example 2

Find the interest rate r when an amount of $882 is received at the end of 2 years on an $800 investment.

Solution

By Equation (3), $A = P(1+r)^2$; in this case, $A = 882$, $P = 800$. Substituting,

$$882 = 800(1+r)^2$$

Note that $(1.05)^2 = 1.1025$.

$$\text{or } (1+r)^2 = \frac{882}{800} = 1.1025$$

$$1 + r = \pm\sqrt{1.1025} = \pm 1.05$$

Thus, $$r = -1 + 1.05 = .05, \text{ or } 5\%$$

Example 3

Weighty Scales wishes to manufacture a bathroom scale whose top is made by cutting a small square from each corner of a rectangular sheet of metal 12 inches by 16 inches and then bending down the sides to form an inverted tray. If the face area is to be 140 square inches, what are the dimensions of the top of the scale?

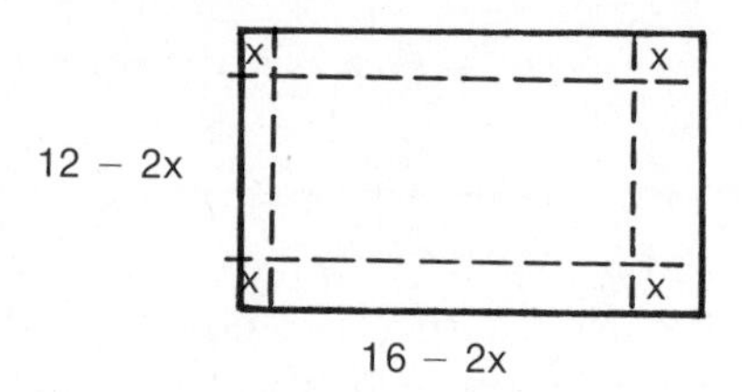

Solution

Here is a diagram showing the situation. The area of the base is 140 square inches. Thus

$$(16 - 2x)(12 - 2x) = 140$$

$$\text{or } 2(8 - x) \cdot 2(6 - x) = 140$$

$(8 - x)(6 - x) = 35$	Dividing by 4
$48 - 14x + x^2 = 35$	Multiplying
or $x^2 - 14x + 13 = 0$	Writing in standard form
$(x - 13)(x - 1) = 0$	Factoring
$x = 13$ or $x = 1$	Solving for x

Since it is physically impossible to cut a 13-by-13-inch square from each corner of the given sheet of metal, the squares to be cut out are 1 inch on a side. Hence, the dimensions of the top are $16 - 2 = 14$ by $12 - 2 = 10$ inches.

We turn now to an application in business. You may have heard of the term "to break even"; in business, the break-even point is the point at which the revenue R of a company equals the cost of the manufactured goods. In symbols, $R = C$. For example, if a company produces x thousands of an item and sells the items for \$2 each, the revenue R in thousands of dollars can be expressed as $R = 2x$. If the cost C in thousands of dollars is $C = x^2 - 4x + 5$ for $x \geq 2$, the break-even point occurs when

The break-even point is when revenue R equals cost C, that is, $R = C$.

$$2x = x^2 - 4x + 5$$

$$\text{or } x^2 - 6x + 5 = 0$$

$$(x - 5)(x - 1) = 0$$

$$x = 5 \quad \text{or} \quad x = 1$$

Thus, the company must produce five-thousand items to break even. Note that $x = 1$ is not used because $x \geq 2$.

Example 4

A company produces x units per day. Each of the items sells for 5 dollars. If the daily cost of production is $C = x^2 - x + 1$, how many units should be produced daily to break even?

Solution

Since each unit sells for \$5, the revenue $R = 5x$. To break even, we need $C = R$, that is,

$$x^2 - x + 1 = 5x \quad \text{or} \quad x^2 - 6x + 1 = 0$$

Thus,

$$x = \frac{6 \pm \sqrt{36 - 4 \cdot 1 \cdot 1}}{2} = \frac{6 \pm \sqrt{32}}{2} = \frac{6 \pm 4\sqrt{2}}{2} = 3 \pm 2\sqrt{2}$$

Since $\sqrt{2}$ is approximately 1.41,

$$x \approx 3 + 2(1.41) = 5.82 \text{ or } x \approx 3 - 2(1.41) = 0.18$$

The company must produce 6 units or 0 units daily.

Since the number of units produced must be an integer, we approximate 5.82 by 6 units. Similarly, $3 - 2(1.41) = 0.18$, which is approximated by 0. Of course, we must check to make sure that if the company produces 6 or 0 units, it breaks approximately even.

PROGRESS TEST

1. An object is dropped from a height of 64 feet. How long does it take for the object to hit the ground?
2. An investor receives $A = \$1089$ after investing $P = \$900$ at compound interest r for two years. If $A = P(1 + r)^2$, the rate r equals ________.
3. A company decides to make a tray by cutting a square from each corner of a rectangular sheet of metal 10 by 20 inches. If the base area of the tray is 144 inches, the dimensions of the base are ________.

4. The revenue in thousands of dollars of a company is given by $R = 2x$, and its costs, (also in thousands of dollars) are $C = 2x^2 - 4x + 4$ where x is the number of units of an item produced). The break-even point for this company is reached when they manufacture ________ units of the item.

EXERCISE 6.6

In Problems 1 through 20, follow the procedures in the preceding examples to solve the given problems.

1. An object is dropped from an altitude of 96 feet. Find the time it takes the object to hit the ground.

2. A plane flying at an altitude of 900 feet drops a bomb. How long does it take before the bomb hits the ground?

3. The distance d (measured in meters) an object travels in time t (seconds) when dropped from a certain height is given by $d = 5t^2$. If an object is dropped from an altitude of 125 meters, how long does it take before the object hits the ground?

4. Use the formula in Problem 3 to find the time it takes an object to drop 50 meters.

5. A wind pressure gauge at Commonwealth Bay, Antarctica, registered 120 pounds per square foot during a gale (Guinness). If the pressure p (in pounds per square feet) from a v-mile-per-hour wind is given as $p = 0.003v^2$, what was the wind velocity?

6. Use the formula in Problem 5 to find the wind velocity at the Golden Gate when a pressure gauge registered a pressure of 7.5 pounds per square foot.

7. A man invested \$100 for two years. At the end of this time, he received \$144. What was his rate of compound interest?

8. A woman borrowed \$100 at compound interest for 2 years. At the end of this time, she paid \$121. What rate of compound interest did she pay?

9. An investor received \$225 on a two-year investment of \$196. If the interest was compounded annually, what was the compound rate?

PROGRESS TEST ANSWERS

1. 2 seconds
2. 10%
3. 18 by 8
4. 2 or 1

10. At what compound interest rate will \$225 grow to \$256 in two years?

11. A company sells its products for \$3 each. If the cost of production is $C = 3x^2 - 5x + 4$ for $x \geq 1$, how many units do they have to produce to break even?

12. Solve Problem 11 if the cost is $x^2 - 4x + 12$.

13. The revenue of a company is given as $R = 3x$, where x is the number of units produced (in thousands). If the cost is $C = 4x^2 - x + 1$, how many units have to be produced before the company breaks even?

14. Solve Problem 13 if the cost is $2x^2 - 2x + 2$ for $x \geq 1$.

15. A company desires to construct a tray by cutting a small square from each corner of a 19-by-14-inch rectangular sheet of metal and bending up the sides. If the area of the base is to be 150 square inches, what are the dimensions of the tray?

16. Repeat Problem 15 if the area is to be 104 square inches.

The following theorem will be used in Problems 17 through 20.

THE PYTHAGOREAN THEOREM. In any right triangle (a triangle with a 90° angle), the sum of the squares of the lengths of the two shorter sides is equal to the square of the length of the longest side (the hypotenuse). This relationship is shown in the accompanying diagram.

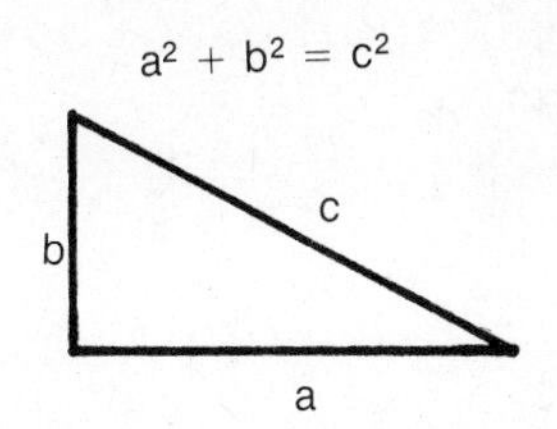

17. The bottom of a ladder is 5 feet from a wall and it reaches a window 12 feet from the ground. How long is the ladder?

18. How long is a wire extending from the top of a 40-foot telephone pole to a point on the ground 30 feet from the foot of the pole?

19. Repeat Problem 18 if the point on the ground is 16 feet away from the pole.

20. How high on the side of a building will a 50-foot ladder reach if its foot is 14 feet from the wall on which the ladder is resting?

SELF-TEST–CHAPTER 6

1. Find the solution set of:

_____________ (a) $x^2 = 4$

_____________ (b) $9x^2 - 4 = 0$

_____________ (c) $(x - 1)^2 = 9$

---

2. Solve:

_____________________ (a) $x^2 - 5x + 6 = 0$

_____________________ (b) $2x^2 - x - 1 = 0$

---

3. Solve by completing the square:

_____________________ (a) $x^2 - 3x + 2 = 0$

_____________________ (b) $2x^2 + 3x - 2 = 0$

---

4. Find the discriminant of the given equation, and determine the character of the roots:

_____________________ (a) $6x^2 + x - 2 = 0$

____

_____________________ (b) $x^2 - x + 1 = 0$

____

---

5. Solve:

_____________________ (a) $x^2 - x - 1 = 0$

_____________________ (b) $x^2 - x = \frac{2}{3}$

---

6. Solve:

_____________________ (a) $\sqrt{x + 2} = x$

__________ (b) $\sqrt{x+1}+5=x$

---

7. Solve:

__________ (a) $\frac{4}{x^2-4}+\frac{1}{x+2}=1$

__________ (b) $\frac{1}{x-3}+\frac{5}{x+3}=0$

---

8. Solve:

__________ (a) $\frac{3}{x+2}=\frac{-1}{x}+2$

__________ (b) $\frac{2}{y-3}=\frac{1}{y-1}+\frac{1}{y}$

---

9. The distance h traveled in t seconds by an object dropped from a certain height is: $h=16t^2$. An object is dropped from a height of 32 feet. How long would it take before the object hits the ground?

---

10. The amount of money A received at the end of two years when P dollars are invested at a compound rate r for two years is expressed as $A=P(1+r)^2$. Find the rate of interest r (written as a per cent) if a person invested $100 for two years and at the end of this time received $121.

CHAPTER 7

HISTORICAL NOTE

The principal feature of this Chapter is an introduction to ideas that belong to the area that mathematicians call analytic geometry. Analytic geometry is a blend of algebra and geometry in which algebra is used to study geometry and geometry to study algebra. This is an exceedingly fruitful blend because each subject sheds much light on the other and, in so doing, forms the basis for much of modern mathematics. The key to an appropriate combination of algebra and plane geometry lies in devising a workable system of associating points of the plane with ordered pairs of numbers.

Historians regard two famous French mathematicians, René Descartes (1596–1650) and Pierre de Fermat (1601–1665), as the originators of the main ideas of modern analytic geometry. As often happens in human affairs, these two people were working independently and almost simultaneously on closely related problems. It is an interesting fact that neither of the two men was a full-time mathematician. Descartes was a soldier and a philosopher who actually devoted only a rather small portion of his time to mathematics. Fermat was a humble and retiring lawyer who apparently enjoyed doing mathematics in his spare time.

Among the legends that attempt to describe the initial flash of genius that led Descartes to apply algebra to geometry is the story that the idea came to him as he was watching a fly crawl on the ceiling near a corner of his room. He suddenly realized that the path of the fly could be mathematically described if one knew a relation connecting the fly's distances from the two walls. If this story is true, it certainly shows that great things can grow from small beginnings.

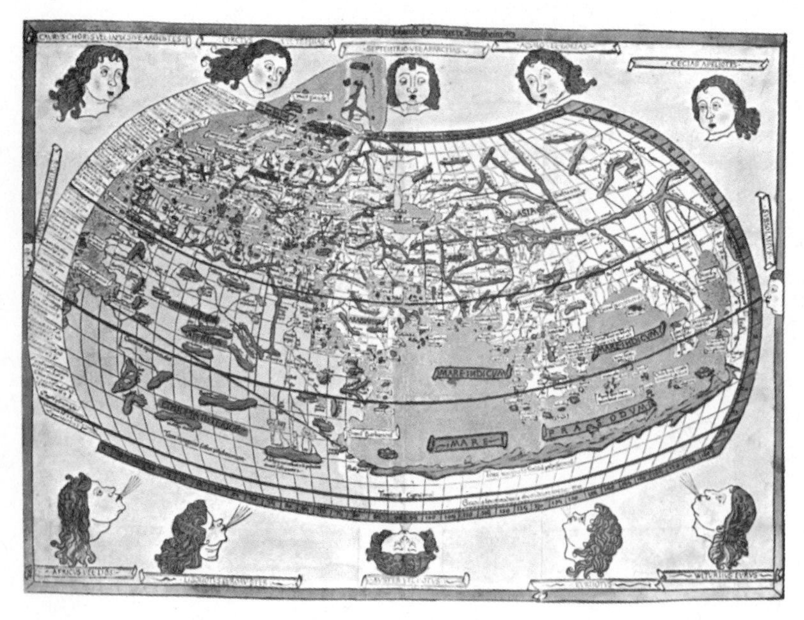

One of the first known maps, showing the latitude and longitude of different points. Similar "grids" are used to locate points on a plane surface. (Courtesy of the British Museum.)

CARTESIAN COORDINATE SYSTEMS

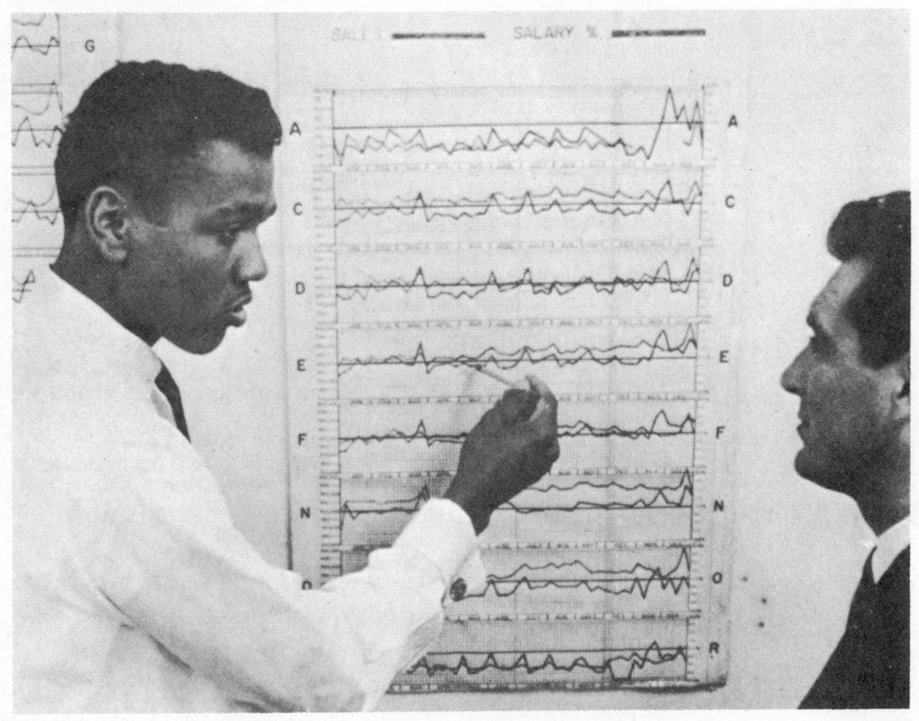

The men are studying a graph detailing sales and salaries. (Courtesy Johnson Publishing Company, Chicago.)

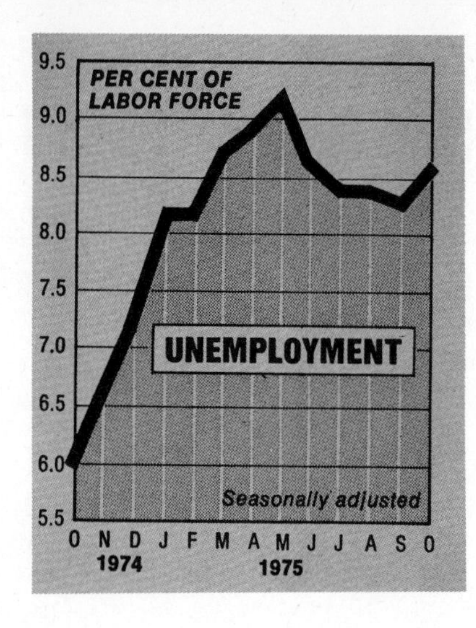

Newsweek—Fenga & Freyer.

OBJECTIVES 7.1

After studying this section, the reader should be able to:

1. **Find the cartesian product of two given sets.**
2. **Graph a given point on the cartesian plane and determine in which quadrant the point lies.**
3. **Graph a linear equation in two variables.**
4. **Find the x- and y-intercepts of a linear equation in two variables.**

7.1 CARTESIAN COORDINATE SYSTEMS

In Section 2.6, we learned that the solution set of a linear equation or inequality could be *represented* or *graphed* on the number line. In this chapter, we extend this idea to linear and quadratic equations in two variables. Since there are two variables involved, these equations can be graphed by using *ordered pairs* of numbers. For example, in the unemployment chart at the beginning of this section, each month is represented along the horizontal axis by its first initial, starting with October, 1974 (O), while the unemployed per cent of the labor force, starting with 5.5, is represented along the vertical axis. From the chart, we see that the per cent of unemployed people in October of 1974 was 6.0, while that in November was 6.5. If we agree to write *ordered pairs* using the month as the first entry and the unemployment rate as the second entry, and put them in parentheses with a comma separating them, we can write ($\mathcal{O}$, 6) and (N, 6.5) to represent these facts. If we choose to represent the months by the numerical order in which they occur (January is 1, February is 2, etc.), the ordered pairs

($\mathcal{O}$, 6) means that in the month of October, the unemployment rate was 6 per cent.

($\mathcal{O}$, 6) and (N, 6.5) can be written as (10, 6) and (11, 6.5), respectively. What would (12, 7.1) represent? This would mean that in December of 1974, the unemployment rate was 7.1 per cent of the labor force. In mathematics, *ordered pairs* of numbers are written as (x, y). The letter x is called the *first component*, or *abscissa*, and y is the *second component*, or *ordinate*. In the ordered pair (2, 3), 2 is the first component and 3 the second component.

In the ordered pair (2, 3), 2 is the abscissa; 3 is the ordinate.

Returning to the unemployment chart, let us now assume that you wish to construct a chart for the first three months of the year—that is, using the set of months $M = \{1, 2, 3\}$. If you know that the unemployment rates in these months are integer per cents, say between 6 and 8 per cent inclusive—i.e., in the set $R = \{6, 7, 8\}$—then the possible set of ordered pairs to be considered is $\{(1,6), (1,7), (1,8), (2,6), (2,7), (2,8), (3,6), (3,7), (3,8)\}$. This set is usually called the *Cartesian* or *cross product* of M and R and is denoted by $M \times R$ (read "M cross R"). The set is constructed by forming all possible ordered pairs for which the first component is an element of M and the second component is an element of R. In symbols, $M \times R = \{(x,y) \mid x \in M \text{ and } y \in R\}$.

Remember, 1 is January, 2 is February, 3 is March.

$M \times R$ is the Cartesian or cross product of M and R.

Example 1

If $A = \{1, 2\}$ and $B = \{3, 4, 5\}$, find
(a) $A \times B$
(b) $B \times A$

Solution

(a) $A \times B = \{(1,3), (1,4), (1,5), (2,3), (2,4), (2,5)\}$
(b) $B \times A = \{(3,1), (3,2), (4,1), (4,2), (5,1), (5,2)\}$
Note that, in general, $A \times B \neq B \times A$.

We are now ready to discuss the Cartesian product of the set R of real numbers and itself, namely, $R \times R$. The set $R \times R$ is an infinite set of ordered pairs of numbers, (x, y), which can be represented geometrically by a plane called the *real plane*. To obtain this representation, we first draw two perpendicular lines, called the x-axis and y-axis, intersecting at a point O, called the *origin* (see Figure 7.1a). Then, using convenient units of length, we make each axis into a number line whose O point is at the origin. In this manner, the real plane is divided into four regions, called *quadrants*, which

This representation is attributed to René Descartes, a French mathematician and philosopher.

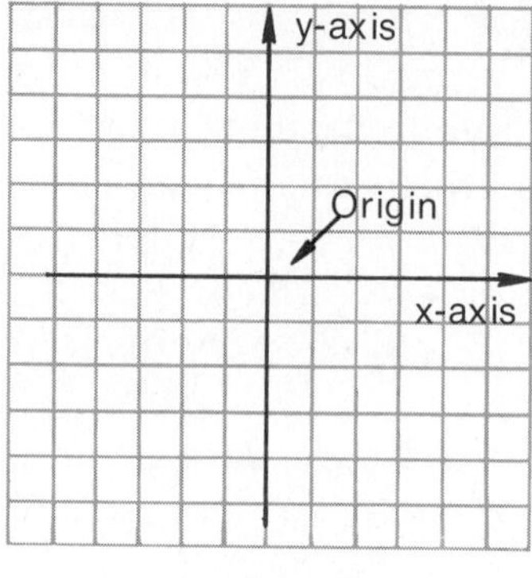

FIGURE 7.1(a)

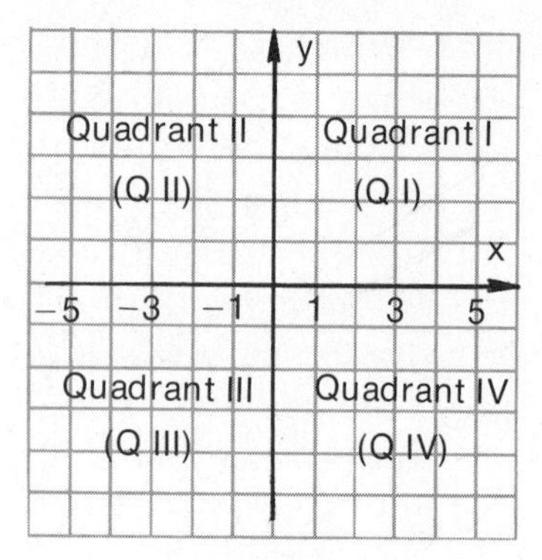

FIGURE 7.1(b)

are numbered as in Figure 7.1b. This configuration is sometimes called the *Cartesian coordinate system*, the *rectangular coordinate system*, or simply the *coordinate plane*. Note that on the x-axis, the *positive* direction is to the *right*, while on the y-axis positive is *up* on the page.

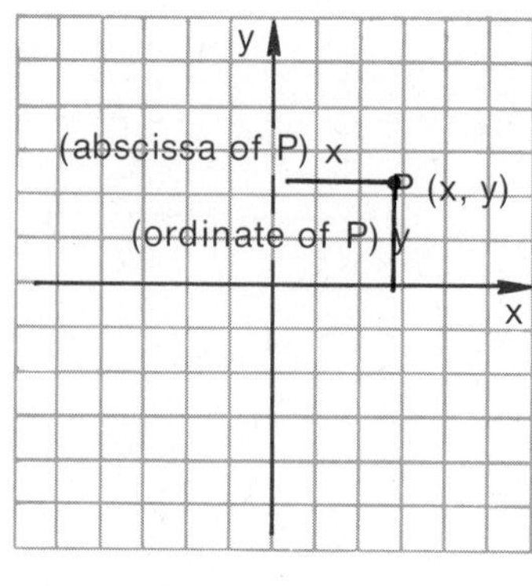

FIGURE 7.2(a)

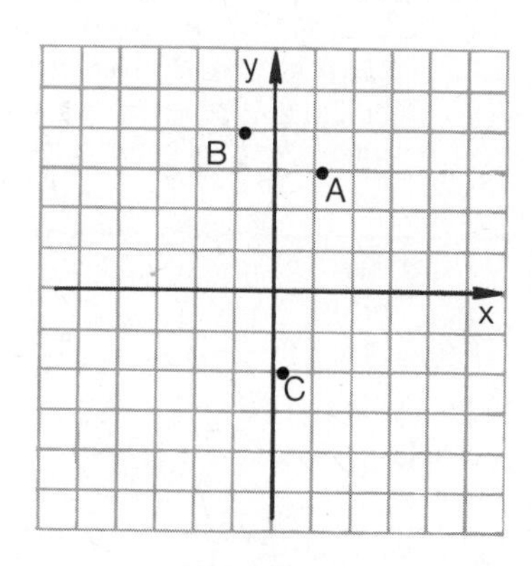

FIGURE 7.2(b)

We associate each point P in the plane with an ordered pair of real numbers (x, y) (see Figure 7.2a). As you can see, there is a one-to-one correspondence between the points in the plane and the rectangular Cartesian coordinates—that is, for each point P in the plane, there is a corresponding unique ordered pair (x, y), and vice versa. Thus, in the ordered pair (x, y), the x-coordinate denotes the directed (perpendicular) distance of the point (x, y) from the vertical axis, to the *right* if x is *positive*, to the *left* if it is *negative*. Similarly, the y-coordinate denotes the directed distance of the point (x, y) from the horizontal axis, *above* the axis if y is *positive*, below if y is *negative*. The components of the ordered pair corresponding to a given point are called the *coordinates* of the point, and the point is called the *graph* of the ordered pair. The *graphs* of the points A(1, 3), B(−1, 4), and C(0, −2) are shown in Figure 7.2b.

Example 2

Graph the ordered pairs A(2, 3), B(−1, 2), C(−2, −1), and D(1, −3).

Solution

The graphs of these points are shown in Figure 7.3.

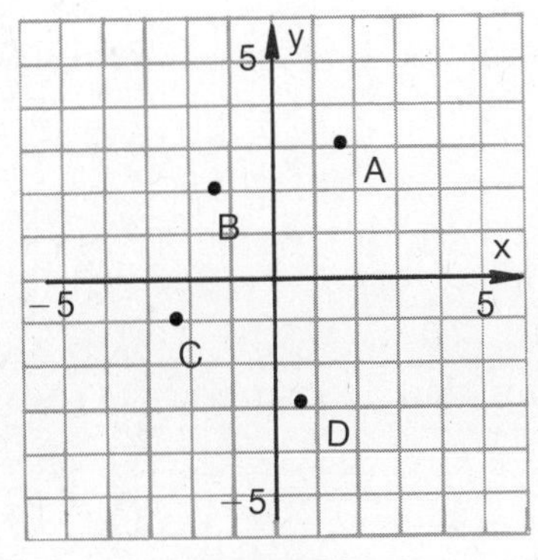

FIGURE 7.3

The equation $2x + 3y = 12$ is an equation in *two* variables, x and y. If in this equation x is replaced by 3 and y by 2, we obtain the *true* statement $2 \cdot 3 + 3 \cdot 2 = 12$. Thus we say that the *ordered pair* (3, 2) is a *solution* of the equation $2x + 3y = 12$ or that it *satisfies* the equation. On the other hand, (2, 3) is *not* a solution of $2x + 3y = 12$, because $2 \cdot 2 + 3 \cdot 3 \neq 12$.

(3, 2) is a solution of $2x + 3y = 12$ because $2 \cdot 3 + 3 \cdot 2 = 12$ is a true statement.

The *solution set* of an equation in two variables can be written using set notation. For example, the solution set of $2x + 3y = 12$ can be written as $\{(x, y) \mid 2x + 3y = 12\}$ or, if we decide to solve for y, as $\left\{(x, y) \mid y = \frac{-2}{3}x + 4\right\}$. The solution set of the equation $2x + 3y = 12$ consists of infinitely many points, so it would be impossible to list all of these points. However, we can find some of these points by substituting values for one of the variables and then computing the corresponding values for the other variable. For example, if we replace x by −3 in the equation $y = \frac{-2}{3}x + 4$, we have

$$y = \frac{-2}{3} \cdot (-3) + 4 = 6$$

If $x = 0$,

$$y = \frac{-2}{3} \cdot (0) \quad + 4 = 4$$

If $x = 3$, $\quad y = \frac{-2}{3} \cdot (3) + 4 = 2$

If $x = 6$, $\quad y = \frac{-2}{3} \cdot (6) + 4 = 0$

If $x = 9$, $\quad y = \frac{-2}{3} \cdot (9) + 4 = -2$

These ordered pairs can be represented in a table as follows:

x	−3	0	3	6	9
y	6	4	2	0	−2

The points (−3,6), (0,4), (3,2), (6,0), and (9,−2) appearing in the table can also be graphed, as shown in Figure 7.4a.

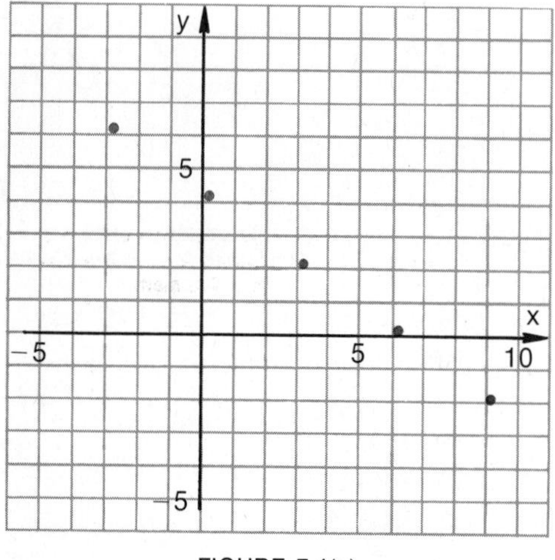

FIGURE 7.4(a)

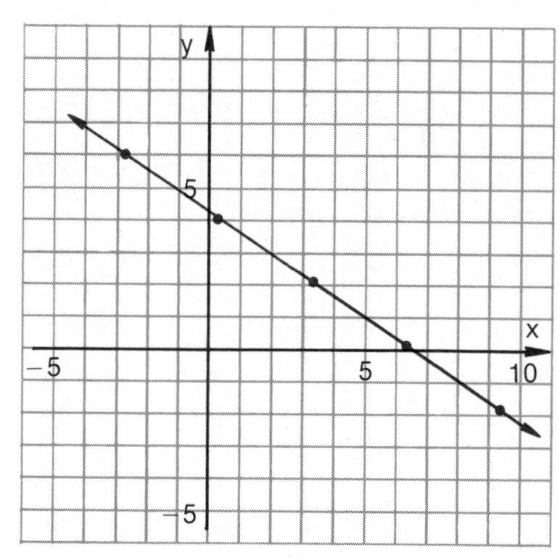

FIGURE 7.4(b)

The points appear to be on a straight line. In fact, it can be proved that every *solution* of $y = \frac{-2}{3}x + 4$ (or $2x + 3y = 12$) corresponds to a point on the line, and vice versa. Thus, the line shown in Figure 7.4b, obtained by joining the points shown in Figure 7.4a, is the *graph* of $y = \frac{-2}{3}x + 4$ (or $2x + 3y = 12$). Note that in Figure 7.4b, the line represents an approximation of the complete graph, which continues without end in both directions, as indicated by the arrows in the figure.

Note that the equations $y = \frac{-2}{3}x + 4$ and $2x + 3y = 12$ are equivalent, and that their graphs are therefore the same.

Example 3

Graph $x + 2y = 8$.

Solution

For $x = -2$, $y = 5$.
For $x = 0$, $y = 4$.
For $x = 2$, $y = 3$.
Thus, we first graph the points $(-2, 5)$, $(0, 4)$, and $(2, 3)$ and then join them with a line, as shown in Figure 7.5.

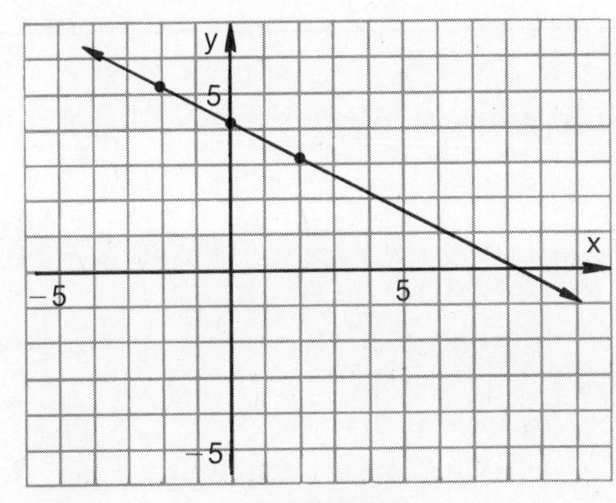

FIGURE 7.5

$Ax + By + C = 0$ is the standard form of a linear equation.

The equation $x + 2y = 8$ has a straight line as its graph. It can be shown that any equation in two variables x and y that is of the form $Ax + By + C = 0$ (A and B not both zero) has a straight line for its graph. For this reason, the equation is called a *linear equation.* We regard the given form, $Ax + By + C = 0$, as the *standard form* of a linear equation.

Any two points will determine a line.

Since any two points will determine a straight line, it is enough to locate two points in order to graph a linear equation. The easiest solutions to compute are those involving zeros, that is, solutions of the form $(x, 0)$ and $(0, y)$. For example, if we let $x = 0$ in the equation $x + 2y = 8$, we obtain $2y = 8$, or $y = 4$. Thus $(0, 4)$ is a point on the graph. Similarly, if we let $y = 0$ in the equation $x + 2y = 8$, we have $x = 8$. Thus, $(8, 0)$ is also on the graph. Since the points $(8, 0)$ and $(0, 4)$ are the points at which the line crosses the x- and y-axes, respectively, the numbers 8 and 4 are called the x- and y-*intercepts.* These two intercepts, as well as the line determined by them (the graph of $x + 2y = 8$), are shown in Figure 7.6.

8 and 4 are the x- and y-intercepts respectively.

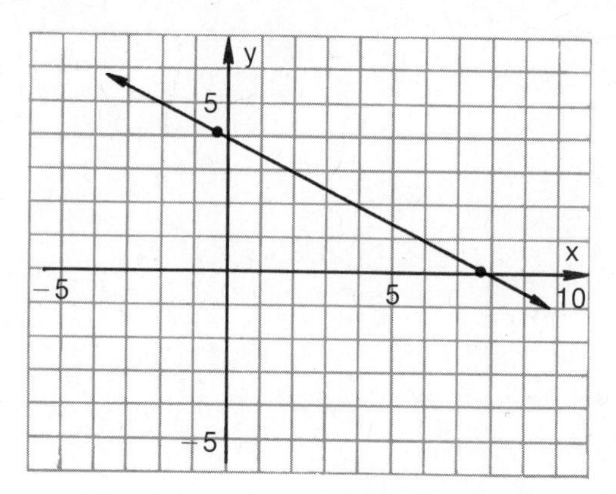

FIGURE 7.6

Example 4 Find the x- and y-intercepts of $y = 3x + 6$ and then graph the line.

Solution To find the x-intercept, we let $y = 0$, obtaining $0 = 3x + 6$ or, $x = -2$. Hence $(-2, 0)$ is on the line.
To find the y-intercept, we let $x = 0$, so $y = 6$. Hence $(0, 6)$ is on the line.
To graph the equation $y = 3x + 6$, we graph the points $(-2, 0)$ and $(0, 6)$ and join them with a line, as shown in Figure 7.7.

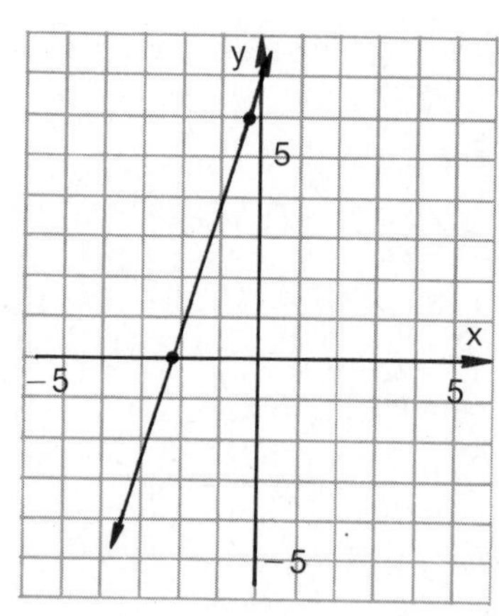

FIGURE 7.7

It is worth noting that the procedure we have discussed works *only* for equations of the form $Ax + By + C = 0$ where *none* of the quantities A, B, and C is 0. Thus, since the equation $2y = 6$ can be written in the form $0 \cdot x + 2y - 6 = 0$, we *cannot* graph $2y = 6$ by finding its intercepts. The equation $2y = 6$ assigns to every value of x a y-value of 3. Thus, for

Note that if $2y = 6$, $y = 3$.

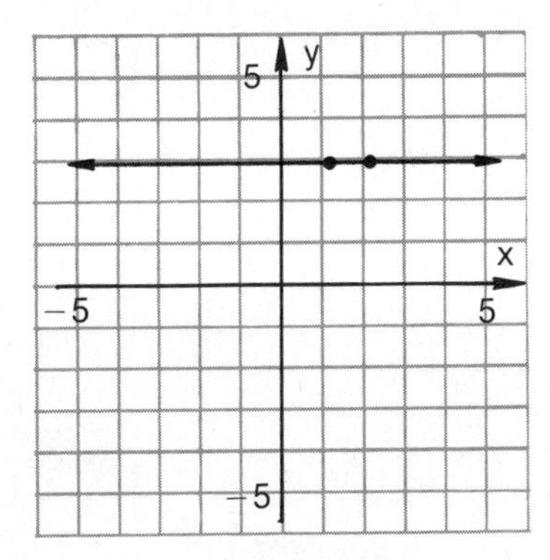

FIGURE 7.8

$x = 1$, $y = 3$, for $x = 2$, $y = 3$, etc. If we graph the points (1, 3) and (2, 3) and connect them with a straight line, we see that the result is a horizontal line, as shown in Figure 7.8.

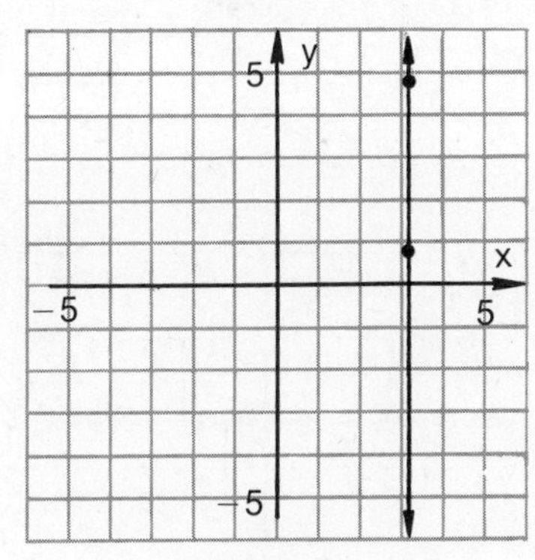

FIGURE 7.9

Similarly, the graph of the equation $2x = 6$ assigns an x value of 3 to every y. Thus, for $y = 1$, $x = 3$, and for $y = 5$, $x = 3$. If we graph the points (3, 1) and (3, 5) we see that the result is a vertical line, as shown in Figure 7.9. In general, the graph of the equation $By = C$ $(B \neq 0)$ is a *horizontal* line for which $y = \frac{C}{B}$; the graph of $Ax = C$ $(A \neq 0)$ is a *vertical* line for which $x = \frac{C}{A}$.

Example 5

Graph the equations:
(a) $2x = 8$
b) $5y = -10$

Solution

(a) Since $2x = 8$ is of the form $Ax = C$, $2x = 8$ is a vertical line for which $x = 4$. If we choose the solutions (4, 1) and (4, 2) and draw a straight line through them, we obtain the graph of $2x = 8$ shown in Figure 7.10.

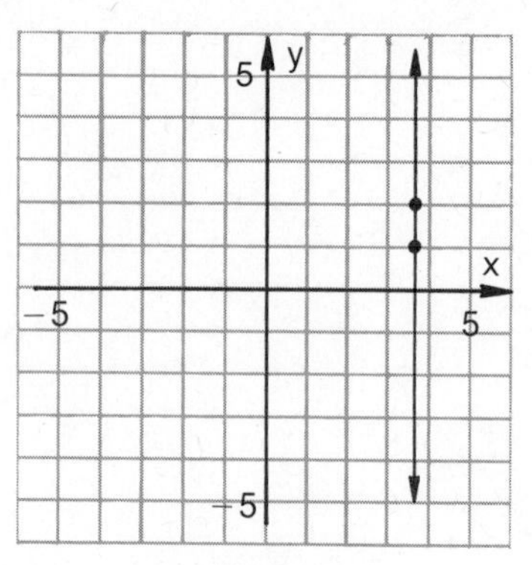

FIGURE 7.10

(b) Since $5y = -10$ is of the form $By = C$, $5y = -10$ is a horizontal line for which $y = -2$. If we choose the solutions $(1, -2)$ and $(2, -2)$ and draw a straight line through them, we obtain the graph of $5y = -10$ shown in Figure 7.11.

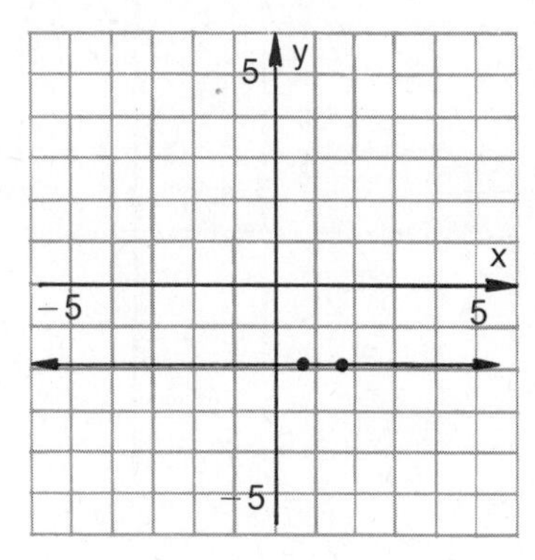

FIGURE 7.11

PROGRESS TEST

1. The Cartesian product of the sets $A = \{a, b, c\}$ and $B = \{d, e\}$ is ________________.
2. Graph the ordered pairs $A(-1, 2)$ and $B(3, -1)$ and name the quadrant containing each point.

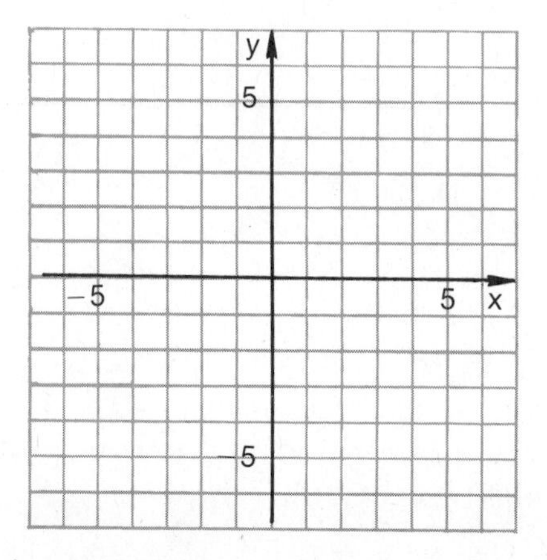

3. Graph the line $3x - y = 6$ on the accompanying rectangular coordinate system.

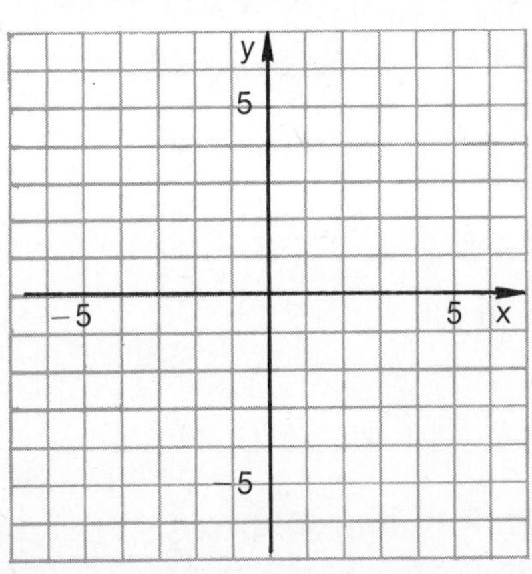

4. The x- and y-intercepts of $2x - 3y = 6$ are, respectively, __________.
5. Use the x- and y-intercepts found in 4 to graph the line $2x - 3y = 6$.
6. The line $3x = -9$ is a __________ line. Graph this line.

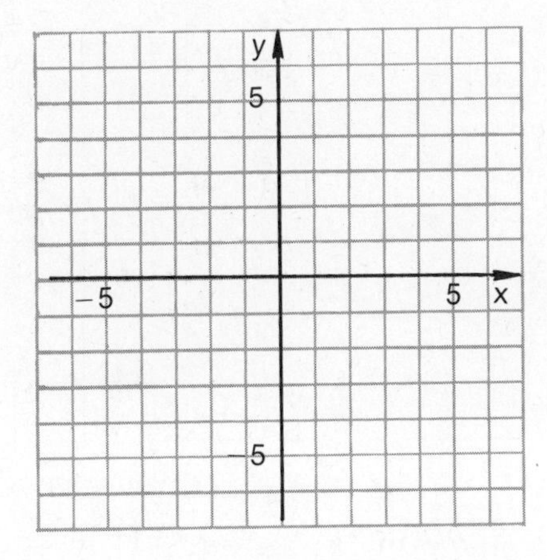

7. The line $2y = -6$ is a __________ line. Graph this line.

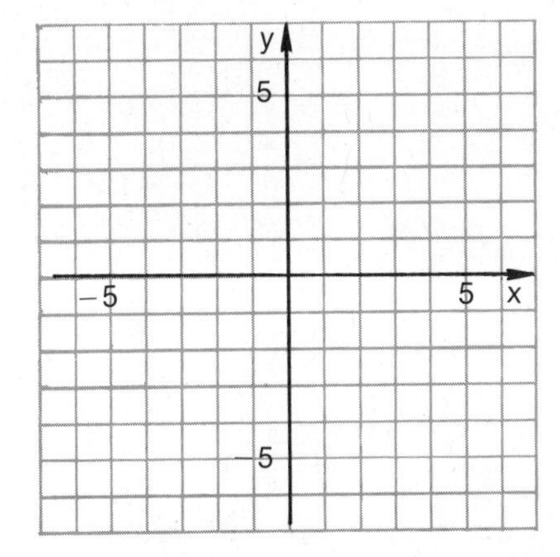

PROGRESS TEST ANSWERS

1. A × B = {(a,d), (b,d), (c,d), (a,e), (b,e), (c,e)}.

2.

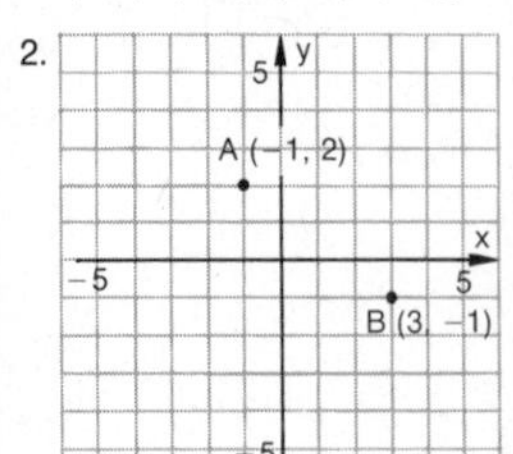

3.

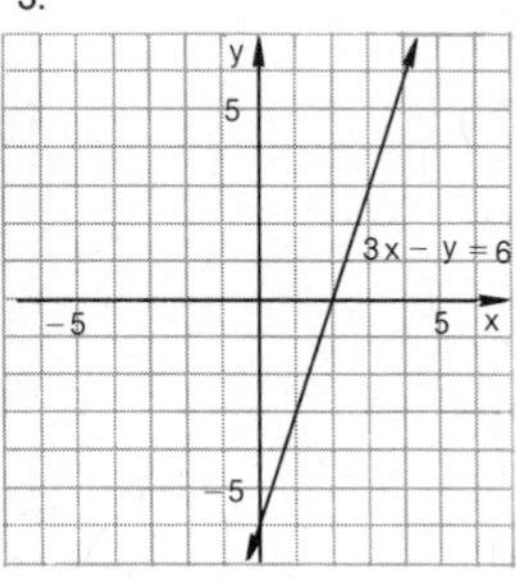

4. 3 and −2.

5.

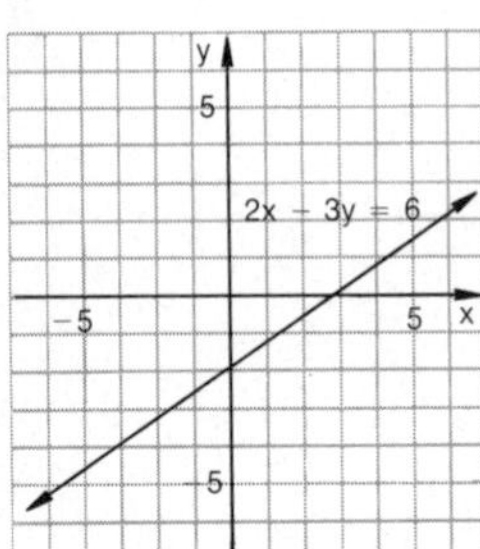

6. vertical

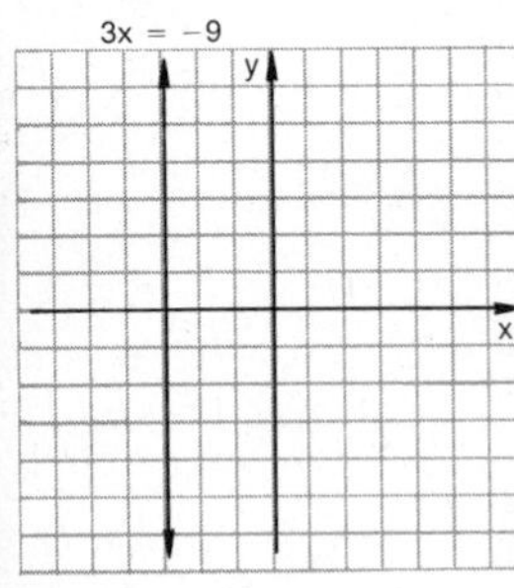

7. horizontal

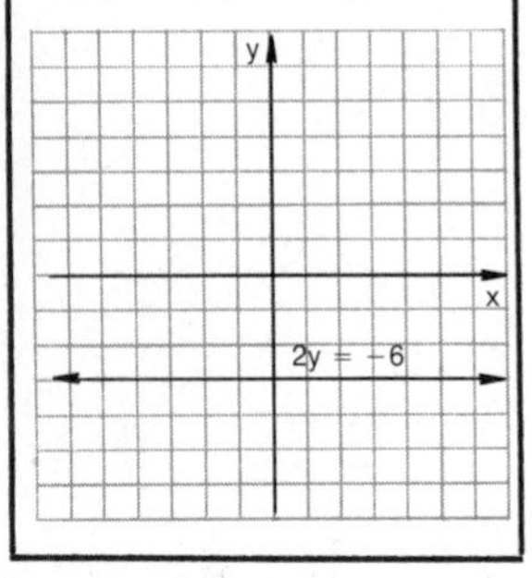

EXERCISE 7.1

In Problems 1 through 4, find A × B.

1. A = {1, 2}, B = {0, 1}

2. A = {−1, 3}, B = {1, 2}

3. A = {1, 2, 3}, B = {a, b}

4. A = {a, b, c}, B = {1, 2, 3}

In Problems 5 through 10, graph the given ordered pairs and state in which quadrant (if any) each lies.

Illustration: (a) $A\left(-3, \frac{-7}{3}\right)$

(b) B(0, −2)

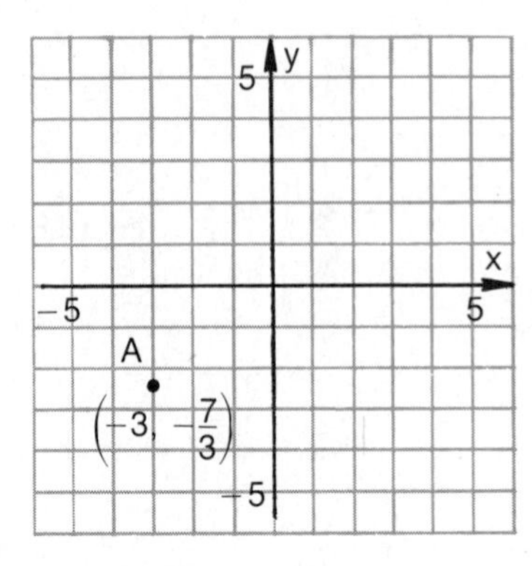

Solution: (a) $A\left(-3, \frac{-7}{3}\right)$ is shown in the graph. The point is in Quadrant III.

(b) B(0, −4) is shown in the graph. The point is on the y-axis, and hence it is *not* in *any* quadrant.

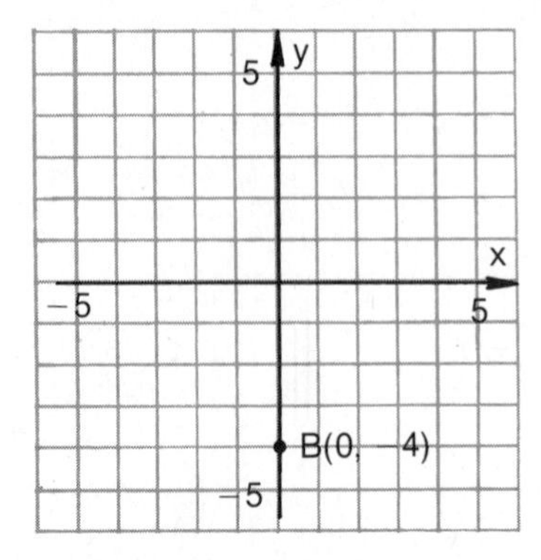

5. C(4, 2)

6. D(−3, 1)

7. E(4, 0)

8. $F\left(-5, -\frac{3}{2}\right)$

9. $G\left(0, \frac{-5}{2}\right)$

10. $H\left(\frac{1}{2}, \frac{-3}{2}\right)$

In Problems 11 through 16, graph the ordered pairs in the given sets.

Illustration: $\{(x, y) \mid y = x + 1, x \in \{-1, 0, 1\}\}$

Solution: For $x = -1$, $y = -1 + 1 = 0$.
For $x = 0$, $y = 0 + 1 = 1$.
For $x = 1$, $y = 1 + 1 = 2$.
We then graph the ordered pairs (−1, 0), (0, 1), and (1, 2) as shown.

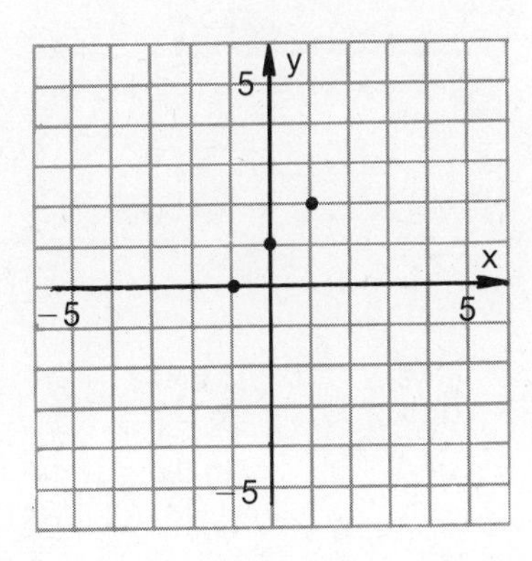

11. $\{(x, y) \mid y = x + 3, x \in \{-2, -1, 0, 1, 2\}\}$

12. $\{(x, y) \mid y = 2x + 1, x \in \{-1, 0, 1\}\}$

13. $\{(x, y) \mid x - y = 4, x \in \{-1, 0, 1\}\}$

14. $\{(x, y) \mid x - 3y = 6, x \in \{1, 3, 5\}\}$

15. $\{(x, y) \mid 2x - y - 3 = 0, x \in \{-1, 0, 1\}\}$

16. $\{(x, y) \mid 2x + y + 2 = 0, x \in \{-1, 0, 1\}\}$

In Problems 17 through 24, graph the given equation.

17. $y = x - 5$

18. $2y = 4x - 2$

19. $2x + 3y = 6$

20. $3x + 2y = 6$

21. $2x - y = 4$

22. $3x - y = 3$

23. $2x + y - 8 = 0$

24. $3x + y - 6 = 0$

In Problems 25 through 30, find the x- and y-intercepts, then graph the equation.

Illustration: $2x - 3y = -6$

Solution: If $x = 0$, $y = 2$. Thus, the y-intercept is 2. If $y = 0$, $x = -3$. Thus, the x-intercept is -3. We graph these intercepts and draw a line through the points as shown.

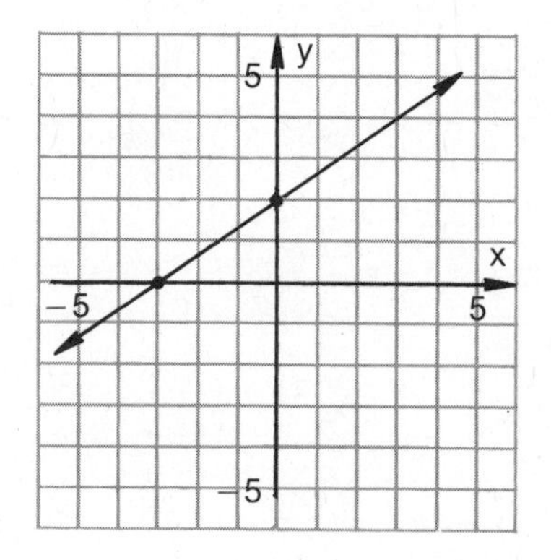

25. $y + 2x = 4$

26. $y + 3x = 6$

27. $x - 5y = -10$

28. $x - 3y = -9$

29. $x - y - 5 = 0$

30. $2x - 3y - 12 = 0$

In Problems 31 through 36, determine if the given line is horizontal or vertical and then graph the line.

Illustration: $-3y = -12$

Solution: $-3y = -12$ is of the form $By = C$, thus it is a *horizontal* line for which $y = 4$, as shown in the graph on the opposite page.

31. $7x = 14$

32. $\frac{3}{2}x = 6$

33. $\frac{-1}{2}y = 4$

34. $\frac{-2}{5}y = 10$

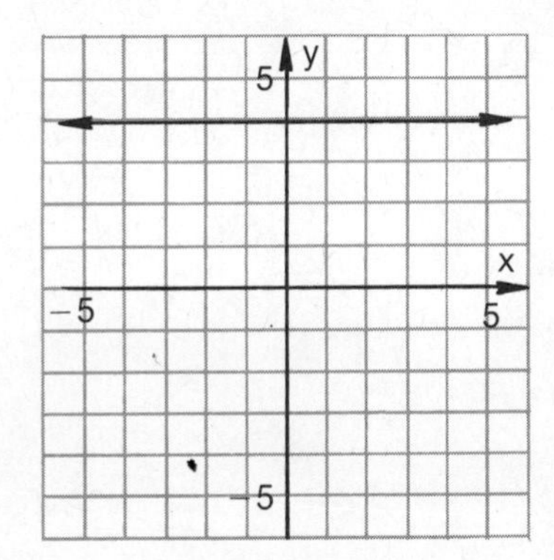

35. $\frac{-3}{4}x = 3$

36. $\frac{-3}{7}x = \frac{6}{7}$

Funky Winkerbean by Tom Batiuk, courtesy of Field Newspaper Syndicate.

OBJECTIVES 7.2

After studying this section, the reader should be able to:

1. **Find the distance between two given points in the plane.**
2. **Find the slope of a line passing through two given points in the plane.**
3. **Determine whether two lines passing through some given points are parallel or perpendicular.**

7.2 THE DISTANCE FORMULA AND THE SLOPE OF A LINE

In the cartoon at the beginning of this section, Les has discovered a formula for the distance between the sun and the earth. In this section we shall develop a formula to find the distance between any two points A and B in the plane. The formula is easy to obtain if we know the Pythagorean theorem, which is stated next.

The Pythagorean Theorem

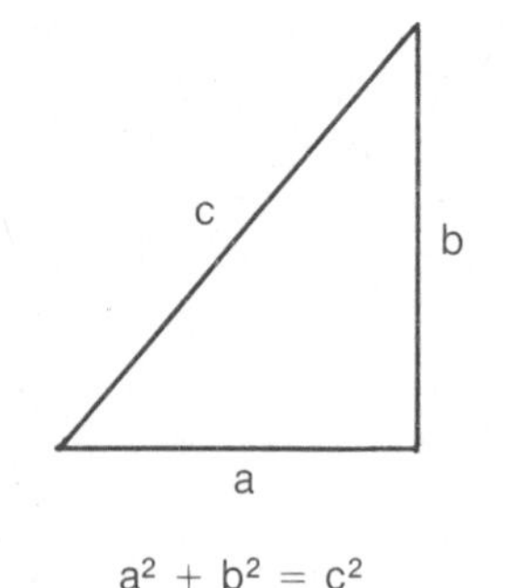

$a^2 + b^2 = c^2$

FIGURE 7.12

PYTHAGOREAN THEOREM

The square of the length of the hypotenuse of any right triangle is equal to the sum of the squares of the lengths of the two legs (see Figure 7.12).

Suppose first that the two points A and B are on the same *horizontal* line, the line $y = b$ shown in Figure 7.13. Then the coordinates of A and B would be, respectively, (x_1, b) and (x_2, b), as in the figure. Because x_1 and x_2 give the distances of A and B from the y-axis, the length of the line segment AB is $|x_2 - x_1|$. If we denote the *length* of the line segment AB by $|AB|$, we have the formula

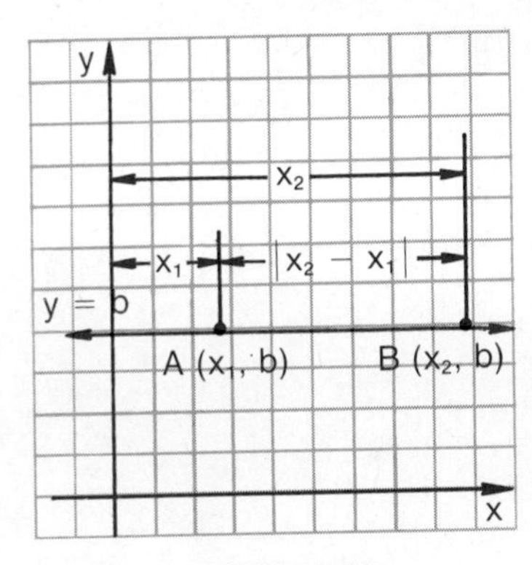

FIGURE 7.13

(1) $|AB| = |x_2 - x_1|$

Similarly, if two points C and D are on the *vertical* line $x = a$, their coordinates are (a, y_1) and (a, y_2). The length of the line segment CD is

(2) $|CD| = |y_2 - y_1|$

Now let us assume that $A(x_1, y_1)$ and $B(x_2, y_2)$ are any two points, as shown in Figure 7.14. As you can see, the line BC is parallel to the y-axis, and AC is parallel to the x-axis.

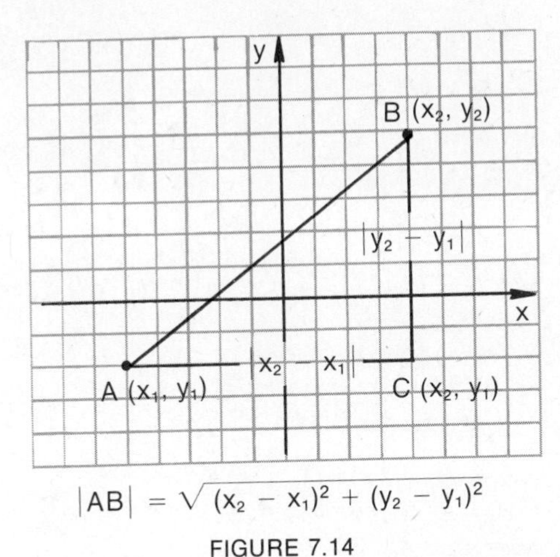

$|AB| = \sqrt{(x_2 - x_1)^2 + (y_2 - y_1)^2}$

FIGURE 7.14

Thus, the lines form the legs of a right triangle with vertices (corners) at A, B, and C. Since A and C are on the same horizontal line, they must have the same y coordinates. Similarly, B and C are on the same vertical line, so they have the same x coordinates. Thus, the coordinates of C must be (x_2, y_1). By using the Pythagorean theorem, we find that the length of the line segment AB is given as:

(3) $|AB| = \sqrt{|AC|^2 + |BC|^2}$

Hence, substituting $|AC| = |x_2 - x_1|$ and $|BC| = |y_2 - y_1|$ into Equation (3), we obtain the distance formula, stated in Theorem 7.1.

THEOREM 7.1

The distance between two points $A(x_1, y_1)$ and $B(x_2, y_2)$ is

$$|AB| = \sqrt{(x_2 - x_1)^2 + (y_2 - y_1)^2}$$

Note that because the numbers $(x_2 - x_1)$ and $(y_2 - y_1)$ are squared, it makes no difference whether they are positive or negative; the result will be the same. This shows that the order in which the given points are taken does not matter.

Example 1

Find the distance between the two given points.
(a) $A(1, 1)$ and $B(5, 4)$
(b) $C(2, 3)$ and $D(-2, 5)$
(c) $E(-2, 1)$ and $F(-2, 3)$

Solution

(a) Using Theorem 7.1 with $x_1 = 1$, $y_1 = 1$, $x_2 = 5$ and $y_2 = 4$, we have $|AB| = \sqrt{(5-1)^2 + (4-1)^2} = \sqrt{(4)^2 + (3)^2} = \sqrt{25} = 5$.

(b) $|CD| = \sqrt{[2-(-2)]^2 + (3-5)^2} = \sqrt{[4]^2 + (-2)^2}$
$= \sqrt{20} = 2\sqrt{5}$

(c) $|EF| = \sqrt{[-2-(-2)]^2 + (1-3)^2} = \sqrt{[0]^2 + (2)^2}$
$= \sqrt{4} = 2$

A second property of a line segment joining two points in the plane is its inclination. For example, if you examine the accompanying graph showing the wholesale price index between the months of June and July and between July and September, you can see that the index rose faster between June and July than between July and September. This can be easily seen because the line segment from June to July is "steeper" than that from July to September.

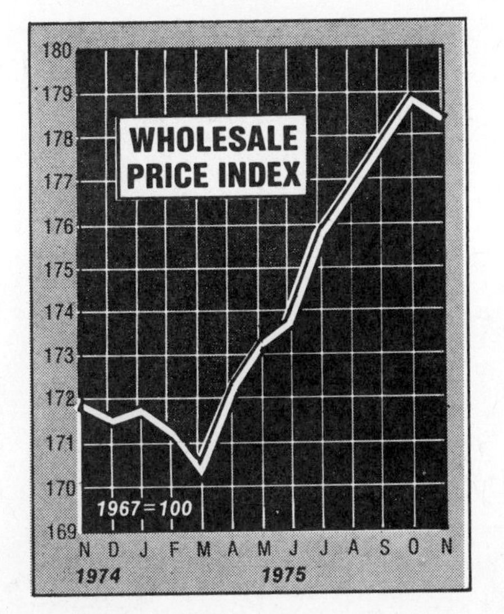

Newsweek—Fenga & Freyer.

In mathematics, there is a method which can be used to measure the "steepness," or inclination, of a line by comparing the *rise* of the segment with a given *run*. Consider the

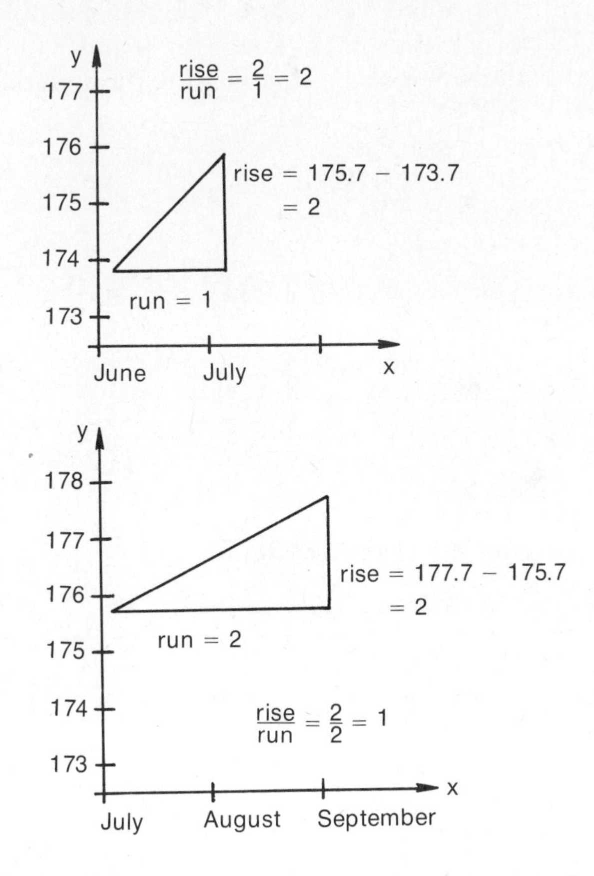

rise and run of the line segments from June to July and July to September shown in the accompanying graphs. The ratio of *rise* to *run* is called the *slope* of the line segment and is denoted by the letter m. Thus,

$$m = \frac{\text{rise}}{\text{run}}$$

We summarize this discussion in the following definition.

DEFINITION 7.1

If $A(x_1, y_1)$ and $B(x_2, y_2)$ are any two distinct points on a line L (which is not parallel to the y-axis), then the slope of L, denoted by m, is given as

$$m = \frac{y_2 - y_1}{x_2 - x_1}$$

It should be noted that when using Definition 7.1, it does not matter which point is taken for A and which for B. For example, the slope of the line passing through the points $A(0, -6)$ and $B(3, 3)$ is

$$m = \frac{3 - (-6)}{3 - 0} = \frac{9}{3} = 3$$

If we choose A to be (3, 3) and B to be (0, −6), the slope is

$$m = \frac{-6 - 3}{0 - 3} = \frac{-9}{-3} = 3$$

Thus, since $\frac{y_2 - y_1}{x_2 - x_1} = \frac{y_1 - y_2}{x_1 - x_2}$, A and B can be interchanged without changing the resulting slope.

Example 2

Find the slope of the line passing through the given points.
(a) $A(-3, 1)$, $B(-1, -2)$
(b) $A(-1, 5)$, $B(-1, 6)$
(c) $A(1, 1)$, $B(2, 3)$

Solution

(a) The slope is $m = \frac{-2 - 1}{-1 - (-3)} = \frac{-3}{2}$.

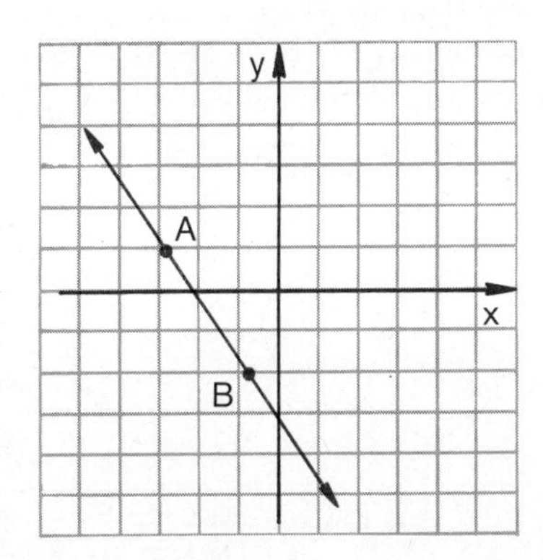

(b) The slope is $m = \dfrac{6-5}{-1-(-1)} = \dfrac{1}{0}$, which is undefined. Notice that the line passing through A(−1, 5) and B(−1, 6) has as its equation $x = -1$, and is a line *parallel* to the y-axis. The fact that the run is zero in lines parallel to the y-axis (that is, *vertical* lines) requires that they be excluded from Definition 7.1; their slope is undefined.

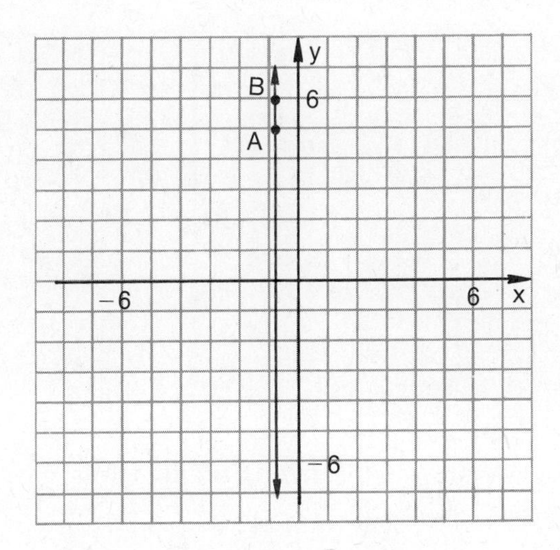

(c) The slope is $m = \dfrac{3-1}{2-1} = \dfrac{2}{1} = 2.$

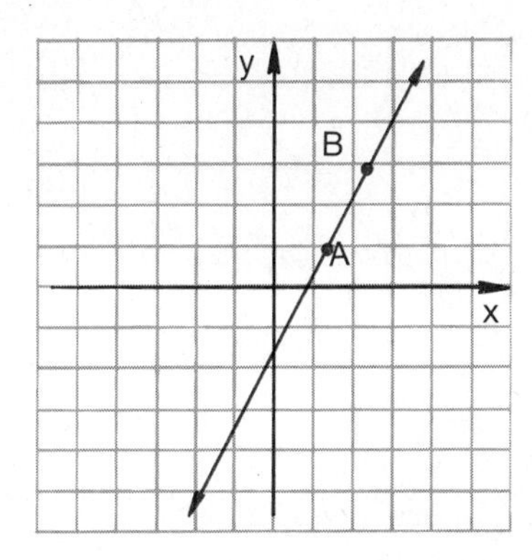

We can see from Example 2 that:

(a) A line that falls from left to right has a *negative* slope.

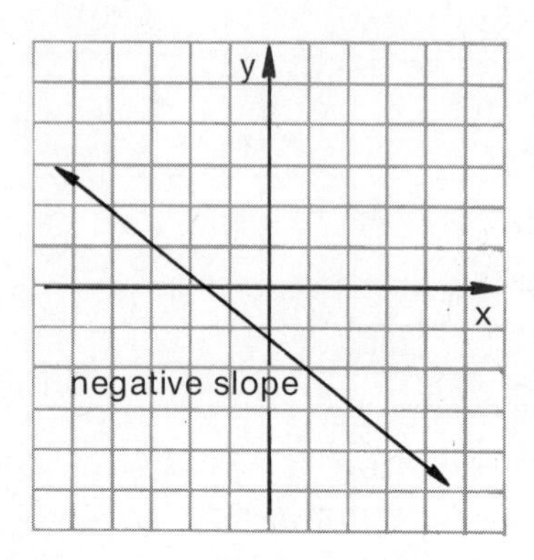

(b) A *vertical* line has no slope (the slope is undefined).

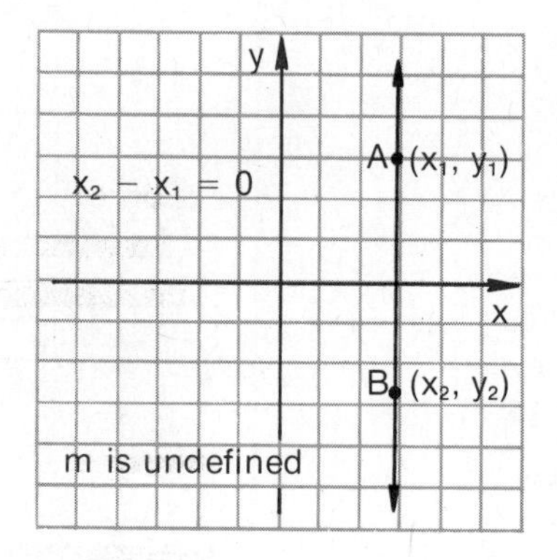

(c) A line that rises from left to right has a positive slope.

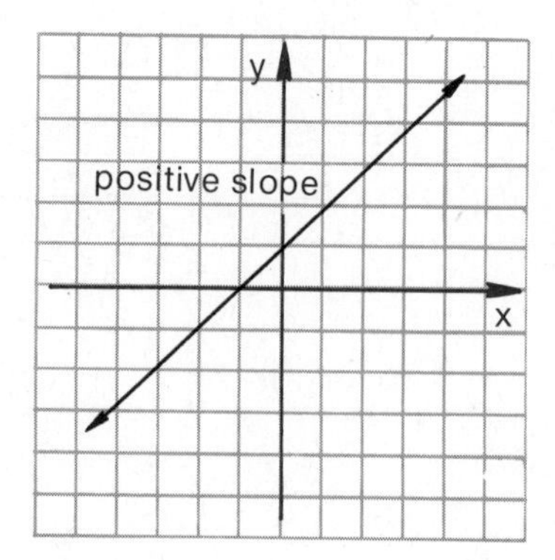

As for lines that are parallel to the x-axis,

(d) A *horizontal* line has slope of 0.

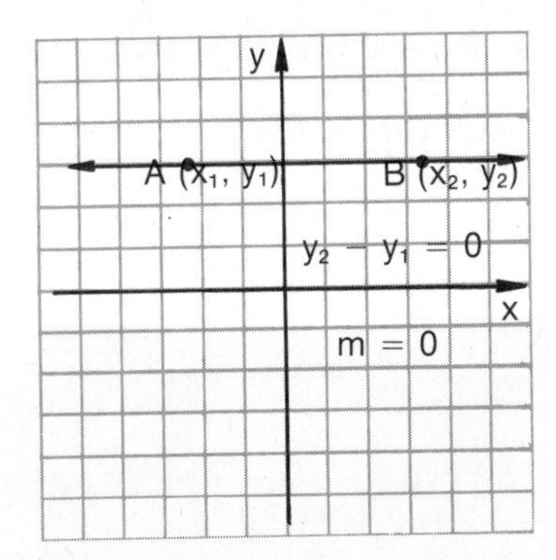

Definition 7.1 can be used to determine when two line segments are parallel. Since two parallel lines must have the same inclination and thus the same slope, we have the following theorem.

THEOREM 7.2

The lines L_1 and L_2 with slopes m_1 and m_2 are parallel if and only if $m_1 = m_2$.

Thus, by Theorem 7.2, the lines passing through points A(4, 2), B(5, −1) and points C(−1, 3), D(0, 0) are parallel because they have equal slopes. The slope of AB is

$$m_1 = \frac{-1-2}{5-4} = \frac{-3}{1} = -3$$

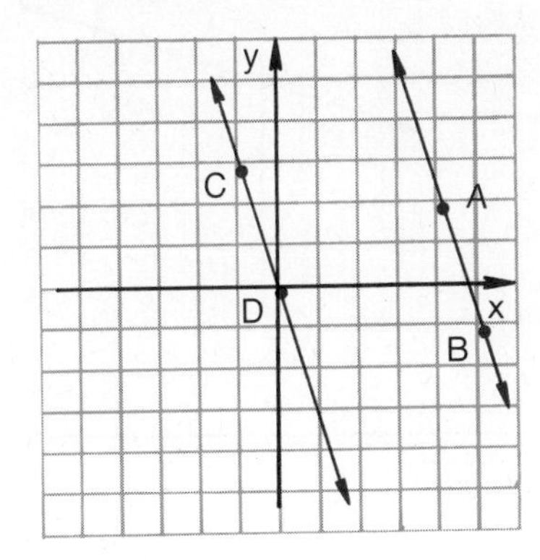

and the slope of CD is

$$m_2 = \frac{0-3}{0-(-1)} = \frac{-3}{1} = -3$$

Finally, it can be shown that two lines with slopes m_1 and m_2 are perpendicular if $m_1 = \frac{-1}{m_2}$, that is, if $m_1 \bullet m_2 = -1$. This fact is stated formally in Theorem 7.3.

THEOREM 7.3

The lines L_1 and L_2 with slopes m_1 and m_2, respectively, are perpendicular if and only if $m_1 \bullet m_2 = -1$.

Thus, the line passing through A(4, 2) and B(5, −1) is perpendicular to the line passing through C(2, 1) and D(5, 2), since the slope of AB is

$$m_1 = \frac{-1-2}{5-4} = \frac{-3}{1} = -3$$

and that of CD is

$$m_2 = \frac{2-1}{5-2} = \frac{1}{3}$$

Thus, $m_1 \bullet m_2 = -1$. These perpendicular lines can be graphed as shown.

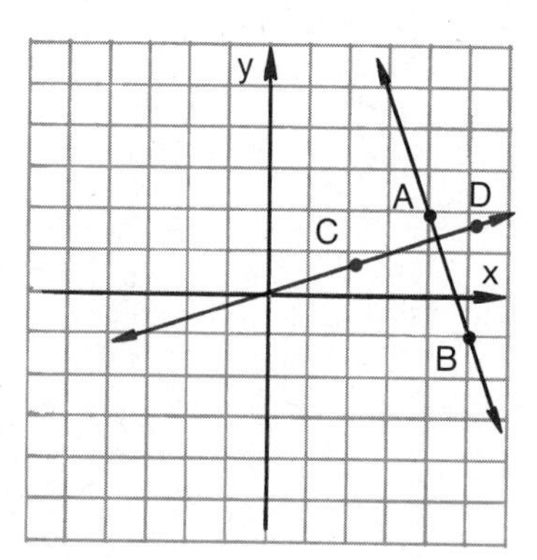

Example 3

A line L_1 has slope $\frac{2}{3}$. Find:

(a) Whether the line passing through A(5, 1) and B(8, 3) is parallel or perpendicular to L_1.
(b) Whether the line passing through C(−1, −4) and D(−3, −1) is parallel or perpendicular to L_1.

Solution

(a) The slope of AB is $m = \frac{3-1}{8-5} = \frac{2}{3}$. Since the slope of AB is identical to that of L_1, the line AB is parallel to L_1.

(b) The slope of CD is $m = \frac{-1-(-4)}{-3-(-1)} = \frac{3}{-2} = \frac{-3}{2}$. Since the slope of L_1 is $\frac{2}{3}$ and $\frac{2}{3} \cdot \frac{-3}{2} = -1$, CD and L_1 are perpendicular.

PROGRESS TEST

1. The distance between the points A(2, −3) and B(8, 5) is ________ units.
2. The slope of the line passing through A(1, −2) and (−3, −4) is ________.
3. The slope of the line passing through A(−1, 2) and B(−1, −2) is ________.
4. The slope of the line passing through A(3, −4) and B(−2, −4) is ________.
5. Line AB passes through A(0, 3) and B(1, 5). A line parallel to AB must have a slope of ________.
6. A line perpendicular to the line AB in Problem 5 must have a slope of ________.

EXERCISE 7.2

In Problems 1 through 10, find the distance between the two given points and find the slope of the line passing through the points. Graph each line on the plane.

Illustration: A(−3, 2) and B(5, 4)

Solution: The distance between the points is

$$|AB| = \sqrt{(x_2 - x_1)^2 + (y_2 - y_1)^2}$$

$$\text{or } |AB| = \sqrt{[5-(-3)]^2 + (4-2)^2}$$

$$= \sqrt{[8]^2 + (2)^2} = \sqrt{68} = 2\sqrt{17}$$

The slope of the line going through the two points is

$$m = \frac{y_2 - y_1}{x_2 - x_1} = \frac{4-2}{5-(-3)} = \frac{2}{8} = \frac{1}{4}$$

The graph is shown at the top of page 378.

PROGRESS TEST ANSWERS

1. 10
2. $\frac{1}{2}$
3. undefined
4. 0
5. 2
6. $\frac{-1}{2}$

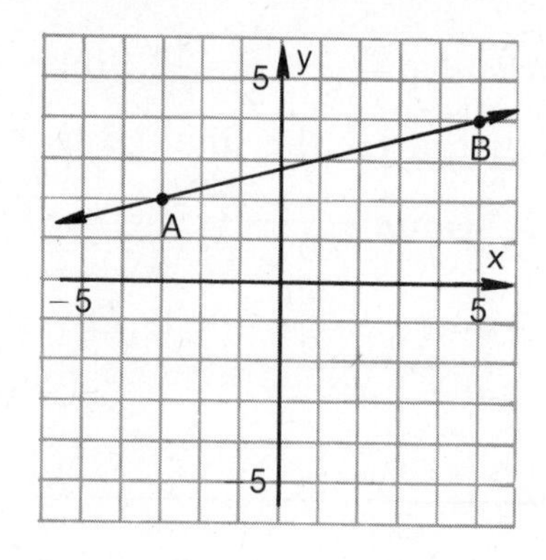

1. A(2, 4), B(−1, 0)

2. A(3, −2), B(8, 10)

3. C(−4, −5), D(−1, 3)

4. C(5, 7), D(−2, 3)

5. E(4, 8), G(1, −1)

6. H(−2, −2), I(6, −4)

7. A(3, −1), B(−2, −1)

8. C(−2, 3), D(4, 3)

9. E(−1, 2), F(−1, −4)

10. G(−3, 2), H(−3, 5)

In Problems 11 through 20, determine whether the lines passing through the two given pairs of points are parallel, perpendicular, or neither.

Illustration: A(2, 3), B(5, 2) and C(4, 2), D(7, 1)

Solution: The slope of AB is $m_1 = \frac{2-3}{5-2} = \frac{-1}{3}$

The slope of CD is $m_2 = \frac{1-2}{7-4} = \frac{-1}{3}$

Since AB and CD have the same slope, then by Theorem 7.2 they are parallel.

11. A(1, 6), B(−1, 4) and C$\left(1, \frac{-7}{2}\right)$, D$\left(\frac{7}{2}, -1\right)$

12. $A(0, 4)$, $B(1, -1)$ and $C(0, 1)$, $D\left(\frac{7}{2}, -1\right)$

13. $A(2, 0)$, $B(4, 5)$ and $D\left(\frac{7}{2}, 0\right)$, $E(1, 1)$

14. $A(1, 1)$, $B(-1, 2)$ and $E(1, -1)$, $F(0, -3)$

15. $A(-1, 1)$, $B(1, 2)$ and $C(1, -1)$, $D(0, -1)$

16. $A(1, 1)$, $B(3, 3)$ and $C(1, -1)$, $D(0, 2)$

17. $A(1, -1)$, $B\left(2, \frac{-1}{2}\right)$ and $C(2, -2)$, $D(1, 0)$

18. $A(1, 1)$, $B\left(\frac{1}{5}, 0\right)$ and $C(1, 1)$, $D\left(0, \frac{9}{5}\right)$

19. $A(0, 1)$, $B(14, -1)$ and $C\left(0, \frac{3}{2}\right)$, $D\left(\frac{7}{2}, 1\right)$

20. $A(2, -2)$, $B(1, -7)$ and $C(1, -3)$, $D(0, -8)$

In Problems 21 through 25, the given points are the vertices of a triangle. Use the distance formula and the Pythagorean theorem to determine if the triangle is a right triangle. Also determine if the triangle is isosceles (two sides equal) or scalene (no sides equal).

21. $(2, 2)$, $(0, 5)$, $(-20, 12)$

22. $(0, 6)$, $(-3, 0)$, $(9, -6)$

23. $(2, 2)$, $(0, 5)$, $(-19, -12)$

24. $(0, 0)$, $(6, 0)$, $(3, 3)$

25. $(2, 2)$, $(-4, -14)$, $(-20, -8)$

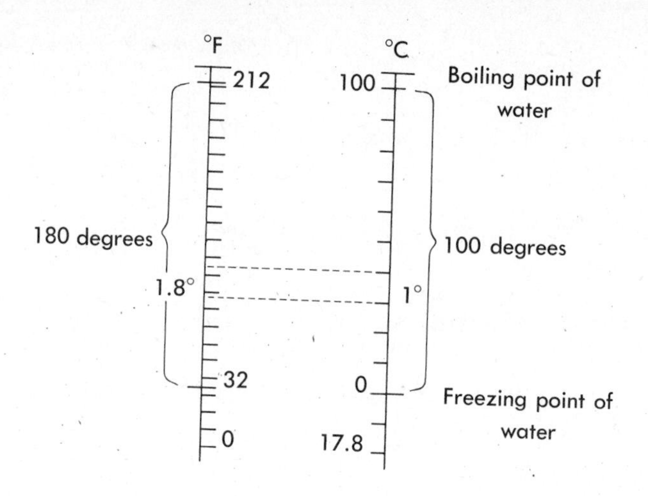

OBJECTIVES

After studying this section, the reader should be able to find the equation of a line, in standard form, given:

a. Two points through which the line passes.
b. The slope of the line and one point through which the line passes.
c. The slope and the y-intercept of the line.
d. One point through which the line passes and the fact that the line is parallel or perpendicular to a second line whose equation is specified.

7.3 EQUATIONS OF A LINE

The figure at the beginning of this section shows two thermometers comparing the Celsius (Centigrade) and Fahrenheit scales. Do you know how to convert temperatures in the Celsius scale to Fahrenheit? One way of doing this is simply to graph ordered pairs of numbers in which the

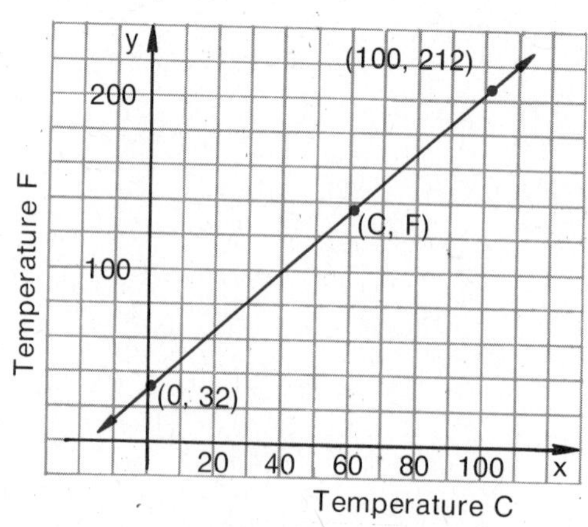

FIGURE 7.15

first coordinate represents the Celsius temperature and the second the corresponding Fahrenheit temperature. Since at 0° Celsius the corresponding Fahrenheit temperature is 32°, one such point is (0, 32). Another one is (100, 212). Thus, the graph of the equation can be easily found, as shown in Figure 7.15. As you know, the graph of a line can be obtained by using *any* two given points on the line. To obtain the *equation* of the line—that is, to find F in terms of C—we select a point on the line and assign it coordinates (C, F). The slope of the line going through (0, 32) and (100, 212) is

The points (0, 32) and (100, 212) correspond to the freezing and boiling points of water, respectively.

$$m = \frac{212 - 32}{100 - 0} = \frac{180}{100} = \frac{9}{5}$$

The slope of the line going through (0, 32) and (C, F) is

$$m = \frac{F - 32}{C - 0} = \frac{F - 32}{C}$$

Since these slopes are the same, we write

$$\frac{F - 32}{C} = \frac{9}{5}$$

$$F - 32 = \frac{9}{5}C \qquad \text{Multiplying both sides by C}$$

$$F = \frac{9}{5}C + 32 \qquad \text{Adding 32 to both members}$$

What we did here was to work with the graph of a line to find its equation (in Section 7.1, we worked from an equation, the standard $Ax + By + C = 0$, to graph a line). In general, if a line goes through two points $P_1(x_1; y_1)$ and $P_2(x_2, y_2)$,

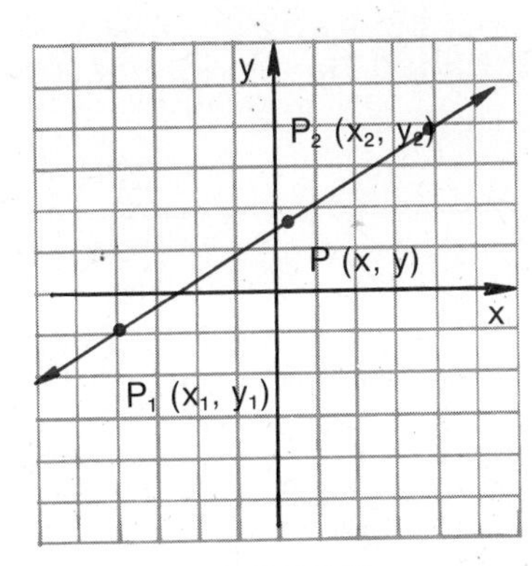

FIGURE 7.16

as shown in Figure 7.16, an equation for this line can be found as follows:

1. Select a general point $P(x, y)$ on the line.
2. The slope of the line P_1P_2 is

$$m = \frac{y_2 - y_1}{x_2 - x_1}, \quad x_2 \neq x_1$$

3. The slope of the line P_1P is

$$m = \frac{y - y_1}{x - x_1}, \quad x \neq x_1$$

4. Since the slopes are equal,

$$\frac{y - y_1}{x - x_1} = \frac{y_2 - y_1}{x_2 - x_1},$$

$$\text{or } y - y_1 = \frac{y_2 - y_1}{x_2 - x_1} \cdot (x - x_1)$$

We summarize this discussion in the following theorem.

THEOREM 7.4

The two-point form.

An equation of a line going through the points (x_1, y_1) and (x_2, y_2) is given as

$$(1) \quad y - y_1 = \frac{y_2 - y_1}{x_2 - x_1} \cdot (x - x_1), \; x_2 \neq x_1$$

Equation (1) is called the *two-point* form of an equation of a line.

Example 1

Find an equation of the line going through the points (5, 2) and (6, 4) and write the equation in standard form.

Solution

Letting $(x_1, y_1) = (5, 2)$ and $(x_2, y_2) = (6, 4)$, we obtain

$$y - 2 = \frac{4 - 2}{6 - 5} \cdot (x - 5)$$

$$\text{or } y - 2 = 2(x - 5)$$

$y - 2 = 2(x - 5) = 2x - 10$, or, in standard form, $2x - y - 8 = 0$.

Thus, in standard form an equation of the line is $2x - y - 8 = 0$.

If we replace $\dfrac{y_2 - y_1}{x_2 - x_1}$ by m in Equation (1) of Theorem 7.4, we obtain

$$(2) \qquad y - y_1 = m(x - x_1)$$

The point-slope form.

Equation (2) is called the *point-slope* form of the equation. The point-slope form enables us to find an equation of a line when a point $P(x_1, y_1)$ and the slope m are given. We shall use this equation in Example 2.

Example 2

Find the equation of the line with slope $m = -2$ and passing through the point (3, 5).

Solution

Here $m = -2$, $(x_1, y_1) = (3, 5)$. Substituting in Equation (2), we get $y - 5 = -2(x - 3)$, or $2x + y - 11 = 0$.

If, in Equation (2), the point $P(x_1, y_1)$ is on the y-axis, then $x_1 = 0$. If we let $y_1 = b$, then $P(x_1, y_1) = P(0, b)$. Thus, the point-slope form of the equation is $y - b = m(x - 0)$. Solving for y, we obtain

$$(3) \qquad y = mx + b$$

The slope-intercept form.

Equation (3) is called the *slope-intercept* form of the equation of the line. Note that in Equation (3), m is the *slope* and b is the *y-intercept* of the line.

Example 3

A line has slope 5 and y-intercept 3. Find the slope-intercept form of the equation of this line.

Solution

Using Equation (3) with $m = 5$ and $b = 3$, we find the equation to be $y = 5x + 3$.

Equation (3) is especially convenient because we can find the slope and y-intercept of a given line by just writing the equation of the line in the form $y = mx + b$. For example, to find the slope and y-intercept of the line $2x + 5y = 7$, we solve for y, obtaining

$$y = \frac{-2}{5}x + \frac{7}{5}.$$

Thus, from Equation (3), we can see that m (the slope) is $\frac{-2}{5}$, and b (the y-intercept), is $\frac{7}{5}$.

Example 4 Find the slope and y-intercept of the line $6x + 3y = 12$.

Solution Solving for y, we obtain $y = -2x + 4$. Thus, the slope is -2 and the y-intercept is 4.

We are already familiar with particular aspects of the slopes of parallel lines and perpendicular lines (Theorems 7.2 and 7.3). We can now use the fact that two parallel lines have identical slopes to find an equation of a line that is parallel to a given line. Likewise, the fact that two perpendicular lines have slopes that are negative reciprocals of each other can be used to find an equation of a line that is perpendicular to a given line.

Example 5 Find the equation of the line that passes through the point (2, 1) and is

(a) parallel to the line $y - x = 1$.
(b) perpendicular to the line $y - x = 1$.

Solution

(a) We first write the equation $y - x = 1$ in the slope-intercept form: $y = x + 1$. We can clearly see that the slope of this line is 1. If we wish to construct another line parallel to $y = x + 1$ and passing through (2, 1), we simply use the point-slope form, with $(x_1, y_1) = (2, 1)$ and $m = 1$. This gives us

Since the new line must be parallel to $y = x + 1$, which has a slope of $m = 1$, the slope of the new line must be 1.

$$y - 1 = 1(x - 2)$$
$$\text{or } y = x - 1$$

(b) The line $y - x = 1$ has slope 1. A line perpendicular to this line must have a slope $\frac{-1}{1} = -1$. Using the point-slope form again, we find that an equation of a line perpendicular to $y - x = 1$ and passing through (2, 1) is

$$y - 1 = -1(x - 2)$$
$$\text{or } y = -x + 3$$

The sketch of all the lines involved is shown in Figure 7.17.

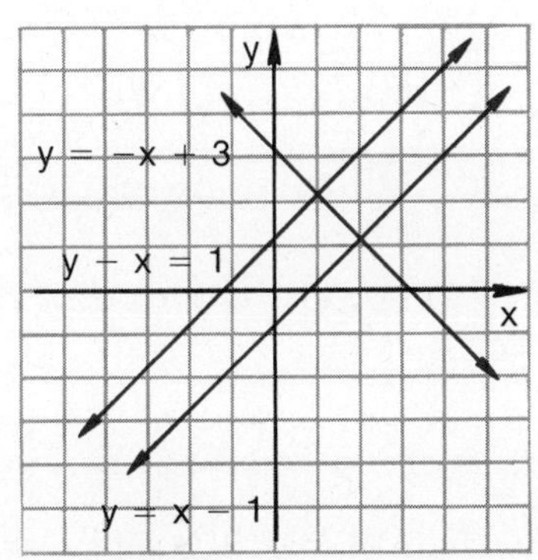

FIGURE 7.17

PROGRESS TEST

1. The two-point form of the equation of the line passing through (3, 1) and (0, 4) is ____________________.

2. The point-slope form of the equation of the line with slope -7 and passing through $(-1, -2)$ is ____________________.

3. The slope-intercept form of the equation of the line with slope $\frac{1}{3}$ and y-intercept $\frac{-3}{4}$ is ____________________.

4. The slope and y-intercept of the line with equation $2y = -3x - 2$ are ________ and ________ respectively.

5. The equation of the line passing through (1, 1) and parallel to $2y = 6x + 5$ is ____________________.

6. The equation of the line passing through (1, 1) and perpendicular to $2y = 6x + 5$ is ____________________.

EXERCISE 7.3

In Problems 1 through 20, a line will be described as (1) passing through two points [e.g., "Through (1, −1) and (2, 2)"], (2) having a particular slope and passing through a specified point [e.g., "With slope 2, through (−3, 5)"], (3) having a particular slope and y-intercept, or (4) passing through a point and parallel or perpendicular to a specified line. Use the given information to find the equation of the line and write it in the standard form $Ax + By + C = 0$.

Illustration: (a) Through (3, 5) and (1, 7)

(b) With slope $\frac{-1}{5}$, through (1, −3)

(c) With slope −4, y-intercept 4

Solution: (a) Using Equation (1), we have

$$y - 5 = \frac{7-5}{1-3} \cdot (x - 3)$$

$$\text{or } y - 5 = -1(x - 3)$$

Writing this equation in standard form, we obtain $x + y - 8 = 0$.

(b) Using Equation (2), we obtain $y + 3 = \frac{-1}{5}(x - 1)$. Writing this equation in standard form (by first multiplying each term by 5), we have $x + 5y + 14 = 0$.

(c) From Equation (3), with $m = -4$ and $b = 4$, we get $y = -4x + 4$. Writing this equation in standard form, we have $4x + y - 4 = 0$.

PROGRESS TEST ANSWERS

1. $y - 1 = \frac{4-1}{0-3}(x - 3)$, or $y - 1 = -1(x - 3)$
2. $y + 2 = -7(x + 1)$
3. $y = \frac{1}{3}x - \frac{3}{4}$
4. $\frac{-3}{2}, -1$
5. $y - 1 = 3(x - 1)$, or $y = 3x - 2$
6. $y - 1 = -\frac{1}{3}(x - 1)$, or $y = -\frac{1}{3}x + \frac{4}{3}$

1. Through (1, −1) and (2, 2)

2. Through (−3, −4) and (−2, 0)

3. Through (3, 2) and (2, 3)

4. Through (3, 0) and (0, 5)

5. With slope 2, through (−3, 5)

6. With slope $\frac{1}{2}$, through (2, 3)

7. With slope -3, through $(-1, -2)$

8. With slope $\frac{-1}{3}$, through $(2, -4)$

9. With slope 5, y-intercept 2

10. With slope $\frac{1}{4}$, y-intercept 2

11. With slope $\frac{-1}{5}$, y-intercept $\frac{-1}{3}$

12. With slope $\frac{-1}{7}$, y-intercept $\frac{-1}{9}$

13. Through $(1, -2)$ and parallel to $y = 2x + 1$

14. Through $(-1, -2)$ and parallel to $2y = -4x + 5$

15. Through $(-5, 3)$ and parallel to $2y + 6x = 8$

16. Through $(-3, -5)$ and parallel to $3y - 6x = 12$

17. Through $(1, 1)$ and perpendicular to $2y = x + 6$

18. Through $(2, 3)$ and perpendicular to $3y = -x + 5$

19. Through $(-2, -4)$ and perpendicular to $y - x = 3$

20. Through $(-3, 5)$ and perpendicular to $2y - x = 5$

In Problems 21 through 26, find the slope-intercept form of the given line.

Illustration: $3x - y = 4$
Solution: Solving for y, we have $y = 3x - 4$, the slope-intercept form.

21. $x - 3y = 5$

22. $4x + 5y = 20$

23. $2y = 6 - 5x$

24. $2y = -3x + 6$

25. $x = 4 - 8y$

26. $2x = 6 - 4y$

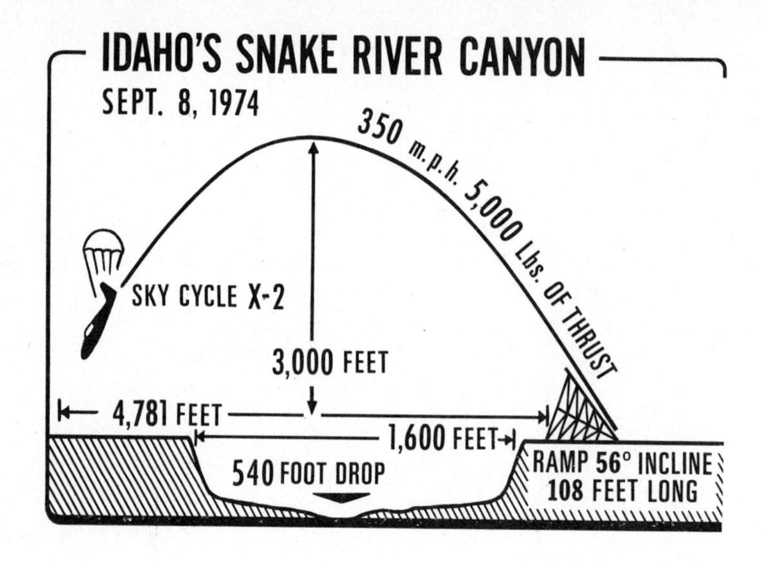

OBJECTIVES 7.4

After studying this section, the reader should be able to:

1. **Graph a parabola.**
2. **Find the vertex of a parabola.**

7.4 THE PARABOLA

The sketch at the beginning of this section shows the intended path of Evel Knievel on his attempted jump across the Snake River Canyon. Do you know the name of the curve his sky-cycle was to follow? The curve is called a *parabola.* In mathematics, the graph of any quadratic equation of the form

$$y = ax^2 + bx + c, \quad a \neq 0$$

Equation for a parabola.

is a *parabola.* The simplest quadratic equation is

$$(1) \qquad y = x^2$$

As in the case of linear equations, the *solution set* of Equation (1) consists of all ordered pairs which satisfy the equation, and the graph of the equation can be obtained by finding the graph of the solution set. The table below gives a few solutions of (1).

x	−3	−2	−1	0	1	2
y	9	4	1	0	1	4

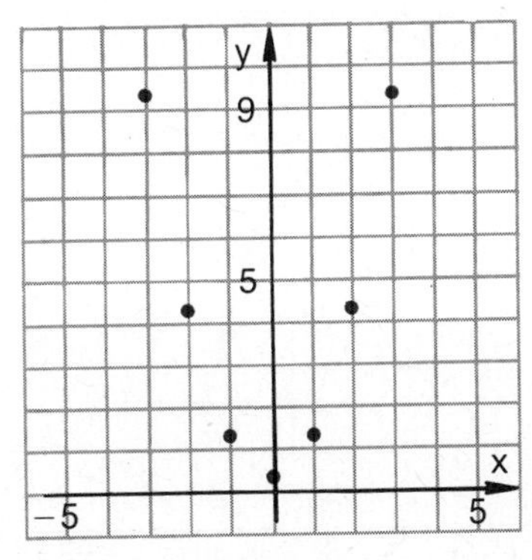

FIGURE 7.18

These solutions are obtained by assigning arbitrary values to x (such as −3, −2, −1, 0, 1, 2) and then finding the corresponding y values. We have graphed these points in Figure 7.18. If these points are connected with a smooth curve, the result is the graph of the parabola $y = x^2$ shown in Figure 7.19.

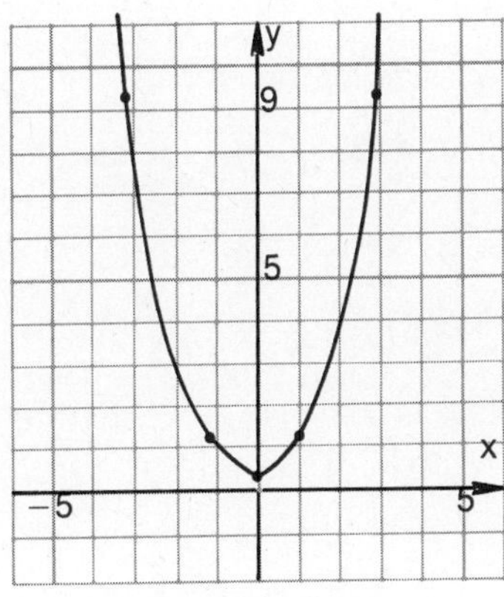

FIGURE 7.19

Example 1

Graph the parabola $y = -x^2 + 1$.

Solution

We make a table by assigning values to x and finding the corresponding values of y. These points are then graphed and connected with a smooth curve, as shown in Figure 7.20.

x	−2	−1	0	1	2
y	−3	0	1	0	−3

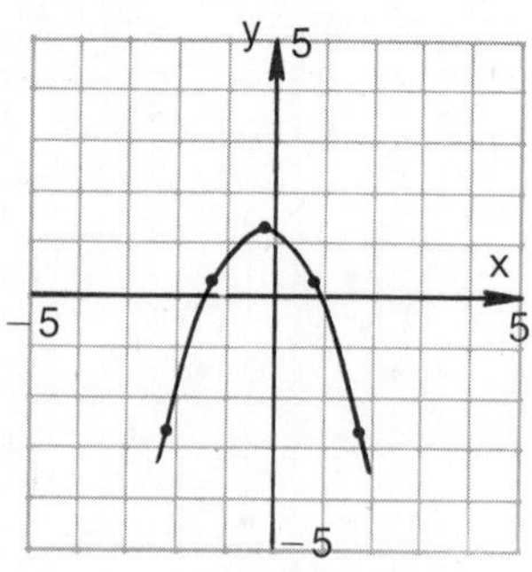

FIGURE 7.20

As you can see, the parabola $y = x^2$ opens *upward*, while the parabola $y = -x^2 + 1$ opens *downward*. In general, the parabola $y = ax^2 + bx + c$

(i) opens upward if a is positive, or

(ii) opens downward if a is negative.

Moreover, the *vertex* (the highest or lowest point on the parabola) has as its x-coordinate $\frac{-b}{2a}$.

Example 2

Graph the parabola $y = 2x^2 + 4x - 1$ and find its vertex.

Solution

Here, $a = 2$ and $b = 4$. Thus, the vertex is at the point where $x = \frac{-4}{2 \cdot 2} = -1$. At this point, $y = 2(-1)^2 + 4(-1) - 1 = -3$. Therefore, the vertex is at $(-1, -3)$. Since a is positive, the parabola opens upward. A sketch of the curve is shown in Figure 7.21.

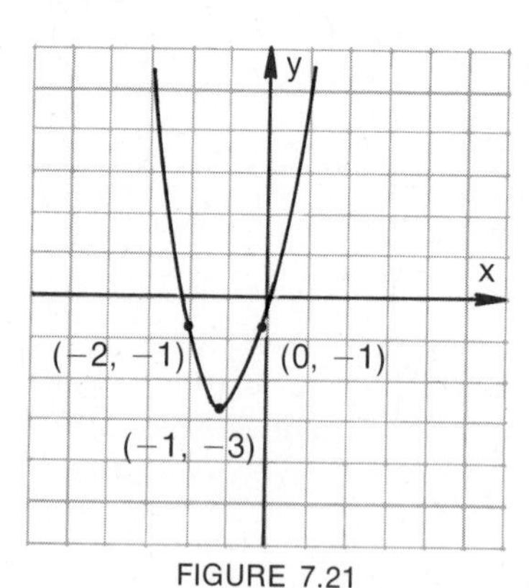

FIGURE 7.21

In Example 2, it is easy to find the point at which the parabola intersects the y-axis, since at this point $x = 0$. Can we find the exact points at which the parabola intersects the x-axis? At these points, $y = 0$, so $2x^2 + 4x - 1 = 0$. Using the quadratic formula to solve for x, we find

$$x = \frac{-4 \pm \sqrt{16 - 4(2)(-1)}}{4} = \frac{-4 \pm \sqrt{24}}{4} = \frac{-2 \pm \sqrt{6}}{2}$$

Thus, the points at which the parabola crosses the x-axis have x-coordinates of $\frac{-2 \pm \sqrt{6}}{2}$ and $\frac{-2 - \sqrt{6}}{2}$.

From the preceding discussion, it is evident that the graph of the parabola can be obtained by using the following procedure:

1. Find the vertex of the parabola by first letting $x = \frac{-b}{2a}$ and then finding the corresponding y values. The line $x = \frac{-b}{2a}$ is called the *axis* of the parabola.
2. Let $y = 0$, and find the corresponding x values (if they exist) by using the quadratic formula or the method of factoring.
3. Let $x = 0$, and find the y-intercept.
4. Connect all the points found in steps 1, 2, and 3, keeping in mind that if a is positive, the parabola will open upward, and that if a is negative, the parabola will open downward.

Example 3

Sketch the graph of $y = x^2 + 3x + 2$.

Solution

We follow the preceding four steps.

1. Since $a = 1$ and $b = 3$, the vertex is at the point where $x = \frac{-3}{2 \cdot 1} = \frac{-3}{2}$. At this point, $y = \left(\frac{-3}{2}\right)^2 + 3\left(\frac{-3}{2}\right) + 2 = \frac{-1}{4}$.

   Hence the vertex is at $\left(\frac{-3}{2}, \frac{-1}{4}\right)$.
2. For $y = 0$, we get

$$x^2 + 3x + 2 = 0$$
$$\text{or } (x + 1)(x + 2) = 0$$

   That is, when $y = 0$, $x = -1$ or $x = -2$.
3. If $x = 0$, $y = 2$.

4. Since a is positive, the parabola opens upward. The sketch is shown in Figure 7.22.

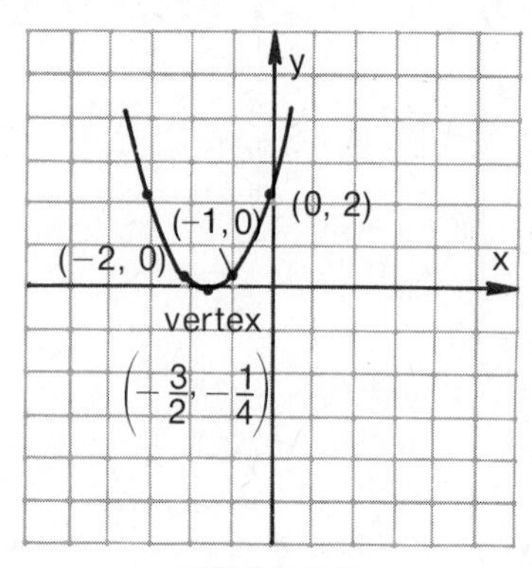

FIGURE 7.22

Note that in Step 2 of the procedure for sketching a parabola the x values corresponding to $y = 0$ may or may not be real. This is because a quadratic equation in one variable may have *one* real solution, *two* real solutions, or *no* real solution. If the equation has *no* real solution, the graph *does not* touch the x-axis; if there is *one* solution, the graph touches the x-axis at just *one* point; and if there are *two* real solutions, the graph crosses the x-axis at two distinct points. All of these cases are shown in Figure 7.23.

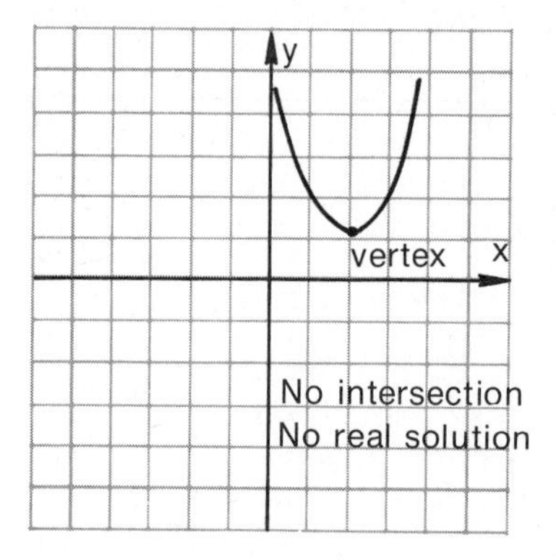

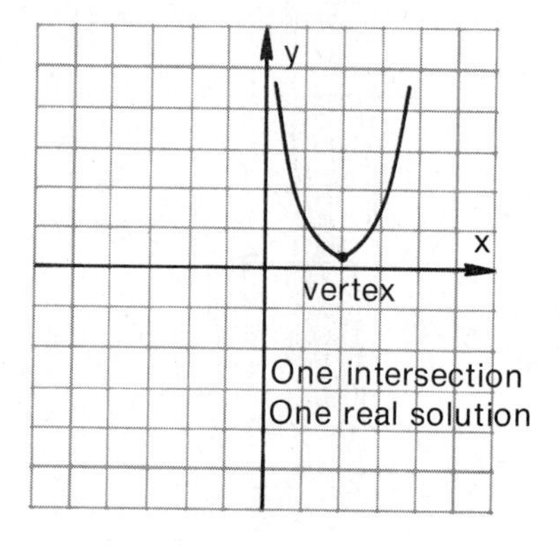

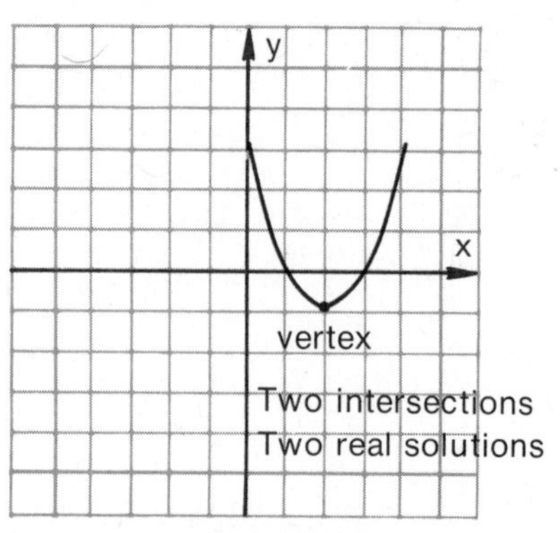

FIGURE 7.23

PROGRESS TEST

1. Graph $y = x^2 - 1$.

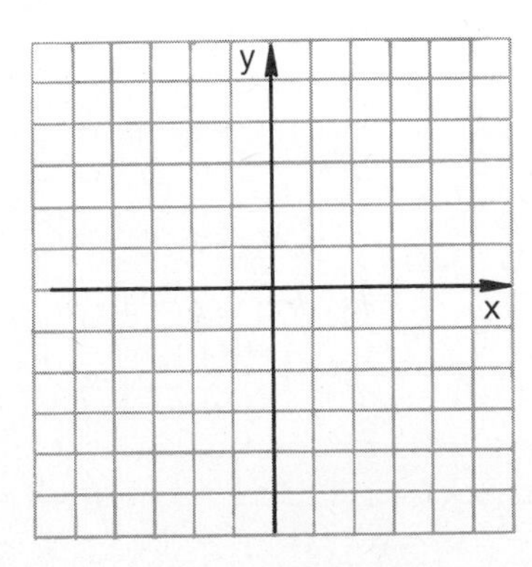

2. The vertex of the parabola $y = x^2 + 6x + 8$ has as its x-coordinate ________ and as its y-coordinate ________.
3. Sketch the graph of $y = x^2 - 2x - 3$.

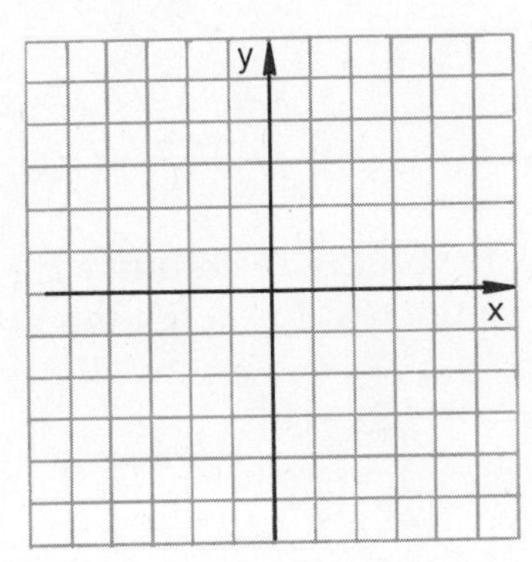

EXERCISE 7.4

In Problems 1 through 10, let x be an integer and $-2 \le x \le 2$. Find solutions for the given equations and use these solutions to construct the graph.

Illustration: $y = x^2 + 3$

Solution:

For $x = -2$, $y = 4 + 3 = 7$, thus $(-2, 7)$ is a solution.
For $x = -1$, $y = 1^2 + 3 = 4$, thus $(-1, 4)$ is a solution.
For $x = 0$, $y = 0 + 3 = 3$, thus $(0, 3)$ is a solution.
For $x = 1$, $y = 1^2 + 3 = 4$, thus $(1, 4)$ is a solution.
For $x = 2$, $y = 2^2 + 3 = 7$, thus $(2, 7)$ is a solution.
The graph of these points (as well as the completed graph) is shown.

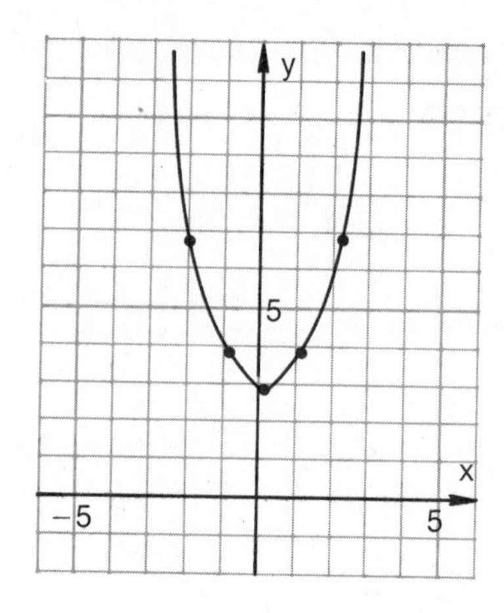

1. $y = x^2 + 1$
2. $y = x^2 + 2$
3. $y = x^2 - 2$
4. $y = x^2 - 3$
5. $y = -x^2 + 2$
6. $y = -x^2 + 3$
7. $y = -x^2 - 1$
8. $y = -x^2 - 2$
9. $y = x^2 + 2x + 1$
10. $y = x^2 + 2x - 1$

In Problems 11 through 20, use the four-step procedure given in the text to sketch the graph of the given equation.

11. $y = x^2 + 2x + 1$
12. $y = x^2 + 2x + 3$
13. $y = x^2 + 6x + 8$
14. $y = x^2 + 4x + 3$
15. $y = -x^2 + 6x - 8$
16. $y = -x^2 + 4x - 3$
17. $y = x^2 - 8x + 12$
18. $y = x^2 - 3x + 2$
19. $y = -x^2 - 4x + 5$
20. $y = -x^2 - 6x - 8$

In Problems 21 through 30, let y be an integer and $-2 \leq y \leq 2$. Find solutions for the given equations and use these solutions to construct the graph.

Illustration: $x = y^2 - 1$

Solution:
For $y = -2$, $x = 3$, thus $(3, -2)$ is a solution.
For $y = -1$, $x = 0$, thus $(0, -1)$ is a solution.
For $y = 0$, $x = -1$, thus $(-1, 0)$ is a solution.
For $y = 1$, $x = 0$, thus $(0, 1)$ is a solution.
For $y = 2$, $x = 3$, thus $(3, 2)$ is a solution.

PROGRESS TEST ANSWERS

1.

2. $x = \frac{-6}{2 \cdot 1} = -3$, $y = (-3)^2 + 6(-3) + 8 = -1$

3. The vertex is at $x = \frac{2}{2} = 1$, $y = 1^2 - 2 - 3 = -4$
When $y = 0$, $x^2 - 2x - 3 = (x - 3)(x + 1) = 0$. Thus, $x = 3$ or $x = -1$.
When $x = 0$, $y = -3$.
The sketch is shown.

These points in the graph of $x = y^2 - 1$ are shown.

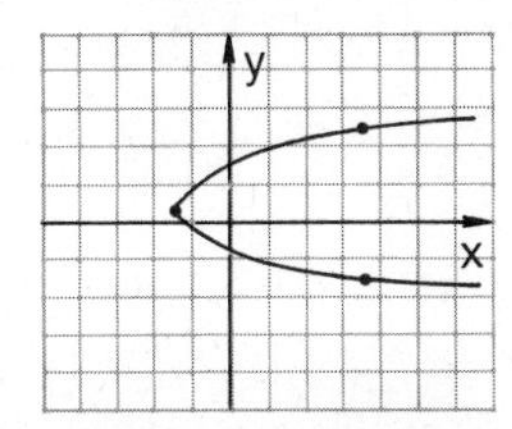

21. $x = y^2 + 1$

22. $x = y^2 + 2$

23. $x = y^2 - 2$

24. $x = y^2 - 3$

25. $x = -y^2 + 2$

26. $x = -y^2 + 3$

27. $x = -y^2 - 1$

28. $x = -y^2 - 2$

29. $x = -2y^2 - 6$

30. $x = -4y^2 - 4$

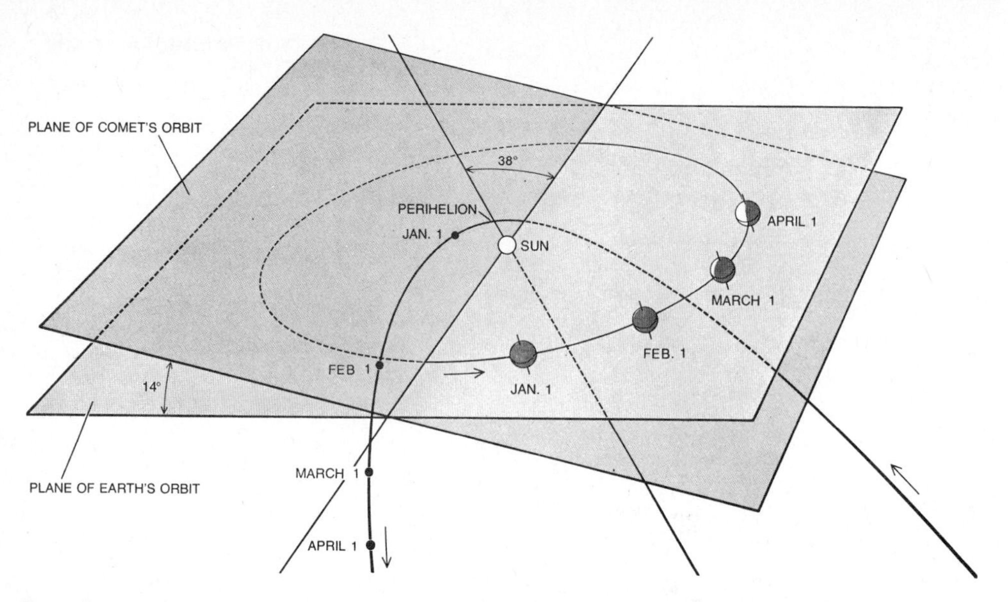

OBJECTIVES 7.5

After studying this section, the reader should be able to:

1. **Find the graph of a circle, given its equation.**
2. **Find the graph of an ellipse, given its equation.**
3. **Find the graph of a hyperbola, given its equation.**

7.5 CIRCLES, ELLIPSES, AND HYPERBOLAS

The diagram at the beginning of this section shows the orbit of the comet Kohoutek with respect to the orbit of the earth. The comet's orbit is an *ellipse*, while the earth's orbit is nearly a perfect *circle*. The ellipse and the circle are called *conic sections* or *conics*, as are the parabola and the hyperbola. As their name implies, conic sections can be obtained by cutting a cone, as with a plane (see Figure 7.24).

We shall discuss here the equations and graphs of the circle, the ellipse, and the hyperbola. First consider the equation $x^2 + y^2 = 25$. Assigning a few arbitrary values to x and solving for y, we find the following ordered pairs: (0, 5), (0, −5), (5, 0), (−5, 0), (3, 4), (3, −4), (−3, 4), and (−3, −4). We graph these points as in Figure 7.25 (a), and then connect the points with a smooth curve. The result is a circle of radius 5 centered at the origin, as shown in Figure 7.25 (b).

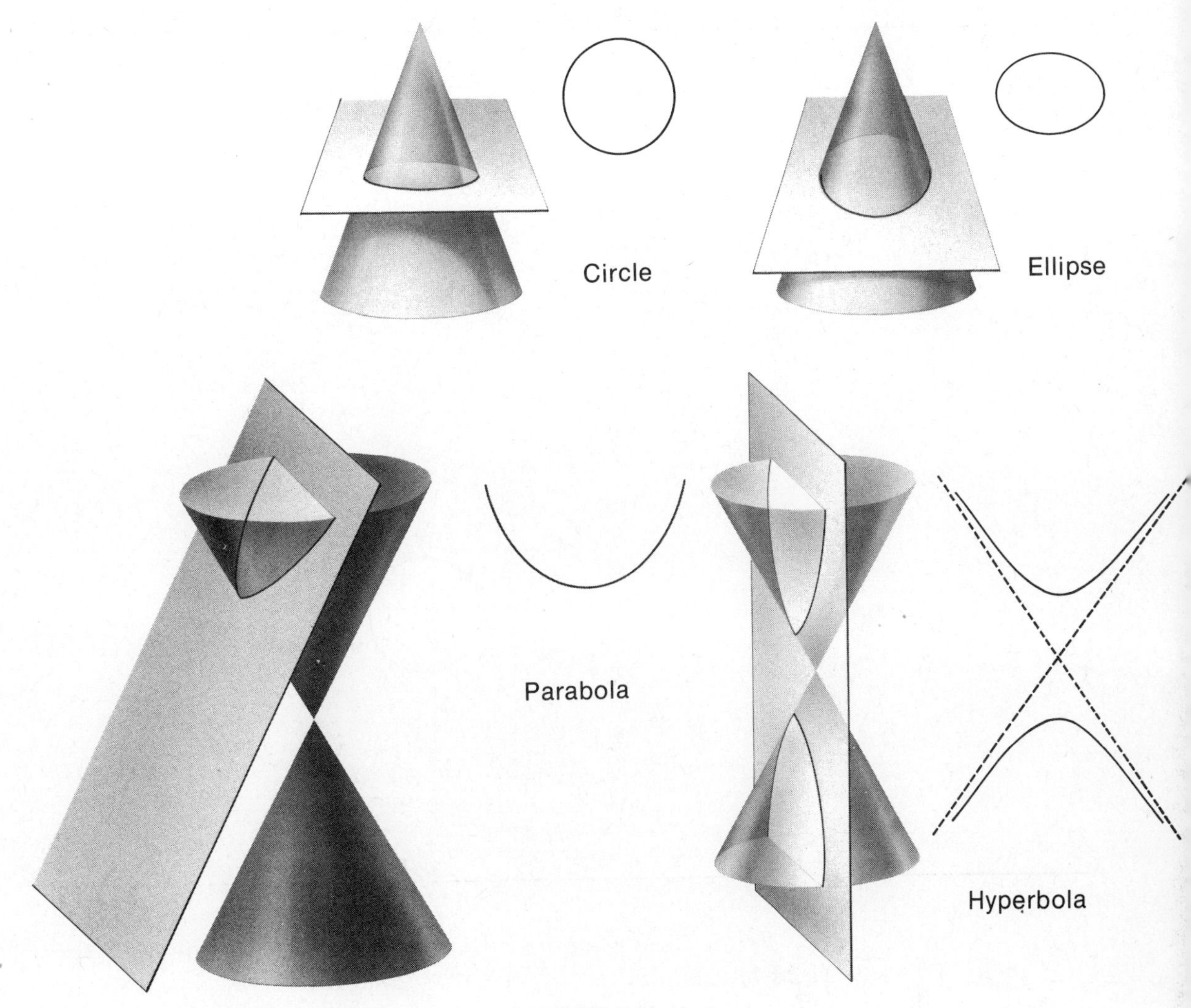

FIGURE 7.24

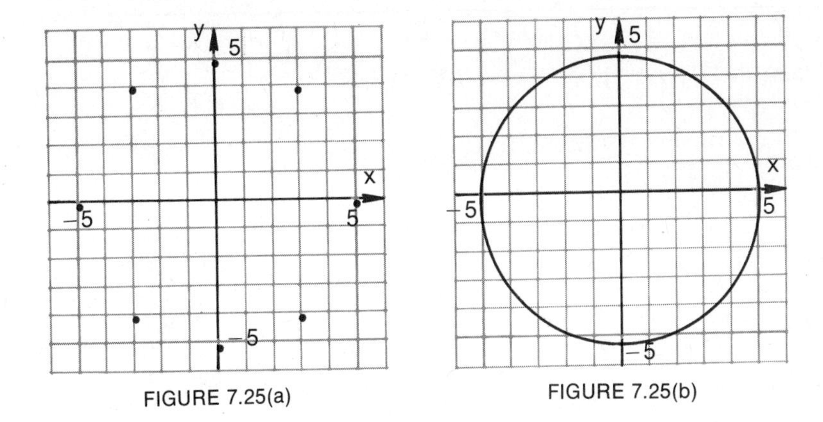

FIGURE 7.25(a)

FIGURE 7.25(b)

Example 1 Graph the equation $x^2 + y^2 = 4$.

Solution The easiest points to obtain are those for which x or y is 0. For $x = 0$, $y = \pm 2$. For $y = 0$, $x = \pm 2$. Thus, (0, 2), (0, −2), (2, 0), and (−2, 0) are on the graph. These four points—as well as the completed graph, showing a circle of radius 2 centered at the origin—are shown in Figure 7.26.

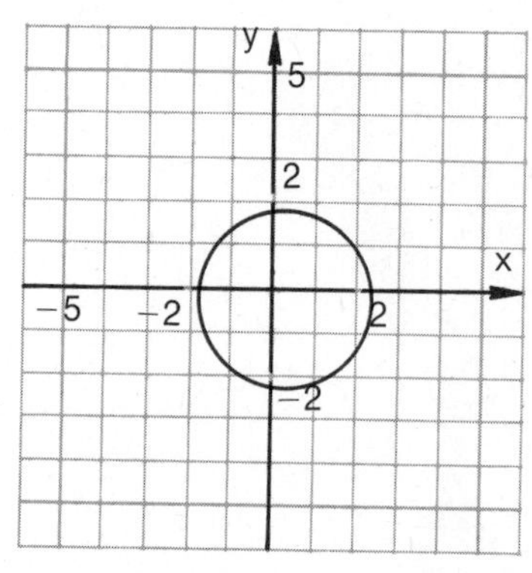

FIGURE 7.26

From the preceding discussion, you can see that the right-hand members of $x^2 + y^2 = 25$ and $x^2 + y^2 = 4$ determine the radius of the circle. In general,

> $x^2 + y^2 = r^2$ is a circle of radius $r > 0$ centered at the origin.

Example 2 Graph the equation $4x^2 + 4y^2 = 1$.

Solution To write the equation in the form $x^2 + y^2 = r^2$, we divide each term by 4, obtaining $x^2 + y^2 = \left(\frac{1}{2}\right)^2$. The graph of this equation is a circle of radius $\frac{1}{2}$ centered at the origin, as shown in Figure 7.27.

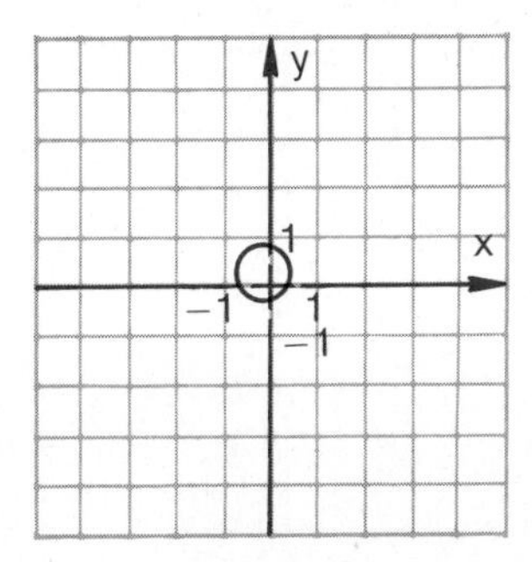

FIGURE 7.27

The second type of equation we shall discuss is that of the ellipse. Consider the equation $4x^2 + 9y^2 = 36$. If we assign some arbitrary values to x and find the corresponding y values, we obtain the following ordered pairs: (0, 2), (0, −2), (3, 0), and (−3, 0). The graph of these points, as well as the completed graph of an ellipse centered at the origin, is shown in Figure 7.28. We now provide a theorem from analytic geometry that will be useful in determining if the graph of a given equation is an ellipse.

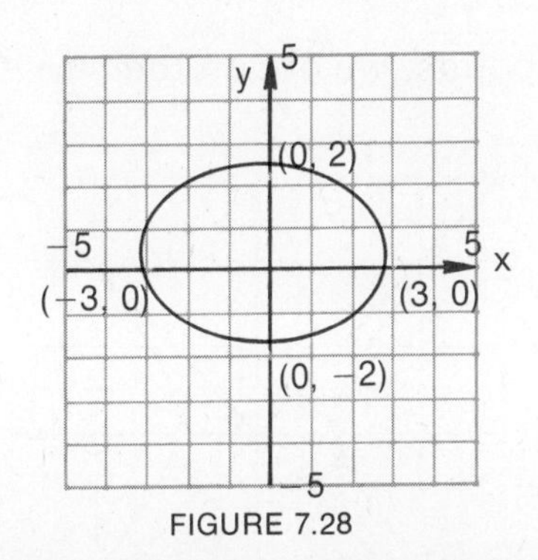

FIGURE 7.28

The graph of an equation of the form $Ax^2 + By^2 = C$, $A \neq B$, A, B, C > 0, is an ellipse with its center at the origin and with its x-intercepts at $\pm\sqrt{\frac{C}{A}}$ and y-intercepts at $\pm\sqrt{\frac{C}{B}}$.

Example 3

Graph the equation $4x^2 + 25y^2 = 100$.

Solution

Letting $x = 0$, we find $25y^2 = 100$, or $y = \pm 2$. For $y = 0$, $4x^2 = 100$, or $x = \pm 5$. These points, as well as the completed graph, appear in Figure 7.29.

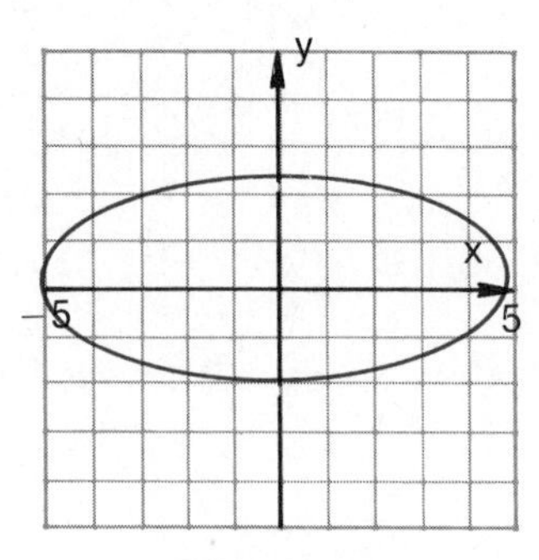

FIGURE 7.29

We now come to the hyperbola. Consider the graph of the curve $x^2 - y^2 = 4$:

When $x = 0$, $y^2 = -4$; thus there are no corresponding real y values.
When $y = 0$, $x = \pm 2$; thus $(\pm 2, 0)$ are on the curve.
When $y = \pm 1$, $x^2 = 5$ and $x = \pm\sqrt{5}$; thus $(\pm\sqrt{5}, \pm 1)$ are on the curve.

When $y = \pm 2$, $x^2 = 8$ and $x = \pm\sqrt{8}$; thus $(\pm\sqrt{8}, \pm 2)$ are on the curve.

Graphing these points, we obtain the curve shown in Figure 7.30.

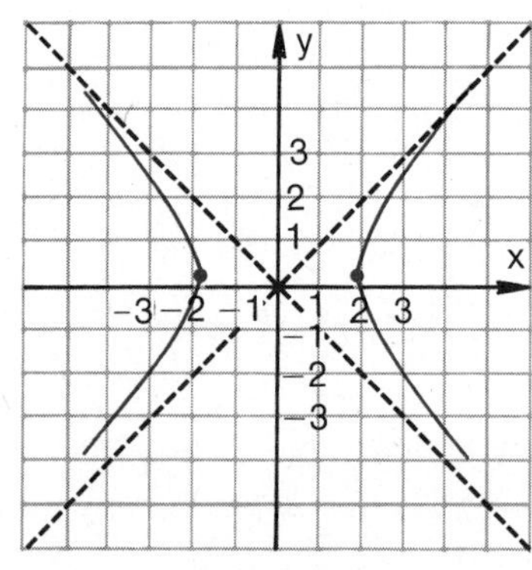

FIGURE 7.30

Note that in Figure 7.30 the graph of the hyperbola approaches the lines $y = x$ and $y = -x$, whose graph is shown dotted on the figure. The lines $y = x$ and $y = -x$ are called the *asymptotes* of the parabola $x^2 - y^2 = 4$. The asymptotes can be very helpful in sketching, since they provide the region within which the hyperbola must lie.

In general, we can state the following:

The graph of an equation of the form $Ax^2 - By^2 = C$, A, B, C,

A, B, C > 0, is an hyperbola with vertices at $\left(\pm\sqrt{\frac{C}{A}}, 0\right)$ and with asymptotes $y = \pm\sqrt{\frac{A}{B}}x$.

Example 4

Graph the equation $x^2 - y^2 = 16$.

Solution

Here the vertices are at $\left(\pm\sqrt{\frac{16}{1}}, 0\right)$, or $(\pm 4, 0)$, and the asymptotes are $y = \pm\sqrt{\frac{1}{1}}\,x$, or $y = \pm x$.

For $y = \pm 1$, $x = \pm\sqrt{17}$; thus $(\pm\sqrt{17}, \pm 1)$ are on the curve.

For $y = \pm 2$, $x = \pm\sqrt{20}$; thus $(\pm\sqrt{20}, \pm 2)$ are on the curve.

For $y = \pm 3$, $x = \pm\sqrt{25}$; thus $(\pm 5, \pm 3)$ are on the curve.

All these points, as well as the graph of the hyperbola, are shown in Figure 7.31.

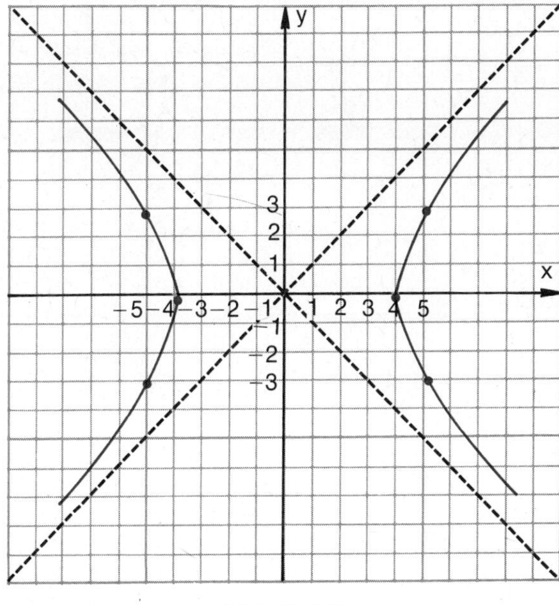

FIGURE 7.31

PROGRESS TEST

1. Graph $x^2 + y^2 = 16$.

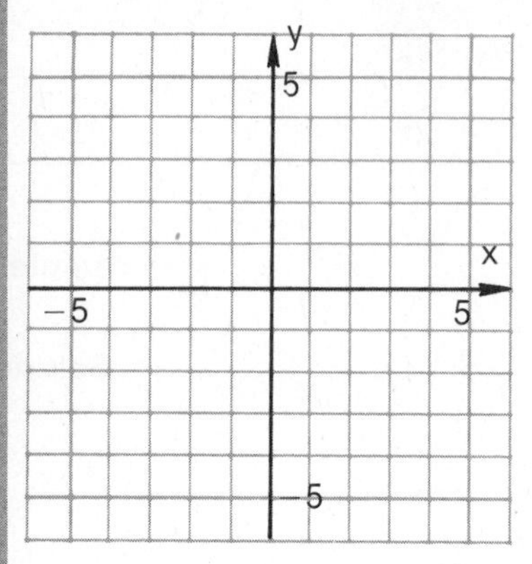

2. Graph $5x^2 + 5y^2 = 125$.

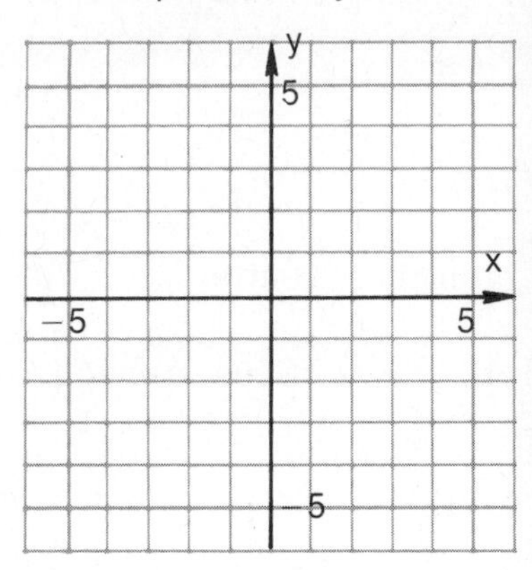

3. Graph $4x^2 + 16y^2 = 64$.

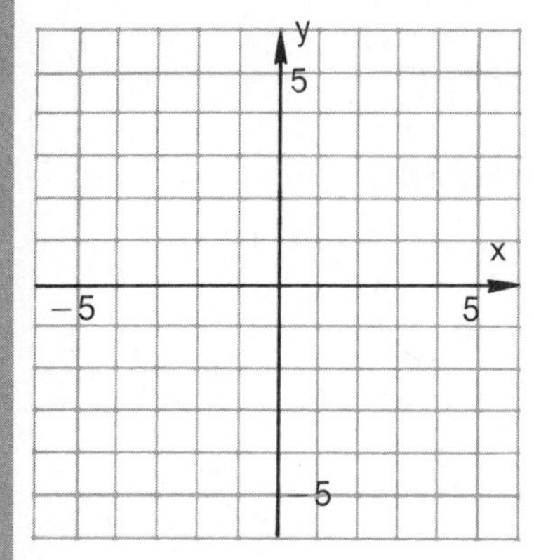

4. Graph $x^2 - y^2 = 9$.

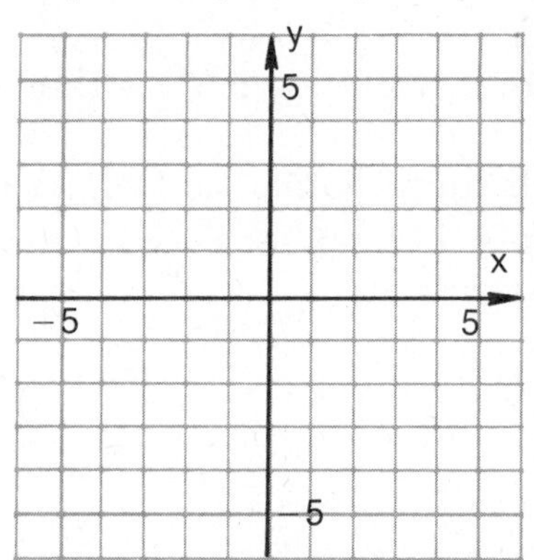

PROGRESS TEST ANSWERS

1.

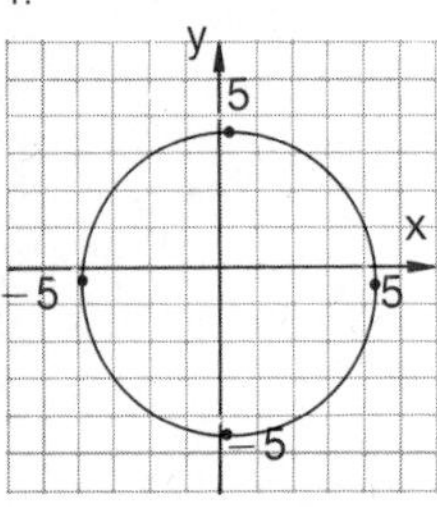

2.

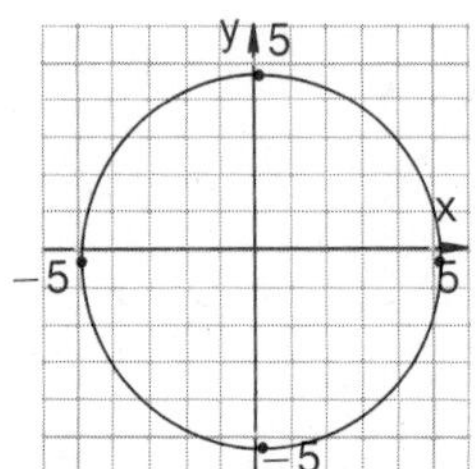

3.

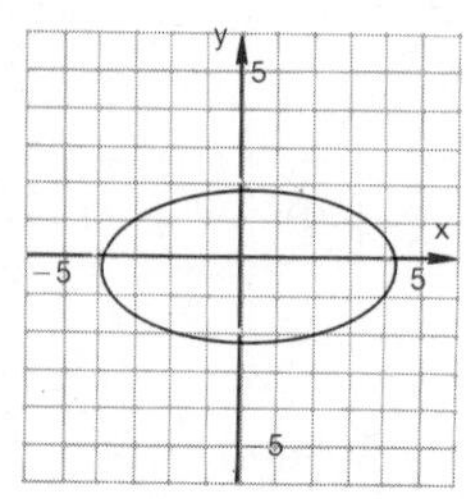

4.

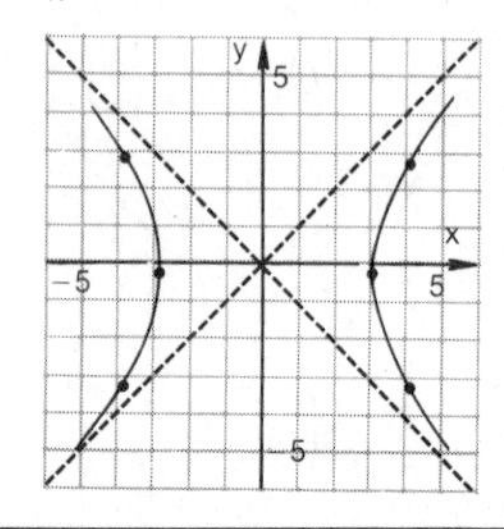

EXERCISE 7.5

In Problems 1 through 20, classify and graph the given equation.

Illustration: (a) $9x^2 = 36 - 4y^2$
(b) $x^2 = 8 + 2y^2$

Solution: (a) Rewriting $9x^2 = 36 - 4y^2$ in the form $9x^2 + 4y^2 = 36$, we see that the equation is that of an ellipse. For $x = 0$, $y = \pm 3$, thus $(0, \pm 3)$ are on the curve. For $y = 0$, $x = \pm 2$; thus $(\pm 2, 0)$ are on the curve. The graph of the equation is shown in the accompanying figure.

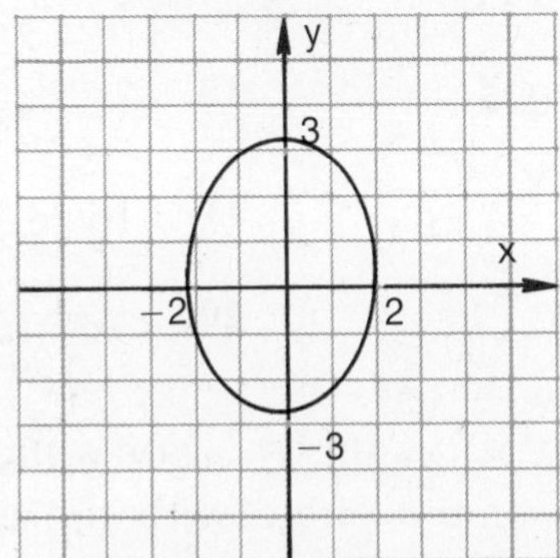

(b) Rewriting the equation in the form $x^2 - 2y^2 = 8$, we see that the equation is that of a hyperbola with vertices at $\left(\pm\sqrt{\frac{8}{1}},\ 0\right)$ and asymptotes $y = \pm\sqrt{\frac{1}{2}}\,x$. For $y = 0$, $x = \pm\sqrt{8}$; thus $(\pm\sqrt{8},\ 0)$ are on the curve. For $y = \pm 1$, $x = \pm\sqrt{10}$, thus $(\pm\sqrt{10}, \pm 1)$ are on the curve. For $y = \pm 2$, $x = \pm 4$, thus $(\pm 4, \pm 2)$ are on the curve. The graph is shown in the accompanying figure.

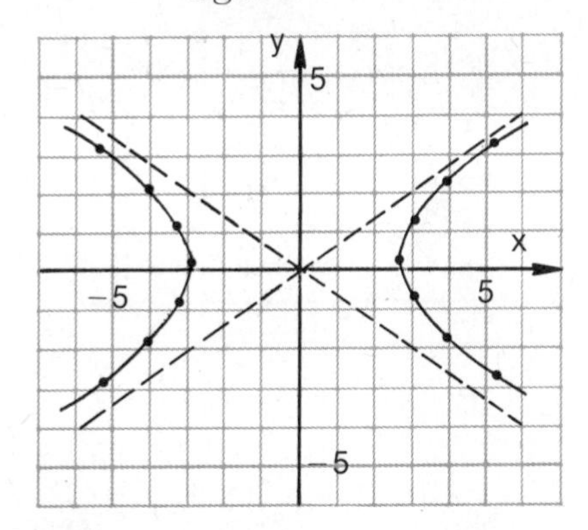

1. $x^2 + y^2 = 36$
2. $x^2 + y^2 = 9$
3. $9x^2 + 9y^2 = 36$
4. $8x^2 + 8y^2 = 32$
5. $3x^2 = \frac{1}{3} - 3y^2$
6. $7x^2 = \frac{7}{4} - 7y^2$
7. $9x^2 + 16y^2 = 144$
8. $9x^2 + 25y^2 = 225$
9. $x^2 + 4y^2 = 16$
10. $x^2 + 9y^2 = 36$
11. $3x^2 = 12 - 4y^2$
12. $4x^2 = 36 - y^2$
13. $x^2 - y^2 = 1$
14. $x^2 - y^2 = 144$
15. $4x^2 - y^2 = 1$
16. $x^2 - 2y^2 = 8$
17. $x^2 - 5y^2 = 25$
18. $x^2 - 4y^2 = 16$
19. $16x^2 = 4y^2 + 1$
20. $9x^2 = 36y^2 + 324$

In Problems 21 through 25, use the distance formula to find the equation of the circle with the given center and radius.

Illustration: Center at $(-2, 3)$, radius of 4.

Solution: The distance from a point (x, y) on the circle to the center of the circle (see diagram) is $\sqrt{[x-(-2)]^2 + (y-3)^2}$. But this distance must also be 4 (the length of the radius). Thus,

$$\sqrt{(x+2)^2 + (y-3)^2} = 4, \text{ or}$$
$$(x+2)^2 + (y-3)^2 = 16.$$

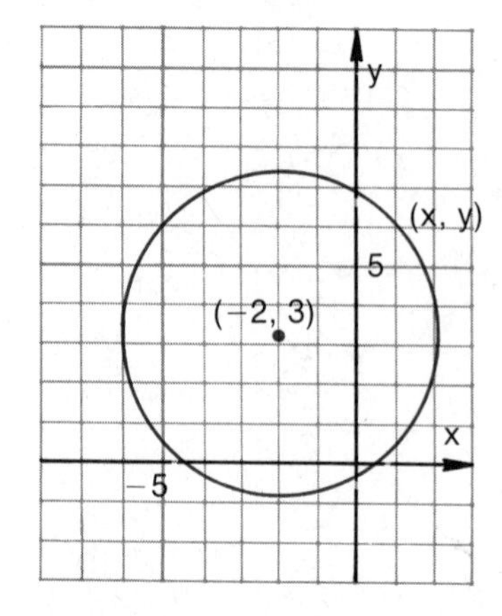

21. Center at $(-1, 3)$, radius 2.

22. Center at $(-2, 4)$, radius 4.

23. Center at $(-3, -2)$, radius 3.

24. Center at $(-5, -2)$, radius 5.

25. Center at $(5, -1)$, radius 6.

REDUCE YOUR BUILDINGS' OPERATING COSTS.

Courtesy of The Kinney Safety Systems, Inc.

OBJECTIVES 7.6

After studying this section the reader should be able to solve word problems involving quadratic equations by using the graph of the given equation.

7.6 APPLICATIONS

The accompanying ad, taken from the Wall Street Journal, explains how to reduce building operating costs. It is a common practice for businesses to hire consultants to advise them on methods to improve performance, reduce costs, and maximize profits. For example, suppose Gadget Manufacturing hires a consultant to analyze building operating costs. After some work, the consultant discovers that the cost C (in hundreds of dollars) of operating a particular building for x hours can be expressed as $C = -x^2 + 6x$ for $0 \leq x \leq 3$. How many hours should the building be used daily to minimize costs, and what is the number of hours of operation which results in the highest operating cost? To solve these problems, we first notice that $C = -x^2 + 6x$ is a parabola opening downward. For $C = 0$,

$$-x^2 + 6x = 0$$

$$\text{or } -x(x - 6) = 0$$

$$x = 0 \text{ or } x = 6$$

$x = 6$ is not used because $0 \leq x \leq 3$.

The vertex of the parabola is at the point where $x = \dfrac{-6}{2(-1)} =$ 3, and at this point $C = -(3)^2 + 6(3) = 9$; that is, the vertex is at (3, 9). Using this information, we can sketch the graph of the parabola as shown in Figure 7.32. From this sketch it is clear that the maximum cost occurs when the building is operated for three hours and the cost of operation is 9 (hundred dollars).

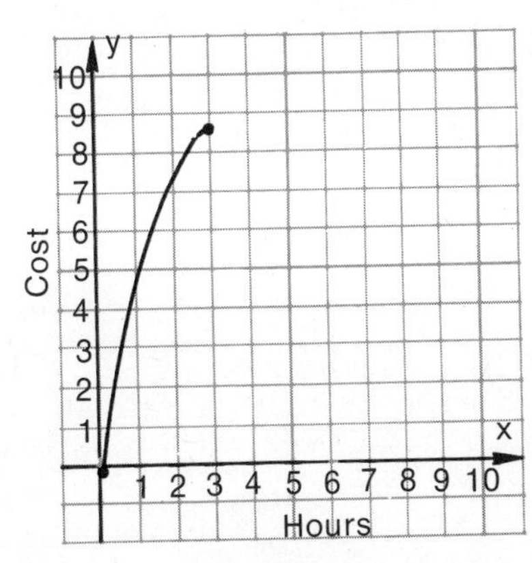

FIGURE 7.32

Example 1

Caliente Enterprises estimates that its profit P (in hundreds of dollars) after making x (thousand) tacos can be expressed as

$$P = -x^2 + 4x$$

(a) Graph the equation.
(b) Find the number of tacos that have to be made to obtain maximum profit.

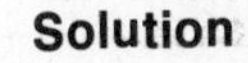

(a) When $P = 0$, $-x^2 + 4x = 0$
or $-x(x - 4) = 0$

Thus, $x = 0$ or $x = 4$.
The vertex of the parabola $P = -x^2 + 4x$ is at the point where $x = \dfrac{-4}{2(-1)} = 2$, and at this point $C = -(2)^2 + 4(2) = 4$; that is, the vertex is at (2, 4). The graph appears in Figure 7.33.

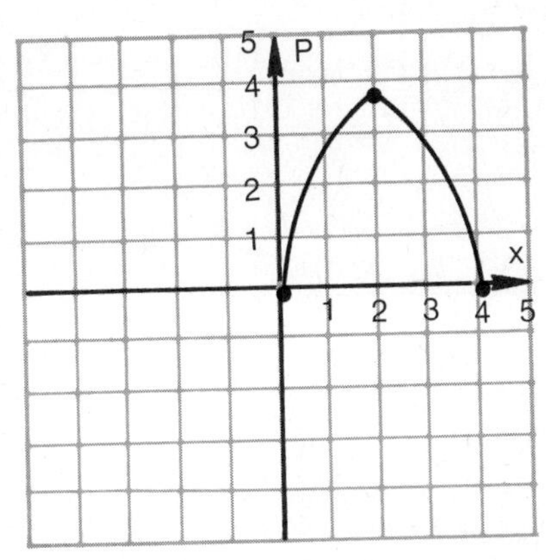

FIGURE 7.33

A linear equation is an equation that can be written in the form $Ax + By + C = 0$, A and B not both zero.

(b) Since the maximum point of the parabola $P = -x^2 + 4x$ is at the vertex (2, 4), the number of tacos that have to be made is 2 (thousand). When this is done, the profit is 4 (hundred) dollars.

Linear equations have many applications in economics. For example, if the total cost y of producing x units of an item is assumed to be linear, then this cost can be written as

m is the marginal cost.

$$y = mx + b$$

b is the fixed cost.

In this equation, m, the slope, represents the cost of producing one unit of the product. This cost is usually called the *marginal cost.* The number b represents the *fixed cost*—that is, the cost incurred before production starts (machines, leasing of buildings, etc.). Thus, if a man invests $2,000 to start manufacturing an item whose production cost is $2 per unit, the total cost y can be expressed as $y = 2x + 2{,}000$. Similarly, if the total cost y of manufacturing certain items changes from $1,200 to $1,350 as the number x of units increases from 400 to 700, the equation $y = mx + b$ can be applied. We first take the points (400, 1200) and (700, 1350) to find

$$m = \frac{150}{300} = \frac{1}{2}$$

Then, using the point (400, 1200) and the point-slope form, we find

$$y - 1200 = \frac{1}{2}(x - 400)$$

$$\text{or } y = \frac{1}{2}x + 1000$$

Example 2

If production is increased from $x = 300$ units of a product to $x = 450$ units, then the cost y increases from $800 to $850. If x and y are linearly related, find:
(a) the cost m per unit.
(b) the fixed cost b.

Solution

(a) Since the cost of producing 300 units is \$800, the point (300, 800) is on the line. The point (450, 850) is also on the line. The slope m, then, is

$$m = \frac{850 - 800}{450 - 300} = \frac{50}{150} = \frac{1}{3}$$

Hence the cost per unit is $\$\frac{1}{3}$, or $33\frac{1}{3}$ cents.

(b) Using the point-slope formula with $m = \frac{1}{3}$ and $(x_1, y_1) =$ (450, 850),

$$y - 850 = \frac{1}{3}(x - 450)$$

$$\text{or } y = \frac{1}{3}x + 700$$

Thus, the fixed cost is \$700.

Example 3

The pollution index p at Main City remains at a constant level, 50 (parts per million), from 6 P.M. until 8 P.M. and then drops linearly to 20 (parts per million) at 8 A.M.; that is, it is 20 (parts per million) 12 hours later.

(a) Draw a graph of the pollution index p with respect to the time t.
(b) Find an equation relating p and t.
(c) Use the equation found in (b) to find the pollution index p at 2 A.M.

Solution

(a) We draw a set of coordinate axes with p as the vertical axis and t as the horizontal axis, starting at 8 P.M., as shown in Figure 7.34.

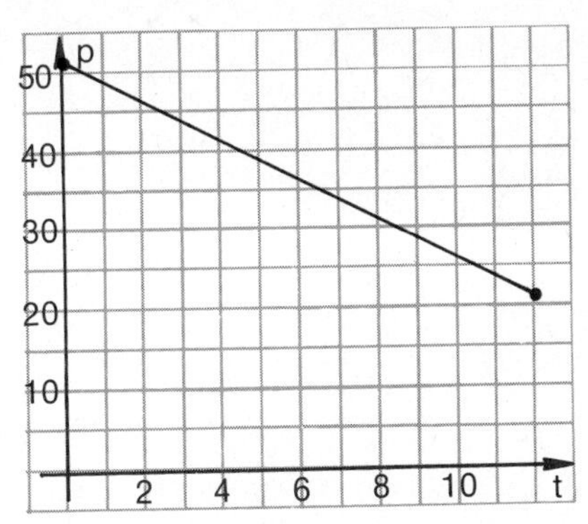

FIGURE 7.34

The numbers on the t axis are the number of hours after 8 P.M.

(b) Since the points (0, 50) and (12, 20) are on the line, the slope is

$$m = \frac{50 - 20}{0 - 12} = \frac{30}{-12} = \frac{-5}{2}$$

Using the point-slope formula with $m = \frac{-5}{2}$ and $(x_1, y_1) =$ (0, 50),

$$p - 50 = \frac{-5}{2}(t - 0)$$

$$p = \frac{-5}{2}t + 50$$

(c) At 2 A.M., $t = 6$ (it is 6 hours past 8 P.M.). At this time, the pollution index p is

$$p = \frac{-5}{2}(6) + 50 = -15 + 50 = 35$$

PROGRESS TEST

1. The profit p in thousands of dollars of a certain business is given as $p = -x^2 + 2x$, where x is the number of items produced (in thousands). Graph this equation.

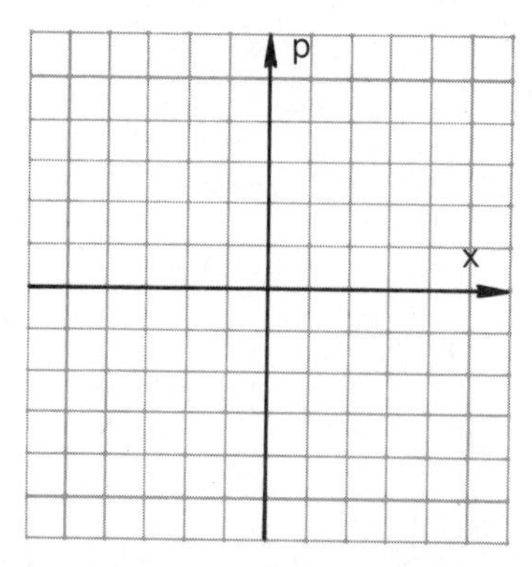

2. The number of items that have to be sold to obtain maximum profits is _________.
3. The total cost y of manufacturing a product changes from \$1,100 to \$1,150 as the number of units increases from 400 to 600. The cost per unit is _________.
4. In problem 3, the fixed cost is _________.

EXERCISE 7.6

In Problems 1 through 8, solve and sketch the graph.

Illustration: The total cost C (in thousands of dollars) of producing x (thousand) units of a certain item is given as $C = x^2 - 2x + 2$. Find the number of units that have to be produced to minimize the cost and sketch the graph of C.

Solution: The equation $C = x^2 - 2x + 2$ is a parabola opening upward. The vertex of this parabola occurs when $x = \frac{-(-2)}{2(1)} = 1$. At this point $C = (1)^2 - 2(1) + 2 = 1$. Thus, the vertex is at (1, 1). This means that when 1 (thousand) units are produced, the cost is 1 (thousand) dollars, the minimum cost possible. To finish the sketch of the parabola, we note that when $x = 0$, $C = 2$. The completed sketch is shown.

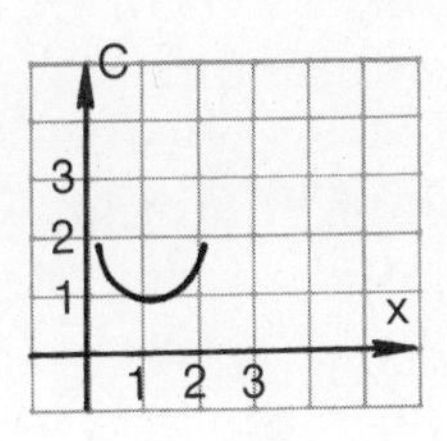

1. The total cost C of producing x (thousand) units of a certain item is given as $C = x^2 - 4x + 5$. Find the number of units that have to be produced to minimize the cost, and sketch the graph of C.

2. Repeat Problem 1, if $C = x^2 - 4x + 6$.

3. The total profit P (in thousands of dollars) obtained by producing x (thousands) units of a certain item is given as $P = -x^2 + 4x - 1$.

(a) Graph P.
(b) Find the number of units that have to be produced to maximize the profits.

4. Repeat Problem 3 if $P = -x^2 + 5x$.

5. The pollution index p (in tens) for a six-hour period in a

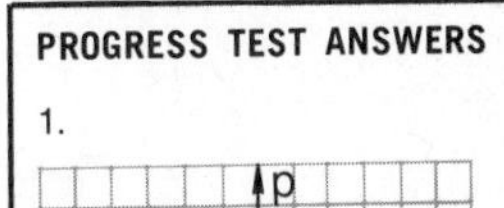

PROGRESS TEST ANSWERS

1.

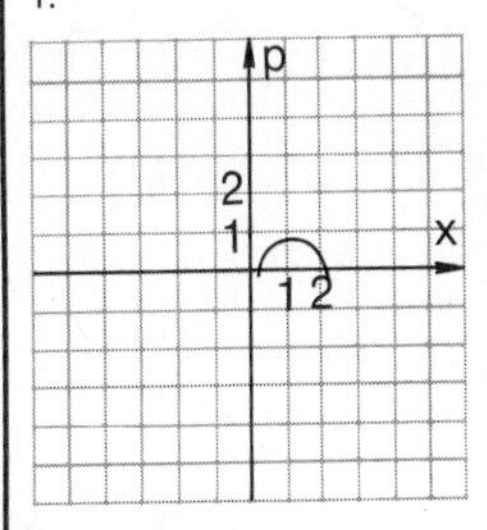

2. 1 (thousand)

3. $\$\frac{1}{4}$ = 25 cents

4. $1,000

certain city is given as $p = t^2 - 6t + 100$, where t is the number of hours elapsed.

(a) Graph p using t as the horizontal axis.
(b) In how many hours will the pollution index be at its lowest?
(c) What will be the lowest pollution index during the six-hour period?

6. The demand D for a certain product is dependent on the number x (in thousands) of units produced, and is given as $D = x^2 - 2x + 3$.

(a) Graph D.
(b) Find the number of units that have to be produced so that the demand is at its lowest.

7. Repeat Problem 6 using $D = x^2 - 4x + 4$.

8. The price P in dollars of a product is dependent upon the amount x (in thousands) of units of the product available (the supply), and is given as $P = x^2 - 6x + 12$.

(a) Graph P.
(b) How many units have to be produced so that the price is at its minimum?

In Problems 9 through 12, $y = mx + b$ represents the total cost of producing x items at a cost of m dollars per item and a fixed cost b.

Illustration: The cost of manufacturing ten cars is \$22,000. Twenty cars can be manufactured for \$40,000. Find both the fixed cost and the marginal cost of manufacturing one car.

Solution: Since the cost in thousands of dollars of producing $x = 10$ cars is $y = 22$, (10, 22) is on the line. Also (20, 40) is on the line. The slope m is $\frac{40 - 22}{20 - 10} = \frac{18}{10} = \frac{9}{5}$. Using this slope and the point (10, 22), we find that the cost y can be expressed as

$$y - 22 = \frac{9}{5}(x - 10)$$

$$y = \frac{9}{5}x + 4$$

Thus, the marginal cost for one car is

$m = \frac{9}{5} = 1.8$ thousand dollars. The fixed cost is 4 thousand dollars.

9. The Princess Dress Shop can produce 10 dresses for $50 and 20 dresses for $90. Find the cost per dress and the fixed cost.

10. A man invests $1000 in preparing his manuscript for publication. Printing costs are $3 per book. If x is the number of books printed,

(a) What is y, the total cost?
(b) What is the total cost of producing 600 books?

11. The total cost of producing a product changes from $5,000 to $7,000 as the number x of units produced changes from 2,000 to 3,000.
(a) What is the fixed cost for the product?
(b) What is the cost per unit?

12. Repeat Problem 11 if the total cost changes from $5,000 to $6,000.

13. The demand D for Moor beer is linearly dependent on the price. When the price P is 30 cents per can, the company sells 10,000 cans per week. However, if the price increases to 40 cents per can, only 5,000 cans are sold.

(a) If $D = mx + b$, find m and b.
(b) Graph the line $D = mx + b$.
(c) In this problem, the slope m represents the change in sales if the price is changed by one cent. Find the effect of lowering or raising the price by one cent.

14. Repeat Problem 13 if it is known that the company sells 20,000 cans when the price is 30 cents per can but only 15,000 cans when the price is 40 cents per can.

15. Knotty Shoelaces finds that it can sell 3,000 pairs of shoelaces if the price is 20 cents per pair. If the price is raised to 30 cents, only 1,000 pairs are sold. The demand D is related to the price x by $D = mx + b$.

(a) Find m and b.
(b) Graph $D = mx + b$.
(c) The slope m represents the change in sales if the price changes by one cent. Find the effect of lowering or raising the price by one cent.

SELF-TEST–CHAPTER 7

1. Graph the given line on the accompanying rectangular coordinate system.

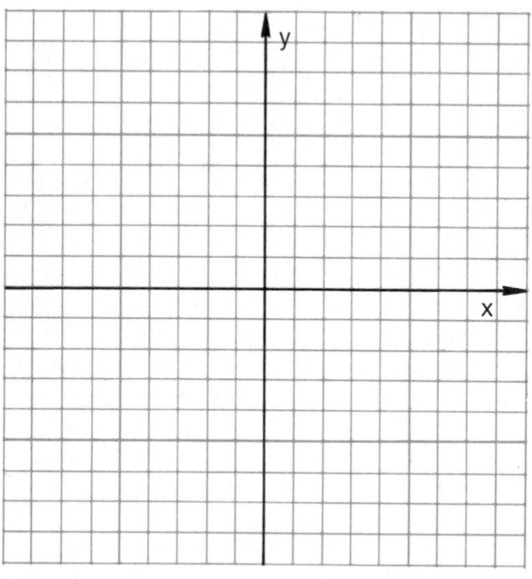

(a) $2x - y = 6$

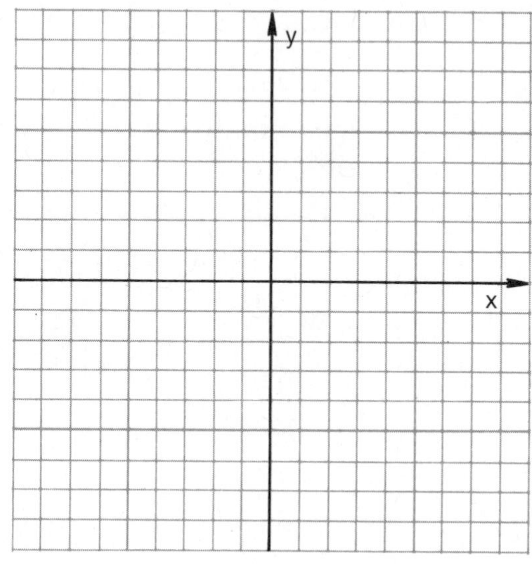

(b) $2x = 6$

---

2. Find the distance between the two given points and the slope of the line passing through them:

   ________ (a) (3, 4) and (6, 9)

   ____

   ________ (b) (−2, 1) and (−2, 3)

   ____

---

3. Write in standard form the equation of the line passing through the points:

   ____________ (a) (1, 2) and (3, 4)

   ____________ (b) (4, 2) and (2, 6)

---

4. Write in standard form the equation of the line with the given slope and passing through the given point.

_______________ (a) $m = -2$, $(3, 4)$

_______________ (b) $m = 0$, $(1, 3)$

---

5. Write in standard form the equation of the line with the given slope and y-intercept.

_______________ (a) $m = -1$, y-intercept 4

_______________ (b) $m = 2$, y-intercept 0

---

6. Write in standard form the equation of the line passing through (1, 2) and

_______________ (a) parallel to the line $y = 2x + 1$

_______________ (b) perpendicular to $y = 2x + 1$

---

7. Graph the given parabola in the accompanying rectangular coordinate system.

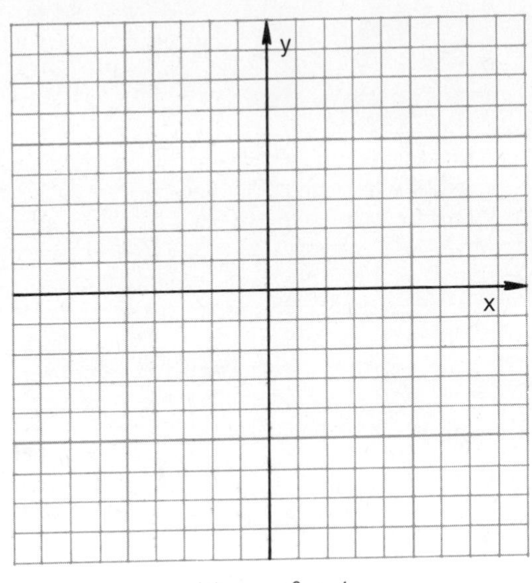

(a) $y = x^2 + 1$

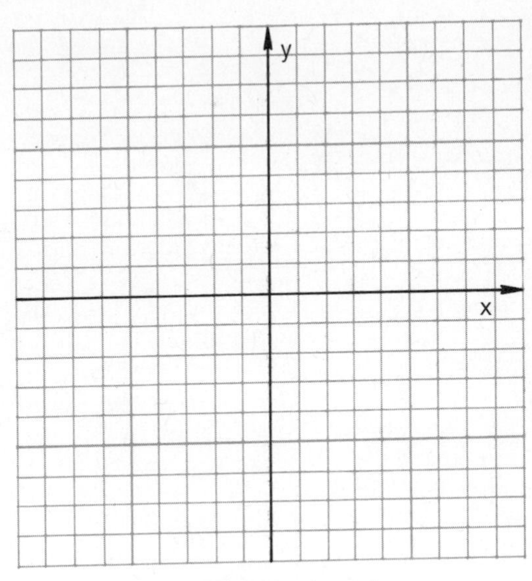

(b) $y = 2x^2 + 4x - 1$

---

8. Graph the given equation.

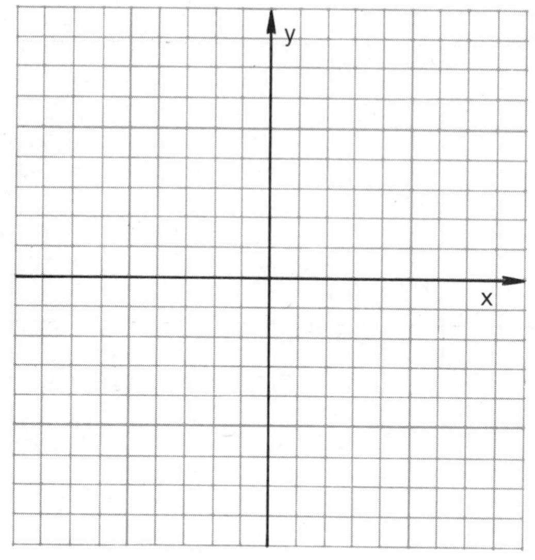

(a) $x^2 + y^2 = 9$

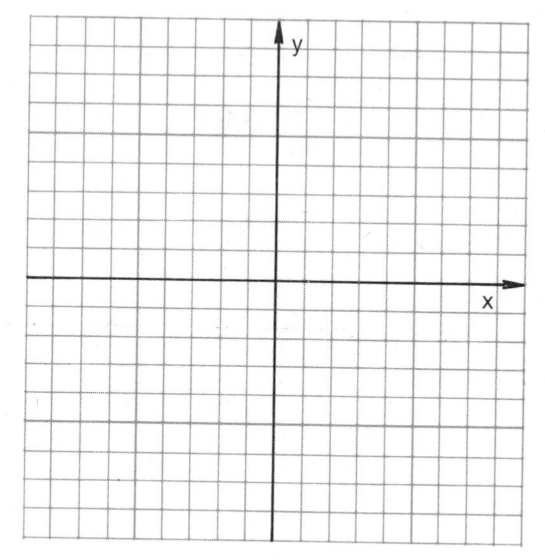

(b) $25x^2 + 25y^2 = 100$

9. Graph the given equation.

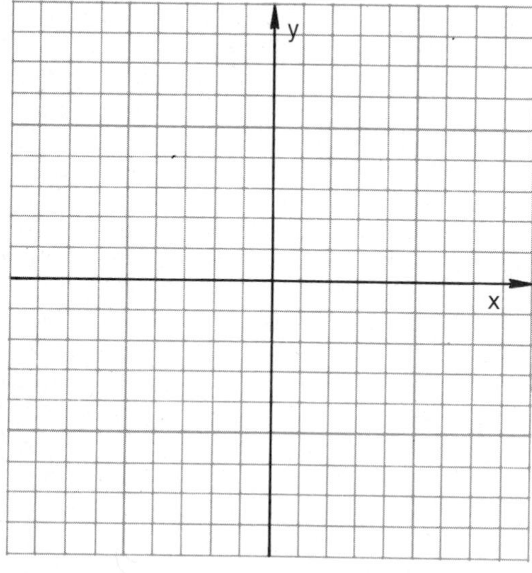

(a) $25x^2 + 4y^2 = 100$

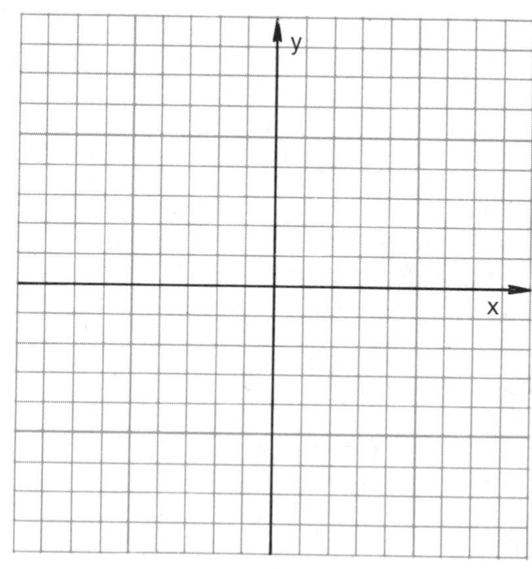

(b) $x^2 - y^2 = 9$

10. ClanKem Manufacturing estimates that their profit P (in hundreds of dollars) after producing x (thousand) units can be expressed as $P = -x^2 + 2x$.

(a) Graph the equation.

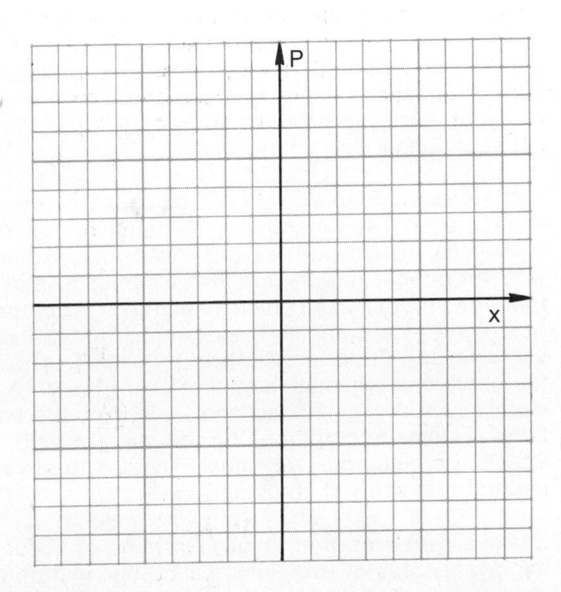

_______ (b) Find the number of units that have to be produced to obtain maximum profits.

CHAPTER 8

HISTORICAL NOTE

Many practical problems involve two or more unknowns and can be solved by means of a system of equations. Approximately 3500 years ago, the Babylonians had already introduced problems that led to such systems of equations. Many of these problems involved only two unknowns, but in at least one case the Babylonians posed a problem that involved ten equations in ten unknowns.

In the present chapter, we shall be concerned with systems of two linear equations in two unknowns. Systems of linear equations have been studied for many centuries, but it was not until after the time of Descartes, who introduced much modern symbolism, that it became possible to treat the problem in an efficient manner. One of the first men to reduce the procedure to a completely systematic method was the German mathematician Carl Friedrich Gauss (1777–1855), who has been called one of the greatest mathematicians that ever lived. The Gauss elimination procedure is still widely used and is, in fact, the basic method for solving large systems of equations on the modern high speed digital computer. The Gauss method is very closely related to the addition and subtraction method that you will study in Section 8.2.

Besides problems that involve systems of linear equations, there is an important class of problems, called linear programming problems, that involve a system of linear inequalities. Such problems are of fundamental importance in scheduling production of items in manufacturing plants, in scheduling deliveries from warehouses to retail outlets in large firms like Sears, and in allocating supplies for the U.S. Air Force.

Carl Friedrich Gauss, one of the greatest mathematicians who ever lived, discovered the elimination method for solving a system of equations, the method that has been named for him. (Courtesy of the Library of Congress.)

SYSTEMS OF EQUATIONS

Freight transport inevitably involves problems associated with systems of equations. (Photo courtesy of Flying Tigers.)

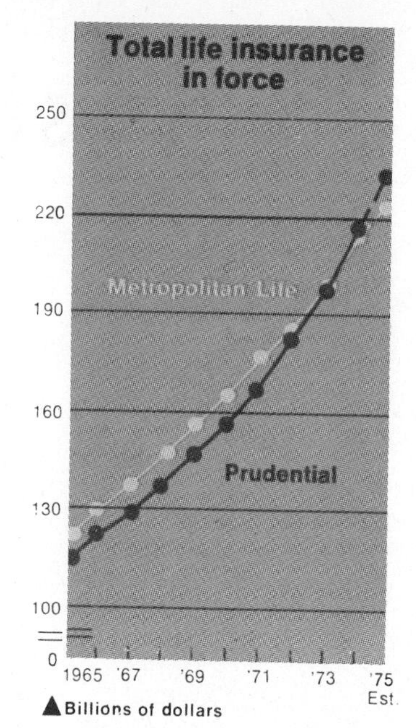

Business Week, Feb. 9, 1976, p. 55.

OBJECTIVES 8.1

After studying this section, the reader should be able to:

1. **Find the solution (or the solution set) of a system of two linear equations in two variables by using the graphical method.**
2. **Determine if a system of two linear equations in two variables is inconsistent or dependent by using the substitution method.**
3. **Find the solution set of a system of two linear equations in two variables by using the substitution method.**

8.1 THE GRAPHICAL SOLUTION OF A SYSTEM OF TWO LINEAR EQUATIONS IN TWO VARIABLES

The graph at the beginning of this section shows the amount (in billions of dollars) of total life insurance in force for the years 1965 to 1975. Can you tell from the graph in what year Metropolitan Life and Prudential had the same amount of insurance in force? It is clear that this happened in 1973. This is because the value representing the total life insurance in force for each of the companies is the same (approximately 200 billion) in the year 1973. If we agree to denote the year by x and the total life insurance in force by y, the point (1973, 200) represents the point at which the two curves *intersect*. In mathematics we can graph a pair of linear equations and find an ordered pair which is a *solution* to both equations by finding the point (if there is one) at which the lines *intersect*. The coordinates of this point will be the *common solution* of both equations. For example, the total life insurance in force, y (in billions of dollars), for company A can be written as

The common solution to a pair of linear equations is the point that satisfies both equations.

$$y = 10x + 110$$

This equation is an approximation of Metropolitan Life's total life insurance in force.

where x is the number of years elapsed after 1965; the total life insurance for company B can be given as

$$y = 5x + 120$$

The year in which both companies have the same amount of life insurance in force can be found by graphing these two equations, as shown in Figure 8.1. Their point of intersection shows when both companies had the same amount of life insurance in force, 130 billion dollars. As you can see from the graph, the point (2, 130) is the *only* ordered pair that is a *solution* of both equations. This point is the *common solution* of the two equations.

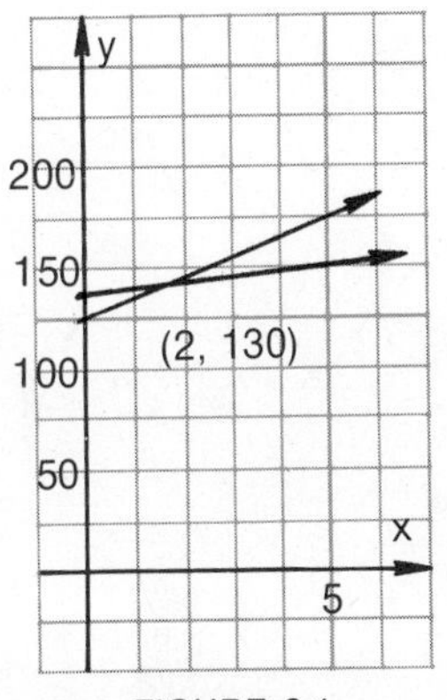

FIGURE 8.1

Example 1

Use the graphical method to find the common solution of the equations $2x - y = 2$ and $y = x - 1$.

Solution

We first graph $2x - y = 2$. When $x = 0$, $y = -2$; thus $(0, -2)$ is on the line. When $y = 0$, $x = 1$; thus $(1, 0)$ is on the line. Drawing a line through these two points, we obtain the graph of $2x - y = 2$, shown in color in Figure 8.2. Similarly, to graph $y = x - 1$, we first let $x = 0$, obtaining $y = -1$. Thus $(0, -1)$ is on the line. If $y = 0$, $x = 1$, and the point $(1, 0)$ is on the line. The line connecting these points for $y = x - 1$ is shown in black in Figure 8.2. The point of intersection of the two lines gives the common solution: (1, 0).

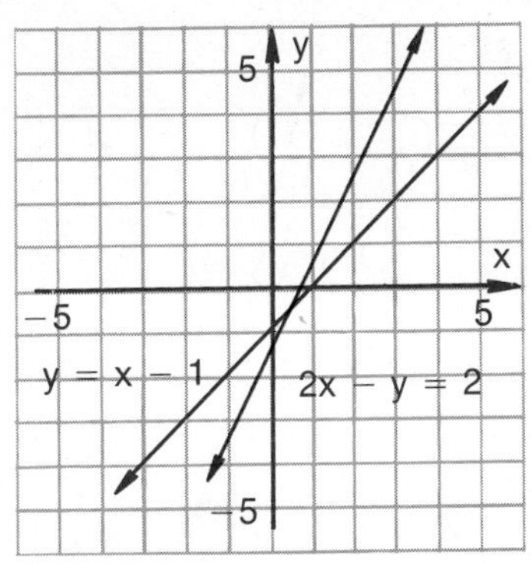

FIGURE 8.2

Example 2 Use the graphical method to find the common solution (if there is one) of the equations $x + y = 3$ and $y = -x - 3$.

Solution To graph $x + y = 3$, we first let $x = 0$, obtaining $y = 3$ and the point (0, 3) on the graph. If $y = 0$, $x = 3$, and (3, 0) is our second point on the graph. The graph of $x + y = 3$, together with the points (0, 3) and (3, 0), is shown in color in Figure 8.3. We graph $y = -x - 3$ similarly. If $x = 0$, $y = -3$, giving us $(0, -3)$ on the graph. When $y = 0$, $x = -3$, and hence $(-3, 0)$ is on the graph. The completed graph for the equation $y = -x - 3$ is shown in black in Figure 8.3. As you can see, it appears that these two lines are parallel and do not have any common solution. This means that there is *no* ordered pair (x, y) that is a solution for both equations.

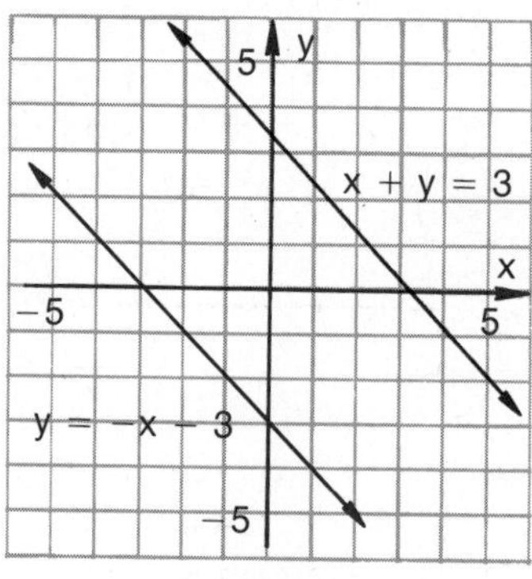

FIGURE 8.3

Example 3 Use the graphical method to find the common solution of $2x + \frac{1}{2}y = 2$ and $y = -4x + 4$.

Solution We first graph $2x + \frac{1}{2}y = 2$. If $x = 0$, $y = 4$; thus (0, 4) is on the

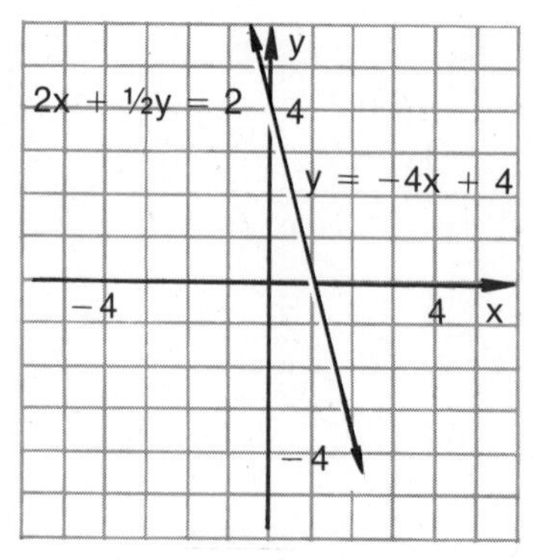

FIGURE 8.4

graph. If $y = 0$, $x = 1$, so $(1, 0)$ is on the graph. We draw a line through these two points, obtaining the graph of $2x + \frac{1}{2}y = 2$, shown in color in Figure 8.4. To graph $y = -4x + 4$, we first let $x = 0$. Then $y = 4$, so $(0, 4)$ is on the graph. Similarly, if $y = 0$, $x = 1$, and hence $(1, 0)$ is on the graph. As you can see, the graphs of the equations $2x + \frac{1}{2}y = 2$ and $y = -4x + 4$ are the same line.

From Examples 1, 2, and 3, it is clear that there are three possibilities when graphing a system of two linear equations.

1. The graphs intersect at one and only one point (Example 1).
2. The graphs are parallel lines (Example 2).
3. The graphs are the same line (Example 3).

The graphs intersect at one and only one point.

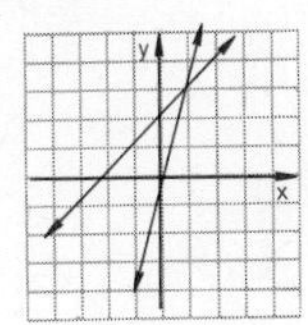

Algebraically, these facts mean that there are three alternatives for the solution set of a system of two linear equations:

1. The common solution set for the equations consists of exactly *one point*,
2. The common solution set for both equations is the *null set* (such systems are called *inconsistent*), or
3. The two equations have the same solution set (such systems are called *dependent*).

The graphs are parallel lines.

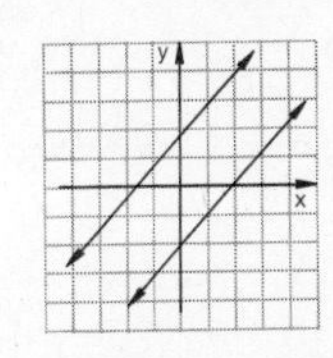

Can we now show algebraically that the system $2x - y = 2$ and $y = x - 1$ used in Example 1 has one and only one solution? We can. This affirmative answer depends on the fact that any ordered pair (x, y) in the solution set of a system of equations must be a solution for each equation. Thus, we can apply the substitution axiom and replace one of the variables in one of the equations by its equivalent from the other equation. For obvious reasons, this method is called the *substitution method.* For example, if

The graphs are the same line.

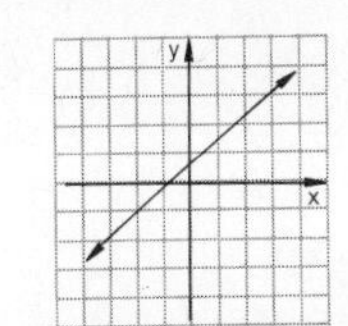

If a system of two linear equations has no solution, the system is <u>inconsistent.</u>

(1) $2x - y = 2$ and
(2) $y = x - 1$, then
$2x - (x - 1) = 2$ (substituting $y = x - 1$ in Equation 1)

This means that if there is a common solution (x, y), then $x = 1$ is the <u>only</u> possibility.

Thus,

$$x + 1 = 2$$
$$x = 1$$

Substituting $x = 1$ in Equation (2), we obtain $y = 0$. Hence, the solution of $2x - y = 2$ and $y = x - 1$ is $(1, 0)$, as was shown graphically in Example 1.

Example 4 Use the substitution method to solve the system

(1) $3y + x = 9$
(2) $x = 3y$

Solution Substituting $x = 3y$ in Equation (1), we obtain

$$3y + 3y = 9$$
$$6y = 9, \text{ or } y = \frac{3}{2}$$

Substituting $y = \frac{3}{2}$ in (2), we have

$$x = 3\left(\frac{3}{2}\right) = \frac{9}{2}$$

Check: $3\left(\frac{3}{2}\right) + \frac{9}{2} = 9$, and $\frac{9}{2} = 3 \cdot \left(\frac{3}{2}\right)$.

Thus, the solution of $3y + x = 9$ and $x = 3y$ is $\left(\frac{9}{2}, \frac{3}{2}\right)$.

How do we recognize that a system such as $x + y = 3$ and $y = -x - 3$ is inconsistent? We can do this with the substitution method as follows. The system is

(1) $x + y = 3$
(2) $y = -x - 3$

We substitute $y = -x - 3$ in Equation (1) to obtain

$$x + (-x - 3) = 3,$$
$$\text{or } -3 = 3$$

Since this is a contradiction, the system $x + y = 3$ and $y = -x - 3$ is inconsistent; it has *no* solution. In other words, the solution set is $\emptyset$.

Example 5 Solve the system

(1) $x + y = 5$
(2) $y = -x$

Solution

Substituting $y = -x$ in equation (1), we get

$$x + (-x) = 5,$$
$$\text{or } 0 = 5$$

Since this is a contradiction, the system $x + y = 5$ and $y = -x$ has *no* solution.

PROGRESS TEST

1. Using the graphical method, find the solution for the system $-x + y = 1$ and $y = -x - 1$.

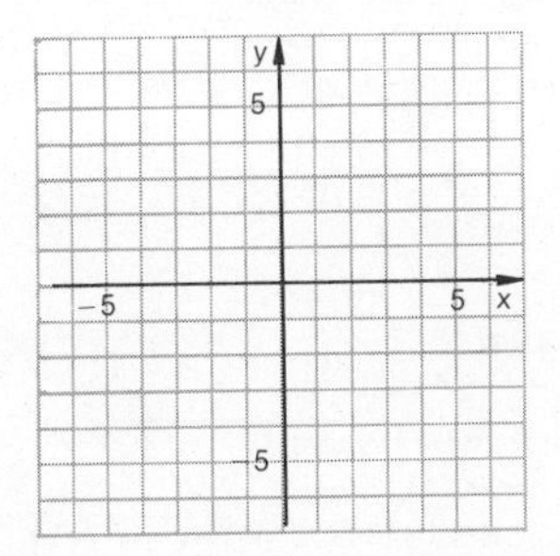

2. The graphical method shows that $x + 2y = 4$ and $x = -2y + 4$ are ____________.

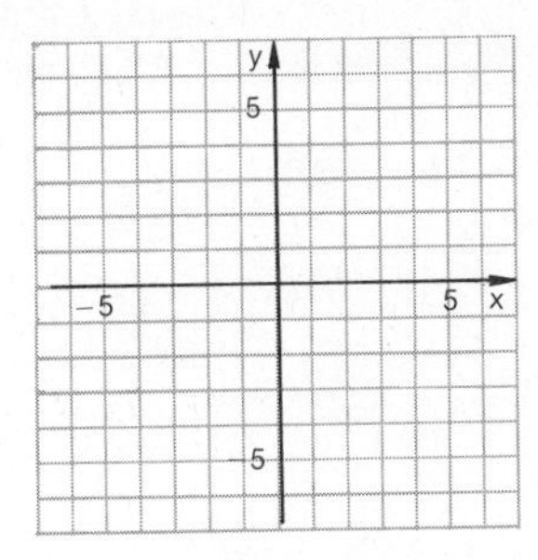

3. By the substitution method, the solution of $2x - 3y = 6$ and $y = x + 1$ is ____________.

EXERCISE 8.1

In Problems 1 through 10, use the graphical method to solve the system of equations.

1. $y = x - 4$
$y = 2x$

2. $y = x - 6$
$y = 3x$

3. $y = x - 5$
$y = 6x - 10$

4. $y = x - 7$
$y = 2x - 3$

5. $2x = -y + 4$
$y = -2x + 4$

6. $2x = -y + 2$
$y = -2x + 2$

7. $y = 3x + 9$
$-y = -2x - 6$

8. $y = x - 1$
$-y = -3x - 1$

9. $2x - y = -2$
$y = 2x + 4$

10. $2x + y = -2$
$y = -2x + 4$

Substitution Problems

11–20. Use the substitution method to solve Problems 1 through 10.

Illustration: Solve the system
(1) $y = x - 6$
(2) $y = 3x$

You should check this solution in the given equation.

Solution: Substituting $y = 3x$ in Equation (1), we have
$3x = x - 6$
$2x = -6$, or $x = -3$
Substituting $x = -3$ in (2), we obtain $y = -9$. Thus, the solution of $y = x - 6$ and $y = 3x$ is $(-3, -9)$.

Word Problems

PROGRESS TEST ANSWERS

1. The solution is (–1, 0).

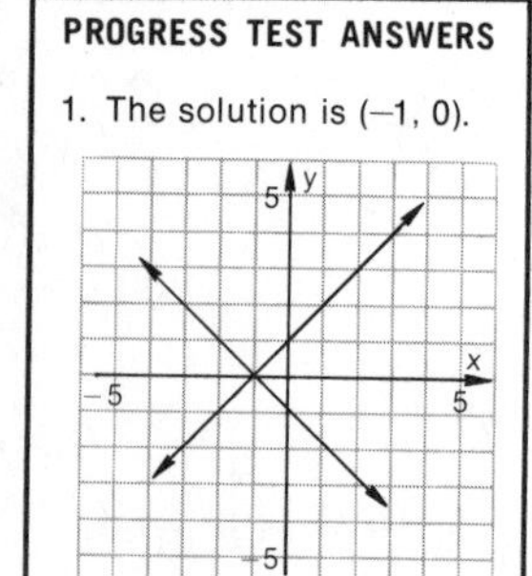

2. dependent

3. (–9, –8)

21. The supply y of a certain item is given in the equation $y = 5x + 40$, where x is the number of days elapsed. Assume the demand is $y = 10x$.

(a) Graph the system $y = 5x + 40$ and $y = 10x$.
(b) In how many days will the supply equal the demand?

22. The supply y of a certain item is $y = 5x + 20$. Assume the demand is $y = 10x$.

(a) Graph the system $y = 5x + 20$ and $y = 10x$.
(b) In how many days will the supply equal the demand?

23. The population x of Town A is 20,000 and is decreasing at the rate of 300 persons per year, so that $x = 20{,}000 -$

$300t$, where t(years) is the time elapsed. The population x of Town B is 10,000 and is increasing at the rate of 100 persons per year, so that $x = 10{,}000 + 100t$. In how many years will the populations of the two towns be the same?

24. Repeat Problem 23 with the decrease in population in Town A as 200 and the increase in Town B as 300 per year.

OBJECTIVES 8.2

After studying this section, the reader should be able to:

1. **Solve a system of two linear equations in two variables by using the addition or subtraction method.**
2. **Determine if a system of two linear equations in two variables is inconsistent.**

8.2 SOLVING LINEAR SYSTEMS OF EQUATIONS IN TWO VARIABLES BY THE ADDITION OR SUBTRACTION METHOD

Note that the graphical method will be somewhat inaccurate, because when $x = 0$, for example, we have to approximate $y = 8.50$.

In the preceding section, we used a graphical method to find the common solution of a pair of equations. This method depends on how accurately we can graph the equations and read the coordinates of the point of intersection. Because of the labor involved in graphing equations, and because of the possible inaccuracy of this method, we now develop other, algebraic methods of solving systems of equations. Consider the system in the advertisement at the beginning of this section: National introduces the 12-hour New York businessman's rate, \$8.50 plus 13¢ per mile. If x represents the number of miles traveled and y the cost in dollars, then

$$(1)\quad y = 0.13x + 8.50$$

Now, suppose that another company offers a deal in which we pay \$10 a day plus 10¢ per mile. If, as before, x is the number of miles traveled and y is the cost in dollars, then

$$(2)\quad y = 0.10x + 10$$

How many miles do we have to travel for the cost of both companies' deals to be the same? To answer this question, we must find an ordered pair (x, y) that satisfies both Equations (1) and (2), that is, $\{(x, y) \mid y = 0.13x + 8.50\} \cap \{(x, y) \mid y = 0.10x + 10\}$. Accordingly, we first write Equations (1) and (2) in the forms:

$$(3)\ -0.13x + y = 8.50$$
$$(4)\ -0.10x + y = 10$$

If we subtract Equation (3) from Equation (4) (or, equivalently, multiply Equation 3 by −1 and then add), we obtain

Remember that $(-0.10x + y) - (-0.13x + y) = -0.10x + y + 0.13x - y = 0.03x$.

$$\begin{array}{ll} (4) & -0.10x + y = 10 \\ (3') & \underline{\ \ 0.13x - y = -8.50} \\ (5) & \ \ 0.03x \qquad = 1.50 \end{array}$$

The cost for National is $y = 0.13(50) + 8.50 = 6.50 + 8.50 = \$15$.

Thus, $x = 50$. Accordingly, when the number of miles traveled is 50, the cost is the same for both companies. The method of addition or subtraction is dependent on the following fact:

The cost for the other company is $y = 0.10(50) + 10 = 5 + 10 = \$15$.

> Any ordered pair (x, y) that satisfies
> (6) $a_1x + b_1y = c_1$ and
> (7) $a_2x + b_2y = c_2$
> will also satisfy
> (8) $A(a_1x + b_1y) + B(a_2x + b_2y) = Ac_1 + Bc_2$, where A and B are any two nonzero real numbers.

This means that we can multiply (6) by A and (7) by B, and then add the results to obtain (8).

Example 1

Solve the given system by the addition or subtraction method:

$$(1)\ 2x + 3y = 8$$
$$(2)\ x - y = -1$$

Solution

We proceed by steps:

1. Write the system:
   $$(1)\ 2x + 3y = 8$$
   $$(2)\ \ x - \ y = -1$$

2. Multiplying Equation (2) by 3, we obtain the system
   $$(1)\ \ 2x + 3y = 8$$
   $$(2')\ \underline{3x - 3y = -3}$$

3. Adding (1) and (2′),
   $5x = 5,$
   or $x = 1$

4. Substituting $x = 1$ in Equation (2), we get $1 - y = -1$; that is, $y = 2$.

5. The common solution is (1, 2), as can easily be checked since
   (1) $2(1) + 3(2) = 8$ and
   (2) $1 - 2 = -1$

In many cases it is not enough to multiply one of the equations by a number to eliminate one of the variables. For example, to find the common solution of

(1) $2x + 3y = -1$
(2) $3x + 2y = -4,$

we multiply *both* equations by numbers chosen so that the coefficients of one of the variables are opposites in the resulting equations. Thus,

1. (1) $2x + 3y = -1$
   (2) $3x + 2y = -4$

2. Multiplying Equation (1) by 2 and Equation (2) by -3, we obtain

Note that the coefficients of y are now opposites; 6 and −6.

$$\begin{array}{ll}(1') & 4x + 6y = -2 \\ (2') & \underline{-9x - 6y = 12}\end{array}$$

3. Adding,
   $-5x = 10$
   or $x = -2$

4. Substituting $x = -2$ in (1′), we get
   $-8 + 6y = -2$
   $6y = 6$, or $y = 1$

Check this!

5. Thus, the common solution of the system is $(-2, 1)$.

Example 2

Solve the system

(1) $3x + 2y = 6$
(2) $6x + 4y = 10$

Solution

We proceed by steps as before:

1. (1) $3x + 2y = 6$
   (2) $6x + 4y = 10$

2. Multiplying Equation (1) by -2, we obtain
   (1′) $-6x - 4y = -12$
   (2) $\underline{6x + 4y = 10}$

3. Adding, we get $0 = -2$.

4. Since we have obtained a contradiction, this system is *inconsistent* and has no solution.

5. There is no ordered pair (x, y) that is a common solution to both equations; thus the solution set is $\emptyset$.

This means that the graphs of the lines $3x + 2y = 6$ and $6x + 4y = 10$ do not intersect and therefore must be parallel.

PROGRESS TEST

1. The solution of the system $x + y = 5$ and $x - y = 1$ is ________.
2. The solution of the system $5x + 3y = 1$ and $2x - y = -4$ is ________.
3. The solution of the system $3x + 5y = 2$ and $6x + 10y = 5$ is ________.

EXERCISE 8.2

In Problems 1 through 30, solve the given system of equations. If there is no solution or if the system is dependent, so state.

Illustration: (1) $\dfrac{x}{6} + \dfrac{y}{2} = 1$

(2) $\dfrac{x}{9} + \dfrac{y}{3} = \dfrac{2}{3}$

PROGRESS TEST ANSWERS

1. (3, 2)
2. (−1, 2)
3. This system is inconsistent and has no solution.

Solution:

1. We first multiply Equation (1) by 6, the LCD of $\frac{x}{6}$ and $\frac{y}{2}$, and Equation (2) by 9, the LCD of $\frac{x}{9}$, $\frac{y}{3}$, and $\frac{2}{3}$. We obtain
   (1′) $x + 3y = 6$
   (2′) $x + 3y = 6$
2. Multiplying Equation (2′) by -1, we have
   (1′) $x + 3y = 6$
   (3) $-x - 3y = -6$
3. Adding, $0 = 0$.
4. Since we obtain $0 = 0$, a true statement, the equations are *dependent*, that is, any ordered pair satisfying (1) will also satisfy (2), and the system has infinitely many solutions. To find some of these solutions, we let $x = 6$ in Equation (1), obtaining $\frac{6}{6} + \frac{y}{2} = 1$ or $1 + \frac{y}{2} = 1$. Thus, $y = 0$, and (6, 0) is a solution for the system. For $x = 12$, $y = -2$, so $(12, -2)$ is also a solution.
5. It should be noted here that when we obtained Equations 1′ and 2′, we came up with *identical* equations. This fact alone would be sufficient to tell us the equations are dependent.

1. $x + y = 8$
$x - y = 2$

2. $x + y = 3$
$x - y = 1$

3. $x + 4y = 2$
$x - 4y = -2$

4. $x - 5y = 15$
$x + 5y = -5$

5. $-x - 2y = -2$
$x - 2y = -2$

6. $x + 3y = -7$
$-x + 2y = -3$

7. $2x + y = 7$
$3x - 2y = 0$

8. $2x + y = 4$
$3x - 2y = -1$

9. $2x - 2y = 6$
$x + y = 2$

10. $3x - 2y = 0$
$x + y = -5$

11. $3x + 5y = 1$
$-6x - 10y = 2$

12. $5x - 2y = 4$
$-10x + 4y = 1$

13. $2x + y = 8$
$3x - y = 7$

14. $x - 3y = -2$
$x + 3y = 4$

15. $2x + 5y = 9$
$3x + 2y = 8$

16. $3x + 5y = 26$
$5x + 3y = 22$

17. $6x + 5y = 12$
$9x - 4y = -5$

18. $5x + 4y = 6$
$4x - 3y = 11$

19. $2x - 3y = 16$
$x - y = 7$

20. $3x - 2y = 35$
$x - 5y = 42$

21. $18x - 15y = 1$
$10x - 12y = 3$

22. $6x - 9y = -2$
$3x - 5y = -6$

23. $\frac{x}{3} + \frac{y}{6} = \frac{2}{3}$

$\frac{2}{5}x + \frac{y}{4} = \frac{1}{5}$

24. $\frac{x}{6} + \frac{y}{3} = \frac{1}{2}$

$\frac{3}{5}x + \frac{y}{4} = \frac{17}{20}$

25. $\frac{5}{6}x + \frac{y}{4} = 7$

$\frac{2}{3}x - \frac{y}{8} = 3$

26. $\frac{1}{5}x + \frac{2}{5}y = 1$

$\frac{1}{4}x - \frac{1}{3}y = \frac{-5}{12}$

27. $\frac{2}{x} + \frac{3}{y} = \frac{-1}{2}$

$\frac{3}{x} - \frac{2}{y} = \frac{17}{12}$

28. $\frac{4}{x} + \frac{2}{y} = \frac{26}{21}$

$\frac{2}{x} - \frac{1}{y} = \frac{-1}{21}$

(Hint: Multiply the first equation by 2, the second one by 3.)

(Hint: Multiply the second equation by 2.)

29. $\frac{2}{x} - \frac{1}{y} = 0$

$\frac{3}{x} + \frac{5}{y} = \frac{13}{4}$

(Hint: Multiply the first equation by 5.)

30. $\frac{1}{x} - \frac{3}{y} = \frac{-13}{10}$

$\frac{5}{x} + \frac{2}{y} = 2$

(Hint: Multiply the first equation by -5.)

In Problems 31 through 35, find the solution.

Illustration: 130 dance-tickets were sold for \$340. If tickets were sold to men for \$3 and to women for \$2, how many of each were sold?

Solution:

1. We are asked for the number of tickets sold to men and the number sold to women.
2. Let m be the number of tickets sold to men and w be the number of tickets sold to women.
3. Since 130 tickets were sold,
   (1) $m + w = 130$
   Also, since the m tickets sold for \$3 each and the w tickets for \$2 each, and the total sales came to \$340,
   (2) $3m + 2w = 340$
4. We now solve
   (1) $m + w = 130$,
   (2) $3m + 2w = 340$.
   Multiplying Equation (1) by -2, we obtain the system

$$\begin{array}{lrcr} (1') & -2m - 2w & = & -260 \\ (2) & 3m + 2w & = & 340 \\ \hline & m & = & 80 \end{array} \quad \text{Adding}$$

   Substituting $m = 80$ in (1) we have $80 + w = 130$, or $w = 50$.

5. The number of tickets sold to men is 80 and to women 50. The verification is left to the student.

31. According to the Guinness Book of World Records, the sum of the ages of the two oldest cats is 70 years. The difference of their ages is 2 years. What are their ages?

32. According to the Guinness Book of World Records, the sum of the heights of the shortest and tallest persons on record (measured in inches) is 130. If the difference in their heights was 84 inches, how tall was each?

33. The height of the Empire State Building and its antenna is 1,472 feet. The difference in height between the building and the antenna is 1,028 feet. How tall is the antenna and how tall is the building?

34. The height of the Eiffel Tower and its antenna is 1,052 feet. The difference in height between the tower and the antenna is 920 feet. How tall is the antenna and how tall is the tower?

35. In 1970, the combined weight of the McCreary Brothers was 1,300 pounds (Guinness). Their weight difference was 20 pounds. What was the weight of each of the McCreary Brothers?

Photo by Jim Goetz.

OBJECTIVES 8.3

After studying this section, the reader should be able to:

1. **Solve a system of equations in three variables by using the technique discussed in the text.**
2. **Determine if a system of equations in three variables is inconsistent or dependent.**

8.3 SOLVING SYSTEMS OF EQUATIONS IN THREE VARIABLES

The man in the photograph has coffees worth \$1.30, \$1.40, and \$1.50 per pound. If he calls these coffees A, B, and C and decides to make 50 pounds of a mixture containing x pounds of A, y pounds of B, and z pounds of C,

$$(1)\quad x + y + z = 50$$

Assume that he decides to have twice as much Brand B as C. Thus

$$(2)\quad y = 2z$$

Finally, if he sells the 50 pounds at \$1.42 per pound, the total price will be \$71.00, or

$$(3)\quad 1.30x + 1.40y + 1.50z = 71$$

Now we rewrite Equations (1), (2), and (3) in the form

$$(4)\quad x + y + z = 50$$

$$(5)\quad y - 2z = 0$$ Subtracting 2z from both members of (2)

$$(6)\quad 13x + 14y + 15z = 710$$ Multiplying each member of (3) by 10

We have obtained a system of linear equation in three unknowns. To solve this system, we first note that Equation (2) *does not* contain x. Because of this fact, we select the pair of Equations (4) and (6) and eliminate x by multiplying (4) by −13, obtaining the new system

$$\begin{array}{lrcl} (7) & -13x - 13y - 13z &=& -650 \\ (6) & 13x + 14y + 15z &=& 710 \\ \hline (8) & y + 2z &=& 60 \end{array}$$ Adding

The system consisting of Equations (5) and (8) is a system of two equations in two unknowns. To eliminate z from this system, we add Equations (5) and (8), obtaining

$$\begin{array}{lrcl} (5) & y - 2z &=& 0 \\ (8) & y + 2z &=& 60 \\ \hline (9) & 2y &=& 60 \\ & \text{or } y &=& 30 \end{array}$$ Adding

Substituting $y = 30$ in (5), we obtain $30 - 2z = 0$, or $z = 15$. Finally, since $x + y + z = 50$, $x + 30 + 15 = 50$, or $x = 5$. Thus, the solution for the system is $x = 5$, $y = 30$, and $z = 15$. This solution can be written as an *ordered triple* of numbers: $(x, y, z) = (5, 30, 15)$. This solution is the intersection of the solution sets of the three equations in the system—that is, $\{(x, y, z) \mid x + y + z = 50\} \cap \{(x, y, z) \mid y = 2z\} \cap \{(x, y, z) \mid 1.30x + 1.40y + 1.50z = 71\} = \{(5, 30, 15)\}$.

As you can see from this example, to solve a system of three linear equations in three unknowns, we can proceed as follows:

1. Select a pair of equations and eliminate one variable from this pair.
2. Select a *different* pair of equations and eliminate the same variable chosen in Step 1.
3. Solve the pair of equations resulting from Steps 1 and 2 (use the procedure outlined in Section 8.2).
4. Substitute the values found in Step 3 in the simplest of the original equations, and then solve for the third variable.
5. Check by substituting the values in each of the original equations.

Example 1

Solve the system

(1) $x + y + z = 12$
(2) $2x - y + z = 7$
(3) $x + 2y - z = 6$

Solution

1. Adding (1) and (3) in order to eliminate z, we obtain

(4) $2x + 3y = 18$

2. Adding (2) and (3), again to eliminate z, we have

(5) $3x + y = 13$

3. We now have the system

(4) $2x + 3y = 18$
(5) $3x + y = 13$

Multiplying Equation (5) by −3, we have

$$\begin{array}{lrcr} (4) & 2x + 3y & = & 18 \\ (6) & -9x - 3y & = & -39 \\ \hline & -7x \qquad & = & -21 \\ & \text{or } x & = & 3 \end{array} \qquad \text{Adding}$$

Substituting $x = 3$ in (5), we get $9 + y = 13$, or $y = 4$.

4. In Equation (1), $x + y + z = 12$. We know now that $x = 3$ and that $y = 4$; substituting these values in (1), we have $3 + 4 + z = 12$. Solving, we find $z = 5$.
5. The solution of the system is (3, 4, 5), as can be easily verified:

(1) $3 + 4 + 5 = 12$
(2) $2(3) - 4 + 5 = 7$
(3) $3 + 2(4) - 5 = 6$

Example 2

Solve the system

(1) $x + 3y - z = 1$
(2) $x - y + z = 4$
(3) $3x + y + z = 3$

Solution

Adding first (1) and (2) and then (1) and (3), we obtain

(4) $2x + 2y = 5$
(5) $4x + 4y = 4$

Multiplying (4) by -2 to eliminate x, we have

$$\begin{array}{ll} (6) & -4x - 4y = -10 \\ (5) & \underline{\ \ 4x + 4y = \ \ \ \ 4} \\ (7) & \qquad\quad 0 = -6 \qquad \text{Adding} \end{array}$$

Since it is impossible for 0 to be equal to -6, this system has *no solution*; it is an *inconsistent system.*

Example 3

Solve the system

$$\begin{array}{ll} (1) & x - 2y + 3z = 4 \\ (2) & 2x - y + z = 1 \\ (3) & x + y - 2z = -3 \end{array}$$

Solution

1. Multiplying (2) by -2 and adding to (1) (to eliminate y), we get

$$\begin{array}{ll} (1) & \quad x - 2y + 3z = 4 \\ (4) & \underline{-4x + 2y - 2z = -2} \\ (5) & -3x \qquad\quad + z = 2 \end{array}$$

2. Adding (2) and (3), again to eliminate y, we obtain

$$\begin{array}{ll} (2) & 2x - y + \ z = 1 \\ (3) & \underline{\ x + y - 2z = -3} \\ (6) & 3x \qquad - z = -2 \end{array}$$

3. We now have the system

$$\begin{array}{ll} (5) & -3x + z = 2 \\ (6) & \underline{\ \ 3x - z = -2} \\ (7) & \qquad\ 0 = 0 \end{array}$$

   Thus, the system is *dependent* and has infinitely many solutions. One such solution is obtained if we let $x = 0$ in (5), obtaining $z = 2$.
4. Substituting $x = 0$ and $z = 2$ in (2), we have

$$\begin{array}{ll} (2) & 2 \cdot 0 - y + 2 = 1 \\ & \qquad - y + 2 = 1 \qquad \text{or } y = 1 \end{array}$$

   So (0, 1, 2) is one of the solutions for the system, as can be easily checked.

PROGRESS TEST

1. The solution of

$$\begin{aligned} x + y - 2z &= 13 \\ x - 3y - z &= -3 \\ x - y + 4z &= -17 \end{aligned}$$

is ______________________.

2. The solution of

$$\begin{aligned} 2x + y + z &= 2 \\ -x + 2y - z &= 4 \\ x + 8y - z &= 8 \end{aligned}$$

is ______________________.

3. The solution of

$$\begin{aligned} x + 2y + z &= -10 \\ x + y - z &= -3 \\ 5x + 7y - z &= -29 \end{aligned}$$

is ______________________.

EXERCISE 8.3

In Problems 1 through 20, use the five-step procedure detailed in the text to solve the given system of equations. If the system has no solution or is dependent, so state.

Illustration:
(1) $x + y = 2$
(2) $y + z = -2$
(3) $x + z = 12$

Solution:

1. We eliminate y from Equations (1) and (2) by first multiplying Equation (2) by -1, and adding the new equation to (1), obtaining

$$\begin{array}{lrcl} (4) & -y - z &=& 2 \\ (1) & x + y \quad &=& 2 \\ \hline & x \quad - z &=& 4 \end{array}$$

2. We now have the system

$$\begin{array}{lrcl} (3) & x + z &=& 12 \\ (5) & x - z &=& 4 \\ \hline \end{array}$$

PROGRESS TEST ANSWERS

1. $(2, 3, -4)$.
2. inconsistent; there is no solution.
3. dependent; there are infinitely many solutions. One such solution is $\left(0, \frac{-13}{3}, \frac{-4}{3}\right)$.

3. (6) $2x \quad = 16$ or $x = 8$ Adding

   Substituting $x = 8$ in (5), we have $8 - z = 4$, or $z = 4$.
4. Substituting $x = 8$ in (1), we obtain $8 + y = 2$, or $y = -6$. Thus, the solution is $(8, -6, 4)$.
5. The verification is left to the student.

1. $x + y + z = 12$
$x - y + z = 6$
$x + 2y - z = 7$

2. $x + y + z = 13$
$x - 2y + 4z = 10$
$3x + y - 3z = 5$

3. $x + y + z = 4$
$2x + 2y - z = -4$
$x - y + z = 2$

4. $2x - y + z = 3$
$x + 4y - z = 6$
$3x + 2y + 3z = 16$

5. $2x - y + z = 3$
$x + 2y + z = 12$
$4x - 3y + z = 1$

6. $x - 3y - 2z = -12$
$2x + y - 3z = -1$
$3x - 2y - z = -5$

7. $x - 2y - 3z = 2$
$x - 4y - 13z = 14$
$-3x + 5y + 4z = 2$

8. $2x + 2y + z = 3$
$-x + y - z = 5$
$3x + 5y + z = 8$

9. $2x + 4y + 3z = 3$
$10x - 8y - 9z = 0$
$4x + 4y - 3z = 2$

10. $9x + 4y - 10z = 6$
$6x - 8y + 5z = -1$
$12x + 12y - 15z = 10$

11. $x - 2y - z = 3$
$2x - 5y + z = -1$
$x - 2y - z = -3$

12. $x - 3y + 6z = -8$
$3x - 2y - 10z = 11$
$5x - 6y - 2z = 7$

13. $2x + y + z = 5$
$-x + 2y - z = 3$
$3x + 4y + z = 10$

14. $x - 3y + z = 2$
$x + 2y - z = 1$
$-7x + y + z = -10$

15. $x + y = 5$
$y + z = 3$
$x + z = 7$

16. $x + 2y = -1$
$2y + z = 0$
$x + 2z = 11$

17. $x - 2y = 0$
$y - 2z = 5$
$x + y + z = 8$

18. $2y + z = 9$
$z - 2y = 1$
$x + y + z = 1$

19. $5x - 3z = 2$
$2z - y = -5$
$x + 2y - 4z = 8$

20. $5x - 2z = 1$
$3z - y = 6$
$x + 2y - z = -1$

21. Find a condition on a, b, and c so that the system
$-4x + 3y = a$
$5x - 4y = b$
$-3x + 2y = c$
has a solution.

22. Show that the system

(1) $2x + 4z = 6$
(2) $3x + y + z = -1$
(3) $2y - z = -2$
(4) $x - y + z = -5$

does not have a solution. (Hint: Solve the system consisting of Equations (1), (2), and (4) and then show that the solution does not satisfy Equation (3).)

23. Find the solution set of

$2x + 4z = 6$
$3x + y + z = -1$
$2y - z = -2$
$x - y - 2z = -5$

24. Find a value of k so that the system

(1) $5x - y + 2z = 2$
(2) $3x + y - 3z = 7$
(3) $x + 5y + z = 5$
(4) $x + ky - z = 9$

has a solution. (Hint: Solve the system consisting of Equations (1), (2), and (3). Then substitute the values of x, y, and z in Equation (4) and solve for k.)

25. Repeat Problem 24 for the system

(1) $2x + 4z = 6$
(2) $3x + y + z = -1$
(3) $2y - z = -2$
(4) $x - y + kz = -5$

The Bettmann Archive.

OBJECTIVES 8.4

After studying this section, the reader should be able to:

1. **Evaluate a two-by-two determinant.**
2. **Use Cramer's rule to find the solution of a system of two equations in two unknowns.**
3. **Use Cramer's rule to determine if a system of two equations in two unknowns is inconsistent or dependent.**

8.4 SOLVING SYSTEMS OF TWO EQUATIONS IN TWO UNKNOWNS BY DETERMINANTS

The reproduction at the top of this page shows Gottfried Wilhem Leibniz, the originator of the theory of determinants. In the year 1693, Leibniz studied and used determinants to solve systems of simultaneous equations. A *determinant* is a square array of numbers of the form

$$\begin{vmatrix} a_1 & b_1 \\ a_2 & b_2 \end{vmatrix}$$

The numbers a_1, a_2, b_1, and b_2 are called the *elements* of the determinant. As you can see, this determinant has *two* rows and *two* columns. For this reason,

$$\begin{vmatrix} a_1 & b_1 \\ a_2 & b_2 \end{vmatrix}$$

is called a two-by-two (2×2) determinant. The *value* of this determinant is defined to be $a_1b_2 - b_1a_2$. In symbolic form,

$$\begin{vmatrix} a_1 & b_1 \\ a_2 & b_2 \end{vmatrix} = a_1b_2 - a_2b_1$$

Note that the value of

$$\begin{vmatrix} a_1 & b_1 \\ a_2 & b_2 \end{vmatrix}$$

is obtained by finding the difference of the "cross-multiplications" shown.

Thus,

$$\begin{vmatrix} 3 & -1 \\ 4 & 1 \end{vmatrix} = 3 \cdot 1 - 4 \cdot (-1) = 3 + 4 = 7$$

and

$$\begin{vmatrix} 3 & -1 \\ -2 & 0 \end{vmatrix} = 3 \cdot 0 - (-2)(-1) = 0 - 2 = -2$$

Determinants can be used to solve linear systems of two equations in two unknowns. For example, consider the system

(1) $a_1x + b_1y = c_1$ $\quad a_1, b_1, c_1$ real numbers
(2) $a_2x + b_2y = c_2$ $\quad a_2, b_2, c_2$ real numbers

Multiplying Equation (1) by $-a_2$ and Equation (2) by a_1, we obtain

(3) $-a_1a_2x - a_2b_1y = -a_2c_1$
(4) $\quad a_1a_2x + a_1b_2y = \quad a_1c_2$

Adding, we obtain

(5) $(a_1b_2 - a_2b_1)y = a_1c_2 - a_2c_1$.

Thus, if $a_1b_2 - a_2b_1 \neq 0$, then

(6) $$y = \frac{a_1c_2 - a_2c_1}{a_1b_2 - a_2b_1}$$

Remember that

$$\begin{vmatrix} a_1 & c_1 \\ a_2 & c_2 \end{vmatrix} = a_1c_2 - a_2c_1$$

and that

$$\begin{vmatrix} a_1 & b_1 \\ a_2 & b_2 \end{vmatrix} = a_1b_2 - a_2b_1$$

The numerator of (6) is the value of

$$\begin{vmatrix} a_1 & c_1 \\ a_2 & c_2 \end{vmatrix},$$

which we denote by D_y; the denominator of (6) is the value of

$$\begin{vmatrix} a_1 & b_1 \\ a_2 & b_2 \end{vmatrix},$$

which we denote by D. Thus,

$$(7)\quad y = \frac{\begin{vmatrix} a_1 & c_1 \\ a_2 & c_2 \end{vmatrix}}{\begin{vmatrix} a_1 & b_1 \\ a_2 & b_2 \end{vmatrix}} = \frac{D_y}{D}$$

Using a similar procedure, we can show that

$$(8)\quad x = \frac{\begin{vmatrix} c_1 & b_1 \\ c_2 & b_2 \end{vmatrix}}{\begin{vmatrix} a_1 & b_1 \\ a_2 & b_2 \end{vmatrix}} = \frac{D_x}{D}$$

We summarize this discussion in the following theorem, which is usually called Cramer's rule.

THEOREM 8.1

The system

$$(1)\quad a_1x + b_1y = c_1$$
$$(2)\quad a_2x + b_2y = c_2$$

has as its solution $x = \frac{D_x}{D}$, $y = \frac{D_y}{D}$,

where $D_x = \begin{vmatrix} c_1 & b_1 \\ c_2 & b_2 \end{vmatrix}$,

$D_y = \begin{vmatrix} a_1 & c_1 \\ a_2 & c_2 \end{vmatrix}$,

$D = \begin{vmatrix} a_1 & b_1 \\ a_2 & b_2 \end{vmatrix}$,

and $D \neq 0$.

Note that both x and y have as their denominators the determinant of the coefficients of the unknowns in (1) and (2). Moreover, the elements of the numerators of x and y are identical to the denominator, *except* that the column containing the coefficients of the unknown being found is replaced by the column of terms c_1 and c_2 on the right of the equations.

The solution of the system

$$(1)\quad 2x + 3y = 13$$
$$(2)\quad 5x - 4y = 21$$

is

$$x = \frac{\begin{vmatrix} 13 & 3 \\ 21 & -4 \end{vmatrix}}{\begin{vmatrix} 2 & 3 \\ 5 & -4 \end{vmatrix}}$$

$$= \frac{(13)(-4) - (21)(3)}{(2)(-4) - (5)(3)} = \frac{-52 - 63}{-8 - 15} = \frac{-115}{-23} = 5$$

$$y = \frac{\begin{vmatrix} 2 & 13 \\ 5 & 21 \end{vmatrix}}{\begin{vmatrix} 2 & 3 \\ 5 & -4 \end{vmatrix}} = \frac{(2)(21) - (5)(13)}{(2)(-4) - (5)(3)} = \frac{42 - 65}{-8 - 15} = \frac{-23}{-23} = 1$$

That is, (5, 1) is the solution of the given system.

Note: It is particularly important to check solutions found in such a mechanical fashion.

Example 1

Use Cramer's rule to solve the system

(1) $x + y = 5$
(2) $3x - y = 3$

Solution

By Cramer's rule,

$$x = \frac{\begin{vmatrix} 5 & 1 \\ 3 & -1 \end{vmatrix}}{\begin{vmatrix} 1 & 1 \\ 3 & -1 \end{vmatrix}} = \frac{-5 - 3}{-1 - 3} = \frac{-8}{-4} = 2$$

$$y = \frac{\begin{vmatrix} 1 & 5 \\ 3 & 3 \end{vmatrix}}{\begin{vmatrix} 1 & 1 \\ 3 & -1 \end{vmatrix}} = \frac{3 - 15}{-1 - 3} = \frac{-12}{-4} = 3$$

This solution can easily be verified by substituting the values for x and y in the original equations:

(1) $2 + 3 = 5$
(2) $3(2) - 3 = 3$

In many cases, the denominator of x and y in Theorem 8.1 is 0. When this happens,

(a) The system is *inconsistent* if both numerators are *not* 0.
(b) The system is *dependent* if both numerators *are* 0.

Example 2

Use Cramer's rule to solve the following system, if possible.

(1) $3x + 6y = 2$
(2) $x + 2y = 4$

Solution

By Cramer's rule,

$$x = \frac{\begin{vmatrix} 2 & 6 \\ 4 & 2 \end{vmatrix}}{\begin{vmatrix} 3 & 6 \\ 1 & 2 \end{vmatrix}} = \frac{4 - 24}{6 - 6} = \frac{-20}{0}$$

$$y = \frac{\begin{vmatrix} 3 & 2 \\ 1 & 4 \end{vmatrix}}{\begin{vmatrix} 3 & 6 \\ 1 & 2 \end{vmatrix}} = \frac{12 - 2}{6 - 6} = \frac{10}{0}$$

If a system is inconsistent, there is no ordered pair (x, y) that satisfies both equations.

As you can see, the denominator is zero, but the numerators are not zero. Thus the system is *inconsistent* and has no solution.

Example 3

Use Cramer's rule to solve the following system, if possible.

(1) $x - y = -6$
(2) $-x + y = 6$

Solution

By Cramer's rule,

$$x = \frac{\begin{vmatrix} -6 & -1 \\ 6 & 1 \end{vmatrix}}{\begin{vmatrix} 1 & -1 \\ -1 & 1 \end{vmatrix}} = \frac{-6 - (-6)}{1 - (-1)(-1)} = \frac{-6 + 6}{1 - 1} = \frac{0}{0}$$

$$y = \frac{\begin{vmatrix} 1 & -6 \\ -1 & 6 \end{vmatrix}}{\begin{vmatrix} 1 & -1 \\ -1 & 1 \end{vmatrix}} = \frac{6 - (-1)(-6)}{1 - (-1)(-1)} = \frac{6 - 6}{1 - 1} = \frac{0}{0}$$

Since in both cases the numerator and denominator are 0, the system is *dependent* and has infinitely many solutions. Moreover, any solution of (1) is a solution of (2). For example, if we let $x = 0$ in Equation (1), then $y = 6$, and $(0, 6)$ is a solution of the system. Similarly, if $y = 0$ in Equation (1), then $x = -6$, and $(-6, 0)$ is another solution of the system.

If a system is dependent, there is an infinite number of ordered pairs (x, y) that satisfy both equations.

The general solution of this system can be expressed as the ordered pair $(k - 6, k)$ for each real value of k.

PROGRESS TEST

1. Using Cramer's rule, the solution of
   $-x - y = -6$
   $5x - 4y = 12$
   is $x =$ ____________ and $y =$ ____________.
2. Using Cramer's rule, the solution of
   $x - 2y = 6$
   $2x - 4y = 5$
   is $x =$ ____________ and $y =$ ____________.
3. Using Cramer's rule, the solution of
   $-x + 2y = -5$
   $2x - 4y = 10$
   is $x =$ ____________ and $y =$ ____________.

EXERCISE 8.4

In Problems 1 through 10, evaluate the given determinant.

Illustration: $\begin{vmatrix} 3 & -1 \\ 4 & 2 \end{vmatrix}$

Solution: $\begin{vmatrix} 3 & -1 \\ 4 & 2 \end{vmatrix} = (3)(2) - (4)(-1) = 6 + 4$
$= 10$

1. $\begin{vmatrix} 1 & 1 \\ 0 & 2 \end{vmatrix}$

2. $\begin{vmatrix} 2 & -1 \\ 4 & 3 \end{vmatrix}$

3. $\begin{vmatrix} -3 & -2 \\ 5 & 1 \end{vmatrix}$

4. $\begin{vmatrix} 2 & -1 \\ -3 & 1 \end{vmatrix}$

5. $\begin{vmatrix} -2 & 0 \\ 5 & -3 \end{vmatrix}$

6. $\begin{vmatrix} 5 & 2 \\ -10 & -4 \end{vmatrix}$

7. $\begin{vmatrix} \frac{1}{2} & \frac{-1}{4} \\ \frac{1}{2} & \frac{3}{4} \end{vmatrix}$

8. $\begin{vmatrix} \frac{1}{5} & \frac{1}{10} \\ \frac{1}{2} & \frac{1}{4} \end{vmatrix}$

9. $\begin{vmatrix} \frac{3}{5} & \frac{1}{2} \\ \frac{-1}{4} & \frac{-1}{2} \end{vmatrix}$

10. $\begin{vmatrix} \frac{4}{5} & \frac{-1}{3} \\ \frac{-1}{2} & \frac{1}{2} \end{vmatrix}$

In Problems 11 through 31, use Cramer's rule to solve the given system. If the system is inconsistent or dependent, so state.

Illustration: (1) $x = y + 1$
(2) $y = -x + 7$

Solution: We *first* write the system in the equivalent form:

$x - y = 1$ Subtracting y from both members in (1)

$x + y = 7$ Adding x to both members in (2)

The solution of the system is

$$x = \frac{\begin{vmatrix} 1 & -1 \\ 7 & 1 \end{vmatrix}}{\begin{vmatrix} 1 & -1 \\ 1 & 1 \end{vmatrix}} = \frac{1+7}{1+1} = \frac{8}{2} = 4$$

$$y = \frac{\begin{vmatrix} 1 & 1 \\ 1 & 7 \end{vmatrix}}{\begin{vmatrix} 1 & -1 \\ 1 & 1 \end{vmatrix}} = \frac{7-1}{1+1} = \frac{6}{2} = 3$$

11. $x + y = 5$
$3x - y = 3$

12. $x + y = 9$
$x - y = 3$

13. $x + y = 9$
$x - y = -1$

14. $2x + y = -1$
$x - 2y = -13$

PROGRESS TEST ANSWERS

1. $x = \frac{\begin{vmatrix} -6 & -1 \\ 12 & -4 \end{vmatrix}}{\begin{vmatrix} -1 & -1 \\ 5 & -4 \end{vmatrix}} = \frac{36}{9} = 4,$

$y = \frac{\begin{vmatrix} -1 & -6 \\ 5 & 12 \end{vmatrix}}{\begin{vmatrix} -1 & -1 \\ 5 & -4 \end{vmatrix}} = \frac{18}{9} = 2$

2. $x = \frac{\begin{vmatrix} 6 & -2 \\ 5 & -4 \end{vmatrix}}{\begin{vmatrix} 1 & -2 \\ 2 & -4 \end{vmatrix}} = \frac{-14}{0},$

$y = \frac{\begin{vmatrix} 1 & 6 \\ 2 & 5 \end{vmatrix}}{\begin{vmatrix} 1 & -2 \\ 2 & -4 \end{vmatrix}} = \frac{-7}{0}$

The system is inconsistent and has no solution.

3. $x = \frac{\begin{vmatrix} -5 & 2 \\ 10 & -4 \end{vmatrix}}{\begin{vmatrix} -1 & 2 \\ 2 & -4 \end{vmatrix}} = \frac{20-20}{4-4} = \frac{0}{0},$

$y = \frac{\begin{vmatrix} -1 & -5 \\ 2 & 10 \end{vmatrix}}{\begin{vmatrix} -1 & 2 \\ 2 & -4 \end{vmatrix}} = \frac{-10+10}{4-4} = \frac{0}{0}$

The system is dependent.

15. $4x + 9y = 3$
$3x + 7y = 2$

16. $5x + 2y = 32$
$3x + y = 18$

17. $x - y = -1$
$x - 2y = -6$

18. $x - 2y = -13$
$3x - 2y = -19$

19. $2x + 3y = -13$
$6x + 9y = -39$

20. $4x + 5y = -2$
$12x + 15y = -6$

21. $x - y = 1$
$x - 2y = 4$

22. $x - 2y = 4$
$4x - 5y = 7$

23. $x + 3y = 6$
$2x + 6y = 5$

24. $x - 2y = 3$
$-x + 2y = 6$

25. $x = 7y - 3$
$2x + 3y = 23$

26. $x = 3y + 1$
$2x + 3y = 20$

27. $y = -3x + 17$
$2x - y = 8$

28. $y = -2x + 14$
$3x - y = 11$

29. $\frac{x}{2} - \frac{y}{3} = \frac{-1}{6}$
$\frac{x}{3} + \frac{y}{4} = \frac{-7}{12}$

30. $\frac{x}{3} - \frac{y}{5} = \frac{4}{3}$
$\frac{x}{4} - \frac{y}{3} = \frac{1}{12}$

In Problems 31 through 40, show that the given equation is true for all real values of the variables.

31. $\begin{vmatrix} 0 & a \\ 0 & b \end{vmatrix} = 0$

32. $\begin{vmatrix} 0 & 0 \\ a & b \end{vmatrix} = 0$

33. $\begin{vmatrix} a_1 & b_1 \\ a_2 & b_2 \end{vmatrix} = -\begin{vmatrix} b_1 & a_1 \\ b_2 & a_2 \end{vmatrix}$

34. $\begin{vmatrix} a_1 & a_2 \\ b_1 & b_2 \end{vmatrix} = -\begin{vmatrix} b_1 & b_2 \\ a_1 & a_2 \end{vmatrix}$

35. $\begin{vmatrix} a_1 & kb_1 \\ a_2 & kb_2 \end{vmatrix} = k\begin{vmatrix} a_1 & b_1 \\ a_2 & b_2 \end{vmatrix}$

36. $\begin{vmatrix} ka_1 & b_1 \\ ka_2 & b_2 \end{vmatrix} = k \begin{vmatrix} a_1 & b_1 \\ a_2 & b_2 \end{vmatrix}$

37. $\begin{vmatrix} a & ka \\ b & kb \end{vmatrix} = 0$

38. $\begin{vmatrix} ka & a \\ kb & b \end{vmatrix} = 0$

39. $\begin{vmatrix} a_1 & a_2 \\ b_1 & b_2 \end{vmatrix} = \begin{vmatrix} a_1 & a_2 \\ a_1 + b_1 & a_2 + b_2 \end{vmatrix}$

40. $\begin{vmatrix} a_1 & a_2 \\ b_1 & b_2 \end{vmatrix} = \begin{vmatrix} a_1 & a_2 \\ ka_1 + b_1 & ka_2 + b_2 \end{vmatrix}$

Photo by Jim Goetz.

OBJECTIVES 8.5

After studying this section, the reader should be able to:

1. **Solve a system of three linear equations in three unknowns by using determinants.**
2. **Determine if a system of three linear equations in three unknowns is inconsistent or has no unique solution.**
3. **Expand a three-by-three determinant along any row or column.**

8.5 SOLVING SYSTEMS OF THREE EQUATIONS IN THREE UNKNOWNS BY DETERMINANTS

The photograph at the beginning of this section shows a Sunoco pump, which blends gasoline to the specific octane requirements of various cars. A customer bought a certain quantity (x gallons) of 220 gas, a certain quantity (y gallons) of 240 gas, and a certain quantity (z gallons) of 260 gas. Her total purchase was 6 gallons. Thus,

(1) $x + y + z = 6$

When she next got gas, she decided to buy twice as many gallons of 240 gas and the same amounts of the other two as she had bought the first time. This time her total purchase was 7 gallons—that is,

(2) $x + 2y + z = 7$

On her next trip to the station, she decided to double the

amount of 260 gas and to buy the same amounts of the other two as she bought on the first purchase. This time her total purchase amounted to 9 gallons, so that

$$(3)\ x + y + 2z = 9$$

Can we find how many gallons of each gas she bought the first time? To do this, we must solve the system

$$(1)\ x + y + z = 6$$
$$(2)\ x + 2y + z = 7$$
$$(3)\ x + y + 2z = 9$$

To solve this system, we first remark that Cramer's rule can be extended to system of three equations in three unknowns. This method requires a knowledge of determinants with three rows and three columns. (These are called 3 × 3 – read "three by three" – determinants.) The three-row, three-column determinant is denoted by

$$\begin{vmatrix} a_1 & b_1 & c_1 \\ a_2 & b_2 & c_2 \\ a_3 & b_3 & c_3 \end{vmatrix} \quad \text{where the a's, b's, and c's are real numbers}$$

The value of this determinant is then defined in terms of second-order determinants as shown:

$$\begin{vmatrix} a_1 & b_1 & c_1 \\ a_2 & b_2 & c_2 \\ a_3 & b_3 & c_3 \end{vmatrix} = a_1 \begin{vmatrix} b_2 & c_2 \\ b_3 & c_3 \end{vmatrix} - b_1 \begin{vmatrix} a_2 & c_2 \\ a_3 & c_3 \end{vmatrix} + c_1 \begin{vmatrix} a_2 & b_2 \\ a_3 & b_3 \end{vmatrix}$$

For example, the value of the determinant

$$\begin{vmatrix} 2 & -3 & 4 \\ -1 & 2 & 1 \\ 1 & 1 & -2 \end{vmatrix}$$

is $(2) \begin{vmatrix} 2 & 1 \\ 1 & -2 \end{vmatrix} - (-3) \begin{vmatrix} -1 & 1 \\ 1 & -2 \end{vmatrix} + (4) \begin{vmatrix} -1 & 2 \\ 1 & 1 \end{vmatrix}$

$$= 2(-4 - 1) + 3(2 - 1) + 4(-1 - 2)$$
$$= -10 + 3 - 12$$
$$= -19$$

We state Cramer's rule for a system of three variables in the next theorem.

THEOREM 8.2

The system

(1) $a_1x + b_1y + c_1z = d_1$
(2) $a_2x + b_2y + c_2z = d_2$
(3) $a_3x + b_3y + c_3z = d_3$

(i) has the unique solution

$$x = \frac{D_x}{D}, \quad y = \frac{D_y}{D}, \quad z = \frac{D_z}{D}$$

where $D = \begin{vmatrix} a_1 & b_1 & c_1 \\ a_2 & b_2 & c_2 \\ a_3 & b_3 & c_3 \end{vmatrix}$,

$D_x = \begin{vmatrix} d_1 & b_1 & c_1 \\ d_2 & b_2 & c_2 \\ d_3 & b_3 & c_3 \end{vmatrix}$,

$D_y = \begin{vmatrix} a_1 & d_1 & c_1 \\ a_2 & d_2 & c_2 \\ a_3 & d_3 & c_3 \end{vmatrix}$,

and $D_z = \begin{vmatrix} a_1 & b_1 & d_1 \\ a_2 & b_2 & d_2 \\ a_3 & b_3 & d_3 \end{vmatrix}$,

if $D \neq 0$;

(ii) is *inconsistent* and has no solution if $D = 0$ and any one of D_x, D_y, D_z is different from zero;

(iii) has *no unique* solution if $D = 0$ and $D_x = D_y = D_z = 0$. (In this case, the system either has no solution or infinitely many solutions. This situation is studied more fully in advanced algebra.)

We now use Cramer's rule to solve the system

(1) $x + y + z = 6$
(2) $x + 2y + z = 7$
(3) $x + y + 2z = 9$

According to the theorem, we must calculate:

$$D = \begin{vmatrix} 1 & 1 & 1 \\ 1 & 2 & 1 \\ 1 & 1 & 2 \end{vmatrix} = (1)\begin{vmatrix} 2 & 1 \\ 1 & 2 \end{vmatrix} - (1)\begin{vmatrix} 1 & 1 \\ 1 & 2 \end{vmatrix} + (1)\begin{vmatrix} 1 & 2 \\ 1 & 1 \end{vmatrix}$$

$$= 3 - 1 - 1 = 1$$

$$D_x = \begin{vmatrix} 6 & 1 & 1 \\ 7 & 2 & 1 \\ 9 & 1 & 2 \end{vmatrix} = (6)\begin{vmatrix} 2 & 1 \\ 1 & 2 \end{vmatrix} - (1)\begin{vmatrix} 7 & 1 \\ 9 & 2 \end{vmatrix} + (1)\begin{vmatrix} 7 & 2 \\ 9 & 1 \end{vmatrix}$$

$$= 18 - 5 - 11 = 2$$

$$D_y = \begin{vmatrix} 1 & 6 & 1 \\ 1 & 7 & 1 \\ 1 & 9 & 2 \end{vmatrix} = (1)\begin{vmatrix} 7 & 1 \\ 9 & 2 \end{vmatrix} - (6)\begin{vmatrix} 1 & 1 \\ 1 & 2 \end{vmatrix} + (1)\begin{vmatrix} 1 & 7 \\ 1 & 9 \end{vmatrix}$$

$$= 5 - 6 + 2 = 1$$

$$D_z = \begin{vmatrix} 1 & 1 & 6 \\ 1 & 2 & 7 \\ 1 & 1 & 9 \end{vmatrix} = (1)\begin{vmatrix} 2 & 7 \\ 1 & 9 \end{vmatrix} - (1)\begin{vmatrix} 1 & 7 \\ 1 & 9 \end{vmatrix} + (6)\begin{vmatrix} 1 & 2 \\ 1 & 1 \end{vmatrix}$$

$$= 11 - 2 - 6 = 3$$

From the theorem,

$$x = \frac{D_x}{D} = \frac{2}{1} = 2, \quad y = \frac{D_y}{D} = \frac{1}{1} = 1, \quad z = \frac{D_z}{D} = \frac{3}{1} = 3$$

Thus, the woman mentioned at the beginning of this section bought two gallons of 220 gasoline, one gallon of 240, and 3 gallons of 260 gasoline on her first purchase.

Example 1 Use Cramer's rule to solve the system

(1) $x + y + 2z = 7$
(2) $x - y - 3z = -6$
(3) $2x + 3y + z = 4$

Solution By Cramer's rule,

$$x = \frac{D_x}{D}, \quad y = \frac{D_y}{D}, \quad \text{and } z = \frac{D_z}{D}.$$

$$D = \begin{vmatrix} 1 & 1 & 2 \\ 1 & -1 & -3 \\ 2 & 3 & 1 \end{vmatrix}$$

$$= (1)\begin{vmatrix} -1 & -3 \\ 3 & 1 \end{vmatrix} - (1)\begin{vmatrix} 1 & -3 \\ 2 & 1 \end{vmatrix} + (2)\begin{vmatrix} 1 & -1 \\ 2 & 3 \end{vmatrix}$$

$$= 8 - 7 + 10 = 11$$

$$D_x = \begin{vmatrix} 7 & 1 & 2 \\ -6 & -1 & -3 \\ 4 & 3 & 1 \end{vmatrix}$$

$$= (7)\begin{vmatrix} -1 & -3 \\ 3 & 1 \end{vmatrix} - (1)\begin{vmatrix} -6 & -3 \\ 4 & 1 \end{vmatrix} + (2)\begin{vmatrix} -6 & -1 \\ 4 & 3 \end{vmatrix}$$

$$= 56 - 6 - 28 = 22$$

$$D_y = \begin{vmatrix} 1 & 7 & 2 \\ 1 & -6 & -3 \\ 2 & 4 & 1 \end{vmatrix}$$

$$= (1)\begin{vmatrix} -6 & -3 \\ 4 & 1 \end{vmatrix} - (7)\begin{vmatrix} 1 & -3 \\ 2 & 1 \end{vmatrix} + (2)\begin{vmatrix} 1 & -6 \\ 2 & 4 \end{vmatrix}$$

$$= 6 - 49 + 32 = -11$$

$$D_z = \begin{vmatrix} 1 & 1 & 7 \\ 1 & -1 & -6 \\ 2 & 3 & 4 \end{vmatrix}$$

$$= (1)\begin{vmatrix} -1 & -6 \\ 3 & 4 \end{vmatrix} - (1)\begin{vmatrix} 1 & -6 \\ 2 & 4 \end{vmatrix} + (7)\begin{vmatrix} 1 & -1 \\ 2 & 3 \end{vmatrix}$$

$$= 14 - 16 + 35 = 33$$

Therefore, $x = \frac{D_x}{D} = \frac{22}{11} = 2$, $y = \frac{D_y}{D} = \frac{-11}{11} = -1$, and $z = \frac{D_z}{D} = \frac{33}{11} = 3$. This solution can easily be verified by substituting our values for x, y, and z in the original equations:

(1) $2 + (-1) + 2 \cdot 3 = 7$
(2) $2 - (-1) - 3 \cdot 3 = -6$
(3) $2(2) + 3(-1) + 3 = 4$

Because it is easy to make numerical errors in evaluating determinants, the check is quite important.

Example 2

Use Cramer's rule to solve the system

(1) $x + y - z = 2$
(2) $2x + y + z = 4$
(3) $-x - y + z = 3$

Solution

By Cramer's rule, if there is a unique solution, it is given by

$$x = \frac{D_x}{D}, \quad y = \frac{D_y}{D}, \quad \text{and } z = \frac{D_z}{D}$$

However,

$$D = \begin{vmatrix} 1 & 1 & -1 \\ 2 & 1 & 1 \\ -1 & -1 & 1 \end{vmatrix}$$

$$= (1)\begin{vmatrix} 1 & 1 \\ -1 & 1 \end{vmatrix} - (1)\begin{vmatrix} 2 & 1 \\ -1 & 1 \end{vmatrix} + (-1)\begin{vmatrix} 2 & 1 \\ -1 & -1 \end{vmatrix}$$

$$= 2 - 3 + 1 = 0$$

and

$$D_x = \begin{vmatrix} 2 & 1 & -1 \\ 4 & 1 & 1 \\ 3 & -1 & 1 \end{vmatrix}$$

$$= (2)\begin{vmatrix} 1 & 1 \\ -1 & 1 \end{vmatrix} - (1)\begin{vmatrix} 4 & 1 \\ 3 & 1 \end{vmatrix} + (-1)\begin{vmatrix} 4 & 1 \\ 3 & -1 \end{vmatrix}$$

$$= 4 - 1 + 7 = 10 \neq 0$$

Hence, the system is inconsistent and has no solution.

We should remark at this point that the evaluation of a 3×3 determinant can be done by introducing the idea of a *minor* of an element in a determinant.

DEFINITION 8.1

In the determinant

$$\begin{vmatrix} a_1 & b_1 & c_1 \\ a_2 & b_2 & c_2 \\ a_3 & b_3 & c_3 \end{vmatrix},$$

the *minor* of an element is the determinant that remains after deleting the row and column in which the element appears.

For example, in the determinant $\begin{vmatrix} a_1 & b_1 & c_1 \\ a_2 & b_2 & c_2 \\ a_3 & b_3 & c_3 \end{vmatrix}$,

The minor of a_1 is $\begin{vmatrix} b_2 & c_2 \\ b_3 & c_3 \end{vmatrix}$

The minor of b_1 is $\begin{vmatrix} a_2 & c_2 \\ a_3 & c_3 \end{vmatrix}$

The minor of c_1 is $\begin{vmatrix} a_2 & b_2 \\ a_3 & b_3 \end{vmatrix}$

the minor of a_1 is $\begin{vmatrix} b_2 & c_2 \\ b_3 & c_3 \end{vmatrix}$,

the minor of b_1 is $\begin{vmatrix} a_2 & c_2 \\ a_3 & c_3 \end{vmatrix}$,

and the minor of c_1 is $\begin{vmatrix} a_2 & b_2 \\ a_3 & b_3 \end{vmatrix}$

Earlier, we presented the value of a 3×3 determinant as

This is the expansion of the determinant by minors along the first row.

$$(4) \quad \begin{vmatrix} a_1 & b_1 & c_1 \\ a_2 & b_2 & c_2 \\ a_3 & b_3 & c_3 \end{vmatrix} = a_1 \begin{vmatrix} b_2 & c_2 \\ b_3 & c_3 \end{vmatrix} - b_1 \begin{vmatrix} a_2 & c_2 \\ a_3 & c_3 \end{vmatrix} + c_1 \begin{vmatrix} a_2 & b_2 \\ a_3 & b_3 \end{vmatrix}$$

The 2×2 determinants in this equation are clearly minors of the elements in the first row of the 3×3 determinant on the left; the right side is called the *expansion* of the 3×3 determinant by *minors* along the first row. An expansion with numbers is carried out in Example 3.

Example 3

Expand

$$\begin{vmatrix} 1 & 1 & 1 \\ 1 & 2 & 1 \\ 1 & 1 & 2 \end{vmatrix}$$

by minors along the first row.

Solution

$$\begin{vmatrix} 1 & 1 & 1 \\ 1 & 2 & 1 \\ 1 & 1 & 2 \end{vmatrix} = (1) \begin{vmatrix} 2 & 1 \\ 1 & 2 \end{vmatrix} - (1) \begin{vmatrix} 1 & 1 \\ 1 & 2 \end{vmatrix} + (1) \begin{vmatrix} 1 & 2 \\ 1 & 1 \end{vmatrix}$$

$$= 3 - 1 - 1 = 1$$

In our work so far, the given determinant has been evaluated by minors along the first row. Actually, it is possible to expand a determinant by the minors of *any row* or *any column*. To do this, it is necessary to define the *sign array* of a determinant. For a three by three (3×3) determinant, the sign array is the following arrangement of alternating signs:

$$(5) \quad \begin{matrix} + & - & + \\ - & + & - \\ + & - & + \end{matrix}$$

To obtain the expansion of

$$\begin{vmatrix} a_1 & b_1 & c_1 \\ a_2 & b_2 & c_2 \\ a_3 & b_3 & c_3 \end{vmatrix}$$

along a particular row or column, we simply write in front of each term in the expansion the corresponding sign from the array. For example, to expand

$$\begin{vmatrix} 1 & 1 & 1 \\ 1 & 2 & 1 \\ 1 & 1 & 2 \end{vmatrix}$$

along the *second row*, we write

Note that we used the signs of the second row of (5), $-+-$, for the first, second, and third terms, respectively.

$$\begin{vmatrix} 1 & 1 & 1 \\ 1 & 2 & 1 \\ 1 & 1 & 2 \end{vmatrix} = -(1)\begin{vmatrix} 1 & 1 \\ 1 & 2 \end{vmatrix} + (2)\begin{vmatrix} 1 & 1 \\ 1 & 2 \end{vmatrix} - (1)\begin{vmatrix} 1 & 1 \\ 1 & 1 \end{vmatrix}$$

$$= -1(1) + 2(1) - 1(0) = 1$$

Example 4

Expand the given determinant along the third column.

(a) $\begin{vmatrix} 0 & 1 & 1 \\ 1 & 2 & -1 \\ 1 & -1 & 3 \end{vmatrix}$

(b) $\begin{vmatrix} 1 & 1 & 0 \\ 0 & -1 & 1 \\ 2 & -1 & -3 \end{vmatrix}$

Solution

(a) $\begin{vmatrix} 0 & 1 & 1 \\ 1 & 2 & -1 \\ 1 & -1 & 3 \end{vmatrix}$

Here we use the signs of the third column in (5): $+-+$.

$$= +(1)\begin{vmatrix} 1 & 2 \\ 1 & -1 \end{vmatrix} - (-1)\begin{vmatrix} 0 & 1 \\ 1 & -1 \end{vmatrix} + (3)\begin{vmatrix} 0 & 1 \\ 1 & 2 \end{vmatrix}$$

$$= (1)(-3) + (1)(-1) + (3)(-1) = -3 - 1 - 3 = -7$$

(b) $\begin{vmatrix} 1 & 1 & 0 \\ 0 & -1 & 1 \\ 2 & -1 & -3 \end{vmatrix}$

$$= +(0)\begin{vmatrix} 0 & -1 \\ 2 & -1 \end{vmatrix} - (1)\begin{vmatrix} 1 & 1 \\ 2 & -1 \end{vmatrix} + (-3)\begin{vmatrix} 1 & 1 \\ 0 & -1 \end{vmatrix}$$

$$= 0 - (1)(-3) + (-3)(-1) = 6$$

PROGRESS TEST

1. What is the solution of the following system?

(1) $x + y + z = 6$
(2) $x - y + z = 2$
(3) $x + y - 2z = -3$

2. What is the solution of the following system?

$$(1)\ x - y - z = 2$$
$$(2)\ x + 2y + z = 6$$
$$(3)\ -x + y + z = 4$$

3. Expand

$$\begin{vmatrix} 1 & -2 & 3 \\ 1 & 0 & 0 \\ 2 & 4 & -1 \end{vmatrix}$$

along the third column.

EXERCISE 8.5

In Problems 1 through 10, evaluate the given determinant.

Illustration:

$$\begin{vmatrix} 1 & 2 & 0 \\ -1 & 2 & 3 \\ 3 & -2 & -1 \end{vmatrix}$$

Solution:
Expanding along the first row, we have

$$\begin{vmatrix} 1 & 2 & 0 \\ -1 & 2 & 3 \\ 3 & -2 & -1 \end{vmatrix}$$

$$= (1)\begin{vmatrix} 2 & 3 \\ -2 & -1 \end{vmatrix} - (2)\begin{vmatrix} -1 & 3 \\ 3 & -1 \end{vmatrix} + 0\begin{vmatrix} -1 & 2 \\ 3 & -2 \end{vmatrix}$$

$$= (1)[(2)(-1) - (-2)(3)] - 2[(-1)(-1) - (3)(3)] + (0)[(-1)(-2) - (3)(2)]$$

$$= (-2 + 6) - 2(1 - 9) + 0$$

$$= 4 + 16$$

$$= 20$$

1. $\begin{vmatrix} 1 & 3 & 2 \\ 2 & 4 & 1 \\ 3 & 6 & 5 \end{vmatrix}$

2. $\begin{vmatrix} 1 & 3 & 5 \\ 2 & 0 & 10 \\ -3 & 1 & -15 \end{vmatrix}$

3. $\begin{vmatrix} 1 & 2 & 3 \\ 4 & 5 & 6 \\ 7 & 8 & 9 \end{vmatrix}$

4. $\begin{vmatrix} 1 & 1 & 1 \\ 2 & 3 & 1 \\ 2 & 4 & 1 \end{vmatrix}$

5. $\begin{vmatrix} 2 & 1 & 3 \\ 1 & 2 & -1 \\ 3 & 1 & 5 \end{vmatrix}$

6. $\begin{vmatrix} -1 & 1 & -1 \\ -2 & 2 & -6 \\ 3 & -3 & 4 \end{vmatrix}$

PROGRESS TEST ANSWERS

1. $x = 1$
   $y = 2$
   $z = 3$
2. The system is inconsistent and has no solution, since $D = 0$ and $D_x = 6 \neq 0$.
3. $+3\begin{vmatrix} 1 & 0 \\ 2 & 4 \end{vmatrix} - 0\begin{vmatrix} 1 & -2 \\ 2 & 4 \end{vmatrix} + (-1)\begin{vmatrix} 1 & -2 \\ 1 & 0 \end{vmatrix}$
   $= 3(4) - 0 + (-1)(2)$
   $= 12 - 2 = 10$

7. $\begin{vmatrix} 1 & 1 & 6 \\ 1 & 1 & 4 \\ 1 & -1 & 2 \end{vmatrix}$

8. $\begin{vmatrix} 1 & 4 & 0 \\ 1 & -3 & 1 \\ 0 & 8 & -1 \end{vmatrix}$

9. $\begin{vmatrix} 0 & -1 & 2 \\ 2 & 1 & -3 \\ 1 & -3 & 1 \end{vmatrix}$

10. $\begin{vmatrix} -3 & 2 & -4 \\ 1 & -1 & 3 \\ 1 & 2 & 10 \end{vmatrix}$

In Problems 11 through 20, solve the given system using Cramer's rule. If the system is inconsistent or has no unique solution, so state.

Illustration: (1) $x + y = 2$
(2) $y + z = -2$
(3) $x + z = 12$

Solution:
To avoid confusion, we first rewrite the system in the form.

$$\begin{aligned} &(1)\ x + y = 2 \\ &(2)\ y + z = -2 \\ &(3)\ x + z = 12 \end{aligned}$$

By Cramer's rule,

$$x = \frac{D_x}{D}, \quad y = \frac{D_y}{D}, \quad z = \frac{D_z}{D} \quad \text{where}$$

$$D = \begin{vmatrix} 1 & 1 & 0 \\ 0 & 1 & 1 \\ 1 & 0 & 1 \end{vmatrix}$$

$$= (1)\begin{vmatrix} 1 & 1 \\ 0 & 1 \end{vmatrix} - (1)\begin{vmatrix} 0 & 1 \\ 1 & 1 \end{vmatrix} + (0)\begin{vmatrix} 0 & 1 \\ 1 & 0 \end{vmatrix}$$

$$= 1 + 1 = 2$$

$$D_x = \begin{vmatrix} 2 & 1 & 0 \\ -2 & 1 & 1 \\ 12 & 0 & 1 \end{vmatrix}$$

$$= (2)\begin{vmatrix} 1 & 1 \\ 0 & 1 \end{vmatrix} - (1)\begin{vmatrix} -2 & 1 \\ 12 & 1 \end{vmatrix} + (0)\begin{vmatrix} -2 & 1 \\ 12 & 0 \end{vmatrix}$$

$$= 2 + 14 = 16$$

$$D_y = \begin{vmatrix} 1 & 2 & 0 \\ 0 & -2 & 1 \\ 1 & 12 & 1 \end{vmatrix}$$

$$= (1)\begin{vmatrix} -2 & 1 \\ 12 & 1 \end{vmatrix} - (2)\begin{vmatrix} 0 & 1 \\ 1 & 1 \end{vmatrix} + (0)\begin{vmatrix} 0 & -2 \\ 1 & 12 \end{vmatrix}$$

$$= -14 + 2 = -12$$

$$D_z = \begin{vmatrix} 1 & 1 & 2 \\ 0 & 1 & -2 \\ 1 & 0 & 12 \end{vmatrix}$$

$$= (1)\begin{vmatrix} 1 & -2 \\ 0 & 12 \end{vmatrix} - (1)\begin{vmatrix} 0 & -2 \\ 1 & 12 \end{vmatrix} + (2)\begin{vmatrix} 0 & 1 \\ 1 & 0 \end{vmatrix}$$

$$= 12 - 2 - 2 = 8$$

Thus, $x = \frac{D_x}{D} = \frac{16}{2} = 8$, $y = \frac{D_y}{D} = \frac{-12}{2} = -6$, $z = \frac{D_z}{D} = \frac{8}{2} = 4$. This solution can easily be verified by substituting $x = 8$, $y = -6$, $z = 4$ in the original equations.

11. $x + y + z = 6$
$2x - 3y + 3z = 5$
$3x - 2y - z = -4$

12. $x + y + z = 13$
$3x + y - 3z = 5$
$x - 2y + 4z = 10$

13. $6x + 5y + 4z = 5$
$5x + 4y + 3z = 5$
$4x + 3y + z = 7$

14. $3x + 2y + z = 4$
$4x + 3y + z = 5$
$5x + y + z = 9$

15. $x - 2y + 3z = 15$
$5x + 7y - 11z = -29$
$-13x + 17y + 19z = 37$

16. $2x - y + z = 3$
$x + 2y + z = 12$
$4x - 3y + z = 1$

17. $5x + 3y + 5z = 3$
$3x + 5y + z = -5$
$2x + 2y + 3z = 7$

18. $x + y = 5$
$y + z = 3$
$x + z = 7$

19. $2y + z = 9$
$-2y + z = 1$
$x + y + z = 1$

20. $x - y = 3$
$y - z = 3$
$x + z = 9$

In Problems 21 through 30, show that for any value of the variables, the given statement is true.

21. $\begin{vmatrix} a & b & 0 \\ c & d & 0 \\ e & f & 0 \end{vmatrix} = 0$

22. $\begin{vmatrix} a & b & c \\ d & e & f \\ 0 & 0 & 0 \end{vmatrix} = 0$

23. $\begin{vmatrix} a & b & c \\ 1 & 2 & 3 \\ a & b & c \end{vmatrix} = 0$

24. $\begin{vmatrix} 1 & a & a \\ 2 & b & b \\ 3 & c & c \end{vmatrix} = 0$

25. $\begin{vmatrix} 1 & 2 & 3 \\ 3 & 1 & 2 \\ k & 2k & 3k \end{vmatrix} = k \begin{vmatrix} 1 & 2 & 3 \\ 3 & 1 & 2 \\ 1 & 2 & 3 \end{vmatrix}$

26. $\begin{vmatrix} 1 & 2 & 3k \\ 3 & 2 & k \\ 0 & 1 & 2k \end{vmatrix} = k \begin{vmatrix} 1 & 2 & 3 \\ 3 & 2 & 1 \\ 0 & 1 & 2 \end{vmatrix}$

27. $\begin{vmatrix} kb_1 & b_1 & 1 \\ kb_2 & b_2 & 2 \\ kb_3 & b_3 & 3 \end{vmatrix} = 0$

28. $\begin{vmatrix} b_1 & b_2 & b_3 \\ kb_1 & kb_2 & kb_3 \\ 1 & 2 & 3 \end{vmatrix} = 0$

29. $\begin{vmatrix} 1 & 1 & 1 \\ 2 & a & a \\ 3 & b & b \end{vmatrix} = 0$

30. $\begin{vmatrix} 0 & 0 & 0 \\ a & b & c \\ d & e & f \end{vmatrix} = 0$

SUPPLY AND DEMAND SCHEDULES FOR WHEAT

	(1) POSSIBLE PRICES ($ per bu.)	(2) QUANTITY DEMANDED (million bu. per month)	(3) QUANTITY SUPPLIED (million bu. per month)	(4) PRESSURE ON PRICE
A	$5	9	18	Downward
B	4	10	16	Downward
C	**3**	**12**	**12**	**Neutral**
D	2	15	7	Upward
E	1	20	0	Upward

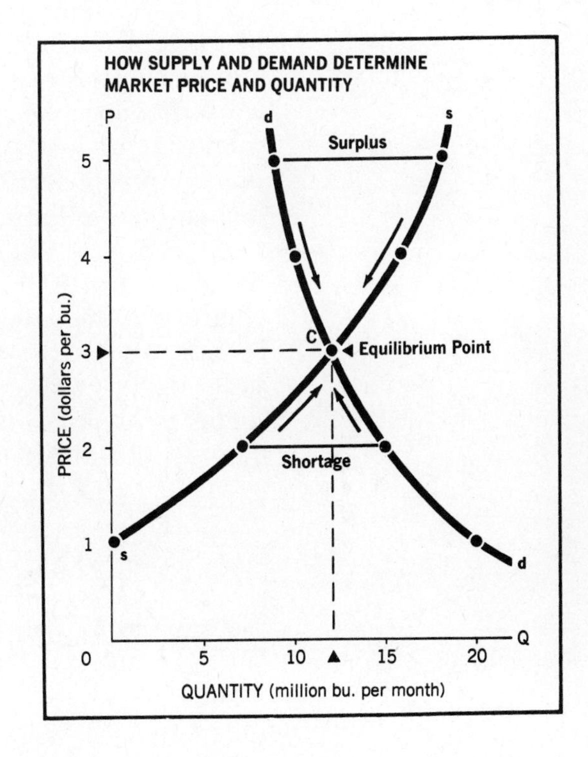

In the graph, the demand curve is labeled d and the supply curve s. (From *Economics*, 10th Ed., by Paul A. Samuelson. Copyright 1976. Used with permission of McGraw-Hill Book Company.)

OBJECTIVES 8.6

After studying this section, the reader should be able to find the solution of a given system of equations involving linear and/or second-degree equations by using the substitution method, and then check the solution by graphing the system.

8.6 SYSTEMS INVOLVING SECOND DEGREE EQUATIONS

The graph at the beginning of this section shows how supply and demand determine the market price and the quantity of wheat available for sale. As you can see from the table

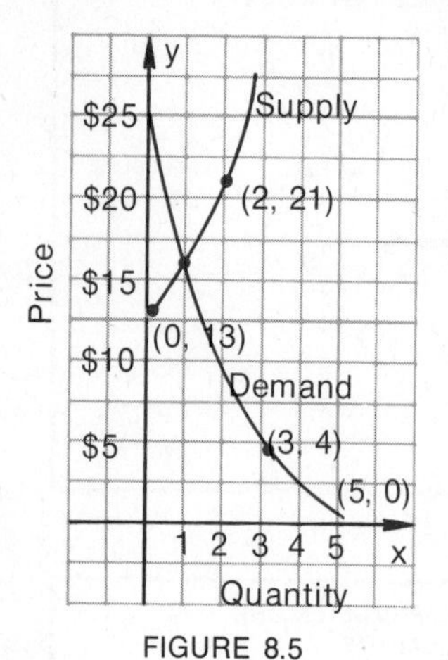

FIGURE 8.5

and the graph, as the price per bushel of wheat *decreases*, the quantity demanded by the consumer *increases*, as shown by the demand curve. If the price per bushel *increases*, sellers are more willing to supply wheat, and thus the quantity of wheat available *increases*, as shown by the supply curve on the graph. The point C of intersection of the two curves is called the *equilibrium point*. At this point, the price of a bushel of wheat is \$3, and the amount demanded by the consumers, 12 million bushels per month, exactly equals the amount supplied by producers.

As we mentioned in Section 8.2, the graphical method we have just used to find the equilibrium point depends on how accurately we graph the equations involved. Thus, as we did before, we shall concentrate on algebraic methods of solving equations. One of the most useful of these methods in that of substitution. For example, suppose that (1) the demand curve for a certain product is given by the equation $y = (x - 5)^2$, where x is the number of units produced and y is the price, and (2) the supply curve is given as $y = x^2 + 2x + 13$, where y is the price and x is the number of units available for sale. To find the *equilibrium point*, we sketch both curves, as shown in Figure 8.5. As you can see, the equilibrium point seems to be at the point (1, 16). To check this guess, we use the substitution method. Consider the system of equations

$$\begin{aligned}&(1)\ y = (x - 5)^2\\&(2)\ y = x^2 + 2x + 13\end{aligned}$$

We wish to find an ordered pair (x, y) which is a solution of *both* (1) and (2). Thus, by the Substitution Axiom, we may write

$$\begin{aligned}(x - 5)^2 &= x^2 + 2x + 13 &&\\ x^2 - 10x + 25 &= x^2 + 2x + 13 &&\text{Expanding}\\ -10x + 25 &= 2x + 13 &&\text{Subtracting } x^2\\ -12x &= -12 &&\text{Subtracting } 2x \text{ and } 25\\ x &= 1 &&\text{Dividing by } -12\end{aligned}$$

If $x = 1$ in Equation (2), then

$$y = (1)^2 + 2(1) + 13 = 16$$

Hence, the solution for the given system is (1, 16), as can easily be verified.

Example 1

Find the solution set of the given system by the substitution method. Check the solution by sketching the graphs of the equations.

(1) $x^2 + y^2 = 25$
(2) $x + y = 5$

Solution

We first rewrite Equation (2) in the equivalent form $y = 5 - x$, obtaining

(1) $x^2 + y^2 = 25$
(3) $y = 5 - x$

Any ordered pair (x, y) which is a solution of both (1) and (2) is also a solution of (1) and (3). Thus, by the Substitution Axiom, we replace y in Equation (1) by $5 - x$, which will produce

(4) $x^2 + (5 - x)^2 = 25$	Substituting in (1)
$x^2 + 25 - 10x + x^2 = 25$	Expanding
$2x^2 - 10x = 0$	Simplifying; subtracting 25
$x^2 - 5x = 0$	Dividing by 2
$x(x - 5) = 0$	Factoring
$x = 0$ or $x - 5 = 0$	
$x = 0$ or $x = 5$	Solving for x

We now place $x = 0$ and $x = 5$ in Equation (3) to obtain the corresponding y values:

$$y = 5 - 0 = 5$$
$$\text{and}\quad y = 5 - 5 = 0$$

Thus, when $x = 0$, $y = 5$, and when $x = 5$, $y = 0$. Therefore, the solution set for the system is $\{(0, 5), (5, 0)\}$.
Note that if we had substituted $x = 0$ and $x = 5$ in Equation (1) rather than in Equation (3), we would have obtained

$$0^2 + y^2 = 25 \text{ and } 5^2 + y^2 = 25$$

That is, $y = \pm 5$ and $y = 0$. In this case, the solutions obtained would have been (0, 5), (0, −5), (5, 0). However, (0, −5) is *not* a solution of Equation (3), since $-5 \neq 5 - 0$. Therefore,

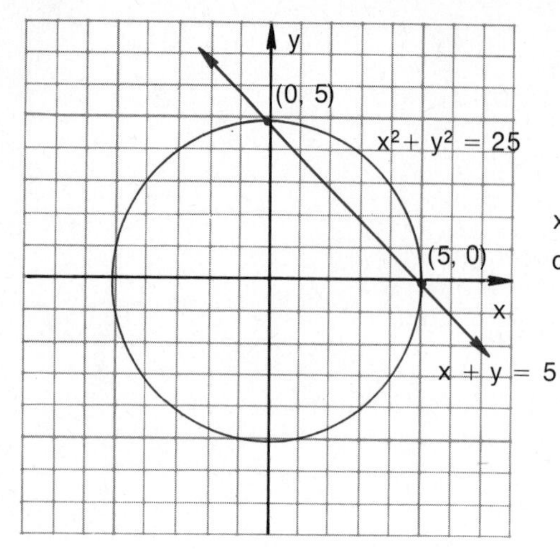

$x^2 + y^2 = 25$ is a circle of radius 5

FIGURE 8.6

the solution set is only $\{(0, 5), (5, 0)\}$. As you can see, if the degrees of the equations are different, one component of a solution should be substituted in the *lower degree* equation to find the ordered pairs satisfying *both* equations. For this reason, we double-check our work by graphing the given system (see Figure 8.6) to verify that our solutions are correct.

Example 2

Find the solution of the given system by the substitution method. Check the solution by sketching the graphs of the equations.

(1) $x^2 + y^2 = 9$
(2) $x + y = 5$

Solution

Rewriting (2) in the form $y = 5 - x$, we obtain the equivalent system

(1) $x^2 + y^2 = 9$
(3) $y = 5 - x$

Substituting $y = 5 - x$ in (1), we get

$$x^2 + (5 - x)^2 = 9$$

$$x^2 + 25 - 10x + x^2 = 9 \quad \text{Expanding}$$

$$2x^2 - 10x + 25 = 9 \quad \text{Simplifying}$$

$$2x^2 - 10x + 16 = 0 \quad \text{Subtracting 9}$$

$$x^2 - 5x + 8 = 0 \quad \text{Dividing by 2}$$

Using the quadratic formula, we get

$$x = \frac{5 \pm \sqrt{25 - 4 \bullet 8}}{2} = \frac{5 \pm \sqrt{-7}}{2} = \frac{5 \pm \sqrt{7}i}{2}$$

Substituting these values in (3), we obtain

$$y = 5 - \frac{(5 + \sqrt{7}i)}{2} \text{ and } y = 5 - \frac{(5 - \sqrt{7}i)}{2}.$$

That is,

$$y = \frac{5}{2} - \frac{\sqrt{7}}{2}i \text{ and } y = \frac{5}{2} + \frac{\sqrt{7}}{2}i$$

Hence, the solution set of the system is

$$\left\{\left(\frac{5 + \sqrt{7}\,i}{2}, \frac{5 - \sqrt{7}\,i}{2}\right), \left(\frac{5 - \sqrt{7}\,i}{2}, \frac{5 + \sqrt{7}\,i}{2}\right)\right\}$$

The graphs of the two equations are shown in Figure 8.7. As you can see, the graphs *do not intersect*. When the solution of a system of equations are imaginary numbers, there are no points of intersection for the graphs. This is because the coordinates of points in the real plane are *real* numbers.

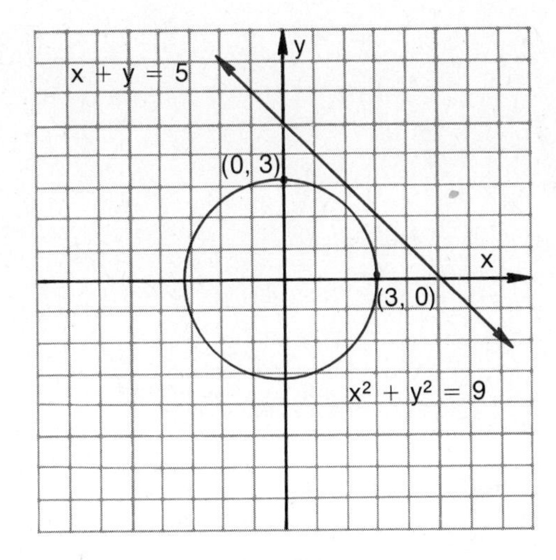

FIGURE 8.7

PROGRESS TEST

1. The demand curve for x units of an item is expressed as $y = (x - 4)^2$. If the supply curve is given as $y = x^2 + x + 7$, the equilibrium point occurs at ______.

2. The solution set of the system
   (1) $x^2 + y^2 = 16$
   (2) $x + y = 4$
   is ______.

3. The solution set of the system
   (1) $x^2 + y^2 = 3$
   (2) $x + y = 3$
   is ______.

EXERCISE 8.6

In Problems 1 through 22, find the solution set of the given system. Check the solution by graphing the system.

Illustration: (1) $y = (x - 1)^2$
(2) $y - x = 1$

Solution: Solving for y in (2), we obtain the equivalent system.
(1) $y = (x - 1)^2$
(3) $y = x + 1$
Substituting $y = x + 1$ in Equation (1), we have

$$x + 1 = (x - 1)^2$$

$$x + 1 = x^2 - 2x + 1 \quad \text{Expanding}$$

$$x^2 - 3x = 0 \quad \text{Subtracting 1 and x}$$

$$x(x - 3) = 0 \quad \text{Factoring}$$

Thus, $x = 0$ or $x = 3$

PROGRESS TEST ANSWERS

1. (1, 9)
2. {(0, 4), (4, 0)}
3. $\left\{\left(\frac{3 + \sqrt{3}\,i}{2}, \frac{3 - \sqrt{3}\,i}{2}\right), \left(\frac{3 - \sqrt{3}\,i}{2}, \frac{3 + \sqrt{3}\,i}{2}\right)\right\}$

Substituting these values in (3), we obtain

$$y = 0 + 1 = 1 \text{ and } y = 3 + 1 = 4$$

Therefore the solution set of the system is $\{(0, 1), (3, 4)\}$, as shown in the graph.

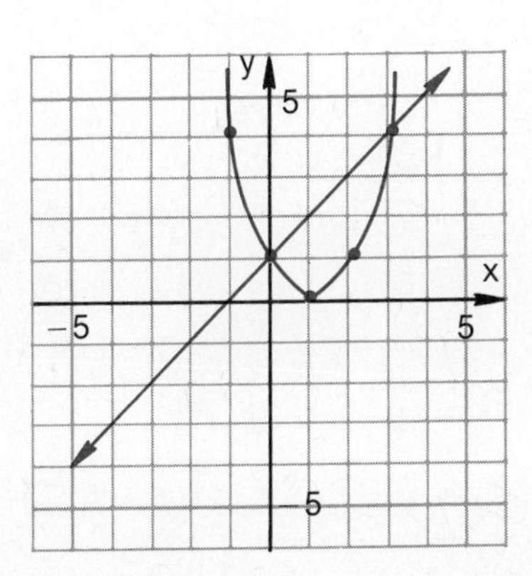

1. $x^2 + y^2 = 36$
   $x + y = 6$

2. $x^2 + y^2 = 49$
   $x + y = 7$

3. $x^2 + y^2 = 25$
   $y - x = 5$

4. $x^2 + y^2 = 36$
   $y - x = 6$

5. $x^2 + y^2 = 25$
   $y - x = 1$

6. $x^2 + y^2 = 8$
   $x - y = 0$

7. $y = x^2 - 5x + 4$
   $x - y = 1$

8. $y = x^2 - 4x$
   $y - 2x = 0$

9. $y = -x^2 + 7x - 6$
   $x - y = 6$

10. $y = -x^2 + 4$
    $y - x = 4$

11. $4x^2 + 9y^2 = 36$
    $y - x = 3$

12. $4x^2 + y^2 = 36$
    $y + 2x = 6$

13. $9x^2 + y^2 = 36$
    $y - 3x = 6$

14. $x^2 + 4y^2 = 16$
    $y + \frac{1}{2}x = 2$

15. $y = 4 - x^2$
    $y = x^2 - 4$

16. $y = 9 - x^2$
    $y = x^2 - 9$

17. $x^2 + y^2 = 36$
$5y = x^2$

18. $x^2 + y^2 = 16$
$6y = x^2$

19. $4x^2 + y^2 = 36$
$y - x = 2$

20. $x^2 + y^2 = 4$
$y - x = 5$

21. $x^2 + y^2 = 9$
$y - x = 6$

22. $x^2 + y^2 = 16$
$y - x = 7$

SELF-TEST—CHAPTER 8

1. Use the accompanying rectangular coordinate system to find the solution, if it exists, of the given system.

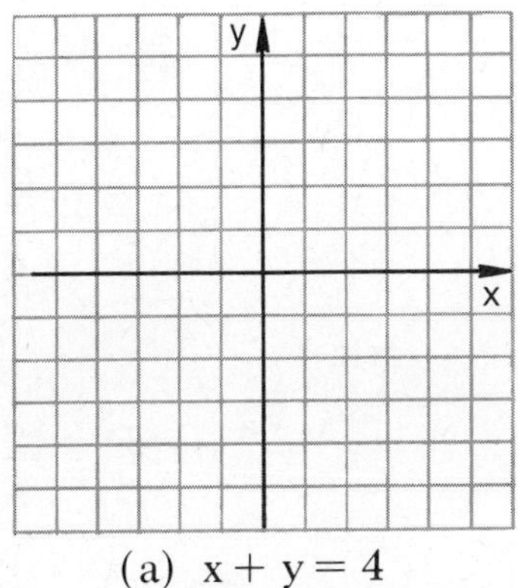

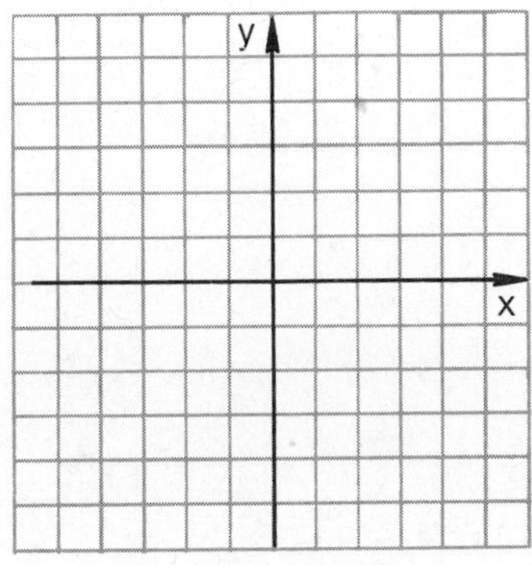

(a) $x + y = 4$
$2x - y = 5$

(b) $x - y = 3$
$2y = 2x - 1$

---

2. Find, if possible, the solution of the given system:

______________ (a) $2x + y = 1$
$y = 2x + 5$

______________ (b) $-x - y = 6$
$y = -x$

---

3. Find, if possible, the solution of the given system:

______________ (a) $2x + y = -4$
$3x - y = -1$

______________ (b) $3x + 5y = 2$
$9x + 15y = 5$

---

4. Find the solution of the system:

______________ $x + y + z = 3$
$x - y + 2z = 3$
$x + y - z = -1$

---

5. Find the value of:

_______________ (a) $\begin{vmatrix} 2 & 1 \\ 1 & 3 \end{vmatrix}$

_______________ (b) $\begin{vmatrix} 8 & 1 \\ -1 & 3 \end{vmatrix}$

---

6. Consider the system

$$2x + y = 6$$
$$x + 3y = -2$$

In terms of determinants, the solution is $x = \frac{D_x}{D}$ and $y = \frac{D_y}{D}$.

_______________ (a) Find $\frac{D_x}{D} = x$

_______________ (b) Find $\frac{D_y}{D} = y$

---

7. Consider the system

$$x + y = 6$$
$$5x - 4y = 12$$

In terms of determinants, the solution is $x = \frac{D_x}{D}$ and $y = \frac{D_y}{D}$.

_______________ (a) Find $\frac{D_x}{D} = x$

_______________ (b) Find $\frac{D_y}{D} = y$

_______________ (c) Is this system dependent or inconsistent?

---

8. Evaluate

_______________ (a) $\begin{vmatrix} 1 & 1 & 2 \\ 1 & -1 & -3 \\ 2 & 3 & 1 \end{vmatrix}$

_______________ (b) $\begin{vmatrix} 1 & 1 & 1 \\ 2 & 3 & 1 \\ 2 & 4 & 1 \end{vmatrix}$

---

9. By the substitution method, we find that the solution set of the following system is ______________________.

$$x^2 + y^2 = 9$$
$$x + y = 3$$

---

10. Consider the system

$$y = (x + 1)^2$$
$$y = 3x + 1$$

Use the given rectangular system to find the solution of the system graphically.

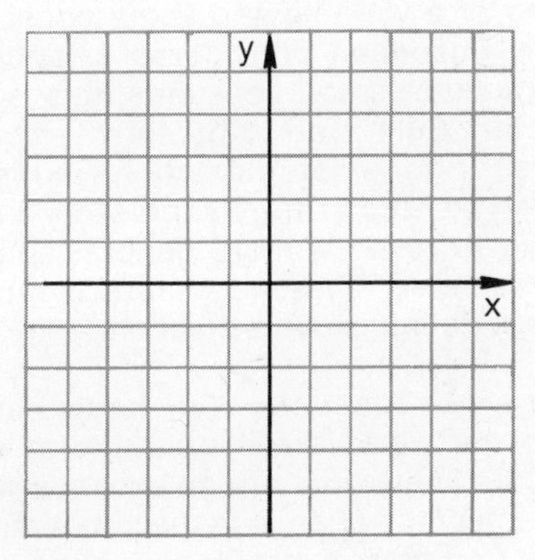

CHAPTER 9

HISTORICAL NOTE

In this chapter we shall study the idea of a function. The term "function" is common to many branches of mathematics, but it is of special importance in algebra and calculus. The word itself, in its Latin equivalent, seems to have been introduced by Leibniz in 1694, although the concept of a function had been developing since the Middle Ages. For example, in Descartes' geometry, curves were represented by equations. Developing this idea further, Leibniz used the term function to denote any quantity connected with a curve, such as the coordinates of a point on the curve, the radius of a curve, or the slope of a curve.

By 1718 Johann Bernoulli had come to regard a function as an expression consisting of a variable and some constants. Somewhat later, Leonhard Euler of Switzerland considered a function as an equation or formula involving variables and constants (this is how functions are presented in this chapter). Euler's concept remained basically unchanged until Fourier considered trigonometric functions in his work in physics. The functions considered by Fourier were expressed in terms of so-called trigonometric series and involved a more general relationship between the variables being considered. These ideas led Lejeune Dirichlet to the formulation of the current definition of a function.

The analysis of functions is of particular importance because so many things in the world are expressible in the language of functions. For example, the metal in a bridge expands when heated, and this expansion is a function of the change in the temperature; the amount of postage paid to send a first-class letter is a function of the weight of the letter. Other important functions in the space age are the speed of a satellite, expressed as a function of the diameter of its orbit, and an astronaut's need for oxygen, expressed as a function of his physical stress.

RELATIONS AND FUNCTIONS

Courtesy of National Aeronautics and Space Administration.

NASA Photo No. 71-H-1439 AS15-85-11471.

Pintos or similar class car.	**Mavericks** or similar class car.	**Monarchs Granadas** or similar class car.
$13.95 any day plus gas. No charge for mileage!	$14.95 a day plus 15¢ a mile and gas.	$16.95 a day plus 17¢ a mile and gas.

Compliments of Hertz, Inc.

OBJECTIVES 9.1

After studying this section, the reader should be able to:

1. **Find the domain and range of a given relation.**
2. **Find the graph of a given relation and then determine its domain and range.**

9.1 RELATIONS

In the ad at the beginning of this section, the daily price for renting a Maverick or a similar class car is $14.95 plus 15¢ per mile. As you can see, the daily cost of renting the car is *related to* the number of miles driven. For example, if a person drives 1 mile, the cost y is

$$y = \$14.95 + \$0.15 = \$15.10$$

If two miles are driven, the cost is

$$y = 14.95 + (0.15)2 = \$15.25$$

If we agree to measure to the nearest mile the distance traveled, we can make a table using two sequences of numbers, one for the number x of miles driven and the other for the daily cost y, as shown in the table below:

No. of miles (x)	Cost (y)
0	14.95
1	15.10
2	15.25
3	15.40
—	—
—	—
—	—

The numbers in the first column are called values of the *independent* variable because they are chosen independently of the second number. The numbers in the second row are called values of the *dependent* variable because they *depend* on the values of the numbers in the first column. The numbers in our table can also be written as *ordered pairs,* as shown below:

(Miles, Cost)
(0, 14.95)
(1, 15.10)
(2, 15.25)
(3, 15.40)
—
—
—

These ordered pairs can be written as the set

$$S = \{(0, 14.95), (1, 15.10), (2, 15.25), \ldots\}$$

The set S is considered a *relation,* which we define formally as follows.

DEFINITION 9.1

A set of ordered pairs of real numbers is a *relation.*

In the relation S, the set of all first components—that is, the set of values of the *independent* variable—is called the *domain* of the relation. The set of values of the *dependent* variable is called the *range* of the relation. The domain D in the relation S is $\{0, 1, 2, 3, \ldots\}$, and the range $R = \{14.95, 15.10, 15.25, 15.40, \ldots\}$.

The domain of a relation S is the set of all first components of the ordered pairs in S; the range of S is the set of all second components of the ordered pairs in S.

Example 1

Find the domain and the range of the relation:

$$A = \{(1, 2), (2, 3), (3, 4)\}$$

elements in the domain

A = (1, 2), (2, 3), (3, 4)

elements in the range

Solution

The domain of A is the set of first components:

$$D = \{1, 2, 3\}$$

The range of A is the set of second components:

$$R = \{2, 3, 4\}$$

The relation we defined by listing the ordered pairs can also be expressed by finding the relationship between x and y by means of an equation. Since the daily cost y of renting a car is \$14.95 plus \$0.15 for each mile x driven, the total daily cost y can be written as

$$y = 14.95 + 0.15x, \text{ where } x = 0, 1, 2, 3, \text{ etc.}$$

The relation defined by this equation is usually written as

$$\{(x, y) \mid y = 14.95 + 0.15x\}$$

Remember that a relation is a set of ordered pairs.

Since a relation is a set of ordered pairs, relations can be graphed in the Cartesian plane. This procedure is illustrated in the next example.

Example 2

Find the graph, the domain, and the range of the relation:

$$\{(x, y) \mid y = 2x - 4\}$$

Solution

The graph of the relation is the graph of the equation $y = 2x - 4$, shown in Figure 9.1. The *domain* of this relation is the set of all real numbers, since any real number x can be used as the first component. Similarly, the *range* of y is the set of all real numbers. Some of the ordered pairs in the relation are $(0, -4)$, $(1, -2)$, $(2, 0)$, and so on.

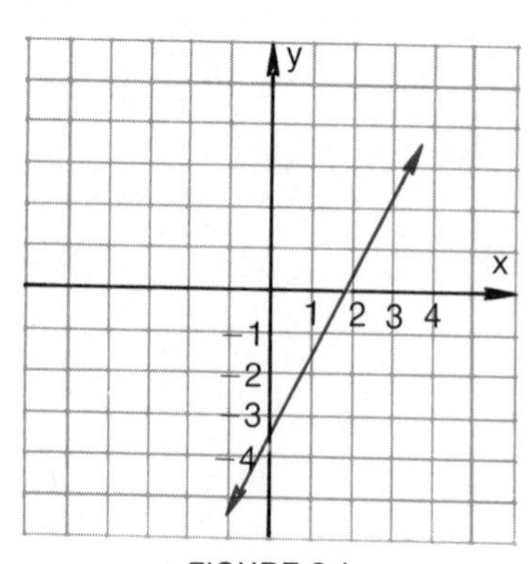

FIGURE 9.1

Example 3

Find the graph, the domain, and the range of the following relations:

(a) $A = \{(x, y) \mid x^2 + y^2 = 4\}$

(b) $B = \{(x, y) \mid y = \sqrt{4 - x^2}\}$

(c) $C = \{(x, y) \mid y = -\sqrt{4 - x^2}\}$

Solution

(a) The graph of $x^2 + y^2 = 4$ is a circle of radius 2 centered at the origin, as shown in Figure 9.2. From the figure, it is clear that x and y must be between -2 and 2, inclusive. Thus the domain of A is $D = \{x \mid -2 \le x \le 2\}$ and the range is $R = \{y \mid -2 \le y \le 2\}$.

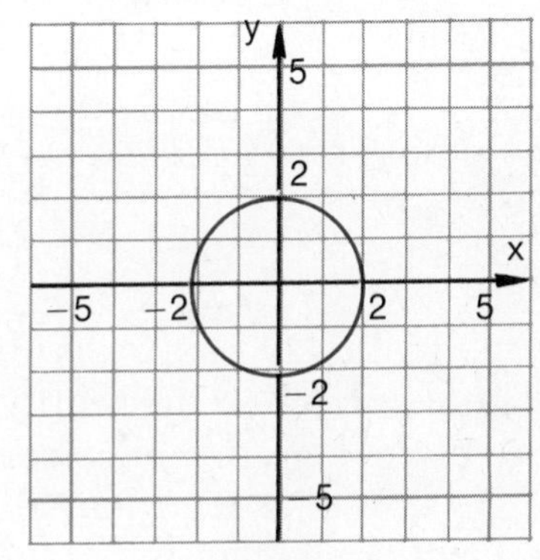

FIGURE 9.2

(b) The graph of $y = \sqrt{4 - x^2}$ is shown in Figure 9.3. The domain of B is $D = \{x \mid -2 \le x \le 2\}$, and the range is $R = \{y \mid 0 \le y \le 2\}$.

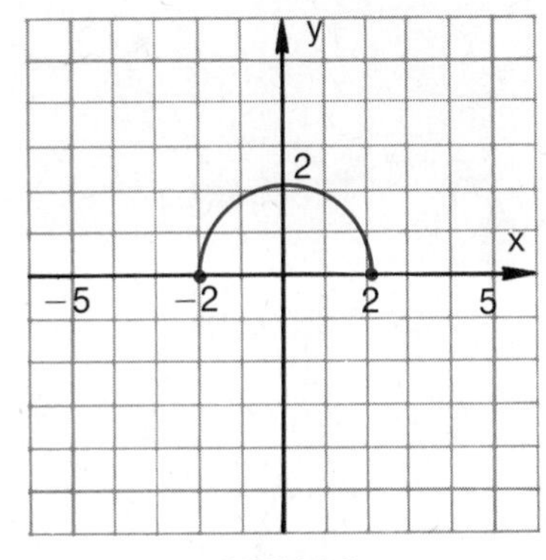

FIGURE 9.3

(c) The graph of $y = -\sqrt{4 - x^2}$ is shown in Figure 9.4. The domain of C is $D = \{x \mid -2 \leq x \leq 2\}$, and the range is $R = \{y \mid -2 \leq y \leq 0\}$.

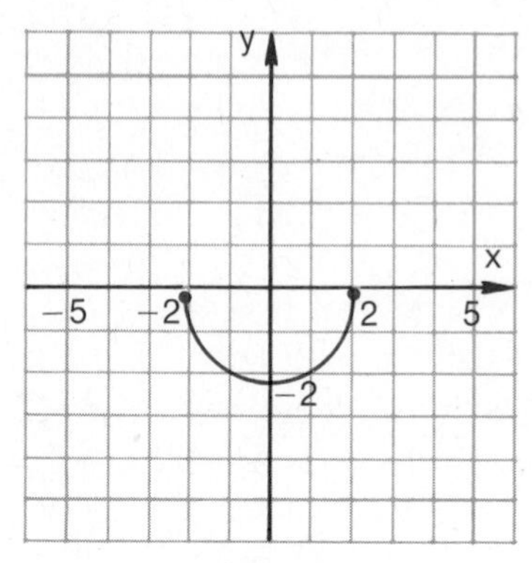

FIGURE 9.4

Example 4

Find the graph, the domain, and the range of the relation:

$S = \{(x, y) \mid 2y < 6\}$

Solution

If we simplify the inequality $2y < 6$ by dividing through by 2, we obtain $y < 3$. The graph of the inequality $y < 3$ consists of all the points in the plane *below* the horizontal line $y = 3$, as shown in Figure 9.5. Note that the inequality $y < 3$ is satisfied by any ordered pair (x, y) *regardless of the x value* as long as $y < 3$. In Figure 9.5, the line $y = 3$ is drawn dashed because points on this line do not satisfy the inequality $y < 3$. From the figure, it is clear that x can be any real number, while y can be any real number smaller than 3. Thus, the domain of S is the set of all real numbers, while the range is $R = \{y \mid y < 3\}$.

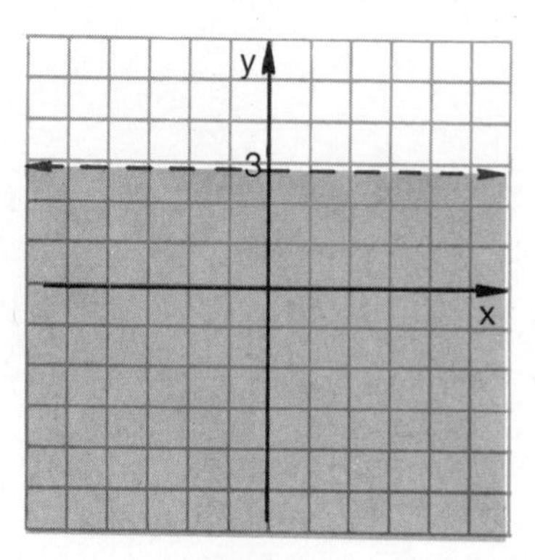

FIGURE 9.5

Example 5

Find the graph, the domain and the range of the relation:
$S = \{(x, y) \mid 2x + 5y \leq 10\}$

Solution

We first graph the relation when $2x + 5y = 10$. Letting $x = 0$, we obtain $y = 2$, and $(0, 2)$ is on the line. When $y = 0$, $x = 5$, and $(5, 0)$ is on the line. Joining $(0, 2)$ and $(5, 0)$ with a line will give us the graph shown in Figure 9.6.

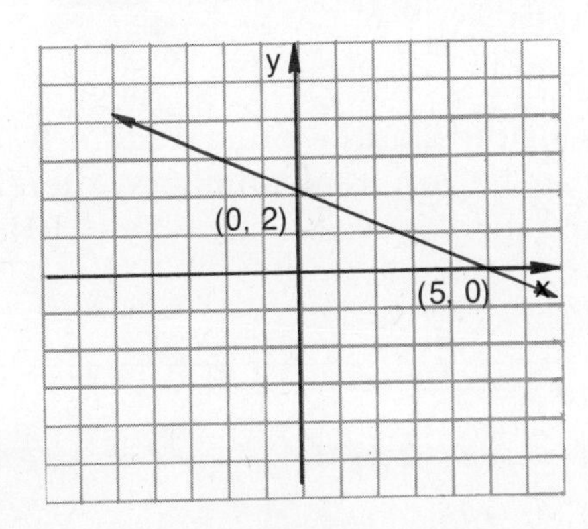

FIGURE 9.6

Where will the points satisfying $2x + 5y < 10$ be? For points such as $(1, 1)$ and $(3, 0)$, lying *below* the line, we have

$$2 \cdot 1 + 5 \cdot 1 < 10$$
$$\text{and } 2 \cdot 3 + 5 \cdot 0 < 10$$

On the other hand, for points such as $(2, 3)$ and $(4, 2)$, lying *above* the line, we have

$$2 \cdot 2 + 5 \cdot 3 > 10$$
$$\text{and } 2 \cdot 4 + 5 \cdot 2 > 10$$

These results lead us to guess that $2x + 5y < 10$ for all points *below* the line $2x + 5y = 10$, and that $2x + 5y > 10$ for all points *above* the line. In fact, it can be shown that the entire solution set has to be *above* or *below* the line in Figure 9.6. We can quickly find out which side of the line to use by substituting the x and y coordinates of any point *not* on the line into $2x + 5y \leq 10$. If the point satisfies the inequality, we shade all the points on that side. If it does not, we shade the points on the other side of the line. When the line does not cross the origin, the easiest point to use is $(0, 0)$. When this point is substituted in the inequality $2x + 5y < 10$, we obtain $2 \cdot 0 + 5 \cdot 0 < 10$, a true statement. Thus, we shade the points below the line, as shown in Figure 9.7. Note that the points *on* the line also satisfy $2x + 5y \leq 10$; thus the line $2x + 5y = 10$

To find which side of the line we should shade, we select a point. If the point satisfies the given inequality, we shade all the points on that side. If our chosen point does not satisfy the inequality, we shade the points on the other side of the line.

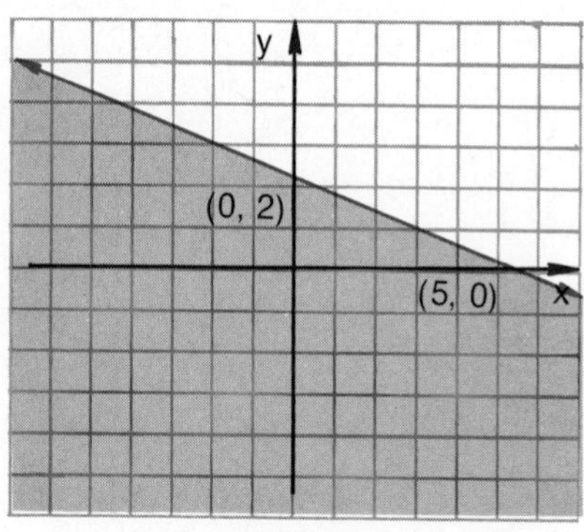

FIGURE 9.7

is drawn solid in the figure. From Figure 9.7 you can see that x and y can be any real numbers; therefore both the domain and the range of $\{(x, y) \mid 2x + 5y \leq 10\}$ are the set of real numbers.

PROGRESS TEST

1. Find the domain and the range of the relation $\{(-1, 2), (-3, 5), (2, 7)\}$.
2. Graph the relation $\{(x, y) \mid y = -x + 3\}$.

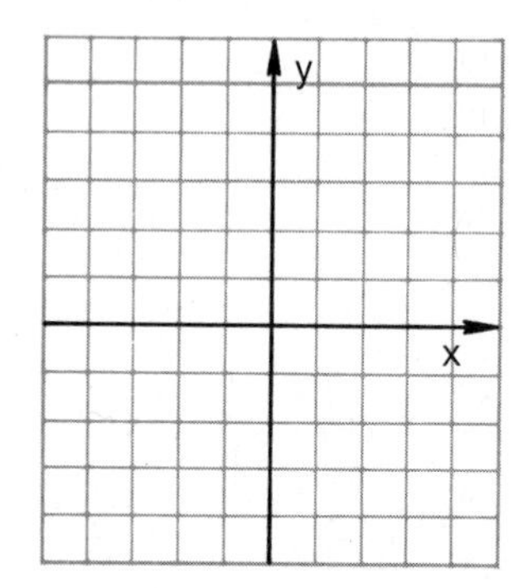

3. The domain and range of the relation given in 2 are, respectively, ______________ ____________ and ______________________.
4. Graph the relation $\{(x, y) \mid y = -\sqrt{9 - x^2}\}$.

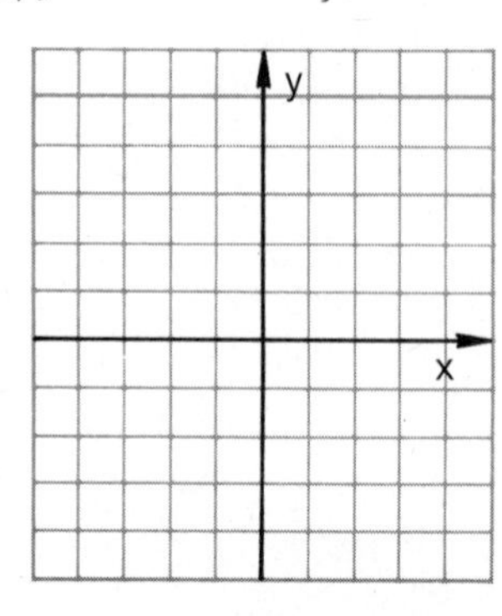

5. The domain and range of the relation in 4 are, respectively, ______ ______ and ______.

EXERCISE 9.1

In Problems 1 through 10, find the domain and range of the given relations.

Illustration: $\{(-5, 3), (2, 4), (5, 3)\}$

Solution: The domain is the set of all *first* components, that is, $\{-5, 2, 5\}$. The range is the set of all second components, that is, $\{3, 4\}$.

1. $\{-3, 0), (-2, 1), (-1, 2)\}$
2. $\{(-1, -2), (0, -1), (1, 0)\}$
3. $\{(3, 0), (4, 0), (5, 0)\}$
4. $\{(0, 1), (0, 2), (0, 3)\}$
5. $\{(1, 2), (1, 3), (2, 2), (2, 3)\}$
6. $\{(2, 1), (1, 2), (3, 4), (4, 3)\}$
7. $\{(1, -1), (3, -1), (5, -1), (7, -1)\}$
8. $\{(-3, 2), (4, 3), (5, 7), (9, 8)\}$
9. $\{(2, 1), (2, 0), (2, -1), (2, -2)\}$
10. $\{(-3, -1), (-3, 0), (-3, 1), (-3, 2)\}$

In Problems 11 through 40, graph the given relation and describe the domain and range of each.

Illustration: $\{(x, y) \mid y = x^2 + 1\}$

Solution: The graph of $y = x^2 + 1$ is a parabola with its vertex at $(0, 1)$, as shown in the accompanying figure. The domain of

PROGRESS TEST ANSWERS

1. The domain is $\{2, -1, -3\}$
   The range is $\{2, 5, 7\}$
2.

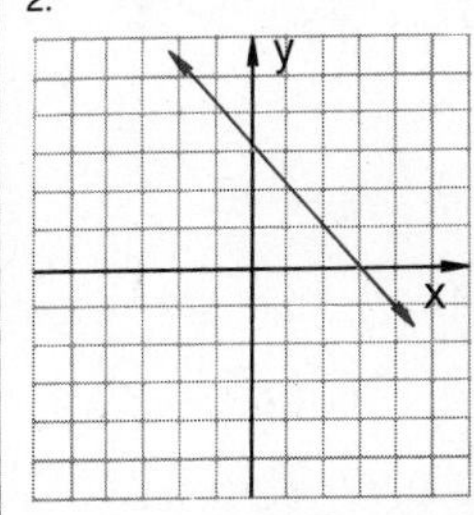

3. the set of real numbers; the set of real numbers
4.

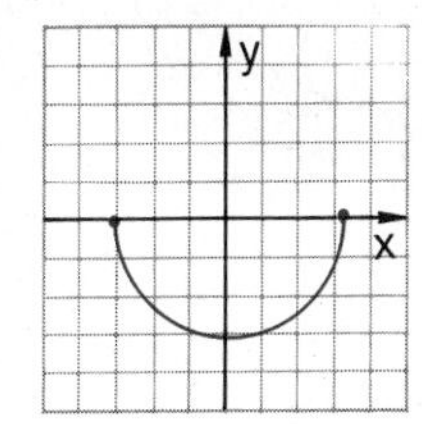

5. $\{x \mid -3 \le x \le 3\}$; $\{y \mid -3 \le y \le 0\}$

this function is the set of real numbers, and the range is the set of real numbers greater than or equal to 1: $\{y \mid y \geq 1\}$.

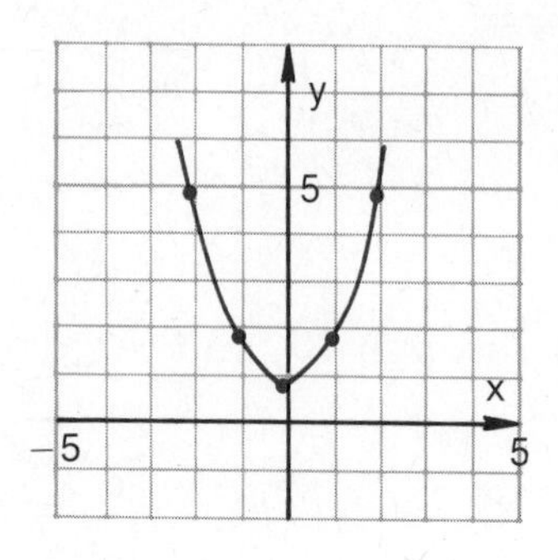

11. $\{(x, y) \mid x^2 + y^2 = 25\}$

12. $\{(x, y) \mid x^2 + y^2 = 16\}$

13. $\{(x, y) \mid y = \sqrt{25 - x^2}\}$

14. $\{(x, y) \mid y = -\sqrt{25 - x^2}\}$

15. $\{(x, y) \mid x = \sqrt{25 - y^2}\}$

16. $\{(x, y) \mid x = -\sqrt{25 - y^2}\}$

17. $\{(x, y) \mid y = x^2 - 1\}$

18. $\{(x, y) \mid y = x^2 + 3\}$

19. $\{(x, y) \mid y = -x^2\}$

20. $\{(x, y) \mid y = -x^2 + 1\}$

21. $\{(x, y) \mid x = y^2\}$

22. $\{(x, y) \mid x = y^2 + 1\}$

23. $\{(x, y) \mid y = -x + 5\}$

24. $\{(x, y) \mid y = -2x + 4\}$

25. $\{(x, y) \mid y = x + 2\}$

26. $\{(x, y) \mid y = 2x - 2\}$

27. $\{(x, y) \mid x < 3\}$

28. $\{(x, y) \mid 3x < 9\}$

29. $\{(x, y) \mid 2y < 8\}$

30. $\{(x, y) \mid 3y < 6\}$

31. $\{(x, y) \mid y < x + 2\}$

32. $\{(x, y) \mid y < 2x - 2\}$

33. $\{(x, y) \mid y < -x + 5\}$

34. $\{(x, y) \mid y < -2x + 4\}$

35. $\{(x, y) \mid y > x + 2\}$

36. $\{(x, y) \mid y > 2x + 4\}$

37. $\{(x, y) \mid y > -x + 1\}$

38. $\{(x, y) \mid y > -2x + 2\}$

39. $\{(x, y) \mid 2y > -6x + 4\}$

40. $\{(x, y) \mid 3y > 6x - 9\}$

In Problems 41 through 50, write an equation for the relationship between the given quantities and state the domain for the resulting relation.

Illustration: The area A of a circle and the length r of its radius.

Solution: $A = \pi r^2$
The domain is $\{r \mid r \geq 0\}$.

41. The circumference C of a circle and the length r of its radius.

42. The area A of a square and the length r of its side.

43. The perimeter P of a square and the length s of its side.

44. The distance D traveled by a car moving at a constant rate r for a period of two hours.

45. The distance D traveled by a car moving at 50 miles per hour and the time t (hours) traveled.

46. The daily cost y of renting a car costing $10 per day plus 15 cents per mile for x miles traveled.

47. The daily cost y of renting a car costing $15 per day plus 10 cents per mile for x miles traveled.

48. The distance D traveled by a car that gets 20 miles to the gallon when y gallons of gas have been used.

49. The number c of chirps made by a cricket that in one minute chirps four times the difference between the temperature t in degrees Fahrenheit and 40.

50. The temperature t in degrees Fahrenheit when a cricket chirps c times in a minute (see Problem 49).

Photo by Fotoart, Tampa, Florida.

OBJECTIVES 9.2

After studying this section, the reader should be able to:

1. **Determine if a given relation is a function and if so indicate its domain and range.**
2. **Graph a given function and then determine its domain and range.**
3. **Evaluate a given function at a given point.**

9.2 FUNCTIONS

The photograph at the beginning of this section illustrates the fact that the water pressure is *related to* or is a *function* of the depth. As you can see, the higher pressure at the lower holes of the can makes water squirt out in a flat trajectory, while the lower pressure at the upper holes produces only a weak stream. As a matter of fact, the pressure p of the water (in pounds per square foot) at a depth of x feet (x a positive integer) is given approximately in the formula

The pressure at any depth x is unique.

(1) $p = 62.5x$

From this formula, you can see that at *any* depth x there is *one and only one* pressure. For example, at 1 foot below the surface, the pressure is 62.5 pounds per square foot. At 2 feet below the surface, the pressure is $(62.5)(2) = 125$ pounds per square foot; at 3 feet below the surface, the pressure is $(62.5)(3) = 187.5$ pounds per square foot.

Remember, x is a positive integer.

feet below surface (x)	pressure p (in pounds per square foot)
1	62.5
2	125
3	187.5
—	—
—	—
—	—

When this information is written as a set of ordered pairs, the result is a *function.*

DEFINITION 9.2

A function is a set of ordered pairs (x, y) (a relation) such that no two distinct ordered pairs have the same first component. The set of all possible values of x is the domain of the function, and the set of all possible values of y is the range of the function.

If a relation is a function, any vertical line can intersect its graph at no more than one point.

On the Cartesian plane, this means that any vertical line can intersect the graph of a function at one point only. If a vertical line intersects a graph at more than one point, the graph is not that of a function.

The ordered pairs defined by the equation $p = 62.5x$ can be obtained by making a table as shown.

feet below surface (x)	pressure (p)	ordered pairs
1	62.5	(1, 62.5)
2	125	(2, 125)
3	187.5	(3, 187.5)
—	—	—
—	—	—
—	—	—

Again, keep in mind that x is a positive integer.

The pairs defined by this table are {(1, 62.5), (2, 125), (3, 187.5), . . .}. The graph of this set of points can be intersected by a vertical line at one point only, and thus the set of ordered pairs is indeed a function. The domain of this function is {1, 2, 3, . . .} and the range is {62.5, 125, 187.5, . . .}. Of course, if we decide to measure the distance below the surface in arbitrary portions of a foot, then the domain and the range are the set of all positive real numbers.

Example 1

Determine if the following relations are functions. If so, indicate the domain and range.
(a) $\{(1, 2), (1, 3), (1, 4)\}$
(b) $\{(2, 1), (3, 1), (4, 1)\}$

Solution

(a) The relation in (a) is *not* a function, because the distinct pairs $(1, 2)$, $(1, 3)$, and $(1, 4)$ have the same first component, 1.
(b) The relation $\{(2, 1), (3, 1), (4, 1)\}$ is a function, since no two distinct pairs have the same first component. The domain of the function is $\{2, 3, 4\}$, and the range is $\{1\}$.

Sometimes we wish to find the value of a function for a certain value of the variable. To do this, we use a special notation. For example, if we wish to emphasize the fact that the pressure of the water is a function of the depth x, we write Equation (1) in the form

(2) $f(x) = 62.5x$ (read "f of x equals 62.5x")

We regard Equation (2) as a rule or formula that tells us how the ordered pairs of the function are to be obtained. Thus, $f(1) = (62.5)(1) = 62.5$, $f(2) = (62.5)(2) = 125$, etc.

In general, when using functional notation, we think of the place occupied by the independent variable as a blank to be filled by numbers in the domain. For example, if a function f is defined by the rule $f(x) = 2x - 1$, then this rule is thought of as

$$f(\quad) = 2(\quad) - 1$$

A note about notation: other letters, such as g and h, are sometimes used to denote a function.

Hence,

$$f(1) = 2(1) - 1 = 1$$

$$f(2) = 2(2) - 1 = 3$$

$$f(\sqrt{3}) = 2(\sqrt{3}) - 1$$

$$f\left(\frac{a}{b}\right) = 2\left(\frac{a}{b}\right) - 1$$

$$f(a + b) = 2(a + b) - 1$$

In every case, the number or expression that occurs in the place of x in f(x) is substituted for x in the rule that defines the function.

Example 2

Let $g(x) = x^2 - 1$. Find $g(-1)$, $g(0)$, and $g(1)$ and complete the table.

x	−1	0	1
y = g(x)			

Solution

Since $g(x) = x^2 - 1$,

$$g(-1) = (-1)^2 - 1 = 0$$
$$g(0) = 0^2 - 1 = -1$$
$$g(1) = 1^2 - 1 = 0$$

The completed table is

x	−1	0	1
y = g(x)	0	−1	0

Example 3

Graph the given relations and determine if they are functions. If so, specify the domain and range.

(a) $\{(x, y) \mid y = 2\}$
(b) $\{(x, y) \mid x = 3\}$
(c) $\{(x, y) \mid y = x + 1\}$

Solution

(a) The graph of $y = 2$ is shown in Figure 9.8. The relation is, of course, a function, since no two distinct ordered pairs have the same first component. The domain of the function defined by $y = 2$ is the set of all real numbers, since for any real number x, y is 2, whereas the range is $\{2\}$.

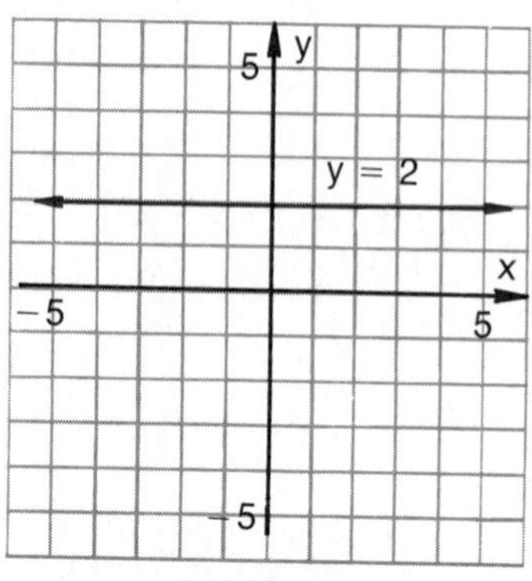

FIGURE 9.8

(b) The graph of $x = 3$ is shown in Figure 9.9. The relation is *not* a function, since at least two different ordered pairs have the same first components. For example, (3, 0) and (3, 1) are two different ordered pairs with the same first component, 3.

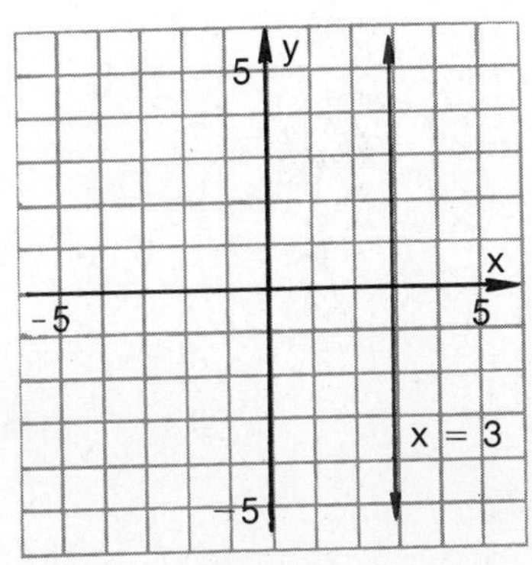

FIGURE 9.9

(c) The graph of $y = x + 1$ is shown in Figure 9.10. Since no two different ordered pairs have the same first components, the relation is a function. Both the domain and range are the set of all real numbers.

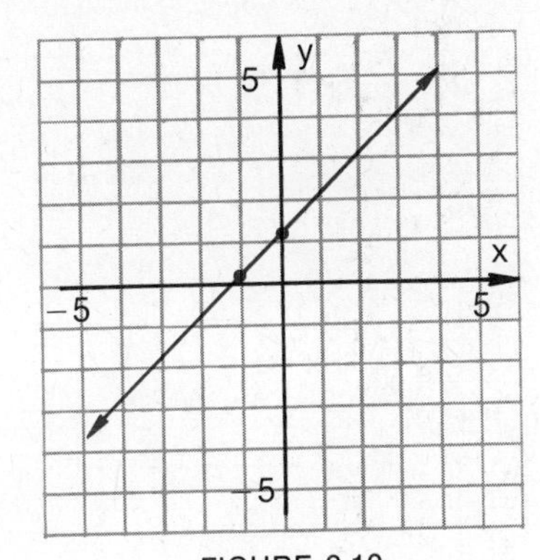

FIGURE 9.10

The functions in Examples 3a and 3c can also be written using functional notation. Thus, we may think of these functions as defined by

(a) $f(x) = 2$

(c) $g(x) = x + 1$

We use this notation in the next example.

We can use f(x) or g(x) instead of y.

Example 4

Consider the function defined by $f(x) = |x|$, the *absolute value* function. Find:

(a) $f(0)$
(b) $f(-1)$
(c) $f(1)$
(d) the domain of f
(e) the range of f
(f) the graph of the function

Solution

(a) Since $f(x) = |x|$, $f(0) = |0| = 0$
(b) $f(-1) = |-1| = 1$
(c) $f(1) = |1| = 1$
(d) Since there is no restriction on x, the domain of x is the set of real numbers.
(e) Since $f(x) = |x|$, a non-negative real number, the range of f is the set of non-negative real numbers.
(f) The graph of $f(x) = |x|$ is shown in Figure 9.11, where we have shown the points (0, 0), (−1, 1), and (1, 1) found in parts (a), (b), and (c).

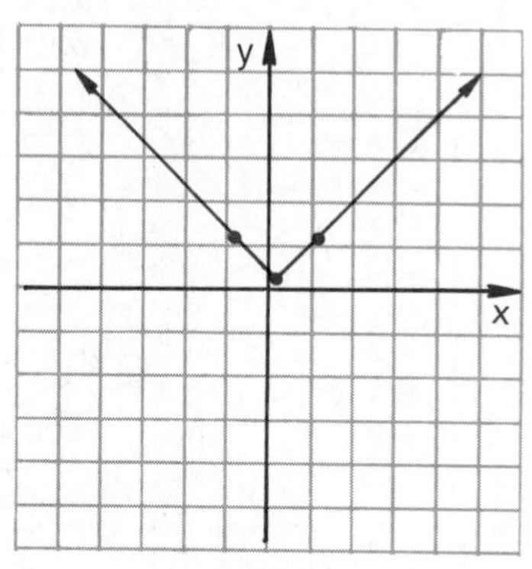

FIGURE 9.11

PROGRESS TEST

1. The relation {(−1, −1), (−1, 0), (−1, 1)} ___________ (is, is not) a function.
2. The relation {(1, 1), (2, 1), (3, 1)} ___________ (is, is not) a function.
3. If $h(x) = -x^2$, $h(-1)$ equals ___________.

4. Graph the function defined by $f(x) = x - 2$ and determine its domain and range.

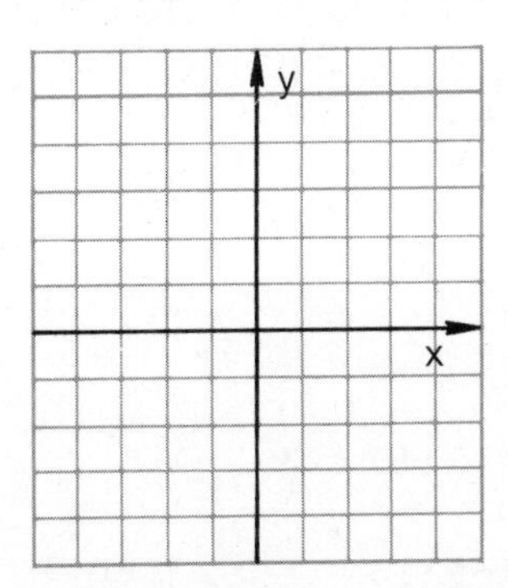

EXERCISE 9.2

In Problems 1 through 6, determine if the given relation is a function. If it is, describe its domain and range.

1. $\{(-2, 0), (-1, 1), (0, 2)\}$

2. $\{(-3, 4), (-2, 3), (-1, 2)\}$

3. $\{(1, 0), (2, 0), (3, 0)\}$

4. $\{(0, -1), (0, -2), (0, -3)\}$

5. $\{(1, 1), (1, 2), (2, 2), (2, 3)\}$

6. $\{(0, 0), (1, 1), (2, 2), (3, 3)\}$

In Problems 7 through 20, graph the given relation and determine if it is a function. If it is, give the domain and range.

Illustration: $\{(x, y) \mid x = y^2\}$

Solution: The equation $x = y^2$ can also be written as $y = \pm\sqrt{x}$. The graph is a parabola

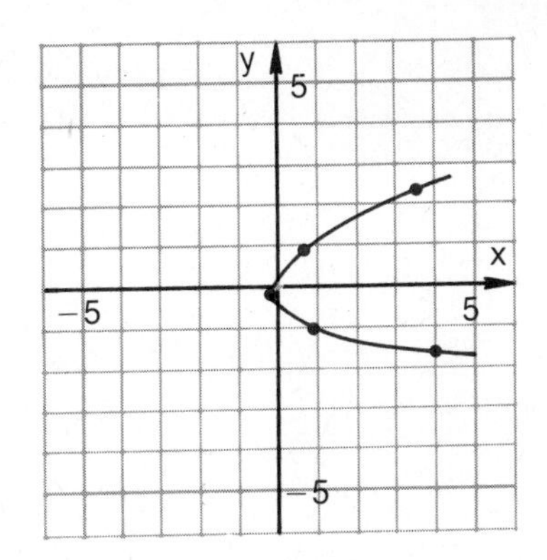

PROGRESS TEST ANSWERS

1. is not
2. is
3. $h(-1) = -(-1)^2 = -1$.
4. Both the domain and the range are the set of all real numbers.

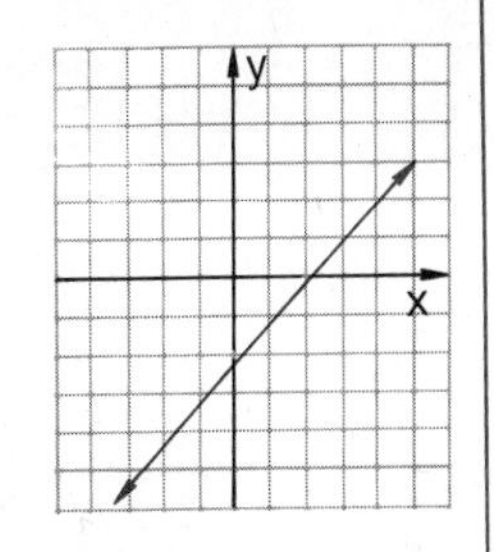

whose vertex is at (0, 0) as shown. Because there are two y values for each $x > 0$, a vertical line can intersect the graph at two points, and the relation is not a function.

7. $\{(x, y) \mid y = x^2 + 3\}$

8. $\{(x, y) \mid y = x^2 - 5\}$

9. $\{(x, y) \mid y = x^2 + 1\}$

10. $\{(x, y) \mid y = x^2 - 1\}$

11. $\{(x, y) \mid y = -x^2\}$

12. $\{(x, y) \mid x = -y^2\}$

13. $\{(x, y) \mid y = x\}$

14. $\{(x, y) \mid y = -3\}$

15. $\{(x, y) \mid y = -x + 1\}$

16. $\{(x, y) \mid y = -2x + 3\}$

17. $\{(x, y) \mid y = |x| + 1\}$

18. $\{(x, y) \mid y = |x| + 2\}$

19. $\{(x, y) \mid |y| = x + 1\}$

20. $\{(x, y) \mid y = |x| - 2\}$

In Problems 21 through 30, find the specified values for the given function.

Illustration: $f(x) = \sqrt{25 - x^2}$; $f(0)$, $f(3)$, $f(4)$

Solution:

$$f(0) = \sqrt{25 - 0^2} = \sqrt{25} = 5$$

$$f(3) = \sqrt{25 - 3^2} = \sqrt{25 - 9} = \sqrt{16} = 4$$

$$f(4) = \sqrt{25 - 4^2} = \sqrt{25 - 16} = \sqrt{9} = 3$$

21. $f(x) = \sqrt{9 - x^2}$; $f(0), f(1), f(3)$

22. $f(x) = -\sqrt{9 - x^2}$; $f(0), f(1), f(3)$

23. $g(x) = |x| + 2$; $g(0), g(1), g(-1)$

24. $h(x) = |x| + 1$; $h(0), h(1), h(-1)$

25. $f(x) = \frac{1}{x}$; $f(1), f(-1), f(2)$

26. $f(x) = \frac{1}{x^2}$; $f(1), f(-1)$

27. $g(x) = \frac{2x - 1}{3}$; $g(0), g(2), g(-2)$

28. $r(x) = \frac{x + 4}{x - 2}$; $r(0), r(-2)$

29. $s(t) = \frac{t^2 + 1}{t}$; $s(1), s(-1)$

30. $u(t) = \frac{t^2 - 1}{t}$; $u(1), u(-1)$

In Problems 31 through 36, find (a) $f(x+h)$, (b) $f(x+h) - f(x)$, and (c) $\frac{f(x+h) - f(x)}{h}$ for the given function.

Illustration: $f(x) = x^2$

Solution:

(a) $f(x+h) = (x+h)^2 = x^2 + 2xh + h^2$

(b) $f(x+h) - f(x) = x^2 + 2xh + h^2 - x^2$
$= 2xh + h^2$

(c) $\frac{f(x+h) - f(x)}{h} = \frac{2xh + h^2}{h} = 2x + h$

31. $f(x) = x^2 + 1$

32. $f(x) = x^2 - 2$

33. $f(x) = 2x + 1$

34. $f(x) = 3x - 1$

35. $f(x) = x^2 + 2x$

36. $f(x) = x^2 - 2x$

Courtesy of General Dynamics Corporation.

OBJECTIVES 9.3

After studying this section, the reader should be able to:

1. **Classify a given polynomial as a linear, quadratic, or cubic function.**
2. **Find the graph of a linear or quadratic polynomial and indicate its domain and range.**
3. **Find the domain of a given rational function.**

9.3 POLYNOMIAL FUNCTIONS

The photo at the beginning of this section shows the flight path of an airplane taking off from an airport. The picture is taken by a special camera that makes two exposures per second. As you can see, the altitude a of the plane depends on the time t after takeoff; in symbols, with a and t measured in suitable units,

(1) $a = f(t)$

If it is known that the airplane is climbing at a constant rate of k units per second, Equation (1) can be rewritten as

A polynomial function f is a function for which f(x) is a polynomial in x of degree n.

(2) $f(t) = a = kt$

The function given in Equation (2) is an example of a *polynomial function.* In general, if f is a function for which f(x) is a polynomial of degree n, then f is called a *polynomial function of degree n.* If the degree of the polynomial is *0*, the function is called a *constant* function; if the degree is *1*, the function is called a *linear* function; if the degree is *2*, the function is called a *quadratic* function; and if the degree is 3, the function is called a *cubic* function. For example,

$f(x) = 4$ defines a *constant* function — The degree of 4 is 0.
$f(x) = -2x + 3$ defines a *linear* function — The degree of $-2x+3$ is 1.
$f(x) = 3x^2 + x + 5$ defines a *quadratic* function — The degree of $3x^2+x+5$ is 2.
$f(x) = 2x^3 - 7x$ defines a *cubic* function — The degree of $2x^3-7x$ is 3.

We shall now discuss *linear* functions. As you recall from Chapter 7, the equation

$$(3)\quad y = mx + b$$

has as its graph a *line* with slope m and y-intercept b. A function defined by a comparable formula,

$$(4)\quad f(x) = mx + b$$

(m and b real constants), is called a *linear function.* One particular linear function is the identity function, defined by

$$(5)\quad f(x) = x,$$

the graph of which is shown in Figure 9.12.

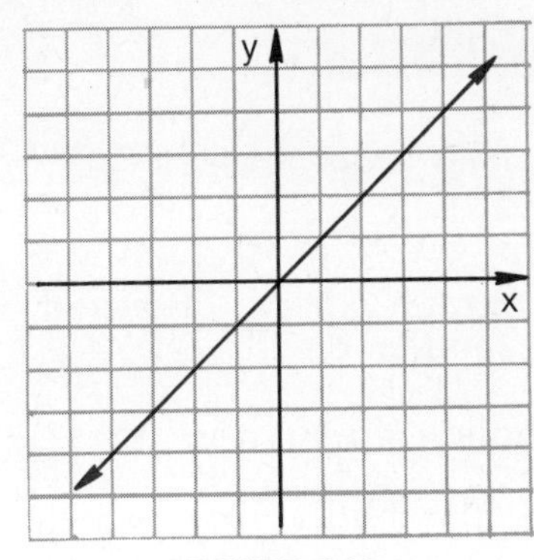

FIGURE 9.12

Example 1

Find the domain and range of the linear function defined by $f(x) = -2x + 3$, and sketch the graph.

Solution

Since there are no restrictions on x, the domain of f is the set of real numbers. The range of f is also the set of real numbers. To determine the graph of a straight line, it is enough to determine two points on the line. When $x = 0$, $f(x) = 3$; thus

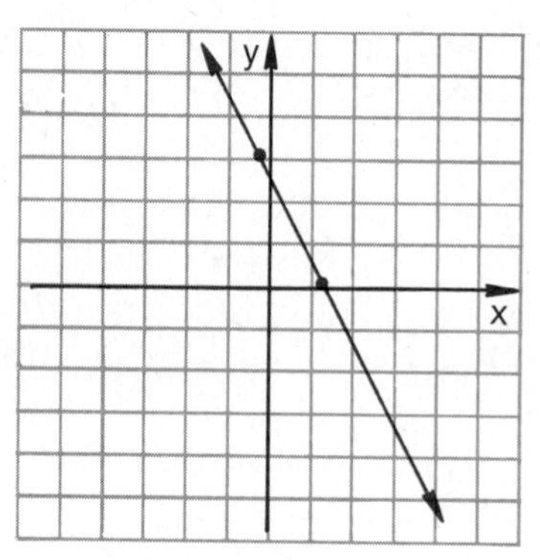

FIGURE 9.13

(0, 3) is on the line. When $f(x) = 0$, $x = \frac{3}{2}$; hence $\left(\frac{3}{2}, 0\right)$ is on the line. These two points are then joined with a line to form the graph of $f(x) = -2x + 3$, as shown in Figure 9.13.

The second type of function we shall study—the *quadratic function*—has already been mentioned in Chapter 6. A quadratic function is a polynomial function of degree 2. For example, $f(x) = -x^2$, $g(x) = 2x^2 + 3$, and $h(x) = -5x^2 - x + 8$ define quadratic functions. The general quadratic function is defined by

$$(6)\quad f(x) = ax^2 + bx + c \quad (a, b, c \text{ real numbers}, a \neq 0)$$

As you recall, the graph of a quadratic polynomial is a *parabola,* and this graph can be obtained by plotting a few points and recalling that the *vertex* (the lowest or highest point) of the parabola occurs where $x = \frac{-b}{2a}$. To simplify the task of graphing parabolas, we suggest the following procedure.

1. Locate the x-intercepts (if there are any) by letting $f(x) = 0$ and solving for x.
2. Locate the y-intercept by letting $x = 0$ and solving for y (recall that $y = f(x)$).
3. Locate the vertex by letting $x = \frac{-b}{2a}$, and then solve for y.
4. Sketch the parabola using the points obtained in Steps 1, 2, and 3.

Example 2

Find the domain of the quadratic function defined by

$$f(x) = x^2 + 2x - 3$$

Sketch the graph and then describe the range.

Solution

The domain is the set of real numbers. We now follow our four steps for graphing a parabola.

1. To find the x-intercepts, we let $f(x) = 0$, obtaining $x^2 + 2x - 3 = 0$, or $(x + 3)(x - 1) = 0$. Thus $x = -3$ or $x = 1$.
2. To find the y-intercept, we let $x = 0$; thus

$$f(0) = 0^2 + 2 \cdot 0 - 3 = -3$$

3. The vertex is at the point where $x = \dfrac{-2}{2} = -1$. At this point, $y = f(-1) = (-1)^2 + (2)(-1) - 3 = -4$.
4. We graph the points obtained in Steps 1, 2, and 3, and sketch the parabola shown in Figure 9.14.

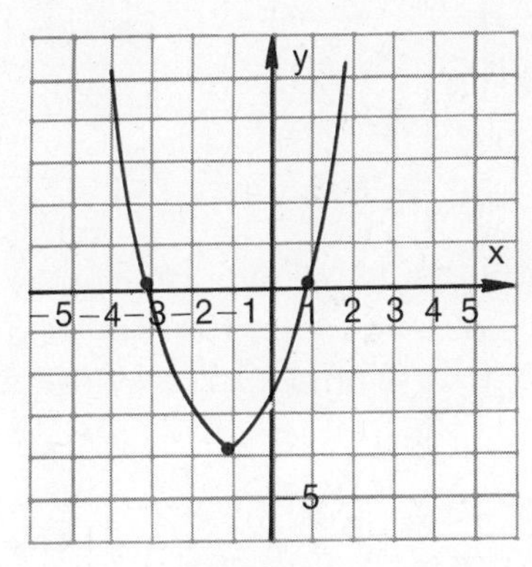

FIGURE 9.14

As you can see from Example 2, the graph of a quadratic function can be used to determine the nature of the roots of the associated quadratic equation, $ax^2 + bx + c = 0$. This is done by inspecting the x-intercepts of the graph of the function given in Equation (6). If $ax^2 + bx + c = 0$ has two real roots and $a > 0$, the graph of $f(x) = ax^2 + bx + c = 0$ intersects the x-axis at two distinct points and opens upward, as shown in Figure 9.15. In case $a < 0$, the parabola opens downward, as shown in Figure 9.16. If the equation has only one root and $a > 0$, the graph is tangent to the x-axis and opens upward, as shown in Figure 9.17. When the equation has one root and $a < 0$, the parabola is tangent to the x-axis

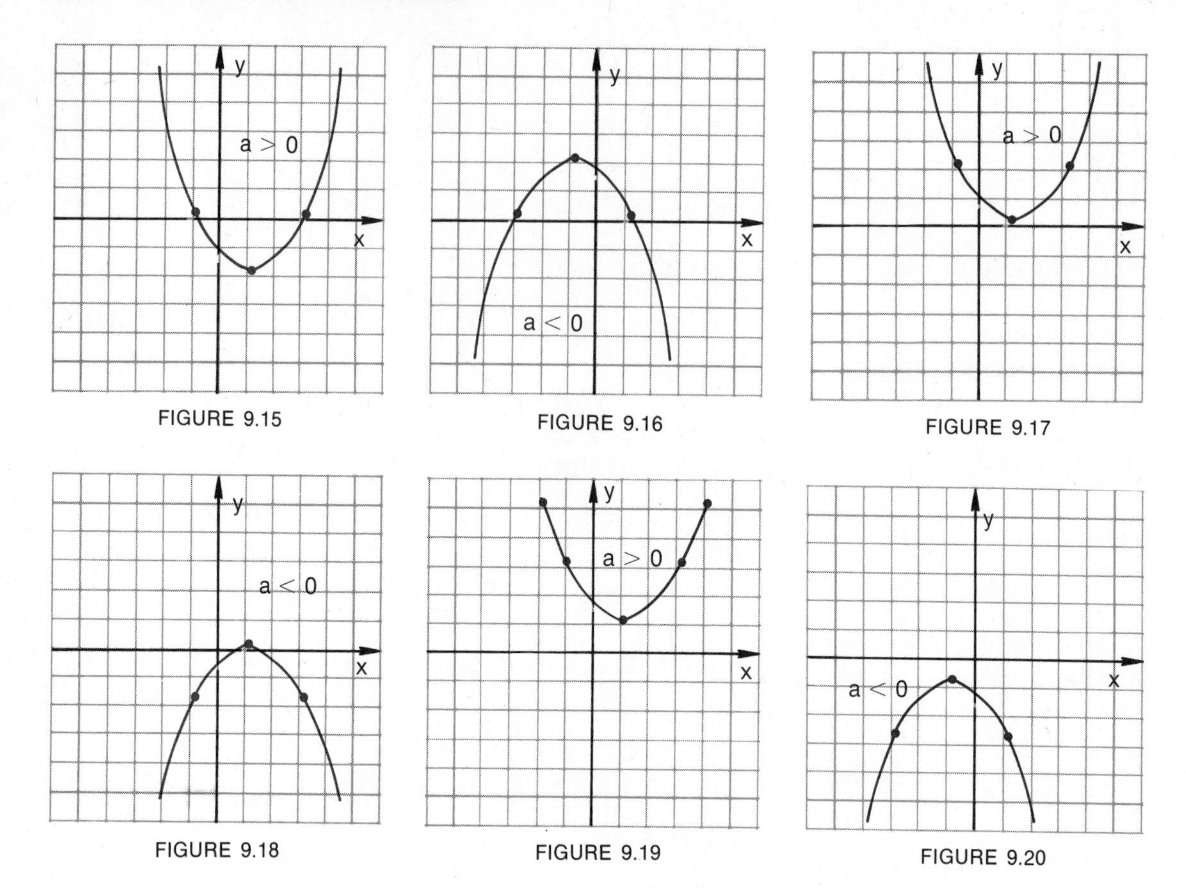

FIGURE 9.15

FIGURE 9.16

FIGURE 9.17

FIGURE 9.18

FIGURE 9.19

FIGURE 9.20

and opens downward, as shown in Figure 9.18. Finally, if the equation has no real roots, the graph does not intersect the x-axis, as shown in Figure 9.19, where $a > 0$, and Figure 9.20, where $a < 0$.

Example 3

Sketch the graph of $f(x) = 2x^2 + 3x + 1$ and determine from the graph the nature of the roots of the quadratic equation $2x^2 + 3x + 1 = 0$.

Solution

The vertex of this parabola is where $x = \frac{-3}{2 \cdot 2} = \frac{-3}{4}$. At this point,

$$y = f\left(\frac{-3}{4}\right) = 2\left(\frac{-3}{4}\right)^2 + 3\left(\frac{-3}{4}\right) + 1$$

$$= \frac{9}{8} - \frac{9}{4} + 1 = \frac{-1}{8}$$

Thus, the vertex is at $\left(\frac{-3}{4}, \frac{-1}{8}\right)$. Since in this case $a > 0$, the parabola opens upward; since the vertex is below the x-axis, there must be *two* distinct real roots (i.e., two x-intercepts) for the equation $2x^2 + 3x + 1 = 0$. A sketch of the graph is shown in Figure 9.21. Note that to determine the nature of the roots of $2x^2+3x+1=0$, we *do not* have to solve the given equation. Since we know that the parabola opens upward, it is enough to find the vertex of $f(x) = 2x^2 + 3x + 1$ and determine if this vertex is *below*, *above*, or *on* the x-axis.

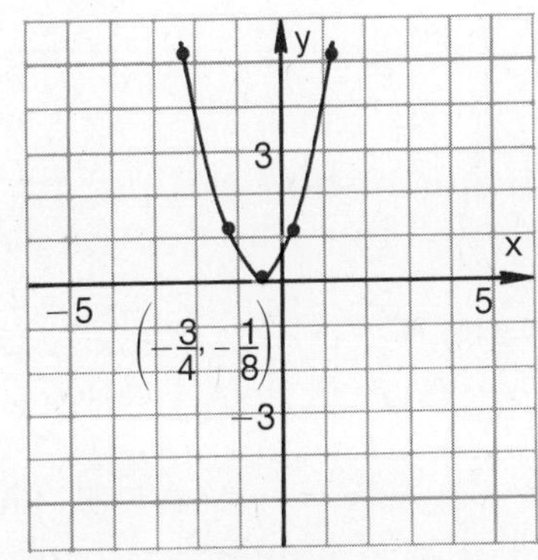

FIGURE 9.21

Polynomial functions can be manipulated in a variety of ways. For example, we can have the quotient of two polynomial functions, called a *rational function*. The function g defined by

A rational function is a function which can be expressed as the quotient of two polynomial functions.

$$g(x) = \frac{x^2 - 7x + 1}{x^2 - 1}$$

is a rational function. This function is undefined when $x = 1$ or -1; accordingly, the domain of g is the set of real numbers *except* 1 and -1. The polynomial and the rational functions constitute a special class of a more general type of function that can be obtained by performing a finite number of algebraic operations (additions, subtractions, multiplications, divisions, extracting roots, and raising to powers) on the identity function and the constant function. The latter functions are in turn special cases of an even more general class called *algebraic functions*, the definition of which is outside the scope of this book. The functions defined by

A function formed by a finite number of algebraic operations on the identity and constant functions is a special kind of algebraic function.

$$f(x) = \frac{(x^2 - 1)^2}{\sqrt{x^2 + 1}}$$

$$\text{and } g(x) = \frac{\sqrt[3]{(x+1)^2} - 7}{(x-1)^3}$$

are examples of *algebraic functions.*

Example 4 Find the domain of the rational function defined by

$$h(x) = \frac{x^2 + x - 1}{(x^2 - 4)(x^2 + x - 12)}$$

Solution Since the denominator of h(x) *can not* be zero, the values that should be excluded from the domain are those for which

$$x^2 - 4 = 0 \quad \text{or} \quad x^2 + x - 12 = 0$$
$$(x+2)(x-2) = 0 \quad (x+4)(x-3) = 0$$

That is, $x = -2$, $x = 2$ or $x = -4$, $x = 3$. The domain of h is therefore the set of all real numbers *except* -2, 2, -4, and 3.

PROGRESS TEST

1. The domain and range of $g(x) = -x + 7$ are ____________________ and ____________________, respectively.
2. Sketch the graph of the function defined by $f(x) = x^2 + 1$.

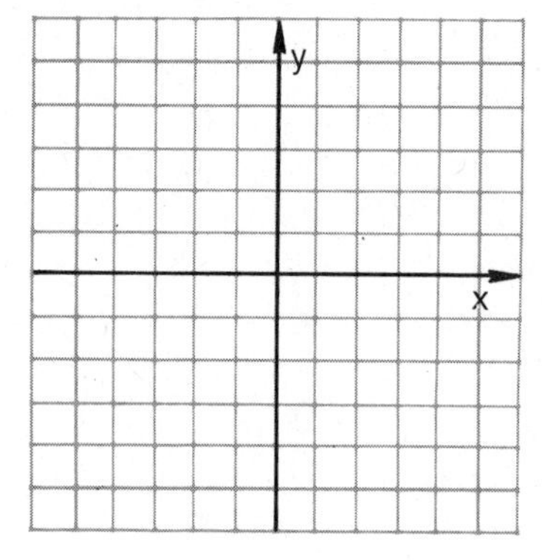

3. The domain and range of the function in Problem 2 are ____________________ __________ and ____________________, respectively.

4. Follow the four-step procedure in the text to graph the function $f(x) = x^2 - 2x - 3$.

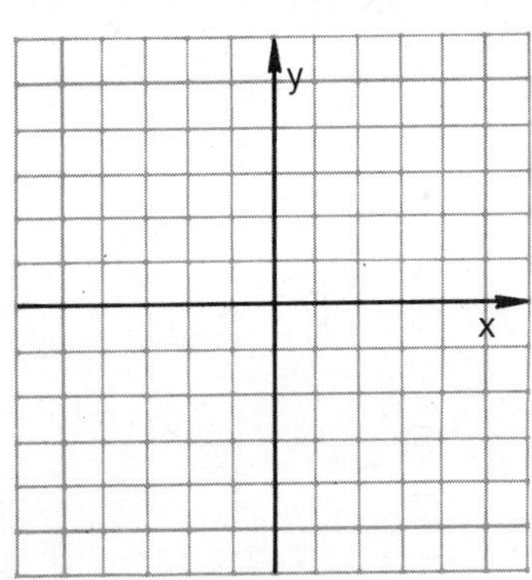

5. The domain of the rational function $g(x) = \frac{x^2 + x + 5}{x^2 - 9}$ is ______________.

EXERCISE 9.3

In Problems 1 through 20, sketch the graph of the given function and give its domain and range.

Illustration: (a) $f(x) = -3x + 6$
(b) $g(x) = -x^2 + 1$

Solution: (a) The graph of $f(x) = -3x + 6$ is a line. If $x = 0$, $y = 6$; hence $(0, 6)$ is on the line. When $f(x) = 0$, $-3x + 6 = 0$, and $x = 2$; thus $(2, 0)$ is on the line. Joining these two points, we obtain the graph of $f(x) = -3x + 6$ shown in Figure 9.22. Both the domain and range of $f(x)$ are the set of all real numbers.

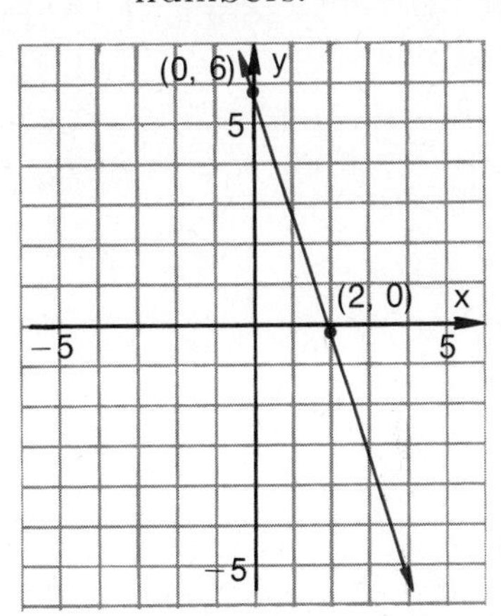

FIGURE 9.22

PROGRESS TEST ANSWERS

1. the set of all real numbers; the set of all real numbers
2.

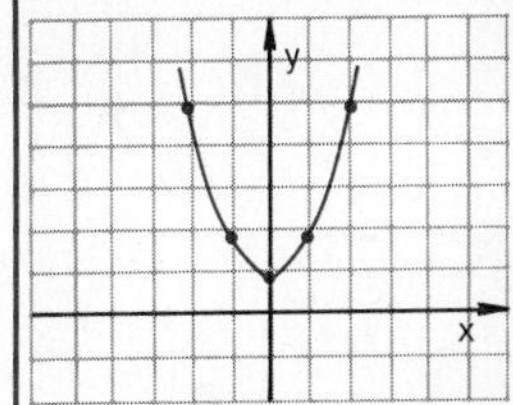

3. the set of all real numbers; the set of all real numbers greater than or equal to 1
4.

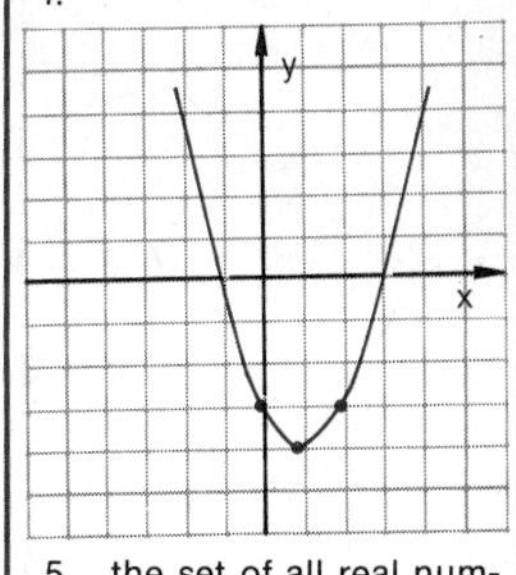

5. the set of all real numbers except 3 and −3.

(b) The graph of $g(x) = -x^2 + 1$ is a parabola opening downward. To sketch the graph, we use the four-step procedure given in the text.

1. We let $g(x) = 0$, obtaining $-x^2 + 1 = 0$, or, equivalently, $x^2 = 1$; solving, $x = \pm 1$, the x-intercepts.
2. We let $x = 0$, obtaining $y = g(x) = -0^2 + 1 = 1$, the y-intercept.
3. The vertex is where $x = \dfrac{-b}{2a} = \dfrac{0}{-2} = 0$. At this point, $y = 1$.
4. We graph the points found in steps 1, 2, and 3 as shown in Figure 9.23.

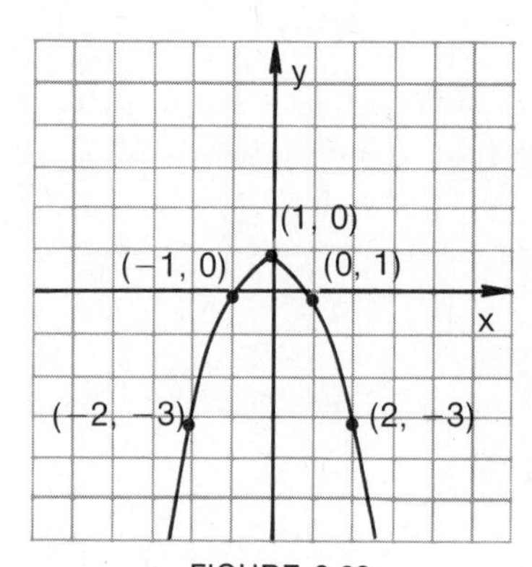

FIGURE 9.23

The domain of the function g(x) is the set of real numbers, and the range is the set of real numbers less than or equal to 1, that is, $\{y \mid y \leq 1\}$.

1. $f(x) = 2x + 6$

2. $f(x) = 3x + 6$

3. $g(x) = -2x - 6$

4. $g(x) = -3x - 6$

5. $h(x) = -3x + 6$

6. $h(x) = -4x + 4$

7. $f(x) = x^2 + 3$

8. $f(x) = x^2 + 4$

9. $g(x) = x^2 - 4$

10. $g(x) = x^2 - 9$

11. $h(x) = -x^2 + 4$

12. $h(x) = -x^2 + 9$

13. $f(x) = -2x^2 + 2$

14. $f(x) = -3x^2 + 12$

15. $g(x) = x^2 + 2x - 8$

16. $g(x) = x^2 - 2x - 8$

17. $h(x) = -x^2 + 4x$

18. $h(x) = -x^2 + x$

19. $f(x) = x^2 + 4x$

20. $f(x) = -x^2 + 2x - 1$

In Problems 21 through 30, sketch the graph of the given quadratic function and determine the nature of the roots of the equation.

Illustration: $f(x) = -x^2 - 2x + 8$

Solution: The graph of $f(x) = -x^2 - 2x + 8$ is a parabola opening downward. We use the four-step procedure to sketch it.

1. For $f(x) = 0$, $-x^2 - 2x + 8 = 0$, that is, $-(x^2 + 2x - 8) = 0$. Solving for x, we have $-(x + 4)(x - 2) = 0$. Thus, $x = -4$ or $x = 2$, the x-intercepts.
2. If we let $x = 0$, $f(x) = y = -0^2 - 2 \cdot 0 + 8 = 8$, the y-intercept.
3. The vertex is where $x = \dfrac{-(-2)}{(2)(-1)} = -1$. At this point, $y = f(-1) = -(-1)^2 - 2(-1) + 8 = 9$.
4. We graph the points obtained in steps 1, 2, and 3, and then sketch the parabola, as in Figure 9.24.

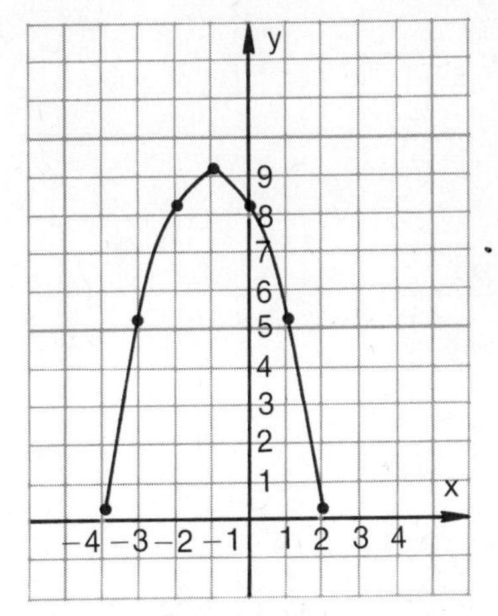

FIGURE 9.24

As you can see from the graph, the equation $-x^2 - 2x + 8 = 0$ has two real roots.

21. $f(x) = -x^2 + 4$

22. $f(x) = -x^2 + 9$

23. $h(x) = x^2 - 4$

24. $h(x) = x^2 - 9$

25. $g(x) = -x^2 + 2x + 8$

26. $g(x) = x^2 - 4x$

27. $u(x) = x^2 - x$

28. $u(x) = x^2 - 4x + 4$

29. $v(x) = x^2 + 4$

30. $v(x) = x^2 + 1$

In Problems 31 through 40, follow the procedure of Example 4 and find the domain of the given rational function.

31. $f(x) = \frac{1}{x}$

32. $f(x) = \frac{1}{x - 1}$

33. $g(x) = \frac{2x}{x^2 - 1}$

34. $g(x) = \frac{-5x}{x^2 - 9}$

35. $h(x) = \frac{3x}{x(x^2 - 4)}$

36. $h(x) = \frac{-7x^2}{x^2(x^2 - 16)}$

37. $u(x) = \frac{5x^2 + 1}{(x^2 - 4)(x^2 + 3x + 2)}$

38. $u(x) = \dfrac{2x^2 + x + 1}{(x^2 - 3x + 2)(x^2 - 1)}$

39. $v(x) = \dfrac{3x^2 + x + 1}{(x^2 + x - 2)(x^2 - 16)}$

40. $v(x) = \dfrac{5x^3 + x^2 + x + 1}{(x - 1)(x^2 - 4)(x^2 - 2x - 3)}$

FOR HOME DELIVERY
OF YOUR
TAMPA TRIBUNE
CALL 272-7422

And, if you are presently a subscriber and have been missed for some reason – phone 272-7422

THE TAMPA TRIBUNE

Published every morning by The Tribune Company at 202 S. Parker Street, Tampa, Florida 33606, an affiliate of Media General, Inc. Second class postage paid at Tampa, Florida.

TRIBUNE HOME DELIVERY RATES

	Daily & Sunday	Daily	Sunday
1 Week	$ 1.10	$ .80	$ .50
4 Weeks	4.40	3.20	2.00
13 Weeks	14.30	10.40	6.50
26 Weeks	28.60	20.80	13.00
52 Weeks	57.20	41.60	26.00

OBJECTIVES 9.4

After studying this section, the reader should be able to:

1. **Find the inverse of a relation expressed as a set of ordered pairs.**
2. **Find the inverse of a relation given by a formula.**
3. **Determine if a given function is one-to-one.**
4. **Determine if a function defined by a formula has another function as an inverse (by using Theorem 9.2).**

9.4 THE INVERSE OF A FUNCTION

The table at the beginning of this section shows the daily and Sunday Tribune house delivery rates for 1, 4, 13, 26, and 52 weeks, respectively. If a customer decides to represent by x the number of weeks he wishes to subscribe to the Tribune, with y as the cost, the daily and Sunday home delivery rate can be expressed as the function.

$$f = \{(1, 1.10), (4, 4.40), (13, 14.30), (26, 29.60), (52, 57.20)\}$$

In this function, the ordered pair (4, 4.40) means that 4 weeks' delivery costs $4.40. The business manager of the Tribune may be more concerned with the *inverse* function:

f^{-1} is read as "the inverse of f" or "f inverse."

$$f^{-1} = \{(1.10, 1), (4.40, 4), (14.30, 13), (28.60, 26), (57.20, 52)\}$$

In this function, the ordered pair (4.40, 4) means that the business manager has received $4.40, and hence must deliver the paper to the subscriber for 4 weeks. As you can see, the function f^{-1} is obtained by *interchanging* the components of each ordered pair of the function f. We make this idea precise in the following definition.

DEFINITION 9.3

If S is a relation, then the *inverse* of S, denoted by S^{-1}, is the relation obtained by interchanging the components of each ordered pair in S.

If follows from Definition 9.3 that the domain of S^{-1} is the range of S, and that the range of S^{-1} is the domain of S. For example, the domain and range of the function f given at the beginning of this section are {1, 4, 13, 26, 52} and {1.10, 4.40, 14.30, 28.60, 57.20}, respectively; the domain and range of f^{-1} are {1.10, 4.40, 14.30, 28.60, 57.20} and {1, 4, 13, 26, 52}, respectively.

Example 1

Let S = {(1, 2), (3, 4), (5, 4)}. Find:
(a) the domain and range of S
(b) S^{-1}
(c) the domain and range of S^{-1}
(d) the graphs of S and S^{-1} on the same coordinate axes

Solution

(a) The domain of S is {1, 3, 5}. The range is {2, 4}.
(b) S^{-1} = {(2, 1), (4, 3), (4, 5)}
(c) The domain of S^{-1} is {2, 4}; the range is {1, 3, 5}.
(d) The graphs of S (in color) and S^{-1} (in black) are shown in Figure 9.25. As you can see, the two graphs are symmetric with respect to the line y = x (shown dotted).

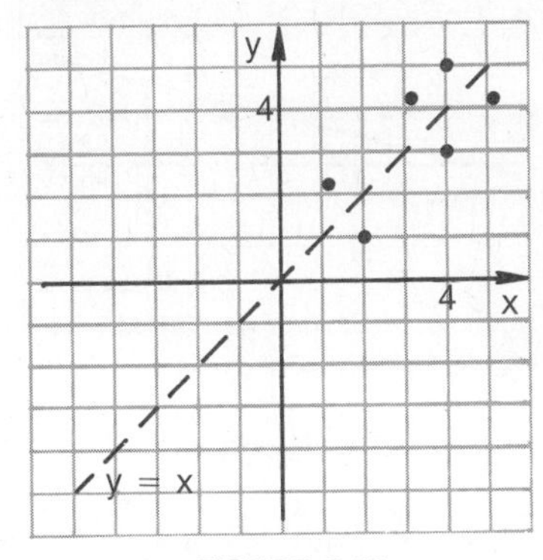

FIGURE 9.25

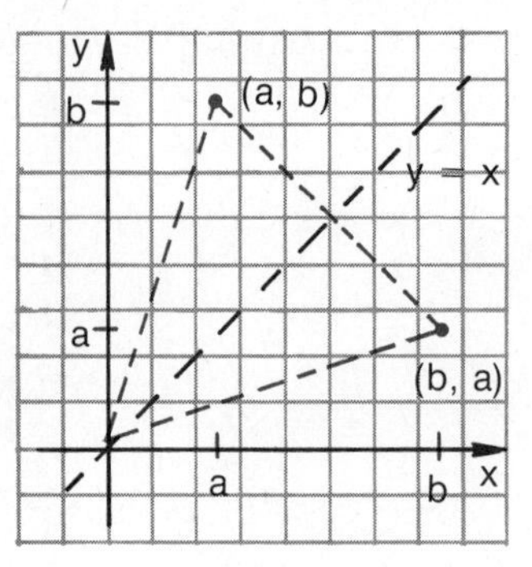

FIGURE 9.26

The symmetry noted in Example 1(d) is always found in the graphs of a relation and its inverse. Figure 9.26 shows that the ordered pairs (a, b) and (b, a) are symmetric to each other (with respect to the line y = x). Now consider the relation

$$(1)\ y = 4x - 4.$$

The inverse of this relation is obtained by interchanging the x and y coordinates, that is, by writing $x = 4y - 4$. Solving for y, we get

$$(2)\ y = \frac{1}{4}(x + 4)$$

The graphs of (1) (in black) and its inverse (2) (in color) are shown in Figure 9.27. Clearly, the graphs are symmetric to each other with respect to the line y = x, shown dotted.

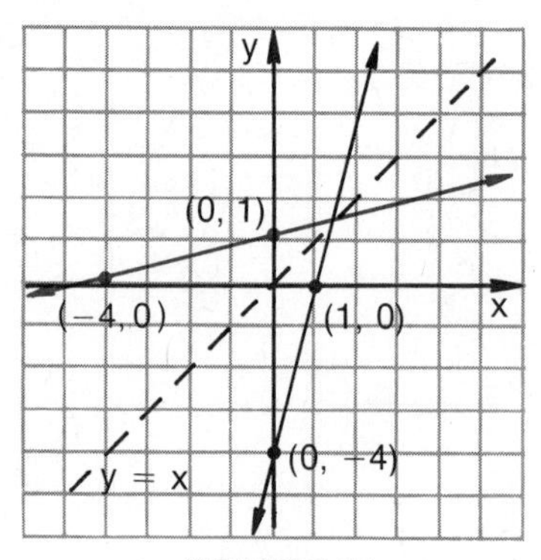

FIGURE 9.27

Example 2 Let $f(x) = y = 4x - 2$. Find $f^{-1}(x)$.

Solution

Since $y = 4x - 2$, we interchange the variables x and y, obtaining $x = 4y - 2$. Solving for y, $y = \frac{x + 2}{4}$.

Thus, the inverse of f(x) is $f^{-1}(x) = \frac{x + 2}{4}$.

Since every function is a set of ordered pairs (a relation), every function has an inverse. Is this inverse always a function? We can very quickly see that it is not. For example, if S is the function defined by $S = \{(1, 2), (3, 2)\}$, the inverse of S is $S^{-1} = \{(2, 1), (2, 3)\}$, which is *not* a function, since two distinct ordered pairs have the same first component, 2. On the other hand, if G is the function defined by $G = \{(3, 4), (5,6)\}$, the inverse is $G^{-1} = \{(4, 3), (6, 5)\}$, which *is* a function. The reason that the inverse of S is not a function is that S has two ordered pairs with the same *second* component. A function in which no two distinct ordered pairs have the same second component is called a *one-to-one* function. The inverse of such a function is always a function. We summarize this discussion in the following theorem.

A one-to-one function is a function which associates with each element of the range a unique element in the domain.

THEOREM 9.1

If the function f is one-to-one, then f^{-1} (the inverse of f) is also a function.

Theorem 9.1 tells us that to determine if the inverse of a function is a function, we must ascertain if the original function is one-to-one. In order to do this, we must return to the definition of a one-to-one function—a one-to-one function *can not* have two ordered pairs with the same second component; thus, if f is one-to-one and $f(x_1) = f(x_2)$, it must be true that $x_1 = x_2$. The converse of this is our criterion: if $f(x_1) = f(x_2)$ implies that $x_1 = x_2$, then f is one-to-one. On the Cartesian plane, this means that any horizontal line can intersect the graph of a one-to-one function at one point only. If a horizontal line intersects the graph at more than one point, the graph is not that of a one-to-one function. This discussion is summarized in the following definition.

Here we have two ordered pairs $(x_1, f(x_1))$ and $(x_2, f(x_2))$ which are equal, so the second components are equal.

DEFINITION 9.4

The function f is one-to-one if whenever $f(x_1) = f(x_2)$ then $x_1 = x_2$.

With this definition, Theorem 9.1 can be restated in the form given below.

THEOREM 9.2

The function f has a function as its inverse if $f(x_1) = f(x_2)$ implies $x_1 = x_2$.

For example, $f(x) = 3x - 6$ has a function as its inverse, since if $f(x_1) = f(x_2)$, we have

$$3x_1 - 6 = 3x_2 - 6$$
$$\text{or } 3x_1 = 3x_2$$
$$\text{that is, } x_1 = x_2$$

Recall that $f(x_1) = 3x_1 - 6$, and that $f(x_2) = 3x_2 - 6$.

Example 3 Find the inverse of $f(x) = y = x^2$. Is the inverse a function?

Solution Since $y = x^2$, we interchange the x and y variables to obtain $y^2 = x$, or $y = \pm\sqrt{x}$. To find if this inverse is a function, we must check if when $f(x_1) = f(x_2)$, $x_1 = x_2$. If $f(x_1) = f(x_2)$, we have

$$x_1^2 = x_2^2$$

However, this *does not* mean that $x_1 = x_2$, since we may have $x_1 = -x_2$. For example, $(2)^2 = (-2)^2$, but $2 \neq -2$. Hence, the inverse of $f(x) = x^2$ is *not* a function, as can be seen in Figure 9.28, which shows the graph of $f(x) = x^2$ in black and the graph of the inverse relation $\{(x, y) \mid y = \pm\sqrt{x}\}$ in color.

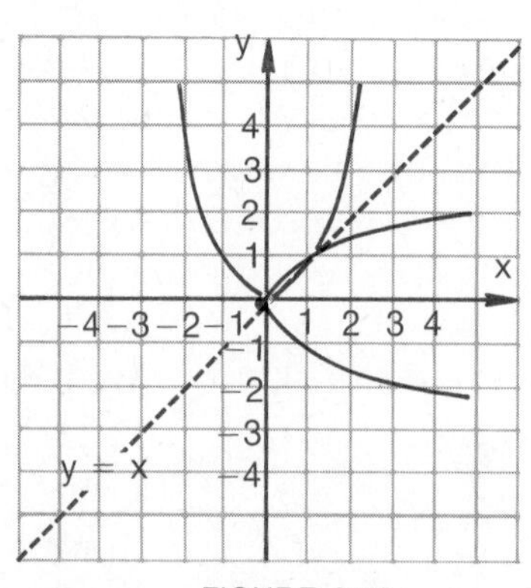

FIGURE 9.28

PROGRESS TEST

1. The inverse of $f = \{(7, 1), (1, 7), (3, 7)\}$ is ______________________.
2. In Problem 1, the inverse relation __________ (is, is not) a function.
3. If $f(x) = y = 2x - 2$, $f^{-1}(x) =$ __________.
4. If $f(x) = 2x - 5$, is f a one-to-one function?
5. Is the inverse of $f(x) = 2x - 5$ a function?

EXERCISE 9.4

In Problems 1 through 10, a relation S is given. Find S^{-1}, draw the graph of S and S^{-1} on the same coordinate axes, and determine if S^{-1} is a function.

Illustration: (a) $S = \{(1, 5), (3, 2)\}$
(b) $\{(x, y) \mid y = 3x\}$

Solution: (a) $S^{-1} = \{(5, 1), (2, 3)\}$. The graphs of S (in black) and S^{-1} (in color) are shown in Figure 9.29. Since S has no two ordered pairs with the same second component, S^{-1} is a function.

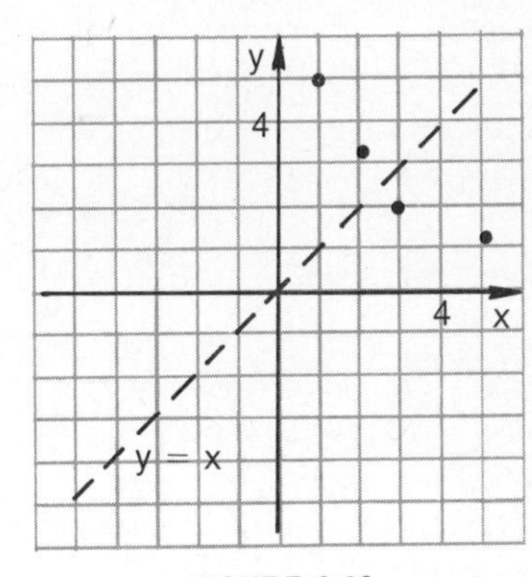

FIGURE 9.29

(b) Given $y = 3x$, we interchange x and y, obtaining $x = 3y$, or $y = \frac{1}{3}x$, the inverse of $f(x) = 3x$. The graphs of

PROGRESS TEST ANSWERS

1. $\{(1, 7), (7, 1), (7, 3)\}$
2. is not
3. $\frac{x + 2}{2}$
4. Yes. If $f(x_1) = f(x_2)$, then $2x_1 - 5 = 2x_2 - 5$ or $2x_1 = 2x_2$ that is, $x_1 = x_2$
5. Yes. Since f is one-to-one (see Question 4), the inverse is a function.

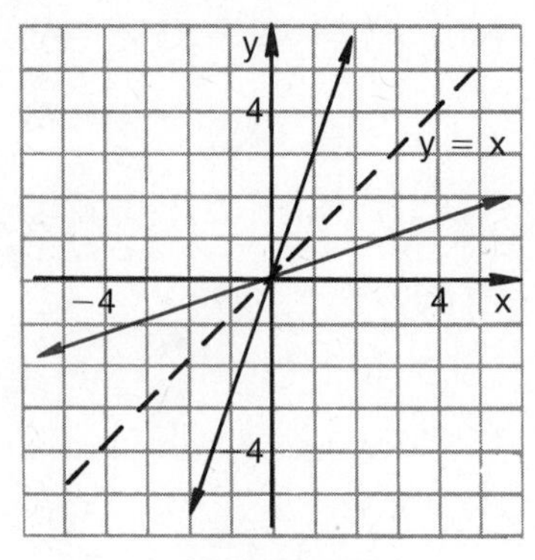

FIGURE 9.30

$f(x) = 3x$ (in black) and $f^{-1}(x) = \frac{1}{3}x$ (in color) are shown in Figure 9.30. To check if f^{-1} is a function, we let

$$f(x_1) = f(x_2)$$
$$\text{or } 3x_1 = 3x_2$$
$$\text{that is, } x_1 = x_2$$

Thus, according to Theorem 9.2, f^{-1} is a function.

1. $\{(1, 3), (2, 4), (5, 6)\}$

2. $\{(-1, 5), (-3, 7), (-4, 7)\}$

3. $\{(x, y) \mid y = 3x + 6\}$

4. $\{(x, y) \mid y = 2x + 4\}$

5. $\{(x, y) \mid y = 2x - 4\}$

6. $\{(x, y) \mid y = 3x - 6\}$

7. $\{(x, y) \mid y = 2x^2\}$

8. $\{(x, y) \mid y = x^2 + 1\}$

9. $\{(x, y) \mid y = x^2 - 1\}$

10. $\{(x, y) \mid y = x^2 - 4\}$

In Problems 11 through 16, a function f is given. Find: (a) the range of f, (b) the domain of the inverse relation, and (c) if the inverse relation is a function.

Illustration: $y = f(x) = \sqrt{x - 1}$

Solution: (a) Since the radical sign indicates the non-negative square root of $x - 1$, the range is the set of all non-negative real numbers.

b) The domain of the inverse relation is the range of f—that is, the set of all non-negative real numbers.

(c) If $f(x_1) = f(x_2)$,

$$\sqrt{x_1 - 1} = \sqrt{x_2 - 1}$$

$$\text{or } x_1 = x_2$$

Thus, f^{-1} is a function.

11. $f(x) = \sqrt{x + 1}$

12. $f(x) = \sqrt{x + 3}$

13. $f(x) = x^2 - 1$

14. $f(x) = x^2 - 4$

15. $f(x) = x^2 + 4$

16. $f(x) = x^2 + 1$

17. The equation $f(x) = \sqrt{9 - x^2}$ defines a function with domain $\{x \mid -3 \leq x \leq 3\}$.
(a) What is the range of f?
(b) Find the inverse relation and its domain.
(c) Is the inverse relation a function?

18. Repeat Problem 17 for the function defined by $f(x) = \sqrt{4 - x^2}$, the domain of which is $\{x \mid -2 \leq x \leq 2\}$.

19. Repeat Problem 17 for the function defined by $f(x) = \sqrt{16 - x^2}$, the domain of which is $\{x \mid -4 \leq x \leq 4\}$.

20. Prove that any linear function defined by $f(x) = ax + b$, with $a \neq 0$, has a function as its inverse.

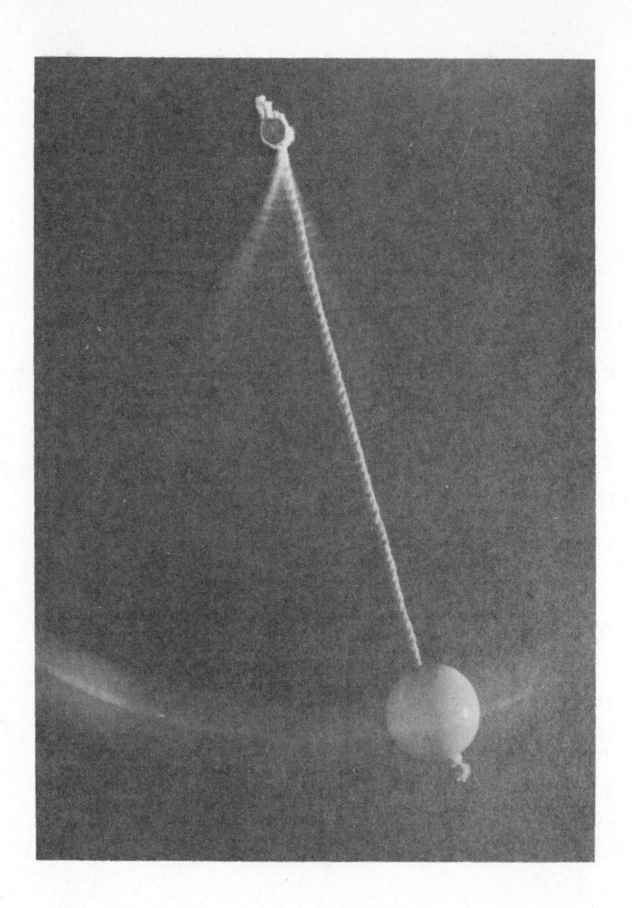

OBJECTIVES 9.5

After studying this section, the reader should be able to:

1. **Find an equation of variation for variables that are directly, inversely, or jointly proportional.**
2. **Find the constant of proportionality in a given variation problem.**
3. **Solve word problems involving variation.**

9.5 VARIATION

The photograph at the beginning of this section shows a swinging pendulum. As the length of the string increases, the time it takes the pendulum to make a full swing (back and forth) also increases. Can we find a formula relating the time T and the length L of the string? Galileo Galilei discovered that the time T (in seconds) for one complete swing varies (changes) in the same manner as the square root of the length of the pendulum. In the language of mathematics, science, and technology, we say that the time T (in seconds) for one complete swing of the pendulum varies directly with the square root of the length of the pendulum. In symbols,

$$T = k\sqrt{L}, \text{ where k is a positive constant}$$

Direct variation is defined formally as follows:

DEFINITION 9.5

A variable y *varies directly* with a variable x if $y = kx$, where k is a positive constant.

The statements

y varies with x
y varies directly with x
and *y is proportional to x*

k is called a proportionality constant.

are all equivalent and are translated as

$$y = kx$$

The constant k is called the *constant of proportionality* (or the constant of variation). For example, the interest earned when investing money at an annual rate of 10% is directly proportional to the principal invested—that is, Interest equals k times the Principal, or

(1) $I = kP$

Since the rate of annual interest is 10%, (1) can be written as

(2) $I = \frac{10}{100}P$; that is, $k = \frac{10}{100}$

Example 1

The photograph on this page shows Masuriya Din, who sports the longest moustache on record (Guinness). Assume that the number N of inches that a moustache grows is directly proportional to the time t in months.

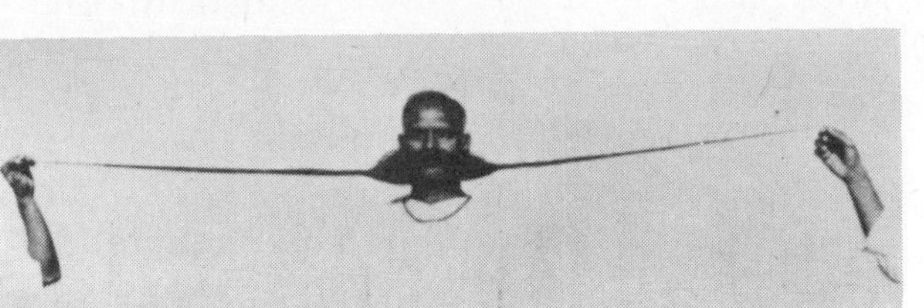

Photo credit: Sterling Publishing Company, Inc., New York.

(a) Find an equation of variation.
(b) If a moustache grows 4 inches in 1 year, find the value of k in part (a).
(c) Estimate the length of Masuriya Din's moustache after he had grown it for 14 years (1949 to 1962, inclusive).

Solution

(a) Since N is directly proportional to t,
$N = kt$

(b) If a moustache grows 4 inches in 1 year, $N = 4$ when $t = 1$. Substituting $N = 4$ and $t = 1$ in the formula $N = kt$, we obtain, $4 = k \cdot 1$, or $k = 4$.

(c) We must find N when $t = 14$. Since in part (b) we found that $k = 4$, $N = 4t$. When $t = 14$, $N = 4 \cdot 14 = 56$ inches. (By the way, Masuriya's moustache actually grew a total of 102 inches between 1949 and 1962.)

Example 2

If a gas such as air is enclosed in a container of fixed volume, the pressure P of the gas on the walls of the container varies directly with the absolute temperature* T of the gas. We will use P measured in pounds per square inch and T in degrees absolute.

(a) Find an equation of variation.

(b) If the air pressure on the walls of the container is 5 pounds per square inch when the temperature is 460° absolute, find the constant k of part (a).

(c) Find the air pressure if the temperature is raised to 552° absolute.

Solution

(a) Since P is directly proportional to T,
$P = kT$

(b) When the air pressure $P = 5$, the temperature $T = 460$. Thus,

$$5 = k \cdot 460 \text{ or } k = \frac{5}{460} = \frac{1}{92}$$

(c) Since we know from part (b) that $k = \frac{1}{92}$, we may write

$$P = \frac{1}{92}T$$

We must find P when $T = 552$. Substituting $T = 552$ in the equation $P = \frac{1}{92}T$, we obtain

$$P = \frac{1}{92} \cdot 552 = 6 \text{ pounds per square inch}$$

*Absolute temperature is measured from *absolute zero*, which occurs at approximately −460° F.

Not all variations in mathematics and science are direct. For example, Robert J. Ringer, the author of the best-selling book *Winning Through Intimidation*, claims that the results R a person obtains are *inversely proportional* to the degree i to which he is intimidated. Here, the phrase *inversely proportional to* means that as the degree to which a person is intimidated *decreases*, the results *increase*. We now give the exact definition of the phrase *inversely proportional to*.

DEFINITION 9.6

A variable y *varies inversely* with a variable x if $y = \frac{k}{x}$, where k is a constant.

The statement *y is inversely proportional to x* is often used to mean *y varies inversely with x*.

With our definition, the claim that the results R a person obtains are inversely proportional to the degree i to which he is intimidated can be written as

$$R = \frac{k}{i}, \quad k \text{ a constant}$$

Example 3

At a constant temperature, the pressure P exerted by a gas varies inversely with the volume V. Assume P is measured in pounds per square foot and V in cubic feet.

(a) Find an equation of variation.

(b) If a pressure of 1760 pounds per square foot is exerted by 2 cubic feet of air in a cylinder fitted with a piston, find the k of part (a).

(c) If the piston is pushed out until the pressure is 704 pounds per square foot, what is the volume of the gas? (Assume no temperature change.)

Solution

(a) Since P varies inversely with V,

$$P = \frac{k}{V}$$

(b) When the pressure $P = 1760$, the volume $V = 2$. Substituting these values in the equation $P = \frac{k}{V}$, we obtain

$$1760 = \frac{k}{2}, \quad \text{or } k = 3520$$

(c) From part (b), $k = 3520$; with this value, our equation is

$$P = \frac{3520}{V}$$

We must find V when $P = 704$. Substituting $P = 704$ in the equation $P = \frac{3520}{V}$, we have

$$704 = \frac{3520}{V},$$

or, equivalently, $704V = 3520$. Dividing,

$$V = \frac{3520}{704} = 5 \text{ cubic feet}$$

Besides the direct and inverse variations we have discussed so far, there can be variation involving a third variable. A variable z can vary *jointly* with the variables x and y. For example, labor costs c vary jointly with the number of workers w used and the number of hours h that they work. The formal expression of joint variation is given below.

The statement "z is proportional to x and y" means the same as "z varies jointly with x and y."

DEFINITION 9.7

A variable z *varies jointly* with the variables x and y if $z = kxy$, k a constant.

The statement *z is proportional to x and y* is sometimes used to mean z *varies jointly with the variables x and y.*

(Applying our definition, the fact that labor costs c vary jointly with the number w of workers used and the number h of hours worked can be expressed as $c = kwh$, k a positive constant.)

Example 4

The lifting force P exerted by the atmosphere on the wings of an airplane varies jointly with the wing area A in square feet and the square of the plane's speed V in miles per hour. If the lift is 1200 pounds for a wing area of 100 square feet and a speed of 75 miles per hour,
(a) find an equation of variation.
(b) find k.

Solution

(a) Since P varies jointly with the area A and the square of the velocity V, $P = kAV^2$.

(b) When the lift $P = 1200$, we know that $A = 100$ and $V = 75$. Substituting these values in the equation $P = kAV^2$, we obtain

$$1200 = k \cdot 100 \cdot (75)^2$$

That is,

$$k = \frac{1200}{100 \cdot 75^2} = \frac{12}{75^2} = \frac{4}{1875}$$

PROGRESS TEST

1. The number N of trees saved by recycling is directly proportional to the height h of a stack of recyclable newspaper. The equation of variation for this relationship is ________.
2. If in Problem 1 it is estimated that a 4-foot stack of newspapers saves 1 tree, $k =$ ________.
3. Referring to Problems 1 and 2, the number of trees saved by a stack of newspapers 6 feet high is ________.
4. The f number on a camera lens and shutter varies inversely with the diameter of the aperture a (opening) provided the distance is set at infinity. The equation of variation for this relationship is ________.
5. If the aperture diameter for the lens in Problem 4 is $\frac{1}{2}$ inch when the f number is 8, $k =$ ________.
6. When the aperture diameter of the camera in Problems 4 and 5 is $\frac{1}{8}$ inch, the f number is ________.
7. The wind force F on a vertical surface varies jointly with the area A of the surface and the square of the wind velocity V. The equation of variation for this relationship is ________.

EXERCISE 9.5

In Problems 1 through 20, find the equation of variation using k as the constant of proportionality.

Illustration: The pressure P of a gas varies directly with its absolute temperature T and inversely with its volume V.

PROGRESS TEST ANSWERS

1. $N = kh$
2. $\frac{1}{4}$
3. $1\frac{1}{2}$
4. $f = \frac{k}{a}$
5. 4
6. 32
7. $F = kAV^2$

Solution: Since P is directly proportional to T and inversely proportional to V, $P = \frac{kT}{V}$.

1. The tension T on a spring varies directly with the distance s it is stretched.

2. The distance s a body falls in t seconds is directly proportional to the square of t.

3. The weight W of a dam varies directly with the cube of its height h.

4. The kinetic energy KE of a moving body is proportional to the square of its velocity v.

5. The weight W of a human brain is directly proportional to the body weight B.

6. The annual interest I received on a savings account varies jointly with the principal P (the amount in the account) and the interest rate r paid by the bank.

7. The cost C of a building varies jointly with the number w of workers used to build it and the cost of materials m.

8. The amount of oil A used by a ship traveling at a uniform speed varies jointly with the distance s and the square of the speed v.

9. The power P in an electric circuit varies jointly with the resistance R and the square of the current I.

10. The volume V of a rectangular container of fixed length varies jointly with its depth d and width w.

11. In a circuit with constant voltage, the current I varies inversely with the resistance R of the circuit.

12. For a wire of fixed length, the resistance R varies inversely with the square of its diameter D.

13. The intensity of illumination I from a source of light varies inversely with the square of the distance d from the source.

14. The force of attraction F between two spheres of mass m_1 and m_2, respectively, varies directly with the

product of the masses and inversely with the square of the distance d between their centers.

15. The illumination I in foot-candles upon a wall varies directly with the intensity i in candlepower of the source of light and inversely with the square of the distance d from the light.

16. The strength S of a horizontal beam of rectangular cross section and of length L varies jointly with the breadth b and the square of the depth d and inversely with the length L.

17. The electrical resistance R of a wire of uniform cross section varies directly with its length L and inversely with its cross-sectional area A.

18. The electrical resistance R of a wire varies directly with the length L and inversely with the square of its diameter d.

19. The weight W of a body varies inversely with the square of its distance d from the center of the earth.

20. z varies directly with the cube of x and inversely with the square of y.

In Exercises 21 through 30, solve the given problem.

Illustration: The horsepower HP that a rotating shaft can safely transmit varies jointly with the cube of its diameter d and the number of revolutions R it makes per minute. It is known that a 2-inch shaft rotating at 1200 revolutions per minute can safely transmit 288 HP.

(a) Find k.

(b) What horsepower can a 1.5 inch shaft safely transmit at 1800 revolutions per minute?

Solution: (a) From the information given, $HP = kd^3R$; when $d = 2$ and $R = 1200$, $HP = 288$, since a 2-inch shaft rotating at 1200 revolutions per minute can transmit 288 HP. Substituting these values in the equation $HP = kd^3R$, we obtain $288 = k(2)^3(1200)$; that is, $k = \dfrac{288}{(8)(1200)} = \dfrac{3}{100}$.

(b) Since we know from part (a) that $k = \frac{3}{100}$, the formula $HP = kd^3R$ can be written as $HP = \frac{3}{100}d^3R$. We are asked to find HP when $d = 1.5$ and $R = 1800$. Substituting these values in the formula, we have

$$HP = \frac{3}{100}(1.5)^3(1800)$$

$$= (3)(1.5)^3(18) = 182.25$$

Thus, the 1.5 inch-shaft can safely transmit 182.25 horsepower at 1800 revolutions per minute.

21. If the body in Problem 2 falls 16 feet in one second, (a) find k, and (b) state how far the body will fall in t seconds.

22. If in Problem 5 (a) it is known that a 120-pound person has a brain weighing 3 pounds, find k, and (b) determine the weight of the brain of a person weighing 200 pounds.

23. If in Problem 7 of the Progress Test (p. 523) it is known that the wind force on 1 square foot of surface is 1.8 pounds when the wind is blowing at 20 mph, (a) find k, and (b) ascertain the force exerted on a 2-foot-square surface when the wind velocity is 45 miles per hour.

24. If the ship in Problem 8 uses 500 barrels of oil in traveling 200 miles at 20 mph, (a) find k, and (b) determine how many barrels of oil are used in traveling 250 miles at 16 mph.

25. If in Problem 16 it is known that a 2-by-4-inch beam 6 feet long and resting on the 2-inch side (b in our equation) will safely support 800 pounds, first find k, and then calculate the safe load that the beam will support when resting on the 4-inch side.

26. The pressure P of a gas varies directly with its absolute temperature T and inversely with its volume V. A gas whose volume is 500 cubic feet and whose temperature is 320° absolute has a pressure of 180 pounds per square inch. The gas is allowed to expand until its pressure is 15 pounds per square inch and its temperature is 300° absolute. What is its final volume?

27. The pressure P exerted by a liquid at a point varies directly with the depth d of the point below the surface of the liquid. If a liquid exerts a pressure of 120 pounds per square foot at a depth of 30 feet, (a) find k, and (b) calculate the pressure that would be exerted at a depth of 50 feet.

28. The distance d in miles a person can see to the horizon from a point h feet above the surface of the earth varies approximately with the square root of the height. If for a height of 600 feet the horizon is 30 miles distant, how far is the horizon from a point which is 864 feet high? (Hint: $\sqrt{864} \approx 29.39$ and $\sqrt{600} \approx 24.49$.)

29. If in the preceding problem the original height is quadrupled, how much farther can the observer see?

30. The force of attraction between two spheres varies directly with the product of their masses and inversely with the square of the distance between their centers. If for a given distance the force of attraction between two spheres is 40 dynes, what is the attractive force when the distance is quadrupled?

SELF-TEST–CHAPTER 9

1. Consider the relation $\{(-2, 2), (-3, 2), (2, 4)\}$. Find:

   ______________ (a) The domain of this relation.

   ______________ (b) The range of this relation.

   ______________ (c) If the relation is a function.

---

2. Use the accompanying rectangular coordinate system to graph the function $f(x) = x^2 + 1$.

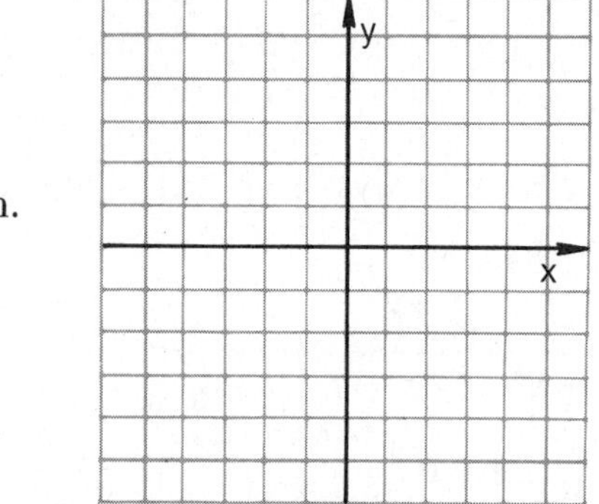

______________ (a) Find the domain of this function.

______________ (b) Find the range of this function.

---

3. Use the accompanying rectangular coordinate system to sketch the graph of $f(x) = 2x^2 + 3x + 1$. After you sketch the graph, discuss the nature of the roots.

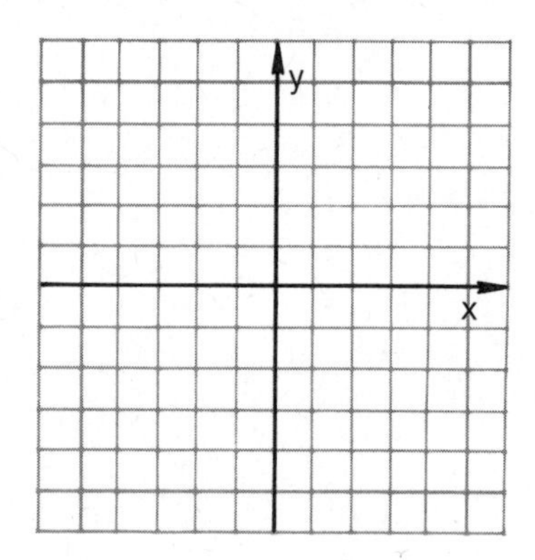

---

4. Consider the relation $f = \{(3, 1), (1, 3), (2, 7)\}$.

   ______________ (a) Find the inverse of this relation.

   ______________ (b) Find the domain of the inverse.

   ______________ (c) Find the range of the inverse.

   ______________ (d) Is f a function?

   ______________ (e) Is f^{-1}, the inverse, a function?

---

5. Consider the function $f(x) = 3x - 3$.

_______________ (a) Is f one-to-one?

_______________ (b) Find $f^{-1}(x)$.

_______________ (c) Is f^{-1} a function?

---

6. Consider the function $f(x) = 2x - 4$.

(a) Graph f(x).

(b) Graph $f^{-1}(x)$.

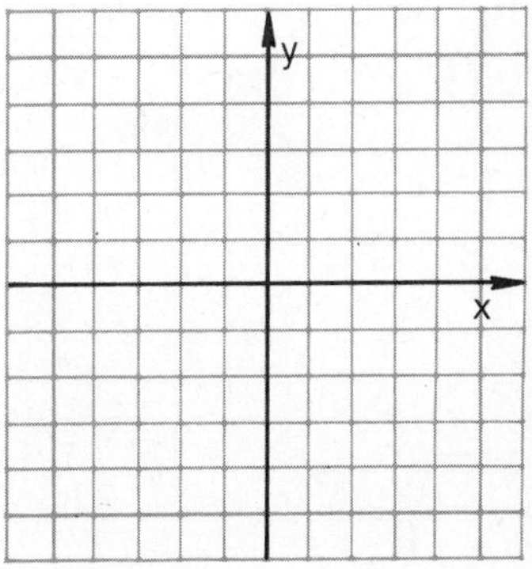

---

7. A variable y varies directly with the square of another variable x. If k is the constant of proportionality,

_______________ (a) Find the equation of variation.

_______________ (b) If it is known that when x is 2 units, y is 8 units, find k.

---

8. The intensity of illumination I from a light source varies inversely with the square of the distance d from the source.

_______________ (a) Find the equation of variation if it is known that a light source has an intensity of 100 candle power at 5 meters.

_______________ (b) What is the intensity of the source at a distance of 1 meter?

---

9. A variable z varies jointly with x and the square of y. Use k as the constant of proportionality.

_______________ (a) Find the equation of variation.

_______________ (b) If z is 6 when x and y are 1 and 4, respectively, find k.

---

10. The total cost C of producing a product is directly proportional to the cost per unit u, and the number n of units produced.

_______________ (a) If it costs $1,000 to produce 5 units whose individual cost is $100, find the constant of proportionality.

_______________ (b) How much will it cost to produce 10 units costing $200 each?

CHAPTER 10

HISTORICAL NOTE

Beginning in the seventeenth century, there was an increased demand by both science and commerce for computations to be performed quickly and accurately. These demands were met by four inventions: the Hindu-Arabic notation, decimal fractions, logarithms, and computers. In this chapter we shall consider the third of these inventions, logarithms. A logarithm is simply the exponent of a number and indicates to what power the number must be raised in order to produce another given number. By being acquainted with logarithms it is possible to save a great amount of time (and to avoid sheer boredom) when performing arduous multiplications and divisions. This was exactly the intent of the inventor of logarithms, John Napier. His ideas about logarithms were published in 1614 in a book entitled *Mirifici logarithmorum canonis descriptio* ("A description of the wonderful law of logarithms").

Napier was an eccentric and consequently many (probably unfounded) anecdotes about him have survived. One of these stories relates how Napier became irritated by his neighbor's pigeons eating his grain. He threatened to impound the birds if his neighbor did not restrict them, but the neighbor refused to cooperate, saying that if Napier could catch the pigeons, he could keep them. The next day the surprised neighbor observed his pigeons staggering about on Napier's lawn and Napier calmly putting them into a large sack. He had rendered them drunk by scattering some brandy-soaked corn about his lawn!

On another occasion, Napier announced that his black rooster was psychic and would identify the servant who had been stealing from the estate. The servants were sent one by one to a dark room and were ordered to pat the rooster on the back. Unknown to the servants, Napier had coated the bird with coal, so that when the servants touched it their hands would become black. Of course, the guilty servant, fearing to touch the rooster, returned with clean hands.

EXPONENTIAL AND LOGARITHMIC FUNCTIONS

Napier invented a set of rods that could be used to multiply numbers without using the multiplication tables. (Photo of Napier's rods courtesy of International Business Machines Corporation.)

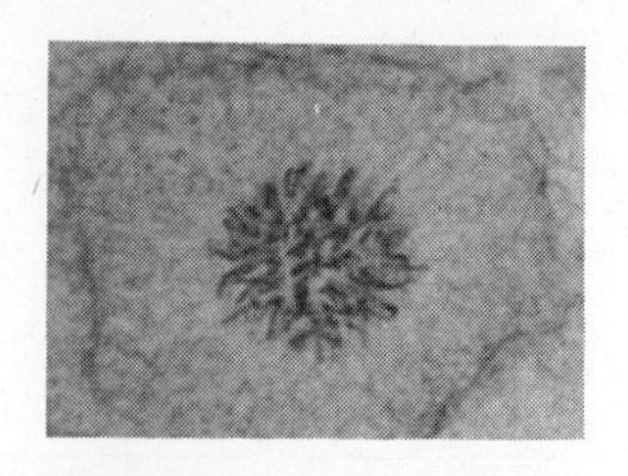
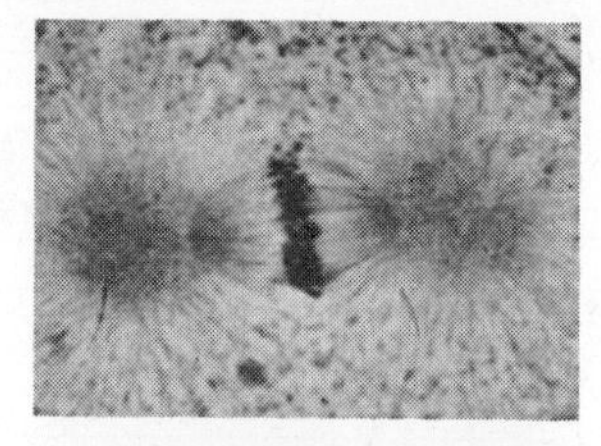
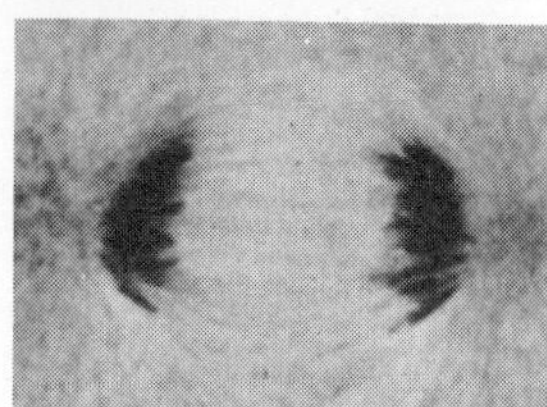
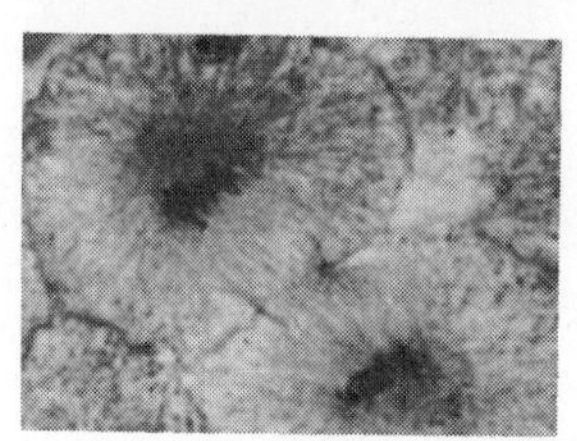

Photo credit: General Biological Supply.

OBJECTIVES 10.1

After studying this section, the reader should be able to:

1. **Graph functions of the form a^x and a^{-x}.**
2. **Graph the exponential function e^x and e^{-x}.**
3. **Determine if a given exponential function is increasing or decreasing by studying its graph.**

10.1 EXPONENTIAL FUNCTIONS

The photograph at the beginning of this section shows a cell reproducing by a process called *mitosis*. In mitosis, a single cell or bacterium divides and forms two identical daughter cells. Each daughter cell then doubles in size and divides. As you can see, the number of bacteria present is a function of time. If we start with one cell and assume that each cell divides after 10 minutes, then the number of bacteria present after t minutes is given by

$$(1)\quad f(t) = 2^{t/10}$$

For example, at the beginning of the process—that is, when $t = 0$—the number of bacteria is

$$f(0) = 2^0 = 1$$

After 10 minutes we have 2 bacteria, and each of these divides into 2 more after the next 10 minutes. Thus, we have $2 \cdot 2 = 2^2$ bacteria at the end of 20 minutes.

After 10 minutes, the number present is

$$f(10) = 2^{10/10} = 2^1 = 2$$

After 20 minutes, the number is

$$f(20) = 2^{20/10} = 2^2 = 4$$

The function $f(t) = 2^{t/10}$ is an example of an *exponential function*, so called because the variable t is an exponent. The following are also examples of exponential functions

(2) $f(x) = 2^x$

(3) $f(t) = \left(\frac{1}{2}\right)^t$

(4) $y = (0.2)^x$

In general, an exponential function is a function defined by

Definition of exponential function.

$$f(x) = b^x, \quad b > 0, b \neq 1$$

In this definition, b is a constant called the *base*, and the *exponent x* is the *variable*.

The exponential function defined by $f(t) = 2^{t/10}$ can be graphed and used to predict the number of bacteria present after a period of time t. To make this graph, we first construct a table giving the value of the function for certain convenient times, as shown.

t	0	10	20	30
$f(t) = 2^{t/10}$	1	2^1	2^2	2^3

The corresponding points can then be graphed and joined with a smooth curve, as shown in Figure 10.1.

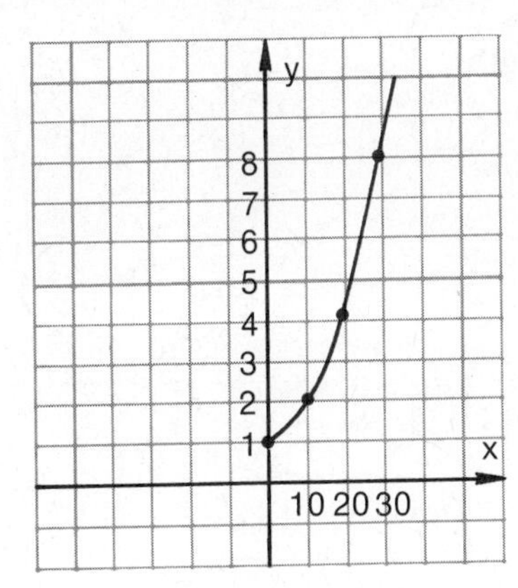

FIGURE 10.1

We must remark here that $2^{t/10}$ has been defined for all rational values of $\frac{t}{10}$ but not for *all* real values. For example, for $t = 20$, we obtain 2^2, which has been defined. Likewise, for $t = -30$, 2^{-3} is also defined. But what does $2^{\sqrt{2}/10}$ mean? At this time we cannot answer this question completely. We can only point out that if x and y are rational numbers such that $x > y$, then

$$(5)\quad b^x > b^y \quad \text{for } b > 1$$

Since a decimal approximation of $\frac{\sqrt{2}}{10} = \frac{1.4142\ldots}{10} = 0.14142\ldots$ can be obtained accurate to any number of decimal places, we can conclude from (5) that because

$$0.1 < \frac{\sqrt{2}}{10} < 0.2, \text{ then } 2^{0.1} < 2^{\sqrt{2}/10} < 2^{0.2}$$

Also, because

$$0.14 < \frac{\sqrt{2}}{10} < 0.15, \text{ then } 2^{0.14} < 2^{\sqrt{2}/10} < 2^{0.15}$$

and because

In advanced mathematics it is proved that $|b^m - b^n|$ becomes arbitrarily small when $|m - n|$ is taken small enough.

$$0.141 < \frac{\sqrt{2}}{10} < 0.142, \text{ then } 2^{0.141} < 2^{\sqrt{2}/10} < 2^{0.142}$$

It is clear that we can continue this process indefinitely, and that by so doing the difference between the number at the left and the number at the right of the inequality can be made as small as we wish. Consequently, we assume that there is only one number, $2^{\sqrt{2}/10}$, that will satisfy each inequality when the process is carried out indefinitely. Since this argument can be applied to any irrational exponent x, we see that b^x can be defined for any real value of x. With this technicality resolved, we can proceed to graph exponential functions.

Example 1

Graph the given functions on the same coordinate system:
(a) $f(x) = 2^x$
(b) $f(x) = \left(\frac{1}{2}\right)^x$

Solution

(a) We first make a table with convenient values for x, and then find the corresponding values for f(x), as shown

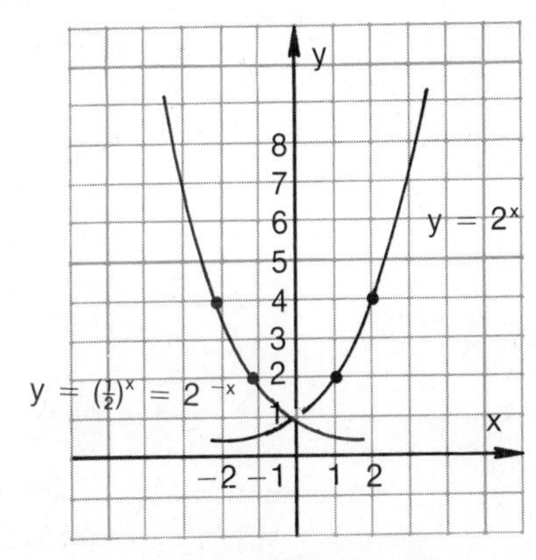

FIGURE 10.2

below. We graph the points and connect them with a smooth curve, as shown in black in Figure 10.2.

x	−2	−1	0	1	2
$f(x) = 2^x$	$2^{-2} = \frac{1}{4}$	$2^{-1} = \frac{1}{2}$	$2^0 = 1$	$2^1 = 2$	$2^2 = 4$

(b) We first note that $f(x) = \left(\frac{1}{2}\right)^x = (2^{-1})^x = 2^{-x}$. We then use suitable values of x to find the values of the function, as shown in the table below. Note that, for example, for $x = -2$, $f(-2) = 2^{-(-2)} = 2^2$, and for $x = 2$, $f(2) = 2^{-2} = \frac{1}{4}$.

x	−2	−1	0	1	2
$f(x) = \left(\frac{1}{2}\right)^x$	2^2	2^1	1	$\frac{1}{2}$	$\frac{1}{4}$

The graph of $f(x) = \left(\frac{1}{2}\right)^x = 2^{-x}$ is shown in color in Figure 10.2.

Note that the graph of $y = \left(\frac{1}{2}\right)^x$ goes *down* to the *right* and thus $\left(\frac{1}{2}\right)^x$ is a *decreasing* function. On the other hand, the graph of $y = 2^x$ goes up to the right and hence 2^x is an *increasing* function.

If the graph of a function goes down to the right, the function is decreasing. If the function goes up to the right, the function is increasing.

In our definition of $f(x) = b^x$, it was required only that the base b be a positive number other than 1. However, in mathematics, science, and engineering there is a *particular* value of b which is of great importance. This value is an

≈ means approximately equal to.

irrational number denoted by e. The value of e is approximately 2.718282, expressed as $e \approx 2.718282$. The reasons for using e as a base are made clear in more advanced mathematics courses, but for now we shall note only that e can be approximated by $\left(1+\frac{1}{n}\right)^n$ where n is sufficiently large. For example,

$$\text{when } n = 100,\ \left(1+\frac{1}{n}\right)^n = 2.704814$$

$$\text{when } n = 1{,}000,\ \left(1+\frac{1}{n}\right)^n = 2.716924$$

$$\text{when } n = 10{,}000,\ \left(1+\frac{1}{n}\right)^n = 2.718146$$

$$\text{when } n = 100{,}000,\ \left(1+\frac{1}{n}\right)^n = 2.718255$$

$$\text{when } n = 1{,}000{,}000,\ \left(1+\frac{1}{n}\right)^n = 2.718282$$

From this point on, we shall define the *exponential function* by

$$f(x) = e^x \quad \text{or} \quad \exp(x) = e^x \quad \text{or} \quad \{(x, y) \mid y = e^x\}$$

Example 2

Use the values in the given tables to graph (on the same coordinate system) $f(x) = e^x$ and $f(x) = e^{-x}$.

x	-2	-1	0	1	2
e^x	0.1353	0.3679	1	2.7183	7.3891

x	-2	-1	0	1	2
e^{-x}	7.3891	2.7183	1	0.3679	0.1353

Solution

Plotting the given values, we obtain the graphs of $f(x) = e^x$ and $f(x) = e^{-x}$ shown in Figure 10.3.

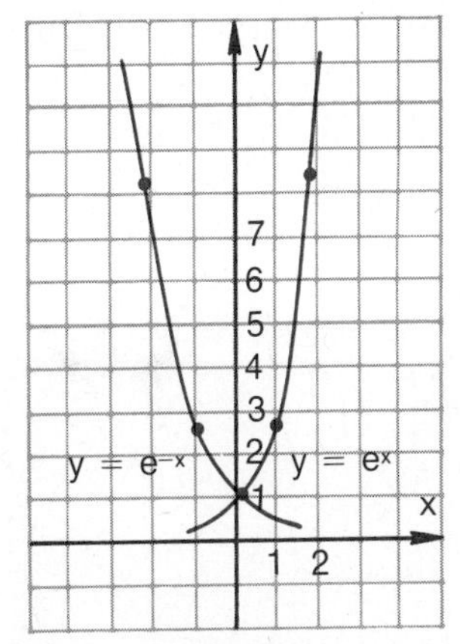

FIGURE 10.3

PROGRESS TEST

1. Graph the function $f(x) = 3^x$.

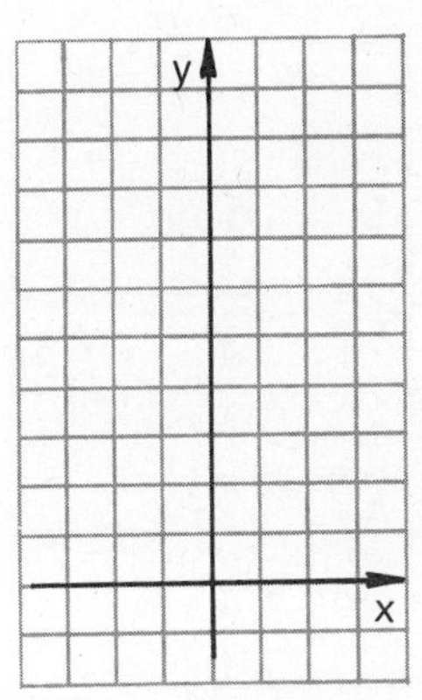

2. Is the function $f(x) = 3^x$ increasing or decreasing?

3. Graph the function $f(x) = 3^{-x}$.

4. Is the function $f(x) = 3^{-x}$ increasing or decreasing?

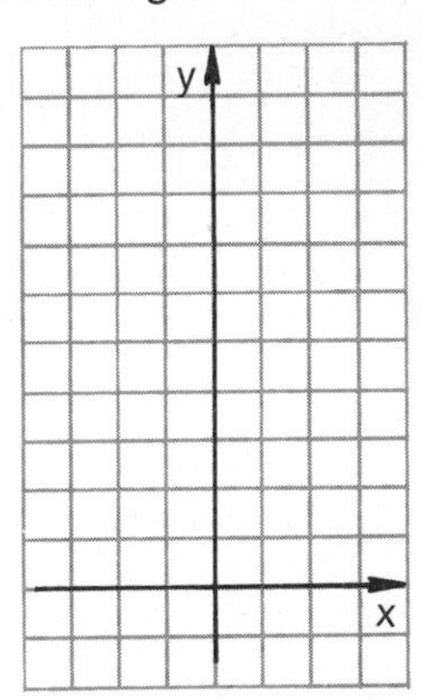

EXERCISE 10.1

In Problems 1 through 6, find the second component of the number pair that satisfies the given equation.

Illustration: $y = 4^{-x}$
(a) (−1,)
(b) (0,)
(c) (1,)

Solution: (a) For $x = -1$, $y = 4^{-(-1)} = 4$.
Thus, $(-1, 4)$ satisfies the equation.

(b) For $x = 0$, $y = 4^0 = 1$.
Thus, $(0, 1)$ satisfies the equation.

(c) For $x = 1$, $y = 4^{-1} = \frac{1}{4}$.

Thus, $\left(1, \frac{1}{4}\right)$ satisfies the equation.

1. $y = 5^x$
(a) (−1,)
(b) (0,)
(c) (1,)

2. $y = \left(\frac{1}{5}\right)^x = 5^{-x}$
(a) (−1,)
(b) (0,)
(c) (1,)

3. $y = 6^x$
(a) (−1,)
(b) (0,)
(c) (1,)

4. $y = \left(\frac{1}{6}\right)^x = 6^{-x}$
(a) (−1,)
(b) (0,)
(c) (1,)

5. $y = 10^x$
(a) (−1,)
(b) (0,)
(c) (1,)

6. $y = \left(\frac{1}{10}\right)^x = 10^{-x}$
(a) (−1,)
(b) (0,)
(c) (1,)

In Problems 7 through 12, use the corresponding information from problems 1 to 6 to graph the given functions.

Illustration: $y = 4^{-x}$

Solution: From the previous illustration, $(-1, 4)$, $(0, 1)$, and $\left(1, \frac{1}{4}\right)$ satisfy the given equa-

PROGRESS TEST ANSWERS

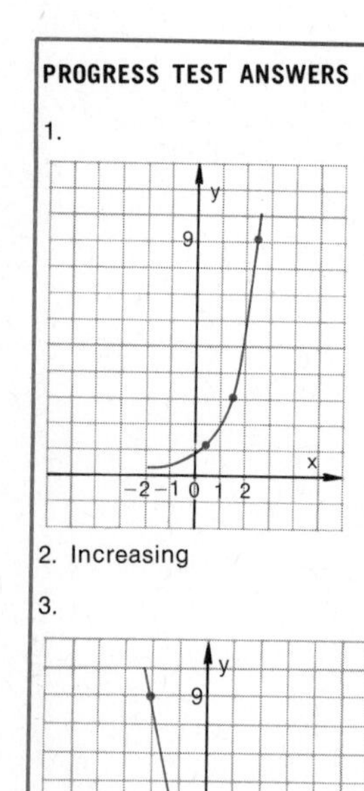

4. Decreasing.

tion. Plotting these points and joining them with a curve, we obtain the graph of $y = 4^{-x}$, as shown in Figure 10.4.

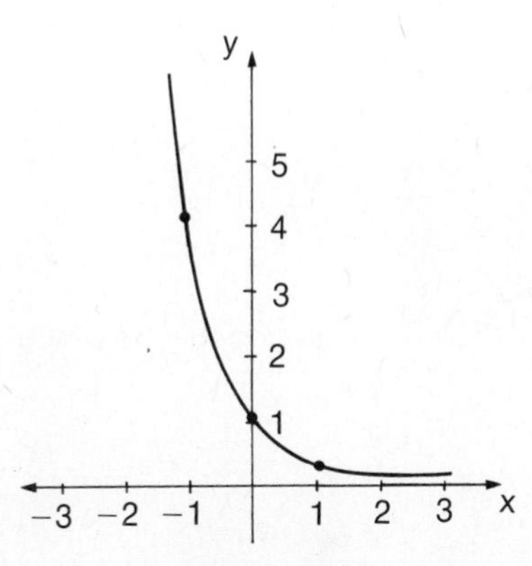

FIGURE 10.4

7. $y = 5^x$

8. $y = \left(\frac{1}{5}\right)^x = 5^{-x}$

9. $y = 6^x$

10. $y = \left(\frac{1}{6}\right)^x = 6^{-x}$

11. $y = 10^x$

12. $y = \left(\frac{1}{10}\right)^x = 10^{-x}$

In Problems 13 through 20, graph the given function and state whether the function is increasing or decreasing. (For problems 17 to 20 you can use the table in Example 2.)

13. $y = \left(\frac{1}{2}\right)^{-x}$

14. $y = \left(\frac{1}{3}\right)^{-x}$

15. $y = \left(\frac{1}{5}\right)^{-x}$

16. $y = \left(\frac{1}{10}\right)^{-x}$

17. $y = e^{2x}$

18. $y = e^{3x}$

19. $y = e^{-2x}$

20. $y = e^{-3x}$

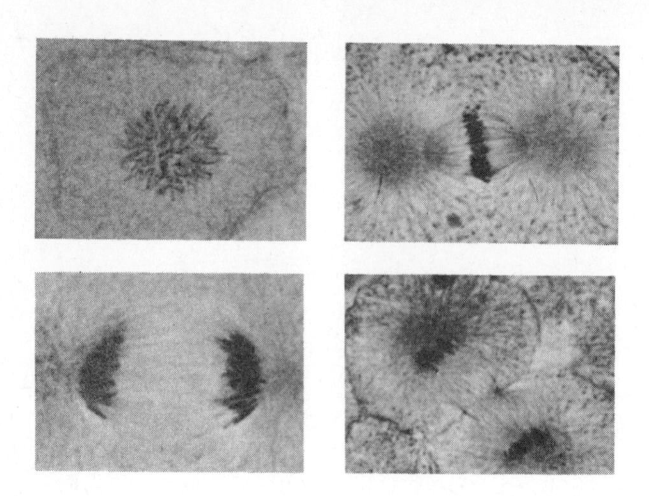

Photo credit: General Biological Supply.

OBJECTIVES 10.2

After studying this section, the reader should be able to:

1. **Write an equation involving exponents in its equivalent logarithmic form.**
2. **Write a logarithmic equation in exponential form.**
3. **Find the logarithm of a number to a specified base by writing the given logarithmic equation in exponential form.**

10.2 LOGARITHMS

The photo at the beginning of the section again shows a cell reproducing by *mitosis*. If we now assume that these cells divide every hour, the number of cells present is still a function of the time, as shown in the table below.

Time (in hours)	0	1	2	3	4	5	6	7	8...
Number (of cells)	1	2	4	8	16	32	64	128	256...

A brief inspection of this table shows that each number in the second row can be obtained by applying the corresponding number in the first row as an exponent to the base 2. For example, in the fourth column of the table, the number 3 in the first row is paired with 8 in the second row, and $2^3 = 8$. Similarly, $2^5 = 32$, $2^7 = 128$, and so on. Thus, if y stands for a number in the second row and x for the corresponding number in the first row, we can write

(1) $y = 2^x$

where y represents a number and x is the corresponding exponent if the base is 2. In somewhat different words, we say "x is the *logarithm* of y to the base 2." It is important to notice that the statement "x is the logarithm of y to the base 2" means simply that x is the exponent that must be applied to the base 2 to yield the number y. Thus, since $8 = 2^3$, the *logarithm* of 8 to the base 2 is 3. Notice that in formula (1), $x = 3$ is the *exponent* that must be applied to the *base* 2 to give the number $y = 8$. For the sake of brevity, we shall write "the logarithm of 8 to the base 2 is 3" as $\log_2 8 = 3$. We can clearly see from our table that

Here we are substituting the word logarithm for exponent.

This is so because $2^3 = 8$.

$$\log_2 16 = 4 \quad \text{because} \quad 2^4 = 16$$
$$\log_2 2 = 1 \quad \text{because} \quad 2^1 = 2$$
$$\log_2 32 = 5 \quad \text{because} \quad 2^5 = 32$$

Example 1

Use the table to find:
(a) $\log_2 64$
(b) $\log_2 256$

Solution

(a) $\log_2 64$ is the *exponent* of 2 in the equation $2^6 = 64$, that is, $\log_2 64 = 6$ because $2^6 = 64$.
(b) $\log_2 256$ is the *exponent* of 2 in the equation $2^8 = 256$, that is, $\log_2 256 = 8$ because $2^8 = 256$.

Clearly, we can find the logarithm of many numbers to a given base by writing the corresponding exponential relationship. For example,

$$\log_{10} 100 = 2 \quad \text{because} \quad 10^2 = 100$$
$$\log_3 27 = 3 \quad \text{because} \quad 3^3 = 27$$
$$\log_{27} 9 = \frac{2}{3} \quad \text{because} \quad 27^{2/3} = 9$$

In general, we have the following definition:

DEFINITION 10.1

$\log_b x = y$ means $x = b^y$.
(Thus, $\log_4 64 = 3$ because $4^3 = 64$, and $\log_5 625 = 4$ because $5^4 = 625$.)

Example 2 Write the equation $5^3 = 125$ in logarithmic form.

Solution By Definition 10.1, $x = b^y$ means $\log_b x = y$. Thus, $125 = 5^3$ means $\log_5 125 = 3$

Example 3 Write the equation $\log_{32} 64 = \frac{6}{5}$ in exponential form and check its accuracy.

Solution By Definition 10.1, $\log_b x = y$ means $x = b^y$. Hence $\log_{32} 64 = \frac{6}{5}$ means $32^{6/5} = 64$. Because $32^{6/5} = (\sqrt[5]{32})^6 = 2^6 = 64$, the equation $\log_{32} 64 = \frac{6}{5}$ is correct.

Example 4 Find:
(a) $\log_{27} 81$
(b) $\log_{10} 0.01$

Solution (a) $\log_{27} 81 = y$ means $81 = 27^y$, or, equivalently,

$$\begin{aligned} 3^4 &= (3^3)^y \\ &= 3^{3y} \end{aligned}$$

Since $3^4 = 3^{3y}$, $4 = 3y$, and $y = \frac{4}{3}$. Thus,

$$\log_{27} 81 = \frac{4}{3}$$

(b) $\log_{10} 0.01 = y$ means $0.01 = 10^y$, or, equivalently,

$$\begin{aligned} 10^{-2} &= 10^y \\ y &= -2 \end{aligned}$$

Thus, $$\log_{10} 0.01 = -2$$

PROGRESS TEST

1. $\log_2 128$ equals ______.

2. When expressed in logarithmic form, the equation $5^{-2} = \frac{1}{25}$ is written as ______________________________________.

3. When written in exponential form, the equation $\log_{216} 36 = \frac{2}{3}$ is written as ______________________________.

4. $\log_3 \frac{1}{27}$ equals ______.

EXERCISE 10.2

In Problems 1 through 8, follow the procedure of Example 1 to find the value of each of the following logarithms.

1. $\log_2 256$
2. $\log_9 81$
3. $\log_{11} 121$
4. $\log_7 49$
5. $\log_8 128$
6. $\log_9 243$
7. $\log_2 \frac{1}{64}$
8. $\log_{10} 1{,}000$

In Problems 9 through 20, follow the procedure of Example 2 to write each of the following equations in logarithmic form.

9. $2^7 = 128$
10. $3^4 = 81$
11. $81^{1/2} = 9$
12. $16^{1/2} = 4$
13. $10^3 = 1{,}000$
14. $10^{-3} = 0.001$
15. $216^{1/3} = 6$
16. $64^{1/6} = 2$
17. $10^{0.47712} = 3$
18. $10^{1.30103} = 20$
19. $N = b^5$
20. $M = a^{-4}$

PROGRESS TEST ANSWERS

1. 7, since $2^7 = 128$
2. $\log_5 \frac{1}{25} = -2$
3. $(216)^{2/3} = 36$
4. -3, since $3^{-3} = \frac{1}{27}$

In Problems 21 through 28, follow the procedure of Example 3 to write each of the following equations in exponential form.

21. $\log_9 729 = 3$

22. $\log_7 343 = 3$

23. $\log_2 \frac{1}{256} = -8$

24. $\log_5 \frac{1}{125} = -3$

25. $\log_{81} 27 = \frac{3}{4}$

26. $\log_{625} 5 = \frac{1}{4}$

27. $\log_{10} 300 = 2.47712$

28. $\log_{10} 2000 = 3.30103$

In Problems 29 through 50, use Definition 10.1 to find the value of the letter in each of the following equations.

Illustration: (a) $\log_{16} y = 1.75$

(b) $\log_z 9 = \frac{1}{3}$

Solution: (a) By Definition 10.1, $\log_b x = y$ means $x = b^y$. Thus, $\log_{16} y = 1.75$ means $y = (16)^{1.75} = (16)^{1+3/4} = (16)^{7/4}$. Since $16 = 2^4$ and $y = (16)^{7/4}$, $y = (2^4)^{7/4} = 2^7 = 128$.

(b) By Definition 10.1, $\log_z 9 = \frac{1}{3}$ means $z^{1/3} = 9$, or $\sqrt[3]{z} = 9$. Thus, z must be a number whose cube root is 9, that is, $z = 9^3 = 729$. You can also argue that since $z^{1/3} = 9$, and we want z^1, we must cube both sides of the equation: $[z^{1/3}]^3 = 9^3$, or $z = 729$.

29. $\log_3 x = 27$

30. $\log_4 x = 5$

31. $\log_{25} z = \frac{5}{2}$

32. $\log_9 z = \frac{7}{2}$

33. $\log_{16} x = 1.75$

34. $\log_{10} R = -5$

35. $\log_{10} N = 5$

36. $\log_{32} V = 0.8$

37. $\log_{36} Q = -2.5$

38. $\log_{27} w = \frac{5}{3}$

39. $\log_{64} x = \frac{-7}{6}$

40. $\log_x 64 = 2$

41. $\log_z 32 = 5$

42. $\log_z 625 = 4$

43. $\log_a 128 = 7$

44. $\log_a 15 = \frac{1}{2}$

45. $\log_k 7 = \frac{1}{3}$

46. $\log_m 2 = \frac{1}{6}$

47. $\log_b 27 = \frac{-3}{4}$

48. $\log_b 1000 = \frac{3}{2}$

49. $\log_c 0.1 = \frac{-1}{2}$

50. $\log_c 0.01 = \frac{-1}{2}$

Photo credit: Steve McCutcheon, Alaska Pictorial Service.

OBJECTIVES 10.3

After studying this section, the reader should be able to:

1. **Evaluate the logarithm of a given number by using the three theorems given in the text.**
2. **Evaluate the logarithms of some appropriate numbers whose values are specified.**
3. **Transform the left member of a logarithmic equation into the right member of the equation by using the three theorems given in the text.**

10.3 PROPERTIES OF LOGARITHMS

The photograph at the beginning of the section shows part of the damage done by the 1964 earthquake in Anchorage, Alaska. If I_0 denotes the minimum intensity of an earthquake (used for comparison purposes), this earthquake was $10^{8.4}$ times as intense as the minimum—that is, the intensity of the 1964 earthquake was $I_0 \cdot 10^{8.4}$. Actually, the intensity or magnitude of earthquakes is measured on the Richter scale, a scale that operates on a logarithmic basis. On this scale, the magnitude R of an earthquake is expressed as

$$R = \log_{10} \frac{I}{I_0}$$

Thus, on the Richter scale, the Anchorage earthquake had a magnitude

To find $\log_{10} 10^{8.4}$, we have to recall that $\log_{10} 10^{8.4} = y$ means that $10^y = 10^{8.4}$; that is, $y = 8.4$.

$$R = \log_{10} \frac{I_0 \bullet 10^{8.4}}{I_0} = \log_{10} 10^{8.4} = 8.4$$

This problem would have been easier to solve if we had known a theorem which is a restatement of one of the laws of exponents. The theorem supplies this equation:

$$\log_b M^r = r(\log_b M)$$

Applying the equation, we get

$$\log_{10} 10^{8.4} = 8.4(\log_{10} 10) = (8.4)(1) = 8.4$$

We now state this theorem and two others which will help in computations involving logarithms. *These theorems are valid for positive real numbers b, M, and N and all real numbers r.*

THEOREM 10.1

$\log_b MN = \log_b M + \log_b N$

PROOF:

Let $x = \log_b M$ and $y = \log_b N$. Then
$b^x = M$ and $b^y = N$.
Multiplying $MN = b^x \bullet b^y = b^{x+y}$.
Thus, $\log_b MN = x + y = \log_b M + \log_b N$.

THEOREM 10.2

$\log_b \frac{M}{N} = \log_b M - \log_b N$

PROOF:

Let $x = \log_b M$ and $y = \log_b N$. Then
$b^x = M$ and $b^y = N$.

Dividing, $\frac{M}{N} = \frac{b^x}{b^y} = b^{x-y}$.

Thus, $\log_b \frac{M}{N} = x - y = \log_b M - \log_b N$.

THEOREM 10.3

$\log_b M^r = r(\log_b M)$
(The proof of this theorem is left to the student.)

Example 1

In 1626, Peter Minuit bought the island of Manhattan from the Indians for goods worth 60 Dutch guilders, or approximately \$24. If this money were invested at 5% compounded annually, at the end of one year the money would be worth \$24(1.05); at the end of two years, $\$24(1.05)^2$; at the end of three years, $\$24(1.05)^3$. By the time of the bicentennial, 1976, the money would be worth $\$24(1.05)^{350}$. If it is known that $\log_{10} 24 = 1.3802$ and that $\log_{10} 1.05 = 0.0212$, find $\log_{10} 24(1.05)^{350}$.

Solution

By the way, the amount $24(1.05)^{350}$ is approximately $\$6.26 \times 10^8$.

$$\begin{aligned}
\log_{10} 24(1.05)^{350} &= \log_{10} 24 + \log_{10}(1.05)^{350} && \text{By Theorem 10.1}\\
&= \log_{10} 24 + 350(\log_{10} 1.05) && \text{By Theorem 10.3}\\
&= 1.3802 + 350(0.0212)\\
&= 1.3802 + 7.4200\\
&= 8.8002
\end{aligned}$$

Example 2

Show that

$$\log_b \sqrt{\frac{MN}{PQ}} = \frac{1}{2}(\log_b M + \log_b N - \log_b P - \log_b Q)$$

Solution

$$\begin{aligned}
\log_b \sqrt{\frac{MN}{PQ}} &= \log_b \left(\frac{MN}{PQ}\right)^{1/2} && \text{Since } \sqrt{a} = a^{1/2}\\
&= \frac{1}{2}\left(\log_b \frac{MN}{PQ}\right) && \text{By Theorem 10.3}\\
&= \frac{1}{2}(\log_b MN - \log_b PQ) && \text{By Theorem 10.2}\\
&= \frac{1}{2}(\log_b M + \log_b N - \log_b P && \text{By Theorem 10.1}\\
&\quad - \log_b Q)
\end{aligned}$$

PROGRESS TEST

If it is known that $\log_{10} 2 = 0.301$, $\log_{10} 3 = 0.477$, and $\log_{10} 5 = 0.699$, then

1. $\log_{10} 6 =$ __________
2. $\log_{10} \frac{5}{2} =$ __________
3. $\log_{10} 5^4 =$ __________
4. $\log_{10} 36 =$ __________
5. $\log_{10} \frac{16}{75} =$ __________

EXERCISE 10.3

1. The worst earthquake ever recorded occurred in the Pacific Ocean near Colombia. The intensity of this earthquake was $10^{8.9}$ as great as that of an earthquake of minimum intensity I_0. What was the magnitude of this earthquake on the Richter scale?

2. The San Francisco earthquake of 1906 was $10^{8.3}$ times as intense as an earthquake of minimum intensity I_0. What was the magnitude of the San Francisco earthquake on the Richter scale?

3. The Los Angeles earthquake of 1971 had an intensity $10^{6.7}$ times as great as that of an earthquake of minimum intensity I_0. What was its magnitude on the Richter scale?

4. The Maiano, Italy, earthquake of 1976 had an intensity $10^{6.9}$ times as great as that of an earthquake of minimum intensity I_0. What was its magnitude on the Richter scale?

In Problems 5 through 20, find the value of each of the logarithms listed if it is known that $\log_{10} 2 = 0.301$, $\log_{10} 3 = 0.477$, and $\log_{10} 5 = 0.699$. In all of these problems, the base is 10.

Illustration: $\log \frac{\sqrt{24}}{5}$

PROGRESS TEST ANSWERS

1. $\log_{10} 6 = \log_{10} 3 + \log_{10} 2$
$= 0.477 + 0.301$
$= 0.778$

2. $\log_{10} \frac{5}{2} = \log_{10} 5 - \log_{10} 2$
$= 0.699 - 0.301$
$= 0.398$

3. $\log_{10} 5^4 = 4(\log_{10} 5)$
$= 4(0.699)$
$= 2.796$

4. $\log_{10} 36 = \log_{10} (2 \cdot 3)^2$
$= 2(\log_{10} 2 \cdot 3)$
$= 2(\log_{10} 2 + \log_{10} 3)$
$= 2(0.301 + 0.477)$
$= 1.556$

5. $\log_{10} \frac{16}{75} = \log_{10} 16 - \log_{10} 75$
$= \log_{10} 2^4 - \log_{10} 3 \cdot 5^2$
$= 4(\log_{10} 2) - \log_{10} 3 - \log_{10} 5^2$
$= 4(0.301) - 0.477 - 2(\log_{10} 5)$
$= -0.671$

Solution:

$$\log \frac{\sqrt{24}}{5} = \log \sqrt{24} - \log 5 \qquad \text{By Theorem 10.2}$$

$$= \log (24)^{1/2} - \log 5 \qquad \text{Since } \sqrt{a} = a^{1/2}$$

$$= \frac{1}{2} \log 24 - \log 5 \qquad \text{By Theorem 10.3}$$

$$= \frac{1}{2} \log 2^3 \cdot 3 - \log 5 \qquad \text{Since } 24 = 2^3 \cdot 3$$

$$= \frac{1}{2} [\log 2^3 + \log 3] - \log 5 \qquad \text{By Theorem 10.1}$$

$$= \frac{1}{2} [3(\log 2) + \log 3] - \log 5 \qquad \text{By Theorem 10.3}$$

$$= \frac{1}{2} [3(0.301) + 0.477] - 0.699$$

$$= \frac{1}{2} [1.380] - 0.699$$

$$= 0.690 - 0.699$$

$$= -0.009$$

5. $\log 36$

6. $\log 54$

7. $\log 135$

8. $\log 180$

9. $\log \frac{24}{5}$

10. $\log \frac{81}{64}$

11. $\log 0.75$

12. $\log 0.045$

13. $\log 18^3$

14. $\log (7.2)^3$

15. $\log \sqrt{75}$

16. $\log \sqrt[3]{240}$

17. $\log \sqrt[4]{48}$

18. $\log \sqrt[5]{7.5}$

19. $\log \sqrt{\frac{128}{1250}}$

20. $\log \sqrt[3]{\frac{64}{125}}$

In Problems 21 through 30, use the properties of logarithms to transform the left member into the right member in each of the given equations.

Illustration: $\log \frac{26}{7} - \log \frac{5}{21} + \log \frac{5}{26} = \log 3$

Solution:

$\log \frac{26}{7} - \log \frac{5}{21} + \log \frac{5}{26}$

$= [\log 26 - \log 7]$ By Theorem 10.2
$\quad - [\log 5 - \log 21]$
$\quad + [\log 5 - \log 26]$

$= \log 26 - \log 7 - \log 5 + \log 21$
$\quad + \log 5 - \log 26$

$= \log 21 - \log 7$

$= \log 3 \cdot 7 - \log 7$ Since $21 = 3 \cdot 7$

$= \log 3 + \log 7 - \log 7$ By Theorem 10.1

$= \log 3$

21. $\log \frac{26}{7} - \log \frac{15}{63} + \log \frac{5}{26} = \log 3$

22. $\log 9 - \log 8 - \log \sqrt{75} + \log \sqrt{\frac{25}{27}} = -3 \log 2$

23. $\log b^3 + \log 2 - \log \sqrt{b} + \log \frac{\sqrt{b^3}}{2} = 4 \log b$

24. $\log k^2 - \log k^{-2} - \log \sqrt{k} - \log k^{-1} = \frac{9}{2} \log k$

25. $\log k^{3/2} + \log r - \log k - \log r^{3/4} = \frac{1}{4} (\log k^2 r)$

26. $\log a - \frac{1}{6} \log b - \frac{1}{2} \log a + \frac{1}{3} \log b = \frac{1}{6} \log a^3 b$

27. $\log \left(y - \frac{1}{y^2}\right)^3 = 3 \log (y-1) + 3 \log (y^2 + y + 1) - 6 \log y$

28. $\log \frac{x^2 (x+5)^{3/2}}{x-5} = 2 \log x + \frac{3}{2} \log (x+5) - \log (x-5)$

29. $\log \frac{(x^2 - 4)\sqrt{x^2 + 2x + 4}}{(x^3 - 8)^2} = \log (x+2) - \log (x-2) -$
$\frac{3}{2} \log (x^2 + 2x + 4)$

30. $\log \left[\frac{1}{12(z-3)^3} - \frac{1}{12(z+3)^2}\right] = \log z - 2 \log (z+3) -$
$2 \log (z-3)$

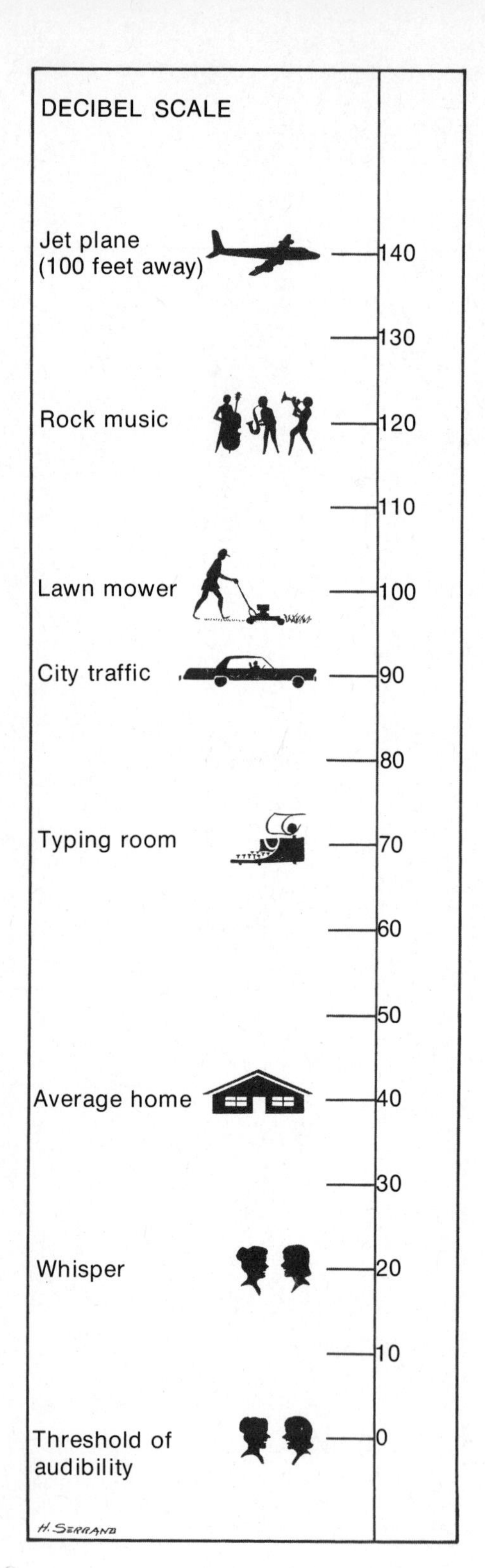

A decibel is a unit based on the faintest sound a person can hear. The scale is logarithmic, so that a sound 10 times as intense as another is 1 Bel louder.

OBJECTIVES 10.4

After studying this section, the reader should be able to:

1. Write a given number in scientific notation.

2. **Find the characteristic and the mantissa of a given number by using a table of logarithms.**
3. **Give the characteristic of a logarithm using the −10 notation.**
4. **Find the antilogarithm of a given number.**

10.4 COMPUTATION WITH COMMON LOGARITHMS

The chart at the beginning of this section shows a scale for measuring the loudness of sounds. This scale is called the decibel scale. The loudness L of a sound of intensity I measured in decibels is expressed as

$$L = 10 \log_{10} \frac{I}{I_0}$$

where I_0 is the minimum intensity detectable by the human ear. For example, the sound of a riveting machine 30 feet away is 10^{10} times as intense as the minimum intensity I_0, and hence its loudness in decibels is expressed as

$$L = 10 \log_{10} \frac{10^{10} \not{I_0}}{\not{I_0}} = 10 \log_{10} 10^{10} = (10 \cdot 10) \log_{10} 10$$

$$= 100 \text{ decibels}$$

As you can see, many of the applications of logarithms use the base 10. This system is called *the common system of logarithms*. In mathematics, it is customary to omit the base 10 in log notation for common logarithms:

Common logarithms are logarithms using base 10.

$$\log_{10} a = \log a$$

Throughout this chapter, *if no base appears with the log, assume that the base is 10.*

Before discussing the use of a common logarithm table, we shall mention the fact that any ordinary number written in ordinary decimal notation can be written as a product of a number between 1 and 10 and an appropriate power of 10. When a number is written in such a fashion, the number is said to be written in *scientific notation*. Thus,

A number is written in scientific notation as a product of a number between 1 and 10 and a power of 10.

$$3967 = 3.967 \times 10^3$$
$$384.2 = 3.842 \times 10^2$$
$$9.85 = 9.85 \times 10^0$$
$$0.74 = 7.4 \times 10^{-1}$$
$$0.00456 = 4.56 \times 10^{-3}$$

Example 1

On July 15, 1972, Pioneer 10 entered the asteroid belt, a region 280 million kilometers wide located between Mars and Jupiter. Write the 280-million in scientific notation.

Solution

280 million can be written as 280,000,000. In scientific notation,

$$280{,}000{,}000 = 2.8 \times 10^8$$

Since $\log 1 = 0$ and $\log 10 = 1$, the logarithm of any number, say N, between 1 and 10 must be a number between 0 and 1; that is, log N is a decimal with no whole number part. Thus, to find the logarithm of the number 158, we first write the number in scientific notation as 1.58×10^2, and then proceed as follows:

$$\begin{aligned} \log 158 = \log 1.58 \times 10^2 &= \log 1.58 + \log 10^2 \\ &= \log 1.58 + 2 \log 10 \\ &= \log 1.58 + 2 \end{aligned}$$

Similarly,

$$\begin{aligned} \log 15.8 = \log 1.58 \times 10 &= \log 1.58 + \log 10 \\ &= \log 1.58 + 1 \end{aligned}$$

$$\begin{aligned} \log 0.158 = \log 1.58 \times 10^{-1} &= \log 1.58 + \log 10^{-1} \\ &= \log 1.58 - 1 \end{aligned}$$

Thus, if it is known that $\log 1.58 = 0.1987$, we have:

$$\begin{aligned} \log 158 &= \log 1.58 + 2 \\ &= 0.1987 + 2 \end{aligned}$$

$$\begin{aligned} \log 15.8 &= \log 1.58 + 1 \\ &= 0.1987 + 1 \end{aligned}$$

$$\begin{aligned} \log 0.158 &= \log 1.58 - 1 \\ &= 0.1987 - 1 \end{aligned}$$

In these examples, the decimal part, 0.1987, of the common logarithm is called the *mantissa*. The exponent of 10, which is the integer part of the logarithm, is called the *characteristic*.

Example 2

Find the logarithm of 1580, and its characteristic.

Solution

$$\begin{aligned}\log 1580 = \log (1.58 \times 10^3) &= \log 1.58 + \log 10^3 \\ &= 0.1987 + 3 \\ &= 3.1987\end{aligned}$$

Since $1580 = 1.58 \times 10^3$, the characteristic is 3.

In order to keep the mantissa of a logarithm positive, we do not combine the mantissa and characteristic when the latter is negative. Thus, if

$$\begin{aligned}\log 2 &= 0.3010 \\ \log 0.002 &= 0.3010 - 3\end{aligned}$$

This difference is ordinarily left in indicated form (rather than as -2.6990, which is the actual negative value). This is done in order to make easy reference to tables that give only positive mantissas. Moreover, there is a convenient way of writing a negative logarithm so that the non-negative mantissa is indicated. For example, if we write $7 - 10$ in place of -3 in the example above, we have

$$\begin{aligned}\log 0.002 &= 0.3010 - 3 \\ &= 0.3010 + (7 - 10) \\ &= 7.3010 - 10\end{aligned}$$

Of course, we could have also written

$$\begin{aligned}\log 0.002 &= 0.3010 + (2 - 5) \\ &= 2.3010 - 5 \\ \text{or } \log 0.002 &= 0.3010 + (6 - 9) \\ &= 6.3010 - 9\end{aligned}$$

but the more conventional notation is $7.3010 - 10$. The student should examine the following table to observe how characteristics may be written in the -10 form.

Number	Characteristic of log
0.25	9 − 10
0.0378	8 − 10
0.000069	5 − 10

TABLE 10.1 Some Common Logarithms

x	0	1	2	3	4	5	6	7	8	9
3.0	.4771	.4786	.4800	.4814	.4829	.4843	.4857	.4871	.4886	.4900
3.1	.4914	.4928	.4942	.4955	.4969	.4983	.4997	.5011	.5024	.5038
3.2	.5051	.5065	.5079	.5092	.5105	.5119	.5132	.5145	.5159	.5172
3.3	.5185	.5198	.5211	.5224	.5237	.5250	.5263	.5276	.5289	.5302
3.4	.5315	.5328	.5340	.5353	.5366	.5378	.5391	.5403	.5416	.5428

Now suppose we wish to find

$$\log 324$$

Remember that the characteristic of the log is the power of 10.

Since $324 = 3.24 \times 10^2$, the characteristic of log 324 is 2. The mantissa of the logarithm can be found in the accompanying table, which is part of Appendix 1 (pp. 688 to 689). To read log 3.24, we first locate the row headed 3.2, then move across to the column headed 4. Combining the number there with the characteristic, we have

$$\log 324 = 0.5105 + 2 = 2.5105$$

Example 3

Using the table, find the logarithm of:
(a) 3150
(b) 0.00325

Solution

(a) Since $3150 = 3.150 \times 10^3$, the characteristic is 3. The mantissa is found along the row headed 3.1, under column 5; the value there is 0.4983. Thus,

$$\begin{aligned}\log 3150 &= 0.4983 + 3\\ &= 3.4983\end{aligned}$$

(b) Since $0.00325 = 3.25 \times 10^{-3}$, the characteristic is -3, which is customarily written as $(7 - 10)$. The mantissa, 0.5119, appears in the table in the row headed 3.2 and under column 5. Thus,

$$\begin{aligned}\log 0.00325 &= 0.5119 - 3\\ &= 7.5119 - 10\end{aligned}$$

In Example 3(a) we found that $\log 3150 = 3.4983$. If we wished to find x if $\log x = 3.4983$, we could argue that the sequence of digits in x is determined from the mantissa 0.4983. We then refer back to Table 10.1 and follow down the first column of mantissas until the number nearest but less than 0.4983 occurs. This number is 0.4914. An examination of this section of the table shows 0.4914 is in the row labeled 3.1; we move along this row to find the number 0.4983, and see that it is in column 5. Thus,

$$\log 3.15 = 0.4983$$

Because the characteristic of log x is 3,

$$\log x = 3.15 \times 10^3$$

or, equivalently,

$$\log x = \log 3150$$

Thus,

$$x = 3150$$

The number x that corresponds to a given logarithm is often referred to as the *antilogarithm.* For example,

$$\log x = 3.4983$$

may be written as

$$\text{antilog } 3.4983 = x$$

If we wanted to find, say, antilog 2.4249, the procedure would be to follow down the first column of mantissas in Appendix 1 until the number nearest but less than 0.4249 occurs. This number is 0.4150; moving from there, we find that 0.4249 is in the row labeled 2.6 and under column 6. Hence,

$$\text{antilog } 0.4249 = 2.66$$

Since the characteristic of 2.4249 is 2,

$$\text{antilog } 2.4249 = 2.66 \times 10^2 = 266$$

Example 4

Find (a) antilog 3.4330 and (b) antilog 6.4330 − 10.

Solution

(a) Both logarithms have the mantissa 0.4330, which from the table is found to have an antilogarithm of 2.71. Since the characteristic of 3.4330 is 3,

$$\text{antilog } 3.4330 = 2.71 \times 10^3 = 2710$$

(b) Since the antilog of 0.4330 is 2.71,

$$\text{antilog } 6.4330 - 10 = 2.71 \times 10^{-4} = 0.000271$$

The ideas we have studied are extremely useful in computations pertaining to "compound interest." Investments at compound interest earn interest each period they are invested. This interest is then added to the principal and earns interest itself. The formula that gives the total amount A at the end of n conversion periods if P dollars are invested at a rate r is

$$A = P(1 + r)^n$$

Thus, if $1000 is invested at 5% interest compounded annually for 10 years, the amount A at the end of 10 years is expressed as

$$A = 1000(1 + 0.05)^{10}$$

To find this amount, we proceed as follows:

$$\begin{aligned}
\log A &= \log 1000(1 + 0.05)^{10} \\
&= \log 10^3 + \log (1 + 0.05)^{10} && \text{Since } 1000 = 10^3 \\
&= 3 \log 10 + 10 \log (1.05) && \text{By Theorem 10.3} \\
&= 3 + 10(0.0212) && \text{Since } \log 10 = 1 \text{ and } \log 1.05 = 0.0212 \\
&= 3.212
\end{aligned}$$

Thus, A = antilog 3.212. From Appendix 1, the mantissa, 0.212, corresponds (approximately) to 1.63, and the characteristic of 3.212 is 3. Thus,

$$A = 1.63 \times 10^3 = 1630$$

Thus, the amount at the end of 10 years is approximately $1630.

Example 5

Find the approximate amount obtained at the end of 20 years if \$500 is invested at 5% interest compounded annually.

Solution

The amount is given by $A = P(1 + r)^n$, where $P = 500$, $r = 0.05$, and $n = 20$. Thus,

$$\begin{aligned}
A &= 500(1 + 0.05)^{20} \\
\log A &= \log 500(1 + 0.05)^{20} \\
&= \log 500 + \log (1 + 0.05)^{20} \\
&= \log (5 \times 10^2) + 20 \log (1 + 0.05) \\
&= \log 5 + \log 10^2 + 20 \log 1.05 \\
&= 0.6990 + 2 + 20(0.0212) \\
&= 2.6990 + 0.4240 \\
&= 3.1230 \\
A &= \text{antilog } 3.1230 \\
&= 1.33 \times 10^3 \text{ (approximately, since the antilog } 0.1239 = 1.33) \\
&= \$1330
\end{aligned}$$

PROGRESS TEST

1. The speed of Pioneer 10 was 131,000 kilometers per hour. When written in scientific notation, 131,000 = __________.
2. log 131,000 = __________.
3. log 0.005 = __________.
4. antilog 5.4048 = __________.
5. antilog 7.6571 − 10 = __________.
6. The approximate amount A obtained at the end of 10 years if \$100 is invested at 6% compounded annually is __________.

EXERCISE 10.4

1. Pioneer 10, a U.S. spacecraft, was launched March 2, 1972, on a 992,000,000-kilometer journey past Jupiter. Write 992,000,000 in scientific notation.
2. The star Epsilon Eridani is 0.000025 times as bright as the sun. Write 0.000025 in scientific notation.
3. Our solar system is 30,000 light years from the center of the Milky Way galaxy. Write 30,000 in scientific notation.

PROGRESS TEST ANSWERS

1. 1.31×10^5
2. $0.1173 + 5 = 5.1173$
3. $\log .005 = \log (5 \times 10^{-3}) = \log 5 - 3 = \log 0.6990 - 3 = 7.6990 - 10$
4. $2.54 \times 10^5 = 254{,}000$
5. $4.54 \times 10^{-3} = 0.00454$
6. $A = 100(1 + 0.06)^{10} = 1.79 \times 10^2 = \$179$

4. Light travels 186,000 miles per second. Write 186,000 in scientific notation.

5. Barnard's Star is 0.0004 times as bright as the sun. Write 0.0004 in scientific notation.

6. Infrared light has an average wavelength of 0.000001407 meter. Write this number in scientific notation.

In Problems 7 through 20, find the characteristic of the logarithm of the given number.

Illustration: (a) 5,128.33
(b) 0.4135

Solution: (a) Since $5,128.33 = 5.12833 \times 10^3$, the desired characteristic is 3.
(b) Since $0.4135 = 4.135 \times 10^{-1}$, the desired characteristic is -1.

7. 4.1402
8. 520.82
9. 21,076
10. 53.964
11. 0.37235
12. 0.08779
13. 1868.37
14. 0.00457
15. 0.12038
16. 784.11
17. 93.816
18. 4.57×10^4
19. 6.7432×10^{-5}
20. 3.3648×10^{-4}

In Problems 21 through 30, use the -10 notation to give the characteristic of the logarithm of the given number.

Illustration: 2.039×10^{-4}

Solution: The characteristic of the logarithm of 2.039×10^{-4} is -4, which can be written as $6 - 10$.

21. 0.83927
22. 0.00284
23. 0.06926
24. 0.00004
25. 0.07525
26. 0.31713
27. 0.05079
28. 1.8467×10^{-2}
29. 3.9052×10^{-1}
30. 6.1109×10^{-3}

In Problems 31 through 44, use Appendix 1 to find the logarithm of the given number.

31. 74.48

32. 952.0

33. 1837

34. 3.046

35. 0.04371

36. 50.18

37. 0.1283

38. 3.632

39. 164,200

40. 257.5

41. 0.008606

42. 0.01004

43. 0.06737

44. 9.444×10^{-4}

In Problems 45 through 50, use Appendix 1 to find the value of N for the given logarithm.

45. $\log N = 1.2676$

46. $\log N = 0.44091$

47. $\log N = 0.7671$

48. $\log N = 9.8305 - 10$

49. $\log N = 8.0346 - 10$

50. $\log N = 7.3115 - 10$

In Problems 51 through 54, find the approximate amount accumulated at the end of the given time if the specified sum is invested at the given compound annual rate for the specified period of time.

	PRINCIPAL	RATE	TIME
51.	\$100	7%	10 years
52.	\$1,000	6%	20 years
53.	\$500	5%	30 years
54.	\$200	10%	10 years

In chemistry, the pH (hydrogen potential) of a solution

is defined by $pH = -\log_{10}[H^+]$, where H^+ is the hydrogen ion concentration of the solution in moles per liter. In Problems 55 through 60, find the pH of a solution whose hydrogen ion concentration is as given.

Illustration: $[H^+] = 8 \times 10^{-6}$

Solution: Substituting in our equation for pH,

$$
\begin{aligned}
pH &= -\log[8 \times 10^{-6}] \\
&= -(\log 8 + \log 10^{-6}) \\
&= -(\log 8 - 6) \\
&= -0.9031 + 6 \qquad \text{From} \\
&= 5.0969 \qquad \text{Appendix 1}
\end{aligned}
$$

55. $[H^+] = 7 \times 10^{-7}$

56. $[H^+] = 1.5 \times 10^{-9}$

57. eggs whose $[H^+]$ is 1.6×10^{-8}

58. tomatoes whose $[H^+]$ is 6.3×10^{-5}

59. milk whose $[H^+]$ is 4×10^{-7}

60. $[H^+] = 5 \times 10^{-8}$

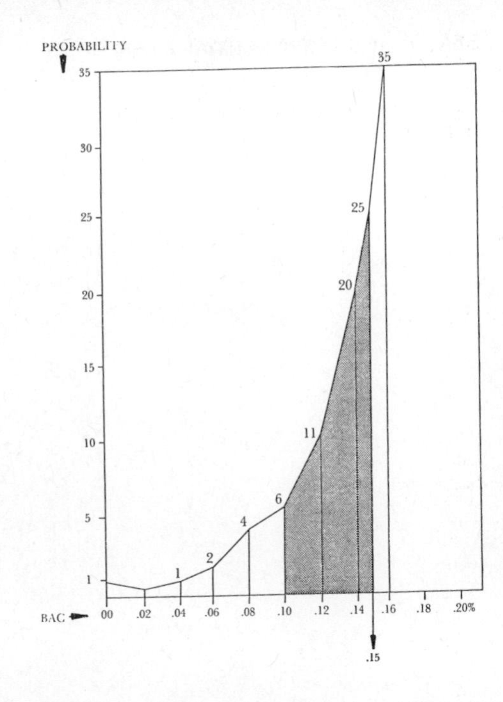

Indiana University Institute for Research in Public Safety.

OBJECTIVES 10.5

After studying this section, the reader should be able to:

1. **Solve an exponential equation using the properties and theorems discussed in the text and the appropriate logarithmic tables.**
2. **Solve word problems involving exponential equations.**

10.5 EXPONENTIAL EQUATIONS

The chart at the beginning of this section shows the probability of having an accident as a function of blood alcohol level (BAC). The exact formula relating the probability P (in per cent) of having an accident when the per cent of alcohol in the blood is b is expressed as

$$(1)\ P(b) = e^{kb}$$

As you can see from the chart, this probability is 25% when the alcohol level in the blood is .15%. Can we find k? To do this, we let $P(b) = 25$, and $b = 0.15$ in Equation (1). We then have

An exponential equation is an equation with variables used as exponents.

$$(2)\ 25 = e^{0.15k}$$

Equation (2) is an *exponential* equation—that is, an equation in which a variable occurs in an exponent. To solve this equation, we make the following assumptions.

$\log_e N$ is the natural logarithm of N and is written as ln N.

A_1 If $M = N$, then $\log_a M = \log_a N$, provided M, N, $a > 0$, and $a \neq 1$.

A_2 If $\log_a M = \log_a N$, then $M = N$.

In particular, if $M = N$, then $\log_e M = \log_e N$.

The number $\log_e N$ is called the *natural logarithm* of N and is abbreviated ln N, to use typical notation,

$$(3)\ \log_e x = \ln x$$

With this notation, the two assumptions above can be written in the form:

A_1' If $M = N$, then $\ln M = \ln N$ (given that the provisions noted in assumption A_1 still hold).

A_2' If $\ln M = \ln N$, then $M = N$.

We also restate in terms of natural logarithms the theorems we proved for general logarithms:

THEOREM 10.1

$\ln MN = \ln M + \ln N$

THEOREM 10.2

$\ln \frac{M}{N} = \ln M - \ln N$

THEOREM 10.3

$\ln M^r = r \cdot \ln M$

Moreover, we have

THEOREM 10.4

$\ln e = 1$

THEOREM 10.5

$\ln e^k = k$

THEOREM 10.6

$\ln 1 = 0$

We are now ready to solve Equation (2). Using A_1' in Equation (2), we have

$$\ln 25 = \ln e^{0.15k}$$

$$= 0.15k \cdot \ln e \qquad \text{By Theorem 10.3}$$

$$= 0.15k \qquad \text{Since } \ln e = 1$$

Solving $\ln 25 = 0.15k$, we obtain

$$k = \frac{\ln 25}{0.15}$$

The value of ln 25 can be found in Appendix 2; the value is 3.2189 (rounding off 3.21888). We then have

$$k = \frac{3.2189}{0.15} = 21.5 \qquad \text{(rounding to the nearest tenth)}$$

Actually, $\frac{3.2189}{.15} = 21.459333$ . . . , but we round off at 21.5.

Example 1

If $P(b) = e^{21.5b}$, at what blood level will the probability of having an accident be 100%?

Solution

We wish to find b, when $P(b) = 100$. That is, we need to solve the equation $100 = e^{21.5b}$. We have

$$\ln 100 = \ln e^{21.5b} \qquad \text{By } A_{1'}$$

$$= 21.5b \cdot \ln e \qquad \text{By Theorem 10.3}$$

$$= 21.5b \qquad \text{Since } \ln e = 1$$

$$b = \frac{\ln 100}{21.5} \qquad \text{Solving } \ln 100 = 21.5b \text{ for b}$$

$$= \frac{4.6052}{21.5}$$

Since $100 = 10^2$, $\ln 100 = 2 \ln 10$ or $2 \times 2.30259 = 4.6052$ (See Appendix 2)

$$= 0.214$$

Thus, when the alcohol level is approximately 0.21%, the probability of an accident is 100%. Of course, when your blood alcohol content is 0.21%, you may not even be able to drive!

Exponential equations can also be used to calculate the growth rate of a population. For example, in 1974 the population of the world was approximately 3.9 billion, and the yearly growth rate was 2%. The growth equation is

$$(4)\ P(t) = 3.9e^{0.02t}, \text{ where t is the time in years}$$

To estimate the world population in 1984, we let $t = 10$ in (4), obtaining

$$P(10) = 3.9e^{0.02(10)} = 3.9e^{0.2}$$

From Appendix 3, you can see that $e^{0.2} = 1.2214$.

We must now look up the value of $e^{0.2}$, found in Appendix 3. The value is 1.2214, so we have

$$3.9(1.2214) \approx 4.8 \text{ billion}$$

It is possible to find the time T it takes for the world population to double. Since

$$(5)\ P(t) = P_0e^{kT}$$

where P_0 is the population of the world at a given time, we wish to find T when

$$(6)\ 2P_0 = P_0e^{kT}$$

We do this in the next example.

Example 2 Solve for T in the equation $2P_0 = P_0e^{kT}$.

Solution

$$2P_0 = P_0e^{kT}$$

$$2 = e^{kT} \qquad \text{Dividing by } P_0$$

$$\ln 2 = \ln e^{kT} \qquad \text{Assumption } A_1'$$

$$= kT \ln e$$

$$= kT$$

Thus, $\ln 2 = 0.6931 = kT$, or

$$T = \frac{0.6931}{k}$$

Based on estimates, the annual world population growth for 1976 was $k = 2\%$. Thus,

$$T = \frac{0.6931}{0.02} = 34.7$$

Hence, the population will double by about the year 2011.

Exponential equations also occur in the field of biology. For example, if B is the number of bacteria present in a culture after t minutes, then under ideal conditions

(7) $B = ke^{0.05t}$

If we know that the initial number of bacteria is 1000, we can find how many bacteria will be present after 60 minutes. Since we know that at $t = 0$, $B = 1000$, we have

$$1000 = ke^{0}$$

Thus, $k = 1000$, and we rewrite Equation (7) as

(8) $B = 1000e^{0.05t}$

After 60 minutes,

$$B = 1000e^{0.05(60)} = 1000e^{3} = 1000(20.086) = 20{,}086 \text{ bacteria}$$

From Appendix 3, you can see that $e^3 = 20.086$.

We shall use Equation (8) in the next example.

Example 3

Use Equation (8) to find the time T it takes to have 50,000 bacteria present.

Solution

We have to find T, when $B = 50{,}000$, that is, solve

$$50{,}000 = 1{,}000e^{0.05T}$$

$$50 = e^{0.05T} \qquad \text{Dividing by 1000}$$

$$\ln 50 = \ln e^{0.05T} \qquad \text{Using } A_1'$$

$$= 0.05T \ln e \qquad \text{Theorem 10.3}$$

$$= 0.05T$$

Thus,

$$T = \frac{\ln 50}{0.05} = \frac{3.9120}{0.05} = 78.2 \text{ minutes}$$

PROGRESS TEST

1. If $P(b) = e^{21.5b}$, the alcohol level b at which the probability of an accident is 50% is ______________.
2. If in Example 2 we know that the annual population growth for Europe is 1%, the time it would take the European population to double would be ______________ years.
3. If the equation for the number B of bacteria present after time T is $B = 1000e^{0.05T}$, the time it would take to have 25,000 bacteria present is ______________ minutes.

EXERCISE 10.5

In Problems 1 through 6, use the procedure of Example 1 to find at which blood level the probability of having an accident is as given:

1. 60%

2. 70%

3. 75%

4. 80%

5. 90%

6. 95%

In Problems 7 through 10, assume that the number of bacteria present in a culture after t minutes is given by $B = 1000e^{0.04t}$. Find the time it takes to have:

7. 2000 bacteria

8. 5000 bacteria

9. 25,000 bacteria

10. 50,000 bacteria

In Problems 11 through 14, follow the procedure in the Illustration and Solution below to find the half life of 100 grams of the indicated element.

Illustration:
Cesium 137, whose decay rate is 2.3% per year.
Solution:
The half life of a substance can be found by using the formula

$$N(t) = N_0e^{-kt}$$

PROGRESS TEST ANSWERS

1. .18%
2. $\frac{0.6931}{0.01} = 69.31$
3. 64.4

where t is the time in years, k the decay rate, and N_0 the initial amount of material present. We need to find the time t it takes to end up with half of our original amount of material. We start with 100 grams of cesium 137, which has a decay rate of 2.3% per year; thus

$$k = 2.3\% = 0.023$$

$$N_0 = 100$$

We wish to find t when $N(t) = 50$; that is, we need to solve

$$50 = 100e^{-0.023t}$$

$\frac{1}{2} = e^{-0.023t}$ Dividing both members by 100

$\ln \frac{1}{2} = \ln e^{-0.023t}$ By A_1'

$= -0.023t \ln e$

$= -0.023t$ Since $\ln e = 1$

$\ln \frac{1}{2} = \ln 1 - \ln 2$ Theorem 10.2: $\ln \frac{M}{N} = \ln M - \ln N$

$= -0.023t$

$0 - 0.6931 = -0.023t$ From Appendix 2

$\frac{-0.6931}{-0.023} = t$ Solving for t

Our solution is $t = \frac{-0.06931}{-0.023} = 30.13$ years.

11. Plutonium, whose decay rate is 0.003% per year.

12. Krypton, whose decay rate is 6.3% per year.

13. A radioactive isotope whose decay rate is 5.2% per year.

14. A radioactive isotope whose decay rate is 0.002 per year.

Word Problems

15. If the initial weight of an animal is w_0, its weight after t days of withholding food is $w = w_0e^{-0.0071}$.
(a) What per cent of its weight does the animal lose daily?
(b) After 20 days have passed, what per cent of the weight of the animal remains?

16. Repeat Problem 15 if $w = w_0e^{-0.008t}$.

17. If $P_0 = 14.7$ lb/in^2 (pounds per square inch) is the atmospheric pressure at sea level, the atmospheric pressure at an altitude of a feet is given by $P = P_0e^{-0.00005a}$.
(a) Find the pressure at 5,000 feet.
(b) Find the pressure at 10,000 feet.

18. If in Problem 17 the atmospheric pressure P is measured in inches of mercury, $P = 30(10)^{-0.09a}$, where a is the altitude in miles
(a) What is the atmospheric pressure at sea level?
(b) What is the atmospheric pressure at 5 miles above sea level?

19. The home range H of an animal is the region to which it confines its movements. When the body weight of the animal is w, $H = w^{1.41}$ units. If a small animal weighs e pounds, find H.

20. Repeat Problem 19 for an animal weighing 10 pounds.

SELF-TEST—CHAPTER 10

1. On the accompanying rectangular coordinate system, graph the function $f(x) = 2^x$.

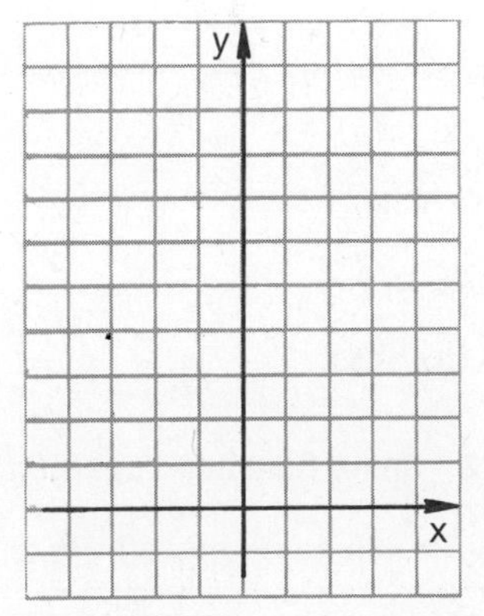

---

2. On the accompanying rectangular coordinate system, graph the function $f(x) = \left(\frac{1}{2}\right)^x$.

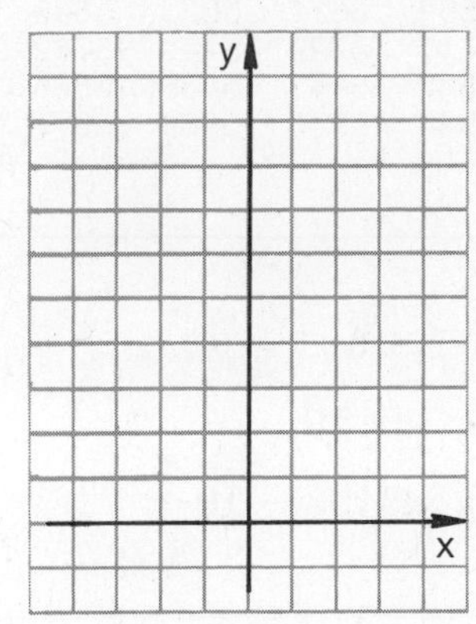

---

3. Find:

_______________ (a) $\log_2 16$

_______________ (b) $\log_3 81$

_______________ (c) $\log_2 \frac{1}{8}$

---

4. If it is known that $\log_{10} 2 = 0.301$, $\log_{10} 3 = 0.477$, and $\log_{10} 5 = 0.699$, find:

_______________ (a) $\log_{10} 36$

_______________ (b) $\log_{10} 10$

_______________ (c) $\log_{10} \frac{5}{2}$

---

5. Write the given numbers in scientific notation

______________ (a) 287,000

______________ (b) 0.000123

---

6. Use the accompanying table to find:

______________ (a) log 335

______________ (b) log 0.00335

TABLE 10.2 Some Common Logarithms

x	0	1	2	3	4	5	6	7	8	9
3.0	.4771	.4786	.4800	.4814	.4829	.4843	.4857	.4871	.4886	.4900
3.1	.4914	.4928	.4942	.4955	.4969	.4983	.4997	.5011	.5024	.5038
3.2	.5051	.5065	.5079	.5092	.5105	.5119	.5132	.5145	.5159	.5172
3.3	.5185	.5198	.5211	.5224	.5237	.5250	.5263	.5276	.5289	.5302
3.4	.5315	.5328	.5340	.5353	.5366	.5378	.5391	.5403	.5416	.5428
3.5	.5441	.5453	.5465	.5478	.5490	.5502	.5514	.5527	.5539	.5551
3.6	.5563	.5575	.5587	.5599	.5611	.5623	.5635	.5647	.5658	.5670
3.7	.5682	.5694	.5705	.5717	.5729	.5740	.5752	.5763	.5775	.5786
3.8	.5798	.5809	.5821	.5832	.5843	.5855	.5866	.5877	.5888	.5899
3.9	.5911	.5922	.5933	.5944	.5955	.5966	.5977	.5988	.5999	.6010

---

7. Use the table in Problem 6 to find:

______________ (a) antilog 3.5159

______________ (b) antilog 5.5198 − 10

---

8. The amount of money A accrued at the end of n years when a certain amount P is invested at a compounded annual rate r is given by $A = P(1 + r)^n$. If a person invests $100 at 5% interest compounded annually, find the approximate amount accrued at the end of 20 years. (Hint: $\log_{10} 1.05 = 0.0212$, and $\log_{10} 2.655 \approx 0.4240$.)

---

9. Find x in the given equations:

______________ (a) $\log_3 x = 243$

______________ (b) $\log_{64} x = \frac{7}{6}$

______________ (c) $\log_x 0.01 = \frac{-1}{2}$

---

10. The number of bacteria present in a culture after t minutes is given as $B = 1{,}000\,e^{kt}$.

_______________ (a) If there are 2,117 bacteria after 3 minutes, find k. (Use Table 10.3.)

TABLE 10.3

x	e^x	e^{-x}
0.55	1.7333	0.5769
0.60	1.8221	0.5488
0.65	1.9155	0.5220
0.70	2.0138	0.4966
0.75	2.1170	0.4724

_______________ (b) How many bacteria would there be after 4 minutes? (Hint: $e \approx 2.7$.)

CHAPTER 11

HISTORICAL NOTE

We have already mentioned the name of Leonhard Euler several times in this book. Euler (pronounced "oiler") is one of the greatest scientists that Switzerland has ever produced. His father, a good mathematician himself, taught the boy mathematics but insisted on his attending the University of Basel to study Hebrew and theology. There, Leonhard was befriended by the famous Bernoulli family of mathematicians, who persuaded his father that the boy was destined to become a great mathematician rather than a Calvinist pastor. In 1727, Euler accepted the chair of mathematics at the new St. Petersburg Academy formed by Peter the Great. In St. Petersburg, Euler became a voluminous writer on mathematics, attaining an enormous productivity which continued throughout his life, even though in about 1768 he had the misfortune of becoming totally blind.

Euler's work is an outstanding example of the eighteenth century practice of manipulating formulas involving infinite processes. For example, if the binomial theorem (which we study in this chapter) is applied formally to $(1-2)^{-1}$, we find that $-1 = 1 + 2 + 4 + \ldots$, a result which caused Euler no small wonderment! Also, if the series

$$x + x^2 + x^3 + \ldots = \frac{x}{1-x}$$

$$\text{and } 1 + \frac{1}{x} + \frac{1}{x^2} + \ldots = \frac{x}{x-1}$$

are added, we obtain $\ldots \frac{1}{x^2} + \frac{1}{x} + 1 + x + x^2 + \ldots = 0$. In this chapter you will learn that the "sum" of these infinite series can only be obtained under certain special conditions.

Ideas about sequences and series are, of course, much older than the eighteenth century. For example, the *Liber Abaci*, a book published by Leonardo da Pisa (Fibonacci) in the Middle Ages, posed the following problem: What is the number of pairs of rabbits at the beginning of any month if a single pair of newly born rabbits is put into an enclosure at the beginning of a given month and if each pair breeds a new pair at the beginning of the second month following birth and an additional pair at the beginning of each month thereafter? The solution is dependent upon the ability to find the sum of the terms in a certain sequence, and the problem is solved in the text.

SEQUENCES, SERIES, AND THE BINOMIAL THEOREM

The spirals in the daisy shown here are seen as two distinct sets radiating clockwise and counterclockwise, with each set always made up of a predetermined number of spirals. Most daisies have 21 and 34, adjacent numbers in the Fibonacci sequence discussed in this chapter. (Photo credit: Rutherford Platt.)

Leonardo Fibonacci

David Eugene Smith Collection, Columbia University Libraries.

OBJECTIVES 11.1

After studying this section, the reader should be able to:

1. **Find a specified term in a given sequence.**
2. **Find a specified term in a given sequence if the formula for the nth term of the sequence is given.**
3. **Find the sequence associated with a given function when the formula for the nth term is given.**

11.1 SEQUENCES

The engraving reproduced at the beginning of this section shows Leonardo Fibonacci, one of the greatest mathematicians of the Middle Ages and author of a book called the *Liber Abaci.* In this book, Fibonacci proposed the following problem:

> Let us suppose you have a one-month-old pair of rabbits and assume that in the second month, and every month thereafter, they produce a new pair. If each new pair does the same, and none of the rabbits die, can we find out how many pairs of rabbits there will be at the beginning of each month?

Here is the solution:

Beginning month number	1	2	3	4	5	6	7	...
Number of pairs	1	1	2	3	5	8	13	...

A sequence is a set of numbers arranged according to some given law.

The numbers 1 1 2 3 5 8 13 . . . form a *sequence*, called the *Fibonacci sequence.* In mathematics, a *sequence* is a set of numbers arranged according to some given law. For example, the digits 1,2,3,... 9 are a sequence. The numbers in a sequence are called the *terms* of the sequence, and we refer to them as the *first term*, the *second term*, the *third term*, and so on. It is customary to denote the terms in a sequence by using subscripts. Thus, s_1, s_2, and s_3 will denote the first, second, and third terms of a sequence. In the sequence of counting numbers 1, 2, 3, 4, . . .

$$s_1 = 1$$
$$s_2 = 1 + 1 = 2$$
$$s_3 = 2 + 1 = 3$$
$$s_4 = 3 + 1 = 4$$

As you can see from this pattern, the nth term, called the *general term*, is n. Thus,

$$s_{31} = 31$$
$$s_{49} = 49$$

and $\quad s_n = n$

Of course, it is possible to construct a sequence whose first four terms are 1, 2, 3, 4 and whose thirty-first term is different from 31. However, we assume the simplest sequence is 1, 2, 3, 4, 5, etc.

Example 1

Consider the sequence 2, 4, 6, 8, . . . Find:
(a) s_2 and s_4
(b) the tenth term—that is, s_{10}
(c) The nth term—that is, s_n

Solution

(a) In this case,

$$s_1 = 2$$
$$s_2 = 4$$
$$s_3 = 6$$
$$s_4 = 8$$

Hence, $s_2 = 4$, and $s_4 = 8$.

(b) You can clearly see that

$$s_1 = 2 \cdot 1$$
$$s_2 = 2 \cdot 2$$
$$s_3 = 2 \cdot 3$$
$$s_4 = 2 \cdot 4$$

and so on. Thus, $s_{10} = 2 \cdot 10 = 20$.
(c) From the pattern in (b), $s_n = 2n$.

Example 2

In the publishing industry, large printed sheets are folded to make the pages of a book. If sheets are folded once, we have 2 pages, or a *folio.* If sheets are folded twice, we have 4 pages, or a *quarto.* If sheets are folded three times, we have 8 pages, or an *octavo.* Further folding of sheets produces page units called "16 mo," "32 mo," etc.

(a) Write the sequence associated with the number of folds.

(b) If a sheet is folded 6 times, how many pages will result? If the sheet is folded n times, how many pages will result?

Solution

(a) $s_1 = 2$
$s_2 = 4$
$s_3 = 8$
$s_4 = 16$
Thus, the sequence is 2, 4, 8, 16, . . .

(b) We can rewrite the terms of the sequence in the form

$$s_1 = 2^1$$
$$s_2 = 2^2$$
$$s_3 = 2^3$$
$$s_4 = 2^4$$

and so on. Thus, if a sheet is folded *six* times, the number of pages will be $s_6 = 2^6 = 64$. If the sheet is folded n times, the number of pages will be $s_n = 2^n$.

In Examples 1 and 2, we found the general term of a given sequence. However, if only a *finite* number of successive terms are given with no rule specified to obtain the general term, then a *unique* general term cannot be obtained. For example,

$$(1)\quad s_n = 2n + 3$$

and $$(2)\quad s_n = (2n + 3) + \frac{1}{6}(n - 1)(n - 2)(n - 3)$$

will both yield

$$s_1 = 5$$
$$s_2 = 7$$
$$s_3 = 9$$

for the first three terms. However, (1) gives 11 and (2) gives

12 for the fourth term. This shows that the first three terms do not determine the fourth term. This procedure can be generalized to show that no finite number of terms can determine the next term uniquely.

No finite number of terms determines the next term uniquely.

In the following example, we show how to obtain the first few terms of a sequence when the general term is given.

Example 3

Find s_1, s_2, s_3, and s_4 if

$$s_n = \frac{n(n-1)}{2}$$

Solution

$$s_1 = \frac{1(1-1)}{2} = 0$$

$$s_2 = \frac{2(2-1)}{2} = 1$$

$$s_3 = \frac{3(3-1)}{2} = 3$$

$$s_4 = \frac{4(4-1)}{2} = 6$$

Thus, the first four terms are 0, 1, 3, and 6.

We should mention that sometimes *function notation* is used to denote the terms in a sequence. For instance, in Example 3 we could write

$$s(n) = \frac{n(n-1)}{2}$$

With this notation,

$$s(1) = \frac{1(1-1)}{2} = 0$$

$$s(2) = \frac{2(2-1)}{2} = 1$$

and so on.

Some books define a *sequence* as a function whose domain is a set of successive positive integers. The elements in the

A sequence is a function whose domain is the positive integers.

range of s written in the order s(1), s(2), s(3), . . . are said to form a sequence. With this notation, the function associated with the sequence 2, 4, 6, 8, . . . of Example 1 is $s(n) = 2n$, $n \in \{1, 2, 3, \ldots\}$. Similarly, the function $s(n) = 2^n$, $n \in \{1, 2, 3, \ldots\}$, defines the terms of the sequence 2, 4, 8, 16, . . . given in Example 2.

Example 4 Consider the function $s(n) = 2n - 1$, $n \in \{1, 2, 3, \ldots\}$. Find the sequence associated with this function.

Solution

For $n = 1$, $s(1) = 2 \cdot 1 - 1 = 1$
For $n = 2$, $s(2) = 2 \cdot 2 - 1 = 3$
For $n = 3$, $s(3) = 2 \cdot 3 - 1 = 5$
For $n = 4$, $s(4) = 2 \cdot 4 - 1 = 7$

Thus, the sequence is 1, 3, 5, 7, . . .

PROGRESS TEST

1. The seventh term in the sequence 3, 6, 9, . . ., where $s_n = 3n$, is ______.
2. The nth term in the sequence 5, 10, 15, . . . is ______.
3. If $s(n) = \frac{n(n+1)}{2}$, s_1 equals ______.
4. The sequence associated with $s(n) = \frac{n(n+1)}{2}$, $n \in \{1, 2, 3, \ldots\}$, is ______.

EXERCISE 11.1

In Problems 1 through 20, find a tenth and an nth term to fit the given sequence.

Illustration: $-8, -4, 0, 4, \ldots$

Solution:

term number	1	2	3	4	5	6
term	$-8 = 4(-2)$	$-4 = 4(-1)$	$0 = 4(0)$	$4 = 4(1)$	$8 = 4(2)$	$12 = 4(3)$

PROGRESS TEST ANSWERS

1. 21
2. 5n
3. 1
4. 1, 3, 6, . . .

From the table,

term number	4	5	6
term	$4 = 4(4-3)$	$8 = 4(5-3)$	$12 = 4(6-3)$

Thus, a suitable tenth term is $4(10 - 3) = 28$, and the nth term is $4(n - 3)$.

1. 1, 2, 3, 4, . . .

2. 5, 6, 7, 8, . . .

3. 7, 10, 13, 16, . . .

4. 3, 10, 17, 24, . . .

5. 55, 50, 45, 40, . . .

6. 33, 30, 27, 24, . . .

7. $\frac{1}{2}, \frac{1}{3}, \frac{1}{4}, \frac{1}{5}, \ldots$

8. $\frac{1}{2}, \frac{2}{3}, \frac{3}{4}, \frac{4}{5}, \ldots$

9. $-1, 1, -1, 1, \ldots$

10. $-1, 3, -5, 7, \ldots$

11. $x, x^2, x^3, x^4, \ldots$

12. $x^2, x^4, x^6, x^8, \ldots$

13. $-x, x^3, -x^5, x^7, \ldots$

14. $-x^2, x^4, -x^6, x^8, \ldots$

15. $x, \frac{x^2}{2}, \frac{x^3}{3}, \frac{x^4}{4}, \ldots$

16. $\frac{x}{5}, \frac{x^2}{10}, \frac{x^3}{15}, \frac{x^4}{20}, \ldots$

17. $-x, x, -x, x, \ldots$

18. $x, -x, x, -x, \ldots$

19. $\frac{x}{2}, \frac{x^2}{4}, \frac{x^3}{8}, \frac{x^4}{16}, \ldots$

20. $\frac{x}{2}, \frac{x^3}{4}, \frac{x^5}{8}, \frac{x^7}{16}, \ldots$

In Problems 21 through 34, find the first three terms of a sequence with the given general term.

Illustration: (a) $s_n = 1 - \frac{1}{n}$

(b) $s(n) = (-1)^n 3^n$

Solution: (a) For $n = 1$, $s_1 = 1 - \frac{1}{1} = 0$

For $n = 2$, $s_2 = 1 - \frac{1}{2} = \frac{1}{2}$

For $n = 3$, $s_3 = 1 - \frac{1}{3} = \frac{2}{3}$

(b) For $n = 1$, $s(1) = (-1)^1 3^1 = -3$
For $n = 2$, $s(2) = (-1)^2 3^2 = 9$
For $n = 3$, $s(3) = (-1)^3 3^3 = -27$

21. $s_n = 2n - 3$

22. $s_n = 2n + 3$

23. $s_n = \frac{n(n-2)}{2}$

24. $s_n = \frac{n(n+2)}{2}$

25. $s(n) = 1 + \frac{1}{n}$

26. $s(n) = 1 + \frac{2}{n}$

27. $s_n = n^2$

28. $s_n = -n^2$

29. $s(n) = \frac{n}{2n+1}$

30. $s_n = \frac{n}{2n-1}$

31. $s_n = (-1)^n$

32. $s_n = (-1)^{2n}$

33. $s(n) = (-1)^n 2^n$

34. $s(n) = (-1)^n 5^n$

35. A "Super Ball" is made of highly compressed synthetic rubber, and when dropped on a hard floor will make a sequence of bounces. Each successive bounce is about $\frac{9}{10}$ as high as its previous bounce. If a Super Ball is dropped from a height of 10 feet, how high will it bounce

(a) on the first bounce?
(b) on the second bounce?
(c) on the nth bounce?

36. There is a legend that the King of Persia offered the inventor of chess anything he wished as a reward for his invention. The man asked that 1 grain of wheat be placed on the first square of the chessboard, 2 grains on the second, 4 grains on the third, and so on. How many grains would there be

(a) on the fifth square?
(b) on the ninth square?
(c) on the nth square?

37. A colony of bacteria has 100 members and doubles every hour. How many bacteria are there

(a) after 2 hours?
(b) after 4 hours?
(c) after n hours?

38. A free-falling body falls approximately 16 feet the first second, 48 the next second, 80 the third second and so on. How far does it fall the 8th second? The nth second?

39. A man earned $10 Monday and doubled his salary every day thereafter. How much money did he earn on Saturday?

40. A racer moves 6 meters in the first second of a certain race and 25 centimeters more than in the previous second each second thereafter. How far does she go on the 8th second? The nth second? (Hint: 100 centimeters = 1 meter.)

Photo by Parachutes Incorporated, Orange, Massachusetts.

OBJECTIVES 11.2

After studying this section, the reader should be able to:

1. **Find a specified term in a given arithmetic progression.**
2. **Find the common difference d in a given arithmetic progression.**
3. **Find the sum of the first n terms in a given arithmetic progression.**
4. **Find the number of terms in an arithmetic progression for which the first term, the common difference, and the sum are specified.**

11.2 ARITHMETIC PROGRESSIONS

The photograph at the beginning of this section shows a skydiver plunging toward earth. Do you know how far he will fall the first five seconds? A free-falling body falls about 16 feet in the first second, 48 in the next, 80 in the third, and so on. In fact, the number of feet traveled each successive second is given by the sequence

(1) 16, 48, 80, 112, 144, . . .

An arithmetic progression is a progression in which each term after the first is formed by adding a common difference to the preceding term.

The sequence in (1) is an example of an *arithmetic progression*. An *arithmetic progression* is a sequence in which each term after the first is obtained by *adding* a quantity called the *common difference* to the preceding term. Thus, if a sequence is formed by the rule

(2) $s_1 = a_1$
and $s_{n+1} = s_n + d$

the sequence is an arithmetic progression. For example, the sequence given in (1) is an arithmetic progression, since $s_1 = 16$ and $s_{n+1} = s_n + 32$. That is, each term after the first is obtained by adding 32 to the preceding term, as shown here:

(3) 16, 16 + 32 = 48, 48 + 32 = 80, 80 + 32 = 112, . . .

It is customary to denote the first term of an arithmetic progression by a_1, the common difference by d, and the nth term by a_n. The complete sequence is then defined by the *recursive definition* shown in Equation (2). Thus, in the sequence 7, 12, 17, 22, . . ., $a_1 = 7$ (the first term) and $a_{n+1} = a_n + 5$. Hence d = 5. As for the *general term*, a_n, the formula is easily found to be $a_n = 2 + 5n$.

(2) is now written as $s_1 = a_1$
$a_{n+1} = a_n + d$

Here we are rewriting equation (2) by using a_{n+1} and a_n instead of s_{n+1} and s_n.

Example 1

When an arithmetic sequence is given, you can always find d by taking any two successive terms and finding their difference. For example, if we take 13 and 16, d = 16 − 13 = 3.

Consider the arithmetic progression 7, 10, 13, 16, ...
(a) Find a_1
(b) Find d
(c) Find a_n

Solution

(a) $a_1 = 7$ is the first term of the progression.
(b) We rewrite the progression as follows:

$a_1 = 7$
$a_2 = 7 + 3 = 10$
$a_3 = 10 + 3 = 13$
$a_4 = 13 + 3 = 16$, and so on.

As you can see, each successive term is obtained by adding 3 to the preceding term—that is, $a_{n+1} = a_n + 3$. Thus d = 3.

(c) The general term is obtained by starting with 4 and adding 3 n times. Thus, $a_n = 4 + 3n$.

You can check that $a_n = 4 + 3n$ yields the progression 7, 10, 13, 16, . . . :
For n = 1, $a_1 = 4 + 3 = 7$
For n = 2, $a_2 = 4 + 3 \cdot 2 = 10$
For n = 3, $a_3 = 4 + 3 \cdot 3 = 13$
and so on.

Equation (2) shows that an arithmetic progression can be completely specified if we know a_1 (the first term), d (the common difference), and n (the number of terms). Thus, if $a_1 = 5$, $d = 4$, and $n = 5$, we have the progression

$$5, 9, 13, 17, 21$$

In general, the first n terms of an arithmetic progression may be written in the form

$$(4) \quad a_1, a_1 + d, a_1 + 2d, a_1 + 3d, \ldots a_1 + (n-1)d$$

Thus, the nth term of an arithmetic progression is

$$(5) \quad a_n = a_1 + (n-1)d$$

Now we go back to our original problem of trying to determine how far the skydiver mentioned at the beginning of this section falls after 5 seconds. As you recall, he falls

$$(1) \quad 16, 48, 80, 112, 144, \ldots$$

feet in the first, second, third, fourth, and fifth seconds respectively. Thus, we need to find

$$16 + 48 + 80 + 112 + 144,$$

the sum of the first five terms of the sequence in (1). The sum of the first five terms of a sequence is denoted by

$$S_5 = a_1 + a_2 + a_3 + a_4 + a_5$$

To generalize, the sum of the first n terms can be written (using Equation 4) as:

$$(6) \quad S_n = a_1 + (a_1 + d) + (a_1 + 2d) + \ldots + (a_n - d) + a_n$$

The terms on the right side of Equation (6) can be rewritten in reverse order, giving us

$$(7) \quad S_n = a_n + (a_n - d) + \ldots + (a_1 + 2d) + (a_1 + d) + a_1$$

Adding (6) and (7), we obtain

$$\begin{aligned}(8) \quad 2S_n &= (a_1 + a_n) + (a_1 + \not{d} + a_n - \not{d}) + \ldots \\ &\quad + (a_n - \not{d} + a_1 + \not{d}) + (a_n + a_1) \\ &= (a_1 + a_n) + (a_1 + a_n) + \ldots + (a_1 + a_n) + (a_1 + a_n) \\ &= n(a_1 + a_n)\end{aligned}$$

$a_1 + a_n$ is added n times, so we obtain $n(a_1 + a_n)$.

Thus,

$$(9)\quad 2S_n = n(a_1 + a_n)$$

Dividing, we get

$$(10)\quad S_n = \frac{n(a_1 + a_n)}{2}$$

We are now able to determine how far the skydiver dropped after 5 seconds. We need to find

$$S_5 = \frac{5(a_1 + a_5)}{2}$$

where $a_1 = 16$ and $a_5 = 144$. Substituting,

$$S_5 = \frac{5(16 + 144)}{2} = \frac{5(160)}{2} = 400 \text{ feet}$$

Example 2

A club raffles a certain item by selling 100 sealed tickets numbered 1, 2, 3, . . . , 100. These tickets are selected at random by purchasers, who pay the number of cents equal to the number on the ticket. How much money does the club receive?

Solution

The number of cents that the club receives is

$$S_{100} = 1 + 2 + 3 + \ldots + 100$$

Here $a_1 = 1$, $n = 100$, and $a_n = 100$. Thus, using Equation (10),

$$S_{100} = \frac{100(1 + 100)}{2} = \frac{100(101)}{2} = 50(101) = 5{,}050$$

Expressed in dollars, the club receives \$50.50.

Example 3

The sum of the first ten terms of an arithmetic progression is 205 and the tenth term is 34. Find:

(a) a_1, the first term

(b) d, the common difference.

Solution

(a) From Equation (10), with $S_{10} = 205$,

$$205 = \frac{10(a_1 + 34)}{2}$$

$10(a_1 + 34) = 410$	Multiplying both members by 2
$10a_1 + 340 = 410$	By the Distributive Law
$10a_1 = 70$	Subtracting 340 from both members
$a_1 = 7$	Dividing both members by 10

(b) From Equation (5), $a_n = a_1 + (n - 1)d$. We know that the tenth term, a_{10}, equals 34, and that $a_1 = 7$. Substituting in (5), we obtain

$$34 = 7 + (10 - 1)d$$
$$\text{or } 34 = 7 + 9d$$

Solving for d,

$$d = 3$$

Thus, the common difference is 3.

Example 4

Find the number of terms in the arithmetic progression for which $a_1 = 13$, $d = 10$, and $S_n = 384$.

Solution

By Equation (10) with $S_n = 384$ and $a_1 = 13$, we have

$$384 = \frac{n(13 + a_n)}{2}$$

$$\text{or } 768 = n(13 + a_n)$$

From Equation (5),

$$\begin{aligned} a_n &= a_1 + (n - 1)d \\ &= 13 + (n - 1)10 \\ &= 13 + 10n - 10 \\ &= 3 + 10n \end{aligned}$$

Returning to $768 = n(13 + a_n)$, we can substitute $a_n = 3 + 10n$:

$$768 = n(13 + 3 + 10n)$$
$$= n(16 + 10n)$$

Multiplying and arranging in standard form,

$$10n^2 + 16n - 768 = 0$$

$$5n^2 + 8n - 384 = 0 \quad \text{Dividing each member by 2}$$

$$(5n + 48)(n - 8) = 0 \quad \text{Factoring}$$

Since n must be a positive integer, the solution is $n = 8$.

PROGRESS TEST

The arithmetic progression

(1) 9, 12, 15, 18, . . .

will be used in Problems 1 through 4.

1. In arithmetic progression (1), a_1 equals ______.
2. In arithmetic progression (1), d equals ______.
3. In arithmetic progression (1), a_n equals ______.
4. The sum of the first 20 terms in arithmetic progression (1) equals ______.
5. The sum of the first ten terms of an arithmetic progression is 255 and the tenth term is 48. In this progression $a_1 =$ ______ and $d =$ ______.
6. The number of terms in an arithmetic progression for which $a_1 = 7$, $d = 6$, and $S_n = 340$ is ______.

EXERCISE 11.2

In Problems 1 through 10, an arithmetic progression is given. Find: (a) a_1, the first term, (b) d, the common difference, and (c) a_n, the nth term.

Illustration: $-8, 2, 12, 22, \ldots$

Solution: (a) $a_1 = -8$
(b) Since each term is obtained by adding 10 to the preceding one, $d = 10$.
(c) $a_n = -8 + (n - 1)(10) = -18 + 10n$

PROGRESS TEST ANSWERS

1. 9
2. 3
3. $6 + 3n$
4. $S_n = \frac{20(9 + 66)}{2} = 750$
5. 3; 5
6. 10

1. $5, 8, 11, 14, \ldots$

2. $5, 10, 15, 20, \ldots$

3. $11, 6, 1, -4, \ldots$

4. $43, 32, 21, 10, \ldots$

5. $3, -1, -5, -9, \ldots$

6. $0.6, 0.2, -0.2, -0.6, \ldots$

7. $8, 0, -8, -16, \ldots$

8. $\frac{2}{3}, \frac{5}{6}, 1, \frac{7}{6}, \ldots$

9. $\frac{-5}{6}, \frac{-1}{3}, \frac{1}{6}, \frac{2}{3}, \ldots$

10. $\frac{-1}{4}, \frac{1}{4}, \frac{3}{4}, \frac{5}{4}, \ldots$

In Problems 11 through 20, certain elements of an arithmetic progression are given. Find the elements whose values are not given.

Illustration: $a_1 = 7$, $n = 10$, $d = 6$; a_{10}, S_{10}

Solution: By Equation (5) with $a_1 = 7$ and $n = 10$,

$$\begin{aligned} a_{10} &= 7 + (10 - 1)6 \\ &= 7 + 54 \\ &= 61 \end{aligned}$$

By Equation (10) with $a_1 = 7$, $a_n = 61$, and $n = 10$,

$$S_{10} = \frac{10(7 + 61)}{2} = 5(68) = 340$$

11. $a_1 = 7$, $n = 15$, $d = 6$; a_{15}, S_{15}

12. a_1, d, S_8 in the sequence $4, 10, 16, 22, \ldots$

13. a_1, d, S_{10} in the sequence $3, -1, -5, -9, \ldots$

14. $a_1 = -2,\ d = -5,\ a_n = -72;\ n,\ S_n$

15. $a_1 = 3,\ a_6 = 8;\ d,\ S_6$

16. $a_1 = -1,\ a_{10} = -4;\ d,\ S_{10}$

17. $a_1 = 6,\ S_{14} = -280;\ d,\ a_{14}$

18. $a_1 = 15,\ a_n = -25,\ S_n = -85;\ d,\ n$

19. $d = 40,\ S_{40} = 40;\ a_1,\ a_{40}$

20. $a_1 = 4,\ d = 2,\ a_n = 30;\ n,\ S_n$

21. The ninth and thirty-third terms of an arithmetic progression are 48 and 288 respectively. Find the fifty-first term.

22. The sixteenth and forty-sixth terms of an arithmetic progression are 61 and 151. Find the thirty-second term.

23. A certain property valued at $30,000 will depreciate $1380 the first year, $1340 the second year, $1300 the third year, and so on. Based on this information, what will be the worth of the property twenty years from now?

24. Strikers at a plant were ordered to return to work and told they would be fined $50 the first day they failed to do so, $60 the second day, $70 the third day and so on. If their union paid a $680 fine, in how many days did they go back to work?

25. Show that the sum of the first n natural numbers is $\frac{n(n+1)}{2}$.

26. Show that the sum of the first n odd natural numbers is n^2.

27. Show that the sum of the first n even natural numbers is $n^2 + n$.

28. Show that the sum of the first n natural numbers divisible by 8 is $4n(n+1)$.

Photo credit: Stock, Boston, Inc.

OBJECTIVES 11.3

After studying this section, the reader should be able to:

1. **Find a specified term in a given geometric progression.**
2. **Find the common ratio r in a given geometric progression.**
3. **Find the nth term and the sum of a given geometric progression.**
4. **Find the sum (if possible) of a given infinite geometric series.**
5. **Use series to write a given repeating decimal as the ratio of two integers.**

11.3 GEOMETRIC PROGRESSIONS

The photograph at the beginning of this section shows a giant chessboard. The game of chess is said to have originated in Persia. Legend has it that the Shah (or king) of Persia offered the inventor of the game anything he wanted as a reward. The inventor asked that one grain of wheat be placed on the first square of the chessboard, two grains on the second, four on the third, and so on. The sequence enumerating the number of grains in each square is given by

(1) 1, 2, 4, 8, 16, . . .

and the inventor was to receive the sum of the first 64 terms in this sequence (there are 64 squares on a chessboard). As you can see, each *term* after the first in this sequence is obtained by *doubling* (*multiplying* by 2) the preceding term.

Thus (1) is *not* an arithmetic progression. This sequence is called a *geometric progression.*

Definition of a geometric progression.

A *geometric progression* is a sequence in which each term after the first is obtained by *multiplying* the preceding term by a number called the *common ratio,* denoted by the letter r. The result of multiplying each term by a number is to produce a fixed ratio between any two consecutive terms. If a geometric progression is given, r can be found by taking the ratio of two successive terms. For example, in the sequence in (1),

$$r = \frac{16}{8} = 2 \quad \text{or} \quad \frac{8}{4} = 2 \quad \text{or} \quad \frac{2}{1}$$

As in the case of an arithmetic progression, a geometric progression can be completely described by specifying the following three numbers:

a_1, the first term
r, the common ratio
n, the number of terms

Since in a geometric progression each term after the first is obtained by multiplying the preceding term by r, the first n terms in such a sequence are

$$(2)\quad a_1, a_1r, a_1r^2, a_1r^3, \ldots, a_1r^{n-1}$$

The nth term of the geometric progression is expressed as

$$(3)\quad a_n = a_1r^{n-1}$$

Example 1

Consider the geometric progression

$$1, \frac{1}{10}, \frac{1}{100}, \frac{1}{1000}, \ldots$$

Find:
(a) a_1
(b) r
(c) a_n

Solution

(a) a_1 is the first term, 1.

(b) r is the common ratio of any two successive terms. Thus

$$r = \frac{\frac{1}{10}}{1} = \frac{1}{10} \qquad \left(\text{or } \frac{\frac{1}{100}}{\frac{1}{10}} = \frac{1}{10}\right)$$

(c) By Equation (3) with $a_1 = 1$ and $r = \frac{1}{10}$,

$$a_n = 1 \cdot \left(\frac{1}{10}\right)^{n-1} = \frac{1}{10^{n-1}}$$

If we return to the inventor of chess and decide to find out how many grains of wheat he was to obtain, we first rewrite the progression in (1) as

(4) $1, 2, 2^2, 2^3, \ldots$

Then our answer will be the sum

$$S_{64} = 1 + 2 + 2^2 + 2^3 + \ldots + 2^{64}$$

To evaluate this sum, we recall that the first n terms of a geometric sequence are given by Equation (2); thus

(5) $S_n = a_1 + a_1r + a_1r^2 + \ldots + a_1r^{n-2} + a_1r^{n-1}$

Multiplying both sides of (5) by r, we obtain

(6) $rS_n = a_1r + a_1r^2 + a_1r^3 + \ldots + a_1r^{n-1} + a_1r^n$

Subtracting Equation (6) from Equation (5) yields

(7) $S_n - rS_n = a_1 - a_1r^n$

Factoring Equation (7) gives us

$$S_n(1 - r) = a_1(1 - r^n)$$

Now, if $r \neq 1$, we obtain

(8) $S_n = \dfrac{a_1(1 - r^n)}{1 - r}$

Thus, to find the sum in (5), we use Equation (8) with $n = 64$, $a_1 = 1$, and $r = 2$, obtaining

$$S_{64} = \frac{1(1 - 2^{64})}{1 - 2} = \frac{1 - 2^{64}}{-1} = 2^{64} - 1$$

a number which is larger than 18 quintillion (18,000,000,000,000,000,000). It is interesting to know that this is considerably more wheat than is grown in the entire world in one year, and it is probably the reason for the rest of the legend—that when the Shah's wise men found out how much wheat was needed, the Shah ordered the inventor of chess to lose his head!

Recall that $\frac{a}{-b} = \frac{-a}{b}$. Thus $\frac{1 - 2^{64}}{-1} = -1(1 - 2^{64}) = 2^{64} - 1$. To find an approximation for 2^{64}, we let $x = 2^{64}$. Then $\log x = \log 2^{64} = 64 \log 2 \approx 64(.301) \approx 19.264$. The antilog of 0.264 is approximately 1.84, and the characteristic is 19. Thus, $x \approx 1.84 \times 10^{19}$.

Example 2

Consider the geometric sequence 4, 8, 16, 32, ... Find:
(a) r
(b) a_n
(c) S_n

Solution

(a) r is the common ratio between successive terms: $\frac{8}{4} = 2$.

(b) $a_n = a_1 r^{n-1}$ (Equation 3). Letting $a_1 = 4$ and $r = 2$ in this equation, we have

$$a_n = 4(2)^{n-1}$$

(c) $S_n = \frac{a_1(1 - r^n)}{1 - r}$ (Equation 8). Letting $a_1 = 4$ and $r = 2$, we obtain

$$S_n = \frac{4(1 - 2^n)}{1 - 2} = -4(1 - 2^n) = -2^2(1 - 2^n) = 2^{n+2} - 4$$

In the preceding examples, we have found the sum of the first n terms of a geometric progression. We now consider what happens if n is allowed to increase without bound. We indicate that the number of terms is *infinite*—that is, unlimited—by writing

$$(9)\quad a_1 + a_1 r + a_1 r^2 + \ldots + a_1 r^{n-1} + \ldots$$

The expression given in (9) is called an *infinite geometric series.* We can find the sum of the first n terms of this series by using the formula

$$S_n = \frac{a_1(1 - r^n)}{1 - r} = \frac{a_1}{1 - r}(1 - r^n)$$

Now, if r is less than 1 in absolute value, that is, $|r| < 1$, or, equivalently, $-1 < r < 1$, then r^n becomes smaller and smaller as n increases. For example, if $r = \frac{1}{10}$,

$$r^2 = \left(\frac{1}{10}\right)^2 = \frac{1}{100}$$

$$r^3 = \left(\frac{1}{10}\right)^3 = \frac{1}{1000}$$

$$\text{and } r^4 = \left(\frac{1}{10}\right)^4 = \frac{1}{10{,}000}$$

Clearly,

$$r^n = \left(\frac{1}{10}\right)^n$$

can be made as small as we please by making n sufficiently large. Thus, if $|r| < 1$, the factor $(1 - r^n)$ in $S_n = \frac{a_1}{1 - r}(1 - r^n)$ can be made as close to 1 as we please by making n large enough. Consequently, as n is taken larger and larger, the sum

$$S_n = \frac{a_1}{1 - r}(1 - r^n)$$

is more and more closely approximated by

$$\frac{a_1}{1 - r}$$

This discussion is summarized in the following definition.

DEFINITION 11.1

For $|r| < 1$, $S_\infty = a_1 + a_1 r + a_1 r^2 + \ldots + a_1 r^{n-1} + \ldots$

$$= \frac{a_1}{1 - r}$$

We call S_∞ the *sum* of the series.
If $|r| \geq 1$, the sum of n terms does not get closer and closer to any fixed number as n increases without bound. In such cases, we say that S_∞ does not exist.

For $r = \frac{1}{2}$ and $a_1 = 1$, we can give a graphical interpretation of the behavior of S_n as n increases without bound. We consider the series

$$(10)\quad 1 + \frac{1}{2} + \frac{1}{4} + \frac{1}{8} + \ldots + \left(\frac{1}{2}\right)^{n-1} + \ldots$$

For this series,

$$S_1 = 1$$

$$S_2 = 1 + \frac{1}{2} = 1\frac{1}{2}$$

$$S_3 = 1 + \frac{1}{2} + \frac{1}{3} = 1\frac{3}{4}$$

$$S_4 = 1 + \frac{1}{2} + \frac{1}{3} + \frac{1}{4} = 1\frac{7}{8}$$

The sums are shown in Figure 11.1. As you can see, each step cuts in half the remaining distance to the point marked 2.

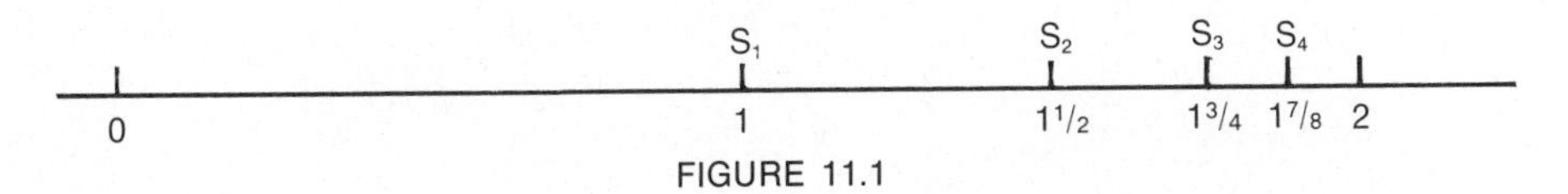

FIGURE 11.1

Thus, by making n sufficiently large, we can make the value of S_n as close as we please to 2. For the series in (10),

$$S_\infty = \frac{a_1}{1 - r} = \frac{1}{1 - \frac{1}{2}} = 2$$

Example 3

Find the sum of the geometric series

$$2 - \frac{4}{3} + \frac{8}{9} - \ldots.$$

Solution For the given series,

$$a_1 = 2 \text{ and } r = \frac{\frac{-4}{3}}{2} = \frac{-2}{3}$$

Hence,

$$S_\infty = \frac{2}{1 - \frac{(-2)}{3}} = \frac{2}{\frac{5}{3}} = \frac{6}{5}$$

An important application of Definition 11.1 occurs in connection with repeating decimals such as

$0.\overline{3}$ means the 3 repeats endlessly.

0.333 . . . , usually written as $0.\overline{3}$,

0.414141 . . . , usually written as $0.\overline{41}$, and

0.142857142857 . . . , usually written as $0.\overline{142857}$

If we wish to express the decimal 0.333 . . . as a ratio of two integers, we first write

$$(11)\quad 0.333\ldots = 0.3 + 0.03 + 0.003 + \ldots$$

or, equivalently,

$$(12)\quad \frac{3}{10} + \frac{3}{100} + \frac{3}{1000} + \ldots$$

which is a geometric progression with

$$r = \frac{\frac{3}{100}}{\frac{3}{10}} = \frac{1}{10}$$

Since r is less than 1, we can use Definition 11.1, with $a_1 = \frac{3}{10}$ and $r = \frac{1}{10}$ to find the sum in (11). We obtain

$$S_\infty = 0.333\ldots = \frac{3}{10} + \frac{3}{100} + \frac{3}{1000} + \ldots = \frac{\frac{3}{10}}{1 - \frac{1}{10}} = \frac{1}{3}$$

Example 4

Write 0.414141 . . . as a ratio of two integers.

Solution

$$0.414141 = 0.41 + 0.0041 + 0.000041 + \ldots$$

$$= \frac{41}{100} + \frac{41}{10{,}000} + \frac{41}{1{,}000{,}000} + \ldots$$

This is a geometric progression with

$$a_1 = \frac{41}{100} \text{ and } r = \frac{\frac{41}{10{,}000}}{\frac{41}{100}} = \frac{100}{10{,}000} = \frac{1}{100}$$

(which is less than 1)

Substituting these values in Definition 11.1, we have

$$S_\infty = \frac{\frac{41}{100}}{1 - \frac{1}{100}} = \frac{\frac{41}{100}}{\frac{99}{100}} = \frac{41}{99}$$

Thus, $0.414141\ldots = \frac{41}{99}$.

PROGRESS TEST

The geometric progression (1) $1, \frac{1}{2}, \frac{1}{4}, \frac{1}{8}, \ldots$ will be used in Problems 1 through 4.

1. In progression (1), a_1 equals ________.
2. In progression (1), r equals ________.

3. In progression (1), a_n equals ______.
4. In progression (1), S_n equals ______.
5. In the geometric progression $1, \frac{1}{3}, \frac{1}{9}, \frac{1}{27}, \ldots$, S_∞ equals ______.
6. When written as a ratio of integers, 0.666 . . . equals ______.
7. When written as a ratio of integers, 0.434343 . . . equals ______.

EXERCISE 11.3

In Problems 1 through 10, a geometric progression is given. Find: (a) a_1; (b) r; (c) a_n; (d) S_n, (e) S_∞ (if possible).

Illustration: $1, \frac{1}{4}, \frac{1}{16}, \frac{1}{64}, \ldots$

Solution: (a) a_1 is the first term, 1.

(b) $r = \frac{\frac{1}{4}}{1} = \frac{1}{4}$

(c) $a_n = a_1 r^{n-1} = 1\left(\frac{1}{4}\right)^{n-1} = \frac{1}{4^{n-1}}$

(d) $S_n = \frac{a_1(1 - r^n)}{1 - r}$

$$= \frac{1\left[1 - \left(\frac{1}{4}\right)^n\right]}{1 - \frac{1}{4}} = \frac{4\left[1 - \left(\frac{1}{4}\right)^n\right]}{3}$$

(e) $S_\infty = \frac{a_1}{1 - r} = \frac{1}{1 - \frac{1}{4}} = \frac{4}{3}$

PROGRESS TEST ANSWERS

1. 1
2. $\frac{\frac{1}{2}}{1} = \frac{1}{2}$
3. $a_n = a_1 r^{n-1} = 1 \cdot \left(\frac{1}{2}\right)^{n-1} = \left(\frac{1}{2}\right)^{n-1}$
4. $S_n = \frac{a_1(1 - r^n)}{1 - r} = \frac{1 \cdot \left[1 - \left(\frac{1}{2}\right)^n\right]}{1 - \frac{1}{2}} = \frac{1 - \frac{1}{2^n}}{\frac{1}{2}} = 2\left(1 - \frac{1}{2^n}\right) = 2 - \frac{1}{2^{n-1}}$
5. $S_\infty = \frac{a_1}{1 - r} = \frac{1}{1 - \frac{1}{3}} = \frac{3}{2}$
6. $\frac{2}{3}$
7. $\frac{43}{99}$

1. 3, 6, 12, 24, . . .

2. $\frac{1}{3}, 1, 3, 9, \ldots$

3. 8, 24, 72, 216, . . .

4. $\frac{1}{5}, \frac{1}{10}, \frac{1}{20}, \frac{1}{40}, \ldots$

5. $16, -4, 1, \frac{-1}{4}, \ldots$

6. $3, -1, \frac{1}{3}, \frac{-1}{9}, \ldots$

7. $\frac{-3}{5}, \frac{3}{2}, \frac{-15}{4}, \frac{75}{8}, \ldots$

8. $60, -6, \frac{6}{10}, \frac{-6}{100}, \ldots$

9. $\frac{-3}{4}, \frac{-1}{4}, \frac{-1}{12}, \frac{-1}{36}, \ldots$

10. $\frac{-5}{6}, \frac{-1}{3}, \frac{-2}{15}, \frac{-4}{75}, \ldots$

In Problems 11 through 20, certain elements of a geometric progression are given. Find the elements whose values are not given.

Illustration: $a_1 = 8$, $S_3 = 248$; a_3, r

Solution:

We use Equation (8):

$$S_n = \frac{a_1(1 - r^n)}{1 - r}$$

Substituting,

$$248 = \frac{8(1 - r^3)}{1 - r}$$

$$31 = \frac{1 - r^3}{1 - r} \qquad \text{Dividing both members by 8}$$

$$\frac{(1 - r)(1 + r + r^2)}{1 - r} = 31 \qquad \text{Factoring and simplifying}$$

$$r^2 + r - 30 = 0 \qquad \text{Subtracting 31 from both members}$$

$$(r + 6)(r - 5) = 0 \qquad \text{Factoring}$$

$$r = -6 \text{ or } r = 5$$

Thus, there are two possible values for r: -6 or 5.
For $r = -6$, $a_3 = 8(-6)^2 = 288$.

For $r = 5$, $a_3 = 8(5)^2 = 200$.
Note that for $r = -6$,

$$S_3 = \frac{8[1 - (-6)^3]}{1 - (-6)} = \frac{8[217]}{7} = 8 \cdot 31 = 248$$

and for $r = 5$,

$$S_3 = \frac{8[1 - (5)^3]}{1 - 5} = \frac{8[-124]}{-4} = 248$$

So the information obtained is correct.

11. $a_1 = 1,\ S_3 = 1\frac{3}{4};\ a_3,\ r$

12. $a_1 = 4,\ S_3 = 7;\ a_3,\ r$

13. $a_1 = 3,\ S_3 = 21;\ a_3,\ r$

14. $a_1 = \frac{1}{2},\ S_3 = \frac{39}{50};\ a_3,\ r$

15. $r = 2,\ S_8 = 1785;\ a_1,\ a_8$

16. $a_6 = \frac{-16}{27},\ r = \frac{-1}{3};\ a_1,\ S_6$

17. $a_1 = -4,\ a_n = 108,\ S_n = 80;\ r,\ n$

18. $a_1 = \frac{3}{4},\ a_n = 192,\ S_n = 255\frac{3}{4};\ r,\ n$

19. $a_1 = \frac{16}{125},\ r = \frac{5}{2},\ a_n = 12\frac{1}{2};\ n,\ S_n$

20. $a_1 = 7,\ r = 2,\ a_n = 896;\ n,\ S_n$

In Problems 21 through 26, a geometric series is given. Find S_∞ if possible.

Illustration: (a) $\frac{-3}{4} + \frac{-1}{4} + \frac{-1}{12} + \ldots$

(b) $16 + 24 + 36 + \ldots$

Solution: (a) $S_\infty = \frac{a_1}{1 - r}$. In this case,

$$a_1 = \frac{-3}{4} \text{ and } r = \frac{\frac{-1}{4}}{\frac{-3}{4}} = \frac{1}{3}$$

Since $|r| < 1$,

$$S_\infty = \frac{\frac{-3}{4}}{1 - \frac{1}{3}} = \frac{\frac{-3}{4}}{\frac{2}{3}} = \frac{-9}{8}$$

(b) In this case, $a_1 = 16$ and $r = \frac{24}{16} = \frac{3}{2}$.

Since $|r| \nless 1$, S_∞ does not exist.

21. $6 + 3 + 1\frac{1}{2} + \ldots$

22. $12 + 4 + 1\frac{1}{3} + \ldots$

23. $(-6) + (-3) + \frac{(-3)}{2} + \ldots$

24. $(-8) + (-4) + (-2) + \ldots$

25. $2 + (-1) + \frac{1}{2} + \ldots$

26. $9 + (-3) + (1) + \ldots$

27. $4 + (-8) + (16) + \ldots$

28. $\frac{-1}{10} + \frac{1}{5} + \left(\frac{-2}{5}\right) + \ldots$

29. $(-4) + (-2) + (-1) + \ldots$

30. $(-5) + (-10) + (-20) + \ldots$

In Problems 31 through 40, write the given decimal as an equivalent fraction.

Illustration: $3.222\ldots$

Solution: $3.222\ldots = 3 + 0.2 + 0.02 + 0.002 + \ldots$

$$= 3 + \frac{2}{10} + \frac{2}{100} + \frac{2}{1000} + \ldots$$

We now find

$$S_\infty = \frac{2}{10} + \frac{2}{100} + \frac{2}{1000} + \cdots.$$

$$= \frac{a_1}{1 - r}$$

Here $a_1 = \frac{2}{10}$ and $r = \frac{1}{10}$; hence

$$S_\infty = \frac{\frac{2}{10}}{1 - \frac{1}{10}} = \frac{\frac{2}{10}}{\frac{9}{10}} = \frac{2}{9}$$

Thus, $3.222\ldots = 3 + \frac{2}{9} = \frac{29}{9}$

31. 0.444 . . .

32. 0.555 . . .

33. 0.313131 . . .

34. 0.454545 . . .

35. 0.272727 . . .

36. 4.060606 . . .

37. 2.3161616 . . .

38. 0.4303030 . . .

39. 0.140140140 . . .

40. 1.123123123 . . .

41. The population of a certain town increases at the rate of 4% per year. If the present population is 200,000, what will be the population in 5 years?

42. The population of a town has increased geometrically from 59,049 to 100,000. If this growth occurred in 5 years, what is the growth rate for the town?

43. The number of bacteria in a culture increased from 64,000 to 125,000 in 6 days. Find the daily rate of increase, if this rate is assumed to be constant.

44. When dropped on a hard surface, a Super Ball takes a series of bounces, each bounce being about $\frac{9}{10}$ as high as the preceding bounce. If a Super Ball is dropped from a height of 10 feet, find the approximate distance the ball travels before coming to rest.

45. Repeat Problem 44 if the ball is dropped from a height of 20 feet.

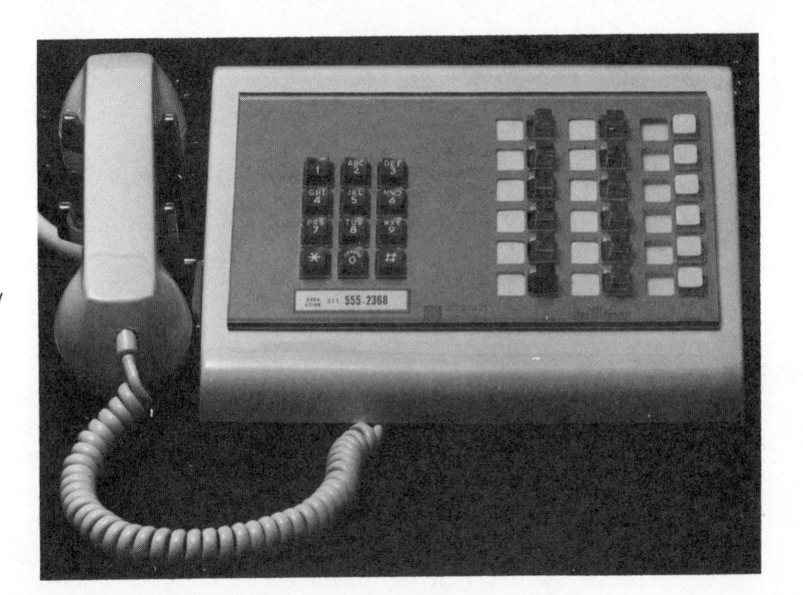

Photo credit: A. T. & T. Company Photo Center.

OBJECTIVES 11.4

After studying this section, the reader should be able to:

1. **Find the value of expressions involving numbers written using factorial notation.**
2. **Evaluate $\binom{n}{r}$ when n and r are natural numbers and n is larger than or equal to r.**
3. **Expand a binomial by using the binomial theorem given in the text.**
4. **Find the rth term of a binomial expansion.**

11.4 THE BINOMIAL THEOREM

The photograph at the beginning of this section shows a telephone with 12 black buttons for extensions. Do you know how many different conversations between two parties are possible if there is a total of 12 telephones, one at each extension? To answer this question, we need to know how many *combinations* of 2 objects are possible when 12 objects are available. In mathematics, the number of combinations of r objects that can be formed when n objects are available is denoted by

$$(1)\quad \binom{n}{r}$$

$\binom{n}{r}$ means the number of combinations of r objects that can be formed when n objects are available.

In the case of the 12 telephones, we need to find

$$(2)\quad \binom{12}{2}$$

To facilitate defining $\binom{n}{r}$, we first introduce a special symbol, n! (read "n factorial"), defined as:

$$(3) \quad n! = n(n-1)(n-2) \ldots (1), \quad n = 1, 2, 3, \ldots$$

Thus,

$$5! = 5 \cdot 4 \cdot 3 \cdot 2 \cdot 1 = 120$$

$$7! = 7 \cdot 6 \cdot 5 \cdot 4 \cdot 3 \cdot 2 \cdot 1 = 5{,}040$$

This notation can also be used to denote the product of consecutive integers beginning with integers different from 1. For example,

Note that in writing 7 • 6 • 5 in factorial notation, we write 7! (the first number in 7 • 6 • 5) in the numerator, and 4! (one less than the last number in 7 • 6 • 5) in the denominator.

$$7 \cdot 6 \cdot 5 = \frac{7!}{4!}$$

since

$$\frac{7!}{4!} = \frac{7 \cdot 6 \cdot 5 \cdot \not{4} \cdot \not{3} \cdot \not{2} \cdot \not{1}}{\not{4} \cdot \not{3} \cdot \not{2} \cdot \not{1}}$$

Similarly,

$$9 \cdot 8 \cdot 7 \cdot 6 = \frac{9!}{5!}$$

and

$$10 \cdot 9 \cdot 8 \cdot 7 = \frac{10!}{6!}$$

Also,

$$\begin{aligned} n! &= n[(n-1)(n-2)(n-3) \ldots (1)] \\ &= n(n-1)! \end{aligned}$$

Thus, we can write

$$(4) \quad n! = n(n-1)!$$

With this notation,

$$9! = 9 \cdot 8!$$
$$13! = 13 \cdot 12!$$

If we let n = 1 in Equation (4), we obtain

$$1! = 1 \cdot 0!$$

Hence, we define 0! as follows:

$$0! = 1$$

Example 1

Find
(a) 4!
(b) $\dfrac{8!}{5!}$

Solution

(a) By Equation (3), $4! = 4 \cdot 3 \cdot 2 \cdot 1 = 24$.
(b) By Equation (3), $8! = 8 \cdot 7 \cdot 6 \cdot 5 \cdot 4 \cdot 3 \cdot 2 \cdot 1$ and $5! = 5 \cdot 4 \cdot 3 \cdot 2 \cdot 1$. Thus,

$$\frac{8!}{5!} = \frac{8 \cdot 7 \cdot 6 \cdot \cancel{5} \cdot \cancel{4} \cdot \cancel{3} \cdot \cancel{2} \cdot \cancel{1}}{\cancel{5} \cdot \cancel{4} \cdot \cancel{3} \cdot \cancel{2} \cdot \cancel{1}} = 336$$

We are now ready to use factorial notation to define the symbol given in Equation (1). We do this as follows:

DEFINITION 11.2

$$\binom{n}{r} = \frac{n!}{r!(n-r)!}$$

Thus,

$$\binom{5}{2} = \frac{5!}{2!(5-2)!} = \frac{5!}{2!\,3!} = \frac{5 \cdot 4 \cdot \cancel{3!}}{2!\,\cancel{3!}} = 10$$

$$\binom{5}{5} = \frac{5!}{5!(5-5)!} = \frac{5!}{5!\,0!} = 1$$

The number of conversations possible with the 12 telephones is

$$\binom{12}{2} = \frac{12!}{2!\,10!} = \frac{12 \cdot 11 \cdot \cancel{10!}}{2!\,\cancel{10!}} = 66$$

Example 2

Evaluate $\binom{6}{4}$.

Solution

By Definition 11.2,

$$\binom{6}{4} = \frac{6!}{4!\,2!} = \frac{6 \cdot 5 \cdot \cancel{4!}}{\cancel{4!}\,2!} = \frac{6 \cdot 5}{2} = 15$$

The notation $\binom{n}{r}$ is sometimes used in algebra to obtain the expansion of the binomial

$$(a + b)^n$$

From our previous work, we know the following results:

$$(a + b)^1 = a + b$$
$$(a + b)^2 = a^2 + 2ab + b^2$$
$$(a + b)^3 = a^3 + 3a^2b + 3ab^2 + b^3$$
and $$(a + b)^4 = a^4 + 4a^3b + 6a^2b^2 + 4ab^3 + b^4$$

Before attempting to write the formula for $(a + b)^n$, we make the following observations on aspects which appear to be common to the four expansions:

1. The first term is a^n and the last term is b^n.
2. The exponents of a *decrease* by 1 from term to term, while those of b *increase* by 1 from term to term.
3. The coefficients of each term can be written using the $\binom{n}{r}$ notation, where n is the exponent to which the binomial is being raised and r is the exponent of a. Thus, in the expansion $(a + b)^3 = a^3 + 3a^2b + 3ab^2 + b^3$, $n = 3$, and the coefficient of $3a^2b$, which is 3, can be written as $\binom{3}{2}$, where 2 represents the exponent of a. Similarly, the coefficient of b^3, which is 1, can be represented by $\binom{3}{0}$, where 0 is the exponent of a.

Note that since 0 is the exponent of a, and $a^0 = 1$, a^0b^3 is simply written as b^3 in the expansion.

Example 3

Given that $(a + b)^4 = a^4 + 4a^3b + 6a^2b^2 + 4ab^3 + b^4$, write each coefficient in the expansion using the $\binom{n}{r}$ notation.

Solution

$$(a + b)^4 = a^4 + 4a^3b + 6a^2b^2 + 4ab^3 + b^4$$

$$= \binom{4}{4} a^4 + \binom{4}{3} a^3b + \binom{4}{2} a^2b^2 + \binom{4}{1} ab^3 + \binom{4}{0} b^4$$

The student can easily verify that

$$\binom{4}{4} = 1, \binom{4}{3} = 3, \binom{4}{2} = 6, \binom{4}{1} = 4, \text{ and } \binom{4}{0} = 1$$

From Example 3, it is evident that the formula for $(a + b)^n$ can be written using the $\binom{n}{r}$ notation, where, as before, n represents the exponent of the binomial being expanded and r is the exponent of the a. When we also recall that the exponents of a *increase* by 1 from term to term while those of b *decrease* by 1 from term to term, we write the binomial expansion as in the next equation:

This general result is proved in more advanced algebra courses.

$$(5)\quad (a + b)^n = \binom{n}{n} a^n + \binom{n}{n-1} a^{n-1}b^1 + \binom{n}{n-2} a^{n-2}b^2 + \ldots + \binom{n}{0} b^n$$

Thus,

$$(a + b)^5 = \binom{5}{5} a^5 + \binom{5}{4} a^4b^1 + \binom{5}{3} a^3b^2 + \binom{5}{2} a^2b^3 + \binom{5}{1} a^1b^4 + \binom{5}{0} b^5$$
$$= a^5 + 5a^4b^1 + 10a^3b^2 + 10a^2b^3 + 5ab^4 + b^5$$

Note that to avoid confusion, we first write the expansion completely, and *then* substitute the coefficients

$$\binom{5}{5} = 1, \binom{5}{4} = 5, \binom{5}{3} = 10, \binom{5}{2} = 10, \binom{5}{1} = 5, \text{ and } \binom{5}{0} = 1$$

in the expansion.

Example 4

Expand $(x - 2y)^4$.

Solution

Here $n = 4$, and we have x instead of a and $-2y$ instead of b. Making these substitutions in (5), we obtain

$$(x-2y)^4 = \binom{4}{4} x^4 + \binom{4}{3} x^3(-2y)^1 + \binom{4}{2} x^2(-2y)^2 + \binom{4}{1} x^1(-2y)^3 + \binom{4}{0} (-2y)^4$$

$$= x^4 + 4x^3(-2y)^1 + 6x^2(-2y)^2 + 4x(-2y)^3 + (-2y)^4$$

$$= x^4 + 4x^3(-2y) + 6x^2(4y^2) + 4x(-8y^3) + (16y^4)$$

$$= x^4 - 8x^3y + 24x^2y^2 - 32xy^3 + 16y^4$$

PROGRESS TEST

1. $8!$ equals __________.
2. $\frac{10!}{7!}$ equals __________.
3. $\binom{7}{3}$ equals __________.
4. $(a+2b)^3$ equals ____________________.

EXERCISE 11.4

In Problems 1 through 14, evaluate the given expression.

Illustration: (a) $9!$

(b) $\frac{11!}{9!}$

(c) $\binom{11}{2}$

Solution: (a) $9! = 9 \cdot 8 \cdot 7 \cdot 6 \cdot 5 \cdot 4 \cdot 3 \cdot 2 \cdot 1 = 362{,}880$

(b) $\frac{11!}{9!} = \frac{11 \cdot 10 \cdot \cancel{9!}}{\cancel{9!}} = 110$

(c) $\binom{11}{2} = \frac{11!}{2!\,9!} = \frac{11 \cdot 10 \cdot \cancel{9!}}{2!\,\cancel{9!}}$

$$= \frac{110}{2} = 55$$

PROGRESS TEST ANSWERS

1. $8! = 8 \cdot 7 \cdot 6 \cdot 5 \cdot 4 \cdot 3 \cdot 2 \cdot 1 = 40{,}320.$
2. $\frac{10 \cdot 9 \cdot 8 \cdot 7!}{7!} = 720$
3. $\binom{7}{3} = \frac{7!}{3!\,4!} = \frac{7 \cdot 6 \cdot 5 \cdot 4!}{3!\,4!} = 35$
4. $a^3 + 6a^2b + 12ab^2 + 8b^3$

1. $3!$

2. $6!$

3. $10!$

4. $2!$

5. $\frac{6!}{4!}$

6. $\frac{11!}{10!}$

7. $\frac{9!}{6!}$

8. $\frac{3!}{0!}$

9. $\binom{6}{2}$

10. $\binom{6}{5}$

11. $\binom{11}{1}$

12. $\binom{11}{0}$

13. $\binom{4}{0}$

14. $\binom{7}{3}$

In Problems 15 through 20, simplify the given expression.

Illustration: $\frac{(n+2)!}{(n-1)!}$

Solution:

$$\frac{(n+2)!}{n-1)!} = \frac{(n+2)(n+1)(n)(n-1)(n-2)(n-3)\ldots(1)}{(n-1)(n-2)(n-3)\ldots(1)}$$

$$= (n+2)(n+1)(n)$$

15. $\frac{(n+3)!}{n!}$

16. $\frac{(n-1)!}{(n-3)!}$

17. $\frac{(n+2)!}{(n-1)!}$

18. $\frac{(n+2)(n+3)!}{(n+4)!}$

19. $\frac{(2n+4)!}{(2n+2)!}$

20. $\frac{(2n+1)!}{(2n-1)!}$

In Problems 21 through 30, follow the procedure of

Example 4 and use Equation (5) to expand the given binomial.

21. $(a+3b)^3$

22. $(a-3b)^3$

23. $(x+4)^4$

24. $(x-4)^4$

25. $(2x-1)^5$

26. $(2x+1)^5$

27. $(2x-3y)^3$

28. $(2x+3y)^3$

29. $\left(\frac{1}{x}+\frac{1}{y}\right)^3$

30. $\left(\frac{1}{x}-\frac{1}{y}\right)^3$

In Problems 31 through 35, use the fact that the rth term of the binomial expansion of $(a+b)^n$ is given by

$$\binom{n}{n-r+1} a^{n-r+1} b^{r-1}$$

Illustration: Find the fifth term in the expansion of $(x+3)^8$.

Solution: Here $n=8$, $r=5$, $x=a$ and $b=8$. Substituting in the formula above, we have

$$\binom{8}{8-5+1} x^{8-5+1}(3)^{5-1}$$

$$=\binom{8}{4} x^4(3)^4$$

$$=\frac{8 \bullet 7 \bullet 6 \bullet 5 \bullet 4!}{4!4!} x^4(3)^4 = 70x^4(81)$$

$$=5{,}670x^4$$

31. Find the fourth term in the expansion of $(x + 3)^8$

32. Find the fourth term in the expansion of $(x - 2)^8$

33. Find the fifth term in the expansion of $(2x - y)^6$.

34. Find the sixth term in the expansion of $(3x - 2y)^7$.

35. Find the third term in the expansion of $(3x - 4y)^3$.

SELF-TEST–CHAPTER 11

1. Consider the sequence 3, 5, 7, ...

_______________ (a) Find the eighth term in this sequence.

_______________ (b) Find s_4.

---

2. Consider the sequence associated with $s(n) = 2n + 1$, $n \in \{1, 2, 3, \ldots\}$.

_______________ (a) Find the first five terms of the sequence.

_______________ (b) Find $s(81)$

---

3. Consider the arithmetic progression 6, 9, 12, 15 . . . .

_______________ (a) Find a_1

_______________ (b) Find the common difference d.

---

4. The first term of an arithmetic progression is 11 and its common difference is 5. Find:

_______________ (a) a_5

_______________ (b) a_n

---

5. Find the sum of:

_______________ (a) $2 + 4 + 6 + \ldots + 100$

_______________ (b) $2 + 4 + 6 + \ldots + 2n$ (that is, S_n, the sum of the first n terms of the given sequence)

---

6. The sum of the first 6 terms of an arithmetic progression is 48 and the sixth term is 13. Find:

_______________ (a) a_1, the first term

_______________ (b) d, the common difference

---

7. Consider the geometric sequence $\frac{1}{2}, \frac{1}{4}, \frac{1}{8}, \ldots \left(\frac{1}{2}\right)^n, \ldots$.

_______________ (a) Find r, the common ratio.

_______________ (b) Find, if possible, S_∞.

---

8. The first term of a geometric progression is 1. The sum of the first three distinct terms is 3. Find:

_______________ (a) a_3

_______________ (b) r, the common ratio

---

9. Find:

_______________ (a) $6!$

_______________ (b) $\frac{8!}{6!}$

_______________ (c) $\binom{6}{2}$

---

10. Expand the binomial $(a-2b)^3$ and show all the terms in simplified form.

ANSWERS

EXERCISE 1.1

1. correct
3. Not correct
5. {1, 2, 3, 4, 5, 6, 7}
7. {. . . , −1, 0, 1, 2, 3, 4, 5, 6, 7}
9. {M, i, s, p}
11. The set of whole numbers less than 2.
13. The set of integers between −4 and 0.
15. The set consisting of the first and last letters in the English alphabet.
17. The set of natural numbers less than 6.
19. {x | x is an odd natural number less than 6}
21. {x | x is a natural number less than 6}
23. {x | x is a letter in the word "woman"}
25. $\in$
27. $\in$
29. $\in$
31. =
33. $\neq$
35. (a) $P \neq Q$ (b) $\{1, 3, 5\} = P$
37. { }
39. $\{\sqrt{2}, \pi\}$
41. $\left\{\frac{1}{3}\right\}$
43. {−4}

EXERCISE 1.2

1. Correct
3. Incorrect
5. Incorrect
7. {2, 3}
9. {1}
11. {1, 2, 3, 4}
13. {3}
15. {3}
17. {1, 2}, {1, 3}, {1, 4}, {2, 3}, {2, 4}, {3, 4}

19. (a) No. If a number is divisible by 2, it is not necessarily divisible by 4.
(b) Yes. If a number is divisible by 4, it is also divisible by 2.

21. $\mathcal{U} = \{1, 2, 3, 4, 5\}$

23. (a) $\mathcal{U}$ (b) $\emptyset$ (c) $\emptyset$

25. (a) $D \cap S$ (b) $F \cap T$ (c) $M \cap D$

27. (a)

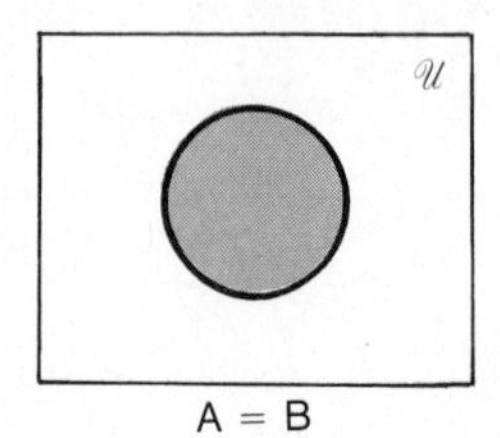

(b)

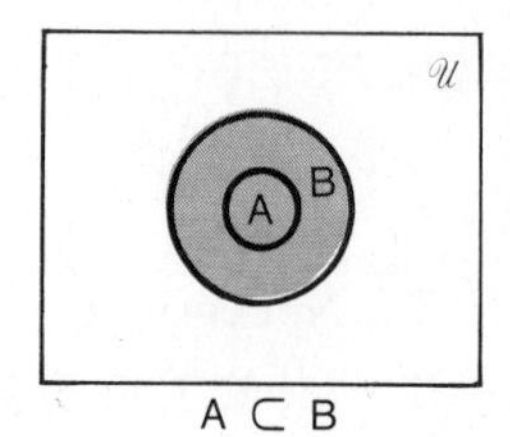

29. (a) Florida (b) Texas

31. (a) $2^2 = 4$ (b) $2^4 = 16$ (c) 2^n

EXERCISE 1.3

1. −8 −7 −6 −5 −4 −3 −2 −1 0 1 2 3 4 5 6 7 8

3. −1 0 ¾ 1

5. −3¼
−5 −4 −3 −2 −1 0 1 2 3 4 5

7. 4⅓
−5 0 1 2 3 4 5

9. $-2\frac{1}{4}$

11. 1

13. $2\frac{1}{5}$

15. $4\frac{1}{2}$

17. $0.333\ldots = 0.\overline{3}$

19. $0.\overline{571428}$

21. $0.8333\ldots = 0.8\overline{3}$

23. $\frac{11}{100}$

25. $\frac{73}{99}$

27. $\frac{781}{333}$

29. (a) $\frac{182}{99}$
(b) $\frac{15}{7}$
(c) $\frac{23}{7}$

31. Rational

33. Irrational

35. $\sqrt{9} = 3$, which is rational

37. Rational
39. Rational
41. Rational
43. $\frac{313}{99}$
45. $\frac{7}{6}$

EXERCISE 1.4

1. F-3
3. F-1
5. F-4
7. E-4
9. E-4
11. F-6
13. F-5
15. F-3
17. E-3
19. F-7
21. F-8
23. E-2
25. E-4
27. Th. 1.3
29. Th. 1.2

EXERCISE 1.5

1. 20
3. -0.1
5. 2
7. -0.8
9. $\frac{1}{5}$
11. 9
13. $0.\overline{33}$
15. 5
17. $3 + (-7)$
19. $2x + (-3z)$
21. $-3a + (-2b)$
23. $-x + 7y$
25. 8
27. -3
29. -0.9
31. 5
33. $\frac{2}{3}$
35. 0
37. -0.7
39. -0.3
41. 0.5
43. -0.6
45. 0
47. 3
49. 3
51. 16,000 feet
53. (a) $476 - (-216) = 692$ years
    (b) $1939 - (-323) = 2{,}262$ years
55. 20 kilometers

EXERCISE 1.6

1. -40
3. -12
5. 50
7. 60
9. 40
11. 30
13. -40
15. 16
17. -125
19. -1296
21. 32
23. 243
25. -2
27. -4
29. 2
31. 0
33. Not defined
35. -2
37. 9
39. 3
41. $-3 = (-3)(1)$
43. $-16 = (4)(-4)$
45. $-56 = (8)(-7)$
47. $5 \cdot \left(\frac{1}{7}\right)$
49. $3 \cdot \left(\frac{1}{8}\right)$

51. $41 \cdot \left(\frac{1}{43}\right)$

53. $-7 \cdot \left(\frac{1}{11}\right)$

55. $-2 \cdot \left(\frac{1}{7}\right)$

57. $4 \cdot \left(\frac{1}{-3}\right)$

59. $5 \cdot \left(\frac{1}{-6}\right)$

61. 54

63. 38,325 pints

65. (a) 54 (b) 72 (c) 99

67. 160

69. $15

EXERCISE 1.7

1. (a) 34 (b) 70
3. (a) 27 (b) 3
5. 53
7. 34
9. 9
11. −13
13. 24
15. 4
17. 3
19. 0
21. 1
23. −55
25. 3
27. −6
29. 1
31. 10
33. 3
35. 18
37. 6
39. 110

SELF-TEST—CHAPTER 1

1. (a) {5, 6, 7, 8, 9} (b) {1, 3, 5, 7, 9} (c) {T, a, 1, h, s, e}
2. (a) {x | x is a natural number less than 5}
   (b) {x | x is a vowel in the English alphabet}
3. (a) { } or ∅
   (b) {1, 2, 3, 4, 5, 6, 8, 10}
   (c) {1, 2, 3, 4, 5, 7, 9}
4. (a) Irrational (b) Rational, $\frac{31}{99}$ (c) Irrational
   (d) Rational, $\frac{157}{50}$ (e) Rational, $\frac{41}{100}$
5. (a) Associative Law of Addition
   (b) Negative or Opposite
   (c) Distributive Law
6. (a) 4 (b) 8 (c) −2
7. (a) 6 (b) 6 (c) −1
8. (a) 24 (b) −56 (c) 16 (d) −27 (e) −4
9. (a) 8 (b) −2
10. (a) 6 (b) 1

EXERCISE 2.1

1. Two terms
   The coefficient of x is 1
   The coefficient of 2 is 2
3. Two terms
   The coefficient of 2x is 2
   The coefficient of $-x^2$ is -1
5. One term
   The coefficient of $3(x+1)$ is 3
7. Two terms
   The coefficient of $(x+2)^2$ is 1
   The coefficient of 2 is 2
9. One term
   The coefficient of $a(b+c)$ is a
11. 16x
13. $-4y$
15. 0
17. 6xy
19. $x-1$
21. $3x+2$
23. $6x+3$
25. $5x-8$
27. $6x+2$
29. $-x-1$
31. 2h
33. $2.89h+70.64$
35. $p+83$
37. $a-t$
39. $5h-190$
41. $\frac{100M}{C}$

EXERCISE 2.2

1. All real numbers except -8
3. All real numbers except 3
5. All real numbers except 1 and -1
7. 3 is a solution
9. -1 is a solution
11. 3 is a solution
13. $\frac{-1}{5}$ is not a solution
15. 3 is not a solution
17. $x=1$
19. $x=2$
21. An identity
23. $x=1$
25. No solution
27. No
29. No

EXERCISE 2.3

1. $\{8\}$
3. $\{2\}$
5. $\{1\}$
7. $\{6\}$
9. $\left\{\frac{10}{3}\right\}$
11. $\{4\}$
13. $\{1\}$
15. $\left\{-\frac{11}{80}\right\}$
17. $\left\{\frac{1}{8}\right\}$
19. $\{0\}$
21. $\emptyset$
23. $\emptyset$
25. $I=\frac{E}{R}$
27. $h=\frac{V}{\pi r^2}$

29. $L = \frac{P - 2w}{2}$

31. $x = \frac{y - b}{m}$

33. $W = \frac{Fgr}{v^2}$

35. $V_2 = \frac{P_1V_1}{P_2}$

37. $r = \frac{s - a}{s}$

39. $a = \frac{-bf}{f - b} = \frac{bf}{b - f}$

EXERCISE 2.4

1. 10, 11, 12
3. 51, 52, 53
5. 16 L.P.'s and 104 singles
7. 16 feet
9. 253 pounds
11. 213 pounds
13. 80 miles
15. 48
17. Width, 12 feet; Length, 5000 feet
19. Width, 518 feet; Length, 716 feet

21.

	% Anti-freeze	×	Qts. of Solution	=	Qts. of Antifreeze
Original	0.50		30 − x		0.50(30 − x)
Water	0		x		0
New Sol.	0.30		30		9

$$0.50(30 - x) = 9$$
$$x = 12$$

12 quarts of water should be added.

23.

	Price	× Amount	=	Total Price
Adoration	370	x		370x
Another	160	1 − x		160(1 − x)

$$370x + 160(1 - x) = 300$$

$\frac{2}{3}$ ounces of Adoration, $\frac{1}{3}$ ounces of the other.

25.

	Price	× Amount	=	Total Price
Copper	65	x		65x
Zinc	40	50 − x		40(50 − x)
Brass	50.75	50		2537.5

$$65x + 40(50 - x) = 2537.5$$
$$x = 21.5$$

21.5 pounds of copper, 28.5 pounds of zinc.

27.

	%	× Amt. of Solution	= Total Salt
10%	0.10	x	0.10x
20%	0.20	15	3
16%	0.16	(x + 15)	0.16(x + 15)

$$0.10x + 3 = 0.16(x + 15)$$
$$x = 10$$

10 gallons

29.

	% of Carbohydrate	× Amount	= Pure Carbohydrate
20%	0.20	x	0.20x
60%	0.60	100	60
50%	0.50	x + 100	0.50(x + 100)

$$0.20x + 60 = 0.50(x + 100)$$
$$x = 33\frac{1}{3}$$

$33\frac{1}{3}$ grams

31.

	% of Vermouth	× Amount	= Vermouth
40%	0.40	x	0.40x
20%	0.20	32 − x	0.20(32 − x)
30%	0.30	32	9.60

$$0.40x + 0.20(32 - x) = 9.60$$
$$x = 16$$

16 ounces

33.

	% of Peroxide	× Amount	= Pure Peroxide
90%	0.90	x	0.90x
98%	0.98	3000 − x	0.98(3000 − x)
96%	0.96	3000	2880

$$0.90x + 0.98(3000 - x) = 2880$$
$$x = 750$$

750 gallons

35.

	% of Acid	× Amount	= Pure Acid
0%	0	x	0
8%	0.08	100 − x	0.08(100 − x)
6%	0.06	100	6

$$0.08(100 - x) = 6$$
$$x = 25$$

25 cubic centimeters

EXERCISE 2.5

1.

	R	× T	= D
H	90	T − 2	90(T − 2)
C	60	T	60T

$$90(T - 2) = 60T$$
$$T = 6$$
$$T - 2 = 4$$

It takes 4 hours.

3.

	R	× T	= D
Jet	(5R − 150)	1/2	1/2(5R − 150)
Sm. plane	R	2	2R

$$1/2(5R - 150) = 2R$$
$$5R - 150 = 4R$$
$$R = 150 \text{ mph}$$

150 mph for the small plane
600 mph for the jet

5.

	R	× T	= D
P_1	480	T	480T
P_2	480	T − 1	480(T − 1)

$$480T + 480(T - 1) = 1632$$
$$T = \frac{2112}{960} = 2.2 \text{ hrs.}$$

They meet at 11:12.

7.

	R	×	T	=	D
G	240		T		240T
C	320		7 − T		320(7 − T)

$$320(7 - T) = 240T$$
$$T = 4$$

$240 \cdot 4 = 960$ miles

9. Let T be the time she walked.

$$4T + 40(1.5 - T) = 15$$
$$4T + 60 - 40T = 15$$
$$T = \frac{45}{36} \text{ hrs., or } 1\frac{1}{4} \text{ hrs.}$$

She rode $\frac{1}{4}$ of an hour.

11.

	P	×	R	=	I
2½ Inv.	P		0.025		0.025P
3¼ Inv.	15,000 − P		0.035		0.035(15,000 − P)

$$0.025P + 0.035(15{,}000 - P) = 435$$
$$P = \$9{,}000$$

$9,000 was invested at 2.5%
$6,000 was invested at 3.5%

13.

	P	×	R	=	I
5% Inv.	8000		0.05		400
10% Inv.	P		0.10		0.10P
TOTAL Inv.	8000 + P		0.06		0.06(8000 + P)

$$400 + 0.10P = 0.06(8000 + P)$$
$$P = 2000$$

She should invest $2000.

15.

	P	×	R	=	I
15% Ring	P		0.15		0.15P
10% Ring	6000 − P		0.10		0.10(6000 − P)
TOTAL	6000		0.11		660

$$0.15P + 0.10(6000 - P) = 660$$
$$P = 1200$$

$1200 for one kind, $4800 for the other.

17.

	P	×	R	=	I
5% Inv.	P		0.05		0.05P
6% Inv.	20,000 − P		0.06		0.06(20,000 − P)

$$0.05P - 10 = 0.06(20{,}000 - P)$$
$$P = \$11{,}000$$

\$11,000 are invested at 5%
\$9,000 are invested at 6%

19.

	P	×	R	=	I
12% Inv.	$\left(\frac{1}{3}\right)P$		0.12		$0.12 \cdot \left(\frac{1}{3}\right)P$
8% Inv.	$\left(\frac{2}{3}\right)P$		0.08		$0.08 \cdot \left(\frac{2}{3}\right)P$

$$0.12 \cdot \left(\frac{1}{3}\right)P + 0.08 \cdot \left(\frac{2}{3}\right)P = 784$$
$$P = 8400$$

EXERCISE 2.6

1. $\{x \mid x \le 2\}$

3. $\{x \mid x > 2\}$

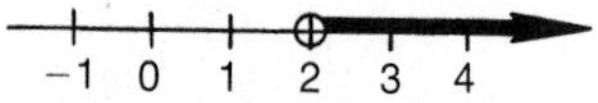

5. $\{x \mid x < 3\}$

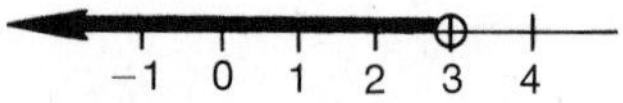

7. $\{x \mid x \ge 2\}$

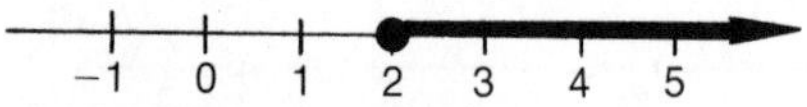

9. $\{x \mid x < 3\}$

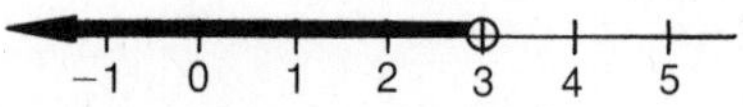

11. $\{x \mid x < 6\}$

13. $\{x \mid x < 4\}$

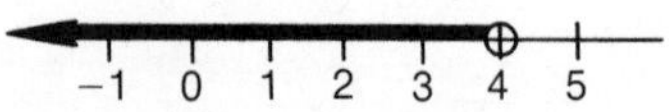

15. $\left\{x \,\middle|\, x < \frac{9}{2}\right\}$

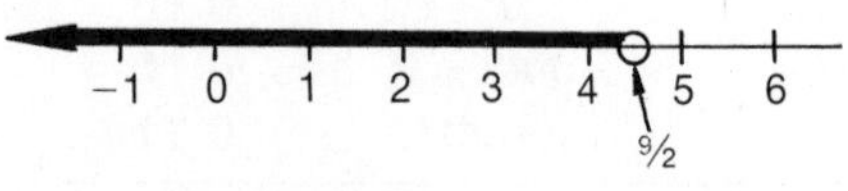

17. $\left\{x \,\middle|\, -\frac{1}{2} \le x \le 2\right\}$

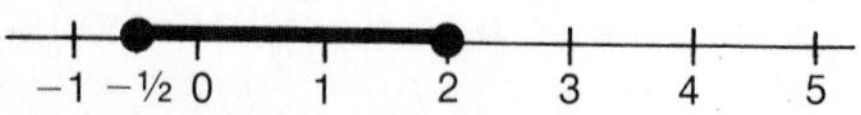

19. $\{x \mid -2 \le x \le 4\}$

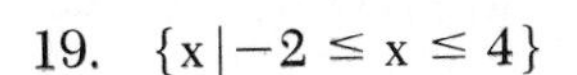
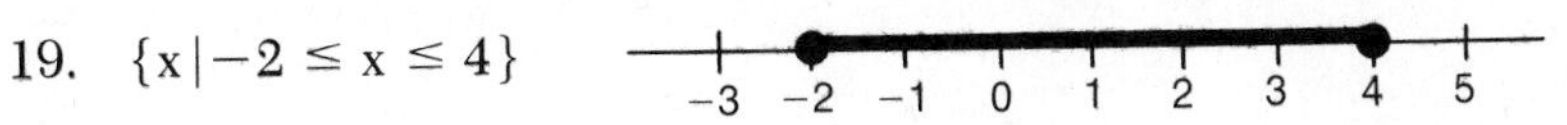

21. $\{x \mid -7 < x < -3\}$

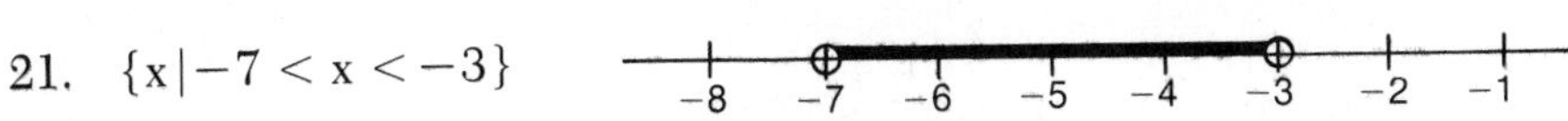

23. $\{x \mid -2 \le x \le 1\}$

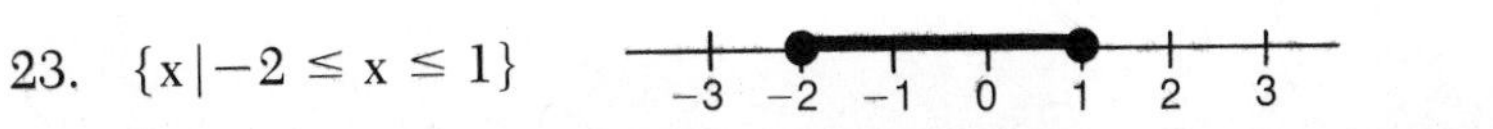

25. $\{x \mid x \le -1 \text{ or } x \ge 3\}$

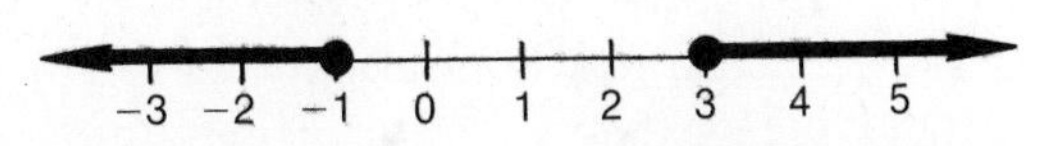

27. $\{x \mid x \le -1 \text{ or } x \ge 2\}$

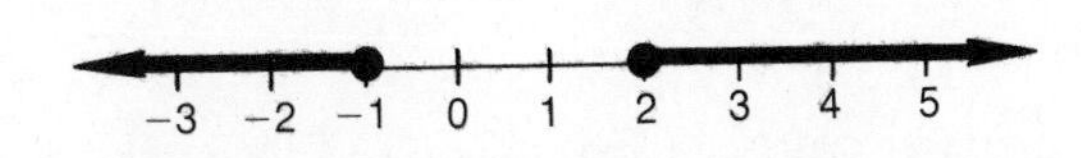

29. $\left\{x \,\middle|\, x < -1 \text{ or } x > \frac{5}{2}\right\}$

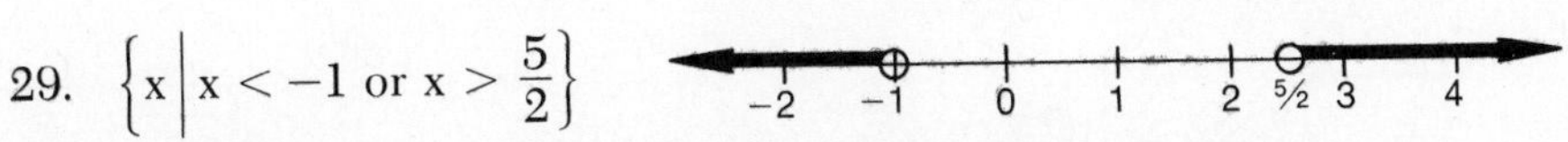

31. 94 or better
33. $2,500
35. $2,500

SELF-TEST – CHAPTER 2

1. (a) $4x - 4$
   (b) 6
2. (a) All real numbers except −2
   (b) All real numbers except −2
3. (a) $x = 8$
   (b) $x = \frac{3}{2}$
   (c) $x = -30$
4. (a) ∅
   (b) $\{6\}$
5. (a) 415 (b) 265
6. 32
7. $500
8. 300 miles

9. (a) $x < -\frac{1}{2}$

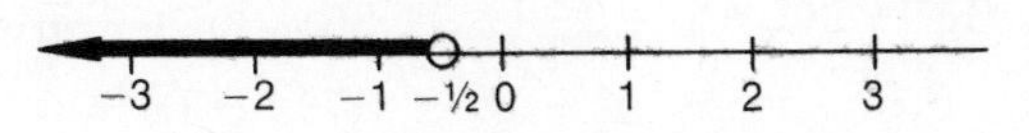

(b) $-3 \le x \le 2$

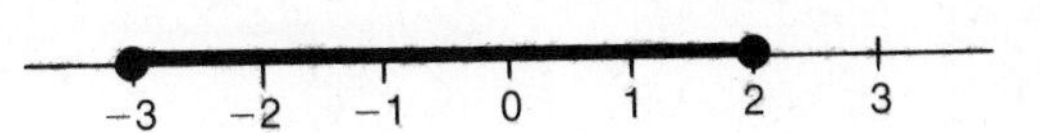

10. (a)

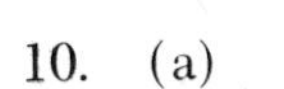

(b)

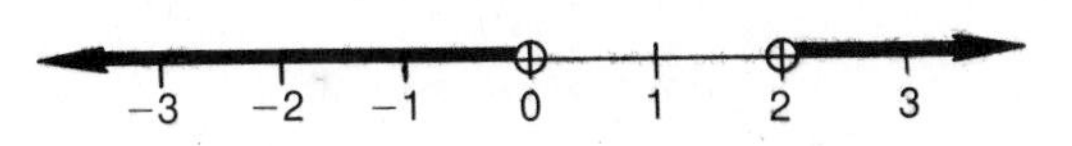

EXERCISE 3.1

1. Monomial. Degree 4
3. Binomial. Degree 3
5. Monomial. Degree 3
7. Trinomial. Degree 5
9. Zero polynomial. No degree
11. $x^3 + y^4$
13. $x^4 + y^4 + z^4$
15. $3x^2 - 4y^3 + 5(x + y)^2$
17. $z^3 = (-2)^3 = -8$
19. $(x - 2y + z)^2 = [2 - 2 \cdot (-1) - 2]^2$
    $= [2 + 2 - 2]^2$
    $= 4$
21. -1
23. 16
25. $R(-1) = -1$, $S(1) = 2$
    So, $R(-1) - S(1) = -3$

EXERCISE 3.2

1. $6x^2 - 5$
3. $x^2 - 2x - 4$
5. $3x^2 + 4x - 9$
7. $-5y^2 - 13y - 3$
9. $4x^3 - 12x^2 + 9x - 6$
11. $7y^2 + 12y - 11$
13. $3v^3 - 2v^2 + v - 7$
15. $-5u^3 - 5u^2 - 3u + 10$
17. $4x^3 + 2xy - 1$
19. $x^3 - x^2 - 4$
21. $-2a^2 + 5a$
23. $4y$
25. $3x^2 + 7y$
27. Commutative Law
29. Distributive Law
31. Associative Law
33. Commutative Law
35. Distributive Law
37. $P(0) = 0^2 - 2 \cdot 0 + 3 = 3$
    $Q(0) = 2 \cdot 0^2 + 3 \cdot 0 - 1 = -1$
    $P(0) + Q(0) = 3 + (-1) = 2$
39. $P(x) - P(x) = 0$

EXERCISE 3.3

1. $20a^2$
2. $12x^4y^2$
5. $-24u^3v^4w^4$
7. x^{2n+1}
9. x^{14}
11. a^6b^3
13. $16a^4b^8$
15. $108x^2y^6$
17. $-576x^8y^5$
19. $2^{2n}x^{4n}y^{3n}z^{3n}$
21. a^5
23. $2xy$
25. $3xy^3z^2$
27. $-x^6y$
29. x^4y^5
31. $3a^5b^4c^6$

EXERCISE 3.4

1. $12x^2 - 6x$
3. $-3x^3 + 9x^2$
5. $-24x^3 + 16x^2 - 8x$
7. $9x^2 + 12x + 3$
9. $5x^2 + 11x - 12$
11. $3a^2 + 14a - 5$
13. $2y^2 + 7y - 15$
15. $x^2 - 8x + 15$
17. $6x^2 - 7x + 2$
19. $4x^2 + 4ax - 15a^2$
21. $x^2 + 15x + 56$
23. $4a^2 + 10ab + 4b^2$
25. $16u^2 + 8uv + v^2$
27. $4y^2 + 4yz + z^2$
29. $9a^2 - 6ab + b^2$
31. $a^2 - b^2$
33. $25y^2 - 4y^2$
35. $b^2 - 9a^2$
37. $3x^3 + 9x^2 + 6x$
39. $-3x^3 + 12x^2 - 9x$
41. $x^3 + 6x^2 + 9x$
43. $-2x^3 + 4x^2 - 2x$
45. $4x^2y^2 - y^4$
47. $-3x^2 - 3x$
49. $4a + 3$

EXERCISE 3.5

1. $4(x + 4)$
3. $8(a - 2)$
5. $-6(x - 7)$
7. $-3(a + 5)$
9. $6x^2(y + 2)$
11. $3xy(2x - 3y + 2)$
13. $2a^2b^2(4a^2b^2 - 3 + 2ab)$
15. $6a^3(3a^3y^4 - 4b^3 + 2ab)$
17. $-5ur^2s(2u^2 + 5us + 3r)$
19. $3xy(x - 2y + 4)$
21. $(x + 1)(x + y)$
23. $(2x - 1)(y + z)$
25. $(2x - y)(5x - 3)$
27. $2(3x - y)(a - 3b)$
29. $5(a - 1)(a^2 + 1)$
31. $(x + 1)(x^n + y^n)$
33. $x^{n+1}(a + b + c)$
35. $x^n(x^3 - x^n)$

EXERCISE 3.6

1. $(x + 2)(x + 3)$
3. $(a + 2)(a + 5)$
5. $(x + 4)(x - 3)$
7. $(x + 2)(x - 1)$
9. $(x + 1)(x - 2)$
11. $(x + 2)(x - 5)$
13. $(a - 7)(a - 9)$
15. $(y - 2)(y - 11)$
17. $(x + 1)^2$
19. $(y + 11)^2$
21. $(2x + 1)^2$
23. $(3x + 5y)^2$
25. $(6a + 4)^2$
27. $(y - 1)^2$
29. $(x - 7)^2$
31. $(7a - 2x)^2$
33. $(4x - 3y)^2$
35. $(y + 8)(y - 8)$

37. $(a+c)(a-c)$
39. $(8+b)(8-b)$
41. $(6a+7b)(6a-7b)$
43. $(x+y)(x-y)$
45. $(9x+1)(x+4)$
47. $(3a+1)(a-2)$
49. $(2y+5)(y-4)$
51. $(4x-3)(x-2)$
53. $(2x+3)(3x-4)$
55. $(3a+2)(7a-1)$

EXERCISE 3.7

1. $(x+5)(x^2-5x+25)$
3. $(a+1)(a^2-a+1)$
5. $(2x+y)(4x^2-2xy+y^2)$
7. $(x-1)(x^2+x+1)$
9. $(5a-2b)(25a^2+10ab+4b^2)$
11. $(a+2b+c)(a+2b-c)$
13. $(2x-y+1)(2x-y-1)$
15. $(3y-2x+5)(3y-2x-5)$
17. $(4a+x+3y)(4a-x-3y)$
19. $(y+a-b)(y-a+b)$
21. $(x+2y)(x-2y+1)$
23. $(2x-3y)(2x+3y+1)$
25. $(x^2+4)(x+2)(x-2)$
27. $(4x^2+9)(2x+3)(2x-3)$
29. $(x^3+1)^2=[(x+1)(x^2-x+1)]^2$

SELF-TEST—CHAPTER 3

1. (a) Binomial, degree 3
   (b) Trinomial, degree 5
   (c) Monomial, degree 8
2. (a) 3 (b) 6 (c) 9
3. (a) $5x^2+x$ (b) x^2y+8xy
4. (a) $3x^2-6xy$ (b) $10x^2+17x+3$ (c) x^2-y^2
   (d) $-ab^2c^2$ (e) x^{2n}
5. (a) $4x^7y^9$ (b) $-4a^2b^2c^4$
6. (a) $2(x+3)$ (b) $(x+4)(x-4)$ (c) $(4x-1)(x-1)$
   (d) $(x+y)(x^2-xy+y^2)$ (e) $(x-y)(x^2+xy+y^2)$
7. (a) $(ax+1)(x+1)$ (b) $(y-3x)(y^2+3xy+9x^2)$
   (c) $(x^2+4)(x+2)(x-2)$ (d) $-5(x-2)$ (e) $4x(x+2)$
8. (a) 7 (b) 9 (c) 4
9. (a) terms (b) 4
10. (a) x^2+2x-2 (b) x^2-6

EXERCISE 4.1

1. $\frac{4xy^2}{6y^3}$
3. $\frac{-a}{y-x}$
5. $\frac{x(x-y)}{x^2-y^2}$
7. $\frac{x(x+y)}{x^2-y^2}$
9. $\frac{x}{3y-2x}$
11. $\frac{y}{2}$
13. $\frac{-x}{x-5}=\frac{x}{5-x}$
15. $\frac{-2x}{5y}$
17. $\frac{x+y}{x-y}$
19. $\frac{1}{x-2}$
21. $\frac{x^3}{y^3}$
23. 3
25. $\frac{1}{3x+2y}$
27. $\frac{(x-y)^2}{x+y}$
29. $y-1$
31. $\frac{x+y}{x-y}$
33. $\frac{y-5}{y+6}$
35. -1
37. $-(x+3)$
39. $-(y^2+2y+4)$
41. $x-2$
43. $y+4+\frac{1}{y+1}$
45. $x+3$
47. $y-10-\frac{1}{y-3}$
49. $x+1$

EXERCISE 4.2

1. $\frac{3x}{5}$
3. $\frac{5x}{3}$
5. $\frac{3+2x}{5(x+2)}$
7. $\frac{x+2}{2(x+1)}$
9. $\frac{2x+5}{3(x-1)}$
11. 210
13. 180
15. $12x^2y$
17. a^2-b^2
19. $x^2(x+y)^2$
21. $\frac{2x-5}{(x-2)(x-3)}$
23. $\frac{2a+6}{(a+4)(a+2)}$
25. $\frac{2}{(a+2)(a+4)}$
27. $\frac{2a-1}{5a-7b}$
29. $\frac{2y^2+6y+3}{y^2-9}$
31. $\frac{-y^2+5y}{y^2-4}$
33. $\frac{10x^2-13xy-6y^2}{4x^2-9y^2}$
35. $\frac{16y^2}{(x-5y)(x-3y)}$
37. $\frac{2a+3}{(a-2)(a+5)}$
39. $\frac{6x^2-28xy}{(x-2y)(x+2y)(x-3y)}$
41. $\frac{2x-3}{(2-x)(2+x)}$
43. $\frac{6x+4}{(x-3)(x+4)(x+5)}$
45. 0
47. $\frac{7x^2-9xy+14y^2+x+2y}{x^3+8y^3}$
49. $\frac{a^3-2a^2+8a-6}{a^3+27}$

EXERCISE 4.3

1. $\frac{3}{10}$
3. $\frac{2x}{3}$
5. $\frac{2y}{21z^4x^2}$
7. $\frac{4}{x+1}$
9. 1
11. $\frac{-5}{x(x+y)}$
13. $\frac{3y-2}{y-1}$
15. 1
17. $1-x$
19. $\frac{a+b}{a-b}$
21. $\frac{27}{50}$
23. $\frac{5x}{3}$
25. $\frac{9ad}{c}$
27. $\frac{3x}{x+1}$
29. $\frac{4(y+5)}{3(y+2)}$
31. $\frac{a+b}{a-b}$
33. $\frac{-(4a^2+2a+1)}{2u^2w^2}$
35. $\frac{-(1+x)}{5x}$
37. $\frac{2y+3}{3y-5}$
39. $\frac{2x-3}{5-2x}$

EXERCISE 4.4

1. $\frac{3}{4}$
3. $\frac{a}{c}$
5. $\frac{z}{xy}$
7. $\frac{2z}{5y}$
9. $\frac{1}{3}$
11. $\frac{ab-a}{b+a}$
13. $\frac{x}{xy-2}$
15. $\frac{x-y}{x^2y^2}$
17. $\frac{9}{5}$
19. $\frac{2a^2-a}{2a+1}$
21. $\frac{x^2}{2x-1}$
23. $\frac{30}{43}$

EXERCISE 4.5

1. $\{6\}$
3. $\{-5\}$
5. $\left\{\frac{1}{3}\right\}$
7. $\{-4\}$
9. $\left\{\frac{26}{9}\right\}$
11. $\left\{\frac{-1}{12}\right\}$
13. $\emptyset$
15. The set of all real numbers except $\frac{-1}{3}$
17. $\{-11\}$
19. $\{7\}$
21. $\{2\}$
23. $\{-3\}$
25. $\left\{\frac{-3}{2}\right\}$
27. $\emptyset$
29. $\{0\}$

EXERCISE 4.6

1. $\frac{T}{3} + \frac{T}{5} = 1$

$$5T + 3T = 15$$
$$8T = 15$$
$$T = \frac{15}{8} \text{ hours}$$

3. $\frac{T}{9} + \frac{T}{10} = 1$

$$10T + 9T = 90$$
$$19T = 90$$
$$T = \frac{90}{19} \text{ hours}$$

5. Let T be the time it takes the other press. In 1 hour, this press does $\frac{1}{T}$ of the job, and in 2 hours, $\frac{2}{T}$. The other press takes $\frac{1}{2}$ as much, so it does $\frac{2}{\frac{1}{2}T}$ in 1 hour. Together, they do

$$\frac{2}{T} + \frac{2}{\frac{1}{2}T} = 1$$
$$\frac{2}{T} + \frac{4}{T} = 1$$
$$T = 6$$

7. $\frac{30}{R+5} = \frac{20}{R-5}$

$$30(R-5) = 20(R+5)$$
$$30R - 150 = 20R + 100$$
$$10R = 250$$
$$R = 25$$

9. $\frac{700}{R-50} = \frac{900}{R+50}$

$$700(R+50) = 900(R-50)$$
$$700R + 35{,}000 = 900R - 45{,}000$$
$$80{,}000 = 200R$$
$$R = 400$$

11. $\frac{200}{115-w} = \frac{260}{115+w}$

$$200(115+w) = 260(115-w)$$
$$200w + 260w = (260 \cdot 115) - (200 \cdot 115)$$
$$460w = 60 \cdot 115$$
$$w = \frac{60 \cdot 115}{460} = 15$$

13. $\frac{1{,}000}{R} = \frac{950}{R - \frac{1}{4}}$

$1{,}000R - 250 = 950R$
$50R = 250$
$R = 5$

15. $\frac{T}{9} - \frac{T}{21} = 1$

$7T - 3T = 63$
$4T = 63$
$T = \frac{63}{4} = 15\frac{3}{4}$

17. $\frac{T}{7} + \frac{T}{21} = 1$

$3T + T = 21$
$4T = 21$
$T = \frac{21}{4}$

19. $\frac{T}{60} + \frac{T}{90} = 1$

$3T + 2T = 180$
$T = 36$ seconds

21. $\frac{T}{9} - \frac{T}{6} = -1$

$2T - 3T = -18$
$-T = -18$
$T = 18$

SELF-TEST—CHAPTER 4

1. (a) $\frac{8}{4}$ or 2 (b) $\frac{-a}{a-7}$ or $\frac{a}{7-a}$ (c) $\frac{-a}{b}$ (d) $\frac{x^2}{y}$ (e) $\frac{3}{x}$
2. (a) $\frac{3}{a-b}$ (b) $\frac{a}{b-a}$ (c) $\frac{1}{b-a}$
3. (a) $\frac{2+y}{3}$ (b) $\frac{1}{x+y}$ (c) $\frac{2}{a^2bc}$
4. (a) $x^3 - 3x^2 + 10x - 28$, remainder 83
   (b) $3x^2 + 6x + 8$, remainder 15
   (c) $x^2 - 2x + 4$, 0 remainder
5. (a) 120 (b) $24x^2y$ (c) $a^2 - b^2$ (d) $2(a-1)$
   (e) $2(x+1)(x-1)^2$
6. (a) $\frac{a-2}{3}$ (b) $\frac{2}{b}$ (c) $\frac{-a}{(a+3)(a-3)}$ (d) $\frac{b+3}{(b+2)(b-2)}$
   (e) $\frac{x^2 - x + 1}{x-1}$
7. (a) $\frac{3ax^2}{4}$ (b) $\frac{x}{y}$ (c) $\frac{3x(2x+3)}{(x-1)(1-2x)}$
8. (a) $\frac{ac}{b^2}$ (b) $56ax$ (c) $\frac{a-7}{a+3}$
9. (a) $\frac{c}{b}$ (b) $\frac{1}{10}$ (c) $\frac{2a^2 - a}{2a+1}$
10. (a) $\{-2\}$ (b) $\left\{-\frac{1}{5}\right\}$

EXERCISE 5.1

1. x^7
3. x^1
5. $x^{-3} = \frac{1}{x^3}$
7. $x^{-11} = \frac{1}{x^{11}}$
9. a^6
11. $\frac{1}{a^6}$
13. y^{12}
15. $y^{-6} = \frac{1}{y^6}$
17. $y^{-12} = \frac{1}{y^{12}}$
19. x^{20}
21. $4x^2y^4$
23. $\frac{x^2}{4y^4}$
25. $\frac{a^2}{b^6}$
27. $\frac{-27a^3}{8b^6}$
29. a^8b^4
31. $\frac{x^{14}}{y^7}$
33. $x^{27}y^6$
35. $\frac{8x}{9}$
37. 4
39. $\frac{-3}{y}$
41. 1
43. x^{3n-1}
45. y^{3n}
47. 1
49. $x^{-7n} = \frac{1}{x^{7n}}$
51. $\frac{1}{256}$
53. $\frac{36}{13}$
55. 16
57. $1 + \frac{x}{y}$
59. $\frac{b^3 + a^3}{a^2b^2}$

EXERCISE 5.2

1. 2
3. 2
5. -2
7. $-\frac{1}{4}$
9. 2
11. 2
13. 3
15. not a real number
17. 3
19. 3
21. $-\frac{1}{2}$
23. not a real number
25. 9
27. 25
29. $\frac{1}{4}$
31. 16
33. 16
35. $+16$
37. $\frac{1}{16}$
39. 7
41. $x + x^{4/3}$
43. $y^{7/4} - y^{5/4}$
45. $\frac{1}{x}$
47. $\frac{x^2}{y}$
49. x^3y^{11}

EXERCISE 5.3

1. 5
3. -4
5. $2\sqrt{2}$
7. $2\sqrt{3a}$
9. $3ab\sqrt{2ab}$
11. $2xy\sqrt[3]{2}$
13. $2x\sqrt[3]{5xy}$
15. $xy\sqrt[4]{xy^3}$
17. $-3a^2b^3\sqrt[5]{b^2}$
19. $\frac{\sqrt{13}}{7}$
21. $\frac{\sqrt{17}}{2x}$
23. $\frac{\sqrt[3]{3}}{4x}$
25. $\frac{\sqrt{6}}{3}$
27. $\frac{-\sqrt{14}}{7}$
29. $\frac{\sqrt{10a}}{2a}$
31. $\frac{\sqrt{10ab}}{8ab}$
33. $\frac{-\sqrt{6ab}}{2a^2b^2}$
35. xy
37. $\frac{-\sqrt[3]{21}}{3}$
39. $\frac{\sqrt[3]{12x}}{4x}$
41. $3^{1/2} = \sqrt{3}$
43. $(2a)^{1/2} = \sqrt{2a}$
45. $(5x^3y)^{1/2} = \sqrt{5x^3y} = x\sqrt{5xy}$
47. $(7x^5y^3)^{1/2} = x^2y\sqrt{7xy}$
49. $(2ab)^{1/2} = \sqrt{2ab}$

EXERCISE 5.4

1. $15\sqrt{2}$
3. $9\sqrt{5a}$
5. $\sqrt{2}$
7. $-5a\sqrt{2}$
9. $-26\sqrt{3}$
11. $7\sqrt[3]{5}$
13. $-12\sqrt[3]{3}$
15. $3\sqrt[3]{3}$
17. $4\sqrt[3]{3a}$
19. $\frac{9\sqrt[3]{2} - 2\sqrt[3]{3}}{6}$
21. $15 - 3\sqrt{2}$
23. $2 + 3\sqrt{2}$
25. $14\sqrt{15} + 30$
27. $6\sqrt{15} - 15$
29. $-8\sqrt{21} + 20\sqrt{14}$
31. $55 + 13\sqrt{15}$
33. $42 + 21\sqrt{2}$
35. $-441 + \sqrt{35}$
37. -1
39. -23
41. $5 - 2\sqrt{6}$
43. $a^2 + b - 2a\sqrt{b}$
45. $\frac{3\sqrt{2} + \sqrt{6}}{2}$
47. $\frac{2(3 + \sqrt{2})}{7}$

49. $a(3+\sqrt{5})$

51. $\frac{(3a+2b)(3-\sqrt{2})}{7}$

53. $\frac{(\sqrt{a}+b)^2}{a-b^2}$

55. $\frac{(\sqrt{a}+\sqrt{2b})^2}{a-2b}$

EXERCISE 5.5

1. $5i$
3. $5i\sqrt{2}$
5. $24i\sqrt{2}$
7. $-12i\sqrt{2}$
9. $3+8i\sqrt{7}$
11. $6+4i$
13. $-2-6i$
15. $-5-6i$
17. $-2+5i$
19. $-7+4i$
21. $8-i$
23. $10+5i$
25. $-3-2\sqrt{2}i$
27. $-1+2\sqrt{2}i$
29. $-7+3\sqrt{5}i$
31. $12+6i$
33. $-12+20i$
35. $-4+6i$
37. $-3+3\sqrt{3}i$
39. $-6+9i$
41. $28+12i$
43. $-20+20i$
45. $3+11i$
47. 13
49. $24+7i$
51. 31
53. $-3i$
55. $6i$
57. $\frac{2}{5}+\frac{1}{5}i$
59. $\frac{-6}{5}+\frac{3}{5}i$
61. $-1+2i$
63. $\frac{17}{13}-\frac{6}{13}i$
65. $\frac{-3}{2}i$
67. $\left(\frac{12+\sqrt{10}}{18}\right)+\left(\frac{4\sqrt{5}-3\sqrt{2}}{18}\right)i$
69. $\left(\frac{3+\sqrt{6}}{12}\right)+\left(\frac{3\sqrt{2}-\sqrt{3}}{12}\right)i$

SELF-TEST—CHAPTER 5

1. (a) a^9b^3 (b) $\frac{1}{x}$
2. (a) $x^{12}y^8$ (b) $\frac{y^{10}}{x^{15}}$
3. (a) $-x^3$ (b) $\frac{z^3}{y^2}$
4. (a) -2 (b) $\frac{1}{8}$

5. (a) $xy^2\sqrt[4]{x^2}$ (b) $9a^2b^3$

6. (a) $\frac{2(\sqrt{x}+\sqrt{y})}{x-y}$ (b) $\frac{-1}{2}+\frac{3}{2}i$

7. (a) $\frac{a^6}{b^4}$ (b) $x^{3/5}+x$

8. (a) $6+4\sqrt{2}$ (b) $-2+3\sqrt{3}$

9. (a) $x^{5/4}$ (b) $x^{-5/6}=\frac{1}{x^{5/6}}$

10. (a) $1+i$ (b) $3+3i$ (c) $2+3i$
(d) $3+i$ (e) $\frac{-5}{5}-\frac{5}{5}i=-1-i$

EXERCISE 6.1

1. $\{8,-8\}$
3. $\{11i,-11i\}$
5. $\{13,-13\}$
7. $\{2i,-2i\}$
9. $\left\{\frac{7}{6},-\frac{7}{6}\right\}$
11. $\left\{\frac{9}{2}i,-\frac{9}{2}i\right\}$
13. $\left\{\frac{5\sqrt{3}}{3},-\frac{5\sqrt{3}}{3}\right\}$
15. $\left\{\frac{6\sqrt{5}}{5}i,-\frac{6\sqrt{5}}{5}i\right\}$
17. $\left\{\frac{10\sqrt{3}}{3},-\frac{10\sqrt{3}}{3}\right\}$
19. $\left\{\frac{9\sqrt{13}}{13}i,-\frac{9\sqrt{13}}{13}i\right\}$
21. $\{-3,-7\}$
23. $\{-2+5i,-2-5i\}$
25. $\{6+3\sqrt{2},6-3\sqrt{2}\}$
27. $\{1+2\sqrt{7}i,1-2\sqrt{7}i\}$
29. $\{1+5\sqrt{2},1-5\sqrt{2}\}$
31. $\{5+4\sqrt{2},5-4\sqrt{2}\}$
33. $\{9+8i,9-8i\}$
35. $\{-1+4\sqrt{2},-1-4\sqrt{2}\}$
37. $\{2+5\sqrt{2},2-5\sqrt{2}\}$
39. $\{5+3\sqrt{3}i,5-3\sqrt{3}i\}$

EXERCISE 6.2

1. $\{-1,-2\}$
3. $\{1,-4\}$
5. $\left\{\frac{1}{2},\frac{1}{3}\right\}$
7. $\{0,3\}$
9. $\{8,-8\}$
11. $\{9,-9\}$
13. $\{0,-6\}$
15. $\{0,3\}$
17. $\{3,9\}$
19. $\{-1,-5\}$
21. $\{5,-3\}$
23. $\left\{-\frac{2}{3},-1\right\}$
25. $\left\{1,\frac{1}{2}\right\}$
27. $\left\{1,-\frac{1}{2}\right\}$
29. $\{2,-6\}$
31. $\left\{1,\frac{1}{2}\right\}$
33. $\{-2,-4\}$

EXERCISE 6.3

1. $\{-1, -5\}$
3. $\{-3, -5\}$
5. $\{-3 + i, -3 - i\}$
7. $\{4, 6\}$
9. $\{3, 7\}$
11. $\{4 + i, 4 - i\}$
13. $\left\{-1 + \frac{\sqrt{2}}{2} i, -1 - \frac{\sqrt{2}}{2} i\right\}$
15. $\{-1 + 5i, -1 - 5i\}$
17. $\left\{\frac{2}{5}, \frac{3}{5}\right\}$
19. $\left\{\frac{1 + 2i}{2}, \frac{1 - 2i}{2}\right\}$
21. 1 sec; 5 sec.
23. 4 sec.
25. No

EXERCISE 6.4

1. $\{1, -2\}$
3. $\{-2 + \sqrt{3}, -2 - \sqrt{3}\}$
5. $\left\{\frac{3 + \sqrt{17}}{2}, \frac{3 - \sqrt{17}}{2}\right\}$
7. $\left\{1, \frac{5}{7}\right\}$
9. $\left\{\frac{-4 + 3i}{5}, \frac{-4 - 3i}{5}\right\}$
11. $\left\{-2, -\frac{3}{2}\right\}$
13. $\left\{1, \frac{3}{2}\right\}$
15. $\left\{-3, -\frac{1}{2}\right\}$
17. $\frac{5}{16}$. Two real and unequal roots.
19. 0. A single real root.
21. 5%
23. 6%
25. 9%

EXERCISE 6.5

1. $\{0\}$
3. $\{6\}$
5. $\{8\}$
7. $\{11\}$
9. $\{9\}$
11. $\{0, -1\}$
13. $\{1\}$
15. $\{0\}$
17. $\{1\}$

19. $\{4\}$
21. $\{4\}$
23. $\left\{1, \frac{1}{5}\right\}$
25. $\left\{0, -\frac{5}{2}\right\}$
27. $\{7, -3\}$
29. $\{0\}$
31. $\{2, 1\}$
33. $\left\{-1, \frac{-16}{5}\right\}$
35. $\left\{\frac{1}{2}, \frac{-7}{5}\right\}$

EXERCISE 6.6

1. $\sqrt{6}$ sec.
3. 5 sec.
5. 200 miles/hours
7. 20%
9. $7\frac{1}{7}\%$
11. 2 units
13. $\frac{1}{2}$. We have to approximate to 0 or 1.
15. 15 in. by 10 in.
17. 13 ft.
19. $\sqrt{1856} \approx 43.1$ ft.

SELF-TEST—CHAPTER 6

1. (a) $\{2, -2\}$ (b) $\left\{\frac{2}{3}, \frac{-2}{3}\right\}$ (c) $\{4, -2\}$
2. (a) $x = 2$ or $x = 3$ (b) $x = 1$ or $x = \frac{-1}{2}$
3. (a) $x = 1$ or $x = 2$ (b) $x = \frac{1}{2}$ or $x = -2$
4. (a) 49, two real roots (b) -3, two complex roots
5. (a) $x = \frac{1 \pm \sqrt{5}}{2}$ (b) $x = \frac{1}{2} \pm \frac{\sqrt{33}}{6}$
6. (a) $x = 2$ (b) $x = 8$
7. (a) $x = 3$ (b) $x = 2$
8. (a) $x = \pm 1$ (b) $y = \frac{3}{5}$
9. $\sqrt{2}$ seconds
10. 10%

EXERCISE 7.1

1. $\{(1, 0), (1, 1), (2, 0), (2, 1)\}$
3. $\{(1, a), (1, b), (2, a), (2, b), (3, a), (3, b)\}$
5. The point is in quadrant I.

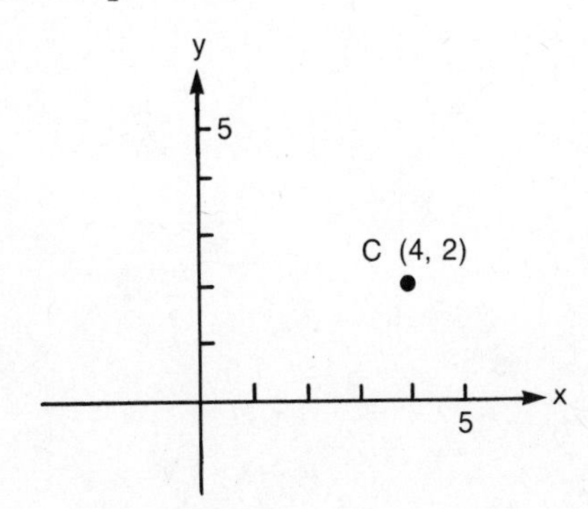

7. The point is on the x-axis and therefore is not in any quadrant.

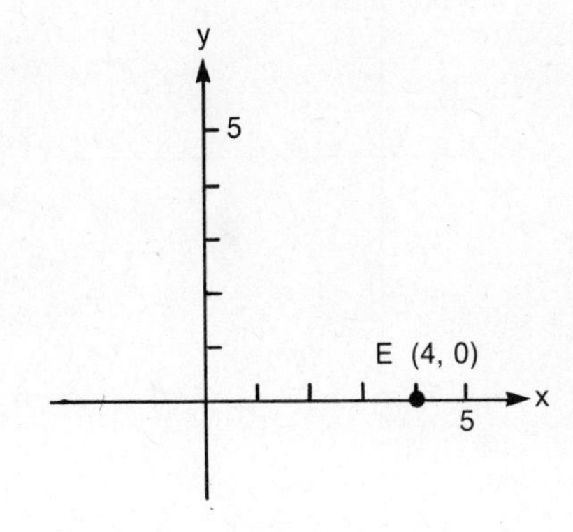

9. The point is on the y-axis and therefore is not in any quadrant.

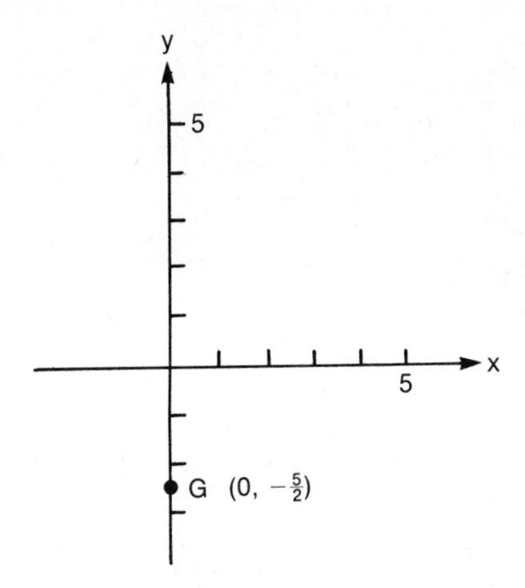

11. $\{(-2, 1), (-1, 2), (0, 3), (1, 4), (2, 5)\}$

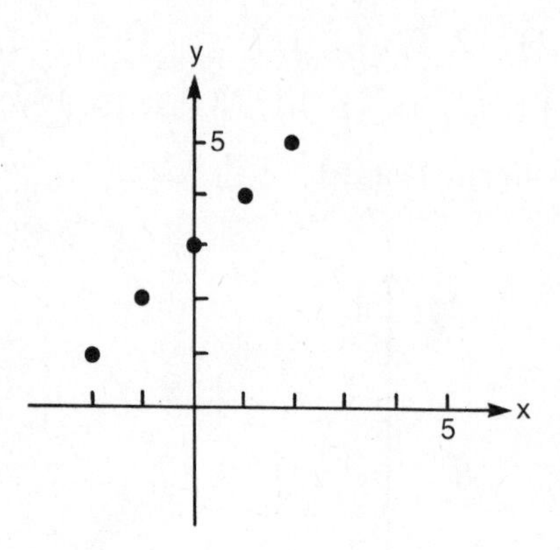

13. $\{(-1, -5), (0, -4), (1, -3)\}$

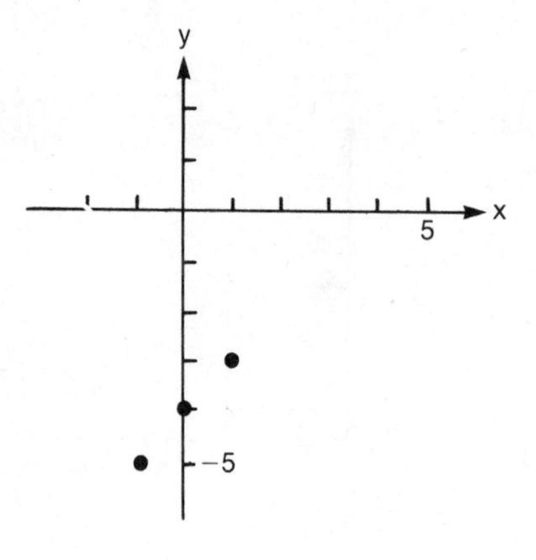

15. $\{(-1, -5), (0, -3), (1, -1)\}$

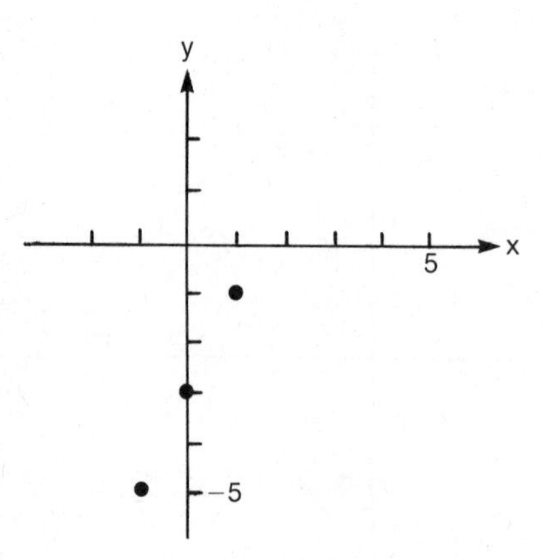

17.

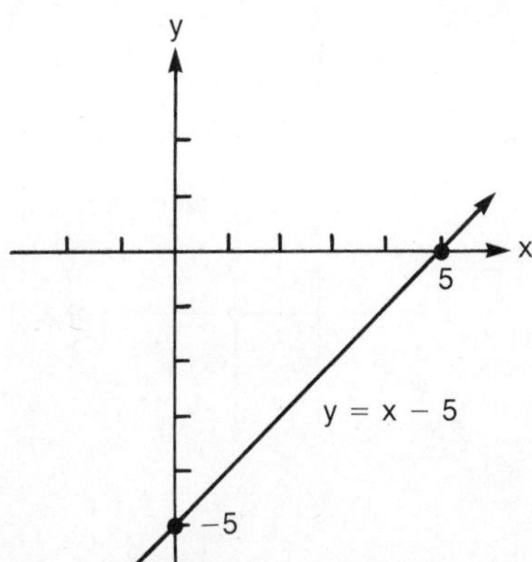

19.

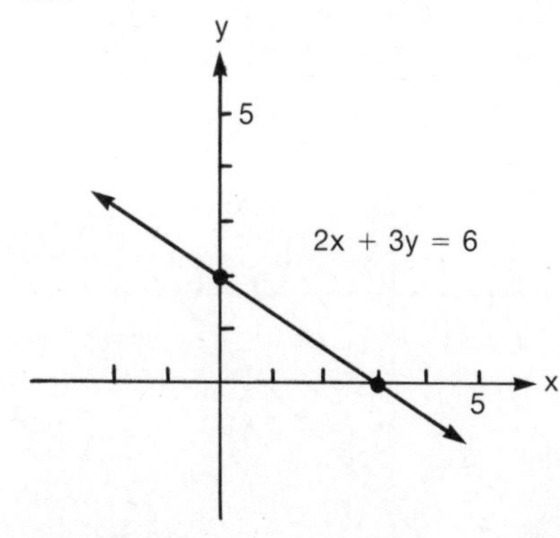

21.

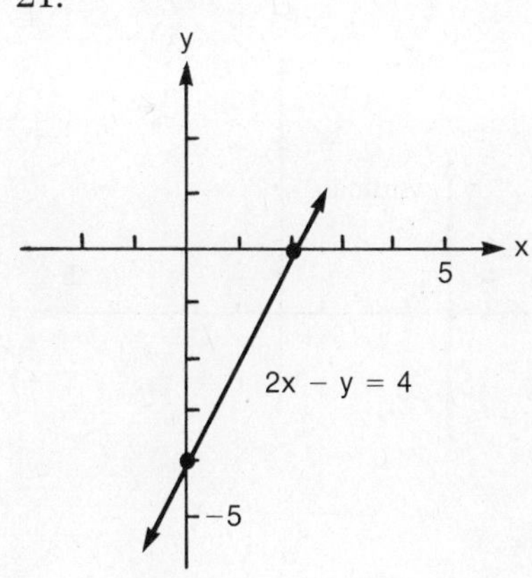

23.

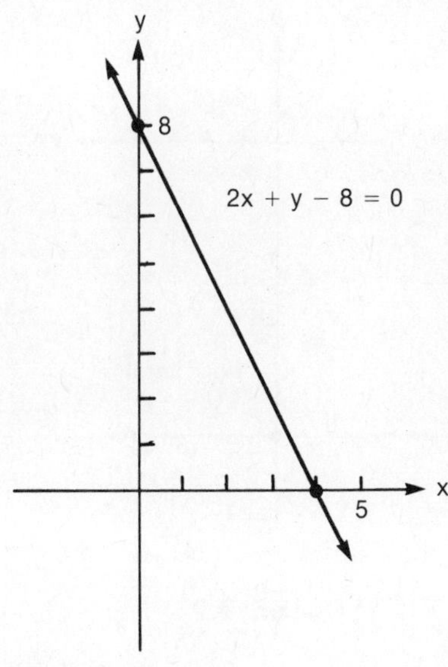

25.

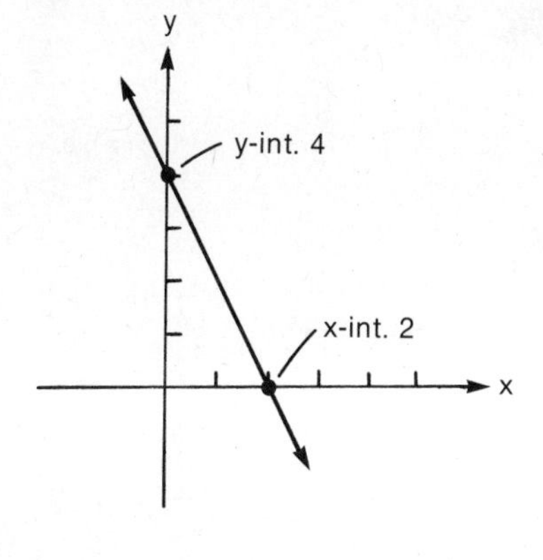

27.

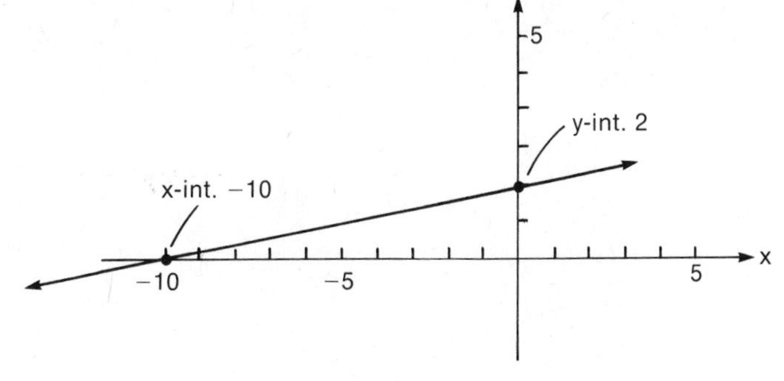

29.

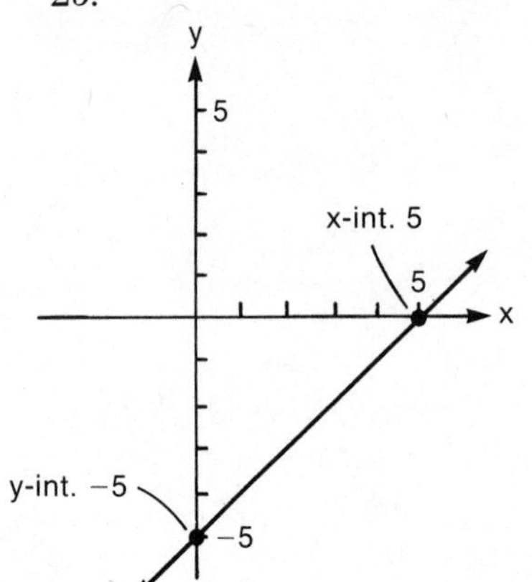

31.

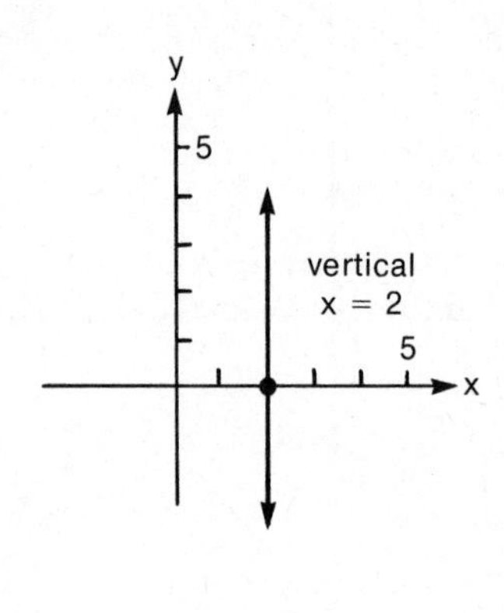

33.

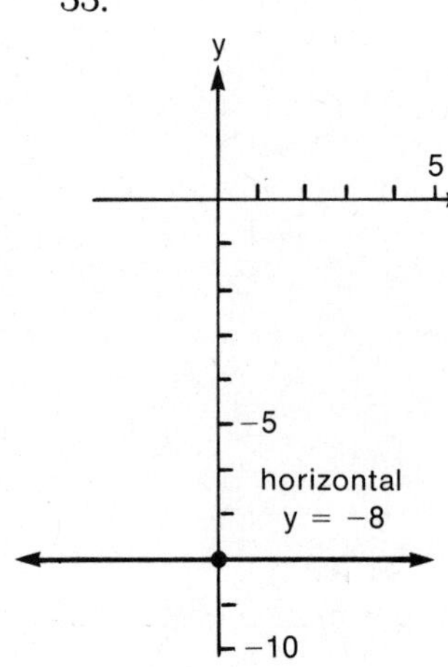

35.

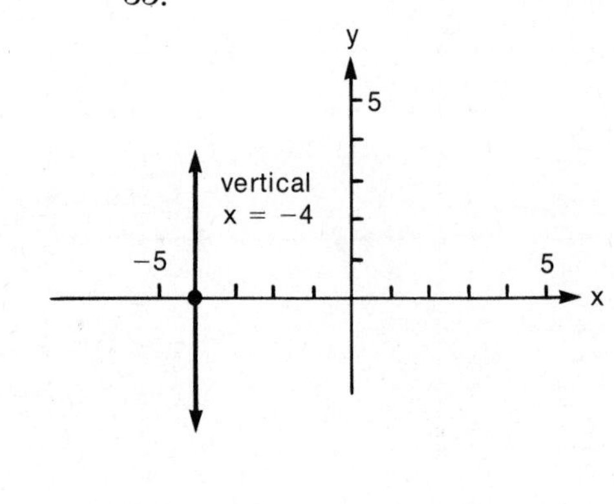

EXERCISE 7.2

1. distance = 5; slope = $\frac{4}{3}$

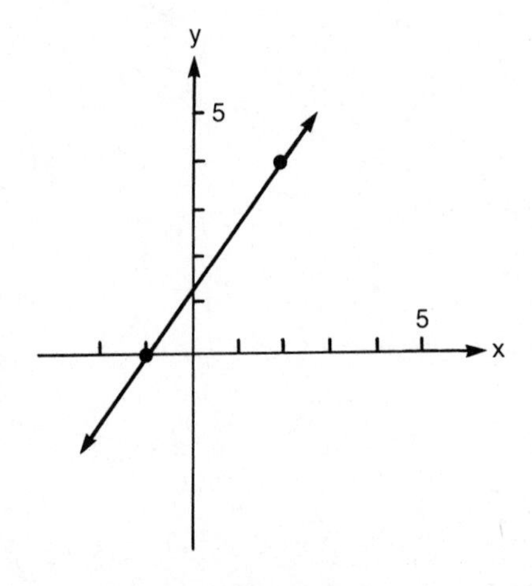

3. distance $= \sqrt{73}$; slope $= \dfrac{8}{3}$

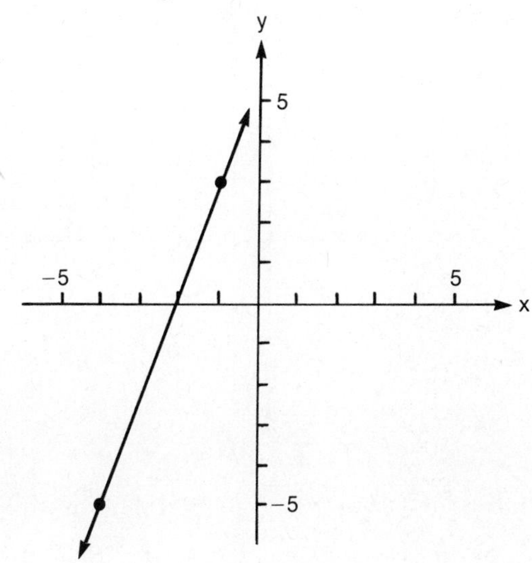

5. distance $= 3\sqrt{10}$; slope $= 3$

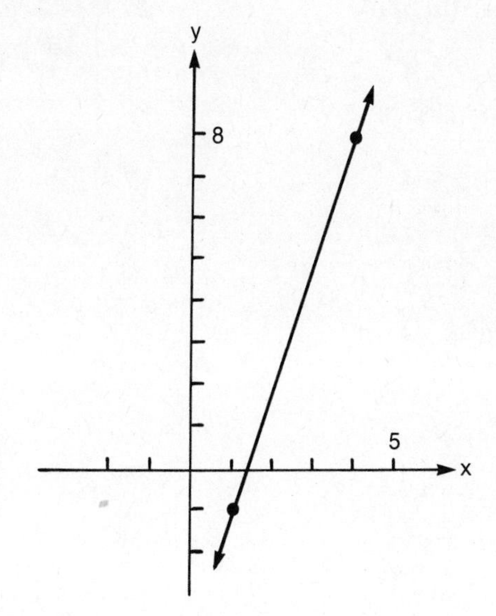

7. distance $= 5$; slope $= 0$

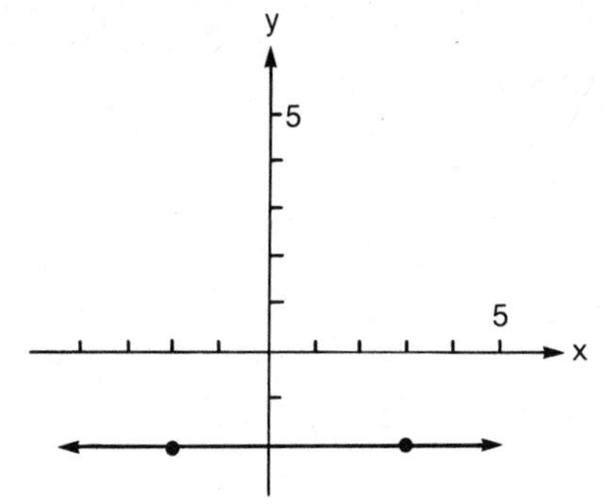

9. distance = 6; slope is undefined

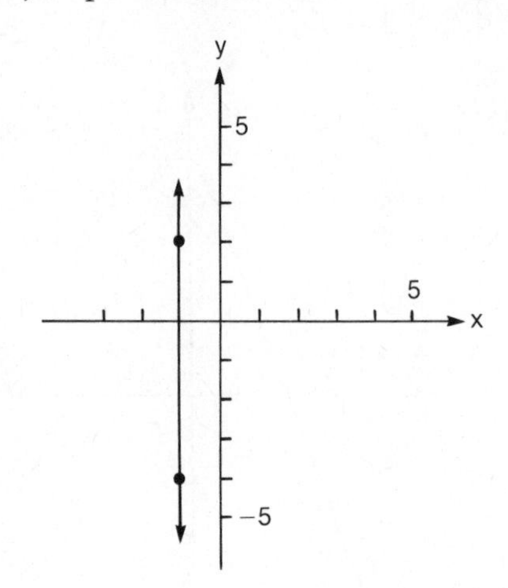

11. Parallel
13. Perpendicular
15. Neither
17. Perpendicular
19. Parallel
21. Scalene
23. Scalene
25. Isosceles

EXERCISE 7.3

1. $-3x + y + 4 = 0$
3. $x + y - 5 = 0$
5. $2x - y + 11 = 0$
7. $3x + y + 5 = 0$
9. $5x - y + 2 = 0$
11. $3x + 15y + 5 = 0$
13. $2x - y - 4 = 0$
15. $3x + y + 12 = 0$
17. $2x + y - 3 = 0$
19. $x - y - 2 = 0$
21. $y = \frac{1}{3}x - \frac{5}{3}$
23. $y = -\frac{5}{2}x + 3$
25. $y = -\frac{1}{8}x + \frac{1}{2}$

EXERCISE 7.4

1. $(-2, 5), (-1, 2), (0, 1), (1, 2), (2, 5)$

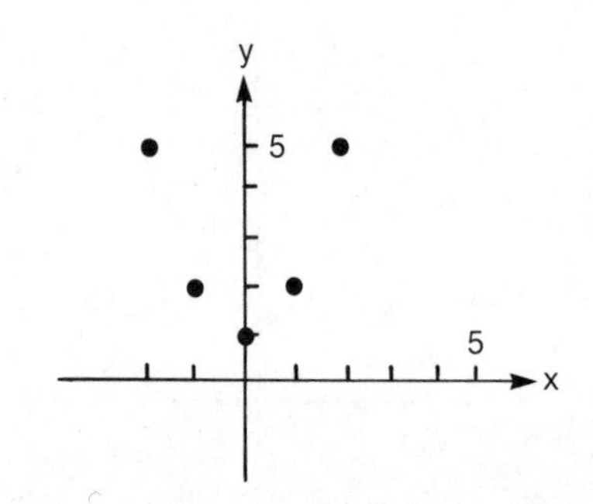

3. $(-2, 2)$, $(-1, -1)$, $(0, -2)$, $(1, -1)$, $(2, 2)$

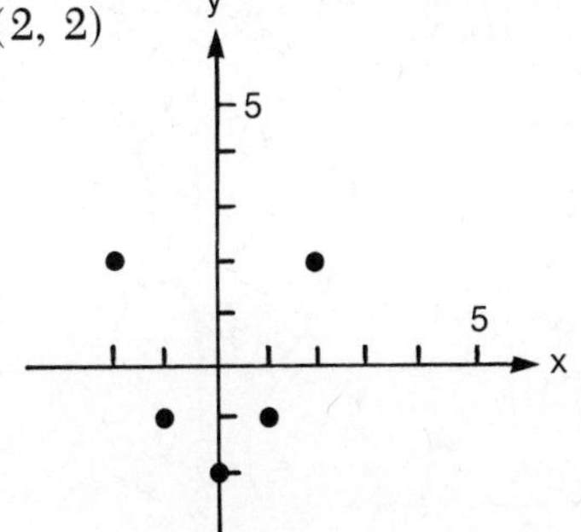

5. $(-2, -2)$, $(-1, 1)$, $(0, 2)$, $(1, 1)$, $(2, -2)$

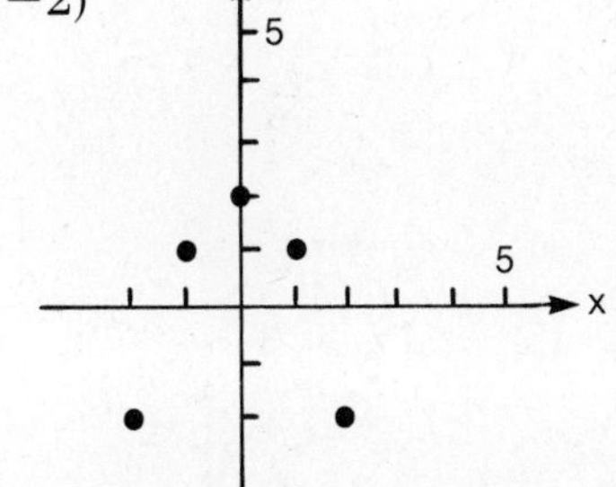

7. $(-2, -5)$, $(-1, -2)$, $(0, -1)$, $(1, -2)$, $(2, -5)$

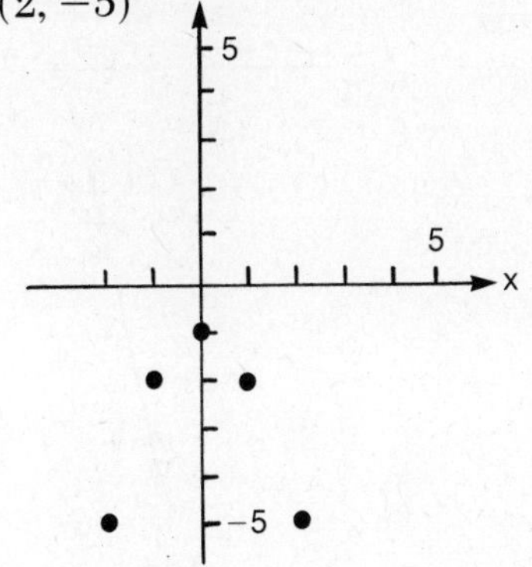

9. $(-2, 1)$, $(-1, 0)$, $(0, 1)$, $(1, 4)$, $(2, 9)$

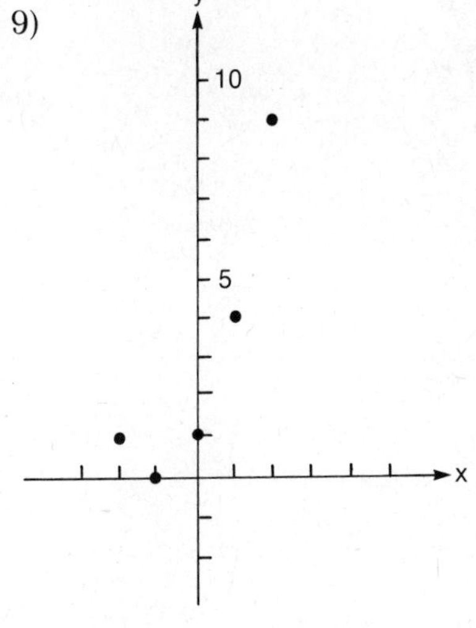

11.

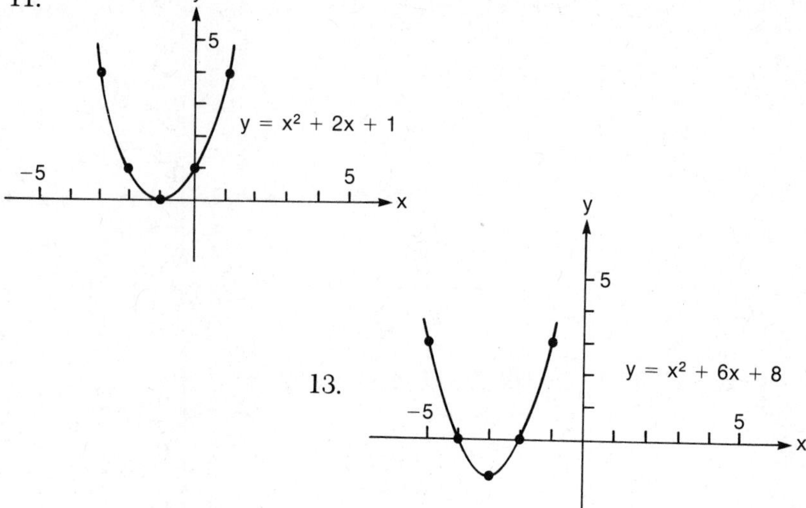
y
5
−5
5
x
y = x² + 2x + 1
13.
y
5
−5
5
x
y = x² + 6x + 8

15.

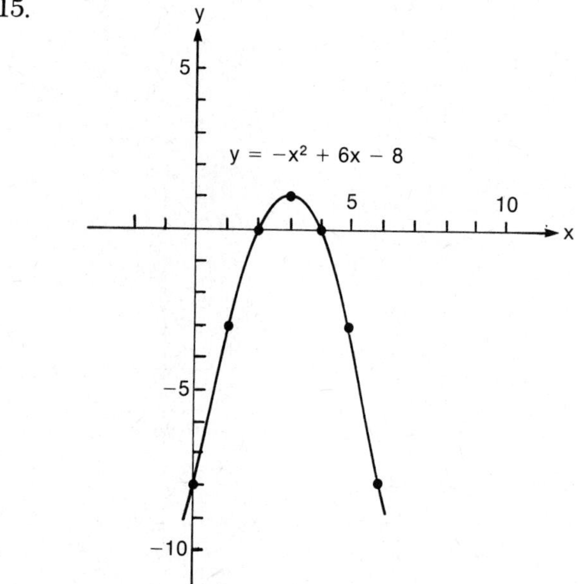
y
5
y = −x² + 6x − 8
5
10
x
−5
−10

17.

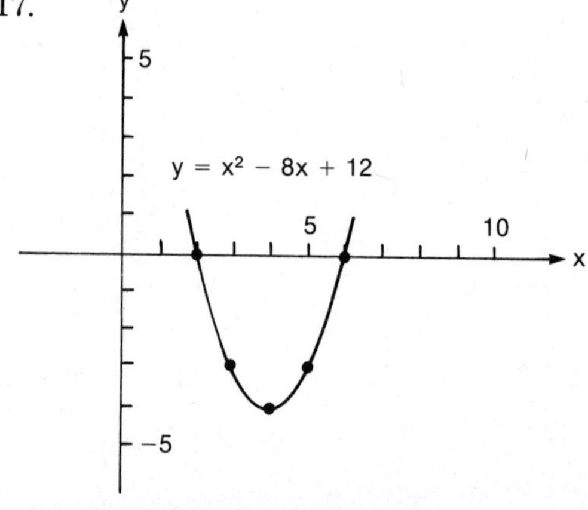
y
5
y = x² − 8x + 12
5
10
x
−5

19.

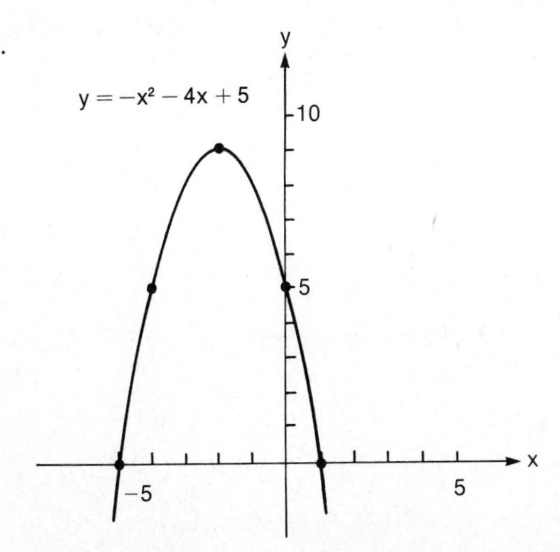

21. (5, −2), (2, −1), (1, 0), (2, 1), (5, 2)

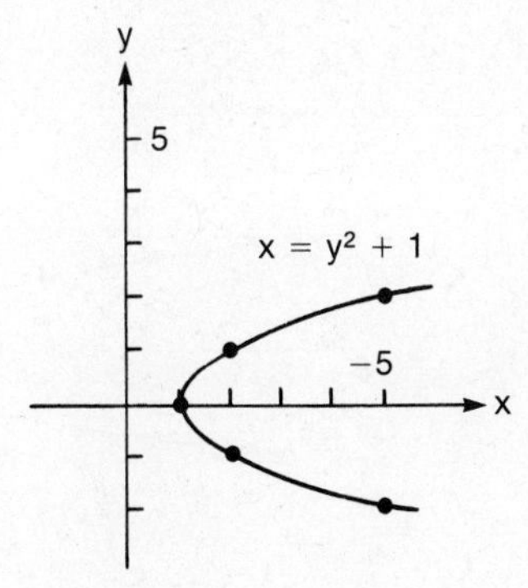

23. (2, −2), (−1, −1), (−2, 0), (−1, 1), (2, 2)

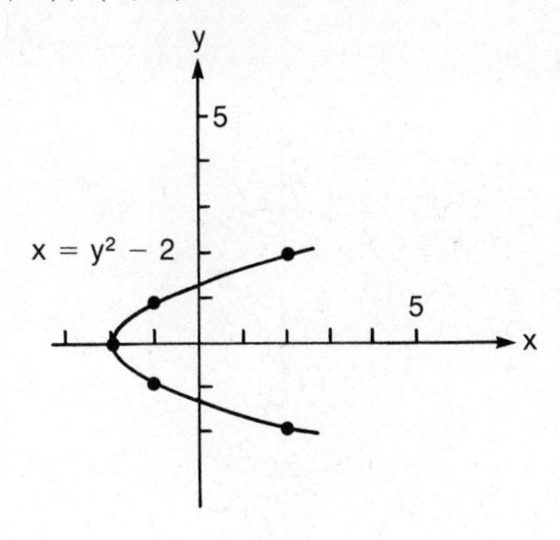

25. (−2, −2), (1, −1), (2, 0), (1, 1), (−2, 2)

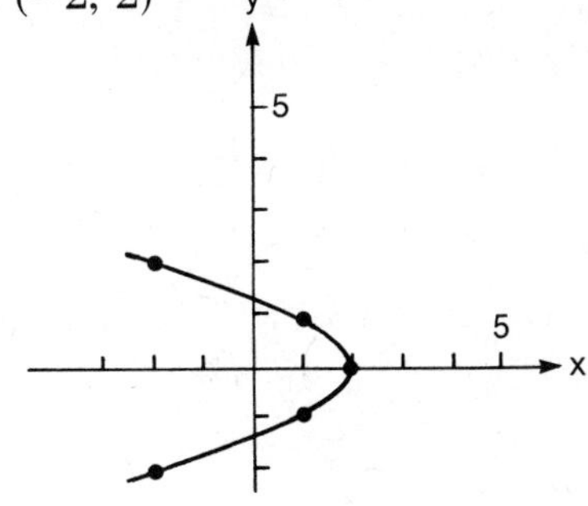

27. $(-5, -2), (-2, -1), (-1, 0), (-2, 1), (-5, 2)$

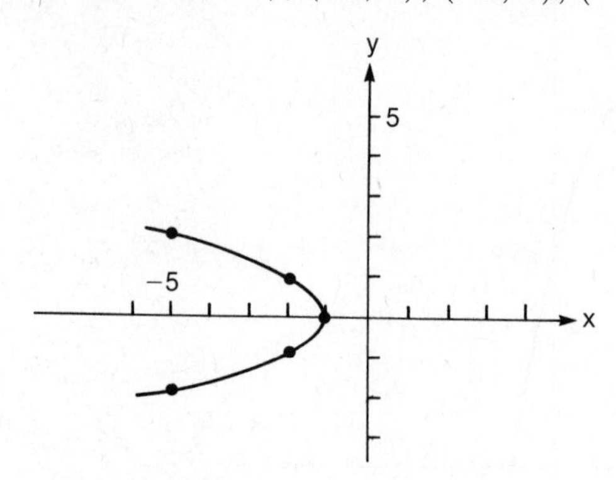

29. $(-14, -2), (-8, -1), (-6, 0), (-8, 1), (-14, 2)$

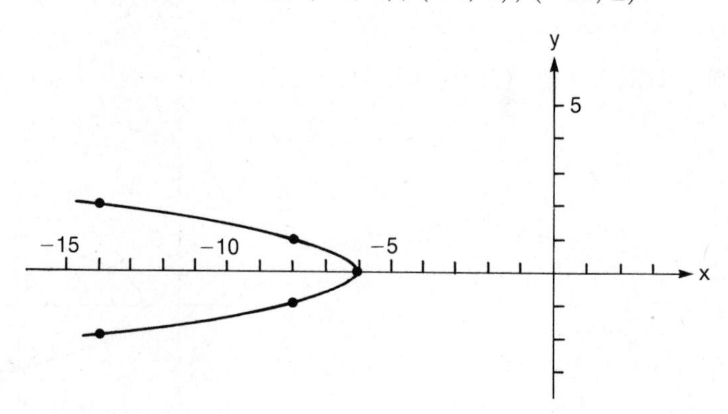

EXERCISE 7.5

1.

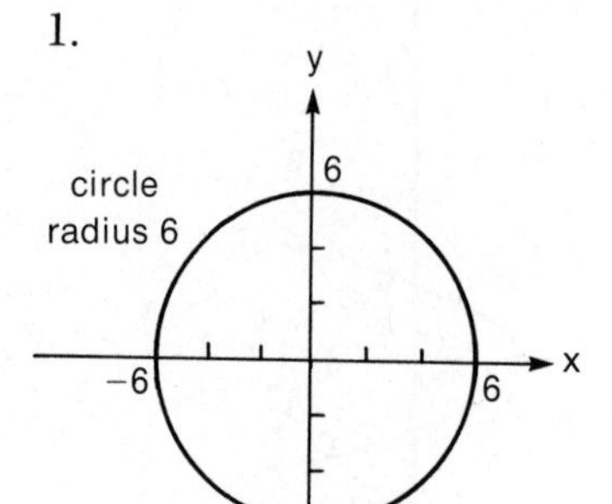

3.

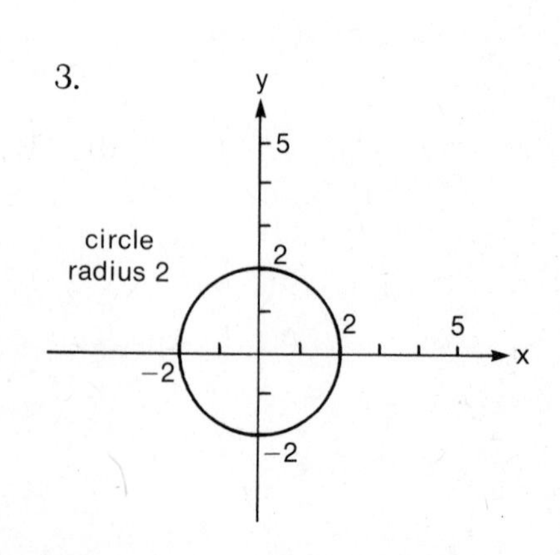

5.

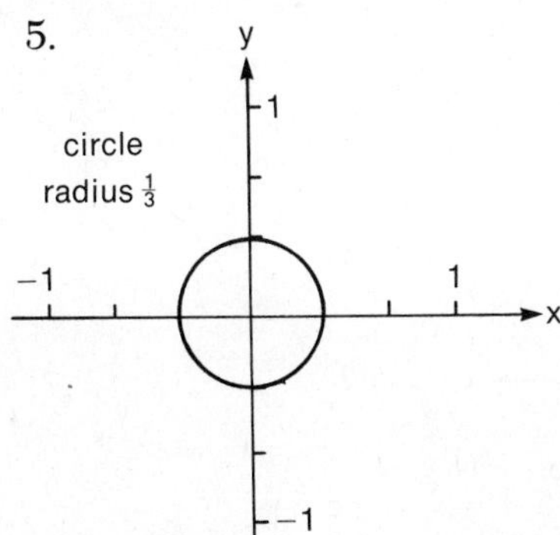
y
1
circle
radius $\frac{1}{3}$
−1
1
x
−1

7.

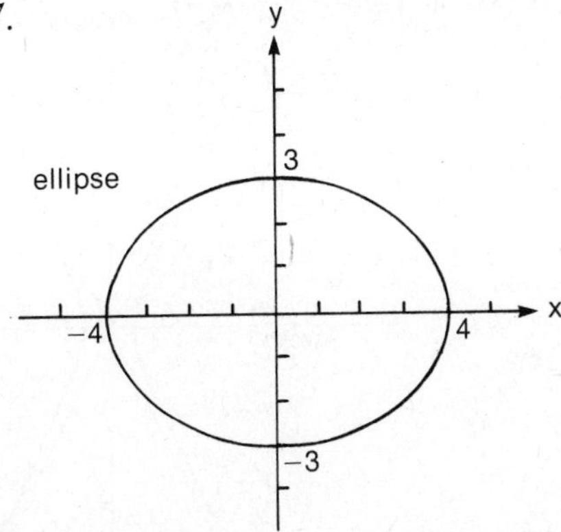
y
3
ellipse
−4
4
x
−3

9.

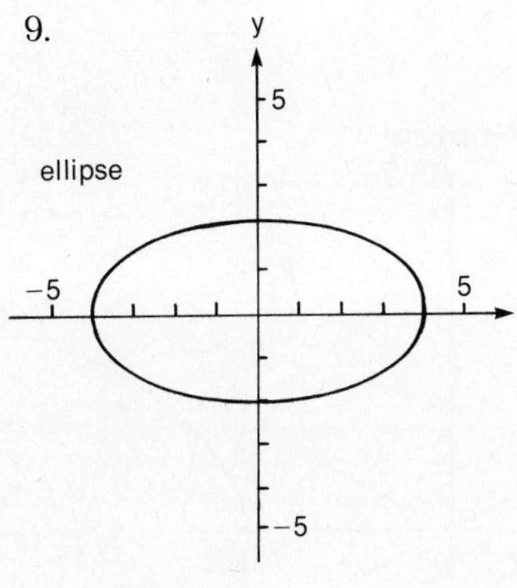
y
5
ellipse
−5
5
x
−5

11.

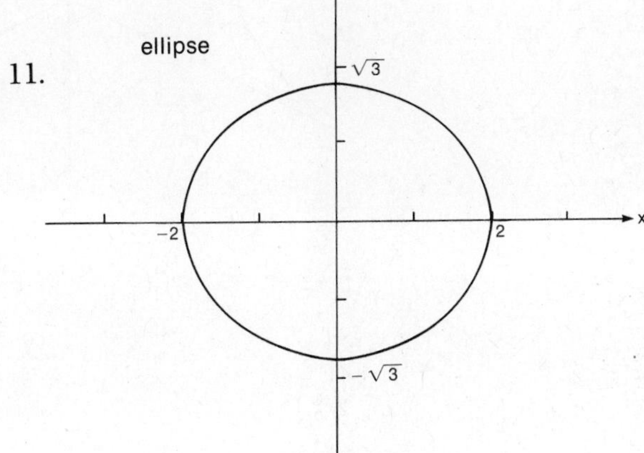
y
ellipse
$\sqrt{3}$
−2
2
x
$-\sqrt{3}$

13.

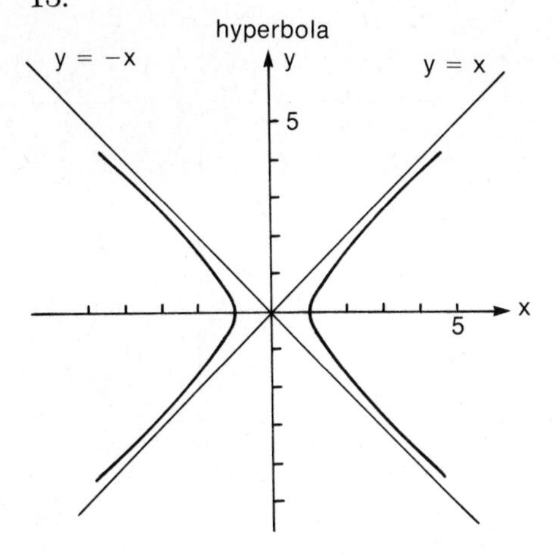
hyperbola
y = −x
y
y = x
5
x
5

15.

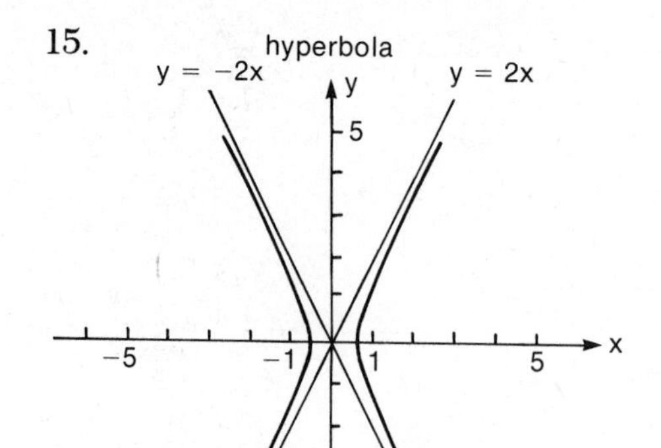

17.

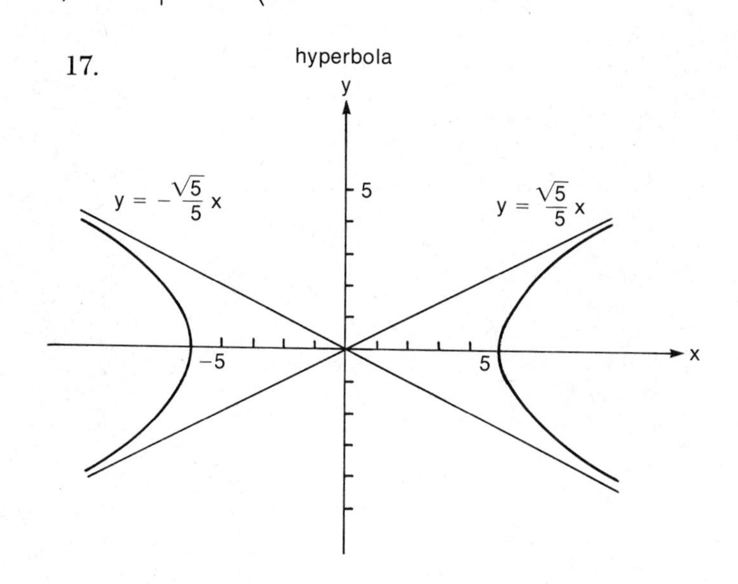

19.

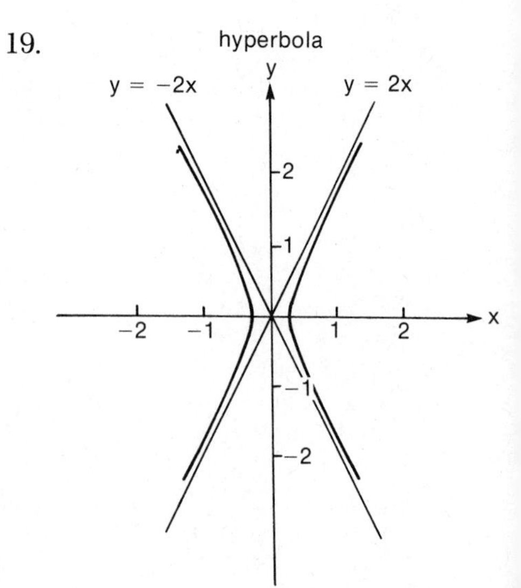

21. $(x+1)^2 + (y-3)^2 = 4$

23. $(x+3)^2 + (y+2)^2 = 9$

25. $(x-5)^2 + (y+1)^2 = 36$

EXERCISE 7.6

1. 2,000 units

3. (a)

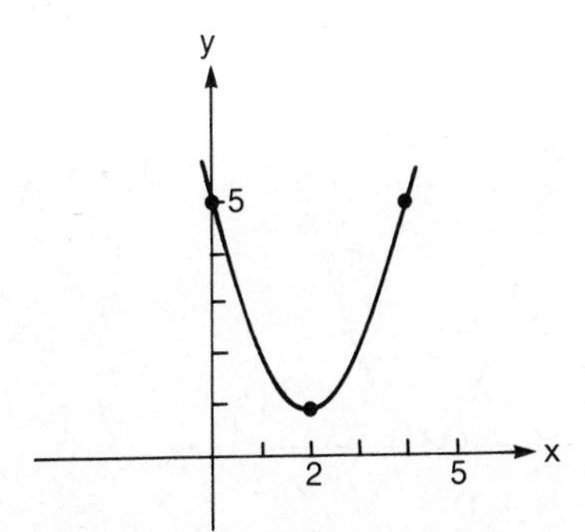

(b) 2,000 units

5. (a)

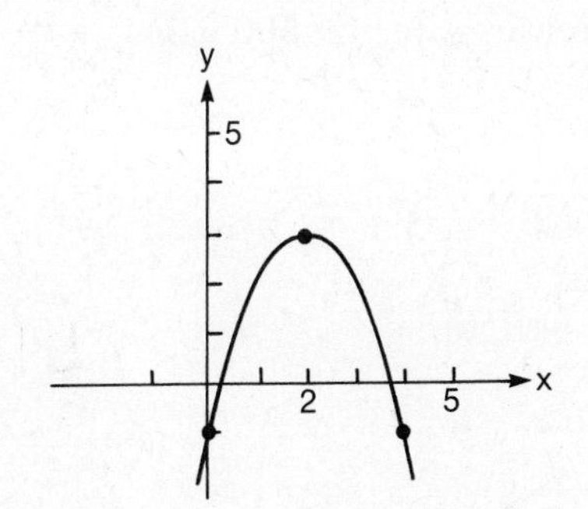

(b) 3 hours
(c) 10

7. (a)

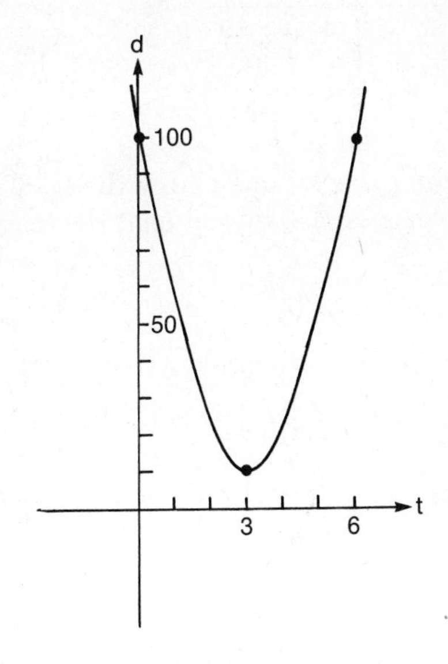

(b) 2,000 units

9. \$4 per dress and \$10 of fixed cost

11. (a) \$1,000
(b) \$2

13. (a) $m = -\frac{1}{2}$; $b = 25$

(b)

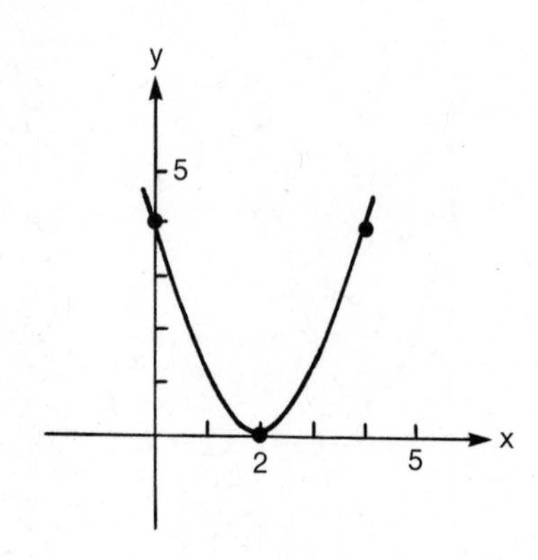

(c) A drop of one cent should sell 500 more cans. An increase of one cent decreases sales by 500 cans.

15. (a) $m = -2$; $b = 7$

(b)

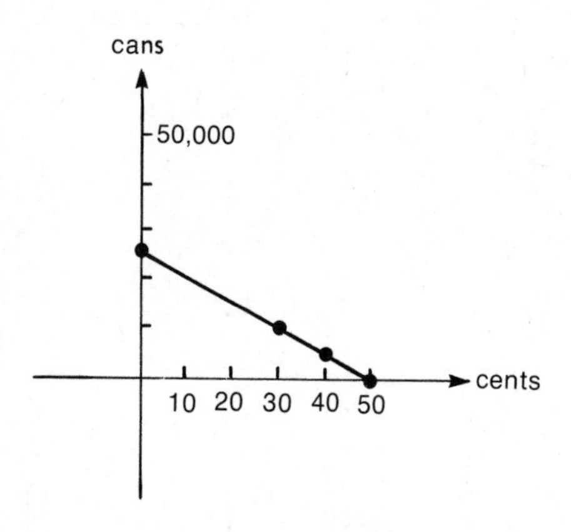

(c) Lowering the price by one cent will sell 200 more pairs of shoelaces. An increase of one cent decreases sales by 200 pairs.

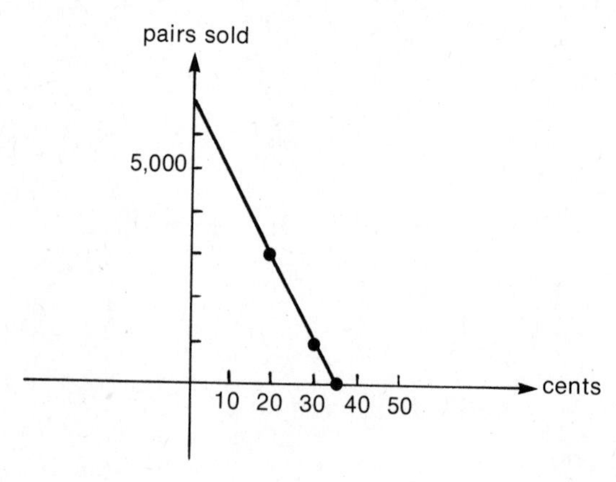

SELF-TEST—CHAPTER 7

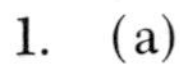

1. (a)

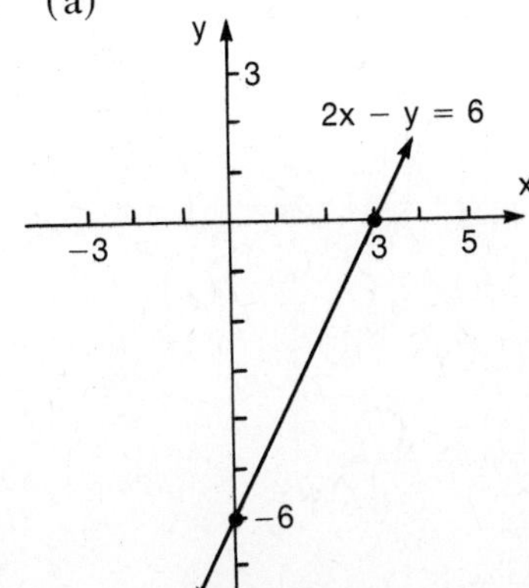

(b)

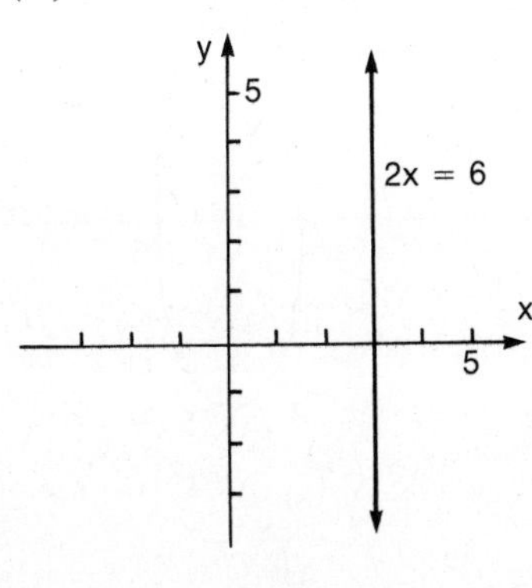

2. (a) distance $= \sqrt{34}$; slope $= \frac{5}{3}$
   (b) distance = 2; slope is undefined
3. (a) $x - y + 1 = 10$ (b) $2x + y - 10 = 0$
4. (a) $2x + y - 10 = 0$ (b) $y - 3 = 0$
5. (a) $x + y - 4 = 0$ (b) $2x - y = 0$
6. (a) $2x - y = 0$ (b) $x + 2y - 5 = 0$
7. (a)

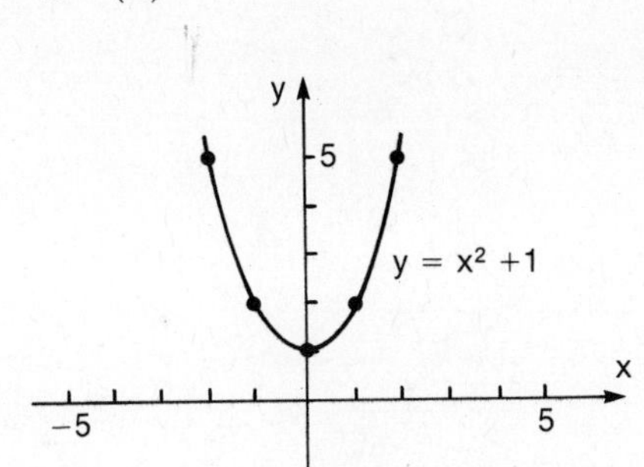

(b)

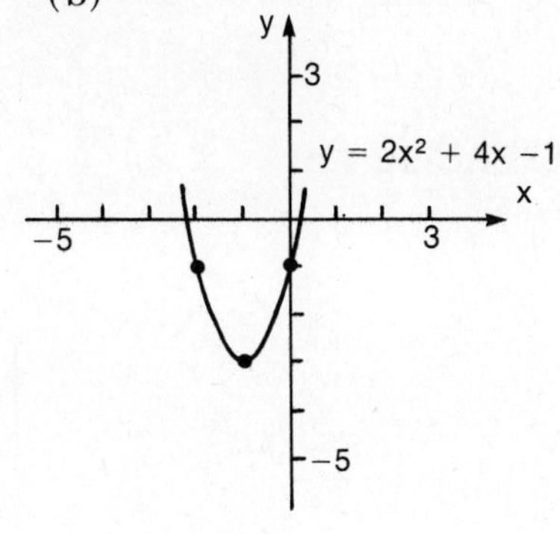

8. (a)

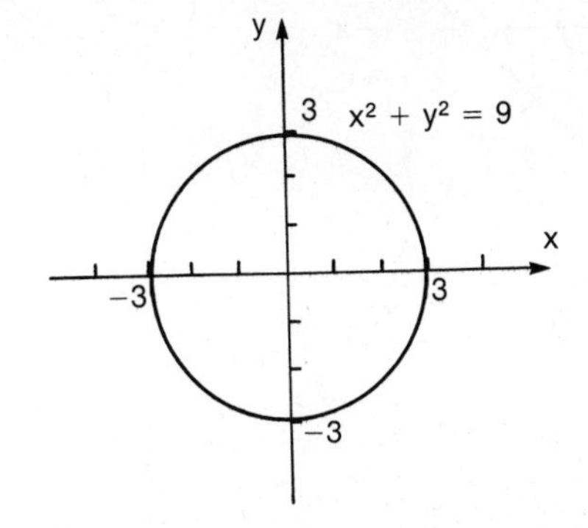

(b)

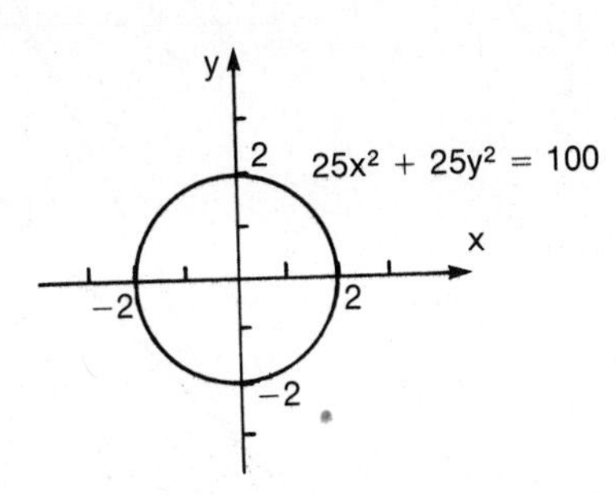

9. (a)

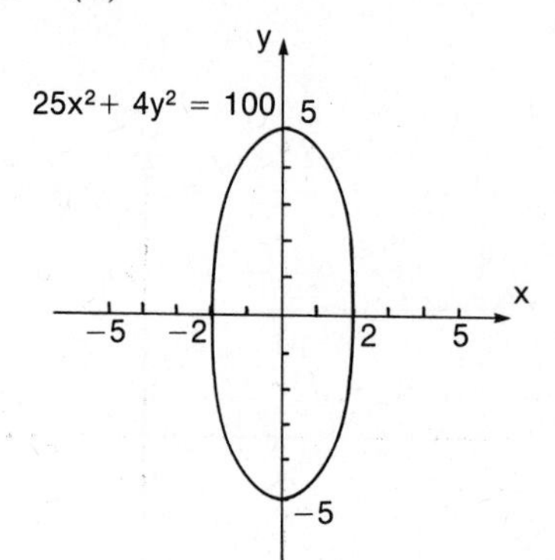

(b)

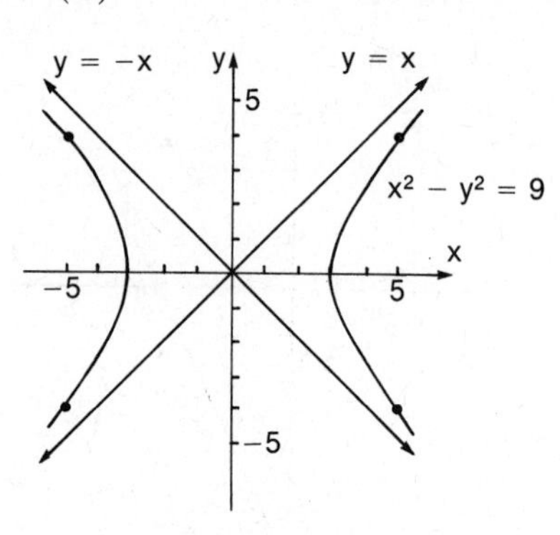

10. (a)

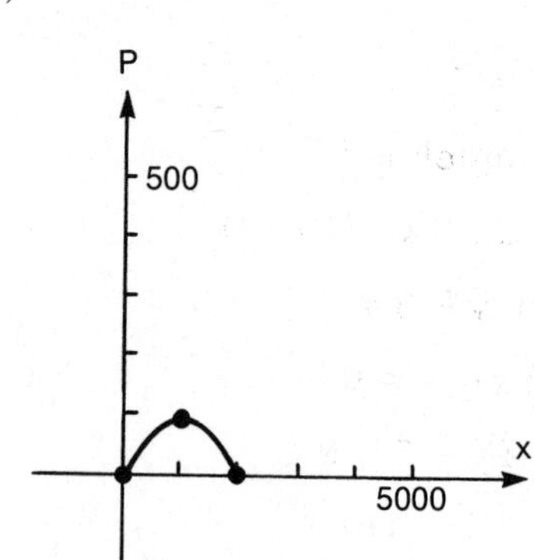

(b) one thousand

EXERCISE 8.1

1.

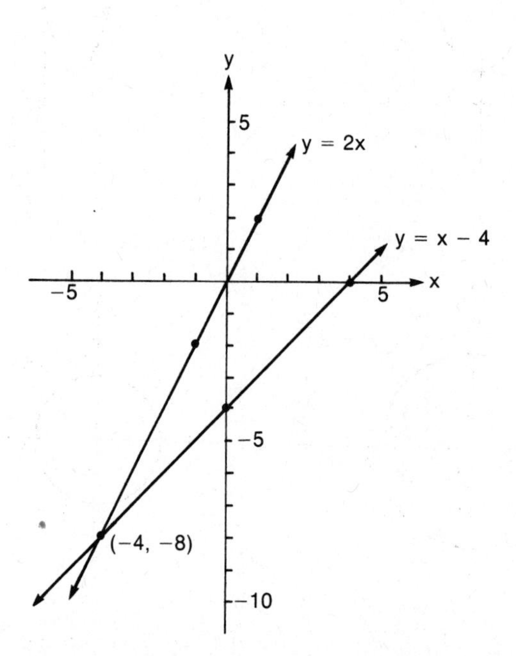

3.

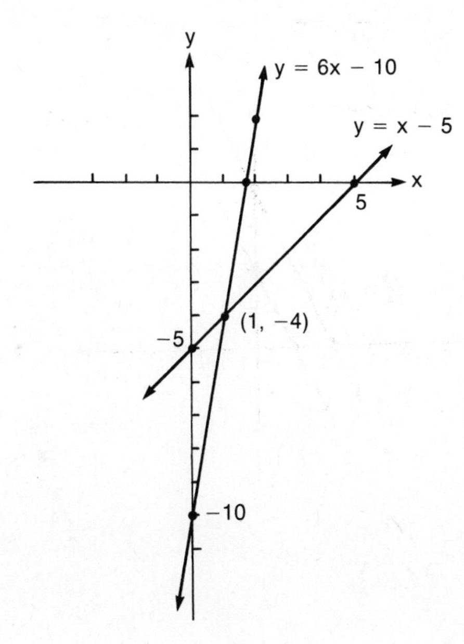

5.

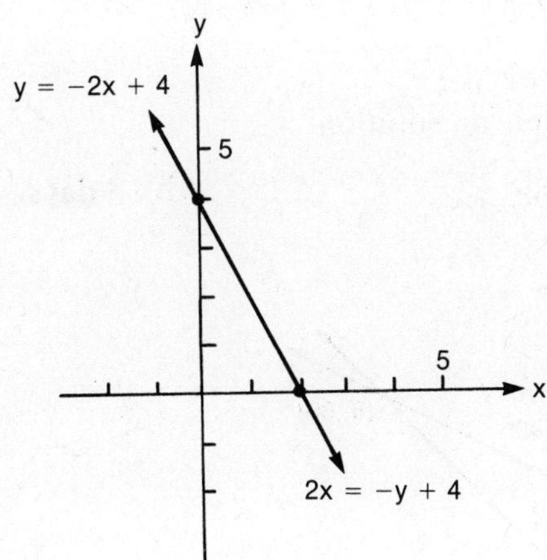

7.

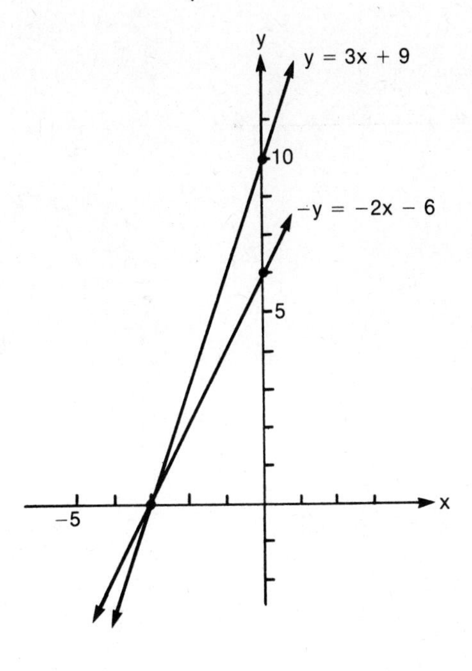

9.

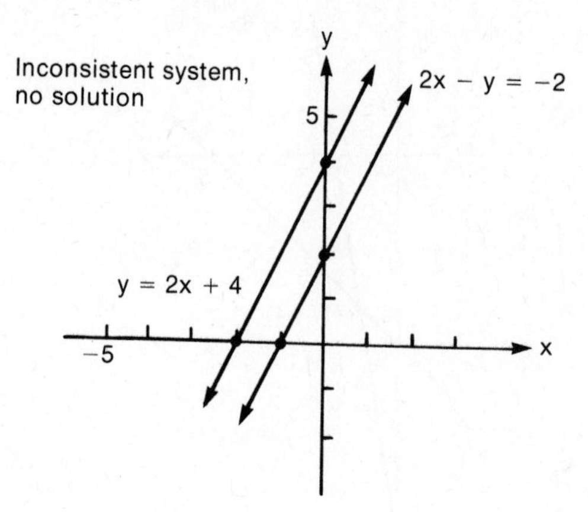

11. $(-4, -8)$

13. $(1, -4)$

15. Dependent

17. $(-3, 0)$

19. Inconsistent, no solution

21. (a) (b) 8 days

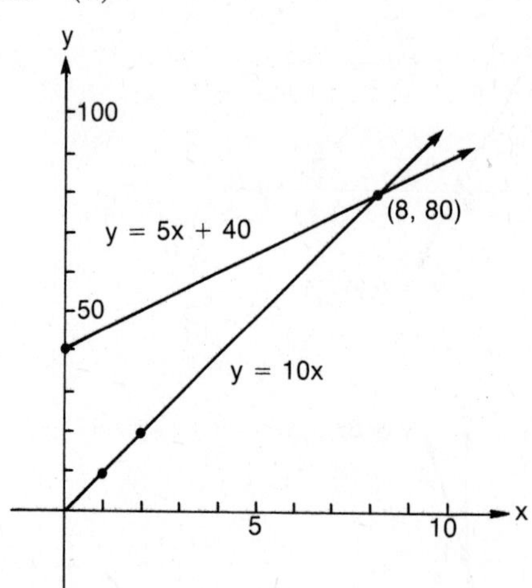

23. 25 years

EXERCISE 8.2

1. $(5, 3)$

3. $\left(0, \frac{1}{2}\right)$

5. $(0, 1)$

7. $(2, 3)$

9. $\left(\frac{5}{2}, -\frac{1}{2}\right)$

11. No solution

13. (3, 2)

15. (2, 1)

17. $\left(\frac{1}{3}, 2\right)$

19. (5, −2)

21. $\left(-\frac{1}{2}, -\frac{2}{3}\right)$

23. (8, −12)

25. (6, 8)

27. (4, −3)

29. (4, 2)

31. 36 years, 34 years

33. 1250 ft. for the building, 222 ft. for the antenna

35. 660 pounds, 640 pounds

EXERCISE 8.3

1. (5, 3, 4)

3. (−1, 1, 4)

5. (3, 4, 1)

7. Inconsistent (no solution)

9. $\left(\frac{1}{2}, \frac{1}{4}, \frac{1}{3}\right)$

11. Inconsistent (no solution)

13. Inconsistent (no solution)

15. $\left(\frac{9}{2}, \frac{1}{2}, \frac{5}{2}\right)$

17. (6, 3, −1)

19. (−2, −3, −4)

21. $2a + b = c$

23. (−1, 0, 2)

25. $k = -2$

EXERCISE 8.4

1. 2

3. 7

5. 6

7. $\frac{1}{2}$

9. $\frac{-7}{40}$

11. (2, 3)

13. (4, 5)

15. (3, −1)

17. (4, 5)

19. Dependent

21. (−2, −3)

23. Inconsistent, no solution

25. $\left(\frac{152}{17}, \frac{29}{17}\right)$

27. (5, 2)

29. (−1, −1)

EXERCISE 8.5

1. −7
3. 0
5. −1
7. −4
9. −9
11. (1, 2, 3)
13. (3, −1, −2)
15. (3, 0, 4)
17. (−5, 1, 5)
19. (−6, 2, 5)

EXERCISE 8.6

1. {(0, 6), (6, 0)}

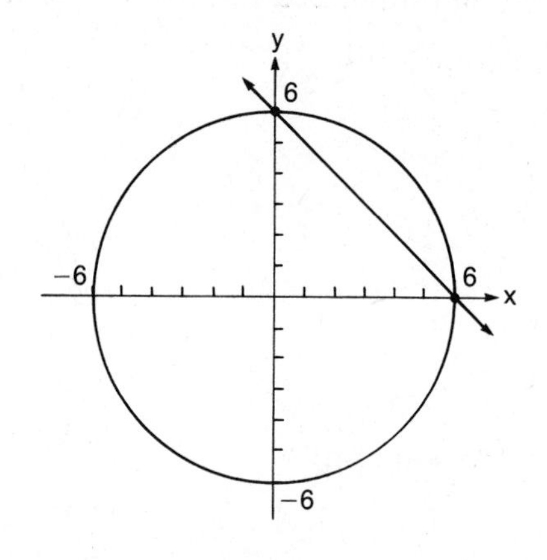

3. {(0, 5), (−5, 0)}

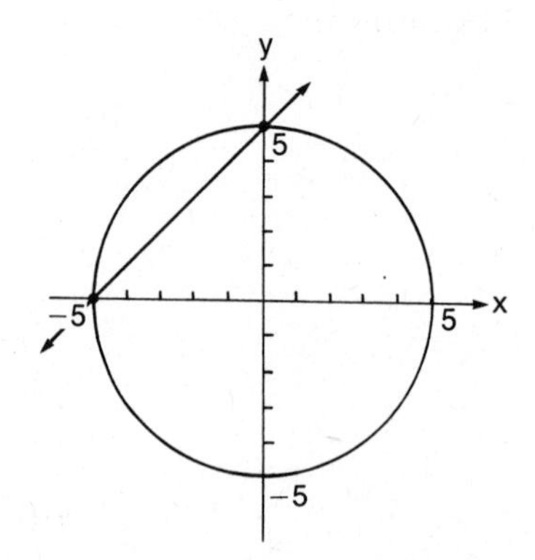

5. $\{(3, 4), (-4, -3)\}$

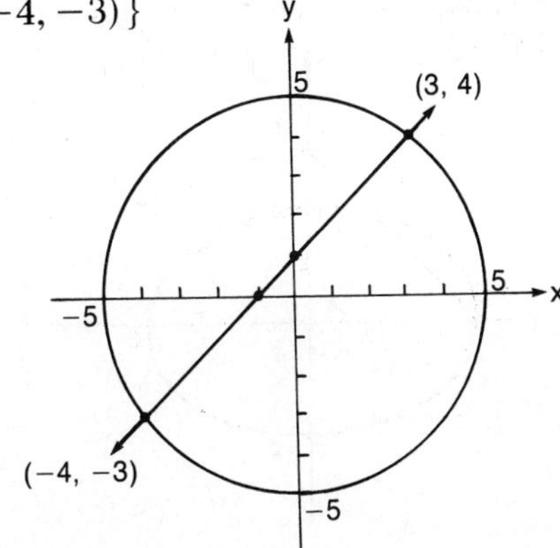

7. $\{(1, 0), (5, 4)\}$

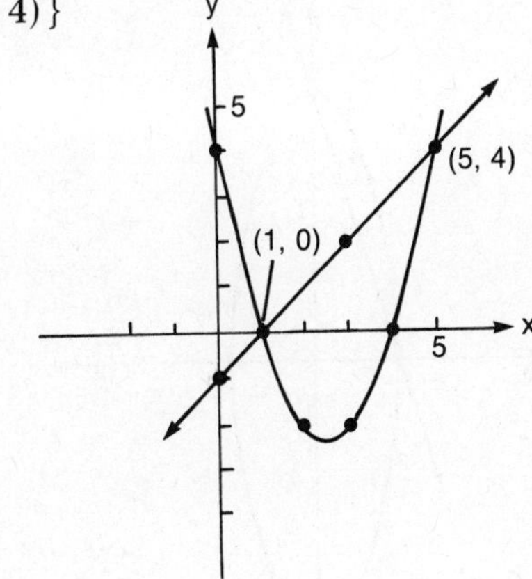

9. $\{(6, 0), (0, -6)\}$

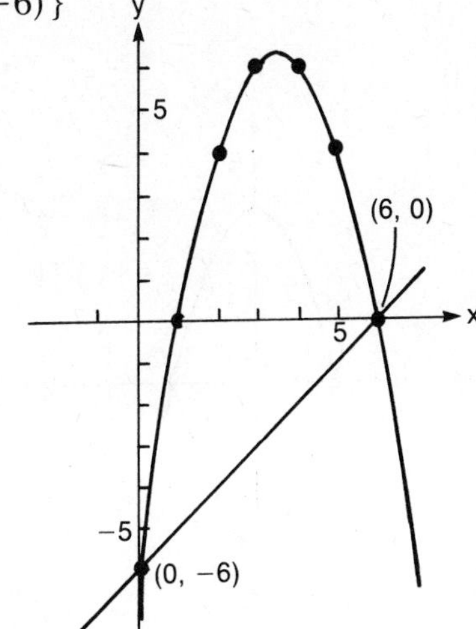

11. $\left\{(-3, 0), \left(-\frac{15}{13}, \frac{24}{13}\right)\right\}$

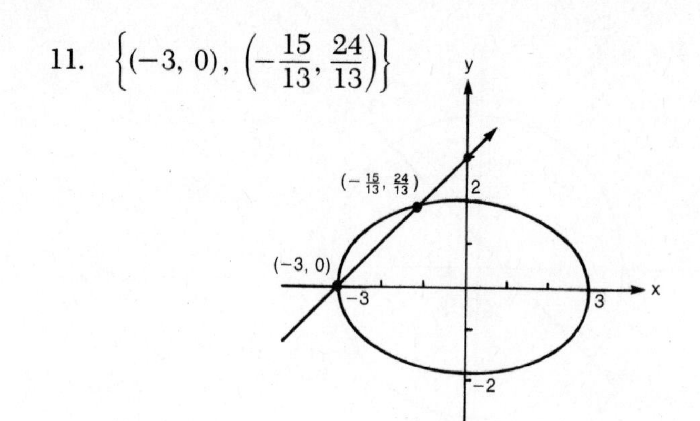

13. $\{(0, 6), (-2, 0)\}$

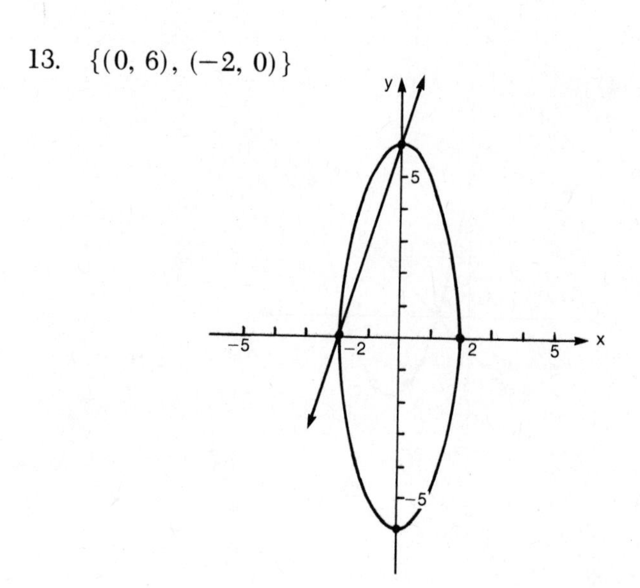

15. $\{(2, 0), (-2, 0)\}$

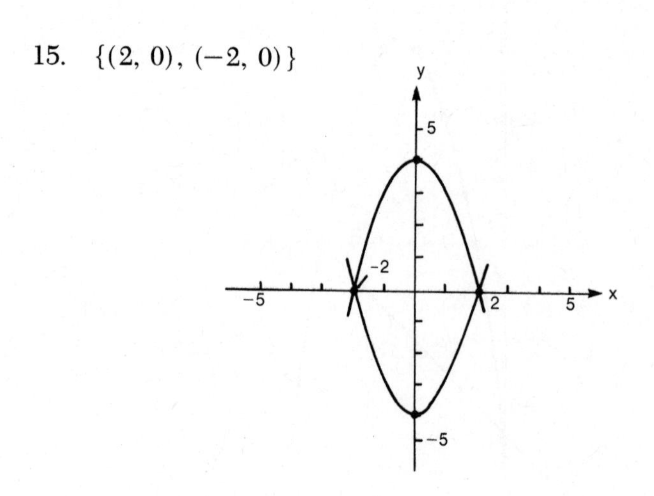

17. $\{(2\sqrt{5}, 4), (-2\sqrt{5}, 4)\}$

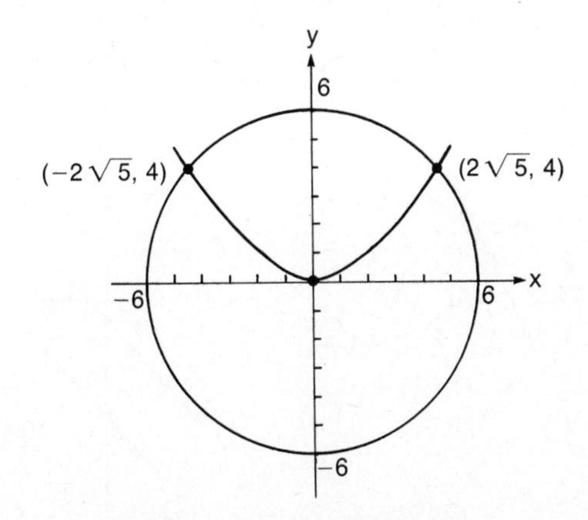

19. $\left\{\left(\frac{-2+2\sqrt{41}}{5}, \frac{8+2\sqrt{41}}{5}\right), \left(\frac{-2-2\sqrt{41}}{5}, \frac{8-2\sqrt{41}}{5}\right)\right\}$

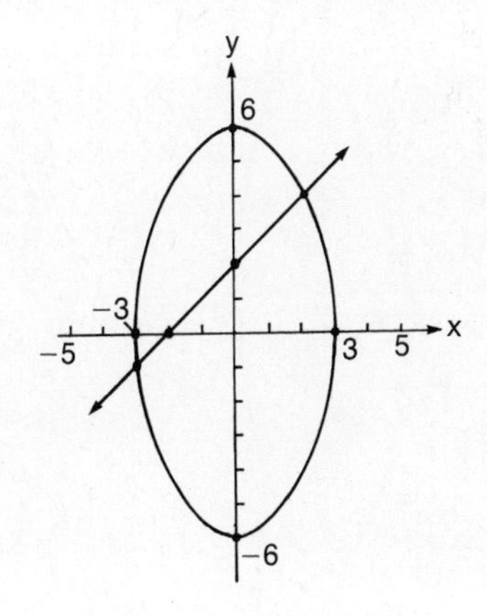

21. No real solutions

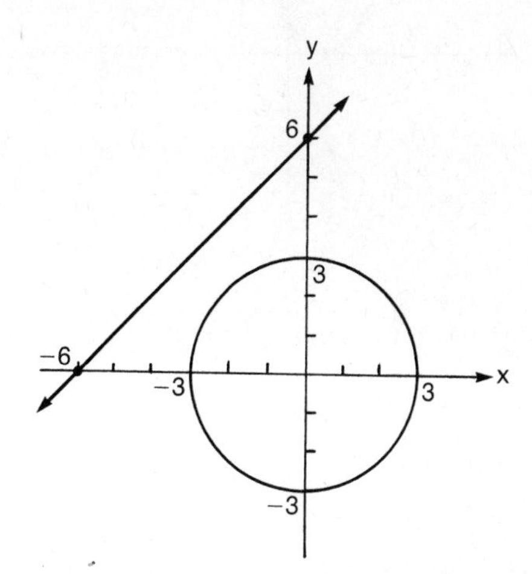

SELF-TEST—CHAPTER 8

1. (a) (b)

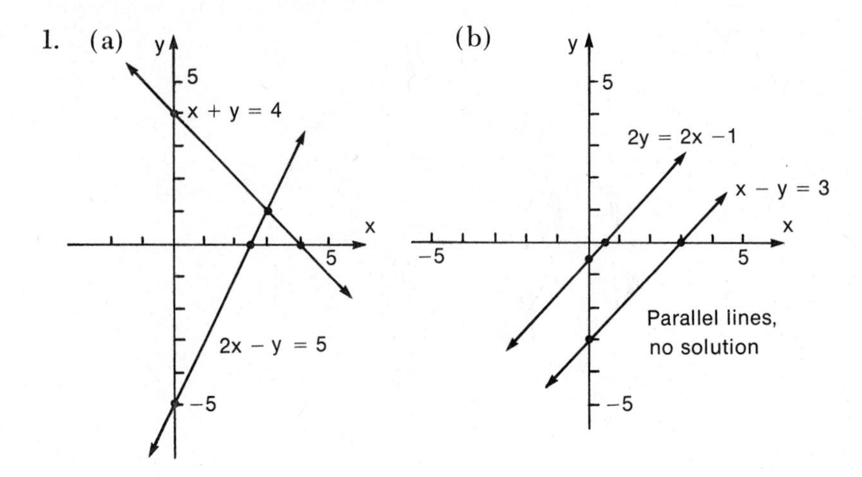

2. (a) $x = -1$, $y = 3$ (b) No solution
3. (a) $x = -1$, $y = -2$ (b) No solution
4. $x = 0$, $y = 1$, $z = 2$
5. (a) 5 (b) 25
6. (a) $x = 4$ (b) $y = -2$
7. (a) $x = 4$ (b) $y = 2$ (c) Neither
8. (a) 11 (b) 1
9. $\{(0, 3), (3, 0)\}$
10.

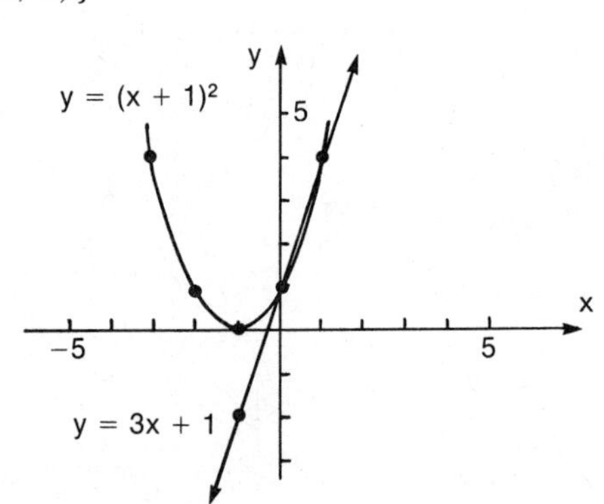

EXERCISE 9.1

1. $D = \{-3, -2, -1\}$, $R = \{0, 1, 2\}$
3. $D = \{3, 4, 5\}$, $R = \{0\}$
5. $D = \{1, 2\}$, $R = \{2, 3\}$
7. $D = \{1, 3, 5, 7\}$, $R = \{-1\}$
9. $D = \{2\}$, $R = \{1, 0, -1, -2\}$

11. $D = \{x \mid -5 \leq x \leq 5\}$, $R = \{y \mid -5 \leq y \leq 5\}$

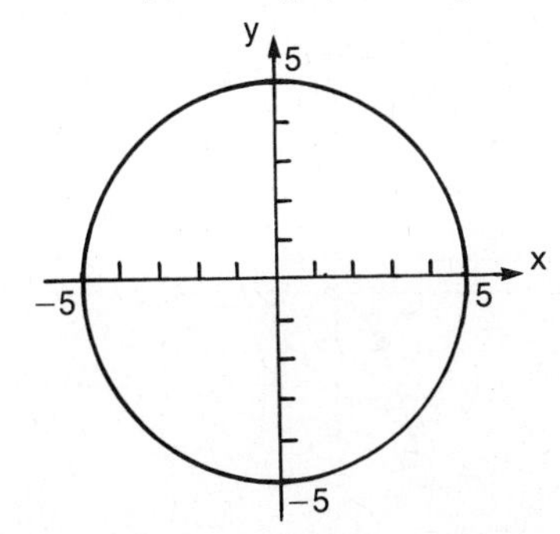

13. $D = \{x \mid -5 \leq x \leq 5\}$, $R = \{y \mid 0 \leq y \leq 5\}$

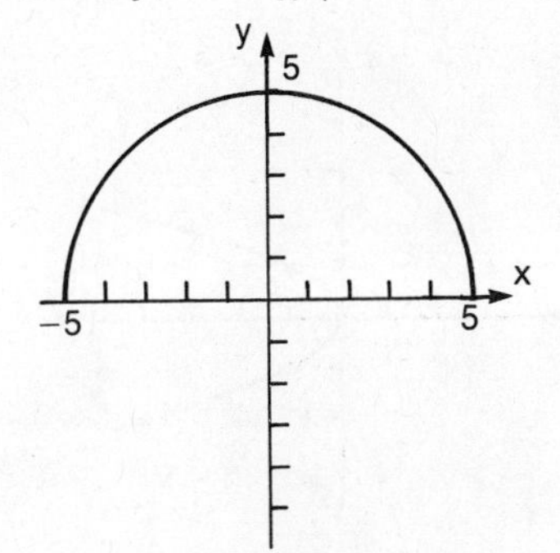

15. $D = \{x \mid 0 \leq x \leq 5\}$, $R = \{y \mid -5 \leq y \leq 5\}$

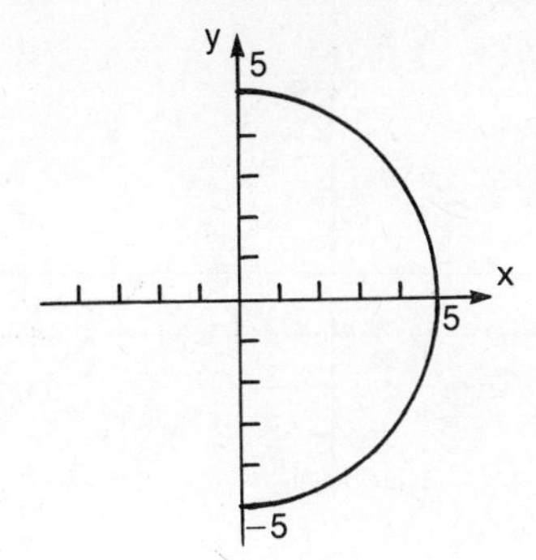

17. $D = \{x \mid x \text{ is a real number}\}$, $R = \{y \mid y \geq -1\}$

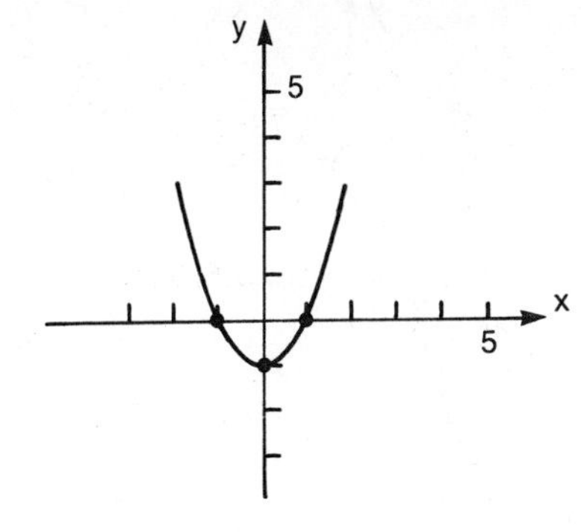

19. D = {x | x is a real number}, R = {y | y ≤ 0}

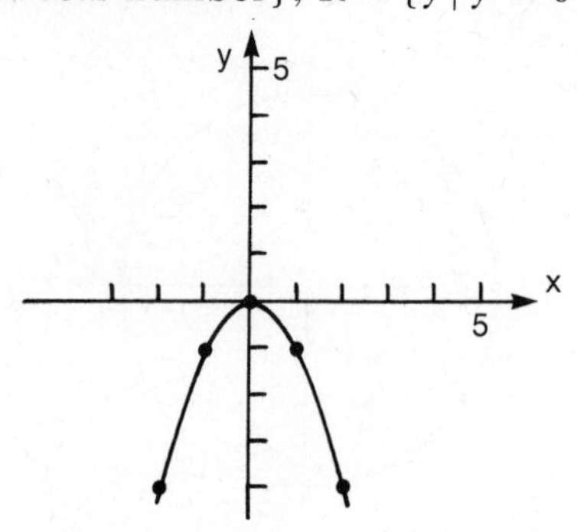

21. D = {x | x ≥ 0}, R = {y | y is a real number}

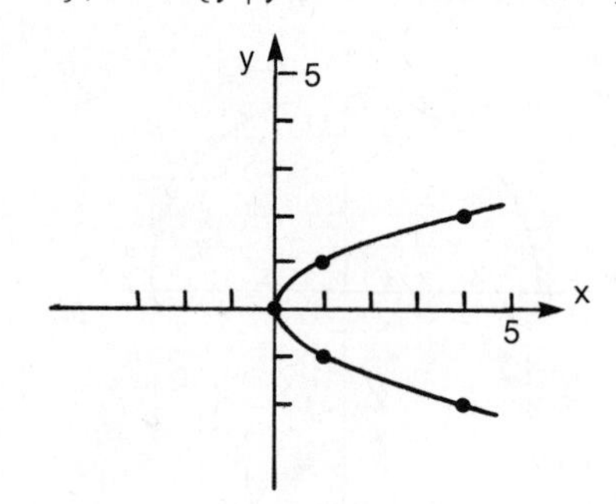

23. D = {x | x is a real number}, R = {y | y is a real number}

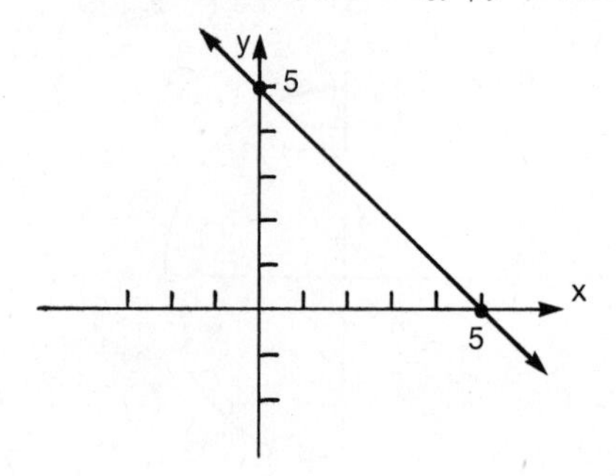

25. D = {x | x is a real number}, R = {y | y is a real number}

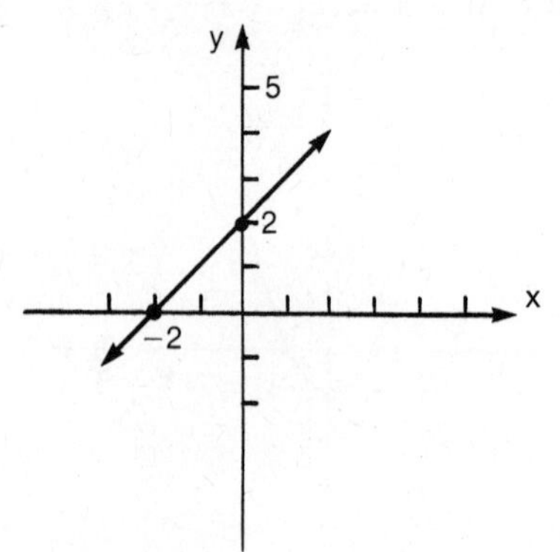

27. $D = \{x \mid x < 3\}$, $R = \{y \mid y \text{ is a real number}\}$

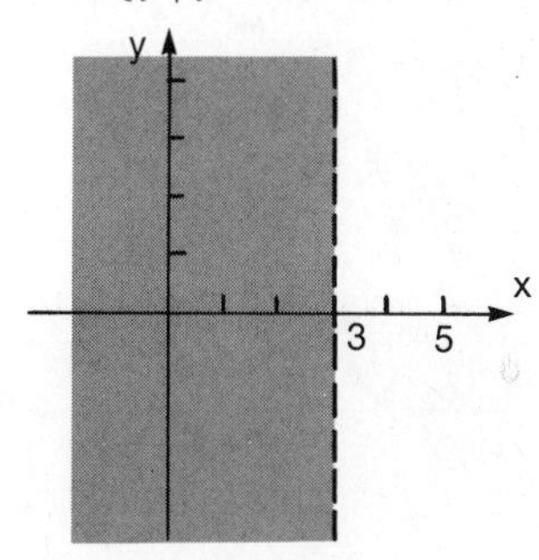

29. $D = \{x \mid x \text{ is a real number}\}$, $R = \{y \mid y < 4\}$

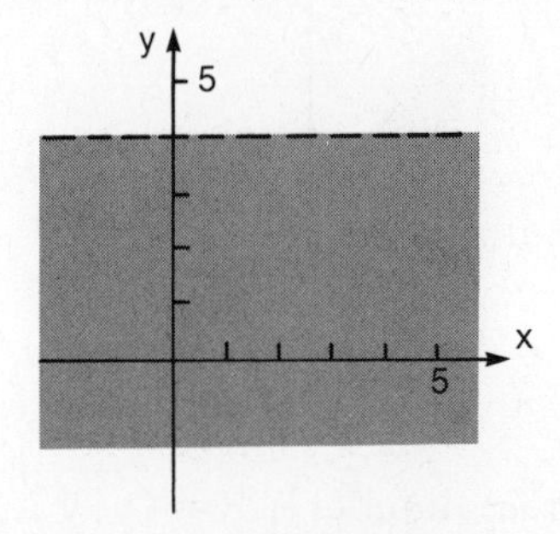

31. $D = \{x \mid x \text{ is a real number}\}$, $R = \{y \mid y \text{ is a real number}\}$

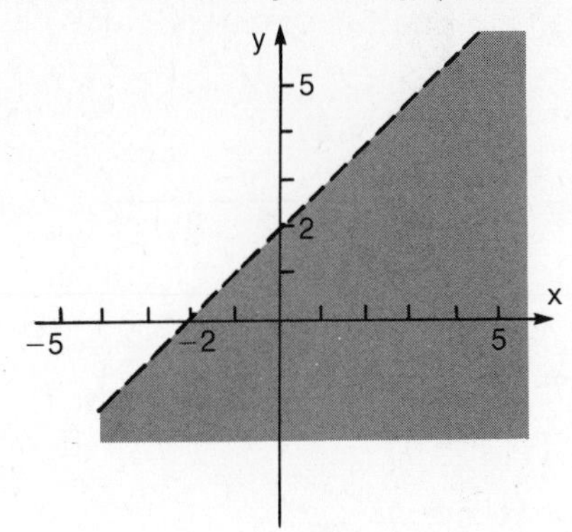

33. $D = \{x \mid x \text{ is a real number}\}$, $R = \{y \mid y \text{ is a real number}\}$

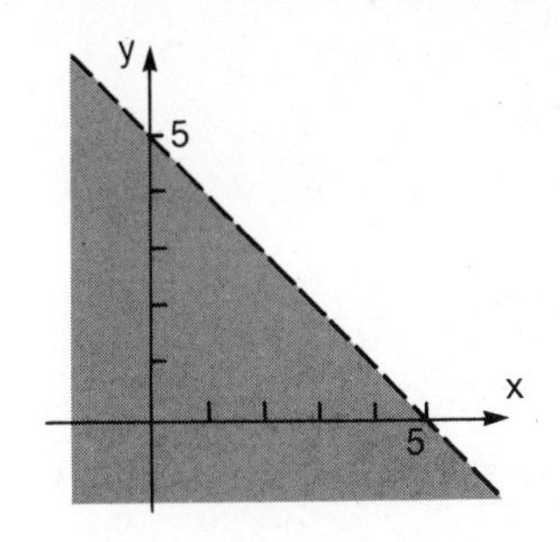

35. $D = \{x \mid x \text{ is a real number}\}$, $R = \{y \mid y \text{ is a real number}\}$

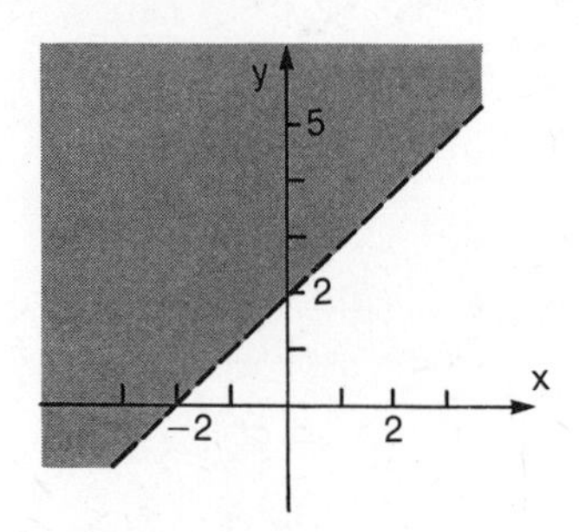

37. $D = \{x \mid x \text{ is a real number}\}$, $R = \{y \mid y \text{ is a real number}\}$

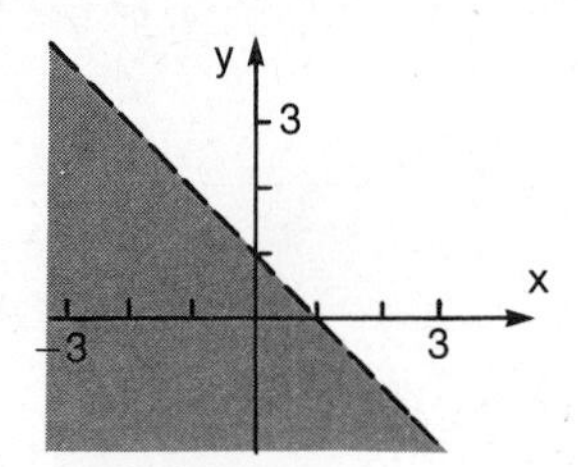

39. $D = \{x \mid x \text{ is a real number}\}$, $R = \{y \mid y \text{ is a real number}\}$

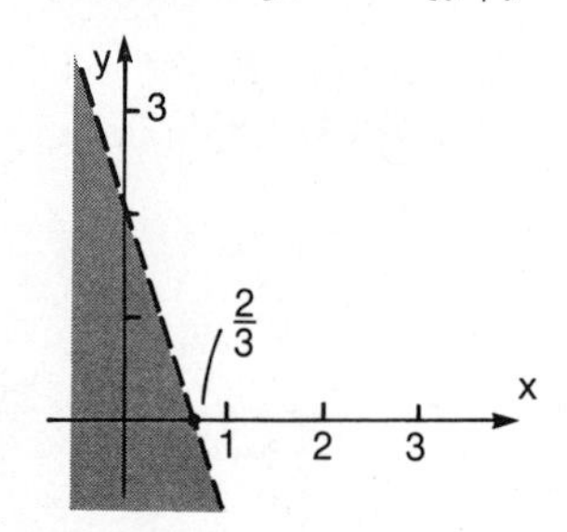

41. $C = 2\pi r$, $D = \{r \mid r \geq 0\}$

43. $P = 4s$, $D = \{s \mid s \geq 0\}$

45. $D = 50t$, $D = \{t \mid t \geq 0\}$

47. $y = 0.10x + 15$, $D = \{x \mid x \geq 0\}$

49. $C = 4(t - 40)$, $D = \{t \mid t \geq 40\}$

EXERCISE 9.2

1. Function
   $D = \{-2, -1, 0\}$, $R = \{0, 1, 2\}$

3. Function
   $D = \{1, 2, 3\}$, $R = \{0\}$

5. Not a function

7. Function
$D = \{x \mid x \text{ is a real number}\}$, $R = \{y \mid y \geq 3\}$

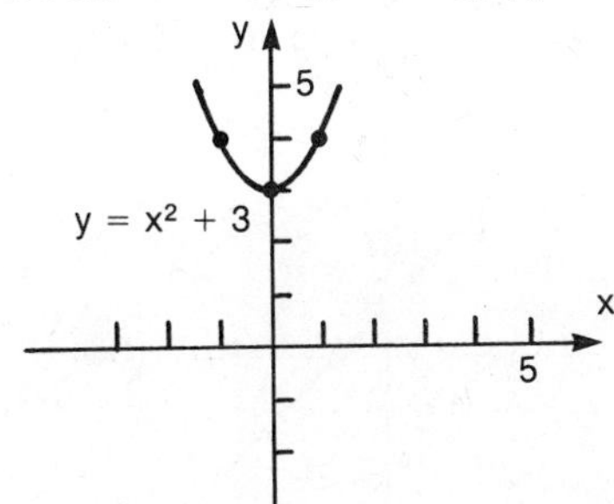

9. Function
$D = \{x \mid x \text{ is a real number}\}$, $R = \{y \mid y \geq 1\}$

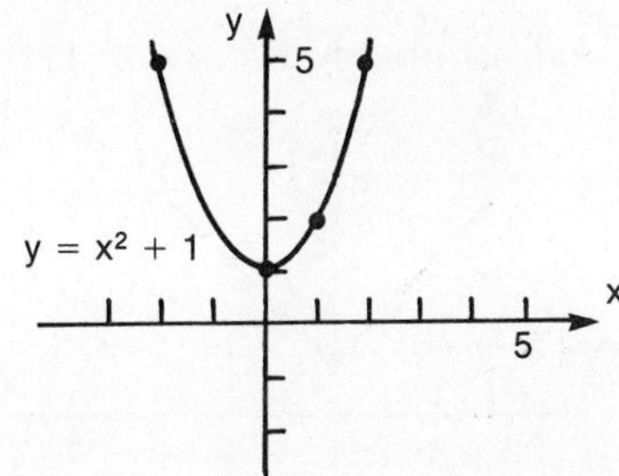

11. Function
$D = \{x \mid x \text{ is a real number}\}$, $R = \{y \mid y \leq 0\}$

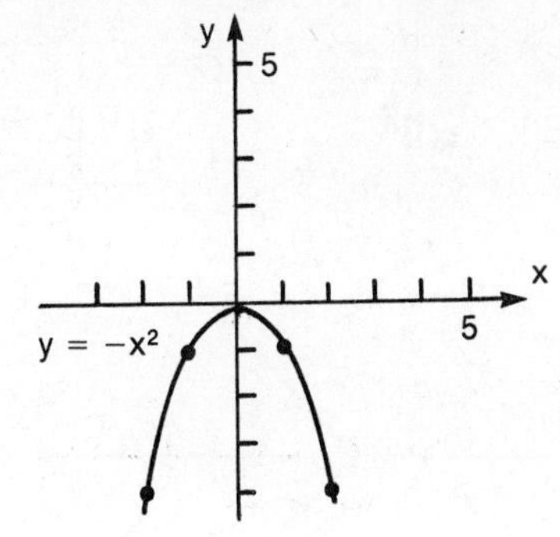

13. Function
$D = \{x \mid x \text{ is a real number}\}$, $R = \{y \mid y \text{ is a real number}\}$

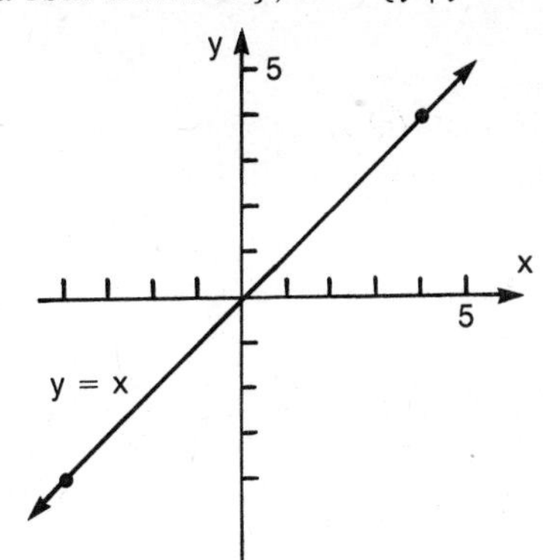

15. Function
$D = \{x \mid x \text{ is a real number}\}$, $R = \{y \mid y \text{ is a real number}\}$

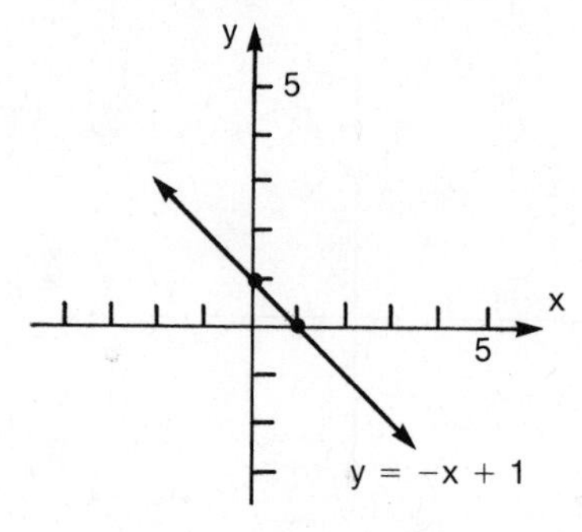

17. Function
$D = \{x \mid x \text{ is a real number}\}$, $R = \{y \geq 1\}$

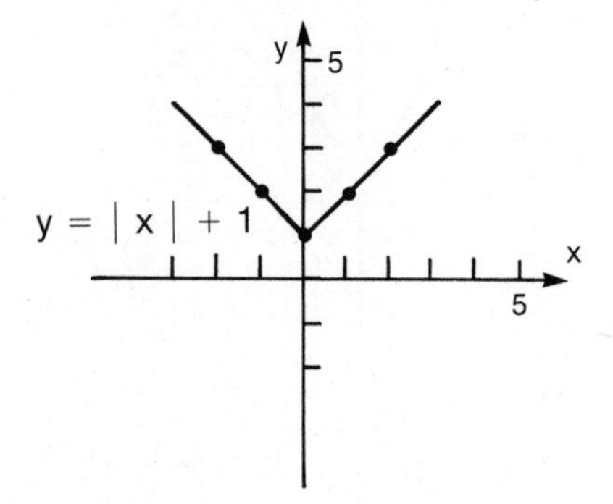

19. Not a function

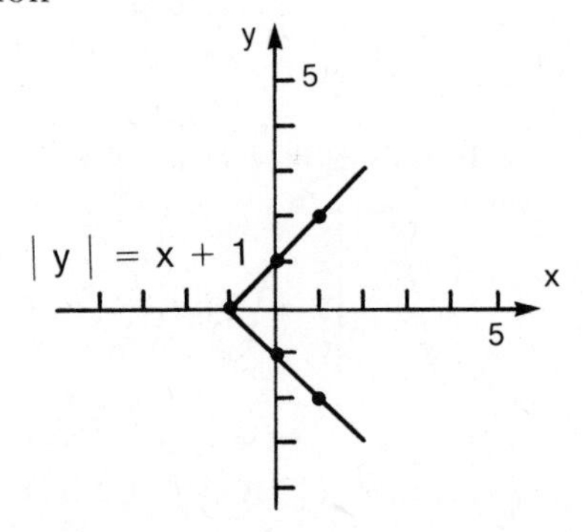

21. $f(0) = 3$, $f(1) = 2\sqrt{2}$, $f(3) = 0$

23. $g(0) = 2$, $g(1) = 3$, $g(-1) = 3$

25. $f(1) = 1$, $f(-1) = -1$, $f(2) = \frac{1}{2}$

27. $g(0) = -\frac{1}{3}$, $g(2) = 1$, $g(-2) = -\frac{5}{3}$

29. $s(1) = 2$, $s(-1) = -2$

31. (a) $x^2 + 2xh + h^2 + 1$
(b) $2xh + h^2$
(c) $2x + h$

33. (a) $2x + 2h + 1$
(b) $2h$
(c) 2

35. (a) $x^2 + 2xh + h^2 + 2x + 2h$
(b) $2xh + h^2 + 2h$
(c) $2x + h + 2$

EXERCISE 9.3

1. $D = \{x \mid x \text{ is a real number}\}$, $R = \{y \mid y \text{ is a real number}\}$

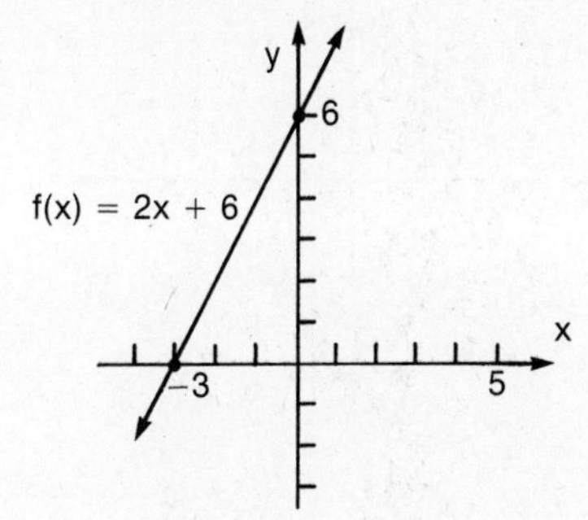

3. $D = \{x \mid x \text{ is a real number}\}$, $R = \{y \mid y \text{ is a real number}\}$

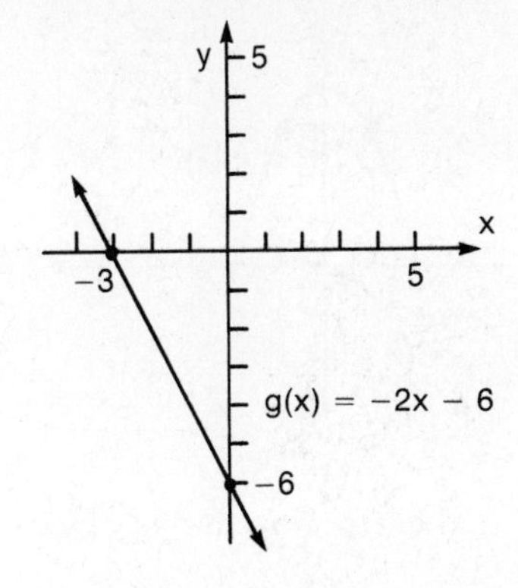

5. $D = \{x \mid x \text{ is a real number}\}$, $R = \{y \mid y \text{ is a real number}\}$

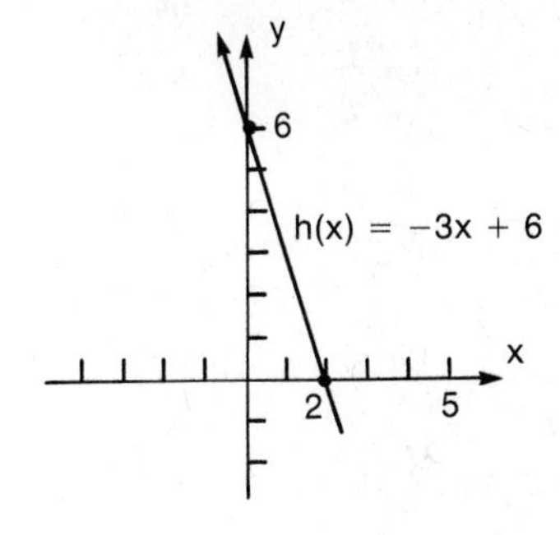

7. D = {x | x is a real number}, R = {y | y ≥ 3}

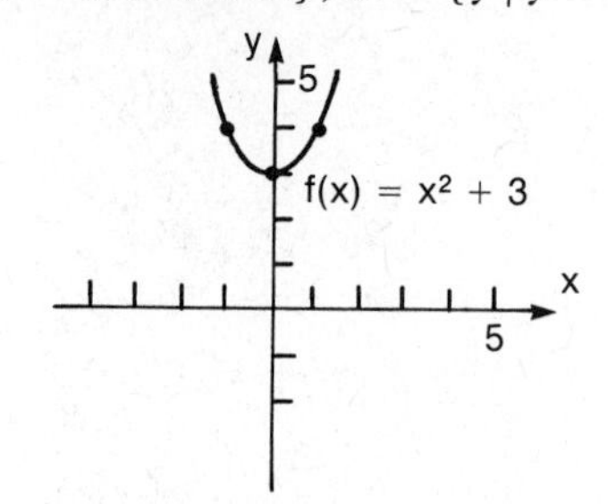

9. D = {x | x is a real number}, R = {y | y ≥ −4}

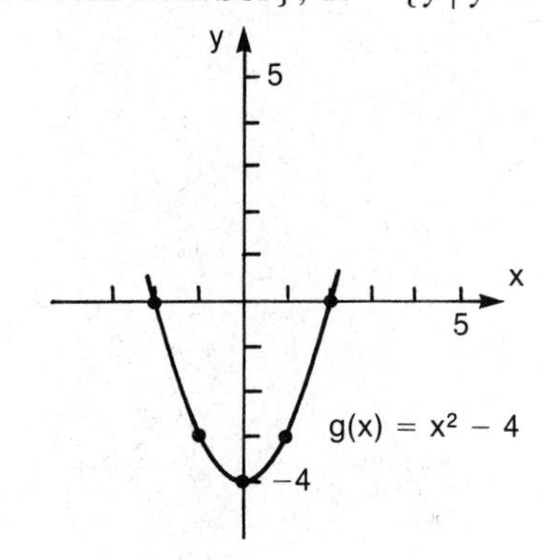

11. D = {x | x is a real number}, R = {y | y ≤ 4}

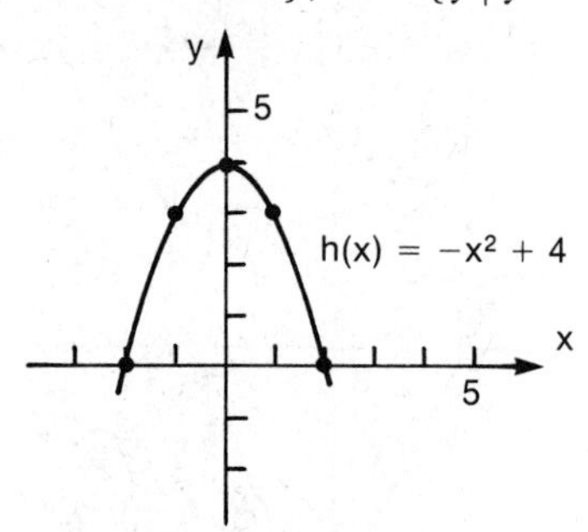

13. D = {x | x is a real number}, R = {y | y ≤ 2}

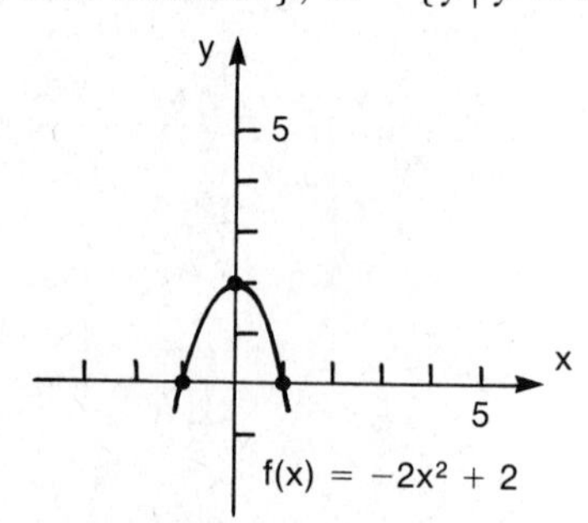

15. D = {x | x is a real number}, R = {y | y ≥ −9}

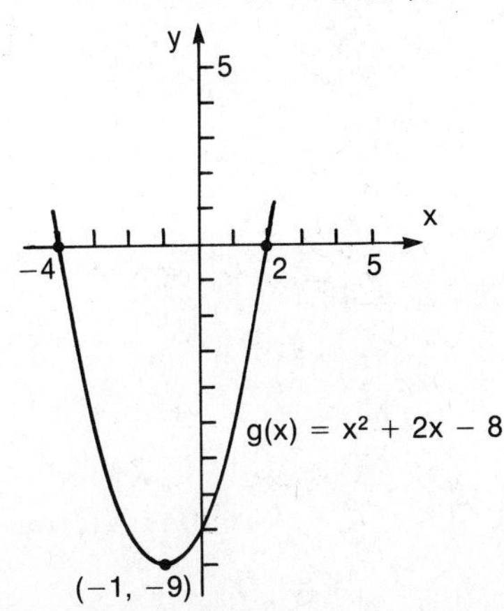

17. D = {x | x is a real number}, R = {y | y ≤ 4}

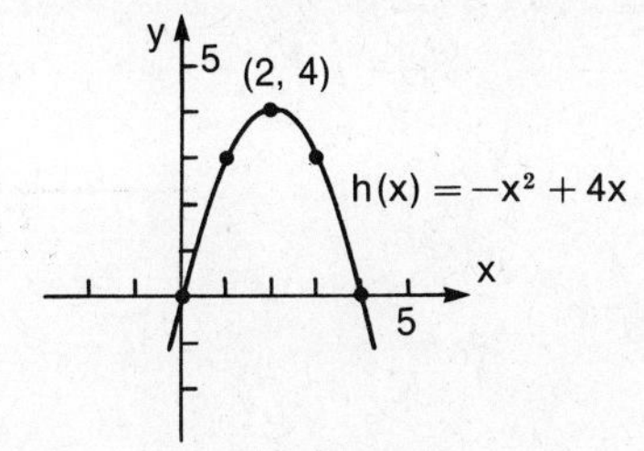

19. D = {x | x is a real number}, R = {y | y ≥ −4}

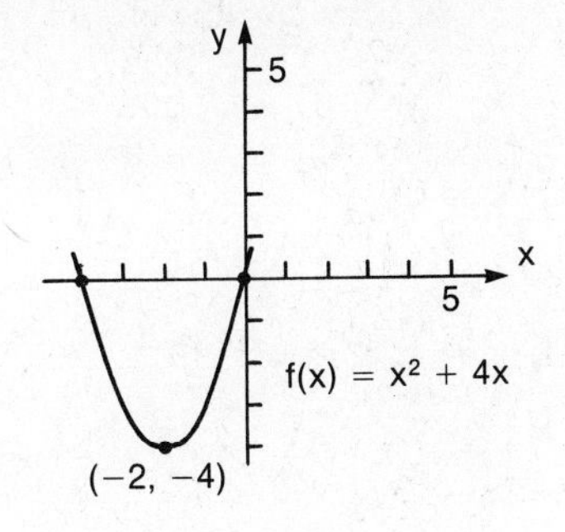

21. Two real roots

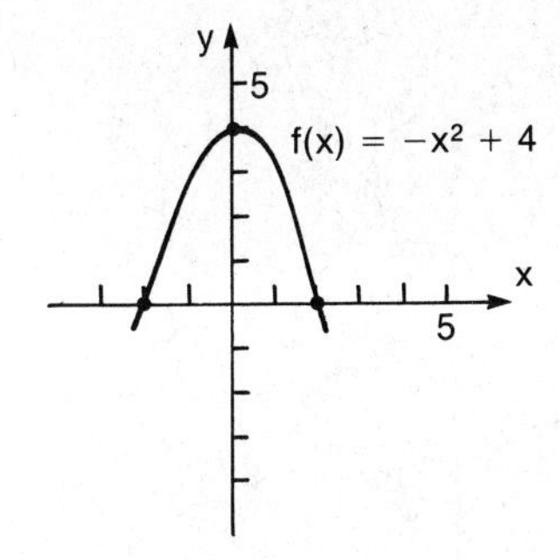

23. Two real roots

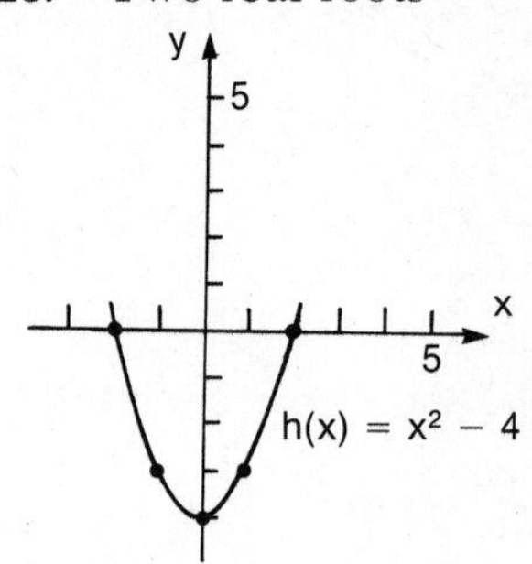

27. Two real roots

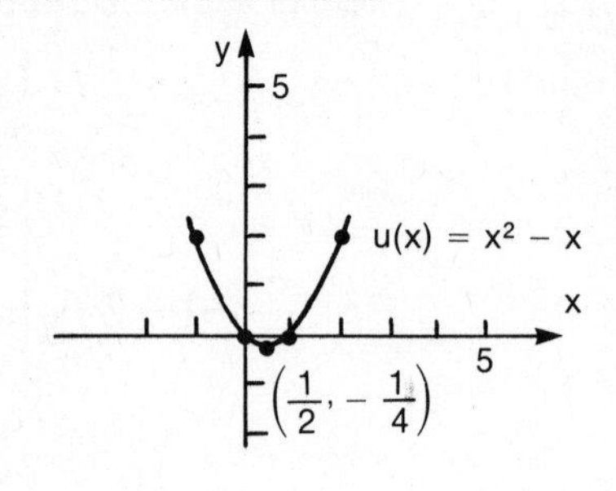

25. Two real roots

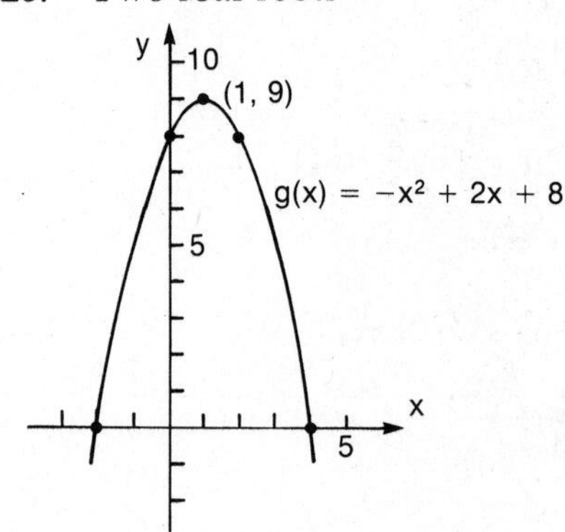

29. No real roots

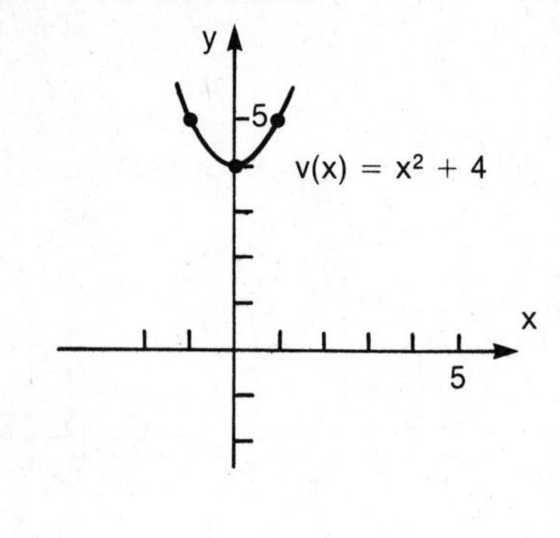

Note: In Problems 31 to 40, x is a real number.

31. $D = \{x \mid x \neq 0\}$

33. $D = \{x \mid x \neq \pm 1\}$

35. $D = \{x \mid x \neq \pm 2, x \neq 0\}$

37. $D = \{x \mid x \neq \pm 2, x \neq -1\}$

39. $D = \{x \mid x \neq \pm 4, x \neq 1, x \neq -2\}$

EXERCISE 9.4

1. $S^{-1} = \{(3, 1), (4, 2), (6, 5)\}$
   S^{-1} is a function

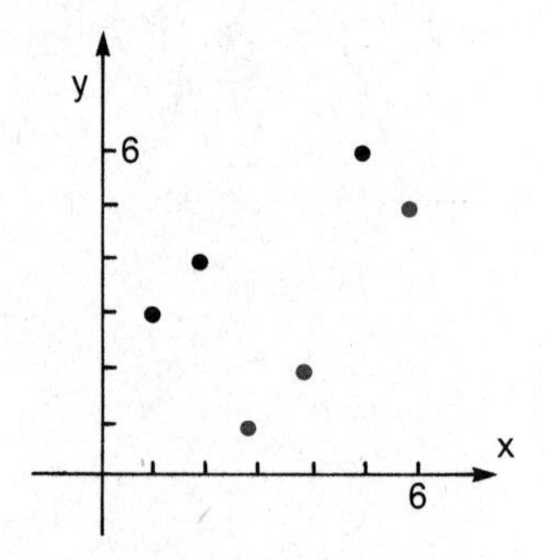

3. $S^{-1} = \left\{(x, y) \,\middle|\, y = \frac{x-6}{3}\right\}$
S^{-1} is a function

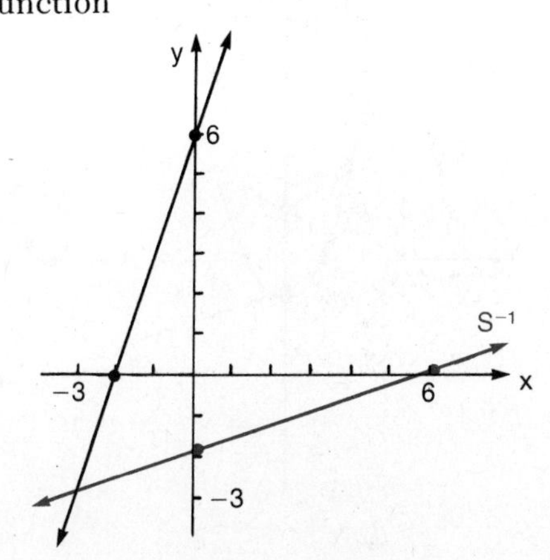

5. $S^{-1} = \left\{(x, y) \,\middle|\, y = \frac{x+4}{2}\right\}$
S^{-1} is a function

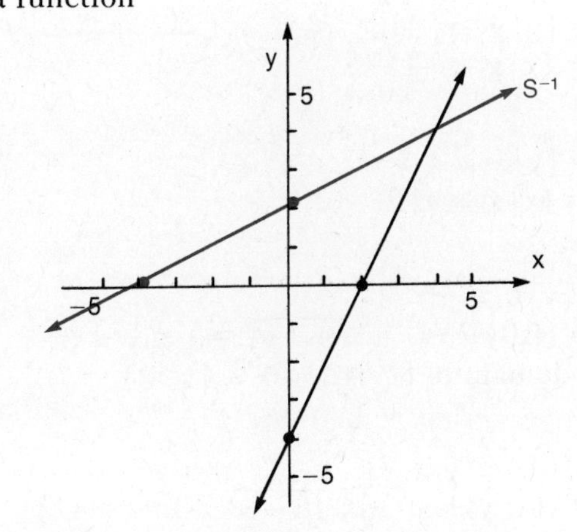

7. $S^{-1} = \left\{(x, y) \,\middle|\, y = \pm\sqrt{\frac{x}{2}}\right\}$
S^{-1} is not a function

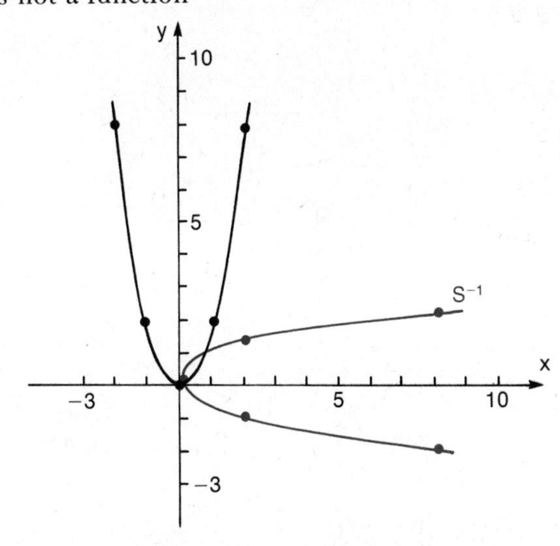

9. $S^{-1} = \{(x, y) \mid y = \pm\sqrt{x+1}\}$
S^{-1} is not a function

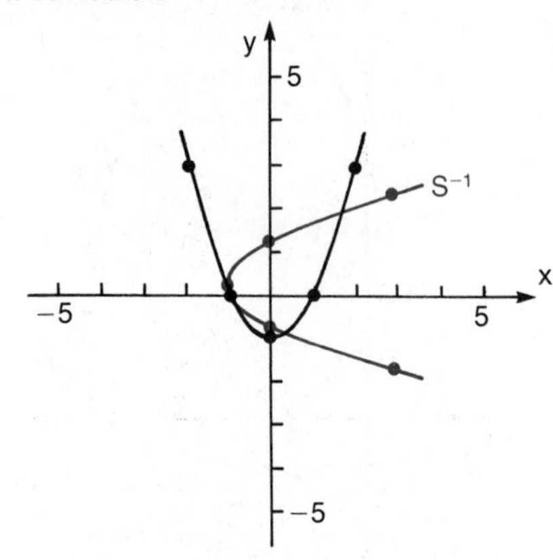

11. (a) R is the set of non-negative real numbers
(b) D is the set of non-negative real numbers
(c) Yes

13. (a) $R = \{y \mid y \geq -1\}$
(b) $D = \{x \mid x \geq -1\}$
(c) No

15. (a) $R = \{y \mid y \geq 4\}$
(b) $D = \{x \mid x \geq 4\}$
(c) No

17. (a) $R = \{y \mid 0 \leq y \leq 3\}$
(b) $f^{-1} = \{(x, y) \mid x = \pm\sqrt{9 - y^2}, -3 \leq y \leq 3\}$
The domain of f^{-1} is $\{x \mid 0 \leq x \leq 3\}$
(c) No

19. (a) $R = \{y \mid 0 \leq y \leq 4\}$
(b) $f^{-1} = \{(x, y) \mid x = \pm\sqrt{16 - y^2}, -4 \leq y \leq 4\}$
The domain of f^{-1} is $\{x \mid 0 \leq x \leq 4\}$
(c) No

EXERCISE 9.5

1. $T = ks$
3. $W = kh^3$
5. $W = kB$
7. $C = kwm$
9. $P = kRI^2$
11. $I = \frac{k}{R}$
13. $I = \frac{k}{d^2}$
15. $I = \frac{ki}{d^2}$
17. $R = \frac{kL}{A}$
19. $W = \frac{k}{d^2}$
21. $k = 16$; $s = 16t^2$
23. (a) $\frac{9}{2000} = 0.0045$
(b) 18.225 lbs

25. (a) $k = 150$
    (b) $s = 400$ lbs

27. $k = 4$; 200 lbs per square foot

29. Twice as far.

SELF-TEST—CHAPTER 9

1. (a) $\{-2, -3, 2\}$ (b) $\{2, 4\}$ (c) Yes
2. (a) $\{x \mid x \text{ is a real number}\}$ (b) $\{y \mid y \geq 1\}$

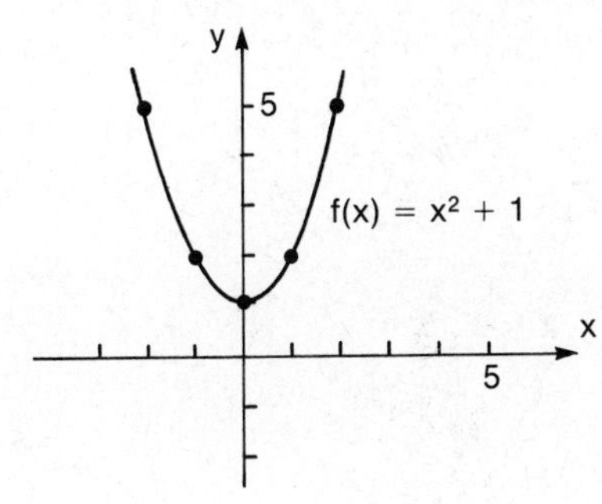

3. Two real roots

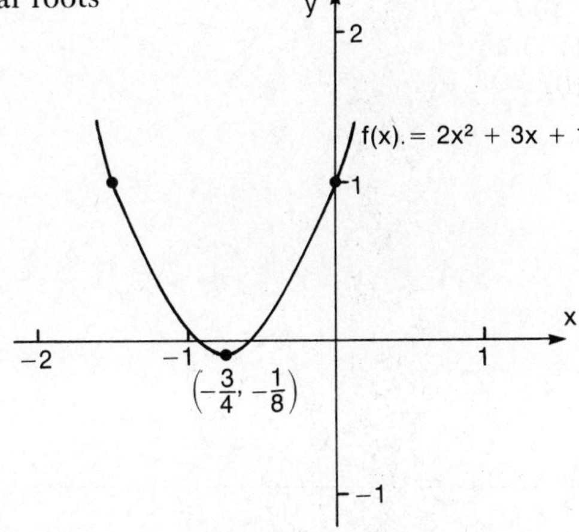

4. (a) $\{(1, 3), (3, 1), (7, 2)\}$ (b) $\{1, 3, 7\}$
   (c) $\{3, 1, 2\}$ (d) Yes (e) Yes
5. (a) Yes (b) $f^{-1}(x) = \dfrac{x + 3}{3}$ (c) Yes
6.

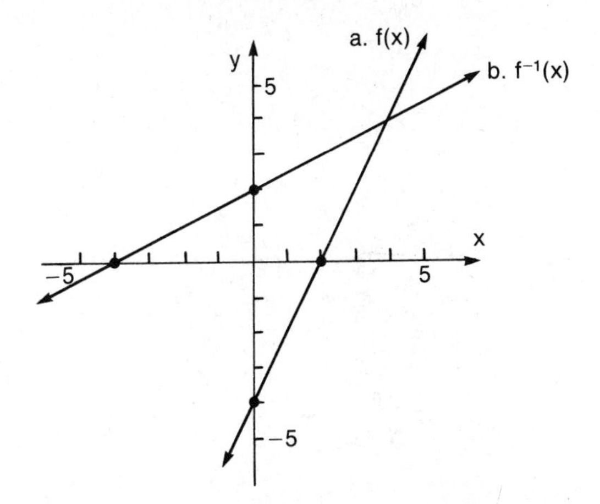

7. (a) $y = kx^2$ (b) $k = 2$

8. (a) $I = \dfrac{2500}{d^2}$ (b) $I = 2500$

9. (a) $z = kxy^2$ (b) $k = \dfrac{3}{8}$

10. (a) $k = 2$ (b) $4000

EXERCISE 10.1

1. (a) $\left(-1, \dfrac{1}{5}\right)$
   (b) (0, 1)
   (c) (1, 5)

3. (a) $\left(-1, \dfrac{1}{6}\right)$
   (b) (0, 1)
   (c) (1, 6)

5. (a) $\left(-1, \dfrac{1}{10}\right)$
   (b) (0, 1)
   (c) (1, 10)

7.

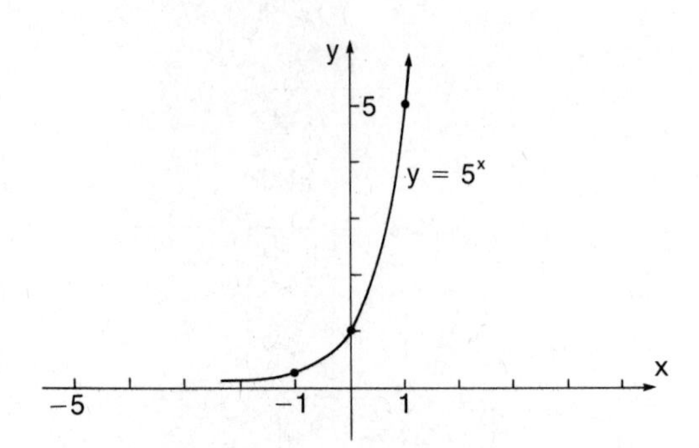

9.

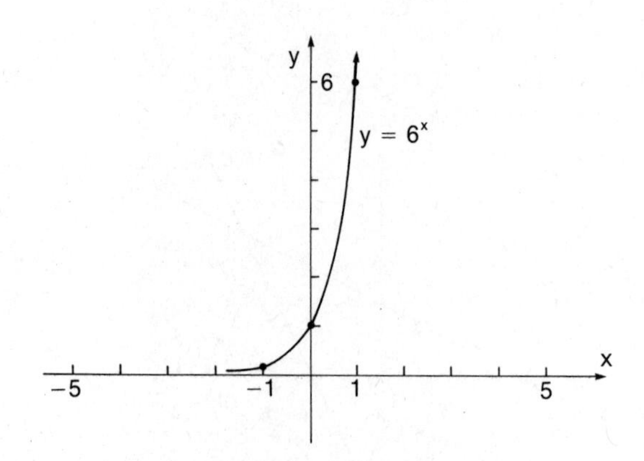

11.

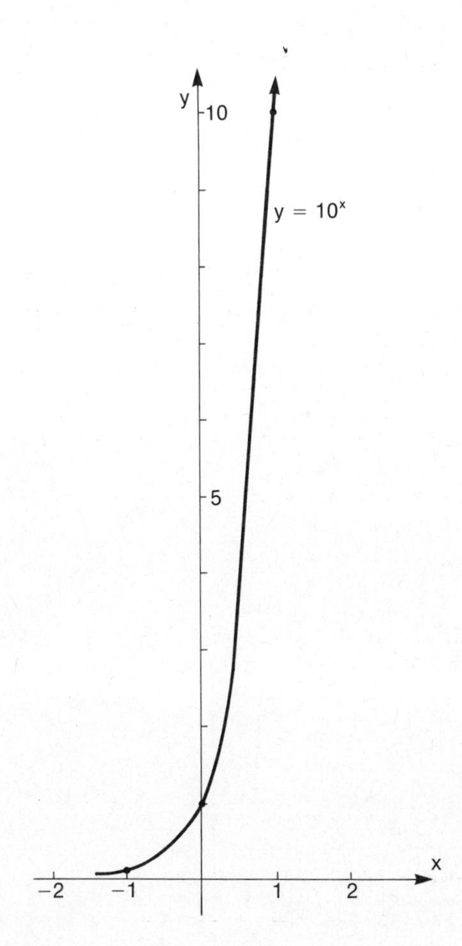

13. Increasing

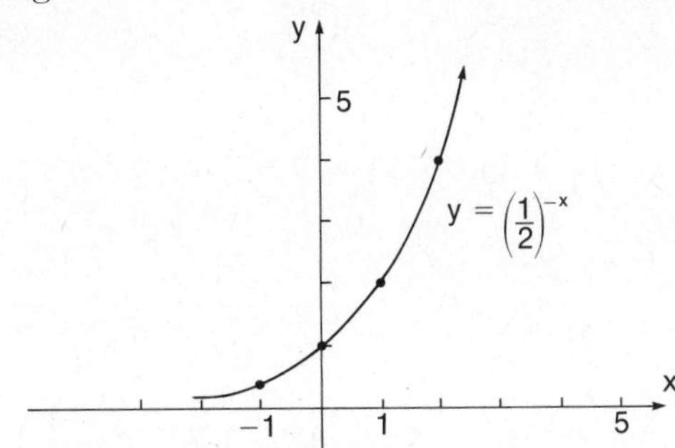

15. Increasing

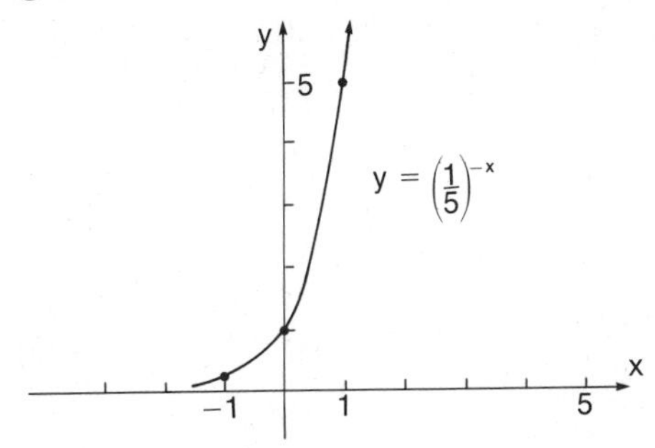

17. Increasing

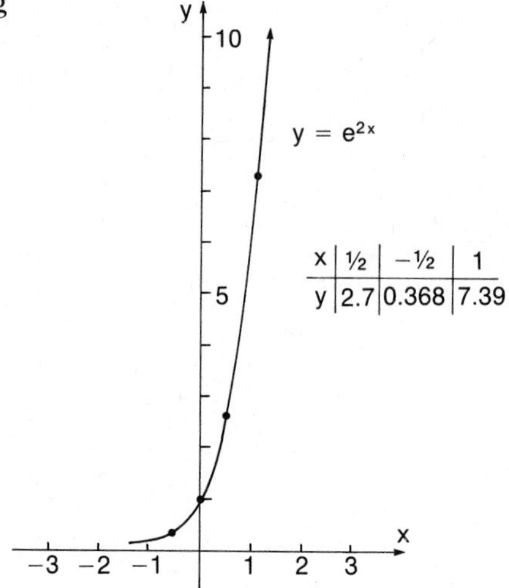

19. Decreasing

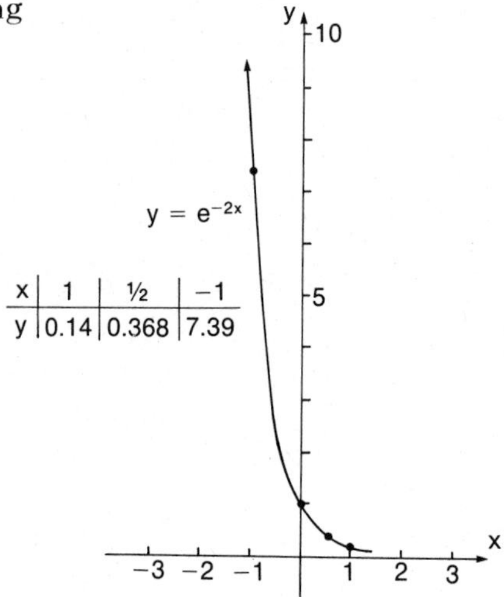

EXERCISE 10.2

1. 8
3. 2
5. $\frac{7}{3}$
7. -6
9. $\log_2 128 = 7$
11. $\log_{81} 9 = \frac{1}{2}$
13. $\log_{10} 1000 = 3$
15. $\log_{216} 6 = \frac{1}{3}$
17. $\log_{10} 3 = 0.47712$
19. $\log_b N = 5$
21. $9^3 = 729$
23. $2^{-8} = \frac{1}{256}$
25. $81^{3/4} = 27$
27. $10^{2.47712} = 300$

29. $x = 3^{27}$

31. $z = 5^5$

33. $x = 2^7$

35. $N = 10^5$

37. $Q = \left(\frac{1}{6}\right)^5$

39. $x = \left(\frac{1}{2}\right)^7$

41. $z = 2$

43. $a = 2$

45. $k = 7^3$

47. $b = \frac{1}{81}$

49. $c = 100$

EXERCISE 10.3

1. 8.9
3. 6.7
5. 1.556
7. 2.130
9. 0.681
11. −0.125
13. 3.766
15. 0.938
17. 0.420
19. −0.495

Problems 21 to 30—even *and* odd alike—are left for the student to solve.

EXERCISE 10.4

1. 9.92×10^8
3. 3×10^4
5. 4×10^{-4}
7. 0
9. 4
11. −1
13. 3
15. −1
17. 1
19. −5
21. 9 − 10
23. 8 − 10
25. 8 − 10
27. 8 − 10
29. 9 − 10
31. 1.8720
33. 3.2641
35. 8.6404 − 10
37. 9.1082 − 10
39. 5.2154
41. 7.9348 − 10
43. 8.8285 − 10
45. 18.518
47. 5.849
49. 0.0108
51. Approximately $197 ($196.72)
53. Approximately $2161 ($2160.97)
55. 6.2
57. 7.8
59. 6.4

EXERCISE 10.5

1. 0.19%
3. 0.20%
5. 0.209%
7. 17.33 minutes
9. 80.5 minutes
11. 23,105 years
13. 13.35 years
15. (a) 0.7%
    (b) about 87%
17. (a) 11.45 pounds per square inch
    (b) 8.92 pounds per square inch
19. 4.09

SELF-TEST—CHAPTER 10

1.

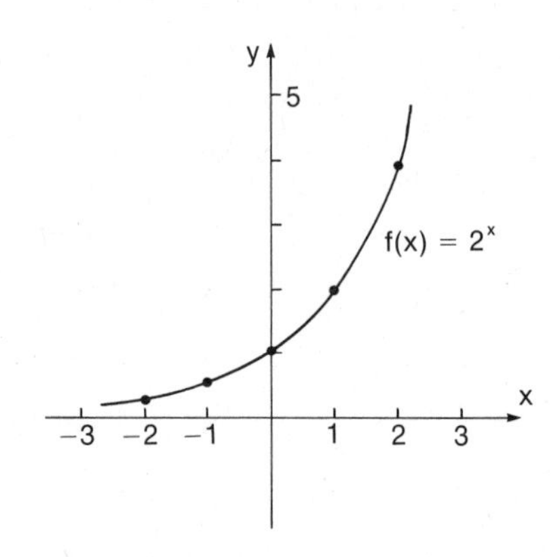

2.

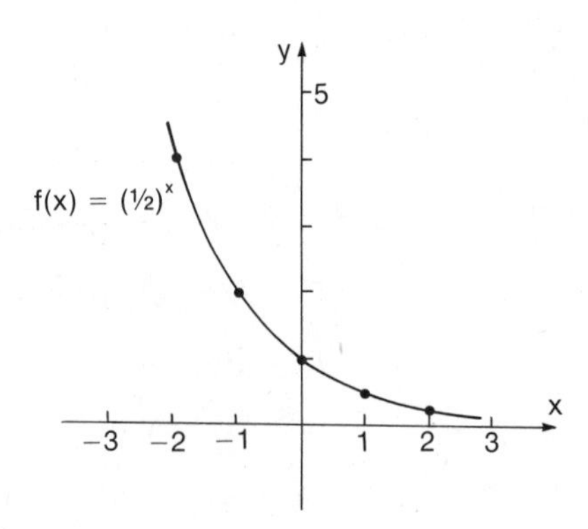

3. (a) 4 (b) 4 (c) -3

4. (a) 1.556 (b) 1 (c) 0.398

5. (a) 2.87×10^5 (b) 1.23×10^{-4}

6. (a) 2.5250 (b) $0.5250 - 3$ or $7.5250 - 10$

7. (a) 3,280 (b) 0.0000331

8. $265.50

9. (a) $x = 3^{243}$ (b) $x = 128$ (c) $x = 10{,}000$

10. (a) $k = \frac{1}{4}$ (b) $1000e \approx 2{,}700$

EXERCISE 11.1

1. tenth term: 10
   nth term: n

3. tenth term: 34
   nth term: $3n + 4$

5. tenth term: 10
   nth term: $5(12 - n)$

7. tenth term: $\frac{1}{11}$
   nth term: $\frac{1}{n + 1}$

9. tenth term: 1
   nth term: $(-1)^n$

11. tenth term: x^{10}
    nth term: x^n

13. tenth term: x^{19}
    nth term: $(-1)^n x^{2n-1}$

15. tenth term: $\frac{x^{10}}{10}$
    nth term: $\frac{x^n}{n}$

17. tenth term: x
    nth term: $(-1)^n x$

19. tenth term: $\frac{x^{10}}{2^{10}}$
    nth term: $\frac{x^n}{2^n} = \left(\frac{x}{2}\right)^n$

21. $-1, 1, 3$

23. $-\frac{1}{2}, 0, \frac{3}{2}$

25. $2, \frac{3}{2}, \frac{4}{3}$

27. 1, 4, 9

29. $\frac{1}{3}, \frac{2}{5}, \frac{3}{7}$

31. $-1, 1, -1$

33. $-2, 4, -8$

35. (a) 9 ft (b) $\frac{81}{10}$ ft (c) $\frac{9^n}{10^{n-1}}$

37. (a) 400 (b) 1600 (c) $2^n \bullet 100$

39. $320

EXERCISE 11.2

1. (a) 5
 (b) 3
 (c) $2 + 3n$

3. (a) 11
 (b) -5
 (c) $16 - 5n$

5. (a) 3
 (b) -4
 (c) $7 - 4n$

7. (a) 8
 (b) -8
 (c) $16 - 8n$

9. (a) $\frac{-5}{6}$
 (b) $\frac{1}{2}$
 (c) $\frac{-4}{3} + \frac{1}{2}n$

11. $a_{15} = 91$
 $S_{15} = 735$

13. $a_1 = 3$
 $d = -4$
 $S_{10} = -150$

15. $d = 1$
 $S_6 = 33$

17. $a_{14} = -46$, $d = -4$

19. $a_1 = -779$, $a_{40} = 781$

21. $a_{51} = 468$

23. \$19,800

Note: Problems 24 to 28—even *and* odd—are left for the student to solve.

EXERCISE 11.3

1. (a) 3
   (b) 2
   (c) $3 \cdot 2^{n-1}$
   (d) $3(2^n - 1)$
   (e) Not possible

3. (a) 8
   (b) 3
   (c) $8 \cdot 3^{n-1}$
   (d) $4(3^n - 1)$
   (e) Not possible

5. (a) 16
   (b) $-\frac{1}{4}$
   (c) $(-4)^{3-n}$
   (d) $\frac{64}{5}\left[1 - \left(-\frac{1}{4}\right)^n\right]$
   (e) $\frac{64}{5}$

7. (a) $\frac{-3}{5}$
   (b) $\frac{-5}{2}$
   (c) $\frac{-3}{5}\left(-\frac{5}{2}\right)^{n-1}$
   (d) $\frac{-6\left[1 - \left(-\frac{5}{2}\right)^n\right]}{35}$
   (e) Not possible

9. (a) $-\frac{3}{4}$
   (b) $\frac{1}{3}$
   (c) $\left(\frac{-3}{4}\right)\left(\frac{1}{3}\right)^{n-1}$
   (d) $-\frac{9}{8}\left[1 - \frac{1}{3^n}\right]$
   (e) $-\frac{9}{8}$

11. $a_3 = \frac{1}{4}$, $r = \frac{1}{2}$
    $a_3 = \frac{9}{4}$, $r = \frac{-3}{2}$

13. $a_3 = 12$, $r = 2$
    $a_3 = 27$, $r = -3$

15. $a_1 = 7$, $a_8 = 896$

17. $r = -3$, $n = 4$

19. $n = 6$, $S_n = \dfrac{\frac{16}{125}\left[\left(1 - \frac{5}{2}\right)^n\right]}{1 - \frac{5}{2}}$

21. $S_\infty = 12$

23. $S_\infty = -12$

25. $S_\infty = \frac{4}{3}$

27. Does not exist

29. $S_\infty = -8$

31. $\frac{4}{9}$

33. $\frac{31}{99}$

35. $\frac{3}{11}$

37. $\frac{2293}{990}$

39. $\frac{140}{999}$

41. $200{,}000(1.04)^5$

43. $r = \left(\frac{5}{4}\right)^{3/5}$, or about 11.2%

45. 380 ft.

EXERCISE 11.4

1. 6
3. 3,628,800
5. 30
7. 50
9. 15
11. 11
13. 1
15. $(n+3)(n+2)(n+1)$
17. $(n+2)(n+1)(n)$
19. $(2n+4)(2n+3)$
21. $a^3 + 9a^2b + 27ab^2 + 27b^3$
23. $x^4 + 16x^3 + 96x^2 + 256x + 256$
25. $32x^5 - 80x^4 + 80x^3 - 40x^2 + 10x - 1$
27. $8x^3 - 36x^2y + 54xy^2 - 27y^3$
29. $\frac{1}{x^3} + \frac{3}{x^2y} + \frac{3}{xy^2} + \frac{1}{y^3}$
31. $1512x^5$
33. $60x^2y^4$
35. $144xy^2$

SELF-TEST—CHAPTER 11

1. (a) 17 (b) 9
2. (a) 3, 5, 7, 9, 11 (b) 163
3. (a) 6 (b) 3
4. (a) 31 (b) $11 + 5(n-1)$
5. (a) 2550 (b) $\frac{n(2+2n)}{2} = n(1+n)$
6. (a) 3 (b) 2
7. (a) $\frac{1}{2}$ (b) 1
8. (a) 4 (b) -2
9. (a) 720 (b) 56 (c) 15
10. $a^3 - 6a^2b + 12ab^2 - 8b^3$

APPENDICES

APPENDIX 1. FOUR-PLACE LOGARITHMS

	0	1	2	3	4	5	6	7	8	9
1.0	.0000	.0043	.0086	.0128	.0170	.0212	.0253	.0294	.0334	.0374
1.1	.0414	.0453	.0492	.0531	.0569	.0607	.0645	.0682	.0719	.0755
1.2	.0792	.0828	.0864	.0899	.0934	.0969	.1004	.1038	.1072	.1106
1.3	.1139	.1173	.1206	.1239	.1271	.1303	.1335	.1367	.1399	.1430
1.4	.1461	.1492	.1523	.1553	.1584	.1614	.1644	.1673	.1703	.1732
1.5	.1761	.1790	.1818	.1847	.1875	.1903	.1931	.1959	.1987	.2014
1.6	.2041	.2068	.2095	.2122	.2148	.2175	.2201	.2227	.2253	.2279
1.7	.2304	.2330	.2355	.2380	.2405	.2430	.2455	.2480	.2504	.2529
1.8	.2553	.2577	.2601	.2625	.2648	.2672	.2695	.2718	.2742	.2765
1.9	.2788	.2810	.2833	.2856	.2878	.2900	.2923	.2945	.2967	.2989
2.0	.3010	.3032	.3054	.3075	.3096	.3118	.3139	.3160	.3181	.3201
2.1	.3222	.3243	.3263	.3284	.3304	.3324	.3345	.3365	.3385	.3404
2.2	.3424	.3444	.3464	.3483	.3502	.3522	.3541	.3560	.3579	.3598
2.3	.3617	.3636	.3655	.3674	.3692	.3711	.3729	.3747	.3766	.3784
2.4	.3802	.3820	.3838	.3856	.3874	.3892	.3909	.3927	.3945	.3962
2.5	.3979	.3997	.4014	.4031	.4048	.4065	.4082	.4099	.4116	.4133
2.6	.4150	.4166	.4183	.4200	.4216	.4232	.4249	.4265	.4281	.4298
2.7	.4314	.4330	.4346	.4362	.4378	.4393	.4409	.4425	.4440	.4456
2.8	.4472	.4487	.4502	.4518	.4533	.4548	.4564	.4579	.4594	.4609
2.9	.4624	.4639	.4654	.4669	.4683	.4698	.4713	.4728	.4742	.4757
3.0	.4771	.4786	.4800	.4814	.4829	.4843	.4857	.4871	.4886	.4900
3.1	.4914	.4928	.4942	.4955	.4969	.4983	.4997	.5011	.5024	.5038
3.2	.5051	.5065	.5079	.5092	.5105	.5119	.5132	.5145	.5159	.5172
3.3	.5185	.5198	.5211	.5224	.5237	.5250	.5263	.5276	.5289	.5302
3.4	.5315	.5328	.5340	.5353	.5366	.5378	.5391	.5403	.5416	.5428
3.5	.5441	.5453	.5465	.5478	.5490	.5502	.5514	.5527	.5539	.5551
3.6	.5563	.5575	.5587	.5599	.5611	.5623	.5635	.5647	.5658	.5670
3.7	.5682	.5694	.5705	.5717	.5729	.5740	.5752	.5763	.5775	.5786
3.8	.5798	.5809	.5821	.5832	.5843	.5855	.5866	.5877	.5888	.5899
3.9	.5911	.5922	.5933	.5944	.5955	.5966	.5977	.5988	.5999	.6010
4.0	.6021	.6031	.6042	.6053	.6064	.6075	.6085	.6096	.6107	.6117
4.1	.6128	.6138	.6149	.6160	.6170	.6180	.6191	.6201	.6212	.6222
4.2	.6232	.6243	.6253	.6263	.6274	.6284	.6294	.6304	.6314	.6325
4.3	.6335	.6345	.6355	.6365	.6375	.6385	.6395	.6405	.6415	.6425
4.4	.6435	.6444	.6454	.6464	.6474	.6484	.6493	.6503	.6513	.6522
4.5	.6532	.6542	.6551	.6561	.6571	.6580	.6590	.6599	.6609	.6618
4.6	.6628	.6637	.6646	.6656	.6665	.6675	.6684	.6693	.6702	.6712
4.7	.6721	.6730	.6739	.6749	.6758	.6767	.6776	.6785	.6794	.6803
4.8	.6812	.6821	.6830	.6839	.6848	.6857	.6866	.6875	.6884	.6893
4.9	.6902	.6911	.6920	.6928	.6937	.6946	.6955	.6964	.6972	.6981
5.0	.6990	.6998	.7007	.7016	.7024	.7033	.7042	.7050	.7059	.7067
5.1	.7076	.7084	.7093	.7101	.7110	.7118	.7126	.7135	.7143	.7152
5.2	.7160	.7168	.7177	.7185	.7193	.7202	.7210	.7218	.7226	.7235
5.3	.7243	.7251	.7259	.7267	.7275	.7284	.7292	.7300	.7308	.7316
5.4	.7324	.7332	.7340	.7348	.7356	.7364	.7372	.7380	.7388	.7396
5.5	.7404	.7412	.7419	.7427	.7435	.7443	.7451	.7459	.7466	.7474
5.6	.7482	.7490	.7497	.7505	.7513	.7520	.7528	.7536	.7543	.7551
5.7	.7559	.7566	.7574	.7582	.7589	.7597	.7604	.7612	.7619	.7627
5.8	.7634	.7642	.7649	.7657	.7664	.7672	.7679	.7686	.7694	.7701
5.9	.7709	.7716	.7723	.7731	.7738	.7745	.7752	.7760	.7767	.7774

APPENDIX 1. FOUR-PLACE LOGARITHMS—*continued*

	0	1	2	3	4	5	6	7	8	9
6.0	.7782	.7789	.7796	.7803	.7810	.7818	.7825	.7832	.7839	.7846
6.1	.7853	.7860	.7868	.7875	.7882	.7889	.7896	.7903	.7910	.7917
6.2	.7924	.7931	.7938	.7945	.7952	.7959	.7966	.7973	.7980	.7987
6.3	.7993	.8000	.8007	.8014	.8021	.8028	.8035	.8041	.8048	.8055
6.4	.8062	.8069	.8075	.8082	.8089	.8096	.8102	.8109	.8116	.8122
6.5	.8129	.8136	.8142	.8149	.8156	.8162	.8169	.8176	.8182	.8189
6.6	.8195	.8202	.8209	.8215	.8222	.8228	.8235	.8241	.8248	.8254
6.7	.8261	.8267	.8274	.8280	.8287	.8293	.8299	.8306	.8312	.8319
6.8	.8325	.8331	.8338	.8344	.8351	.8357	.8363	.8370	.8376	.8382
6.9	.8388	.8395	.8401	.8407	.8414	.8420	.8426	.8432	.8439	.8445
7.0	.8451	.8457	.8463	.8470	.8476	.8482	.8488	.8494	.8500	.8506
7.1	.8513	.8519	.8525	.8531	.8537	.8543	.8549	.8555	.8561	.8567
7.2	.8573	.8579	.8585	.8591	.8597	.8603	.8609	.8615	.8621	.8627
7.3	.8633	.8639	.8645	.8651	.8657	.8663	.8669	.8675	.8681	.8686
7.4	.8692	.8698	.8704	.8710	.8716	.8722	.8727	.8733	.8739	.8745
7.5	.8751	.8756	.8762	.8768	.8774	.8779	.8785	.8791	.8797	.8802
7.6	.8808	.8814	.8820	.8825	.8831	.8837	.8842	.8848	.8854	.8859
7.7	.8865	.8871	.8876	.8882	.8887	.8893	.8899	.8904	.8910	.8915
7.8	.8921	.8927	.8932	.8938	.8943	.8949	.8954	.8960	.8965	.8971
7.9	.8976	.8982	.8987	.8993	.8998	.9004	.9009	.9015	.9020	.9026
8.0	.9031	.9036	.9042	.9047	.9053	.9058	.9063	.9069	.9074	.9079
8.1	.9085	.9090	.9096	.9101	.9106	.9112	.9117	.9122	.9128	.9133
8.2	.9138	.9143	.9149	.9154	.9159	.9165	.9170	.9175	.9180	.9186
8.3	.9191	.9196	.9201	.9206	.9212	.9217	.9222	.9227	.9232	.9238
8.4	.9243	.9248	.9253	.9258	.9263	.9269	.9274	.9279	.9284	.9289
8.5	.9294	.9299	.9304	.9309	.9315	.9320	.9325	.9330	.9335	.9340
8.6	.9345	.9350	.9355	.9360	.9365	.9370	.9375	.9380	.9385	.9390
8.7	.9395	.9400	.9405	.9410	.9415	.9420	.9425	.9430	.9435	.9440
8.8	.9445	.9450	.9455	.9460	.9465	.9469	.9474	.9479	.9484	.9489
8.9	.9494	.9499	.9504	.9509	.9513	.9518	.9523	.9528	.9533	.9538
9.0	.9542	.9547	.9552	.9557	.9562	.9566	.9571	.9576	.9581	.9586
9.1	.9590	.9595	.9600	.9605	.9609	.9614	.9619	.9624	.9628	.9633
9.2	.9638	.9643	.9647	.9652	.9657	.9661	.9666	.9671	.9675	.9680
9.3	.9685	.9689	.9694	.9699	.9703	.9708	.9713	.9717	.9722	.9727
9.4	.9731	.9736	.9741	.9745	.9750	.9754	.9759	.9763	.9768	.9773
9.5	.9777	.9782	.9786	.9791	.9795	.9800	.9805	.9809	.9814	.9818
9.6	.9823	.9827	.9832	.9836	.9841	.9845	.9850	.9854	.9859	.9863
9.7	.9868	.9872	.9877	.9881	.9886	.9890	.9894	.9899	.9903	.9908
9.8	.9912	.9917	.9921	.9926	.9930	.9934	.9939	.9943	.9948	.9952
9.9	.9956	.9961	.9965	.9969	.9974	.9978	.9983	.9987	.9991	.9996

APPENDIX 2. NATURAL LOGARITHMS: 1.00 to 5.09

N	0	1	2	3	4	5	6	7	8	9
1.0	0.00000	0995	1980	2956	3922	4879	5827	6766	7696	8618
1.1	9531	*0436	*1333	*2222	*3103	*3976	*4842	*5700	*6551	*7395
1.2	0.1 8232	9062	9885	*0701	*1511	*2314	*3111	*3902	*4686	*5464
1.3	0.2 6236	7003	7763	8518	9267	*0010	*0748	*1481	*2208	*2930
1.4	0.3 3647	4359	5066	5767	6464	7156	7844	8526	9204	9878
1.5	0.4 0547	1211	1871	2527	3178	3825	4469	5108	5742	6373
1.6	7000	7623	8243	8858	9470	*0078	*0682	*1282	*1879	*2473
1.7	0.5 3063	3649	4232	4812	5389	5962	6531	7098	7661	8222
1.8	8779	9333	9884	*0432	*0977	*1519	*2058	*2594	*3127	*3658
1.9	0.6 4185	4710	5233	5752	6269	6783	7294	7803	8310	8813
2.0	9315	9813	*0310	*0804	*1295	*1784	*2271	*2755	*3237	*3716
2.1	0.7 4194	4669	5142	5612	6081	6547	7011	7473	7932	8390
2.2	8846	9299	9751	*0200	*0648	*1093	*1536	*1978	*2418	*2855
2.3	0.8 3291	3725	4157	4587	5015	5442	5866	6289	6710	7129
2.4	7547	7963	8377	8789	9200	9609	*0016	*0422	*0826	*1228
2.5	0.9 1629	2028	2426	2822	3216	3609	4001	4391	4779	5166
2.6	5551	5935	6317	6698	7078	7456	7833	8208	8582	8954
2.7	9325	9695	*0063	*0430	*0796	*1160	*1523	*1885	*2245	*2604
2.8	1.0 2962	3318	3674	4028	4380	4732	5082	5431	5779	6126
2.9	6471	6815	7158	7500	7841	8181	8519	8856	9192	9527
3.0	9861	*0194	*0526	*0856	*1186	*1514	*1841	*2168	*2493	*2817
3.1	1.1 3140	3462	3783	4103	4422	4740	5057	5373	5688	6002
3.2	6315	6627	6938	7248	7557	7865	8173	8479	8784	9089
3.3	9392	9695	9996	*0297	*0597	*0896	*1194	*1491	*1788	*2083
3.4	1.2 2378	2671	2964	3256	3547	3837	4127	4415	4703	4990
3.5	5276	5562	5846	6130	6413	6695	6976	7257	7536	7815
3.6	8093	8371	8647	8923	9198	9473	9746	*0019	*0291	*0563
3.7	1.3 0833	1103	1372	1641	1909	2176	2442	2708	2972	3237
3.8	3500	3763	4025	4286	4547	4807	5067	5325	5584	5841
3.9	6098	6354	6609	6864	7118	7372	7624	7877	8128	8379
4.0	8629	8879	9128	9377	9624	9872	*0118	*0364	*0610	*0854
4.1	1.4 1099	1342	1585	1828	2070	2311	2552	2792	3031	3270
4.2	3508	3746	3984	4220	4456	4692	4927	5161	5395	5629
4.3	5862	6094	6326	6557	6787	7018	7247	7476	7705	7933
4.4	8160	8387	8614	8840	9065	9290	9515	9739	9962	*0185
4.5	1.5 0408	0630	0851	1072	1293	1513	1732	1951	2170	2388
4.6	2606	2823	3039	3256	3471	3687	3902	4116	4330	4543
4.7	4756	4969	5181	5393	5604	5814	6025	6235	6444	6653
4.8	6862	7070	7277	7485	7691	7898	8104	8309	8515	8719
4.9	8924	9127	9331	9534	9737	9939	*0141	*0342	*0543	*0744
5.0	1.6 0944	1144	1343	1542	1741	1939	2137	2334	2531	2728
N	**0**	**1**	**2**	**3**	**4**	**5**	**6**	**7**	**8**	**9**

Since the first two figures are the same for several lines, they are given only on the first line on which they occur. The point at which these first two figures change is indicated by the asterisk.

APPENDIX 2. NATURAL LOGARITHMS: 5.00 to 9.09

N	0	1	2	3	4	5	6	7	8	9
5.0	1.6 0944	1144	1343	1542	1741	1939	2137	2334	2531	2728
5.1	2924	3120	3315	3511	3705	3900	4094	4287	4481	4673
5.2	4866	5058	5250	5441	5632	5823	6013	6203	6393	6582
5.3	6771	6959	7147	7335	7523	7710	7896	8083	8269	8455
5.4	8640	8825	9010	9194	9378	9562	9745	9928	*0111	*0293
5.5	1.7 0475	0656	0838	1019	1199	1380	1560	1740	1919	2098
5.6	2277	2455	2633	2811	2988	3166	3342	3519	3695	3871
5.7	4047	4222	4397	4572	4746	4920	5094	5267	5440	5613
5.8	5786	5958	6130	6302	6473	6644	6815	6985	7156	7326
5.9	7495	7665	7834	8002	8171	8339	8507	8675	8842	9009
6.0	9176	9342	9509	9675	9840	*0006	*0171	*0336	*0500	*0665
6.1	1.8 0829	0993	1156	1319	1482	1645	1808	1970	2132	2294
6.2	2455	2616	2777	2938	3098	3258	3418	3578	3737	3896
6.3	4055	4214	4372	4530	4688	4845	5003	5160	5317	5473
6.4	5630	5786	5942	6097	6253	6408	6563	6718	6872	7026
6.5	7180	7334	7487	7641	7794	7947	8099	8251	8403	8555
6.6	8707	8858	9010	9160	9311	9462	9612	9762	9912	*0061
6.7	1.9 0211	0360	0509	0658	0806	0954	1102	1250	1398	1545
6.8	1692	1839	1986	2132	2279	2425	2571	2716	2862	3007
6.9	3152	3297	3442	3586	3730	3874	4018	4162	4305	4448
7.0	4591	4734	4876	5019	5161	5303	5445	5586	5727	5869
7.1	6009	6150	6291	6431	6571	6711	6851	6991	7130	7269
7.2	7408	7547	7685	7824	7962	8100	8238	8376	8513	8650
7.3	8787	8924	9061	9198	9334	9470	9606	9742	9877	*0013
7.4	2.0 0148	0283	0418	0553	0687	0821	0956	1089	1223	1357
7.5	1490	1624	1757	1890	2022	2155	2287	2419	2551	2683
7.6	2815	2946	3078	3209	3340	3471	3601	3732	3862	3992
7.7	4122	4252	4381	4511	4640	4769	4898	5027	5156	5284
7.8	5412	5540	5668	5796	5924	6051	6179	6306	6433	6560
7.9	6686	6813	6939	7065	7191	7317	7443	7568	7694	7819
8.0	7944	8069	8194	8318	8443	8567	8691	8815	8939	9063
8.1	9186	9310	9433	9556	9679	9802	9924	*0047	*0169	*0291
8.2	2.1 0413	0535	0657	0779	0900	1021	1142	1263	1384	1505
8.3	1626	1746	1866	1986	2106	2226	2346	2465	2585	2704
8.4	2823	2942	3061	3180	3298	3417	3535	3653	3771	3889
8.5	4007	4124	4242	4359	4476	4593	4710	4827	4943	5060
8.6	5176	5292	5409	5524	5640	5756	5871	5987	6102	6217
8.7	6332	6447	6562	6677	6791	6905	7020	7134	7248	7361
8.8	7475	7589	7702	7816	7929	8042	8155	8267	8380	8493
8.9	8605	8717	8830	8942	9054	9165	9277	9389	9500	9611
9.0	9722	9834	9944	*0055	*0166	*0276	*0387	*0497	*0607	*0717
N	0	1	2	3	4	5	6	7	8	9

APPENDIX 2. NATURAL LOGARITHMS: 9.00 to 10.09

N	0	1	2	3	4	5	6	7	8	9
9.0	9722	9834	9944	*0055	*0166	*0276	*0387	*0497	*0607	*0717
9.1	2.2 0827	0937	1047	1157	1266	1375	1485	1594	1703	1812
9.2	1920	2029	2138	2246	2354	2462	2570	2678	2786	2894
9.3	3001	3109	3216	3324	3431	3538	3645	3751	3858	3965
9.4	4071	4177	4284	4390	4496	4601	4707	4813	4918	5024
9.5	5129	5234	5339	5444	5549	5654	5759	5863	5968	6072
9.6	6176	6280	6384	6488	6592	6696	6799	6903	7006	7109
9.7	7213	7316	7419	7521	7624	7727	7829	7932	8034	8136
9.8	8238	8340	8442	8544	8646	8747	8849	8950	9051	9152
9.9	9253	9354	9455	9556	9657	9757	9858	9958	*0058	*0158
10.0	2.3 0259	0358	0458	0558	0658	0757	0857	0956	1055	1154
N	**0**	**1**	**2**	**3**	**4**	**5**	**6**	**7**	**8**	**9**

APPENDIX 2. NATURAL LOGARITHMS: 10 to 99

10	2.30259	**25**	3.21888	**40**	3.68888	**55**	4.00733	**70**	4.24850	**85**	4.44265
11	2.39790	26	3.25810	41	3.71357	56	4.02535	71	4.26268	86	4.45435
12	2.48491	27	3.29584	42	3.73767	57	4.04305	72	4.27667	87	4.46591
13	2.56495	28	3.33220	43	3.76120	58	4.06044	73	4.29046	88	4.47734
14	2.63906	29	3.36730	44	3.78419	59	4.07754	74	4.30407	89	4.48864
15	2.70805	**30**	3.40120	**45**	3.80666	**60**	4.09434	**75**	4.31749	**90**	4.49981
16	2.77259	31	3.43399	46	3.82864	61	4.11087	76	4.33073	91	4.51086
17	2.83321	32	3.46574	47	3.85015	62	4.12713	77	4.34381	92	4.52179
18	2.89037	33	3.49651	48	3.87120	63	4.14313	78	4.35671	93	4.53260
19	2.94444	34	3.52636	49	3.89182	64	4.15888	79	4.36945	94	4.54329
20	2.99573	**35**	3.55535	**50**	3.91202	**65**	4.17439	**80**	4.38203	**95**	4.55388
21	3.04452	36	3.58352	51	3.93183	66	4.18965	81	4.39445	96	4.56435
22	3.09104	37	3.61092	52	3.95124	67	4.20469	82	4.40672	97	4.57471
23	3.13549	38	3.63759	53	3.97029	68	4.21951	83	4.41884	98	4.58497
24	3.17805	39	3.66356	54	3.98898	69	4.23411	84	4.43082	99	4.59512

APPENDIX 3. EXPONENTIAL FUNCTIONS: 0.00 to 1.50

x	e^x	e^{-x}	x	e^x	e^{-x}	x	e^x	e^{-x}
0.00	1.0000	1.000000	**0.50**	1.6487	0.606531	**1.00**	2.7183	0.367879
0.01	1.0101	0.990050	0.51	1.6653	.600496	1.01	2.7456	.364219
0.02	1.0202	.980199	0.52	1.6820	.594521	1.02	2.7732	.360595
0.03	1.0305	.970446	0.53	1.6989	.588605	1.03	2.8011	.357007
0.04	1.0408	.960789	0.54	1.7160	.582748	1.04	2.8292	.353455
0.05	1.0513	0.951229	**0.55**	1.7333	0.576950	**1.05**	2.8577	0.349938
0.06	1.0618	.941765	0.56	1.7507	.571209	1.06	2.8864	.346456
0.07	1.0725	.932394	0.57	1.7683	.565525	1.07	2.9154	.343009
0.08	1.0833	.923116	0.58	1.7860	.559898	1.08	2.9447	.339596
0.09	1.0942	.913931	0.59	1.8040	.554327	1.09	2.9743	.336216
0.10	1.1052	0.904837	**0.60**	1.8221	0.548812	**1.10**	3.0042	0.332871
0.11	1.1163	.895834	0.61	1.8404	.543351	1.11	3.0344	.329559
0.12	1.1275	.886920	0.62	1.8589	.537944	1.12	3.0649	.326280
0.13	1.1388	.878095	0.63	1.8776	.532592	1.13	3.0957	.323033
0.14	1.1503	.869358	0.64	1.8965	.527292	1.14	3.1268	.319819
0.15	1.1618	0.860708	**0.65**	1.9155	0.522046	**1.15**	3.1582	0.316637
0.16	1.1735	.852144	0.66	1.9348	.516851	1.16	3.1899	.313486
0.17	1.1853	.843665	0.67	1.9542	.511709	1.17	3.2220	.310367
0.18	1.1972	.835270	0.68	1.9739	.506617	1.18	3.2544	.307279
0.19	1.2092	.826959	0.69	1.9937	.501576	1.19	3.2871	.304221
0.20	1.2214	0.818731	**0.70**	2.0138	0.496585	**1.20**	3.3201	0.301194
0.21	1.2337	.810584	0.71	2.0340	.491644	1.21	3.3535	.298197
0.22	1.2461	.802519	0.72	2.0544	.486752	1.22	3.3872	.295230
0.23	1.2586	.794534	0.73	2.0751	.481909	1.23	3.4212	.292293
0.24	1.2712	.786628	0.74	2.0959	.477114	1.24	3.4556	.289384
0.25	1.2840	0.778801	**0.75**	2.1170	0.472367	**1.25**	3.4903	0.286505
0.26	1.2969	.771052	0.76	2.1383	.467666	1.26	3.5254	.283654
0.27	1.3100	.763379	0.77	2.1598	.463013	1.27	3.5609	.280832
0.28	1.3231	.755784	0.78	2.1815	.458406	1.28	3.5966	.278037
0.29	1.3364	.748264	0.79	2.2034	.453845	1.29	3.6328	.275271
0.30	1.3499	0.740818	**0.80**	2.2255	0.449329	**1.30**	3.6693	0.272532
0.31	1.3634	.733447	0.81	2.2479	.444858	1.31	3.7062	.269820
0.32	1.3771	.726149	0.82	2.2705	.440432	1.32	3.7434	.267135
0.33	1.3910	.718924	0.83	2.2933	.436049	1.33	3.7810	.264477
0.34	1.4049	.711770	0.84	2.3164	.431711	1.34	3.8190	.261846
0.35	1.4191	0.704688	**0.85**	2.3396	0.427415	**1.35**	3.8574	0.259240
0.36	1.4333	.697676	0.86	2.3632	.423162	1.36	3.8962	.256661
0:37	1.4477	.690734	0.87	2.3869	.418952	1.37	3.9354	.254107
0.38	1.4623	.683861	0.88	2.4109	.414783	1.38	3.9749	.251579
0.39	1.4770	.677057	0.89	2.4351	.410656	1.39	4.0149	.249075
0.40	1.4918	0.670320	**0.90**	2.4596	0.406570	**1.40**	4.0552	0.246597
0.41	1.5068	.663650	0.91	2.4843	.402524	1.41	4.0960	.244143
0.42	1.5220	.657047	0.92	2.5093	.398519	1.42	4.1371	.241714
0.43	1.5373	.650509	0.93	2.5345	.394554	1.43	4.1787	.239309
0.44	1.5527	.644036	0.94	2.5600	.390628	1.44	4.2207	.236928
0.45	1.5683	0.637628	**0.95**	2.5857	0.386741	**1.45**	4.2631	0.234570
0.46	1.5841	.631284	0.96	2.6117	.382893	1.46	4.3060	.232236
0.47	1.6000	.625002	0.97	2.6379	.379083	1.47	4.3492	.229925
0.48	1.6161	.618783	0.98	2.6645	.375311	1.48	4.3929	.227638
0.49	1.6323	.612626	0.99	2.6912	.371577	1.49	4.4371	.225373
0.50	1.6487	0.606531	**1.00**	2.7183	0.367879	**1.50**	4.4817	0.223130

APPENDIX 3. EXPONENTIAL FUNCTIONS: 1.50 to 3.00

x	e^x	e^{-x}	x	e^x	e^{-x}	x	e^x	e^{-x}
1.50	4.4817	0.223130	**2.00**	7.3891	0.135335	**2.50**	12.182	0.082085
1.51	4.5267	.220910	2.01	7.4633	.133989	2.51	12.305	.081268
1.52	4.5722	.218712	2.02	7.5383	.132655	2.52	12.429	.080460
1.53	4.6182	.216536	2.03	7.6141	.131336	2.53	12.554	.079659
1.54	4.6646	.214381	2.04	7.6906	.130029	2.54	12.680	.078866
1.55	4.7115	0.212248	**2.05**	7.7679	0.128735	**2.55**	12.807	0.078082
1.56	4.7588	.210136	2.06	7.8460	.127454	2.56	12.936	.077305
1.57	4.8066	.208045	2.07	7.9248	.126186	2.57	13.066	.076536
1.58	4.8550	.205975	2.08	8.0045	.124930	2.58	13.197	.075774
1.59	4.9037	.203926	2.09	8.0849	.123687	2.59	13.330	.075020
1.60	4.9530	0.201897	**2.10**	8.1662	0.122456	**2.60**	13.464	0.074274
1.61	5.0028	.199888	2.11	8.2482	.121238	2.61	13.599	.073535
1.62	5.0531	.197899	2.12	8.3311	.120032	2.62	13.736	.072803
1.63	5.1039	.195930	2.13	8.4149	.118837	2.63	13.874	.072078
1.64	5.1552	.193980	2.14	8.4994	.117655	2.64	14.013	.071361
1.65	5.2070	0.192050	**2.15**	8.5849	0.116484	**2.65**	14.154	0.070651
1.66	5.2593	.190139	2.16	8.6711	.115325	2.66	14.296	.069948
1.67	5.3122	.188247	2.17	8.7583	.114178	2.67	14.440	.069252
1.68	5.3656	.186374	2.18	8.8463	.113042	2.68	14.585	.068563
1.69	5.4195	.184520	2.19	8.9352	.111917	2.69	14.732	.067881
1.70	5.4739	0.182684	**2.20**	9.0250	0.110803	**2.70**	14.880	0.067206
1.71	5.5290	.180866	2.21	9.1157	.109701	2.71	15.029	.066537
1.72	5.5845	.179066	2.22	9.2073	.108609	2.72	15.180	.065875
1.73	5.6407	.177284	2.23	9.2999	.107528	2.73	15.333	.065219
1.74	5.6973	.175520	2.24	9.3933	.106459	2.74	15.487	.064570
1.75	5.7546	0.173774	**2.25**	9.4877	0.105399	**2.75**	15.643	0.063928
1.76	5.8124	.172045	2.26	9.5831	.104350	2.76	15.800	.063292
1.77	5.8709	.170333	2.27	9.6794	.103312	2.77	15.959	.062662
1.78	5.9299	.168638	2.28	9.7767	.102284	2.78	16.119	.062039
1.79	5.9895	.166960	2.29	9.8749	.101266	2.79	16.281	.061421
1.80	6.0496	0.165299	**2.30**	9.9742	0.100259	**2.80**	16.445	0.060810
1.81	6.1104	.163654	2.31	10.074	.099261	2.81	16.610	.060205
1.82	6.1719	.162026	2.32	10.176	.098274	2.82	16.777	.059606
1.83	6.2339	.160414	2.33	10.278	.097296	2.83	16.945	.059013
1.84	6.2965	.158817	2.34	10.381	.096328	2.84	17.116	.058426
1.85	6.3598	0.157237	**2.35**	10.486	0.095369	**2.85**	17.288	0.057844
1.86	6.4237	.155673	2.36	10.591	.094420	2.86	17.462	.057269
1.87	6.4883	.154124	2.37	10.697	.093481	2.87	17.637	.056699
1.88	6.5535	.152590	2.38	10.805	.092551	2.88	17.814	.056135
1.89	6.6194	.151072	2.39	10.913	.091630	2.89	17.993	.055576
1.90	6.6859	0.149569	**2.40**	11.023	0.090718	**2.90**	18.174	0.055023
1.91	6.7531	.148080	2.41	11.134	.089815	2.91	18.357	.054476
1.92	6.8210	.146607	2.42	11.246	.088922	2.92	18.541	.053934
1.93	6.8895	.145148	2.43	11.359	.088037	2.93	18.728	.053397
1.94	6.9588	.143704	2.44	11.473	.087161	2.94	18.916	.052866
1.95	7.0287	0.142274	**2.45**	11.588	0.086294	**2.95**	19.106	0.052340
1.96	7.0993	.140858	2.46	11.705	.085435	2.96	19.298	.051819
1.97	7.1707	.139457	2.47	11.822	.084585	2.97	19.492	.051303
1.98	7.2427	.138069	2.48	11.941	.083743	2.98	19.688	.050793
1.99	7.3155	.136695	2.49	12.061	.082910	2.99	19.886	.050287
2.00	7.3891	0.135335	**2.50**	12.182	0.082085	**3.00**	20.086	0.049787

INDEX

A

B

C

D

E

F

P

Q

R

W

X

Y

Z

W9-AZG-776

HANDBOOK OF RESEARCH (

Handbook of Research on Family Business

Edited by

Panikkos Zata Poutziouris

Associate Professor in Entrepreneurship and Family Business, Cyprus International Institute of Management, Cyprus and Visiting Fellow, Manchester Business School, UK

Kosmas X. Smyrnios

Professor and Director of Research, School of Management, RMIT University, Melbourne, Australia

Sabine B. Klein

Assistant Professor in Family Business and Academic Director of the European Family Business Center, European Business School, Germany and Visiting Research Fellow, INSEAD, France

IN ASSOCIATION WITH IFERA – THE INTERNATIONAL FAMILY ENTERPRISE RESEARCH ACADEMY

Edward Elgar
Cheltenham, UK • Northampton, MA, USA

Published by
Edward Elgar Publishing Limited
Glensanda House
Montpellier Parade
Cheltenham
Glos GL50 1UA
UK

Edward Elgar Publishing, Inc.
William Pratt House
9 Dewey Court
Northampton
Massachusetts 01060
USA

A catalogue record for this book
is available from the British Library

Library of Congress Cataloguing in Publication Data

Handbook of research on family business/edited by Panikkos Zata Poutziouris,
Kosmas X. Smyrnios, Sabine B. Klein.
p. cm. — (Elgar original reference in association with IFERA)
Includes bibliographical references and index.
1. Family-owned business enterprises. I. Poutziouris, Panikkos Zata.
II. Smyrnios, Kosmas X., 1954– . III. Klein, Sabine B., 1962– .
IV. Series.
HD62.25.H36 2006
338.6—dc22

2005037395

ISBN-13: 978 1 84542 410 7 (cased)
ISBN-10: 1 84542 410 7 (cased)

Printed and bound in Great Britain by MPG Books Ltd, Bodmin, Cornwall

Contents

Contributors

Joseph H. Astrachan (PhD) is the director of the Cox Family Enterprise Center, holds the Wachovia Eminent Scholar Chair of Family Business, is Professor of Management and Entrepreneurship in the Coles College of Business at Kennesaw State University and is a distinguished research chair at Loyola University Chicago (USA). He is a founding board member of IFERA and serves as editor of *Family Business Review* and as an editorial board member of several other academic journals.

Åsa Björnberg holds the Institute for Family Business (IFB UK) Research Fellowship in association with LIFBRI at London Business School (UK). Her background is in Organizational and Clinical psychology, and her research interests centre on family development/functioning in relation to leadership, culture and performance of family firms. Åsa Björnberg also works as a personal development coach and an academic translator.

Kristin Cappuyns is Research Associate at IESE Business School, Barcelona Spain. She has co-authored numerous research papers, books and case studies on the subject of family business in different disciplines including governance, business management, business ethics and values systems. She is a founding board member of International Family Enterprise Research Academy (IFERA).

Guido Corbetta is AIdAF-Alberto Falck Professor of Strategic Management in Family Firms, and Director of the Entrepreneurship and Entrepreneurs Research Centre (EntER) at Bocconi University in Milan, Italy. His research interests are mostly in the area of family business, strategic management and entrepreneurship. He has a 20 years experience with family companies, where he has been often consultant and member of the board of directors.

Justin Craig received his PhD from Bond University in Australia. He is an assistant professor of entrepreneurship at Oregon State University.

Sharon M. Danes (PhD) is Professor of Family Social Science, University of Minnesota (USA) has over 125 refereed research articles, book chapters, and outreach publications emphasizing the intersection of economic and social decision-making. Research focus is the impact of the interconnectedness between the family and business systems as it affects the viability of family businesses.

Rik Donckels is managing director of Cera and the former director of the Small Business Research Institute at the Catholic University Brussels (Belgium).

Kimberly A. Eddleston (PhD) is an Assistant Professor at Northeastern University (USA), where she holds the Riesman Research Professorship and Tarica-Edwards Fellowship.

She was recently selected as a Family Owned Business Institute Research Scholar by the Family Owned Business Institute of the Seidman College of Business at Grand Valley State University. She received her PhD in Management from the University of Connecticut. Her research has appeared in journals such as the *Academy of Management Journal*, *Academy of Management Executive*, *Academy of Management Perspectives*, *Human Resource Management Review*, *Journal of Occupational and Organizational Psychology*, *Entrepreneurship Theory and Practice*, *Journal of Business Venturing*, and *Journal of Applied Psychology*.

Margaret A. Fitzgerald (PhD) is an Associate Professor in the Department of Child Development and Family Science at North Dakota State University in Fargo (USA). She teaches courses in financial planning and public policy. Dr Fitzgerald's research is in the area of family business and is focused on gender and management issues, copreneurs and business social responsibility.

Miguel Angel Gallo is Emeritus Professor of General Management in the Department of General Management at IESE, Barcelona (Spain), where he has served as a full professor from 1975 to September 2003, and Chairman of the Family Business Chair at IESE from its foundation (1987) until September 2003. His areas of specialization include strategic management, organizational design, boards of directors and family businesses. He is fellow of the Strategic Management Society and Honorary President of IFERA. He is also a co-founder and partner of Family Business Consulting Group International and Chairman of FBCG Spain. He is a member of the board of directors of several family businesses in Spain, Portugal and Mexico.

Ercilia García-Álvarez (PhD), is full professor in management at the Universitat Rovira i Virgili Tarragona (Spain), Fellow of the Family Firm Institute and member of the board of the Qualitative Research Network, European Sociological Association. Her research in family business has received awards from the Family Business Network (2000) and Family Firm Institute (2001). She has published in many international journals such as: *Family Business Review*, *Journal of Business Research*, *Field Methods* and the *European Sociological Review*.

Alberto Gimeno Sandig (PhD) is Professor of Business Policy Department at ESADE, Barcelona, Spain. He is Program Director for both the Senior Executive Program and Family Enterprise program. Alberto is a member of the Body of Knowledge of Family Firm Institute (FFI), as well as of the Family Business Network (FBN) and of the International Family Enterprise Research Academy (IFERA). He is former professor at the World Economic Forum and lectures in both national and international family enterprise forums.

Luca Gnan is Associate Professor of Organizational Design and Behaviour at the University of Tor Vergata, Rome (Italy) and is a Faculty Member of the Strategic Management Department at SDA Bocconi School of Management, Milan (Italy).

Silvia Gómez Ansón (PhD) is Associate Professor of Finance at the University of Oviedo (Spain). She graduated in Business Administration at the Complutense University of

Madrid, has a master in International Economics at the University of Konstanz (Germany) and obtained her PhD at the University of Oviedo. Her research interests include corporate governance, family firms and corporate finance.

Toshio Goto is on the Faculty of Integrated Engineering at the Graduate School for the Creation of New Photonic Industries, Hamamatsou (Japan). His background is in business strategy and his research focuses on strategies for sustainable growth, and especially for that of family businesses.

Timothy G. Habbershon (EdD) is the founding Director of the Institute for Family Enterprising at Babson College in Wellesley, Massachusetts, USA. He is also an Assistant Professor of Entrepreneurship and holds the Presidents Term Chair in Family Enterprising. His articles on family-based entrepreneurship have appeared in the *Journal of Business Venturing* and the *Family Business Review* and he has a regular column in *Business Week*'s *Small Business Magazine*. In addition he is the founder and principal of the Telos Group, a consulting firm specializing in transition and growth strategies for family firms.

Annika Hall (PhD) is a research fellow and lecturer at Jönköping International Business School (Sweden), specializing in the fields of organization theory, strategy and family business. She is a board member of International Family Enterprise Research Academy (IFERA).

George W. Haynes (PhD) is an Associate Professor in the Department of Health and Human Development at Montana State University (USA). Dr Haynes teaches small business management and research methods courses and has been actively engaged in family business finance, employee wellness and substance abuse treatment demand research.

Ramona K.Z. Heck (PhD) is the Peter S. Jonas Distinguished Professor of Entrepreneurship in the Department of Management of the Zicklin School of Business at Baruch College, The City University of New York (USA). Dr Heck teaches and conducts research related to family businesses and the owning family's internal social and economic dynamics, the effects of the family on the family business viability over time, the economic impact of family businesses on communities, minority business ownership and gender issues within family firms.

Carole Howorth (PhD) is a Senior Lecturer in Entrepreneurship at Lancaster University Management School (UK) where she researches and teaches on family business and entrepreneurship. Her research has been published in national and international journals. Prior to entering academia Carole was owner-manager of two family businesses.

Frank Hoy (PhD) is director of the entrepreneurship programme and the Family and Closely Held Business Forum at the University of Texas at El Paso (USA). From 1991 to 2001, he served as dean of the College of Business Administration at UTEP. Prior to that he held the Carl R. Zwerner Professorship of Family-Owned Business at Georgia State University. Dr Hoy is a past president of the United States Association for Small Business and Entrepreneurship and a past editor of *Entrepreneurship Theory and Practice*.

Wilfred V. Huang (PhD) is Professor of Management Information Systems at the College of Business, Alfred University-New York (USA), where he holds the Raymond Chair in Family Business.

Cynthia R. Jasper (PhD) is Professor and Chair of the Department of Consumer Science at the University of Wisconsin-Madison (USA). She is interested in decision-making within family businesses, especially pertaining to retirement and estate planning, and business management issues.

Zhao Jing (PhD) is the Professor of Management Information Systems at the College of Management, China University of Geosciences, Wuhan, China.

Theodoros A. Kalkanteras is a graduate of the MBA International Programme of the Athens University of Economics and Business, Greece.

Franz W. Kellermanns (PhD) is an Assistant Professor of Management in the College of Business and Industry at the Mississippi State University (USA). He was recently selected as a Family Owned Business Institute Research Scholar by the Family Owned Business Institute of the Seidman College of Business at Grand Valley State University. He received his PhD from the University of Connecticut. His current research interests include strategy process and entrepreneurship with a focus on family firms. His research has appeared in journals such as the *Journal of Management*, *Journal of Business Venturing*, *Entrepreneurship Theory and Practice* and the *Academy of Management Learning and Education*. He is the co-editor of the recent book *Innovating Strategy Process* in the Strategic Management Society Book Series.

Andrew Keyt is the Executive Director of the Loyola University Chicago Family Business Center (USA) which is widely recognized as a leading think tank in issues unique to business owning families. In addition, he is the President and Founder of a private consulting firm, Keyt Consulting. Having served as a manager in two family-owned firms, and as member of his own family partnership, Keyt has experienced the challenges of family business at first hand. As a consultant to family firms he specializes in dealing with family conflict and communication, working with adult sibling/cousin teams, succession planning strategic planning and emergency management transition. A cum laude graduate of Kenyon College (BA), Keyt completed a Masters in Family Systems Theory from Northwestern University with a concentration on family business.

Sabine B. Klein (PhD) is the Academic Director of the European Family Business Center and Assistant Professor in Family Business at the European Business School at Oestrich-Winkel, Germany. She is founding member of IFERA and, since 2003, President of IFERA. Her research has been awarded several prizes. She is serving on the review board of several journals and academic conferences.

Gaston J. Labadie (PhD) is Dean and Professor of Human Behaviour and Organizational Behaviour at Universidad ORT Uruguay. He is the research director of Study Group in

Economics, Organization and Social Policies (GEOPS) and member of the editorial boards of *Management Research* and *Latin American Business Review*.

Johan Lambrecht is Director of the Research Centre for Entrepreneurship at European University College Brussels (EHSAL) and Catholic University Brussels and Professor at EHSAL – Brussels (Belgium).

Suzanne Lane is Program Director of the Loyola University Chicago Family Business Center (USA) and has been extensively involved in the Center's programming and research initiatives. She specializes in working with President/CEOs, board directors, and senior management in areas such as strategic planning, leadership transitions, board development and corporate governance. Susanne currently serves on many boards of advisers for non-profit organizations throughout Chicago.

Isabelle Le Breton-Miller is President of OER, Inc. in Montreal, a strategic and organizational management consultancy, and Senior Research Associate at the University of Alberta. Her recent book (with Danny Miller) is *Managing for the Long Run* (Harvard Business School Press, 2005), which has been chosen by JP Morgan Chase as one of the 10 'must read' books of 2005. It is to be translated into six languages. Her practice and continuing research focuses on how firms can better design their organizations to manage for the long run.

Jordi López-Sintas (PhD) is full professor in Business Economics at the Universitat Autònoma de Barcelona (Spain) and elected director of the Humanities Research Centre (CERHUM). He is also convenor of the track 'Combining qualitative and quantitative methods' at the Qualitative Research Network, European Sociological Association, and in 2006 he became Director of the Advanced Seminar in Qualitative Research (SAIC). His research in family business has received awards from the Family Business Network (2000) and Family Firm Institute (2001). He has published in many international journals such as: *Family Business Review*, *Journal of Business Research*, *Field Methods* and the *European Sociological Review*.

Myriam Lyagoubi (PhD) is associate professor of corporate finance at EM Lyon Business School (France). She has been conducting research in the field of family business for several years. In 2003, she was awarded the F.B.N. Miguel Angel Gallo Award for the most innovative paper of the year.

Ian C. MacMillan is the Executive Director of the Sol C. Snider Entrepreneurial Center and Dhirubhai Ambani Professor of Entrepreneurial Management, Wharton School, University of Pennsylvania (USA). He has published numerous articles and books on organizational politics, new ventures and strategy formulation. His articles have appeared in the *Harvard Business Review*, the *Sloan Management Review*, the *Journal of Business Venturing* and others. He is co-author with Rita McGrath of the best-selling books *The Entrepreneurial Mindset*, which focuses on how managers and entrepreneurs can create a continuous stream of growth opportunities for their firms, and *MarketBusters*, which focuses on strategies firms can use to dramatically change and grow their existing businesses.

Gaia Marchisio (PhD) is Assistant Professor of Management at the Michael J. Coles College of Business and faculty associate of the Coles College Cox Family Enterprise Center, both at Kennesaw State University (USA). Her research primarily concerns family business, corporate entrepreneurship, and strategic management, with particular interest in: fostering entrepreneurship in family business; family businesses' strategic planning process; and going public and family offices.

Pietro Mazzola is the director of the Master in Investor Relations e Financial Analysis holds at IULM University in Milan (Italy) in collaboration with the Italian Stock Exchange, he is Full Professor of Management at IULM University and is Senior Faculty Member at the Strategic and Entrepreneurship Management Department at SDA Bocconi, Bocconi University School of Management, Milan.

Kristi McMillan is the Associate Director of the Cox Family Enterprise Center at Kennesaw State University, which she joined in 1994. Ms McMillan has a Master of Science in Conflict Management and is co-author of the acclaimed book *Conflict and Communication in the Family Business*.

Leif Melin (PhD) is Professor of Strategy and Organization at Jönköping International Business School (Sweden). He is the founding Director of the Center for Family Enterprise and Ownership (CeFEO). He has published widely in international journals and book volumes and he serves on editorial boards for several academic journals, such as *Organization Studies* and *Strategic Organization*.

Xavier Mendoza Mayordomo (PhD) is Dean of the ESADE Business School, Barcelona, Spain, and Professor of the Business Policy Department and the Institute of Public Management. He is the Academic Vice-Chair of the Supervisory Board de la European Academy of Business in Society and member of the Editorial Advisory Committee of the *Corporate Governance*.

Susana Menéndez-Requejo (PhD) is a Professor of Finance at the University of Oviedo (Spain). She is the Director of the Family Business Chair at the same university. Her research interests are in the areas of Corporate Finance (capital structure, corporate governance) and Family Firms.

Danny Miller is President of Paradox Learning Resources and Chaired Professor in Strategy and Family Enterprise at HEC Montreal and the University of Alberta. He has authored six books and over 100 articles, and has held professorships at McGill University and the Columbia Business School. He consults with numerous Fortune 500 companies, and has directed major thought leadership projects for several international management consulting firms. His practice and current research concerns how firms can develop sustainable competitive advantage by expanding their time horizons and changing their strategies, metrics and incentives.

Alessandro Minichilli is a post-doctoral fellow at Bocconi University in Milan (Italy) where he received is PhD in Business Administration and Management. He is lecturer in

Corporate Governance and Business Administration. His research mostly deals with boards of directors in large companies, with a focus on a behavioural perspective on board activity. He is also concerned with the development of evaluation systems for the corporate boards in quoted companies.

Sandra L. Moncrief-Stuart holds dual Masters Degrees in Marriage and Family Therapy and Social Work. Sandra is currently an individual and family therapist in Michigan. Previously, Sandra worked with Joe Paul and the Aspen Family Business Group designing and refining consulting methodologies and working with multi-generational family businesses.

Daniela Montemerlo is Associate Professor of Business Administration at the University of Insubria, Professor of Strategic Management in Family Business at Bocconi University and Senior Faculty Member of the Strategic Management Department at SDA Bocconi School of Management, Milan (Italy). She is a founding board member and Fellow of IFERA.

Ken Moores (PhD) is the Director of Bond University's Australian Centre for Family Business – a centre he established in 1994 and in which he served as Foundation Director from 1994 to 1998. Professor Moores pioneered research and recognition of family business in Australia and has achieved wider recognition for his work including his 2003 book, *Learning Family Business: Paradoxes and Pathways* (co-authored with Mary Barrett). Professor Moores served as Vice-Chancellor and President of Bond University from 1997 to 2003.

Nigel Nicholson is Professor of Organizational Behaviour at London Business School (UK) where he is also the director of the Leadership in Family Business Research Initiative (LIFBRI). This major new initiative aspires to make London Business School one of the world's leading centres for the study of family business. He has published 18 books and monographs and over 180 articles on many aspects of business psychology, leadership and organization.

Mattias Nordqvist (PhD) is a research fellow and co-director of the Center for Family Enterprise and Ownership (CeFEO) at Jönköping International Business School (Sweden). He is also Research Associate and Visiting Scholar for Family Enterprising at the Arthur M. Blank Center for Entrepreneurship at Babson College (USA). His main interests are strategizing, governance and entrepreneurial processes within the context of family businesses.

Joe Paul specializes in family business leadership and the resolution of family issues that interfere with asset development. He authored several family business assessment instruments, is a Fellow and Director Emeritus of the Family Firm Institute, and a partner in the Aspen Family Business Group and Global Family Business Advisors.

David Pistrui (PhD) is the Professor of Business at the Illinois Institute of Technology – Chicago (USA) where he holds the Coleman Foundation Chair in Entrepreneurship.

Panikkos Zata Poutziouris (PhD) is the Associate Professor in Entrepreneurship and Family Business at the Cyprus International Institute of Management and Visiting Fellow for Family Business Initiatives at Manchester Business School (UK). Whilst on the Faculty of University of Manchester-MBS he served on the Advisory Board of the Institute for Family Business (UK) and on the Board of Directors of the Institute for Small Business Entrepreneurship (UK). Currently he is the founding Vice-President of the International Family Enterprise Research Academy and serves on the editorial boards of *Family Business Review* and *Journal of Small Business Management*. In 2004, he received the FFI Barbara Hollander Award, in recognition of his work to promote the interests of family business. Panikkos, has carried out research and consultancy projects in the area of strategic financial development of owner-managed companies for numerous financial institutions, government bodies, enterprise support agencies and family firms.

Vassilios D. Pyromalis is a graduate of the MBA International Programme of the Athens University of Economics and Business, Greece.

Michaela E. Rogdaki is a graduate of the MBA International Programme of the Athens University of Economics and Business, Greece.

María Sacristán Navarro (PhD) is Associate Professor of Business Administration at the Rey Juan Carlos University, Madrid (Spain). She graduated in Business Administration and obtained her PhD at the Complutense University of Madrid. Her research interests include corporate governance, family firms and strategic management.

Willem Saris (PhD) is Professor in Methodology of the Social Sciences, University of Amsterdam and ICREA Professor at ESADE-URL. He is a member of the central co-ordination team of the European Social Survey, awarded with the prestigious Descartes Prize in 2005, and is President of the European Survey Research Association (ESRA).

Holly L. Schrank (PhD) is a Professor of Consumer Sciences and Retailing at Purdue University (USA). She focuses on the impact of boundary changes in the family, business and ownership systems of the family business. She also has research interests in impacts of disasters on business.

Melissa Carey Shanker is an accomplished consultant, educator and researcher in the field of family business. Shanker played a key role in the growth and development of the Chicago-based Loyola University Family Business Center (USA), one of the oldest and most respected centres in the world where she designed and directed the innovative Next Generation Leadership Institute, an educational programme designed to develop family business leaders.

Pramodita Sharma (PhD) is a Professor of Management and Associate Dean at the School of Business, Wilfrid Laurier University (Canada). She is the recipient of various international research awards, including the prestigious NFIB Dissertation Award from the Entrepreneurship division of the Academy of Management. She is an Associate Editor of the *Family Business Review* and serves on the editorial boards of various entrepreneurship journals. Dr Sharma serves on the board of the Family Firm Institute, the International Family Enterprise Research Academy and is the Representative-at-Large of the Entrepreneurship division of the Academy of Management.

George P. Sigalas is a graduate of the MBA International Programme of the Athens University of Economics and Business, Greece.

Kosmas X. Smyrnios (PhD) holds the position of Professor and Director of Research in the School of Management at RMIT University, Melbourne, Australia and is Associate Editor of the *Family Business Review*. Kosmas has developed an extensive applied research record with over 70 international refereed publications across the disciplines of marketing, psychology, physics, management and accounting. Kosmas has been involved in a number of prominent national and international research projects. He is a founding board member of the International Family Enterprise Research Academy (IFERA). In 1998 and 2001, he was awarded prizes for the Best International Research Papers at the 9th and 12th World Family Business Network Conference in Paris and Rome, respectively. Kosmas is a recipient of over $1.5 million in research funding, and is frequently called upon to provide expert media commentary on pertinent matters relating to family firms and SMEs.

Lucrezia Songini is a Lecturer in Bocconi University, Milan and senior faculty member of the Accounting and Control Department of the SDA Bocconi School of Management, Milan (Italy) She is professor of Management Accounting in the Università degli Studi del Piemonte Orientale 'Amedeo Avogadro', Novara, in the Business Administration Department in Casale Monferrato.

Kathryn Stafford (PhD) is an Associate Professor at the Ohio State University in the Department of Consumer and Textile Sciences. She teaches a course on Businèss-Owning families and conducts research on the management practices of business-owning families and family businesses.

Lloyd Steier (PhD) is a Professor in Strategic Management and Organization at the University of Alberta School of Business – Canada. He holds a research chair in family enterprise and entrepreneurship and is the academic director of the Centre for Entrepreneurship and Family Enterprise and the Alberta Business Family Institute.

Josep Tàpies (PhD) is professor in the departments of general management and finance and holder of the Chair of Family-Owned Business at IESE, Barcelona (Spain). His areas of specialization include family business, strategic management, private equity, mergers and acquisitions and management buy-outs. He writes and teaches courses in management and governance of family business in the MBA programme, and strategic management in several executive education programmes.

Salvatore Tomaselli (PhD) is Professor of Business Policy at the 'Università di Palermo' (Italy). His areas of specialization include strategic management, organizational design, boards of directors and family businesses. In 1983 and 1999 he received the FBN Award for the best research paper presented at the FBN Annual World Conference. He is a founding board member and Fellow of IFERA and member of the Strategic Management Society. He is also partner of Family Business Consulting Group International and FBCG Spain.

Rosa Nelly Trevinyo-Rodríguez, currently pursuing a PhD at IESE Business School (Spain) has worked for several years as a Professor at Instituto Tecnológico y de Estudios Superiores de Monterrey, Campus Monterrey (ITESM), MÉXICO. She taught courses on the MBA Programme and other undergraduate programmes at ITESM and Universidad Mexicana del Noreste (UMNE) focusing in the management of Family Businesses and in Strategic Management/Valuation analysis.

Lorraine M. Uhlaner (PhD) is Director of the European Family Business Institute at Erasmus University Rotterdam in the Netherlands, sponsored by Arenthals Grant Thornton Accountants and Advisors, Fortis Bank, and Mees Pierson, the private bankers of Fortis Bank. Before joining the Erasmus University Rotterdam she served as full professor in Management at Eastern Michigan University. She is author of a number of journal articles on family business and entrepreneurship, and is co-author of the book, *Dynamic Management of Growing Firms: A Strategic Approach*, published by Prentice-Hall.

George S. Vozikis is the Edward Reighard Chair in Management at California State University, Fresno and the Director of the Institute for Family Business. Prior to joining California State University at Fresno, he taught at the University of Tulsa where he was the Davis D. Bovaird Endowed Chairholder of Entrepreneurial Studies and Private Enterprise, and the Founding Director of the Family-Owned Business Institute, as well as the Tulsa University Innovation Institute.

Harold P. Welsch (PhD) is the Professor of Management at DePaul University, Chicago (USA) where he holds the Coleman Foundation Chair in Entrepreneurship.

Paul Westhead (PhD) is the Professor of Enterprise in the Enterprise Division at Warwick Business School. He is also a Visiting Professor at Bodø Graduate School of Business, Nordland Regional University, Bodø, Norway. His research interests include family firms, habitual entrepreneurs, internationalization of small firms, training programme take-up and benefits, technology-based firms, and Science Parks.

Mary Williams is Professor of Management and Department Head of the MIS and Decision Sciences Department at Widener University in Chester, Pennsylvania, USA. She has presented research in entrepreneurship at the Academy of Management Meetings, the International Atlantic Economic Meetings, and the National Business and Economic Society Meetings. Her entrepreneurship research appears in the *Journal of Business Venturing*, *Frontiers of Entrepreneurship Research*, and *Research in Entrepreneurship and Management*. Her research in family business appears in the *Journal of Family and Economic Issues* and the *Family Business Review*.

Mary Winter (PhD) has recently retired from the position of Professor of Human Development and Family Studies and Associate Dean for Research and Graduate Education at the College of Family and Consumer Sciences, Iowa State University. Her research interests have focused on the responses of ordinary families to extraordinary circumstances. She has studied the responses of families in Mexico and Poland to changes in their country's economy, and resource development and allocation among US families with a family business.

Shaker A. Zahra (PhD) is the Robert E. Buuck Professor in Entrepreneurial Studies at the Carlson School of Management, University of Minnesota (USA). His research covers corporate, technological and international entrepreneurship. He has published nine books and his research has appeared in several journals such as *Academy of Management Journal*, *Academy of Management Review*, *Academy of Management Executive*, *Strategic Management Journal*, *Journal of International Business Studies*, *Journal of Management*. He has been awarded several grants and has garnered dozens of prestigious honors and awards.

Foreword

I take much pleasure in contributing the Foreword to this outstanding *Handbook of Family Business*. Having dedicated my academic life to the field of family business, I am excited to observe that now, decades after humble beginnings in the 1970s, that this topic is gaining increasing momentum. The growing number of rigorous research projects is promising, and now, academia is taking into account the fact that family enterprises are dominant around the world.

The study of family business is challenging, to say the least. This line of inquiry is interdisciplinary in nature, involving different fields such as finance, organizational behaviour, law, tax, child psychology and ethics, to name a few. In order to master the challenges associated with studying disparate topics like these, one not only requires a broad education but also extensive experience and a disciplined way of thinking.

Family business research was international and interdisciplinary from the outset. Interestingly, the editors of this book mirror this aspect of the field: emanating from three different countries, namely Cyprus, Australia and Germany, and respectively, their principal academic disciplinary backgrounds are entrepreneurial management, psychology and strategy. Panikkos Poutziouris, Kosmas Smyrnios, and Sabine Klein are a team typical of the family business arena.

The editors have collated a selection of notable papers, all of which have been blind peer reviewed. As a senior scholar of this field, this handbook contributes substantially to furthering this area, serving as a watershed for research students and investigators. Consistent with this view, I sincerely hope that this scholarly text is only the start of a long line of tradition founded by IFERA in encouraging researchers to undertake rigorous and relevant projects and to publish them in volumes like the one you are holding in your hands.

Miguel Angel Gallo
Professor Emeritus, IESE Business School
University of Navarra, Barcelona, Spain
Honorary President IFERA

Acknowledgements

The publishers wish to thank the following who have kindly given permission for the use of copyright material:

Blackwell Publishing Ltd for articles: Astrachan, Joseph H. and Melissa Carey Shanker (2003), 'Family businesses' contribution to the US economy: a closer look', *Family Business Review*, **16**(3), 211–19; Sharma, Pramodita (2004), 'An overview of the field of family business studies: current status and directions for the future', *Family Business Review*, **17**(1), 1–36; and Astrachan, Joseph H., Sabine B. Klein and Kosmas X. Smyrnios (2002), 'The F-PEC scale of family influence: a proposal for solving the family business definition problem', *Family Business Review*, **15**(1), 45–58.

Elsevier Ltd for articles: Miller, Danny, Lloyd Steier and Isabelle Le Breton-Miller (2003), 'Lost in time: intergenerational succession, change and failure in family business', *Journal of Business Venturing*, **18**(4), 513–31 and Habbershon, T.G., M. Williams and I.C. MacMillan (2003), 'A unified systems perspective of family firm performance', *Journal of Business Venturing*, **18**(4), 451–65.

Every effort has been made to trace all the copyright holders but if any have been inadvertently overlooked the publishers will be pleased to make the necessary arrangements at the first opportunity.

Introduction: the business of researching family enterprises

Panikkos Zata Poutziouris, Kosmas X. Smyrnios and Sabine B. Klein

The *Handbook of Family Business Research* is a substantial collection of papers manifesting recent advances in the theory and practice of family business research. This compilation is, to a large extent, in response to the extensive growth of the family business discipline as a topic of academic inquiry. The principal objective underlying this volume was to provide readers with a compilation of authoritative and scholarly papers, providing an overview of current thinking and contributing to the further advancement of the field.

Emergence of a family business theme

Family enterprises, irrespective of scale of operation, legal form, industrial activity, and level of socio-political and market development have been the backbone of corporate life, across nations, remaining a cornerstone of socio-economic development. Historically, family firms are, for the most part, enduring institutions. Their importance parallels socio-cultural advances, technological advances, and the so-called new market order associated with globalization.

Family business research, as an academic field of inquiry, is relatively young. The emergence of this topic of research can be attributed largely to the proactive approach of family business practitioners whose early efforts focused on practice-based articles and case studies, as noted by Donnelley (1964) and Barnes and Hershon (1976).

Arguably, in the early days, treatment of this topic from the practitioners' side has often been subjected to journalistic exploitation, whilst from an academic perspective research has tended to be data driven. Despite adoption of increasingly sophisticated empirical and analytic procedures, causal-based research has remained myopic. However, articulation of conceptual and theoretical papers is gaining momentum.

The field of family business research was retarded by a series of factors, including a chronic failure by scholars and practitioners to establish a *consensus* as to what constitutes a family business; adoption of methodologies not supported by theory; survey designs lacking robustness; application of relatively unsophisticated statistical methods; and a dearth of large-scale (longitudinal) databases to enable cross-sectional and time-series econometric analyses, enabling significant comparisons between family and other enterprises (Brockhaus, 1994). Moreover, the focus of researchers was geared towards addressing phenomenological problems rather than systematically exploring and advancing theoretical paradigms. Specifically, early research tended to focus their lines of inquiry on succession, business governance and related structures, and performance, but failed to identify important variables and their interrelationships, leading to sound theoretical conceptualizations. Today, prominent mainstream journals publish family business research owing to its rigour, richness and relevance. Investigators have moved beyond descriptive case study and survey research, employing triangulation approaches driven by theory.

Recent developments

During the new millennium, family firm research has demonstrated significant advances, not only in terms of quality, but also in articulation of new developments and establishment of prominent international bodies. Below, we outline a number of these developments.

(a) Following the early push of a family business agenda by the Family Firm Institute (FFI) which has traditionally been geared to educationalists and practitioners, we saw the emergence of the Family Business Network (FBN) International, establishing close links with family business owner-managers and academics. Since these times, the International Family Enterprise Research Academy (IFERA), a group of academics committed to the advancement of family business as a science-based discipline, has arisen.

(b) The International Family Research Enterprise Academy has been actively campaigning for the advancement of family business through international collaborative projects and work with emerging researchers. Its vision is to be a driving force of an international network that ensures that family business, as a multi disciplinary field, becomes a leading topic of business research. The Academy comprises a global network for scholars (including new doctoral students), committed to family business research, providing investigators with valuable resources; and access to literature, family businesses, business families, and a network of scholars. Through its association with FBN-International in staging annual Global FBN research forums, IFERA has helped to bridge the gap between research, theory and practice, on the one hand, and professionals that service these businesses, business families, owner-managers and the community, on the other. The valuable contribution made by IFERA is amply demonstrated by the quality of refereed conference proceedings and publications (see Poutziouris, 2000; Corbetta and Montemerlo, 2001; Koiranen and Karlsson, 2002; Poutziouris and Steier, 2003; Tomasselli and Melin, 2004).

(c) Another indicator afforded to the recognition of this field is the inclusion of the *Family Business Review* journal in the Social Citation Index by Thomson ISI, recognizing the academic standard of FBR by the scientific community.

(d) The Canadian Theorization Initiative, in response to calls for the deployment of alternative research methodologies essential for theory development and research per se, has published a series of special issues in the *Journal of Business Venturing*, and the *Entrepreneurship Theory and Practice* journal. Similarly, in the North America, the Family Enterprise Research Conference has helped to promote family business at the Academy of Management-Entrepreneurship Division.

Consistent with these developments, research in this field has broken through the glass ceiling and is being published in top-tier journals, particularly in areas focusing on listed family businesses and their performance (Anderson and Reeb, 2003; Anderson et al., 2003; Burkart et al., 2003; Gomez-Mejia et al., 2001; Villalonga and Amit, 2004), agency costs (Morck and Yeung, 2003) and nepotistic altruism (Schulze et al., 2003). These notable investigations, inter alia, have helped elevate the family business theme. Within this backdrop, the *Handbook of Family Business Research* raises the bar further, demonstrating an interdisciplinary and multidimensional field of research, traversing process theory, case studies, application of quantitative and qualitative procedures that employ primary, secondary, narrative and ethnographic methods, along with papers that contribute to the validation of theoretical constructs.

Unequivocally, this volume contributes significantly to the advancement of family business research, theory, and practice. This selection of extant research and conceptual articles:

- helps to validate the protagonist role family firms play in social-economic milieus;
- provides an in depth treatment of operational and definitional issues surrounding what constitutes a family business;
- offers a systematic account of the historical development of the field of family business;
- embraces methodologies encompassing micro and macro perspectives;
- introduces theoretical bases, conceptualizations, and paradigms underpinning family business entrepreneurship, the papers of which challenge the orthodox microeconomic view of *homo-economicus* firms by highlighting the virtues of family influence, *familiness* and social capital; and
- finally, proffers a selection of empirical studies addressing the current family business research agenda.

Structure of the book

The *Handbook of Research on Family Business* involves 33 substantial contributions, five of which are reprints published in top-tier journals. This collection is organized in seven parts: 'Frontiers of Family Business'; 'Theorizing Family Businesses and Business Families'; 'Family Business Research: Metrics and Methodologies'; 'Family Business Themes in Focus'; 'Family Business Succession'; 'Family Business Performance: Global and Trans-cultural Issues'; and 'Family Business Finance'.

Part One 'Frontiers of a Family Business' establishes the academic scene and reviews developments of the family business field.

- Frank Hoy and Pramodita Sharma provide a chronology detailing the evolution of educational programmes and research.
- Pramodita Sharma, based on a review of 217 published studies, reflects on the status of the field and scope of family business.
- Joseph Astrachan and Melissa Shanker present an empirical research framework for assessing the economic role of family firms.

Part Two 'Theorizing Family Businesses and Business Families' is a selection of papers dealing with pertinent theoretical constructs concerning family business entrepreneurship.

- Timothy Habbershon, Mary Williams and Ian MacMillan propose a unified systems perspective of family firm performance, their model of which encapsulates the notion of familiness, and the systemic relationship between resources and capabilities as sources of advantage (or constraint) and performance.

- Ramona Heck, Sharon M. Danes, Margaret Fitzgerald, George Haynes, Cynthia Jasper, Holly Schrank, Kathryn Stafford and Mary Winter, based on the Sustainable Family Business Model, proffer an analytical framework for the study of the family's dynamic role within family business entrepreneurship.
- Nigel Nicholson and Åsa Björnberg introduce the concept of Critical Leader Relationships (CLRs), capturing the neglected theme of shared leadership. These investigators explore possible dyad types in family business CLRs, and propose a Situation–Process–Qualities (SPQ) model of leadership effectiveness as applied to CLRs.
- Lorraine Uhlaner proposes a theoretical framework within which the business family, as a team, can be explored to identify aspects that impact on business strategy and performance.
- Alberto Gimeno Sandig, Gaston Labadie, Willem Saris and Xavier Mendoza Mayordomo advance a theoretical model, incorporating business and family complexity, and identifying internal factors that explain family business performance.

Part Three 'Family Business Research: Metrics and Methodologies' comprises a collection of papers dealing with definitional, methodological, and measurements issues.

- Joseph Astrachan, Sabine Klein and Kosmas Smyrnios report on their pioneering work concerning the Family: Power–Experience–Culture (F-PEC) scale, a systematic measure of family influence on businesses.
- Paul Westhead and Carole Howorth examine variables relating to family firm objectives and ownership, and management structures, leading to an identification of a taxonomy of private family firm types.
- Ken Moores and Justin Craig demonstrate how elements of the Balanced Scorecard, an accepted strategic management and measurement tool, can be adapted to the family business context for use as a strategic tool and as a means of contributing to the professionalization of family firms.
- Sandra Moncrief-Stuart, Joe Paul and Justin Craig present a methodological paper validating the *Aspen Family Business Inventory*, an assessment tool designed specifically for use by consultants working with families in business.

Part Four 'Family Business Themes in Focus' incorporates a series of in-depth papers focusing on functional areas of the family business strategic management.

- Ercilia García-Álvarez and Jordi López-Sintas present evidence from in-depth cross case analysis to delineate a model encompassing value transmission and successors' socialization, for facilitating family business continuity.
- Annika Hall, Leif Melin and Mattias Nordqvist, utilizing a case study approach identifying key parameters associated with everyday, micro and human aspects of organizational life, submit a theoretical framework concerning family business strategizing and direction.
- Lucrezia Songini reflects on the theory and practice of professionalization processes in family firms.

- Miguel Angel Gallo and Salvatore Tomaselli explore the theoretical and practical dimensions of *protocols* in family business strategic planning processes.
- Joseph Astrachan, Andrew Keyt, Suzanne Lane and Kristi McMillan focus on the governance schemes of family and non-family businesses, and tender propositions concerning accountability in firms.
- Rosa Nelly Trevinyo-Rodríguez and Josep Tàpies conceptualize a knowledge transfer model of family firms concentrating on internal and external relationships in family-enterprise next-generation systems.
- Franz Kellermanns and Kimberly Eddleston furnish a conceptual paper targeting management strategies relating to task and relationship conflict in family firms.

Part Five 'Family Business Succession' incorporates a collection of in-depth papers focusing on the popular theme of strategic succession planning:

- Danny Miller, Lloyd Steier and Isabelle Le Breton-Miller, within the context of inductive and case study methodologies, examine the main issues associated with business successions that fail.
- Johan Lambrecht and Rik Donckels develop an explanatory model of business transfer derived from an examination of transitional paths associated with family business succession.
- Pietro Mazzola, Gaia Marchisio and Joseph Astrachan explore the specific benefits of strategic planning processes as a next-generation training tool, particularly during the post-succession stage of the business transfer planning process.
- Vassilios Pyromalis, George Vozikis, Theodoros Kalkanteras, Michaela Rogdaki and George Sigalas advance an integrated framework for evaluating the success of the family business succession process according to gender specificity.

Part Six 'Family Business Performance: Global and Trans-cultural Issues' presents internationally based research initiatives.

- Kristin Cappuyns, through a prism of multiple case studies of Spanish family firms, reports on internationalization via strategic alliances.
- David Pistrui, Wilfred Huang, Harold Welsch and Zhao Jing describe seminal attributes, characteristics and growth orientations of mainland Chinese entrepreneurs, highlighting the contributions of relationships, roles, family and culture in the development of private small and medium enterprises (SMEs).
- Guido Corbetta and Alessandro Minichilli examine the essential features of boards of directors in Italian publicly listed companies.
- Luca Gnan and Daniela Montemerlo compare Italian family versus non-family SMEs in terms of ownership and governance issues.
- Toshio Goto documents the historical development of the Japanese family business economy, providing solid explanations for its longevity.

Part Seven 'Family Business Finance' is a compilation of empirical papers exploring financial and related performance issues pertaining to privately held and quoted family companies.

- Myriam Lyagoubi investigates relationships between family ownership and debt financing behaviour in French privately held and public companies.
- Panikkos Zata Poutziouris, in a study of the UK London main stock market, profiles the UK Family Business PLC economy, and through the lens of a Family Business Index, reports on their performance.
- Susana Menéndez-Requejo evaluates the impact of family governance and ownership on Spanish firm performance.
- María Sacristán Navarro and Silvia Gómez Ansón reviews the influence of family ownership, management and control on the performance of Spanish quoted companies.

'Epilogue'

- Shaker Zahra, Sabine Klein and Joseph Astrachan undertake an element of crystal-ball gazing and provide reflections on the future directions and new frontiers open to this field.

To sum up, this volume features a kaleidoscope of competitive research papers, dealing with conceptual and theoretical frameworks, new methodologies, leading to an identification of best research practice. As is evidenced by this collection, one imperative is the need to continue developing balanced family business entrepreneurship frameworks where new standards of research are achieved in order to strengthen theories relating to family-controlled firms. Another imperative calls for the use of alternative research methods incorporating sources of competitive advantage and dynamic capabilities of family firms, namely, familiness and family influence. This imperative is consistent with the words of Aristotle who is reported as saying that the family is the association established by nature for the supply of people's everyday wants (Aristotle, 400 BC). These words reflect the dual and interrelated nature of family business entrepreneurship.

In closing, we express our gratitude to the IFERA Board who have entrusted us with the co-ordination of this initiative, and of course, to Edward Elgar Publishing Limited for recognizing the emergence of the importance of the theme of family business entrepreneurship. Special thanks are extended to Jo Betteridge, Caroline Cornish and Francine O'Sullivan, and their Elgar editorial team. We appreciate the support of the publishers of the *Journal of Entrepreneurship and Theory and Practice*, *Journal of Business Venturing* and the *Family Business Review* journal for permission to utilize a number of reprints. We are also very grateful to the contributors for their co-operation and zeal to deliver top-quality research papers. Many thanks are also extended to our reviewers who were generous with their time and provided valuable guidance to the authors and editorial team.

Reviewers

Åsa Björnberg, London Business School (UK)
Justin Craig, Oregon State University (USA)
Maria Ercilia Garcia, Universitat Rovira i Virgili de Tarragona, (Spain)
Christopher Graves, Adelaide University (Australia)
Annika Hall, Jönköping International Business School (Sweden)

Thomas Zellweger, St Gallen University (Switzerland)
Ramona Heck, Baruch College (USA)
Carole Howorth, University of Lancaster (UK)
Bakr Ibrahim, Concordia University (Canada)
Peter Jaskiewicz, European Business School (Germany)
Andrew Keyt, Loyola University Chicago (USA)
Eleni Kostea, University of Cyprus (Cyprus)
Eddy Laveren, University of Antwerp (Belgium)
Jordi Lopez, Universitat Autonoma de Barcelona (Spain)
Myriam Lyagoubi, EM LYON (France)
Gaia Marchisio, Kennesaw State University (USA)
Susana Menéndez Requejo, University of Oviedo (Spain)
Nicos Michaelas, Demetra Investment PLC (Cyprus)
Daniela Montemerlo, SDA Bocconi and University of Insubria (Italy)
Mikko Mustakallio, Helsinki Institute of Technology (Finland)
María Sacristán Navarro, University Rey Juan Carlos (Spain)
Mattias Nordqvist, Jönköping International Business School (Sweden)
David Pistrui, Illinois Institute of Technology (USA)
Lucrezia Songini, SDA Bocconi (Italy)
Khaled Soufani, Concordia University (Canada)
Lloyd Steier, University of Alberta (Canada)
Jill Thomas, Adelaide University, (Australia)
Rosa Nelly Trevinyo-Rodríguez, IESE (Spain)
Salvo Tomaselli, Palermo University (Italy)
Lorraine Uhlaner, Erasmus University Rotterdam (Netherlands)
Elina Varamaki, University of Vasa (Finland)
Yong Wang, Wolvehampton University, UK
So-Jin Yoo, NEWI, North East Wales Institute (UK)

On behalf of authors, reviewers, and the editorial team we trust that you find this volume of scholarly papers not only of interest, but also relevant.

References

Anderson, R. and Reeb, D. (2003), 'Founding-family ownership and firm performance: evidence from the S&P 500', *Journal of Finance*, June (3), 1301–28.

Anderson, R., Mansi, S. and Reeb, D. (2003), 'Founding family ownership and the agency cost of debt', *Journal of Financial Economics*, **68**, 263–85.

Barnes, L.B. and Hershon, S.A. (1976), 'Transferring power in the family business', *Harvard Business Review*, July/August, 105–114.

Brockhaus, R.H. (1994), 'Entrepreneurship and family business research: comparisons, critique, and lessons', *Entrepreneurship Theory and Practice*, **19**(1), 25–38.

Burkart, M., Panunzi, F. and Shleifer, A. (2003), 'Family firms', *Journal of Finance*, **58**(5), 2167–201.

Corbetta, G. and Montemerlo, D. (eds) (2001), *The Role of the Family in the Family Business*. Book of Research Forum Proceedings: 12th Annual Family Business Network World Conference, FBN Roma – SDA Bocconi Publication.

Donnelley, R.G. (1964), 'The family business', *Harvard Business Review*, **42**, 93–105.

Gomez-Mejia, L., Nunez-Nickel, M. and Gutierrez, I. (2001), 'The role of family ties in agency contracts', *Academy of Management Journal*, **44**, 81–95.

Koiranen, M. and Karlsson, N. (eds) (2002), *The Future of the Family Business: Values and Social Responsibility*. Research Forum Proceedings, Book of Research Forum Proceedings: 13th Annual Family Business Network World Conference, FBN Helsinki – University of Jyvaskyla Publication.

Morck, R. and Yeung, B. (2003), 'Agency problems in large family business groups', *Entrepreneurship Theory and Practice*, **27**(4), 367–82.
Poutziouris, P. (ed.) (2000), *Family Business: Tradition or Entrepreneurship in the New Economy*, Book of Research Forum Proceedings: 11th Annual Family Business Network World Conference , FBN London – Manchester Business School Publication .
Poutziouris, P. and Steier, L. (eds) (2003), *New Frontiers in Family Business Research: the Leadership Challenge*, Book of Research Forum Proceedings: 14th Annual Family Business Network World Conference, FBN – IFERA Lausanne Publication.
Schulze, W.S., Lubatkin, M.H. and Dino, R.N. (2003), 'Toward a theory of agency and altruism in family firms', *Journal of Business Venturing*, **18**(4), 473–90.
Tomaselli, S. and Melin, L. (eds) (2004), *Family Firms in the Wind of Change*, Research Forum Proceedings, Book of Research Forum Proceedings: 15th Annual Family Business Network World Conference, FBN – IFERA Copenhagen Publication.
Villalonga, B. and Amit, R.H. (2004), 'How do family ownership, control, and management affect firm value?', EFA 2004 Maastricht Meeting, Paper No. 3620.

PART I

FRONTIERS OF A FAMILY BUSINESS

1 Navigating the family business education maze

Frank Hoy and Pramodita Sharma

Introduction

Although we can safely assume that family businesses predate recorded history (Colli, 2003), formal educational and research programs focusing specifically on family-owned firms are recent phenomena (for example, Hoy and Verser, 1994; Wortman, 1994). Litz (1997) provided a persuasive analysis of the reasons for the neglect of family business studies in the business schools citing a 'longstanding pattern of interaction between business firms, business regulators, academic institutions, and individual academic researchers' (p. 56). However, fuelled by a growing awareness of the importance and dominance of family firms in most countries (for example, IFERA, 2003), the interest in family business studies is growing at a rapid pace, as reflected in recent review articles (Bird et al., 2002; Sharma, 2004; Chrisman et al., 2005).

Aims and contributions of chapter

As the field is gaining momentum, it is important to pause momentarily to capture its evolution with an aim to preserve our legacy and provide a common historical basis of our past. This chapter endeavors to provide such a pause to the field of family business studies. While we draw our inspiration from the example of Katz (2003) who captured the evolution of entrepreneurship as a discipline of study and instruction, our focus is much narrower as it is limited to tracking the history of family business studies only. Through this attempt, we hope to preserve our beginnings and provide navigational guidelines to family business scholars and practitioners of tomorrow.

Scope and limitations

To capture the evolution of educational and research programs aimed to create and disseminate knowledge related to family firms, a chronology of key events that have shaped the intellectual trajectory of family business studies was developed and is shared in the next section (Table 1.1). The study of family business lies at the convergence of several research fields including anthropology, family therapy, family studies, organizational studies, sociology and psychology (to name a few). Owing to the limitations of time and space, it is impossible to identify and report all the events and writings that have a bearing on where we stand today. While we have endeavored to provide a comprehensive list of significant events that have helped shape our field of study, this table should only be viewed as the tip of an unexplored iceberg. We invite readers[1] to join us in this effort to refine our chronology as we capture our collective past and lay a foundation to carve the future of family business studies.

Findings

In hindsight, it seems remarkable to us that the early contributions, coming as they did from prestigious business schools such as Indiana University and Harvard, had so

Table 1.1 Chronology of family business studies

1953	Grant H. Calder completes the first doctoral dissertation on family business studies in North America entitled 'Some management problems of the small family controlled manufacturing business', School of Business, Indiana University
1953	Christensen's book, *Management Succession in Small and Growing Enterprises* published by Harvard University Press
1954	*Cases in the Management of Small, Family-Controlled Manufacturing Businesses* published at the Indiana University Press (first family business-specific case book)
1958	English's book, *Financial Problems of the Family Company* published by Sweet and Maxwell, London.
1961	Trow's article 'Executive succession in small companies', published in *Administrative Sciences Quarterly*, **6**
1964	Donnelley's article entitled 'The family business', published in *Harvard Business Review*, **42**(3)
1968	Churchman publishes *The Systems Approach*, Dell Publishers, New York
1968	Alfred Lief's *Family Business: A Century in the Life and Times of Strawbridge & Clothier* published by McGraw-Hill Book Company
1968	Léon Danco holds first interdisciplinary seminar on family business
1971	Levinson's article entitled 'Conflicts that plague family business', published in *Harvard Business Review*, **49**(2)
1972	Ianni and Ianni's book entitled *A Family Business* published by Russell Sage Foundation, New York
1974	Levinson's article entitled 'Don't choose your own successor', published in *Harvard Business Review*, **52**(6)
1975	Léon Danco's *Beyond Survival: A Business Owner's Guide for Success* published by Reston Publishing
1975	Simon A. Hershon completes his dissertation entitled 'The problem of management succession in family businesses', Harvard University
1976	Barnes and Hershon's article entitled 'Transferring power in the family business', published in *Harvard Business Review*, **53**(4)
1978	Streich Chair in Family Business established at Baylor University (Chair holder: Nancy Upton)
1978	Becker and Tillman's book entitled *The Family-Owned Business* published by the Commerce Clearing House, Chicago
1978	Longnecker and Schoen's article entitled 'Management succession in the family business', published in *Journal of Small Business Management*, **16**(3)
1981	Chair of Private Enterprise established at Kennesaw State College (first Chair holder: Craig Aronoff appointed in 1983)
1981	Elaine Kepner presents a workshop on 'Family dynamics and family owned organizations', at the Gestalt Institute of Cleveland conference
1982	Wharton Family Business Program launched at the Wharton Applied Research Center (Founding Director: Peter Davis)

Table 1.1 (continued)

1982	John A. Davis completes his dissertation entitled 'The influence of life stage on father–son work relationship in family companies', Harvard University
1982	John Davis and Renato Tagiuri develop the first compendium of literature on family-owned business entitled 'Bibliography on family business', unpublished document, Harvard Business School
1983	*Organizational Dynamics* publishes a special issue on family business studies (Co-editors: Richard Beckhard and Warner Burke)
1983	Karen L. Vinton completes her dissertation entitled 'The small, family-owned business: a unique organizational culture', unpublished document, University of Utah
1984	Gibb W. Dyer Jr completes his dissertation entitled 'Cultural evolution in organizations: the case of a family owned firm', Sloan School of Management, Massachusetts Institute of Technology
1984	Yale establishes program for the Study of Family Firms (Founding members: Joe Astrachan, Ivan Lansberg, Sharon Rogolsky)
1985	The College of Business, Oregon State University starts the second Family Business Program in the US (Founding Director: Patricia A. Frishkoff)
1985	First 'for credit' family business course offered
1985	Rosenblatt and deMik publish *The Family in Business*, Jossey-Bass
1985	First Family Business Research Conference, University of Southern California (Chair: John Davis)
1986	Founding of the Family Firm Institute Inc. (FFI) (Founding President: Barbara Hollander)
1986	John Ward's book *Keeping the Family Business Healthy* published as part of the Jossey-Bass series on Management of Family Owned Businesses (Consulting Editors: Richard Beckhard, Peter Davis, Barbara Hollander)
1986	Kennesaw State College establishes the Family Business Center and runs the First Family Business Forum (Founding Director: Craig Aronoff)
1986	Andrew Errington edits, *The Farm As a Family Business: An Annotated Bibliography*, Agricultural Manpower Society, University of Reading Farm Management Unit
1987	First Family Business Chair launched in Europe at IESE Business School, University of Navarra, Barcelona (Chair holder: Miguel Ángel Gallo)
1987	Institute for Family Enterprise established at Baylor University (Founding Director: Nancy Upton)
1988	*Family Business Review* began publication (Editor-in-Chief Ivan Lansberg). A Jossey-Bass publication
1988	First FFI research conference hosted by Boston University School of Management, chaired by Marion McCollom Hampton (20 attendees)
1988	Carl R. Zwerner endowed professorship in Family Business established at Georgia State University in the US (Chair holder: Frank Hoy)
1989	Baylor University establishes Institute for Family Business conducts first conference (Chair: Nancy Upton)

Table 1.1 (continued)

1989	'The outstanding outsider and the fumbling family', by Thomas A. Teal and Geraldine E. Willigan, first family business case published in *Harvard Business Review*
1989	FFI established the Best Doctoral Dissertation Award (first award recipient: Colette Dumas) and the Best Unpublished Research Paper Award (first award recipient: Stewart Malone)
1989	Wendy Handler completes her dissertation entitled 'Managing the family firm succession process: the next generation family member's experience', Boston University
1990	Founding of the Family Business Network (FBN)
1992	First FFI Educators Conference hosted by Northeastern University Center for Family Business, Boston
1993	First Mass Mutual Annual Gallup survey of family businesses (the first large sample study of FBs in the United States)
1994	Family Business Division established at the United States Association of Small Business Enterprise (USASBE)
1994	*Entrepreneurship Theory and Practice* publishes a special issue on family business studies (Co-editors: Gibb W. Dyer Jr and Wendy Handler)
1994	FFIs Case Series Project launched (Editor: Jane Hilburt-Davis)
1994	FAMLYBIZ listserv established by Scott Kunkel, University of San Diego
1994	Max Wortman publishes first major review article of family business studies in *Family Business Review*
1995	The first annual Psychodynamics of Family Businesses (PDFB) conference hosted by the Northwestern University (Chair: Ken Kaye)
1995	FFIs Body of Knowledge (BOK) task force created
1996	*A Review and Annotated Bibliography of Family Business Studies* by Sharma, Chrisman and Chua published by Kluwer Academic (first major bibliography of family business studies)
1996	First Family Business major offered at the Texas Tech University
1997	First National Family Business survey (first large study using household sample). Findings reported in *Family Business Review* special issue **7**(3)
1997	Gersick, Davis, Hampton and Lansberg's book entitled *Generation to Generation* published by Harvard Business School Press
1998	Sharma's dissertation receives the NFIB Best Dissertation award (first family business dissertation recognized by the Entrepreneurship division of the Academy of Management)
1999	First Family Business professorship established in Northern Europe at the University of Jyväskylä, Finland (Chair: Matti Koiranen)
2000	Doctoral program for family business studies established at the University of Jyväskylä, Finland
2001	International Family Enterprise Research Academy (IFERA) founded (Founding President: Albert Jan Thomassen). First IFERA conference hosted by INSEAD Fountainbleau (co-organizers: Christine Blondel and Nicholas Rowell; 35 attendees)

Table 1.1 (continued)

2001	First Theories of Family Enterprise (ToFE) conference co-hosted by the universities of Alberta (UoA) and Calgary (UoC) at the School of Business, UoA, Edmonton (co-organizers: Jim Chrisman, Jess Chua and Lloyd Steier)
2003	*Journal of Business Venturing* publishes two special issues on family business studies (Co-editors for **18**(4) issue: Jim Chrisman, Jess Chua and Lloyd Steier, and for **18**(5) issue: Ed Rogoff and Ramona Heck)
2005	International Masters Programme for Family Business established at the University of Jyväskylä, Finland (lectured in English and Finnish)
2005	FITS-project established, a strategic alliance between Finnish (Jyväskylä), Italian (Bocconi), and Swiss (Lugano) universities to offer Doctoral and Post-Doctorals in Family Business.
2005	First Family Enterprise Research Conference (FERC) hosted by the Austin Family Business Program, Portland (co-organizers: Mark Green and Pramodita Sharma; 55 attendees)
2005	Miller and Le Breton-Miller's book entitled *Managing for the Long Run* published by Harvard Business School Press
2005	*Family Business Review* is listed in Social Science Citation Index (SSCI) and Current Contents/Social and Behavioral Sciences (CCBS) (Editor: Joe Astrachan)

little initial stimulative effect on the field. More critical appears to have been the influence of practitioners – family business owners and family business consultants. Popular trade books by authors such as Léon Danco and the launch of university-based outreach programs such as the one at the University of Pennsylvania demonstrated that there was a latent demand in the business community for education, training and development focused on issues peculiar to organizations characterized by family ownership and control. This recognition precipitated more scholarly research in order to have a knowledge base for instruction. To complete the circle, researchers returned to the early contributions for theoretical and empirical justifications for investigations.

Also in retrospect, the formation of associations whose memberships consisted of individuals and organizations with family business constituents or clienteles, that is, the Family Firm Institute (1986) and the Family Business Network (1990), fostered further research investigations for benchmark information and best practices that could be learned and used to serve family enterprises. In 1988, the Family Firm Institute-sponsored *Family Business Review* began offering an outlet for refereed publications. This breakthrough encouraged more academics to produce new knowledge for the field. The academic community is characterized by the axiom, 'publish or perish'. Journal access is an incentive to contribute to a stream of research.

Future

Guided by lessons of history, our critical reflections of the past enable us to speculate about our future (cf. Van Fleet and Wren, 2005). At the time of writing this chapter, it was estimated that there were in excess of 200 family business educational programs based at institutions of higher education worldwide. To this can be added other colleges and universities offering courses addressing family business issues, but short of academic concentrations or outreach programs. All this suggests that a critical mass has been reached with family business studies gaining acceptance as a legitimate field of study.

For over a decade at the annual meeting of the Family Firm Institute, an award has been presented for the best doctoral dissertation on the subject of family business. Every year multiple dissertations are submitted for consideration. This has two implications. The first is that both new faculty entering the field and faculty at their degree-granting institutions value family business issues as significant areas of study. The second is that the continuous streams of research assure that topical issues will be investigated using up-to-date research designs and analytical tools. Further, these dissertations are coming from all regions of the globe, with recent award winners being from Canada, Spain, South Africa, Sweden and the United States, indicating widespread appeal of family business studies for young scholars.

Chronology of family business studies

Methods

Multiple sources of information were used to develop the chronology of key events shaping family business studies (Table 1.1). These include:

- Reviews of primary and secondary historical source documents including many of the earliest books and review articles (for example, Christensen, 1953; Danco, 1975; Dyer Jr, 1986; Sharma et al., 1996; Trow, 1961; Ward, 1986; Wortman Jr, 1994).
- Reviews of websites including of some of the major professional associations focused on family business studies, for example the Family Firm Institute (FFI), the Family Business Network (FBN) and the International Family Enterprise Research Academy (IFERA).
- Reviews of websites of some of the oldest and most well established family business programs across the world. Examples include the Cox Family Enterprise Center at Kennesaw University, the Baylor Institute for Family Business at Baylor University, the Wendel International Centre for Family Enterprise at INSEAD and many more.
- A survey of leading thinkers and professionals in the field. Discussions with many faculty and practitioners who have directly influenced the development of the field. Examples include, Craig Aronoff, Joe Astrachan, Jim Chrisman, Guido Corbetta, John Davis, Nancy Drozdow, Gibb Dyer Jr, Miguel Gallo, Judy Green, Jane Hilburt-Davis, Paul Karofsky, Ken Kaye, Matti Koiranen, Kosmas Smyrnios, Nancy Upton and Karen Vinton, to name a few.
- Finally, the authors immodestly acknowledge their own expertise and contributions to the field. The first author was the inaugural holder of the Carl R. Zwerner Professorship in Family-Owned Businesses, thought to be the third endowed position established dedicated to family business education and research. In addition to writing an award-winning dissertation on family business, the second author has participated in the development of extensive reviews of the literature (Chrisman et al., 2005; Sharma, 2004; Sharma et al., 1996, 1997).

Reflections of the past and current status
A critical analysis of Table 1.1 reveals the field of family business studies can be viewed as having gone through the following stages in its growth.

1950s and 1960s – era of rugged pioneers The decades of the 1950s and 1960s were marked by negative connotations of family businesses both within the circles of higher education as well as within the business community. At this time of perceived negativity towards family business, it was the rugged efforts of entrepreneurial pioneers who ventured into exploring and writing about family firms.

In the broader community, an interesting schizophrenia has been extant. Business founders frequently labeled their firms by the family name. It is not unusual to see signs declaring 'John Doe and Sons' or 'Jones Brothers' (the gender bias implied by these hypothetical examples reflects the historic male dominance of the family firm, particularly regarding succession). Despite these appellations, many company owners objected to being categorized as family enterprises. Their assumption was that a family business was somehow less professional than a non-family firm. The stereotype for many family-owned enterprises could be summarized in the negative connotation, nepotism (Ewing, 1965).

Within the confines of academia, efforts were devoted to develop scientific knowledge based on rigorous empirical investigations and to disseminate it in the classroom. Application of the scientific method translated not only into content, but also into pedagogical processes. In business education specifically, students were taught to be analytical in their approach to management. For many in the education profession, this resulted in encouraging students to focus on objective data and to apply quantitative analysis tools (Bennis and O'Toole, 2005). Emotions and affective behavior were addressed as variables to be controlled by practicing effective managerial techniques in leadership, organization, communication, motivation, and so on (for example, Dyer Jr, 2003). On those rare occasions when family issues might arise in course material or class discussions, students would be taught to segregate what was perceived to be irrelevant variables from business management in order to prevent them from entering into the decision-making process or from disrupting the organizational system (cf. Ghoshal, 2005). While some valiant efforts at conduct of academic research on family firms are visible especially at Harvard and Indiana universities, most of the writings at the time were completed by rugged pioneers such as Donnelley and Trow who were working against the prevailing norms and trends.

1970s – enter practitioner-consultants Practitioner-consultants were the first to begin to fill the educational void, offering training and development programs for family businesses. Disciplines represented by these consultants included law, accounting, psychology, financial planning, general management and others. Early contributors enhanced their ability to market training and consulting by publishing books aimed towards the business owner. For many consultants, the books were elements of marketing strategies to promote themselves as experts in consulting or as professional speakers. Their markets were family business owners and prospective successors to those owners. The lecture circuit often consisted of trade associations that had large numbers of independent businesses among their membership. This early stage marked the identification and cultivation of the owner-manager market segment for education programs.

Early efforts towards building a community of interested scholars in family business research started to emerge, for example, Danco's interdisciplinary seminars. Evidence of scholarly seriousness started to surface, as did the realization of the complexity and interdisciplinary nature of family business studies.

1980s – the decade of institution building By 1980, there were published indications that awareness was arising in the practitioner community regarding the need for education, and for educating more than just the business owner/founder (Poe, 1980). Additionally, popular books about family-owned enterprises brought attention to the interactions between families and the firms that they owned and/or managed (Ward, 1986). A small number of academic scholars who believed that family business warranted study joined the practitioner-consultants in their pursuit of aiding family business owner-managers. These early scholars tended to blend research, consulting and teaching. Research investigations often relied on convenience samples, participant observation or case studies. Thus, the books published by the academics were generally directed toward practitioner markets rather than the academic community. Few courses were offered at American universities specifically oriented towards the ownership and management of family companies.

The most distinctive trend of the 1980s, however, is that of institution building. Family business programs started to emerge in universities. The Wharton Family Business Program at the University of Pennsylvania was the first to be launched, in 1982. Oregon State University started its program in 1985 and was followed by Kennesaw a year later. In 1988, *Nation's Business* magazine identified 20 universities in the United States as having established some form of family business program within or affiliated to their business schools. For the most part, these were outreach or continuing education-type programs designed for business owners in the communities or regions served by the institutions. Family business programs have proliferated at universities. Today there are more than a hundred in the United States plus programs in at least 10 other countries (Family Firm Institute (FFI) and www.ffi.org).

For many universities, the creation of a forum was imposed by external pressure. Successful entrepreneurs and individuals involved in troubled family enterprises pressed universities to go beyond standard business education and offer help and advice to this segment of the business community. When Carl R. Zwerner endowed a professorship in family business in the United States at his alma mater, Georgia State University, he commented that he received an invaluable education in how to manage a business successfully, but the program taught him nothing about how to get along with family members in his companies. In 1988, he called upon Georgia State to infuse family business issues into the curriculum so that future entrepreneurs would not encounter the same obstacles he experienced.

As the forums began to spread across the nation and into other countries, the educational programs were initially targeted towards the same market segment that the practitioner-consultants targeted, that is, business owners and their immediate family members. In many ways, this was highly logical. This was the group that was in immediate need of education. They were confronting family issues daily that facilitated or hindered commercial operations. Additionally, adjunct faculty who gained their experience in consulting or were from a family business background directed many of the new programs. As professional, tenured or tenure-track faculty became involved, they often relied on practitioner-consultants to lead seminars and workshops or to work directly with the family enterprises that were seeking help. The success of these programs led to the identification of internal markets at the universities: undergraduate students and graduate students whose families owned businesses and who had expectations of working in those businesses. It naturally followed that business schools would begin to experiment with introducing family business courses. Many schools anticipated that there would be a latent market beyond the students with family business backgrounds. Other targets included students intending to start their own companies with family business participation, either in employment or investment; students seeking professional degrees such as accounting and expecting to have family enterprises among their clientele; and students who could foresee that they might have opportunities for employment as non-family managers in family-owned businesses.

Another significant event occurred in the 1980s. The Family Firm Institute was founded in 1986 as a nexus for those concerned with family enterprises. The diverse membership of the FFI includes consultants, attorneys, accountants, therapists, financial advisors, academics and others who have family-owned firms as there clientele. In 1988, the FFI launched the *Family Business Review* (*FBR*), the only scholarly journal devoted to family business. Through the *FBR*, researchers have an outlet for the findings of their family enterprise investigations.

1990s and beyond – growth escalates Interest in family business studies has been escalating since the 1990s, with large studies being commissioned to understand the influence of family firms in various nations (for example, Mass Mutual study, National Family Business Survey). Entrepreneurship associations such as the United States Association of Small Business Enterprise (USASBE) started to recognize and establish separate divisions focused on family business studies. Efforts to consolidate knowledge and provide guidance for future directions started to be seen as evidenced by published bibliographies (for example, Sharma et al., 1996) and establishment of the Body of Knowledge (FFI). Emphasis started to increase on doctoral education in family firms especially both in Europe and North America, as rigorous empirical studies on family businesses started to emerge (for example, Bornheim, 1997; Handler, 1989; Sharma, 1997; Thomas, 1999).

New avenues to share research started to be established, for example, the International Family Enterprise Research Academy (IFERA), the Theories of Family Enterprise (ToFE) and the Family Enterprise Research Conference (FERC). Scholars from various disciplines started focusing their research on family firms (Miller and Le Breton-Miller, 2005), special issues of entrepreneurship journals started to appear (*Entrepreneurship Theory and Practice*, **27**(4), **28**(4), **29**(3); *Journal of Business Venturing*, **18**(4), **18**(5)), as some top-ranked journals published family business research (for example, Andersen and Reeb, 2003; Gomez-Mejia et al., 2001; Schulze et al., 2001). The recent selection of *Family Business Review* for coverage in the Social Science Citation Index (SSCI) and the Current Contents/Social and Behavioral Sciences (CCBS) marks a major milestone. An exciting future lies ahead for family business studies!

Going forward: implications for the future

One cause for prospects for family business education to be exciting is the continuing interaction between educators and practitioners, which exists to a greater extent than in most disciplines. In reviewing the milestones in the evolution of family business as a subject of scholarly inquiry, we relied on expert knowledge and opinion of academics. To propose a core body of knowledge for family business education, we turn to the expert opinion of practitioners, specifically the practitioner members of the Family Firm Institute (FFI).

The FFI is an international professional membership organization dedicated to providing interdisciplinary education and networking opportunities for family business advisors, consultants, educators and researchers, and to raising public awareness about trends and developments in the family business field.

Family Firm Institute members are lawyers, therapists, financial professionals, business and management consultants, family business advisors, educators and researchers – the entire spectrum of professionals who advise, study or work with family businesses and family offices. The organization serves as a network for exchanging information and learning techniques for better serving the family business community.

The FFI was founded in 1986 in recognition of the multidisciplinary nature of the family business phenomenon. In 1995, the FFI Board of Directors created a Body of Knowledge (BOK) task force. The task force was charged with identifying knowledge areas of the family business profession and determining the depth of knowledge necessary to demonstrate competence. The task force became a permanent committee of the FFI in 1999.

According to the BOK, a 'body of knowledge for family business practice is a peer-developed distillation of what competent family business advisors, consultants and educators (collectively "family business professionals") must know to work effectively in the field of family business' (The Family Firm Institute, Inc., p. 3). A primary purpose for developing a body of knowledge was to provide 'guidance to educators in business schools as they develop curricula for use with graduate students' (ibid., p. 4).

One of the first steps taken by the task force was to identify four major content areas:

- Behavioral science;
- Financial;
- Law; and
- Management science.

The task force also decided that they should acknowledge different levels of expertise in each category, labeling them as aware, knowledgeable and skilled.

An interesting aspect of the Body of Knowledge formulation process is that it not only identifies content areas for curricula development, but also helps in segmenting markets for family business education. As previously explained, the membership of the Family Firm Institute is heterogeneous, including such diverse backgrounds as law, accounting, psychology, psychiatry, education, and even the occasional practicing family business owner-manager.

Looking at the various professions active in the family business arena, and thinking of the normal disciplinary and degree-level divisions within academic institutions, we can propose a list of prospective audiences for family business education. From the foregoing discussion, we find that the original audience for education was, logically, family business owners and other family members in the firm, especially prospective successors. With the advent of programs at universities, undergraduate and graduate students became target markets. Although these groups could be further segmented, few programs would be large enough to direct courses specifically at sub-groups, such as students already in family enterprises versus students who anticipated a family business clientele at some future time. Thus, educational course offerings have to be sufficiently broad to encompass multiple groups. Distinctions between graduate and undergraduate education are more readily formed. The initiation and growth of the Family Firm Institute demonstrated a demand for education among individuals and firms working with family businesses. As the organization matured, members recognized that there was an additional need for education, training and consulting to non-family employees of family businesses, as well as for family members regarding how to recruit and retain non-family members in a family enterprise.

Table 1.2 contains a matrix listing content areas from FFI's Body of Knowledge, indicating the market segments that might benefit from educational programs, and is intended as a guide or example. Educators would be expected to modify the rows and columns of the table in accordance with the market segments that they identify for their programs.

In this example, it will be noted that most boxes are marked for each of the target market categories. This is not surprising. There is a need for content acquisition across all groups if they are to be effective in fulfilling their roles. Recall also, the BOK acknowledged varying depths of learning. Thus, the matrix could be extended by indicating which groups should become aware, which should become knowledgeable and which should become skilled. One might reasonably assume that the expectations for teaching undergraduate students would fall primarily into the 'aware' level. For business owners, 'knowledgeable' might be appropriate in most content areas. Professionals and consultants should certainly be 'skilled' in their purported expertise, but must be 'aware' or 'knowledgeable' in broad content areas as well.

Conclusions

With apologies to Harriet Beecher Stowe, family business education, like Topsy, has just 'growed.' After being long ignored by academia, the family business market was discovered by consultants and others with family business clienteles. Universities initiated programs in response to market demand more than from inspiration or research-based need determination. Programs and course offerings proliferated primarily as a function of expert opinion more than a careful assessment of constituent needs. One university program stands as a unique model for future program development: Stetson University. At Stetson, their family business program takes a holistic approach and includes a combination of research, teaching and outreach. As momentum is building, we need to maintain our focus on the distinctiveness of family firms based on a reciprocal influence of family on business and vice versa (Astrachan, 2003; Hoy, 2003; Rogoff and Heck, 2003).

Thanks to the opinions of experts in multiple disciplines, it is possible to determine educational content and access course materials. These opinions have value in that they come from individuals who have direct and frequent contact with family businesses. The experts often have experience as members of family firms. Such membership may occur in childhood and youth, may be either as family member or non-family employee, may have resulted from a change in career after being a family business owner-manager, or other such experience. There is reason to expect, however, that the course content proposed from the work of the Family Firm Institute's Body of Knowledge Committee, in particular, is an appropriate starting point for curriculum development.

Table 1.2 Topics and targets of family business education

CORE KNOWLEDGE	Under graduate	Graduate	Owners	Successors	Non-family employees	Professionals	Consultants
BEHAVIORAL SCIENCE							
Individual and family development	X	x	x	x		x	x
Life cycle issues	X	x	x	x		x	X
Family system theories		x				x	X
Conflict management	X	x	x	x	X	X	X
Gender issues	X	x	x	x	X		X
Birth order issues	X	x	x	x			X
Ethics of consulting			x			X	
FINANCIAL							
Roles of financial advisors			x	x	x	X	
Reading financial statements	X	X	X	X			
Business valuation		X	X	X		X	
Accessing liquidity	X	X	X	X		X	
Employee stock ownership		X	X	X	X	X	
Insurance	X	X	X	X		X	
Wealth management	X	X	X	X		X	X
LAW							
Choosing a lawyer	X	X	X	X		X	
Lawyer's mission and role	X	X	X	X	X	X	X
Forms of business organization	X	X		X	X		X
Estate planning	X	X	X	X		X	X
Role of trustees		X	X	X		X	
Contract dispute and resolution	X	X	X	X	X	X	X
Ownership agreements	X	X	X	X		X	X
MANAGEMENT SCIENCE							
Role of management consultants		X	X	X	X	X	X
Strategic planning	X	X	X	X	X	X	X
Basic management theories	X	X		X	X		
Human resource management	X	X		X	X		
Boards of directors	X	X	X	X		X	X
Leadership	X	X	x	X	x		X

In this chapter, we encouraged matching content to market. We examined the evolution of family business education to identify the providers of educational programs and the audiences those providers targeted. Tracking consultants, scholars, associations and others, we find multiple market segments with differing but overlapping needs. Further, we see that the audiences may vary in the levels of knowledge they need to acquire. We propose the BOK materials as a logical starting point for designing educational programs and linking them to different groups of students. Our charge to academic scholars is to contribute to this body of knowledge through rigorous research. While doing so in the buzz of publications and the publish or perish culture, however, it is critical that we do not lose sight of the relevance of research in a professional discipline under the disguise of scientific rigor (Ghoshal, 2005; Warren and O'Toole, 2005; Zahra and Sharma, 2004).

Note

1. Please bring to our attention other key events that have helped shape family business studies by sending a note either to fhoy@utep.edu or pshama@wlu.ca.

References

Anderson, R. and D. Reeb (2003), 'Founding-family ownership and firm performance: evidence from S&P500', *Journal of Finance*, **58**, 1301–28.

Astrachan, J.H. (2003), 'Commentary on the special issue: the emergence of a field', *Journal of Business Venturing*, **18**, 567–72.

Barnes, L.B. and S.A. Hershon (1976), 'Transferring power in the family business', *Harvard Business Review*, **53**(4), 105–14.

Becker, B.M. and F.A. Tillman (1978), *The Family-owned Business*, Chicago, IL: Commerce Clearing House.

Bennis, W.G. and J. O'Toole (2005), 'How business schools lost their way', *Harvard Business Review*, May, 96–104. Reprint # R0505F.

Bird, B., H. Welsch, J.H. Astrachan and D. Pistrui (2002), 'Family business research: the evolution of an academic field', *Family Business Review*, **15**(4), 337–50.

Bornheim, S.P. (1997), 'The evolutionary family business ecology: a grounded theory approach', unpublished dissertation, Universität St. Gallen.

Chrisman, J.J., J.H. Chua and P. Sharma (2005), 'Trends and directions in the development of a strategic management theory of the family firm', *Entrepreneurship Theory and Practice*, **29**(5), 555–75.

Christensen, C.R. (1953), *Management Succession in Small and Growing Enterprises*, Boston, MA: Division of Research, Graduate School of Business Administration, Harvard University Press.

Colli, A. (2003), *The History of Family Business: 1850–2000*, Cambridge: Cambridge University Press.

Danco, L. (1975), *Beyond Survival: A Business Owner's Guide to Success*, Reston, VA: Reston Publishing Company.

Dyer, Jr, W.G. (1986), *Cultural Change in Family Firms*, San Francisco, CA: Jossey-Bass.

Dyer, Jr, W.G. (2003), 'The family: the missing variable in organizational research', *Entrepreneurship Theory and Practice*, **27**(4), 401–16.

Ewing, D.W. (1965), 'Is nepotism so bad?', *Harvard Business Review*, **43**(1), 22.

Gersick, K.E., J.A. Davis, M.M. Hampton and I. Lansberg (1997), *Generation to Generation: Life Cycles of the Family Business;* Cambridge, MA: Harvard Business School Press.

Ghoshal, S. (2005), 'Bad management theories are destroying good management', *Academy of Management Learning and Education*, **4**(1), 75–91.

Gomez-Mejia, L., M. Nuñez-Nickel and I. Gutierrez (2001), 'The role of family ties in agency contracts', *Academy of Management Journal*, **44**, 81–95.

Handler, W.G. (1989), 'Managing the family firm succession process: the next generation family member's experience', PhD dissertation, Boston University.

Hoy, F. (2003), 'Commentary: legitimizing family business scholarship in organizational research and education', *Entrepreneurship Theory and Practice*, **27**(4), 417–22.

Hoy, F. and T.G. Verser (1994), 'Emerging, business, emerging field: entrepreneurship and the family firm', *Entrepreneurship Theory and Practice*, **19**(1), 9–23.

Ianni, F.A.J. and E. Reuss-Ianni (1972), *A Family Business*, New York: Russell Sage Foundation.

International Family Enterprise Research Academy (IFERA) (2003), 'Family businesses dominate', *Family Business Review*, **16**(4), 235–40.

Katz, J.A. (2003), 'The chronology and intellectual trajectory of American entrepreneurship education 1987–1999', *Journal of Business Venturing*, **18**, 283–300.
Litz, R.A. (1997), 'The family firm's exclusion from business school research: explaining the void; addressing the opportunity', *Entrepreneurship Theory and Practice*, Spring issue, **21**, 55–71.
Miller, D. and I. Le Breton-Miller (2005), *Managing for the Long Run: Lessons in Competitive Advantage from Great Family Businesses*, Cambridge, MA: Harvard Business School Press.
Poe, R. (1980), 'The SOBs', *Across the Board*, May, reprinted in C.E. Aronoff and J.L. Ward (eds), *Family Business Sourcebook* (1991), Detroit, MI: Omnigraphics, pp. 38–48.
Rogoff, E.G. and R.K.Z. Heck (2003), 'Editorial: evolving research in entrepreneurship and family business: recognizing family as the oxygen that feeds the fire of entrepreneurship', *Journal of Business Venturing*, **18**(5), 559–66.
Schulze, W., M. Lubatkin, R. Dino and A. Buchholtz (2001), 'Agency relationships in family firms: theory and evidence', *Organization Science*, **12**, 99–116.
Sharma, P. (1997), 'Determinants of the satisfaction of the primary stakeholders with the succession process in family firms', PhD dissertation, University of Calgary.
Sharma, P. (2004), 'An overview of the field of family business studies: current status and directions for future', *Family Business Review*, **17**(1), 1–36.
Sharma, P., J.J. Chrisman and J.H. Chua (1996), *A Review and Annotated Bibliography of Family Business Studies*, Norwell, MA: Kluwer Academic.
Sharma, P., J.J. Chrisman and J.H. Chua (1997), 'Strategic management of the family business: past research and future challenges', *Family Business Review*, **10**(1), 1–35.
Thomas, J. (1999), 'The exercise of leadership in family business: perceptions of owners and managers, family and non-family in the Australian small to medium enterprise sector', unpublished PhD dissertation, University of South Australia.
Trow, D.B. (1961), 'Executive succession in small companies', *Administrative Science Quarterly*, **6**, 228–39.
Van Fleet, D.D. and D.A. Wren (2005), 'Teaching history in business schools: 1982–2003', *Academy of Management Learning and Education*, **4**(1), 44–56.
Ward, J.L. (1986), *Keeping the Family Business Healthy: How to Plan for Continuing Growth, Profitability, and Family Leadership*, San Francisco, CA: Jossey-Bass.
Warren, B.G. and J. O'Toole (2005), 'How business schools lost their way', *Harvard Business Review*, **83**(5), 96–104.
Wortman, Jr, M.S. (1994), 'Theoretical foundations for family-owned businesses: a conceptual and research based paradigm', *Family Business Review*, **7**(1), 3–27.
Zahra, S. and P. Sharma (2004), 'Family business research: a strategic reflection', *Family Business Review*, **17**(4), 331–46.

2 An overview of the field of family business studies: current status and directions for the future

Pramodita Sharma

Whether measured in terms of number of published articles,[1] publication outlets,[2] schools offering family business programs,[3] research support provided by private donors and foundations,[4] or the membership of family firm associations,[5] the interest in family business studies is increasing. As a field of study develops, it is important to intermittently pause to evaluate the progress made and reflect on the directions to pursue in future so as to gain deeper insights into the phenomenon of interest. The purpose of this review is to provide such a reflective moment for the field of family business studies, as the primary scholarly journal of the field, *Family Business Review*, embarks on its new journey with Blackwell Publishing.

The guiding principle of any professional investigation in social sciences is to clarify our understanding of the segment of the social world that is of interest (Lindblom and Cohen, 1979). Scholars and practitioners interested in family firm studies seek to gain new insights and knowledge into the causal processes that underlie these firms (cf. Lewin, 1940). Theory is an efficient tool that guides the development of knowledge because it helps make connections among observed phenomenon, thereby helping build conceptual frameworks that stimulate understanding (Sutton and Staw, 1995). It aids in building connections between the work at hand and preexisting research, thus making use of our cumulative knowledge to reveal a range of alternatives for effective action (Lindblom and Cohen, 1979; Moore, 1962; Weiss, 1977). Kurt Lewin's (1945) often-quoted endorsement of theory, 'there is nothing so practical as a good theory,' suggests the key role of theory in guiding effective practice.

This article is based on a rigorous review of 217 peer-reviewed articles on family business studies. The point of departure for this review was the previous consolidation attempts of the literature by Sharma et al.[6] (1996, 1997). Although an attempt is made to provide an overview of the literature, the size and scope of it precludes detailed descriptions of individual studies or an exhaustive listing of every article that was reviewed. Instead, given the importance of theoretical knowledge in a scholarly inquiry and development of a field of study (Whetten, 1989, 2002), the focus here is primarily on research that is theoretically oriented.[7]

Scholarly research can be undertaken at various levels of analyses: individual, interpersonal/group, organizational, and societal (cf. Low and MacMillan, 1988). Although it is possible to theorize across multiple levels, given the preparadigmatic status of family business studies, with few exceptions, most of the literature is focused on one level rather than the conceptually complex domain of multiple-level theorizing (cf. McKinley et al., 1999). This article organizes the family business literature according to the four levels of analyses, provides an assessment of the status of our current understanding at each level, and presents suggestions for future research.

However, before presenting the highlights of the literature, the following section engages in a discussion of the domain or the scope of the field of family business studies. This is accomplished through a discussion of definitional issues, basis of distinctiveness of the field, and various facets of family firms' performance. Strategies for efficient creation and dissemination of knowledge that should enable the field of family business studies to progress toward being considered a legitimate scholarly field that is theoretically rich and practically useful are shared in the last section.

Domain of the field of family business studies

In an information-overloaded and competitive world of organizational studies, in order to attract intellectual and financial resources, interested scholars and practitioners need to provide convincing reasons for directing research efforts on family business studies (cf. McKinley et al., 1999). Thus far, the main reason provided by scholars for directing scholarly research toward family firms has largely been the observed dominance of these firms on the economic landscape of most nations.[8] Although an effective starting point in generating interest and gaining attention, this approach is not unilaterally sufficient to gain legitimacy for the field. Convincing, theoretically based answers must be provided for questions such as: Are family firms really different from other business organizations? and Why do these firms deserve special research attention? In working toward a response to such questions, there is a need to clarify the definition of family firms, source of distinctiveness of the field, and the different facets of family firm performance. Each of these is discussed below.

Definition

A challenging task in most social sciences,[9] the importance of establishing clear definitions of family firms cannot be denied as these will assist in building a cumulative body of knowledge. Numerous attempts have been made to articulate conceptual and operational definitions of family firms. Various scholars have reviewed existing definitions, made attempts of consolidation of thoughts, and conceptualized another definition of family firms (for example, Chua et al., 1999; Handler, 1989a; Litz, 1995). The focus of most of these efforts has been on defining family firms so that they can be distinguished from nonfamily firms. Although none of these articulations has yet gained widespread acceptance, most seem to revolve around the important role of family in terms of determining the vision and control mechanisms used in a firm, and creation of unique resources and capabilities (for example, Chrisman et al., 2003a; Habberson et al., 2003).

Reflecting on the well-established fact that a large majority of firms in most countries have a significant impact of 'family' in them (for example, Astrachan et al., 2003; Corbetta, 1995; Klein, 2000), scholars question the homogeneity of these firms (Sharma, 2002). Empirical research has revealed that these firms are only rarely an either-or scenario (Tsang, 2002). Instead, they vary in terms of degrees of family involvement. Attempts to capture the varying extent and mode of family involvement in firms have been directed in three general directions: articulation of multiple operational definitions of family firms (for example, Astrachan and Shanker, 2003; Heck and Stafford, 2001; Westhead and Cowling, 1998); development of scales to capture various types of family involvement (Astrachan et al., 2002b); and development of family firm typologies (Sharma, 2002).

Using three modes of family involvement, Astrachan and Shanker (2003) provide three operational definitions of family firms.[10] Their broad definition uses the criteria of family's retention of voting control over the strategic direction of a firm. In addition to retention of such control by the family, the mid-range definition includes firms with direct family involvement in day-to-day operations. The most stringent of definitions classifies firms as family firms only if the family retains voting control of the business and multiple generations of family members are involved in the day-to-day operations of the firm. Using these definitions, these researchers estimate between 3 to 24.2 million family firms in the United States that provide employment to 27–62 per cent of the workforce, and contribute 29–64 per cent of the national GDP.

Astrachan et al. (2002b) have presented a validated ready-to-use scale for assessing the extent of family influence on any business organization. This continuous scale is comprised of three subscales: power, experience, and culture (F-PEC scale). Particularly impressive in this study is the power scale, which articulates the interchangeable and additive influence of family power through ownership, management, and/or governance. The experience scale measures the breadth and depth of dedication of family members to the business through the number of individuals and generations of family members involved in the business. Family's commitment to the business and values are used for the culture scale. These scholars encourage researchers to move away from a bi-polar treatment of firms as family or nonfamily firms toward exploring the mediating and moderating effects of family involvement in their studies.

Using the well-established overlapping three-circles model (Lansberg, 1988), each circle representing family membership, ownership, and managerial roles of internal family firm stakeholders, Sharma (2002) has proposed a typology that identifies 72 distinct nonoverlapping categories of family firms according to the extent of family involvement in terms of ownership and management. This 'collectively exhaustive,' 'mutually exclusive,' and 'stable' system of classification meets some of the key criteria of a good classification system (Chrisman et al., 1988, p. 416; McKelvey, 1975, 1982). Subjecting this classification system to empirical tests should help identify the types of family firms that prevail in each nation at any point in time. Research has revealed that the national fiscal laws (for example, inheritance and capital gains taxes) influence the type of family firm that prevails in a country, as firm leaders make attempts to minimize tax payment and retain the fruits of their labor within their family and business (for example, Burkart et al., 2003; Foster and Fleenor, 1996; Wells, 1998). An important task that lies ahead for the field is to subject the theoretical taxonomy developed to empirical tests in different nations. Such an effort should assist in the development of empirical taxonomies that can operationally distinguish family from nonfamily firms and between different types of family firms (cf. Miller, 1996).

Distinctiveness

In 1994, after conducting a thorough review of the family business literature, Wortman commented: 'no one really knows what the entire field is like or what its boundaries are or should be' (Wortman, 1994, p. 4). Perhaps the lack of definitional clarity at the time compounded the difficulties of pinpointing the source of distinctiveness of the field (cf. Hoy, 2003). As progress is being made on the development of definitions of family firm based on the varying extent and nature of family involvement in a firm, some clarity on the domain and distinctiveness of the field of family business studies is being experienced.

Encouraged by suggestions based on previous reviews of the literature (Sharma et al., 1997; Wortman, 1994), one stream of effort aimed at finding the source of distinctiveness in family firm studies was directed toward comparative studies of family and nonfamily firms (for example, Anderson and Reeb, 2003; Coleman and Carsky, 1999; Gudmundson et al., 1999; Lee and Rogoff, 1996; Littunen, 2003; Westhead et al., 2001; Zahra et al., in press). This research revealed mixed results, with family and nonfamily firms being different on some dimensions (for example, entrepreneurial activities undertaken, performance, perception of environmental opportunities and threats) but not on others (for example, strategic orientation, sources of debt financing). Although these efforts have aided in improving our understanding of these firms, no set of distinct variables separating family and nonfamily firms has yet been revealed. Clearly, there is a need to conduct a meta-analysis of this research stream to determine what these efforts have collectively disclosed in terms of distinctions between family and nonfamily firms.

Scholars have suggested broad-based conceptual models of sustainable family businesses that take into account the reciprocal relationship between family and business systems (Stafford et al., 1999). These models are aimed toward the simultaneous development of functional families and profitable firms. Others have encouraged the adoption of a 'family embeddedness perspective' by including the characteristics of family systems in research studies (Aldrich and Cliff, 2003; Chrisman et al., 2003c; Zahra et al., in press). Large-scale carefully designed empirical studies have revealed that the success of family firms depends on the effective management of the overlap between family and business, rather than on resources or processes in either the family or the business systems (Olson et al., 2003). A seeming convergence is appearing that it is the reciprocal impact of family on business that distinguishes the field of family business studies from others (for example, Astrachan, 2003; Dyer, 2003; Habbershon et al., 2003; Rogoff and Heck, 2003; Zahra, 2003).

To crystallize the source of distinctiveness of this field of study from other related fields, it is important to understand its distinction from and linkages with other fields of study. Noteworthy efforts toward this end have been made in the recent special issues of *Entrepreneurship Theory and Practice* (**27**(4)) and *Journal of Business Venturing* (**18**(4) and **18**(5)). These issues are directed toward exploring the linkages between family business studies and other disciplines or theories developed in other fields.[11] Such efforts must be continued in the future as they propel the field toward establishing its niche and identity in the domain of organizational studies.

Family firm performance

Recognition of the intertwinement of family and business in family firms has led to a definition of high-performing family firms that takes into consideration performance on both family and business dimensions (Mitchell et al., 2003). It is generally accepted that these firms aim to achieve a combination of financial and nonfinancial goals (Davis and Taguiri, 1989; Olson et al., 2003; Stafford et al., 1999). Research has revealed significant variations in perceptions of family firm stakeholders regarding even the most fundamental issues (for example, Poza et al., 1997; Sharma, 1997). An important direction for the future is to understand the extent of a lignment in the definition of success used by the key players of family firms. The tenets of stakeholder theory may prove useful in gaining such an understanding (Freeman, 1984). Perhaps, an alignment of stakeholders' perspective on

FAMILY DIMENSION

BUSINESS DIMENSION	Positive	Negative
Positive	I *Warm hearts* *Deep pockets* High emotional and financial capital	II *Pained hearts* *Deep pockets* High financial but low emotional capital
Negative	III *Warm hearts* *Empty pockets* High emotional but low financial capital	IV *Pained hearts* *Empty pockets* Low financial and emotional capital

Figure 2.1 Performance of family firms

what 'success' means to them could be an important predictor of success of family firms, as such an alignment can lead to agreement on appropriate mode and extent of involvement of key family and nonfamily members in the firm. On the contrary, a mismatch in the definitions of success or goals that different stakeholders strive to achieve for the family firm could point toward a tenacious source of conflict (Astrachan and McMillan, 2003).

If family firm performance refers to high performance in terms of family and business dimensions, at any point in their lifecycle, family firms may be successful on either one or both these dimensions. Using a two by two matrix (Figure 2.1) four variations of the performance of family firms can be conceptualized based on whether a positive performance is experienced on one or both dimensions (cf. Davidsson, 2003; Sorenson, 1999). Although good performance on the family dimension indicates firms with high cumulative emotional capital, good business performance indicates firms with high cumulative financial capital.

Warm hearts–deep pockets Firms in Quadrant I of Figure 2.1 are the successful family firms; they experience profitable business as well as family harmony. In other words, they enjoy high cumulative stocks of both financial and emotional capital that may help sustain the family and business through turbulent economic and emotional times. Staying in this quadrant over a sustained period of time would be the most desirable performance combination for family firms. Haniel in Germany, Cranes papers, S.C. Johnson, J.M. Smucker, Cargill, and Nordstorm in the United States, Kikkoman in Japan, Beaudoin, Thomson, and Molsons in Canada, and Antinori, Ferragamo, and Torrini in Italy are examples of such firms.

Pained hearts–deep pockets Quadrant II firms are characterized by business success but also are tension prone or exhibit failed family relationships. This scenario has been observed in many large family firms, such as McCains in Canada and Pritzkers in the United States, that continue to expand globally and experience increased profits, but the family relationships have been strained by discontent and conflict (for example, Pitts, 2001). Such firms carry high stocks of financial capital but are low on family emotional capital. Relational issues have been found central to the sustainability and success of family firms as good relationships can overcome bad business decisions but the opposite is more difficult to achieve (Olson et al., 2003; Ward, 1997). This is because, unlike with unrelated parties, relationships among family members are densely linked, wherein the tremors of one bad relationship are felt throughout the tight web of other relationships (Astrachan, 2003). Emotional capital and stability provide the fuel to reap the benefits of other types of capital (Puhakka, 2002). Thus, the long-term survival of firms in this quadrant is dependent on them developing support mechanisms aimed at mending family relationships and moving toward Quadrant I.

Warm hearts–empty pockets Quadrant III firms enjoy strong relationships among family members, though their businesses are low performers. In other words, they are endowed with high levels of emotional capital but low financial capital. The Southam family of the Southam newspaper chain in Canada, Agnelli's of Italy, and Ford Motor Company in the past few years exemplify such firms. The strength of the glue among family relationships can aid these firms to endure poor business performance for some time. However, over longer periods of time, accumulated resources are likely to deplete, causing stress in family relationships as well. Although the nature of intervention required to turn these firms toward Quadrant I is different than that required by firms in Quadrant II, a move toward Quadrant I will be needed for long-term sustainability of such firms.

Pained hearts–empty pockets Quadrant IV firms are failed firms that perform poorly on both the family and business end. Although failure on the business dimension can be used as a learning experience that may even enable these family members to launch another venture in future (Davidsson, 2003), failure on the family dimension is likely to create long-term, far-reaching tremors that may take several years to fade, if they do so at all. Although the most desirable position for these firms would be Quadrant I, they may have to follow the path through Quadrant II or III to reach that happy state.

Care must be exercised in the path followed and strategies used to move toward a more favorable quadrant such that firms avoid tripping into the next worst quadrant instead. For example, firms in Quadrant II may be enticed to pay family members with hopes of achieving family harmony and moving into Quadrant III, while their aim is ultimately to achieve a position in Quadrant I. However, over time, they may find themselves unable to pay those fees for sustaining family harmony, which in turn may land them in Quadrant IV instead of I.

The above description of possible outcomes of family business performance is a simplification as only two dimensions, each with only two extreme positions, are considered. Further refinements in conceptualization of family firm performance will be essential to ensure clarity in dependent variables used in theory development and empirical research. Comprehensive scales that measure the performance of family firms along various business and family dimensions will need to be developed and validated. Olson et al. (2003) have provided a very good start in this direction. Future efforts can modify these scales to develop the equivalent of a 'Family Business Score Card' (cf. Kaplan and Norton, 1996).

Research also needs to be directed toward understanding why family firms find themselves in a particular quadrant, the factors that influence movement from one quadrant to the next, and pathways followed to move to a quadrant with superior performance on one or both dimensions.

Some preliminary evidence of the role of family involvement on firm performance and key strategic decisions such as CEO pay have been revealed through recent research conducted on publicly listed family firms. Using accounting and market measures, Anderson and Reeb (2003) found that firm performance increases until a family owns about a third of the business, after which it tends to decrease. Gómez-Mejia et al. (2003) find family member CEOs receive significantly lower pay than nonfamily CEOs, although family leaders are better protected from systematic (industrywide) or unsystematic (unique to business) risks. This study confirms similar findings by McConaughy (2000).

Important extensions of this research stream would be to conduct research to understand the role of family involvement along other dimensions, such as power, through a combination of governance, management, and ownership, family culture and structure, and experience in terms of number of family members and generations involved on firm performance in publicly and privately held family firms (Astrachan et al., 2002b; Heck, in press).

Summation

The discussion thus far reveals an increasing interest in the field of family business studies. The intertwinement and reciprocal relationships between the family and business systems is being recognized as the key feature distinguishing this field of study from others. Efforts are underway to develop conceptual and operational definitions of family firms. Instead of one definition, a range of definitions that capture varying extent and mode of family involvement in these firms are being used. Some preliminary efforts are underway to develop general purpose classification systems that distinguish family firms from nonfamily firms and between different types of family firms. A framework to understand firm performance along business and family harmony dimensions is presented.

Levels of analysis in family business studies

In this section, the family business literature is organized according to its focus on the four levels of analyses: individual, interpersonal/group, organizational, and societal. At each level, a review of the prevailing literature is presented so as to highlight the topics that have received attention, provide an assessment of the prevailing understanding, and give suggestions for future research.

Individual level
Stakeholders have been defined as 'any group or individual who can affect or is affected by the achievement of firm's objectives' (Freeman, 1984, p. 47). Freeman (1984) identified 16 generic stakeholders[12] and distinguished between primary (those who affect a firm's objectives) and secondary (those affected by a firm's objectives) stakeholders. However, he did not include 'family members' as a distinct generic category. In an extension of this concept into the family firm context, Sharma (2001) distinguished between internal and external family firm stakeholders. Those involved with the firm either as employees[13] (receive wages), and/or owners (shareholders), and/or family members are referred to as *internal* stakeholders. On the other hand, stakeholders not linked to a firm either through employment, ownership, or family membership, but that have the capacity to influence the long-term survival and prosperity of a firm, are referred to as *external* stakeholders. At the individual level of analysis, family business studies have devoted varying attention to four categories of internal stakeholders: founders, next-generation members, women, and nonfamily employees. Research related to each is discussed below.

Founders Due to their anchoring role in a firm, organizational leaders have been recognized as having a significant influence on culture, values, and performance of their firms (Collins and Porras, 1994; Schein, 1983). Family business literature recognizes the influential position of founders. Due to their long tenures and the centrality of their position in their family and firm, founders exert considerable influence on the culture and performance of their firms during and beyond their tenure (Andersen et al., 2003; García-Álvarez et al., 2002; Kelly et al., 2000; McConaughy, 2000). Efforts have been made to understand the leadership styles adopted by these leaders and their relationship with other family and nonfamily members (Aldrich and Cliff, 2003; Lubatkin et al., 2003; Sorenson, 2000).

As compared to nonfamily executives, tenures of family business leaders have been found to be longer. In a sample of publicly traded American firms, McConaughy (2000) found the tenure of family business leaders to be almost three times longer than that of nonfamily executives (17.6 years vs. 6.43 years). These long tenures have been attributed to these CEOs facing higher cognitive costs and psychological barriers to exit their firms (Gómez-Mejia et al., 2003, Lansberg, 1988). Although some have been reported to experience loneliness and boredom in their positions (Gumpert and Boyd, 1984; Malone and Jenster, 1992), others remain energetic and rejuvenated throughout their tenure (Keynon-Rouvinez, 2001). It can be speculated that a combination of individual traits, family structure and values, future goals for the enterprise and the envisioned role of the founder in it, and contextual factors such as the state of economy or industry growth, would influence the disposition of founders during the course of their tenure. Perhaps those involved in mentoring of future leaders or philanthropic activities may continue to feel an excitement in their work lives. It would be useful to understand the reasons for the observed differences in the energy and excitement levels of founders with respect to their jobs and firms.

Using the social network theory (Brass, 1995), Kelly et al. (2000) have developed the concept of founder centrality within a family firm and its influence both during and after the tenure of a founder. They suggest three dimensions of centrality – betweenness (central to the flow of information), closeness (direct linkages with top management group), and connectivity (ability to influence the most connected members). A variety of hypotheses are proposed, such as that high founder centrality should lead to (1) an alignment of perceptions between founder and other family and nonfamily executives, (2) better firm performance along the dimensions of success that are important to a founder, and (3) a stronger influence of the founder on the firm after his or her tenure ends.

Using 13 cases of Spanish family firms, Garciá-Álvarez et al. (2002) observe that the founders' view of the role of business in their family influences the mode and process of socialization they use for next-generation family members, thereby influencing the culture of the firm beyond their tenure. Those who regard their business as a means to support the family, value the feeling of family and limit the growth of their firm. They communicate higher values of group orientation to their successors, who were found to join the firm at a young age, lower position, and with low levels of formal education. On the other hand, founders viewing business as an end in itself encourage successors to achieve high levels of formal education and experience outside the business before joining the family firm at senior levels.

Research conducted on publicly traded firms by Anderson and Reeb (2003) and Anderson et al. (2003) reveals a positive role of founder on firm performance in terms of accounting profitability measures, market performance, and cost of debt financing for family firms. This performance compares favorably with performance of family descendants as well as outsiders as CEOs, suggesting that founders bring unique value to the firm.

Family business leaders have been observed to adopt five leadership styles: participative, autocratic, laissez-faire, expert, and referent (Sorenson, 2000). Participative leaders, who value the input from and consistently evaluate family and nonfamily employees, were found to achieve high performance both on family and business dimensions. However, this study did not reveal conclusive findings on the affects of other leadership styles on performance related to family or business dimensions, suggesting a need for further research on this topic. Research on personality traits and attitudes regarding appropriate power distance between family and nonfamily members may inform why founders adopt different leadership styles. Moreover, a clearer understanding of the long-term goals of founders in terms of performance on family and business dimensions may influence their management and leadership styles and any observed differences in these styles over the course of their long tenure. Stage of life through which an individual, family, and the business are going may further influence the observed leadership style of founders.

The relationship of founders with other family members has received some attention. Two recent conceptual efforts are noteworthy.

1. In a crisp articulation of the linkages between the fields of entrepreneurship and family business studies, Aldrich and Cliff (2003) suggest that families aid founders to recognize the opportunities around which to create a venture and lend support to ensure its birth and sustenance over time. Other research has provided empirical support for the integral role played by family in providing both the financial and nonfinancial support to founders for creating new ventures (Astrachan et al., 2003; Erikson et al., 2003).

2. Using behavioral economics and organizational justice theories, Lubatkin et al. (2003) propose that the extent of self-control exerted by founders differenti-ates 'far-sighted' founders from those suffering from 'myopic altruism.' Although far-sighted founders are able to withhold immediate gratification of each and every need of family members in favor of actions that enhance long-term value for the family and the firm, myopic altruists find it difficult to take such actions, thereby violating rules of procedural

and distributive justice, leading to their being perceived as unjust by family and nonfamily members.

As is evident from this account, at this point in time, a significant amount of research focused on family firm founders is in theory development stages. Although this is a good starting point toward developing clearer insights regarding the role of family firm founders, these conceptual ideas need to be subjected to empirical tests to gauge their validity and generalizability for practice. It would be useful to understand the role of family composition, values and beliefs, and individual personality and dispositional traits of founders on their position in the business; how different types of family involvement during founding and later life stages of a firm influence founder and firm performance along different dimensions; and factors that lead to far- versus near-sightedness of founders in terms of careful planning that leads to sustainable family firms. Although, so far, this literature has focused on individual founders or controlling owners, research needs to be directed toward understanding founding teams of the same or different genders and ethnic backgrounds, given the predicted trend toward team leadership in family firms (Astrachan et al., 2002a).

Next generation Handler (1989b) successfully directed the attention of the field toward the importance of focusing on next-generation family members and understanding their perspectives. Following her suggestions, research has focused in three general directions: desirable successor attributes from the perspective of leaders; performance enhancing factors; and reasons these family members decide to pursue a career in their family firms.

Exploratory research conducted both in Western and Eastern cultures revealed 'integrity' and 'commitment to business' as the two most desirable next-generation attributes from the viewpoint of the firm leaders (Chrisman et al., 1998; Sharma and Rao, 2000). Other attributes found important are ability to gain respect of nonfamily employees, decision-making abilities and experience, interpersonal skills, intelligence, and self-confidence. The attributes considered important by the leaders are relatively versatile in their applicability to different situations and cultures. However, it would be important to understand why attributes are rated higher or lower, if there are differences based on current and future performance objectives of a firm. Despite the central position of firm leaders, given their emotional involvement with the next generation and their bounded and parenting rationalities[14] (Ling, 2002; Simon, 1957), leaders may not be in the best position to accurately assess either the list of desirable successor attributes or the extent to which members of the next generation possess them. Future research in this area will benefit from collecting such data from multiple respondents in family firms and drawing on the psychology literature to develop mechanisms to assess amounts of each attribute possessed by the next generation of leaders. Moreover, whether possession of different attributes leads to high performance on financial and/or nonfinancial dimensions must be studied.

Due to their long tenures, family firm leaders possess a significant amount of idiosyncratic or tacit knowledge related to the firm (Lee et al., 2003). It has been suggested that the performance of the next generation is likely to be based on the effectiveness with which this knowledge, and social networks, are transferred across generations (Cabrera-Suárez et al., 2001; Steier, 2001). Research on effectiveness of knowledge transfer between a source and recipient has unequivocally revealed the importance of the absorptive capacity of the recipient and the nature of the relationship between the source and the recipient (Cohen and Levinthal, 1990; Szulanski, 1995). Defined as the ability to acquire, assimilate, transform, and exploit new knowledge, recipients' absorptive capacity has been found to be dependent on the existing stocks of knowledge and skills (Szulanski, 2000; Zahra and George, 2002).

Mirroring this research, the family business literature has revealed that the level of preparedness of the next generation and its relationship with the senior generation have a significant influence on the next generation's performance (for example, Goldberg, 1996; Morris et al., 1997). A supportive relationship characterized by mutual respect enables the smooth transition of knowledge, social capital, and networks across generations (Steier, 2001). With this research, preliminary steps have been taken to understand some of the factors that enhance firm performance and knowledge transfer from one generation to the next. In the future it would be useful to understand whether the mode of preparedness of the next generation should vary based on the goals of family firms, and the interests, attitudes, and psychological traits of involved family members. Effort should also be directed to understanding the contextual factors that impede or enhance transfer of knowledge across generations.

As compared to their peers who come from nonfamily business settings, junior-generation members of family firms were found to have lesser clarity about their abilities, talents, goals, and career interest (Eckrich and Loughead, 1996). Although this observation reveals a difference in vocational clarity of family members, it would be interesting to understand why this lack of clarity prevails, and its implication for individual disposition and firm performance. Perhaps it is a product of the socialization processes that inculcate a sense of obligation among juniors to pursue a career in their firms (García-Álvarez et al., 2002). Knowing they have no real choices to make, they subdue consideration of what their own interests might be. However, these are only some speculative explanations and need to be subjected to careful theoretical development and empirical testing.

Stavrou (1998) observed several reasons for next generation decisions to join the family firm. Why were these differences observed? Would differences in motivating factors impact the performance outcomes of these family members? Drawing on organizational commitment literature, Sharma and Irving (2002) have developed a theoretical model to understand the behavioral and performance implications of next-generation family members based on their reasons for pursuing a career in their family firms. Behavioral and performance variations are expected depending on whether juniors join their family firms because they want to, from a sense of obligation, due to involved opportunity costs, or from a sense of need. Research directed to assess the validity of this model would improve our understanding of the motivational factors directing next-generation members toward family firms. Also fruitful would be to understand the role of environmental context (family, industry, and business) on the motivations to join and performance of next-generation family members.

Women Research on this topic suggests that a majority of women in family firms continue to remain in the background, frequently occupying the role of a household manager and taking on the primary responsibility for the household and child-rearing tasks (Cole, 1997; Fitzgerald and Muske, 2002). Although they may seem to occupy a subdued role, such a positioning provides them with a unique vantage point that aids in the development of a rich understanding of the prevailing issues and relationship dynamics (Dumas, 1998; Lyman et al., 1995). They can also provide the emotional reservoir to be drawn on for efficient conduct of the business and management of relationships among family members (cf. Puhakka, 2002). If used astutely, their observations, intuition, and emotional capital can make a difference between the success and failure of a family firm, though formal research has still not reached these topics.

Based on interviews with 11 spouses of successful family firms, Poza and Messer (2001) describe six different types of roles adopted by these women: jealous spouse, chief trust officer, partner or co-preneur, vice-president, senior advisor, and free agent. In another similar attempt, Curimbaba (2002) interviewed 12 potential heiresses of Brazilian family firms to report that they occupy either a professional, invisible, or anchor role in their firms. Although these studies based on small convenience samples provide an indication of the varying types of roles that women in family firms tend to adopt, they do not explain the reasons that prompt their adoption or the implications these role adoptions have on firm performance. This leaves an opportunity to conduct theoretically oriented large-sample studies to understand the role of females in family firms.

The last few years have witnessed a number of female leaders taking over the reins of their family businesses, in some instances alongside their male relatives while in others outcompeting them. Some examples include Marcy Syms (Syms Corporation), Gina Gallo (E&J Gallo Wineries), and Abigail Johnson (Fidelity Investments) in the United States and Gail Regan (CARA operations) and Martha Billes (Canadian Tire) in Canada. Despite this trend, no systematic research has yet been directed toward understanding the contextual and individual factors that buoy these women into leadership positions, their performance goals in terms of family and business dimensions, or the leadership and managerial styles adopted by them, pointing toward an interesting and ripe area for serious study.

Nonfamily employees In terms of number of individuals involved and the impact on the success and growth of family firms, nonfamily employees are an important stakeholder group (Chrisman et al., 1998; Gallo, 1995; Ibrahim et al., 2001). Moreover, these individuals may possess idiosyncratic knowledge of the firm that may be prove valuable in mentoring of future-generation leaders, or filling in the leadership role should a need arise (Lee et al., 2003). In larger firms, nonfamily executives have been found to play a critical role in strategic decision making (Chua et al., 2003). However, it is only recently that some efforts are being directed to understand the complexity of their role and their perceptions.

Using transaction costs and social cognition theories, Mitchell et al. (2003) have theoretically demonstrated that in comparison to employees in nonfamily settings, family business employees need to manage dramatically complex cognitions even for performing simple transactions. This conceptualization provides a theoretical explanation of why some individuals may prefer not to work in family firms.

Lubatkin et al. (2003) use behavioral economics and distributive justice theories to suggest that nonfamily employees' perceptions of fairness, in terms of resource allocation exhibited by controlling owners, will be dependent on the extent of self-control exhibited by these individuals. If they are perceived to make decisions that gratify immediate needs of family members as opposed to promoting long-term value for the family firm, they will be perceived as unjust. Such perceptions are likely to lead to dissatisfaction of nonfamily employees and reduce the likelihood of high performance or long tenures of these employees.

We have hardly scratched the surface of understanding this stakeholder group. The theoretical models proposed need empirical verification. Clearly, there is a need to devote more attention to understanding the perspective of nonfamily employees, issues that are important to them, and that would lead to superior performance of these individuals along various dimensions.

Interpersonal/group level
Significant research attention has been devoted to this level of analysis. Three topics related to interpersonal or group levels that have been investigated are nature and types of contractual agreements, sources of conflict and management strategies, and intergenerational transitions.

Nature and type of contractual agreements In the field of family business studies, interest in this topic was kindled when two sets of scholars – Gómez-Mejia et al. (2002) and Schulze et al. (2001) – began to question the applicability of the central tenets of agency theory in the context of family firms. As their works received acceptance in mainstream journals of organizational studies, they attracted immediate and widespread attention in the field, leading to a number of subsequent conceptual developments and empirical studies (for example, Burkart et al., 2003; Chrisman et al., 2002a; Gómez-Mejia et al., 2003; Greenwood, 2003; Ling, 2002; Lubatkin et al., 2003; Schulze et al., 2003a, 2003b; Steier, 2003).

Built on the central tenets proposed by Adam Smith (1796), Berle and Means (1932), and Max Weber (1947) agency theory was conceived and popularized in organizational studies by Jensen and Meckling (1976) and Ross (1973).[15] It is based on the idea that the separation of ownership and management in firms leads to a principal–agent relationship in which the managers (agents) may not make decisions that are in the best interest of owners (principals). Thus, suggestions were made to develop mechanisms to align these interests (Jensen and Meckling, 1976). It was expected that an alignment of ownership and management within a family would alleviate the agency problems in family firms because individual family members would engage in altruistic behaviors wherein they subjugate their self-interests for the collective good of the family.

Two different perspectives exist on the reasoning that motivates family members to engage in other-regarding behavior as opposed to self-regarding acts. The economist perspective is that 'altruism is self-reinforcing and motivated by self-interest because it allows the individual to simultaneously satisfy altruistic (other-regarding) preferences and egotistic (self-regarding) preferences' (Schulze et al., 2001, p. 102). From this perspective, family members are viewed as utility maximizers who are rooted in economic rationality. Research related to majority and minority family member shareholders points toward examples of family members who may be motivated partially or exclusively by their self-interests rather than other-regarding family-oriented behavior (Morck and Yeung, 2003; Morck et al., 1988).

However, an alternate viewpoint, rooted in theological perspective, is offered by stewardship theory (Davis et al., 1997; Greenwood, 2003). Following the views of McGregor (1960), Maslow (1970), and Argyris (1973), this theory uses a humanistic and self-actualizing model of humankind, wherein an individual views himself or herself as a 'steward whose behavior is ordered such that pro-organizational collectivistic behaviors have higher utility than individualistic, self-serving behaviors' (Davis et al., 1997, p. 24). From this viewpoint, the other-regarding or selfless behavior exhibited by controlling family business owners is motivated by their collectivistic rationality that there is greater utility in cooperative behavior (cf. Hofstede, 1980, 2001). Regardless of the reasoning underlying the other-regarding behavior in family firms, these views led to a belief that there was no need for a formal governance mechanism in such instances of aligned management and ownership, as it would be an unnecessary expense that would deter firm's financial performance.

Scholars working in the context of family firms argue that while the agency costs caused by the separation of ownership and management may be reduced to some extent in family firms, other types of problems arise, revealing darker implications of altruism (Gómez-Mejia et al., 2002; Schulze et al., 2001). When dealing with members of one's own family, problems of 'myopic altruism' may arise, wherein controlling owners may experience a lack of self-control due to which they have difficulty restraining their impulse to gratify every need and wish of their family (Lubatkin et al., 2003). In other instances, being boundedly rational, these owners may not be award of the behaviors that would lead to highest expected outcomes for their children (cf. Simon, 1957). Ling (2002) argues that even when parents attempt to engage in self-control, their fundamental ideological beliefs and values will constrain and determine the governance choices made by them (cf. Todd, 1985). All these underlying causes may lead to adverse selection or entrenchment in family firms, leading to placing family members in positions for which they are not best qualified (Burkart et al., 2003). Moreover, family members may engage in shirking or free-riding behaviors to the detriment of firm performance (Gómez-Mejia et al., 2002; Schulze et al., 2001).

Steier (2003) has argued that variants of agency contracts among family members occur within a continuum of positivealtruistic and economically-oriented rationalities among family members. As both positive and negative aspects of altruism have received some empirical support, it is being suggested that family firm leaders engage in self-control and adopt governance mechanisms that would aid in curbing the negative tendencies of altruism even when owners and managers belong to the same family (Gómez-Mejia et al., 2002).

This research stream has been successful in elaborating the boundary conditions of agency theory and extending its theoretical range (Greenwood, 2003). Although the major focus of this research has been on relationships between family members who are owners and family employees, Chrisman et al. (2002) observe that it has opened rich avenues for scholarly examination of agency relationships among various internal and external stakeholder groups of family firms. Some of these relationships have received previous scholarly attention. For example, Morck et al. (1988) and Morck and Yeung (2003) have documented the potential agency costs to minority shareholders in firms that have an entrenched dominant shareholding, while Myers (1977) and Smith and Warner (1979) focus on agency costs in owner-lender relationships. The literature on venture capital financing can be informative in further developing an understanding of this latter relationship.

Future research on this topic would benefit from taking into consideration the conceptualization of success prevailing in a family firm. For example, if maintenance of performance on family dimensions is important, it may be in the long-term strategic interest to keep members of one generation (even if they are underperformers) engaged in a business, with the hope of improving the training and fostering the interest of future generations in the business. In other words, some generations may primarily act as bridge – or connector – generations, maintaining family harmony and financial performance at par or subpar levels until more competent or prepared family members become available. Clearly, the time horizons under consideration in this instance are significantly longer than in cases discussed currently in agency theory. Perhaps systematic study of dynastic family firms may be informative in understanding how these firms have sustained through multiple generations with varying levels of alignment of skills, abilities, and interests of family members of different generations with tasks undertaken by the firm. As the field engages in this inquiry, it would be fruitful to examine the differences between explicit and psychological contracts among the different stakeholders (Argyris, 1960; Kotter, 1973; Levinson, 1962), and take into consideration the role of the family's culture, beliefs, and value systems on the nature and effectiveness of contracts among different stakeholder groups in family firms.

Sources of conflict and management strategies An embeddedness of the family and business systems, which in their original forms are based on fundamental sociological differences, makes family firms a ripe context for misunderstandings and conflict (Boles, 1996; Miller and Rice, 1988; Swartz, 1989). Conflict has been described as 'awareness on the part of the parties involved of discrepancies, incompatible wishes, or irreconcilable desires' (Jehn and Mannix, 2001, p. 238).

Based on work-groups conflict literature (for example, Jehn, 1995,1997), three types of conflicts have been conceptualized: task (disagreement on what tasks should be accomplished), process (disagreement on how to accomplish the tasks), and relationship (based on interpersonal incompatibilities about values, attitudes, and so on). Cross-sectional studies in this literature (for example, Jehn, 1995; Shah and Jehn, 1993) have revealed that relationship conflict is detrimental to individual and group performance, reducing the likelihood that members of a group will work together in the future. A moderate level of task conflict has been found to increase group performance in cognitively complex tasks as it allows groups to benefit from different opinions and avoid group thinking (Janis, 1982). Process conflict has been associated with lower levels of productivity and group morale (Jehn, 1997). Most of these studies, however, have been cross-sectional in nature, focusing on static levels of conflict and ignoring temporal issues. Even in a case where an attempt was made to understand patterns of conflict over time (for example, Jehn and Mannix, 2001), the study was conducted on graduate students who, at best, have to work together on projects for the relatively limited duration of their program of study.

The family business literature is just beginning to develop conceptual models to understand the nature, causes, and implications of different types of conflict. Scholars recognize the positive and negative aspects of conflict, comparing it to 'social friction' (Astrachan and Keyt, 2003; Mitchell et al., 2003). Cosier and Harvey (1998) have proposed that process and task conflicts can be beneficial because they promote creativity and innovation. Preliminary evidence of cross-generational innovation would support this notion (Litz and Kleysen, 2001). Building on this idea, Kellermanns and Eddleston (2002) suggest that task and process conflicts interact with relationship conflict to influence firm performance. These researchers also theorize that the relationship between conflict and performance is moderated by the ownership structure of the firm.

Another stream of literature has attempted to understand how conflicts may be resolved and the impact of adopted resolution strategies on financial and nonfinancial dimensions of firm performance. Sorenson (1999) examined the five conflict management strategies of competition, collaboration, compromise, accommodation, and avoidance used by family firms. Although collaboration strategies lead to positive outcomes on both family and business dimensions, the avoidance and competition strategies performed poorly on both dimensions. Compromise and accommodation were better for the family-related outcomes but not for the business-related ones. Astrachan and McMillan (2003) and Habbershon and Astrachan (1996) have suggested that systems for regular collective encounter among family business stakeholders aid in the development of shared cognitive maps and beliefs. In turn, these shared perceptions enable prediction and pro-active management of conflict, thus increasing the effectiveness of intervention strategies, should these be used.

Due to the relative stability of membership over time and multiplicity in variety of interactions that take place among family members, family firms offer a natural setting to understand the root causes and temporal dimensions of conflict (Astrachan and McMillan, 2003; Grote, 2003). Moreover, effective resolution of conflict is likely to influence firm performance in terms of financial and nonfinancial dimensions. Thus, this is an extremely important area for future research effort as we need to understand the root causes of each type of conflict, whether there is any linkage between the overall goals of a firm and the frequency of the nature of conflict experienced, and resolution strategies that are more helpful in different types of conflict situations. It is possible to conceive, for example, that relationship conflict may be caused by allocation rules of distributive justice (equity, equality, or need based[16]) that prevail in a family (Lubatkin et al., 2003); or differences in fundamental norms guiding a family's values about the nature of relationship among siblings (for example, whether one sibling is regarded above others in terms of inheritance of parental property or all are considered equal) (for example, Todd, 1985); or disagreements in terms of choices made along other dimensions of life such as mate selection (for example, Kaye, 1999).

Intergenerational transition Since the inception of this field of study, significant research efforts have been devoted to the topic of succession (Handler, 1994). The interest continues (for example, Burkart et al., 2003; Le Breton-Miller et al., in press; Lee et al., 2003; Sharma et al., 2003a). Earlier reviews revealed the importance of this topic and described efforts devoted to describing the phenomenon of succession process and observed best practices (Bird et al., 2002; Sharma et al., 1996; Wortman, 1994).

A majority of family firm leaders have been found to be desirous of retaining family control past their tenure (for example, Astrachan et al., 2002a). Although the initial reactions to such preference were relegated to the propensity of family firms toward nepotism, recent conceptual thinking suggests such preference to be a rational and efficient choice when: (1) the prevailing legal system accords low shareholder protection, such that separation of ownership and control becomes inefficient, (2) the family gains significant nonpecuniary and reputational benefits from retaining the leadership within the family, and (3) the competitive advantages of a firm are based in idiosyncratic knowledge that can only be transferred efficiently to family members or the most-trusted outsiders (Burkart et al., 2003; Lee et al., 2003).

Both incumbents and successors play critical roles in this process, although they attribute more importance to the others' role (Sharma et al., 2003a). Significant differences in perceptions about the process have been identified repeatedly (Handler, 1989b; Poza et al., 1997; Sharma, 1997), pointing toward the importance of engaging in processes that lead to development of collective beliefs (Habbershon and Astrachan, 1996). Using the theory of planned behavior from the social psychology literature, Sharma et al.'s (2003b) study of 118 family firm leaders revealed that the presence of a trusted successor willing to take over the leadership of a firm was the spark that controls the succession planning process. This suggests a need to engage the next-generation family members in succession planning, as it is their careers and lives that are involved in this decision. The pursuit of understanding the extent of interest of next-generation family members in their firms and the best mode for getting these individuals involved in the firm must continue (Fiegener et al., 1996).

Attempts have been made to reveal various dimensions and phases of the succession process, differentiating between successful and unsuccessful successions, and identifying the factors that contribute to effective successions (Cadieux et al., 2002; Davis and Harveston, 1998; Gersick et al., 1997; Harveston et al., 1997; Morris et al., 1997; Murray, 2003; Poza et al., 1997; Sharma et al., 2001; Sharma et al., 2003a). Most of these studies subject theoretically developed models to empirical tests, thereby improving our understanding of the succession process (Rogoff and Heck, 2003). This process has been revealed to be a multistaged phenomenon with trigger events or markers distinguishing one stage from the other (Cadieux et al., 2002; Gersick et al., 1997; Keating and Little, 1997; Lansberg, 1999; Murray, 2003). For example, using an analogy of a relay race, Dyck et al. (2002) suggest the importance of sequence (appropriateness of successors' skills and experiences), timing, technique (details by which succession will be achieved), and communication between the predecessor and successor. It is generally agreed that this process extends over time and needs to be carefully planned (Davis and Harveston, 1998; Harveston et al., 1997; Sharma et al., 2003a). Recently, Le Breton-Miller et al. (in press) developed an excellent integrative model for successful successions that describes the succession process while taking into account the contextual variables within the family, industry, and society. Although research needs to be directed toward subjecting this model to empirical testing, this effort is successful in providing a comprehensive conceptual framework to understand the succession process in family firms.

Parallel efforts have been directed toward understanding the reasons that successions fail when failure is defined as successor dismissal or firm bankruptcy (Dyck et al., 2002; Miller et al., 2003). Based on their study of 16 failed successions, D. Miller et al. (2003) note that at the heart of failed successions is the misalignment between an organizational past and future. Three observed patterns of this alignment are conservative (attachment to the past), rebellious (wholesale rejection of the past), and wavering (incongruous blending of the past and present). Each pattern leads to different performance implications. These studies present conceptual models that are simply waiting for large-scale empirical testing. In related efforts, it would be useful to carefully consider the effect of performance objectives, family values and beliefs, and other contextual variables that might influence the effectiveness of a succession process.

In the past few years, questions have been asked about whether continuity of a family business is always a good thing (Drozdow, 1998; Kaye, 1996). Although experience and intuition point toward a negative answer to this question, systematic conceptual development of this issue has not yet been undertaken. Some have made suggestions for adopting broader definitions of 'success' of succession (Kaye, 1996) and differentiating between elements of a family business that should and should not be transferred across generations. Research based on a resource-based view of the firm suggests the importance of transferring the tacit embedded knowledge (Cabrera-Suárez et al., 2001), networks and social capital (Steier, 2001), passion (Andersson et al., 2002), and innovative spirit (Litz and Kleysen, 2001) across generations, as such transfers would lead to competitive advantages for family firms. Future research needs to be directed toward understanding effective ways of transferring these resources across generations, as well as exploring the extent of importance of their transfer in different types of family firms located in varied cultures (Dyer, 1988). Literatures on diffusion of innovation and knowledge transfer could be informative in this regard (for example, Rogers, 1962; Szulanski, 1995, 2000).

Organizational level

At the organizational level of analysis, efforts have been largely directed toward the identification and management of resources in family firms. Resource-based theory of the firm has been used to inform the research directed toward identification and management of the unique resources in family firms. Habbershon and Williams (1999) suggest that it is the 'familiness' or the idiosyncratic internal resources built into a firm as a result of the involvement of family that makes family firms distinctive. Further, they argue that 'familiness' can used either as a source of strategic competence (distinctive) or encumbrance (constrictive) by family firms.

In an article that conceptually consolidates different types of capital, Sirmon and Hitt (2003) distinguish between five types of capital resources and the characteristics that distinguish family from nonfamily firms. These sources of capital are human, social, survivability, patient, and governance structures. They suggest that to gain competitive advantage, family firms need to evaluate, acquire, shed, bundle, and leverage their resources efficiently. The interaction between family and business in these firms provides some advantages and challenges to pursuing these activities.

To truly understand the strategic decision processes of family firms, it is important to incorporate the role of family beliefs and culture. In one related attempt, Sharma and Manikutty (2003) have presented a conceptual model for understanding the interactive role of prevailing community culture and family beliefs on resource-shedding decisions in family firms. They hypothesize varying levels of inertia to divest unproductive business units by family firms depending on the values held by the owning family and the culture that prevails in the community where the family business is located. However, empirical testing of the conceptual models developed in this research stream is necessary.

Research directed toward understanding the sources of financial capital used by family firms has consistently revealed a 'pecking order' with highest preference given to internal financing, followed by debt and equity financing (Coleman and Carsky, 1999; Erikson et al., 2003; Morck and Yeung, 2003; Poutziouris, 2002; Romano et al., 2000). External financing is generally avoided because it is a source of accountability. However, research focused on understanding the cost of debt financing in publicly traded family versus nonfamily firms has revealed a lower cost of debt financing in family firms (Anderson et al., 2003). The rationale offered for this finding is that the bondholders perceive lower conflict of interest with family firms due to their long-term orientation and undiversified portfolios. This perception, in turn, leads to a reduction in the cost of debt financing (Anderson et al., 2003).

Clearly, more attention needs to be directed toward the firm level. For example, there is a need to understand the mechanisms family firms use to develop, communicate, and reinforce desired vision and organizational culture over extended tenures of leaders and across generations; strategies used to maintain long-term relationships with external stakeholders and other organizations; ethical dilemmas faced and resolution strategies used; and human resource strategies used, especially as these firms provide limited leadership opportunities for nonfamily executives.

Societal/environmental level

A majority of the research efforts directed toward understanding the role of family firms at the societal level have focused on establishing the extent of economic importance of these firms in various nations such as Germany (Klein, 2000), the Gulf region (Davis et al., 2000), Italy (Corbetta, 1995), Spain (Gallo, 1995), Sweden (Morck and Yeung, 2003), and the United States (Astrachan and Shanker, 2003; Heck and Stafford, 2001). As a consistently high influence of family firms has been found in most nations where such studies have been undertaken, perhaps it is time to get to the question of why these firms endure, try to understand the impact of fiscal systems on the formations that persist in different environments, and take a look at the role of these firms in their communities. Theories such as institutional theory and population ecology might be used in such endeavors.

Summation

Overall, the majority of research on family firms in the past decade or so has been directed toward the individual or group levels, with only scant recent interest in the organizational level. Topics such as organizational vision and culture development, marketing strategies used, human resource practices, interorganizational relationships, and so forth remain unstudied. Further, the impact of family firms at the societal level has largely been ignored, except for the documentation of a large number of these firms in different nations.

At the individual level, founders and next-generation members have received the most attention, with only some attention shown women and nonfamily employees. The long terms and significant influence of founders on their firms during and after their tenures is well established. However, the reciprocal impact of family on founders and the firms is only just beginning to gain attention. Although different leadership styles have been observed, there is still lack of clarity on styles that may be more effective given different organizational goals and personality traits of founders or leadership teams. The focus has largely remained on individual founders; issues related to team founding and leadership await attention. From the perspective of leaders, committed next-generation family members with high integrity are desirable successors, even though such individuals might remain unclear about their abilities, skills, or career interests. Women are found to play multiple roles. Nonfamily employees face a complex environment in family firms. Our understanding of either of these stakeholders is preliminary at this stage, showing a need for more systematic research attention in future.

At the interpersonal level, agency theory has dominated the research related to the nature of contractual agreements between family owners and family employees. These efforts have revealed that an alignment of ownership and management within a family may not reduce the overall agency costs because, although some costs are alleviated, new types of problems arise. The dark side of altruism has been revealed, displaying human limitations in terms of accurate understanding of how actions taken today might influence the future of a firm, or the impact of one's control impulses in decisions related to family members. Research on the nature of conflict and resolution strategies has highlighted different types of conflicts and varying degrees of effectiveness of resolution mechanisms, although this stream is still in its infancy. In terms of the leadership transition process, it is now clear that this process is a long one and marked with trigger events. Both the departing and incoming leaders play a critical role, although their perceptions on key dimensions may vary significantly. Comprehensive conceptual models of the succession process have been developed, and are awaiting empirical testing.

Moving forward: strategies for knowledge creation and dissemination

The ultimate aim of the field of family business studies is to improve the functioning of family firms. This aim can be achieved by gaining deeper understanding of the forces that underlie these firms. Creation and dissemination of usable knowledge is a painstaking effort that requires strategic thinking. Not only must we efficiently use our collective intellectual resources, we must continuously attract and retain good thinkers who will devote their energies toward gaining insights into the world of family firms. In this section, I present some thoughts related to gaining efficiencies in the tasks of knowledge creation and dissemination. Strategies that can be adopted at the individual and community level to expedite our collective understanding of family firms are discussed.

Knowledge creation

Three aspects related to creating knowledge about family firms merit some consideration: What topics deserve attention? How can we effectively organize our ideas around questions of interest? How can we design effective scientific investigations?

Choosing research topics and questions Asking the right questions is the first critical step in finding the right answers. A major difficulty in a new field of study is to determine the projects that must be undertaken and intelligently formulate the research questions. Review papers and directions for future research listed in research articles provide some suggestions to individual scientists. However, the level of thoughtfulness and sophistication with which projects are chosen can be greatly enhanced by a mutually interactive process among scientists, and between scientists and practitioners, in which the system achieves a rationality superior to that of any individual in it.

Some efforts along this dimension are underway. For example, the International Family Enterprise Research Academy's (IFERA) annual researchers meeting and the scholars program at the Family Owned Business Institute (FOBI) at Seidman School of Business, Grand Valley State University have initiated efforts to aid critical evaluation of research proposals related to family business studies. Such efforts can avoid uncoordinated efforts of isolated individuals and help make good choices for research projects. Moreover, such meetings enable development of coordinated efforts among scholars, as exemplified by the F-PEC study that involves nine scholars in four nations. Another good example of research collaboration is the 1997 National Family Business Survey that involved 25 scholars from 17 institutions across the United States and Canada (Winter et al., 1998), with follow-up US reinterviews in 2000 (Winter et al., in press). Further progress can be made by involving family firm practitioners and the scientific community more closely in relation to issues faced by both communities.

Development and organization of conceptual thoughts After developing well-thought-out research questions, the next step in scientific investigation involves development and organization of conceptual thoughts. Theory is an efficient mechanism to build conceptual frameworks that stimulate understanding and provide a strong foundation for conducting systematic research (Sutton and Staw, 1995). In explaining a phenomenon, a good theory identifies the variables that are important and why, specifies how and why the variables are interrelated, and identifies conditions under which they should or should not be related (Campbell, 1990). A high priority needs to be placed on continuously building and improving our theoretical and conceptual knowledge base as over time such efforts aid in generating usable knowledge (Lindblom and Cohen, 1979).

In the academic realm, theoretical knowledge is the distinctive intellectual capital that provides legitimacy to a field of study (Elsback et al., 1999). It assists in developing a language to express the reality around us, build on each others' work, and attract and retain the attention of scholars from other disciplines to contribute to the field (McKinley et al., 1999).

Similar to this pursuit in most other social sciences, the ultimate aim of the field of family business studies is to develop 'theory/ies of family firms' that take into account the reciprocal relationships between family and business systems. A starting point for achieving this ultimate objective is to reexamine the current theories in the family and organizational fields to test the extent of their validity when these two systems are intertwined (Figure 2.2). Such filtering process will ensure that the theories developed are valuable and robust so that they will apply to a vast majority of organizations in the world (Chrisman et al., 2003c).

Research directed toward an examination of agency theory in the context of family firms (for example, Gómez-Mejia et al., 2002, 2003; Schulze et al., 2001, 2003a, 2003b) is an excellent example of how a reexamination of an accepted theory in the domain of family firms has revealed the limited scope of the original theory, suggested extensions that aid in its elaboration and refinement, and at the same time aided the field of family business studies to gain deeper insights and rapid legitimacy in the broader academic arena.

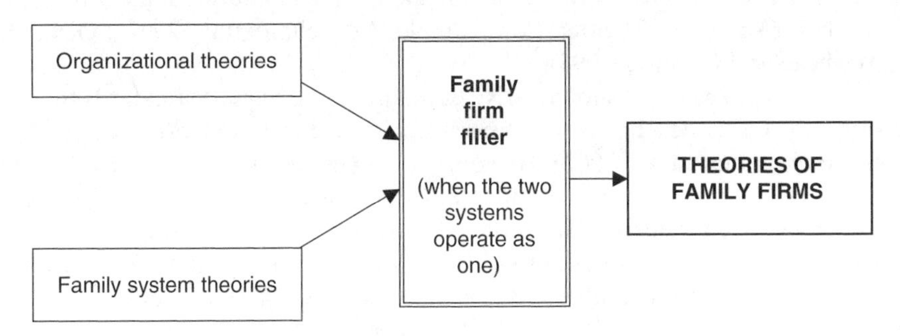

Figure 2.2 Toward the development of theories of family firms

As a community of scholars, we must direct efforts toward training researchers interested in the field to learn the craft of theory building and writing for scholarly publishing. Research conferences such as the 'Theories of Family Enterprise Conferences' organized by the Universities of Alberta, Calgary, and Wharton, are an effective way to promote and support scholarship in the area. Such efforts should be continued and similar initiatives undertaken to promote interaction between senior and newer scholars. At an individual level, we need to make continuous efforts to educate ourselves on the nuances of theory building, and share our research findings in varied venues.

Designing effective studies Research designs must be chosen based on the research question of interest and the prevailing level of understanding on the issue. In the 'full cycle' of research, there is an important role of both qualitative/process and quantitative/variance approaches to develop rich generalizable theories (Cialdini, 1980; Mohr, 1982). Although the process theories use events and states to tell a story about how outcomes are achieved, the variance theories use independent variables as necessary and sufficient causes of variation in dependent variables (Elsbach et al., 1999). In an emerging field of study as standards are being laid down, it is critical to understand and adopt widely accepted tenets and guiding principles of the chosen approach (for example, Langley, 1999; Whetten, 2002). Research on succession in family firms has made good progress through its pursuit of rigor in both qualitative (for example, Handler, 1989a) and quantitative (for example, Sharma, 1997) methods.

As significant differences in perceptions of leaders and other family members have been revealed (for example, Poza et al., 1997; Sharma, 1997), it is necessary for researchers to use multirespondent data-collection methods to capture different prevailing perspectives. Topics of interest in family firm studies such as firm performance along financial and non-financial dimensions across generations, sources of conflict, efficacy of resolution strategies, and succession process extend over long periods of time, and suggest a need for longitudinal studies or cross-sectional studies repeated over time.

As we work toward building cumulative knowledge on family business studies, it is extremely important to share in detail the methods used, definitions of variables of interest and their operationalization, and research instruments (Handler, 1989a). The trend in the field is in the right direction as good descriptions of methods are beginning to emerge (for example, Cole, 1997; Danes et al., 2002; Keating and Little, 1997; Mustakallio et al., 2003; Poza and Messer, 2001; Smyrnios et al., 2003), and definitions and research instruments are being shared more often (for example, Astrachan et al., 2002a; Olson et al., 2003; Westhead and Cowling, 1998).

In our role as reviewers, we should strive to maintain the highest standards by encouraging research that is developed on a theoretically strong foundation and is methodologically sound (Bird et al., 2002). Although our judgments should be based on high quality, our reviews should be developmental so that we can support each other toward the conduct of good research. In short, the field can benefit from using rigorous designs both in qualitative and quantitative research methods, longitudinal or repeated cross-sectional studies using multiple respondents, and adoption of a culture of sharing and mutual support.

Dissemination of knowledge

Efficient dissemination of acquired knowledge is at least as important as its acquisition. Scholars of an emerging field of study need to find effective ways to share their ideas both within the academic community and among practitioners. Below, strategies that can enable effective dissemination in both these communities are presented for consideration.

Dissemination within the academic community It is largely through academic journals and conferences that research is communicated within the academic community. Although we are fortunate in this field to have this journal [*Family Business Review*] devoted exclusively to family business studies, in order to attract more scholars to the field and generate widespread interest and credibility, it is important to continue our efforts to disseminate our research in a variety of journals and conferences, and invite scholars from other fields of study to conferences devoted to family business studies in an effort to increase awareness.

At the community level, efforts can be made to organize meetings where successful senior scholars are placed in mentoring roles for junior family business researchers so as to generate interest in a wider community of academics and aid the new scholars to learn the craft of publishing (Whetten, 2002).

Dissemination among practitioners For improving the functioning of family firms, it is critical that created knowledge be effectively disseminated into the communities of practice. However, transfer of knowledge is a challenging task as significant losses are experienced in the transmission process, leading to a time lag between knowledge creation and its use (Lindblom and Cohen, 1979). Adding to the challenge is a general human resistance to change and the adoption of new ideas (Lewin, 1943).

Research has revealed that knowledge transfer can be expedited by involving the recipients in the generation of research questions and studies, ensuring communication mechanism are adapted to the absorptive capacity of users, and developing an intimate trusting relationship with the users (Szulanski, 1995).

At individual levels, researchers must continue to make efforts to communicate research findings in a manner conducive to practitioners. Usage of diagrams and models has been suggested to help in providing structure to thoughts and communication of complex ideas (Whetten, 2002). Enlightened practitioners must make efforts to keep abreast of new research findings. Collaborative efforts between scholars and academics can be hugely beneficial in generating and disseminating usable knowledge. Associations such as the Family Firm Institute can play a significant role in aiding the knowledge transfer process by providing opportunities for increased interaction between scholars and practitioners, and making efforts to develop objective standards of achievement, without which no learning can take place.

Summation

Overall, this review has revealed a positive trend in the field toward more sophisticated research that is based on rich theory-based conceptualizations of various phenomenon of interest. Although such efforts should continue in future, it is equally important to subject the theoretical models developed to carefully designed empirical (qualitative or quantitative) studies. Only through continuous theory development and testing can we find ourselves closer to the creation of usable knowledge.

In conclusion, the state of the field of family business studies can be described using Jim Collins's analogy (2001) of a huge heavy metal flywheel mounted horizontally on an axle. The aim of interested scholars is to turn this wheel of understanding family firms fast and long. At first, through persistent efforts of early researchers in the field such as Beckhard, Danco, Dyer, Hollander, Lansberg, Levinson, and Ward, it inched slowly and imperceptibly. As more individuals joined the field, the wheel gained momentum. At this point in time, the wheel seems to be turning slowly using its own weight. Before it gains unstoppable momentum, it is worthwhile to take stock of its current direction and take care that future efforts are made in a desirable manner so as to ensure that the wheel of understanding family firms moves expeditiously. This is my attempt at stock taking for the field of family business studies.

Acknowledgements
I am grateful for the detailed suggestions and comments provided by Joe Astrachan and Ramona Heck that helped improve this paper significantly. Thanks also to Jim Chrisman and Jess Chua for their encouragement and support.

Notes

1. An ABI inform search indicates the number of articles on 'family business' in peer-reviewed scholarly journals has shown a dramatic increase: 33 articles up to 1989; 110 from 1990 to 1999 (an average of 11 articles per year); and 195 articles in the four-year period from 2000 to 2003 (almost 49 articles per year indicating over fourfold increase!).
2. Family business research has begun to emerge in mainstream journals such as *Academy of Management Journal*, *Academy of Management Review*, *Journal of Finance*, and *Organizational Science* (for example, Anderson and Reeb, 2003; Anderson et al., 2003; Burkart et al., 2003; Gómez-Mejia et al., 2002, 2003; Lee et al., 2003; Schulze et al., 2001, 2003a). Special issues of family business research of some of the top-ranked entrepreneurship journals such as *Entrepreneurship Theory and Practice* (**27**(4)), and *Journal of Business Venturing* (**18**(4) and **18**(5)) have been published in the year 2003.
3. The AACSB (Association to Advance Collegiate Schools of Business) website http://www.aacsb.edu/members/communities/interestgrps/familybusndoc.asp lists nearly 50 accredited schools with family business programs. Aronoff and Ward (1995) report that more than 70 universities, including leading schools such as Harvard, North-Western, Notre Dame, the University of California in Los Angeles (UCLA), Wharton, INSEAD, and IMD, have active family business programs.
4. Some notable examples include the Coleman, Cox, Kauffman, Mass Mutual, and Raymond Foundations in the United States; Lombard Odier Hentsch and Cie in Europe; and the Tanenbaum foundation in Canada.
5. Examples include the Family Firm Institute (FFI), which was founded in 1986, had about 500 members in 1992, and now has nearly 1200 members; Canadian Association of Family Enterprises (CAFÉ), which was established in 1983 with 15 founding members and now boasts more than 2400 members representing almost 900 family firms across Canada.
6. In addition to these broad-based literature reviews, Bird et al. (2002); Chrisman et al. (2003b); Dyer and Sanchez (1998), Handler (1994), and Wortman (1994) have presented more focused reviews of a selection of the family business literature.
7. Although not included in this article, some insightful experiential and prescriptive articles include Hubler (1999); Kaye (1999); Krasnow (2002); McCann (2003); Mendoza and Krone (1997); and Murphy and Murphy (2001).
8. Evidence of the prevalence of family firms has been provided by Klein (2000) in Germany; Morck and Yeung (2003) in Sweden; and Astrachan and Shanker (2003) and Heck and Stafford (2001) in the United States.
9. For example, a struggle for resolving the issue of definitions continues in the literatures of entrepreneurship (Shane and Venkataraman, 2000), corporate entrepreneurship (for example, Sharma and Chrisman, 1999), and leadership (Yukl, 1989).
10. Two other examples of multiple operational definitions include:
    1. The 1997 and 2002 National Family Business Surveys that use four different definitions of family firms based on the level of involvement of family in the business. These studies indicate a significant influence of family firms in the United States with over 8.6 million families (one out of every 10 households)

in the United States owning family firms. Collectively, these businesses generated between $1.3–10.4 trillion in gross revenues in 1996, depending on the definition used (Heck and Stafford, 2001; Heck and Trent, 1999).

2. Westhead and Cowling (1998) present seven operational definitions to classify firms in the United Kingdom according to varying levels of family ownership, managerial involvement, and CEO perception of the firm being a family business or not, and conclude that these firms are a numerically important group of businesses in the United Kingdom.

11. For example, Aldrich and Cliff (2003) explore linkages of family business studies with entrepreneurship; Stewart (2003) does it with anthropology; Sirmon and Hitt (2003) and Zahra (2003) link it to strategic management.
12. Owners, employees, unions, customers, consumer advocates, competitors, suppliers, media, environmentalists, governments, local community organizations, political groups, financial community, trade associations, activist groups, and special interest groups (Freeman, 1984, pp. 25, 55).
13. The term 'employees' is used broadly and includes all levels of the employed workforce in a firm.
14. Bounded rationality refers to the limited ability of human beings to process information and understand the environment around them (Simon, 1957). In the context of family firms, it is the limited understanding of parents as to the attributes of the next generation that would lead to highest expected returns (referred to as parenting rationality by Ling (2002)).
15. For a more detailed account of development of this stream of research, see Chrisman et al. (2003b).
16. In equity, allocations are commensurate with contributions; in equality, equal allocations are made regardless of contributions; and allocations are based on need to survive with dignity in need-based systems (Cohen, 1987; Gilliland, 1993; Lubatkin et al., 2003).

References

Aldrich, H.E. and Cliff, J.E. (2003), 'The pervasive effects of family on entrepreneurship: toward a family embeddedness perspective', *Journal of Business Venturing*, **18**(5), 573–96.

Anderson, R. and Reeb, D. (2003), 'Founding family ownership and firm performance: evidence from the S&P 500', *Journal of Finance*, **58**, 1301–28.

Anderson, R., Mansi, S.A. and Reeb, D. (2003), 'Founding family ownership and the agency cost of debt', *Journal of Financial Economics*, **68**, 263–85.

Andersson, T., Carlsen, J. and Getz, D. (2002), 'Family business goals in tourism and hospitality sector: case studies and cross-case analysis from Australia, Canada, and Sweden', *Family Business Review*, **15**(2), 89–106.

Argyris, C. (1960), *Understanding Organizational Behavior*, Homewood, IL: Dorsey Press.

Argyris, C. (1973), 'Some limits of rational man organizational theory', *Public Administration Review*, **33**, 253–67.

Aronoff, C.E. and Ward, J.L. (1995), 'Family-owned businesses: a thing of the past or the model for the future', *Family Business Review*, **8**(2), 121–30.

Astrachan J.H. (2003), 'Commentary on the special issue: the emergence of a field', *Journal of Business Venturing*, **18**(5), 567–72.

Astrachan, J.H. and Keyt, A.D. (2003), 'Commentary on: the transacting cognitions of non-family employees in the family business setting', *Journal of Business Venturing*, **18**(4), 553–8.

Astrachan, J.H. and McMillan, K.S. (2003), *Conflict and Communication in the Family Business*, Marietta, GA: Family Enterprise.

Astrachan, J.H. and Shanker, M.C. (2003), 'Family businesses' contribution to the U.S. economy: a closer look', *Family Business Review*, **16**(3), 211–19.

Astrachan, J.H., Allen, I.E. and Spinelli, S. (2002a), *Mass Mutual/Raymond Institute American Family Business Survey*, Springfield, MA: Mass Mutual Financial Group.

Astrachan, J.H., Klein, S.B. and Smyrnios, K.X. (2002b), 'The F-PEC scale of family influence: a proposal for solving the family business definition problem', *Family Business Review*, **15**(1), 45–58.

Astrachan, J.H., Zahra, S.A. and Sharma, P. (2003), 'Family-Sponsored Ventures', presented at the First Annual Global Entrepreneurship Symposium, New York. Available at <http://www.emkf.org/pdf/UN_ family_sponsored_report.pdf>.

Berle, A. and Means, G. (1932), *The Modern Corporation and Private Property*, New York: Macmillan.

Bird, B., Welsch, H., Astrachan, J.H. and Pistrui, D. (2002), 'Family business research: the evolution of an academic field', *Family Business Review*, **15**(4), 337–50.

Boles, J.S. (1996), 'Influences of work–family conflict on job satisfaction, life satisfaction, and quitting intentions among business owners: the case of family operated businesses', *Family Business Review*, **9**(1), 61–74.

Brass, D.J. (1995), 'A social network perspective on human resources management', in *Research in Personnel and Human Resources Management* (vol. 13, pp. 39–79), Greenwich, CT: JAI Press.

Burkart, M., Panunzi, F. and Shleifer, A. (2003), 'Family firms', *Journal of Finance*, **58**(5), 2167–201.

Cabrera-Suárez, K., De Saa-Pérez, P. and García-Almeida, D. (2001), 'The succession process from a resource- and knowledge-based view of the family firm', *Family Business Review*, **14**(1), 37–46.
Cadieux, L., Lorrain, J. and Hugron, P. (2002), 'Succession in women-owned family businesses: a case study', *Family Business Review*, **15**(1), 17–30.
Campbell, J.P. (1990), 'The role of theory in industrial and organizational psychology', in M.D. Dunnette and L.M. Hough (eds), *Handbook of Industrial and Organizational Psychology* (pp. 39–73), Palo Alto, CA: Consulting Psychologists Press.
Chrisman, J.J., Chua, J.H. and Litz, R.A. (2002), 'Do family firms have higher agency costs than non-family firms?', presented at the second annual Theories of Family Enterprises Conference, University of Pennsylvania.
Chrisman, J.J., Chua, J.H. and Litz, R.A. (2003a), 'Commentary: a unified perspective of family firm performance: an extension and integration', *Journal of Business Venturing*, **18**(4), 467–72.
Chrisman, J.J., Chua, J.H. and Sharma, P. (1998), 'Important attributes of successors in family businesses: an exploratory study', *Family Business Review*, **11**(1), 19–34.
Chrisman, J.J., Chua, J.H. and Sharma, P. (2003b), 'Current trends and future directions in family business management studies: toward a theory of the family firm', written as part of Coleman Foundation White Paper Series. Available at <http://www.usasbe.org/knowledge/whitepapers/index.asp>.
Chrisman, J.J., Chua, J.H. and Steier, L. (2003c), 'Editorial: an introduction to theories of family business', *Journal of Business Venturing*, **18**(4), 441–8.
Chrisman, J.J., Hofer, C.W. and Boulton, W.R. (1988), 'Toward a system for classifying business strategies', *Academy of Management Review*, **13**(3), 413–28.
Chua, J.H., Chrisman, J.J. and Sharma, P. (1999), 'Defining the family business by behavior', *Entrepreneurship Theory and Practice*, **23**(4), 19–39.
Chua, J.H., Chrisman, J.J. and Sharma, P. (2003), 'Succession and non-succession concerns of family firms and agency relationships with nonfamily managers', *Family Business Review*, **16**(2), 89–107.
Cialdini, R.B. (1980), 'Full-cycle social psychology', in L. Bickman (ed.), *Applied Social Psychology Annual* (vol. 1, pp. 21–47), Beverly Hills, CA: Sage.
Cohen, R.L. (1987), 'Distributive justice: theory and research', *Social Justice Research*, **1**, 19–40.
Cohen, W.M. and Levinthal, D. (1990), 'Absorptive capacity: a new perspective on learning and innovation', *Administrative Science Quarterly*, **35**(1), 128–52.
Cole, P.M. (1997), 'Women in family business', *Family Business Review*, **10**(4), 353–71.
Coleman, S. and Carsky, M. (1999), 'Sources of capital for small family-owned businesses: evidence from the national survey of small business finances', *Family Business Review*, **12**(1), 73–85.
Collins, Jim (2001), *Good to Great: Why Some Companies Make the Leap and Others Don't*, New York: Harper Business Press.
Collins, J.C. and Porras, J.I. (1994), *Built to Last: Successful Habits of Visionary Companies*, New York: HarperCollins.
Corbetta, G. (1995), 'Patterns of development of family businesses in Italy', *Family Business Review*, **8**(4), 255–65.
Cosier, R.A. and Harvey, M. (1998), 'The hidden strengths in family business: functional conflict', *Family Business Review*, **11**(1), 75–9.
Curimbaba, F. (2002), 'The dynamics of women's roles as family business managers', *Family Business Review*, **15**(3), 239–52.
Danes, S.M., Reuter, M.A., Kwan, H. and Doherty, W. (2002), 'Family FIRO model: an application to family business', *Family Business Review*, **15**(1), 31–43.
Davidsson, P. (2003), 'The domain of entrepreneurship research: some suggestions', in J. Katz and S. Shepherd (eds), *Advances in Entrepreneurship, Firm Emergence and Growth* (vol. 6, pp. 315–72), Oxford: Elsevier/JAI Press.
Davis, J.A. and Taguiri, R. (1989), 'The influence of life-stage on father–son work relationships in family companies', *Family Business Review*, **2**(1), 47–74.
Davis, J.A., Pitts, E.L. and Cormier, K. (2000), 'Challenges facing family companies in the Gulf region', *Family Business Review*, **13**(3), 217–37.
Davis, J.H., Schoorman, F.D. and Donaldson, L. (1997), 'Toward a stewardship theory of management', *Academy of Management Review*, **22**(1), 20–47.
Davis, P. and Harveston, P.D. (1998), 'The influence of family on the family business succession: a multi-generational perspective', *Entrepreneurship Theory and Practice*, **22**(3), 31–53.
Drozdow, N. (1998), 'What is continuity?', *Family Business Review*, **11**(4), 337–47.
Dumas, C. (1998), 'Women's pathways to participation and leadership in family-owned firms', *Family Business Review*, **11**(3), 219–28.
Dyck, B., Mauws, M., Starke, F.A. and Miske, G.A. (2002), 'Passing the baton: the importance of sequence, timing, technique, and communication in executive succession', *Journal of Business Venturing*, **17**, 143–62.
Dyer Jr, W.G. (1988), 'Culture and continuity in family firms', *Family Business Review*, **1**(1), 37–50.

Dyer Jr, W.G. (2003), 'The family: the missing variable in organizational research', *Entrepreneurship Theory and Practice*, **27**(4), 401–16.
Dyer Jr, W.G. and Sanchez, M. (1998), 'Current state of family business theory and practice as reflected in *Family Business Review* 1988–1997', *Family Business Review*, **11**(4), 287–95.
Eckrich, C.J. and Loughead, T.A. (1996), 'Effects of family-business management and psychological separation on the career development of late adolescents', *Family Business Review*, **IX**(4), 369–86.
Elsbach, K.D., Sutton, R.I. and Whetten, D.A. (1999), 'Perspectives on developing management theory, circa 1999: moving from shrill monologues to (relatively) tame dialogues', *Academy of Management Review*, **24**, 627–33.
Erikson, T., Sørheim, R. and Reitan, B. (2003), 'Family angels vs. other informal investors', *Family Business Review*, **16**(3), 163–71.
Fiegener, M.K., Brown, B.M., Prince, R.A. and File, K.M. (1996), 'Passing on strategic vision: favored modes of successor preparation by CEOs of family and non-family firms', *Journal of Small Business Management*, **34**(3), 15–26.
Fitzgerald, M.A. and Muske, G. (2002), 'Copreneurs: an exploration and comparison to other family businesses', *Family Business Review*, **15**(1), 1–16.
Foster, J.D. and Fleenor, P. (1996), 'The estate tax drag on family businesses', *Family Business Review*, **9**(2), 233–52.
Freeman, E. (1984), *Strategic Management: A Stakeholder Approach*, Boston, MA: Pitman.
Gallo, M.A. (1995), 'Family businesses in Spain: tracks followed and outcomes reached by those among the largest thousand', *Family Business Review*, **8**(4), 245–54.
García-Álvarez, E., López-Sintas, J. and Saldaña-Gonzalvo, P. (2002), 'Socialization patterns of successors in first- and second-generation family businesses', *Family Business Review*, **15**(3), 189–203.
Gersick, K.E., Davis, J.A., Hampton, M.M. and Lansberg, I. (1997), *Generation to Generation: Life Cycles of the Family Business*, Cambridge, MA: Harvard Business School Press.
Gilliland, S.W. (1993), 'The perceived fairness of selection systems: an organizational justice perspective', *Academy of Management Review*, **18**, 694–734.
Goldberg, S.D. (1996), 'Research note: effective successors in family-owned businesses: significant elements', *Family Business Review*, **9**(2), 185–97.
Gómez-Mejia, L.R., Larraza-Kintana, M. and Makri, M. (2003), 'The determinants of executive compensation in family-controlled publicly traded corporations', *Academy of Management Journal*, **44**(2), 226–37.
Gómez-Mejia, L.R., Núñez-Nickel, M. and Gutiérrez, I. (2002), 'The role of family ties in agency contracts', *Academy of Management Journal*, **44**(1), 81–95.
Greenwood, R. (2003), 'Commentary on: toward a theory of agency and altruism in family firms', *Journal of Business Venturing*, **18**(4), 491–4.
Grote, J. (2003), 'Conflicting generations: a new theory of family business rivalry', *Family Business Review*, **16**(2), 113–24.
Gudmundson, D., Hartman, E.A. and Tower, C.B. (1999), 'Strategic orientation: differences between family and non-family firms', *Family Business Review*, **12**(1), 27–39.
Gumpert, D.E. and Boyd, D.P. (1984), 'The loneliness of the small business owner', *Harvard Business Review*, **62**(6), 18–24.
Habbershon, T.G. and Astrachan, J.H. (1996), 'Research note: perceptions are reality: how family meetings lead to collective action', *Family Business Review*, **10**(1), 37–52.
Habbershon, T.G. and Williams, M.L. (1999), 'A resource-based framework for assessing the strategic advantages of family firms', *Family Business Review*, **12**, 1–25.
Habbershon, T.G., Williams, M.L. and MacMillan, I. (2003), 'A unified systems perspective of family firm performance', *Journal of Business Venturing*, **18**(4), 451–65.
Handler, W.C. (1989a), 'Methodological issues and considerations in studying family businesses', *Family Business Review*, **2**(3), 257–76.
Handler, W.C. (1989b), 'Managing the family firm succession process: the next generation family members' experience', doctoral dissertation, School of Management, Boston University.
Handler, W.C. (1994), 'Succession in family businesses: a review of the research', *Family Business Review*, **7**(2), 133–57.
Harveston, P.D., Davis, P.S. and Lynden, J.A. (1997), 'Succession planning in family business: the impact of owner gender', *Family Business Review*, **10**(4), 373–96.
Heck, R.K.Z. (in press), 'Commentary on: entrepreneurship in family vs. non-family firms: a resource-based analysis of the effect of organizational cultures', *Entrepreneurship Theory and Practice*.
Heck, R.K.Z. and Stafford, K. (2001), 'The vital institution of family business: economic benefits hidden in plain sight', in G.K. McCann and N. Upton (eds), *Destroying Myths and Creating Value in Family Business* (pp. 9–17), Deland, FL: Stetson University.
Heck, R.K.Z. and Trent, E. (1999), 'The prevalence of family business from a household sample', *Family Business Review*, **12**(3), 209–24.
Hofstede, G. (1980, 2001), *Culture's Consequences*, Beverly Hills, CA: Sage.

Hoy, E (2003), 'Commentary: legitimizing family business scholarship in organizational research and education', *Entrepreneurship Theory and Practice*, **27**(4), 417–22.
Hubler, T. (1999), 'Ten most prevalent obstacles to family-business succession planning', *Family Business Review*, **12**(2), 117–21.
Ibrahim, A.B., Soufani, K. and Lam, J. (2001), 'A study of succession in a family firm', *Family Business Review*, **14**(3), 245–58.
Janis, I.L. (1982), *Victims of Groupthink*, 2nd edn, Boston, MA: Houghton-Mifflin.
Jehn, K.A. (1995), 'A multimethod examination of the benefits and detriments of intragroup conflict', *Administrative Science Quarterly*, **40**, 256–82.
Jehn, K.A. (1997), 'A qualitative analysis of conflict types and dimensions in organizational groups', *Administrative Science Quarterly*, **42**, 530–57.
Jehn, K.A. and Mannix, E.A. (2001), 'The dynamic nature of conflict: a longitudinal study of intragroup conflict and group performance', *Academy of Management Journal*, **44**(2), 238–51.
Jensen, M.C. and Meckling, W.H. (1976), 'Theory of the firm: managerial behavior, agency costs, and ownership structure', *Journal of Financial Economics*, **3**, 305–60.
Kaplan, R.S. and Norton, D.P. (1996), *The Balanced Scorecard: Translating Strategy into Action*, Boston, MA: Harvard Business School Press.
Kaye, K. (1996), 'When the family business is a sickness', *Family Business Review*, **9**(4), 347–68.
Kaye, K. (1999), 'Mate selection and family business success', *Family Business Review*, **11**(2), 107–15.
Keating, N.C. and Little, H.M. (1997), 'Choosing the successor in New Zealand family farms', *Family Business Review*, **10**(2), 157–71.
Kellermanns, F.Z. and Eddleston, K.A. (2002), 'Feuding families: when conflict does a family firm good', paper presented at the Academy of Management meetings, Denver, CO.
Kelly, L.M., Athanassiou, N. and Crittenden, W.F. (2000), 'Founder centrality and strategic behavior in family-owned firm', *Entrepreneurship Theory and Practice*, Winter, 27–42.
Keynon-Rouvinez, D. (2001), 'Patterns in serial business families: theory building through global case studies', *Family Business Review*, **14**(3), 175–87.
Klein, S.B. (2000), 'Family businesses in Germany: significance and structure', *Family Business Review*, **13**(3), 157–81.
Kotter, J.P. (1973), 'The psychological contract: managing the joining-up process', *California Management Review*, **15**(3), 91–8.
Krasnow, H.C. (2002), 'What to do when talking fails: strategies for minority owners to turn stock certificates into money', *Family Business Review*, **15**(4), 259–68.
Langley, A. (1999), 'Strategies for theorizing from process data', *Academy of Management Review*, **24**(4), 691–710.
Lansberg, I. (1988), 'The succession conspiracy', *Family Business Review*, **1**(2), 119–43.
Lansberg, I. (1999), *Succeeding Generations: Realizing the Dream of Families in Business*, Boston, MA: Harvard Business School Press.
Le Breton-Miller, I., Miller, D. and Steier, L. (in press), 'Towards an integrative model of effective FOB succession', *Entrepreneurship Theory and Practice*.
Lee, M. and Rogoff, E.G. (1996), 'Comparison of small businesses with family participation versus small businesses without family participation: an investigation of differences in goals, attitudes, and family/business conflict', *Family Business Review*, **9**(4), 423–37.
Lee, D.S., Lim, G.H. and Lim, W.S. (2003), 'Family business succession: appropriation risk and choice of successor', *Academy of Management Review*, **28**(4), 657–66.
Levinson, H. (1962), *Men, Management, and Mental Health*, Cambridge, MA: Harvard University Press.
Lewin, K. (1940), 'Formalization and progress in psychology', in G.W. Lewin (ed.), *Resolving Social Conflicts: Selected Papers on Group Dynamics by Kurt Lewin*, New York: Harper and Brothers.
Lewin, K. (1943), 'Defining the "field at a given time" ', in D. Cartwright (ed.), *Field Theory in Social Science: Selected Theoretical Papers by Kurt Lewin*, New York: Harper and Brothers.
Lewin, K. (1945), 'The research center for group dynamics at Massachusetts Institute of Technology', *Sociometry*, **8**, 126–35.
Lindblom, C.E. and Cohen, D.K. (1979), *Usable Knowledge: Social Science and Social Problem Solving*, New Haven, CT and London: Yale University Press.
Ling, Y. (2002), 'Parenting rationality and the diversity in family firm governance', paper presented at the Academy of Management meetings, Denver, CO.
Littunen, H. (2003), 'Management capabilities and environmental characteristics in the critical operational phase of entrepreneurship – a comparison of Finnish family and nonfamily firms', *Family Business Review*, **16**(3), 183–97.
Litz, R.A. (1995), 'The family business: toward definitional clarity', in *Proceedings of the Academy of Management* (pp. 100–104), Briarcliff Manor, NY: Academy of Management.

Litz, R.A. and Kleysen, R.F. (2001), 'Your old men shall dream dreams, your young men shall see visions: toward a theory of family firm innovation with help from the Brubeck family', *Family Business Review*, **14**(4), 335–51.
Low, M.B. and MacMillan, I.C. (1988), 'Entrepreneurship: past research and future challenges', *Journal of Management*, **14**(2), 139–61.
Lubatkin, M.H., Ling, Y. and Schulze, W.S. (2003), 'Explaining agency problems in family firms using behavioral economics and justice theories', paper presented at the Academy of Management meetings, Seattle, WA.
Lyman, A., Salangicoff, M. and Hollander, B. (1995), 'Women in family business: an untapped resource', *SAM Advanced Management Journal*, Winter, 46–9.
Malone, S.C. and Jenster, P.V. (1992), 'The problem of plateaued owner manager', *Family Business Review*, **5**(1), 21–41.
Maslow, A.H. (1970), *Motivation and Personality*, New York: Harper and Row.
McCann, G. (2003), 'Where do we go from here? Strategic answers for university-based family business programs', *Family Business Review*, **16**(2), 125–44.
McConaughy, D.L. (2000), 'Family CEO vs. nonfamily CEOs in the family controlled firm: an examination of the level and sensitivity of pay to performance', *Family Business Review*, **13**(2), 121–31.
McGregor, D. (1960), *The Human Side of the Enterprise*, New York: McGraw-Hill.
McKelvey, B. (1975), 'Guidelines of the empirical classification of organizations', *Administrative Science Quarterly*, **20**, 509–25.
McKelvey, B. (1982), *Organizational Systematics: Taxonomy, Evolution, Classification*, Berkeley, CA: University of California Press.
McKinley, W., Mone, M.A. and Moon, G. (1999), 'Determinants and development of schools in organizational theory', *Academy of Management Review*, **24**, 634–48.
Mendoza, D.S. and Krone, S.P. (1997), 'An interview with Judy G. Barber: prenuptial agreements, intimacy, trust, and control', *Family Business Review*, **10**(2), 173–84.
Miller, D. (1996), 'Configurations revisited', *Strategic Management Journal*, **17**, 505–12.
Miller, D., Steier, L. and Le Breton-Miller, I. (2003), 'Lost in time: intergenerational succession, change and failure in family business', *Journal of Business Venturing*, **18**(4), 513–51.
Miller, E.J. and Rice, A.K. (1988), 'The family business in contemporary society', *Family Business Review*, **1**(2), 193–210.
Mitchell, R.K., Morse, E.A. and Sharma, P. (2003), 'The transacting cognitions of non-family employees in the family businesses setting', *Journal of Business Venturing*, **18**(4), 533–51.
Mohr, L.B. (1982), *Explaining Organizational Behavior*, San Francisco, CA: Jossey-Bass.
Moore Jr, B. (1962), *Political Power and Social Theory*, New York: Harper and Brothers.
Morck, R. and Yeung, B. (2003), 'Agency problems in large family business groups', *Entrepreneurship Theory and Practice*, **27**(4), 367–82.
Morck, R., Shleifer, A. and Vishny, R. (1988), 'Management ownership and market valuation: an empirical analysis', *Journal of Financial Economics*, **20**, 293–316.
Morris, M.H., Williams, R.O., Allen, J.A. and Avila, R.A. (1997), 'Correlates of success in family business transitions', *Journal of Business Venturing*, **12**, 385–401.
Murphy, D.L. and Murphy, J.E. (2001), 'Protecting the limited liability feature of your family business: evidence from the US court system', *Family Business Review*, **14**(4), 325–34.
Murray, B. (2003), 'The succession transition process: a longitudinal perspective', *Family Business Review*, **16**(1), 17–33.
Mustakallio, M., Autio, E. and Zahra, S.A. (2003), 'Relational and contractual governance in family firms: effects on strategic decisions making', *Family Business Review*, **15**(3), 205–22.
Myers, S. (1977), 'The determinants of borrowing', *Journal of Financial Economics*, **5**, 147–75.
Olson, P.D., Zuiker, V.S., Danes, S.M., Stafford, K., Heck, R.K.Z. and Duncan, K.A. (2003), 'Impact of family and business on family business sustainability', *Journal of Business Venturing*, **18**(5), 639–66.
Pitts, G. (2001), *In the Blood: Battles to Succeed in Canada's Family Businesses*, Toronto: Doubleday Canada Trade Paperback.
Poutziouris, P.Z. (2002), 'The views of family companies on venture capital: empirical evidence from the UK small to medium-size enterprising economy', *Family Business Review*, **14**(2), 277–91.
Poza, E.J. and Messer, T. (2001), 'Spousal leadership and continuity in the family firm', *Family Business Review*, **14**(1), 25–36.
Poza, E.J., Alfred, T. and Maheshwari, A. (1997), 'Stakeholder perceptions of culture and management practices in family firms – a preliminary report', *Family Business Review*, **10**(2), 135–55.
Puhakka, V. (2002), 'Entrepreneurial business opportunity recognition: Relationships between intellectual and social capital, environmental dynamism, opportunity recognition behavior, and performance', unpublished doctoral dissertation, Universitas Wasaensis, Vaasa.
Rogers, E.M. (1962), *Diffusion of Innovation*, New York: Free Press.

Rogoff, E.G. and Heck, R.K.Z. (2003), 'Editorial: evolving research in entrepreneurship and family business: recognizing family as the oxygen that feeds the fire of entrepreneurship', *Journal of Business Venturing*, **18**(5), 559–66.

Romano, C.A., Tanewski, G.A. and Smyrnios, K.X. (2000), 'Capital structure decision making: a model for family business', *Family Business Review*, **16**, 285–310.

Ross, S. (1973), 'The economic theory of agency: the principal's problem', *American Economic Review*, **63**, 134–9.

Schein, E.H. (1983), 'The role of the founder in creating organizational culture', *Organizational Dynamics*, Summer, 14.

Schulze, W.S., Lubatkin, M.H. and Dino, R.N. (2003a), 'Exploring the agency consequences of ownership dispersion among the directors of private family firms', *Academy of Management Journal*, **46**(2), 174–94.

Schulze, W.S., Lubatkin, M.H. and Dino, R.N. (2003b), 'Toward a theory of agency and altruism in family firms', *Journal of Business Venturing*, **18**(4), 473–90.

Schulze, W.S., Lubatkin, M.H., Dino, R.N. and Buchholtz, A.K. (2001), 'Agency relationships in family firms: theory and evidence', *Organizational Science*, **12**(2), 99–116.

Shah, P. and Jehn, K. (1993), 'Do friends perform better than acquaintances? The interaction of friendship, conflict, and task', *Group Decisions and Negotiation*, **2**, 149–66.

Shane, S. and Venkataraman, S. (2000), 'The promise of entrepreneurship as a field of research', *Academy of Management Review*, **25**(1), 217–26.

Sharma, P. (1997), 'Determinants of the satisfaction of the primary stakeholders with succession process in family firms', doctoral dissertation, University of Calgary.

Sharma, P. (2001), 'Stakeholder management concepts in family firms', in *Proceedings of 12th Annual Conference of International Association of Business and Society* (pp. 254–59).

Sharma, P. (2002), 'Stakeholder mapping technique: toward the development of a family firm typology', paper presented at the Academy of Management meetings, Denver, CO.

Sharma, P. and Chrisman, J.J. (1999), 'Toward a reconciliation of the definitional issues in the field of corporate entrepreneurship', *Entrepreneurship Theory and Practice*, **23**(3), 11–27.

Sharma, P. and Irving, G. (2002), 'Four shades of family business successor commitment: Motivating factors and expected outcomes', best unpublished paper award winner at the annual conference of Family Firm Institute, Dallas, TX.

Sharma, P. and Manikutty, S. (2003), 'Shedding of unproductive resources in family firms: role of family structure and community culture', best unpublished paper award winner at the annual conference of Family Firm Institute, Toronto.

Sharma, P. and Rao, S.A. (2000), 'Successor attributes in Indian and Canadian family firms: a comparative study', *Family Business Review*, **13**(4), 313–30.

Sharma, P., Chrisman, J.J. and Chua, J.H. (1996), *A Review and Annotated Bibliography of Family Business Studies*, Norwell, MA: Kluwer Academic.

Sharma, P., Chrisman, J.J. and Chua, J.H. (1997), 'Strategic management of the family business: past research and future challenges', *Family Business Review*, **10**(1), 1–35.

Sharma, P., Chrisman, J.J. and Chua, J.H. (2003a), 'Predictors of satisfaction with the succession process in family firms', *Journal of Business Venturing*, **18**(5), 667–87.

Sharma, P., Chrisman, J.J. and Chua, J.H. (2003b), 'Succession planning as planned behavior: some empirical results', *Family Business Review*, **16**(1), 1–15.

Sharma, P., Chrisman, J.J., Pablo, A. and Chua, J.H. (2001), 'Determinants of initial satisfaction with the succession process in family firms: a conceptual model', *Entrepreneurship Theory and Practice*, **25**(3), 1–19.

Simon, H.A. (1957), *Models of Man*, New York: Wiley.

Sirmon, D.G. and Hitt, M.A. (2003), 'Managing resources: linking unique resources, management, and wealth creation in family firms', *Entrepreneurship Theory and Practice*, **27**(4), 339–58.

Smith, A. (1796), *An Inquiry into the Nature and Causes of the Wealth of Nations* (2nd US edn, vol. 2), Philadelphia, PA: Thomas Dobson.

Smith, C. and Warner, J. (1979), 'On financial contracting: an analysis of bond covenants', *Journal of Financial Economics*, **7**, 117–61.

Smyrnios, K.X., Romano, C.A., Tanewski, G.A., Karofsky, P.I., Millen, R. and Yilmaz, M.R. (2003), 'Work–family conflict: a study of American and Australian family businesses', *Family Business Review*, **16**(1), 35–51.

Sorenson, R.L. (1999), 'Conflict management strategies used in successful family businesses', *Family Business Review*, **12**(2), 133–46.

Sorenson, R.L. (2000), 'The contribution of leadership style and practices to family and business success', *Family Business Review*, **13**(3), 183–200.

Stafford, K., Duncan, K.A., Dane, S. and Winter, M. (1999), 'A research model of sustainable family businesses', *Family Business Review*, **12**(3), 197–208.

Stavrou, E.T. (1998), 'A four factor model: a guide to planning next generation involvement in the family firm', *Family Business Review*, **11**(2), 135–42.
Steier, L. (2001), 'Next-generation entrepreneurs and succession: an exploratory study of modes and means of managing social capital', *Family Business Review*, **14**(3), 259–76.
Steier, L. (2003), 'Variants of agency contracts in family financed ventures as a continuum of familial altruistic and market rationalities', *Journal of Business Venturing*, **18**(5), 597–618.
Stewart, A. (2003), 'Help one another, use one another: toward an anthropology of family business', *Entrepreneurship Theory and Practice*, **27**(4), 383–96.
Sutton, R.I. and Staw, B.M. (1995), 'What theory is *not*', *Administrative Science Quarterly*, **40**, 371–84.
Swartz, S. (1989), 'The challenges of multidisciplinary consulting to family-owned businesses', *Family Business Review*, **2**(4), 329–39.
Szulanski, G. (1995), 'Unpacking stickiness: an empirical investigation of the barriers to transfer best practice inside the firm', *Academy of Management Journal*, **38**, 437–41.
Szulanski, G. (2000), 'The process of knowledge transfer: a diachronic analysis of stickiness', *Organizational Behavior and Human Decision Processes*, **82**(1), 9–27.
Todd, E. (1985), *The Explanation of Ideology, Family Structures, and Social Systems*, Oxford: Basil Blackwell.
Tsang, E.W.K. (2002), 'Learning from overseas venturing experience: the case of Chinese family businesses', *Journal of Business Venturing*, **17**, 21–40.
Ward, J.L. (1997), 'Growing the family business: special challenges and best practices', *Family Business Review*, **10**(4), 323–37.
Weber, M. (1947), *The Theory of Social and Economic Organization*, A.M. Henderson and T. Parsons (eds), Glencoe, IL: Free Press.
Weiss, C.H. (1977), *Using Social Science Research in Public Policy Making*, Lexington, MA: D.C. Heath.
Wells, P. (1998), 'Essays in international entrepreneurial finance', doctoral dissertation, Harvard University.
Westhead, P. and Cowling, M. (1998), 'Family firm research: the need for a methodology rethink', *Entrepreneurship Theory and Practice*, Fall, 31–56.
Westhead, P., Cowling, M. and Howorth, C. (2001), 'The development of family companies: management and ownership imperatives', *Family Business Review*, **14**(4), 369–85.
Whetten, D.A. (1989), 'What constitutes a theoretical contribution?', *Academy of Management Review*, **14**, 490–95.
Whetten, D.A. (2002), 'Modelling-as-theorizing: a systematic methodology for theory development', in D. Partington (ed.), *Essential Skills for Management Research* (pp. 45–71), Thousand Oaks, CA: Sage Publications.
Winter, M., Danes, S.M., Koh, S.K., Fredericks, K. and Paul, J.J. (in press), 'Tracking family businesses and their owners over time: sample attrition, manager departure, and business demise', *Journal of Business Venturing*.
Winter, M., Fitzgerald, M.A., Heck, R.K.Z., Haynes, G.W. and Danes, S.M. (1998), 'Revisiting the study of family businesses: methodological challenges, dilemmas, and alternative approaches', *Family Business Review*, **11**(3), 239–52.
Wortman Jr, M.S. (1994), 'Theoretical foundations for family-owned businesses: a conceptual and research based paradigm', *Family Business Review*, **7**(1), 3–27.
Yukl, G. (1989), 'Managerial leadership: a review of theory and research', *Journal of Management*, **15**(2), 251–89.
Zahra, S.A. (2003), 'International expansion of US manufacturing family businesses: the effect of ownership and involvement', *Journal of Business Venturing*, **18**(4), 495–512.
Zahra, S.A. and George, G. (2002), 'Absorptive capacity: a review, reconceptualization, and extension', *Academy of Management Review*, **27**(2), 185–203.
Zahra, S.A., Hayton, J.C. and Salvato, C. (in press), 'Entrepreneurship in family vs. non-family firms: a resource based analysis of the effect of organizational culture', *Entrepreneurship Theory and Practice*.

3 Family businesses' contribution to the US economy: a closer look

Joseph H. Astrachan and Melissa Carey Shanker

Introduction

How has the available research on family businesses' economic impact changed since 1995, amid greater attention by the White House on estate tax reform, seamless access to information on the Internet, and more university-based family business centers than ever? Unfortunately, there is not much new research.

There is still little doubt that family-owned and operated businesses are large contributors to the US economy. However, just how to determine the exact extent of their impact continues to be difficult. A vast study of all of the family business literature and research since our original findings has convinced us that our definition-based formulas for estimating family businesses' overall economic impact is still the most accurate information available. With greater access to government information, we were able to refine our original data greatly and apply these new figures to our existing formulas, resulting in a new and improved framework with which to evaluate just how important family businesses are to the US economy.

Defining a family business: the ultimate challenge

Given the private nature of most family businesses, accurate information about them is not readily available. The even greater challenge in quantifying family businesses' collective impact is that there is no concise, measurable, agreed upon definition of a family business. Experts in the field use many different criteria to distinguish these businesses, such as percentage of ownership, strategic control, involvement of multiple generations, and the intention for the business to remain in the family.

All of these criteria can be important characteristics for describing a family business, depending on where the business is in its life cycle. In our research, we created a range of possible family business definitions from a broad, inclusive definition to a narrow and more exclusive one. The level of inclusiveness depends on the perceived degree of family involvement in the business.

Our broad definition, the outer circle of the 'bull's-eye,' is the most inclusive and requires only that there be some family participation in the business and that the family have control over the business' strategic direction (see Figure 3.1). This definition covers the gamut of possibilities, from a large public company that has descendants from the original founding family as stockholders or on the board to an independent building contractor whose daughter manages his books and whose grandson performs occasional manual labor for him.

Our middle definition narrows the field by requiring that the business owner intends to pass the business on to another member of his or her family and that the founder or descendant of the founder plays a role in running the business (see Figure 3.1). The latter requirement separates out those businesses where the original family may have a stake in

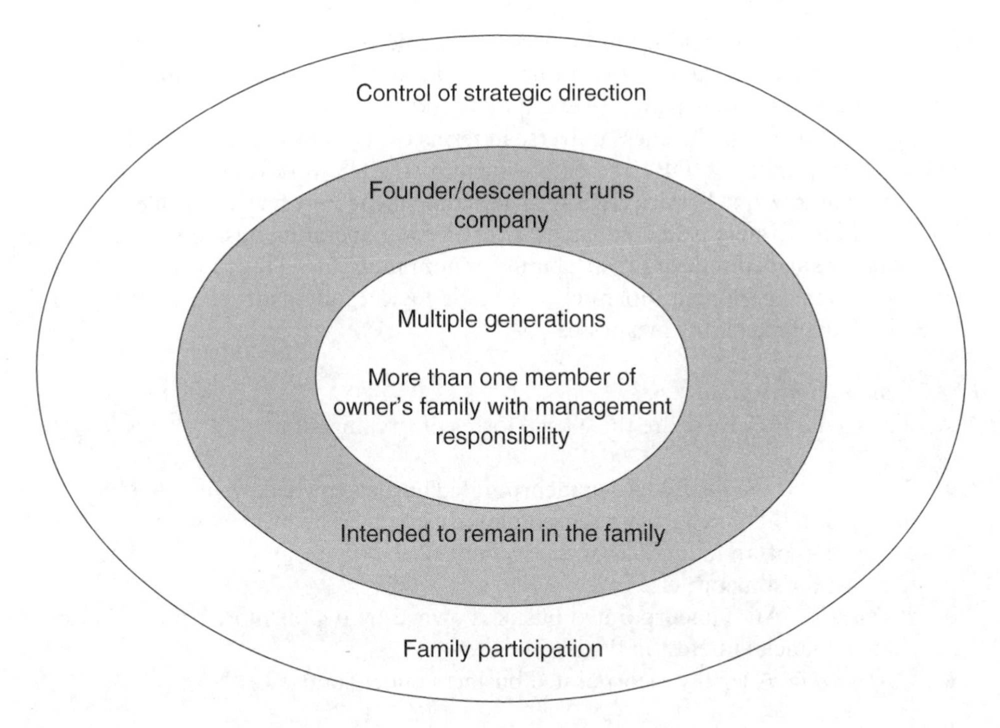

Figure 3.1 *Family business universe*

the business or a role on its board but very little interaction in day-to-day operations. The other requirement – intention – is a tricky concept to quantify, but, we believe, an important delineator. We believe that if an entrepreneur's long-term vision for his or her business is to build something for his or her children, then the planning and strategic decisions he or she makes will be different from those of a nonfamily business.

A family business in the center ring, our narrowest definition, may involve a grandparent/founder as chairman, two or three siblings in top management, one sibling with ownership but no day-to-day responsibilities, and younger cousins in entry-level positions (see Figure 3.1). In this scenario, multiple generations have a significant impact on the business. Although a common example of a family business at this stage of its life cycle, the founder no doubt had a similar profile to many of the 'entrepreneurs' included in the other circles of the bull's-eye just a generation or two earlier.

Although we agree with the importance of the intangible aspects that make family businesses unique, they make quantifying family businesses' economic impact more elusive. Unlike the impartial measurements used to identify other business types, that is, sales, number of employees, etc., the characteristics most often used to define family businesses are difficult, if not impossible, to collect. This is the primary reason why more research in this area has not been conducted.

Quantifying family businesses

Our research creates a framework for understanding the size of the family business universe based on possible criteria used to define one. A loose definition will ultimately include more businesses and result in larger economic contributions. A narrower definition results in a more homogeneous group of businesses, but less total economic impact. We have used existing information to extrapolate and make educated estimates on the size and impact of the family business universe in terms of its total size, contributions to the gross domestic product (GDP), and employment of the US workforce.

The Internal Revenue Service (IRS) provides one of the very few accessible sources of information on privately held companies. Every legally operating business in the United States, large or small, public or private, family or nonfamily, files a tax return with the IRS. Looking at each component separately, we made logical judgments about each group's propensity to include family businesses.

Legal form of organization

According to the IRS, there are three legal forms of organization:

- *Individual proprietorship*. An unincorporated business owned by an individual. Also included in this category are self-employed persons. The business may be the only occupation of an individual or the secondary activity of an individual who works full time for someone else.
- *Partnership*. An unincorporated business owned by two or more persons having a shared financial interest in the business.
- *Corporation*. A legally incorporated business under state laws.

In 2000, 17.9 million sole proprietor businesses, 2 million partnerships, 5.5 million corporations, and 1.8 million farms filed for a total of approximately 27.2 million tax returns (see Figure 3.2).

Model for broad family business definition

It can be argued that a sole proprietorship (an unincorporated business owned by a single person, with no paid employees) is a type of family business; many scholars have incorporated this idea into their *family business* definitions. In support of this theory, it is likely that a high number of family members are helping out in such enterprises.

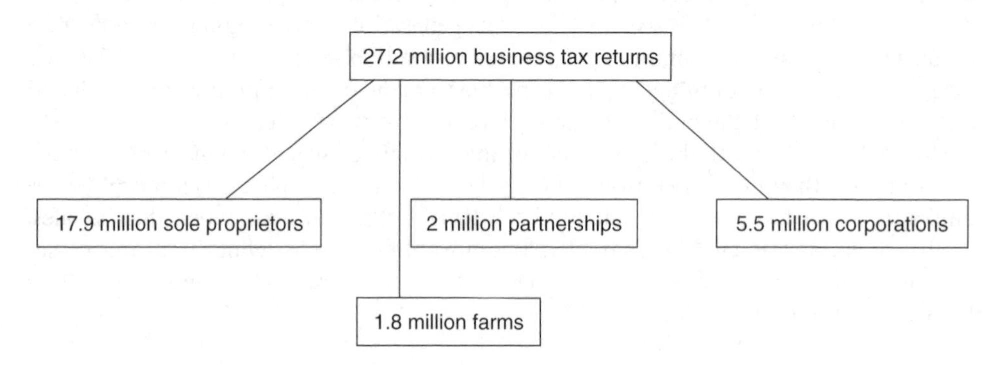

Source: US Department of Treasury, Internal Revenue Service (2000).

Figure 3.2 Breakout of total business tax returns in 2000

A study by Kirchoff and Kirchoff (1987) finds that smaller family businesses very often use both paid and nonpaid family labor, especially when starting out. (It is also likely that if family members are paid, they receive cash that is not reported.) A family farm is an example of a business in which family members work for the business but are unlikely to be listed as paid employees. In keeping with this theory, we have included all of the 17.9 million sole proprietorships and 1.8 million farms in our broadest family business universe, based on the belief that although only one family member is officially running the business, the family dynamics involved in businesses of this type qualify it as a family business in our broadest definition.

Partnerships and corporations may be somewhat less likely to exhibit this level of direct family involvement. Although the majority of partnerships and corporations are private, research done on public companies gives us a rare glimpse at empirically based data on the total number of family businesses present in public corporations. Burch (1972) finds that 47 per cent of the 1965 *Fortune* 500 were family businesses, McConaughy (1994) finds that 21 per cent of the *Business Week* 1000 list qualified, and Jetha (1993) finds that 37 per cent of the 1992 *Fortune* 500 businesses qualified as family businesses under his definition. Although the definitions and methodologies used were different, this research seems to say that one-third to one-half of the very largest public companies are family businesses.

Unfortunately, this research on public companies represents only a small sliver of US businesses. Even in terms of public companies, the majority are not traded on the big exchanges but rather via the over-the-counter market and 'pink sheets,' where smaller and closely held companies as well as high-tech start-ups go to gain access to capital. In many cases, these 'public' companies remain in the owner's or family's strategic control and are, therefore, more likely to be family businesses that fit our broad definition, despite being considered 'public' companies.

With what the research on public companies tells us, and based on what we know to be true about other large family businesses, we estimate that 60 per cent of all public and private partnerships and corporations are family businesses under our broad definition. This is a conservative estimate based on years of experience working with these types of family entities.

Therefore, the model we use to determine the total number of family businesses and their impact on the economy for our broadest definition is 100 per cent of sole proprietors and farms and 60 per cent of partnerships and corporations.

Narrowing the field

Although the IRS information is helpful, it is not necessarily an accurate picture of the total number of businesses, especially in terms of sole proprietors, because many people have full-time jobs elsewhere, operate multiple businesses, and/or file multiple tax returns. To get a better understanding of the type of sole proprietors that exist and their propensity to be family run, we referred to two sources of data from the US Census Bureau (US Department of Commerce, 1997) to help us meet the more specific criteria of our middle and narrow definitions.

Middle definition: 'intention'

The intention to pass on the business is an elusive but important distinction for a family business. The National Federation of Independent Business gives us one way to think about this. According to its research, which is based on special runs of the 1997 Census data, of the 21 million businesses listed in the 1997 Census, only 12 million represented the owner's 'principal occupation' (US Department of Commerce, 1997). Although intention is impossible to measure, we believe that if the business is not the primary focus of the family and the primary source of income, it is unlikely that the time and effort has been given to plan to keep it in the family for the future generation. With this in mind, the middle definition universe will consider only 'principal occupation' sole proprietors. We have assumed that all partnerships and corporations represent the owner's principal occupation and, thus, subtracted the total number of these tax returns from the 12 million 'principal occupation businesses.' This move left us with approximately 4.5 million 'principal occupation' sole proprietors to consider.

Thus, when we apply our family business formula to the middle universe, we use the same framework as the broad model, but now limit the number of sole proprietors from 17.9 million to the 4.5 million that declare that their sole proprietorship is their 'principal occupation.'

Our narrow definition: multiple generations and family managers

According to the US Census Bureau, nearly three-quarters of all US business firms have no payroll, or are 'nonemployers.' Using Census data to help us narrow our IRS figures made sense because the primary source for the Census' nonemployer statistics is administrative records of the IRS. Nonemployer figures consist primarily of sole proprietorship businesses filing IRS Form 1040 Schedule C, although a very small percentage of the data is derived from filers of partnership and corporation tax returns that report no paid employees. These data undergo complex processing, editing, and analytical review at the Census Bureau to distinguish nonemployers from employers.

The Census states that the majority of these nonemployer businesses are very small, typically the second or third business in a household, and many are not the primary source of income for their owners. The Census also estimates that nonemployers account for only 3 per cent of business receipts (US Department of Commerce, 1997). We believe this number is quite understated due to the fact that many transactions take place in cash in these small businesses and are typically not reported. Either way, nonemployers, that is, sole proprietors, represent the majority of businesses in the US economy, but their makeup varies widely. The US Census estimates that 16.5 million nonemployer firms existed in 2000 (see Table 3.1).

The remaining 5.6 million businesses are employers and are made up primarily of partnerships and corporations (see Table 3.1). Also included in this figure are about one million self-employed businesses owners that have paid employees.

At first glance, it may be confusing that the Census figures the total number of US firms to be 22 million vs. the IRS's 27 million (see Table 3.1). The reason for this is that the IRS

Table 3.1 Employers and nonemployers in 2000

All firms	22 182 499
Nonemployers (firms with no payroll)	16 529 955
Employers	5 652 544

is counting all business tax returns, not individual business establishments. As mentioned earlier, many sole proprietors operate multiple businesses and file multiple tax returns. In addition, whenever two different sources of information are used, differing methodologies will result in different data.

So, how does the Census' nonemployer information pertain to our family business definitions? Our narrow definition requires, among other things, that 'more than one member of the owner's family have significant management responsibility.' This implies that the firm is an employer and that family members are on the payroll.

Therefore, in this definition, we will consider only the 5.6 million employer businesses in our formulas. Although not included in the Census' employer statistics, we add farms because they are traditionally family-run operations that 'employ' many family members' efforts and meet the criteria of our narrowest definition.

In addition to requiring that family members be employed in the business, our narrow definition also states that multiple generations must be involved. Two earlier research studies help us here. John Ward's 1987 research on succession finds that approximately one-third of post-start-up family businesses survive and reach the second generation of ownership (Ward, 1987). In addition, a Mass Mutual family business study, which surveyed 1002 family businesses, supports Ward's statistic by finding that 35 per cent of the businesses it contacted had multiple generations working in the business (Arthur Andersen, 1995).

Therefore, the narrow definition requires that we eliminate a majority of the 27.2 million tax returns in 2000. We can consider only the 5.6 million employer businesses, and then, the existing research tells us that within the family business universe, only 35 per cent of those family businesses employ multiple generations of the same family. Thus, the formula for our narrowest family business definition will include only 35 per cent of the businesses included in our broad definition, or 35 per cent of sole proprietors, and 21 per cent (35 per cent of the estimated 60 per cent used in our broad definition) of partnerships and corporations.

These family business formulas for the broad, middle, and narrow definitions explained above are applied below to determine the total number of family businesses in the US economy, their contribution to GDP, and the number of workers they employ.

Total number of family businesses in the United States

The outer ring of our bull's-eye (broad) definition finds a total of 24.2 million family businesses in the United States, or 89 per cent of all 2000 business tax returns.

According to our middle definition, 10.8 million family businesses operate in the United States, representing 39 per cent of all 2000 tax returns but 89 per cent of businesses that the owner claims are his or her 'principal source of income.'

Our high-family-involvement definition finds 3 million family firms in the United States, representing 11 per cent of all 2000 tax returns but 54 per cent of all 'employer' businesses.

The results of each definition obviously show vast differences. The three rings of the bull's-eye in Figure 3.3 show how definitions can affect the size of the family business universe. We are not suggesting that the center ring is the real or best family business universe – only the most narrowly defined. This illustrates that the criteria used to define a family business play heavily on the overall perception of family businesses' contributions to our economy.

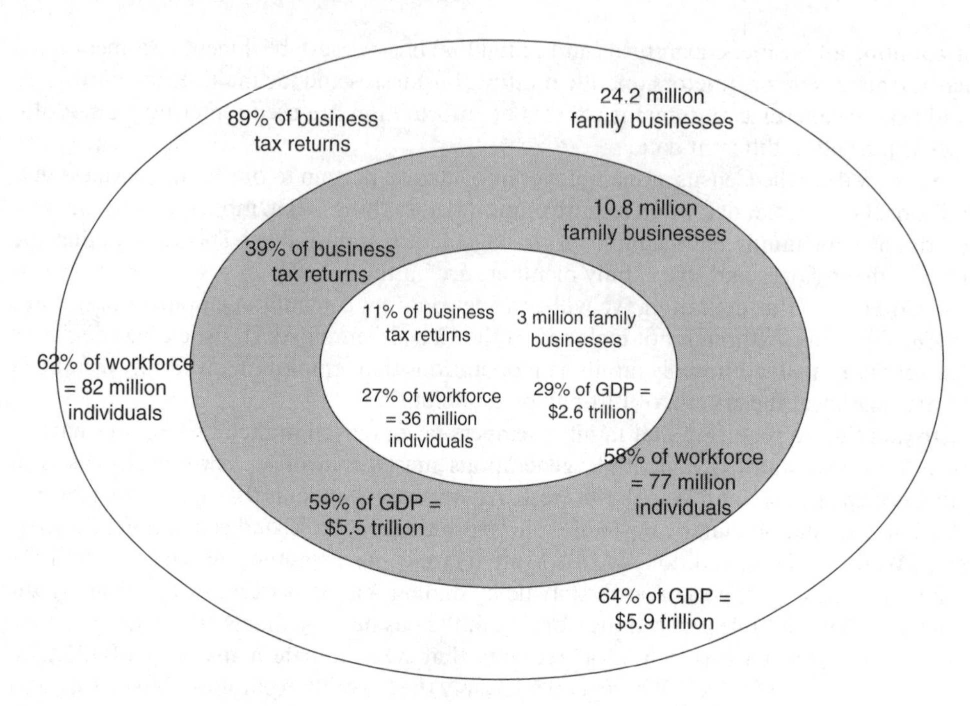

Figure 3.3 Defining family business: the family business bull's-eye

Family businesses' contribution to GDP

The US gross domestic product is the market value of the goods and services produced by labor and property located in the United States. The Bureau of Economic Analysis (US Small Business Administration, 2000a) provides us with the breakdown of real GDP by sector (see Table 3.2).

The Small Business Administration (SBA) then breaks down the business sector once more into small business (fewer than 500 employees) and big business (more than 500 employees) (US Small Business Administration, 2000b). In our research, the SBA's breakdown of the GDP into small and large businesses allows us to relate these figures back to our IRS-based formulas and make educated estimates about the number of small and large family businesses represented in this data.

Nonemployer firms are not included in the SBA's GDP calculations because of their relatively small (estimated 3 per cent of sales and receipts) contribution to the total gross product and the difficulty in collecting information about these very small businesses. Of the approximately 5.6 million employer firms, only about 17 000 – less than 1 per cent – were deemed big businesses, that is, those employing 500 or more in 2000. The remaining 5.5+ million small businesses most likely consist of all three types of IRS subgroups: sole proprietors, partnerships, or corporations. Because we are looking only at employer firms, the number of sole proprietors that contributed to the GDP will be limited, but we included farms in this calculation as well. (See the appendix for specifics.)

Table 3.2 Real gross domestic product by sector (2000)

Total	$9220 billion
Business (including farms)	$7879
Small business	$4097
Big business	$3782
Households and institutions	$389
General government	$959
(Residual)	−$6.9

Source: US Small Business Administration, Bureau of Economic Analysis (2000a).

Table 3.3 Division of total US workforce in 2000

US workforce in 2000	No. of employees (in thousands)	%
Total workforce	133 101	100
Government	20 680	15
Small business	55 729	42
Big business	54 976	41
Farm	1716	1

Source: US Department of Labor, Bureau of Labor Statistics, ERS-William Edmondson (farm) (2000).

Broad, middle, and narrow definitions

As illustrated in Figure 3.3, dividing the small and large businesses into IRS categories and applying the broad definition model established above (i.e., 100 per cent of sole proprietors and farms and 60 per cent of partnership and corporations) results in family businesses contributing 64 per cent of the GDP, or $5907 billion ($5.9 trillion). Applying the formula for the middle definition results in family businesses contributing 59 per cent of the GDP, or $5481 billion ($5.5 trillion). Applying our narrow definition to this data results in family businesses contributing 29 per cent of the US GDP, or $2566 billion ($2.6 trillion).

Family business employment

The US government employs 15 per cent of all American workers. Other public and private enterprises employ the remaining 112 million workers. Approximately 56 million work for businesses employing fewer than 500 employees (72 per cent in businesses with fewer than 20 employees), and about 55 million work for big businesses employing more than 500 workers. Table 3.3 illustrates the division.

Using the narrow and broad family business models, we can estimate the number of small and larger businesses that are family run and then extrapolate the number of workers employed by each category from the workforce information provided above.

Using the broadest *family business* definition, family businesses employ 62 per cent of the US workforce, or approximately 82 million individuals. Our narrow definition would result in 27 per cent of the workforce, or 36 million people.

Conclusion

No matter what criteria are used, family businesses represent a substantial portion of the US economy and have a massive impact on the economy as a whole. This research provides a range of estimates based on the degree of family involvement in a business. The difficulty in more accurately quantifying this impact stems from the lack of a universally agreed-upon definition for a family business and the fact that many of the criteria most important in defining a family business are difficult, if not impossible, to collect.

Clearly, more research is needed to study family businesses' importance to the US economy, particularly as our leaders argue over the right tax policy to stimulate our lagging economy.

References

Arthur Andersen Center for Family Business (1995), *American Family Business Survey*, St Charles, IL.

Burch, P. (1972), *Managerial Revolution Reassessed: Family Control in America's Largest Corporations*, Lexington, MA: Lexington Books.

Jetha, H. (1993), 'The industrial Fortune 500 study', unpublished research, Loyola University, Chicago.

Kirchoff, B.A. and Kirchoff, J.J. (1987), 'Family contributions to productivity and profitability in small business', *Journal of Small Business Management*, **25**, 25–31.

McConaughy, D. (1994), 'Founding-family-controlled corporations: an agency-theoretic analysis of corporate ownership and its impact upon performance, operating efficiency and capital structure', doctoral dissertation, University of Cinncinati.

US Department of Commerce (1997), *1997 Economic Census*, US Census Bureau, Washington, DC.

US Department of Labor, Bureau of Labor Statistics (2000, August), *Employment in Perspective. Earnings and Job Growth*, Washington, DC: Government Printing Office.

US Department of Treasury, Internal Revenue Service (2000), *Statistics of Income*, Washington, DC: US Government Printing Office.

US Small Business Administration, Bureau of Economic Analysis (2000a), *The Nation*, Washington, DC: US Government Printing Office.

US Small Business Administration (2000b), *The State of Small Business*, Washington, DC: US Government Printing Office.

Ward, J. (1987), *Keeping the Family Business Healthy: How to Plan for Continuing Growth, Profitability and Family Leadership*, San Francisco, CB: Jossey-Bass.

Appendix

Definition	Characteristics of business
Broad	• Family controls strategic direction • Family participates in business
Middle	• Founder/descendant runs business • Business is intended to remain in family
Narrow	• Multiple generations participate in business • More than one member of owner's family has management responsibility

PART II

THEORIZING FAMILY BUSINESSES AND BUSINESS FAMILIES

4 A unified systems perspective of family firm performance

Timothy G. Habbershon, Mary Williams and Ian C. MacMillan

1 Introduction

Achieving strategic competitiveness is difficult in today's turbulent and complex market place. These difficulties are compounded when firms do not have a clear understanding of what affects their performance. Recognizing the antecedents to firm performance allows leaders to exploit their organizational resources and capabilities and to make the requisite strategic choices to pursue future opportunities. The heart of the strategic management process is to achieve the performance outcomes that allow firms, including family-influenced firms, to be competitive over time.

To date, the family firm literature has generally emphasized improving family relationships without a strong strategic management focus on firm performance (Sharma et al., 1997). Anecdotal descriptions of organizational behavior are often substituted as strategy models, and attempts to define a family firm or to delineate between the performance requirements of so-called family firms and nonfamily firms have left family and business leaders confused at best (Chua et al., 1999; Gudmundson et al., 1999). More often, the response is to discount, ignore, or isolate the family factors from the business and resort to traditional strategy models for the business. The end result is that these leaders fail to account for major systemic influences that impact their performance outcomes. In short, they do not have an adequate performance model.

Theory and practice indicate that in family-influenced firms, there are complex arrays of systemic factors that impact strategy processes and firm performance outcomes. Habbershon and Williams (1999) have suggested that these unique systemic family influences can be captured through the resources and capabilities of the organization. The idiosyncratic firm level bundle of resources and capabilities resulting from the system interactions is referred to as the 'familiness' of the firm.

In this chapter, we pursue the thinking of Habbershon and Williams (1999) and more specifically develop a unified systems model of family firm performance that demonstrates how the systemic interactions of the family unit, business entity, and individual family members are linked to performance outcomes. The performance model blends systems theory thinking with strategic management theory in order to show how family influences can lead to a potential competitive advantage.

The first section presents the current thinking from the field of family business studies on the family business as a strategic entity and evaluates it from a strategic management perspective. The second section builds the performance model for family-influenced firms. It begins with a general utility function of value creation for the family business social system and moves to a more specific wealth creation function for a subset of firms we refer to as 'enterprising families.' Enterprising families are those committed to transgenerational wealth creation, which is shown to be a function of a family-based advantage (advantage$_f$). The advantage$_f$ is found when the enterprising families system generates 'distinctive familiness' (resources$_f$ and capabilities$_f$) that can be exploited for generating advantage-based rents. The chapter concludes by presenting the defining function for the enterprising families system, demonstrating that family-influenced firms hold the potential for positive and synergistic outcomes. By striving to fulfill the defining function of the system, family and business leaders gain a fuller understanding of the antecedents to firm performance and are better able to explore their advantage for transgenerational wealth creation.

2 The family business as a strategic entity

Discussions of strategy, planning, growth, or the performance of family firms frequently reference the tensions and contradictions that arise between the family system and the business system. Whether it is the global adaptation of family businesses in China (Yeung, 2000), the financial decision making patterns of family firms (Romano et al., 2000), their strategic orientation towards market opportunities (Gudmundson et al., 1999), or the formulation and implementation of strategy (Harris et al., 1994), the tensions among the needs, desires, goals, and practices of the family versus the business are discussed as strategic factors affecting firm level outcomes.

For nearly 2 decades, the two or three overlapping circles models (Figures 4.1a and b) have been the standard theoretical models for picturing family and business as interlinking systems that explain the competitive tensions in strategy making.

These models have been used to distinguish the family business system as a distinct strategic entity (Hollander and Elman, 1988; Swartz, 1989), to describe the strategically relevant attributes and constituencies in the systems (Tagiuri and Davis, 1983, reprinted as a classic, 1996), to discuss the family business' unique strategy making processes (Carlock and Ward, 2001), and to explain how each of the subsystems move through stages over time (Gersick et al., 1999). Hoy and Vesser (1994) asserted that the critical strategic management issues for family firms (founder transition, business continuation, succession, tax planning, and owner-manager life cycles) are located in the nexus of the overlapping areas of the circles.

The overlapping circles are useful organizational behavior models for describing the complex individual and organizational phenomena associated with the overlapping subsystems and for identifying the stakeholder perspectives, roles, and responsibilities. From a strategic management perspective, however, the overlapping circles models and the ensuing theory have limitations for identifying performance outcomes and explaining how these interactions materially influence firm level outcomes (Chua et al., 1999).

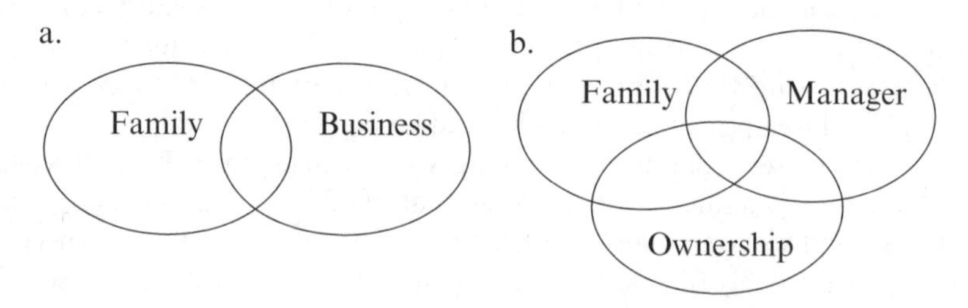

Figure 4.1 Overlapping circles models

Because the overlapping circles models descriptively picture a static degree of interaction (overlap) between the family and business, they perpetuate a trade-offs approach to strategy. The prevailing view in the overlapping circles models is that the family and business are two complex social systems that, when combined, differentiate family businesses from other organizations by the degree to which the systems boundaries overlap (Ibrahim and Ellis, 1988; McCollom, 1990; Stafford et al., 1999; Whiteside and Brown, 1991). The dominant perspective is that the business system is interpenetrated by the family system, resulting in constraints on the performance outcomes of the business (Stafford et al., 1999; Whiteside and Brown, 1991). This 'dual systems approach' (Swartz, 1989) emphasizes managing the boundaries between two qualitatively different social systems in order to develop coping strategies for addressing the inherent contradictions (Davis and Stern, 1980; Lansberg, 1983). There are those who stress the equal power and importance of the family and the business (Carlock and Ward, 2001; Hollander, 1984; Stafford et al., 1999; Ward, 1987), but strategy development is still presented as a satisficing process that balances the competing interests of the subsystems or that manages the changing needs and interests of the constituency groups represented in the overlapping circles of the system through time (Carlock and Ward, 2001; Gersick et al., 1999).

The dualistic stereotyping of the subsystem functioning – family as emotional based and business as task based – creates an exaggerated notion of overlap and subsystem boundaries (Whiteside and Brown, 1991). It establishes an a priori classification of inputs and actions that predisposes the assessment of strategic processes and outcomes. Using a dual systems approach, strategy making for family businesses focuses on a series of internal negative trade-offs to manage the overlap between family and business rather than a process for finding the systemic synergy that can lead to strategic competitiveness for the firm. We avoid the limitations of the dual systems approach by introducing a unified systems perspective of performance in the family business system.

3 A unified system performance model for family-influenced firms

A unified systems model of family firm performance focuses not only on describing stakeholder constituencies and conditions, but also shows how the parts of the system interact to generate idiosyncratic antecedents to firm performance. Using a deductive method, we begin with a general performance proposition in which the outcome of interest is maximization of the utility function of the family business social system.

3.1 The family business social system

Senge (1990) described systems thinking as 'a discipline for seeing wholes . . . interrelationships rather than things . . . patterns of change rather than static snapshots' (p. 68). Similarly, Ackoff (1994) defined a system as a whole that cannot be divided into independent parts. A social system model must, therefore, show how the systemic influences of the system are a product of the continuous interaction of the parts if the utility function of the system is said to represent the system as a whole.

The family business social system is a 'metasystem' comprised of three broad subsystem components: (1) the controlling family unit – representing the history, traditions, and life cycle of the family; (2) the business entity – representing the strategies and structures utilized to generate wealth; and (3) the individual family member – representing the interests, skills, and life stage of the participating family owners/managers (see Figure 4.2a).

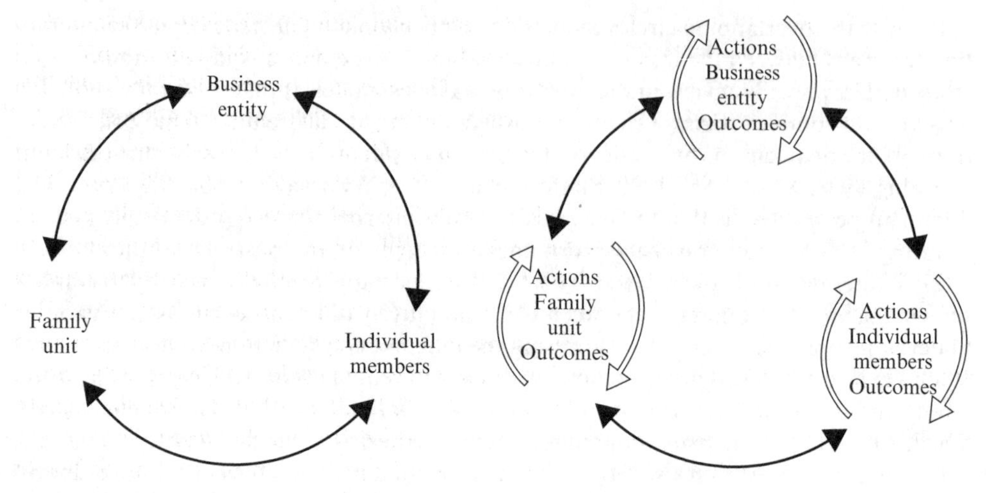

Figure 4.2 Unified systems models

While we acknowledge that on the surface the unified systems model does not seem useful in analysing specific stakeholder interactions and relationships, we deliberately keep the model broad in order to capture the 'systemic influences'. Each subsystem does, however, lend itself to a more in-depth stakeholder analysis. By defining the arguments in the utility function of the system according to the stakeholder(s) of interest, the performance analysis of the system becomes stakeholder specific.

The point of the models (a and b) is to show the circular feedback processes with continuous influence rather than picturing degrees of subsystem overlap and isolated points of influence as in the overlapping circles model described earlier. To capture systemic strategic influences, it is necessary to show how events in one of the parts of the system ultimately are both a cause and effect in the other subsystem components.

Figure 4.2b shows how the subsystems have their own action and outcome interactions that continuously feed back into the metasystem. These subsystem interaction loops represent the stakeholder interests of the subsystem – goals, traditions, life cycle stage, values, and so on – that generate subsystem performance or success measures. In this regard, the utility function of the metasystem is, either implicitly or explicitly, dependent on the subsystems and the interactions in and between the subsystems. The model also recognizes that influences from external stakeholders or the environment that enter the family business social system through a subsystem component are included in the metasystem. It is impossible to bracket off the influences of one subsystem from the other, or to speak as if one part of the system can be pulled apart from the other.

We now take the first step in building our performance model by looking at arguments that may be included in the metasystem utility function: the income levels of shareholders, the number of family members active in the business, the role of family members in the business, family reputation, short-run profit, long-run profit, market share, return on investment, the level of community involvement, philanthropy interests, dividend levels, the quickest sale of the business, and/or long-run wealth accumulation, and so on. The arguments would be any outcome that has value to any members of a subsystem. It is not our intent to further define or judge the value of the possible utility arguments noted above, but rather to hint at the type of arguments that might be defined as creating transgenerational value to a familial coalition who is in control of defining and pursuing the 'vision' of the metasystem. The model is highly general and is no different than that for a public company stating that their goal is to maximize shareholder value. We are intentionally stating the obvious – that the metasystem defines its own utility function based upon subsystem components and the systemic influences between and within the subsystem components. We do so in order to make the point that defining a utility function for the family business social system must include systemic activities.

In proposition form, the utility function for transgenerational value creation in the metasystem is

$$\text{Utility} = f(\text{arguments that positively affect transgenerational value})$$

Once the defining function of the metasystem is outlined, the utility function will focus on a more narrow set of arguments specific to this paper.

3.2 The defining function of a family business social system

Critical to our argument for a unified systems approach is Ackoff's (1994) criterion for fulfilling the definition of a social system. He stated that the 'whole' must have one or more defining functions that cannot be carried out by the parts taken separately. Ackoff (1994) describes a social system and its outcomes very differently from the dualistic assumptions and satisficing resolutions of a competing family and business associated with the overlapping circles models. He provides a description of a system and its definition of the defining function as follows:

> A system is a whole that cannot be divided into independent parts . . . . The whole has one or more defining functions . . . The defining function of a system cannot be carried out by any one part of the system taken separately . . . . When an essential part of a system is separated from the system of which it is a part, that part loses its ability to carry out its defining function . . . . Synergy is the increase in the value of the parts of a system that derives from their being parts of the system – that is, from their interactions with other parts of the system. Such an increase in value can occur only if the parts can do something together that they cannot do alone . . . . A social system should serve the purposes of both its parts and the system of which it is a part. It should enable its parts and its containing systems to do things they could not otherwise do. (Ackoff, 1994, pp. 21–31)

The implications of Ackoff's (1994) views are significant for a systemic approach to family business strategy and performance assessment. First, the system must have a defining function that is identifiable, positive, and cannot be generated by the subsystem components taken separately. A healthy family, a profitable business, and a fulfilled individual are positive outcomes, but they could each exist without the systemic interactions of the metasystem and cannot in and of themselves be called defining functions.

Second, systemic interactions must create synergy that increases the value of the component parts and the system as a whole. The system should be able to synergistically do something that the parts cannot do separately. If the synergy must increase the value of the parts in the system and has an outcome that is synergistically positive, then viewing the family and the business as competing entities that are more effective when they are kept apart cannot be said to form a synergistic system. It is difficult to imagine making an argument that a family and a business comprise a social system if the system becomes more synergistic as it is pulled apart.

Third, it is possible for a 'collection of parts' to be what Ackoff (1994) refers to as an 'unsystemic aggregation' (p. 25). He specifically mentions a holding company that cannot identify a defining function other than the common ownership of the entities. This analysis implies that a family business (with their subsystem components) does not inherently possess the attributes of an effective social system. If the system does not generate positive synergistic outcomes that can be called a defining function, then it must be considered an unsystemic aggregation of parts (family, business, and individuals). A family business by any definition should not de facto be considered a social system. It must meet a definitional hurdle that includes generating a positive synergistic outcome that fulfills its defining function as a system. Much of the family business literature that attempts to explain the negative outcomes of family firms may be trying to explain unsystemic behavior as if it is normative and systemic – families, firms, and individuals that interact without positive and synergistic systemic outcomes (aggregations) versus those in which a positive value-added defining function can be identified (social systems).

In the general unified systems performance model we have developed so far, it is difficult to identify a defining function since we have 'permitted' the system to pursue the vague goal of value creation. We now move beyond the general model and focus on a subset of family-influenced firms, specifically those pursuing wealth rather than value creation outcomes. We refer to these families as 'enterprising families' and to the metasystem as the 'enterprising families system.'

3.3 Enterprising families

By focusing on a specific subset of family-influenced firms – namely 'enterprising families' – we are able to define the arguments in the utility function as those that potentially influence transgenerational wealth creation. In the enterprising families system, a vision forged by the controlling familial coalition directs the enterprising activities of the family unit, business entity, and individual family members so as to pursue the maximum potential wealth for current and future generations of family members. The defining function of the enterprising family social system is that which synergistically and positively enables them to create transgenerational wealth.

Our focus on wealth creation stems from our conviction that firms that do not pursue an advantage in wealthy creation will in the long run have their strategic competitiveness eroded and will be selected out of the market place. Only those who are creating wealth can refer to themselves as transgenerational enterprising families since they are the ones who can guarantee their continued existence.

Business leaders create wealth for their organizations by fulfilling the primary objective of business: generating above-average returns in the market (Rowe, 2001). Generating above-average returns is obtained when a firm achieves strategic advantage and successfully exploits that advantage over other firms. Firms that do not have a competitive advantage or are in unattractive industries earn average returns at best.

We argue that in an enterprising families system, the challenge is to cultivate distinctive family-based resources and capabilities that hold the potential for rent-creating advantage. As long as these distinctive resources can be developed in ways that lead to competitive advantage, the results will be above-average returns and transgenerational wealth creation.

Wealth creation in an enterprising family is a function of performance in the form of rent and rent generation potential. We thus state a narrow form of the first proposition and introduce the second proposition for the performance model of enterprising families:

Utility = *f*(transgenerational wealth potential)

Transgenerational wealth of the enterprising family = *f*(rent generation potential)

This proposition has a number of implications for developing our unified systems model. First, the assumption underlying the model is that the system is directed towards transgenerational wealth creation. Second, to generate wealth, we must articulate a performance model such that the business entity subsystem captures rents, and thereby is the engine for wealth creation. Third, since our stated interest is in family-influenced wealth creation and not just wealth creation, we must articulate a performance model such that it captures the distinctive systemic influences of the family unit and individual family member subsystems on the performance outcomes of the business entity. To build such a system performance model, we utilize the strategic framework of the resource-based view of the firm.

3.4 The resource-based view of the firm

A systemic performance model must be able to account for a broad array of organizational influences and connect them to performance. Within the field of strategic management, the resource-based view is an established theoretical model that links these organizational influences to firm level resources, capabilities, and rent performance outcomes (Henderson and Cockburn, 1994; King and Zeithaml, 2001; Miller and Shamsie, 1996; Yeoh and Roth, 1999). Resource-based strategy scholars have incorporated a balanced process perspective that integrates the influences from psychology, organizational development, evolutionary economics, entrepreneurship, and systems dynamics (Macintosh and Maclean, 1999). With its emphasis on path-dependent behavior (Teece et al., 1997), deeply embedded resources and capabilities (Makadok, 2001), and idiosyncratic firm level advantages (Barney, 1991), it is a useful model for thinking about how systemic family influence creates the potential for advantage and corresponding performance outcomes. The resource-based model assumes that each organization is a collection of idiosyncratic resources and capabilities that differentiate firm performance across time and is the source of their returns (Hitt et al., 2001). This leads to our third enterprising families performance model proposition:

Rent generation = *f*(resources and capabilities)

The definitional distinction between a resource and a capability highlights the systemic nature of the resource-based approach. Broadly speaking, resources refer to all of a firm's assets and organizational attributes (Barney, 1991) including knowledge and processes controlled by them. Examples of how organizational processes can be related to performance include assessing the long-term impact of outsider assistance on the growth of new ventures (Chrisman and McMullan, 2000), researching the impact of technological innovativeness on small firms (Hadjimanolis, 2000), determining the effects of different human resource policies on firm outcomes (Olalla, 1999), understanding how the cognitive and emotional biases of decision makers impact the way in which they allocate resources (McGrath and Dubini, 1998), and determining how resource picking and capability building enable managers to create economic rents for firms (Makadok, 2001).

Makadok (2001) defined a capability as a 'special type of resource – specifically, an organizationally embedded nontransferable firm specific resource whose purpose is to improve the productivity of other resources' (p. 389). Amit and Schoemaker (1993) distinguished resources and capabilities by conceptualizing resources as factor stocks that are deployed through a firm's capabilities. Teece et al. (1997) argued that capabilities must be built rather than bought, and Makadok (2001) made the distinction between 'resource-picking' and 'capability-building' (p. 389). Miller and Shamsie (1996) distinguished between 'systemic' resources that are embedded in the organization and 'discrete' resources that are more readily transferable.

Research has shown that resources and capabilities create 'chains' of interactions that are directly and indirectly (Yeoh and Roth, 1999) linked to firm performance, competitive advantage, and firm wealth creation. For example, social capital has been shown to enhance knowledge acquisition (Yli-Renko et al., 2001), alliance formation (Chung et al., 2000), and interunit linkages (Tsai, 2000). Similarly, learning has been shown to affect the ability of organizations to build alliances (Khanna et al., 2000) and to positively change other capabilities (Helfat, 2000). Identifying these systemic links in the resource and capability chain is an important step in understanding firm level performance outcomes.

Due to the systemic interaction of the family unit, business entity, and individual family members, family-influenced firms are unusually complex, dynamic, and rich in intangible resources and capabilities. Many of the potential advantages associated with family firms are found in their path-dependent resources, idiosyncratic organizational processes, behavioral and social phenomena, or leadership and strategy making capabilities (Habbershon and Williams, 2000). These systemic influences lead to the idiosyncratic resources and capabilities unique to the enterprising family and which we in turn can link to their performance outcomes (for a more complete literature review of the resource-based view and its link to the family business literature, see Habbershon and Williams, 1999).

3.5 The 'familiness' of a firm

The systemic influences generated by the interaction of the subsystems – family unit, business entity, and individual family members – create an idiosyncratic pool of resources and capabilities. These resources and capabilities have deeply embedded defining characteristics that we refer to as the 'family factor' (f factor) and connote as resources$_f$ and capabilities$_f$. Figure 4.3 shows how each of the three subsystems can generate family-based systemic resources$_f$, and capabilities$_f$, that become inputs into the metasystem performance model.

Any of the resources and capabilities that could be associated with a given firm might have an f factor influence, either positive ($f+$) or negative ($f-$). We refer to positive f factor influences as 'distinctive' and note that they hold the potential to provide an advantage. We refer to negative f factor influences as 'constrictive' and note that they hold the potential to constrain competitiveness. For example, family-influenced firms may have unique potential for trust$_{f+-}$, cost of capital$_{f+-}$, HR policies$_{f+-}$, leadership development$_{f+-}$, alliance building strategies$_{f+-}$, decision making$_{f+-}$, and so on, depending upon the specific context of the systemic influences of the family business system.

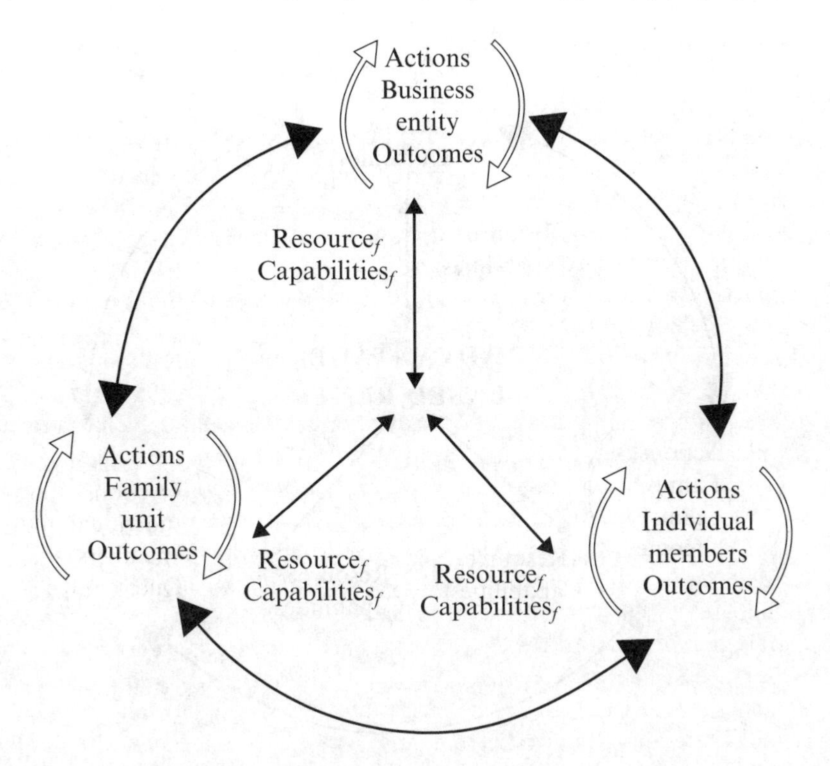

Figure 4.3 *Resources$_f$ and capabilities$_f$.*

In light of the model in Figure 4.3, we present our fourth enterprising families performance model proposition:

$$\text{Resources}_f \text{ and capabilities}_f = f(\text{systemic influences of an enterprising families system})$$

The 'familiness' of the firm can thus be referred to as the summation of the resources$_f$ and capabilities$_f$ (Σf) in given firm. This idiosyncratic familiness bundle of resources and capabilities provides a potential differentiator for firm performance and explains the nature of family influence on performance outcomes. Hence:

$$\text{Familiness} = \Sigma\ (\text{resources}_f \text{ and capabilities}_f)$$

As noted above, it is 'distinctive familiness' (Σf^+) that holds the potential for providing firms with a competitive advantage. We now present the fifth enterprising families performance model proposition:

$$\text{Advantage}_f = f(\text{distinctive familiness})$$

We conclude, therefore, that rent-generating performance for the family form of business organization is a function of those advantages$_f$ that stem from the distinctive familiness of

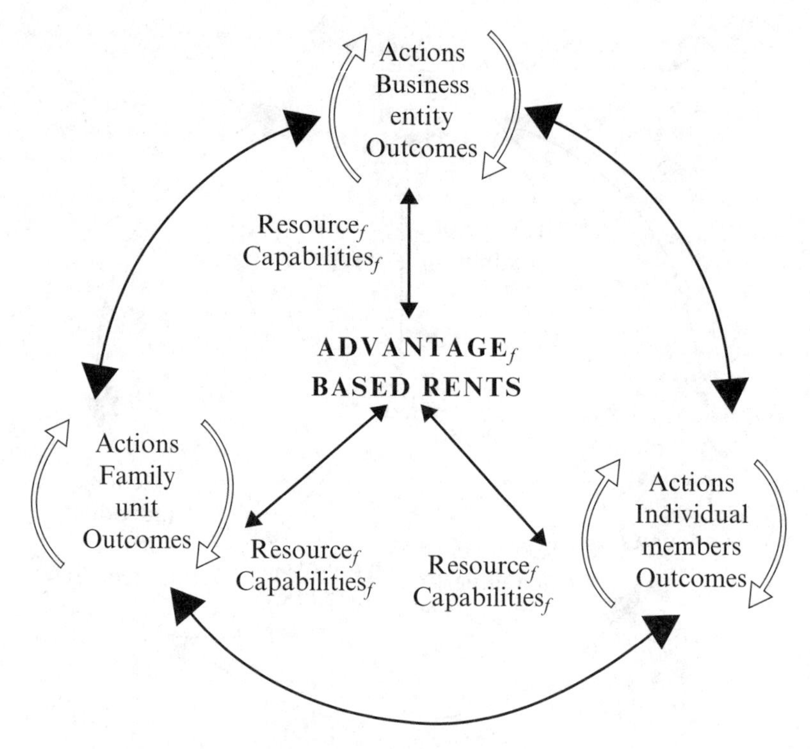

Figure 4.4 Unified systems model of firm performance

a particular firm. The final proposition for our enterprising families' performance model is presented as the following:

$$\text{Rent generating performance}_f = f(\text{advantage}_f)$$

Figure 4.4 presents the completed unified systems performance model for enterprising families. It shows how the resources$_f$ and capabilities$_f$ generated by the interaction of the subsystems lead to advantage$_f$ and the possibility of generating supernormal rents. Wealth creation is thus tied to the systemic influences in the system as they create an idiosyncratic bundle of distinctive familiness resources for the firm.

3.6 The defining function of the enterprising families system

We are now ready to present the defining function of the system for enterprising families. The defining function of a system whose purpose is to create wealth – an enterprising families system – must address competitive advantage and above-average returns since that is the source of sustained wealth creation. Based upon the reasoning in our performance model, an advantage for family-influenced firms is rooted in the systemic influences we have called the 'f factor +' of resources$_f$ and capabilities$_f$ or its distinctive familiness. The defining function of an enterprising families system must, therefore, include the commitment and ability to generate distinctive familiness leading to an advantage$_f$.

Going a step further, the defining function of an enterprising families system must also identify the strategic intent of the system as a whole or define the utility function for the system in a cohesive fashion. Chua et al. (1999) get at this point in their theoretical definition of a family business. They argued that strategic intent is captured in the vision of an organization, and stated that a family business is defined by the vision of the dominant coalition of one or more families who have the intent to sustain it across generations. Combining their theoretical premise concerning strategic intent with our systemic model for wealth creation, we present the defining function of enterprising families system as

> The systemic vision of the familial coalition that leads them to pursue distinctive familiness for the purpose of transgenerational wealth creation.

Our defining function for the enterprising families system fulfills Ackoff's definitional requirements. (1) It includes the systemic interaction of the family unit, the business entity, and family members as individuals in order to fulfill the defining function and can therefore be said to be both positive and synergistic. (2) Following on this, the value of the whole can be said to be greater than the sum of its parts because the parts cannot create the same outcome when they operate as individual subsystems. (3) The model also requires the system to be unified about its purpose (utility maximization) and to clarify the antecedent functions in the performance model (distinctive familiness) in relation to the performance outcomes (transgenerational rents) of the system.

4 Conclusions

Wernerfelt (1997) challenged resource-based scholars to gain a more specific understanding of the nature of different resources and capabilities rather than discussing them only in terms of their effects. Our '*f* factor' approach describes the nature or source of the resources and capabilities in family-influenced firms and adds depth to the explanatory power of the antecedents in the performance model.

The unified systems performance model does not make an a priori judgment about the degree or nature of the *f* factor influence. In this regard, the model applies to all types of firms – from the copreneurial couple, to the multigenerational owned and managed firm, to the family-controlled public company. It allows the researcher to identify and describe the *f* factor influence as antecedents to the resources and capabilities and link them to specific advantages and performance of the firm. The model overcomes the constraints of debating the definition of a 'family business' or of establishing boundary conditions that limits the investigation. It focuses on the degree and nature of the systemic family influences with a clear line to the impact on outcomes.

The unified systems performance model for enterprising families addresses a number of other critical issues for exploring the nature of family influence on business and wealth creation. First, it isolates the performance of the business entity as the appropriate outcome measure for a system intended to create wealth (Sharma et al., 1997). By defining the arguments in the utility function of the metasystem as those that create wealth, it removes the confusion concerning the role of the business entity subsystem in relation to the other subsystems.

Second, it identifies the systemic conditions and constituencies generated by the system as influences or inputs into the wealth creation process. Rather than downplaying the role of the family in relation to the business or creating dualistic tensions, our model demonstrates how the influence of the family and individual family members can be evaluated more positively and synergistically in relation to the business entity.

Third, it clarifies the use of dependent and independent variables in the systemic model. For example, generation of economic rents is the dependent variable, and subsystem and metasystem outcomes are the independent variables. Further, it is possible to isolate any one of the subsystems and develop stakeholder models that assess the systemic influences of that subsystem on any appropriate outcome measure. As is the circular nature of systems, each of the outcome measures (dependent variables) for a given subsystem can also serve as systemic inputs (independent variables) into the outcome measures of any other subsystem or for the metasystem as a whole.

Fourth, the model calls enterprising family leaders to be metasystems leaders. By this, we mean they need to intentionally manage the interaction of the family unit, business entity, and individual family members as an important source of their resource and capabilities pool. Ignoring the systemic family influences or attempting to bracket them off will negatively impact performance outcomes. Family leaders should clarify the defining function of their family business system as a vision that pursues distinctive familiness for the purpose of transgenerational wealth creation.

Lastly, the model calls for researchers to more aggressively study the degree and nature of family influence on firms and wealth creation. The resource-based unified systems perspective provides an accepted strategic management framework for identifying the antecedent *f* factor resources$_f$ and capabilities$_f$ to performance outcomes and for developing models to empirically test the relationships derived from maximizing the utility function of enterprising families. As the *f* factor resources$_f$ and capabilities$_f$ are identified and tested, a fuller understanding of the 'familiness' of firms will be developed.

References

Ackoff, R.L. (1994), *The Democratic Corporation*, New York: Oxford University Press.

Amit, R. and Shoemaker, P. (1993), 'Strategic assets and organizational rent', *Strategic Management Journal*, **14**, 33–46.

Barney, J. (1991), 'Firm resources and sustained competitive advantage', *Journal of Management*, **17**(1), 99–120.

Carlock, R.S. and Ward, J.S. (2001), *Strategic Planning for the Family Business: Parallel Planning to Unify the Family and Business*, New York: Palgrave.

Chrisman, J. and McMullan, W.E. (2000), 'A preliminary assessment of outsider assistance as a knowledge resource. The longer-term impact of new venture counseling', *Entrepreneurship Theory and Practice*, **24**(3), 37–53.

Chua, J.H., Chrisman, J.J. and Sharma, P. (1999), 'Defining the family business by behavior', *Entrepreneurship Theory and Practice*, **23**(4), 19–39.

Chung, S., Singh, H. and Lee, K. (2000), 'Complementarity, status similarity and social capital as drivers of alliance formation', *Strategic Management Journal*, **21**(1), 1–22.

Davis, P. and Stern, D. (1980), 'Adaptation, survival, and growth of the family business: an integrated systems perspective', *Human Relations*, **34**(4), 207–24.

Gersick, K.E., Lansberg, I., Desjardins, M. and Dunn, B. (1999), 'Stages and transitions: managing change in the family business', *Family Business Review*, **12**(4), 287–97.

Gudmundson, D., Hartman, E.A. and Tower, C.B. (1999), 'Strategic orientation: differences between family and non-family firms', *Family Business Review*, **12**(1), 27–39.

Habbershon, T. and Williams, M.L. (1999), 'A resource-based framework for assessing the strategic advantages of family firms', *Family Business Review*, **12**(1), 1–22.

Habbershon, T.G. and Williams, M.L. (2000), 'A model for understanding the competitiveness of family-controlled companies', in P. Poutziouris (ed.), *Tradition or Entrepreneurship in the New Economy*, Manchester: Manchester Business School, pp. 94–115.

Hadjimanolis, A. (2000), 'A resource-based view of innovativeness in small firms', *Technology Analysis and Strategic Management*, **12**(2), 263–81.

Harris, D., Marinez, J.I. and Ward, J.L. (1994), 'Is strategy different for the family-owned business?', *Family Business Review*, **7**(2), 159–74.

Helfat, C.E. (2000), 'Guest editor's introduction to the special issue: the evolution of firm capabilities', *Strategic Management Journal*, **21**, 955–9.

Henderson, R. and Cockburn, I. (1994), 'Measuring competence? Exploring firm effects in pharmaceutical research', *Strategic Management Journal*, **15**, 63–84.

Hitt, M.A., Ireland, R.D. and Hoskisson, R.E. (2001), *Strategic Management: Competitiveness and Globalization*, Cincinnati, OH: South-Western College Publishing.

Hollander, B.S. (1984), 'Toward a model for family-owned business', paper presented at the Meeting of the Academy of Management, Boston.

Hollander, B.S. and Elman, N.S. (1988), 'Family-owned businesses: an emerging field of inquiry', *Family Business Review*, **1**(2), 145–63.

Hoy, F. and Vesser, T.G. (1994), 'Emerging business, emerging field: entrepreneurship and the family firm', *Entrepreneurship Theory and Practice*, **19**(1) Fall, 9–24.

Ibrahim, A.B. and Ellis, W.H. (1988), *Family Business Management: Concepts and Practice*, Dubuque, IA: Kendall/Hunt.

Khanna, T., Gulati, R. and Nohria, N. (2000), 'The economic modeling of strategy process: "clean models" and "dirty hands"', *Strategic Management Journal*, **21**(7), 781–90.

King, A.W. and Zeithaml, C.P. (2001), 'Competencies and firm performance: examining the causal ambiguity paradox', *Strategic Management Journal*, **22**, 75–98.

Lansberg, I. (1983), 'Managing human resources in family firms: the problem of institutional overlap', *Organizational Dynamics*, **12**(1), 39–46.

Macintosh, R. and Maclean, D. (1999), 'Conditioned emergence: a dissipative structures approach to transformation', *Strategic Management Journal*, **20**(4), 297–316.

Makadok, R. (2001), 'Toward a synthesis of the resource-based and dynamic-capability views of rent creation', *Strategic Management Journal*, **22**, 387–401.

McCollom, M.E. (1990), 'Problems and prospects in clinical research on family firms', *Family Business Review*, 3(3), 245–62.

McGrath, R.G. and Dubini, P. (1998), 'Salient options: strategic resource allocation under uncertainty', working paper series, Snider Entrepreneurial Research-Center, The Wharton School, University of Pennsylvania.

Miller, D. and Shamsie, J. (1996), 'The resource based view of the firm in two environments: the Hollywood film studios from 1936 to 1965', *Academy of Management Journal*, **39**(3), 519–43.

Olalla, M.F. (1999), 'The resource-based theory and human resources', *International Advancement Economic Research*, **5**(1), 84–92.

Romano, C.A., Tanewski, G.A. and Smyrnios, K.X. (2000), 'Capital structure decision making: a model for family business', *Journal of Business Venturing*, **16**, 285–310.

Rowe, W.G. (2001), 'Creating wealth in organizations: the role of strategic leadership', *Academy Management Executive*, **15**(1), 81–94.

Senge, P.M. (1990), *The Fifth Discipline*, New York: Doubleday.

Sharma, P., Chrisman, J.J. and Chua, J.H. (1997), 'Strategic management of the family business; past research and future challenges', *Family Busines Review*, **10**(1), 1–34.

Stafford, K., Duncan, K.A., Dane, S. and Winter, M. (1999), 'A research model of sustainable family businesses', *Family Business Review*, **12**(3), 197–208.

Swartz, S. (1989), 'The challenges of multidisciplinary consulting to family-owned businesses', *Family Business Review*, **2**(4), 329–39.

Tagiuri, R. and Davis, J.A. (1996), 'Bivalent attributes of the family firm', *Family Business Review*, **9**(2), 199–208.

Teece, D.J., Pisano, G. and and Shuen, A. (1997), 'Dynamic capabilities and strategic management', *Strategic Management Journal*, **18**(7), 509–33.

Tsai, W. (2000), 'Social capital, strategic relatedness and the formation of intraorganizational linkages', *Strategic Management Journal*, **21**(9), 925–39.

Ward, J.L. (1987), *Keeping the Family Business Healthy: How to Plan for Continuing Growth, Profitability and Family Leadership*, San Francisco, CA: Jossey-Bass.

Wernerfelt, B. (1997), *A Resource-based View of the Firm. Resources, Firms and Strategies: A Reader in Resource-Based Perspective*, Oxford: Oxford University Press, pp. 3–18.

Whiteside, M.F. and Brown, F.H. (1991), 'Drawbacks of a dual systems approach to family firms: can we expand our thinking', *Family Business Review*, **4**(4), 383–95.

Yeoh, P.L. and Roth, K. (1999), 'An empirical analysis of sustained advantage in the U.W. pharmaceutical industry: impact of firm resources and capabilities', *Strategic Managment Jounal*, **20**, 637–53.

Yeung, W. (2000), 'Limits to the growth of family-owned business? The case of Chinese transnational corporations from Hong Kong', *Family Business Review*, **13**(1), 55–70.

Yli-Renko, H., Autio, E. and Sapienza, H.J. (2001), 'Social capital, knowledge acquisitions, and knowledge exploitation in young technology-based firms', *Strategic Management Journal*, **22**(6), 587–613.

5 The family's dynamic role within family business entrepreneurship

Ramona K.Z. Heck, Sharon M. Danes, Margaret A. Fitzgerald, George W. Haynes, Cynthia R. Jasper, Holly L. Schrank, Kathryn Stafford and Mary Winter

This chapter offers a conceptual and analytical review of the Sustainable Family Business Model (Stafford et al., 1999). The SFB Model is a comprehensive and flexible model that enhances the understanding of the dynamic role of family within family business entrepreneurship through its systems orientation. It explores the entrepreneurship of the business within the social context of the family. Unlike many other models that take an individual approach to the study of the family business, it emphasizes the overlap of the family and business systems while recognizing the unique characteristics of each of the systems. The chapter presents the theoretical perspective and major premises of the SFB Model of family business. One of the features of the model is its recognition that processes differ in times of stability versus times of change; it includes the Family Fundamental Relationship Orientation Model (FIRO Model) as a working model that explicates the reconstruction that is needed during times of changes for family businesses to remain resilient and, thus, sustainable over time. The chapter offers specific illuminations of the SFB Model's major components/concepts. The methodological issues of the 1997/2000 National Family Business Surveys (1997/2000 NFBSs) are discussed relative to their comparative advantage for studying the owning family and its business. The chapter offers an analytical model for study of the family business based on the SFB Model. Additionally, it more fully identifies the salient family factors that influence the business and vice versa as well as placing the family business within its community context. Suggestions for further applications and future research using the SFB Model are offered.

Introduction

Throughout history and world wide, families and business have always existed to a large extent in tandem (Morck and Yeung, 2002; Narva, 2001). The economic necessity of earning a living and supporting a family is often the underlying motivation for starting and growing a business (Winter et al., 1998). Among other motivators, lifestyle and wealth accumulation goals play an important role in whether a particular family member or members choose to start a business in conjunction with their family. At the same time that the business provides income to the family, the family may serve as a critical supply of paid and unpaid labor, as well as contribute additional resources such as money, space, equipment and other factors of production in the business.

A comprehensive and flexible conceptual model such as the Sustainable Family Business Model (SFB Model) (Stafford et al., 1999) can enhance our understanding of the dynamic role of the family in family business entrepreneurship. Astrachan (2003) has commented that the SFB Model both conceptually and empirically 'exemplifies what is at the heart of the family business field: the study of the reciprocal impact of family on business' (p. 570). The SFB Model will be delineated in this chapter along with several empirical applications, demonstrating its integration of the family, business, and community.

The nature of family businesses

Regardless of definition, the majority of businesses in the United States are family businesses (Heck and Trent, 1999). Among United States family businesses represented in the 1997 and 2000 National Family Business Surveys (1997/2000 NFBSs), a nationally representative sample of households with a family business, about two-thirds are owned by families, have a family member manager and at least two family members working in the business (Heck and Trent, 1999). A little over half of these family businesses are legally organized as sole proprietorships, however, nearly 80 per cent of all family businesses are one-owner businesses, irregardless of their legal organization. However, spouses, parents, adult children, and other relatives often play a major decision-making role in the business. Ownership, clearly, does not completely depict the nature of the family business. The majority of family businesses have two or more members of the residential family working in the business and one-quarter involve extended family members who may be paid or unpaid. Only 13 per cent are firms in the hand of the descendant generation of owner-managers, of which 23 per cent expecting ownership change in the next five years.

In tracking these same family businesses over a three-year period using the 1997/2000 NFBSs, researchers have found that the most important factor in continuity is the respondents' assessment of business success (Winter et al., 2004). Successful family businesses continue, are sold or gifted when the owner-manager leaves the business. A business closure should not be viewed as a business or managerial failure but rather as the demise of the business. That movement away from the value-laden language of 'failure' is advised because although some changes may be failures, others should be viewed as ordinary business or family developments. The business owner may have accomplished all that he or she wanted in the business or may just want to do something different. Other life course developments may dictate business discontinuance such as a new regulation that would dictate adjustments in the business that are beyond the financial capacity of the business.

The reciprocal relationship between families and business often evolves from the natural dynamics of both entities being in close proximity to each other, particularly in the case of the home-based business (Heck et al., 1995). These dynamics are important because approximately one-half of family businesses in the 1997 NFBSs were home based in nature (Fredericks and Winter, 2003) and made major contributions to both rural and urban economies (Heck and Stafford, 2001; Rowe et al., 1999). Families and the family home, in fact, often serve as transparent incubators for the germination of business ideas and endeavors both in and out of the home as well as the storefront or factory. The birthplace of entrepreneurial ventures is often in the home.

Even when a more restrictive and limited definition is used, family businesses comprise an important group of businesses publicly and privately (Anderson and Reeb, 2003; Astrachan and Shanker, 2003). Renewed interest in the role of the family in business performance is long overdue. The family is now seen as the competitive advantage in long-run business success because it facilitates the development of future leadership and enhances annual shareholder return, return on assets, annual revenue growth and income growth. Five key factors for superior family firm performance may include: (a) leadership development from within the family, (b) quick decision-making ability, (c) employee loyalty, (d) investing in growth by family owners, and (e) no absentee landlords via family boards. Despite these logical keys to superior performance, surprising little study has been made of the family's role.

Ownership and size considerations
Family businesses exist in a wide variety of ownership structures and in sizes from micro to very large-scale operations based on the empirical examination of the nationally representative 1997/2000 NFBSs data. However, business ownership is only one defining aspect because ownership is merely a legal structure choice that is often arbitrary and fails to truly represent the involvement and management of the family business (Heck and Trent, 1999). The SFB Model permits the examination of ownership as a control variable or operating dimension. Who owns the business lies within the overlap of the family and the business systems.

A major advantage of the SFB Model is its adaptability to a full range of business sizes, measured in either the number of employees or gross annual receipts. The United States Small Business Administration (SBA) uses the number of employees and average annual receipts to determine if non-banking businesses qualify as small businesses for each North American Industry Classification System (NAICS) code. In general, the United States SBA classifies businesses with 500 or fewer employees as a small business; however some businesses can have as many as 1500 employees (such as telecommunications resellers) or sales of $30 million or less (such as facility support services) and still retain their small business classification (see www.sba.gov/size for more details). The largest employer in the 1997/2000 NFBSs had 350 employees and gross sales of $86 million (1996 dollars) was the highest gross annual receipts. Although the SFB Model has been initially tested with the NFBSs samples, it can be utilized with any business which operates with some family involvement, regardless of size. In fact, the level of family involvement and the impact of that involvement can be investigated by studying the extent of the overlap between the family and the business as depicted in the SFB Model.

Previous research
Few researchers have noted the connections between entrepreneurship and the family (Gartner, 2001; Upton and Heck, 1997). Moreover, entrepreneurship research literature has given little attention to the interrelatedness of families and businesses (for example, Davidsson and Wiklund, 2001; Shane and Venkataraman, 2000; Timmons, 1999). Entrepreneurship research, in particular, rarely acknowledges the underlying family dynamics of the owning family and its effects on the business. Some recent attention has been given to family ownership and its relationship to the ongoing performance of extant businesses (Anderson and Reeb, 2003). The effects of the business on the family have been entirely omitted by most entrepreneurship researchers (Aldrich and Cliff, 2003; Rogoff and Heck, 2003).

In contrast, the SFB Model has been conceptually developed (Stafford et al., 1999) and tested empirically (Olson et al., 2003), a model which fully encompasses the family perspective relative to the business enterprise and vice versa. In this model, family and business are equal and overlapping systems that move simultaneously towards mutual sustainability. Families and businesses, according to Olson and her co-authors (2003), tend to move in parallel, with success in one leading to success in the other. Similarly, problems or changes in one result in problems or changes in the other.

Trends in entrepreneurship research

Early research in entrepreneurship often utilized psychological attribute models which are now considered to be limiting (Ibrahim and Ellis, 2004). The rational economic model has been employed to examine entrepreneurship and family businesses by integrating the business profit function of the business into the monetary constraint of the utility maximization model of the household (Lopez, 1986). The SFB Model was the first to recognize the importance of profit maximization in the entrepreneurial venture or the family business by recognizing the interdependence of resources, and extends the theoretical argument beyond the objective measures employed in the profit function. Networking theory states that individuals get information, resources and social support through the network of social relationships (Aldrich, 1999). More recent entrepreneurship research places the entrepreneur(s) within a social context that is, in part, the family (Aldrich and Cliff, 2003).

Trends in family business research

In general, family business research has consistently employed a systems approach, but attention has been focused on the business system without adequate consideration of the family system. For instance, many times the legal and economic aspects of succession are preformed or studied within family businesses, but little, if any attention is given to implication of changes in management and the disruptions those changes create in the owning family dynamics. More recently, family business researchers have more equally focused on both the family and business with mutual sustainability between them as embodied in the SFB Model (Stafford et al., 1999). Researchers developing this theoretical model have then empirically examined the performance of the family and the business systems underlying the family business entity by using this more comprehensive view of the two systems and their overlap (or interface) (Olson et al., 2003).

Subsequently, the family dimensions that relate to development of a new business, including recognition of an opportunity and the mobilization of resources, are being considered and investigated (Aldrich and Cliff, 2003; Steier, 2003). A few entrepreneurship researchers (who typically take a business focused approach) are taking a renewed and more in-depth view of the family within emerging businesses (for example, Aldrich and Cliff, 2003). In particular, Aldrich and Cliff (2003) have identified sociohistorical changes in the family such as household counts, size and composition as well as roles and relationships within the family that impinge on the family and its possible motivations to create new businesses. These researchers note that previous literature in areas of emergence and opportunity recognition, along with the decisions that create the new venture, does not include the family as either a conceptual or empirical dimension. Current networking research has given some attention to the effect of family on the mobilization of resources in creating new business ventures. For example, Aldrich and Cliff (2003) have offered a 'family embeddedness perspective' that suggests family system characteristics such as transitions, resources, and norms lead to the creation of new businesses. The processes of creating new ventures include recognition of an opportunity, the decisions surrounding the business launch, locating and mobilizing resources and founding. These processes then produce outcomes such as survival, objective performance and subjective success through a sequential and progressive path that provides feedback to the family system.

Overview of previous conceptual frameworks

The family is a critical element in the mix of resources that the entrepreneur needs at every stage of a venture (Rogoff and Heck, 2003). Over time, the mutual sustainability between the business and the family becomes an important goal of each system relative to both objective and subjective outcomes. This section will offer an overview of family systems theory and business system frameworks. These two conceptual orientations have already provided important contributions to the study of family businesses. Those contributions need to be acknowledged as foundations for the argument about using the SFB Model that integrates the concepts of these unique approaches into the systemic study of family businesses, including the overlap of the business and family systems.

Both entrepreneurship and family business researchers view the business system as important and examine traditional topics such as strategy, management, production, labor, performance at business stages such as start-up, growth, maturity and exit (Rogoff and Heck, 2003). Although the family business is often delineated by ownership, management involvement and multiple generations, entrepreneurship more narrowly focuses on new venture opportunities and emergence. Moreover, family business research uses family systems theory while entrepreneurship is rooted in economics, management, strategy, finance, psychology and sociology. Systems theory is key to understanding the linking/overlapping family and business systems (Stafford et al., 1999). Family systems theory uses a group-level perspective and acknowledges the significance of developmental change at both the individual and group levels. This theory locates behavior in systems of relationships and encompasses more than the notion that families are collections of individuals who have dyadic relationships with each other. Therefore, families are not just simply a group of individuals, but these individuals through family interactions and transactions create unique system or group attributes that are more than the sum of individuals and their attributes. These unique interactions may well be at the heart of what makes family businesses different.

Family systems theory suggests that the family group will attempt to maintain a consistent mode of operation, an emotional equilibrium. When change occurs as the result of either the developmental processes or external events, it disturbs the equilibrium and activates regulating mechanisms. Within the family system, both intellectual and emotional reactions to change occur. The theory examines effects of members on one another and on the system itself. Family systems theory also illuminates such processes as the mechanisms of loss and replacement, the family life cycle (span), subsystem interactions (spouses, parent–child; siblings), and communication strategies such as triangulating (Kepner, 1983). Systems theory is equally useful in understanding the business. Systems theory has been especially important to business researchers in their attempts to understand the importance of culture in underpinning formal institutions, which, in turn, underpin societal business systems (Whitley, 1999).

Several paradigms have focused on the role of the business within the context of the family business. The concept of success has been an integral part of these frameworks. Success, however, can also be defined beyond financial success of the business to factors that may be of value to the family system such as independence, family security, and being able to choose where the family will live (Kuratko et al., 1997). In other words, success within a business helps families meet goals other than just those connected to financial well-being or making a profit. Conversely, success within the owning family helps business meet its goals (Danes et al., 2002; Danes et al., 1999). Two different approaches have evolved to examine determinants of success within the business. One approach is to examine a business as part of the larger economy. The other approach is to examine the role of the individual as entrepreneur within the business.

The business within the economy approach is exemplified by the work of Davidsson (1991) and Greenburger and Sexton (1987). Davidsson (1991) contended that business growth is the result of the sum of three factors – ability, need, and opportunity; however in studying these factors, he found that only explained 25 per cent of the variance in business growth was explained. Greenberger and Sexton (1987) developed a business success model focusing on the role of the entrepreneur and how that role changed depending on the extent of business success. They contend that business success depends on the capabilities of individuals within the business but, also, on such aspects as organizational vision and empowerment of subordinates. Clearly, this approach has resulted in limited understanding. Entrepreneurs start businesses for lifestyle reasons, too (Davis-Brown and Salamon, 1987; Winter et al., 2004). Although growth and profits are important to business owners, they may want more time with family and friends, more leisure time, or more control over the time they spend at work.

The individual within the business approach is exemplified by the work of Becker (1993) and others (Ehrenberg and Smith, 1997; Zuiker, 1998), who note that individuals within family businesses bring their unique human capital to a business. These authors contend that this unique human capital often determines the success of the business. In these studies human capital was defined as the amount of skills, knowledge, intelligence, and health that an individual brings to a business. This capital can be used to gain both monetary and nonmonetary resources for the individual as well as for the business. This approach further assumes that the individual, as entrepreneur, is economically mobile and that as the amount of human capital grows, so do the rewards associated with it.

Utilizing both family systems theory and business system frameworks enables researchers to study the owning family and the business entity as well as the overlap between these two major systems. Exploring the family business from a broader vista offers richer understandings. The SFB Model offers a means to explore this richness and provide more satisfying answers to questions about the nature of family business.

The Sustainable Family Business Model

The Sustainable Family Business (SFB) Model (Figure 5.1) is a theoretical model that draws from family systems theory, giving equal recognition to family and business systems and to the interplay between them which is necessary for the achievement of mutual sustainability (Stafford et al., 1999). Both the family and the business are social systems which are purposive and rational. These two social systems transform available resources and constraints via interpersonal and resource transactions into achievements. Achievements in this model can be both objective or subjective (Olson et al., 2003). The model further recognizes that family and business are both affected by environmental and structural change, and that responses are different when changes occur.

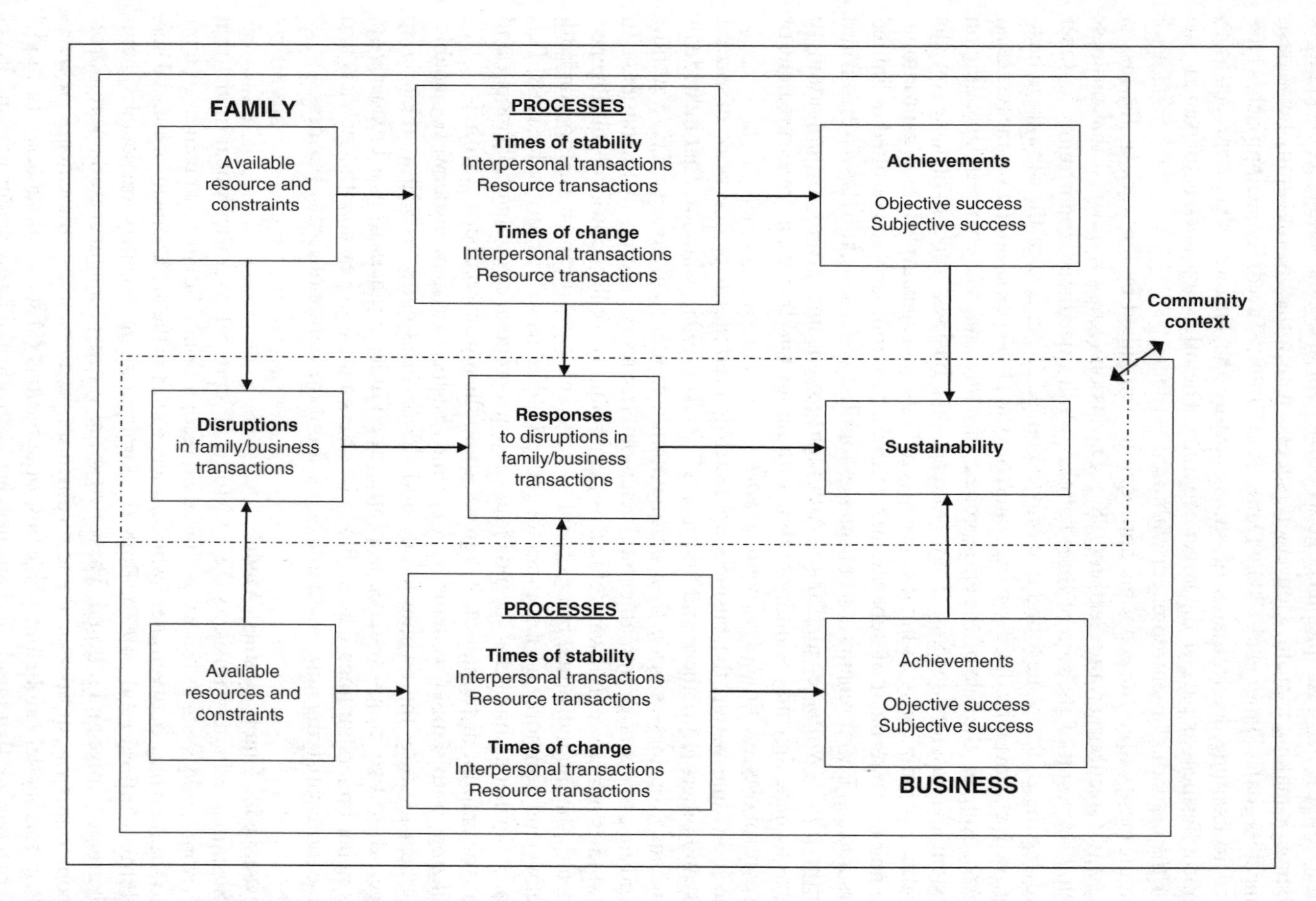

Figure 5.1 Sustainable Family Business (SFB) Model

Resources can be economic (for example, assets and liabilities), demographic (for example, number of children, number of family and nonfamily employees), functional (for example, skills, number of hours worked in the business), or psychosocial (for example, perceptions, attitudes, beliefs). The SFB Model is a dynamic theory that incorporates change as a major premise (Figure 5.1). The model is flexible enough to address either family or business issues independently and in conjunction with each other. The flexibility of the SFB Model facilitates the study of family businesses with all their complexity and diversity in such things as size, stage of the family and business cycles, mix of family and nonfamily employees, types of businesses, and ways of legally structuring the business. It emphasizes the sustainability of the family business system as a holistic entity and treats the family and business systems equitably. The SFB Model implies that the sustainability of a family business is a function of both business success and family functionality (Stafford et al., 1999). The model also rests on the notion that an individual in either system may affect both systems (Heck and Trent, 1999). The diagram of the SFB Model indicates that the boundaries of the family and business are permeable; the degree of overlap in the systems varies as do the desired goals and realized achievements of each system.

Family as social context for entrepreneurship

Much of the traditional business performance literature is plagued by the underlying assumption that individuals make economic decisions in a social vacuum (Aldrich and Zimmer, 1986). In contrast, the SFB Model locates entrepreneurship of the business within the social context of the family, indicating that this social network is the milieu out of which the family business initiates, grows, and encounters transitions. The appropriateness of locating entrepreneurship within the family context is substantiated by Aldrich's work (1999) on social networks of family-owned firms. The current literature on family-owned business success and performance primarily emphasizes business system issues. For example, a number of authors have indicated that family tasks and values often are placed in opposition to those of the business (Danes et al., 2005; Danes and Olson, 2003). There is a tendency to consider the family as a system that impedes the functioning of the business, and the family is seen as the part of the situation that must be managed (Danes et al., 2002; Ward, 1997).

The SFB Model permits a neutral approach and does not assume that family is in competition with or in conflict with business. Rather, the SFB Model recognizes that disruptions created by change are normal and occur at the intersection of family and business. It further suggests that management of conflict evolving from disruptions may serve to project the family or business into needed constructive change that fosters sustainability rather than assuming conflict progresses into purely destructive conflict that would negatively influence survivability and sustainability (Danes and Morgan, 2004; Danes et al., 1999).

Resource and interpersonal transactions

The SFB Model suggests that at various times, resource transactions (for example, utilization or transformation of time, energy and money) and interpersonal transactions (for example, communication, personal relationships, conflict management) from either the business or family may facilitate or inhibit the sustainability of family businesses. For instance, interpersonal dynamics among family business members have sometimes been depicted as an obstacle to successful multigenerational transmission of family businesses

(Lansberg and Astrachan, 1994; Rodriguez et al., 1999). Yet, the family is also likely to be a source of support that can help a family business overcome adversity and social change (Simon and Hitt, 2003). In particular, Werbel and Van Auken (forthcoming) suggest that family members may be likely to provide financial resources through outside sources of earned income, emotional support in the form of encouragement, and instrumental support in the form of knowledge or physical assistance in helping the family business to survive (Procidano and Heller, 1983).

A unique feature of family businesses is that family members often work in the business system. This human capital (as depicted in Figure 5.1) can be a resource or constraint depending on the life cycle stage of either the family or the business. For example, during the early years of a family venture, the family often provides the firm with a steady supply of trustworthy human resources (Ward, 1997). In fact, Chrisman et al. (2002) stated that new family firms might not face the same liability of newness because of the support provided by family members. In their formative years, family firms often benefit from the overlap of family and business systems because the informal nature of family relations frequently carries over into the firm and serves to foster commitment and a sense of identification with the founder's dream (Haynes et al., 1999; Van Auken, 2003; Van Auken and Neeley, 2000; Winborg and Landström, 2000). Human resources are also critical in later years relative to expanding the business or in identifying and training a potential successor. The human resource pool of the owning family may be limited or eventually outstripped by the demands of a growing business. Succession remains a tremendous challenge to family businesses as they transition between generations, and it clearly demonstrates the importance of family as a source of critical resources.

Processes during times of change

The SFB Model further recognizes that different processes occur in each system during times of stability and times of change (Danes et al., 2002; Stewart and Danes, 2001). Ward (1997) indicated that the long-term sustainability of any family business depends on its ability to anticipate and respond to change. Modified patterns of interaction are needed for a family business to remain healthy when responding to changes that occur during normative transitions or non-normative crises in either the family or the business system (Danes, 1999; Danes et al., 2002). During times of change, the patterns of interpersonal and resource interactions are adjusted at the intersection of the family and business systems to sustain that family business (Danes et al., 2002; Stafford et al., 1999).

Objective and subjective success

Financial measures of business success have long been the gold standard against which family businesses have been measured. However, subjective indicators such as motivations (for example, maintaining personal freedoms), rewards (for example, meeting challenges), goals (for example, increasing family security by building the business) and perceptions of success also are important assessments for measuring success (Cooper and Artz, 1995; Kuratko et al., 1997; Stafford et al., 1999). The SFB Model recognizes that family business achievements are evaluated in both objective and subjective terms. Utilizing both types of success measures in family business research leads to an understanding of the entire context in which business owners choose to invest their time and money, whether they choose to stay in business, how they work with customers and employees, and how they recognize and solve problems (Olson et al., 2003). Objective and subjective measures are each an important part of a complete outcome assessment of outcomes or achievements (Cooper and Artz, 1995; Cooper et al., 1988).

Family firm within the community context

The SFB Model (Figure 5.1) recognizes that the firm is part of a larger system by placing the family business within its community context. Members of the family and business system may interact with the community; the impetus for the manner and degree to which that interaction with the community occurs is rooted in the meanings that family business members give to that activity. The owning family provides a fertile environment of values, attitudes, and beliefs that serve as inputs into the family firm culture. One of the attitudes from the family system that often transfers into the business through its family employees is a social responsibility to the community. The interaction between the family business and its community context is critical because success of the family business depends on whether the firm is managed in harmony with the local community culture (Astrachan, 1988). A positive symbiosis between the family business and its community host is more productive for both the firm and the community compared with a situation where there is not a good match between the two cultures.

Managing change within family businesses

In the SFB Model, change and the disruptions resulting from change are positioned within the overlap of the family and business systems. Change is positioned at the intersection of these two systems because dynamic and interdependent relationships that are needed to successfully maneuver through change require the resources and interpersonal interactions of both systems (Danes and Morgan, 2004). During times of change, resources that contribute to creative problem-solving around the disruptions are garnered from each system to manage their response to change. Other models in the family business field do not explicitly recognize that processes are different in times of change versus times of stability (Figure 5.1). Nor do they acknowledge explicitly that processes within each system must be reconstructed during times of change to incorporate the adaptations that each system has made to accommodate the disruptions.

A sustainable family business depends upon its ability to remain resilient in times of normative, planned change as well as during non-normative, unexpected change (Connor, 1992; Danes, 1999). Normative changes are those that are planned and expected such as changing the product line. The non-normative changes are those that are unexpected and that create a crisis situation such as when a critical, valued employee leaves unexpectedly on short notice. The family business has resources and processes at the intersection of the family and business systems that may aid in adjusting to changes. How the family members utilize those resources and processes during times of disruption caused by change may facilitate or inhibit the sustainability of the family business (Danes, 1999; Danes et al., 2002).

Because sustainability of the family business depends, in part, upon how it adapts to change, it is important to understand in detail the resource and interpersonal transactions that increase the probability of traversing the process of change. The Family FIRO (Fundamental Interpersonal Relationship Orientation) Model (Danes et al., 2002) (Figure 5.2) is a working model about the human dynamics of change and can be used for assessment and problem-solving during the change process. It that provides a means of

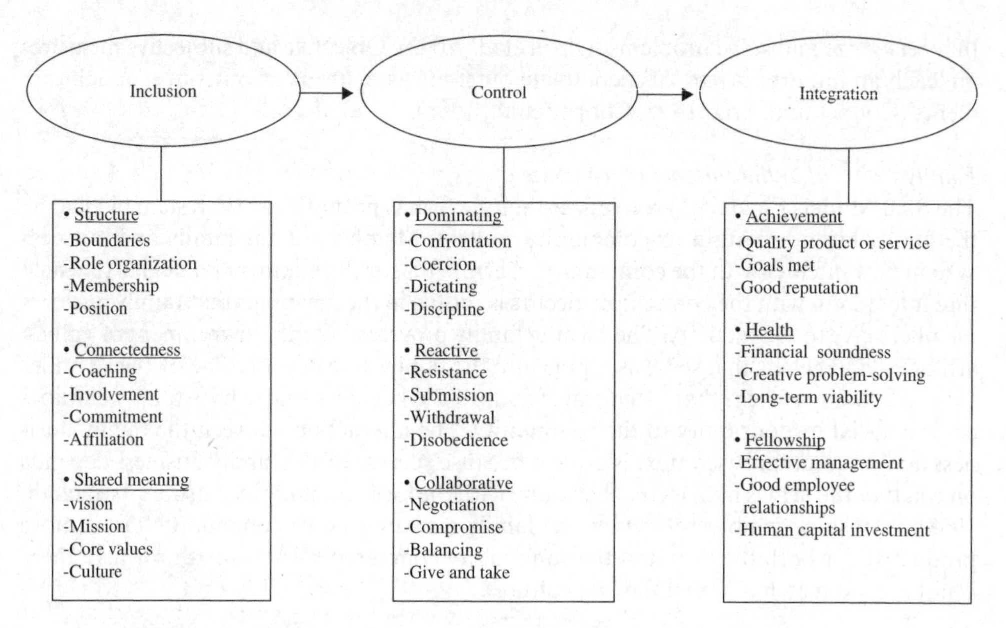

Figure 5.2 Fundamental Interpersonal Relationship Orientation (FIRO) Model

examining more precisely the resource and interpersonal transactions that occur during change located at the intersection of family and business systems within the SFB Model.

The Family FIRO Model of human dynamics of change
The Family FIRO Model more precisely identifies the human dynamics of change and provides for setting priorities across a full range of family business changes. Danes et al. (2002) empirically tested the Family FIRO Model on family businesses, and they found support for the developmental sequence of its three dimensions: inclusion, control, and integration (Figure 5.2 includes the subcategories of each dimension and the components of each subcategory). Danes et al. (2002) found support for a major proposition of the Family FIRO Model in that a sense of *inclusion* in a family business and the manner in which *control* issues were managed had important influences on the *integration* (successful achievements) of the family business system. The Family FIRO Model is a systems theory-based model as is the SFB Model. Thus, the Family FIRO dimensions of inclusion, control, and integration fit hand-in-glove with the major concepts of the SFB Model (available resources and constraints, processes, and achievements).

Understanding what composes each dimension of the Family FIRO Model and the relationship of the dimensions is outlined first followed by an example that explicates the developmental nature of the dimensions. The *inclusion* dimension of the Family FIRO Model is not just about structure. It is not only about who is part of the family business (either explicitly or implicitly), but is also about how and to what extent family members are connected to the family business, and the degree to which there is shared meaning among family business members. The components of the subcategories of the inclusion dimension more precisely identify critical resources and constraints that affect the way people respond to change. During times of change, the subcategories of inclusion (*struc-*

ture, *connectedness*, and *shared meaning*) will need to be reassessed and reconstructed within the family business to maneuver through any change successfully. If the subcategories of structure, connectedness, and shared meaning from the inclusion dimension are not reconstructed after a change, conflict and stress result, affecting the incorporation of the change and the long-term achievements of the family business system.

The control dimension of the Family FIRO Model reflects the responses to change (as depicted in the SFB Model) among family business members. Control refers to interactions that concern influence and power that people hold during times of change. The responses within the change process include tensions resulting from the disruption that evolves out of change. Those tensions may either be creative tensions or distracting tensions. Creative tensions (for example, collaborative) project the current reality within the business toward the vision that family business members hold for the business. These tensions create energy and excitement that motivate family business members to engage in behaviors that lead to business success. Distracting tensions (for example, dominating or reactive) pull the vision back toward the current reality and prevent the business from moving successfully toward its goals.

In order for the family business to remain sustainable in times of change (defined as achievement, health, and fellowship within the integration dimension of the Family FIRO Model), the inclusion dimension components need to be revisited and reconstructed to minimize the level of distracting tensions and maximize the level of creative tensions. The control dimension of the model suggests that conflict is a normal part of healthy change (Danes and Amarapurkar, 2000; Danes and Morgan, 2004). However, because of the variability in how and whether families who own businesses reconstruct the subcategories of inclusion after experiencing the disruptions of change, an interesting paradox seems to exist regarding family conflict over business issues (Danes et al., 1999). On the one hand, a certain level of tension acts as a creative mechanism and can increase the achievement, health, and fellowship (integration) of the family business (Danes and Morgan, 2004; Danes et al., 2000b). Very high levels of distracting tension, on the other hand, can have the opposite effect: reduced achievement, health and fellowship (Danes and Olson, 2003; Danes et al., 1999, 2000a).

For example, a husband and wife might be embroiled in a disagreement about expanding the family business. The underlying issue might be the question of who has more authority to make the decision. If the wife does not feel that she has a full voice, she may assert her position more strongly or disagreeably than she would otherwise; in the same way, if the husband feels that his legitimate leadership is being threatened, he may be more stubborn than otherwise. The result can be a stalemate and poorer family business functioning. In the lens of the Family FIRO Model, the couple would be wise to begin with an airing of perceptions from each member of the couple, and desires about who should participate, and how, in the decision-making process. Clarity there, which may take repeated conversations, can often lead to a more collaborative decision-making resulting in better outcomes.

If a couple concentrates on the conflict (control dimension of the Family FIRO model) rather than addressing the issues representative of the inclusion dimension, the planning for change will most likely not occur (Danes et al., 2002). In the long term, conflict that remains unaddressed could not only stalemate a decision, affecting the health of the business, but it could mire the system over time (Ward, 1997). Investing in family processes and responses to disruptions caused by change could potentially yield large returns. For instance, one study has shown that reducing family tension alone would substantially increase annual firm revenues (Olson et al., 2003).

Human and economic capital adjustments to change

Business disruptions can require the business and family to pool resources to sustain the family business. Business disruptions caused by natural disasters, labor actions, cyber or virus attacks and other major disruptions, can be especially serious and place the business and family at risk (Keating, 2001). These types of disruptions not only impact the business and family, but they have very serious ramifications for the host community. If the business community finds itself unable to recover from these disruptions, these communities face the dual challenges of decreased business and agency services and increased social and economic needs. Businesses leave because they are unprofitable and agencies depart because they are serving too few people, while those remaining in the community grapple with the challenges of a declining community. Internal to the family business, tensions resulting from the disruptions of change affect the interdynamics between spouses who own family businesses. The tensions that occur between spouses at the intersection of the family and business systems often center around resources such as the allocation of finances, the distribution of time across the family and business, or the energy and commitment provided to either or both systems (Danes and Morgan, 2004; Danes et al., 1999).

Business and family finances are linked within family-owned businesses, especially in the early stages of the business (Aldrich and Cliff, 2003; Haynes et al., 1999; Steier and Greenwood, 2000). Owing to the economic bonds of marriage, a spouse becomes a critical stakeholder in the family business. In addition, entrepreneurs often consult with and are influenced subtly by their spouses (Aldrich and Cliff, 2003). Regardless of a spouse's degree of direct participation in the business, spousal attributes, such as spousal commitment, permeate family relationships and can affect business performance by influencing the entrepreneur's attitudes, resources, and motivation toward the business.

In family businesses, spousal commitment to the business may underscore the nature of the marital relationship and may affect the health and achievements of the family business. Few studies have examined the role or impact of spousal relationships on family business performance (Foley and Powell, 1997; Poza and Messer, 2001). Even though some studies have examined the impact of family business on the marital relationship (Miller et al., 1999), little is known about the simultaneous trade-offs to achieve both family and business goals. Strong spousal commitment can be a source of competitive advantage and facilitate the success of family businesses (Harris et al., 1994). Spouses can provide active support in the form of personal or temporal resources (Danes and Olson, 2003; Fitzgerald and Muske, 2002). Whether the spouse makes direct work contributions or not, the extent to which the spouse listens, offers ideas, and makes suggestions about the business can positively influence the quality of the owner's decision-making.

Beyond the interdynamics between spouses who own family businesses, the intermingling of finances between the family and business systems is crucial to the achievements of the family business. In general, the small business finance research has overlooked the intermingling of family and businesses finances in business-owning families. It has concentrated on models of profit maximization and risk tolerance preferences of the owner-manager or the financial and regulatory structure of corporate financial markets. Furthermore, the family business research has frequently used samples from the upper bounds of what constitutes a 'small' business (Haynes and Avery, 1997). Investigating the intermingling of finances between the family and business, Haynes et al. (1999) found that two-thirds of family businesses intermingled household and business finances, indicating that those finances are inextricably intertwined. Business to family intermingling was more likely to occur when the location of the business was in a rural or small town as opposed to an urban area or if the business borrowed money or operated as a C or S corporation.[1] Sole proprietorships were more likely to use family resources in the business than other types of businesses, as were those who borrowed money, were younger owners and were owners without children.

The intermingling of financial resources is not necessarily completely positive and without cost. A potential negative is the inability to capture these interchanges in the financial records of either the family or the business. The lack of such data, while confusing for the family's financial picture, may be catastrophic to a business. Simply put, the business may not know if it is making a profit and may be jeopardizing its long-term future. The need to establish and maintain separate financial accounts is crucial to business management (Burns and Bolton, 2001). Separation of business and personal records remains a key in helping the business plan and in responding to bankers and governmental entities. Another potential negative includes the inability to repay a debt at the time needed; when borrowing business capital, the lending agency assumes that the money will be used in the business venture. Intermingling the financial resources makes it virtually impossible to predict the impact that the loan transaction has on business financial results and success.

Community context

The original SFB Model (Stafford et al., 1999) has been enhanced to guide the evaluation of the economic and social contributions of family businesses to their communities, and the impact of the community context on the family and its business. The original model provided a framework for assessing the resources, constraints, processes and achievements of the family and business and the interplay between the two systems. The enhanced SFB Model provides a framework to examine the interconnection between family businesses and their community host by acknowledging the interconnectedness of the family, the business, and the community.

According to Post (1996), a business has responsibilities to its consumers, employees, owners/shareholders, environment, and community. The enhanced SFB Model enables researchers to assess the economic and social responsibility of the business to the community, as well as the contributions of the community to its businesses and families. The economic contributions of the family business might include provision of merchandise or services, taxes paid and jobs created. Social contributions could include involvement in the community, leadership positions in civic or other local organizations, financial or technical assistance provided to the community, donations to local schools or youth programs, other philanthropic contributions, and the contributions of time, space and products businesses make. One specific process of interest is the altruism of the family and business systems amid changes (perturbations to) in the family or business systems or in the community.

Altruistic behavior as defined by economists is when one individual's utility function includes the utility function of at least one other person (Feldstein, 1975) Altruistic behavior can be pure, where one person's actions are for the benefit of another person(s), or impure, where one person's actions benefit themselves and at least one other person. For instance, a business owner might donate money to the local food bank to purchase food for a holiday where the benefit is received by the person receiving the food (an example of pure altruism). He/she might donate money to public radio because he/she likes listening to public radio and without his/her contribution, and many others, public radio would not be available to anyone, including himself/herself (an example of impure altruism). In the former case, the business owner receives no particular benefit, but in the latter, he/she does receive a benefit.

Profit motives can be intertwined with altruistic motives. For instance, business owners may donate to a local school because they realize that keeping the school in the community will keep families and their children in the community and increase the sales volume of their business. However, business owners may have a very pure motivation for being altruistic because they realize that without their giving, the community will decline and eventually may fail. In addition, altruism may engender information asymmetries in the family and business (Schulze et al., 2000). One of the challenges of the SFB Model is to provide some guidance in unbundling profit from altruistic motives and distinguishing between pure and impure altruism.

The community is important to the examination of family business. The community provides employees, educational institutions, security, and partners that a business needs to operate. On the other hand, family businesses success is closely intertwined with that of the community. Business owners and their families not only benefit from what the community offers families and businesses, but family businesses bring jobs, income and wealth to the community. Families and businesses also donate their time, talents and resources to the community to help sustain it. This enhanced SFB Model is important because family businesses are vital to communities and communities are vital to family business. Lack of attention paid to the influence of public policy on economic and social contributions made by these family businesses jeopardizes not only the businesses but also the communities where they reside and operate.

Empirical dimensions related to testing the SFB Model

Studying the complex dynamics in family-owned businesses at the intersection of family and business systems requires systemic theories, longitudinal data, multiple informants, multiple indicators of concepts (Menard, 2002; Zahra and Sharma, 2004), and the inclusion of both objective and subjective indicators of concepts. The methodology for the 1997 NFBS (Winter et al., 1998) and its follow-up, the 2000 NFBS (Winter et al., 2004), conducted by the Family Business Research Group (FBRG), included all of these critical facets. The FBRG is a multi-state research effort supported by the Cooperative States Research, Education, and Extension Service (CSREES) of the United States Department of Agriculture, the Agriculture Experiment Stations at 17 different universities, and the Social Sciences and Humanities Research Council of Canada for the University of Manitoba.

Household sampling procedures

Unlike other family business studies (for example, Astrachan and Koenko, 1994; Mass Mutual, 1995), the sampling frame for the 1997 NFBS consisted of households rather than businesses. Using a probability sample of all 50 states in the United States, not only allowed for the study of the complex dynamics within family businesses, but the methodology of the study created the opportunity to discuss with more accuracy than in the past, the prevalence of various definitions of family businesses.

More the 14 000 household telephone numbers were called in 1997 to ascertain whether someone in the household was either the owner of a family business who was involved in the day-to-day operations of that business, or was the manager of a family business that he/she expected to inherit. A total of 1536 households included someone who met the criteria. However, even further restrictions were placed on the sample because of its focus on the interaction of business and family in a family business setting (see Winter et al., 1998 for details). Qualifying families identified by a screening questionnaire, were recontacted for two different 30-minute telephone interviews, one for the business manager and one for the family manager. The family manager was defined as the 'person who actually manages the household, that is, the one who takes care of most of the meal preparation, laundry, cleaning, scheduling family activities, and oversees child care.' (Winter et al., 1998, p. 244) When the family manager and the business manager were the same individual, a 45-minute combined interview was administered. The final sample size was 708 households that owned a family business.

To provide data about family businesses over time, researchers reinterviewed the 1997 NFBS sample three years later. In households in which two different people were interviewed, attempts were made to reinterview each individual. When only one individual in a household was interviewed in 1997, only that individual was reinterviewed in 2000. To enhance the possibility that the 1997 individuals could be located in 2000, every six months between survey waves, the research group mailed a one-page summary of research results to each household interviewed initially. Address correction was requested with each mailing, and addresses in the database were updated after each mailing. Only 61 of the 708 households could not be located for the 2000 survey; an additional 94 households were contacted, but either refused to be interviewed or were deemed ineligible for the study in 2000. Data were gathered in 2000 from 553 households, more than three-quarters of the 708 households surveyed in 1997.

Tracking changes over time can be done in a variety of ways. Panel studies, defined as gathering data from the same individuals at different points in time, are among the most fruitful when attempting to understand factors that influence characteristics and behaviors over time. It is unlikely, however, that all the initial sample members will be reinterviewed through the panel period. Attrition can be a serious problem if the characteristics of individuals who remain in the sample differ significantly from those people who were not reinterviewed because statistical inference based on information from those still in the sample may no longer be representative of the target population. One of the most important conclusions of the post-stratification analysis of the 2000 NFBS data is that attrition had affected the representativeness of the sample, but in a way that could be corrected by including measures of business stability in future analyses. Age of the manager of the business, number of employees, and gross income are all variables that should be included in analyses using that data because those reinterviewed in 2000 were in established businesses.

Operational definitions of the family and family business

Definitional issues are complex in the study of family businesses. The FBRG researchers focused on families as defined by the United States Bureau of the Census, as a group of people related by blood, marriage or adoption, who share a common dwelling. The 1997/2000 NFBSs excluded single-person households and nonfamily households and included opposite-sex and same-sex cohabitors. Some researchers have adopted Hollander and Elman's (1988) definition of a family business as a business that is owned and managed by one or more family members. This definition, which can be implemented using the NFBSs, is less restrictive than others are (Heck and Trent, 1999) because it is not dependent on the involvement of multiple family members. A less restrictive definition permits a more restrictive definition if desired, yet allows empirical analyses of the data assessing the effect of type and degree of family involvement (Heck and Trent, 1999; Winter et al., 1998).

Importance of multiple perspectives
Asking one person to speak for the family (household) is not unusual in social science research. In fact, in national panel studies in the United States such as the Panel Study of Income Dynamics (PSID) and the Survey of Income and Program Participation (SIPP), and the United States population census, one person is asked to report for the family (household) (Hill, 1992). Having only one person speak for the family is appropriate when the information is straightforward and factual. However, this approach is strongly criticized when a study focuses on attitudes, beliefs, and options (Conger and Elder, 1994; McDonald, 1980; Qualm, 1981; Safilios-Rothschild, 1969) such as were those included in the 1997/2000 NFBSs.

With the exception of Rosenblatt et al. (1985), family members are generally not questioned about the interface between the family and the business. A single respondent may distort the reporting of what a family business is like and how it operates and interacts with the owning family (Winter et al., 1998). In a study of at-home income generation, Heck et al. (1992) documented that business and family experiences and perceptions for home-based workers were distinct, even when the same person managed both the business and the family.

Because the primary focus of the 1997/2000 NFBSs was the interaction between the business and the family, the views of both the household manager and the business manager were needed. Questions appropriate to the respondent's role were included in each interview schedule. In addition, a small group of identical system goals and conflict questions were asked of each individual, allowing researchers to ascertain the different perceptions of family members on issues such as goals and conflicts related to the family business.

Empirical application and testing of the SFB Model
The household sampling and detailed information available via the 1997/2000 NFBSs enable the empirical testing of the SFB Model as well as myriad yet unstudied research questions relative to the family business. In this section, the empirical applications and overall testing of the SFB Model will be further explored.

An operationalization of the SFB Model is illustrated in Figure 5.3. The empirical measures chosen to represent the various theoretical constructs are based on an extensive review of the previous family business literature and related literatures such as family studies and household management. For example, the process measure of family tension was based on the known writings and research about conflictual issues within family businesses (Danes and Lee, 2004; Danes and Morgan, 2004; Danes et al., 1999). The FBRG researchers also systemically engaged in selected and preliminary field observations with families who owned businesses. In particular, disruption response measures were based

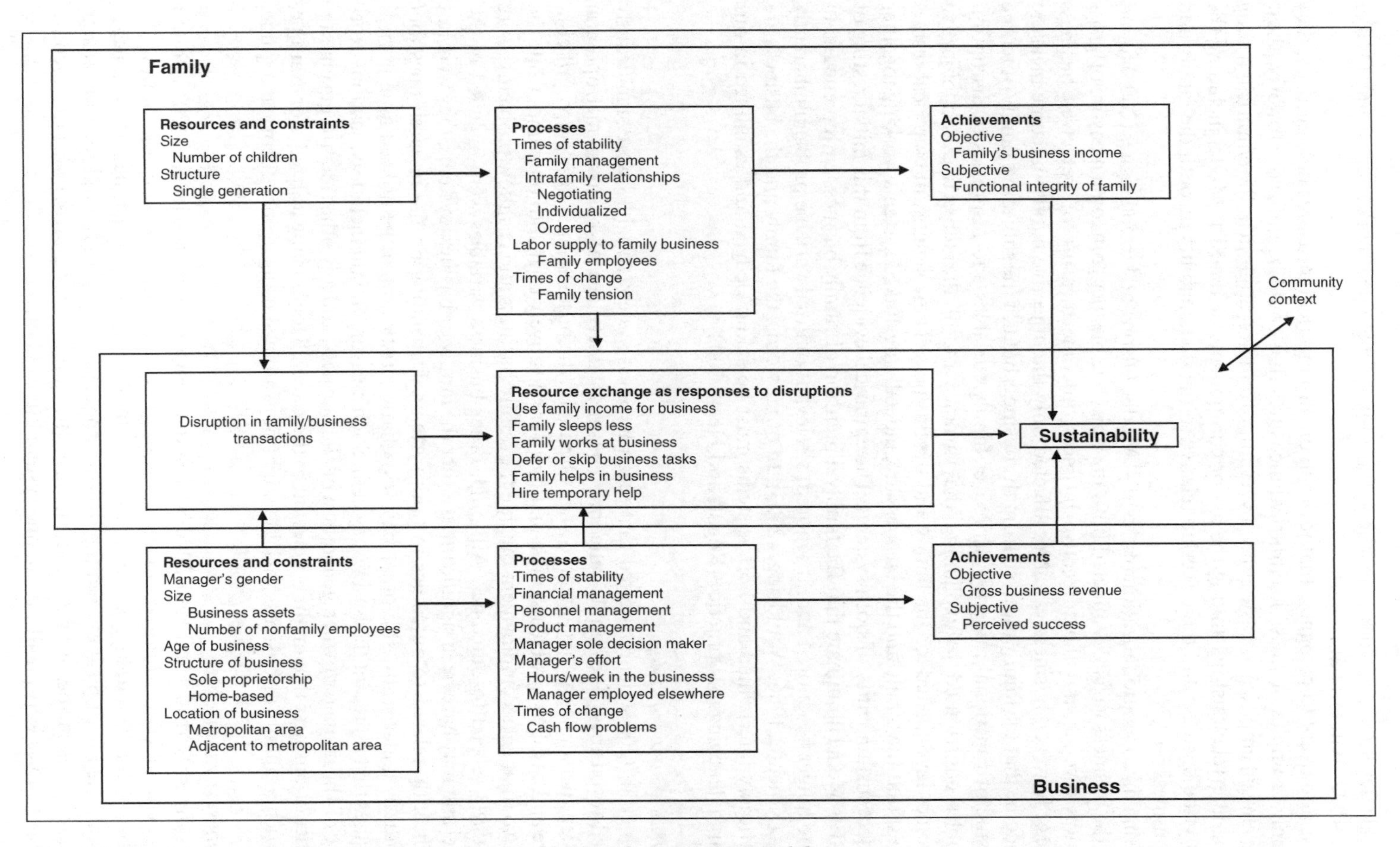

Figure 5.3 An operationalization of the Sustainable Family Business Model

on scale development work from the ground up and included replication testing (Miller et al., 1999; Winter et al., 1993).

It is important to recognize that for each of the empirical measures in Figure 5.3, they are only examples of possible empirical measures that could be utilized. However, these same chosen empirical measures are a very important first attempt at the empirical testing of the interrelationships among the major components of the SFB Model including the family, the business and their overlap relative to the sustainability of both the family and the business.

Using this operationalization of the SFB Model, Olson and her colleagues (2003) found that the business variables affected both the objective business outcome measure of gross business revenue and the subjective business outcome measure of perceived business success; however, business variables affected only the objective family outcome measure of income that the family received from the business but not the subjective family outcome measure of functional integrity. Moreover family variables and responses to disruption variables contributed to explaining the variance of all the subjective and objective outcomes investigated for the family and the business. These significant empirical results suggest that the family and the business are, indeed, interlinked systems working together to affect sustainability of both systems. The empirical evidence from this initial testing of the SFB Model illuminate that the family dimension of family business entrepreneurship can no longer be ignored and, if ignored, a relevant omitted variable problem inherently exists (Kmenta, 1986; Wooldridge, 2003). In other words, the family and the business are inextricably intertwined and both systems must be acknowledged and examined in our research, teaching and practice (Rogoff and Heck, 2003).

Business system variables

Olson et al. (2003) demonstrated that seven business system variables were significantly related to business and family outcomes (business assets, age of business, metropolitan location, personnel management, owner's hours of the business, manager employed elsewhere and cashflow problems). The first five were positively related to outcomes while the last two (manager's employment outside the family business and cashflow problems) were negatively related to outcomes. Although other business variables were included in the model and may have been significant in either the family or business systems, they were not significantly related to outcomes in both systems. Depending on the research question/ hypothesis under study, still other business characteristics might be utilized as empirical measures that represent the business system. Such measures might include education/ experience of the manager/owner, generation of the business, and any other identifying characteristics of business that might be relevant to the research focus being utilized. For example, a particular family or business variable might be examined using agency theory as a guide.

Family system variables

Using the SFB Model and 1997/2000 NFBSs data, Olson and her colleagues (2003) found that the number of children, an individualized style of interaction in the family, the number of family employees, and the level of family tensions all significantly and negatively affected both business and family outcomes except for family employees increasing gross revenues in the business and, in turn, the family's share of the business income. Other variables were significant in one system but not both.

Again, various research questions or foci might necessitate employing other empirical measures in representing the family system dimension. Such variables as the family life cycle stage might further describe the family. Gender and age combinations of children in the family as well as the more complex nature of today's blended families might warrant additional detail in representing exactly who comprises the family unit. Also, there are many other theories of family interaction that would require additional measures to represent interacting styles or the nature of family relationships. For example, the notion of power and status within the family might require additional measures of perceptions and attitudes about the power of various family members. Power or status of various family members has received little empirical attention relative to succession issues within the family business as a whole (Chrisman et al., 1998). Other researchers have begun to explore new empirical measures for family influence including measures of power and their relationships to outcomes such as success, failure, strategy and operations (Astrachan et al., 2002).

Responses to disruption variables
Although the overlap between the family and the business likely illustrates the most straightforward way in which the family and business systems are interlocked, this area has only recently been studied empirically (Danes and Amarapurkar, 2001; Danes and Morgan, 2004; Danes and Lee, 2004; Danes and Olson, 2003; Danes et al., 2002, 2005). Olson et al. (2003) used a series of empirical measures that represent various ways in which either the family or the business made adjustments to achieve outcomes in both systems. Family income and help in the business were related negatively to outcomes in both systems while hiring temporary help increased outcomes in both systems.

Other measures might be employed to represent the interaction between the family and the business. For example, measures are needed for modified patterns of interaction related to structure, shared meanings, and connectedness during times of change, for modes of conflict during times of stability and times of change, for physical location of the work area (particularly in home-based businesses) and for the degree of intrusion of the family on the business and vice versa. Continued development is needed in family business research in operationalizing processes that occur in times of stability versus times of change. This development is needed to more precisely understand the dynamics that occur to create equilibrium of the family business system in the short run and sustainability and health in the longrun. For example, more research is needed about conflict management in a range of family businesses rather than only those with major dysfunction, in order to be proactive with family businesses in both training and consultation. Research about processes that occur at the intersection of family and business systems and the impact of the outcomes of those processes on the sustainability of the family business over time has only just begun.

In summary, common omitted variables in family business research are family variables. The SFB Model (see Figure 5.3) provides some guidance in choosing family variables to include in these analytical models. For instance, a model of family business success would include characteristics of the business owner (such as age, gender, education, experience and others), the business (size, age, legal organization, location and others) and the family (number of children, duration of residence, family functionality, family tensions and others). While business success may be a function of the experience of the owner, it is impacted by disruptions within the family system. The inclusion of family variables recognizes the intermingling of resources between the family and business, and reduces the misspecification bias in the model.

Further research needs/questions

The SFB Model developed elsewhere (Stafford et al., 1999) and further delineated in this chapter can be applied throughout the study of family businesses. We know so little about how families actually interact to start, maintain, grow and exit businesses. Such study might explore the family system relative to its intricate nature and dynamics, the unique business system in relation to its owning family and, of course, the overlap between both. Further applications and possible extensions of the SFB Model are discussed below.

The family system

Not only do businesses go through stages of development, families do too. The theories of the stages of family life span are well established in the literature and the interconnections among the stages of development for the family and the business need to be further explored. The SFB Model could be used to examine differences among both family and business developmental stages. Do the family stages vary relative to time within each stage? How do these family developmental stages match up to the needs and demands? Do younger families of young entrepreneurs interact with their businesses differently or view risk differently than more mature families? Do families with either or both spouse experiencing retirement view business opportunities differently? How do couples engage in both their spousal relationship and possible copreneurial adventures to the benefit of each person and for each effort?

In today's world, blended families and their relationships to business ownership have been slow in changing (Aldrich and Cliff, 2003). Blended families often face particular challenges in merging their families as well as combining family and business. Using the SFB Model, one could examine how the demands of a blended family affect the demands of starting and operating a business. The geographical location of the family and its members can have an important effect on new business ventures and ongoing business activities. Choosing a business for lifestyle reasons (for example, moving out of urban areas or having a homebased business) will continue to be critical in the location of new businesses. Ethnicity and culture offer important new opportunities for family business research. Is ethnicity or culture of the family related to how it meets the demands of the family and the business? Applications of the SFB Model to business-owning ethnic families would allow researchers to explore such possible differences. Furthermore, do families worldwide operate in similar fashion relative to the delineations represented in SFB Model? Or do family systems manifest differently in various cultures? How well does the SFB Model apply to the extended family structure found in many countries around the world? How does family culture influence business culture?

The business system

The SFB Model might be utilized to study differences relative to the business system. Do businesses of different sizes and structures function differently relative to their interactions with their families and vice versa? How do these interactions change with changes in business legal structure? How can the family and business interactions in a smaller business differ from those of multiple-owner businesses, or those with outside shareholders? Do families and businesses combine in different ways by industry type? For example, does the industry type relate to the prevalence or likelihood of a family business? Do family businesses function differently in different industries? The SFB Model could also be utilized to examine the business system over time as it interacts with the family and the community. As the business system adapts and grows over time, how does this affect its interactions with the family system and the community? Do businesses in various stages of development act differently towards their owning families and their communities?

Family and business overlap

Certainly the nature of the SFB Model makes it particularly effective in studying the interface between the family and the business. What internal dynamics develop between the family and the business during the start-up phase or in growth or slowdown periods? How do the family and business interact during the exiting or closing of a family business? The effect of family breakups and dissolutions on business, and vice versa, deserve much further study. In addition, the family business literature devotes much attention to the succession process within the family business but primarily from the point of view of business sustainability. The SFB Model offers an important advantage and opportunity for studying the succession process from both the business and family perspectives. Not only can the basic family and business systems be studied, but also the subsequent disruptions and responses to succession can be examined.

The community

The extended SFB Model examines the interactions between the family, business and community. Further research is needed to better understand the impact of selected public programs and policies on both the family and business entities. This research will help business owners, community leaders, politicians and others develop programs to sustain and grow family businesses and communities. Three important to research questions are as follows:

1. What is the contribution of family businesses to communities?
2. What is the impact of public policies on the family business in communities?
3. What is the impact of public policies on the contribution of family businesses to communities?

This research agenda could be accomplished using existing data from the 1997/2000 NFBSs and a new wave of panel data. The new wave of panel data would focus on a more comprehensive inventory of the contributions of family businesses to the community and a retrospective assessment of the influence of public programs and policies on the family business and communities. This study is important because family businesses are vital to communities, and lack of attention paid to the influence of public policy on economic and social contributions made by these family businesses jeopardizes these families, businesses and the communities where they reside.

The importance of the family business to the community is often paramount. Typically, the family business enhances its community or communities via generating revenues and employment; it also may provide leadership, time, space, products and services, and financial support. A comprehensive view of the family business allows researchers to further understand the internal dynamics and how the community can support and benefit from such enterprises. The SFB Model affords the ability to study the owning family in relation to the business and acknowledges that both the family and the business make economic contributions, among other contributions, to their communities. In addition, the SFB Model affords the ability to examine community inputs and policies as they affect the family and the business.

Final summary

The SFB Model offers a comprehensive view of the family business and suggests that such businesses are far more complicated than was previously realized. Only careful delineation of the components of the family and the business in the context of the community will afford researchers a sufficiently rich understanding of the internal dynamics of the family business and how they relate to external environments such as their communities. Future research must continue to take a broader view while encompassing the detailed richness of the basic components of systems and subsystems that make up this larger system, namely, the family business.

Acknowledgement

This chapter reports results from the Cooperative Regional Research Project, NE-167R, 'Family Businesses: Interaction in Work and Family Spheres', partially supported by the Cooperative States Research, Education and Extension Service, US Department of Agriculture, and the Experiment Stations at University of Hawaii at Manoa, University of Illinois, Purdue University (Indiana), Iowa State University, Michigan State University, University of Minnesota, Montana State University, University of Nebraska, Cornell University (New York), Baruch College (New York), North Dakota State University, Ohio State University, Pennsylvania State University, Texas A&M University, Utah State University, University of Vermont, University of Wisconsin-Madison and the Social Sciences and Humanities Research Council of Canada (for University of Manitoba).

This chapter also reports results from the Cooperative Regional Research Project, NE-167 'Family Business Viability in Economically Vulnerable Communities', partially supported by the Cooperative States Research, Education, and Extension Service (CSREES); US Department of Agriculture; Baruch College, the experiment stations at the University of Arkansas, University of Hawaii at Manoa, University of Illinois, Purdue University (Indiana), Iowa State University, University of Minnesota, Cornell University (New York), North Dakota State University, The Ohio State University, Oklahoma State University, Utah State University, and University of Wisconsin-Madison.

Note

1. US Small Business Administration definition: the main differences between S and C lie in the fact that a C corporation is taxed a Federal Corporate Income tax. An eligible domestic corporation can avoid double taxation (once to the shareholders and again to the corporation) by electing to be treated as an S corporation. Generally, an S corporation is exempt from federal income tax other than tax on certain capital gains and passive income. On their tax returns, the S corporation's shareholders include their share of the corporation's separately stated items of income, deduction, loss and credit, and their share of non-seperately stated income or loss.

References

Aldrich, H.E. (1999), *Organizations Evolving*, London: Sage Publications.

Aldrich, H.E. and J.E. Cliff (2003), 'The pervasive effects of family on entrepreneurship: toward a family embeddedness perspective', *Journal of Business Venturing*, **18**(5), 573–96.

Aldrich, H.E. and C. Zimmer (1986), 'Entrepreneurship through social networks', in D.L. Sexton and R.W. Smilor (eds), *The Art and Science of Entrepreneurship*, Cambridge, MA: Ballinger, pp. 3–23.

Anderson, R.C. and D.M. Reeb (2003), 'Founding-family ownership and firm performance: evidence from the S&P 500', *Journal of Finance*, **58**(3), 1301–28.

Astrachan, J.H. (1988), 'Family firm and community culture', *Family Business Review*, **1**(2), 165–89.

Astrachan, J.H. (2003), 'Commentary on the special issue: the emergence of a field', *Journal of Business Venturing*, **18**(5), 567–72.

Astrachan, J.H. and T.A. Kolenko (1994), 'A neglected factor explaining family business success: human resource practices', *Family Business Review*, **7**(3), 251–62.

Astrachan, J.H. and M.C. Shanker (2003), 'Family businesses' contribution to the U.S. economy: a closer look', *Family Business Review*, **16**(3), 211–19.

Astrachan, J.H., S.B. Klein and K.X. Smyrnios (2002), 'The F-PEC scale of family influence: a proposal for solving the family business definition problem', *Family Business Review*, **15**(1), 45–58.

Becker, G.S. (1993), *Human Capital: A Theoretical and Empirical Analysis with Special Reference to Education* (3rd edition), Chicago, IL: University of Chicago Press.

Burns, M.M. and C.B. Bolton (2001), *Business Savvy for Today's Entrepreneur*, Stillwater, OK: New Forums Press.

Chrisman, J.J., J.H. Chua and P. Sharma (1998), 'Important attributes of successors in family businesses: an exploratory study', *Family Business Review*, **11**(1), 19–34.

Chrisman, J.J., J.H. Chua and L.P. Steir (2002), 'The influence of national culture and family involvement on entrepreneurial perceptions and performance at the state level', *Entrepreneurship Theory and Practice*, **26**(4), 113–29.

Conger, R.D. and G.H. Elder Jr (eds) (1994), *Families in Troubled Times*, New York: Aldine de Gruyter.

Conner, D.R. (1992), *Managing at the Speed of Change*, New York: Villard.

Cooper, A.C. and K.W. Artz (1995), 'Determinants of satisfaction for entrepreneurs', *Journal of Business Venturing*, **10**(6), 439–57.

Cooper, A.C., C.Y. Woo and W.C. Dunkelberg (1988), 'Entrepreneurs' perceived chances for success', *Journal of Business Venturing*, **3**, 97–108.

Danes, S.M. (1999), *Change: Loss, Opportunity, and Resilience*, University of Minnesota Extension Service publication #FO-7421-S, St. Paul, MN: UMES.

Danes, S.M. and S. Amarapurkar (2001), 'Business tensions and success in farm family businesses', *Family Economics and Resource Management Biennial*, **4**, 178–90.

Danes, S.M. and Y.G. Lee (2004), 'Tensions generated by business issues in farm business-owning couples', *Family Relations*, **53**, 357–66.

Danes, S.M. and E.A. Morgan (2004), 'Family business-owning couples: an EFT view into their unique conflict culture', *Contemporary Family Therapy*, **26**(3), 241–60.

Danes, S.M. and P.D. Olson (2003), 'Women's role involvement in family businesses, business tensions, and business success', *Family Business Review*, **16**(1), 53–68.

Danes, S.M., N. Fitzgerald and K.C. Doll (2000a), 'Financial and relationship predictors of family business goal achievement', *Financial Counseling and Planning*, **11**(2), 43–53.

Danes, S.M., H.R. Haberman and D. McTavish (2005), 'Gendered discourse about family business', *Family Relations*, **54**, 116–30.

Danes, S.M., R. Leichtentritt, M.E. Metz and C. Huddleston-Casas (2000b), 'Effects of conflict styles and conflict severity', *Journal of Family and Economic Issues*, **21**, 259–86.

Danes, S.M., M.A. Rueter, H.-K. Kwon and W. Doherty (2002), 'Family FIRO model: an application to family business', *Family Business Review*, **15**(1), 31–43.

Danes, S.M., V.S. Zuiker, R. Kean and J. Arbuthnot (1999), 'Predictors of family business tensions and goal achievement', *Family Business Review*, **12**(3), 241–52.

Davidsson, P. (1991), 'Continued entrepreneurship: ability, need, and opportunity as determinants of small firm growth', *Journal of Business Venturing*, **6**(6), 405–29.

Davidsson, P and J. Wiklund (2001), 'Levels of analysis in entrepreneurship research: current research practice and suggestions for the future', *Entrepreneurship Theory and Practice*, **25**(3), 81–99.

Davis-Brown, K. and S. Salamon (1987), 'Farm families in crisis: an application of stress theory to farm family research', *Family Relations*, **36**, 368–73.

Ehrenberg, R.G. and R.S. Smith (1997), *Modern Labor Economics: Theory and Practice* (6th edition), Reading, MA: Addison-Wesley.

Feldstein, M. (1975), 'The income tax and charitable contributions: part II – the impact on religious, educational and other organizations', *National Tax Journal*, **28**, 81–100.

Fitzgerald, M.A. and G. Muske (2002), 'Copreneurs: an explanation and comparison to other family businesses', *Family Business Review*, **15**(1), 1–16.
Foley, S. and G. Powell (1997), 'Reconceptualizing work–family conflict for business/marriage partners: a theoretical model', *Journal of Small Business Management*, **35**, 36–47.
Fredericks, K. and M. Winter (2003), 'A profile of home-based entrepreneurs', in R.K.Z. Heck, A.N. Puryear and P.A. Tombline (eds), *A Toolkit for Home-based Entrepreneurs*, New York: Baruch College, Lawrence N. Field Center of Entrepreneurship and Small Business, pp. 1–8.
Gartner, W.B. (2001), 'Is there an elephant in entrepreneurship? Blind assumptions in theory development', *Entrepreneurship Theory and Practice*, **25**(3), 27–39.
Greenberger, D.B. and D.L. Sexton (1987), 'A comparative analysis of the effects of the desire for personal control on new venture initiations', *Frontiers of Entrepreneurship Research. 1987: Proceedings of Seventh Annual Babson College Entrepreneurship Research Conference*, Wellesley, MA: Babson College, pp. 239–53.
Harris, D., J.I. Martinez, and J.L. Ward (1994), 'Is strategy different for the family-owned businesses', *Family Business Review*, **7**(2), 159–74.
Haynes, G.W. and R.J. Avery (1997), 'Family businesses: can the family and the business finances be separated? Preliminary results', *Entrepreneurial and Small Business Finance*, **5**(1), 61–74.
Haynes, G.W., R. Walker, B.S. Rowe and G.-S. Hong (1999), 'The intermingling of business and family finances in family-owned businesses', *Family Business Review*, **12**(3), 225–39.
Heck, R.K.Z. and K. Stafford (2001), 'The vital institution of family business: economic benefits hidden in plain sight', in G.K. McCann and N. Upton (eds), *Destroying Myths and Creating Value in Family Business*, Deland, FL: Stetson University, pp. 9–17.
Heck, R.K.Z. and E.S. Trent (1999), 'The prevalence of family business from a household sample', *Family Business Review*, **12**(3), 209–24.
Heck, R.K.Z., A.J. Owen and B.R. Rowe (eds) (1995), *Home-based Employment and Family Life*, Westport, CT: Auburn House.
Heck, R.K.Z., M. Winter and K. Stafford (1992), 'Managing work and family in home-based employment', *Journal of Family and Economic Issues*, **13**(2), 187–212.
Hill, M.S. (1992), *The Panel Study of Income Dynamics: A User's Guide*, Newbury Park, CA: Sage Publications.
Hollander, B.S. and N.S. Elman (1988), 'Family-owned businesses: an emerging field of inquiry', *Family Business Review*, **1**(2), 145–64.
Ibrahim, A.B. and W.H. Ellis (2004), *Family Business Management: Concepts and Practice* (2nd edition), Dubuque, IA: Kendall/Hunt.
Keating, R. (2001), *Small Business Survival Index 2001: Ranking the Policy Environment for Entrepreneurship Across the Nation*, Washington, DC: Small Business Survival Committee.
Kepner, E. (1983), 'The family and the firm: A coevolutionary perspective', *Organizational Dynamics*, (Summer), reprinted in C.E. Aronoff and J.L. Ward (eds) (1991), *Family Business Sourcebook*, Detroit: Omnigraphics, pp. 474–88.
Kmenta, J. (1986), *Elements of Econometrics* (2nd edition), New York: Macmillan, pp. 443–6.
Kuratko, D.F., J.S. Hornsby and D.W. Naffziger (1997), 'An examination of owners' goals in sustaining entrepreneurship', *Journal of Small Business Management*, **35**(1), 24–33.
Lansberg, I.S. and J.H. Astrachan (1994), 'Influence of family relationships on succession planning and training: the importance of mediating factors', *Family Business Review*, **7**(1), 37–57.
Lopez, R. (1986), 'Structural models of the farm household that allow for interdependent utility and profit maximization decisions', in I. Singh, L. Squire and J. Straus (eds), *Agricultural Household Models: Extensions, Applications and Policy*, Baltimore, MD and London: Johns Hopkins University Press, pp. 306–25.
Mass Mutual (1995), *Family Business: 1995 Research Findings*, Springfield, MA: Mass Mutual.
McDonald, G.W. (1980), 'Family power: the assessment of a decade of theory and research, 1970–1979', *Journal of Marriage and the Family*, **42**, 841–54.
Menard, S. (2002), *Longitudinal Research*, Thousand Oaks, CA: Sage Publications.
Miller, N.J., M.A. Fitzgerald, M. Winter and J. Paul (1999), 'Exploring the overlap of family and business demands: household and family business managers adjustment strategies', *Family Business Review*, **12**(3), 253–68.
Morck, R. and B. Yeung (2002), 'Family control and the rent-seeking society', unpublished paper presented at the 2002 Theories of the Family Enterprise Conference, Wharton Business School, Philadelphia, PA.
Narva, R.L. (2001), 'Heritage and tradition in family business: how family-controlled enterprises connect the experience of their past to the promise of their future', in G.K. McCann and N. Upton (eds), *Destroying Myths and Creating Value in Family Business*, Deland, FL: Stetson University, pp. 29–38.
Olson, P.D., V.S. Zuiker, S.M. Danes, K. Stafford, R.K.Z. Heck and K.A. Duncan (2003), 'The impact of the family and the business on family business sustainability', *Journal of Business Venturing*, **18**(5), 639–66.
Post, J. (1996), 'The new social contract', in J.W. Houck and O.F. Williams (eds), *Is the Good Corporation Dead? Social Responsibility in a Global Economy*, Lanham, MD: Rowan and Littlefield, pp. 49–64.

Poza, E.J. and T. Messer (2001), 'Spousal leadership and continuity in the family firm', *Family Business Review*, **14**(1), 25–35.
Procidano, M.E. and K. Heller (1983), 'Measures of perceived social support from friends and from family: three validation studies', *American Journal of Community Psychology*, **11**(1), 1–24.
Qualm, D. (1981), 'Random measurement error as a source of discrepancies between the reports of wives and husbands concerning marital power and task allocation', *Journal of Marriage and the Family*, **43**, 521–36.
Rodriguez, S.N., G.J. Hildreth and J. Mancuso (1999), 'The dynamics of families in business: how therapists can help in ways consultants don't', *Contemporary Family Therapy*, **21**(4), 453–68.
Rogoff, E.G. and R.K.Z. Heck (2003), 'Evolving research in entrepreneurship and family business: recognizing family as the oxygen that feeds the fire of entrepreneurship', *Journal of Business Venturing*, **18**(5), 559–66.
Rosenblatt, P.C., L. De Mik, R.M. Anderson and P.A. Johnson (1985), *The Family in Business*, San Francisco, CA: Jossey-Bass.
Rowe, B.R., G.W. Haynes and K. Stafford (1999), 'The contribution of home-based business income to rural and urban economies', *Economic Development Quarterly*, **13**(1), 66–77.
Safilios-Rothschild, C. (1969), 'Family sociology or wives' family sociology: a cross-cultural examination of decision-making', *Journal of Marriage and the Family*, **32**, 290–301.
Schulze, W.S., M.H. Lubatkin, and R.N. Dino (2000), 'Altruism and agency in family firm', *Academy of Management's Best Paper Proceedings*, New York: Academy of Management.
Shane, S. and S. Venkataraman (2000), 'The promise of entrepreneurship as a field of research', *Academy of Management Review*, **25**(1), 217–26.
Simon, D. and M. Hitt (2003), 'Managing resources: linking unique resources, management, and wealth creation in family firms', *Entrepreneurship Theory and Practice*, **27**(4), 339–58.
Stafford, K., K.A. Duncan, S.M. Danes and M. Winter (1999), 'A research model of sustainable family businesses', *Family Business Review*, **12**(3), 197–208.
Steier, L. (2003), 'Variants of agency contracts in family-financed ventures as a continuum of familial altruistic and market rationalities', *Journal of Business Venturing*, **18**(5), 597–618.
Steier, L. and R. Greenwood (2000), 'Entrepreneurship and the evaluation of angel financial networks', *Organizational Studies*, **21**(1), 163–92.
Stewart, C.C. and S.M. Danes (2001), 'The relationship between inclusion and control in resort family businesses: a developmental approach to conflict', *Journal of Family and Economic Issues*, **22**, 293–320.
Timmons, J.A. (1999), *New Venture Creation*, Boston, MA: Irwin McGraw-Hill.
Upton, N.B. and R.K.Z. Heck (1997), 'The family business dimension of entrepreneurship', in D.L. Sexton and R.W. Smilor (eds), *Entrepreneurship: 2000*, Chicago, IL: Upstart, pp. 243–66 .
Van Auken, H. (2003), 'An empirical investigation of bootstrap financing among small firms', *Journal of Small Business Strategy*, **14**(2), 22–36.
Van Auken, H. and L. Neeley (2000), 'Pre-launch preparations and the acquisition of capital', *Journal of Developmental Entrepreneurship*, **5**, 169–82.
Ward, J.L. (1997), *Keeping the Family Business Healthy: How to Plan for Continuing Growth, Profitability, and Family Leadership*, Marietta, GA: Business Owner Resources.
Werbel, J.D. and H. Van Auken (forthcoming), 'Family dynamics and family business financial performance: the case for spousal commitment', *Family Business Review*.
Whitley, R. (1999), *Divergent Capitalisms*, Oxford: Oxford University Press.
Winborg, J. and H. Landström (2001), 'Financial bootstrapping in small business: examining small business managers' resource acquisition behaviors', *Journal of Business Venturing*, **16**(3), 235–54.
Winter, M., S.M. Danes, S.-K. Koh, K. Fredericks and J. J. Paul (2004), 'Tracking family businesses and their owners over time: panel attrition, manager departure, and business demise', *Journal of Business Venturing*, **19**, 535–59.
Winter, M., M.A. Fitzgerald, R.K.Z. Heck, G.W. Haynes and S.M. Danes (1998), 'Revisiting the study of family businesses: methodological challenges, dilemmas, and alternative approaches', *Family Business Review*, **11**(3), 239–52.
Winter, M., H. Puspitawati, R.K.Z. Heck and K. Stafford (1993), 'Time-management strategies used by households with home-based work', *Journal of Family Economic Issues*, **14**, 69–92.
Wooldridge, J.M. (2003), *Introductory Econometrics: A Modern Approach* (2nd edition), Mason, OH: SouthWestern, A Division of Thomson Learning.
Zahra, S.A. and P. Sharma (2004), 'Family business research: a strategic reflection', *Family Business Review*, **17**(4), 331–46.
Zuiker, V.S. (1998), *Hispanic Self-employment in the Southwest Rising above the Threshold of Poverty*, New York: Garland.

6 Critical leader relationships in family firms

Nigel Nicholson and Åsa Björnberg

No man is an island. (John Donne, English poet, 1624)

Critical leader relationships (CLRs): a neglected topic

In this chapter we introduce a new concept to capture a theme that is neglected in the literature on leadership, yet which is familiar in the field of family business as a phenomenon of great significance: shared leadership. There is no dedicated literature on the subject, and the concept is almost unheard of in non-family firms. Nor is it much discussed in the family business literature, despite the fact that a recent survey found around one in seven US family firms considered themselves to be co-led (Raymond Institute, 2003).

Five academic literatures are relevant to the analysis of CLRs: leadership, leader–member exchange (LMX) research, studies of intimate relationships, research into executive boards and top management teams, and the general management practitioner literature. Only in the last of these does one find writing directly bearing on the topic. Our thinking draws on all of these in various ways, as follows.

From the leadership literature comes the idea that individual differences in personality dispositions are critical to business performance (Hogan et al., 1994; Miller and Toulouse, 1986) and as a cause of 'derailment' or failure (Van Velsor and Leslie, 1995). The literature also underscores the idea that personality effects, often represented as leadership style consequences, are contingent on contextual factors (Conger, 2005). Leader–member exchange theory and research is a fairly circumscribed subfield in which leader–follower relationships are an explicit focus, showing that qualities of interaction emerge through the exchange and enactment of specific behaviours by both parties (Schriesheim et al., 1999; Yukl, 2002). But it is only in the literature on marriage, the family and intimate relationships that one finds a focus on the outcomes of the personal identities of parties to a dyad, though even here this focus is quite rare (Cooper and Sheldon, 2002). Among these, the results generally support the idea that homophily (similarity of characteristics) forms the basis for attraction and partner choice (Luo and Klohnen, 2005; McPherson et al., 2001), though other patterns have also been noted (Costa and McCrae, 1992). Homophily has also been found to be both a cause and an effect of friendship (Gibbons and Olk, 2003). We await systematic research in the management sphere on this theme, though within the top management teams literature, there is work on the demographic profiles of these groups (Barsade et al., 2000; Bunderson and Sutcliffe, 2002).

A more sophisticated approach to individual differences has been suggested by Moynihan and Peterson (2002). They develop the notion of personality configurations as a key to understanding top team dynamics and decision-making, contingent on context. This 'contingent configuration' approach reasons that factors such as organizational culture and task type moderate the relationship between group-member personality and performance.

Although there is scarcely a literature to draw upon in terms of dyadic leadership configurations, two practitioner-oriented publications do offer encouragement however. The first was a book by David Heenan and Warren Bennis (1999) who forcefully made the case that leadership is often practised in too much isolated singularity, when co-leadership would enhance their effectiveness. At the same time they showed how in many cases the common view over attributes leadership to the single designated chief, overlooking the key role of a leader's principal partnership. The co-leadership theme was subsequently taken up in an article by O'Toole and colleagues (O'Toole et al., 2002) who discuss the preconditions for its success or failure. We shall revisit their ideas later, but first there is need to clarify what this concept means and how it manifests itself.

The use of the concept of 'co-leadership' in these works and in family business discussions tends to be rather loose and undifferentiated. We wish to advance the case that this potentially important phenomenon requires a more precise treatment if we are to advance knowledge about it. Accordingly we propose the adoption of a new concept to aid the study of leadership, what we shall call critical leader relationships (CLR henceforth), to cover the range of ways in which leaders cooperate with other individuals.

We concur with Heenan and O'Toole and colleagues that the reality of leadership is that individuals often have key partnerships, alliances and adjacent roles that contribute not just additively but interactively with the leader's personal qualities to deliver unique business outcomes. In the family business field one can see that CLRs are a significant force, and many differing in form from the models that may be found elsewhere – often close and personal, and sometimes traversing the boundaries of the firm. Indeed, this may be one of the most important contributory causes to the 'bivalence' of family firms (Tagiuri and Davis, 1996) – their tendency to outperform non-family businesses, while simultaneously being vulnerable to untimely demise.

Leadership in context

It is generally accepted that the personality, skills and style of a leader make a difference to the performance of even the largest businesses (Hogan et al., 1994; Miller and Toulouse, 1986). They do so principally in relationship to the outer and inner challenges of leadership. The outer challenge is the leader's role in scanning the environment, scouting for talent and resources, networking with external stakeholders and formulating a mission for the business to pursue. The inner challenge is the leader's role in setting up the systems and structures that make a business efficient and effective, and attending to the values, practices, and relationships that build a distinctive and healthy culture. Overarching both is the leader's symbolic and emblematic position, as a figurehead and focal point for the identity of the business. When all three are vested in one person there are the recognized dangers of an excess of unitary power, which when accompanied by charisma can drift into demagoguery (Conger, 1990), contrary to the heroic and 'romantic' ideal of leadership (Meindl, 1993).

Within the literature there are two correctives to the singular leader concept. One is the idea of institutional 'substitutes for leadership'(Kerr and Jermier, 1978) – systems, procedures, rules and the machinery of bureaucracy – that can buffer or even minimize the impact of leaders. The other is the idea of situational leadership, which focuses on contingencies that require the leader to adapt his or her style in order to be effective. These contingencies include the level of maturity or self-managing capability of followers, the leader's powers and position in the hierarchy, and the nature of the strategic challenge the leader faces (Yukl, 2002). The most popular current framework for partitioning these factors resolves to a typology of three main types, each with merits for a class of circumstances: transactional, transformational and charismatic leadership (Bass, 1985; Conger and Kanungo, 1987).

The literature in these areas does put the leader in context but maintains the individualistic presumption. We have no wish to completely discard this assumption, since the personal characteristics of the leader are important and irreducible, and often leaders do act and make decisions in relative isolation. However, we submit that this might be more the exception than the rule; more often leaders are framing their decisions after intense deliberations with one or more other parties. Our thesis here is that in many firms – family and non-family alike – there is one critical relationship axis that can be identified that tends to eclipse others. We do not rule out the possibility of multiple CLRs, or the existence of leaders who give over areas of decision-making to collective bodies, such as corporate boards and top management teams, but we assert here that the dyad as a unit of analysis has special significance (Gonzalez and Griffin, 2002; Thompson and Walker, 1982).

We appeal to the psychology literature to support this assertion – namely that human capacity for intense or continuous relationships is limited (Hendrick and Hendrick, 2000). Moreover, we tend to enact these relationships either serially, or in parallel in distinctive domains, to avoid the likelihood of role and relationship conflicts (Jackson and Schuler, 1985). The applied psychology literature has devoted a lot of attention on this presumption to dyadic leader–member exchange (Dienesch and Liden, 1986). In management practice, and particularly in family businesses, dyadic leadership is a well-known phenomenon. Many leaders have a 'number two', and the key partnerships of the chief executive with the person responsible for the finances (chief finance officer – CFO typically) and the chief operating officer (COO), are familiar in the corporate landscape, as is the partnership between chairman and chief executive officer (CEO) in many larger businesses.

Our general case, therefore, is that business success and failure may be as much attributable to a web of relationships as to the character of the person nominated as leader, especially in family business, where co-leadership is often the declared model. We believe this topic needs to be unpacked, for some leadership is more 'shared' than others, and the form of the sharing matters.

CLRs: an analytical framework

First let us offer a precise definition of a CLR: *a leader's partnership that provides advice or support on a regular basis for the most important work-related decisions the leader has to make*. We do not offer a definition of a leader. In this chapter most of our attention is devoted to the person or persons designated or recognized as the chief executive authority of a business. However, the analysis we offer could apply to various other leadership positions – such as divisional head of a large business, or temporary leader of a project.

In this chapter, we distinguish between *CLR forms* and *dyad types*. CLR forms refer to different models of CLR that may be adopted, in terms of variation in power distance and role differentiation. Dyad types refer to the various combinations of roles and kinship relationships that may occur in family firm CLRs.

CLR forms

Our analysis is based upon observations we have made formally and informally in the field – that is, the reports of informants in the family business arena, and our ongoing studies of family firms – and in our case analyses of archival material. From these sources we observe that CLR partnerships exist on a continuum with varying degrees of power distance and role differentiation. Distinctive forms of CLRs can be identified as points on the continuum. We perceive the main forms to be the following.

Autonomous This refers to cases where there is a single acknowledged leader of a business who generally makes decisions alone, either with or without receiving advice and information. Owner-managers of new enterprises may be in this position, but so also may leaders in high power-distance organizations, where they have no single person they choose to consult with regularly. In some cases the leader may refer only to a group structure – a board or executive team – for advice and guidance, without making regular recourse to any single partner.

Assistant The leader here relies on one person more than others to execute decisions and anticipate their feasibility. This is where the leader has a preferred chief of staff or a key personal assistant on whom s/he depends for decision-making. Most of the influence flows from top down, but support is primarily bottom up, from the number two to the number one. Power distance and role differentiation are both high.

Reciprocal These are CLRs where there is a clear separation and division of labour, usually hierarchical, as in junior–senior partnerships where a leader makes policy with the counsel of a close subordinate. The chairman–managing director (MD) relationship has this character where the former is the acknowledged leader of the business and the MD or CEO has bounded responsibility for business operations. CEO's relationship with CFOs, COOs and other designated domain specific leaders fall into this category. Influence and support are bi-directional usually, but asymmetrical within domains. Power distance and role differentiation are moderate to high.

Balanced This is where co-leadership is declared, but power and responsibility are not differentially distributed across roles. One may be nominally superior to the other, but their contributions are mutual as co-leaders of a business. Support flows both ways, and influence from party to party according to their domains of expertise and authority. The chairman–CEO relationship has this character where both share an interest in the external and internal environments and both consider themselves to be leaders of the business, yet they retain distinct domains of responsibility. Power distance and role differentiation are medium to low.

Equal This covers cases where leadership is genuinely shared between two people (rarely more). Influence and support are notionally equal and mutual, though this will depend upon the personality dynamics of the relationship. They may share the same title, but not necessarily. If they do not, then they still treat each other as co-leaders, have equal rights of access to the main domains of decision-making and are both are able to substitute for each other. Power distance and role differentiation are both low.

Transitional In some cases the distribution of power and authority is in flux, with a leader transferring authority to a protégé. Here support and influence are mainly top down to start with, but changing as the process unfolds, if it is able to do so for this process is notoriously fraught with difficulty in family firms, where seniors fail to withdraw and successors are not adequately mentored. Power distance and role differentiation are changing; reducing where this is a straightforward transfer of leadership.

These CLR forms may be found in all types of business, but they are not equally distributed across types of firm. Family businesses present a stark contrast in some forms of preferred CLR. This is not widely understood outside the family business area. In the UK a BBC television programme has featured a former corporate chief executive, each week trouble-shooting a family business case, under the programme title, *I'll Show Them Who's Boss*. The underlying presumption was that no business can prosper without secure and monolithic leadership, and generally hostile to the co-leadership forms that many family businesses adhere to.

CLRs in family businesses: six possible dyad types

In family-business, unique combinations of CLR partnerships occur – what we shall call 'dyad types' combining different relations within the family and family with non-family. These dyad types can exist inside (intrafirm) or outside the firm (extrafirm). Each of these dyad types tends toward different degrees of power distance more than others, as described above. A priori, one can distinguish six different dyad types in family firms:

1. Intergenerational – usually father–son, sometimes father–daughter, and rarely mother–child; in the extended family, they can also occur between uncle/aunts and nephews/nieces.
2. Intragenerational – often between cousins and siblings.
3. Copreneurial – husband and wife partnerships; can apply to buddies who found a business, which then is maintained as dual family business.
4. Intrafirm family/non-family – takes many forms, most common is family chairman and non-family CEO but many other variations occur.
5. Extrafirm family/non-family – most commonly family leaders with non-family external advisers; also applies where the leadership is non-family but bound into a CLR with non-executive family owner as a source of influence and advice.
6. Intrafamilial informal – often between a family leader and a spouse who is not involved in the day-to-day running of the business, can also apply to siblings in a similar situation.

In Table 6.1, examples that we have come across in family businesses are used to illustrate different CLR forms and dyad types.

Our proposal raises some important questions, to which research needs to seek answers, to do with what factors shape the character of CLRs: (1) what are the 'inputs' or determinants of CLRs? and (2) which CLR forms and partnership combinations of CLR work best, that is, what are their 'outcomes', especially their effectiveness? The two questions are linked. We shall consider the first question here, then review the dyad types in family business CLRs in greater detail. We will conclude the chapter with a discussion of critical leader relationships and leadership effectiveness.

Table 6.1 Illustrations of CLRs in family firms

CLR form: Assistant *Dyad type:* Intrafamilial informal	In this family business, support is provided to the chairman by a spouse. The spouse is consulted in every important decision the chairman has to make. The spouse has no formal role in the family business, and support is given in an informal way
CLR form: Reciprocal *Dyad type:* Intrafirm family – non-family	A chairman/joint managing director in a family business works very closely with the non-family MD. This relationship is described as 'joint MD-ship', and is a replica of the CEO's father's leadership configuration, who had a similar partnership with the previous non-family MD. The degree of trust is very high, but there is a division of responsibility and job roles – the non-family MD is in charge of operations. However, there is some overlap in decision-making. The CLR form can be described as 'Reciprocal'
CLR form: Balanced *Dyad type:* Copreneurial	This husband and wife team had a 'proverbial coin-toss' about who was going to hold the title of president. They do not use titles on their business cards, and lead the company as a couple. Decisions are made together, and influence varies according to area of expertise in a balanced way
CLR form: Equal *Dyad type:* Intrafirm intragenerational	Two twin brothers lead this family firm in a truly equal fashion. They have no designated job titles or roles. Their areas of responsibility overlap, and they have had to create a reporting system with subordinates in order to avoid mixed messages as far as possible
CLR form: Transitional *Dyad type:* Extrafirm family – non-family	In this family firm, a professional coach has been brought in to provide personal development coaching and mentoring to support a successor in his emerging leadership role. This is expected to last over a limited period of time, and can thus be considered as 'Transitional'

Determinants of family firm CLRs

The inputs to CLRs consist of an intricate set of interweaving factors: the personal qualities of the individuals concerned; their relatedness; the surrounding milieu and structural imperatives. We shall look at these in turn, but it is important to keep in mind that in many cases relationships are a matter of choice. We might not be able to choose to whom we are related but we can choose which of our relatives we wish to treat as friends, confidants and helpers. We can also strive to create conditions that sustain the kinds of relationships we prefer.

Relatedness This is an especially important consideration in family business. Elsewhere Nicholson (2005) has supplied a neo-Darwinian analysis of the kind of force this exerts. On the one hand, the genetic bond pulls people together by ties of dependence and affection (Neyer and Lang, 2003). At the same time there are inherent conflicts. The parent–child relationship typically involves a pull between the interests of parents seeking to shape and control their offspring's destiny, and children seeking to grasp autonomy and resources to enable them to determine their own destiny (Trivers, 1974). This drama is played out in different ways. Parents and children often have insights that enable them to exercise restraint or to depart from the stereotypical model. Parent–child partnerships are a familiar phenomenon in the family business scene, and they may enjoy long periods of stability; but mostly it is an evolving relationship that requires continual mutual adjustment, which may prove difficult to implement.

Siblings are also locked into an ambivalent bond of love and rivalry, according to the literature (Dunn and Plomin, 1990; Hertwig et al., 2002; Sulloway, 2001). Age gaps and gender differences mitigate the rivalry, though it depends upon the other factors discussed below: the character of the individuals and contextual factors that determine whether the bond will sustain a working relationship. In the case of cousins, the Darwinian perspective predicts that the bond will be genetically and emotionally weaker. Here the struggle may be less to withstand conflict so much as to maintain enough interest to cooperate.

Finally, there is the bond of affinal partners (marriage or its equivalence). Here the genetic bond is absent, except through the equal (genetically) shared interest in offspring. Asymmetries in offspring relationships (step-relationships) are a source of potential divergence in marital relationships, and the absence of offspring makes the association as fragile as any friendship or love alliance (Danes and Olsen, 2003; Fitzgerald and Muske, 2002; Foley and Powell, 1997). In extended family networks, the non-blood ties of marriage that link branches can be a fracture line (Gersick et al., 1997). The relevance of genetic relations to CLRs is that they represent different 'pulls' of potential commitment to collaborate. They are, of course, moderated by other factors such as age gaps, family size and available alternative CLRs.

Personal identity Personal chemistry, as it is popularly termed, is the essence of the CLR. This occurs at a number of levels – ability, values, experience and knowledge, and personality. Do opposites attract or will people team up on the basis of complementarities? Both seem to occur but, surprisingly, research has only scratched the surface of these dynamics. Much of the work in this area comes from the literature on love and marriage, which generally favours the 'birds of a feather' or 'homophily' hypothesis (Gibbons and Olk, 2003; Johnson et al., 2004), though all kinds of personality combinations can be found (Costa and McCrae, 1992). In friendship and working association, the tendency for homophily is more pronounced (McPherson et al., 2001). Moreover, whatever differences there are in traits, there is evidence that emotional convergence will occur over time in close relationships (Anderson et al., 2003).

Behaviour genetics points out, however, that because non-additive genetic influences determine much of the variation in personality and other individual differences, the correlations between the characteristics of close family members – parent and children, and siblings – is close to zero, except in the case of identical twins (where associations tend to be upwards of 0.5, depending upon the measured trait) (Loehlin et al., 1998; Lykken et al., 1992). The consequence is that there is more variation or difference where family members are yoked together than when people can associate by the principle of 'elective affinity' – working with people whom they have chosen to partner because of 'good chemistry'.

There are clearly dangers in this – principally of 'cloning' – CLRs with people who share the leader's biases. This may just double up the leader's blindspots when what they really need is challenge, or the support of someone with a different perspective. The ideal balance in a CLR would be where there is sufficient similarity in qualities that enable empathy and exchange to take place, and contrast in those that are most instrumental in mastering the challenges facing the leader. We shall return to this theme at the end of this chapter.

Milieu and structure This refers to the ambient culture and networks within which relationships form. This is a function of firm, family and national culture (Shams and Björnberg, 2006). Biases towards different dyad types and forms of CLRs can be expected in different cultures, depending on such factors as power distance and prevailing cultural norms to do with family life. For instance, in Latin societies, high levels of emotional cohesion could act as a buffer against sibling rivalry in intragenerational CLRs. Similarly, in Japan, reciprocal CLR forms may be more common owing to a high degree of power distance. Factors such as these govern the availability of candidates for a CLR and the prevailing norms of conduct that would support a CLR.

All families develop their own cultures (Björnberg and Nicholson, 2005; Nicholson and Björnberg, 2004), and these too will support or discourage the possibility of a family based CLR. Likewise, the prevailing organizational culture: local norms govern the CLRs that may be considered feasible. Shared leadership may be a feature of the founding period – where relatives, buddies or marriage partners are working together – and continue thereafter. In other businesses the leader can be a more remote figure. The causes can be both structural and cultural; with high power distance, linear hierarchies are reinforced by local norms of leader remoteness. In some cases the form of the CLR is determined surrounding organizational or governance systems. By virtue of the strictures of boards a chairman and a CEO may be locked into a partnership that neither of them might otherwise willingly choose. Tradition may also play a part. In many businesses the finance director has a critical role and is often a party to a CLR. But structures are mutable, and one of the first acts of many incoming leaders is to revise structures to support the CLRs they prefer. Changing the culture may be more difficult.

CLRs in family firms – a closer look at the partnership combinations and their specific challenges

Table 6.2 summarizes what might be predicted for the most common intersections between CLR forms and dyad types. These derive more from our archive of case reports and the observations of commentators than from systematic research, of which there has been little or none. Predictions at this stage are tentative, although initial results from current and ongoing pilot work is providing strong support for the viability of the overall conception. More systematic empirical investigations seek to underpin the validity of the approach.

Intergenerational CLRs

The key feature of the intergenerational CLR is that it grows. It has antecedents in immaturity and dependency, and it has a future in succession or transition. This trajectory has several implications. One is that early roots of the relationship need to have been healthy and nurtured. A second is that some mentoring should take place to bring the junior party into partnership with the senior (Wright and Wright, 1987). A third implication is that the mature form of the relationship will transit into another form, if the parties are sensitive to the need for it to evolve. The relationship is never likely to be of the equal or balanced

Table 6.2 Predominant CLR forms and family dyad types – likely patterns

Power distance/ Role differentiation	Forms of CLR	Dyad types in Family Business CLRs					
		1) Inter-generational	2) Intra-generational	3) Copreneurial	4) Intrafirm family/non-family	5) Extrafirm family/non-family	6) Intrafamilial informal
High	Autonomous	x	x	Possible	x	x	Common
	Assistant	Short-term	Rare	Common	Possible	Common	Common
	Reciprocal	Possible	Common	Common	Common	Possible	Rare
	Balanced	Short-term	Common	Possible	Common	Rare	Rare
	Equal	Rare/unstable	Possible	Rare/unstable	Rare	x	x
Low	Transitional	Common	Rare	Rare	Possible	x	Rare

forms, but more likely to move through stages of assistant–reciprocal–transitional forms. There is one notable piece of research on this CLR – Davis and Tagiuri's (1989) examination of father–son relationships in family firms. They found it most problematic when the age gap was modest; the parent not ready to let go at the time when the son is becoming impatient for leadership responsibility.

Intergenerational research reports that the two most common types of conflicts between adult children and their parents concern communication and interaction and habit and lifestyle choices (Clarke et al., 1999). In family firms, clashes can also arise owing to differences in education and expectation of work–life balance. Depending on the contingencies of the relationship, such as personality and patterns of communication, conflicts can present an opportunity for renewal if it surfaces as a creative debate about goals and values (Lansberg, 1999). Who better to challenge the parent-leader than his or her child? Or, under certain circumstances, who worse?

In the absence of specifically relevant research it is not possible to make firm predictions about the role of gender in intergenerational CLRs, but research on dyads in nuclear families has provided indicative evidence that different combinations are not distinct from each other so much as some dyads have unique, distinguishing features (Russell and Saebel, 1997). Anecdotal evidence suggests that father–daughter are generally less problematic than father–son relationships, possibly owing to a lesser need for differentiation and thus competition, but the field awaits systematic research to test this and other ideas about family CLRs.

Intragenerational CLRs

Many family businesses develop in the second generation around the sibling bond. The so-called sibling partnership (Gersick et al., 1997) usually rests upon one ascendant sibling, with others content in junior roles or as disinterested shareholders. True equality, as in intergenerational CLRs, seems to be a rarity, though honour may be served by having titles that denote equality. More usually there will be mutual acceptance of the ascendancy of one sibling over the others. Where strong norms of primogenitor prevail in the local culture, incongruent hierarchy – the superiority of a younger over an older sibling – may be difficult to institute, but necessary to achieve a good fit with the different qualities of the parties (Barnes, 1988). Successful family firms are those where conflict can be positive force (Harvey and Evans, 1994; Kellermans and Eddleston, 2004), which it has the unique opportunity to be because the strength of the kinship tie allows more adaptable and therefore dynamic problem-solving than would otherwise be possible.

Cousin leaders have a different set of challenges to face. At this stage of the life cycle, the firm and the family are often more complex, and ownership is more dispersed. Cousins do not share the same parents, and loyalty to the company does not rest on a personal bond to the founder, who is removed by at least one generation. Thus, cousin relationships tend to be less intense and more purely political than the greater emotional bond of siblings (Gersick et al., 1997; Ward, 2004). For example, different family branches may present a source of cousin rivalry, in which one cousin will struggle to keep the firm in his/her branch of the family (Lansberg, 1999).

It is rare for intragenerational CLRs to take on the assistant form. However, the exception to this rule is when one sibling acts as a surrogate parent. Reasons for same-generation siblings or cousins assuming a parental stance include wide age gaps and/or the premature death of a parent. The degree of conflict or rivalry depends on whether this is an accepted role division in the family or not.

Similarity in age and possible similarity in family stages may lead to competing needs, but also to increased understanding. As such, reciprocal and balanced forms of intragenerational relationships are quite common. These often build on division of labour, as the following quote from a team of sisters in a two-generation family business illustrates: 'Elina will be responsible for the financial and marketing issues of the company, whereas I will be responsible for issues related to research and development . . . but at the same time the final decision about every aspect concerning the business will be made by both of us' (Sakellariou, 2004, p. 26).

The way in which the succession process is handled is of great importance to avoid conflicts in intragenerational CLRs. The transitional intragenerational CLR is more of a rarity, with the exception of the surrogate parenting form or alliances between widely age-separated cousins.

The copreneurial CLR

Couples and partners in business – 'copreneurs' – face a range of challenges, compounded by the duality of professional and romantic involvement, as well as gender and dominance dimensions in the relationship. Research confirms that it is often hazardous because of uncertainties and concerns about inequities in division of labour, rewards, rights and responsibilities (Danes and Olsen, 2003; Fitzgerald and Muske, 2002; Foley and Powell, 1997). In effect, this amounts to the relationship becoming overloaded with areas where cooperation is required, with insufficient opportunities for separate development.

Research shows that copreneurs are significantly less egalitarian than dual-earner marriages (Marshack, 1994). Although the spouses may co-own and co-lead the business, a more traditional division of labour is observed among copreneurs, with the woman normally bearing the major responsibility for childcare and household management. The implication is that the lack of a clear physical boundary between work and home requires copreneurs to rely on gender roles as conceptual boundaries. It is argued that the stereotypical gender-roles that copreneurs often become locked into contribute to why many firms are unable to transfer management to the second generation. This gene-drain results from a lack of socialization of the young women in the family and the alienating effects on children as a result of watching the founder-owner work excessively long hours (Marshack, 1994).

Very little is known about same sex-copreneurs and the degree to which they enact traditional gender roles, but anecdotal evidence suggests that male gay relationships are gendered, and that traditional gender roles apply.

Copreneurial CLRs are the most flexible in terms of the different CLR forms they assume, since they are not constrained by familial bonds. Commonly, they assume the assistant or reciprocal form, in which one party accepts a significant degree of subordination. Also, it is possible for them to be autonomous, or at the other end of the spectrum, balanced. Copreneur CLRs of the truly equal form are rare and unstable, for the reasons we have identified. Transitional copreneurs CLRs are also rare, except where there is progressive withdrawal by one party into domestic duties or other interests, leaving the spouse to find new partnerships or lead alone.

Intrafirm family/non-family CLRs

This is probably the most common and important CLR in the family firm, for the partnership between a family and a non-family leader has two special qualities. One is that it is usually an indispensable element in the development of the family firm that aspires to be anything more than small and domestic. In the growing firm, the introduction of a non-family leader is inevitable except in the most populous and talented extended families. Second, the non-family leader has a special role of bringing professionalism and the intelligence of wider experience into the business. Various role combinations and CLR forms are possible, with either the family or non-family leader in the ascendant leadership role. The most common, however, is that the family member occupies the more general and senior position, and the non-family member is in a reciprocal relationship associated through an operational or finance specialism. Balanced and equal leadership may also occur, where there is genuine power sharing, and spheres of responsibility are adjacent and overlapping, as occurs in many chairman/CEO relationships. In some governance systems a family or non-family CLR partner may be in the role of non-executive director. This is especially likely in larger and more mature businesses. However, family–non-family CLRs usually stop short of truly equal sharing of the leadership role, according to anecdotal evidence, certainly occurring less readily than a transitional CLR where power is progressively transferred, generally from family to non-family. The flow can reverse, as when non-family can also mentor family next-generation would-be leaders into their role.

Common report suggests that the search for a good fit of the non-family leader is difficult, often because they will be bound into such a CLR. The demands for loyalty are intense, as is the need for psychological qualities that will enable the non-family leaders to manage the emotional complexities of the role. Least suitable may be self-made entrepreneurs, a group who are said to find it most troublesome to create reciprocal mentorship-style CLRs, owing to their autonomous personality style (Kets de Vries, 1996; Liberman, 2004).

Extrafirm family/non-family CLRs

Critical leadership relationships across a firm's boundaries between family and non-family are mainly between leaders and long-standing and trusted family friends or professional advisers, acting as coaches, mentors or catalysts to change. Family members in this CLR may include non-executive family members or family owners. In some cases, lesser frequency of interaction means more detachment, and a greater possibility to provide an outside point of view. One challenge for the outside party in a professional advisory relationship is that the client identity may change. In terms of the CLR, the leader is the client, but in many cases the focus can switch to different members of the family or the whole family. Managing these multiple roles requires a special blend of flexibility, sensitivity and ethical awareness on the part of the adviser, especially if he or she has more than one close relationship with family members (Bork et al., 1996).

CLR forms are restricted for this dyad type. The external non-family party often acts as a sounding board, with the family leader making the ultimate decisions – that is, the leader his or her their autonomy and the CLR is no more than an intermittent arrangement. Otherwise the non-family party may be in an assistant role. It is possible for the relationship to assume a more reciprocal form, with two-way influence and support, but truly balanced relationships of this form are not possible in the absence of executive responsibility. Adviser–client relationships, since they involve a purchase of services, are more unilateral, whereas friendships require a more mutual approach, and thus are more emotionally involving. Reciprocal and balanced CLRs of this type are more feasible when the relationship is informal (such as friendships) or excludes day-to-day running of the business (such as non-executive directorships).

Intrafamilial informal CLRs
Intrafamilial informal CLRs occur when a family leader relies on another family member, who is not in a formal executive role, for support and advice. This is probably one of the least understood but most important of CLRs. Its most common manifestation is usually the family leader and non-executive spouse, who is the key source of influence, but without ever being publicly recognized as such. Wives have been termed 'chief emotional officer' to capture the essence of the role. Although important, it is not a sustainable model in the growing business. This may be a founder's most preferred CLR during the early years of the firm's development, before other more formal and directly involved partners become available. The nature of an informal wife–husband CLR is often characterized by a high frequency of interaction. When the informality persists but the boundaries of the non-founder spouse involvement are blurred, this can lead to tension over decision-making, tasks or values and beliefs (Danes and Olson, 2003). For example, women who work in the business, but are not designated owners, often do so without pay or a job description. Nevertheless, the intrafamilial informal CLR has a lesser degree of complexity compared with that of the copreneurs, given the absence of the professional involvement of one party. Other varieties of this CLR may be found where a family leader seeks or is subject to the authority of retired family founders, or family owners not directly involved in the business. If these are sufficiently powerful they can be yoked into reciprocal or even balanced CLRs, but not for long. Only family members directly engaged in executive roles will be likely targets for any kind of significant CLR. The transitional form will only apply when a family member not formally involved is being inducted into the business.

Our review leads to a few observations. One is that 'equal' CLRs are a rarity, probably much more uncommon than role titles would lead one to believe. These titles may be formulated to conceal the reality of unequal relationships where this would be incongruent with family norms or external expectations, that is, they are face-saving or political labels. Second, some types of family CLR are more flexible, or can take more forms than others. Family–non-family partnerships (including copreneurs) seem to have more possible CLR forms than family–family CLRs, though the latter may have greater depth and power when they are effectively mobilized. This last point leads us into the question of what does determine their effectiveness. This we shall now review.

CLRs and leader effectiveness
Before considering the issue of CLR effectiveness, we briefly comment on what the leadership literature says about the effectiveness of the individual leader. Three broad conclusions can be drawn. First, that fit between leader characteristics and situational demands is a key determinant, hence the need for leaders to adapt their style over time to changing circumstances, or for organizations to switch leaders, seeking new leader characteristics for new challenges if the current leader is unable to adapt. This is the

contingency view of leadership, which has a long pedigree in the literature (House, 1971). Second, there is a literature on individual differences that does identify certain traits (such as the personality dimension 'conscientiousness') with high performance, and others as risk factors, impelling leader 'derailment' – sometimes called the 'dark side' of leadership (Conger, 1990; Van Velsor and Leslie, 1995). Third, there is a literature than can be broadly characterized as processual; to do with the skills that every leader should master in order to secure success. These include relationship-building, effective communications with followers, sound methods of decision-making, and ability to influence (through vision, story-telling, and trust) (Bass, 1998; Collins, 2001; Denning, 2004).

These three aspects of leadership effectiveness – leadership situations, processes and qualities – would seem equally applicable to CLR effectiveness. The two works referred to at the outset of this chapter point in this direction. In the Heenan and Bennis (1999) book it is clear that what they call 'co-leadership' is much more than what we have called the reciprocal model, and their recommendations are really focused on what it takes to be a great number two. Their conclusions amount to the need for a second in command to be resilient, cooperative, committed, insightful and flexible – necessary to sustain teamwork with often dominant and difficult bosses. This is fine as far as it goes, but these worthwhile case-based reflections sharpen the feeling that the field needs a more systematic analytical approach to the interdependencies.

The article by O'Toole et al. (2002) charting 'the promise and pitfalls of shared leadership', moves a step closer to this, by implicitly embracing several of the CLR forms we have identified earlier, but in a somewhat undifferentiated fashion. Sally (2002) lists a number of qualities, structures and rules from the Republican Roman model of shared leadership to describe how effective co-leadership can be achieved in modern society. Like Heenan and Bennis their focus is mainly what on we have designated 'reciprocal' co-leadership. Their conclusion is that shared leadership is needed when 'the challenges a corporation faces are so complex that they require a set of skills too broad to be possessed by any one individual' (O'Toole et al., 2002, p. 68). This is a good but partial answer to the question, and in the remainder of the article the authors do go beyond their own formulation to discuss many of the elements we shall try to capture, using the SPQ framework, summarized in Figure 6.1.

The leadership situation

The *complexity* of the challenge facing the leader is one important dimension, as O'Toole et al. suggest. The imperative here is for the CLR to achieve integration on the 'two heads are better than one' principle. *Breadth* of domain is another important dimension of situational demand. This includes the familiar bilateral model of one taking responsible for the outer challenge of leadership (the external environment) and the other leader keeping house – overseeing the internal community. A third dimension of particular relevance to family businesses – a chief reason for cousins to be yoked together – is *constituencies*. By this we mean sets of factional interests that may be non-aligned or relatively independent. Forms of CLR can be prudential against these turning into factions. Fourth and finally are *subcultures*. These may or may not coincide with hierarchical or lateral domains of organizational structure, but it may be important to have involved in CLRs individuals who can speak knowingly to, and be trusted by, different populations.

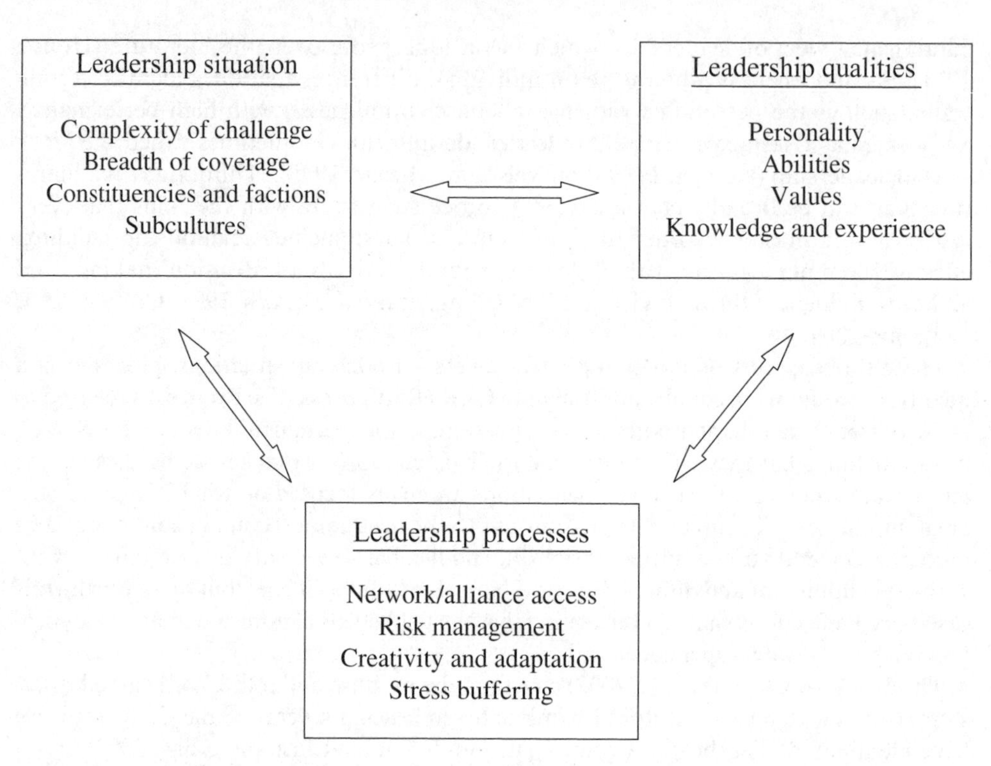

Figure 6.1 The SPQ (situation–process–qualities) model of leadership effectiveness applied to CLRs

Leadership processes

The list of processes relevant to leadership situations is potentially as long as the varieties of positions in which leadership is possible. Without the guidance of prior research on CLRs we are forced to speculate on which of these are likely to be key. Again we focus on four. First is access to *networks and alliances*. This is a key aspect of partnerships – and of course the partnership model is par excellence a CLR model. Parties pool their contacts with potentially great economic benefit to the business. Second is *risk management*. In environments were the costs of failure are potentially high and the probabilities uncertain, a close partner in decision-making may be a vital backstop to protect against catastrophic misjudgement. Usually risk management is widely shared within businesses where it is a factor, but leaders have a constant need for devil's advocates and critical restraints against the driven qualities of many leaders. Third is *creative and flexible adaptation*. A CLR can be the source of challenged assumptions, fresh perspective and open ideas, against the predominant tendency of leadership to be task-driven and convergent. Fourth CLRs are a buffer against *stress*. It is often said that it is lonely at the top, and wherever the buck stops there is pressure to be borne. Stress has unpredictable effects – driving some people to conservatism and other to high risk-taking. A CLR is not a guarantee against this but there is an extensive literature about the buffering effects of social support, and a CLR is likely to be powerful in this direction.

Leadership qualities
The four areas of principal relevance are *personality*, *abilities*, *values* and *knowledge/ experience*. One can argue that ideally in a CLR one requires homophily on *values* and diversity in *abilities*, so that centrifugal and centripetal forces are in balance. Within the domain of *personality* the answer is likely to be much more complex – what has been called a contingent configural approach (Moynihan and Peterson, 2002). For example, low emotionality might be something that both parties need in highly conflictual settings, but in other circumstances diversity in emotional responsiveness might be more appropriate. We lack space to develop all relevant hypotheses here, but various triangulations of personality, situation, and CLR form would seem to be implicated. Finally, in the domain of *knowledge and experience*, again one can assert that some common ground can help to ensure an easy flow of communication and trust, and some diversity in order to comprehend the diversity of the leadership situation.

It will be apparent that to take these ideas further requires a more fully worked out set of predictions around the complex interdependencies of S, P and Q factors, and the forms of CLR that may be enacted. For example, one can hypothesize, and indeed the literature supports the idea, that family–non-family shared leadership would be best suited to leadership situations where family and non-family interests are both powerful, and leadership situations where diverse alliances are needed.

Looking at the SPQ diagram one can see that CLRs are a means, as Heenan and Bennis (1999) and O'Toole et al. (2002) point out, for the contents of the Q box and the P box to be enlarged to encompass more of what is in the S box, that is, a broader repertoire of attributes and skills to be deployed.

Conclusion
To conclude, this analysis has several implications.

1. A CLR can actually increase the range of challenges an organization can face and master.
2. Duplication of qualities and processes is only desirable where the magnitude of the challenge requires weight of numbers.
3. Self-awareness by both parties to a CLR is essential if they are to know what they have to attend to.
4. Insights into each other and a stream of relevant communication between parties to a CLR are essential maintain alignment.
5. Regular attention needs to be paid to how the CLR should adapt its form to a changing environment, and respond to changes in the relationship through the personal development or needs of either party. CLRs have their life span and there will come a time to disband them without rancour.

And specifically in relation to family firms. We can add:

6. Leaders and other agents need to think dispassionately about which form of CLR fits the situation and the qualities of the people, rather than creating, for example, a sham equal CLR model that will not in reality be sustainable.

7. The more flexible the family firm's approach to types of CLR that can be sustained, the more likely the firm is to capture the positive premium of 'familiness'.

Further research is needed to shed light on this clearly important and neglected topic. There is a need for descriptive studies on the extent of CLR forms as they appear in firms of all types, and for research to assess whether some are more sustainable in the family firm context than others. In our work we are deploying a psychometric approach to the analysis of qualities of CLRs, and the consequences of particular combinations, taking a contingent approach to identify the forms of CLR and leadership challenges that particular dyadic personality configurations are able to deal with least problems and most effectiveness.

References

Anderson, C., Keltner, D. and John, O.P. (2003), 'Emotional convergence: implications for individuals, relationships, and cultures', *Journal of Personality and Social Psychology*, **84**, 1054–68.

Barnes, L.B. (1988), 'Incongruent hierarchies: daughters and younger sons as company CEOs', *Family Business Review*, **1**, 9–21.

Barsade, S.G., Ward, A.J., Turner, J.D. and Sonnenfeld, J.A. (2000), 'To your heart's content: a model of effective diversity in top management teams', *Administrative Science Quarterly*, **45**, 802–36.

Bass, B.M. (1985), *Leadership and Performance Beyond Expectation*, New York: Free Press.

Bass, B.M. (1998), *Transformational Leadership: Industrial, Military and Educational Impact*, Mahwah, NJ: Lawrence Erlbaum Associates.

Björnberg, Å. and Nicholson, N. (2005), 'Family climate: the development of a new measure for use in family business research', London Business School working paper.

Bork, D., Jaffe, D.T., Lane, S.H., Dashew, L. and Heisler, Q.G. (1996), *Working with Family Businesses*, San Francisco, CA: Jossey-Bass.

Bunderson, J.S. and Sutcliffe, K.M. (2002), Comparing alternative conceptualizations of functional diversity in management teams: process and performance effects, *Academy of Management Journal*, **45**, 875–93.

Clarke, E.J., Preston, M., Raksin, J. and Bengtson, V.L. (1999), 'Types of conflicts and tensions between older parents and adult children', *The Gerontologist*, **39**, 261–70.

Collins, J. (2001), *Good to Great*, New York: HarperCollins.

Conger, J. (2005), 'Leadership, contingencies', in N. Nicholson, P. Audia and M. Pillutla (eds), *The Blackwell Encyclopedia of Management: Organizational Behavior* (2nd edition), Oxford: Blackwell.

Conger, J.A. (1990), 'The dark side of leadership', *Organizational Dynamics*, **19**(2), 44–55.

Conger, J.A. and Kanungo, R.N. (1987), 'Toward a behavioral theory of charismatic leadership in organizional settings', *Academy of Management Review*, **12**, 637–47.

Cooper, M.L. and Sheldon, M.S. (2002), 'Seventy years of research on personality and close relationships: substantive and methodological trends over time', *Journal of Personality*, **70**, 783–812.

Costa, P.T. and McCrae, R.R. (1992), *Revised NEO Personality Inventory (NEO PI-R) and NEO Five-Factor Inventory. Professional Manual*, Lutz, Florida: Psychological Assessment Resources, Inc.

Danes, S.M. and Olson, P.M. (2003), 'Women's role involvement in family businesses, business tensions, and business success', *Family Business Review*, **16**, 53–68.

Davis, J.A. and Tagiuri, R. (1989), 'The influence of life-stage on father–son work relationship in family companies', *Family Business Review*, **2**, 47–74.

Denning, S. (2004), 'Telling tales', *Harvard Business Review*, **82**(5), 122–8.

Dienesch, R.M. and Liden, R.C. (1986), 'Leader–member exchange model of leadership: a critique and further development', *Academy of Management Review*, **11**, 618–34.

Dunn, J. and Plomin, R. (1990), *Separate Lives: Why Siblings Are So Different*, New York: Basic Books.

Fitzgerald, M.A. and Muske, G. (2002), 'Copreneurs: an exploration and comparison to other family businesses', *Family Business Review*, **15**, 1–15.

Foley, S. and Powell, G.N. (1997), 'Reconceptualizing work–family conflict for business/marriage partners: a theoretical model', *Journal of Small Business Management*, **35**, 36–47.

Gersick, K.E., Davis, J.A., Hampton, M.M. and Lansberg, I. (1997), *Generation to Generation*, Cambridge, MA: Harvard Business School Press.

Gibbons, D. and Olk, P.M. (2003), 'Individual and structural origins of friendship and social position among professionals', *Journal of Personality and Social Psychology*, **84**, 340–51.

Gonzalez, R. and Griffin, D.W. (2002), 'Modeling the personality of dyads and groups', *Journal of Personality*, **23**, 901–24.

Harvey, M. and Evans, R.E. (1994), 'Family business and multiple levels of conflict', *Family Business Review*, **7**, 331–48.
Heenan, D.A. and Bennis, W. (1999), *Co-leaders: The power of Great Partnerships*, New York: Wiley.
Hendrick, S. and Hendrick, C. (2000), *Close Relationships: A Sourcebook*, Thousand Oaks, CA: Sage.
Hertwig, R., Davis, J.N. and Sulloway, F.J. (2002), 'Parental investment: how an equity motive can produce inequality', *Psychological Bulletin*, **128**, 728–45.
Hogan, R., Curphy, G.J. and Hogan, J. (1994), 'What we know about leadership: effectiveness and personality', *American Psychologist*, **49**, 493–504.
House, R.J. (1971), 'A path–goal theory of leadership effectiveness', *Administrative Science Quarterly*, **16**, 321–39.
Jackson, S.E. and Schuler, R.S. (1985), 'A meta-analysis and conceptual critique of research on role ambiguity and role conflict in organizations', *Organizational Behavior and Human Decision Processes*, **32**, 16–78.
Johnson, W., McGue, M., Krueger, R.F. and Bouchard, T.J. (2004), 'Marriage and personality: a genetic analysis', *Journal of Personality and Social Psychology*, **86**, 285–94.
Kellermans, F.W. and Eddleston, K.A. (2004), 'Feuding families: when conflict does a family firm good', *Entrepreneurship Theory and Practice*, **29**(3), 209–28.
Kerr, S. and Jermier, J.M. (1978), 'Substitutes for leadership: their meaning and measurement', *Organizational Behavior and Human Performance*, **22**, 375–403.
Kets de Vries, M.F.R. (1996), *Family Business: Human Dilemmas in Family Firms*, London and Boston, MA: International Thompson Business Press.
Lansberg I. (1999), *Succeeding Generations*, Boston, MA: Harvard Business School Press.
Liberman, V. (2004), 'Mentoring the lonely CEO', *Across the Board*, **41**(6), 59–60.
Loehlin, J.C., McCrae, R.R., Costa, P.T. and John, O.P. (1998), 'Heritabilities of common and measure-specific components of the Big Five personality factors', *Journal of Research in Personality*, **32**, 431–53.
Luo, S. and Klohnen, E.C. (2005), 'Assortative mating and marital quality in newlyweds: a couple-centered approach', *Journal of Personality and Social Psychology*, **88**, 304–26.
Lykken, D.T., McGue, M., Tellegen, A. and Bouchard, T.J. (1992), 'Emergenesis: genetic traits that may not run in families', *American Psychologist*, **47**, 1565–77.
Marshack, K.J. (1994), 'Copreneurs and dual-career couples: are they different?', *Entrepreneurship Theory and Practice*, **19**(1), 49–94.
McPherson, M., Smith-Lovin, L. and Cook, J.M. (2001), 'Birds of a feather: homophily in social networks', *Annual Review of Sociology*, **27**, 415–44.
Meindl, J.R. (1993), 'Reinventing leadership: a radical, social psychological approach', in J.K. Murnighan (ed.), *Social Psychology in Organizations*, Englewood Cliffs, NJ: Prentice-Hall, pp. 89–118.
Miller, D. and Toulouse, J.-M. (1986), 'Chief executive personality and corporate strategy and structure in small firms', *Management Science*, **32**, 1389–409.
Moynihan, L.M. and Peterson, R.S. (2002), 'A contingent configuration approach to understanding the role of personality in organizational groups', in B.W. Staw and R. Sutton (eds), *Research in Organizational Behavior*, Greenwich, CT: JAI Press.
Neyer, F.J. and Lang, F.R. (2003), 'Blood is thicker than water: kinship orientation across adulthood', *Journal of Personality and Social Psychology*, **84**, 310–21.
Nicholson, N. (2005), 'Gene politics: a new view of family business', paper to the Academy of Management Annual Congress, Hawaii.
Nicholson, N. and Björnberg, Å. (2004), 'Evolutionary psychology and the family firm: structure, culture and performance', in S. Tomaselli and L. Melin (eds), *Family Firms in the Wind of Change*, Research Forum Proceedings, IFERA, Lausanne.
O'Toole, J., Galbraith, J. and Lawler, E.E. (2002), 'When two (or more) heads are better than one: the promise and pitfalls of shared leadership', *California Management Review*, **44**, 65–83.
Raymond Institute (2003), *Focus on the American Family Business Survey*, the Raymond Report, 11th edn, 17 March.
Russell, A. and Saebel, J. (1997), 'Mother–son, mother–daughter, father–son, and father–daughter: are they distinct relationships?', *Developmental Review*, **17**, 111–47.
Sakellariou, C. (2004), 'Family business succession: a family's perspective. A case study of Cosmetia S.A', MSc dissertation, London School of Economics and Political Science.
Sally, D. (2002), 'Co-leadership: lessons from Republican Rome', *California Management Review*, **42**(4), 84–99.
Schriesheim, C.A., Castro, S.L. and Cogliser, C.C. (1999), 'Leader–member exchange (LMX) theory: a comprehensive review of theory, measurement and data-analytic practices', *Leadership Quarterly*, **10**, 63–113.
Shams, M. and Björnberg, Å. (2006), 'Issues in family business: an international perspective', in P.R. Jackson and M. Shams (eds), *Developments in Work and Organizational Psychology: Implications for International Business*, Oxford: Elsevier.

Sulloway, F.J. (2001), 'Birth order, sibling competition, and human behavior', in J.H. Fetzer (series ed.) and H.R. Holcomb III (vol. ed.), *Conceptual Challenges in Evolutionary Psychology, Innovative Research Strategies: Studies in Cognitive Systems*, vol. 27, Dordrecht: Kluwer, pp. 39–83.
Tagiuri, R. and Davis, J.A. (1996), 'Bivalent attributes of the family firm', *Family Business Review*, **9**, 199–208.
Thompson, L. and Walker, A.J. (1982), 'The dyad as the unit of analysis: conceptual and methodological issues', *Journal of Marriage and the Family*, **44**, 889–900.
Trivers, R.L. (1974), 'Parent–offspring conflict', *American Zoologist*, **14**, 249–64.
Van Velsor, E. and Leslie, J.B. (1995), 'Why executives derail: perspectives across time and cultures', *Academy of Management Executive*, **9**, 62–73.
Ward, J.L. (2004), *Perpetuating the Family Business*, New York: Palgrave Macmillan.
Wright, C.A. and Wright, S.D. (1987), 'The role of mentors in the career development of young professionals', *Family Relations*, **36**, 204–8.
Yukl, G. (2002), *Leadership in Organizations*, Upper Saddle River, NJ: Prentice-Hall.

7 Business family as a team: underlying force for sustained competitive advantage

Lorraine M. Uhlaner

Introduction

One of the challenges in past research on family owned and managed firms has been to identify the family characteristics that matter, in prediction of firm strategies and business performance. The purpose of this chapter is to introduce a new concept, that of business family, and to identify aspects of the business family likely to impact business strategy and performance. The proposed framework draws on a wide range of research from different social science and business disciplines. The basic premise is that business families can be viewed as a specific type of team. Similar to teams more generally, the effective business family shares values and norms, has clear roles and procedures, and is able to resolve conflicts effectively among its members. A model and propositions are generated that relate different aspects of the business family, including business family cohesiveness, performance norms and characteristics of business family effectiveness, with business performance.

For the purpose of this chapter, the Dynamic System Planning (DSP) Model, a general systems theory based model of organization effectiveness, is used to identify different dimensions of business performance. Also important in the development of the framework are the notions of family orientation and business orientation, viewed in this chapter as two types of business family performance norms.

The first section of this chapter presents background from the existing literature, including definitions of key terms such as the family, family business, business family, teams and groups. Parallels are drawn between the concept of business family and teams which form the basis for propositions and a framework later in the chapter. The second section of the chapter presents other relevant background from the family business literature including family and business philosophy or norms (Carlock and Ward, 2001; Ward, 1987), as well as a description of the DSP Model in the context of the family business. The third section of the chapter presents a model that focuses on the relationship between different aspects of the business family (its norms, cohesiveness and degree of effectiveness as a team) and business performance. The final two sections of the chapter provide guidelines for future research and conclusions.

Background

Definition of key concepts: family, family business versus business family, group and team

Among the many definitions provided in the family therapy literature, *family* has been defined as 'people who have a shared history and shared future, bound by blood, legal and/or historical ties' (Carter and McGoldrick, 1999, p. 1). In the context of family business research, however, one can more clearly define *family* as the group of people related either by blood or marriage to the founder or founders of the business. A characteristic that sets families apart from many other social systems is the fact that membership can be biologically determined, and as such, membership in the family can be permanent – at least with respect to blood relationships (Borwick, 1986).

For purposes of the present discussion, *family business* is defined as a firm – regardless of company size, sector, or legal structure (though most typically privately held) – in which the majority of the ownership resides in the hands of one family and in which at least two members of the same family either own and/or manage the firm together. In the present context, the *business family* refers to the subgroup of individuals from the family, as defined above, who either own or work in the same business enterprise. This can include family members in paid or unpaid positions, including governance roles on the family council or board of directors.

A parallel can be drawn between the concepts of family and business family and the social-psychological concepts of a *group* and *team*, respectively. The social psychology concepts of group and team date back at least to the early 1950s (Cartwright and Zander, 1953, 1968). For the purposes of present discussion, *group* is defined as a 'collection of two or more persons who interact with one another in such a way that each person influences and is influenced by the others' (Wagner and Hollenbeck, 1995, p. 310). *Team* is defined as 'a collection of individuals who are interdependent in their tasks, who share responsibility for outcomes, who see themselves and who are seen by others as an intact social entity embedded in one or more larger social systems . . . and who manage their relationships across organizational boundaries' (Cohen and Bailey, 1997, p. 241). Furthermore, *group norms* are defined as acceptable standards of behaviour that are shared by a group's members.

The family can be viewed as a relatively stable group: family members interact with influence, and identify themselves with one another. The business family, in contrast, is a type of team. At a moment in time, the business family forms in order to share responsibility for the success of one or more business enterprises. In contrast to passive owners in large publicly held firms, the business family is likely to view itself as an intact social entity and is furthermore embedded in other social systems (the core family, the ownership system and the business enterprise). The distinction between family and business family is that of choice. People usually cannot choose the family of their birth (although, of course, partners form new families by choice). However, individuals entering into a business family have at some point made a conscious choice to do so.

The concept of team is found in only a few past references in the family business research literature. For instance, Stevenson recommends that families pursue a transition over time from a dependency relationship between child and entrepreneur parent to one of more equal footing, noting that 'the ultimate goal of a family business is interdependence – where the family is able to work together as a team' (Astrachan, 1996, p. 212). Filbeck and Smith (1997) also refer to the family team, but primarily in the context of management.

Team development and team effectiveness – past research

This subsection provides relevant background from team research, including stages of group development (Tuckman, 1965), team effectiveness (Hackman, 1990; Mohrman et al., 1995), and characteristics of the team found to determine team effectiveness (Huszczo, 1996).

One of the earliest models of group formation was proposed by Tuckman (1965), who identifies four stages in group development, which he defines as *forming* (formation of the group), *storming* (a stage of conflict where cliques and factions may form and team members vie for position), *norming* (where agreement and consensus begins to form, and which is characterized by close relationships and cohesiveness) and *performing* (where the team works effectively with a shared vision). In later research, Tuckman and Jensen (1977) add a fifth stage, *mourning*, to reflect the stage when a team dissolves. In the family firm, Gersick and colleagues emphasize the shift that takes place when family members begin to enter the firm (entering the business stage), parallel to the forming stage, and when family work together productively (working together stage) which can be seen as parallel to the norming and performing stages (Gersick et al., 1997). Of course, as with teams more generally, not all business families manage to reach Tucker's fourth stage of performing, which requires a shared vision among business family members. However, the model provides some clues as to the changes that take place in developing an effective team, including the decision to work together, development of consensus about the purpose of the group, roles and procedures, and a shared vision.

According to more recent team research, team effectiveness is multifaceted. Hackman (1990) proposes three criteria including team performance, quality of team process and team satisfaction. *Team performance* is the extent to which the group's productive output (that is, its product, service, or decision) meets the standards of quantity, quality, and timeliness of the people who receive, review and/or use that output. *Quality of team process* is the degree to which in the process of working together, the team enhances the capability of members to work together interdependently in the future (that is, the degree to which the whole is greater than the sum of the parts). Finally, *team satisfaction* is the degree to which the group experience contributes to the growth and personal well-being of team members (Mohrman et al., 1995).

Other team-focused research identifies characteristics of teams which may determine team performance (Cohen and Bailey, 1997; Stewart, 2006). Huszczo (1996) provides a helpful list of these characteristics including: a shared vision, purpose, goals and values; talented members – not only a full range of abilities or competencies, but also their proper development and utilization; clear responsibilities – (role) expectations that are well established; reasonable and efficient operating procedures to carry out work and make decisions; constructive interpersonal relationships and effective conflict management; appropriate rewards – both monetary but nonmonetary as well; and constructive external relationships (Huszczo, 1996, p. 16).

One can draw parallels between the team and family business literature. For instance, family business researchers identify a shared family vision, climate of trust, and open honest communication as prerequisites for successful interaction among family members in the family business (Carlock and Ward, 2001; Habbershon et al., 2003; Ward, 1987). In an empirical study, Mustakallio (2002) reports a statistically significant linkage between shared vision among family members and a family's commitment to business decisions. But much more research is needed to confirm the importance of these other aspects to business and family performance.

Group cohesiveness and team performance – past research

Group cohesiveness is the degree to which members feel attracted to one another and the group as a whole, sometimes also referred to as *esprit de corps* (Wagner and Hollenback, 1995). Some variables that enhance cohesiveness include shared values and interests of group members, agreement on group goals, frequency of interaction, and isolation from other groups. Research carried out by Rokeach and Rokeach (1989) supports the conclusion that parents and family influence values and that people's values have their roots in early childhood (Schermerhorn et al., 2000). Cohesiveness of the family is also likely to be enhanced in many families through other determinants such as frequency of interaction, and isolation from outsiders (common in the modern Western family), although the strength of such determinants may diminish as frequency diminishes and as grown children move away from parents and each other.

Bowen's family systems theory, rooted in general systems theory (Bowen, 1981; Von Bertalanffy, 1968), provides yet another explanation for the cohesiveness of the family. According to Bowen, families are viewed as social systems, with individuals making up parts of the family system. These individuals develop predictable patterns of interaction which evolve into rules or beliefs about acceptable behaviour (Guttman, 1991; Kerr and Bowen, 1988). Many interactions among family members are seen as the result of the pressure to maintain and/or restore equilibrium or balance in the family system (the concept of homeostasis). Similar to cohesive groups more generally, when a family member deviates too strongly from the prescribed norms, for instance, pressure is exerted by the others to conform. The tension between the opposing forces of individuation (expression of self) and togetherness or loyalty to the family is an outgrowth of this pressure to maintain balance in the family system, and also helps to explain the highly emotional level found within many families (Friedman, 1986; Gersick et al., 1997). Given the strong emotional forces that may be found in the family system, one might also surmise that business families, built upon the family system, tend to be more cohesive, and thus may have more intense relationships and other dynamics than a similar, nonfamily group of owners and/or managers.

Research on cohesiveness may help to explain why the influence of family on the business may be a competitive advantage for some firms but a detriment for others. Research on group cohesiveness supports the conclusion that, rather than improving team productivity, cohesiveness serves the function of reducing variability around group norms. Thus, performance may be higher, if the norm is for high productivity, but can actually be lower if the group norm is for lower productivity (Schermerhorn et al., 2000). Thus the cohesiveness of the business family, and related performance norms associated with that business family, need to be considered together in predicting business performance and other business behaviours.

The 'f' factor

Some strands in recent family business research attempt to identify more clearly the dimensions of the family that may have an impact on the business, variously referring to 'faminess' (Habbershon and Williams, 1999), and the 'the f-factor' (Habbershon and Pistrui, 2004; Habbershon et al., 2003; Uhlaner and Habbershon, 2005). Most of these dimensions identify specific aspects of the family that may influence the performance of the firm, such as family leadership, family vision, family governance, family relationships, family performance objectives, and family strategy (Habbershon and Pistrui, 2004).

In reviewing the different dimensions of the 'f' factor, with the exception of family relationships, the majority of these dimensions pertain to the influence and/or relationship of the family on the business rather than characteristics of the family itself. Thus family leadership refers to the leadership of family in the business. Family vision relates to the vision the family has with respect to the business, and so forth. Thus, in considering these characteristics, they can best be seen (with the possible exception of family relationships) as characteristics of the business family, rather than of the family as a whole.

Family institutions and the business family – background

The family business literature describes a number of family institutions thought to improve the function of the family with respect to the firm, including informal family meetings, formal family meetings, family councils and family plans (Mustakallio et al., 2002; Suaré and Santana-Martin, 2004). Suaré and Santana-Martin define *family governance* as 'the set of institutions and mechanisms whose aim is to order the relationships occurring within the family context and between the family and the business' (Suaré and Santana-Martin, 2004, p. 146). They view these different institutions as means for establishing and communicating group norms and for setting rules governing the behaviour of family members with respect to the firm (that is, who may own shares, who may become a managers, and so forth). One might argue that these are thus essentially norms not so much for the family per se, but the business family. Family plans or constitutions are also a way to clarify the roles, norms, and rules governing behaviour of the business family (see also Neubauer and Lank, 1998). These institutions can be seen as ways to enhance team effectiveness by stimulating a shared vision, clearer responsibilities and suitable operating procedures. Bringing family together through these family institutions can also enhance their relationships and resolve conflicts more effectively.

Some of the family institutions discussed in the literature are clearly geared to the broader family group, such as informal social gatherings for birthdays and so forth. Other institutions might arguably be aimed primarily or exclusively at the business family, or potential members of the business family, that is, those interested in the activity of the firm. Thus a family council is focused on the firm, whether or not all attendees are currently owners or working otherwise in the firm. A family protocol or family constitution, similarly, also relates to the business: it generally describes rules, norms, or a mission relevant to the business, rather than the family at large. Even where the institutions may involve those family members outside the business family, such as social gatherings, one can argue that these institutions can be seen as ways to facilitate effective functioning of the business family.

Other determinants of effectiveness in the business family

Drawing on other parallels with the team literature, it is also important to consider other determinants of business family effectiveness including the talents of individual family members and reward systems.

Based on his meta-analysis of team composition characteristics, Stewart (2006) provides statistical support for the positive effect of individual ability and disposition on team performance. Clearly more research is needed to confirm the types of skills and composition of the business family that may be associated not only with more effective functioning of the business family but also with positive influence on business performance. However, drawing from business team research and family business case studies, it is likely to be a relevant variable.

Based again on analogy to the broader research on teams, reward systems provided to family members participating in the business family may also be important determinants of individual and group performance. These rewards may be monetary, with respect to stock dividends for owners or salary to family managers. But one should also consider nonmonetary rewards as well, such as opportunity for social interaction or fulfilment of the need for self-expression and achievement. Money can create a source of conflict and misunderstanding, especially between passive owners and family working within the business, the former often expecting some form of dividend for their shares while the latter often see greater need to keep funds reinvested in the firm to assure its long-term survival and growth (Astrachan, 1993).

Summary: families as groups and business families as teams

To summarize the key points of this section, we propose viewing the family as a group, that is, as two or more individuals who are conscious of each other and part of the same social system. Unlike families in general, who may not have shared objectives, the business family can be viewed as a team, with the common goal of owning and running one or more business enterprises. A quick comparison of the extant literature on family business and the team-building literature suggests a variety of interesting parallels that may provide useful in future research including research on stages of group development, group cohesiveness, group norms and team effectiveness.

Examining the business system: past work from strategic management and organization theory

Before presenting a model that links business families with business performance, this section presents two remaining streams of literature relevant for the present discussion: the notions of family and business orientation and an introduction to the Dynamic System Planning (DSP) Model, in the context of the family firm.

Competing objectives of the business family: family orientation and business orientation

Critical to a family firm's survival and success is its ability to balance and integrate the demands or needs of the family and the business (Carlock and Ward, 2001; Ward, 1987). In one interpretation, family and business are seen as opposite ends of the same spectrum, with balance between family and business seen as a point midway on the continuum. Empirical research by Leenders and Waarts (2003), however, based on an empirical study of 220 Dutch family businesses, supports the more general conclusion that family and business should be treated as independent factors rather than inverse functions of one another. Using terminology from the group and team literatures, these orientations may be viewed as types of *group performance norms*. Thus business families high on *family orientation* place importance on behaviours fulfilling the needs and demands of family, whereas business families high on *business orientation* place importance on behaviours fulfilling the needs and demands of the business.

According to Carlock and Ward (2001), families who can balance family and business systems create a positive environment where the family thrives and the business performs well. However, as Leenders and Waarts warn, and Carlock and Ward (2001) allude to, it is not such a simple task to balance the objectives of both the family and business systems, especially when the orientation towards both is high.

An overview of the DSP Model and its application to family business
This section provides an overview of the Dynamic System Planning (DSP) Model, a general systems theory based framework originally developed to study growing small and medium-sized firms (Hendrickson and Psarouthakis, 1998), and also used to examine family firms (Uhlaner and Habbershon, 2005; Uhlaner and Hunt, 1999). General systems theory applications to organizations can be traced to two streams of applications in the early 1960s – a 'subsystems' approach defined by Katz and Kahn (1966) and an organization problem-solving model developed by Basil Georgopoulos and colleagues (Georgopoulos, 1986). The DSP Model is an offshoot of the latter, identifying seven problems or *issues* that a business must address in order to function effectively. The DSP Model provides an alternative to the value-chain (Porter, 1985) and the balanced scorecard approaches (Kaplan and Norton, 1992), and is easily adapted to different types of firms, regardless of technology, structure, age, phase of development or company size. The DSP Model has been partially validated in several published research studies (Hendrickson, 1992; Hendrickson and l'Abbe Wu, 1993; Hendrickson and Psarouthakis, 1998; Hendrickson and Tuttle, 1997).

The DSP Model is derived from two key assumptions of general systems theory, which in turn, define the seven DSP issues an organization must have the ability to manage properly in order to assure effective performance. The first assumption is that an organization must combat entropy to survive and does so by assuring an efficient transformation of inputs to outputs (and back again to inputs). This one assumption forms the basis for six of the seven DSP issues, including resource acquisition, market strategy, work flow, employee relations, resource allocation, and technical mastery. The second assumption is that organizations are social systems embedded in other social systems. This assumption defines the seventh issue, government and community relations. According to the DSP Model, appropriate management of these seven issues guarantees the *financial viability* of the firm – its profits, (growth of) assets, and long-term survival. In the remainder of this section, each of these seven issues is explained more fully, in the context of the family firm.

Market strategy in the family firm Market strategy refers to a firm's ability to select the right mix and characteristics of products or services, with appropriate target market, pricing, and distribution channels to assure the necessary sales. Effectiveness criteria for market strategy include sales growth and total sales. The business family may impact market strategy in several ways. Market strategy and company direction in general, may change as a new generation takes over. Or certain family values may be instilled in the mission of the business – for instance, making the customer feel like one of the family by providing more personalized service.

Resource allocation in the family firm Resource allocation refers to the firm's ability to assign the various resources so that different departments and individuals in the firm get the resources they need to do their work. Budgeting and inventory control are examples of more formal resource allocation strategies with the 'squeaky wheel' approach (the one complaining the loudest getting the resources), and other informal means often used in smaller and/or less professionally run firms. Effectiveness criteria for resource allocation include adequate cash flow and, more generally, a sense by employees and management that resources are available where and when needed. In the family business, managers from the business family may have easier access to resources within the company. This phenomenon may be accentuated in firms lacking a formal budgeting process. Family firms may also assign a family member – even one not thoroughly trained for the position – as corporate treasurer or bookkeeper, in order to keep a close eye on the family's assets.

Resource acquisition in the family firm Resource acquisition relates to a firm's ability to obtain needed inputs: money, people, materials, and external information. Examples of resource acquisition strategies include investor relations, recruitment, purchasing, and research and development. Effectiveness criteria for resource acquisition include adequate staffing, finances, materials and information. Regarding family influence, suppliers may have family ties. Family members are frequently an important source of capital (Hendrickson and Psarouthakis, 1998). And, of course, family members form a frequent recruitment pool for both start-up and managerial positions (Hendrickson and Psarouthakis, 1998).

Work flow and the family firm Work flow relates to the firm's ability to structure and coordinate work appropriately among its employees. Examples of strategies for structuring the firm include specialization, centralization of decision-making and departmentalization. Hierarchy of authority, informal meetings or the team approach are examples of coordination strategies. Effectiveness criteria for work flow include smoothness with which tasks are carried out. This is often difficult to determine except when things go wrong. Thus role conflict and resulting role stress may provide evidence of poor structure. Signs of coordination problems may include slow or inaccurate fulfilment of customer orders or other work 'slipping through the cracks' (Georgopoulos, 1986; Hendrickson and Psarouthakis, 1998). Information flow and the division of decision-making responsibilities are other aspects of effective work flow. Information may also flow differently as a result of family dynamics, passing outside the formal chain of command and with the possible consequence, for example, that a more senior nonfamily manager may be left out of the loop. Decision-making responsibility may also be distributed differently as a result of family dynamics.

Employee relations and the family firm Employee relations relates to the firm's ability to motivate and satisfy employees, and to instil a positive corporate culture. Strategies may include incentive programmes and techniques to build awareness of company values, such as a company handbook or company-wide meetings. Effectiveness criteria for employee relations can thus include both attitudinal measures (employee satisfaction, motivation, commitment to the firm) as well as behavioural measures such as level of absenteeism and employee turnover. Although nepotism in allocation of rewards comes immediately to mind as a negative impact of the business family on the firm, there may also be a positive influence. For instance, business family members may have a stronger sense of loyalty and commitment and willingness to make short-term sacrifices for the long-term good of the firm. Supporting evidence based on research of the Dutch economy suggests family businesses provide an important stabilizing force in the economy because they are less likely to close their doors during periods of economic downturn (van Engelenburg and Kommers, 2001).

Technical mastery and the family firm Technical mastery is the ability of the firm to produce goods and services with adequate quality, efficiency and timeliness. Technical innovation may also be considered an aspect of this issue. Effectiveness criteria for technical mastery include technical quality, technical innovation (that is, new products and processes) and overall productivity. Training programmes and investments in new technologies are two examples of technical mastery strategies. In companies where tacit knowledge is handed down from one generation to the next, family businesses often have an advantage in preserving technical know-how within the firm. On the other hand, family businesses who fail to encourage the younger generation to get outside work experience and training before joining the firm may lag behind firms who recruit outsiders.

Government and community relations and the family firm The seventh and final DSP issue relates to a firm's ability to get along with everyone other than with those stakeholders linked to the transformation of inputs to outputs. Examples of such stakeholders include the government, community groups, and family members outside the business. Effective criteria for government and community relations include goodwill and a good company reputation in the community (local, regional, or business) as well as the absence of conflict with outside groups. Positive handling of the government and community relations issue is often evidenced by the absence of problems – thus a lack of lawsuits from former employees, consumer action groups or the government. Positive evidence may include conformance with environmental protection laws, positive recognition from the general community for its reputation towards employees or customers, and level of harmony with family members outside the business family.

A proposed model for examining the influence of the business family on the firm

This section presents a rudimentary framework for examining the impact of the business family on the firm, considering such variables as performance norms, effectiveness, and cohesiveness of the business family. Other variables, such as talents of individuals and commitment of the business family to firm goals, are included in the framework, although propositions involving these variables are beyond the scope of this chapter. Business performance is a multidimensional concept which includes effectiveness of each of the DSP issues, financial viability, and the quality of the overall vision for the firm by the top management team. The overall model is presented in Figure 7.1.

The intervening variables of ownership and management composition represent the percentage of ownership and management, respectively, which come from the business family, as opposed to nonfamily. The business family is thought to have a greater impact the more highly it is represented in either the ownership or management groups. Note also that that the model suggests that members of the family may impact business performance, but this is most likely indirect by way of the business family. Thus, the spouse of a son who works in the firm, but who is not herself employed by the firm or assigned to some other role such as a family council representative, may influence the son's behaviour, but is assumed to have a limited direct effect on the business itself.

A thorough development of the model, including related propositions, is beyond the scope of this chapter. However, an initial set of propositions and suggested relationships are presented in the remainder of this section of the chapter to give some direction for future research.

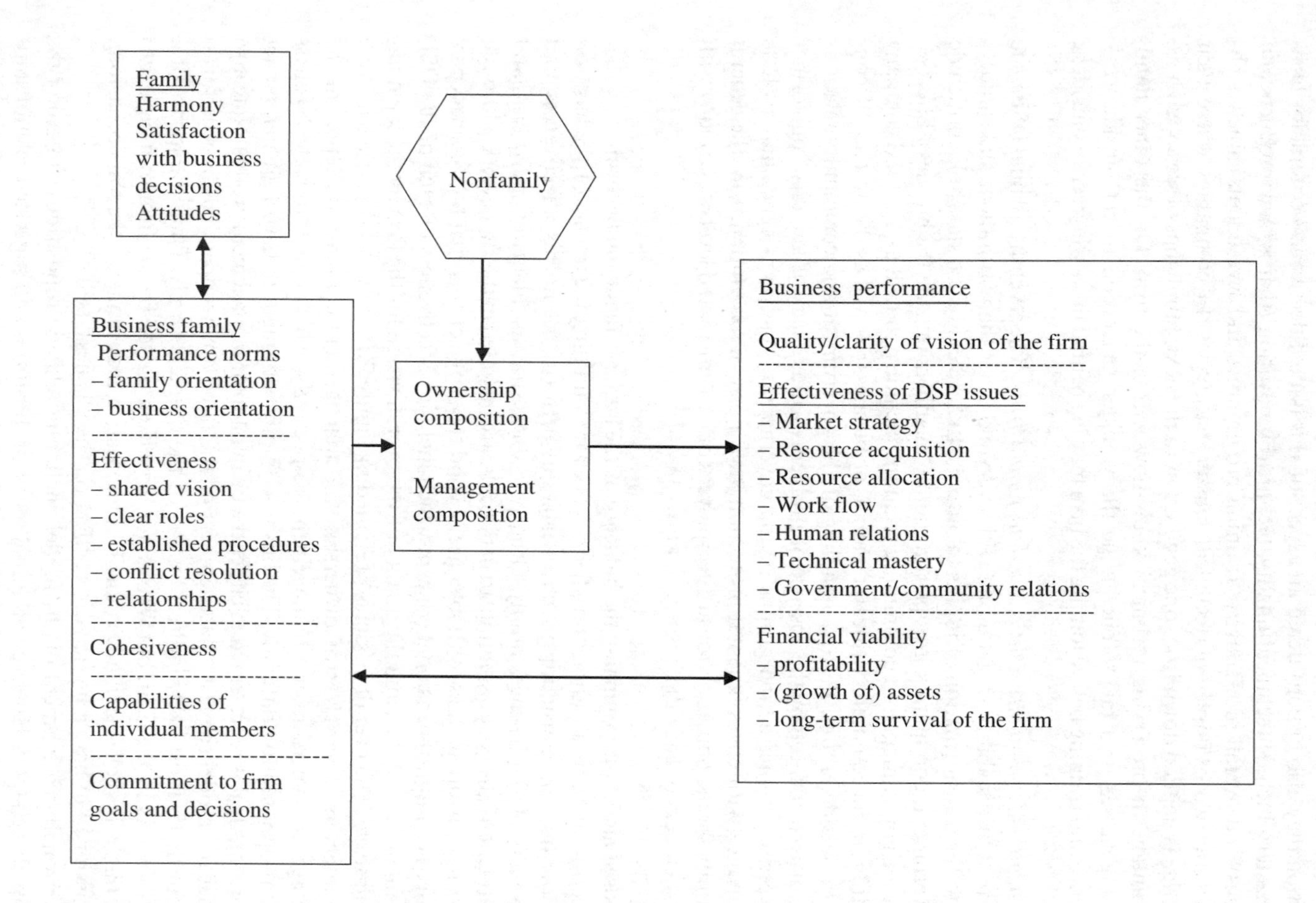

Figure 7.1 The business family and business performance

Performance norms, group cohesiveness and business performance
The first two propositions draw upon research on performance norms and group cohesiveness. Consistent with group research findings, the assumption is that a tight-knit or cohesive business family per se does not in and of itself guarantee better business performance. However group cohesiveness of the business family is an important moderating variable in predicting the relationship between group performance norms and business performance.

Treating business orientation and family orientation as types of performance norms, one can thus state the following two propositions:

> *Proposition 1* Especially under conditions of high business family cohesiveness, the greater the emphasis on business orientation by the business family, the greater the business performance.
>
> *Proposition 2* Especially under conditions of high business family cohesiveness, the greater the family orientation, the better the family performance, that is, the greater the family harmony the greater the satisfaction of family members with business decisions affecting the family.

Note that by definition, members of the business family are all part of the larger family system. Thus, aspects of the business family logically will influence the family and vice versa. Therefore it would seem logical that performance norms related to family would influence family behaviour.

The relationship between business family effectiveness and business performance
Examining other research from the groups and teams literature, and viewing the business family as a type of team, the proposed framework also predicts that business families with characteristics consistent with those of effective teams more generally (that is, which have a shared vision or norms, clear roles, clear procedures for making decisions, a high level of individual talent and good quality relationships), are likely to influence business performance in a more positive manner than business families which operate as less effective teams. Thus, proposition 3 is stated as follows:

> *Proposition 3* The more effectively a business family functions together as a team, the better the business performance is expected to be.

The underlying rationale of proposition 3, to restate briefly the arguments presented thus far in this chapter, is that business families can be viewed as a type of team, and thus the research accumulated to date about teams in general applies to business families. However, further research of course is needed to validate this notion, since past research suggests that for different types of teams, the same determinants of effectiveness do not always apply (Stewart, 2006).

Based on the rationale of logic, one might also predict that the greater the overlap between the business family and total ownership of the firm, the more influence the business family is likely to have. A business family only owning a very small proportion of the shares will have less influence than one owning most or all of the shares. Similarly, the greater the overlap with the total management team, the more influence the business family will have on management. Restating proposition 3, keeping these ideas into account, one can state the following:

> *Proposition 3a* Under the condition that the majority or all the business family *owns* the firm, the greater the influence the business family's performance is likely to have on business performance.

And similarly:

> *Proposition 3b* Under the condition that the majority or all the business family *manages* the firm, the greater the influence the business family will have on business performance.

It is beyond the scope of this chapter to elaborate upon which aspects of business performance are more likely to be influenced by greater business family ownership versus that of business family management. However, this distinction may be important as may the difference in the way business family owners or managers function together as a subgroup of the total business family. For the remainder of this section, we speak in general terms of business families that run a business, whether or not individual business family members are legal owners, managers, and/or members of a business or family-related governance board (for instance, a family council or board of directors) but future research may benefit from examining finer nuances among subgroups, especially within large family firms.

Business family and market strategy

Past research on the influence of family ownership and management on market strategy leads to equivocal conclusions. On the one hand, some researchers conclude that family businesses are less likely to pursue growth-oriented strategies than nonfamily firms, (Donkels and Fröhlich, 1991; Gomez-Mejia et al., 1987). But, on the other hand, especially in larger quantitative studies, results are less clear. For instance, Jorissen and colleagues find that differences in strategy disappear when controlling for size and sector differences (Jorissen et al., 2002). Daily and Thompson (1994) also find no differences in strategic posture and firm growth when comparing family owned/managed, entrepreneur owned/managed, and professionally managed firms in a large sample of heating and air conditioning wholesalers.

These results suggest that differences in growth orientation and sales growth, more generally, cannot be explained merely by differences in ownership or management composition (family versus nonfamily). An alternative is to consider some of the other variables mentioned thus far, including cohesiveness and effectiveness of the business family and norms of the business family. In parallel with propositions 1 and 3 for instance, one might state:

> *Proposition 4* Under conditions of high business family cohesiveness, the stronger the business orientation, the better sales growth is likely to be.

On the other hand, following the format of proposition 2 above, cohesive business families with a strong family orientation may place a stronger emphasis on continuity and independence of the firm, even if this comes at the expense of sales growth and profits, to assure that wealth remains in the family over the long term. Note that the effect of family orientation is not thus necessarily a negative influence on the business. For instance, emphasis on continuity may improve the odds of survival of the firm. Other possible consequences resulting from a strong family orientation may be less helpful however. For instance, a strong family orientation may result in a market strategy that accommodates the preferences of the founder or other family, even if that strategy is not sound. For instance, geographic expansion may be based on the need to create an opportunity for a son or son-in-law wanting to move to a particular city, rather than a rational decision based on market analysis in that region. And companies often continue particular products or services, or even the company itself, in honour or homage to the founder, even when it no longer makes business sense to do so. Again, these latter decisions are expected more likely in situations where family orientation is high, especially in combination with low business orientation.

Business family and resource allocation

As with market strategy, past research that merely compares family versus nonfamily firms is not adequate to predict differences in resource allocation strategies, especially once size and sector are controlled for (Jorissen et al., 2002). In parallel to the logic of the previous section, one might predict that in addition to ownership and management composition, one might consider the performance norms (business and family) and cohesiveness of the business family. Thus, one might state the following propositions:

Proposition 5 Firms run by cohesive business families with a stronger business orientation are more likely to use more 'rational' or formal allocation strategies that focus on business needs, whereas

Proposition 6 Firms run by cohesive business families placing a strong emphasis on family orientation (especially in the absence of a strong business orientation) are more likely to allocate resources favouring family interests.

Examples of more rational allocation strategies might include use of formal budgeting and inventory control, professional cash flow management techniques and reliance on a capital expenditure plan – in short, allocation as much as possible on business needs.

On the other hand, allocation to favour family over nonfamily might include allocation (arbitrarily) of funds to family-run over nonfamily-run departments, or large issuance of dividends to family owners at the expense of accommodating the capital needs of the firm. These are, of course, propositions that would need to be tested empirically.

Business family and resource acquisition

Case studies on family firms suggest that family firms also vary in the resources obtained by the firm. Thus, whereas some firms might focus more clearly on rational choices, such as the most qualified people or best materials for the price, other companies may make decisions based on family or historical considerations. Once again, building on the propositions 1 and 2, one might thus state the following:

Proposition 7 Firms run by cohesive business families with a stronger business orientation are more likely to select people and material that are most efficient or suitable for the business, even when this runs counter to family demands or needs.

Thus, in firms run by cohesive business families with a strong business orientation, suppliers would be selected on rational selection criteria (best price, best quality for the price)

rather than favouring suppliers that have friendships or long-standing relationships with the firm. And the most qualified applicants would be hired, even if this meant passing over family members. Such firms may also be more likely to choose to expand the firm using financial resources from outside investors, even if this dilutes family interests.

On the other hand, where family orientation performance norms predominate, one might state the following proposition:

> *Proposition 8* Firms run by a cohesive business family with a stronger family orientation (especially where business orientation is weak) are more likely to take family considerations into account in selecting resources for the firm, including people hired, suppliers selected or equity sources used for raising funds.

Thus, it may be that where family comes first, the firm's managers may hire family over nonfamily, even when less qualified (especially where business orientation is low), remain with suppliers that have been loyal to the family even when better suppliers emerge over time and retain family control by refusing nonfamily investors, choosing instead either to borrow from banks or to accept equity from family only, regardless of business opportunity. On the other hand, cohesive business families with a strong family orientation may benefit more highly from access to investment funds or loans from other family members.

In the cases where both business and family orientation are high, one can see that competing goals may lead to conflicting resource acquisition strategies. In this latter case, it may be that business families that function as effective teams have worked out such conflicts by setting up a family constitution or other family institutions to resolve such conflicts, or by creating mechanisms for improving the qualifications pool of potential family applicants.

Business family and work flow

Once again using the analogy from propositions 1, and the research on group cohesiveness, norms and group performance, one might presume that firms run by cohesive business families with a stronger business orientation are also more rationally organized. Thus one can state as follows:

> *Proposition 9* Firms run by cohesive business families with a stronger business orientation are more likely to assign roles to those employees most capable of doing the work, decisions are handled by the person most qualified to do so, and information and decisions flow according to a formal chain of command, regardless of family status.

On the other hand, in firms run by cohesive families with a strong family orientation, especially where business orientation is weak, family demands may take precedence over business demands to the extent that assignment of key roles, access to information and decision-making responsibility is given to family even when they may be less qualified to carry out the job, leading to poorer work flow. Thus, one might state:

> *Proposition 10* Firms run by cohesive business families with a stronger family orientation (especially where business orientation is weak) may favour family over

> nonfamily in assignment of roles and responsibilities, sharing of information and decision-making responsibility even when this runs counter to the formal chain of command, or even in place of a formal chain of command.

It may be a mistake to assume, however, that family orientation always runs counter to effective business performance. For instance, a cohesive business family that functions as an effective team may be able to work smoothly together making mutual adjustments needed to informally coordinate efforts.

Business families and employee relations
Regarding employee relations, continuing the logic from previous sections, one might state the following:

> *Proposition 11* Firms run by cohesive business families with stronger business orientation, are more likely to reward employees according to business criteria (position, seniority or performance level) and therefore result in a more motivated and satisfied workforce.

On the other hand, family orientation (especially with a balanced business orientation) may have a positive impact in that it may lead to stronger organization commitment by family members (Uhlaner and Hunt, 1999). Family values may also have more influence, for better or worse, on the business culture. One might also assume that business families that function as an effective team will also share stronger commitment to the firm's success.

Business families and technical mastery
Regarding technical mastery, companies run by cohesive business families with a stronger business orientation may be more apt to use professional training or to hire more professionally trained technical staff, and management would take advantage of networks with outsiders and use of such programmes as ISO 9000 to assure higher quality. Jorrissen and colleagues find little difference in networking activities once controlling for size and sector in family and nonfamily firms (Jorissen et al., 2002), suggesting once again, the need for examining variables other than ownership and management composition. Stating this in the form of a proposition:

> *Proposition 12* Firms run by cohesive business families with stronger business orientation, are more likely to use professional practices to assure better quality and productivity, and more likely to obtain outside information to improve technical know-how.

On the other hand, business families with a stronger family orientation (especially combined with a weak business orientation) may permit tolerance of low-quality work by less competent family members (Uhlaner and Hunt, 1999). However, companies run by cohesive families with a strong family orientation may also benefit with improved technical mastery. For instance, case studies suggest that in family firms, a firm's reputation for quality is a direct reflection of the family itself (Uhlaner and Hunt, 1999). Also, long family apprenticeships, often beginning at a young age, can help to carry on knowledge that assures quality of the product.

Business families and government and community relations
Carrying the logic of Proposition 1 to the final DSP issue, one might state:

> *Proposition 13* Firms run by cohesive business families with stronger business orientation, are more likely to use professional personnel to stay abreast of legal requirements for employment, the environment, and so forth.

On the other hand, family orientation may have a positive effect on social responsibility as well. For instance, in one study based on a relatively small set of Dutch firms (Uhlaner et al., 2004), family firms often view their employees, and even suppliers, as an extension of the family. Furthermore, particularly in firms bearing the family name, concern for the environment seems stronger: soiling the environment reflects badly on the family name. The same study also finds that family firms are most likely to show their philanthropy outside the firm by supporting organizations linked with family members (such as a family member's church, school, or sport club). On the other hand, one might predict that companies run by cohesive business families with a stronger business orientation approach these issues in a more professionalized manner – staying abreast of rules governing employment, the environment and other regulations.

Summary: strong business family as competitive advantage (or not)
A review of the existing body of family business literature lacks clear predictions about the business strategies and tactics that are more or less likely in the family firm. Because of this, research from the group and team literature, especially findings related to group cohesiveness and performance norms, were applied in the present discussion to each of the DSP issues to illustrate the differences in predictions that may occur as a result of differences in norms. Thus, the type of performance norms (business versus family orientation) together with the cohesiveness of the business family, may help to predict differences in strategy followed in different family firms. Although it would appear that the influence of business orientation is generally positive, with respect to the firm, whereas the influence of family orientation is negative, this is not always the case. For most of the DSP issues (with the exception of resource allocation), examples are provided to show how family orientation may have a positive effect on business performance, including, for instance, the emphasis on the continuity and independence of the firm (market strategy), mutual adjustment and team coordination (work flow), and stronger commitment to the firm (employee relations). Positive effects of strong family orientation are also considered for other DSP issues: research suggests that for instance, family orientation may also be linked to greater social responsibility especially for stakeholders within the firm or in the immediate community (government and community relations), availability of financial resources as well as people willing to work (at a lower wage) (resource acquisition) and preservation of know-how from long-term apprenticeships (technical mastery). However, emphasis on family orientation in the absence of balance towards business orientation may serve to harm the firm's performance, and eventually (the potential for) family wealth derived from the firm.

Programme for future research
A variety of methods might be used to test the propositions and other ideas suggested thus far in this chapter. Qualitative research may be helpful in clarifying the way in which

the concept of team should be viewed in the family business. It is proposed here that a distinction is made between the concepts of family and the business family, but a review of existing cases as well as new research may help to clarify whether this distinction makes sense in practice. As a part of this exploration, it may also be helpful to clarify whether it is useful to consider the business family members who are owners separately from those that work in the firm, rather than combining them as was proposed in this chapter. However, future research needs to clarify this concept further and, in particular, to consider whether only current or also potential family owners and workers should be considered part of the business family.

Qualitative research, especially the use of structured interviews of several business family members in each of several firms, (using a multiple case study approach) would provide an initial means to clarify the way in which techniques and concepts from the team-building literature should map onto that of the business family and family business. In short, initial qualitative research may help to clarify the suitability of the concept of team and how it applies to the family firm.

However, without quantitative research, it will be difficult to test the propositions and framework adequately. Large samples are needed that reflect not only from a range of company sizes and sectors, but also reflect different performance norms and cohesiveness of the business family. Validation of the team characteristics described by Huszczo (1996) is needed in the context of business families to see whether these characteristics do indeed predict different aspects of team effectiveness as defined by Hackman (1990). It is important either to control for the ownership and management composition, furthermore, either by sampling adequately from family businesses where ownership and management by family members range widely, or expressly to choose only those firms where the majority of owners and managers are family members and draw conclusions for that population only. With large samples and random sampling to assure variation on the relevant variables, data analysis techniques such as multiple regression analysis and/or other multivariate techniques such as structural equations can be applied to test these propositions more adequately.

Conclusions

This chapter reviews a variety of concepts from the social science and business literature to propose a preliminary framework for understanding the influence of the business family on the family owned and managed firm. Past research is unclear about the nature of the family system and, in particular, which characteristics might help to explain under which conditions family orientation can be viewed as a strategic advantage or disadvantage in the family owned and managed firm.

This chapter suggest that rather than examine the family as a whole, research should concentrate on the subgroup of people within the family, referred to as the business family, and that by viewing the business family as a team, interesting linkages can be made to conclusions already drawn from the group and team research literature. In particular, similar to teams, family business writers have suggested that successful family businesses need to have a shared vision, clear expectations, protocols or procedures for interaction (including but not limited to various family institutions such as informal social gatherings, family councils and family plans), positive reward systems for both family owners and employees, and capable individuals.

This chapter also highlights some of the past research on family business and strategic management. A finding that is important is that business families may vary in their group norms: with varying emphasis on the demands or needs of family and firm. The Dynamic System Planning Model, a model derived from general systems theory, is presented as a means to define different aspects of the business performance that may be influenced by the business family. A model and propositions are suggested that might link different aspects of the business family to business and family performance, derived from assumptions built from the group and team research literature, and the added assumption that the business family can be viewed as a type of team. Past research on group cohesiveness, group performance norms and team effectiveness are used to make predictions about business and family performance. In particular, differences in norms regarding family and business orientation may help to explain differences in business strategy and performance outcomes, especially when controlling for cohesiveness of the business family.

The field of family business research is still a relatively young, emerging field compared with more established fields such as social psychology, sociology, economics family therapy, and more established specialties in business (such as organization behaviour and strategy) and can benefit from reviewing key findings from those fields. At the very least, in spite of many remaining unanswered questions, it is hoped that this chapter might serve to stimulate further thinking regarding the integration of the vast body of research on groups and teams with research in the field of family business, in order to provide better insight into determinants of effectiveness of the family firm. Future research will have to determine whether or not the business family is really a force for competitive advantage, but the answer is likely to require a fairly complex combination of factors, rather than a simple comparison of the composition of ownership and management between family- and nonfamily-owned and managed firms.

References

Astrachan, J.H. (1996), 'Preparing the next generation for wealth: a conversation with Howard H. Stevenson', in R. Beckhard (ed.), *The Best of FBR: A Celebration*, Brookline, MA: Family Firm Institute, pp. 211–15.

Borwick, I. (1986), 'The family therapist as business consultant', in L.C. Wynne, S.H. McDaniel and T.T. Weber (eds), *Systems Consultation: A New Perspective for Family Therapy*, New York: Guilford, pp. 423–40.

Bowen, M. (1981), *Family Therapy in Clinical Practice*, Northvale, NJ: Jason Aronson.

Carlock, R.S. and J.L. Ward (2001), *Strategic Planning for the Family Business: Parallel Planning to Unify the Family and Business*, New York: Palgrave.

Carter, B. and M. McGoldrick (1999), 'Overview: the expanded family life cycle', in B. Carter and M. McGoldrick (eds), *The Expanded Family Life Cycle: Individual, Family, and Social Perspectives*, Boston, MA: Allyn and Bacon, pp. 1–26.

Cartwright, D. and A. Zander (1953), *Group Dynamics: Research and Theory*, New York: Harper and Row.

Cartwright, D. and A. Zander (1968), 'The nature of group cohesiveness', in D. Cartwright and A. Zander (eds), *Group Dynamics*, New York: Harper and Row, pp. 91–109.

Cohen, S.G. and D.E. Bailey (1997), 'What makes teams work: group effectiveness research from the shop floor to the executive suite', *Journal of Management*, **23**(3), 239–90.

Daily, C. and S. Thompson (1994), 'Ownership structure, strategic posture, and firm growth', *Family Business Review*, **7**(3), 237–50.

Donkels, R. and E. Fröhlich (1991), '"Are family businesses really different?", European experiences from STRATOS', *Family Business Review*, **4**(2), 149–60.

Engelenburg, R. van and L. Kommers (2001), *Business Transfers in the Netherlands*, Den Haag: RZO.

Filbeck, G. and L.L. Smith (1997), 'Team building and conflict management: strategies for family businesses', *Family Business Review*, **10**(4), 339–52.

Friedman, E. (1986), 'Emotional process in the marketplace: the family therapist as consultant with work systems', in L.C. Wynne, S.H. McDaniel and T.T. Weber (eds), *Systems Consultation: A New Perspective for Family Therapy*, New York: Guilford, pp. 398–422.

Georgopoulos, B.S. (1986), *Organization Structure, Problem Solving and Effectiveness: A Comparative Study of Hospital Services*, San Francisco, CA: Jossey-Bass.

Gersick, K.E., J.A. Davis, M.M. Hampton and I. Lansberg (1997), *Generation to Generation: Life Cycles of the Family Business*, Boston, MA: Harvard Business School Press.

Gomez-Mejia, L.R., H. Tosi and T. Hinkin (1987), 'Managerial control, performance and executive compensation', *Academy of Management Journal*, **30**(1), 51–70.

Guttman, H.A. (1991), 'Systems theory, cybernetics, and epistemology', in A.S. Gurman and D.P. Kniskern (eds), *Handbook of Family Therapy*, New York: Brunner/Mazel, pp. 41–62.

Habbershon, T.G. and J. Pistrui (2004), 'A model for strategic dialogue, establishing congruency in your mindset and methods', paper presented at FBN-IFERA (International Family Enterprise Research Academy) conference, Jonkoping, Sweden.

Habbershon, T.G. and M.L. Williams (1999), 'A resource-based framework for assessing the strategic advantages of family firms', *Family Business Review*, **12**(1), 1–22.

Habbershon T.G., M.L. Williams and I. MacMillan (2003), 'Familiness: a unified systems theory of family business performance', *Journal of Business Venturing*, **18**(4), 451.

Hackman, J.R. (ed.) (1990), *Groups that Work (and Those that Don't): Creating Conditions for Effective Teamwork*, San Francisco, CA: Jossey-Bass.

Hendrickson, L.U. (1992), 'Bridging the gap between organization theory and the practice of managing growth: the dynamic system planning model', *Journal of Organizational Change Management*, **5**(3), 18–37.

Hendrickson, L.U. and N. l'Abbe Wu (1993), 'Technical mastery: basis for strategic manufacturing management', *Productivity*, **34**(2), 199–207.

Hendrickson, L.U. and J. Psarouthakis (1998), *Dynamic Management of Growing Firms: A Strategic Approach*, Ann Arbor, MI: University of Michigan Press.

Hendrickson, L.U. and D. Tuttle (1997), 'Dynamic management of the environmental enterprise: a qualitative analysis', *Journal of Organizational Change Management*, **10**(4), 363–82.

Huszczo, G.E. (1996), *Tools for Team Excellence: Getting Your Team into High Gear and Keeping it There*, Palo Alto, CA: Davies-Black.

Jorissen, A., E. Laveren, R. Martens and A. Reheul (2002), 'Differences between family and nonfamily firms: the impact of different research samples with increasing elimination of demographic sample differences', *Conference Proceedings, RENT XVI 16th workshop*, Barcelona, Spain.

Kaplan, R.S. and D.P. Norton (1992), 'The balanced scorecard: measures that drive performance', *Harvard Business Review*, **70**(1), 71–9.

Katz, D. and R.L. Kahn (1966), *The Social Psychology of Organizations*, New York: John Wiley and Sons.

Kerr, M.E. and M. Bowen (1988), *Family Evaluation*, New York: W.W. Norton.

Leenders, M. and E. Waarts (2003), 'Competition and evolution of family business: the role of family and business orientation', *European Management Journal*, **21**(6), 686–97.

Mohrman, S.A., S.G. Cohen and A.M. Mohrman (1995), *Designing Team-Based Organizations: New Forms for Knowledge Work*, San Francisco, CA: Jossey-Bass.

Mustakallio, M., E. Autio and S.A. Zahra, (2002), 'Relational and contractual governance in family firms: effects on strategic decision making', *Family Business Review*, **15**(3), 205–22.

Mustakallio, M.E. (2002), 'Contractual and relational governance in family firms: effects on strategic decision-making quality and firm performance', doctoral dissertations 2002/2, Helsinki, Finland.

Neubauer, F. and A.G. Lank (1998), *The Family Business: Its Governance for Sustainability*, London: Macmillan.

Porter, M. (1985), *Competitive Advantage: Creating and Sustaining Superior Performance*, New York: Free Press.

Rokeach, M. and S.J.B. Rokeach (1989), 'Stability and change in American value priorities, 1968–1981', *American Psychologist*, May, 775–84.

Schermerhorn, J.R., J.G. Hunt and R.N. Osborn (2000), *Organizational Behavior*, New York: John Wiley and Sons.

Stewart, G.L. (2006), 'A meta-analytic review of relationships between team design features and team performance', *Journal of Management*, **32** (1), forthcoming.

Suaré, K.C. and D.J. Santana-Martin (2004), 'Governance in Spanish family business', *International Journal of Entrepreneurial Behavior and Research*, **10**(1/2), 141–63.

Tuckman, B.W. (1965), 'Development sequences in small groups', *Psychological Bulletin*, **63**(6), 384–99.

Tuckman, B.W. and Jensen, M.A.C. (1977), 'Stages of small group development revisited', *Group and Organizational Studies*, **2**(4), 419–27.

Uhlaner, L.M. and T. Habbershon (2005), 'Family influence, strategy, and innovation in family firms: application of the DSP-"f" model in an exploratory four-country investigation', paper presented at FBN-IFERA (International Family Enterprise Research Academy) conference, Barcelona, Spain.

Uhlaner, L.M. and J. Hunt (1999), 'The strategic leadership of family businesses: application of the dynamic system planning model', *Proceedings, International Council for Small Business 1999 World Conference*, Naples, Italy.
Uhlaner, L.M., H.J.M. van Goor-Balk and E. Masurel (2004), 'Family business and corporate social responsibility in a sample of Dutch firms', *Journal of Small Business and Enterprise Development*, **11**(2), 186–94.
Von Bertalanffy, L. (1968), *General Systems Theory*, Harmondsworth: Penguin.
Wagner, J.A. and J.R. Hollenbeck (1995), *Management of Organizational Behavior*, Englewood Cliffs, NJ: Prentice-Hall.
Ward, J. (1987), *Keeping the Family Business Healthy*, San Francisco, CA: Jossey-Bass.

8 Internal factors of family business performance: an integrated theoretical model*

Alberto Gimeno Sandig, Gaston J. Labadie, Willem Saris and Xavier Mendoza Mayordomo

Introduction

This chapter puts forward a theoretical model that identifies the internal factors explaining family business performance. Studies of family companies have mainly compared the performance of these firms with the performance of non-family companies. In our view, the results of such studies are contradictory given the enormous variety of both family and non-family companies, and the partial scope of the models and tests performed. Hence, there is a need for an integrated model that systematically considers the internal factors that influence the performance of family businesses and that is not based on flawed comparisons with other kinds of firms. This work presents such a model, examining theories and evidence on how the family condition and its dimensions influence performance.

We define family firms as those in which one or several families exercise influence, identified in the literature by the term *familiness* (Habbershon and Williams, 1999). We propose an approach to family companies based upon two dimensions: the family and the company. Such an approach is similar to the one suggested in early family business literature (Davis and Stern, 1980; Lansberg, 1983; Rosenblatt et al., 1985; Vilanova, 1985). We model the contingency and structural factors that have an impact on family and business management (the factors chosen being based upon the literature review). In so doing, we introduce the relationships among these constructs and number them, so that a cumulative graphic representation of the model is developed in 'stages', with the full systemic model represented in the final section. A more extensive justification of the theoretical relations proposed in this chapter can be found in Gimeno et al. (2005).

Business and family complexity

The lifecycle of the business has been widely studied. Perhaps the studies that had the greatest impact were those by Chandler (1962), Greiner (1972) and Churchill and Lewis (1983). While these authors agree in proposing an evolutionary sequence, there is no consensus regarding the dimensions and variables for defining this evolution. In the family business literature, Gersick et al. (1997) define the company dimension ('Company Axis', in their terminology) as a sequential evolution, in which there are three main stages: start-up, expansion-formalization, and maturity.

This chapter takes a different approach, employing a concept of company development that is based on *complexity*. This concept was originally developed in the Fundamental Sciences, and is now widely employed in the Social Sciences, albeit mainly in a metaphorical sense (Luhmann, 1996; Morin, 1995). More recently, it has been applied in Economics and in Management Sciences (Arthur, 1999; Krugmann, 1996).

Thus, the development of a company is not just a question of growth. Development also relates to the firm's increasing complexity in general,[1] whether it involves processes, products, markets, degree of internationalization, technology, and so on.

The complexity concept is also useful for understanding the business family. The complexity of family dynamics profoundly influences the company, as the family business literature has stated extensively, since the family holds a controlling stake in the firm.

Therefore, a proper analysis of management of family businesses requires an understanding of the role played by the family as a social system, and the way in which it exerts its influence on the firm by virtue of ownership.

The term 'family' refers to a social unit with different degrees of belonging. At its core is a group that is organized around couples and their offspring (the 'nuclear family'), which operates within a larger, less organized group of relations (the 'extended family') (Parsons, 1994). Business families exhibit special functions that go beyond the typical family functions (Rosenblatt et al., 1985). In addition to the nurturing and socialization functions,[2] these families also have to ensure the future of the companies they own. This broadening of functions creates ambivalent situations (Lansberg, 1983), affecting the relationships between family members (Davis and Taguiri, 1989).

The additional function of business families create greater demands regarding the way the family works. This is why a business family needs additional communication channels so that, according to Olson et al. (1983), the family can maintain the necessary cohesion and ability to adapt.

The family influences the company by virtue of vested interests. Indeed, a family company is defined by the fact that a business family can exercise influence over a firm primarily because of its ownership stake (Davis, 1983). In other words, family ownership is what determines the family nature of a company.

Gersick et al. (1997) propose an evolutionary pattern in business families, basing their ideas on a study of family transitions and lifecycle (Levinson, 1978, 1996). Ward (1991) argues that there are three clearly defined stages in the ownership of family companies in which the evolution of a family business is characterized by the progressive dilution of shareholdings and family relationships. This three stages have usually been identified with a given generation.[3] This relationship between generation and the nature of ownership would not occur if there were just a single heir, or if the succession policy involved concentration of shareholders.

Family influence has also been considered to go beyond the idea that family businesses can be defined by ownership and management dimensions (Heck and Trent, 1999), to also taking in control and family involvement aspects (Heck and Stafford, 2001).

Daily et al. (2003) proposed a resource theory to study ownership. They stressed the need to identify who owns each resource, considering both resources and capabilities (Grant, 1991). Thus, company owners only own certain resources (mainly assets). There are other resources (particularly capabilities) which they do not own (for example, knowledge, commitment, habits, relationships, and so on.), even though they benefit from the residual profits generated by such resources. This perspective contributes to understand the phenomena that arises as company ownership becomes more complex. Part of these resources can be transferred simply by shares changing hands, but transferring other resources is much more difficult. This is especially true when dealing with firm transitions from an organization run by a single controlling entrepreneur to one run by siblings. In this case, the entrepreneur is usually the owner of significant resources that are directly associated with him or her. The transfer of these resources is only partially possible.

The lifecycles of the family and ownership can be linked by employing the concept of family complexity. A family with three generations will be much more complex merely because more people are involved. A company founder may also have siblings and cousins, but that does not mean that the firm constitutes a 'sibling partnership' or a 'cousin consortium' (since, although those relations are part of the family, they do not constitute part of the company). We are, therefore, solely interested in cases where the family is linked to the company and influences its behavior and the complexity of the family subset holding material or psychological ownership of the firm. Accordingly, when referring to the 'family', we mean the business family: specifically, that sub set of the extended family that currently determines company behavior or will do so in the future.

Family complexity is defined by the number of family members and the kind of relationships established among them, the number of generations alive at a given point in time, and so on. Greater family complexity implies bigger differences between family members in terms of personality, abilities, interests, personal circumstances, education, and so on. The influence of each family member is also contingent on his or her position relative to the nuclear family. The combination of the individual lifecycles of the various family members also constitutes a complexity factor, as Davis and Taguiri (1989) have indicated.

Family complexity can be modified by the family itself through its inheritance policy.[4] The modern Western practice to apply egalitarian criteria to all family members of the same generation will increase the complexity of the family company over time. As Ward and Dolan (1998) observed, family complexity increases with each succeeding generation. This underlines the first proposition of the model:

Proposition 1 The complexity of family companies grows over time.

Company development is also shaped by time. Various authors propose the existence of a time pattern in company lifecycles. Time affects positively the complexity of the company through a process of accumulation of resources that leads to further business growth, internationalization, number of processes, amount of knowledge, and so on. This implies the second proposition:

Proposition 2 Business complexity is affected by the passage of time.

Thus, time affects complexity at both the family and the company levels (Figure 8.1).

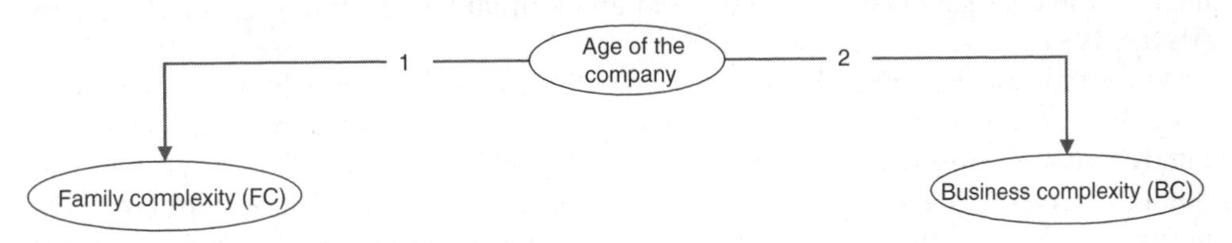

Figure 8.1 Effect of time on family and business complexity

Family business performance

In order to be able to evaluate management behavior, it is necessary to have an evaluation standard (Simon, 1983). In the case of family companies, one must consider whether the dominant values are those of the business family, those of the researcher, or those of a third party. Given that one of the main objectives of a system is its own survival (Luhmann, 1996), the issue of survival capability is a useful criterion for evaluating company management. This criterion can act as a *metapoint of view* (Morin, 1995).

Gimeno et al. (1997) have argued that a company's survival depends on two dimensions: economic performance and threshold of performance, the latter being understood as 'the level of performance below which the dominant organizational constituents would act to dissolve the company' (p. 750). Gimeno et al. place special emphasis on the expectations of the owners as factors to explain why companies continue to march on, even when their economic performance is patently unsatisfactory.

Various authors (Morris et al., 1997; Sharma et al., 2001) associate successful succession planning of family companies with two similar dimensions proposed by Gimeno et al. (1997), namely, company results in the post-succession stage, and family satisfaction with the succession process as a whole. The impact of non-economic performance factors on company objectives and values has been widely documented. Such factors include: maintaining a good work atmosphere and avoiding clashes; providing offspring: work, prestige and social recognition; financial security; a pleasant lifestyle; sense of justice, social development; and parental affection.

These objectives and values (which are clearly different from those related to economic performance) can be grouped under a 'Satisfaction' dimension. This represents family members' satisfaction with the state of family–company relationships. This satisfaction will therefore be related to family members' expectations with respect to the company.

The performance of family firms, consequently, depends on two qualitatively different dimensions, namely, economic-financial performance and satisfaction with family–company relations.

There are many studies that describe the 'disorder' that may prevail within family-owned companies. Succession problems have been described as one of the main factors that tend to weaken family companies (Bird et al., 2002), whether because of the psychological profile of a powerful entrepreneur (Kets de Vries, 1993), the dynamic relationship between father and sons (Matthews et al., 1999), the loss of leadership (Lansberg, 1999), or lack of planning (Carlock and Ward, 2001; Lansberg, 1988; Ward, 1988a). An increase in the number of shareholders leads to differences regarding objectives and values (Ward, 1997), which makes it difficult to build a dominant coalition capable of effectively leading the company. This leads to a general loss of confidence among the various players (McCollom, 1992). The extension of family equity implies a loss of shareholder commitment to the company (Thomas, 2002) and an eventual lose of entrepreneurial capacity (Payne, 1984).

Astrachan and Kolenco (1994) stress that the energy absorbed by systems to keep working increases as the company grow. Hence, the growth in family complexity can lead to family companies losing competitiveness or 'closing in' on themselves from a systems viewpoint (McConaughy and Phillips, 1999). The loss of performance as ownership complexity increases is also explained by an agency theory approach. Agency costs would not exist in a company in which the manager wholly owned the enterprise.[5] Agency costs occur once the

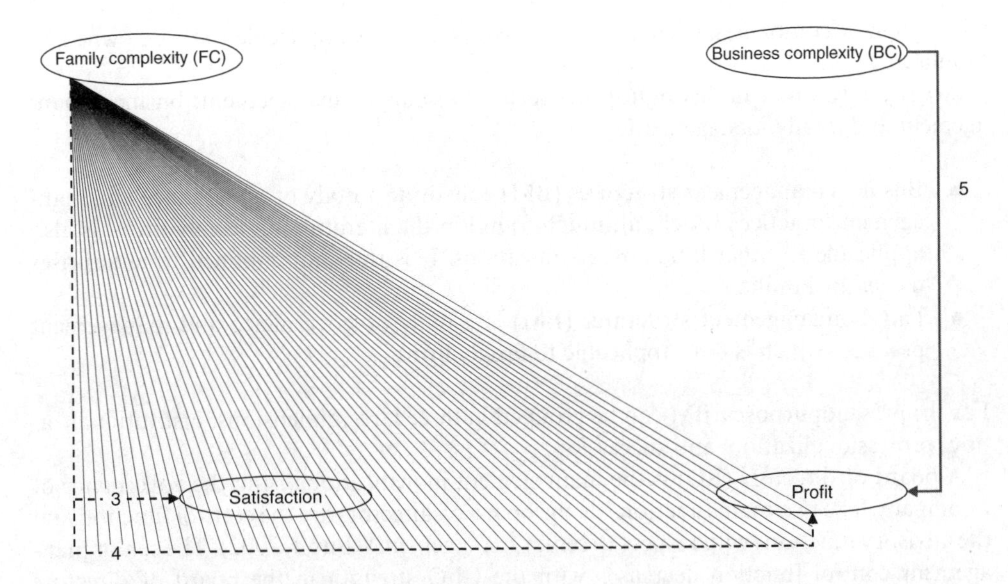

Figure 8.2 Impact of complexity on performance

'entrepreneur-controller' leaves management even if the entrepreneur retains full ownership over the firm (Ang et al., 2000). Schulze et al. (2001) and Gómez-Mejía et al. (2001) argue that family companies also incur in agency costs at the 'entrepreneur-controller' stage, given that 'altruistic' decisions lead these entrepreneurs to make poor choices that are based on non-economic preferences.

This ultimately implies the following effects:

Proposition 3 The complexity of family shareholders has a negative impact on family satisfaction (conflict and succession problems).

Proposition 4 The complexity of family shareholders has a negative impact on financial results (loss of resources and capabilities, agency costs, and succession).

The literature has also linked a company's development and size to its performance, arising from factors such as: market share (Buzzell et al., 1975; company size (Scherer, 1980); exploitation of barriers to entry (Porter, 1982); economies of scale (Besanko et al., 1996); learning curve (Ghemawat, 1985); development of the resource base (Sobel, 1984) and financial capacity (Palich and Cardinal, 2000).

This suggests the following relation:

Proposition 5 The business complexity has a positive impact on a firm's economic performance (Figure 8.2).

Business management

Management of a family firm involves two kinds of governance structures and practices. Some of these management aspects are not unique to family companies, while others only make sense for family firms. For instance, a board of directors is a useful body for a very wide range of companies, whereas a family council is only applicable to those owned by families.

Accordingly, one can distinguish between two kinds of management: business management and family management:

- Business management structures (BM) constitute a body of governance and management practices which, although found in the literature on family firms, is also applicable to other kinds of organizations. It is thus not confined to enterprises owned by families.
- Family management structures (FM) is a body of governance and management practices which is only applicable to family firms.

For the present purposes, BM can be grouped in three large dimensions: institutionalization, professionalization, and succession.

A board of directors (BD) is the main element in institutionalizing the governance of a company. The two main functions of the board of directors are the control function and the advisory function (with respect to senior management) (Birgit, 2002). The senior management control function decreases with the CEO strength in the board of directors (Warther, 1998) and increases with the role of external advisors (Byrd and Hickman, 1992). The advisory function is affected by information the CEO shares with the board of directors, and depends on the CEO's perception of job risk (Holderness et al., 1999). A CEO who feels his job is at risk will be less willing to share information with the board.

There is an agency problem in family companies (Schulze et al., 2001) that stems from the influence of family loyalties and ties on the decisions taken by family managers. This makes it even more important that the board makes adequate succession plans (Ward, 1991), develops the management team (Lansberg, 1999), and facilitates communication between family members (LaChapelle, 1998).

Management is a professional discipline because of its content (the description of managers' tasks), its instruments (how managers behave in their jobs), and its contingency aspects (what impinges upon managers' actions) (Squires, 2001). This implies stressing the importance of regularities in administrative decision-making.

Another approach to the management role, stresses the importance of the entrepreneurial behavior by creating discontinuities (Schumpeter, 1985). The differences between both schools (professional and entrepreneurial) are reflected in the various aspects of business management, such as information, knowledge, decision-making processes, and organizational structure. The professional manager relies much more on explicit knowledge (Nonaka and Takeuchi, 1995), and develops abstraction and codification processes (Boisot, 1995) so that knowledge can be passed on and shared. By contrast, entrepreneurial management relies more in tacit knowledge (Dreyfus and Dreyfus, 1986). There are also differences in the strategic decision-making processes. Professional management tends to rely on planned strategies (Porter, 1996), while entrepreneurial management tends to develop emergent strategies (Mintzberg and McHugh, 1985). Professional management and entrepreneurial management also tend to institute different organizational structures. The variable that most clearly defines an organizational structure is its level of centralization (Chandler, 1977). Professional management tends to develop more decentralized firms, whereas entrepreneurial management is associated with centralized companies.

A family business is often founded by an individual who creates and develops a company in accordance with his or her capabilities and needs. This entrepreneur develops considerable knowledge as a result of both his business activities and life experience. His or her power within the company means he or she is not required to codify his or her knowledge so that it can be shared with the rest of the management team. Centralization of decision-making allows for the alignment of decisions, in such a way that there is no need for pooling knowledge or planning a strategy in a group context. However, it is difficult to sustain centralized management practice in the long run. Tacit knowledge cannot be transferred and therefore tends to disappear when the founder leaves the organization. A centralized organization is highly dependent on the entrepreneur. This individual considerably influences the organization's hierarchy, processes, coordination systems, and work habits. Centralization and accumulation of knowledge give family companies considerable flexibility and adaptability; however, at the same time, this creates a structure highly dependent on the leader/founder (Kelly et al., 2000). Accordingly, the company is strongly influenced by the founder's lifecycle. As a result, various authors recommend establishing formal strategic planning processes (Carlock and Ward, 2001; Malone and Winter, 1989; Ward, 1988b, 1991), and 'professionalization' of the family management structure (Dyer, 1989; Hofer and Charan, 1984).

Professional management affects succession, as well. The succession of the CEO in family companies exhibits specific features. This is particularly true during the succession of the founder, given: the lack of previous experience, internal rules and guidelines; the founder's considerable power (often having combined the roles of CEO and controlling shareholder for years) and, usually, the board's inability to properly control the succession process itself. The literature on family business has widely covered the issue of succession, in which a double transition takes place with regard to management and ownership. Family companies seem particularly vulnerable during the succession process. This phenomenon has often been considered the culprit for the disappearance of such firms. That is why different authors prescribe proper planning as a way to overcome this weakness (Carlock and Ward, 2001; Malone, 1989; Ward, 1988b).

The management literature assumes the existence of a BD, which helps to ensure management continuity. This is one of the reasons why the literature on family companies emphasizes this governing body (Lansberg, 1999; Neubauer and Lank, 1999; Ward, 1991). A family company enjoys a high level of BM when it has institutionalized its decision-making processes (through an effective BD), and professionalized its management practices (that is, there are systems for explicitly rendering tacit knowledge; the strategy is well planned and the company's organization is as decentralized as possible) and the succession is prepared.

The various dimensions considered by BM are interrelated. A greater degree of BM implies less CEO discretion,[6] ultimately owing to: the supervisory functions exercised by the BD; a less ad hoc approach to running the company; a more professionalized management team and greater anticipation of the factors affecting decisions. The BD must lead the company towards professionalization and ensure that the firm anticipates the main factors determining succession. The board, in turn, can only execute its monitoring and advisory roles if the company develops explicit knowledge and communicates planned strategies. This same organization also needs to ensure management professionalization and continuity, which has the effect of reducing the company's dependence on the CEO.

Impact of business management

There are various studies that attempt to identify the relationship between BM and the firm's economic performance. The basic argument is that good corporate governance (Poza, 1988; Ward, 1991; Neubauer and Lank, 1999; Cowling, 2003; Kaplan, 1997), professionalization of management practices (Ward, 1988a; Porter, 1982; Gimbert, 1998), and planning for the succession (Covin and Slevin, 1990; Eisenhardt and Bourgeois, 1988; Lansberg, 1988, 1999) all improve companies' financial results.

Therefore, the following effect is expected:

Proposition 6 Enhancing BM directly benefits a firm's economic performance.

Company complexity also has an impact on BM. Various authors (Chandler, 1962; Greiner, 1972) have argued that BM eases a firm's transition to the next stage in its development. Inevitably, this will bring new challenges that must be overcome in order to avoid future crises. This implies that a company's growth stage influences the way a company is managed.

This argument, in turn, implies the following proposition:

Proposition 7 An increase in company complexity has a direct positive impact on BM.

The resource based theory states that a company is capable of achieving a competitive position through its various resources and capabilities (Collis and Montgomery, 1995; Grant, 1991; Wernerfelt, 1984). The development of BM is a way of transferring part of the company's resources that fosters the development of the firm. These authors consider acquisition, maintenance, and exploitation of resources to be the basic activities of managing a company. In this context, BM is the development of the firm's resources and capabilities in order to achieve corporate growth.

This suggests the following relationship:

Proposition 8 The development of BM has a direct positive effect, increasing company complexity in the process.

Other authors (Miller and Friesen, 1984; Scott and Bruce, 1987) consider that new demands on the company arise during each stage of development. These demands must then be addressed by appropriate internal changes. Each change helps to propel the company forward to a new stage of its development (7 and 8; see Figure 8.3).

Family management

As stated earlier, our model groups the different management practices that are only applicable to family businesses, within the construct of family management (FM). Those management practices that constitute FM can be grouped in four general management dimensions. Three of these coincide with those also found in BM (institutionalization, communication and succession), and one is specific to family firms (family–company differentiation). The fact that FM and BM share the same management dimensions does not mean that the management variables are the same; but, rather, that they can be grouped in the same category despite being different. For example, the 'succession' dimension

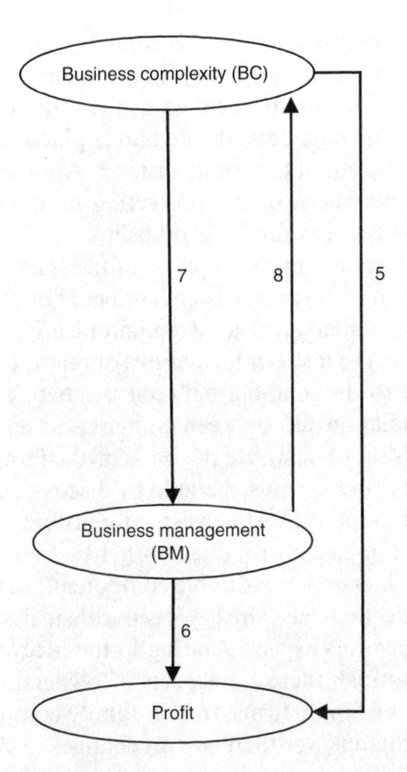

Figure 8.3 Impact of business management

covers aspects relating to management succession in the BM construct, although it refers to ownership succession in the FM construct.

Institutional theory is a suitable framework for analysing the governance of family companies. For those authors proposing this theoretical approach (Meyer and Rowan, 1977; Zucker, 1983), organizational behavior is determined by a set of values and ideas that originate in an institutional context. These institutional settings foster systems for repeating patterns and hence create organizational stability (Barley and Tolbert, 1997).

One of the most widely accepted governance structures, regarding family influence on the company, is the family council (FC). Another issue consistently emphasized by the authors is the need for formal rules[7] to limit family influence on the company. These rules tend to refer to the division of property and economic rights among family members. Hence, there is a sphere in which the family defines its relations as a business family (that is, FC), a sphere for corporate governance (that is BD), and a system of rules that provides a functional behavioral framework (Baulenas and Gimeno, 2000). Thus, institutional workings are not simply the product of these boards; rather, such bodies have attained sufficient power and legitimacy within the organizational social system itself, so that family–company relations are institutionalized, and provide an instrument that limits the family's destabilizing potential.

Family–company differentiation is another dimension of family management. As previously explained, business and family are two interrelated systems (Davis and Stern, 1980). This does not stop them having different roles, rules, and functions (Lansberg, 1983; Chandler, 1990). Therefore, clearly defined limits between company and family are needed. The differentiation between family and company mainly involves economic relations and human resource policies regarding family members (Vilanova, 1985). In the human resource field, this differentiation concerns the criteria applied to the incorporation and promotion of family members, the demands placed upon them, and authority and hierarchical relations based on corporate criteria. Applying special criteria, such as equality between various members, or the protection of the least able member, would signal a low differentiation between family and business.

The relational aspect of communication plays a fundamental role in the literature on family firms. Family companies have often been hampered by communication issues (Kets de Vries, 1993). Conflicts, understood as a communicative phenomenon, arise from differences either with respect to tasks (information) or relations (Watzlawick et al., 1981). It is especially important to differentiate between the two levels in family companies, owing to the strong interrelationship between content and relations. Logically, relations are of great importance, as they constitute the basis of the family system. Content is also important, since a business family must be able to discuss business matters effectively. From a communication standpoint, FM consists of clarifying matters, so that relational aspects and content do not interfere with one another.

Succession in family businesses has a double component: management succession and ownership succession. The first has already been considered in the context of BM. Management succession involves aspects relating to the lifecycle of the first CEO or his or her performance. In contrast, there are aspects of ownership succession that are specific to, and characteristic of family firms. In non-family companies, there are situations involving shares changing hands, yet there are no changes in the structure of ownership arising from shareholders' lifecycles.

An increase in family complexity involves an increase in the number of shareholders and a dilution in their relationships with one another. When the 'nuclear family' threshold is surpassed, the family starts to lose cohesiveness. One implication is the need to construct the identity of the business family in the form of a 'common dream' (Lansberg, 1999) or the 'entrepreneurial family group' (Habbershon and Pistrui, 2002). This common identity is a mechanism to help unify the family unit beyond just nuclear families and the branches they comprise.

Family management succession means preparing the ownership transmission from the legal and financial point of view but also creating family cohesion based upon the concept of value creation.

Impact of family management

Family management practices help to regulate the role of the family in the business and thus allows the family to: self-impose a system of limits to manage differences that would otherwise lead to conflict (Levinson, 1971); help resolve conflicts when they arise (Sorensen, 1999); limit nepotism (Vinton, 1998); increase harmony and satisfaction (Malone and Winter, 1989); and, maintain family commitment to the company's entrepreneurial mission (Habbershon and Pistrui, 2002).

Family management often leads to an increase in the family's welfare by enhancing its capacity to identify, accept, and resolve differences. It also reduces the potential for disorder in the company. A natural by-product is increased company competitiveness and therefore better financial performance.

All this supports proposition 9 and 10:

Proposition 9 FM has a direct positive impact on satisfaction with family–company relations.

Proposition 10 FM has a direct positive impact on the firm's economic performance.

As noted earlier, effects (1) and (2) suggest that an increase in family complexity reduces company performance in terms of satisfaction and business results. The effects identified in the preceding paragraph (9) and (10) also suggest that FM has a direct positive impact on satisfaction and business results. Consequently, FM can counteract the negative effects produced by an increase in family complexity.

It therefore implies that:

Proposition 11 Business families tend to develop FM as they become more complex.

The family influences the way the company operates (Astrachan, 1988; Chua et al., 1999), in either a positive or negative fashion (Steier et al., 2004). The agency problem in family companies (Gómez-Mejía et al., 2001; Schulze et al., 2001) may indicate a negative influence stemming from altruism.

Family management limits the power of the founder and family managers in relation to non-managers, thus allowing greater company professionalization (Hofer and Charan, 1984).

This supports the following proposition:

Proposition 12 The development of FM should have a direct positive impact on BM.

The combination of these effects on FM is shown graphically in Figure 8.4:

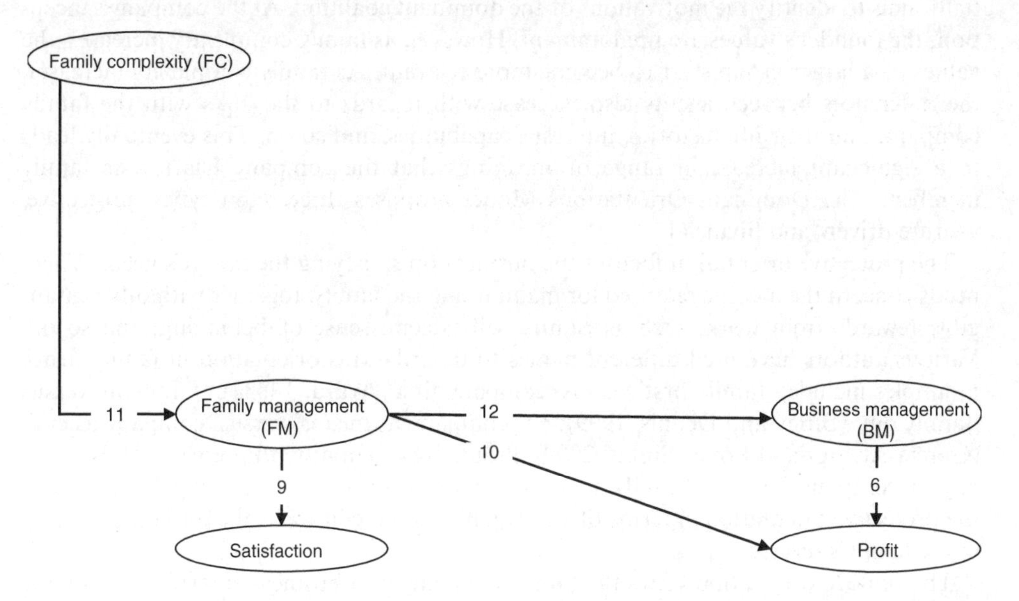

Figure 8.4 The impact of family management on business complexity

The ideological dimension of family business

Understanding the behavior found in family firms requires considering the individuals' perception of reality, as influenced by their position within the family and the business project. Terms such as *world vision, paradigm, beliefs, culture, mental model, organizational culture, cognitive map, mental category*, and *cognitive program* are evidence of this phenomena. They have been applied to the family through the family paradigm concept, which has been defined as a set of shared family experiences that are integrated in the form of constructs. Reiss (1981, p. 382) expresses it as: 'A shared construct specifies that the family acts in this fashion because it is collectively convinced that its social setting is the best in the world.'

Both the founder's culture and its transmission mechanisms influence the management of a family company, whether in a BM context, or in relation to FM.

The values of entrepreneurs and firms have been widely studied, whereas there is limited literature concerning the values of business families and/or their members. The ideological components of both systems have been considered as separate phenomena. Therefore, families are addressed in terms of 'family paradigms', and companies in terms of 'organizational cultures'. The literature hardly addresses the issue of 'family business culture'.

The few studies that do exist analyse the value systems in family companies regarding philosophy, values, and attitudes. This information has been graphed on an axis running from greater to lesser family involvement, with the middle point representing a balance between the two systems (Birley and Dennis, 1999; Ward, 1988a). Chua et al. (1999) stress the importance of a family company value system.

An effective family company performance model needs to incorporate how reality is interpreted by each member. This model must add numerous perspectives, beyond that of just the founder, since eventually there will be a shift in values as a dominant coalition is formed.

Within the integrated model, we use the Dominant Orientations Model produced by Otálora et al. (1990) and Gimeno (1999) for the ideological component. The model essentially aims to identify the motivations of the dominant coalition. At the company's inception, the founder's values are predominant. However, as family complexity increases, the values of a larger group start to become more relevant. As family complexity increases, the differences between agents also increase with regards to the links with the family company, and their life histories, interests, capabilities, and so on. This eventually leads to a significant increase in range of meanings that the company has for its family members. The Dominant Orientations Model proposes three main types: protective, venture-driven, and financial.

The protective orientation focuses the business on satisfying the family's needs. These needs concern the income required for maintaining the family, together with other intangible rewards from work, such as: status, self-esteem, sense of belonging, and so on. Various authors have used different names to describe this orientation in family firms. Examples include: 'family first' versus 'company first' (Ward, 1988a); 'family in' versus 'family out' (Birley and Dennis, 1999); or, 'company as means' versus 'company as end' (García-Álvarez and López-Sintas, 2000). Protective orientation thus corresponds basically to 'company first' and 'family in', or 'company as means'. The principal objective of the protective orientation in terms of the organizational culture, is the quest for stability in the family's service.

The venture orientation views the family company as a business enterprise led by the family, or at least its most enterprising members. It implies a vision of company growth and a willingness to act. This vision is not subject to family interests. Profitability is not an end in itself but rather simply a means to attain growth and development. The company's purpose is to provide work and support for family members. Given that the aim is company development, there should be an associated increase in business complexity. Habbershon and Pistrui (2002), in stating the role of a business family, actually characterize what we define as a family adopting a venture orientation.

In a company with a financial orientation, the aim is to maximize profits and optimize assets. The value of the company lies in its capacity to generate economic value for the family. In this case, the central objective is the optimization of financial variables, such as profitability, liquidity, risk, and so on. This orientation supports the idea that the company must have a principal, as required by agency theory.

These three orientations correspond to radial categories that cannot be defined from characteristics shared by each member in the category. As a result, the orientations are characterized by a variation of the central model (Lakoff, 1987). With regard to radial categories, family companies could belong to all the categories, but in varying degrees.

Dominant orientations influence BM decisions. A purely functional board of directors is superfluous in a company whose *raison d'être* is family service. Family capacities define the management practices, which, obviously, leads to less professionalization. Under the protective orientation, succession planning is irrelevant since continuity is not one of the family's objectives. All this implies that the protective orientation has a net negative impact on BM.

The prevalence of a venture orientation means that the family entrepreneur runs the company in accordance with his own egocentric management style, which makes it illogical for the BD to orient the company towards greater professionalization and to exercise control over the managers (Cannella and Lubatkin, 1993). The family entrepreneur creates management models that represent alternatives to the professional ones. Furthermore, it would be unreasonable for the BD to demand performance that goes beyond the personal demands the entrepreneur places upon himself (Fredrickson et al., 1988). In this case, succession is relatively straightforward. Effectively, this requires simply repeating the model, with a new 'entrepreneur' picking up the reins from his or her predecessor during the transition process. The greatest challenge is finding the right person to succeed the entrepreneur (Gimeno and Baulenas, 2003). This implies that a venture orientation has a negative impact on FM.

The financial orientation also influences BM in positive terms, stressing company performance. The BD focus on results and CEOs who fail to deliver them are likely to be replaced (Mizruchi, 1983). Management professionalization is a logical approach under this orientation as it reduces risk, and provides more control of the CEO and the company. It reduces the firm's dependency on the CEO too. This orientation should encourage succession planning in order to sustain financial performance. This means that financial orientation should have a positive effect on BM.

In summary, we expect the following causal relations (see Figure 8.5):

Proposition 13 Protective orientation has a negative effect on BM.

Proposition 14 Venture orientation has a negative effect on BM.

Proposition 15 Financial orientation has a positive effect on BM.

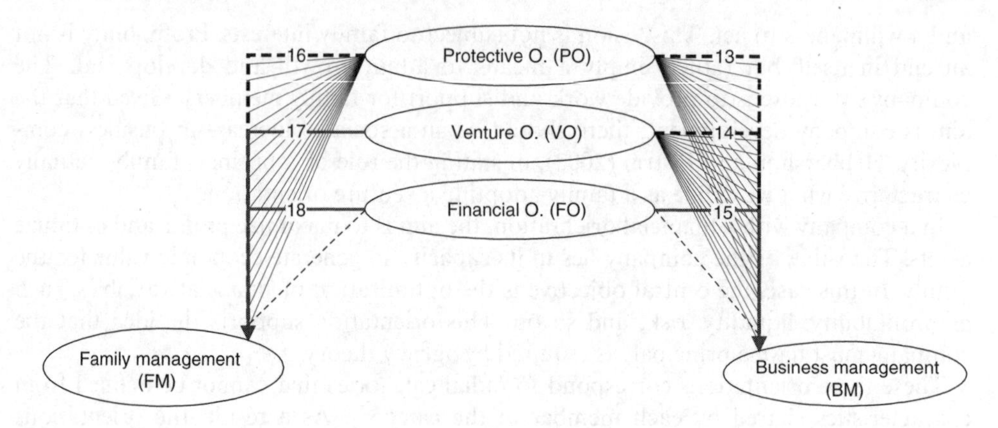

Figure 8.5 *Impact of ideology on business management and family management*

The ideological dimension should also affect family management. Self-imposed limits appear logical, depending on the family's dominant orientation. However, such limits are unwarranted if the family has a protective orientation since this entails the company being at the family's service. Consequently, a protective orientation specifically does not foster institutional systems, differentiation between family and company, and development of pertinent communication skills to facilitate family relations. This last concept is fundamental to establish limits on the family's influence. This implies a negative effect of protective orientation on FM.

On the other hand, it is sensible for the family to establish limits to intervention within the company in the case of the venture orientation. There are more important objectives, such as the business project, to which the family is willing to subordinate its interests. Developing management instruments that limit the family's ability to divert resources is consistent with an venture orientation. These policies help ensure that resources are employed sensibly. The venture orientation hence fosters FM.

The financial orientation also has an impact on FM. Company financial interests delineate a clear differentiation between the family and company spheres. The family organizes itself in such a way as to maximize returns, accepting numerous self-imposed limits, if necessary. The company does not cater to special interests involving, for instance, occupational status or family satisfaction. In this case, there is a logical need to clearly define spheres delimiting the family and company. This therefore implies that the financial orientation has a positive impact on FM.

Hence:

Proposition 16 Protective orientation has a negative effect on FM.

Proposition 17 Venture orientation has a positive effect on FM.

Proposition 18 Financial orientation has a positive effect on FM.

Summary of model

The various effects of internal factors influencing family business behavior are summarized in the model shown in Figure 8.6.

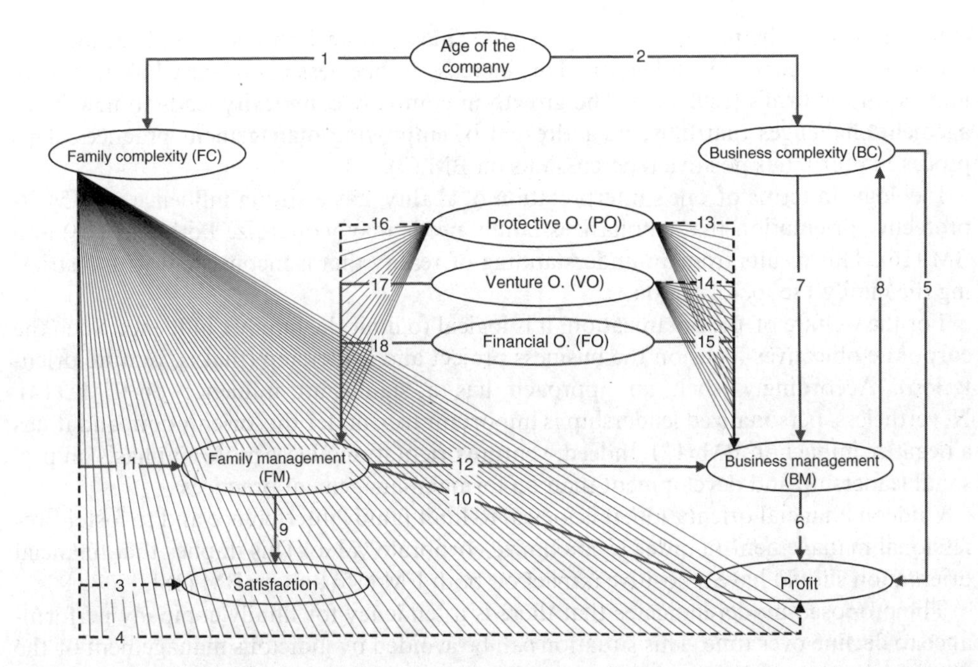

Figure 8.6 Family Business Performance Model

This is a contingent model of family business management. Under the proposed model, the complexity of family shareholdings increases with time (1) as the number of family members grow. This is accompanied by an increase in company complexity (2) due to the accumulation of resources over time.

An increase in the family shareholdings complexity over time has an indirect negative impact on the performance of a family company. More family complexity leads to a loss of family satisfaction (3) owing to growing differences between family shareholders. It also has a negative impact on the company's performance (4), given that the business family's commitment diminishes.

An increase in family complexity must impact FM (11) because the family must spend more effort in creating order in its relationships within the company. Consequently, the family imposes stricter interfamily limits as family complexity grows.

The management of family–company relations (that is, FM) has a direct positive impact on two dimensions of company performance, namely, family satisfaction (9) and profits (10). Family management improves the firm's capacity in dealing with family differences without harming interfamily relations. This typically fosters a much greater family commitment to the firm.

Family management also has a direct impact on BM. Indeed, the family's ability to impose limits on itself and to confine family rules and roles to the family sphere allows the company to apply its own rules and roles. Limiting family encroachment substantially facilitates corporate management and creates a more professional work environment. Thus, FM has a positive impact on BM (12).

Professional management of the company improves its economic results (6) and increases its complexity (8), as well. Therefore management, in the process of improving company results, helps the company expand, incorporate knowledge and technology, widen its product range, and so on. This increase in business complexity has a positive impact on the firm's results (5). The growth in company complexity leads to new management challenges that must be addressed by improving management practices. This process, in turn, has positive repercussions on BM (7).

Ideology, in terms of one's interpretation of reality, has a strong influence on FM. A protective orientation that emphasizes family needs may jeopardize both FM (13) and BM (16). This results from an understanding of reality that is inconsistent with restricting the family's scope-of-action.

For the welfare of the organization, it is logical to limit the family's influence when the corporate objectives focus on the business project and leadership (that is, venture orientation). Accordingly, such an approach has a significant influence on FM (14). Nevertheless, personalized leadership is important too, under this orientation, and it has a negative impact on BM (17). Indeed, venture orientation places more emphasis on personal leadership and development than on formal management practices.

Under a financial orientation, it is logical to limit family interference and to foster professional management in order to maximize corporate value. This implies that financial orientation should have a positive impact on both FM (15) and on BM (18).

The proposed model indicates that there is a tendency for family company performance to decline over time. This situation can be avoided by judicious management of the relations between family and company. This, in turn, would improve business management and, ultimately the firm's results. The business family's ideological framework may either help or hamper management of both family–company relations and the company itself.

Concluding remarks

The model provides an integrated framework for analysing the performance of family firms and for interpreting pre-existing findings and relationships posed among some of the variables included in the model. The model presented is a typical example of a structural equation model. Specific estimation and testing procedures have been developed (Jöreskog, 1969) for these kinds of models. To validate this model, the various concepts have to be measured for a sample of family businesses. The effects hypothesized in this chapter can then be estimated and the model as a whole can be tested. This not only provides information about the direct effects specified in the model but also about indirect effects and spurious relationships. If the model does not fit the data, there is scope for refinements. The authors have already collected data for the empirical testing of the Family in Business Performance Model.

Notes

* Acknowledgement: we are thankful to Marcel Planellas, Pedro Parada, Joan Manel Batista, Joan Sureda, Eugenia Bieto, Iván Lansberg, Fernando Casado, Emil Herbolzheimer, Panikkos Poutziouris and two anonymous reviewers for their valuable comments.

1. Complexity implies the emergence of new system properties arising from the dynamic relationships between its constituent elements (Morel and Ramanujam, 1999). A more complex system is both more adaptable to its environment, but also has greater potential to become disordered (Luhmann, 1996). Thus, a more complex company will have more elements (staff, products, suppliers, clients, technologies, countries, cultures, and so on), which will create richer behavior patterns, but which will also require systems to impose order and limit the company's propensity to become chaotic.

2. Families perform two main functions in human society: nurturing and socialization (Minuchin, 1979; Reiss, 1981). The first consists of supporting the physical and emotional development of family members, while the second involves passing on values, rules, and social position to those members.
3. Controlling owner, sibling partnership and cousin consortium.
4. For example, the primogeniture model widely used in farming societies in which, unlike in the modern world, management and technology played little or no role in either creating or destroying value.
5. Often referred to as a 'zero agency-cost firm'.
6. Hambrick and Finkelstein (1987) define managerial discretion as the freedom managers have in making decisions and shaping the company's organization.
7. This formal definition of rules has been given various names: shareholders' agreement, protocol, family constitution, capital principles.

References

Ang, J. Cole, R.A. and Wuh Lin, J. (2000), 'Agency costs and ownership structure', *Journal of Finance*, **4**(1), 81–106.

Arthur, B. (2000), 'Cognition: the black box of economics', in D. Colander (ed.), *The Complexity Vision and the Teaching of Economics*, Northampton: Edward Elgar Publishing.

Arthur, W.B. (1999), 'Complexity and the economy', *Science*, **284**(5411), 107–9.

Astrachan, J. (1988), 'Family firm and community culture', *Family Business Review*, **1**(2), 165–89.

Astrachan, J.H. and Kolenko, T.H. (1994), 'Neglected factors explaining family business success: human resource practices', *Family Business Review*, **7**(3), 251–62.

Barley, S. and Tolbert, P. (1997), 'Institutionalisation and structuration: studying the links between action and institution', *Organization Studies*, **18**(1), 93–117.

Baulenas, G. and Gimeno, A. (2000), 'Comunicación, Contextos, Reglas y Órganos de Gobierno', *Congreso Nacional de Investigación sobre la Empresa Familiar*, Organismo Público Valenciano de Investigación, Valencia: Organismo Público Valenciano de Investigación.

Besanko, D., Dranove, D. and Shanley, M. (1996), *Economics of Strategy*, New York: John Wiley and Sons.

Bird, B., Welsch, H., Astrachan, J.H. and Pistrui, D. (2002), 'Family business research: the evolution of an academic field', *Family Business Review*, **15**(4), 337–50.

Birgit, R. (2002), 'Four essays in corporate governance', University of Chicago, ed. Proquest doctoral dissertations.

Birley, S. and Dennis, N. (1999), 'The family and the business', *Long Range Planning*, **32**(6), 598–609.

Boisot, M.H. (1995), *Information Space: A Framework for Learning in Organizations, Institutions and Culture*, London: Routledge.

Buzzell, B., Gale, B. and Sultant, R. (1975), 'Market share: key to profitability', *Harvard Business Review*, January–February, 97–106.

Byrd, J. and Hickman, K. (1992), 'Do outside directors monitor managers?', *Journal of Financial Economics*, **32**(2), 195–221.

Cannella, A. and Lubatkin, M. (1993), 'Succession as a socio-political process: internal impediments to outsider selection', *Academy of Management Journal*, **32**(6), 763–94.

Carlock, R. and Ward, J. (2001), *Strategic Planning for the Family Business*, New York: Palgrave.

Chandler, A.D. (1962), *Strategy and Structure: Chapters in the History of the Industrial Enterprise*, Cambridge, MA: MIT Press.

Chandler, A.D. (1977), *The Visible Hand*, Cambridge, MA: Harvard University Press.

Chandler, A.D. (1990), *Scale and Scope: The Dynamics of Industrial Capitalism*, Cambridge, MA: Harvard University Press.

Chua, J.H., Chrisman, J.J. and Sharma, P. (1999), 'Defining family business by behaviour', *Entrepreneurship Theory and Practice*, **23**(4), 19–30.

Churchill, N. and Lewis, V. (1983), 'The five stages of small business growth', *Harvard Business Review*, **61**(3), 30–50.

Collis, D.J. and Montgomery, C.A. (1995), 'Competing on resources: strategy in the 1990s', *Harvard Business Review*, **73**(3), 118–29.

Covin, L.G. and Slevin, T.J. (1990), 'New ventures strategic posture, structure and performance: an industry life cycle analysis', *Journal of Business Venturing*, **5**(2), 123–35.

Cowling, M. (2003), 'Productivity and corporate governance in smaller firms', *Small Business Economics*, **20**(4), 335–45.

Daily, C.M., Dalton, D.R. and Rajagopalan, N. (2003), 'Governance through ownership, centuries of practice, decades of research', *Academy of Management Journal*, **46**(2), 151–8.

Davis, J. and Taguiri, R. (1989), 'The influence of life stage on father–son work relationships in family companies', *Family Business Review*, **11**(1), 47–74.

Davis, P. (1983), 'Realizing the potential of the family business', *Organizational Dynamics*, **12**(1), 47–57.
Davis, P. and Stern, D. (1980), 'Adaptation, survival, and growth of the family business: an integrated systems perspective', *Human Relations*, **34**(4), 207–24.
Dreyfus, H.I. and Dreyfus, S.E. (1986), *Mind over Machine: The Power of Human Intuition and Expertise in the Era of Computer*, New York: Free Press.
Dyer, W.G. Jr (1989), 'Integrating professional management into a family owned business', *Family Business Review*, **2**(3), 221–35 .
Eisenhardt, K.M. and Bourgeois, L.J. (1988), 'Politics of strategic decision making in high-velocity enviroments: towards a mid-range theory', *Academy of Management Journal*, **31**(4), 737–70.
Fredrickson, J.W., Hambrick, D.C. and Baumrin, S. (1988), 'A model of CEO dismissal', *Academy of Management Review*, **13**(2), 255–70.
García-Álvarez, E. and López-Sintas, J. (2000), 'Tradition or entrepreneurship in family business: is that the question?', *Proceedings of the FBN 11th Annual World Conference*: *Tradition or Entrepreneurship in the New Economy*, Ed. Panikkos Poutziouris, London.
Gersick, K.E., Davis, J.A., Hampton, M.M. and Lansberg, I. (1997), *Empresas Familiares: Generación a Generación*, México: McGraw-Hill.
Ghemawat, P. (1985), 'Building strategy on experience curve', *Harvard Business Review*, (63), 143–52.
Gimbert, X. (1998), *El Enfoque estratégico de la empresa: principios y esquemas básicos*, Bilbao: Deusto.
Gimeno, A. (1999), 'The prevalent expectations of family business owners: an extension of the family, a corporation or a business project', *Proceedings of the IV CEMS Academic Conference*, Barcelona.
Gimeno, A. and Baulenas, G. (2003), 'Modelos de empresa familiar: identidad y estructura', *Iniciativa Emprendedora*, (40), 95–109.
Gimeno, A., Labadie, G.J., Mendoza, X. and Saris, W. (2005), 'El desempeño en la empresa familiar, in M. Garrido and J.M. Fugardo (eds), *El patrimonio Familiar, Profesional y Empresarial. Sus Protocolos*, Barcelona: Bosch.
Gimeno, J., Folta, T.B., Cooper, A.C. and Woo, C.Y. (1997), 'Survival of the fittest? Entrepreneurial human capital', *Administrative Science Quarterly*, **42**(4), 750–84.
Gómez-Mejía, L.R., Núñez-Nickel, M. and Gutiérrez, I. (2001), 'The role of family ties in agency contracts', *Academy of Management Journal*, **44**(1), 81–96.
Grant, R.M. (1991), 'The resource-based theory of competitive advantage: implications for strategy formulation', *California Management Review*, **33**(3), 114–36.
Greiner, L. (1972), 'Evolution and revolution as organizations grow', *Harvard Business Review*, July–August, 37–46.
Habbershon, T.G. and Pistrui, J. (2002), 'Enterprising families domain: family-influenced ownership groups in pursuit of transgenerational wealth', *Family Business Review*, **15**(3), 223–37.
Habbershon, T.G. and William, M.L. (1999), 'A resource-based framework for assessing the strategic advantages of family firms', *Family Business Review*, **12**(1), 1–25.
Hambrick, D.C. and Finkelstein, S. (1987), 'Managerial discretion: a bridge between polar views on organizations', in L.L. Cummings and B.M. Staw (eds), *Research in Organizational Behaviour*, Greenwich, CT: JAI Press, vol. 9, pp. 369–406.
Heck, R.K.Z. and Stafford, K. (2001), 'The vital institution of family business: economic benefits hidden in plain sight', in G.K. McCann and N. Upton (eds), *Destroying Myths and Creating Value in Family Business*, Deland, Fl: Stetson University, pp. 9–17.
Heck, R.K.Z. and Trent, E.S. (1999), 'The prevalence of the family business from a houshold sample', *Family Business Review*, **11**(3), 239–52.
Hofer, C.W. and Charan, R. (1984), 'The transition to professional management: mission impossible?', *American Journal of Small Business*, **1**(1), 1–11.
Holderness, C., Kroszner, R. and Sheehan, D. (1999), 'Were the good old times that good? Changes in managerial stock ownership since the Great Depression', *Journal of Finance*, **54**(2), 435–69.
Jöreskog, K.G. (1969), 'A general approach to confirmatory maximum likelihood factor analysis', *Psychometrika*, (34), 183–202.
Kaplan, S. (1997), 'Corporate governance and corporate performance: a comparison of Germany, Japan and the U.S.', in D.H. Chew (ed.), *Studies in International Corporate Finance and Governance Systems. Comparison of the U.S., Japan and Europe*, Oxford: Oxford University Press.
Kelly, L., Athanassiou, N. and Crittenden, W.F. (2000), 'Founder centrality and strategic behavior in the *family*-owned firm', *Entrepreneurship Theory and Practice*, **25**(2), 27–43.
Kets de Vries, M. (1993), 'The dynamics of family controlled firms: the good and the bad news', *Organizational Dynamics*, **21**(3), 59–62.
Krugman, P. (1996), *The Self-Organizing Economy*, Oxford: Basil Blackwell.
LaChapelle, K. (1998), 'The trust catalyst in the family-owned business', *Family Business Review*, **11**(1), 1–18.
Lakoff, G. (1987), *Woman, Fire and Dangerous Things*, Chicago, IL: University of Chicago Press.

Lansberg, I. (1983), 'Managing human resources in family firms: the problem of institutional overlap', *Organizational Dynamics*, **12**(1), 39–46.
Lansberg, I. (1988), 'The succession conspiracy', *Family Business Review*, **1**(2), 11–19.
Lansberg, I. (1999), *Succeeding Generations*, Boston, MA: Harvard Business School Press.
Levinson, H. (1971), 'Conflicts that plague family businesses', *Harvard Business Review*, March–April, 134–5.
Levinson, D. (1978), *Seasons of a Man's Life*, New York: Basic Books.
Levinson, D. (1996), *Seasons of Woman's Life*, New York: Knopf.
Luhmann, N. (1996), *Introducción a la Teoría de Sistemas*, Mexico, DF: Universidad Iberoamericana, AC.
Malone, S.C. and Winter, M. (1989), 'Selected correlates of business continuity planning in the family business', *Family Business Review*, **2**(4), 341–53.
McCollom, M. (1992), 'The ownership trust and succession paralysis in the family business', *Family Business Review*, **5**(2), 145–60.
McConaughy, D.L. and Phillips, G.M. (1999), 'Founders versus descendants: the profitability, efficiency, growth characteristics and financing in large, public, founding-family-controlled firms', *Family Business Review*, **42**(2), 123–31.
Meyer, J. and Rowan, B. (1977), 'Institutionalized organizations: formal structure as myth and ceremony', *American Journal of Sociology*, **83**(2), 340–63.
Miller, D. and Friesen, P.H. (1984), 'A longitudinal study of the corporate life cycle', *Management Science*, **30**(10), 1161–83.
Mintzberg, H. (1979), *The Structuring of Organizations*, New York: Prentice-Hall.
Mintzberg, H. and McHugh, A. (1985), 'Strategy formation in an ad-hocracy', *Administrative Science Quarterly*, (30), 160–97.
Minuchin, S. (1979), *Familias y terapia familias*, Barcelona: Gedisa.
Mizruchi, M.S. (1983), 'Who controls whom? An examination of the relationship between management and boards of directors in large American corporations', *Academy of Management Review*, (8), 426–35.
Morel, B. and Ramanujam, R. (1999), 'Through the looking glass of complexity: the dynamics of organizations as adaptative and evolving systems', *Organization Science*, **10**(3), 278–93.
Morin, E. (1995), *Sociología*, Barcelona: Editorial Tecnos.
Morris, M.H., Williams, R.O., Allen, J.A. and Ávila, R.A. (1997), 'Correlates of success in family business transitions', *Journal of Business Venturing*, (12), 358–401.
Neubauer, F. and Lank, A.G. (1999), *La Empresa Familiar*, Barcelona: Ediciones Deusto.
Nonaka, I. and Takeuchi, H. (1995), *The Knowledge-creating Company*, Oxford: Oxford University Press.
Olson, D.H., McCubbin, H.C., Barnes, H., Larsen, A., Muxen, M. and Wilson, M. (1983), *Families: What Makes Them Work*, Beverly Hills, CA: Sage Publications.
Otálora, G., Vilanova, A. and Gimeno, A. (1990), 'Los modelos de capital dominante', working document, ESADE.
Palich, L.E. and Cardinal, L. (2000), 'Curvilinearity in the diversification-performance linkage: an examination of over three decades', *Strategic Management Journal*, **21**(2), 155–75.
Parsons, T. (1994), 'La estructura social de la familia', in R.N. Anshen (ed.), *La Familia*, Barcelona: Edicions 62.
Payne, P.L. (1984), 'Family business in Britain: an historical and analytical survey', in A. Akio Okochi and Y. Shigeaki (eds), *Family Business in the Era of Industrial Growth; Its Ownership and Management*, Tokyo: University of Tokyo Press.
Porter, M. (1982), *Estrategia Competitive*, Mexico: CECSA.
Porter, M. (1996), 'What is strategy', *Harvard Business Review*, **74**(6), 61–78.
Poza, E. (1988), 'Managerial practices that support entrepreneurship and continued growth', *Family Business Review*, **1**(4), 339–59.
Reiss, D. (1981), *The Family's Construction of Reality*, Cambridge, MA: Harvard University Press.
Rosenblatt, P.C., de Mik, L., Anderson, R.M. and Johnson, P.A. (1985), *The Family in Business: Understanding and Dealing with the Challenges Entrepreneurial Families Face*, San Francisco, CA: Jossey-Bass.
Scherer, F.M. (1980), *Industrial Market Structure and Economic Performance*, Chicago, IL: Rand McNally College Publishing.
Schulze, W.S., Lubatkin, M.H., Dino, R.N. and Buchholtz, A.K. (2001), 'Agency relationships in family firms: theory and evidence', *Organization Science*, **12**(2), 99–116.
Schumpeter, J.A. (1985), *The Theory of Economic Development*, London: Oxford University Press.
Scott, B.R. and Bruce, R. (1987), 'Five stages of growth in small business', *Long Range Planning*, **20**(3), 45–52.
Sharma, P., Christman, J., Pablo, A. and Chua, J.H. (2001), 'Determinants of initial satisfaction with the succession process in family firms: a conceptual model', *Entrepreneurship Theory and Practice*, **25**(3), 17–36.
Simon, H.A. (1983), *Reason in Human Affairs*, Oxford: Basil Blackwell.
Sobel, R. (1984), *The Rise and Fall of the Conglomerate Kings*, New York: Stein and Day.
Sorensen, R.L. (1999), 'Conflict management strategies used by successful family firms', *Family Business Review*, **12**(4), 325–39.

Squires, G. (2001), 'Management as a professional discipline', *Journal of Management Studies*, **38**(4), 473–88.
Steier, L., Crisman, J. and Chua, J. (2004), 'Entrepreneurial management in family firms: an introduction', *Entrepreneurship Theory and Practice*, **28**(4), 295–303.
Thomas, J. (2002), 'Freeing the shackles of family business ownership', *Family Business Review*, **15**(4), 321–31.
Vilanova, A. (1985), 'El Modelo de los Cuatro Niveles', unpublished work, ESADE.
Vinton, K.L. (1998), 'Nepotism: an interdisciplinary model', *Family Business Review*, **11**(4), 297–303.
Ward, J. (1988a), *Keeping the Family Business Healthy*, San Francisco, CA: Jossey-Bass.
Ward, J. (1991), *Creating Effective Boards for Private Enterprises: Meeting the Challenges of Continuity and Competition*, San Francisco, CA: Jossey-Bass.
Ward, J.L. (1988b), 'The special role of strategic planning for family businesses', *Family Business Review*, **1**(2), 105–17.
Ward. J. (1997), 'Growing the family business: special challenges and best practices', *Family Business Review*, **10**(4), 323–38.
Ward, J. and Aronoff, C.E. (1994), 'Managing family-business conflict', *Nation's Business*, **82**(11), 54–5.
Ward, J. and Dolan, C. (1998), 'Defining and describing family business ownership configurations', *Family Business Review*, **11**(2), 305–10.
Warther, M. (1998), 'Board effectiveness and board dissent: a model of the board's relationship to management and shareholders', *Journal of Corporate Finance*, (4), 53–70.
Watzlawick, P., Beavin, J. and Jackson, D. (1981), *Teoría de la Comunicación Humana: Interacciones, Patologías y Paradojas*, Barcelona: Herder.
Wernerfelt, B. (1984), 'A resource-based view of the firm', *Strategic Management Journal*, **5**(2), 171–80.
Zucker, L. (1983), 'Organizations as institutions', in S.B. Bacharach (ed.), *Research in the Sociology of Organizations*, Greenwich, CT: JAI Press, pp. 1–47.

PART III

FAMILY BUSINESS RESEARCH: METRICS AND METHODOLOGIES

9 The F-PEC scale of family influence: a proposal for solving the *family business* definition problem[1]

Joseph H. Astrachan, Sabine B. Klein and Kosmas X. Smyrnios[2]

The definition problem in family business research

Although in 1989, Handler said that 'defining the family firm is the first and most obvious challenge facing family business researchers' (p. 258), more then 10 years later, the challenge remains. To date, there is 'no widely accepted definition of a family business' (Littunen and Hyrsky, 2000, p. 41). Instead, various definitions are reported in the literature.

An analysis of the literature suggests three principal ways in which to consider the plethora of definitions: content, purpose, and form. Most definitions and classifications focus on content (for example, Handler, 1989; Heck and Scannell Trent, 1999; Litz, 1995). However, definitions cited earlier in the literature mostly concern ownership (for example, Berry, 1975; Lansberg et al., 1988), ownership and management involvement of an owning family (Barnes and Hershon, 1976; Burch, 1972), and generational transfer (Ward, 1987). In contrast, more recent definitions concentrate on family business culture (Dreux IV and Brown, 1999; Litz, 1995).

A definition of *family business* can either serve a distinct research purpose (for example, Dean, 1992) or assist in differentiating family from nonfamily firms (Klein, 2000a). Moreover, definitions can be employed for structural purposes, such as subdividing a sample into various categories (Daily and Thompson, 1994). Definitions can also be employed for explanatory purposes. For instance, Harris, Martinez, and Ward (1994) use a multifaceted definition to develop a theory about the evolution of family-owned businesses from founder-managed firms to cousin-run enterprises.

Somewhat problematically, however, a number of investigators avoid the use of clear definitions, maintaining that classification of family business is done on a case-to-case basis. Lack of definitional clarity can be attributed to difficulties associated with differentiating family from nonfamily enterprises (Wortman, 1995).

Operationalization and specificity of definitions has improved in recent times. However, one concern remains: A definition of *family* is often missing. This notable absence poses problems, particularly in an international context where families and cultures differ not only across geographical boundaries, but also over time. One way of overcoming this problem, especially in empirical research, is to specify levels and types of relationships as well as kinship ties of involved persons. Another way is to provide from the outset a clear and concise definition of what is meant by *family*.

Another, though less frequent, concern relates to difficulties associated with categorizing companies that are influenced by two or more unrelated families. For example, two families – Miele and Zinkann, who are descendants of unrelated founders – own and manage Miele in Germany. Although two families influence this company, the influence of one family balances the other. Thus, the influence of multiple-family ownership is not necessarily additive. Given this situation, we suggest in such circumstances that the influence of each family must be considered within any measure that assesses family influence.

To be functional, a definition must be unambiguous and transparent in such a way that it can be quantified. For example, Lea's (1998) definition is very difficult to operationalize:

> A business is a family business when it is an enterprise growing out of the family's needs, built on the family's abilities, worked by its hands and minds, and guided by its moral and spiritual values; when it is sustained by the family's commitment, and passed down to its sons and daughters as a legacy as precious as the family's name. (p. 1)

Furthermore, a definition should measure what it purports to measure and assist in providing reliable (replicable) research results.

In an early attempt to view family businesses as nonmonolithic, Shanker and Astrachan (1996) classify definitions by degree of family involvement. Their three-tier categorization ranges from broad (little direct family involvement), to middle (some family involvement), to narrow (a lot of family involvement). In contrast, Klein (2000b) prepared a modular classification in which different criteria are regarded as independent rather than additive.

Definitions that differ only slightly make it difficult not only to compare across investigations but also to integrate theory. Smyrnios, Tanewski, and Romano (1998) point out that 'complexities associated with arriving at a sound definition of a family firm raised a number of methodological concerns related to sampling issues, appropriate group comparisons, and establishing appropriate measures used to derive statistics' (p. 51). This complexity can raise confusion and call into question the credibility of family business research (Habbershon and Williams, 1999). It is our view that a *family business* definition should be clear about to which dimensions it refers. Moreover, a definition should be transparent and unambiguous. Perhaps most important, a definition should be modular, and its operationalization should lead to reliable and valid results.

A detailed review of definitions employed in studies reveals that there is no clear demarcation between family and nonfamily businesses and that no single definition can capture the distinction between the two types of entities. Artificially dichotomizing family vs. nonfamily firms when no such clear-cut dichotomy exists creates more problems than it attempts to solve. In this paper, we propose that there are discrete and particular qualities or characteristics of a business that are more appropriately measured on a continuous rather than dichotomous scale. We also suggest measures that can be used to tap different qualities of businesses. These measures make it possible to differentiate levels of family involvement. In addition, these measures provide a framework integrating different theoretical and methodological approaches to the study of family business.

From the one definition toward a continuum of family business

Utilizing the 'family universe bull's-eye,' Shanker and Astrachan (1996) outline a continuum ranging from high to low levels of family involvement. One difficulty associated with this approach is that different aspects of family involvement are directly found on the continuum itself. For example, Shanker and Astrachan suggest that a business with much family involvement has at least one family member in a management position and multiple generations work in and own the company. As this scheme comprises three categories of family involvement, finer distinctions that could be useful in understanding family business behavior appear without recognition.

A relevant issue, therefore, is not whether a business is family or nonfamily, but the extent and manner of family involvement in and influence on the enterprise. In our view, there are three important dimensions of family influence that should be considered: power, experience, and culture. These three dimensions, or subscales, comprise the F-PEC, an index of family influence. This index enables comparisons across businesses concerning levels of family involvement and its effects on performance as well as other business behaviors.

The F-PEC also allows researchers to utilize data derived from subscales and total scores as independent, dependent, mediating, or moderating variables. Interestingly, during the late 1930s, Lazarsfeld (1937, p. 127f, quoted after Schnell et al., 1995, p. 161) identified three reasons for developing a scale: functional reduction, arbitrary numerical reduction, and pragmatic reduction. With respect to the F-PEC, pragmatic reduction is perhaps the most important reason for its development.

As well as pragmatic implications, the F-PEC will herald objectivity and standardization of measurement across investigations. F-PEC development is based on main themes derived from an in-depth content analysis of various definitions of *family business*. Scales of the F-PEC provide an overall measure of family influence. A discussion of the three subscales of the F-PEC follows.

The power dimension of family influence: ownership, governance, and management participation

A family can influence a business via the extent of its ownership, governance, and management involvement (see Figure 9.1). A measure should not only take these issues into account, but also legal, political, and economic considerations associated with different countries. For example, in the case of board structures and compositions, most western countries, including the United States, involve a one-level board system. Germany, Switzerland, and the Netherlands have a two-level system in which a board member of one board (management or governance) is, by law, not permitted to be a concurrent member of both levels of governance. The F-PEC power subscale takes into account the percentage of family members on each board level as well as the percentage of members who are named through family members on the management and governance boards.

The involvement of family members as leaders of family firms has been a matter of interest for researchers and practitioners since the early 1970s (for example, Danco, 1975).

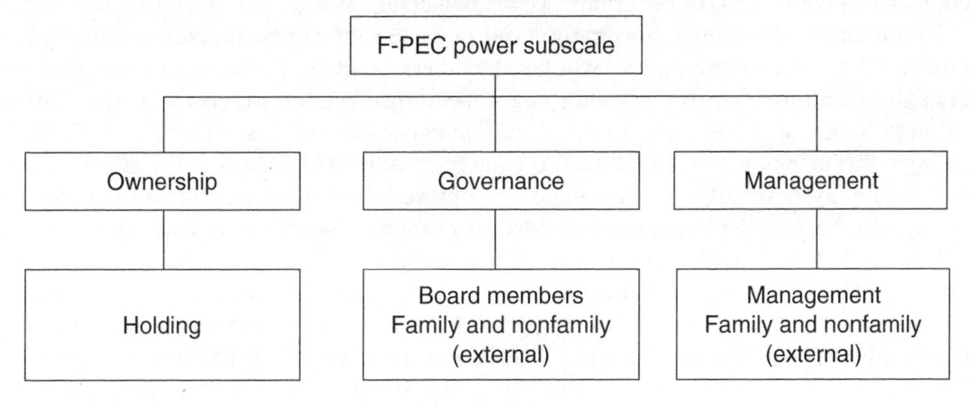

Figure 9.1 Dimensions of the F-PEC power subscale

This interest has focused on a number of different topics, including legitimate leadership (Kehr, 1996), performance (Monsen, 1996), principal-agent theory (Aronoff and Ward, 1995), and governance structure (Neubauer and Lank, 1998). Although these topics are important, the F-PEC is not concerned with whether a nonfamily CEO would serve the business better, whether a family CEO will reduce control costs, or whether a family CEO is highly motivated (Aronoff and Ward, 1995). The F-PEC power subscale assesses the degree of overall influence or power either in the hands of family members or in those named by the family. This level of influence via ownership, management, and governance is, therefore, viewed as interchangeable as well as additive.

In line with this view, Klein (2000a, 2000b) integrates ownership, governance, and management involvement of the family into a definition in which the level of influence in another could balance a lack of influence in one of these three domains. Although the Klein definition provides only a discrete determination (family vs. nonfamily), it does combine several criteria into one continuum and, thus, shows a number of precursor characteristics appropriate for the development of an index or scale. Discussing how this continuum functions, Klein (2000a) states that 'influence in a substantial way is considered if the family either owns the complete stock or, if not, the lack of influence in ownership is balanced through either influence through corporate governance or influence through management' (p. 158).

Notwithstanding, Klein did not comment on the importance of indirect influences for international comparisons. This issue is important as tax and legal structures across national boundaries encourage different forms of ownership. In some countries, for example, it is an advantage to own a company through other entities (for example, trusts, companies, or holding companies), and understanding the actual levels of family ownership and governance control can be difficult to decipher. For instance, it can be difficult to assess the extent of influence of a family who owns a business through a holding company. Faccio and Lang (2002, p. 10) take into account the indirect influence of a stakeholder through 'the product of two ownership stakes along the chain' of owning companies or family members. An example of this ownership chain includes a family that owns 100 per cent of a holding company that itself owns 100 per cent of the company. Obviously, this family has 100 per cent influence through ownership. However, a family that owns 50 per cent of a holding company that itself owns 50 per cent of the stock of a company has only a 25 per cent influence via ownership.

Family influence through governance and management can be measured as the proportion of family representatives who are members of the governance or management boards. In contrast, indirect influence might mean members of a board who are named through family members but are not family members themselves. A family's influence through this means, although indirect, is usually considerable. To assess this direct influence optimally, a weighting system must be employed. In mixed cases, the proportion of family members on the board will be added to a weighted proportion of members.

Consider the following example: two of five board members are family, two are nominated or elected by family members, and one is representative of a minor nonfamily shareholder. Our weighting system suggests that this board comprises 44 per cent of family influence to the overall power subscale. This proportion is calculated by aggregating 40 per cent of family influence (that is, two of five members are family) and 4 per cent of indirect influence (two of five multiplied by 0.1).

The experience dimension of family influence: generation in charge

This section discusses the family business experience subscale in relation to succession and the number of family members who contribute to the business. A number of authors (for example, Barach and Ganitsky, 1995; Birley, 1986; Heck and Scannell Trent, 1999; Ward, 1987, 1988) state that an enterprise can be viewed only as a family business when a transfer to the next generation is intended. Other authors (for example, Daily and Thompson, 1994) consider that at least one generational transfer should have occurred. For others (for example, Klein, 2000b), a founder-run entity can be regarded as a specific case of a family business. Despite these differences in viewpoints, all authors agree that each succession adds considerable valuable business experience to the family and the company.

It could be argued that the level of experiences gained from the succession process is greatest during the shift from first to second generations. During the first generation of ownership, many new rituals are installed. Thus, second and subsequent generations of ownership contribute proportionally less value to this process. As shown in Figure 9.2, family business experience of succession is regarded as involving an exponential continuum. Accordingly, dimensions involving a generation of family ownership and who is on the management and governance boards are weighted according to a nonlinear algorithm.

The number of family members associated with the business also contributes to the experience dimension. As a case in point, the wife of the family CEO can influence the business in a substantial way. Poza and Messer (2001) state that 'CEO spouses play a key, even if often invisible, role in most family-controlled corporations' (p. 25). Furthermore, discussions between owner-parents and their young adult children on business topics can enrich the business in a substantial way.

In some families, the contribution of the young generation over time is even more visible. One example is the Schmidt family in Germany. The youngest son of the Schmidt family, which owns and manages a bank in Southern Germany, in 1994 founded Consors – a subsidiary dealing with online brokerage. Today, Consors is one of the biggest online banks in Europe and has been listed on the Frankfurt stock exchange since 1999. The contribution of the son to the family business by founding his own business as a start-up in a similar

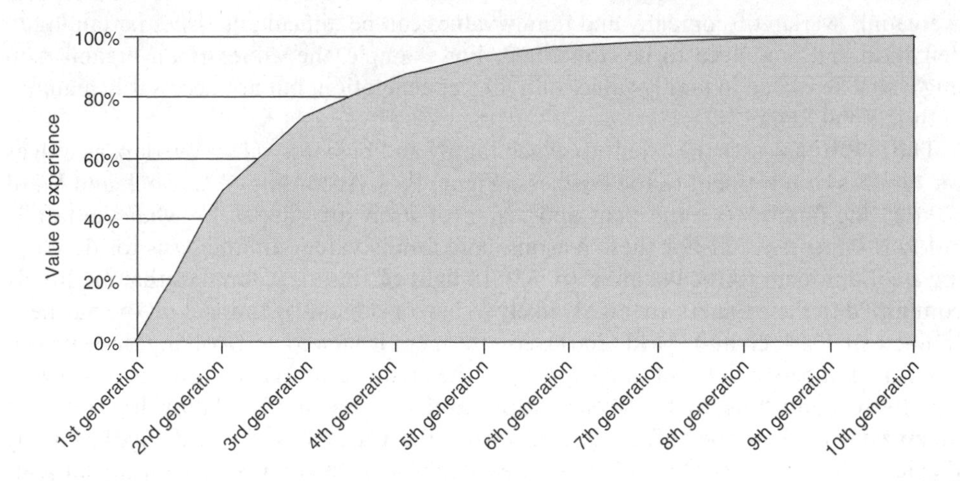

Figure 9.2 The experience of succession curve

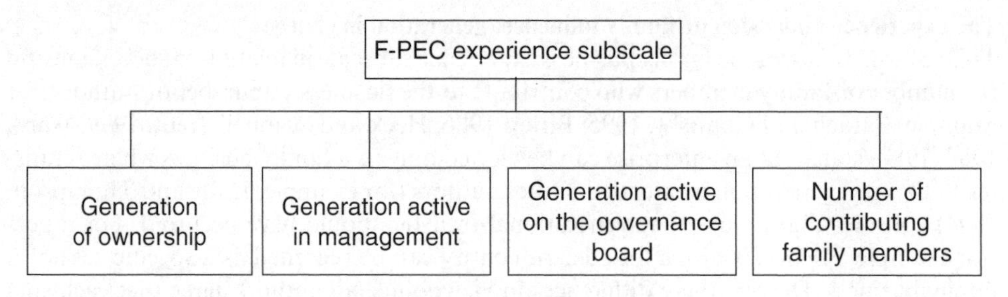

Figure 9.3 Dimensions of the F-PEC experience subscale

field is undeniable, even given the recent difficulties facing the Schmidt bank itself and, therefore, Consors as well. The family gained substantial experience as a result of their son's entrepreneurial input. Therefore, the number of family members dedicated to the business is viewed as an important indicator of how much experience the business receives from the family. Figure 9.3 shows the dimensions of the F-PEC experience subscale.

The culture dimension of family business: family and business values

Gallo (2000) considers business culture an important family enterprise element. According to his perspective, a firm can be considered a family business when family and business share assumptions and values. Other researchers define a family firm in terms of how the CEO, its managers, or its owners view the business. For example, it is reasonable to assume that owners or managers who regard their enterprise as a family business are highly likely to be attentive to issues and opinions of family members, as well as meeting the needs of family members.

However, anchoring values in an organization takes time. Klein (1991) finds that core values of key personnel (that is, individuals who have led an organization for more than 10 years) usually form part of the culture of their organization. The values of these significant individuals can be seen embedded in internal political matters, the ways in which conflicts are handled, and the degree of centralization vs. decentralization. Notwithstanding, evaluating overlap of company and family values can be difficult, as issues pertaining to definition and time need to be considered. For example, the values of an organization might well be rooted in family values of a former generation, but not necessarily manifest in the current family.

The F-PEC assesses the extent to which family and business values overlap, as well as the family's commitment to the business (Figure 9.4). According to Carlock and Ward (2001), 'the family's commitment and vision of itself are shaped by what the family holds as important . . . For these reasons, core family values are the basis for developing a commitment to the business' (p. 35). In light of this view, families that are highly committed to the business are highly likely to have a substantial impact on the business. In line with Carlock and Ward (2001), commitment is viewed as involving three principal factors: a personal belief and support of the organization's goals and visions, a willingness to contribute to the organization, and a desire for a relationship with the organization. A number of items comprising the Carlock and Ward (2001) Family Business Commitment Questionnaire are integrated into the F-PEC culture subscale (see the appendix).

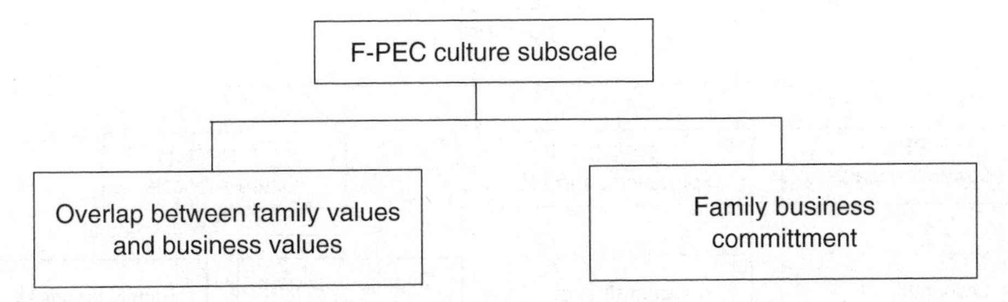

Figure 9.4 Dimensions of the F-PEC culture subscale

The F-PEC scale

As discussed earlier, the F-PEC comprises three subscales: power, experience, and culture. The F-PEC measures the extent of family influence on any enterprise. In marked contrast to previous work in this field, the F-PEC is not concerned with arriving at a precise or all-encompassing definition of *family business* or with differentiating this type of enterprise from its counterparts. However, development of a standardized instrument, like the F-PEC, enables sound comparisons across investigations and use of measures of family influence as either dependent, independent, moderating, or mediating variables. Figure 9.5 shows subscales along with their dimensions making up the F-PEC scale.

Procedures for determining the psychometric properties of the F-PEC scale

A team of experts, including academic researchers, family business owners, and practitioners, developed all items forming part of the F-PEC. Development of the scale proceeded through focus group discussions and pilot testing on a number of family business owners. Data relating to the F-PEC were analysed utilizing principal components and maximum likelihood factor analytic procedures and structural equation modeling techniques. Items demonstrating ambiguity, redundancy, and lack of discriminatory power were eliminated.

Dimensions of the F-PEC measure

Following suggestions by Gorsuch (1983), McDonald (1985), and Pedhazur and Pedhazur Schmelkin (1991), both factor analytic methods were used to assess the stability, number, and simplicity of factor structures. A cutoff score of $r = 0.40$ was considered reasonable for inclusion of a variable in interpretation of a factor (Stevens, 1986; Lambert et al., 1991). Items that did not meet the above-mentioned item loading criterion and those items that lacked discriminatory power were deleted.

Psychometric properties

Internal reliability (consistency) coefficients (Cronbach's alpha) for the F-PEC subscales and overall scale were also determined. Cronbach's alpha assessed the degree to which items making up a factor are intercorrelated or share similarities in their measurement of a particular construct, such as culture.

Items that make up the three subscales of the index were then evaluated for unidimensionality and reliability. A unidimensional factor comprises items that share a similar trait or construct. Congeneric measurement models were produced by allowing each item to

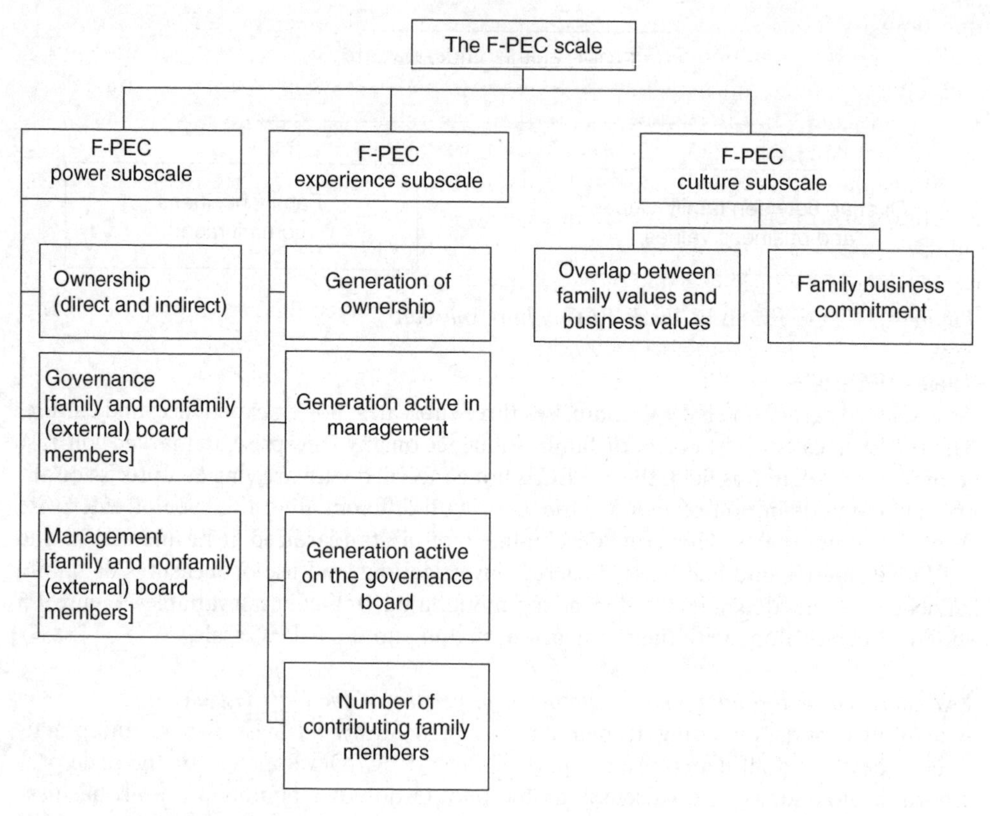

Figure 9.5 The F-PEC scale

respond to its underlying concept (Jöreskog and Sörbom, 1989). Goodness of fit of a measure was used to assess the degree to which observed data scores are predicted by an estimated model. Results should indicate whether items adequately fit hypothesized models and whether items have acceptable reliabilities (Hair et al., 1995).

External validity
To demonstrate external validity (that is generalizability), the F-PEC was tested on large sample groups (for example, $n > 500$) in different countries, including the United States, Germany, Australia, and Britain as well as in Europe. Cross-cultural comparisons also involved subjecting the F-PEC to the rigorous statistical procedures outlined previously.

Discussion
The F-PEC index of family influence on the business provides researchers, for the first time, with a tested standardized instrument that allows integration of different theoretical positions as well as comparisons of different types of data. Once the F-PEC's reliability and validity are demonstrated, it will encourage researchers to conduct more international research on a solid basis, as well as encourage researchers from outside the family business field to include family business issues in their research. The time so far spent on definition problems might be invested in either pure research of fundamental

questions to develop a theoretical framework of family businesses and/or in empirical studies – both crossnational and an in-depth understanding of special items. In the long run, international studies might lead to a better understanding of national peculiarity, thereby enriching the discussion of researchers, consultants, and family business members so that they can learn from each other and from other nationalities.

We also hope that practitioners will regain trust in research results, which might help encourage and finance further research. Questions concerning family businesses that consultants, family business members, and companies dealing with family businesses raise could lead to direct projects that don't depend on first having to define *family business*. At the same time, it would be possible to compare the obtained data with already-existing data that were gained on the same basis.

Apart from research implications, the F-PEC will help teachers and scholars of the family business field to understand the possible ways through which family members and families as an entity gain, loose, or maintain influence on their business. This will help in the development of agendas for both university courses and executive courses, emphasizing the management of this family influence on the business in a way that balances family and business needs. Such knowledge helps both the family and the business to perform even better.

We believe that the F-PEC is only the beginning and will help to establish the family business as an independent research field attracting high-standard researchers and dedicated practitioners.

Notes

1. This paper is seeded in the thoughts of the first named author, though these early ideas were fully realized only following the second meeting of the International Family Enterprise Academy in Amsterdam in 2000. Since this time, a number of discussions have been held on this topic and researchers around the world (for example, Germany, United States, Australia) have begun an international collaboration on this research. The operationalization of family vs. nonfamily enterprises has been a matter of concern from the very beginning of family business research. In most studies, the categorization of firms has culminated in the use of the classification as an independent variable. This approach, while important, has contributed to several problems, such as the lack of comparability of empirical data, confusion over what is meant by the term *family business*, and unconstructive discussion among researchers.
2. We would like to thank the participants of the 2000 (Amsterdam, the Netherlands) and 2001 (INSEAD, Fontainebleau) International Family Enterprise Research Academy for their valuable, provocative, and challenging thoughts and comments.

References

Aronoff, C.E. and Ward, J.L. (1995), 'Family-owned businesses: a thing of the past or a model for the future?', *Family Business Review*, **8**(2), 121–30.

Barach, J.A. and Ganitsky, J.B. (1995), 'Successful succession in family business', *Family Business Review*, **8**(2), 131–55.

Barnes, L.B. and Hershon, S.A. (1976), 'Transferring power in the business', *Harvard Business Review*, July–August, 105–14.

Berry, B. (1975), 'The development of organization structure in the family firm', *Journal of General Management*, **III**(1), 42–60.

Birley, S. (1986), 'Succession in the family firm: the inheritor's view', *Journal of Small Business Management*, **24**(3), 36–43.

Burch, P. (1972), *Managerial Revolution Reassessed: Family Control in America's Largest Corporations*, Lexington, MA: Lexington Books.

Carlock, R.S. and Ward, J.L. (2001), *Strategic Planning for the Family Business – Parallel Planing to Unify the Family and Business*, Houndsmill, NY: Palgrave.

Daily, C.M. and Thompson, S.S. (1994), 'Ownership structure, strategic posture, and firm growth: an empirical examination', *Family Business Review*, **7**(3), 237–50.

Danco, L. (1975), *Beyond Survival – A Business Owner's Guide for Success*, Cleveland, OH: The University Press.
Dean, S.M. (1992), 'Characteristics of African-American family-owned businesses in Los Angeles', *Family Business Review*, **5**(4), 373–95.
Dreux, D.R., IV and Brown, B.M. (1999), *Marketing Private Banking Services to Family Businesses*, available: http://www.genusresources.com/Mark.Priv.Bank.Dreux_5.html
Faccio, M. and Lang, L.H.P. (2002), 'The ultimate ownership of Western European Corporations', *Journal of Financial Economics*, **65**(3), September, 365–95.
Gallo, M.A. (2000), 'Conversation with S. Klein at the IFERA meeting held at Amsterdam University, April.
Gorsuch, R.L. (1983), *Factor Analysis*, 2nd edn, Hillsdale, NJ: Lawrence Erlbaum.
Habbershon, T.G. and Williams, M.L. (1999), 'A resource-based framework for assessing the strategic advantages of family firms', *Family Business Review*, **13**(1), 1–25.
Hair, J.F., Anderson, R.E., Tatham, R.L. and Black, W.C. (1995), *Multivariate Data Analysis*, Englewood Cliffs, NJ: Prentice-Hall.
Handler, W.C. (1989), 'Methodological issues and considerations in studying family businesses', *Family Business Review*, **2**(3), 257–76.
Harris, D., Martinez, J.I. and Ward, J.L. (1994), 'Is strategy different for the family-owned business?', *Family Business Review*, **7**(2), 159–74.
Heck, R.K.Z. and Scannell Trent, E. (1999), 'The prevalance of family business from a household sample', *Family Business Review*, **12**(3), 209–24.
Jöreskog, K.G. and Sörbom, D. (1989), *LISREL 7: A Guide to the Program and Application*, Chicago, IL: SPSS Inc.
Kehr, H. (1996), *Die Legitimation von Führung*, published (on microfiche) doctoral dissertation, Ludwig-Maximilians-Universität München.
Klein, S.B. (1991), *Der Einfluß von Werten auf die Gestaltung von Organisationen*, Berlin: Duncker and Humblot.
Klein, S.B. (2000a), 'Family businesses in Germany: significance and structure', *Family Business Review*, **13**(3), 157–81.
Klein, S.B. (2000b), *Familienunternehmen – Theoretische und empirische Grundlagen*, Wiesbaden: Gabler.
Lambert, Z., Wildt, A. and Durand, R. (1991), 'Approximating confidence intervals for factor loadings', *Multivariate Behavioral Research*, **26**(3), 421–34.
Lansberg, I., Perrow, E.L. and Rogolsky, S. (1988), 'Family business as an emerging field', *Family Business Review*, **1**(1), 1–8.
Lazarsfeld, P.F. (1937), 'Some remarks on the typological procedures in social research', *Zeitschrift für Sozialforschung*, 119–39.
Lea, J. (1998), 'What is a family business? More than you think', available: http://www.bizjournals.com/triangle/stories/1998/11/02/smallb3.html
Littunen, H. and Hyrsky, K. (2000), 'The early entrepreneurial stage in Finish family and nonfamily firms', *Family Business Review*, **13**(1), 41–54.
Litz, R.A. (1995), 'The family business: toward definitional clarity', *Family Business Review*, **8**(2), 71–81.
McDonald, R.P. (1985), *Factor Analysis and Related Methods*, Hillsdale, NJ: Lawrence Erlbaum.
Monsen, J. (1996), 'Ownership and management: the effect of separation on performance', in C.E., Aronoff, J.H. Astrachan and J.L. Ward (eds), *Family Business Sourcebook II* (pp. 26–33), Marietta, GA: Business Owner Resources.
Neubauer, F. and Lank, A.G. (1998), *The Family Business – 1st Governance for Sustainability*, London: Macmillan.
Pedhazur, E. and Pedhazur Schmelkin, L. (1991), *Measurement, Design, and Analysis: An Integrated Approach*, Hillsdale, NJ: Lawrence Erlbaum.
Poza, E.J. and Messer, T. (2001), 'Spousal leadership and continuity in the family firm', *Family Business Review*, **14**(1), 25–36.
Schnell, R., Hill, P.B. and Esser, E. (1995), *Methoden der empirischen Sozialforschung* (5th completely revised edition), München, Wien: R. Oldenbourg.
Shanker, M.C. and Astrachan, J.H. (1996), 'Myths and realities: family businesses' contribution to the US economy – a framework for assessing family business statistics', *Family Business Review*, **9**(2), 107–19.
Smyrnios, K.X., Tanewski, G.A. and Romano, C.A. (1998), 'Development of a measure of the characteristics of family business', *Family Business Review*, **11**(1), 49–60.
Stevens, J. (1986), *Applied Multivariate Statistics for the Social Sciences*, 3rd edn, Mahwah, NJ: Lawrence Erlbaum.
Ward, J.L. (1987), *Keeping the Family Business Healthy: How to Plan for Continuing Growth Profitability and Family Leadership*, San Francisco, CA: Jossey-Bass.
Ward, J.L. (1988), 'The special role of strategic planning for family businesses', *Family Business Review*, **1**(2), 105–17.
Wortman, M.S. (1995), 'Critical issues in family business: an international perspective of practice and research', *Proceedings of the 40th International Council for Small Business Research Conference*, Sydney: NCP Printing, University of Newcastle, NSW, Australia.

Appendix: F-PEC questionnaire

Definitions

- Family is defined as a group of persons including those who are either offspring of a couple (no matter what generation) and their in-laws as well as their legally adopted children.
- Ownership means ownership of stock or company capital. When the percentage of voting rights differs from percentage of ownership, please indicate voting rights.
- Management board refers to the company board that manages or runs an entity(ies).
- Persons named through family members represent the ideas, goals, and values of the family.

Part 1: The Power Subscale

1. Please indicate the proportion of share ownership held by family and nonfamily members:

   (a) Family ___________%
   (b) Nonfamily ___________%

2. Are shares held in a holding company or similar entity (e.g., trust)? 1. ❑ Yes 2. ❑ No

   If YES, please indicate the proportion of ownership:

   (a) Main company owned by: (i) direct family ownership: ____%
   (ii) direct nonfamily: ____ ownership: ____%
   (iii) holding company: ____%

   (b) Holding company owned by: (i) family ownership: ____%
   (ii) nonfamily ownership: ____%
   (iii) 2nd holding company: ____%

   (c) 2nd holding company owned by: (i) ____ family ownership: ____%

3. Does the business have a governance board? 1. ❑ Yes 2. ❑ No
   If YES:
   (a) How many board members does it comprise? _________ members
   (b) How many board members are family? _________ family members
   (c) How many nonfamily (external) members nominated by the family are on the board? _______ nonfamily members

4. Does the business have a management board? 1. ❑ Yes 2. ❑ No
   If YES:
   (a) How many persons does it comprise? ____________members
   (b) How many management board members are family? _______ family members
   (c) How many nonfamily board members are chosen through them? _________nonfamily members

Definitions

- The founding generation is viewed as the 1st generation
- Active family members involve those family members who contribute substantially to the business. These individuals might hold official positions in the business as shareholders, board members or employees.

Part 2: The Experience Subscale

1. What generation owns the company? _______ generation
2. What generation(s) manage(s) the company? _______ generation
3. What generation is active on the governance board? _______ generation
4. How many family members participate actively in the business? _______ members
5. How many family members do not participate actively in the business but are interested? _______ members
6. How many family members are not (yet) interested at all? _______ members

Part 3: The Culture Subscale

Please rate the extent to which:

1. Your family has influence on your business. *Not at all* *To a large extent* *1......... 2.......... 3......... 4.......... 5*
2. Your family members share similar values. *Not at all* *To a large extent* *1......... 2.......... 3......... 4.......... 5*
3. Your family and business share similar values. *Not at all* *To a large extent* *1......... 2.......... 3......... 4.......... 5*

Please rate the extent to which you agree with the following statements:

4. Our family members are willing to put in a great deal of effort beyond that normally expected in order to help the family business be successful. *Strongly Disagree* *Strongly Agree* *1......... 2.......... 3......... 4.......... 5*
5. We support the family business in discussions with friends, employees, and other family members. *Strongly Disagree* *Strongly Agree* *1......... 2.......... 3......... 4.......... 5*
6. We feel loyalty to the family business. *Strongly Disagree* *Strongly Agree* *1......... 2.......... 3......... 4.......... 5*
7. We find that our values are compatible with those of the business. *Strongly Disagree* *Strongly Agree* *1......... 2.......... 3......... 4.......... 5*
8. We are proud to tell others that we are part of the family business. *Strongly Disagree* *Strongly Agree* *1......... 2.......... 3......... 4.......... 5*

9. There is so much to be gained by participating with the family business on a long-term basis.	*Strongly Disagree Strongly Agree* *1......... 2.......... 3......... 4.......... 5*
10. We agree with the family business goals, plans and policies.	*Strongly Disagree Strongly Agree* *1......... 2.......... 3......... 4.......... 5*
11. We really care about the fate of the family business.	*Strongly Disagree Strongly Agree* *1......... 2.......... 3......... 4.......... 5*
12. Deciding to be involved with the family business has a positive influence on my life.	*Strongly Disagree Strongly Agree* *1......... 2.......... 3......... 4.......... 5*
13. I understand and support my family's decisions regarding the future of the family business.	*Strongly Disagree Strongly Agree* *1......... 2.......... 3......... 4.......... 5*

We thank you very much for your support!

10 Identification of different types of private family firms

Paul Westhead and Carole Howorth

Introduction

The importance and powerful influence of family in all aspects of entrepreneurship and business has recently been highlighted (Aldrich and Cliff, 2003; Rogoff and Heck, 2003). Despite their significant contribution to the economy, research into private family firms is relatively neglected. Perhaps this is because they are believed to be less interesting owing to a lack of agency problems, little separation of ownership and control and shared goals. Many view private family firms as a homogenous group, but casual observation suggests that they differ with regard to their ownership and management structures and the extent to which family objectives dominate. Currently, there is limited understanding of how, why, or the extent to which, private family firms differ. This has implications for the future of family firms research, because the use of an overarching family firm definition, and the failure to recognize contrasts between 'types' of family firms may impact on the validity and generalizability of findings. Also, most studies fail to explore the link between the 'type' of family firm and its performance, which is important for the development and efficacy of practitioner and public policy support.

The stereotypical family firm with good relationships and low information asymmetries may occur less frequently than expected. Indeed, family firms may not be the corporate governance panacea predicted by agency models (Schulze et al., 2001). Limitations of agency theory in the family business and entrepreneurship context (Arthurs and Busenitz, 2003; Astrachan, 2003; Greenwood, 2003; Randoy and Goel, 2003) indicate it only provides a partial explanation of private family firm dynamics (Howorth et al., 2004). Agency theory focuses on firm-level ownership and management, but the family system increases complexity (Neubauer and Lank, 1998). Ownership and management are, generally, part of a performance-based system, while the family is a relationship-based system. Where a system is performance based and rational economic objectives can be assumed, agency theory can be applied. In a relationship-based system, where non-financial objectives prevail and behaviour may not be economically rational, the explanatory power of agency theory is more limited, especially where goal congruence exists (Arthurs and Busenitz, 2003). Nevertheless, separation of ownership and control does occur in some family businesses, indicating that agency theory has some explanatory power. However, complementary theories are required which incorporate the relationship aspects and long-term interactions of family firms.

We utilize agency and stewardship perspectives to formulate a conceptual framework highlighting differences between private family firms. Variables relating to family firm objectives and ownership and management structures are used to empirically identify a taxonomy of private family firm 'types'.

The following research question is the focus of this chapter:

Can 'types' of private family firms be identified?

The following discussion reports several stages in the classification of private family firms. In profiling 'types' of family firms, we characterize the 'average' private family firm and the variants. Seven 'types' of family firms are identified. Data on the size and performance differences between the detected 'types' of family firms are provided. Superior (and weaker) performance is found to be associated with 'outlier' family firm 'types'. Implications relating to the findings are discussed.

Theoretical perspectives

The popular image of a family firm negates agency problems, owing to close matching of owners' and managers' motivations and attitudes, and little information asymmetry. Stewardship theory suggests managers' and employees' motives are aligned to those of the organization (Davis et al., 1997). Further, stewardship theory does not assume financial goals or economic rationality, and thus may be more applicable to family firms. Unlike agency theory, stewardship theory can encompass pure altruism, which is selfless and not motivated by economic returns. Thus, in the stereotypical family firm, stewardship theory would expect behaviour that puts the organization first, a strong psychological ownership of the family firm and a high occurrence of altruism. Altruism stems from loyalty and commitment between family members (Schulze et al., 2003). Where altruism is high, communication and cooperation will be high but information asymmetries will be low. The flip side is that offspring may take advantage of parents' altruism leading to free-riding, shirking, squandering and increased information asymmetry. Biases and filters may also compound information flows in an altruistic environment. Thus, in some family firms there may be a high level of information asymmetry, in contrast to the stereotypical family firm.

Westhead et al. (2002) suggest that some private family firms do not conform to the stewardship perspective. Family firms exhibit a variety of motivations, more and less complex management and ownership structures, and differing levels of company performance (Westhead and Howorth, 2004). For some, financial objectives are important (Smyrnios and Romano, 1994). A number of family firms may, for example, introduce agency control mechanisms such as performance related pay (Schulze et al., 2003) and non-executive directors (NEDs) (Westhead et al., 2001).

In determining where an agency or stewardship perspective is appropriate we need to consider the assumptions of the perspectives. Agency theorists assume a focus on financial objectives and individuals that are self-serving (Astrachan, 2003). Conversely, stewardship theorists assume a focus on non-financial objectives and individuals that are organization-serving. In considering family firms and the applicability of agency or stewardship perspectives, we focus on the dominant coalition, that is, the group of individuals (owners and managers) who have power and influence within the firm. Where the dominant coalition is organization-serving they will put the needs of the family and/or the firm before their own desires, in line with a stewardship perspective. Where the dominant coalition is self-serving, individuals needs and desires will dominate and agency mechanisms may be in place to align owners' and managers' interests. Figure 10.1 presents these variations as a basis for identifying 'types' of private family firms. The stereotypical family firm is expected to conform to the classic stewardship perspective of organization-serving and focus on non-financial objectives, such as maintaining family control and avoiding debt, rather than on establishing ownership and management structures that encourage superior firm performance. This latter family firm resides in quadrant C in

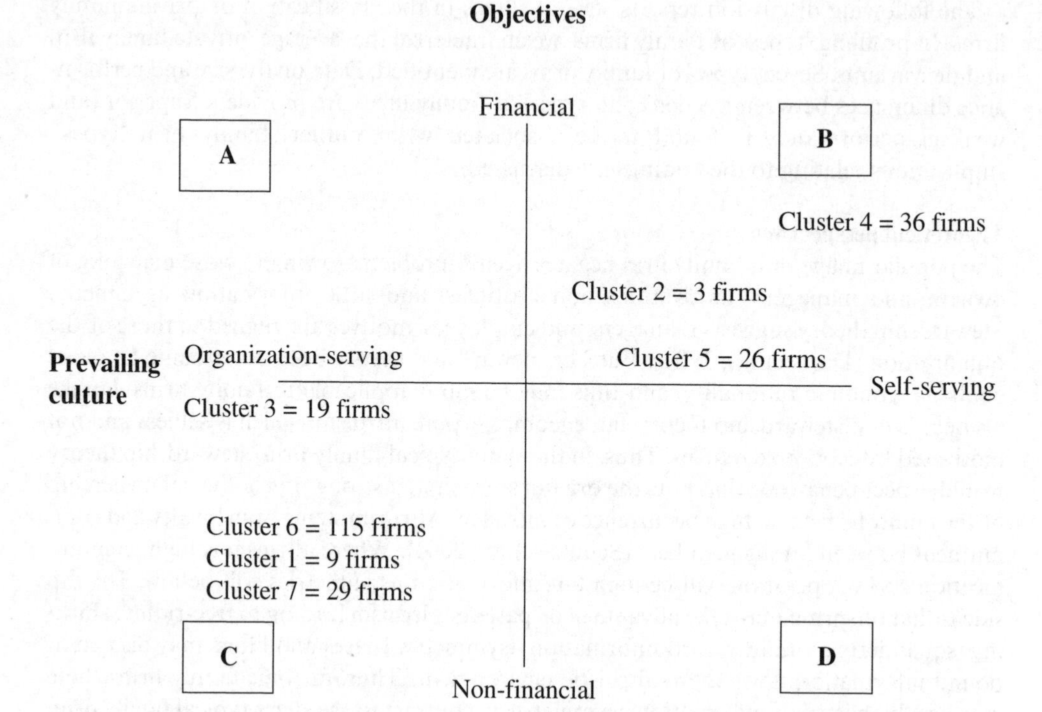

Figure 10.1 Differentiating factors between 'types' of private family firms

Figure 10.1. Firms that exhibit the classic agency perspective of the firm reside in quadrant B. Variations in the extent to which the prevailing culture within a firm is organization- or self-serving, and/or in the importance of financial objectives, provide potential for firms to reside in quadrants A and D. It is expected that family firms will exhibit varying degrees of organization-serving and self-serving motivations with altruistic and self-serving behaviours, at times, occurring simultaneously (Jensen, 1994). There will also be differing emphases on financial and non-financial objectives, and often a combination of the two. Therefore, the lines within the framework in Figure 10.1 represent continua rather than mutually exclusive extremes.

Operationalization

In operationalizing this framework, family firms can be asked directly about the extent to which they focus on various objectives. It is more difficult to identify the extent to which they are organization- or self-serving. However, where self-serving behaviour prevails, firms will more closely resemble the agency perspective of the firm and are more likely to implement agency control mechanisms, such as the appointment of NEDs and larger boards of directors. Thus, differences between private family firms are expected to manifest themselves in variations in their ownership and management structures. We, therefore, focus on ownership and management structures, as well as company objectives. These three factors may be interrelated, and this can shape the formation of 'types' of private family firms.

Objectives
Family and business motives are intertwined in family firms. Family firms may differ with regard to the reported importance of particular family objectives (Chua et al., 2003). Important objectives cited by owners of family firms include: survival of the family business as a going concern (Westhead and Cowling, 1997); continued independent ownership of the firm (Binder Hamlyn, 1994); transfer of ownership to the next generation (Gersick et al., 1997); maintaining financial independence (Donckels and Fröhlich, 1991); and employment of family members (Westhead, 1997). There may also be variations in the extent to which financial objectives are emphasized (Feinberg, 1975; Smyrnios and Romano, 1994).

Nine company objectives variables were utilized (Table 10.1). In identifying 'types' of family firms, we expect that there will be variations in the extent to which family firms emphasize financial and non-financial (or family) objectives. Family firms that focus more on financial objectives may reflect the agency perspective of the firm. The latter firms may be associated with greater separation of company ownership and control, and they may be more likely to employ agency control mechanisms.

Ownership and management
Family influence on ownership and management is not yet fully understood (Rogoff and Heck, 2003). The stereotypical family firm may have an advantage owing to the lack of agency problems (Randoy and Goel, 2003). However, there may be variations in family ownership and control (Daily and Dollinger, 1992; Greenwood, 2003). Even where the family retains a majority stake in the business, dilution of ownership may enable non-family members to shape business development and objectives.

As family firms progress from one generation to the next, the structural form of ownership and management will change (Howorth and Ali, 2001). Where firms are characterized by increasing complexity of ownership, shareholdings will become more diverse. Individual shareholders with less control will require extra monitoring to ensure that their interests are secured. This could impact on the objectives of the firm and the management structure employed.

Variables relating to the number of ordinary shareholders, the proportion of ordinary shares owned by the largest family group, and first versus multi-generation family firms were ascertained (Table 10.1). In identifying 'types' of family firms, we expect variations in the diversity of shareholdings and the proportion of family ownership. Family firms with diverse shareholdings, and diluted family ownership, are expected to be associated with financial objectives and agency control mechanisms. However, multi-generation family firms may place a great deal of emphasis on family history and there may be pressures to put the family and/or firm first (Rose, 2000) and therefore they are expected to be associated with increased emphasis on family objectives.

Family managers and directors can reinforce the power of family owners by exerting pressure to ensure that 'family agendas' are considered. This can be curbed by external influences (for example, compliance with legal requirements and agency control mechanisms attached to the provision of external finance). Owners of growing family firms may expand the management team organically by promoting competent existing family members to key management positions, ensuring family members manage and own the business. Some owners may recruit non-family professional managers to 'professionalize'.

Table 10.1 Cluster characteristics of family companies by company ownership, control and objectives variables

Variables (a)	Variables related to principal components	Clusters							Global mean	Std dev
		1	2	3	4	5	6	7		
Number of directors (a)	1	2.11*	8.67**	2.21	3.78*	3.08	2.50	3.41	2.91	1.53
Number of people in the management team (a)	1	3.56	24.00**	2.53*	6.61*	4.65	3.48	5.28	4.49	3.68
Proportion of directors from the largest family group (a)	1	85.19	33.73**	89.47	62.35*	66.70*	89.71	82.99	81.31	25.31
Proportion of management team from family (a)	1	51.69	12.58**	81.93*	35.56*	44.59	69.84	55.38	59.65	33.29
Board employed a non-executive director (b) (c)	1	0.00	0.67**	0.05	0.56**	0.08	0.06	0.31	0.17	0.38
Proportion of shares owned by family (a)	1	91.78	75.33*	96.90	77.44*	85.73	94.64	89.07	90.20	15.77
A prime objective is to accumulate family wealth (a) (d)	2	4.00	4.33*	2.90*	2.83*	3.39	3.97	4.38*	3.70	0.95
A prime objective is to maintain/enhance owners lifestyle (a) (d)	2	4.11	4.33*	3.63	2.89*	3.42	4.02	4.48*	3.81	1.03
A prime objective is to ensure survival of the business (a) (d)	3	4.89	4.33*	4.32*	4.83	3.89**	4.84	4.86	4.69	0.55
A prime objective is employees job security	3	4.44	3.33**	3.37**	4.31	3.54*	4.44	4.17	4.19	0.77
A prime objective is to ensure independent ownership (a) (d)	3	4.44	5.00**	3.68*	4.06	2.92**	4.29	4.79*	4.13	0.87
Second or more generation family firm (b) (c)	4	0.56	0.33	0.21	0.78*	0.46	0.30	0.76*	0.45	0.50
A prime objective is to pass business to next generation (a) (d)	4	3.00	4.00*	3.05	3.36	2.69*	3.39	4.10*	3.36	1.04
A prime objective is increase market value of business (a) (d)	5	4.22	4.67*	2.79**	4.19	4.15	4.32	3.24*	4.03	0.89
A prime objective is reputation and status in locality (a) (d)	5	4.56*	3.00*	3.05*	4.42*	3.62	4.14	2.97*	3.90	1.01

Number of ordinary shareholders (logs to base 10) (a)	6	3.66**	0.69	0.37	0.70	0.44	0.42	0.59	0.61	0.72
It is important that day-to-day operations are the responsibility of family members (a) (d)	6	2.56*	3.67	3.47	2.69*	2.81	3.50	3.62	3.28	1.03
Number of companies in the cluster		9	3	19	36	26	115	29		

Notes:
(a) Kruskal-Wallis coefficient statistically significant at the 0.001 level of significance for the seven clusters.
(b) A chi-square statistic could not be calculated because more than 20 per cent of the observed categories had less than five expected observations. Chi-square coefficient statistically significant at the 0.001 level of significance for five clusters (excluding the respondents in clusters 1 and 2).
(c) Measured on a scale where 1 = 'yes' and 0 = 'no'.
(d) Measured on a scale where 1 = 'strongly disagree', 2 = 'disagree', 3 = 'neutral', 4 = 'agree', and 5 = 'strongly agree'.
** Cluster mean which deviates by more than a standard deviation from the respective global mean.
* Cluster mean which deviates by more than half a standard deviation from the respective global mean.

However, difficulties may arise from the conflict of the economic objectives of non-family members and the non-economic objectives of family. Relatively few family firms employ NEDs (Westhead et al., 2001). In the majority of family firms the CEO is a family member (Kelly et al., 2000), which may lead to greater emphasis on family objectives. Larger boards of directors and more outside directors may be associated with agency problems. Many family firms have small boards, dominated by family directors, but this may be associated with size of the firm (Daily, 1995).

Six variables relating to the size and composition of the board and management team were ascertained (Table 10.1). In identifying 'types' of family firms, we expect variations in the size and composition of the board of directors and the management team. Larger boards and more outside (non-family) representation on the board and within the management team are expected to be associated with family firms that focus on financial objectives, more diverse shareholdings and smaller proportions of family ownership. Outside representation may include (but is not restricted to) the employment of one or more NEDs.

Size and performance

There may also be variations linking the above variables to firm performance. As yet, there is no consensus surrounding the links between the ownership and control of a family firm and its performance (James, 1999). Firm performance is associated with a complex array of factors including, but not limited to, family influences. The favouring of family members may lead to more able non-family members seeking employment outside the family business and a group of less ambitious non-family managers remaining, which may retard firm development. On the other hand, family firms may report improved performance due to high levels of commitment, long-term objectives, and close relationships between owners, managers and employees (James, 1999).

Family firms with more 'outside' involvement in their management structure may report improved levels of performance arising from expert advice, specialist skills and resources not possessed internally. Non-executive directors can play an important role in corporate governance (that is, reducing potential agency costs) and pressure to perform. They can defuse personal conflicts, strengthen commitment to business continuity, and provide an objective view of succession. Daily and Dollinger (1993) found that professionally managed firms were larger in terms of number of employees than family owned and managed firms.

Firm performance may be associated with company objectives (Feinberg, 1975). Family firms who do not focus on wealth creation in the long term may struggle to compete, threatening their continued existence. Also, family firm performance may be retarded because owners focus on maintaining or enhancing their lifestyle (Westhead, 1997). Some family firms have no plans to grow, or may only grow at a pace consistent with meeting the needs of the family. Growth to employ family members may in the short term, lead to sub-optimal investment and lower firm profitability (James, 1999). Professional managers however, may be more self-serving in line with the agency perspective of the firm.

Thus, company objectives and the ownership and management structures of family firms may be linked with firm performance. Also, the three factors may be interrelated. Identification of 'types' of private family firm may differentiate between firms reporting superior, or weaker, performance. We expect superior family firm performance to be associated with 'types' of family firms with more outside involvement on the board of directors and in management, employment of one or more NEDs and less emphasis on family objectives.

Multiple measures of firm size and performance were ascertained. This is especially important for family firms where wealth creation may not be the major objective (Chrisman et al., 2003). Financial data can be difficult to interpret, particularly for small unquoted companies, where compensation strategies of owners can have a profound influence on recorded profits. Absolute scores on financial performance criteria can also be affected by industry-related factors. Moreover, financial results may not match the perception of performance. Firm size was explored in terms of absolute gross sales revenues (£s), people (full-time and part-time) employed, and absolute gross sales revenues as a proportion of people employed. Six additional performance indicators were collected relating to absolute change in gross sales revenues (£s), 1991–94; absolute change in number of people employed, 1991–94; whether the business exported outside the UK; percentage of gross sales exported; profitability of the business in 1994; and a weighted average performance score (WEIGHTED) (that is controlling for industry differences between firms) (Naman and Slevin, 1993; Westhead and Cowling, 1997).

Research design

Information from a stratified random sample of family and non-family independent unquoted companies in the UK was collected in 1995 (Westhead, 1997). In total, 427 valid questionnaires were obtained. A noteworthy 48 per cent valid response rate was reported. Response bias tests were conducted. Chi-square and 't' test analyses indicated that the respondents to the survey were not significant different from the non-respondents with regard to industry; location of the business by standard region as well as urban and government designated 'assisted' area location; age of the company; employment size of the company; and the sales revenue of the company.

A company was regarded as a family firm if more than 50 per cent of ordinary voting shares were owned by members of the largest single family group and the company was perceived to be a family business (Westhead and Cowling, 1998). In total, 272 private companies (64 per cent) were regarded as family companies. In 98 per cent of the surveyed companies the CEO provided data. In total, 146 first generation family firms (54 per cent) and 126 multi-generation family firms (46 per cent) were identified. Respondents that filed missing information returns to any of the selected 18 variables were excluded from any further analysis. In total, 237 respondents provided complete data sets for the selected variables.

An R-mode principal components analysis (PCA) was utilized to transform and orthonormalize the original data, identifying six underlying constructs. Cluster analysis then explored the standardized component score matrix (that is, six orthogonal variables for each firm). Seven 'types' of private family firms were identified. The predictive validity of the taxonomy was assessed by discriminant analysis. Bivariate analysis was conducted to compare the 'types' of family firms, and to examine size and performance differences.

Results

An R-mode PCA was utilized to detect the links between the 18 selected variables. The variable relating to whether the CEO is a member of the single dominant family group that owns the business had a low communality, and was removed from the model. All the assumptions of the model were satisfied with regard to the inclusion of the remaining 17 variables. The varimax rotated model identified six components with eigenvalues greater than unity, and the model accounted for 62 per cent of the total variance. Each variable had a component loading of 0.50 or higher on only one component. Adequate convergent validity was evident. Labelling of components was based on loadings that were statistically significant at the 0.05 level. The components were labelled as follows: closely and family owned and controlled (Component 1); family lifestyle objectives (Component 2); security and survival objectives (Component 3); family succession objectives (Component 4); wealth maximization objectives (Component 5); and concentrated shareholding and family control objective (Component 6). Trends isolated in the rotated component structure appeared to have meaningful expression in that they highlight contrasting financial and non-financial objectives, as well as the concentration of family ownership and management. Component 1 scores will be positive for family firms that align with the stewardship perspective, and negative for firms that align with the agency perspective. High positive Component 5 scores represent a focus on financial objectives, while high positive scores on Component 2 represent a focus on non-financial objectives. High positive scores on the other components represent a focus on non-financial objectives.

The 237 firms by six components matrix of component scores formed the data matrix for an agglomerative hierarchical QUICK CLUSTER analysis. The seven-cluster solution is interpretable, and occurs before the distances at which clusters are combined become too large. The seven clusters highlight different permutations of company objectives, as well as company ownership and management structures. One-way analysis of variance tests showed significant differences between clusters on all components. Because the QUICK CLUSTER analysis explored a component score matrix that accounted for 62 per cent of the variance in the 'raw' data, Kruskal-Wallis and chi-square statistics were also calculated between each of the seven clusters with respect to the original 'raw' data relating to the 17 variables analysed by the PCA. Table 10.1 shows that statistically significant contrasts were recorded among the seven clusters with respect to all 17 'raw' variables.

Descriptive labels for each of the seven clusters were based on differences between cluster means and global means for each of the 17 'raw' variables (Table 10.1). Further, the cluster mean for each of the earlier identified principal components was calculated. The results from the two analyses are reasonably consistent and allow 'types' of private family firms to be labelled as follows:

Cluster 1 – Diverse shareholdings but family management not emphasized. The nine firms in cluster 1 have diverse shareholdings, they attach less importance to family management but place greater emphasis on the status of the firm. Cluster 1 firms do not differ significantly from the average with regard to the proportion of family ownership or control, and the importance of other objectives.

Cluster 2 – Diluted control and focus on family lifestyle objectives. The three firms in cluster 2 are significantly different from other firms across a wide range of variables. Firms have larger boards of directors and management teams, more outside (non-family) representation on the board and management team, less family ownership, and they are more likely to employ a NED. Family wealth and lifestyle are emphasized with less emphasis given to long-term security. Independent ownership of the firm and increasing market value are important objectives.

Cluster 3 – Family owned and controlled with less emphasis on wealth maximization and security objectives. The 19 firms in cluster 3 are family owned and controlled and they are not focusing on family wealth maximization or security company objectives. Increasing market value and employee job security are less likely to be emphasized.

Cluster 4 – Diluted ownership and control with less focus on family lifestyle objectives. The 36 firms in cluster 4 have diluted ownership and control and they are more likely to be multi-generation firms. Less emphasis is placed on family lifestyle objectives. Firms have more directors and managers but a smaller percentage of them are family members. Cluster 4 firms are more likely to employ a NED. Further, family members own a smaller percentage of shares.

Cluster 5 – Dilution of family control and less focus on security or inter-generational ownership transfer objectives. The 26 firms in cluster 5 highlight less family control and less focus is being placed on the survival of the business, employee job security, independent ownership, or inter-generational succession objectives.

Cluster 6 – Family owned and controlled firms with a focus on family objectives. The 115 firms in cluster 6 are family owned and controlled and they focus on family objectives. We can infer that they are 'average' (or stereotypical) family firms.

Cluster 7 – Multi-generation family firms with a focus on family objectives. The 29 firms in cluster 7 are generally multi-generation family firms focusing on family lifestyle and inter-generational ownership transfer objectives. Independent ownership is more likely to be emphasized and less focus is given to increasing market value.

The seven 'types' of private family firms were allocated to the quadrants suggested by the earlier theoretical discussion. In doing this, each type was compared with regard to their emphasis on financial or non-financial objectives and the presence of agency control mechanisms as discussed earlier (see 'Operationalization'). Family firm 'types' with agency control mechanisms such as larger boards of directors and use of non-executive directors tended to emphasize financial objectives in line with the agency perspective. Figure 10.1 shows that the family firm 'types' appear to fall into either quadrant B or C. There was no evidence of family firms reporting objectives and ownership/management structures which would situate them in quadrants A or D. In line with agency theory, this suggests that family firms predominantly focusing upon financial objectives report a self-serving culture. Moreover, in line with stewardship theory, family firms predominantly focusing upon non-financial objectives generally report an organization-serving culture. This provides some confirmation of the validity of these two theoretical perspectives.

The appropriateness of the seven-cluster taxonomy was tested using discriminant analysis. Classification results from the final model showed that the seven-cluster solution is optimal. Approximately 92 per cent of firms were correctly classified to the cluster identified by the cluster analysis.

Table 10.2 summarizes the differences between the seven 'types' of family firms with regard to their demographic characteristics. A larger proportion of cluster 3 firms were engaged in services, while a larger proportion of cluster 5 firms were engaged in construction. No marked differences were detected in location of the firms. Fewer firms in

Table 10.2 Cluster characteristics of family companies by company demographics and performance (a)

	Cluster														Chi-square stat significance level
	1		2		3		4		5		6		7		
	No.	%	No.	%	No.	%	No.	%	No.	%	No.	%	No.	%	
1. Main industrial activity															(b)
Agriculture, forestry and fishing	1	11.1	1	33.3	1	5.3	2	5.6	1	3.8	15	13.0	5	17.2	
Manufacturing	1	11.1	0	0.0	1	5.3	2	5.6	2	7.7	8	7.0	1	3.4	
Construction	1	11.1	0	0.0	2	10.5	6	16.7	6	23.1	17	14.8	5	17.2	
Services	6	66.7	2	66.7	15	78.9	26	72.2	17	65.4	75	65.2	18	62.1	
2. Location															(b)
Rural	2	22.2	1	33.3	6	31.6	9	25.0	2	7.7	38	33.0	7	24.1	
Urban (c)	7	77.8	2	66.7	13	68.4	27	75.0	24	92.3	77	67.0	22	75.9	
3. CEO is a member of the single dominant family group that owns the business															(b)
No	1	11.1	1	33.3	2	10.5	12	33.3	4	15.4	7	6.1	3	10.3	
Yes	8	88.9	2	66.7	17	89.5	24	66.7	22	84.6	108	93.9	26	89.7	
4. Business exported sales outside the UK in 1994															(b)
No	6	66.7	3	100.0	13	72.2	25	71.4	20	76.9	78	68.4	20	71.4	
Yes	3	33.3	0	0.0	5	27.8	10	28.6	6	23.1	36	31.6	8	28.6	
5. For the financial year ending in 1994 the business operated at:															(b)
Loss or break-even	1	11.1	0	0.0	2	11.1	4	11.4	8	30.8	17	15.0	5	17.2	
Profit	8	88.9	3	100.0	16	88.9	31	88.6	18	69.2	96	85.0	24	82.8	

Notes:
(a) Derived from postal questionnaire survey.
(b) A chi-square statistic could not be calculated because more than 20 per cent of the observed categories had less than five expected observations.
(c) Business located in an area with 10 000 or more people.

clusters 2 and 4 reported that their CEO was a family member. Table 10.3 shows that the firms in clusters 7 and 4 were, on average, older.

The seven 'types' of family firms were compared with regard to their size and performance. Table 10.3 shows that firms in clusters 2, 4 and 7 reported significantly higher sales revenues (SALES94), while firms in clusters 1 and 3 reported significantly lower sales. Firms in clusters 2 and 4 were larger in employment size (EMPLOY94), whilst firms in clusters 1, 3 and 6 were smaller. Further, a weakly significant difference with regard to the absolute change in gross sales revenues (SALESCH) was detected. Firms in cluster 2 reported larger increases in sales, while firms in cluster 1 reported the smallest increases in sales.

Correlation analysis showed that the WEIGHTED variable was significantly associated with five of the size and performance variables. The WEIGHTED variable can, therefore, be viewed as a reasonable surrogate measure of overall firm size and performance. Firms in clusters 2 and 1 reported significantly higher WEIGHTED scores, while firms cluster 3 reported the lowest scores.

No marked differences were detected between the 'types' of firms with regard to the propensity to export or to report a profit in 1994 (Table 10.2). Further, no significant differences between the 'types' of firms were detected with regard to sales revenues to employees (SALESR94); change in number of employees (EMPLOYCH); and percentage of sales exported (%EXPORT) (Table 10.3). However, significant pairwise differences with regard to absolute employment change were noted. Cluster 2 firms, on average, reduced their employment size by 51 people and cluster 3 firms only reported increases of 0.16 people.

Conclusions and implications

In total, 172 private family firms (73 per cent) exhibit behavior in line with the stewardship perspective (that is, quadrant C in Figure 10.1), and a further 65 firms (27 per cent) exhibit behavior in line with the agency perspective (that is, quadrant B). None of the 'types' of firms were located in quadrants A or D. This indicates the appropriateness and complementarity of the two theoretical perspectives, and provides evidence of conceptual validity. Family firms focusing on financial objectives generally report ownership and management structures in line with agency theory. Conversely, family firms focusing on non-financial objectives generally appear to have more of an organization-serving culture in line with stewardship theory. Agency theory, therefore, may not be universally relevant (Arthurs and Busenitz, 2003) to discuss the behaviour of all private family firms, but it does have some validity. The agency perspective should, therefore, be used alongside complementary theories.

Variations were observed in the size and performance of the seven 'types' of family firms. The 115 firms in cluster 6 (that is, 'average' family firms) report average performance. Cluster 2 firms report superior sales growth and WEIGHTED scores. Firms in clusters 4 and 7 are larger, but do not report superior sales or WEIGHTED scores. Firms in clusters 3 and 1 are smaller in employment and sales size. Cluster 3 firms report the lowest WEIGHTED scores, indicating dissatisfaction with their performance. Firms in cluster 1 report little sales growth, but they are generally satisfied with their performance as suggested by their higher WEIGHTED scores.

Policy-makers and practitioners are seeking to encourage the supply of entrepreneurs and new ventures and improve the performance of existing firms. This study has shown

Table 10.3 Cluster characteristics of family companies by company age, size and performance

	Clusters															
	1		2		3		4		5		6		7		No. of Obs	Kruskal-Wallis sig
Varibles	Mean	Median	Mean	Median	Mean	Median	Mean	Median	Mean	Median	Mean	Median	Mean	Median		
AGE (i, ii, iii, iv)	36.67	15.00	42.33	37.00	27.79	25.00	54.08	47.50	36.89	26.50	29.46	19.00	61.55	43.00	237	0.0000
SALES94 (v, vi, vii)	988	670	58 666	35 000	1021	400	9625	2356	3908	1650	1394	783	4171	1710	216	0.0000
EMPLOY94 (viii, ix, x)	18.11	11.00	321.67	280.00	11.32	6.00	130.77	42.00	40.89	24.50	19.97	12.00	45.00	21.00	233	0.0000
SALESR94	73 443	40 476	260 882	96 429	131 876	58 452	98 343	53 571	88 805	65 517	88 766	65 553	121 209	73 529	215	0.6290
SALESCH (xi, xii, xiii)	286	182	15 333	13 000	329	29	1264	212	827	121	302	92	616	223	212	0.0549
EMPLOYCH (xiv, xv)	3.33	0.00	−51.00	−51.00	0.16	0.00	14.11	1.00	1.12	−1.00	2.37	0.00	4.39	0.00	231	0.7298
%EXPORT	3.33	0.00	0.00	0.00	10.67	0.00	8.31	0.00	5.39	0.00	5.49	0.00	2.50	0.00	233	0.9298
WEIGHTED (xvi, xvii)	13.56	15.33	18.72	20.67	9.93	9.67	12.49	12.25	11.19	10.92	12.58	12.33	11.43	11.33	233	0.0336

Notes:
Total absolute gross sales revenues financial year ending 1994 (£000s) (SALES94); total employees (full-time and part-time) end of 1994 (EMPLOY94); total absolute gross sales revenues financial year ending 1994 (£s) as proportion of total employees (SALESR94); total absolute gross sales revenue change 1991 to 1994 (£000s) (SALESCH); total absolute employment change 1991 to 1994 (EMPLOYCH); percentage of sales exported in 1994 (%EXPORT); weighted average performance score was calculated for each company (Naman and Slevin, 1993), based upon the importance respondents attached to six selected performance indicators (that is, sales revenues level, sales revenue growth rate, cash flow, return on shareholder equity, gross profit margin and net profits from operations, each rated on a scale ranging from 1 'very little importance' to 5 'extremely important') and the level of satisfaction their business had achieved with regard to each of these indicators (that is each reported on a scale ranging from 1 'highly dissatisfied' to 5 'highly satisfied'). This scale has a Cronbach's Alpha of 0.85 (WEIGHTED).
(i) At the 0.1 level significant difference between clusters 2 & 3.
(ii) At the 0.05 level significant difference between clusters 5 & 6.
(iii) At the 0.01 level significant difference between the following pairs of clusters: 1 & 4, 1 & 5, 2 & 4, 2 & 5, 3 & 5 and 4 & 5.
(iv) At the 0.001 level significant difference between the following pairs of clusters: 1 & 6, 1 & 7, 2 & 6, 2 & 7, 3 & 4, 3 & 6, 3 & 7, 4 & 6, 4 & 7, 5 & 7 and 6 & 7.
(v) At the 0.05 level significant difference between the following pairs of clusters: 1 & 2, 1 & 3 and 5 & 6.
(vi) At the 0.01 level significant difference between the following pairs of clusters: 3 & 5, 5 & 7 and 6 & 7.
(vii) At the 0.001 level significant difference between the following pairs of clusters: 1 & 4, 1 & 5, 1 & 6, 1 & 7, 2 & 3, 2 & 4, 2 & 5, 2 & 6, 2 & 7, 3 & 4, 3 & 6, 3 & 7, 4 & 6 and 4 & 7.

(viii) At the 0.05 level significant difference between the following pairs of clusters: 1 & 2 and 6 & 7.
(ix) At the 0.01 level significant difference between clusters 1 & 3 and 2 & 3.
(x) At the 0.001 level significant difference between the following pairs of clusters: 1 & 4, 1 & 5, 1 & 6, 1 & 7, 2 & 4, 2 & 5, 2 & 6, 2 & 7, 3 & 4, 3 & 5, 3 & 6, 3 & 7, 4 & 6, 4 & 7, 5 & 6 and 5 & 7.
(xi) At the 0.1 level significant difference between the following pairs of clusters: 1 & 5, 1 & 6 and 1 & 7.
(xii) At the 0.05 level significant differences between the following pairs of clusters: 1 & 2, 1 & 3, 1 & 4, 2 & 4, 2 & 5, 2 & 6 and 2 & 7.
(xiii) At the 0.001 level significant difference between clusters 2 & 3.
(xiv) At the 0.1 level significant difference between clusters 1 & 2.
(xv) At the 0.05 level significant difference between clusters 2 & 3.
(xvi) At the 0.1 level significant difference between clusters 3 & 6.
(xvii) At the 0.05 level significant difference between the following pairs of clusters: 1 & 3, 1 & 4, 1 & 5, 1 & 6, 1 & 7, 2 & 3, 2 & 4, 2 & 5, 2 & 6, 2 & 7 and 3 & 4.

that information relating to company objectives as well as ownership and management (that is, the 17 variables explored here) would be a useful means of identifying particular types of family firms in order to allocate assistance and resources. While most family firm 'types' emphasize a range of family and financial objectives, the worst-performing family firms appear to focus exclusively on family objectives. To retain knowledge within the entrepreneurial pool and to foster wealth creation and job generation, policy-makers and practitioners need to encourage more owners of weaker-performing family firms to focus on financial objectives alongside family objectives.

The most important result from this study is of relevance to all the communities of practice involved with family firms, that is, academics, policy-makers and practitioners. Clearly, it is inappropriate to view private family firms as a single homogeneous entity. Family firms researchers should be aware that the failure to recognize contrasts between 'types' of private family firms might impact on the validity and generalizability of research evidence. Policy-makers and practitioners should consider the specific needs of different 'types' of family firms. The development issues facing the majority of 'average' family firms may not be the same as those cited by other 'types' of family firms. Rather than the provision of 'blanket support' available to all 'types' of family firm owners, there may be a need to provide targeted support. In-depth and repeated dialogues with the owners of various 'types' of family firms will also assist policy-makers and practitioners to allocate appropriate assistance.

The framework developed in this study and the empirical testing provide evidence of the validity of frameworks employing complementary theories. Future research should consider complementary theories rather than a general theory of family firms. The limitations of this study would indicate that more research is needed to further test the validity of the presented taxonomy using a variety of national, cultural, regional and industrial settings. There is clearly a need for multi-paradigm and multidisciplinary studies. Longitudinal studies could explore the causal links between selected variables. This study has highlighted that multivariate regression analysis may fail to consider the 'outlier' firms, which are associated with superior (and weaker) performance relative to 'average' firms. Further research could compare 'average' and 'outlier' private family firms. Clearly, whichever methodology is selected, future research and policy should take into account that family firms are not a homogenous entity and that different 'types' of private family firm do exist.

References

Aldrich, H.E. and Cliff, J.E. (2003), 'The pervasive effects of family on entrepreneurship: toward a family embeddedness perspective', *Journal of Business Venturing*, **18**(5), 573–96.

Arthurs, J.D. and Busenitz, L.W. (2003), 'The boundaries and limitations of agency theory and stewardship theory in the venture capitalist/entrepreneur relationship', *Entrepreneurship Theory and Practice*, **28**(2), 145–62.

Astrachan, J.H. (2003), 'Commentary on the special issue: the emergence of a field', *Journal of Business Venturing*, **18**(5), 567–72.

Binder Hamlyn (1994), *The Quest for Growth: A Survey of UK Private Companies*, London: Binder Hamlyn.

Chrisman, J.J., Chua, J.H. and Zahra, S. (2003), 'Creating wealth in family firms through managing resources: comments and extensions', *Entrepreneurship Theory and Practice*, **27**(4), 359–66.

Chua, J.H., Chrisman, J.J. and Steier, L.P. (2003), 'Extending the theoretical horizons of family business research', *Entrepreneurship Theory and Practice*, **27**(4), 331–38.

Daily, C.M. (1995), 'An empirical examination of the relationship between CEOs and directors', *Journal of Business Strategies*, **12**, 50–68.

Daily, C.M. and Dollinger, M.J. (1992), 'An empirical examination of ownership structure in family and professionally managed firms', *Family Business Review*, **5**(2), 117–36.

Daily, C.M. and Dollinger, M.J. (1993), 'Alternative methodologies for identifying family- versus non family-managed businesses', *Journal of Small Business Management*, **31**, 79–90.

Davis, J., Schoorman, F.D. and Donaldson, L. (1997), 'Toward a stewardship theory of management', *Academy of Management Review*, **22**(1), 20–47.

Donckels, R. and Fröhlich, E. (1991), 'Are family businesses really different? European experiences from STRATOS', *Family Business Review*, **4**(2), 149–60.

Feinberg, R.M. (1975), 'Profit maximization vs. utility maximization', *Southern Economic Journal*, **42**(1), 130–31.

Gersick, K.E., Davis, J., Hampton, M.M. and Lansberg, I. (1997), *Generation to Generation: Life Cycles of the Family Business*, Boston, MA: Harvard Business School Press.

Greenwood, R. (2003), 'Commentary on "Toward a theory of agency and altruism in family firms" ', *Journal of Business Venturing*, **18**(4), 491–4.

Howorth, C. and Ali, Z.A. (2001), 'A study of succession in a family firm', *Family Business Review*, **14**(3), 231–44.

Howorth, C., Westhead, P. and Wright, M. (2004), 'Information asymmetry in management buyouts of family firms', *Journal of Business Venturing*, **19**(4), 509–34.

James, H.S. Jr (1999), 'Owner as manager, extended horizons and the family firm', *International Journal of the Economics of Business*, **6**(1), 41–56.

Jensen, M.C. (1994), 'Self-interest, altruism, incentives, and agency theory', *Journal of Applied Corporate Finance*, **2**(Summer).

Kelly, L.M., Athanassiou, N. and Crittenden, W.F. (2000), 'Founder centrality and strategic behavior in the family-owned firm', *Entrepreneurship Theory and Practice*, **25**(2), 27–42.

Naman, J.L. and Slevin, D.P. (1993), 'Entrepreneurship and the concept of fit: a model and empirical tests', *Strategic Management Journal*, **14**(2), 137–53.

Neubauer, F. and Lank, A.G. (1998), *The Family Business: Its Governance for Sustainability*, Basingstoke: Macmillan.

Randoy, T. and Goel, S. (2003), 'Ownership structure, founder leadership, and performance in Norwegian SMEs: implications for financing entrepreneurial opportunities', *Journal of Business Venturing*, **18**(5), 619–37.

Rogoff, E.G. and Heck, R.K.Z. (2003), 'Evolving research in entrepreneurship and family business: recognizing family as the oxygen that feeds the fire of entrepreneurship', *Journal of Business Venturing*, **18**(5), 559–66.

Rose, M.B. (2000), *Firms, Networks and Business Values: The British and American Cotton Industries since 1750*, Cambridge: Cambridge University Press.

Schulze, W.S., Lubatkin, M.H. and Dino, R.N. (2003), 'Toward a theory of agency and altruism in family firms', *Journal of Business Venturing*, **18**(4), 473–90.

Schulze, W.S., Lubatkin, M.H., Dino, R.N. and Buchholtz, A.K. (2001), 'Agency relationships in family firms: theory and evidence', *Organization Science*, **12**(2), 99–116.

Smyrnios, K. and Romano, C. (1994), *The Price Waterhouse/Commonwealth Bank Family Business Survey 1994*, Sydney: Department of Accounting, Monash University.

Westhead, P. (1997), 'Ambitions, "external" environment and strategic factor differences between family and non-family companies', *Entrepreneurship and Regional Development*, **9**(2), 127–57.

Westhead, P. and Cowling, M. (1997), 'Performance contrasts between family and non-family unquoted companies in the UK', *International Journal of Entrepreneurial Behaviour & Research*, **3**(1), 30–52.

Westhead, P. and Cowling, M. (1998), 'Family firm research: the need for a methodological rethink', *Entrepreneurship Theory and Practice*, **23**(1), 31–56.

Westhead, P. and Howorth, C. (2004), 'Ownership and management structure, company objectives and performance: an empirical examination of family firms', paper presented at the IFERA 4th Annual Research Conference, Jönköping, Sweden.

Westhead, P., Cowling, M. and Howorth, C. (2001), 'The development of family companies: management and ownership issues', *Family Business Review*, **14**(4), 369–85.

Westhead, P., Howorth, C. and Cowling, M. (2002), 'Ownership and management issues in first and multi-generation family firms', *Entrepreneurship and Regional Development*, **14**(3), 247–69.

11 From vision to variables: a scorecard to continue the professionalization of a family firm

Ken Moores and Justin Craig

This chapter builds on previous projects we have conducted that have concentrated on the key areas of corporate governance and strategic planning in family businesses. Whereas our previous projects have enlisted an additive approach (that saw the family perspective *added* to the business), this current research takes on an integrated approach and seeks to integrate issues that influence the family and business systems. Specifically, in this research we use innovation action research (Kaplan, 1998) to illustrate how the Balanced Scorecard that includes reference to family business challenges has been introduced and used to assist family members, board members and management in a third-generation Australian family-owned business by the lead author who is a non-executive director of the business. The process of scorecard development is discussed and the development of the core essence, vision and mission statements, strategic objectives, measures and targets, which can be scrutinized by family business stakeholders to ascertain consistency with the vision of the company, is outlined. A conceptual mapping framework is introduced and propositions that will guide future projects are detailed.

Introduction

Families that work together face many challenges. Much of the friction in family businesses can be attributed to the overlap of the family and business systems. The emotional bonds between family members become intertwined with business issues (Craig and Lindsay, 2002; Lansberg, 1983). As a result, the family business is rarely, if ever, viewed as a total system (Schneider, 1989). Family business is usually seen from either the business perspective, from the family perspective, or as two conflicting systems. A family business, from the business perspective, is a system that is task orientated and competency based (Davis and Stern, 1980). The primary task is the generation of goods and service through organized behaviour for the purpose of making a profit. As a result, social relations are very much influenced and guided by the norms and principles that facilitate the productive process. As such, 'the family business is an enterprise that is based upon the concept of merit and is a system that values the person based upon what s/he does' (Lansberg, 1983, p. 42). Alternatively, from the family perspective, the family business is a kinship system in which members are related by blood or law. This system operates within the environment of the household, is not a place, but rather a 'pattern of appropriate conduct, coherent, embellished and well articulated' (Goffman, 1959, p. 75). In this system, the glue that holds the family together is cooperation and unity, its emotional bonding and affectionate ties that develop between and among its members, as well as a sense of responsibility and loyalty to the group as a system (Schneider, 1989). It is a system largely based on the concept of need. That is, the family's primary social function is to assure the care and nurture of its members. Specifically, 'social relations in the family are structured to satisfy family members' various developmental needs and tend toward valuing the person based upon who he/she is' (Kepner, 1983, p. 60).

Family business research has now evolved to the point where 'to understand the family business we must recognise that the two subsystems (family and business) co-exist and it is their relative powers that make a family business unique' (Sharma et al., 1997, p. 20). Strategic planning and strategy formulation has been seen as a way in which family and business goals can be integrated. However, as Sharma et al. (1997) point out 'family business is more likely to have multiple, complex, and changing goals rather than a singular, simple, and constant goal' (p. 17). These authors also suggest that, although more attention has been paid to the process of strategy formulation and the content of strategy in family businesses, relatively little is still known. Harris et al. (1994) agree that 'the assessment of family business characteristics and their influence on strategy leaves more questions than answers' (p. 171), and Chrisman et al. (2003) contend that this situation is still largely the case.

The purpose of this chapter is to demonstrate using an accepted strategic management and measurement tool, the Balanced Scorecard (BSC), how family and business goals can be integrated. Specifically, we enlist an innovation action research (Kaplan, 1998) process to address the following research question: How can the four perspectives of the Balanced Scorecard be adapted to integrate the potentially divergent family and business goals that exist in family-owned businesses? We show how the BSC can be adapted to the family business context as a measurement and management, as well as a communication, tool that is easily interpretable by those involved in family business (see also, Craig and Moores, 2005). We address a gap in the literature by focusing on how family and business goal integration can be concurrently addressed. Specifically, we highlight a framework that can assist family businesses understand that, as their firm morphs into an increasingly complex business, strategy becomes increasingly important as strategic decisions effect, and need to be communicated to, an increasingly diverse group of family and non-family stakeholders.

First, we review the BSC literature including an outline of the foundation vision and mission statements, which are at the core of the scorecard development process. We then introduce the family business action research site to which the BSC is applied. We proceed to outline the process that illustrates how the four perspectives of the BSC have been introduced to the family business and include the objectives, measures and targets that the family has established to ensure the integration of family and business strategic goals. Finally, we include a conceptual process model and introduce a series of propositions that will drive future projects.

Literature review

Organizations use various systems to measure financial and non-financial indicators. The Balanced Scorecard is one such measurement and management system that has received endorsement from many of the world's most successful organizations. The BSC was developed by Kaplan and Norton (1992) and links the measurement of financial and non-financial indicators to firm strategy.

Originally developed as a performance measurement tool (Kaplan and Norton, 1992), the BSC has evolved into an organizing framework, an operating system, and a strategic management system (Kaplan and Norton, 1996). As exclusive reliance on financial measures in a management system is insufficient, the BSC highlights the difference between lag indicators versus lead indicators. Financial measures are 'lag indicators that report on the outcomes from past actions' (Kaplan and Norton, 2001, p. 18). Examples of lag indicators are return on investment, revenue growth, customer retention costs, new product revenue, revenue per employee, and the like. These lagging outcome indicators need to be complemented (supplemented) by measures of the drivers of future financial performance, that is, lead indicators. Examples of lead indicators are revenue mix, depth of relationships with key stakeholders, customer satisfaction, new product development, diversification preparedness and contractual arrangements.

The BSC also addresses the measurement and management of tangible versus intangible assets. Examples of tangible assets include items such as inventory, property, plant and equipment (Chandler, 1990) while examples of intangible assets are 'customer relationships, innovative products and services, high-quality and responsive operating processes, skills and knowledge of the workforce, the information technology that supports the workforce and links the firm to its customers and suppliers, and the organizational climate that encourages innovative problem-solving and improvement' (Kaplan and Norton, 2001, p. 88). The BSC enables the firm to distinguish four distinct strategically important perspectives: financial, customer, internal processes, innovation and learning. These are individualized by the organization around the vision and the mission, and enable the management team to establish objectives, measures and targets. Theoretically, as Kaplan and Norton point out 'the academic literature, rooted in the original performance management aspects of the scorecard, focuses on the BSC as a measurement system but has yet to examine (in detail) its role as a management system' (Kaplan and Norton, 2001, p. 100) and it is our aim to address this in the context of family-owned businesses.

At the core of the BSC, and an integral step before attempting to build what Kaplan and Norton (2001) refer to as strategy maps, is the necessity to review mission statements: why the company exists, the core values and what the company believes in. A strategic vision can then be developed. The vision 'creates a clear picture of the company's overall goal . . . the strategy identifies the path intended to reach that destination' (Kaplan and Norton, 2001, p. 19). The BSC provides a framework for organizing strategic objectives into four perspectives: (1) financial, (2) customer, (3) internal business processes, and (4) learning and growth.

The financial perspective

Economic growth strategies are usually approached from a revenue growth or productivity perspective. Revenue growth involves either increasing revenue from new markets, new products and new customers, or increasing sales to existing customers. Productivity strategies involve either improving cost structures by expense reduction, or the more effective utilization of assets (Kaplan and Norton, 2001). These widely accepted metrics form the financial perspective of the BSC.

Customer perspective

The unique mix of product, price, service, relationship, and image that the company offers, is at the core of any business strategy, and are introduced in the BSC via the customer perspective. This customer-value proposition defines how the company differentiates itself from competitors and is crucial because it helps an organization 'connect its internal processes to improved outcomes with its customers' (Kaplan and Norton, 2001, p. 19). Value propositions include operational excellence, customer intimacy and product leadership, and sustainable strategies are based on excelling at one of the three while maintaining threshold standards with the other two. Identification of a value proposition allows the company to know which class and type of customer to target. In addition, the customer perspective identifies the intended outcomes from delivering a differentiated value proposition, for example market share in targeted customer segments, account share with targeted customers, acquisition and retention of customers in the targeted segments and customer profitability (Kaplan and Norton, 2001).

Internal process perspective

The internal process perspective captures the critical organizational activities that will determine the means by which the company will achieve the differentiated value proposition and the productivity improvements for the financial objectives (Kaplan and Norton, 2001). These are captured by (1) spurring innovation to develop new products and services and to penetrate new markets and customer segments; (2) increasing customer value by expanding and deepening customer relationships with existing customers; (3) achieving operational excellence by improving supply-chain management, internal processes, asset utilization, resource-capacity management and so on; and (4) becoming a good corporate citizen by establishing effective relationships with external stakeholders. Related financial benefits typically occur in short-term, intermediate and long-term stages.

Innovation and learning perspective

The foundation of any strategy is the innovation and learning perspective. Employee capabilities and skills, technology, and corporate climate are needed to support the strategy. These objectives enable the company to 'align its human resources and information technology with the strategic requirements from its critical internal business processes, differentiated value proposition, and customer relationships' (Kaplan and Norton, 2001, p. 20).

Each of the four perspectives are individualized by the organization around the vision and the mission, and objectives, measures and targets are established accordingly. The BSC can be used to accomplish four important management processes: (1) translating the vision – objectives and measures; (2) communicating and linking – by bringing understanding to employees relating to critical objectives and how they will be measured; (3) business planning – helps organizations integrate plans by using objectives to set targets; and (4) feedback and learning – tests the viability of overall strategy. As such, the BSC benefits shareholders, management and employees. Shareholders are provided with an improved understanding of the workings of the operation, are able to realize that the business is more than just money orientated and a greater focus can be placed on building a stronger operation rather than just financial performance. Management are provided with greater ability to communicate the vision, deal with change and growth, provide more relevant and structured information, create a workforce that has an interest in the strategic direction of the operation and is working towards the achieving the same mission, and are better able to measure true performance. Employees have greater ownership of their position, a belief in the strategic direction of the operation through their input, a say in the development of their skills and an understanding of how their performance is measured.

Method

Research design

The method adopted in this research is a form of action research. People, however, have different meanings and interpretations about action research. The form adopted here is that which Balanced Scorecard creator, Robert Kaplan, labelled 'innovation action research' (Kaplan, 1998). In innovation action research, scholars develop and refine theory (of new management practice) they believe to be broadly applicable to a wide variety of organizations.

Scholars engaged in innovation action research play an active role in implementing their ideas in actual organizations. The concepts must promise sufficient benefits (that is, they represent a solution to a real problem) and be articulated clearly enough that organizations are willing to commit their own resources to an implementation experience. The research cycle in innovation action research involves:

- observing and documenting innovative practice (see, Craig and Lindsay, 2002; Craig and Moores, 2002, 2004; Moores and Barrett, 2003)
- teaching and speaking about the innovation (see, Moores and Craig, 2002, 2003)
- writing articles and cases (see, Craig and Moores, 2005)
- implementing the concept in a new organization (current chapter).

Bracketed references address how we have followed this research cycle.

Case study: O'Reilly's Rainforest Guesthouse

O'Reilly's Rainforest Guesthouse is a family-owned and family-operated business that was established in 1926. It is located in the Lamington National Park in Queensland, Australia. The operation is still known as a 'Guesthouse' to protect the heritage of the business. In today's terms, however, it is better described as an 'eco-tourism resort'. The Guesthouse accommodates up to 180 guests, employs 80 full-time staff (increased from 27 staff in 1990), and enjoys a 70 per cent year round occupancy. The business has a turnover in excess of $10 million (up from $3.7 million in 1994). The business has developed an international reputation in eco-tourism and is the winner of numerous training, family business, and tourism industry awards.

Currently, the business is managed by the third generation. This generation is made up of members from two family groups: four offspring of Peter O'Reilly and 10 offspring of Vince O'Reilly. The CEO is the eldest son of Peter O'Reilly. The family has been pursuing a professionalization agenda for a number of years, especially under the leadership of its current third-generation CEO, who learned business outside the family business in the hospitality industry both in Australia and overseas. Included among its professional measures are both an active board of directors and regular family retreats. The seven-member board of directors meets monthly and is chaired by a non-family independent director and includes two other independent non-family directors. Additionally, the family has been holding family meetings since the early 1990s. These meetings have now evolved into retreats that are typically held over two days and include all family members plus spouses/partners.

Board of director reporting was targeted as an area to improve further in 2004. To that end, an external consultant was engaged to examine the key financial drivers with a view to targeting areas to improve performance, and the reporting thereof. Coinciding with this initiative, senior management were encouraged to align their reports to the board of directors more closely with the key objectives in the current strategic plan. This was to be refined following the adoption of the 2005–08 strategic plan, which was to be considered by the board of directors in the March/April period. In preparation for this planning, it was necessary for family stakeholders to clarify their expectations in terms of the vision and goals that they held for the family firm. The BSC was adopted as the mechanism to frame the new reporting policies.

Beginning the Balanced Scorecard introduction process

The clarification of family stakeholder expectations occurred at the November 2004 family retreat. In anticipation of this meeting, the results of a survey of family members that had been circulated some months earlier were summarized and presented for discussion, confirmation or amendment. Specifically, the responses of 26 family members to the question related to their personal vision and goals for the family business over the next 5 to 10 years were classified according to (1) family as owners, (2) family as employees, (3) family as family, and (4) family as community. The emergent themes distilled from these responses were that the family were in favour of (1) growth of the total business through diversification 'away from the mountain', and (2) the consolidation of its mountain guesthouse operation by enhancing the guest experience. As well, there was agreement that the family were to be encouraged to be involved in the business and that education would be used to develop this interest. The family felt strongly that the business continues to acknowledge the history of its foundation by perpetuating the values of prior generations and developing flexible ownership structures to enable continued generational involvement. Also, family members were especially united in their goal to emphasize environmental responsibility, an acknowledged key ideology of founding generations. Refer to the appendix for summary details of this survey.

These responses were further 'tested' in an open forum and in group discussions during the November 2004 family retreat. The emergent set of family expectations was ratified as the family's vision for the business and circulated to the family and to the board of directors. The key outcomes, as reported to the family and to the board of directors, were:

Family as owners

To grow the family business.

To consolidate the Guesthouse operation.

To diversify and grow the business, thereby providing increased opportunities for family to work and have career paths within the business.

To implement structures (that is ownership, financial, strategic) to aid this growth and diversification.

Family as employees

To encourage family members to pursue career options via employment in the diversified business.

To develop two distinct employment policies for family members – (1) for operations (lower level/internships and vacation employment), and (2) for management.
To emphasize that respect has to be earned and is a vital part of the company framework. A separate induction programme will be required for family members so that they know and appreciate the extra demands and expectations placed upon them in the family business work environment.

Family as family
To encourage all family members (including those not working in the business) to contribute to the perpetuation of the family's values by their representation of the family in various forums.
To further improve communication, especially to celebrate milestones that will be maintained in part via internal newsletters.
To encourage family to stay (holiday/visit regularly) in the mountain resort.
To ensure the development of succession plans and to consider developing more flexibility to enable the identification of exit strategies for those needing them.

Family as community
To be known as a leader of eco-tourism in Australia.
To build further the family's reputation as ethical, honest, fair and supporters of the broader local community.
To be known as an employer of choice.
To recognize the indigenous heritage of the region.

Using this shared and articulated vision for the family and the business, senior family members and non-family managers held their strategic planning meeting in February 2005. The purpose of this meeting was to identify a three-year plan and direction for O'Reilly's in accordance with the family's (owners) expectations and within the board of director's planning parameters.

Subsequently, each manager developed objectives, measures and targets within a Balanced Scorecard framework for their assigned area of responsibility. These were to be consistent with the agreed direction and plan, and be available for team discussion and agreement for submission to the April board of director's meeting. From this agreed three-year plan, a budget for 2005–06 will then be developed to resource the achievement of agreed targets. This budget will accompany the three-year plan in the board submission.

Applying the Balanced Scorecard to the O'Reilly family business

Core essence Whereas vision and mission statements at the centre of the BSC are effectively management tools, in family businesses, there is a need to identify the core essence of the family and therefore the family business. The O'Reilly family's core essence statement was established earlier in the professionalisation process as part of the second to third generation transition, as:

> We treat strangers like friends, friends as family and family as gold.

This core essence statement encapsulates the values that serve as the foundation for the vision and mission. The current senior management team confirmed the key values for the business as:

- to respect, support and trust each other and our guests
- to act fairly and decisively
- to recognize success is due to team efforts and cooperation
- to behave ethically with integrity and honesty
- to maintain our culture and be proud of our heritage
- to support our local community
- to believe in and deliver sustainable tourism practices – providing a net benefit for the social, economic, natural and cultural environments of our area.

Vision statement A company's vision is arguably at its strongest in the founder generation and is at risk of being diluted over time (Gallo, 2000). Members of the O'Reilly family team developed their vision statement (that addressed their *core ideology* and *envisioned future*) by taking cognizance of family expectations, origins and history of the business and parameters suggested by the board as follows:

> To grow the business by applying professional management guided by our core values and the strong ethical business ethics of the founders, to achieve global recognition as a leader within Australia's ecotourism sector.

Mission statement In their aim to make every guest feel special by exceeding expectations, members of the O'Reilly family team developed their mission statement as follows:

> Make every guest feel special.

With the core essence, the vision and mission decided upon, the BSC framework then enabled the O'Reilly family to decide what is required to adhere to these statements in order to remain financially sound, customer focused, professional and innovative.

The financial perspective (FP) From a financial perspective, family businesses have been found to have long-term rather than short-term financial goals (Anderson et al., 2003) and this influences strategic decisions. Family business success has typically not been tied to, or established from, the same performance measures as other business types. Often, ownership transition and efficiency of the family business system rather than wealth-creation and financial performance are used to monitor successful performance (Habbershon and Pistrui, 2002; Sharma et al., 1997; Sorensen, 2000).

The O'Reilly family divided their *financial perspective* objectives into (1) return, (2) growth, and (3) sustainability, with the understanding that the business is a family business and that the incumbent leadership develop strategies that address both current and future generational needs. As illustrated in Table 11.1, the identified measures and targets ensure that, from a financial perspective objective, capital investment is directed to achieve long-term growth and sustainability.

Table 11.1 Financial perspective

BSC code	Objective	Measure	Target
FP1	Return/growth/ sustainability *(with particular emphasis on building a strong business that meets the requirements and expectations of current and future members of the O'Reilly family)*	The capital budget	50% invested in the current business and 50% invested in new growth opportunities with a 20% variance and implemented this current financial year. Minimum investment in dollars to be $250,000 each.
		Return on capital employed (ROCE)	12% (stepped to 10% this current year)

Customer perspective (CP) The O'Reilly family divided their customer perspective objectives into (1) the experience, (2) the differentiation on our value proposition, and (3) market share. Specifically, the customer perspective objectives are to make every guest feel special through the provision of enjoyable and educational experiences strongly linked to nature. This is to be achieved by remaining relevant to customers through competitive management of the destination. The identified measures and targets for these customer perspective objectives are outlined in Table 11.2 and, as can be seen, the stated objectives include adherence to the family aspect of the family business.

Internal process perspective (IPP) It has been suggested that the family, as a family, develops internal processes that facilitate the containment, confrontation and resolution of family problems (Davis and Stern, 1988). Moores and Barrett (2003) suggest that

> (1) managers of family firms should adopt management systems which are adequate for the demands of their external and internal environments, as well as their firm's stage of development, (2) management approaches should form an internally consistent package of strategies, structures and systems, (3) management systems must dynamically evolve as the business grows and matures, (4) professionalism in management is vital for systems development, and (5) without succession plans, professionalization of the firm is seriously inhibited. (p. 148)

Thus, internal processes for family businesses (like all businesses) are necessary to include in strategy development. Arguably, what makes internal processes (particularly changing these processes) more problematic in family businesses, is the influence of the founder or the incumbent generation and the preparation for succession.

The O'Reilly family divided their *internal process perspective* objectives into (1) operational excellence, (2) ecotourism leadership, and (3) good corporate citizenship. Specifically, the family business decided to plan to build further the culture of the organization that was founded on ecotourism leadership and ensure that this feature is communicated both inside and outside the business, thereby enhancing its reputation as a good corporate citizen. The identified measures and targets for these internal process perspective objectives are summarized in Table 11.3. The stated objectives are influenced by the values that have been established by previous generations of the O'Reilly family.

Table 11.2 Customer perspective

BSC code	Objective	Measure	Target
CP1	Provide enjoyable and educational experiences strongly linked to nature *while ensuring exposure to, and acknowledgment of the contribution of previous generations of the O'Reilly family*	Participation rates monitored and to include tracking of awareness of O'Reilly family involvement in the business	$20 pppd of total unbundled guests Ensure all guests meet with at least one member of the O'Reilly family during their stay
		Feedback forms include reference to O'Reilly family involvement in the business	Discovery activities are rated at 6 or above (out of a max score of 7) Answer 'yes' to the question related to meeting at least one O'Reilly family member during their stay
CP2	To ensure that every guest feels special *with consistent acknowledgment of the fact that the resort is a 'guest house' owned and operated by members of the O'Reilly family*	Feedback forms monitored and include questions related to appreciation of the role of the O'Reilly family in the business	'Exceed expectations' on 60% of all returned forms Answer 'yes' to question related to being aware of the involvement of O'Reilly family
		Return and recommendations guests (monitored using Fidelio system)	Return business to make up a minimum of 55% of all guests Consistent reference to the fact that family involvement is a reason for the decision to return
CP3	Remain competitive, innovative and relevant to our customers *while at the same time retaining strong links to the legacy of previous generations of the O'Reilly family*	Earnings before interest and tax (EBIT)	A minimum of 8% growth EBIT per year from 2004/05
		Earnings before interest and tax (EBIT) Break-even for the Guesthouse operation Feedback forms	20% To be at the maximum level of 55% of turnover by 2005/06 95% report their holiday reason was satisfied and that the O'Reilly family factor was appreciated
CP4	To be managers of our destination *and be cognizant that the incumbent generation*	Documentation of capacity, potential issues, risks, future planning and development of	Documented by September '05 with representatives of both O'Reilly families having input

Table 11.2 (continued)

BSC code	Objective	Measure	Target
	should be aware that decisions made now will potentially effect future generations	facilities within the destination	
		Family members to provide input to the process	Implementation started October 2005

Innovation and learning perspective (ILP) Family firms have been shown to place substantial importance upon innovation practices and strategy. Successful family firms have been found to manage and adjust their innovative strategy (Craig et al., 2005). Like innovation, continual learning in the family business is crucial to survival, as highlighted by Moores and Barrett (2003):

> Just as the element of 'family' in family owned businesses influences how they are managed, that is, how the manager deals with the contextual factors such as life cycle stage, context and control, the element of 'family' can be expected to influence how people in family owned businesses learn to manage them. In fact, having to deal with the additional layer of complexity created by the family means that the tasks and priorities involved in learning to manage a family business lead to specific and enduring paradoxes. The family will turn out to be just as important a contingency factor as any of the others in the business context – and often more so. And just as understanding the stage of the business life cycle helps illuminate management priorities in general, it can help in understanding the paradoxes that come with each stage of learning the family business. (p. 32)

The O'Reilly family divided their *innovation and learning* objectives into (1) culture, (2) capabilities, and (3) technology; specifically, to become an employer of choice by developing both a learning culture and internal succession processes with particular reference to developing competent family members to fill the needs of the growing family business. The identified measures and targets for these *innovation and learning* perspective objectives are summarized in Table 11.4.

Discussion

To ensure that the strategy map adhered to the direction and expectations that the family had decided upon in their prior meetings, the Board was able to summarize and tabulate their progress. This provided them with two outcomes. First, it enabled them to ensure that both the family and business systems were integrated and, second, they were able to identify areas that still needed to be addressed. The summary table appears in Table 11.5.

As a consequence of this process, we are able to introduce the first propositions resulting from this innovation action research as follows:

Proposition 1a In multigenerational family firms seeking to integrate their family and business systems, the adoption of a BSC framework will generate outcomes of *family respect* and *business clarity*.

Proposition 1b In multigenerational family firms seeking to integrate their family and business systems, the adoption of a BSC framework will generate outcomes of *family engagement* and *business direction*.

Table 11.3 Internal process perspective

BSC code	Objective	Measure	Target
IPP1	Communicate and build on the culture of our organization *with particular emphasis being placed on the fact that O'Reilly's is a family business*	Feedback forms will include reference to the fact that O'Reilly's is a family business that is committed to building a sustainable and competitive business that has at its core a shared vision	90% of guests from all returned forms say their name was remembered
		Induction process reviewed for family and non-family members	All staff to go through an induction within 6 weeks of starting at O'Reilly's
		Revisit employee manuals to ensure that employees and family members are socialized into the business	
IPP2	Ecotourism leadership *that carries on a significant legacy in that previous members of the O'Reilly family were pioneers of ecotourism in Australia*	Compliance against Environmental Management Plan	Commission an independent audit of the plan every 12 months
		Guide accreditation	Minimum of three guides to be accredited at all times
		Contribution to conservation	Provide $5900 per year towards fauna/flora research
			Active participation in relevant professional associations: EA, WTA, IA
		Innovation	Minimum of two improvements to the delivery, display or interpretation of our natural/cultural environment
		Indigenous respect and sensitivity	Traditional owners or representatives are involved in the development of interpretative material that presents their heritage
IPP3	Build and maintain a profile as a good corporate citizen in the broader community *and ensure that members of the family understand the responsibilities that come with being a member of the O'Reilly family*	Budget community/ charity involvement and philanthropic efforts ensuring input from family members	1% of total room nights and or the equivalent of $50 000. $5000 in current year growing by 5% per year for the life of this plan
		Environmental Management Plan	Monthly implementation

Table 11.4 Innovation and learning perspective

BSC code	Objective	Measure	Target
ILP1	Have a learning culture throughout our company *and, in particular, ensure that family members are given the opportunity to develop an understanding of the O'Reilly family business*	Documentation of training plan that includes detailed strategies to ensure that O'Reilly family members have the opportunity to make a contribution to the family business	Completed by April each year Implementation of the plan – monthly
ILP2	Develop people for internal succession *and continue to ensure that family members are aware of their responsibilities while acknowledging that each generation knows that the challenges that they will face will be different to the ones faced by generations that have preceded them*	Internal appointments include, where suitable, involvement of O'Reilly family members	10 personnel per year are promoted to a new position
		Performance reviews for all employees including objective reviews of family members' contribution.	Full review each year in May with a review in November
		Employee feedback ensuring that family members are treated equally	75% state they are developing skills to assume greater responsibility
		Management growth includes developing family members' competence as managers	40% of all management positions to be filled internally
ILP3	Be seen as an employer of choice *which will include providing an attractive working environment for family members who wish, and have the required competencies, to join the family business*	Recruitment spending	< $ Budget as retention rates increase
		Rating against other employers	75% above average
		Compare with similar family businesses	
		Staff recommend O'Reilly's as a place of employment	75% above average
		Family members demonstrate willingness to be actively involved in family business matters	
		Retention rates	80% turnover 2005/06 decreasing by 5%

Table 11.5 Synthesis of family expectations and BSC perspectives

	Family expectations	BSC Code/Comment
1. Family as owners	1.1 To grow the family business	FP1
	1.2 To consolidate the Guesthouse operation	CP3, CP4
	1.3 To diversify and grow the business, thereby providing increased opportunities for family to work and have career paths within the business	CP1, CP2
	1.4 To implement structures (i.e., ownership, financial, strategic) to aid this growth and diversification	In process
2. Family as employees	2.1 To encourage family members to pursue career options via employment in the diversified business	ILP1
	2.2 To develop two distinct employment policies for family members – (1) for operations (lower level/internships and vacation employment), and (2) for management	ILP2, ILP3
	2.3 To emphasize that respect has to be earned and is a vital part of the company framework	Implied (but continue to monitor)
3. Family as family	3.1 To encourage all family members (including those not working in the business) to contribute to the perpetuation of the family's values by their representation of the family in various forums	ILP1
	3.2 To further improve communication, especially to celebrate milestones that will be maintained in part via internal newsletters	Needs to be formally included
	3.3 To encourage family to stay (holiday/visit regularly) at the mountain resort	Needs to be formally included
	3.4 To ensure the development of succession plans and to consider developing more flexibility to enable the identification of exit strategies for those needing them	Needs to be formally included
4. Family as community	4.1 To be known as a leader of eco-tourism in Australia	IPP2, Included in Vision Statement
	4.2 To build further the family's reputation as ethical, honest, fair and supporters of the broader local community	IPP3
	4.3 To be known as an employer of choice	ILP3
	4.4 To recognize the indigenous heritage of the region	IPP2

Increased family harmony was evident at the May 2005 family meeting at which family members could see that their 'expectations' had been clearly acknowledged in the business plans by the directors and senior management. Specifically, family members articulated without prompting that it was 'refreshing' to see that their input was valued. The CEO also commented that he was less burdened with the responsibility of consistently communicating all facets of the business to family members and was able to focus his energies on the business. We therefore introduce our second proposition as follows:

Proposition 2 In multigenerational family firms seeking to integrate their family and business systems, the adoption of a BSC framework will generate outcomes in which family stakeholders feel *valued* and the business has sharper *focus*.

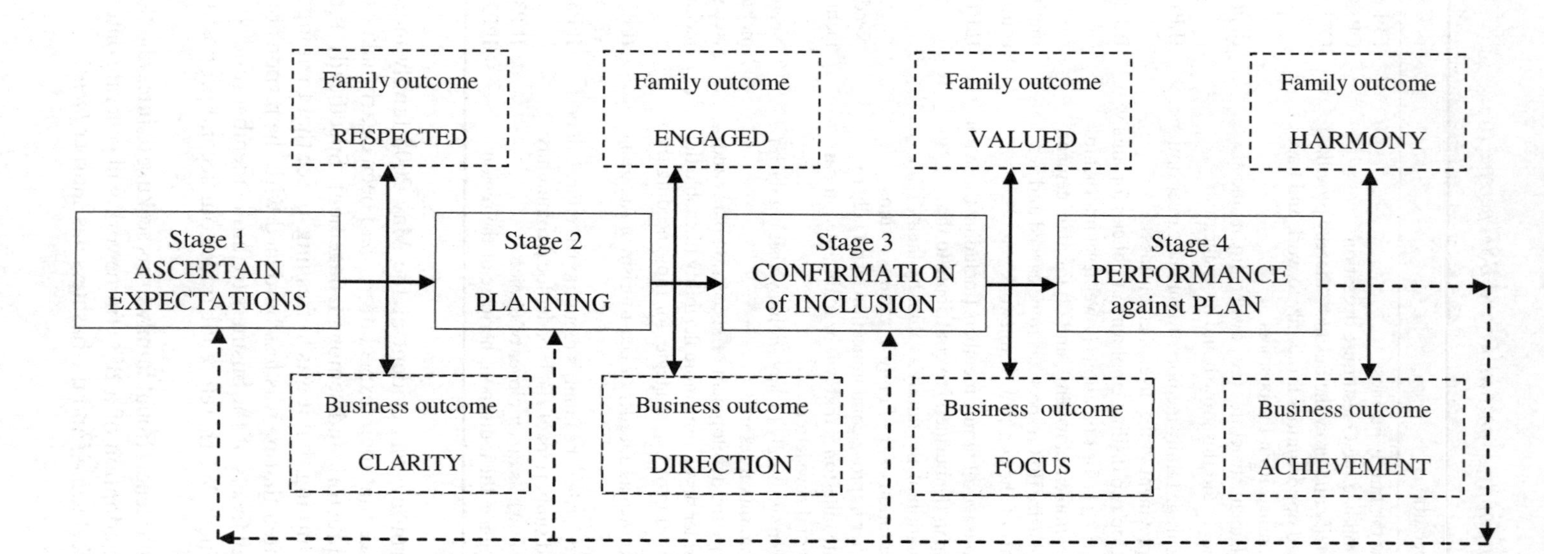

Figure 11.1 Conceptual mapping framework integrating family and business system integration

With the BSC framework now introduced to the business and family members seemingly satisfied and confident with the process, this project is now entering the monitor and evaluation stage. We are not, therefore, able to assess or report significant long-term advantages for both the family and the business systems. However, given the initial response and informal feedback from family and non-family stakeholders, and for the sake of completing the innovation action research cycle, we tentatively introduce the final proposition as follows:

> *Proposition 3* In multigenerational family firms seeking to integrate their family and business systems, the adoption of a BSC framework will generate outcomes of family *harmony* and business *achievement*.

Finally, in Figure 11.1 we introduce a conceptual mapping framework that summarizes the process and the family and business outcomes resulting from family- and business-system integration using the BSC.

Conclusion and future research

The introduction of the Balanced Scorecard framework to the O'Reilly family business enterprise has, thus far, proved beneficial to the various stakeholders, that is, family, directors, and managers. The BSC has enabled the family to be assured that their expectations have been installed as a central consideration in the development of the business strategies and plans. The methodology of scorecard development ensures at least recognition of the visions of key stakeholders. The subsequent development of missions, strategic objectives, measures and targets can then be scrutinized by stakeholders to ascertain their consistency with these visions. The board of directors more regularly undertakes this scrutiny on behalf of the stakeholders in its endeavour to balance the family's needs with those of a business nature. The Balanced Scorecard provides a reporting regime that reflects both these needs. Management can then be encouraged by the scorecard reporting requirements to focus on the operational aspects of the firm's strategy in its reporting to the board of directors.

We will continue to be involved in, monitor and evaluate the introduction of family-business specific initiatives to the O'Reilly family business. This ongoing evaluation under the innovation action research cycle is considered vital as a way to contribute to theory development (Kaplan, 1998). Proponents of the BSC are also encouraged to engage in this theory-building process by documenting their experience of addressing the integration of family-system and business-system driven strategic goals through the introduction of the BSC framework to multigenerational family businesses.

References

Anderson, R., S.A. Mansi and D. Reeb (2003), 'Founding family ownership and the agency cost of debt', *Journal of Financial Economics*, **68**, 263–85.

Chandler, A.D. (1990), *Scale and Scope: The Dynamics of Industrial Capitalism*, Cambridge, MA: Harvard University Press.

Chrisman, J.J., J.H. Chua and P. Sharma (2003), 'Current trends and future directions in family business management studies: toward a theory of the family firm', *Coleman White Paper*, www.usasbe.org/knowledge/whitepapers.

Craig J.B. and N.J. Lindsay (2002), 'Incorporating the family dynamic into the entrepreneurship process', *Journal of Small Business and Enterprise Development*, **9**(4), 416–30.

Craig, J.B. and K.J. Moores (2002), 'How Australia's Dennis Family Corporation professionalized its family business', *Family Business Review*, **15**(1), 59–70.
Craig, J.B. and K.J. Moores (2004), 'The professionalization process in family business: the Dennis family case study', in J.H. Astrachan, P. Poutziouris and K. Soufani (eds), *Family Business Casebook Journal 2004*, Kennesaw: Kennesaw State University Publisher.
Craig, J.B. and K.J. Moores (2005), 'Balanced scorecards to drive the strategic planning of family firms', *Family Business Review*, **18**(2) 105–22.
Craig, J.B., G. Cassar and K. Moores (2005), 'Innovation in family business: a ten-year study', *Family Business Review*, **19**(1), 1–10.
Davis, P. and D. Stern (1980), 'Adaptation, survival and growth of the family business: an integrated systems perspective', *Human Relations*, **34**(4), 207–24.
Davis, P. and D. Stern (1988), 'Adaptation, survival, and growth of the family business: an integrated systems perspective', *Family Business Review*, **1**(1), 210–20.
Gallo, M.A. (2000), Conversation with S. Klein at the IFERA meeting held at Amsterdam University, April, cited in Astrachan, J.H., Klein, S.B. and Smyrnios, K.X. (2002), 'The F-PEC scale of family influence: a proposal for solving the family business definition problem', *Family Business Review*, **15**(1), 31–58.
Goffman, E. (1959), *The Presentation of Self in Everyday Life*, New York: Doubleday Anchor Books.
Habbershon, T.G. and J. Pistrui (2002), 'Enterprising families domain: family-influenced ownership groups in pursuit of transgenerational wealth', *Family Business Review*, **15**(3), 223–38.
Harris, R., J. Martinez and J. Ward (1994), 'Is strategy different for the family-owned business?', *Family Business Review*, **7**(2), 159–74.
Kaplan, R.S. (1998), 'Innovation action research: creating new management theory and practice', *Journal of Management Accounting Research*, **10**, 89–118.
Kaplan, R.S. and D.P. Norton (1992), 'The Balanced Scorecard: measures that drive performance', *Harvard Business Review*, January–February, 71–9.
Kaplan, R.S. and D.P. Norton (1996), 'Using the Balanced Scorecard as a strategic management system', *Harvard Business Review*, January–February, 75–85.
Kaplan, R.S. and D.P. Norton (2001), *The Strategy-Focused Organization*, Boston, MA: Harvard Business School Press.
Kepner, E. (1983), 'The family and the firm: a co-evolutionary perspective', *Organizational Dynamics*, **12**(1), 58–70.
Lansberg, I. (1983), 'Managing human resources in family firms: the problem of institutional overlap', *Organizational Dynamics*, **12**(1), 39–46.
Moores, K.J. and M. Barrett (2003), *Learning Family Business: Paradox and Pathways*, London: Ashgate Publishing.
Moores, K.J. and J.B. Craig (2002), 'A balanced scorecard approach to strategy in family business', paper presented at the Family Business Australia National Conference, Coffs Harbour, October.
Moores, K.J. and J.B. Craig (2003), 'Advancing family business development: balancing founder values with professional management', paper presented at the *International Council for Small Business Conference*, Belfast, June.
Schneider, A.J. (1989), 'How members of family owned businesses do the business of family', unpublished dissertation, UMI Dissertation Services.
Sharma, P., J.J., Chrisman and J.H. Chua (1997), 'Strategic management of the family business: past research and future challenges', *Family Business Review*, Spring, 1–35.
Sorenson, R.L. (2000), 'The contribution of leadership style and practices to family and business success', *Family Business Review*, **13**(3), 183–200.

Appendix: Survey of family member goals and visions

Illustrative comments that exemplify each of these are the following responses:

Family as owners

- The business needs to keep a family orientation, as that is what makes us different to our competitors. A long-term focus of growth is necessary, with emphasis on minimal environmental impact.
- To grow and become more successful. To still be a family company in 10, 50 or 100 years.

Family as owners	• Grow as a family business • Consolidate as a family business • No growth • Great, better, best experience
Family as employees	• Encouragement • Education and interest
Family as family	• Perpetuating values • Ownership structure
Family as community	• Environmental leadership

- For the future, I see the expansion and updating of the company as being fundamental, however, the maintaining of the current culture will be as important.
- Avenues found for diversification of the core business.
- I personally want a successful and profitable business to hand onto the 4th generation. We should consolidate our business and make it the very best it can be, which I believe is the best in the industry.
- As the business continues to grow I think that it should not just look at future projects before first looking after what it has already achieved, making sure that it is all up to the family standards.
- I feel that, though we need to grow and take risks, we should also look at consolidating what we have already.
- To my mind, the guest house and accommodation is big enough considering it is a family business.
- I think we sometimes forget what a great business we have. We should be bursting with pride, and at the same time, striving to do better.
- To become the 'best' resort/guesthouse in the region. Certainly a goal of this magnitude is 5–10 years away but if we strive for this goal by ensuring every single guest encapsulates the 'O'Reilly experience' after every visit, it is not unachievable in the long term. I want people to come to O'Reilly's with no expectations and leave as though they'll never find an experience like it anywhere else.

Family as employees

- All family members should be encouraged to work in the business but it is not an automatic right. They must be able and be prepared to contribute to the business.
- Continue to provide opportunities for future involvement of family members and encourage them to be involved.
- Educate - 3rd and 4th generation in business basics via family courses.
- I have a personal goal to one day be associated with management in the company. Currently doing business at university to aid in achieving this goal!

Family as family

- Communication amongst family members continues to develop strengthening relationships, supporting each other and helping one another realise their dreams.

- Acknowledge history and sacrifice that was made before the current leadership i.e. what they gave.
- Our family business needs to always retain that 'personal' touch and not leave it to staff. Family in the business must lead the way. People (guests) need to be made to feel 'special' – that is part of our Mission Statement.
- Ownership structure with more flexibility to enable continued generational involvement without adversely affecting the business.
- The problems at the moment are to do with exit strategies of the business. I hope that the family does not become too concerned with the money factor, as growth within the company is rising at a level not seen before in the company.

Family as community

- The main goal I would like for the family business is to be more environmentally aware and focused. As the business has expanded we have lost our obligation to preserve the natural environment, which was the key ideology of the founding generations.
- I would also like to see a greater emphasis placed on our environmental responsibilities.

12 Working with families in business: a content validity study of the Aspen Family Business Inventory

Sandra L. Moncrief-Stuart, Joe Paul and Justin Craig

This chapter reports on a research that uses Lawshe's (1975) individual item method and Gregory's (1996) overall assessment method, on order to measure the content validity of the Aspen Family Business Inventory (AFBI), an assessment instrument designed specifically for use by consultants working with families in business. Nineteen experts in the field of family business consulting rated the AFBI's scales and items for relevance, fit, clarity and overall content. In addition to establishing the content validity of the AFBI, this research is designed to familiarize family business consultants and researchers with, and encourage them to use, techniques similar to those introduced in this chapter as a first step in establishing the validity of the instruments that they use in working with families in business.

Introduction

Businesses are usually categorized by the type of service they offer. These categories include retailers, manufacturers, and service providers. Family businesses are a subset of business found in each of these categories that are made up of organizations formed around a family unit. Despite operating within a wide realm of industries and functions, family businesses possess a number of similarities with each other.

Over the past 15 years there has been an increased understanding of the unique dynamics found in family businesses. Understanding family business dynamics involves integrating a variety of disciplines, including family systems theory. In order to provide comprehensive services, professionals working with family businesses often seek consultation with other professionals who specialize in related fields. For example, a family therapist may consult with an insurance agent, or an accountant may consult with a lawyer. This is necessary to ensure that the broad range of family-firm needs are addressed and understood. In assisting a family business under stress, family business consultants must tackle the task of determining where intervention is most needed. Until recently, there has been limited research into, and limited formal assessment tools available to help with, this task.

The Aspen Family Business Inventory was designed as a constructive first step to help understand the special set of dynamics that influence families and to begin dialogue among the key players about these key dynamics. The purpose of this research is to determine the content validity of the AFBI, and use the results to evaluate the instrument's degree of success at capturing crucial elements of family business assessment. The interrelated disciplines of family business will be first discussed in order to provide background into the complex web of processes typically found in family businesses. Then, we will outline how prominent family business experts were asked to rate and critique the relevance of the scales and items that make up the AFBI. Based on this, we report the quantitative measures for content validity that were calculated.

A synthesis of family and family business systems

In the 1950s, family systems theories evolved from what started in the 1930s as family-life education (Thomas, 1992). Family systems theories varied with the practitioner, however, they were all similar in that they focused on a systemic, rather than an individual paradigm. Some family systems theorists use a directive, problem-solving manner while others try to join the family using an emphatic stance. For example, Bowen's (1978) way of being is to act as a calm, neutral coach or interpreter, the one who is capable of getting the members of dyads to speak to each other without triangulating a third family member or therapist into the conflict. Bowen uses insight gained from looking at intergenerational information to assist families in changing. Structural family therapists like Minuchin (1974), join the family and act like one of the members using the same language style of the family members. They help families to re-establish more effective hierarchical structures. The therapists' work revolves around establishing appropriate boundaries and altering coalitions or enacting a problem in the counselling room so that the counsellor can observe the interaction and intervene immediately (Cheston, 2000). Strategic family therapy uses specific tasks that eliminate symptoms and change the way members interact (Haley, 1973).

Regardless, family systems theorists and practitioners contend that individual behaviors and related problems need to be interpreted with respect to the whole order functioning around each person. Moreover, family systems theorists assume that individual health and well-being are not separate from the quality of family relationships. The connection between family interaction processes and the onset and course of individual symptoms is an area of ongoing study within family therapy (for example, Beavers and Hampson, 1990; Fisher and Ransom, 1995; Friedman et al., 1997; Harvey and Bray, 1991).

Family systems theories have gained increasing acceptance in the broader field of psychology. Family system theories have proved successful over a broad range of problematic family and individual issues, including family conflict (De Shazer, 1985), and improvement in family functioning (Doherty et al., 1986; Giblin, 1986). Family system theories have been applied to an increasing range of family situations, and more and more over the past 10 years researchers have applied family systems theories to the dynamics of family businesses.

Family business systems

The study of family business dynamics is in the early phase. There is a growing amount of research material that analyses the psychological component of this segment of corporations. However, Smyrnios et al. (1998) note that the family business literature lacks 'theoretical and empirical integration' (p. 50) (see also, Chrisman et al., 2003). When conceiving the global framework for family business systems, many disciplines including psychology, sociology, economics, law, organizational theory, organizational behavior, social issues, policy, and general systems need to be considered (Wortman, 1994). Though several conceptualization attempts have been made, like family systems theories, there is not an overall accepted model of family business dynamics.

In an attempt to simplify the dynamics of family businesses, Swartz (1989) uses a dual systems model. He believed that the business and the family were separate, but overlapping systems. Countering this, Whiteside and Brown (1991) argue that family businesses are more complex than dual relationships between family members who are also coworkers. They propose that the family firm is a single entity that is a hybrid between a family and a business. Whiteside and Brown (1991) believe that to understand the dimensions of a family business, one must not only be versed in the concepts relating to family business, but must be aware of the unique interaction of the entire family business system. Whiteside and Brown suggest that separating family and business entities can lead to three errors: 'a stereotyping of subsystem functioning, inconsistent and inadequate analysis of interpersonal dynamics, and exaggerated notions of subsystem boundaries and an under analysis of whole system characteristics' (1991, p. 386).

In other research, Poza et al. (1997) contend that 'relationships found between elements of family culture and management practices suggest the utility of approaches that address the whole system' (p. 140) which again suggests that family businesses are a unique mix of systems. Gersick et al. (1997) add another level of complexity to the relationship complexities of family firms. These researchers identify three main players in family business: owners, family members, and employees. An individual may fall into one, two or all three of these roles. Those engaged in family businesses can then be categorized into seven levels of involvement, and each individual interacts with others who have the same, or varying levels of involvement. Gersick et al. believe that in understanding the levels of individual involvement, a leader may more easily sort out personal versus professional differences and conflicts.

Using any model, family business consultants with a psychosocial approach will argue that, when family problems are behind family conflicts, even the best traditional legal, financial or management advice can be difficult to implement. Only once family issues are explored and diffused, can a family firm have the skills needed to move ahead. For example, Davidow, a licensed psychologist, and Narva, a lawyer, worked together in family business consulting (Malkin, 1991). In an interview on their consulting style, Malkin discovered that these consultants admitted to regularly accessing family dynamics while attempting to improve the overall functioning of the business system. Davidow and Narva believed and argued that the functionality of the business needed to be fully explored on many levels. The use of specialized family business consultants is becoming more common as family business look to increase their competitive advantages by first examining and solving the underlying family business system conflicts.

Family systems and family business dynamics

If family dysfunction underlies many family business struggles, proper application of family system theories should improve functionality of the family and the family firm (Deacon, 1996; Kaslow, 1993). As family business consultants, Paul (1995) and Kaslow (1993) applied contextual family system theories to family business processes. The language of contextual theory, which includes the terms ledger, bookkeeping, loyalty, justice, fairness, hierarchy, and merit, fits nicely with business linguistics. This fit grants family business members without psychological aptitude the ability to both understand and resonate with therapy-oriented theory.

Boszormenyi-Nagy, a founder of family therapy, formulated contextual theory. Boszormenyi-Nagy and Krasner (1986) use an integrative stance when they wrote that a 'truly comprehensive grasp of human existence is inevitably composed of both individual and relational realities' (p. 7). Boszormenyi-Nagy's contextual approach takes into account one's 'psychology, transactional patterns and responsibilities, and how they intertwine' (p. 8). Contextual based therapists must work to uncover and understand the unique psychological realities of each family member. Contextual therapists are also person centered, and they attempt to join with each family member by verbalizing active concern for every family member's view.

Paul (1995) argues that contextual theory 'probes the most fundamental dimensions of human relationships by addressing concepts such as intergenerationally derived entitlement, fairness, loyalty, and family obligations in a way that no other theory does' (p. 4). He found that using these concepts of contextual family therapy brings process to family business assessment and consulting. Kaslow (1993) also suggests that contextual theory focuses on 'multigenerational issues about power and control' (p. 8) and is directly applicable to family business succession issues. Both Paul and Kaslow provide anecdotal examples of the application of family business theories to family business situations. However, despite significant advances in recent years, the study of family business dynamics, and applicable family and systems theory is in its infancy.

Challenges facing family businesses

Kaslow (1993) believes that 'providing for one's family and being connected to them financially as well as emotionally is a major motivating factor for most adults' (p. 4). To many, family businesses allow individuals to intertwine both career and family aspirations. However, Kaslow and other researches recognize that many circumstances inhibit the family business and the family from functioning in a healthy manner. Paul (1996) argues that the greatest threat to family businesses is not out in the marketplace, but within the family business itself. Rosenblatt et al. (1985) support this and found that 90 per cent of the 59 firms surveyed reported stress related to managing family business relationships. Other researchers have uncovered trends that underlie this stress. Some of the issues that these researchers highlight as problematic for family firms include: governance structures, succession planning, under-representation of family members, adding outsiders to the system, and problems related to conflict and conflict resolution.

Governance structures

The governance, or power structure, of a family business may take on a variety of forms. The terms authoritarian, democratic, dictatorial, and laissez-faire are all potential descriptions of how power may be utilized and distributed on other members in the family firm system (Deacon, 1996). Clear governance structures illustrate how a company's values help to guide a firm's decisions, but family firm governance structures are often clouded by multiple roles created by family and family business. Governance initiatives are more likely to be successful once there is a clear understanding of all the multiple levels of perspectives of individuals involved in the decision-making process (Magretta, 1998).

Succession planning

Succession planning is a crucial part of planning for, and ensuring the survival of, the family business. Many family firms are unable to survive generational succession intact. For example, Paul (1996) reviewed the results of a study which indicated that the death of the founder was the catalyst for the downfall in 78 per cent of family business failures, and when the founder lives to retirement, it is often difficult for the founder to turn over the reign of leadership (see also, Davis and Harveston, 1999; Moores and Barrett, 2003). McClendon and Kadis (1991) report that only 30 per cent of family businesses succeed in the second generation. For this reason, articles related to succession and succession planning are common in family business literature, and were over-represented in the early family business research agenda.

Kaslow (1993) focuses on problems intertwined with family power and loyalty in successor planning. He believes that most families involved in family businesses could benefit from family therapy interventions and argues that interventions often assist families in moving through succession planning by creating awareness around problematic family and business patterns that could make the process unsuccessful. Kaslow highlights many potential obstacles in succession planning. One obstacle relates to the representation of family members, however, this obstacle's impact is not simply limited to difficulties in succession planning.

Under-representation of family members

The under-representation of family members encompasses issues related to family members who choose to not participate as an employee of the family firm, or who the family firm decisions-makers do not allow to participate. Often, gender-related decisions impact who participates, and who does not, and the literature highlights this trend. Kirschner (1992), and Galiano and Vinturella (1995) review the problems of families who ignore or under-represent the contributions of women family members in business operations. Kirschner (1992) provides two specific case examples where, for the perceived good of the family firm, one child, a daughter, is cut out of family functioning. Acting as a family business consultant, Kirschner was able to both assist in reunifying the split family and improving the related dynamics in the firm. After surveying 10 women in responsible positions in family firms, Galiano and Vinturella (1985) reported another type of gender bias. They concluded that a glass ceiling often exists for female family members wishing to succeed in their family's firms.

The role of outsiders

Just as all family members may not participate in the family firm, all family firm participants may not be family members. The term 'outsiders' is used to describe non-family members who hold positions within a closely held business. Such a term carries strong meaning as it highlights that in many family firms one only truly belongs if one is part of the family system. Despite the exclusionary term, Schwartz and Barnes (1991) argue that family firms are more successful when their governing boards include at least some non-family members. These researchers suggest that non-family board members offer the most help with unbiased perspectives, accountability of management, contact networks, challenging questions, compensation for executives, and long-term perspectives. Debate and conflict continue about the role and acceptance of 'outsiders' in family firms.

Conflict and conflict resolution

In order to succeed, all organizations have a need to identify and resolve conflict. Without conflict resolution skills, functioning systems can fall into chaos. Within family firms, the importance of conflict resolution is heightened as 'family firms are fertile fields for conflict' (Harvey and Evans, 1994, p. 331). And without clear boundaries, business conflicts can easily be channeled into family issues, and vice versa.

Harvey and Evans (1994), Kaye (1991) and Paul (1996) focus on the complexities of sustained conflict in a family business system. Paul reports that family business successors (second generation business leaders) indicate that conflict between family members, or between family member and non-family employees, was the second major source of family business failure. Ylvisaker (1990) agrees that family business 'not only invites an intensifying stress on tensions already evident in the family but intrudes on the differences that arise from the subjectivity of determining life paths' (p. 336). Kets de Vries (1996) summarizes five 'bad news' scenarios that lead to conflictual situations in family firms: incompetent family members chosen to run the business, 'spoiled kid syndrome', increased marital conflict, entrepreneurial aggression, and a tendency towards autocratic rule. Each of these situations underscore the need for positive conflict resolution in problematic interactions in both business and family relationships.

Jaffe (1990) highlights the challenge of managing dual work and family relationships: 'family involvement adds several layers of complexity to a business' (p. 25). He addresses several complexities as they relate to family, marital, parental and employee relationships, communication, and succession. Underscoring the discussion, Jaffe argues that healthy families with healthy businesses are assured a secure future.

This list of family business challenges certainly is not complete. As research continues, more challenges facing family businesses are uncovered and dissected. However, in looking at family business assessment, it is important to understand the base of literature that is currently available. The Aspen Family Business Inventory is one of the first instruments that attempts to bring together the potential challenges facing family firms with the goal of increasing the overall systemic understanding of the family and the family firm.

The Aspen Family Business Inventory

Jaffe (1996) and Paul (1996), working with several other family business consultants, developed the Aspen Family Business Inventory (AFBI), (formerly the Family Business Assessment Inventory). This assessment is based on their and others' experience as family business consultants, reviews of the family business literature, and concepts of contextual therapy.

The AFBI is built on the concept that family businesses are a single hybrid entity. This model suggests that the family and the business interact to form a unique hybrid system. Paul (1996) concedes that this model can become more complex with the addition of outsider employee and management participation, however, this assessment is not designed for use with those who are not part of the family system.

The authors of the AFBI agree that traditional business assessments and traditional family assessments do not effectively capture the dimensions of family firm hybrid system. The AFBI was not designed to test anything about an individual family member. Instead, the individual results are compared with other family member's results. Similarities and differences in responses provide valuable information about individual and family perceptions. This is consistent with Boszormenyi-Nagy and Krasner (1986), who used an integrative stance when they wrote that a 'truly comprehensive grasp of human existence is inevitably composed of both individual and relational realities' (p. 7). The AFBI seeks to uncover these realities and determine how they are both similar and different. Then this information can be used to understand and improve problematic relationships.

Jaffe (1996) and Paul (1996) agree that exploring individual perceptions with all family members, including those not employed in the family business, often leads to improved family relations. Based on contextual theory, the AFBI allows family businesses to explore individual and system perceptions of all family members.

The AFBI is composed of 10 scales with 10 questions each. The first five scales focus on how the family manages its personal relationships, and the next five address how the family manages its relationships in business. A five-point Likert scale is used for all responses that vary from strongly disagree to strongly agree (see Appendix).

Instrument reliability and validity

Any instrument without proven validity and reliability offers questionable results (Gregory, 1996). An instrument that is reliable demonstrates consistency over time. An instrument that is valid provides inferences that are appropriate, meaningful and useful (APA, 1999). The APA *Standards for Educational and Psychological Testing* (1999) state that validity is 'the most important consideration in test evaluation' (p. 9), and psychometric testing should be backed up with data on criterion, construct, and content-related validity. Criterion-related validity refers to an instrument's ability to relate meaningfully with other related and measurable elements. Construct validity is a measure of how well an instrument's items capture the domain of the underlying construct. Content validity is 'determined by the degree to which the question, tasks or items on a test are representative of the universe the test was designed to sample' (Gregory, 1996, p. 108). Devellis (1995) and Haynes et al. (1995) concur that content validity concerns itself with the extent to which a certain set of items reflect a content domain. These definitions give rise to the question of how one can operationalize content, domain and universe of a test in a way that measures instrument validity. Others have also struggled with this question. Validity research reveals that there are no widely accepted forms of determining content validity (Gregory, 1996; Lawshe, 1975). Litwin (1995) recommends that a content validity study be organized to ensure that a survey's contents are complete. One of the most common procedures for determining content validity involves surveying experts about an instrument's relevance to the specific field (Gregory, 1996; Haynes et al., 1995; Lawshe, 1975; Litwin, 1995; Martuza, 1977; Post, 1991; Tilden et al., 1990). Though content validity assessments are important in instrument development, Litwin (1995) cautions that content validity assessments are not meant to be scientific measures. He suggests that content validity assessment does not always require statistical analysis. Other researchers recommend the calculation of content validity ratios or indexes (Gregory, 1996; Lawshe, 1975; Topf, 1986; Waltz et al., 1994).

Instrument development

Gregory (1996) suggests that 'creating a new (instrument) involves both science and art' (p. 129). The American Psychological Association (APA, 1999) focuses on the science of instrument development by providing 25 standards for test instrument development. The standards covered item provisions, item selection, and evaluation of tests for specific purposes. Gregory concurs and highlights the many stages of instrument development, while stressing the importance of defining the purpose, selecting the scale, constructing the items, testing the items, and revising the test as required.

In defining the purpose of the test, a developer must start with a 'hypothesis that implies a domain of content' (Nunnally and Bernstein, 1994, p. 292), and then link concepts and items appropriately (Tilden et al., 1990). Ideally, a focused test definition will result in a clearer picture of the domain of content. Gregory (1996) and Nunnally and Bernstein (1994) agree that in isolating the domain of content, the test developers need to consider many factors including who the test will be directed towards, what the instrument is intended to measure, and how the test will differ from other available inventories.

Item selection procedures vary depending on the nature of the scale's purpose. The APA (1999) recommends that test developers tailor questions to suit the intended population of users. Nunnally and Bernstein (1994), Gregory (1996) and Klakovich (1995) emphasize the importance of creating clear items suitable for the scale. Gregory also raises questions about the homogeneity, range, number and type of items that should be used.

Once items are created, they need to be tested to determine if they are successful in tapping the expected domains. Assessing the reliability and validity of the items accomplishes this. Finally, after the initial assessment is complete, any problematic items should be removed or improved in the next version of the instrument.

Method

Procedure

With the goal of the study being to collect and calculate content validity ratios of the AFBI scales, and the AFBI's individual item's relevance, fit and clarity, 100 family business consultants were mailed a package containing a letter, a copy of the Aspen Family Business Inventory, a questionnaire sheet, and a stamped return envelope. The letter asked the potential subjects to complete a five-page questionnaire about the items appearing on the AFBI. A content validity questionnaire was developed with a focus on content relevance, conceptual fit, and clarity. These areas have been proposed by several content validity researchers (Imle and Atwood, 1988; Messick, 1980; Tilden et al., 1990). Potential subjects were assured that their responses would be kept confidential.

Three weeks after the initial mailing, all potential participants were called and reminded that there was still time to return the response sheet.

Subjects

All participants were members of the Family Firm Institute, a national organization based in the United States with family-firm specialists in a number of fields. Expert participants were chosen from the Family Firm Institute directory. The participants are termed experts on the basis that they are all consultants to the world of family business and, in their related fields, they all have several years experience working with family businesses. The Family Firm Institute directory lists experts in several fields of family business including: insurance, finance, psychosocial, law, and management. Twenty specialists were randomly selected from each of these fields. From the 100 surveys sent, there was a 19 per cent response rate. The family 15 male and 4 female business experts who participated came from a diverse field of specialties. The group included two lawyers, four accountants, four psychologists/therapists, eight business consultants and an insurance agent – all of whom work with family businesses. Several participants are published authors in the family business literature with experience working with family business varying from 1 to 10 years (five respondents), 11 to 20 years (eight respondents), five reported more than 20 years and one respondent did not answer this question. Seven respondents had college degrees, five had master's-level degrees, five had doctoral degrees and two elected not to answer this question.

Results

Nineteen family business experts returned useable response sheets. Content validity was calculated using two methods: Lawshe's (1975) individual item method, and Gregory's (1996) overall assessment method.

Using Lawshe's quantitative approach to content validity, the content validity ratio (CVR) was calculated for relevancy, fit and clarity of each item in the AFBI.

$$\text{CVR} = \frac{(na) - (N/2)}{N/2}$$

where *na* is the number of experts who agree or strongly agree, and *N* is the total number of experts participating. Using this formula, when less than 50 per cent of the experts agree with an item, the CVR will be negative. When all experts agree the CVR is 1.00. According to Lawshe (1975), items with CVRs below 0.49 are not considered to have acceptable content validity. This represents a 75 per cent level of agreement among subjects. Other content validity researchers have suggested a more stringent 80 per cent agreement (Post, 1991).

The inventory's overall content validity was adapted from a formula presented by Gregory (1996) and Topf (1986):

$$\text{Overall content validity}(\%) = \frac{B}{(A+B)}$$

where *B* is the number of items found to be have acceptable content validity, and $A + B$ represents all the items on the instrument. See the appendix for comprehensive results.

Relevance

Scale Every scale's overall CVR was over 0.79, which means agreement on each scale's relevance was 90 per cent or greater. Though all CVR scale relevance scores were within the acceptable range, the Business of the Family sections received significantly lower CVR scores than the Business of the Business sections ($t = 3.797$, $p < 0.001$).

Item With few exceptions, the relevance of individual items was found to be quite good. All individual items in Scales 3, 4, 6, 7, 8, 9 and 10 scored CVRs higher than 0.60, or 80 per cent agreement. Looking at all items in the acceptable range (above 75 per cent agreement), only 8 scored less than a CVR of 0.60, or less than 80 per cent agreement. All of the lower scoring items were found in Scales 1, 2 and 5.

Items in Scale 2 (Quality of Family Life) scored the lowest content validity. Four items (5, 6, 7, and 10) had lower than acceptable content validity scores, and two of these items (7 and 10) also had lower than acceptable scores for fit in that scale.

Fit

Of the 100 items, 96 received acceptable CVR rating for fit in the category. Seven items received scores above 75 per cent, but less than 80 per cent. Two items in Scale 2 (7 and 10), one item in Scale 1 (8), and one item in Scale 4 (4) had lower than acceptable scores.

Clarity

Three of the items did not score within the acceptable range for the CVR for clarity. Six items were at the lower end of acceptability. Five of the unacceptable or lower range of acceptable items were found in Scale 1.

Overall assessment

At the end of the response sheet, subjects were given the opportunity to share written comments about the assessment. Twelve participants chose to do this, and 10 made suggestions for additional scales and/or individual items. In this section, all respondents were asked if the response sheet directions were clear. All respondents indicated that they were.

Written response from the 'Overall assessment' section provided more insight into the thoughts of the family business experts. Participants offered a variety of suggestions, additions and critiques. Scale ideas that respondents indicated were missing from the assessment included: ownership and estate and family board council meeting. Additional items were requested on topics relating to individuals' desire to participate and future participation, roles of non-family business executives, planning and succession issues for non-family business members, career planning and development. Other respondents requested more specific questions on conflict resolution and on family business boundaries.

Further suggestions were related to question wording and assessment style. A participant suggested that items with the word 'all' should instead include the word 'most'. Another suggested that a 'not applicable' answer option should be added, while yet another recommended that paragraph-style questions should supplement the Likert scale items.

Discussion

Overall, the AFBI scored very high in terms of overall category relevance, item relevance, item fit and item clarity. There are several possible explanations for such a positive response. First, the developers of the inventory have a combined experience in the field that totals over 60 years, and each could be described as a pioneer of the family firm field. This experience prepared them to format a comprehensive assessment for this field. Second, there are few available instruments designed for use with family businesses. Several family business experts indicated that they were intrigued with simply the concept of the AFBI. This may have skewed their answers favorably. Without strong theoretical constructs, participants, though experts in the field, may lack clear theoretical frameworks against which to assess the AFBI. And, since there is little written information on family business assessment, the AFBI may have brought expert participants new clarity to their understanding of their field, which, in turn, resulted in high CVR results. Finally, it should be noted that subjects who chose to respond might have done so because they were interested in, and had a positive response to, seeing a copy of the AFBI.

Even with the overwhelming positive response, the validity assessment did produce some interesting findings and trends. The AFBI is divided into two sections: the Business of the Family (Scales 1–5), and the Business of the Business (Scales 6–10). Interestingly, the Business of the Family scale relevance scores were lower than the Business of the Business scales, and all of the Business of the Business items was found to have CVR relevance ratings well above acceptable levels. Seventeen of the 50 questions in the Business of the Family scales had CVR below 0.60, or 80 per cent agreement, with seven of these items' relevance falling into unacceptable levels. This produced a significant difference between the relevance scale CVR scores for Section I versus Section II. This may suggest that family business professionals are able to find more agreement around business operating related issues. Clearly, the family business consultants in this sample found more agreement in the business rather than the family sphere.

Items that scored the lowest CVRs were only tangentially related to both the relationship between family members, and the business related issues. These included five questions from Scale 2 (Quality of Family Life):

5. Our family is active in the community.
6. Everyone is active in fitness and caring for one's health.
8. Family members have outside hobbies and interests.
9. Family gatherings are fun and go well.
10. Family members are involved in charitable activities.

Other items in the same scale that reflected more directly on relationships between family members had higher, and acceptable, CVRs. These results suggest that family business experts place more importance on issues that are more directly related to relationships, or business functioning.

Though there are too few subjects to make significant observations, further analysis of the professional focus of each of the respondents indicated that subgroup family business expert participants with careers in accounting, investment services, or finance were more likely to disagree about the relevance of the unacceptable items in Scale 2. In a larger study it would be possible to make more comparisons between the responses of the subgroups of participants.

The survey responses suggest that, with only a few exceptions, a majority of the experts agreed that most items fit their categories. Two of the items that scored below acceptable CVR ratings for relevance, also scored below CVR rating for fit. These items were both in Scale 2:

6. Everyone is active in fitness and caring for their health.
7. Our family has activities where we all learn together.

The developers of the scale need to evaluate the place of these items in the AFBI. Though an average person may agree that caring for one's health, and families learning together are good things, the experts clearly questioned whether these issues serious impact family business functioning.

For the most part, the AFBI received acceptable CVR ratings for clarity. All items were composed in the one direction meaning high scores produce more favorable results. However, this did not appear to affect the participants' ratings of clarity. The nine items that received lower that 80 per cent agreement on clarity should be reviewed, with a special focus on:

Scale 1

5. Love and affection is shown equally to all children and grandchildren.
8. Family members are not jealous of what other family member have.

Scale 3

6. Our family has clear and separate processes for making decisions about ownership, management and family issues.

There were no items that scored below acceptable CVR levels in all categories (relevance, fit and clarity). This would suggest that experts did not give unacceptable ratings to an item simply because they did not understand the meaning of the item.

The authors of the AFBI clearly divide the inventory into two sections: the Business of the Business, and the Business of the Family. One of the instrument's developers and several other family business researchers have conceded that family business dynamics encompass more elements, especially as the business grows into the successor stage (Gersick et al., 1997; Paul, 1996). To be a successful inventory for all the dynamics of family business, perhaps an entire section should be devoted to the 'Business of the Outsiders'.

The experts also provided interesting input for the overall AFBI. In this section, expert subjects were given full range to express their opinions about this assessment inventory. The area that the AFBI does not cover, and that both the expert subjects and the literature suggest that it should, is the role of non-family member participating in the business. An expansion of questions designed for non-family business participants would enhance the overall assessment's ability to capture the beliefs of all family and business key players.

Limitations

Litwin (1995) recommends that content validity studies include both the review of experts, and review by those for whom the scale is intended. This content validity study would be strengthened by the addition of content validity assessment by members of family businesses. Only 21 per cent of the experts who completed the survey were women. An equal number of male and female participants would have eliminated possible gender bias. Robinson and Phillips (1995) recommend that content validity questionnaires add items not included in the original assessment to see if experts can distinguish between items that are included versus not included. To gain more credibility the AFBI should also be further evaluated for reliability, and construct and criterion-related validity. Traditionally, content validity studies include limited subjects and this study would have been improved had more family business experts participated.

Recommendations

Content validity assessment is the first step in determining the validity of a new instrument. The AFBI faired well in this initial content validity analysis. Future studies looking at other aspects of validity and reliability are under way, including confirmatory factor analysis and internal consistency assessment of the AFBI's items.

Conclusion

This project offers many opportunities for future research in the field of family business. The results from the preliminary content analysis of the Aspen Family Business Inventory suggest that this instrument satisfactorily covers many relevant elements of family business assessment. Moreover, family business consultants and researchers are encouraged to use techniques like those introduced in this project as a first step in establishing the validity of the instruments that they use in working with families in business.

References

American Psychological Association (APA) (1999), *Standards for Educational and Psychological Testing*, Washington, DC: APA.

Beavers, W. and R. Hampson (1990), *Successful Families: Assessment and Intervention*, New York: Norton.

Boszormenyi-Nagy, I. and B.R. Krasner (1986), *Between Give and Take*, New York: Brunner/Mazel.

Bowen, M. (1978), *Family Therapy in Clinical Practice*, New York: Aronson.

Cheston, S.E. (2000), 'A new paradigm for teaching counselling theory and practice', *Counsellor Education and Supervision*, **39**, 254–69.

Chrisman, J.J., J.H. Chua and P. Sharma (2003), 'Current trends and future directions in family business management studies: toward a theory of the family firm', *Coleman White Paper*, www.usasbe.org/knowledge/whitepapers.

Davis, P.S. and P.D. Harveston (1999), 'In the founder's shadow: conflict in the family firm', *Family Business Review*, **12**(4), 311–24.

De Shazer, R. (1985), *Key Solutions to Brief Therapy*, New York: W.W.Norton.

Deacon, S.A. (1996), 'Utilizing structural family therapy and systems theory in the business world', *Contemporary Family Therapy*, **18** (4), 549–65.

Devellis, R.F. (1995), *Scale Development: Theory and Applications*, Newbury Park, CA: Sage Publications.

Doherty, W.J., M.E. Lester and G.K. Leigh (1986), 'Marriage encounter weekends: couples who win and couples who lose', *Journal of Marital and Family Therapy*, **12**(1), 49–61.

Fisher, L. and D. Ransom (1995), 'An empirically derived typology of families: relationships with adult health', *Family Process*, **34**, 161–82.

Friedman, M., W. McDermut, D. Solomon, C. Ryan, G. Keitner and I. Miller (1997), 'Family functioning and mental illness: a comparison of psychiatric and nonclinical families', *Family Process*, **36**, 357–67.

Galiano, A.M. and J.B. Vinturella (1995), 'Implication of gender bias in the family business', *Family Business Review*, **8**(3), 177–88.

Gersick, K.E., J.A. Davis, M.M. Hampton and I. Lansberg (1997), *Generation to Generation: Life Cycles of the Family Business*, Cambridge, MA: Harvard Business School Press.

Giblin, P. (1986), 'Research and assessment in marriage and family enrichment: a meta-analysis study', *Journal of Psychotherapy and the Family*, **2**(1), 79–96.

Gregory, R.J. (1996), *Psychological Testing*, Boston, MA: Allyn and Bacon.

Haley, J. (1973), *Uncommon Therapy*, New York: Norton.

Harvey, D. and J. Bray (1991), 'Evaluation of an intergenerational theory of personal development: family process determinants of psychological health distress', *Journal of Family Psychology*, **4**, 298–325.

Harvey, M. and R.E. Evans (1994), 'Family business and multiple levels of conflict', *Family Business Review*, 7(4), 331–48.

Haynes, S.N., D.C.S. Richard and E.S. Kubany (1995), 'Content validity in psychological assessment: a functional approach to concepts and methods', *Psychological Assessment*, **7**(3), 238–47.

Imle, M.A. and J.R. Atwood (1988), 'Retaining quantitative validity while gaining reliability and validity: development of the transitions to parenthood concerns scales', *Advances in Nursing Science*, **11**(1), 61–75.

Jaffe, D.T. (1990), *Working With the Ones You Love*, Berkeley, CA: Conari Press.

Jaffe, D.T. (1996), *The Family Business Assessment Inventory*, San Francisco, CA: Changeworks Solutions.

Kaslow, F.W. (1993), 'The lore of family businesses', *American Journal of Family Therapy*, **21**(1), 3–16.

Kaye, K. (1991), 'Penetrating the cycle of sustained conflict', *Family Business Review*, **6**(1), 21–44.

Kets de Vries, M.F.R. (1996), 'The dynamics of family controlled firms: the good and the bad news', in C. Aronoff, J. Astrachan and J. Ward (eds), *Family Business Sourcebook II*, (pp. 312–23), Marietta, GA: Business Owners Resources.

Kirschner, S. (1992), 'The myth of the sacrifice of the daughter: implications in family owned business', *The American Journal of Family Therapy*, **20**(1), 13–24.
Klakovich, M. (1995), 'Development and psychometric evaluation of the Reciprocal Empowerment Scale', *Journal of Nursing Measurement*, **3**(2), 127–43.
Lawshe, C.H. (1975), 'A quantitative approach to content validity', *Personnel Psychology*, **28**, 563–75.
Litwin, M.S. (1995), *How to Measure Survey Reliability and Validity*, Thousand Oaks, CA: Sage Publications.
Magretta, J. (1998), 'Governing family owned enterprise: an interview with Finland's Krister Alhstrom', *Harvard Business Review*, January–February, 112–23.
Malkin, R. (1991), 'Integrating family and business perspectives: a conversation with Tom Davidow and Richard Narva', *Family Business Review*, **4**(4), 433–44.
Martuza, V.R. (1977), *Applying Norm Referenced and Criterion Referenced Measurement in Education*, Boston, MA: Allyn and Bacon.
McClendon, R. and L.B. Kadis (1991), 'Family therapists and the family business: a view of the future', *Contemporary Family Therapy*, **13**(6), 641–51.
Messick, S. (1980), 'Test validity and the ethics of assessment', *American Psychologist*, **35**, 1012–27.
Minuchen, S. (1974), *Families and Family Therapy*, Cambridge, MA: Harvard University Press.
Moores, K.J. and M. Barrett (2003), *Learning Family Business: Paradox and Pathways*, London: Ashgate Publishing.
Nunnally, J.C. and I.H. Bernstein (1994), *Psychometric Theory*, 3rd edn, New York: McGraw-Hill.
Paul, J.J. (1995), 'Contextual consultation', unpublished manuscript.
Paul, J.J. (1996), 'Family business survival', *Blueprint for Business Success*, **16**, 1–5.
Post, M.T. (1991), 'Content validity of the Family Systems Stressors Inventory', unpublished manuscript.
Poza, E.J., T. Alfred and A. Maheshwari (1997), 'Stakeholder perceptions of culture and management: practices in family and family firms', *Family Business Review*, **10**(2), 135–55.
Robinson, B.E. and B. Phillips (1995), 'Measuring workaholism: content validity of the Work Addiction Risk Test', *Psychological Reports*, **77**, 657–8.
Rosenblatt, P.C., L. deMik, R.M. Anderson and P.A. Johnson (1985), *The Family in Family Business: Understanding and Dealing with the Challenges Entrepreneurial Families Face*, San Francisco, CA: Jossey-Bass.
Schwartz, M.A. and L.B. Barnes (1991), 'Outside boards and family buinesses: another look', *Family Business Review*, **4**(3), 269–86.
Smyrnios, K., G. Tanewski and C. Romano (1998), 'Development of a measure of the characteristics of family business', *Family Business Review*, **11**(1), 49–60.
Swartz, S. (1989), 'The challenges of multidisciplinary consulting to family owned-businesses', *Family Business Review*, **2**(4), 41–54.
Thomas, M.B. (1992), *An Introduction to Marital and Family Therapy*, New York: Macmillan.
Tilden, V.P., C.A. Nelson and B.A. May (1990), 'Use of quantifiable methods to enhance content validity', *Nursing Research*, **39**, 172–5.
Topf, M. (1986), 'Three estimates of interrater reliability for nominal data', *Nursing Research*, **35**, 253–5.
Waltz, C.F., O.L. Strickland and E.R. Lenz (1994), *Measurement in Nursing Research*, Philadelphia, PA: F.A. Davis.
Whiteside, M.F. and F.H. Brown (1991), 'Drawbacks of the dual system approach to family firms: can we expand our thinking?', *Family Business Review*, **4**(4), 383–95.
Wortman, M.S. (1994), 'Theoretical foundations for family owned business: a conceptual and research based paradigm', *Family Business Review*, **7**(1), 3–27.
Ylvisaker, P.N. (1990), 'Family foundations: high risk, high reward', *Family Business Review*, **3**(4), 331–6.

Appendix: Content validity ratio and percentage agreement of AFBI scales and items

	Relevance		Fit		Clarity	
	CVR	%	CVR	%	CVR	%
Part I: The Business of the Family						
Scale 1: Trust, Fairness and Family Connecting	**0.90**	**95**	**n/a**	**n/a**	**n/a**	**n/a**
1. People in the family trust each others' motives and intentions.	0.90	95	0.67	83	0.78	89
2. Our family gatherings are an emotionally safe place to be.	0.68	84	0.78	89	0.56	78
3. In our family we are open and honest with one another.	0.79	90	0.67	83	0.78	89
4. Past conflicts have been settled without buildup of ongoing resentment or negative feelings.	0.68	84	0.67	83	0.56	78
5. Love and affection is shown equally to all children and grandchildren.	0.58	79	0.56	78	0.44	72
6. We openly express affection for one another.	0.47	74	0.56	78	0.89	94
7. In-laws are fully accepted and feel part of the family.	0.79	90	0.56	78	0.78	89
8. Family members are not jealous of what other family members have.	0.58	79	0.44	72	0.44	72
9. The younger generation of our family seem to be acquiring strong values from the older generation.	1.00	100	0.67	83	0.56	78
10. I trust all of my family members in matters of business.	0.89	95	0.99	100	0.89	94
Scale 2: Quality of Family Life	**0.79**	**90**	**n/a**	**n/a**	**n/a**	**n/a**
1. Family member truly care about each other.	0.79	90	0.68	83	0.64	82
2. Family members clearly enjoy being with each other.	0.79	90	0.78	89	0.88	94
3. Alcohol or substance abuse is not a problem in our family.	0.99	100	0.56	78	0.88	94
4. The family spends time together relaxing in non-business activities.	0.58	79	0.78	89	0.88	94
5. Our family is active in the community.	0.37	68	0.67	83	0.88	94
6. Everyone is active in fitness and caring for their health.	0.37	68	0.44	72	0.77	88
7. Our family has activities where we all learn together.	0.37	68	0.33	67	0.67	83
8. Family members have outside hobbies and interests.	0.58	79	0.67	83	0.77	88
9. Family gatherings are fun and go well.	0.47	74	0.65	82	0.89	94
10. Family members are involved in charitable activities.	0.26	63	0.89	94	0.77	88
Scale 3: Communication and Resolving Conflict	**0.90**	**97**	**n/a**	**n/a**	**n/a**	**n/a**
1. We share our dreams and visions for the future one another.	0.79	95	0.67	83	0.78	89
2. There is a free and open flow of information in the family.	0.90	95	0.89	94	0.55	78

	Relevance		Fit		Clarity	
	CVR	%	CVR	%	CVR	%
3. We communicate well with each other about what we want from the business and the family.	0.79	90	0.89	94	0.78	89
4. We are able to resolve major conflicts and differences with one another.	0.90	95	0.89	94	0.78	89
5. When a family member has a problem with another family member, he or she speaks to that person directly.	0.90	95	0.89	94	0.78	89
6. Our family has clear and separate processes for making decisions about ownership, management and family issues.	0.90	95	0.78	89	0.44	53
7. We can communicate openly about sensitive or uncomfortable issues.	0.79	90	0.89	94	0.78	89
8. We are willing to share bad news.	0.90	95	0.89	94	0.78	89
9. We are responsive to one another's concerns and feelings.	0.90	95	0.89	94	0.67	83
10. We listen to one another.	0.90	95	0.89	94	0.78	89
Scale 4: Balancing Self and Family Interests	**0.90**	**95**	**n/a**	**n/a**	**n/a**	**n/a**
1. Our family shares a sense of purpose that guides our lives.	0.58	79	0.56	78	0.78	89
2. Family members respect each other's privacy.	0.90	95	0.89	94	0.89	94
3. We are tolerant of differences in beliefs and opinions within the family.	0.90	95	0.89	94	0.89	94
4. We have regular family meetings to discuss issues that are important to the family.	0.68	90	0.44	72	0.78	89
5. We do not try to achieve success at another family member's expense.	0.79	90	0.67	83	0.89	94
6. Our family is highly respected in our community.	0.58	79	0.56	78	0.89	94
7. The family provides members with adequate material and emotional resources for their future.	0.58	79	0.78	89	0.67	83
8. We encourage family members to be self reliant.	0.79	90	0.89	94	0.78	89
9. We are as supportive to family members who choose careers outside the business as we are to those we work in the business.	0.90	95	0.89	94	0.89	94
10. Our family has produced psychologically healthy and productive people.	0.58	79	0.66	83	0.56	78
Scale 5: Individual Growth and Development	**0.90**	**94**	**n/a**	**n/a**	**n/a**	**n/a**
1. My family gives me credit for my personal accomplishments and milestones in life.	0.79	90	0.77	88	0.99	100
2. I am being adequately prepared for my future.	0.58	79	0.77	88	0.65	82
3. I know what I want my life to be about.	0.79	90	0.88	94	0.77	88
4. My family encourages me to develop a sense of purpose in life separate from the business.	0.65	82	0.88	94	0.75	88
5. My family has encouraged me to discover my own way.	0.67	83	0.65	82	0.77	88
6. I have been given credit for my contributions to the interests of the family.	0.65	82	0.88	94	0.77	88

	Relevance		Fit		Clarity	
	CVR	%	CVR	%	CVR	%
7. I feel confident about my future.	0.44	72	0.77	88	0.88	94
8. I am basically satisfied with the level of trust and fairness between me and the other family members.	0.99	100	0.50	75	0.65	82
9. I feel that my family understands me.	0.89	94	0.77	88	0.77	88
10. My family generally likes me for who I am.	0.56	78	0.77	88	0.77	88
	Part II:		*The Business of the Business*			
Scale 6: Business Direction and Planning	**0.90**	**95**	**n/a**	**n/a**	**n/a**	**n/a**
1. We have a strong and clear vision for the future of the business.	0.90	95	0.88	94	0.88	94
2. Our employees and customers know what our business stands for.	0.78	89	0.88	94	0.88	94
3. The family agrees on what our business stands for.	0.78	89	0.88	94	0.63	81
4. Our business is about more than making money.	0.90	95	0.77	83	0.63	81
5. Our family's values are in harmony with our business policies and operations.	0.78	89	0.88	94	0.99	100
6. We run our business like a business, with detailed financial reports, plans, clear roles and strategy.	0.90	95	0.88	94	0.77	88
7. As our business has grown, our profits have grown as well.	0.99	100	0.88	94	0.77	88
8. Income is fairly divided between investment in the future of the business, managers' compensation, and distribution to owners.	0.78	89	0.77	83	0.77	88
9. Regular business meetings are held to plan and review progress.	0.89	94	0.88	94	0.88	94
10. We have long and short range business plans, that the family understands and accepts.	0.79	90	0.88	94	0.88	94
Scale 7: Progressive Management	**0.90**	**95**	**n/a**	**n/a**	**n/a**	**n/a**
1. The head of the business doesn't need to be in control of everything.	0.99	100	0.88	94	0.77	88
2. The business hires and retains competent non-family member managers in responsible positions.	0.79	90	0.77	88	0.88	94
3. We listen to the new ideas from our younger generation of family managers.	0.68	89	0.88	94	0.88	94
4. We share company performance data and planning with non-family members.	0.90	95	0.88	94	0.88	94
5. Outside advisors meet with us regularly and are willing to give us 'bad news'.	0.79	90	0.88	94	0.77	88
6. Managers in the business feel comfortable raising difficult issues to the leaders.	0.90	95	0.88	94	0.88	94
7. Inquiry and innovation is encouraged in the company.	0.79	90	0.88	94	0.77	88
8. Roles and responsibilities in family and non-family managers are clear and people are accountable for results.	0.90	95	0.88	94	0.77	88

	Relevance		Fit		Clarity	
	CVR	%	CVR	%	CVR	%
9. We do a good job of delegating through the organization.	0.90	95	0.99	100	0.88	94
10. Employees and customers know what our business stands for.	0.90	95	0.88	94	0.77	88
Scale 8: Family Participation	**0.90**	**95**	**n/a**	**n/a**	**n/a**	**n/a**
1. Family members in our business would be successful in comparable jobs in other companies.	0.77	88	0.53	77	0.88	94
2. Family members in the business are paid at market value for their contributions to the company.	0.90	95	0.77	88	0.88	94
3. We have clear policies about how family members can become employed in the business, which are understood and followed.	0.90	95	0.88	94	0.88	94
4. We clearly and fairly evaluate the performance of family members working in the business.	0.90	95	0.88	94	0.88	94
5. Family employees know where they stand in the business, including both limits and opportunities.	0.99	100	0.99	100	0.88	94
6. Family members and in-laws are encouraged to continually learn about business and develop their leadership ability.	0.90	95	0.88	94	0.65	82
7. Heirs have opportunity to influence the future of the business.	0.79	90	0.77	88	0.65	82
8. Advancement and promotion are based on merit.	0.90	95	0.99	100	0.77	88
9. Heirs gain professional work experience elsewhere before they enter the business.	0.99	100	0.99	100	0.77	88
10. The family trains heirs in their responsibility to the business as owners.	0.79	90	0.88	94	0.77	88
Scale 9: Family Business Boundaries	**0.79**	**90**	**n/a**	**n/a**	**n/a**	**n/a**
1. Family policies concerning the business are fair to all.	0.79	90	0.75	88	0.75	88
2. Family perks are kept to a minimum and distributed equitably and legally.	0.90	90	0.75	88	0.75	88
3. Family managers and heirs share power and have input into the business.	0.68	84	0.63	81	0.63	81
4. Owners have a say in the overall direction of the business, but not in its everyday operations.	0.68	84	0.63	81	0.63	81
5. Women have equal opportunities to participate in the business.	0.79	90	0.63	81	0.88	94
6. There has been discussion and planning for the potential roles of heirs before they enter the business.	0.90	95	0.75	88	0.75	88
7. We have a written dividend policy.	0.79	90	0.75	88	0.88	94
8. Business decisions (hiring, pay) are not made in response to family members concerns.	0.68	84	0.75	88	0.75	88

	Relevance		Fit		Clarity	
	CVR	%	CVR	%	CVR	%
9. Family conflicts are resolved by the family, and do not influence business operations of decisions.	0.90	95	0.88	94	0.88	94
10. The Board listens and monitors family concerns and questions about the business.	0.79	90	0.75	88	0.50	75
Scale 10: Ownership and Management Continuity	**0.90**	**95**	**n/a**	**n/a**	**n/a**	**n/a**
1. There is a sense of development and direction in our company.	0.79	95	0.75	88	0.63	81
2. The company leaders are preparing the organization for the future.	0.90	95	0.88	94	0.63	81
3. We deal promptly with changes as they happen.	0.90	95	0.75	88	0.63	81
4. We have a written succession plan for the next generation in our business.	0.90	95	0.88	94	0.88	94
5. Potential successors are given fair opportunity to demonstrate their talents.	0.90	95	0.88	94	0.88	94
6. Heirs feel that plans for the future of the business are fair.	0.90	95	0.75	88	0.88	94
7. We have a buy-sell agreement that is well understood by the family.	0.79	85	0.75	88	0.88	94
8. The owners estate plan is understood and supported by the next generation.	0.90	95	0.63	81	0.88	94
9. Older family members in the business are responsive to younger members requests for planning and innovation.	0.90	95	0.99	100	0.88	94
10. The older generation has plans for their future activities and involvement after leaving active management.	0.90	95	0.99	100	0.88	94

PART IV

FAMILY BUSINESS THEMES IN FOCUS

13 Founder–successor's transition: a model of coherent value transmission paths

Ercilia García-Álvarez and Jordi López-Sintas

In this chapter we delineate a model that presents the different coherent options of value transmission and successor's socialization that facilitate family business continuity from first to second generation. Our findings are grounded on combined qualitative and quantitative techniques from an extensive research project of in-depth cross-case analysis. Thus, based on our results we now highlight some issues that families and practitioners should take into account to keep the coherence during the succession process. That is, professionals can assist families in preparing the continuity by (1) identifying family value systems, (2) analysing key variables at play in the family-business system, and (3) proposing a coherent option of continuity that both family and business can pursue. Consequently, families and practitioners can benefit from following the paths suggested here as they are coherent combinations among values and characteristics belonging to each family business and different successor's socialization processes that increase the probability of achieving a successful founder–successor's transition.

Introduction

Some years ago Hambrick and Mason (1984) pointed out that organizations are shaped by their top managers. In this vein the CEO's role has been described as the most powerful, representing the ultimate decision-maker and the person with absolute authority (Kesner and Sebora, 1994). Parties external to the firm are likely to view succession as a signal about the institution's future (Beatty and Zajac, 1987), this makes CEO succession a critical event for virtually every organization (Chaganty and Sambharya, 1987; Davis, 1968; Zald, 1969). If this is, the CEO's role in family business' succession will be even more dramatic because his or her decisions can affect both the family and the business spheres that are intimate interlocked in this kind of firm.

Researchers have shown the importance of CEO values in shaping the future of the organization (Hambrick and Mason, 1984; Hofstede, 1983, 1994; Hofstede et al., 1990). This occupant of key role has enduring effects on the organization owing to his or her ultimate power in deciding who will be the successor and to his or her influence in shaping the value structure of that successor (Vancil, 1987). The latter is especially true in family business when the potential successor is a descendant or a group of descendants of the founder, and the CEO is also the founder of the firm (Dumas, 1990; Handler, 1994).

In this chapter we focus our interest on the value structure that founders try to pass to their family successors. Next, we compare this value structure to that of the founder in order to look for a model that explains the match or mismatch between the founder's and successors' value structure. Finally, we discuss the implications of our results.

Literature review

Values and successors

Most of the literature in family business has shown that founders seek continuity of their business through next-generation family members: children first, followed by other family members (Corbetta and Montemerlo, 1999; Iannarelli, 1992; Kets de Vries, 1993; Llano and Olguin, 1986) and, finally, non-family insiders or other alternative solutions (even outsiders) when next-generation successors are unavailable (Ward, 1987). Hence, the founder must pay attention to both business and family to ensure a range of well-prepared potential successors among next-generation family members. In this task, values are important for founders in the relationship between family and business and also represent a key element in handling a future generational CEO succession.

Brunaker (1996), Kets de Vries (1993) and Corbetta and Montemerlo (1999) have emphasized the founder's role in selecting and conveying a set of well-established values to potential successors as a way to facilitate a successful succession process and to achieve the growth and success of the firm. That is, succession can be facilitated by coherence in values' transmission, an argument that leads Santiago (2000, p. 15) to state that the consistency of values between incumbent and successor is more relevant than the existence of formal planning in the succession process. Nevertheless, the achievement of the firm's growth and future success can be affected not only by consistency of values but also by the very nature of the values being transmitted (Dyer, 1986; Gallo and Cappuyns, 1999).

Founder's typologies

So far research has showed that entrepreneurs do not form a homogeneous group. Thus many authors have tried to classify entrepreneurs' heterogeneity by typologies. These classifications pursue different objectives, for example, differentiating entrepreneurs from managers (Collins and Moore, 1964), identifying distinctive types of entrepreneurs (female entrepreneurs versus male entrepreneurs [Fageson, 1993; Kaish and Gilad, 1991]), successful entrepreneurs vs. unsuccessful entrepreneurs (McClelland, 1987), and linking entrepreneurs' mental systems and values to their firms (Donckels and Fröhlich, 1991). This heterogeneity allows authors to build typologies that identify relevant factors in business set-up and management connected to entrepreneurs' behaviour in their firms. The idea was to identify different types of entrepreneurs and to examine the possibility that the type of firm reflected differences among the entrepreneurs.

The pioneering work of Smith (1967) obtains two contrasting types of entrepreneurs, craftsmen and opportunistic, that he links to two different classes of firms. Later research adds to, and modifies, Smith's typology, producing a huge body of literature that commonly describes several groups of entrepreneurs, but does not agree on their labels, variables to study, or methodology (Chell et al., 1991; Collins and Moore, 1970; Donckels and Fröhlich, 1991; Kets de Vries, 1977; Lafuente et al., 1985; Vesper, 1980). However, the literature recognizes the relevance of entrepreneurs values to their business activity, even though, the majority of typologies do not take values into consideration.

One exception is the research conducted recently by García and López (2001) that explicitly took into account founders' values and their heterogeneity. They found that at least two structural dimensions were needed to represent founders' values: the *business value dimension* (firm versus family orientation) and the *psychosocial value dimension* (self-fulfilment versus group orientation). According to the founders' position in these two value dimensions a taxonomy of four groups of founders was constructed, in which the different groups were named as:

1. *Founder of family tradition*, characterized by group value and business as an end orientation, where the firm is something beyond a mere means to earning a living but its development is constrained by founder's orientation towards the family.
2. *Founder achiever*, that is oriented towards the family in the business value dimension (he sees the firm as a means to earn a family living) and toward the group in the psychosocial value dimension.
3. *Founder strategist*, oriented towards the firm (the business is an end in itself) and towards the self-realization in the psychosocial value dimensions, a fact that has its reflection in the steady growth in the size of their firm and the number of additional businesses whose activities are closely related to the original one.
4. *Founder inventor*, whose personal development is based on the possibility to innovate and invent continually inside his firm (oriented towards the self-realization in the psychosocial value dimension), but being that a means to earn a living, to invent (oriented towards the family in the business value dimension).

Family CEO influence on successor's values

Research has shown the existence of heterogeneity between founders (in values and motivations), consequently, founders will probably differ in the value structure that are trying to pass to their successors. Concretely, founders will try to pass to their successors a value structure akin to their own. We cannot say a priori if this legacy will be a strength or weakness for the evolution of the firm. Nevertheless, we can expect that the future of the firm will be affected by successors' decisions in the firm as well as the business and family life-cycle and competitive environment.

On the bright side of the values legacy, Ussmane (1994, pp. 231–2) reported, on her research in Portugal, that potential successors recognized they had learnt business-oriented values and attitudes from their parents that helped them when they joined the family business. In particular, they stated that they learnt critical family business aspects not discussed in business school through a strong personal relationship with the incumbent (Fiegener et al., 1994, p. 324), a learning that can provide the next generation with an entrepreneurial view of the business (Rosa and Cachon, 1989). Shared values are the building blocks of networks (White, 1993, p. 63) that confer a social capital (Bourdieu, 1989) of value to the successor, whether in pursuing the continuity of the family business or setting up their own ventures. Ward (1997, p. 334) pointed out it is this strategic advantage of family business based on a good reputation, trust and long-term goals that the founder should transfer to the next generation in a co-ordinated long-term effort (Steier, 2000).

On the other hand, Hamel and Prahalad (1995) point out that values' transmission can threaten the future competitiveness of the business through fast environmental changes that firms must face at present, where the competitive advantage arises from new combinations rather than simple adaptation of past practices to new business requirements. In the same vein Johannisson (1987) relates succession problems in family businesses to the founder's difficulty in conveying his or her vision to the next generation, that is, to inculcate entrepreneurship as a way to face the fast environmental changes of the firm and to search for new business opportunities.

These two perspectives show us different points of view as we consider the value of the founder's transmission of values to the next generation. The actual value can range from a clear advantage for successors at the start of their business life, to compromising the firm's future success. Consequently, a study of the content of values founders intend to transmit to the next generation can be helpful as we attempt to understand better this specific context.

Based on the literature review, the aim of our work was to find answers to the following questions:

1. What values are founders trying to pass on to potential successors?
2. Do all founders transmit the same values? If not, how do the values systems being transmitted to potential successors vary based on founder type?
3. Do founder values coincide with those they are trying to convey to potential successors?

Research method

Selecting cases

We focused our research on founders with more than 25 years in business and still in full control of their firm's ownership on the verge of transferring the business from one generation to another. We used several steps in building our sample: (1) We identified first-generation family businesses and their founders from *Las primeras 500 empresas de Galicia* (Gómez and Martínez, 1992), which lists firms with sales above 60 million euros in 1989, and presents their main characteristics, ensuring that they were family firms managed by a founder who currently works with the potential successors; (2) we verified that these firms were still family-run in 1996 by checking the several business directories. We obtained a final list of 28 founders who owned firms of varying sizes that had all the required characteristics; this was our theoretical sampling (Glasser and Strauss, 1967). We sent these founders a personal letter and later phoned them to make an appointment for an interview. In the end, 13 male founders agreed to participate once we guaranteed confidentiality. Their firms' characteristics vary in size (for example, number of employees or sales), sector and market (Table 13.1).

Fieldwork

We wanted to obtain the founders' own point of view, so we used an in-depth, semi-structured interview to produce extensive texts (see appendix). We began our sequencing of questions with a description and then asked questions about experience, behaviour, opinions and values. We included several questions to verify information, looking for contradictions (Patton, 1990, ch. 7). All interviews were conducted by García-Álvarez and took place in each founder's office, lasting from two to six hours. Each interview was tape recorded. Both authors handled the transcription and analysis of the 13 interviews to ensure a thorough knowledge of the process. During these visits to founders' firms we used non-participant observation focusing our attention on contrasting and complementing founders' answers during the interviews regarding their firms.

We compiled all published data we could find on these 13 founders, regarding their businesses and family lives, from annual company reports, newspapers, business magazines, business directories and business rankings. We used this information to contextualize each case and to help us during the fieldwork phase. Later, we added secondary data to our analysis to supplement the information obtained from interviews and observations.

Table 13.1 Profile of founders' firms

Founder's age	Years in business (a)	Market	Industry classification	Family Ownership	Employees (1997) (b)	Sales (1997)
65	32	International	Wood	100%	100–200	6–60
>65	30	Local	Food distribution	100%	100–200	6–60
>65	30	National	Chemicals, fertilizers	Majority control	100–200	6–60
65	32	National	Food, transportation, catering	100%	100–200	60–300
>65	30	Regional	Textile distribution	100%	<100	6–60
>65	31	International	Textile-garment construction	100%	100–200	6–60
>65	24	Regional	Food distribution	100%	>200	60–300
>65	41	Regional	Food distribution, transportation	Majority control	>200	>300
>65	25	Regional	Insurance	100%	<100	<6
65	37	National	Textile-garment construction	100%	100–200	6–60
>65	47	International	Plastics	100%	100–200	6–60
>65	51	Regional	Metal recovery	100%	<100	6–60
>65	49	International	Shipyards	100%	100–200	<6

Notes:
(a) With respect to the year 2000.
(b) Business turnover in millions of euros.

Analytical procedure

We analysed our qualitative data with the programme Atlas.ti. This software is one of the more advanced for qualitative data analysis (text, sound and video) and allowed us to extract, compare, explore, and reassemble meaningful pieces from our extensive amounts of data in a flexible and systematic way (Muhr, 1997). In all 13 cases we pursued a multi-variable analysis in three main steps: (1) we initially carried out an in-depth, case-by-case examination by coding for themes until obtaining the final codebook, as a result we got a founders-by-values matrix; (2) we then used quantitative techniques of matrix analysis to look for and display graphically patterns in the coded data; (3) finally, we developed a qualitative back-up of our cross-case patterns.

This iterative qualitative analytical procedure can be described in the following sequential steps. First, we performed the textual analysis which comprised: (1) full transcription of the interviews, (2) adaptation of the transcription form to work with Atlas.ti, (3) creation of textual quotations, (4) revision, (5) descriptive coding, (6) revision, (7) descriptive code reduction, (8) revision. This was followed by the conceptual task, that is: (9) conceptual code reduction, (10) revision, (11) networks and (12) revision. The main objective of these completely qualitative 12 steps was to elaborate a final codebook containing, among other codes, values that founders intend to transmit to their potential successors from our sample.

We then moved the codes-primary document contingency table (the frequency matrix of founders' values for their potential successors) generated by Atlas.ti to SPSS to visually explore the relationship between codes (values) and primary documents (family firm's founders), with the aid of a non-metric multidimensional scaling technique (Kruskal and Wish, 1978). To do this, we produced a derived dissimilarity chi-squared distance (Dillon and Goldstein, 1984, p. 124). The chi-squared distance is similar to the Euclidean distance, with the distinction that each squared difference between profiles is weighted by the correspondent element of the average founder profile (Greenacre, 1993, pp. 24–31).

Finally, we returned to Atlas.ti and built a conceptual matrix focused on cases that included secondary data and evidences from our observation notes. We continued with a textual analysis and finished by building a summary matrix for each group of founders that contains the main variables and links our theoretical memos.

Structure of values that founders intend to transmit to potential successors

We obtained a code frequency table of 28 values for each case. Founders consider the specific role of potential successors and select values to be conveyed on this basis, that is, they build a values system that blends family and business. Founders emphasize the business sphere, as shown by the fact that business orientation was mentioned twice as frequently as the second value, hard work. This main value shows founders' intention to persuade potential successors to devote their professional career to the family firm. However, the relevance of the business arena is conditioned by the explicit requirement of the successor to keep the family together by the value family orientation. Some founders prefer the next generation to be founders as well. This is expressed as autonomy and entrepreneurship, rather than a mere focus on continuing with the family business through growth.

There were 22 other codes that lagged well behind the six main values and expressed how the founder felt the potential successor should behave in general as a person (seriousness, active life, prosperous life, constancy, ethical orientation, rigour, simplicity, self-discipline, ambition, gratitude) or within the firm (determination, feeling of family, innovation, stability, satisfaction, people orientation, positive human relations, negative human relations, long-term orientation).

Heterogeneity of values transmitted to potential successors based on founder type

Based on these different combinations of values which founders wish to convey to the next generation, we decided to identify the underlying structure behind the 28 values mentioned and display the founder's position in the space of values for their successors. Using the value code's frequency table, we applied a non-metrical multidimensional scaling (MDS) technique to the derived chi-squared distance matrix.

The MDS solution (Kruskal and Wish, 1978) obtained by using SPSS's ALSCAL procedure (Schiffman et al., 1981) for two dimensions gives an S-stress value of 0.23 (Young's S-stress formula 1). Further dimensions, though, do not significantly reduce the index of misfit between both data orders, but do increase the complexity of its interpretation.

In Figure 13.1 we plot values for potential successors in the derived Euclidean space. For ease of interpretation, values have been labelled with a number and the meaning of each value number is listed below the figure.

The first dimension, horizontal axis, is associated to the values of family sense (11), business orientation (17), honesty (14), hard work (26), and satisfaction (23) on the left orientation; and ambition (1, just below economic interest, 16), innovation (15), economic interest (16), autonomy (2), and entrepreneurship (7) on the right side. This first axis appears to reflect how founders would like their successor to be or behave: oriented to the

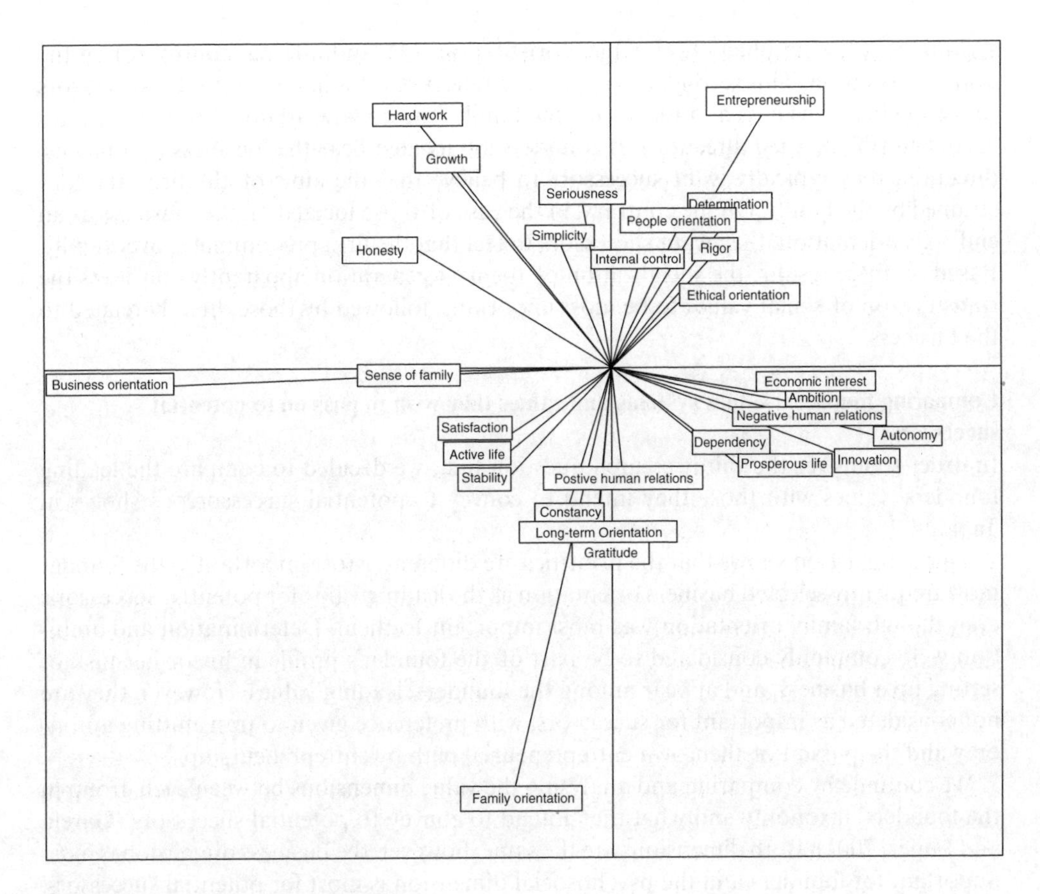

Notes:
1st Horizontal axis: psychosocial dimension
Right orientation, self-fulfilment values.
Left orientation, group orientation values.
2nd Vertical axis: business dimension
Upper orientation, business as an end.
Lower orientation, business as an means.

Figure 13.1 Plot of the values for potential successors in the Euclidean space

group (family) or seeking self-realization. For this reason we labelled it as psychosocial dimension with two orientations, group orientation (to the left) and self-fulfilment orientation (to the right). If founders are located on the 'group orientation values', they try to convey to successors the attitude that their actions are constrained by a desire for group acceptance and that family interest might play an important role in their business decisions. On the other hand, those located on the 'self-fulfilment values' axis orientation try to transmit the value of personal realization, and family probably plays a minor role in business decisions.

The second axis, the vertical one, is associated with gratitude (13, just below long-term orientation, 20), family orientation (19), constancy (15) and positive human relations (4) on the lower orientation; and entrepreneurship (7), growth (8), people orientation (21), seriousness (25), simplicity (24), ethical orientation (18), and internal control (6) on the upper orientation. This second axis appears to reflect how founders want their successors to see the firm: business as a means for the family (the downward direction) or business as an end (the upward direction). If founders are located near the 'business as a means' direction, they typically want successors to believe that the aims of the firm are constrained by the family. On the contrary, in the case of those located on the 'business as an end' axis orientation, they want successors to feel that the firm predominates over family. Based on these results, the socialization of the next generation apparently considers the transmission of social values to be most important, followed by those directly related to the business.

Comparing founders' values systems and values they wish to pass on to potential successors

In order to answer the third research question first, we decided to compare the leading founders' values with those they intend to convey to potential successors, as shown in Table 13.2.

This comparison shows that the priorities are different. Most importantly, the founder most frequently selected business orientation as the leading value for potential successors, even though family orientation was most important to them. Determination and ambition were commonly considered to be part of the founder's profile in his or her task of setting up a business, and appear among the founders' leading values. However, they are not considered as important for successors, with preference given to transmitting autonomy and the pursuit of their own entrepreneurial path by entrepreneurship.

We continue by comparing and analysing the value dimensions between each group in the founders' taxonomy and what they intend to convey to potential successors (García and López, 2001). Both dimensions are the same; however, the business dimension is most important for founders and the psychosocial dimension is most for potential successors. Generally speaking, we observed the highest concordance in value dimensions between founder's type and the values conveyed to the potential successor when both, founder and successor, were male. The highest variation, however, appeared when the successors were a team of women who were the only children of the founder. In teams of potential successors, founders were more likely to attempt to convey group orientation when sons and daughters worked together. Specifically, when reviewing the four groups of founders (García and López, 2001), founders strategists generally agreed most on transmitting

Table 13.2 Leading founders' values versus *leading values to transmit to potential successors*

Leading founders' values	Leading values for potential successors
Hard work	Business orientation
Family orientation	Hard work
Growth	Family orientation
Determination	Autonomy
Ambition	Entrepreneurship
Business orientation	Growth

both business and psychosocial dimensions, and attempted to convey self-fulfilment and business as an end to the potential successor. The family tradition founders showed the most change, with most cases showing a shift from 'business as an end' to 'business as a means' for successors. A group orientation is retained in male or mixed teams of potential successors and changes to self-fulfilment in female teams when they are the only available candidates. Founder achievers keep both social and business dimension, and inventors retain the business dimension, and switch to a social dimension in the case of a team of women.

Proposed model of coherent value transmission and socialization process

We delineate a model that presents the different coherent options of value transmission and successor's socialization that decreases the possibilities of conflict between founder and potential successors, thus facilitating business continuity from first to second generation. As we presented in the previous section, the concordance between founder value dimensions and those transmitted to the founder's potential successors depends on the number and gender of potential successors and on the number of firms the founder owns.

Berger and Luckmann (1966) explained socialization as a complete process of induction of an individual into the objective world of a society. In previous research, we have identified two models of socialization that induce successors into family business (García et al., 2002). These models involve different concepts for founders in terms of the continuity of the family in the company and tend to be linked to a specific value structure. Concretely, the 'founder reproduction socialization model', as the name suggests, is intended to 'clone' the founder in the figure of the successor. Potential successors join the business at an early age with no clear position after finishing secondary studies or three years of college, beginning their career in the business and moving from the shop floor towards managerial positions. The founder supervised the potential successor's training. Founder and offspring share a similar point of view regarding family business and have common business expertise and business networks.

On the other hand, the 'new leader development socialization model' is intended to broaden the possibilities for successors to develop their own approach towards handling their future work in the family business or venturing into a different professional career. Successors typically enter the family business at a later date, for example after finishing their undergraduate or master's degree or working full time outside family business. These descendants started out in management positions related to their academic background, with founders delegating the supervision of these descendants to managers they trusted. The final outcome is both founder and successor have different points of view regarding the family business, with different business expertise and networks.

According to the possible variations on value dimensions and socialization models for potential successors by founder type we can propose a model that summarizes our findings in Figure 13.2. We analyse several cases in which the business is passed on to a single successor and to a team of successors.

Founder's type determines what value dimensions are going to be transmitted to potential successor conditioned by number of firms and number and genre of potential successors. In conveying the psychosocial dimension self-fulfilment value agrees better with the existence of one successor, while group values reinforce the nature of sharing firm's work and responsibility among a team of successors. Relevant for transmitting the business value dimension are the number of firms that founders own and the socialization model their successors follow. Business-as-an-end values are coherent with large-scale business (several firms) and a socialization process that facilitates future successors' role of leadership in this business context. However, business-as-a-means values fits better in cases of medium- to small-scale business (one firm) with successors following a founder reproduction socialization model. Columns show that the variables are the same for each group of dimensions for successors but we can analyse the variations in rows.

In order to facilitate the succession process and not compromise the future success of the firm, founder strategists and family tradition should keep their successors on the left-hand side of Figure 13.2 (business as an end). Thus implies continuing to emphasize among their successors business-as-an-end values and pursuing a new leader development model of socialization. On the other hand, founders achievers and inventors, present four cases of coherence on the right-hand side of the table by transmitting business-as-a-means values and following a reproduction socialization model. Coherence here represents a trade-off between reducing conflict among founder and potential successors and increasing the possibilities of firm's competitiveness and growth. If these types of founders want to provide future challenges for their firms they should consider moving from right to left by changing from business as a means to business as an end and encouraging the new leader development model of socialization for their successors.

The grey diagonal of cells going down to the right-hand corner in Figure 13.2 presents the coincidence between founders' value dimensions and what they intend to transmit to their successors. These cases exemplify values' coherence as pointed out by Santiago (2000, p. 15) and founders' homosocial reproduction, that is a total fit between founders' values and values to be conveyed to potential successors (Hall, 1986; Handler, 1994; Kanter, 1977). These four cases also cohere with the number of firms and the socialization model. In contrast, there are also four cells that represent the total misfit of value dimensions that we rule out from our proposal because is not possible to transmit opposite value dimensions.

The existence of concordance among the founders' values, the values that the founders try to transmit to their potential successors, and successors' socialization process is of high relevance in avoiding distortions between statements (espoused theory) and actions (theory in action). For example one conflictive case appears at the bottom right (just near the corner). Here the founder has a team of potential successors to whom he or she tries to convey group values and business as a means, but successors' socialization models in action are often different. For instance, if the firstborn is male, the founder reproduction model is normally followed. On the other hand, the new leader development model is usually followed with other successors, particularly if they are female. In these cases, the professional adviser encounters a value system that limits the business dimension (business as a means), making the development and effective integration of the entire team of successors more difficult. Moreover, efforts are also needed to manage the coexistence, sometimes, of different socialization models within the team of successors. This aspect is an enormous source of conflict if the founder sets up a hierarchical structure on the basis of birth order and gender rather than skills, especially if the role of each team member in the company is not clarified, and even more if the working relationships and guidelines are not well structured.

	VALUE DIMENSIONS FOUNDER INTEND TO TRANSMIT TO 2nd GENERATION			
	BUSINESS AS AN END		***BUSINESS AS A MEANS***	
FOUNDER TYPE	***SELF-FULFILMENT***	***GROUP VALUES***	***GROUP VALUES***	***SELF-FULFILMENT***
STRATEGIST Business as an end Self-fulfilment	Several firms One successor	Several firms Successors' team	×	One firm One successor
FAMILY TRADITION Business as an end Group values	Several firms One successor	Several firms Successors' team	One firm Successors' team	×
ACHIEVER Business as a means Group values	×	Several firms Successors' team	One firm Successors' team	One firm one son
INVENTOR Business as a means Self-fulfilment	Several firms One successor	×	One firm Successors' team	One firm One son
	NEW LEADER DEVELOPMENT SOCIALIZATION MODEL		***FOUNDER REPRODUCTION SOCIALIZATION MODEL***	

Figure 13.2 Proposed model of coherent value transmission and transition process

However, it cannot be inferred that the founder is certain about the best actions to follow in the firm and the family, nor that potential successors will end up reflecting the value structure that founders try to instil in them (other experiences will influence their resulting value structure, for good and for bad). However, the model of coherent value transmission paths that we have just proposed will serve families and practitioners to increase the probability of achieving a successful founder–successors' transition from first to second family business generation, the most difficult transition according to the evidence.

Conclusions

First, we have shown the legacy of values that founders intend to pass to their potential family successors by building value systems that blend family and business. Founders try to emphasize their entrepreneurial spirit to the next generation (expressed by autonomy and entrepreneurship), instead of only managerial values (for instance, focusing on continuing with the family business through growth). Literature on management acknowledged the influence of CEO values on the future of the organization; however, as far as we are aware, there is no previous research on the influence of incumbent founder's values on the successor's values.

Second, we found that value heterogeneity between founders influences heterogeneity in the values they intend to convey to their successors. We present clear differences in the nature of values to be transmitted to successors, depending on founder type. Thus strategist-type founders emphasize self-fulfilment and business as an end whereas the other three groups of founders – although agreed on business as a means as a value for transmission – opt for different blends of psychosocial values.

Third, we identify that founder's influence is moderated by several structural variables. Within this heterogeneity of values, homosocial reproduction is mediated by two variables: number of firms owned by the founder and the number and gender of potential successors. We compile the variations in a proposed model of value transmission that shows the different legacy of values for each group of founders. This can be used to help identify the values involved, to understand their most immediate effects and to assess both the possibilities for action and the content.

Finally, based on our results we now propose issues that families and practitioners should take into account to keep the coherence during the succession process. In this vein, professionals can assist families in preparing the continuity by (1) identifying family value systems, (2) analysing variables at play in the family-business system, and (3) proposing a coherent option of continuity that both family and business can pursue. The model in our chapter intends to help families and practitioners to follow this path by pointing out the coherent combinations among values and characteristics belonging to each family business and different successor's socialization processes. The final aim is to achieve the best possible generational transfer without adversely affecting the business's competitiveness.

References

Beatty, R.P. and E.J. Zajac (1987), 'CEO change and firm performance in large corporations: succession effects and manager shifts', *Strategic Management Journal*, **8**, 305–17.

Berger, P.L. and T. Luckmann (1966), *The Social Construction of Reality: A Treatise in the Sociology of Knowledge*, Garden City, New York: Anchor Books.

Bourdieu, P. (1989), 'Social space and symbolic power', *Sociological Theory*, **7**(1), 14–25.

Brunaker, S. (1996), 'Introducing second generation family members into the family operated business', doctoral dissertation, Department of Economics, Swedish University of Agricultural Sciences, Uppsala.

Chaganty, R. and R. Sambharya (1987), 'Strategic orientations and characteristics of upper management', *Strategic Management Journal*, **8**, 393–401.

Chell, E., J. Haworth and S. Brearley (1991), *The Entrepreneurial Personality: Concepts, Cases and Categories*, London: Routledge.

Collins, O.F. and D.G. Moore (1964), *The Enterprising Man*, East Lansing: Michigan State University.

Collins, O.F. and D.G. Moore (1970), *The Organization Makers*, New York: Appleton-Century-Crofts.

Corbetta, G. and D. Montemerlo (1999), 'Ownership, governance, and management issues in small and medium-size family businesses: a comparison of Italy and the United States', *Family Business Review*, **12**(4), 361–74.

Davis, S.M. (1968), 'Entrepreneurial succession', *Administrative Science Quarterly*, **13**, 402–16.
Dillon, W.R. and M. Goldstein (1984), *Multivariate Analysis – Methods and Applications*, New York: Wiley.
Donckels, R. and E. Frölich (1991), 'Are family businesses really different? European experiences from Stratos', *Family Business Review*, **4**(2), 149–70.
Dumas, C. (1990), 'Preparing the new CEO: managing the father–daughter succession process in family business', *Family Business Review*, **3**(2), 169–79.
Dyer, W.G. (1986), *Cultural Change in Family Firms: Anticipating and Managing Business and Family Transitions*, San Francisco, CA: Jossey Bass.
Fageson, E.A. (1993), 'Personal values systems of men and women entrepreneurs versus managers', *Journal of Business Venturing*, **8**(5), 409–30.
Fiegener, M.K., M. Brown Bonnie, R.A. Prince and K.M. File (1994), 'A comparison of successor development in family and nonfamily business', *Family Business Review*, **7**(4), 313–29.
Gallo, M.A. and K. Cappuyns (1999), 'Ética de los comportamientos personales en la empresa familiar. Resultados de una encuesta', in Doménech Melé Carné (coord.), *Consideraciones éticas sobre la iniciativa emprendedora y la empresa familiar*, Barcelona: Eunsa, pp. 173–92.
García, E. and J. López (2001), 'A taxonomy of founders based on values: the root of family business heterogeneity', *Family Business Review*, **14**(3), 209–30.
García, E. and J. López (2002), 'Contingency table: a two-way bridge between qualitative and quantitative methods', *Field Methods*, **14**(3), 270–87.
García, E., J. López and P. Saldaña (2002), 'Socialization patterns of successors in first to second-generation family businesses', *Family Business Review*, **15**(2), 189–203.
Glasser, B.G. and A.L. Strauss (1967), *The Discovery of Grounded Theory: Strategies for Qualitative Research*, New York: Aldine de Gruyter.
Gómez, J.L. and J.C. Martínez (1992), *Las primeras 500 empresas de Galicia*, La Coruña: Biblioteca Gallega.
Greenacre, M. (1993), *Correspondence Analysis in Practice*, San Diego, CA: Academic Press.
Hall, B.P. (1986), *The Genesis Effect*, New York: Paulist Press.
Hambrick, D.C. and P. Mason (1984), 'Upper echelons: the organization as a reflection of its top managers', *Academy of Management Review*, **2**, 193–206.
Hamel, G. and C.K. Prahalad (1995), *Compitiendo por el futuro. Estrategia crucial para crear los mercados del mañana*, Barcelona: Editorial Ariel.
Handler, W.C. (1994), 'Succession in family business: a review of the research', *Family Business Review*, **7**(2), 133–57.
Hofstede, G. (1983), 'National cultures in four dimensions: a research-based theory of cultural differences among nations', *International Studies of Management and Organizations*, **13**(1–2), 46–74.
Hofstede, G. (1994), *Cultures and Organizations: Software of the Mind. Intercultural Cooperation and its Importance for Survival*, London: HarperCollins.
Hofstede, G., B. Nevijen and G. Sanders (1990), 'Measuring organizational cultures: a qualitative and quantitative study across twenty cases', *Administrative Science Quarterly*, **35**(2), 286–316.
Iannarelli, C. (1992), 'The socialization of leaders: a study of gender in family business', doctoral dissertation, University of Pittsburgh.
Johannisson, B. (1987), 'Anarchists and organizers: entrepreneurs in a network perspective', *International Studies of Management & Organization*, **17**(1), 49–63.
Kaish, S. and B. Gilad (1991), 'Characteristics of opportunities searches of entrepreneurs vs. executives: sources, interests and general alertness', *Journal of Business Venturing*, **6**(1), 45–61.
Kanter, R.M. (1977), *Men and Woman of the Corporation*, New York: Basic Books.
Kesner, I.F. and T.C. Sebora (1994), 'Executive succession: past, present, and future', *Journal of Management*, **20**(2), 327–72.
Kets de Vries, M.F.R. (1977), 'The entrepreneurial personality: a person at the cross roads', *Journal of Management Studies*, **14**(1), 34–57.
Kets de Vries, M. (1993), 'The dynamics of family controlled firms: the good and the bad news', *Organizational Dynamics*, **21**(3), 59–71.
Kruskal, J.B. and M. Wish (1978), *Multidimensional Scaling*, Sage University Paper series on Quantitative Application in the Social Sciences, 07-011, Beverly Hills, CA, and London: Sage.
Lafuente, A., V. Salas and R. Pérez (1985), 'Tipos de Empresario y de Empresa. El Caso de las Nuevas Empresas Españolas', *Economía Industrial*, November–December, 139–51.
Llano, C. and F. Olguin (1986), 'La Sucesión en la Empresa Familiar', in V.F. Pascual (ed.), *La Empresa Familiar 2*, Barcelona: IESE, Universidad de Navarra.
McClelland, D.C. (1987), 'Characteristics of successful entrepreneurs', *Journal of Creative Behavior*, **21**(3), 219–33.
Muhr, T. (1997), *ATLAS/ti, User's Manual and Reference, Version 4.1*, Berlin: Scientific Software Development.

Patton, M. (1990), *Qualitative Evaluation and Research Methods*, London: Sage.
Rosa, P. and J.-C. Cachon (1989), 'Do graduates from small business backgrounds have more entrepreneurial attitudes than those from employee backgrounds?', *Conference Papers Series, 48/49*, Edinburgh: Scottish Enterprise Foundation.
Santiago, A.L. (2000), 'Succession experiences in Philippine family businesses', *Family Business Review*, **13**(1), 15–40.
Schiffman, S.S., M.L. Reynolds and F.W. Young (1981), *Introduction to Multidimensional Scaling*, Orlando, FL: Academic Press.
Smith, N.R. (1967), *The Entrepreneur and His Firm: The Relationship Between Type of Man and Type of Company*, East Lansing, MI: Michigan State University Press.
Steier, L. (2000), 'Next generation entrepreneurs and succession: modes and means of managing social capital', in P. Poutziouris (ed.), *Tradition or Entrepreneurship in the New Economy? Academic Research Forum Proceedings*, Manchester: Manchester Business School, pp. 329–34.
Ussmane, A.M. (1994), 'A Transferência de Geraçáo na Direcçáo das Empresas Familiares em Portugal', doctoral dissertation, Universidade da Beira Interior, Covilha, Portugal.
Vancil, R.F. (1987), *Passing the Baton: Managing the Process of CEO Succession*, Boston, MA: Harvard Business School Press.
Vesper, K. (1980), *New Venture Strategies*, Englewood Cliff, NJ: Prentice-Hall.
Ward, J.L. (1987), *Keeping the Family Business Healthy*, Boston, MA: Jossey-Bass.
Ward, J.L. (1997), 'Growing the family business: special challenges and best practices', *Family Business Review*, **10**(4), 323–37.
White, H. (1993), 'Values come in styles, which mate to change', in M. Hechter, L. Nadel and R.E. Michod (eds), *The Origin of Values*, New York: Aldine de Gruyter, pp. 63–91.
Zald, M.N. (1969), 'Who shall rule? A political analysis of succession in a large welfare organization', *Pacific Sociological Review*, **8**, 52–60.

Appendix: Translation Protocol Interview

1 To start off, I would like you to describe the main business activities of the company you manage.

Now let's go back to the beginning . . .

2 How did you get the idea of founding a company? Why were you specifically interested in this business?
3 What made you decide to become a businessman?
4 Do you have any business background in your family?
5 What training and experience did you have when you decided to found the company? When you established the company . . .
6 What objectives and ambitions did you want to achieve at the professional level? At the personal level?
7 Could you explain if you had to overcome any difficulties when you started out as a businessman? What were the first few years like?
8 Has the fact that you are doing business in Galicia conditioned the business or does it condition you in any way?
9 Have you had to make any kind of sacrifice to develop your business activity?
10 What qualities or skills do you feel are most important when creating a new business project?
11 From whom or where do you feel you learned the most to carry out business as a businessman?
12 Now I would like you to describe and explain the most important events or incidents for the company and for yourself which have occurred over the years.
13 At present, what is your day like at the company? What kinds of activities do you do normally?

14 Based on what you've just said, what are the tasks you enjoy most? The ones you enjoy least?
15 After all these years, do you feel that you have achieved the objectives you had set when you started your business activity?
16 Do you have any activities outside the company? (business organizations, community activities, hobbies, etc.)

Let's focus more on your company . . .

17 Now I would like you to describe how the responsibilities and tasks of your company are organized and who handles them.
18 How are decisions made?
19 When differences come up in terms of criteria, how is an agreement reached?
20 How do you find out what is going on in the firm?
21 What do you like people working in your company to be like?
22 When you need to hire more staff, how do you select them?
23 Once selected, how are they included in the company?
24 In general, how would you define the staff of this company? (qualified, flexible, etc.)
25 Is there anything you try to instil in the members of your team? If so, how?
26 How do you relate to the staff?
27 How would you define the working style of your company?
28 Do you ever hold any kind of ceremony or celebration within the company?
29 Have you made any kind of innovation over the years? Why?
30 Where did the ideas come from?
31 On what does maintenance of customer and supplier relationships depend?
32 What do you think this company will be like in . . . years?
33 What qualities and/or skills do you feel will be needed to manage it in (time period according to founder's age)?
34 What aspects would you like your company to be known for? What words would you like to be used to define it?

We have talked about you as a businessman and about your company, now let's discuss the family . . .

35 Could you tell me how many members there are in your family, explaining which ones work in the company and which position they hold?
36 Do you have any reservations about including family members in the company? Do you follow any kind of criteria?
37 Do you talk about issues related to the company at home? With whom? And normally when?
38 At times of difficulty or success in your company, did your family participate in it in any way?
39 Is there anything you feel must be instilled in your children? If so, how do you do it?
40 Do you have any expectation(s) in regard to one or more family members continuing with the company in the future? Would you mind if it were someone outside the family?

41 Is your family a conditioning factor in terms of decision-making at the company?
42 (If any children are at the company . . .) In term of hiring one or more of your children in the company, could you explain when it occurred and described how the hiring went? How have they done in the company?
43 What kind of training do they have? (Process)
44 What do you consider most useful for them to learn?
45 What aspects would you like your family to be known for?
46 What is the most important thing in life for you?
47 What opinion do you think people around you have of you as a businessman?
48 Do you feel appreciated and supported as you carry out your business activity? Why? By whom?
49 To finish: if a young person wanted to start working as a businessman/woman in Galicia and came to you for advice, what advice would you give him/her?

14 Understanding strategizing in the family business context

*Annika Hall, Leif Melin and Mattias Nordqvist**

Introduction

This chapter introduces and gives argument for strategizing as a fruitful perspective for researching and understanding the practice of strategy in family firms. Strategy processes are crucial in the development and survival of every family firm. However, the increasing interest from academic research in family businesses is not correspondingly manifested in a large number of studies of strategic processes and outcomes in these firms. So, this chapter's focus on the strategizing perspective also calls for more studies about strategy in the field of family business.

The strategizing perspective is a new stream in the general development of the field of strategic management. Strategizing can be defined as 'the detailed processes and practices which constitute the day-to-day activities of organizational life and which relate to strategic outcomes' (Johnson et al., 2003, p. 14). Advocates for this emerging research perspective have argued for the need to pay more attention to the micro-processes and detailed activities of strategy-making, for instance, to focus more on what people actually do when strategizing, who they are, where they do it, how they do it and why they do it. The mainstream strategy literature has not arrived at this detailed level, but instead stayed at a more general macro level when investigating strategy. The same can be said of the current literature on strategy in family firms.

Strategizing is highly relevant in the context of family businesses – especially for its special attention to social actors and their interaction. In family firms the interaction between family members plays an influential role in strategy formation and the dynamics in which strategies emerge are deeply rooted in family values, emotions and the socio-psychological dimension of ownership (cf. Hall, 2003; Nordqvist, 2005). Consequently, the strategizing perspective outlined in this chapter helps to uncover specific characteristics of strategic activities in family businesses.

In the next section, we briefly review the current literature on strategy in family business research, followed by a more thorough introduction of the strategizing perspective. Then four theoretical perspectives – values, role, arena and legitimacy – are introduced as means of giving strategizing a more precise theoretical direction. A short case description is included, illustrating a period of strategizing activities in a family firm, followed by an analysis of the case using the four theoretical perspectives, and with the purpose of providing an understanding of strategizing in the family business. The chapter concludes with a short summary of the relevance of the strategizing approach in research on family businesses.

A brief review of literature on strategy in family businesses

When reviewing the literature on strategy in family firms it is evident that many scholars draw on a classical approach to strategy (Whittington, 2001), seeing the strategy process as concerned with formal strategic planning, consisting of sequential steps of strategic analysis, choice and implementation (Sharma et al., 1997; Ward, 1988). Strategy processes in family firms are depicted as informal and unstructured, where strategic decisions are made in informal meetings (including around the family kitchen table), with few actors involved, little influence from both external actors and little use of formal strategic tools such as planning techniques. From this it follows that strategy processes in family firms are often characterized as 'irrational', 'unsystematic' and 'non-professional' (Hall, 2002), where informal channels of communication are seen as something negative (Poutziouris et al., 2004). According to scholars in this field of research, a formalized approach to strategy-making is necessary for family firms to thrive and prosper over generations (for example, Carlock and Ward, 2001; Harris et al., 1994). There are also studies influenced by the classical approach focusing on the content of chosen strategies. To give a few examples, research on family business strategy has examined generic strategies (Daily and Dollinger, 1992), strategic posture (Daily and Thompson, 1996), and internationalization (Zahra, 2003). Common to these studies is that they provide an image of family firms' inclination towards specific strategic behaviours without considering why and how these specific strategic contents are predominant in this context.

Another conclusion from the literature influenced by the classical view on strategy is that family members/owners are influential in both the strategy process and the strategic outcome. This is further supported by several authors' interest in comparing family firms and non-family firms, which minimizes the implications of the heterogeneous nature of the family firm population itself. They typically conclude that the family firm is a 'special case of strategic management' (Sandberg, 1992), differentiating it from other business contexts (Chua et al., 2003; Sharma et al., 1997). However, most authors are comfortable with this finding, that is, they do not examine further *how* and *why* the involvement and considerations of family owners influence the strategy process and outcome.

Even so, the interaction between family and business has traditionally been regarded as a problem because of the 'incompatible logics' (Kets de Vries, 1993) of the family system and the business system, and as an obstruction to a healthy development of the family business (for example, Levinson, 1971). However, this conclusion has been increasingly criticized, suggesting an integrated system view as a more fruitful way of understanding family businesses (Fletcher, 2000; Whiteside and Brown, 1991). The integrated view argues that the outcome of two interwoven systems (family and business) is not merely the result of a system overlap, but of a synthesized system with its own dynamic and logic. Understanding the family business means acknowledging the interplay and mutual influence of family and business. On the one hand, family relations, manifested in certain values, traditions, and ways of thinking, might have a strong impact on the business, its culture and long-term perspective. An important characteristic of family business strategy is that it is deeply influenced by family values, goals and relations (Sharma et al., 1997), but also that the business influences the family and its members. The development of the individual family member may be both enhanced and obstructed by the fact that she/he is a member of a family in business (Hall, 2003; Kepner, 1991; Kets de Vries, 1996; Miller and Rice, 1988). From an integrated systems perspective, an understanding of family business strategy must be built on the particular dynamics of the family business, taking into account micro aspects of human interaction. This is essentially what the strategizing perspective sets out to do.

In conclusion, a common factor in the existing literature on family firm strategy is that the research is mainly done at a distance from actual strategic activities. The current literature tends not to arrive at the detailed levels of the everyday activities that constitute the actual strategic development in every family firm. To do this, researchers need to examine strategic processes much closer to the actual situations in which they occur. This means focusing on real-time processes and the micro-level activities in which many different strategists may be involved, that is, to focus on the actual strategizing that takes place. In the next section, we further introduce and discuss the strategizing perspective and its relevance for understanding strategy in family firms.

The strategizing perspective in strategy research

Adopting a strategizing perspective, our focus is on micro-activities that, while often invisible to traditional strategy research, nevertheless, can have significant consequences for organizations and those who work in them (Johnson et al., 2003). In brief, strategizing is what people actually do and that relates to the strategies and strategic outcomes in organizations. The argument is that strategy as activities and practices has been ignored by most scholars in strategic management.

Strategizing builds on the legacy of the process school of strategy that has been of great importance for the development of the strategy field, with several evident contributions. Most important, strategy process research opened up the 'black box' of the organization by the recognition of strategy as an organizational phenomenon with strong cultural and political influences (for example, Pettigrew, 1977, 1979). Studies of autonomous strategic activities were breaking with the traditional top-down view in strategic management research (Burgelman, 1983). Strategy process research demonstrated that strategy is made by human beings, where forces and activities driving or counteracting change emerge from human actions. Populating the arena of strategy-making with human beings paved the way for new theoretical contributions. The socio-cognitive perspective was applied, demonstrating the potential of process studies to capture micro aspects of the strategic thinking and acting made by human beings (for example, Gioia and Chittipeddi, 1991; Hellgren and Melin, 1993). We regard the micro-strategizing perspective as a new movement and means for a necessary renewal of strategy process research. The need for such a renewal gets support from Hambrick (2004) in his claim for theory consolidation in the field of strategy through renewed research on strategy process: 'We need to re-kindle our commitment to understanding complex strategic processes . . . and to reintroduce the human element to our research' (p. 94).

Being highly sympathetic to the process school in the strategy field, the micro-strategy and strategizing perspective aims to take strategy research one step further. As mentioned the emphasis is on the unfolding of strategy through the detailed processes and practices which constitute the day-to-day activities in organizations. Strategy is still about moving an organization forward, from its history into its future. But strategy is not only a macro outcome on the organizational level; it cannot be understood if removed from the micro practice. 'That is what strategy has to be about – not the neat abstractions of the executive suits, but the messy patterns of daily life' (Martin de Holan and Mintzberg, 2004, p. 208). Strategizing accentuates the role of traditions, values and emotions in day-to-day activities and the interpretations and interactions of involved actors. Strategizing emphasizes how daily activities shape the ongoing formation, reproduction and transformation of strategic patterns (Melin et al., 1999), which could be labelled an activity-based view on strategy (Johnson et al., 2003), concerned with the details of organizational work and praxis. The activity-based view 'goes inside the organizations, their strategies and their processes, to investigate what is actually done and by whom. It is concerned with the activities of strategizing' (Johnson et al., 2005, pp. 176–7).

The strategizing perspective means a re-definition of the unit of analysis in strategy research that so far has been dominated by studies and theorizing with the business firm as a whole as the given unit of analysis. This new call for a micro emphasis implies a supplementary demand for understanding and developing strategy theory starting from the micro aspects of strategy process, the everyday activities of the practising strategist(s) as the unit of analysis. As already emphasized, strategizing also implies a stronger focus on the human element of strategy research as human beings have been largely discarded from strategy research as a whole. If considered at all, human actors examined in strategy research have mainly been the managerial elites, and then mainly the single CEO, overemphasizing their general importance in strategy processes. In the actual praxis of strategy, the pattern of human involvement and influence is much more diverse; sometimes it is rather the whole top management team, sometimes the middle managers, sometimes lower-level employees, sometimes consultants and, as in family firms, sometimes it is non-executive family members. Furthermore, quite often it is the interaction and dynamics among all these groups that shape the strategic outcome (Hambrick, 2004). The challenge of strategy researchers is to uncover the people involved and how they actually get on with their strategic activities (Whittington, 2003).

When focusing on micro-activities and human beings, simultaneously we need to contextualize these activities in institutional, social and emotional contexts. In other words, we need to integrate micro activities with macro, societal conditions. Such integration will give new light to a crucial question in strategy research: what activities and processes produce specific strategic outcome on the organizational level? So, even if the micro-strategizing perspective empirically, in the fieldwork, implies almost a phenomenological focus on daily activities, in analyses and theory generation it is intended to be a multilevel approach, integrating the individual and group levels with more macro levels, that is, the organizational level and the societal level. As will be shown, many characteristics of the strategizing perspective fit very well with the typical micro aspects of the family firm's strategy process.

Four theoretical perspectives for understanding strategizing in family businesses

The strategizing view gives preference to certain theoretical perspectives in order to develop a theoretical understanding of strategic activities in the context of family firms. Below, four theoretical perspectives especially relevant for understanding strategizing in family businesses are briefly introduced and discussed. The theoretical perspectives outlined are helpful in understanding how and why strategizing in family businesses unfolds over time, and in the related strategic outcomes. The perspectives – values, role, arena, and legitimacy – all highlight practices, activities and modes of interactions that are inherent in the family businesses context and, consequently, give strategizing in these organizations specific characteristics. We regard these four perspectives as relevant and useful theoretical perspectives for understanding strategizing in the family business context, especially given the specific characteristics of family firms.

Values

The importance of values in family businesses has been stressed by numerous researchers in the field (for example, Aronoff and Ward, 2001; Prokesch, 2002; Schein, 1983). The family tends to be described as perhaps the strongest institution for the transmission of values over generations (Berger and Luckmann, 1966). Through processes of primary socialization, children of business families internalize core family values, thus making them their own (Hall, 2003). Owing to the often close integration between the family and the business, and the common stability of ownership and leadership, these values are spread and translated in the family business, thus constituting a basis for strategizing through interpretations, interactions, and decisions.

Values can be defined as the 'social principles, goals and standards . . . (that) define what the members of an organisation care about' (Hatch, 1997, p. 214), or what is desirable (Shamir, 1991). Also referred to as 'moral or ethical codes' (Hatch, 1997), values underlie judgements about what is right, wrong and desirable. Values are the underpinnings of human action and, as such, they are relevant to the understanding of strategizing activities and practices in the family business context.

Values exert a major – yet often unconscious – influence on organizations, for instance through determining their final goals (Simon, 1947). Organizational processes and practices, such as goal-setting, are based on an ethical component that cannot be objectively, or factually, understood as right or wrong. Given the complexity of the world, values might be regarded as a prerequisite for action. By constituting lenses, or perceptual filters, values reduce the complexity and render the world manageable. The more deep-seated and unconsciously held, the more influential are the values. As they are taken for granted, they are quite unlikely to be questioned or challenged. In this sense, values specify the individual's relations to the environment and thus, they influence strategizing activities and practices. One implication of this for family businesses is the existence of multiple goals typically characterizing these organizations. In addition to profit-making, family businesses tend to have goals related to values stemming from the family such as family unity, concern for individual family members and employees, as well as responsibility for the local community. Hence, strategizing activities and practices tend to be heavily influenced by the special logic provided by the core values of the family.

Role

The three-circle model of the family business (Gersick et al., 1997; Tagiuri and Davis, 1982; Ward, 1987), with the more or less overlapping roles of family member, owner and business manager, points to the fact that roles are complex in family business strategizing. The role perspective is used to show the 'context bound nature of human life' (Burr, 1995, p. 25). Individuals, as members of social settings, occupy several roles, for example, positions in a social context (Ashforth, 2001), with different expectations of their behaviour. These role expectations are important contributions to the role identity, for example, the 'goals, values, beliefs, norms, interactions styles and time horizons that are typically associated with a role' (Ashforth, 2001, p. 6). Since roles have an impact an individuals' vantage points and on how they view the world, these role-specific expectations (that is, the role identity) influence the emotions and actions of the role occupant. However, these expectations are not fixed once and for all. Like identity, the essence of roles is their continuous development: 'the meaning imputed to a given position and the way in which an individual enacts a position are negotiated within structural constraints' (Ashforth, 2001, p. 4).

The integration of different roles is an inherent characteristic of the family business (Tagiuri and Davis, 1982), but highly integrated roles often lead to difficulties in 'decoupling the roles psychologically, disengaging from one in favour of the other' (Ashforth et al. 2000: 481). This may result in 'role blurring', that is, the inability to define the specific content of each role and separate the roles from each other. As pointed out by Ashforth et al. (2000), this is an often occurring dilemma for family members working in family businesses.

Role enactment is closely related to the process of identity and self-construction (Ashforth, 2001; Turner, 1978). Through the enactment of roles, individuals learn about themselves, and create their identities, in the sense that an understanding of who the self is gets mediated through the roles. Continuously learning who they are, individuals seek opportunities to enact the roles they value most and identify with, and they are prone to integrate this role with their other roles. Individuals identifying strongly and exclusively with a particular role might be devastated if forced to exit the role, as they have no alternative role or identity to fall back on. The greater the sacrifice and the investment in time and effort to gain or maintain a role, the more highly valued it tends to be (Turner, 1978). The willingness to leave a role might also be obstructed if alternative roles are ambiguous, for instance, if the content and requirements of these roles are not clear. This might lead to bewilderment and frustration and, hence, a reluctance to exit the role (Roos and Starke, 1981). Role clarification, that is, defining what an occupant of a specific role is supposed to do, is thus a prerequisite for successfully entering a new role. Even under the best of circumstances, role transition processes are, however, challenging, requiring extensive processes of learning and adjustment (Ashforth et al., 2000).

Arena

Strategizing emphasizes interaction between actors and interaction always takes place somewhere, that is, in some type of arena. The arena perspective, here expressed through 'the strategic arena metaphor' (Ericson et al., 2000, 2001; Melin, 1998), is useful for understanding strategizing in the family business context. Ericsson et al. (2001, p. 68) suggest that the 'strategic arena is defined through the dialogues around the issues that are strategic to the individual organization'. The strategic arena includes all possible meeting places that offer an opportunity for communication on strategic issues that can reproduce or change strategies of an organization. The strategic arena can emerge both in formal occasions, such as top management and board meetings, and in more informal situations, such as ad hoc meetings, spontaneous small talk in the hallway, dinners and during travels, and the actors populating the arena may come from different hierarchical levels in the organization as well as from outside the focal organization (Ericson et al., 2000). This means that both formal position and other characteristics such as possessing the right knowledge or social relations with key actors can grant access to the strategic arena. The strategic arena perspective implies viewing the arena as a process that unfolds in different kinds of representations, where the arena is a multiple and changing meeting place for dialogues between actors on strategic issues (Ericson et al., 2000).

The strategic arena metaphor is especially useful and relevant for understanding straegizing, since it does not take its point of departure in the formal organizational structure and hierarchy, but in the situations, settings and venues where strategizing actually occurs (Nordqvist, 2005; Nordqvist and Melin, 2003). The arena is a conception of the actual social context for strategizing activities. In family firms a few actors from the same owner-family often occupy several positions and roles that the strategy and governance literature tend to separate. This overlap creates a blurry organizational structure in which it is not always explicit where, when and by whom different strategic activities are performed. For instance, even if the family firm has top management team meetings and/or an active board with external members, it is not a given that strategizing always occurs in these firm connected arenas. The family context also gives room for relevant strategic arenas, such as a family gathering with family members interacting on strategic issues (Karlsson-Stider, 2001; Nordqvist, 2005). Having said this, the following different aspects of the arena are particularly relevant in the family firm context.

The arena can be described as either *formal* or *informal* depending on the characteristics of the social situation where the arena emerges. A typical example of a formal strategic arena is a periodical board meeting where strategic issues are discussed and settled. A typical example of an informal strategic arena is when actors strategize during a coffee break or in small talk in the hallway.

The arena can be characterized as a *front-stage* or a *back-stage* arena also depending on the characteristics of the social situation where the arena emerges. Following Goffman (1959), a front-stage strategic arena is when one or several actors perform some type of strategic activity with a specific audience in mind. An archetypical example of this is a CEO holding a convincing speech explaining the firm's new strategy to employees. The back-stage strategic arena emerges when and if the CEO after the speech continues to strategize with key actors, addressing issues and using information that was not brought up in the official and communicated version.

The arena can be characterized as *current* or *historical*. This includes a time dimension and refers to whether the strategic arena is a presently or previously important arena. The history and tradition is often important in family firms with actors referring to 'how we used to strategize'. Sometimes this also means a desire to revert to a previous situation.

The strategic arena can be characterized as *closed* or *open*. This means that it can be controlled by the 'power centre' of a dominant actor/coalition, or be more fragmented where many actors are given the possibility of participating in and influencing the outcome of arena dialogues (Melin, 1998).

Regarding what type of arena dominates strategizing in a particular family firm, there is often a blend of arenas in action in strategizing and these arenas can both facilitate and constrain each other. And even if there is a dominant arena where actors strategize at a particular point in time, this arena typically changes over time. Furthermore, different generations in a family firm may prefer different types of arena where they discuss and settle strategic issues.

Legitimacy

The legitimacy perspective contributes to an understanding of strategizing through a focus on how and why firms adopt, implement and act in accordance with different strategic tools and practices not only for their technical and instrumental qualities, but also their symbolic, ritualistic and culturally embedded qualities (Meyer and Rowan, 1977; Suchman, 1995). This means giving attention to institutionalized strategic practices and activities, for instance, how and why different practices and activities in use are related to structures and processes on more macro, societal levels.

Legitimacy can enhance both the stability and comprehensibility of organizational activities. A legitimate organization is often perceived as more worthy, more meaningful, more predictable and more trustworthy (Suchman, 1995, pp. 574–5). According to Suchman (1995) there are three forms of legitimacy: pragmatic, which is based on various stakeholders' self-interests; moral, which is based on normative approval and appropriateness; and cognitive, which is based on comprehensibility and taken-for-grantedness within a specific cultural context. From this broad perspective, legitimacy can be defined as: 'a generalized perception or assumption that the actions of an entity are desirable, proper, or appropriate within some socially constructed system of norms, values, beliefs and definitions' (Suchman, 1995, p. 574). The socially constructed system refers to the shared beliefs of some social group and means that legitimacy depends on a collective audience, yet independent of particular observers. This can be a specific organizational field (DiMaggio and Powell, 1983), which is similar to an industry or a sector, or a population of firms, like family firms (Nordqvist and Melin, 2002), as well as smaller or larger cultural communities such as professions, nations or specific networks of firms and individuals (Scott, 2002).

Family firms face specific challenges regarding how to gain and maintain legitimacy. These challenges have their origin in the overlap between family and the business, and in the supposed negative influence of family on business (for example, Donnelly, 1964; Levinson, 1971) which has caused family businesses to be labelled irrational (Hall, 2002). In business practice, this notion can also be fomented by both internal (for example, employees and shareholders not operatively active in the firm) and external (for example, banks, customers and suppliers) stakeholders who are worried that their interests are not adequately taken into account. As a response to this, family firms may adopt and implement tools, practices and activities to structure the strategic work that signals legitimacy both towards themselves and towards different stakeholders. In other words, the legitimacy perspective suggests that adopting a strategic practice or activity such as, for instance, a strategic planning model or including external members on the board as strategic advisers, can be a way to signal rationality as well as an up-to-date and efficient mode of organizing.

Activities and practices stemming from a search for legitimacy are often spread and translated as 'ideas' and 'best practices'. In relation to family firms, they are of two kinds: first, ideas and best practices successfully implemented by other family firms and, second, and perhaps more commonly, ideas and best practices successfully used by non-family firms. These are often spread through carriers such as popular management books, professional associations (for example, the FBN and the FFI, but also industry-related trade associations), management training courses and business schools, as well as through exchanges among colleagues and networks. In many cases, ideas and practices are not consciously adopted for the sole reason of gaining legitimacy; in many cases the legitimacy reason is largely unconscious among the individuals in question.

Strategizing in a family business – a case illustration

Wiretech Ltd is a second-generation family-owned firm operating in a manufacturing industry. The firm employs around 130 employees and is a subcontractor in a competitive market. Growth has been modest but stable over the past five years. Wiretech is located in a small town, here called Greenbay, in the south of Sweden.

In Wiretech being socially responsible has always been one of the core values of the owning family. The present owners – children of the foundering generation – describe themselves as influenced to a great extent by the values of social responsibility. Through the close interaction between family and business these values were also spread to the business, considerably influencing the practices, processes and modes of interaction characterizing strategizing in the company. Being in business implies, to the owning family, in essence a social commitment.

> This is not a way of making money. I don't run around with the balance sheet in my hip pocket. It's kind of satisfactory that people have a job here, and that it works, and that I enjoy coming here. But in terms of dollars or cents . . . no. I want it to be a secure working place for people living in Greenbay. I want to be able to walk through the village of Greenbay, knowing it's a pleasure to meet. (Steve, CEO and owner)

The family's way of interacting with customers and suppliers is based in respect and friendship. This implies negotiation and employment practices aiming at long-term relations, sometimes at the extent of financial disadvantages.

> No profit maximization. They do business with the ones that conduct well, or have right competencies, or provide quality, that's been more important than prices. (Profit maximization) is nothing big, quite on the contrary, I would say. When I first came here I was astonished by the fact that they worked so hard and made so little profit. (Michael, external manager)

> They care. Many (employees) live close by and they know them quite well. . . . They have a relation to the employees also outside of work. (Thomas, external manager)

> We've never fired anyone. They have not thrown out anyone just because we make less profit. Although, when thinking of the situation we face today, it would perhaps have been needed. (Burt, long-term employee)

After 30 years as CEO of Wiretech, Steve decided to leave this position to bring in an external successor. Apart from the fact that his health at the time was poor, one important reason for the decision was that Steve believed that the company was in need of a new management style. Over the years, Wiretech had been run rather 'spontaneously, with intuition and feeling, not very much planning or calculus . . . but more emotional' (Tom, external manager). For its further development, the company was thought to benefit from a 'professional manager' who could 'make truly wise, rational decisions through an objective perspective' (Steve). Steve had, however, no thoughts of leaving the company completely – 'it would be like depriving oneself from one's beliefs, from one's personality' – but to stay as function manager and board member.

When David, the external CEO came to the company he was very welcome. Both the owner-family and the employees put lots of faith in David's formal competence and long working experience in the publicly held company where he had held several management positions during an extended period of time. As expected, the external CEO brought with him a quite different way of thinking compared with the traditional Wiretech way. Although this had been one of the reasons for hiring David, the family soon ended up with a situation they did not fully approve of. The company was formalized, with clear hierarchies and areas of responsibility. David also started to put into practice quite different decision and selection criteria, such as efficiency, productivity and profitability. In addition, David did not have the same local and relational commitment as the owners. His goal was to expand the company and to increase profit, if necessary by dismissing employees.

Cooperation between the family and the external CEO eventually became more difficult. Steve was not content with the 'rigid constraints of the system which I, as an entrepreneur can never accept' in the form of a formal hierarchy, not allowing him to have a say in practices and processes with which he did not agree. From being used to being 'involved in everything' (Marion, controller and owner), the siblings' influence was severely restricted. 'I wasn't' supposed to interfere, I was told I was too dominant . . . I shouldn't disturb the development' (Steve).

David tried to formalize and, hence, delimit Steve's influence to being mainly that of an owner in the boardroom. Steve, for whom the ability to influence is 'incredibly important' and one of the 'great advantages' of being a family business, did not find this easy. Having been CEO for almost three decades he was unsure of how to exert influence only through the owner role. Even one and a half years after David's appointment, he found the roles 'totally integrated' and 'inseparable'. An ongoing recession helped to worsen the situation, and Steve got all the more irritated and frustrated with circumstances. In spite of this, David was very optimistic, thinking that matters would improve once profits increased.

> When this is the case, my situation in the company will be a whole lot different. That's the key to it. After all, that's why I am here: to make the company grow, and to earn money to the family. I mean, that's rather obvious. . . . Money is what it is all about. (David)

In spite of the optimism, the situation in the company soon passed the point of possible cooperation between the external CEO and the family. Less than two years after his appointment, David left Wiretech, along with other key employees, leaving behind a rather uncertain strategic situation.

Understanding strategizing in the family business

The following interpretation of the case, framed by the theoretical perspectives values, role, arena and legitimacy, is used to provide an understanding of strategizing in the family business. Together these interrelated perspectives help shed light on why and how certain practices and activities are implemented or rejected by interacting strategic actors, and the consequential outcomes for the family business. In doing so, they stress the multidimensionality and complexity of strategizing rendered by the integration of family and business. As is illustrated by the case, micro dimensions of organizational life, often without the strategic actors realizing it themselves, tend to exert a decisive influence on strategy in family businesses. This underlines the usefulness of the strategizing perspective, since it takes these dimensions into account, but often downplayed as irrelevant by traditional strategy approaches.

Values

One important dimension of strategizing in Wiretech is the prevalence of strongly held family values. The case illustrates how deep-seated family values of social responsibility constitute a frame of interpretation, action and, hence, strategizing. Even if David was informed of these values when entering Wiretech, he did not realize the extent to which they would influence his day-to-day work. Coming from a publicly held company, he focused on implementing 'rational' changes in activities and practices without considering the degree of congruence between these changes and the prevailing values. Interestingly, it seems as if Steve and his sister, although being conscious of the values per se, were not aware of the meanings and implications of these for the daily running of the company. Not until the values were severely challenged did the depth of their influence on daily practices and activities become apparent.

As long as family members are the sole managers, they, as a result of primary socialization, are likely to have a common understanding of the company values. The bringing in of external managers implies, however, a radical change to this unanimity, making it essential to take the values seriously. As illustrated by the case, family values are among the most deep-seated and pervasive dimension of strategizing in family businesses. It therefore takes lots of communication and interaction to reach a sufficient degree of mutual understanding of the meaning of the values in the daily strategizing of the organization. That the main strategizing actors in Wiretech never came close to this point is exemplified by David's intentions of rapid expansion, if necessary at the cost of lay-offs, and by the way in which he referred to money-making as the way out of his experienced problems in the company. This stands in sharp contrast to the business goals of the family, of which short-term profit-making has never been given priority over the responsibility felt to its various stakeholders.

Role

A further perspective relevant for understanding strategizing in Wiretech is that of role. Having enacted the role as CEO for almost three decades, Steve had come to identify very closely with it. He refers to himself as an 'entrepreneur' and as an action-oriented 'doer' in the company, used to have a direct influence on all activities and practices. As David succeeded him, this was radically changed. Steve was now supposed to act, think and behave primarily as owner. Until the entrance of David, the roles as owner and CEO had been highly integrated, with priority given to the latter. For this reason, Steve had tremendous difficulties decoupling the roles and disengaging from thinking and acting as CEO. One and a half years after the succession, he still described the roles as blurred. Since no serious efforts were made in specifying the content of the role as owner, it remained ambiguous to Steve and so he never came to identify closely enough with it to use it as a platform for strategizing. As consequence, Steve came to lack an alternative role, or an alternative identity, to fall back on when he exited the role as CEO. This provides an understanding of Steve's reluctance to let go of the role as CEO, based in deep human needs such as having and manifesting a distinct identity, without which any individual would feel bewildered and frustrated. Even though he had formally left the role as CEO, his identity was still based in that role, which explains why Steve continued to seek opportunities to enact it. As David counteracted these efforts, the interaction pattern between David and Steve was increasingly characterized by a struggle for power, not only over the company, but over their own identities (within it). This coloured strategizing in general, not least by creating an even more tense atmosphere in the company with orders and counter-orders given.

Framing the Wiretech case by the concept of role, highlights the importance that individuals about to leave highly valued roles are (made) aware of the need of continuous manifestation of their identity through other roles. Had this been the case in Wiretech, it would most likely have resulted in quite different patterns of interaction between the former and present CEO. Also, it would probably have turned the board and perhaps also a family council into relevant means by which Steve could continue to – and more peacefully and efficiently – influence strategizing in the company.

Arena

The last part of the discussion stresses the arena as a perspective with the potential to provide an enhanced understanding of family businesses strategizing. In family firms, with owners active on a daily basis, formal arenas such as the board might not be the most important for strategizing. One reason for this is highlighted by Steve's difficulties in transferring his involvement in strategizing to formal arenas, such as the board and the management team, regarded by David as the dominant strategic arenas. Never really exiting the role as CEO, Steve continued his everyday interactions in informal arenas, such as small talk and ad hoc conversations with trusted employees. Historically, the typical outcome of this had been quick and intuitive strategic decisions. Now, it merely created confusion and frustration in the company. It also led to tension between informal/formal arenas as well as current/historical arenas, with the former CEO acting primarily on informal and historical arenas, and the present CEO mostly on formal and current ones. What they lacked, however, were common arenas for fruitful discussion. The case, hence, illustrates the difficulties of arriving at an efficient ongoing communication when the character of different arenas for strategizing, and their corresponding roles, is not clearly defined and communicated.

This is further linked to the extent to which the arena is open or closed and the importance of knowledge and social relations for accessing the dominant arena. David lacked the heritage of the firm's history and was never really interested in learning it. Still – indeed, perhaps because of that very reason – he tried to centralize the strategic work and to formalize practices and interaction on formal management arenas. Over time this implied a closing of the dominant strategic arena, especially since David never developed close social relations with other key actors, including family members. Quite radically, this contradicted the strategizing behaviour of the former CEO. Being quite autocratic, Steve was nevertheless always careful to keep his arenas open, in the sense of talking to knowledgeable and trusted employees and other advisers.

The unclear relation between arenas also had impacts on the outcome of strategizing, in the sense that Wiretech's strategies were neither fully reproduced nor fully changed. David worked hard with changing activities and practices, with the intention of altering Wiretech's strategic direction. The spontaneous 'interference' of Steve led, however, to an obstructed change process. In other words, there was a difference between strategizing on back-stage arenas and front-stage arenas.

Legitimacy

The final theoretical perspective turned to for an enhanced understanding of family business strategizing is legitimacy. Upon his arrival the new CEO was welcome and the expectations high. This can be interpreted as being caused by legitimacy claims, although perhaps not exclusively or consciously. In much literature and practice there is a tendency to advocate the recruitment of external, supposedly professional, managers to the top management of family firms that need renewal and strategic change (Hall and Nordqvist, 2005). These managers are often assumed to hold special insights and knowledge that can revitalize a maturing small business (Fletcher, 2002). Recruiting managers with formal management education and significant work experience, often from non-family firms, is seen as a way to break loose from the chains of nepotism, introverted character and close family ties, and instead embrace a universal, objective, rational, analytical and impersonal approach to strategizing. In the Wiretech case, bringing in the new CEO was seen as a way to introduce more 'planning or calculus' and a 'professional manager' who could 'make truly wise, rational decisions through an objective perspective'. Moreover, the new CEO did not just want to introduce new and more formal strategic practices and activities in order to structure the strategizing process formerly based on 'intuition and feeling', but was also expected, by himself and others, to do so. Coming from a publicly held company, with a formal management education David had been socialized into the norms stating the goals and means of a legitimate professional manager (that is, short-term profitability, objectivity, rationality and formalization).

There is, however, a dilemma inherent in the search for legitimacy interpreted in the recruitment of the external CEO in Wiretech. This is related to the difficulties and negative outcomes for strategizing previously observed in this analysis. Since legitimacy, by definition, refers to a generalized perception or assumption, the actions taken to legitimate a certain organization downplay the uniqueness and context-boundedness of that very organization. In other words, in searching for general and universal solutions that are accepted by a wider audience, the particularities and heterogeneity of family-firm strategizing is forgotten. In the case this is illustrated by the importance of family-related core values, history and family members' role identification for the activities, practices and outcomes of strategizing. The introduction of the new CEO in Wiretech implied, over time, dysfunctional and inefficient strategizing. Thus, severe difficulties may sometimes emerge when changing actors, practices and activities of strategizing for reasons of legitimacy, no matter if it is normative, pragmatic or cognitive.

Conclusions and implications

The Wiretech case illustrates how tacit, unconscious dimensions might exert a decisive influence on strategizing practices, and activities and the subsequent outcome in family businesses. During the years of external management of Wiretech – legitimized by management norms assuring its appropriateness – the lack of arenas for communication and mutual understanding of values and roles led to a strategizing situation characterized by a struggle for control over practices and activities. The outcome of this was that Wiretech's strategic direction was neither fully reproduced nor changed. Instead, the overall outcome in terms of strategizing was a strategic standstill, or vacuum, and a notable slowdown of the development of the company coupled with an insecurity among employees as to who was in charge and the future direction of the company. Eventually, this not only resulted in the resigning of David; over time, many other key persons, among them some valuable long-term employees decided to leave the company. To Steve, the worst implication of this was not the problems caused in the business but the 'loss of prestige' by the family, who over the years has been regarded as a highly reliable and socially responsible actor within the local community. The case also illustrates how closely the perspectives of role, values, arena and legitimacy are interrelated. Indeed, they can be seen as four integrated perspectives that jointly constitute a coherent framework for understanding strategizing in the family business context.

This chapter has argued for the relevance of the strategizing approach in strategy research on family businesses. The particular strength of strategizing lies in its focus on everyday, micro, human aspects of organizational life. This focus implies an interest in the activities that really take place in the organization, that is, the daily, continuous practices, processes and interactions through which strategy evolves over time. This micro focus also means that no details of organizational life could, a priori, be disqualified from being strategic. Seemingly unimportant practices such as the daily coffee break or spontaneous hallway meetings could have important strategic consequences. Further, strategizing implies an interest in human aspects of organizational life, such as the interactions and interpretations of strategizing individuals. From this point of view, strategizing is, in essence, 'a theory of social action' (Johnson et al., 2003, p. 11). Taken together, strategizing focuses on 'tacit, deeply embedded, and therefore hard-to-get-at phenomena' (Balogun et al., 2003, p. 199) traditionally neglected – even discarded as irrelevant – by strategy research. As illustrated by the Wiretech case, such phenomena are, however, crucial parts of strategizing activities and outcomes, especially in family businesses, where organizational and family lives are intimately intertwined.

The interest in tacit and hard-to-get-at phenomena as potentially strategic implies the need for an understanding of theoretical perspectives and their strategic relevance. This chapter has provided a theoretical framework for understanding strategizing through four perspectives, all stressing different dimensions of micro, everyday, human aspects of organizational life. The application of values gives meaning to how strategy practices, activities and interactions are based on deep-seated, traditional family values, and how insensitive challenging of these by external actors might lead to decreased efficiency in strategizing. The role perspective emphasizes strategizing as based in human needs such as identity manifestation. The perspective of arena highlights the need for platforms and meeting places for communication, mutual understanding and respect, all aspects crucial for efficient strategizing. The legitimacy perspective is, finally, helpful in understanding the meaning of adopting certain strategic practices as well as the outcome of their implementation.

Together, the applied perspectives provide an understanding of the multidimensionality and complexity of family business strategizing. In a corresponding way, other perspectives could supplement this understanding. Given the special characteristics of family businesses, examples of other relevant perspectives would be power, emotions, identity, culture and sense-making. The application of theory is essential, both for the generation of an understanding of strategizing and for the possibility of generalizing this knowledge analytically (Yin, 1989). The theoretical perspectives not only shed light on the tacit and non-obvious (roles, power, legitimacy, emotions) aspects in a specific case; they also assist in creating a language for understanding other family businesses. In turn, such a language also provides the potential for conscious reflection and action on behalf of family business practitioners and advisers which might, indeed, enhance strategizing in these organization.

Note

* The authors contributed equally to this chapter and are therefore listed in alphabetical order.

References

Aronoff, C.E. and J.L. Ward (2001), *Family Business Values: How to Assure a Legacy of Continuity and Success*, Marinetta, GA: Family Enterprise Publishers.

Ashforth, B.E. (2001), *Role Transition in Organisational Life: An Identity-based Perspective*, Mahwah, NJ: Lawrence Erlbaum Associates.

Ashforth, B.E., G.E. Kreiner and M. Fugate (2000), 'All in a day's work: boundaries and micro role transitions', *Academy of Management Review*, **25**(3), 472–91.

Balogun, J., A.S. Huff and P. Johnson (2003), 'Three responses to the methodological challenges of studying strategizing', *Journal of Management Studies*, **40**(1), 197–224.

Berger, P. and T. Luckmann (1966), *The Social Construction of Reality*, New York: Doubleday.

Burgelman, R. (1983), 'A process model of internal corporate venturing in the diversified major firm', *Administrative Science Quarterly*, **28**, 223–44.

Burr, V. (1995), *An Introduction to Social Constructionism*, London: Routledge.

Carlock, R.S. and J.L. Ward (2001), *Strategic Planning for the Family Business: Parallel Planning to Unify the Family and Business*, Basingstoke: Palgrave.

Chua, J.H., J.J. Chrisman and L.P. Steier (2003), 'Extending the theoretical horizons of family business research', *Entrepreneurship Theory and Practice*, **27**(4), 331–8.

Daily, C.M. and M.J. Dollinger (1992), 'An empirical examination of ownership structure in family and professionally managed firms', *Family Business Review*, **5**(2), 117–36.

Daily, C.M. and S.S. Thompson (1996), 'Ownership structure, strategic posture, and firm growth: an empirical examination', *Family Business Review*, **7**(3), 237–49.

DiMaggio, P.J. and W.W. Powell (1983), 'The iron cage revisited: institutional isomorphism and collective rationality in organizational fields', *American Sociological Review*, **48**, 147–60.

Donnelley, R.G. (1964), 'The family business', *Harvard Business Review*, July–August, 93–105.

Ericson, T., A. Melander and L. Melin (2001), 'The role of the strategist', in H.W. Volberda and T. Elfring (eds), *Rethinking Strategy*, London: Sage Publications.

Ericsson, T., A. Hellqvist, A. Melander and L. Melin (2000), 'Shaping new strategies in professional organizations: the strategic arena approach', paper presented at the EGOS Annual Meeting in Helsinki, Finland.

Fletcher, D. (2000), 'Family and enterprise', in S. Carter and D. Jones-Evans (eds), *Enterprise and Small Business: Principles, Practice and Policy*, London: Prentice-Hall, pp. 155–65.

Fletcher, D. (2002), 'A network perspective of cultural organising and "professional" management in the small, family business', *Journal of Small Business and Enterprise Development*, **9**(4), 400–15.

Gersick, K.E., J.A. Davis, M. McCollom Hampton and I. Lansberg (1997), *Generation to Generation – Life Cycles of the Family Business*, Boston, MA: Harvard Business School Press.

Gioia, D.A. and Chittipeddi, K. (1991), 'Sensemaking and sensegiving in strategic change initiation', *Strategic Management Journal*, **12**(6), 433–48.

Goffman, E. (1959), *The Presentation of Self in Everyday Life*, Harmondsworth: Penguin Books.

Hall, A. (2002), 'Towards an understanding of strategy processes in small family businesses: a multi rational perspective', in D. Fletcher (ed.), *Understanding the Small Family Business*, London: Routledge, pp. 32–45.

Hall, A. (2003), 'Strategizing in the context of genuine relations', JIBS Dissertation Series No. 018, Jönköping International Business School, Jönköping, Sweden.

Hall, A. and M. Nordqvist (2005), 'Professional management in family business contexts: meanings and implications', paper presented at the EURAM 2005 Conference, Munich, Germany.

Hambrick, D.C. (2004), 'The disintegration of strategic management: it's time to consolidate our gains', *Strategic Organization*, **2**(1), 91–8.

Harris, D., J.I. Martinez, and J.L. Ward (1994), 'Is strategy different for the family owned business?', *Family Business Review*, **7**(2), 159–73.

Hatch, M.J. (1997), *Organisation Theory: Modern, Symbolic, and Post-modern Perspecitves*, Oxford: Oxford University Press.

Hellgren, B. and L. Melin (1993), 'The role of strategists' ways-of-thinking in strategic change processes', in J. Hendry and G. Johnson (eds), *Strategic Thinking: Leadership and the Management of Change*, Chichester: John Wiley and Sons.

Johnson, G., L. Melin and R. Whittington (2003), 'Guest editors' introduction: micro strategy and strategizing: towards an activity-based view', *Journal of Management Studies*, **40**(1), 3–22.

Johnson, G., L. Melin and R. Whittington (2005), 'Micro strategy and strategizing: implications for strategy process research', in S.W. Floyd, J. Roos, C.D. Jacobs and F. W. Kellermans (eds), *Innovating Strategy Process*, Oxford: Blackwell, pp. 176–85.

Karlsson-Stider, A. (2001), 'The home – a disregarded managerial arena', in S.-E. Sjöstrand, J. Sandberg and M. Tyrstrup, *Invisible Management: The Social Construction of Leadership*, London: Thomson Learning, pp. 83–104.
Kepner, E. (1991), 'The family and the firm: a coevolutionary perspective', *Family Business Review*, **4**(4), 445–61.
Kets de Vries, M.F.R. (1993), 'The dynamics of family controlled firms: the good and the bad news', *Organisational Dynamics*, **21**(3), 59–71.
Kets de Vries, M.F.R. (1996), *Family Business: Human Dilemmas in Family Firms*, London: International Thompson Business Press.
Levinson, H. (1971), 'Conflicts that plague family businesses', *Harvard Business Review*, **49**(2), 90–98.
Martin de Holan, P. and H. Mintzberg (2004), 'Management as life's essence: 30 years of the nature of managerial work', *Strategic Organization*, **2**(2), 205–12.
Melin, L. (1998), 'Strategisk förändring: om dess drivkrafter och inneboende logik', in B. Czarniawska (ed.), *Organisationsteori på svenska*, Malmö: Liber Ekonomi, pp. 61–85.
Melin, L., T. Ericson and T. Müllern (1999), 'Organizing is strategizing: innovative forms of organizing means continuous strategizing', paper presented at Academy of Management Annual Conference, Chicago.
Meyer, J.W. and B. Rowan (1977), 'Institutionalized organizations: formal structure as myth and ceremony', *American Journal of Sociology*, **83**, 310–63.
Miller, E.J. and A.K. Rice (1988), 'The family business in contemporary society', *Family Business Review*, **1**(2), 193–210.
Nordqvist, M. (2005), 'Understanding the role of ownership in strategizing: a study of family firms', JIBS Dissertation Series No. 029, Jönköping International Business School, Sweden.
Nordqvist, M. and L. Melin (2002), 'The dynamics of family firms: an institutional perspective on corporate governance and strategic change', in D. Fletcher (ed.), *Understanding the Small, Family Firm*, London: Routledge, pp. 94–110.
Nordqvist, M. and L. Melin (2003), 'Understanding strategizing in family firms – exploring the role of strategists and the strategic arena', paper presented at the 48th ICSB World Conference, Belfast, Northern Ireland.
Pettigrew, A. (1977), 'Strategy formulation as a political process', *International Studies of Management and Organizations*', (1/2), 78–87.
Pettigrew, A. (1979), 'On studying organizational cultures', *Administrative.Science Quarterly*, **24**, 570–81.
Poutziouris, P.Z., L.P. Steier and K.X. Smyrnios (2004), 'Guest editorial: a commentary on family business entrepreneurial developments', *International Journal of Entrepreneurial Behaviour and Research*, **10**(1/2), 7–11.
Prokesch, S. (2002), 'Rediscovering family values', in C.E. Aronoff and J.H. Astrachan (eds), *Family Business Sourcebook*, Marinetta, GA: Family Enterprise Publishers, pp. 676–87.
Roos, L.L. Jr and F.A. Starke (1981), 'Organizational roles', in P. Nystrom and W.A. Starbuck (eds), *Handbook of Organizational Design: Adapting Organizations to their Environment*, Oxford: Oxford University Press, pp. 290–308.
Sandberg, W. (1992), 'Strategic management's potential contribution to a theory of entrepreneurship', *Entrepreneurship: Theory and Practice*, Spring , 73–90.
Schein, E.H. (1983), 'The role of the founder in creating organizational culture', *Organizational Dynamics*, **5**(1), 13–28.
Scott, R.W. (2002), *Insitutions and Organisations*, Thousands Oaks, CA: Sage Publications.
Shamir, B. (1991), 'Meaning, self, and motivation in organisations', *Organisation Studies*, **12**(3), 405–24.
Sharma, P., J.J. Chrisman and J.H. Chua (1997), 'Strategic management of the family business: past research and future challenges', *Family Business Review*, **10**(1), 1–35.
Simon, H.A. (1947), *Administrative Behaviour: A Study of Decision-Making Processes in Administrative Organisations*, New York: Macmillan.
Suchman, M.C. (1995), 'Managing legitimacy: strategic and institutional approaches', *Academy of Management Review*, **20**(3), 571–610.
Tagiuri, R. and J.A. Davis (1982), 'Bivalent attributes of the family firm', working paper, Harvard Business School, Cambridge. Reprinted 1996 in *Family Business Review*, **9**(2), 199–208.
Turner, R.H. (1978), 'The role and the person', *American Journal of Sociology*, **84**(1), 1–23.
Ward, J.L. (1988), 'The special role of strategic planning for family businesses', *Family Business Review*, **1**(2), 105–17.
Ward, J.L. (1987), *Keeping the Family Business Healthy: How to Plan for Continuing Growth, Profitability and Family Leadership*, San Francisco, CA: Jossey-Bass.
Whiteside, M.F. and F. Herz Brown (1991), 'Drawbacks of a dual approach to family firms: can we expand our thinking', *Family Business Review*, **4**(4), 383–95.
Whittington, R. (2001), *What is Strategy and Does it Matter?*, London: Thomson Learning.
Whittington, R. (2003), 'The work of strategizing and organizing: for a practice perspective', *Strategic Organization*, **1**(1), 117–26.
Yin, R.K. (1989), *Case Study Research: Design and Methods*, (2nd edn), Newbury Park, CA: Sage Publications.
Zahra, S.A. (2003), 'International expansion of U.S. manufacturing family business: the effect of ownership and involvement', *Journal of Business Venturing*, **18**, 495–512.

15 The professionalization of family firms: theory and practice

Lucrezia Songini

This chapter contributes to the debate on the professionalization of family firms. It is a positioning paper, whose main aim is to highlight the state of the art of literature and studies about the professionalization of family-owned businesses. A detailed review of most significant theoretical streams is presented in order to outline the features, the drivers and the effect of professionalization of family firms. Based on the literature analysis, some research hypotheses are presented, which have been tested on a sample of Italian enterprises. Finally, the implications and limitations of the study are outlined. The structure of the chapter is as follows. First, the main features of family firms are outlined, which can impact on the professionalization. Secondly, the main aspects and features of professionalization are highlighted. Thirdly, the relationships between professionalization and company performance are presented. Fourthly, the main theoretical streams which dealt with professionalization are outlined in order to classify them into two categories: theories which point out the drivers and needs of professionalization and theories which explain mostly the reasons to avoid it. Then, some hypotheses about the professionalization of family firms are presented along with the results of the analysis of a sample of Italian companies.

The features of family firms

The main features of family businesses are the coincidence of shareholder, governance and management roles; an intrinsic fragility related to the very strong and sometimes conflicting relationship between the family and the company interests and contexts (Corbetta, 1995; Gallo, 1993); the presence of three subsystems that need to be coordinated – the family, which considers the firm as a source of financial resources and a way to transfer to the heir and heiress the identity of the family, the shareholders, who are interested in an adequate financial return on their investment, and the managers, who pursue their objectives of getting on careers and earning an adequate remuneration (Tomaselli, 1996). As new generations inherit shares of the company, the number of shareholders increase over time, leading to both a significant degree of fragmentation of ownership structure and the weakening of family ties among shareholders and their identification with the company. Effectiveness of governance models, management systems and their ability to adequately represent a variety of stakeholders (Dumas, 1997; Freeman and McVea, 2001; Frishkoff and Brown, 1997) and to cope with the environmental challenges prove to be drivers of continuity and prosperity for both, the company and the family (Gnan and Montemerlo, 2001).

A general agreement about the definition of family business and its features has not yet emerged among academics. Both theoretical and empirical studies focused on different elements which distinguish family and non-family firms. Some studies pointed out the family's influence on the strategic direction of the firm (Davis and Tagiuri, 1989; Handler, 1989; Pratt and Davis, 1986; Shanker and Astrachan, 1996), while others the intention of the family to maintain the control of the business during generations (Litz, 1995). Both the control of the dominant coalition (Chua et al., 1999) and the presence of unique and peculiar resources and capabilities, such as familiness (Habbershon et al., 2003), were considered features of family-owned businesses too. Chrisman et al. (2003, p. 9) stated that the essence of a family firm consists of: '1. intention to maintain family control of the dominant coalition; 2. unique, inseparable, and synergistic resources and capabilities arising from family involvement and interactions; 3. a vision set by the family controlled dominant coalition and intended for trans-generational pursuance; and 4. pursuance of such a vision'. Astrachan et al. (2002) considered three dimensions of family influence: power (involvement of the family in ownership, governance and management), experience (succession and number of family members who contribute to the business) and culture (overlap between family values and business values, and family business commitment). According to Mustakallio (2002), the various definitions of family business can be summarized into six categories: ownership, management, generational transfer, the family's intention to continue as a family business, family goals and interaction between the family and business.

In this work, we use the term 'family firm' to refer to a company which is both owned and managed by members of one or more families and is perceived as familiar. In particular, a family business is a company where one or more families, with family ties, relationships or solid alliances, own the majority of the capital and are in charge of the governance and management roles (Corbetta, 1995).

The professionalization of family firms: aspects and features

We consider the professionalization of family firms related to the diffusion of the following elements: (1) formal governance mechanisms, such as board of directors, (2) formal strategic planning and control systems (budgeting, reporting, and management accounting), and (3) the involvement of non family members in boards and management, often called professional managers (Dyer, 1996). With regard to governance structures and mechanisms in family firms, they should cope with both the business and the family interests, in order to safeguard shareholder value (Lansberg, 1999). The governance role is accomplished mostly by the board of directors (Gallo and Cappuyns, 1997), which can play different roles: monitoring management on behalf of shareholders, supporting the future development of the business (Demb and Neubauer, 1992; Lorsch and MacIver, 1989; Mace, 1971), planning succession (Harris, 1989). In family businesses the board of directors is critical for its position between family, ownership and the business (Corbetta and Tomaselli, 1996). For some authors, it is even more critical for enterprises in the early stages of development and growth (Danco and Jonovic, 1981; Gallo, 1993; Nash, 1988; Ward, 1991; Ward and Handy, 1988).

With regard to formal strategic planning and control mechanisms, researchers on the professionalization of family firms pointed out that family firms usually are characterized by a lower diffusion of these kinds of mechanisms, as a consequence of widespread entrepreneurship and strong linkages between the family and the enterprise at the ownership, governance and management levels. At the earlier stage of the life cycle of a family firm (the founder or entrepreneurial stage) informal management and control systems are used, little planning and coordination activities are run and decision-making processes are centralized by the entrepreneur. In family businesses, social or relational governance and control mechanisms are strong and long lasting. However, other authors stated that formal mechanisms can help the family-owned businesses to cope with the interests and problems of both the company and the family (Rue and Ibrahim, 1996; Schulze et al., 2003; Ward, 1987, 1988, 1991, 2001). Strategic planning has a peculiar role in family firms, owing to the fact that it can consider the objectives and strategic programmes of both the business and the family (Rue and Ibrahim, 1995, 1996; Sharma et al., 1997; Ward, 1988; Wortman, 1994). For instance, Ward (2001) suggested developing two distinct strategic plans: the company strategic plan, which deals with the company's mission, strategic direction, objectives and programmes, and the family plan, which aims at making explicit personal and professional objectives of family's members and systematically coping with family's issues, such as the future family's involvement and commitment to the company, the uncertainty in the succession process, the rivalry among brothers, sisters and cousins, the uncertainty on the future commitment of the founder, and so on.

Concerning the involvement in the management and boards of non-family members, the professionalization of family-owned businesses does not necessarily imply the involvement of family members in management roles (Gnan and Songini, 2003). Dyer (1996) identifies three different paths: the professionalization of family members; the professionalization of non-family employees, and the employment of new professional managers. The involvement of outsiders/non-family professionals in the governance and management structures and boards can bring objectivity in a family firm's decision-making processes, strategic and succession planning and management (Ibrahim et al., 2001).

To understand the process of professionalization of a family firm, its actual positioning in the life cycle has to be considered too. Actually, the authors agreed on the fact that in the life cycle of a small enterprise a phase emerges, characterized by the increasing complexity of the competitive environment and the company strategy and organizational structure. The separation between power to take decisions (the managers) and the power to control (the owners of the capital) emerges too. This phase is usually positioned in a middle/advanced stage of the life cycle of the company, characterized by the need for a more managerial approach in strategy formulation and management than merely an entrepreneurial one, which asks the company/entrepreneur to delegate the decision-making and management to professional managers and to introduce formal governance, planning and control systems (Deakins et al., 2002; Irvin, 2000; Kroeger, 1974; Perren et al., 1999). In the growth of the enterprise, the division between ownership and managers leads also to the appearance of different interests of shareholders and managers, which are liable to diverge, requiring the introduction of procedures such as governance and control mechanisms, aimed at controlling the activities of managers (Berles and Means, 1932).

The relationship between professionalization and company performance

According to a review of articles dealing with family business management studies, a significant proportion of the studies considered the economic performance of family firms, but only 15 per cent has this topic as primary focus (Chrisman et al., 2003). They focused on performance characteristics such as size, growth, financial structure, productivity and profitability (Gallo, 1995; Gnan and Montemerlo, 2001; Gnan and Songini, 2003; McConaughy and Phillips, 1999; Westhead and Cowling, 1998). It seems that family firms pursue not only economic, but also non-economic goals, with the consequence that the measurement of overall performance becomes much more complex. A study made on the S&P500 (Standard and Poors 500 Stock Index) showed that enterprises influenced by the founding families outperformed those that are not (Anderson and Reeb, 2003). Family firms seem to outperform on a number of aspects, as a consequence of a unique economic vision and lower agency costs (Kirchhoff and Kirchhoff, 1987; Kleiman et al., 1995). A study on the performance and capital structure of large, publicly traded firms controlled by founding families found that these companies have higher profit margins, faster growth rates, higher sales and cash flow per employee, more stable earnings and lower dividend rates (Aronoff and Ward, 1995). Family businesses can have lower costs of family-provided capital, and a longer-term horizon. However, other studies reported that family-owned businesses have lower performance in the long run, owing to the weak family members and succession difficulties (Adams et al., 1996).

The way professionalization is managed can have a significant impact on the performance and survival of the family firm in the long term (Dyer, 1996). Some studies found a correlation between some governance processes, such as that between regular family meetings and board meetings, and business longevity and size (Astrachan and Kolenko 1994). Strategy development and planning can impact on firm performance (Aram and Cowen, 1990) and the growth too (Astrachan and Kolenko, 1994; Ward, 1997). Ward (1988) reported that family businesses tend to have lower performance, owing to a limited use of strategic planning. A study by Chrisman et al. (2002) showed that strategic planning had a greater positive impact on the performance of non-family firms, implying that agency costs are lower in family-owned businesses. Schwenk and Shrader (1993) pointed out that formal planning and control mechanisms are positively correlated with company performance. Gimeno et al. (2004) identified a relationship between business development and complexity and the features of governance structures. Recent studies found that the articulation of agency-costs control mechanisms can have a positive impact on the performance of family firms (Gnan and Songini, 2003). However, the existing evidence about the relationships between the structure of the board, the use of strategic planning and control mechanisms and the enterprise's financial performance is not conclusive (Johnson et al., 1996; Dalton et al., 1998, 1999).

Theoretical streams on the professionalization of family firms

A general agreement about the need, the effects and the path towards professionalization of family-owned businesses has not yet emerged among authors, despite a lot of theories dealt with this issue. In particular, the features, the drivers, the process and the advantages of professionalization of family-owned businesses can be analysed with regard to five different theories: the agency theory, the stewardship theory, the resource-based view (RBV) of the firm theory, the organizational control theory and the company growth theory. These main theoretical streams can be classified into two categories: theories which point out the drivers and need for professionalization of family firms, and theories which explain mostly the reasons to avoid it.

Theories in favour of professionalization of family firms

The agency theory The agency theory considers the formal governance and administrative control systems as a way to align interests and actions of managers and owners

(Jensen and Meckling, 1976; Myers, 1977; Roos, 1973). Owing to the moral hazard and opportunism which the agents could pursue, in a professionally managed firm, control systems are widespread to monitor company operations. The board's main function is to monitor management actions and results on behalf of shareholders (Eisenhardt, 1989; Hillman and Dalziel, 2003; Mizruchi, 1983), dealing with three main roles: output control, behavioural control, strategic control (Fama and Jensen, 1983). Williamson (1981) highlighted that when ownership is concentrated the conflict between managers and owners should disappear. A company performs better to the extent that management and ownership overlap. Family ownership had to be particularly efficient to minimize agency problems owing to the fact that shares are in the hands of agents who can control agency problems, without separating management and control decisions, because of the special relationships with other decision agents (Fama and Jensen, 1983). However, some features of family firms, such as free-riding, ineffective managers, predatory managers, the non-alignment of interests among the non-employed shareholders and the top management team, can increase agency costs (Bruce and Waldman, 1990; Gallo, 1996; Gallo and Lacueva, 1989; Morck et al., 1988). In fact, when one individual or a very few people centralize decision-making process, in order to maintain control within the firm, and do not favour dissention and autonomy among managers, this situation is potentially costly for the family firm, whose performance is likely to suffer (Daily and Dollinger, 1993). Conflicts of interests between family members in different roles can reduce altruism and efficient collaboration and information exchange. Family firms, owing to the scarce financial resources and the reluctance of owners to dilute the company's control, cannot offer the same conditions to their managers as the publicly owned companies. This fact has some consequences: the risk of attracting less competent employees and of employing people with opportunistic behaviours is higher; it is more costly to protect the firm against adverse selection; it is difficult to give shares of the company to better managers. Limited opportunities in a career and lower remuneration decrease the motivations and incentives of competition among people to pursue company goals. The salaries are not aligned with the market levels owing to the tendency to compensate family members without considering their real performance. The altruism in the relationship between father and sons/daughters can hide moral hazard, owing to the fact that parents are likely to be generous with their sons/daughters. Incentives related to the results obtained can avoid opportunistic behaviours by the sons/daughters and favour the pursuit of objectives consistent with those of the company (Bruce and Waldman, 1990). Entrenchment allows managers to extract private benefits from owners, thereby decreasing firm value (Morck et al., 1988). According to some authors, this causes bigger problems in family firms than in non-family ones (Gallo and Vilaseca, 1998; Gomez-Mejia et al., 2001; Morck and Yeung, 2002, 2003). Researchers have shown that altruism and entrenchment have both positive and negative effects on family firm performance. For instance, the implications of management entrenchment are not one-sided. Pollak (1985) states that family enterprises have advantages in incentives and monitoring vis-à-vis non-family firms. Shleifer and Vishny (1997) argue that family ownership and management can add value when the political and legal systems of a country do not provide sufficient protection against the expropriation of minority shareholders' value by the majority shareholder. Burkart et al. (2003) showed that in economies with a strong legal system to prevent expropriation by majority shareholders, the widely held professionally managed firm is optimal. However, where the legal system cannot protect minority shareholders, keeping control and management within the family is optimal.

To promote unity and commitment among shareholders and other family members, who are likely to be future owners of the business, family businesses can adopt different practices (Tagiuri and Davis, 1982), mostly based on governance and control systems. Actually, budgeting, reporting and incentives can help to limit the opportunistic behaviours of agents, because they aim at defining and assigning objectives, monitoring results and compensating adequate performance. Boards of directors can also be useful to communicate to agents the principle's objectives and to control their performance. The involvement of non-family members in the boards and management can help to avoid opportunisms too. Strategic planning is another mechanism for controlling agency costs. For these reasons, many authors said that research in family business can benefit from the agency theory approach (Daily and Dollinger, 1992). In fact, the introduction of formal governance and administrative control systems can help a family firm to cope with its peculiar features, such as free-riding, ineffective managers, predatory managers, and so on. Recent studies founded that the level of professionalization of family firms is mostly related to the adoption of agency cost control mechanisms, that is, formal governance, planning and control systems (Gnan and Songini, 2003; Montemerlo et al., 2004). In particular, the use of agency cost control mechanisms, especially boards of directors and strategic planning, was proposed to increase the company performance of family firms (Schulze et al., 2001, 2003). Other studies pointed out that the more the complexity of ownership and the bigger the size of the enterprise, the greater is the need to adopt a principal–agent relationship and to articulate governance and control structures (Montemerlo et al., 2004). The need for good information systems derives from the agency costs arising from the conflict of interest between owner-managers and lenders, too (Poutziouris et al., 1998). These results point out the need to analyse the role of agency cost control mechanisms in family firms more in depth.

It is worth noting that the agency theory has been criticized, especially by the stewardship theory, for ignoring the effects of good social relationships that might exist among owners and managers, such as in the family firms (Ghoshal and Moran, 1996).

The company growth theory The company growth theory analyses the features, the problems and the advantages of the development and growth process of a firm, aiming at becoming a large enterprise, passing through different stages in its life cycle (Christensen and Scott, 1964; Greiner, 1972; McGuire, 1963; Normann, 1977; Rostow, 1960; Steinmetz, 1969). Specific models to explain the SMEs growth and development towards large size were developed, that pointed out the role of information and administrative control systems to support the enterprise's development process (Churchill and Lewis, 1983; Cooper, 1981; Dodge and Robbins, 1992; Scott and Bruce, 1987). In fact, the use of formal governance and administrative control systems can be related to environmental and firm complexity. According to company growth theory, a small company progresses through distinct stages as it develops (Churchill and Lewis, 1983; Greiner, 1972; Irwin, 2000; Kroeger, 1974; Scott and Bruce, 1987). Successful growth leads to a critical stage, namely professionalization, which requires the owner-manager to change his/her entrepreneurial approach to a more professional one (Deakins et al., 2002; Perren et al., 1999). After this stage of the life cycle, family firms tend to adopt formal control mechanisms and to decentralize decision-making processes (Moores and Mula, 2000). Management accounting, budgeting and reporting are implemented owing to the fact that they can assure a wider delegation of responsibilities (Goffee and Scase, 1987). As the environmental and organizational complexity increases, it also becomes necessary for a company to define more formalized and clear managerial responsibility and to delegate the responsibility of different activities to specialized managers who are in charge of different organizational departments, with the use of appropriate mechanisms such as responsibility accounting, budgeting, and performance evaluation systems. The stage of professionalization is characterized by a more complex relationship between the company and the environment, which requires the adoption of strategic planning and control systems to allow the company to cope also with the evolution of the external environment.

It is worth noting that a family firm is characterized by different life cycles, concerning the family, the business and the ownership (Gersick et al., 1997). Formal planning and control systems can help a family firm to cope with the challenges of family business continuity too (Ward, 1987, 2001). They evolve throughout the life cycle of a family firm (Moores and Mula, 2000). The way the professionalization is handled, with regard to both the involvement in the management of non-family/professional managers (Dyer, 1996) and the adoption of governance, strategic planning and control systems, impact on the success of growth (Gnan and Songini, 2003, 2004).

Theories against professionalization of family firms

The stewardship theory The stewardship theory argues that managers whose needs are based on growth, achievement and self-realization, and who are intrinsically motivated, can make better use of organizational objectives than of personal ones. If they identify with their organizations and are strongly committed to organizational values, they are more likely to pursue organizational objectives. When applied to family firms, the stewardship theory suggests that the coincidence of family and business values and objectives, at least among the first generations, brings individuals to follow collaborative and altruistic behaviours, aimed at pursuing the company goals (Davis et al., 1997). Some studies highlighted that in some situations altruism and kinship obligations can mitigate agency problems (Eaton et al., 2002; Wu, 2001), mainly as a consequence of four elements: they create a unique family firm's history, language and identity; they produce a collective ownership; they reduce information asymmetries among family agents, due to the incentives to communicate and cooperate; they create a unique capability of loyalty and commitment to the firm's long-run performance and strategy (Van den Berghe and Carchon, 2003). As a consequence, in family firms agents tend to pursue the owners' objectives (Jensen and Meckling, 1976), reducing the agency costs. The owner directly manages the company; the interests of owners and managers are aligned, as a consequence of the fact that agents and principles are linked by peculiar relationships, based on familiar ties. Inside family firms an overlap among different roles develops, owing to the fact that people are both family members, owners and managers (Tagiuri and Davis, 1982). This theory indicates that it is the alignment of ownership and control that produces advantages for the family business. The identification of family members with the firm creates a sense of loyalty and trust towards the organization (Menendez-Requejo, 2004). In this context, management is considered a good steward and the role of board is to interact with management to create value, as a mentor, and participate in strategic decision-making with management value (Huse, 2000). In this perspective, the roles of boards are mainly networking, legitimacy, advice and strategic participation. Moreover, in family-owned enterprises the need to account for actions to the owner should not be necessary, explaining the use of less formalized systems which substitute the control ones (Whisler, 1988). The stewardship theory states that in family-owned businesses formal governance and control mechanisms are not necessary, because of the relationships among people. However, these mechanisms can have negative effects on the agent's behaviour. Other authors state that the prevalence of either stewardship or agency relationships depends on some psychological and situational factors. The first considers the degree of identification of the individual with the enterprise and the way power is exercised; the second concerns management philosophy and organizational culture (Craig et al., 2003).

The resource-based view (RBV) of the firm theory The RBV of the firm theory (Barney, 1991) identifies the resources and capabilities that make family enterprise unique and allow it to develop peculiar and family-based competitive advantages. This theory suggests a firm is a family business where the role of a family impacts on its functioning and performance. According to this theory the combination of the family and business systems in a family firm create both economic and non-economic value and 'lead to hard-to-duplicate capabilities or "familiness"' (Chrisman et al., 2003, p. 7). Familiness is a peculiar feature of family enterprise that explains its survival and growth, in addition to two other aspects: emotional involvement of individuals in firm activities and the private language of relatives (Habbershon and Williams, 1999; Habbershon et al., 2003). Studies based on the RBV of the firm theory deal with several topics: which are the peculiar capabilities and competencies of family firms? How do they identify and develop such capabilities? How do they transfer them to new generations and continuously develop new capabilities? The evidence of the RBV of the firm theory with regard to the impact of professionalization is not conclusive. In fact, professionalization can have both positive and negative effects on family-owned businesses.

On the one hand, the overlapping owner and agent relationship can have advantages, such as the reduction of financial reporting, the decrease of regulatory compliance and administrative costs, faster decision-making, longer time horizons, family unity which engenders family commitment and allows longer-term investment return horizon (Poza et al., 2004). On the other hand, some studies pointed out that elements, such as close kinship, ownership and management transfer, conflict of interests and altruism can nullify the value of existing capabilities and obstruct the creation and renewal of new distinctive familiness (Cabrera-Suárez et al., 2001; Steier, 2001a, 2001b, 2003; Stewart, 2003; Wu, 2001). These studies are based mostly on an agency theory approach, which points out that opportunistic behaviours in family enterprises can lead to agency costs that destroy or reduce the specific capabilities of family firms (Gomez-Mejia et al., 2001; Morck and Yeung, 2003; Schulze et al., 2003). Finally, some authors pointed out that boards, managers and planning and control systems can be considered relevant resources for family firms. In particular, the role of strategic planning and control systems consists of identifying the strategic resources of the firm, the strengths and weaknesses of competitors and the opportunities for better using these resources; identifying the capabilities and allocating scarce strategic resources to relevant capabilities; and selecting the strategic decisions that better use resources and capabilities (Grant, 1991). The final goal is to allocate resources in an optimal way to gain a competitive advantage consistently with the resources and capabilities of the firm.

The organizational control theory The way decisions are carried out in an organization is related to the concept of organizational control. In effect, methods of decision-making and interaction between people and organizational units influence control processes. The organizational control theory identifies three different kind of control systems: social control, administrative control, and individual control. A person's behaviour can be influenced by the rules of the groups in which he/she is involved (social control or clan), by the rules, the plans, the programmes and the incentive mechanisms which define organizational objectives, allocate strategic resources, programme and coordinate the management decisions and evaluate performance (bureaucratic or administrative control) or by identification with organizational goals and objectives, values and management philosophies (individual control) (Child, 1972; Galbraith, 1977; Herzberg, 1968; Hopwood, 1974; Johnson and Kaplan, 1987; Mintzberg, 1994; Prahalad and Doz, 1981). With regard to strategic planning, the design school approach suggests that a firm can formulate and implement effective business strategies through the use of models of analysis of the internal and external business environments (Lorange, 1980). However, this approach is often not appropriate for small businesses; instead an incremental, emerging, intuitive and informal approach is more suited to them (Mintzberg, 1994; Normann, 1977; Quinn, 1980). The literature suggests that firms adopt managerial control mechanisms for various purposes, such as the need to cope with increasing firm's and environmental complexity, the need for the entrepreneur to delegate activities, the need to look for external funding or quotation, and so on. These mechanisms can enable any firm, even those managed by stewards rather than agents, to make better strategic decisions in light of its environmental and resource circumstances (Ford, 1988; Schwenk and Shrader, 1993; Ward, 1988). The organizational control theory points out that clan and social control systems are more effective than bureaucratic and administrative control when the strategy formulation, the decision-making processes, and the power in the organization are managed by a few people who share common values and coordinate themselves by informal relationships (Hopwood, 1974; Mintzberg, 1983; Ouchi, 1981). In family firms the social interactions among family members allow the use of informal and cultural mechanisms that substitute or complement the formal administrative systems. Daily and Dollinger (1992) reported that family firms use more informal control processes and systems. Recent research pointed out that informal mechanisms, based on auto-coordination, auto-control and clans are more used in family firms (Uhlaner and Meijaard, 2004).

Concluding remarks, research propositions and hypotheses

The analysis of different theories dealing with the professionalization of family-owned businesses pointed out that the majority agrees on the fact that family firms are characterized by a lower diffusion and use of formal governance, strategic planning and control systems than non-family firms. In particular, stewardship and the RBV of the firm theories state that family firms can be effectively managed without formal managerial mechanisms and the involvement of non-family members as a consequence of the peculiar features of family businesses: familiness, altruism, emotional involvement, common and unique language, values and so on. Organizational control theory applied to family businesses points out that social and individual control systems are more suited to these enterprises, owing to common shared values and languages, informal and kinship relationship, and a small group of people being in charge of ownership, governance and management. However, agency theory and the company growth theory highlight the need for family firms to adopt formal governance, strategic planning and control mechanisms and to involve non-family members in governance and management. Agency theory considers the agency cost control mechanisms necessary for family-owned businesses to cope with their peculiar features that increase the agency costs. However, an inconclusive picture emerged from the analysis of theories, which point out both the advantages and the uselessness of professionalization of family firms. Moreover, to better understand the role, the drivers and the features of the professionalization of family firms, it has to be considered that there is a life cycle in the introduction of different mechanisms and involvement of non-family members, which is related to both the business and the family development. According to company growth theory, a board of directors and strategic planning seem to be introduced in earlier stages than administrative control systems, because they can manage and coordinate the development and growth of both the family and the business during generations. Administrative control systems are implemented in late stages, when the company's large size, strategic and organizational complexity and the competitive environment do not allow a single individual or to a few people to cope with all strategic and management issues and require more delegation. It could be of interest to better compare the diffusion, the features, the objectives and roles of governance, strategic planning and control systems among family firms and non-family enterprises, as well as to analyse the degree of involvement of family and non-family members in ownership, governance and control mechanisms. The drivers of professionalization and the effect on company performance have to be studied too.

According to the evidence of literature analysis, the following propositions and hypotheses are proposed, which deal with the features, the process and the drivers of professionalization of family-owned businesses and its impact on the economic performance:

Proposition 1 Professionalization of family firms is related to the diffusion of governance mechanisms, formal strategic planning and control mechanisms, and the involvement of non-family members in governance and management roles. This proposition is consistent with the evidence of agency and organizational control theories.

Hypothesis 1(H1): The greater the involvement of family members in running the company, the lower the adoption of governance, strategic planning and control mechanisms.

Proposition 2 According to various authors, there is a relationship between the involvement of family members in governance and management roles and the company performance.

Hypothesis 2 (H2): There is a positive relationship between the involvement of family in the board of directors and economic performance of the firm.

Hypothesis 3 (H3): There is a positive relationship between the involvement of family in management roles and economic performance of the firm.

Proposition 3 According to various authors, there is a relationship between the presence of formal governance mechanisms, such as board of directors, strategic planning and control mechanisms, and the company performance.

Hypothesis 4 (H4): There is a positive relationship between the presence of the board of directors and the economic performance of the firm.
Hypothesis 5 (H5): There is a positive relationship between the presence of strategic planning and control mechanisms and the economic performance of the firm.

Proposition 4 According to the company growth theory, there is a sequence in the introduction of governance, strategic planning and control mechanisms, among family enterprises, which are introduced in different stages of the company life cycle: first are implemented governance mechanisms, then strategic planning, and finally administrative control systems. All things being equal, the following hypothesis was defined:

Hypothesis 6 (H6): Board of directors is more widespread in family firms than strategic planning and control mechanisms.

Proposition 5 The increase in the complexity of strategy and organizational structure and the consequent need for the firm to adapt its strategies play a significant role in the widespread of both strategic planning and formal control mechanisms. This is a proposition consistent with the company growth and organizational control theories. All things being equal, the following hypotheses were defined:

Hypothesis 7 (H7): The increase in the strategic business differentiation (strategy complexity) requires a wider adoption of strategic planning and formal control mechanisms.
Hypothesis 8 (H8): The increasing complexity of organizational structure requires a wider adoption of strategic planning and formal control mechanisms.

Research design

Sampling frame

The initial sample consisted of 7964 manufacturing SMEs drawn from AIDA database (by Bureau Van Dijk Electronic Publishing), of Milan province, defined at the four-digit level of the ATECO91 Classification System. Small and medium-sized enterprises are defined as enterprises which:

1. Have fewer than 250 employees, and
2. Have either, an annual turnover not exceeding 40 million euros, or an annual balance sheet total not exceeding 27 million euros,
3. Confirm to the criterion of independence as defined below.

Independent enterprises in which 25 per cent or more of the capital or the voting rights are owned by one enterprise, or jointly by several enterprises fall outside the definition of an SME. The AIDA database contains 1994 to 2001 balance sheet data of about 130 000 incorporated SMEs, representative of the Italian population and operating both in manufacturing and non-manufacturing industries. The Milan province was chosen so that the sample had a high level of internal homogeneity and achieved a broader representativeness of Italian SMEs. The 7964 companies were articulated by range of turnover and industries (Table 15.1).

Table 15.1 Initial sample

Industry	Turnover (million €)					
	0–4	4–8	8–20	20–40	40–80	Total
Chemical	331	123	128	84	48	714
Food	108	50	44	14	16	232
Electronic	847	219	169	72	36	1343
Textile	514	123	128	51	27	843
Mechanic	936	280	229	96	41	1582
Raw material transformation	2117	541	361	141	90	3250
Total	4853	1336	1059	458	258	7964

Procedure

The data collection process proceeded in four phases. First, measurement scales were developed by reviewing relevant literature, by completing five on-site interviews with CEOs from medium-sized firms, with academics and consultants, and by pre-testing the resulting scales with a group of academics and consultants. Next, a single researcher pre-tested the preliminary versions of the resulting questionnaire with some senior executives of SMEs. The third stage consisted of on-site interviews with CEOs or executives in ten SMEs, resulting in the final versions of the questionnaire. In the final stage, the survey was mailed to the companies included in the sampling frame described above. We addressed the surveys to the chief executives of the firms. A single informant was used for each acquisition. Although the use of multiple respondents would have reduced concerns about potential response biases, respondents had to be knowledgeable about the firm and its competitive environment (Campbell, 1955). In a large sample study, identifying and obtaining responses from multiple well-informed respondents is extremely problematic. The key methodological solution in using a single respondent approach is to find the most appropriate respondent. Thus, we qualified our respondents as individuals who held a CEO or equivalent position (president, executive chair and managing director).

Achieved sample

From the initial sample, questionnaires were mailed to 1122 companies, in such a way as to be also representative of the reference population by range of turnover and industries, and for whom we obtained addresses. A total of 166 completed questionnaires were returned, representing a response rate of 15 per cent. This response rate is reasonable given the setting of the survey (small firms), firm diversity, the positions of the respondents (CEO, president, executive chair and managing director), and the sensitivity of the information. Following a check to ensure that these cases all represented family firms, 15 responses were eliminated. Family firms were defined as those companies that met at least one of the following requirements: (1) 51 per cent of equity or more owned by the family; (2) family owns less than 51 per cent but controls the company in partnership with friends, other entrepreneurs, employees; (3) respondents perceive the company to be a family business, whatever the family share, which actually happened in four cases (Greenwald and Associates, 1995). The final data-set included 151 family businesses. Non-response biases were evaluated by comparing the industries represented in the sample with the initial sample used. No differences in the

Table 15.2 Sample description 1: industry

	Frequency	Per cent
Chemical	13	8.7
Food	5	3.2
Electronic	26	17.0
Textile	17	11.2
Mechanic	30	19.8
Raw material transformation	61	40.1
Total	151	100.0

Table 15.3 Sample description 2: turnover

	Frequency	Per cent
From 0 to 4 million €	97	64.4
From 4 to 8 million €	24	16.1
From 8 to 20 million €	19	12.3
From 20 to 40 million €	7	4.8
From 40 to 80 million €	4	2.4
Total	151	100.0

Table 15.4 Sample description 3: employees

	Frequency	Per cent
Up to 15 employees	66	44.5
From 16 to 50 employees	57	38.9
From 50 to 100 employees	11	7.2
Over 100 employees	14	9.4
Total	147	100.0

industries represented were found. Early respondents (first half) were also compared with late respondents (second half), following the Armstrong and Overton procedure (1977). No significant differences were found on key characteristics such as age of the company, size (employees and turnover), market conditions or industry characteristics, suggesting that non-response bias might not be a problem. Overall, the data-set represented a wide range of industries, firms and typology of the firm, as shown in Tables 15.2–15.6.

Descriptive results

Family involvement, board of directors and formal strategic planning and control systems
Consistent with agency and organizational control theories, it was stated that professionalization of family firms is related to the diffusion of governance mechanisms, formal strategic planning and control mechanisms, and the involvement of non-family members in governance and management roles. Accordingly, we hypothesized that (H1): 'The

Table 15.5 Sample description 4: firm typology

	Frequency	Per cent
Family's ownership and governance	120	81.3
Family's ownership	13	8.6
Family's governance	4	2.7
Public company	7	4.8
Co-operative company	4	2.7
Total	148	100.0

Table 15.6 Sample description 5: year of foundation

Valid	150
Mean	1972.5
Median	1982.0
Mode	1989.0
Std. deviation	22.6
Percentiles 25	1956.0
Percentiles 50	1982.0
Percentiles 75	1989.0

Table 15.7 Family involvement versus board of directors and formal strategic planning process

		Board of directors (%)	Formal strategic planning (%)
Family involvement in governance	Up to 25%	42.2	25.1
	From 25% to 50%	100.0	33.7
	From 50% to 75%	100.0	22.1
	Over 75%	67.6	14.7
Family involvement in management	Up to 25%	71.1	19.9
	From 25% to 50%	84.2	36.3
	From 50% to 75%	68.6	1.2
	Over 75%	56.5	24.0

greater the involvement of family members in running the company, the lower the adoption of governance, strategic planning and control mechanisms.'

Table 15.7 reports the extent of family involvement both in governance and in management processes. We cannot state a real strong relationship between the degree of involvement of family members in strategic and managerial decision-making processes and the presence of board of directors and strategic planning (there are only two significant differences, one between the percentages of family involvement in governance for firms with a board of directors – ANOVA $F = 14.683$, $p<0.01$ – and one between the percentages of family involvement in management for firms with a strategic planning – ANOVA $F = 3.087$, $p<0.05$). In other words, the professionalization of family firms

Table 15.8 Family member in charge of governance and management positions

Management positions	% of total
President	66.2
CEO	50.9
Board member	38.7
General manager	29.5
BU director	11.0
Administrative director	31.4
Production director	28.0
Sales and marketing director	23.5
Purchasing director	18.5

seems not to imply the involvement of family members in management positions and activities as a condition for a greater alignment between the interests of owners and managers. Actually, the use of board of directors and strategic planning is related to the presence of managers in charge of the company governance, but they do not necessarily have to be members of the family which owned the firm.

Family involvement and economic performance

According to various authors, there is a relationship between the involvement of family members in governance and management roles and the company performance. It was hypothesized that (H2): 'There is a positive relationship between the involvement of family in the board of directors and economic performance of the firm', and (H3): 'There is a positive relationship between the involvement of family in management roles and economic performance of the firm.'

In Table 15.8 the presence of family in governance and management is highlighted. Family members are involved mostly in the positions of president, CEO, board member and general manager.

Tables 15.9 and 15.10 report statistics for the economic performances. For different levels of family involvement in governance (Table 15.9), significant differences in economic results were found for the two growth indicators: Sales CAGR (Constant Average Growth Rate) (ANOVA $F = 3.655$, $p<0.01$) and Invested capital (Constant Average Growth Rate) (ANOVA $F = 3.681$, $p<0.01$).

For different levels of family involvement in management (Table 15.10), several significant differences (Table 15.11) on economic results for both growth and profitability indicators were found: Sales and Invested capital CAGR (Constant Average Growth Rate), ROS and ROI.

The family involvement in the governance and management indicates the owners' commitment to the family-owned company. This could also explain the good economic performances.

Board of directors, formal strategic planning and economic performance

According to various authors, there is a relationship between the presence of formal governance mechanisms, such as board of directors, strategic planning and control

Table 15.9 Family involvement in governance and economic performances

		Up to 25%	From 25% to 50%	From 50% to 75%	Over 75%
Sales CAGR* (98–00)	N	24	25	23	52
	Mean	0.0255	−0.0206	0.1429	0.0587
	Std. Deviation	0.15056	0.25453	0.14386	0.15387
Invested Capital CAGR* (98–00)	N	24	25	23	52
	Mean	0.0255	−0.0206	0.1429	0.0601
	Std. Deviation	0.15056	0.25453	0.14386	0.15352
Net Assets CAGR* (98–00)	N	24	25	23	52
	Mean	0.0661	0.1573	0.2531	−0.0081
	Std. Deviation	0.10086	0.27865	0.31523	1.20057
ROS (average 98–00)	N	25	26	25	63
	Mean	0.0405	0.0272	0.0539	0.0493
	Std. Deviation	0.04610	0.06100	0.08017	0.06547
ROI (average 98–00)	N	25	26	25	62
	Mean	0.0405	0.0277	0.0526	0.0440
	Std. Deviation	0.04610	0.06038	0.07664	0.04800
ROE (average 98–00)	N	25	25	24	64
	Mean	−0.0011	0.0759	0.1246	0.0312
	Std. Deviation	0.16062	0.30615	0.12838	0.45325

Note: * CAGR = Constant Average Growth Rate.

mechanisms, and the company performance. As a consequence, the following hypotheses were identified: Hypothesis 4 (H4): 'There is a positive relationship between the presence of the board of directors and the economic performance of the firm', and Hypothesis 5 (H5): 'There is a positive relationship between the presence of strategic planning and control mechanisms and the economic performance of the firm'.

Tables 15.12 and 15.13 report the economic performances of the family firms where, respectively, there are the board of directors and the presence of a formal strategic planning process. The results show that the magnitude of economic performances varies significantly across the two situations (presence/absence of board of directors and presence/absence of formal strategic planning).

It is clear that the presence of both a board of directors and strategic planning affects the economic performances of the firms in an asymmetric manner: the growth indicators (Sales, Invested Capital and Net Assets CAGR) are not significantly different between the two situations, whereas profitability indicators are significantly different (both situations present respectively for ROS, $p<0.10$; ROI, $p<0.05$; ROE, $p<0.10$). These results are

Table 15.10 Family involvement in management and economic performances

		Up to 25%	From 25% to 50%	From 50% to 75%	Over 75%
Sales CAGR* (98–00)	N	73	40	12	4
	Mean	0.0898	0.0240	−0.1219	0.0358
	Std. Deviation	0.20824	0.11066	0.10570	0.15316
Invested Capital CAGR* (98–00)	N	73	39	12	4
	Mean	0.0898	0.0256	−0.1219	0.0358
	Std. Deviation	0.20824	0.11004	0.10570	0.15316
Net Assets CAGR* (98–00)	N	73	40	12	4
	Mean	0.1120	−0.0022	0.2327	0.0314
	Std. Deviation	0.35701	1.34428	0.16397	0.14233
ROS (average 98–00)	N	85	41	12	4
	Mean	0.0450	0.0627	−0.0222	0.0760
	Std. Deviation	0.07116	0.04403	0.02817	0.05125
ROI (average 98–00)	N	84	41	12	4
	Mean	0.0407	0.0603	−0.0222	0.0760
	Std. Deviation	0.05840	0.04323	0.02817	0.05125
ROE (average 98–00)	N	85	41	12	4
	Mean	0.0069	0.1279	0.0767	0.1362
	Std. Deviation	0.38873	0.28281	0.10042	0.12234

Note: * CAGR = Constant Average Growth Rate.

consistent with the literature that shows a positive relationship between those mechanisms and economic performance (Schwenk and Shrader, 1993).

Relative diffusion of governance, strategic planning and control mechanisms among family firms

According to the company growth theory, there is not the same degree of diffusion of governance, strategic planning and control mechanisms among family enterprises as a consequence of the presence of a life cycle in their introduction. As a consequence, it was hypothesized that 'Board of directors is more widespread in family firms than strategic planning and control mechanisms' (Hypothesis 6 – H6).

The results point out that a board of directors exists in almost all family enterprises. Seventy-four per cent of family firms in Milan province have a board of directors. Strategic planning is adopted only by 22.6 per cent of the sample companies. Among these enterprises 13.1 per cent have introduced strategic planning within the past five years, while 9.5 per cent have done so more than five years ago. Companies not using strategic planning point out the following main reasons for this: it costs too much

Table 15.11 Family involvement in management and economic performances (ANOVA Tests)

		Sum of Squares	df	Mean Square	F	Sig.
Sales CAGR* (98–00)	Between Groups (Combined)	0.488	3	0.163	5.368	0.002
	Within Groups	3.788	125	0.030		
	Total	4.277	128			
Invested Capital CAGR* (98–00)	Between Groups (Combined)	0.485	3	0.162	5.304	0.002
	Within Groups	3.779	124	0.030		
	Total	4.263	127			
Net Assets CAGR* (98–00)	Between Groups (Combined)	0.620	3	0.207	0.326	0.807
	Within Groups	79.299	125	0.634		
	Total	79.919	128			
ROS (average 98–00)	Between Groups (Combined)	0.071	3	0.024	6.299	0.000
	Within Groups	0.518	138	0.004		
	Total	0.589	141			
ROI (average 98–00)	Between Groups (Combined)	0.072	3	0.024	8.835	0.000
	Within Groups	0.374	137	0.003		
	Total	0.446	140			
ROE (average 98–00)	Between Groups (Combined)	0.443	3	0.148	1.270	0.287
	Within Groups	16.042	138	0.116		
	Total	16.485	141			

Note: * CAGR = Constant Average Growth Rate.

(42.8 per cent), it distracts from day-to-day activities (41.5 per cent), it is too complicate (38.4 per cent), it is useless (37.7 per cent), and it hinders creativity (36.2 per cent). Enterprises which use strategic planning assign to it the following objectives: strategic goals definition (96 per cent), strategy formulation (94.2 per cent), defining a common vision (95 per cent), spreading a common language (68.5 per cent), identifying opportunities (72.9 per cent) and threats (65.8 per cent), points of strength (91.4 per cent) and weaknesses (84.7 per cent). The diffusion of managerial accounting and control systems are generally widespread in the majority of Milan's enterprises. However, incentives, investment analysis and responsibility accounting point to a limited diffusion (Table 15.14).

With regard to the importance assigned to managerial accounting and control systems, Milan's companies consider very important management accounting (81.6 per cent), budget (71.7 per cent) and standard costing (64 per cent). They give less importance to responsibility accounting (42.2 per cent), investment analysis (36.6 per cent) and incentives

Table 15.12 Presence of the board of directors and economic performances

		N	Mean	Std. Deviation
No board of directors	Sales CAGR* (98–00)	33	0.0750	0.14311
	Invested Capital CAGR* (98–00)	33	0.0750	0.14311
	Net Assets CAGR* (98–00)	33	0.1201	0.22026
	ROS (average 98–00)	34	0.0555	0.04387
	ROI (average 98–00)	34	0.0555	0.04387
	ROE (average 98–00)	34	0.0861	0.12822
	Valid N (list wise)	33		
Board of directors	Sales CAGR* (98–00)	92	0.0438	0.19423
	Invested Capital CAGR* (98–00)	92	0.0445	0.19428
	Net Assets CAGR* (98–00)	92	0.0789	0.92402
	ROS (average 98–00)	105	0.0407	0.06958
	ROI (average 98–00)	104	0.0374	0.05887
	ROE (average 98–00)	105	0.0388	0.39043
	Valid N (list wise)	92		

Table 15.13 Presence of a formal strategic planning process and economic performances

		N	Mean	Std. Deviation
No formal strategic planning	Sales CAGR* (98–00)	95	0.0466	0.17781
	Invested Capital CAGR* (98–00)	95	0.0473	0.17779
	Net Assets CAGR* (98–00)	95	0.0617	0.89280
	ROS (average 98–00)	108	0.0416	0.06848
	ROI (average 98–00)	107	0.0382	0.05813
	ROE (average 98–00)	109	0.0356	0.37914
	Valid N (list wise)	95		
Formal strategic planning	Sales CAGR* (98–00)	32	0.0598	0.18039
	Invested Capital CAGR* (98–00)	32	0.0598	0.18039
	Net Assets CAGR* (98–00)	32	0.1575	0.38158
	ROS (average 98–00)	33	0.0579	0.04964
	ROI (average 98–00)	33	0.0583	0.04878
	ROE (average 98–00)	32	0.1036	0.16812
	Valid N (list wise)	32		

Table 15.14 The presence of managerial accounting and control systems

Managerial accounting and control systems	Presence (% of total)
Managerial accounting	67.5
Standard costing	66.5
Budget	81.4
Managerial reporting	71.3
Investment analysis	46.9
Responsibility accounting	45.9
Incentives	28.2

(29.1 per cent). With regard to budget objectives, 76.1 per cent of sample companies give importance to the definition of short-term and operative goals, 72.3 per cent to the assignment of objectives to managers, 58.8 per cent to the motivation of managers and 50.1 per cent to the variance analysis. As far as the diffusion of strategic planning and control systems and their goals is concerned, we can say that control mechanisms mostly focused on short-term horizons and aimed at defining and assigning objectives and measuring performance are widespread (managerial accounting, standard costing, budget, and managerial reporting).

The results indicated that management accounting is the most widespread mechanism, followed by a board of directors and other control mechanisms, such as budget and standard costing. Strategic planning has the lowest diffusion among family enterprises. These results only partially confirmed agency theory, which states that the most widespread agency cost control mechanisms in family-owned businesses are a board of directors and strategic planning.

Strategic and organizational complexity and strategic planning and control mechanisms

According to the company growth and organizational control theories, the increase in the complexity of strategy and organizational structure plays a significant role in the wide spread of both strategic planning and formal control mechanisms.

All things being equal, the following hypotheses were defined: Hypothesis 7 (H7) 'The increase in the strategic business differentiation requires a wider adoption of strategic planning and formal control mechanisms', and Hypothesis 8 (H8) 'The increasing complexity of organizational structure requires a wider adoption of strategic planning and formal control mechanisms'.

With regard to strategy complexity, it was defined as the degree of strategic business differentiation, articulated according to three levels: low differentiation (an enterprise operating in one strategic business area), medium differentiation (a firm operating in not more than three strategic business areas) and high differentiation (an enterprise operating in more than three strategic business areas). Of the sample companies 50.8 per cent had a low differentiation; 36.7 per cent had a medium strategic business differentiation, while 12.6 per cent had a high differentiation. Hypothesis 7 was partially confirmed (Table 15.15). In fact, the relationship between the degree of strategic business differentiation and the presence of strategic planning was not confirmed. However, a relationship was found between the

Table 15.15 The relationships between strategic business differentiation and the presence of strategic planning and control systems

Strategic planning, managerial accounting and control systems	Degree of strategic business differentiation	Mean	Anova test – F
Strategic planning	Low strategic business differentiation (monobusiness enterprise)	0.2323	0.557
	Medium strategic business differentiation (not more than 3 SBA)	0.2585	
	High strategic business differentiation (more than 3 SBA)	0.1633	
	Total	0.2257	
Managerial accounting	Low strategic business differentiation (monobusiness enterprise)	0.6828	1.185
	Medium strategic business differentiation (not more than 3 SBA)	0.7309	
	High strategic business differentiation (more than 3 SBA)	0.5749	
	Total	0.6750	
Budget	Low strategic business differentiation (monobusiness enterprise)	0.7001	4.822**
	Medium strategic business differentiation (not more than 3 SBA)	0.8689	
	High strategic business differentiation (more than 3 SBA)	0.9270	
	Total	0.8137	
Standard costing	Low strategic business differentiation (monobusiness enterprise)	0.5932	2.798
	Medium strategic business differentiation (not more than 3 SBA)	0.6425	
	High strategic business differentiation (more than 3 SBA)	0.8240	
	Total	0.6648	
Managerial reporting	Low strategic business differentiation (monobusiness enterprise)	0.7969	3.149**
	Medium strategic business differentiation (not more than 3 SBA)	0.7186	
	High strategic business differentiation (more than 3 SBA)	0.5589	
	Total	0.7133	
Investment analysis	Low strategic business differentiation (monobusiness enterprise)	0.4598	7.520**
	Medium strategic business differentiation (not more than 3 SBA)	0.3077	
	High strategic business differentiation (more than 3 SBA)	0.7161	
	Total	0.4687	
Responsibility accounting	Low strategic business differentiation (monobusiness enterprise)	0.3148	14.842**

Table 15.15 (continued)

Strategic planning, managerial accounting and control systems	Degree of strategic business differentiation	Mean	Anova test – F
	Medium strategic business differentiation (not more than 3 SBA)	0.3697	
	High strategic business differentiation (more than 3 SBA)	0.8218	
	Total	0.4586	
Incentives	Low strategic business differentiation (monobusiness enterprise)	0.2578	2.884
	Medium strategic business differentiation (not more than 3 SBA)	0.2048	
	High strategic business differentiation (more than 3 SBA)	0.4341	
	Total	0.2823	

Note: ** Sig. ≤ 0.05.

degree of strategic business differentiation and the presence of investment analysis, budget, managerial reporting and responsibility accounting. It seems that in the sample enterprises, strategic goals and programmes are not defined adopting formal strategic planning, confirming that in SMEs an incremental, informal, emerging and intuitive approach in strategy formulation is used, even though the relationship with investment analysis indicates a rational evaluation of strategic alternatives, when strategic complexity increases. However, it seems that the sample enterprises adopt mechanisms based on a short-term horizon when defining their objectives and programmes, instead of strategic and long-term horizons. The increase in strategic complexity requires delegation of objectives and responsibilities, and monitoring and evaluation of the achieved results by formal mechanisms. This is partially consistent with the organizational control theory.

With regard to the complexity of organizational structure, it was hypothesized that it is related to the presence of strategic planning and control mechanisms. The complexity of the organizational structure was defined with regard to four kinds of structures: functional structure (the simplest one), divisional structure (more complex), matrix structure and project structure (the most complex ones). Of the sample enterprises, 64.9 per cent adopt a functional organizational structure, 6.0 per cent a divisional structure, 19.2 per cent a matrix structure and 9.0 per cent a project structure. A relationship between the presence of strategic planning and control mechanism and the complexity of organizational structure was found (Table 15.16). In particular, the use of different mechanisms in different organizational structures, such as managerial accounting, standard costing, managerial reporting, investment analysis and responsibility accounting, can be observed. These are mostly mechanisms aimed at delegating and measuring performance. However, no relationship was found with mechanisms aimed at defining goals and programmes, and reward results, such as strategic planning, budget, and incentives.

The results obtained seem to confirm that as long as organizational and strategic complexity (for example, complexity of organizational structures and strategic differentiation)

Table 15.16 The relationships between the complexity of organizational structure and the presence of strategic planning, managerial and control systems

Strategic planning, managerial accounting and control systems	Complexity of organizational structure	Mean	Anova test – F
Strategic planning	Functional	0.2214	1.396
	Divisional	0.1054	
	Matrix	0.4062	
	Project	0.1833	
	Total	0.2446	
Managerial accounting	Functional	0.5354	5.390**
	Divisional	0.9549	
	Matrix	0.9400	
	Project	0.6378	
	Total	0.6522	
Budget	Functional	0.8467	0.374
	Divisional	0.7642	
	Matrix	0.9106	
	Project	0.8320	
	Total	0.8541	
Standard costing	Functional	0.6899	5.074**
	Divisional	0.3065	
	Matrix	0.9338	
	Project	0.4951	
	Total	0.6899	
Managerial reporting	Functional	0.6205	4.334**
	Divisional	0.7642	
	Matrix	0.9863	
	Project	0.5026	
	Total	0.6924	
Investment analysis	Functional	0.3721	7.740**
	Divisional	1.000	
	Matrix	0.7111	
	Project	0.1263	
	Total	0.4543	
Responsibility accounting	Functional	0.3267	8.309**
	Divisional	0.2537	
	Matrix	0.8219	
	Project	0.7635	
	Total	0.4711	
Incentives	Functional	0.3377	1.071
	Divisional	0.1515	
	Matrix	0.2203	
	Project	0.1352	
	Total	0.2793	

Note: ** Sig. ≤ 0.05.

increase, the enterprises adopt control mechanisms to cope with the new issues of delegation, coordination of managers' actions, and definition of objectives and strategies. This evidence is consistent with the organizational control theory propositions. However, the relationship with managerial control mechanisms, on the one hand, but the non-existing relationship with strategic planning and incentives, on the other hand, confirms the propositions of stewardship theory. In SMEs and family-owned businesses the strategy formulation is carried out by the entrepreneur or a very few people with familiar ties. It is based on an incremental, emerging process. The identification of family members with firm's objectives, loyalty and trust towards the organization do not require formal mechanisms to reward performance.

Conclusions

This chapter is intended to contribute to the debate on the professionalization of family firms. The most relevant theories dealing with this topic have been highlighted and their propositions and evidences compared. An inconclusive picture emerged from the analysis of theories, which revealed advantages and uselessness of professionalization of family firms. Some propositions and hypotheses were derived from the analysed theories and tested on a sample of Italian SMEs family firms.

This study aimed at advancing research on family business in many ways. First, the relationship between professionalization and the use of governance, strategic planning and control mechanisms was identified. The relationship between the involvement of managers and these mechanisms was identified too, even though they have not necessarily to be members of the family. Secondly, the diffusion of governance mechanisms and strategic planning and control systems was presented. It is worth pointing out that the most widespread mechanisms are a board of directors and those control systems focused on short-term horizons and aimed at defining and assigning objectives and measuring performance, such as budget, managerial reporting and managerial accounting. Strategic planning is not as diffused as many authors suggested. It seems that it is not considered by the sample companies as a mechanisms to manage both the family and the company development and growth. Thirdly, family involvement in governance and management roles seems to have an impact on the economic performance of the enterprise. This evidence is consistent with those researches and theories which showed that family enterprises outperform, owing to the family commitment to the company. Fourthly, a board of directors and strategic planning have a relationship with the economic performance of family firms, consistent with the propositions of the agency theory. Finally, strategic and organizational complexity seems to be drivers of professionalization, consistent with the organizational control theory.

From a practitioner standpoint, we can draw some lessons. Family involvement is suggested owing to the positive effect on the company performance. However, it has to be related to the use of governance and strategic planning and control mechanisms, which can positively impact on the performance too. There is a sequence in the introduction of these mechanisms in family firms, which has to be appropriately managed. A bigger role for strategic planning could be useful to cope with both family and company growth issues and to avoid a focus mostly on a short-term horizon.

The limitations of this study can be summarized as follows. It considered only family-owned businesses. However, it could be of interest to compare the diffusion, the features, the objectives and roles of governance, strategic planning and control systems among family firms and non-family enterprises. It focused mostly on governance, strategic planning and control mechanisms, while the involvement of non-family members in ownership, governance and management was not identified. It could be of interest to analyse the role and impact of non-family members in governance and management, especially with regard to some top management positions, such as the CEO, the managing director and the CFO, who can have great autonomy in decision-making. Other drivers of professionalization and their effect on company performance have to be studied too, such as age of the company, number of generations, actual phase in the company life cycle, and size of the enterprise. A similar study applied to large enterprises could be useful to identify the peculiar feature of professionalization consistent with different kinds of firm size. Many studies suggest that large family firms outperform small ones. More attention could be paid to these sources of performance differences.

Although evidence from a large number of firms in a varied set of industries and sizes was presented, this study is subject to the limitations that generally apply to cross-sectional survey-based research: the response rate, although typical, renders the conclusions subject to potential response biases; the fact that the sample is focused on manufacturing firms (although many different manufacturing industries were represented in the sample) limits the possibilities for generalization of results. Studies including non-manufacturing firms could obviously extend the findings.

To conclude, it can be said that the professionalization of family firms to be properly analysed requires adopting a theoretical framework based on the propositions of different theories. Empirical results confirm that to better explain the features, the drivers and the impact of the professionalization of family firms, different theories can contribute, each highlighting specific and peculiar aspects of family-owned businesses. The theories analysed in this work apparently emphasize different features of family-owned businesses, but some convergence can be found among some of their propositions, such as the impact of professionalization and family involvement on company results, the sequence of introduction of governance, strategic planning and control mechanisms, and the relationship between strategic and organizational complexity and professionalization.

References

Adams, J.S., Taschian, A. and Shore, T.H. (1996), 'Ethics in family and non-family owned firms: an exploratory study', *Family Business Review*, **9**(2), 157–70.

Anderson, R. and Reeb, D. (2003), 'Founding-family ownership and firm performance: evidence from S&P500', *Journal of Finance*, **58**(3), 1301–28.

Aram, J.D. and Cowen, S.S. (1990), 'Strategic planning for increased profit in the family owned business', *Long Range Planning*, **23**, 76–81.

Armstrong, J.S. and Overton, T.S. (1977), 'Estimating non-response bias in mail surveys', *Journal of Marketing Research*, **14**, 396–402.

Aronoff, C.E. and Ward, J.L. (1995), 'Family-owned businesses: a thing of the past or a model for the future?', *Family Business Review*, **2**(8), 121–30.

Astrachan, J.H. and Kolenko, T. (1994), 'A neglected factor explaining family business success: human resource practices', *Family Business Review*, **3**(7), 251–62.

Astrachan, J.H., Klein, S.B. and Smyrnios, K.X. (2002), 'The F-PEC scale of family influence: a proposal for solving the family business definition problem', *Family Business Review*, **1**(15), 45.

Barney, J.B. (1991), 'Firms resource and sustained competitive advantage', *Strategic Management Journal*, **17**(1), 99–120.

Berles, A.A. and Means, G.C. (1932), *The Modern Corporation and Private Property*, New York: Macmillan.

Bruce, N. and Waldman, M. (1990), 'The rotten kid meets the Samaritan's dilemma', *Quarterly Journal of Economics*, **105**, 155–65.

Burkart, M., Pannunzi, F. and Shleifer, A. (2003), 'Family firms', *Journal of Finance*, **58**(5), October, 2167–201.

Cabrera-Suárez, K., De Saá-Pérez, P. and Garcia-Almeida, D. (2001), 'The succession process from a resource- and knowledge-based view of the family firm', *Family Business Review*, **14**, 37–46.

Campbell, D.T. (1955), 'The informant in quantitative research', *American Journal of Sociology*, **60**, 339–42.

Child, J. (1972), 'Organization structure and strategies of control: a replication of the Aston study', *Administrative Science Quarterly*, **17**(2), 163–77.

Chrisman, J.J., Chua, J.H. and Sharma, P. (2003), *Current Trends and Future Directions in Family Business Management Studies: Toward a Theory of the Family Firm*, Coleman White Paper Series, Madison, WI: Coleman Foundation and US Association of Small Business and Entrepreneurship.

Chrisman, J., Chua, J. and Steier, L. (2002), 'The influence of national culture and family involvement on entrepreneurial perceptions and performance at the state level', *Entrepreneurship Theory and Practice*, **26**(4), 113–30.

Christensen, C.R. and Scott, B.R. (1964), *Review of Course Activities*, Lausanne: IMEDE.

Chua, J.H., Chrisman, J.J. and Sharma, P. (1999), 'Defining the family business by behaviour', *Entrepreneurship Theory and Practice*, **23**(4), 19–39.

Churchill, N.C. and Lewis, V.L. (1983), 'The five stages of small business growth', *Harvard Business Review*, **3**, 30–50.

Cooper, A. (1981), 'Strategic management: new ventures and small business', *Long Range Planning*, **5**, 39–51.

Corbetta, G. (1995), *Le imprese familiari: Caratteri originali, varietà e condizioni di sviluppo*, Milano: Egea.

Corbetta, G. and Tomaselli, S. (1996), 'Boards of directors in Italian family businesses', *Family Business Review*, **4**(9), 403–21.

Craig, J., Green, M. and Moores, K. (2003), 'Family business leadership: a stewardship and agency life cycle perspective', in P. Poutziouris and L.P. Steier (eds), *New Frontiers in Family Business Research. The Leadership Challenge*, IFERA–FBN Publications, pp. 353–67.

Daily, C.M. and Dollinger, M.J. (1993), 'Alternative methodologies for identifying family- versus non family-managed businesses', *Journal of Small Business Management*, **2**(31), 79–90.

Daily, M. and Dollinger, M.J. (1992), 'An empirical examination of ownership structure in family and professionally managed firms', *Family Business Review*, **2**, 117–36.

Dalton, D.R., Johnson, J.L. and Ellstrande, A.E. (1999), 'Number of directors and financial performance: a meta-analysis', *Academy of Management Journal*, **42**, 674–86.

Dalton, D.R., Daily, C.M., Ellstrande, A.E. and Johnson, J.L. (1998), 'Meta-analytic reviews of board composition, leadership structure, and financial performance', *Strategic Management Journal*, **19**, 269–90.

Danco, L.A. and Jonovic, D.J. (1981), *Outside Directors in the Family Owned Business*, Cleveland, OH: The University Press.

Davis, J. and Tagiuri, R. (1989), 'The influence of life-stage on father–son work relationships in family companies', *Family Business Review*, **2**, 47–74.

Davis, J.H., Schoorman, F.D. and Donaldson, L. (1997), 'Toward a stewardship theory of management', *Academy of Management Review*, **22**(1), 20–47.

Deakins, D., Morrison, A. and Galloway, L. (2002), 'Evolution, financial management and learning in the small firm', *Journal of Small Business and Enterprise Development*, **2**, 7–16.

Demb, A. and Neubauer, F.F. (1992), *The Corporate Board*, New York: Oxford University Press.

Dodge, H.R. and Robbins, J.E. (1992), 'An empirical investigation of the organizational life cycle model for small business development and survival', *Journal of Small Business Management*, **30**(1), 27–37.

Dumas, C. (1997), 'Preparing the new CEO: managing the father–daughter succession process in family business', in C.E. Aronoff, J.H. Astrachan and J.L. Ward (eds), *Family Business Sourcebook II*, Marietta, GA: Business Owner Resources, pp. 434–41.

Dyer, W.G. Jr (1996), 'Integrating professional management into a family owned business', in R. Beckhard (ed.), *The Best of FBR: A Celebration*, Boston: Family Firm Institute, pp. 44–50.

Eaton, C., Yuan, L. and Wu, Z. (2002), 'Reciprocal altruism and the theory of the family firm', paper presented at the Second Annual Conference on Theories of the Family Enterprise: Search for a Paradigm, Philadelphia, December.

Eisenhardt, K. (1989), 'Agency theory: an assessment and review', *Academy of Management Review*, **14**, 57–74.

Fama, E. and Jensen, M. (1983), 'Separation of ownership and control', *Journal of Law and Economics*, **26**, 301–25.

Ford, R.H. (1988), 'Outside directors and the privately-owned firm: are they necessary?', *Entrepreneurship: Theory and Practice*, **13**, 49–57.

Freeman, R.E. and McVea, J.A. (2001), 'Stakeholder approach to strategic management', in M.A. Hitt, R.E. Freeman and J.S. Harrison (eds), *Handbook of Strategic Management*, Oxford: Blackwell.

Frishkoff, P.A. and Brown, B.M. (1997), 'Women on the move in family business', in C.E. Aronoff, J.H. Astrachan and J.L. Ward (eds), *Family Business Sourcebook II*, Marietta, GA: Business Owner Resources, pp. 446–51.

Galbraith, J.R. (1977), *Organization Design*, Reading, MA: Addison Wesley.
Gallo, M. (1993), *Organos de gobierno de la empresa familiar* [Governance bodies in family businesses], Technical note, IESE Research Department DGN-467, Barcelona: IESE.
Gallo, M. (1995), 'The role of family business and its distinctive characteristic behaviour in industrial activity', *Family Business Review*, **8**, 83–97.
Gallo, M. and Vilaseca, A. (1998), 'A financial perspective on structure, conduct, and performance in the family firms: an empirical study', *Family Business Review*, **11**, 35–47.
Gallo, M.A. (1996), 'Accionistas "pasivos" de la Empresa Familiar', in M.A. Gallo (ed.), *La Empresa Familiar 5*, Barcelona: Estudios y Ediciones IESE.
Gallo, M.A and Cappuyns, K. (1997), *Boards of Directors in Family Businesses: Working and Composition. Levels of Usefulness*, Research Paper n. 346 Bis, Barcelona: IESE.
Gallo, M.A. and Lacueva, F. (1989), 'La Crisis Estructural en las Empresas Familiares: Una Observación Internacional del Fenómeno', in V. Font and M.A. Gallo (eds), *La Empresa Familiar 3*, Barcelona: Estudios y Ediciones IESE.
Gersick, K.E. Davis, J.A., McCollom Hampton, M. and Lansberg, I. (1997), *Generation to Generation: Life Cycles of the Family Business*, Boston, MA: Harvard Business School Press.
Ghoshal, S. and Moran, P. (1996), 'Bad for practice: a critique of the transaction cost theory', *Academy of Management Review*, **21**(1), 13–47.
Gimeno, A., Labadie, G. and Saris, W. (2004), 'The effects of management and corporate governance on performance in family business: a model and test in Spanish firms', in S. Tomaselli and L. Melin (eds), *Family Firms in the Wind of Change*, IFERA Publications, pp. 147–71.
Gnan, L. and Montemerlo, D. (2001), 'Structure and dynamics of ownership, governance and strategy: role of family and impact on performance in Italian SMEs', *Research paper. DIR SDA Bocconi*, Milan: SDA Bocconi School of Management.
Gnan, L. and Songini, L. (2003), 'The professionalization of family firms: the role of agency cost control mechanisms', in P. Poutziouris and L.P. Steier (eds), *New Frontiers in Family Business Research: The Leadership Challenge*, IFERA–FBN Publications, pp. 141–72.
Gnan, L. and Songini, L. (2004), 'Glass ceiling or women in command? The role of professionalization in women's family firms', in S. Tomaselli and L. Melin (eds), *Family Firms in the Wind of Change*, IFERA Publications, pp. 172–95.
Goffee, R. and Scase, R. (1987), 'Patterns of business proprietorship among women in Britain', Chapter 5 in R. Goffee and R. Scase (eds), *Entrepreneurship in Europe: The Social Processes*, London: Croom Helm.
Gomez-Mejia, L., Nuñez-Nickel, M. and Gutierrez, I. (2001), 'The role of family ties in agency contracts', *Academy of Management Journal*, **44**, 81–95.
Grant, R.M. (1991), 'The resource-based theory of competitive advantage: implication for strategy formulation', *Califoria Management Review*, **33**(3), 114–35.
Greenwald, M. and Associates (1995), Mass Mutual 1995 Family Business Survey, Springfield, MA: Mass Mutual.
Greiner, L.E. (1972), 'Evolution and revolution as organization growth', *Harvard Business Review*, July–August, 37.
Habbershon, T.G. and Williams, M. (1999), 'A resource-based framework for assessing the strategic advantages of family firms', *Family Business Review*, **12**, 1–25.
Habbershon, T.G., Williams, M. and MacMillan, I. (2003), 'A unified systems perspective of family firm performance', *Journal of Business Venturing*, **18**(4), 451–65.
Handler, W. (1989), 'Methodological issues and considerations in studying family businesses', *Family Business Review*, **2**, 257–76.
Harris, T.B. (1989), 'Some comment on family firm boards', *Family Business Review*, **2**(2), 150–52.
Herzberg, F. (1968), 'One more time: how do you motivate your employees?', *Harvard Business Review*, January–February, 53–62.
Hillman, A.J. and Dalziel, T. (2003), 'Boards of directors and firm performance: integrating agency and resource dependence perspectives', *Academy of Management Review*, **28**(3), 383–96.
Hopwood, A.G. (1974), *Accounting and Human Behaviour*, London: Haymarket.
Huse, M. (2000), 'Boards of directors in SMEs: a review and research agenda', *Entrepreneurship & Regional Development*, **12**, 271–90.
Ibrahim, A., Soufani, K. and Lam, J. (2001), 'A study of succession in a family firm', *Family Business Review*, **14**(3), 245–58.
Irvin, D. (2000), 'Seven ages of entrepreneurship', *Journal of Small Business and Enterprise Development*, **7**(3), 255–60.
Jensen, M.C. and Meckling, W.H. (1976), 'Theory of the firm: managerial behaviour, agency costs and capital structure', *Journal of Financial Economics*, **3**, 305–60.

Johnson, T.H. and Kaplan, R.S. (1987), *Relevance Lost: the Rise and Fall of Management Accounting*, Cambridge, MA: Harvard Business School Press.
Johnson, J.L., Daily, C.M. and Ellstrand, A.E. (1996), 'Boards of directors: a review and research agenda', *Journal of Management*, **22**, 409–38.
Kirchhoff, B.A. and Kirchhoff, J.J. (1987), 'Family contributions to productivity and profitability in small businesses', *Journal of Small Business Management*, **25**(4), 25–31.
Kleiman, B., Petty, J.W. and Martin, J. (1995), 'Family controlled firms: an assessment of performance', *Family Business Annual*, **1**, 1–13.
Kroeger, C.V. (1974), 'Managerial development in the small firms?', *California Management Review*, **1**(XVII), 41–6.
Lansberg, I. (1999), *Succeeding Generations: Realizing the Dream of Families in Business*, Boston, MA: Harvard Business School Press.
Litz, R.A. (1995), 'The family business: toward definitional clarity', *Family Business Review*, **8**(2), 71–81.
Lorange, P. (1980), *Corporate Planning*, New York: McGraw-Hill.
Lorsch, J.W. and MacIver, E. (1989), *Pawns or Potentates*, Boston, MA: Harvard Business School Press.
Mace, M.L. (1971), *Directors: Myth and Reality*, Boston, MA: Division of Research, Graduate School of Business Administration, Harvard University.
McConaughy, D. and Phillips, G. (1999), 'Founders versus descendants: the profitability, efficiency, growth characteristics and financing in large, public, founding-family-controlled firms', *Family Business Review*, **12**, 123–31.
McGuire, J.W. (1963), *Factors Affecting the Growth of Manufacturing Firms*, Seattle, WA: Bureau of Business Research, University of Washington Seattle.
Menendez-Requejo, S. (2004), 'Growth and internationalisation of family businesses', in S. Tomaselli and L. Melin (eds), *Family Firms in the Wind of Change*, IFERA Publications, pp. 284–95.
Mintzberg, H. (1983), *Structures in Five: Designing Effective Organizations*, New York: Prentice-Hall.
Mintzberg, H. (1994), *The Rise and Falls of Strategic Planning*, New York: Prentice-Hall.
Mizruchi, M. (1983), 'Who controls whom? An examination between management and boards of directors in large American corporations', *Academy of Management Review*, **8**, 426–35.
Montemerlo, D., Gnan, L., Schulze, W. and Corbetta, G. (2004), 'Governance structures in Italian Family SMEs', in S. Tomaselli and L. Melin (eds), *Family Firms in the Wind of Change*, IFERA Publications.
Moores, K. and Mula, J. (2000), 'The salience of market, bureaucratic, and clan controls in the management of family firm transitions: some tentative Australian evidence', *Family Business Review*, **13**(2), 91–106.
Morck, R. and Yeung, B. (2002), 'Family control and the rent seeking society', paper presented at Second Annual Conference on Theories of the Family Enterprise, Philadelphia, December.
Morck, R. and Yeung, B. (2003), 'Agency problems in large family business groups', *Entrepreneurship Theory and Practice*, **27**(4), 367–82.
Morck, R., Shleifer, A. and Vishny, R. (1988), 'Management ownership and market valuation: an empirical analysis', *Journal of Financial Economics*, **20**, 293–316.
Mustakallio, M.A. (2002), 'Contractual and relational governance in family firms: effects on strategic decision-making quality and firm performance', doctoral dissertation, Helsinki University of Technology.
Myers, S. (1977), 'Determinants of corporate borrowing', *Journal of Financial Economics*, **5**, 147–75.
Nash, J.M. (1988), 'Boards of privately held companies: their responsibilities and structure', *Family Business Review*, **1**(3), 263–70.
Normann, R. (1977), *Management for Growth*, Chichester: John Wiley and Sons.
Ouchi, W. (1981), *Theory Z: How American Business Can Meet the Japanese Challenge*, Reading, MA: Addison Wesley.
Perren, L., Berry, A. and Partridge, M. (1999), 'The evolution of management information, control and decision-making processes in small, growth-oriented service sector businesses', *Journal of Small Business and Enterprise Development*, **5**(4), 351–62.
Pollak, R.A. (1985), 'A transaction cost approach to families and households', *Journal of Economic Literature*, **23**, 581–608.
Poutziouris, P., Michaelas, N. and Chittenden, F. (1998), *The Financial Affairs of UK SMEs. Family and Private Companies*, working paper, Manchester: Manchester Business School, University of Manchester.
Poza, E.J., Hanlon, S. and Kishida, R. (2004), 'Does the family business interaction factor represent a resource or a cost?', *Family Business Review*, **17**(2), 99–118.
Prahalad, C.K. and Doz, Y.L. (1981), 'An approach to strategic control in MNC's', *Sloan Management Review*, Summer, 5–13.
Pratt, J. and Davis, J. (1986), 'Measurement and evaluation of population of family-owned businesses', *U.S. Small Business Administration Report No. 9202-ASE-85*, Washington, DC: Government Printing Office.
Quinn, J.B. (1980), *Strategies for Change: Logical Incrementalism*, Georgetown: Richard R. Irwin.
Roos, S. (1973), 'The economic theory of agency: the principal's problem', *American Economic Review*, **63**, 134–9.

Rostow, W.W. (1960), *The Stages of Economic Growth*, Cambridge: Cambridge University Press.

Rue, L.W. and Ibrahim, N.A. (1995), 'Boards of directors of family-owned-businesses: the relationship between members involvement and company performance', *Family Business Annual*, **1**,14–21.

Rue, L.W. and Ibrahim, N.A. (1996), 'The status of planning in smaller family owned businesses', *Family Business Review*, **9**, 29–43.

Schulze, W.S., Lubatkin, M.H., Dino, R.M. and Bucholtz, A.K. (2001), 'Agency relationships in family firms: theory and evidence', *Organization Science*, **12**, 99–116.

Schulze, W., Lubatkin, M. and Dino, R. (2003), 'Toward a theory of agency and altruism in family firms', *Journal of Business Venturing*, **18**(4), 473–90.

Schwenk, C.R. and Shrader, C.B. (1993), 'Effect of formal strategic planning on financial performance in small firms: a meta-analysis', *Entrepreneurship: Theory and Practice*, **17**, 53–64.

Scott, M. and Bruce, R. (1987), 'Five stages of growth in small business', *Long Range Planning*, **3**, 45–52.

Shanker, M.C. and Astrachan, J.H. (1996), 'Myths and realities: family businesses' contribution to the US Economy – a framework for assessing family business statistics', *Family Business Review*, **9**, 107–23.

Sharma, P., Chrisman, J.J. and Chua, J.H. (1997), 'Strategic management of the family business, past research and future challenges', *Family Business Review*, **10**, 1–35.

Shleifer, A. and Vishny, R. (1997), 'A survey of corporate governance', *Journal of Finance*, **52**, 737–83.

Steier, L. (2001a), 'Family firms, plural forms of governance, and the evolving role of trust', *Family Business Review*, **14**, 353–67.

Steier, L. (2001b), 'Next-generation entrepreneurs and succession: an exploratory study of modes and means of managing social capital', *Family Business Review*, **14**, 259–76.

Steier, L. (2003), 'Variants of agency contracts in family financed ventures as a continuum of familial altruistic and market rationalities', *Journal of Business Venturing*, **18**(5), 597–618.

Steinmetz, L.L. (1969), 'Critical stages of small business growth: when they occur and how to survive them', *Business Horizons*, February, 29.

Stewart, A. (2003), 'Help one another, use one another: toward an anthropology of family business', *Entrepreneurship Theory and Practice*.

Tagiuri, R. and Davis, J.A. (1982), 'Bivalent attributes of the family firm', *Family Business Review*, **9**, 199–208, reprinted 1996.

Tomaselli, S. (1996), *Longevità e sviluppo delle imprese familiari: problemi, strategie e strutture di governo*, Milano: Giuffrè.

Uhlaner, L.M. and Meijaard, J. (2004), 'The relationship between family orientation, organization context, organization structure and firm performance', in S. Tomaselli and L. Melin (eds), *Family Firms in the Wind of Change*, IFERA Publications, pp. 452–71.

Van den Berghe, L.A.A. and Carchon, S. (2003), 'Agency relations within the family business system: an exploratory approach', *Corporate Governance*, **11**(3), 171–9.

Ward, J.L. (2001), 'The special role of strategic planning for family businesses', *The Best of FBR*, 140–46.

Ward, J.L. (1997), 'Growing the family business: special challenges and best practices', *Family Business Review*, **10**, 323–37.

Ward, J.L. (1988), 'The special role of strategic planning for family businesses', *Family Business Review*, **1**(1), 105–17.

Ward, J.L. (1987), *Keeping the Family Business Healthy*, San Francisco, CA: Jossey-Bass.

Ward, J.L. (1991), *Creating Effective Boards for Private Enterprises*, San Francisco, CA: Jossey-Bass.

Ward, J.L. and Handy, J.L. (1988), 'A survey of board practices', *Family Business Review*, **1**(3), 289–308.

Westhead, P. and Cowling, M. (1998), 'Family firm research: the need for a methodological rethink', *Entrepreneurship Theory and Practice*, **23**(1), 31–56.

Whisler, T.L. (1988), 'The role of the board in the threshold firm', *Family Business Review*, **61**(5), 143–54.

Williamson, O.E. (1981), 'The modern corporation: origins, evolution, attributes', *Journal of Economic Literature*, **19**, 1537–68.

Wortman, M.S. (1994), 'Theoretical foundations for family-owned business: a conceptual and research-based paradigm', *Family Business Review*, **7**, 3–27.

Wu, Z. (2001), 'Altruism and the family firm: some theory', unpublished MA economics thesis, University of Calgary.

16 Formulating, implementing and maintaining family protocols

Miguel Angel Gallo and Salvatore Tomaselli

Introduction

This chapter discusses the main variables that influence the process of writing and implementing a family protocol, and the cause–effect relationships that emerge during that process.

Family protocols are commonly considered by both scholars and practitioners a valuable instrument to govern the relationship between the family and the business; furthermore, most of the authors we encountered point out that the process through which the family goes while writing the protocol is even more important than the content of the final document. On the other hand, most literature we encountered is prescriptive and, with very few exceptions (Corbetta and Montemerlo, 2001; Leon-Guerrero et al., 1998), we have not found any contribution based on empirical surveys that corroborate (or disconfirm) those statements and test the real sturdiness of such tools.

The term 'family protocol', coined by Gallo and Ward (1991), is rooted in the stream of studies on strategic planning in family businesses, developed from the beginning of the 1980s and it is part of the most recent tide of research on corporate governance in family businesses (see, among others, Carlock and Ward, 2001; Corbetta, 1995; Corbetta and Montemerlo, 2000; Gallo, 2000; Gersick et al., 1997; Harris et al., 1994; Lank and Neubauer, 1998; Lansberg, 1999; Tomaselli, 1996; Ward, 1988; Gallo and Ward, 1991). As such studies developed and researchers concentrated their attention on specific aspects, they have coined different terms that, as Lank and Neubauer (1998, p. 89) underlined, assume a variety of forms and cover an ample range of denominations.

A family protocol, as it is intended in this chapter, is a document aimed at maintaining and reinforcing over time and generations unity among family members and their commitment to the success of the family business (Corbetta and Montemerlo, 2000; Gallo, 1994, 2000; Tomaselli, 1996). By writing a family protocol the owning family makes an effort to identify and make explicit and transferable to the subsequent generation and to other stakeholders the main reasons for its own commitment to the business; the philosophy that inspires the family in its relationship with, and control of, the business (ownership, government, management, and so on); the goals pursued by the family and the business; and the rules that govern the relationship between the family and the business (Corbetta and Montemerlo, 2000; Gallo, 1994, 2000; Tomaselli, 1996).

The typical structure of a family protocol we refer to includes two main sections, each of which is usually divided into chapters. The first section states the foundation of the common project of the family concerning the future of the family business, defines the framework for a realistic knowledge of the family business among family members and sets the stage for any kind of rule to be defined and implemented. This section is made up of four pillars: reasons for continuity as a family business and the main values of the family; what kind of relationship the family wants to maintain with the business and, consequently, what type of relationship the family is going to maintain in the future with the business; realistic expectations of family members concerning the family business; and the conditions and circumstances that will determine the end of the family business (either its liquidation, or its division, or its selling out). The glue among these four pillars is the common vision for the future of the family business and the definition of its mission developed by the owning family.

The second section of the family protocol includes norms and rules aimed at fostering trust among family members, by generating consistent and predictable behaviour in family members in their relationship with the business, and vice versa, coherently with the framework set in the previous section. Such norms and rules basically concern: work in the family business (conception of the family concerning work relationships between family members and the family business, prerequisites for working in the family business, selection processes, career paths, and so on); the exercise of power both in the business (top management and governance bodies, control and voting agreements, and so on) and in the relationship between the family and the business (family council, committees, and so on); issues concerning 'money' (buy–sell agreements, evaluation of shares, dividend policies, liquidity funds, common investments, philanthropy, and so on).

As we stated before, this chapter concentrates its attention on the dynamics characterizing the process of writing and implementing a family protocol, as they result from the in-depth observation of some cases.

The survey which this chapter draws on was conducted by the authors from October 1999 to June 2002 to investigate the results of family protocols on a sample of Spanish family businesses. The key questions of our research can be synthesized as follows: what happens after a family protocol has been written and signed by the members of the family? Is it really implemented or does it become a leather-bound book to be placed in the archives? What are the main results (positive or negative) attributable to the formulation and implementation of the family protocol in terms of impact on the degree of unity and commitment? What are the dynamics created in the formulation and implementation (or lack of implementation) processes? What are the elements of the formulation and implementation processes that contributed most to the success of the results? Reasons of space do not allow us to make a complete report on our survey in this chapter, so we will make a brief presentation and discussion of the main results, and concentrate our attention on a case study that discusses the causal relationships characterizing the process of writing and implementing a family protocol.

The chapter is structured as follows: first, we present the theoretical framework of reference of our analysis; secondly, we present the research method we adopted; thirdly, we present our empirical results and discuss them; and fourthly, through the analysis of a case study, we analyse the dynamic relationship characterizing the process of writing and implementing a family protocol.

Theoretical framework

Since our interest was in both structural variables and cause–effect relationship, we adopted for our survey two different theoretical frameworks that, in our opinion, have some complementary characteristics which helped us to analyse in detail the relationships among the different variables.

The first theoretical framework used for the analysis is a version of the model 'structure–conduct–performance' (Caves, 1964) suited for the phenomenon in observation, where we define the variables as follows. The structure is constituted by the structural conditions of the family and the business. The structural dimensions of the family that have finally been taken into account after several trials are: how many and which generations will be immediately or very shortly involved in the existence of the family protocol; how many persons are involved, their age and sex; the economic wealth of these people; the degree to which they are economically dependent on the family business, and the eventual existence of important differences in their economic situations; their level of business knowledge; attitudes in their relations with other family members and the family business; human virtues; the quality of the relations between family members; the level of 'unity' and 'commitment' before starting the formulation and implementation of the family protocol; the existence of one or several family members who act like, and are recognized as, family leaders; and the existence of a common family project in relation to the family business (Lansberg, 1999).

The 'structural' dimensions of the business that were finally taken into account are: absolute size expressed through turnover; size in relation to other businesses in the same industry and in relation to the size of the family; historical growth path of the business. Reasons for the possible crises and periods of heavy development include: diversification of the company into different 'strategic business units', scope and diversity; geographical configuration; level of 'dispersion' of business activities in different geographical areas; level of the technology related to products and processes compared with their business competitors'; how adequate the structuring of responsibilities is to the characteristics of the family business and its strategy; level of inclusion and development of non-family executives (Gallo, 1991a); levels of independence given to executives to fulfil their responsibilities; and level of equilibrium in the development and implementation of the different management systems.

The conduct is constituted by the collective path followed by the participants during the formulation and implementation of the family protocol, in terms of ampleness of the areas covered by the protocol and the speed at which the process advances. The following variables were considered to analyse 'conduct': the main reasons why a family decides to formulate and implant a family protocol should be taken into account, because of their impact on the entire process (considering the difficulty involved in identifying if these reasons are part of the 'structural' dimensions analysed in the previous section, the 'final point' that leads to the start of the formulation process, or derived from the 'conduct', we considered them as the first element in the 'formulation'); the overlapping between the formulation and the implementation; the 'business and family' balance, meaning the level to which the development of the strategy, the professional management systems, the business management and government processes and capacities are 'balanced' with the know-how and qualities of the members of the family; and the 'balance between the government of the family business and the government of the family', which refers to a similar idea but in relation to the implementation of the family protocol. When both bodies and systems of government are capable of handling the power in their relevant areas, balance is good, but if the government of one of the institutions is 'stronger' than the other, then there is an unbalanced situation. The performance is evaluated by the level and quality of implementation of the family protocol and positive and negative results produced by the formulation and implementation of the family protocol in terms of unity and commitment.

Through this framework we wanted to understand the relationship existing among the structural characteristics of the family business, the conduct followed during the formulation and implementation process, and the results achieved through the protocol in terms of unity and commitment.

On the other hand, reasoning about the dynamic relationship among 'structure', 'conduct' and 'performance', and the interaction among their constituents during the process lead us to imagine the existence of circular relationships among different variables. Such an idea induced us to also adopt a second theoretical framework of reference that allowed us to identify and describe, in a systemic perspective, such relationships. For this purpose we used, even though in simplified way, the methodology known as systems dynamics.

Systems dynamics developed initially from the work of Jay W. Forrester. His seminal book *Industrial Dynamics* (Forrester, 1968) is still a significant statement of philosophy and methodology in the field. Since its publication, the span of applications has grown extensively and now encompasses work in corporate planning and policy design, public management and policy, biological and medical modelling, energy and the environment, theory development in the natural and social sciences, dynamic decision-making, complex nonlinear dynamics. This is a methodology for studying and managing complex feedback systems consisting of multiple components of highly heterogeneous characteristics, such as business and other social systems. In fact it has been used to address practically every sort of feedback system. While the word 'system' has been applied to all sorts of situations, feedback is the differentiating descriptor here. Feedback refers to the situation of X affecting Y and Y in turn affecting X, perhaps through a chain of causes and effects. One cannot study the link between X and Y and, independently, the link between Y and X and predict how the system will behave. Only the study of the whole system as a feedback system will lead to correct results.

We considered this methodology useful for several reasons: it increases our level of understanding of problems; it helps to identify the cause and effect relationships and the feedback circuits that determine the dynamic behaviour of a 'system'; it facilitates communications between several components of the 'system', based on a sufficiently objective common premise; it provides an instrument, the simulation model, that allows for a rapid exploration of the effects of alternative policies and behaviours; and it generates a debate on the structure and the dynamic behaviour of the system's variables.

The use of 'dynamic systems analysis' in this chapter does not include the design of a simulation model, and is restricted to the first level of analysis, that is, the definition of diagrams representing the cause and effect circuits between relevant variables in the family protocol formulation and implementation process.

Research method

Since our survey was an exploratory one, aimed at getting an insight on the multiple questions we illustrated in the introduction and also at offering an explanatory model, we decided to adopt a qualitative approach and to use cross-case studies.

Although in the past decade the use of large databases and sophisticated statistics has become the most recurrent research method in the area of family businesses, we consider that qualitative methods and the in-depth analysis of a reduced number of cases can be helpful.

Our sample comprised 12 family businesses, selected among the 31 that during the 1990s had formulated a family protocol with the professional assistance of the IESE's Family Business Chair and advisers related to this. The selected businesses were all big companies from very different industries, ranging from second to fifth generation, and had formulated their protocol at least four years before our survey started, some of them successfully, some unsuccessfully.

The research has been conducted in different phases. In the first phase we selected six family businesses. For each of these we collected, through different (public and private) sources, information useful to define its characteristics, and examined the protocol to know its content. Then we made semi-structured interviews, based on a list of topics that were prepared in advance. On average, each interview lasted around two and half hours. All the interviews were run by Professor Tomaselli, in consideration of the following: he had not participated in the formulation and implementation of any of the protocols, and therefore he did not have any possible bias on the situation of any family business in the sample; he has in-depth knowledge of both the model of protocol drawn and the method of work followed by the Chair, given his his long-lasting relationship with the IESE's Family Business Chair; he has experience in the formulation and implementation of family protocols with different models and in countries other than Spain.

In almost all cases, we interviewed three people: the top person in the family business, the president of the family council or a representative member of the board of directors, and one shareholder, preferably last generation. Subsequently, we discussed the information collected during the interviews with the advisers who had participated in the process of formulation and, eventually, implementation of the protocol. Then we studied the results of our survey using the two different theoretical frameworks, introduced in the previous section.

Based on the results we encountered, we formulated some hypotheses, which we tested through a new sample of six businesses. At the same time, we conducted a two-year follow-up analysis in four of the six businesses of the first sample, to verify if in the mean time new elements had intervened, which could be useful to understand long-term dynamics.

Results

Based on the information collected, we identified the structural dimensions of the family and the businesses that showed to have the highest influence on the formulation and implementation of family protocols. Then, for each identified dimension we assumed as a benchmark the family businesses that showed the best situation, and classified all the other businesses using a five-level Likert scale.

Tables 16.1 and 16.2 report the result of the classification that includes all the businesses in our sample.

With a similar process we examined and classified the elements characterizing the conduct followed during the formulation and implementation process, which we report in Tables 16.3 and 16.4.

Although the dimensions reported in the tables presented here are rather intuitive, we recognize that a comment for each dimension would help the reader to better understand them. On the other hand, reasons of space make it impossible in this chapter. We apologize to the readers for this and intend to give further information in a more extensive report about our research.

Table 16.1 Family dimensions

Company	Economic situation – absence of differences in the family	Knowledge of the business	Attitudes and virtues	Quality of the relations	Presence of leadership	Existence of a common project	Total	Average
ACL	4	2	2	4	4	3	19	3.17
Fonoll	4	3	4	4	3	4	22	3.67
Goiria	4	4	4	5	5	5	27	4.50
Grufor	3	4	2	5	1	1	16	2.67
Oester	3	2	3	4	3	3	18	3.00
Opujol	3	1	2	1	2	1	10	1.67
Rofra	5	4	4	4	5	4	26	4.33
Rogilsa	3	1	3	3	5	4	19	3.17
Sadia	4	5	3	3	2	4	21	3.50
Tarasa	4	4	4	3	4	2	21	3.50
Tonsa	4	1	3	3	2	2	15	2.50
Utiel	5	2	4	2	3	2	18	3.00

Concerning the conduct followed in the formulation and implementation process, an examination of the cases we studied suggested that we formulate a model of analysis based on two dimensions: the 'rhythm', that is the speed at which the process advances, and the 'ampleness', that is the 'quantity' of topics covered by the family protocol. We designed a matrix to classify different typologies of conduct according to the combinations of the two dimensions. For each dimension we identify three levels: minimum (or rather insufficient), enough and good; then we positioned each business of the sample in the matrix (Figure 16.1).

Performance

Since the purpose of the family protocol is to increase the 'unity' in the family business by an ordered, intense and long-lasting 'commitment' of the members of the family to their 'common project', within a framework developed to prepare active shareholders and to prevent and solve conflicts, it is logical that this part of the research examines the results of the formulation and implantation of the family protocol more in terms of the levels of commitment and unity obtained than in relation to the 'economic' success of the business, although commitment and unity have a natural and significant impact on this and an economically successful firm is necessarily part of the 'common project'.

With respect to the main results those who participated in our survey attributed to the family protocol, it is interesting to notice that several interviewees commented as positive the fact of having a 'live' document that raises opportunities to meditate on the future of the company and that can be adjusted over time, as understanding of current circumstances increase.

While the protocol offers to family members positive results such as opportunities to better know each other, to clarify their wills, to discuss problems and to prevent conflicts, in other circumstances it seems that contrary results occur for example, the relatives' resistance to participate actively, loss of unit, conflicts, and so on. In such cases, the same interviewees' admit, the protocol decreases to a 'book of desires'.

Table 16.2 Business dimensions

Company	Growth story	Profitability	Diversification	Geographic configuration–dispersion	Technology	Organization	Executives' authonomy	Management and control systems	Total	Average
ACL	5	5	5	5	5	5	5	5	40	5.00
Fonoll	5	5	4	3	4	4	5	5	35	4.38
Goiria	5	5	5	4	5	5	4	4	37	4.63
Grufor	3	3	4	4	3	2	1	1	21	2.63
Oester	3	4	1	1	3	4	2	1	19	2.38
Opujol	3	4	3	1	3	2	1	1	18	2.25
Rofra	3	5	3	3	5	2	2	3	26	3.25
Rogilsa	4	4	4	4	4	5	4	4	33	4.73
Sadia	2	2	3	4	2	2	4	1	20	2.50
Tarasa	4	4	3	2	3	2	2	2	22	2.75
Tonsa	5	5	5	5	5	5	5	4	39	4.88
Utiel	5	5	5	1	5	2	2	1	26	3.25

Table 16.3 Elements of formulation

Company	Reasons to write a family protocol	Level of participation	Rhythm–continuity	Ampleness of the content before implementation	Equilibrium between family and business	Total	Average
ACL	1	3	1	1	3	9	1.80
Fonoll	5	5	4	4	4	22	4.40
Goiria	5	5	5	5	5	25	5.00
Grufor	1	1	2	1	1	6	1.20
Oester	1	2	4	4	2	13	2.60
Opujol	3	2	2	3	2	12	2.40
Rofra	3	3	2	4	5	17	3.40
Rogilsa	5	5	4	3	4	21	4.20
Sadia	4	4	5	5	4	22	4.40
Tarasa	5	5	4	4	4	22	4.40
Tonsa	1	3	1	1	3	9	1.80
Utiel	4	5	3	1	1	14	2.80

Table 16.4 Elements of implementation

Company	Overlap with formulation	Education	Level of involvement	Existence of chief emotional officer	Equilibrium in the government of the family and the business	Total	Average
ACL	1	1	3	4	3	12	2.40
Fonoll	4	4	5	4	4	21	4.20
Goiria	4	4	5	5	5	23	4.60
Grufor	4	4	5	5	5	23	4.60
Oester	1	1	1	1	1	5	1.00
Opujol	1	3	3	4	3	14	2.80
Rofra	4	5	5	5	5	24	4.80
Rogilsa	2	4	4	3	4	17	3.40
Sadia	4	4	5	1	4	18	3.60
Tarasa	4	5	5	5	4	23	4.60
Tonsa	1	1	2	1	2	7	1.40
Utiel	5	3	5	2	2	17	3.40

We observed such situations in those cases that have not been able to leave position 'A' on the matrix (Figure 16.1) for a long period of time, as testified by the statements made by people interviewed throughout the research:

- 'The family protocol easily becomes a "book of wishes", which, if the business goes well is not needed, and if it does not go well "will fix nothing".'
- 'As we failed to reach a consensus on our main differences of opinion, these differences have got worse and we have lost our unity.'
- 'The formulation of the family protocol did not increase a healthy "family pride".'

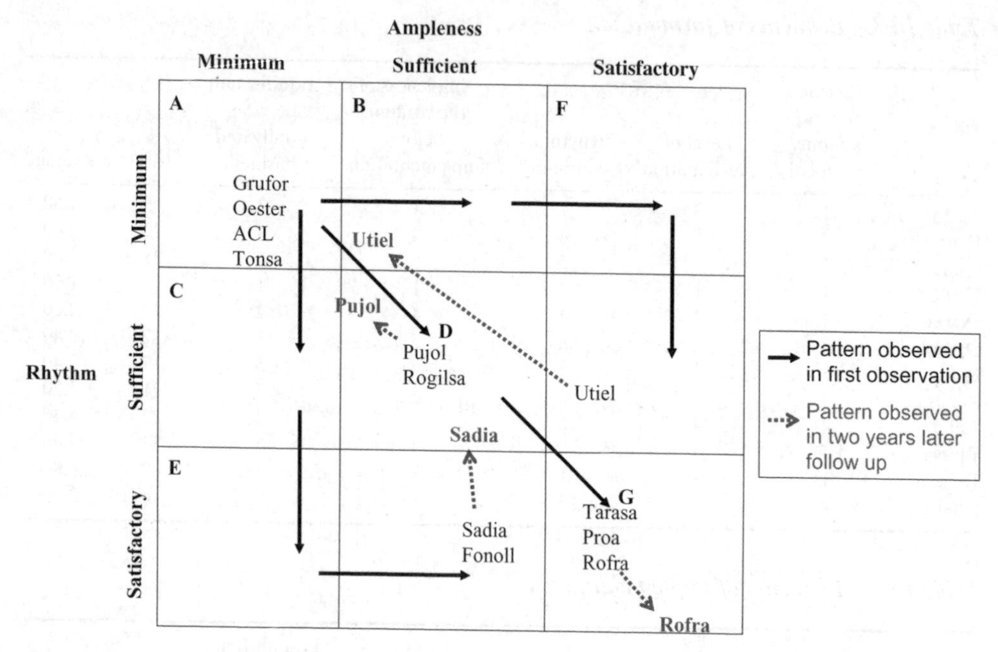

Figure 16.1 Ampleness–rhythm matrix

In all these cases, performance in terms of unity and commitment was poor, whereas in most of the other cases unity and commitment show positive patterns as a result of the implementation of the family protocol.

Discussion

In the previous section we introduced the rhythm–ampleness matrix (Figure 16.1). We consider it worth giving some more information about the interpretation of the different areas of the matrix.

Area 'A': minimum rhythm and ampleness

Obviously, the position 'A' corresponds to the initial position of whatever process of formulation and implementation of a family protocol. Nevertheless, as it is shown by some of the cases we examined, remaining for a long period of time in such position is risky, because it easily translates to a loss of interest by the family in the project, or even to the erosion of trust among family members.

Permanence in quadrant 'A' mainly springs from a lack of commitment by the most influential members of the family in respect of the formulation and implementation of the family protocol. The reasons for such a lack of commitment can differ from one case from another; the same happens with the circumstances and ways through which such a lack of devotion produces paralysing consequences.

Area 'C–E': a protocol with reduced contents advancing at a good rhythm

Permanence in this area can be definitive or it can constitute an intermediary step. In the latter category we find those family businesses which, having experimented with a protocol with the least contents, have started to widen their content, moving towards position G.

The choice between one or other alternative frequently depends on structural conditions or on contingent situations of the family and/or the business. Nevertheless, except in the cases recalled above, the refusal to advance towards position 'G' and the choice of definitive permanence in position 'E' show a lack of interest by the family in trying to further strengthen the family business and to achieve a greater degree of unity and commitment.

Area 'B–F': a protocol with ample contents advancing at a slow rhythm
As for the previous area, such a position can have transitory or definitive character, and it usually depends on contingent situations that the family needs/wants to resolve before proceeding to implementation. However, to remain for too long in position 'F' risks that some members of the family and some non-family executives will become sceptical, doubting the possible success the protocol and even its usefulness. For this reason, as already evidenced with reference to the area 'C–E', in order to avoid the failure of the protocol, it is worth increasing the rhythm as soon as possible, and undertaking the run towards position 'G'.

Area 'D–G': equilibrated ampleness and rhythm
In this area one can find those family businesses that succeed in achieving equilibrium between rhythm and ampleness of contents. These are the situations, that one could define as of greater success, in which the process of formulation and implementation are somehow unified, proceeding with a logic of 'continuous improvement'. As soon as they reach consensus on some parts of the protocol, they proceed to implement these and then continue with further aspects.

Nevertheless, even position 'G' is not a definitive result, as some of the cases we examined show. The change of certain circumstances or the emergence of hidden problems can cause regression, and it is necessary to know how to prevent this.

Our search has shown that, according to what we observed in the two-year follow-up analysis, the majority of businesses find it difficult to advance according to the diagonal 'A–D–G'. Therefore, their conduct seem directed to the continuous search for equilibrium between two extremes: that of minimum ampleness in terms of content of the protocol, obtained however to a sustained rhythm, and that of ampleness of content of the protocol achieved with a slow rhythm.

An example of the first circumstance can be represented by those cases in which in the initial moment it is decided to limit the intervention to put in place a well-functioning board of directors, and to stay in this position for a certain period of time (evolution 'A–C–E'). The second circumstance incurs instead in those cases in which it is decided to cover immediately the different areas of the protocol, while advancing with a minimum rhythm, in order to give family members the time necessary to gain sufficient learning, internalize the different topics, build a wider consent, and so on (evolution 'A–B–F').

According to our results, the two dimension of the structure that present great relationships with the rhythm at which the formulation of the protocol and its implementation advance are: family members' motivation toward the identification of a 'common project', and their commitment to make it become reality; family members' education on the basic concepts of business economics and strategic management and their understanding of the objectives and the content of a family protocol. The results of our research show that when these two dimensions improve, the family business experiences important positive results that have a positive impact on both the formulation and the implementation of the family protocol.

Also the dimensions of the family, the quality of the relationships among its members and the existence of leadership influence the rhythm: small families are able, all other things being equal, to advance more quickly, while larger families have to face greater difficulties not only of an organizational nature, but also concerning the construction of the consent among a greater number of people; good relationships among family members facilitates mutual trust and allows consent to be reached more easily; a recognized leader's presence helps in moments of stagnation or when it is necessary to find a way to get out of difficult situations.

On the business side, the rhythm in the formulation and implementation of the family protocol seems most influenced by the profitability of the family business and well-designed organizational structure and control systems.

Analysing the relationship between structure, conduct and performance, we observed that increased unity in the family seems to be related to the fact that the formulation process offers an opportunity to have a forum of discussion where family members can receive information, discuss different perspectives, better know each other, and gain reciprocal trust. Another variable that has an influence on unity is the compatibility between personal versus group goals.

Commitment seems to be related to the 'quality' of the common project that the family is able to design and to the degree in which family members' behaviours are consistent with such a project. The formulation and implementation of a family protocol seems to influence the determinants of commitment at both family and business level. At family level, the education of family members that is usually part of the process generates an awareness in the family. At the business level an increase in professionalization is usually the premise for higher commitment. Anyhow, it is worth putting on record that introducing commitment to the project into active participation of family members is in most cases a true challenge for those who are in charge of making the protocol work. Even in most successful cases, it is not easy to obtain dedication from family members, especially those who are not active in the family business as managers or directors.

Once we analysed our sample in accordance with the first theoretical framework, we moved on to the dynamics that generate the outcomes we observed. In the next section we will use one of the observed cases to show how we made such dynamic analysis, and the results we obtained.

Analysing cause–effect relationships: the Pujol case

The Pujol family comprised three generations: the founder and his wife, their seven children, and 12 members of the third generation. In 1995, when the family protocol was approved, only the seven siblings constituting the second generation – whose ages ranged from 30 to 45 years – were involved in the family business. In fact, the founder and his wife had not been in charge of the business for many years, given their very advanced age, whereas the 12 members of the third generation were too young to have any involvement in the family business.

The economic situation of these seven siblings was sufficiently good, excellent if compared with the penuries the family experienced years before, given the very humble origin of the family. On the other hand, there were important economic differences among them, owing to the remunerations that perceived for their work. Three of the siblings worked in the family business, whereas the other four worked in diverse activities outside the family business. One of the siblings had a salary that allowed him to live modestly and the others, owing to lack of education and ability, carried out works of low remuneration or of small dedication.

The members of the second generation had very different attitudes and virtues. Four of them were industrious, the three others did not have any job and had been living of the monthly allowances they received from the family company. Some of them fully trusted the siblings who managed the family business, others did not fail to show their disagreement on any kind of decision. In a special way, two of them maintained a permanent discrepancy with the rest in all economic issues.

The three siblings who managed the family business, through their effort and dedication, showed their personal intention to develop of the company, although one of them, Jorge, seemed to worry more about the improvement of his personal status than about the growth and evolution of the family business. The other four siblings looked at the company merely as a source of personal income.

The family business comprised two different and independent companies, one managed by Antonio and Rosa, the other by Jorge. Sales amounted to about 30 million euros for each of the companies, and both companies ranked among the best competitors in their respective industries. However, mainly due to the maturity of their products, for long time the two business had not been growing and developing as necessary and experiencing a sort of stagnation. As a consequence, profitability, that had been very high in the past, was beginning to diminish, although without going down to extremes.

The two companies had a very simple structure of responsibilities. In each of them one family member was the CEO and sales manager, and two non-family managers were in charge of operations and administration areas, supported by a number of non-family middle managers. Management and control systems were rudimentary. When the companies experienced their most important growth, local consultants were contracted to help solve the most pressing problems, but in 1994, when the board of directors was established, there was nothing more than an annual balance sheet, and this was not subject to any kind of auditing.

In 1994, Jorge frequently talked with Antonio and Rosa about the convenience of establishing a board of directors with external and independent consultants that might help to overcome the situation the companies found themselves in. Jorge argued that he was impressed by the recommendations he had frequently heard in conferences and read in the specialized press on boards of directors.

Jorge asked Antonio and Rosa to meet an independent consultant he had in mind as a candidate for being a board member in their family business. Antonio and Rosa agreed to meet the consultant and after doing so they agreed with Jorge's proposal to have him on their business' board. Once Antonio and Rosa agreed the proposal, the three asked the rest of the second generation to express their opinion about the idea of establishing a board of directors with independent members and hiring in the board the consultant they had met. The proposal was almost unanimously accepted, although most of the siblings, especially those who lived apart from the family business, did not know much about the mission of the board, neither did they understand what could be expected from some independent consultants that they found 'expensive'.

The new board of directors was settled, with the participation of four siblings and two independent consultants.

During the first two years of operation of the board, the situation of each of the companies constituting the family business evolved in a different way. The company managed by Antonio, and where Rosa worked, experienced significant improvement. Thanks to the incorporation of excellent non-family managers and the adoption of appropriate management and control systems, the company was able to advance firmly in the implementation of the demanding development strategy designed by the board. However, the second company, managed by Jorge, continued to stagnate: turnover among non-family managers hired by Jorge was very high; the efforts to revitalize the product and overcome its maturity were ineffective; Jorge proved to be unable to react properly to the evolution in distribution channels. All of this occurred in spite of the constant input from the independent directors. In the mean time, one of the independent directors, who had served as a consultant to several families in business, suggested *writing and implementing* a family protocol to regulate the relationship between the family and the business.

In 1995 the formulation of the family protocol began. The four siblings who were on the board were the only family members who had the ability and minimum preparation to give a positive contribution to the process. So it was decided that they would work out a draft with the support of the independent manager, and, once they reached a sufficiently complete draft, the document would be submitted to the rest of the family for discussion.

Supported by the independent director, the four siblings started discussing different aspects of the relationship between the family and the business. The independent director participated in the meetings, took notes and edited a draft that was then discussed, reformed and approved in board meetings.

It took about a year before the document was sufficiently complete in the basic aspects of the relationships between the family and the business, covering the following areas: how to incorporate the family values into the business; what type of business is it going to be in the future; what should family members, as shareholders, expect from the company; what possible reasons could cause the family to want to leave/sell (and so on) the company at a future date; rules for entering the company and promotion therein; how family members should act as owners; how family members should govern the business; and how to maintain unity within the family business.

Then, a family meeting was organized, to which the seven siblings, their spouses and the independent board members were invited. In this meeting the family protocol was presented to the rest of the family. The attitude of most family members was rather passive; they simply listened to the explanations of the independent consultant, without asking any question or raising any issue for discussion.

A few days later, two of the siblings who did not work in the family business, made known their formal intention to sell their share of the business. In short, with the help of lawyers and auditors, the buy–sell negotiations started. About one year later an agreement was reached and the two siblings sold their share to the remaining five siblings.

Once these dissidents were removed, the remaining five members of the second generation, their spouses and the children older than 14 years (a total of 15 people) were invited to participate in an education programme designed for the purpose by the external board members. The external board members encouraged the other directors to hold, every three months, regular meetings among the five siblings and their spouses. These meeting were to be organized similarly to shareholders' general meetings, and were aimed at increasing family members' knowledge of the business and involving non-active siblings as much as possible.

Given the small base of knowledge of the members of the family, these activities had to be run with great patience. It was Rosa who put greatest effort into making all this successful, so that, little by little, she began to be considered unanimously as the 'Chief Emotional Officer of the family'.

The efforts to implement the family protocol began to give good results: the passive attitude of family members was constantly evolving into active participation.

But still the company managed by Jorge did not improve its performance, in spite of the clear and strong arguments raised by the independent board members about the causes of this situation and the numerous plans of action approved by the board. Decisions were not implemented and the company remained as it was.

Everybody doubted Jorge's ability to manage the company out of its stagnation and the independent directors started to state openly their intention to resign from the board; on the other hand, the two non-active siblings did not take a position and did not make known their position in the meetings. The 'unstable balance' could tip in any moment.

Jorge, after failing once again to incorporate a new manager who had been the general manager of a company of considerably greater size, asked a consultant with a very good reputation to take charge of the project to turn the company around. The other board members thought there was little chance that any relevant innovation proposed by the new consultant would be implemented by Jorge, and they began to comment on the possibility of selling the business either to third parties or to Jorge.

In this period, Vicente – one of the siblings who did not work in the family company – asked for 'liquidity' to address the financial needs of his own venture. According to the family protocol, the family company bought some of his shares. The shares were to be maintained in the company's portfolio for an established period of time to allow the seller to buy them back, in case he wanted and had resources to do so. However, against the protocol's statement, Vicente did not allow the auditing company who assisted the group to audit the books of his company to identify the causes of the financial deficit that it periodically experienced.

Taking advantage of this situation, Jorge clearly represented that he wanted to take over the business he was in charge of, in order to be able to manage it 'his own way'. At the same time, he promised Vicente that he would be appointed as sales manager in the company (although it was evident that Vicente had no qualification to carry out such role), so obtaining his support. The independent directors decided to resign to allow family members to resolve their disputes without external interferences in the boardroom and also to prevent the risk of being involved in such dispute. But they continued to give their professional support to the family. The decision to separate was taken. The task of evaluating the companies' equity and preparing the patrimonial separation of the two businesses was given to two experts, but they arrived at very different conclusions and could not reach an express agreement. Consequently, an award by arbitration was requested.

A few months later the award by arbitration was stated. Antonio commented to the two former consultants:

> The award by arbitration has already arrived and it has been accepted by all the siblings. Now the three of us must pay about 9 million euros to Jorge and Vicente, who followed him as sales manager. We have enough time, and with the resources generated by the business we will be able to make it.
>
> The three of us agree we want to set up a new board, and we will ask the persons you suggested to sit in our board. However, you have known us for a long time and we all are totally confident of you, I ask that you help us on two very critical points.
>
> The first consists in that we never again want fratricidal fights. They are an incredible loss of time . . . in the last months I have hardly been able to take care of the business. The displeasures have been so great as to affect even the relationship with my wife and children. Never again!
>
> I don't know the way. I don't know if it is better that one of us buys stocks until having 51% of equity, or we establish 'golden shares', or we establish formal agreements, or what else . . . I do not want to ask for 'power in itself', I don't want to decide everything 'my own way'. I only want we establish a structure of power that can help us prevent situations like the one we have just experienced from happening again in the future. And in case something happens anyhow, we can manage the situation in a professional way, in the best interest of the business.
>
> The second thing I ask you is that you help us formulate a new family protocol. The one we had has been very useful, but it is no longer valid and it is necessary to write a new one, consistent with the present and future situation of the family business.
>
> My wife and children who didn't know anything about the business when we wrote the first protocol, now are informed and they love the idea of continuity. One of my children who has completed his studies as an engineer would like to enter the family business and to follow the tradition of the family in spite of the grievances.
>
> Rosa's children are still very young, but she and her husband want we go through a similar process to the one we followed before, and the owners are committed to the future of the company.
>
> Pedro, our third brother, is 60 years old and still single. It doesn't seem that he will have a family. Can you imagine the situation should he decide to leave his shares equally divided among all his siblings, or should he die without making a will?

Analysis of the Pujol case

In the Pujol case we can find two different phases characterizing very different outcomes. In its early stages the formulation and implementation process of the family protocol favours the development of many virtuous loops that shows the positive influence of family members' education on unity and commitment. The independent directors clearly identified that the low level of business education of most family members could be a problem and, while working towards the professionalization of the companies' management and control systems, they drove a slow process of formulation of the family protocol and facilitated means that could support the education of family members. However, although it could seem the opposite, the lack of education of family members and the fact that only some of them carried out work with any entrepreneurial and managerial content that demanded some effort, have not been the most influential variables in the seeming failure of the family protocol, as is clear if one observes the cause–effect loops.

Figure 16.2 shows three virtuous loops, providing evidence of the synergic effect they have on learning, commitment and unity.

In its early stages the formulation and implementation process of the family protocol favours the development of many virtuous loops that show the positive influence of family members' education on unity and commitment.

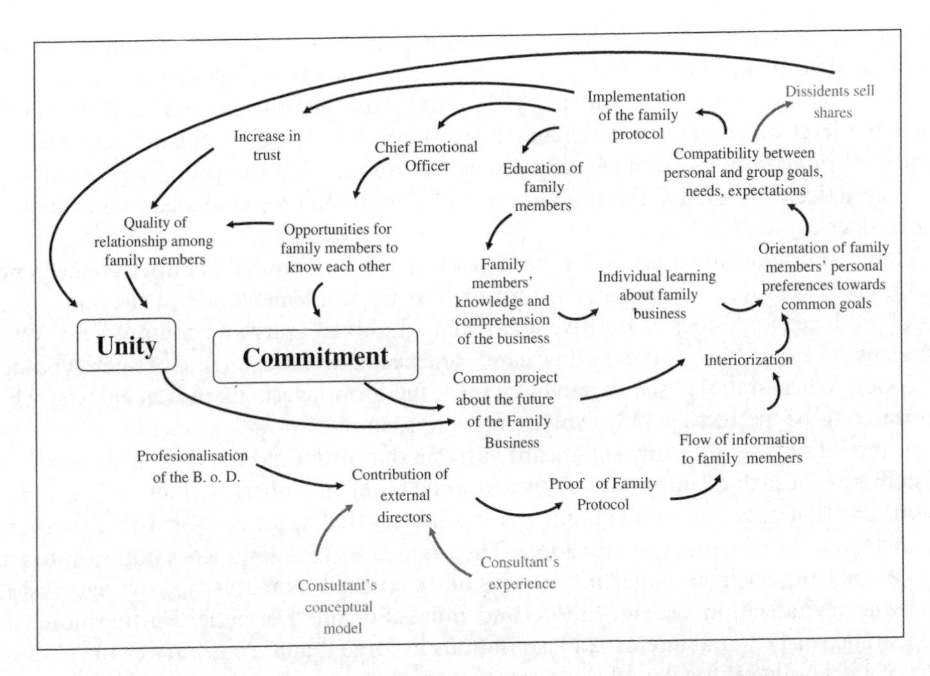

Figure 16.2 *Cause–effect loops and their synergic effect on learning, commitment and unity*

The first loop is founded on the positive effects of the decision to share the information about the family business with all family members. Supported by the independent director, the four siblings who sit on the board write a draft of the family protocol. The consultant, who is an expert in family business, has great influence on the process through his conceptual model and offers significant help in a situation where the business knowledge of family members is poor. The information provided allows family members to internalize the spirit of the family protocol, and shows the family the evident need to orient their personal preferences and expectations towards common objectives. As a result of this, the family members who do not agree with the idea of the family protocol decide to sell their shares in the business. For the other members of the family, the implementation of the family protocol increases their level of mutual confidence, and improves the quality of their personal relationships. All this, reinforced by the departure of the dissident members, increases the level of unity, which in turn leads to an increase in the ability develop a common project on the future of the business, again internalized by the family.

The second virtuous loop is related to the learning generated by the implementation of the family protocol. The family members' training process, which started at the same time as the implementation, causes an increase in their knowledge of the business. This gives rise to a flow of learning related to the peculiar characteristics of family businesses, which helps to improve the orientation of personal preferences towards common objectives and, therefore, creates more compatibility between personal objectives and common needs.

The third virtuous loop is related to the growth in the family's commitment, and its efforts to establish a common project for the future of the business, advancing in the implementation of the family protocol. As a result of the efforts to start the implementation of the family protocol, one member of the family takes on the responsibilities of the 'emotional head' and starts a flow of activities that enables the family to get to know each other better. This 'emotional head' is very important for the young members of the family, since the level of commitment remains high, preventing the centrifugal forces that are typical of large families from creating dispersion, and leading the younger generation to become actively involved in the establishment of a common project related to the future of the business.

The evolution of the case, as it has been described so far, shows that until the emergence of the crisis with Jorge the process of formulation and implementation of the family protocol produced a number of positive loops that reinforced unity and commitment. What happens later on, shows, on the other hand, one peculiar characteristic of such typology of loops. When some of the variables change, the loop inverts its movement and what appeared to be 'perfect' quickly evolves into a 'disaster'.

In the Pujol case, the truly significant variable that produced the degeneration of the system was the lack of important virtues in some family members: virtues such as industriousness that only half of the family have, sincerity that appears nonexistent in several of them, and loyalty that is not in Jorge. The case shows that Jorge does not want to recognize that the business under his responsibility reached its maturity years ago and its strategic revitalization becomes more and more difficult over time. Furthermore, the appearance of personal interest and aspirations in Jorge that are contrary to the philosophy of the family protocol, and the desire to pursue his own personal interest produce an erosion in the reciprocal confidence with the rest of the family. In such circumstances, the rupture of 'unity' appears irreparable, and separation is the best thing that the independent board members can recommend to the rest of the family to preserve their whole patrimony from the risk of major losses and to maintain at least part of the family entrepreneurial project.

The challenge to the family will be in how to articulate the legal and financial aspect of the separation, and arrange it in a way that generates the smallest possible damage to the older generation, and does not cause an intense rupture of the relationships among siblings, as happened with the first separation.

Conclusions

The survey this chapter draws from represents an attempt to deepen our knowledge of family protocols and the dynamics that occur during the formulation and implementation process. We are conscious that the survey presents a number of limitations owing to the reduced number of family businesses in the sample, and the fact that we considered only one country and only one model of family protocol. On the other hand, through cross-case study we have been able to make an in-depth analysis of the process of writing and implementing a family protocol, which probably would not have been possible had we assumed a greater sample or different models and contexts.

We hope that, despite those limitations, the results of our survey can be of help to both academics and practitioners, to better understand the dynamics that occur when writing and implementing a family protocol. It is our opinion that the analysis of the results determined by the introduction of the family protocols and other tools suggested by the academics and practitioners, not only enables the real sturdiness of such tools to be tested, it also facilitates a deeper understanding of the dynamics that characterize family businesses. If the specific of family businesses is represented by the presence of the family, the study of the dynamics that characterize such presence appears essential, at least in our view.

Literature on family protocols shows general agreement in attributing great importance to the formulation process of the protocol. The results of our investigation seem to confirm this statement and identify the reasons why this occurs. In many cases where we observed negative results of the protocol, shortcomings are also identified in the formulation process.

Several academics and advisers underline the importance of a participative process. Our investigation puts in evidence that this is a necessary but not sufficient condition; in addition it is necessary that the most influential members of the family have certain attitudes. Sincerity, shared and real interest in the future of the company, and maturity in the members of the family seem to be the variables that describe the conditions that can favour better (or harm) the successes of the family protocol. These are the attitudes determining that a long and participative process is useful to digest and metabolize knowledge, ideas, concepts, and so on or, conversely, is a useless waste of time. Other important aspects in the process are related to the communication habits among the family. Sincerity, ability to express ones own ideas, ability to resolve conflicts can be affected by good or bad communication habits in the family.

Building on our findings, academics can find inputs for future research aimed, for instance, at corroborating our results in different contexts, testing the models we propose, going one step forward in the line of system dynamics and building a simulation model, analysing our findings referring to different theoretical frameworks (that is, resource-based view, agency theory), linking our model to other models proposed to study family businesses (that is F-PEC Scale; see Klein et al., 2005), deepening the analysis of the relationship between certain structural characteristics of specific family businesses and aspects to be emphasized in the family protocol.

Consultants and family-business members can refer to our results to verify the existence of preliminary conditions to start the project of a family protocol, follow paths that fit the structural characteristic of the family business, and consider in a proper way the different variables that influence the success of the project.

References

Barach, J.A. (1993), 'Keeping the family in the firm', in M.A. Gallo (ed.), *La Empresa Familiar 4*, Publicaciones de la Cátedra de Empresa Familiar, IESE – Universidad de Navarra.

Bianchi, C. (1996), *Modelli contabili e modelli 'dinamici' per il controllo di gestione in un'ottica strategica*, Milano: Giuffrè.

Carlock, R.S. and J. Ward (2001), *Strategic Planning for The Family Business: Parallel Planning to Unify the Family and Business*, New York: Palgrave.

Caves, R. (1964), *American Industry: Structure, Conduct Performance*, New York: Prentice-Hall.

Coda, G. (1995), *L'Orientamento Strategico dell'Impresa*, Turin: UTET.

Corbetta, G. (1995), *Le Imprese Familiari. Caratteristiche Originali, Varietà e Condizioni di Sviluppo*, Milan: Egea.

Corbetta, G. and D. Montemerlo (2000), 'Family protocols: our experience in Italy', *The Family Business Network Newsletter*, (26), May.

Davis, P. (1989), *Succession and Succession Planning*, The Wharton School, University of Pennsylvania, Philadelphia.

Forrester, J. (1968), *Principles of Systems*, Reading, MA: MIT Press.

Gallo, M.A. (1991a), *Family Business: Non Family Managers*, IESE, Research Paper no. 220 bis.

Gallo, M.A. (1991b), 'Las Trampas Profundas de las Empresas Familiares', *Staff Catalunya*, no. 10.

Gallo, M.A. (1994), *Family Protocol*, IESE Technical Note no. 501.
Gallo, M.A. (1997), 'Fortalezas y Debilidades de la Empresa Familiar', *La Empresa Familiar*, Barcelona: Biblioteca IESE de Gestion de Empresas.
Gallo, M.A. (2000), 'Family protocol', *The Family Business Network Newsletter*, (26), May.
Gallo, M.A. and J. Ward. (1991), *Protocolo Familiar*, Nota Técnica de la division di Investigacion del IESE DGN-448, Barcelona.
Gallo, M.A., C. Cappuyns, G. Corbetta, D. Montemerlo and S. Tomaselli (2001), 'Unity and commitment', research paper presented at FBN 12th Conference, Rome.
Gersick, K.E., J.A. Davis, M. McCollom Hampton and I. Lansberg (1997), *Generation to Generation: Life Cycles of the Family Business*, Boston, MA: Harvard Business School Press.
Harris, R., J. Martinez and J. Ward (1994), 'Is strategy different for the family-owned business?', *Family Business Review*, **VII**(2), 159–74.
Jaffe, D.T. (1990), *Working with the Ones You Love: Conflict Resolution & Problem Solving Strategies for a Successful Family Business*, Berkeley, CA: Conari Press.
Jones, W.D. (1982), 'Characteristics of planning in small firms', *Journal of Small Business Management*, **20**(1), 15–19.
Lank, A.G. and F. Neubauer (1998), *The Family Business: its Governance for Sustainability*, London: Macmillan.
Lansberg, I. (1988), 'The succession conspiracy', *Family Business Review*, **1**(2), 119–43.
Lansberg, I. (1999), *Succeeding Generations*, Boston, MA: Harvard Business School Press.
Lansberg, I. and J.H. Astrachan (1994), 'Influence of family relationships on succession planning and training: the importance of mediating factors', *Family Business Review*, **7**(1), 39–59.
Leon-Guerrero, A.Y., J.E. McCann III and J.D. Haley Jr (1998), 'A study of practice utilization in family businesses', *Family Business Review*, **11**(2), 107–20.
Levinson, H. (1990), 'Conflictos que Aquejan a las Empresas Familiares', *La Empresa Familiar 3*, Barcelona: Estudios y Ediciones IESE S.A, pp. 61–78.
Montemerlo, D. (2001), *Il Governo delle Imprese Familiari: Modelli e Strumenti per Gestire i Rapporti tra Proprietà e Impresa*, Milan: Egea.
Mogi, Y. (1993), 'Constitution helps family meet internal challenger', Acts of the Seminar Global Perspectives on Family Businesses, Loyola University Chicago Family Business Center.
Scharma, P., J.J. Christman and J.H. Chua (1997), 'Strategic management of the family business: past research and future challenges', *Family Business Review*, **10**(1), 1–35.
Schein, E.H. (1978), *Career Dynamics*, Reading, MA: Addison Wesley.
Tomaselli, S. (1996), *Longevità e Sviluppo delle Imprese Familiari: Problemi,Strategie e Strutture di governo*, Milan: Giuffrè.
Ward, J. (1987), *Keeping the Family Business Healthy. How to Plan for Continuing Growth, Profitability, and Family Leadership*, San Francisco, CA: Jossey-Bass.
Ward, J. (1988), 'The special role of strategic planning for family business', *Family Business Review*, **1**(2), 105–17.

17 Generic models for family business boards of directors

Joseph H. Astrachan, Andrew Keyt, Suzanne Lane and Kristi McMillan

Many laws and proposals from around the globe are trying to address board practices that have allowed massive corporate failure. While valuable for large public companies, they may be harmful to family-owned businesses because they stem from a *market model* rather than a *control model* of corporate governance, and ignore the issue at the heart of corporate governance problems today: accountability. It is our proposition that the unique characteristics of control-model companies require a different approach to corporate governance than market-model companies. Control-model companies must focus on the unique need of the boards to have the competencies to be held accountable, and hold others accountable, for their actions. This chapter provides recommendations and propositions testable by future research that will lead to greater board accountability and, in turn, positive identifiable results in board and company performance.

Introduction

Governance reform of publicly held corporations is the white-knuckled *topic de jour* among denizens of board rooms, federal oversight bodies, the halls of academia and newsrooms. Indeed, the recent nearly daily onslaught of reports of unethical and criminal conduct, fomented by a greed and lust for power, by certain CEOs, CFOs, board members and other corporate chieftans – from the Enrons to the Global Crossings of America's Blue Chip business landscape – has triggered endless calls for change. Many corporate leaders, acting of their own volition, already have reconfigured the moral compasses by which future executive conduct is to be reckoned (Donaldson, 2003). The US Congress stepped in with Sarbanes–Oxley, a sweeping federal statute providing business leaders with a road map by which they are able to align corporate goals with more effective governance systems.

This is good, fitting and proper. But a critical sub-text has been missing from this often searing debate. Namely, what is the significance of these governance reforms for the publicly held corporation's distant but economically robust brethren – namely, the closely held, family-owned business? Should these family-owned entities be held to the same governance guidelines and standards that apply to those firms making up the ranks of the Fortune 500, for example? Corporate behemoths with a 200 million share stock float who dwarf the family-held entity, run by husbands, wives, sons and daughters with a stock float of a few thousand shares? To put it another way, does one size fit all (Corbetta and Salvato, 2004)?

This is not a trivial question. Recent data, disclosed by Astrachan and Shanker (2003), concludes that family businesses in the US represent the lion's share of all annual US tax return filings: a stunning 89 per cent of the total. These entities generate no less than

64 per cent of the gross domestic product. By another compelling measure, the family-owned business employs a whopping 62 per cent of the nation's workforce. Because of the contribution family-owned companies provide to the US economy, it is imperative that they undertake a concerted effort to maintain and enhance governance standards of their own to assure their continued success.

It is the assertion of this chapter that there are significant differences between these publicly and privately held firms that impact the governance processes necessary to provide accountability. Indeed, many of the most popularized corporate governance practices may be detrimental to family businesses (Corbetta and Salvato, 2004). Many of these recommendations may harm family unity or might be too complex for private firms, and many are only applicable to very large, public companies with dispersed ownership. Popular corporate governance practices are focused towards a market model of corporate governance, found prevalently in the United States and the United Kingdom, which involves companies with a widely dispersed shareholder base and a majority of independent, outside board members. In contrast, the typical family-owned business exhibits characteristics of the control model of corporate governance, found prevalently in continental Europe, Latin America, and Asia, which involves companies with a concentrated shareholder base and family member 'insiders' active in management and on the board.

These popular recommendations may also be problematic in that many promote form over substance, as these so called 'best practices' often become generalized laundry lists that do not lead to identifiable positive results (Atkinson and Salterio, 2002; Robinson, 2002). And although some of the issues most prevalently discussed in family and non-family business literature, such as the significance of independent boards, are important, fixating on such issues tends to overshadow the issue that is at the heart of corporate governance problems around the globe: so, what do we mean by accountability?

Accountability in the context of corporate governance can be thought of as the group processes, characteristics and actions of individual board members that hold management accountable for the decisions taken and for their implementation. This concept of accountability might be operationalized by exploring a series of questions such as: do you have a governance system? Who does it hold accountable? What does it do to hold these people accountable? How are goals set? How are rewards and sanctions set? Once answers to these questions are obtained, we can move to measuring their frequency, intensity and duration, and then correlating them with corporate performance.

Throughout the balance of this chapter, we will offer propositions for accountability in control-model firms, that will operationalize the concept of accountability and look at their impact on performance. As we then look at corporate governance guidelines for family firms, we must focus on the need for the primary governing body – the board – to have these competencies to hold others accountable, and be held accountable, for their actions.

It is our assertion that the propositions contained herein for control-model companies, will lead to great accountability as well as to board and firm performance. These recommendations and propositions will provide a foundation for testing processes of accountability in future research.

To understand the implications of the propositions described here both for family business owners and researchers, we first point to a significant source of bias in popular proscriptions for 'best practices' for board behavior. In the second section of this chapter we describe: (1) the board competencies necessary to ensure shareholder accountability;

(2) ways in which shareholders should exert their rights in order to hold the board accountable; and (3) actions the board must take to hold management accountable for their actions, all of which incorporate propositions for further consideration.

The 'market-model bias' in corporate governance reform

Corporate governance is embedded in the cultural, legal, and financial frameworks of various countries. These frameworks have given rise to two models of corporate governance: market and the control (Table 17.1).

Market model

The market model of corporate governance is common in countries where capital markets are highly liquid and shareholders are widely dispersed, such as in the US, UK, and Ireland. This model involves a large dispersed class of investors with no prior connections to the companies listed on the public exchanges (Coombes and Watson, 2001). The focus of corporate governance reform in countries employing this model is on board structures and practices that ensure that the board is a distinct entity, capable of objectivity and able to act separately from management (Gregory and Simmelkjaer, 2002). It also insists on independent boards and demands a high level of financial and business disclosure. Examples of companies that follow the market model include most public companies in the United States, such as General Electric and AT&T. One difficulty of this model is that it seeks the near impossible mission of eliminating all current and future conflicts of interest.

Table 17.1 Characteristics of the market and control models

Market	Control
Setting	Setting
• Prevalent in UK, US	• Prevalent in continental Europe, Asia, Latin America
• More reliance on public markets	• More reliance on private capital
• High ownership liquidity	• Illiquid ownership
• Shareholders are anonymous investors, not managers	• Concentrated shareholder base often overshadows minority shareholders
• Widely dispersed shareholders	• Shareholders view company as more than an asset and as interested in financial and non-financial returns
• Shareholders only have financial connections to the company	
Elements of governance	Elements of governance
• High level of disclosure	• Secretive
• Focus on short-term strategy	• Focus on long-term strategy
• Independent board members	• Shareholders with control rights in excess of cash flow rights
• Shareholders view company as one of many assets held	• Shareholders have connections to the company other than financial (i.e. managers, board members, family)
• Ownership and management are separate and at arm's length	• Insider board members
	• Ownership and management overlap significantly

Control model

The control model of corporate governance, commonly found in Asia, Latin America, and much of continental Europe, is prevalent where control rights are not fully separated from ownership, and ownership tends to be concentrated. The model sees conflicts of interest as endemic and seeks to institutionalize them or provide sanctions for them rather than eliminate them. An example of this is when a large shareholder, such as a family or institution, maintains a control stake. The purpose of investment for these types of shareholders is not to produce short-term gains as with most market-model corporations (Shleifer and Vishny, 1997); rather, in particular for family businesses, the shareholders tend to maintain a long-term perspective on their investment that benefits current, as well as future, generations.

An example of a control model company is Fiat SA, Italy's third most valuable company (LaPorta et al., 1999), where ultimate control (over 25 per cent) belongs to the Agnelli family, and members of that family are also board members and part of management teams. United States examples of this are the Ford family, which maintains approximately 40 per cent voting power, and the Sulzberger family, owners of the *New York Times*, where the family owns 18 per cent of the company while maintaining voting control over board members through a special class of stock not available to outsiders. For control-model companies, it is natural for the owners to expect to have a board presence, particularly because they are not anonymous (Coombes and Watson, 2001). Typically, shareholders of control-model companies are managers, as well. This often results in shareholders having control rights significantly in excess of their cash flow rights. This concentration of ownership in the control model is an easily identifiable concern that has led to corporate governance reform focusing on the fair treatment of minority shareholders (Gregory and Simmelkjaer, 2002).

The market-model bias

Perhaps because of media attention on dramatic cases of corporate abuse in the US and UK, the ease with which information on publicly listed companies is attained, and the tremendous amount of wealth in the public equity markets, corporate governance recommendations throughout the world have formed a 'market-model bias' towards best practices that lead to increased transparency and financial disclosure through outside, independent boards that attempt to be objective. Although desirable in form, this inclination towards market-model best practices in all situations is flawed in substance owing to its excessive reliance on concepts in agency theory.

Rooted in economics, finance, and Western attitudes towards property, agency theory relies on the problems that occur in public companies when there is a separation of ownership from control, as pointed out by Berle and Means (1932). Agency theory asserts that management and ownership pursue different interests, where top managers may be more interested in their own personal welfare than that of shareholders (Jensen and Meckling, 1976). Agency theory takes a Hobbesian view where managers are assumed to be primarily self-interested and need to be watched more closely to make sure they do not break rules or contractual obligations. Because of its focus on management opportunism, agency theory takes a 'monitoring approach,' in which the board's role of monitoring management and the value of extrinsic motivation (for example, compensation) become of utmost importance (Hillman and Dalziel, 2003). As a result of a need for discipline owing to underlying feelings of distrust for management, market-model proponents have utilized concepts in agency theory to advocate outside independent boards, CEO/chairman duality, and outcome-based compensation, such as stock options. Even though monitoring management and extrinsic motivation are significant, the focus of corporate governance worldwide must be broadened beyond these two concepts, in particular owing to the imbalance in the way risks, penalties, and rewards are shared (Daily et al., 2003; Plender, 2003). As a result, market-model practices are potentially lacking the following key components.

First, market-model practices do not address the board's ability to monitor management, which invokes the 'collaborative approach' of stewardship theory. Stewardship theory proponents, tapping into insights from sociology and psychology, focus on the need to enhance collaboration and decision-making between the board and management by empowering managers (Davis et al., 1997). Somewhat contrary to agency theory's implications, stewardship theory stresses the advisory capacity of the board and operates with the assumption that managers are able to identify personally with the firm, internalize its mission, and obtain satisfaction from intrinsic motivation (Sundaramurthy and Lewis, 2003). Unlike the myopic view of agency theory in the market model, both concepts of agency and stewardship theory can be combined to encourage trust in capabilities, distrust of human limitations, and conflict aimed at tasks and not individuals in order to manage appropriate amounts of monitoring and collaboration (Sundaramurthy and Lewis, 2003). It is the balance between monitoring and collaboration among governance actors that allows for an effective governance system (Demb and Neubauer, 1992).

Second, market-model practices overlook the diverse identities of various types of investors, such as families, who may have different interests, time horizons, and strategies from typical public firm investors (Aguilera and Jackson, 2003). While focusing on the dispersed and inactive shareholder base of public companies, market-model proponents fail to address the interests of companies with concentrated ownership such as control-model companies. Market-model proponents see family business governance as lacking objectivity, because they often have strong influence over management and the board. However, family-owned businesses, regardless of the legal, financial, and cultural frameworks in which they reside, have been able to operate successfully within the control model of corporate governance. Much of the success of family businesses and control-model companies in general is because the investors value non-financial returns and long-term business health. As long as shareholders of family firms receive communication, education, and a sense of shared interests and values exemplified by the board, these shareholders tend not to demand high current returns on their equity (de Visscher et al., 1995). Family business investors believe in the business's long-term potential for future generations, and it is the long-term investment philosophy of family businesses that creates one of their greatest competitive advantages (Habbershon and Williams, 1999).

While for a typical family-owned business (see Figure 17.1), the control model may be a workable form of corporate governance, in order to be successful, the focus of corporate governance must be on accountability, and the financial transparency and reporting promoted by the market-model bias, may not be the answer. After all, the essence of good governance involves being able to hold the corporation and its leaders accountable for delivering on their commitments, while still preserving an atmosphere of trust and unity. Therefore, the next section of this chapter discusses the board competencies to ensure accountability in the control model.

- The family can control effective strategic direction of the business.
- The business contributes significantly to the family's income, wealth, or identity.

Figure 17.1 Primary attributes of the typical family-owned firm

Establishing accountability in family business corporate governance around the world

Corporate governance principles typically specify that the board should either promote the interests of the company, the interests of the shareholders, or both (Gregory and Simmelkjaer, 2002). While in most control-model countries the emphasis is on promoting company interests, the focus of market-model Anglo-American corporations is on promoting the shareholders' immediate interest (Mobius, 2001). This has created an environment in market-model countries emphasizing structure over accountability. It is our assertation that accountability of the board for the activities of the corporation, rather than structure should be the central theme of corporate governance for control-model firms. How that accountability is expressed and to whom it is directed varies somewhat, depending on how the primary objective of the corporation is viewed (Gregory and Simmelkjaer, 2002).

Family firms generally have the benefits of a strong identity and sense of unity that enable them to carry on a long-term view of the business and its sustainability (Kets de Vries, 1993; Taguri and Davis, 1996). On the other hand, sometimes unification of the family through means such as nepotism, resistance to change, and curtailing growth can damage the economic interests of shareholders and cause the failure of the company (Kets de Vries, 1993). Accountability for the family firm involves making decisions that do not sacrifice long-term health for short-term personal or corporate gain.

The following propositions delineate ways in which the typical family-owned business operating under the control model should promote accountability. In addition to being helpful courses of action for family-business owners and managers, the following propositions offer a foundation for further consideration and testing by family-business researchers. It is with a focus on the accountability of the board, being the primary decision-making body, that the separation of management from ownership can occur without losing the identity of the family business.

Family-owned boards must have the competencies to be held accountable

Qualifications

Market-model bias Strategic business competencies and diversity of skills and background are sufficient board qualifications.

Propositions for control-model accountability

Family firms — Strategic business competencies and diversity of skills and background are important, but not sufficient. The most critical qualification is having the ability to hold the company accountable and the discipline to not interfere in company operations.

Proposition 1 — An increase in board accountability will increase board performance.

Much of the corporate governance literature for large public firms and family businesses alike focuses on strategic business competencies and diversity of skills and backgrounds when discussing board qualifications. Many in the corporate governance field believe that an ideal board profile includes active or retired CEOs and other professionals with expertise in such areas as finance, marketing, operations, technology, law, and public policy (Moore, 2002). In addition, a common theme apparent in corporate governance codes worldwide is that the quality, experience, and independence of the board affect its ability to perform its duties (Gregory and Simmelkjaer, 2002). Although these qualifications are important and desirable, they are not sufficient for family firms. Particularly for the typical family business, family dynamics often impinge on what should be a culture of communication and open dissent (Sonnenfeld, 2002). Often conflicts arise because the roles board members play in the family (parent) can emotionally clash with the objective role that is to be carried out in the business (the employer). Therefore, it is the way in which board members interact and communicate with each other and management that is a primary determinant of board success or failure (Sonnenfeld, 2002).

The board of the typical family firm must have the competencies to ensure strategic guidance of the company, effectively monitor management, and be accountable to the company and its shareholders. This concept of a *competency-based board* means that as long as board members conduct a forum of open communication, embody a culture of open dissent, have a basic understanding of business (that is, have an understanding of risks and of the measures that lead to financial and non-financial indicators of success), and collaborate with the management team, board members can be held accountable for their actions and those of management.

Size

Market-model bias — Smaller boards are more manageable.

Propositions for control-model accountability

Family firms — Mid-sized boards promote greater accountability.

Proposition 2 — Boards that consist of seven to 12 members will increase board accountability.

The corporate governance literature is split regarding the appropriate size of a board. Many corporate governance specialists assert that since individual responsibility tends to dissolve in larger groups, smaller boards are more desirable for family businesses (Neubauer and Lank, 1998; Ward, 1991). Many of these authorities recommend that the most effective boards range between five and nine directors (Nash, 1995; Newell and Wilson, 2002).

Others insist that boards that are 'too large' can encounter coordination problems and difficulty developing effective communication and teamwork (Felton et al., 1995).

However, other experts believe that a range of nine to 15 directors is beneficial (Moore, 2002). For example, in many of the European Union member states, the average size of the board is closer to 12 or 13 (Gregory and Simmelkjaer, 2002). In addition, many experts believe that smaller boards may not have enough breadth and can hamper the separation of director and committee assignments (Moore, 2002). Indeed, a larger board can lead to greater accountability as long as each individual board member has the necessary competencies to render good judgment, have their judgment be evaluated by his/her peers, and, in turn, be held accountable for his/her actions. In addition, more board members may also imply more eyes capable of noticing problems and ensuring accountability. Therefore, the most effective board for the typical family firm in the control model consists of seven to 12 people. However, although larger boards lead to greater feedback, and therefore, greater accountability, larger boards may inhibit full participation; therefore, boards should not be too large as to create factions that limit participation and communication. For example, a study conducted by Yermack (1996) found a negative relation between board size and firm value as boards become too large (that is, 12 directors or more).

Independent outsiders

Market-model bias — Boards should minimize the use of insiders and include a significant proportion of independent outsiders.

Propositions for control-model accountability

Model family firms — Independent status is largely irrelevant to achieving accountability. A board of owners may be beneficial.

Proposition 3 — A board of qualified 'insiders' can improve company performance.

The ability to exercise objective judgment of management's performance is critical to the board's ability to monitor management (Gregory and Simmelkjaer, 2002). A general consensus among market-model proponents has developed that this is an issue of board composition, and that boards should include a significant proportion of outsiders (Gregory and Simmelkjaer, 2002). The corporate governance codes in market-model countries, such as the United States, devote considerable attention to the appropriate mix of inside and outside members to ensure that the board is distinct enough from the management team to play a supervisory role and to bring a diversity of opinions to bear on issues facing the company (Gregory and Simmelkjaer, 2002). Additionally, most market-model corporate governance advocates agree that in order to strengthen their autonomy, boards should comprise a substantial majority of 'independent' directors, that is, directors who are free from commercial or personal ties that could impair their ability to probe and challenge management (Felton and Watson, 2002). Many market-model proponents suggest that independent board members should play an important role in areas where the interests of management, the company, and shareholders diverge, such as executive remuneration, succession planning, changes of corporate control, take-over defenses, large acquisitions, and the audit function (OECD, 1999). While some commentators contend that boards should comprise at least half outsiders (Newell and Wilson, 2002), others state an ideal board should consist of only independent outside directors in addition to the CEO and chairman (Ward, 1991).

Family businesses, however, have been criticized for being slow to adopt the concept of outsiders. The *2002 Mass Mutual Financial Group/Raymond Institute: American Family Business Survey* indicates that family members still constitute the majority of board members. Entrepreneurs and family businesses traditionally have resisted bringing in outsiders because they do not want someone directing their actions or revealing family secrets (*2002 American Family Business Survey*; Nash, 1995). We suspect much of the reluctance to adopt a board is due to a desire to avoid accountability. But despite the fact that businesses often need fresh creative perspective, objectivity, and openness – all traits generally considered to be advantages brought about by independent outsiders (Aronoff and Ward, 2002b) – we believe family businesses can achieve these results by adopting a competency-based view and composing their boards accordingly.

Since outsiders can often be easily swayed by compensation, perks, recognition, and potential as well as actual business dealings (with the company as well as with other board members), we would propose that a board occupied by outsiders does not guarantee objectivity. Even though boards consisting primarily of insiders (current or former managers/employees of the firm) or dependent outsiders (directors who have business relationships with the firm and/or family or social ties with the CEO) may be considered to be less effective at monitoring others (Lynall et al., 2003), insiders, in particular, are seen to provide rich firm-specific knowledge and strong commitment to the firm (Sundaramurthy and Lewis, 2003). We believe a board member that has the ability to render an opinion unfettered by other board members is more important than whether or not the individual works inside the business. Additionally, even though several academic studies have provided evidence that suggests that independent outsiders add real value to a company (Felton et al., 1995; Rosenstein and Wyatt, 1990), outside independent board configurations have not been associated with firm performance (Dalton et al., 1998). Independence is a mindset of disinterest that cannot be predicted by the lack of prior relationships of the parties involved. Therefore, in order to promote objectivity and accountability, board members should consist of individuals who base their decision-making on the merits of the decision rather than extraneous influences or considerations, such as personal relationships or financial and personal gain.

Frequency of board meetings

Market-model bias The focus of board meetings is on how many meetings are sufficient.

Propositions for control-model accountability

Model family firms The focus of board meetings is communication, conflict resolution, and accountability. To achieve this, we recommend no more than six, nor less than three meetings per year.

Proposition 4 Three to six board meetings per year will increase board accountability and board performance.

Current corporate governance recommendations suggest various frequencies for board meetings. Many governance professionals in the US claim that six meetings per year in alternate months is a good balance for most companies, supplemented by occasional special meetings (Moore, 2002). The European average is about eight meetings per year (Gregory and Simmelkjaer, 2002). It has been proposed in the family business literature that boards meet formally at least four times per year, supplemented by additional monthly executive committee meetings attended by lead directors, the chairman, and the CEO along with senior management (Ward, 1991).

However, many governance experts have failed to account for the reason behind having board meetings in making recommendations about meeting frequency. Board meetings exist to provide a forum to conduct regular and purposeful communication, ensure accountability, and resolve conflict. In order for this to occur, the board of a typical family business would need to meet anywhere between three to six times per year in order to keep the lines of communication open between the board and management, and between the board and the shareholders. At board meetings, board members require management to give them company information to analyse, and the shareholders, in turn, require the board to give them information and reasons behind, or how to improve, company performance. However, when extraordinary events are occurring within the business, and in turn more accountability is created on behalf of the board, then more meeting times are needed. Although the board must be aware that its role in governance is an active one, its role should not carry over into the role of management. Therefore, having more than six meetings without a crisis probably means the board is operating in a managerial fashion, and not promoting accountability, while holding fewer than three meetings per year is probably promoting form over substance and not providing accountability.

Content and process of board meetings

Market-model bias — The content of board meetings should include all key matters brought forward by senior executives to the board, while the process by which boards make decisions is by consensus.

Propositions for control-model accountability

Model family firms — The content of board meetings should include all key matters brought forward by senior executives to the board, while the process by which boards make decisions is by simple majority vote to create greater accountability. Additionally, all pertinent board information should be encapsulated in a board manual.

Proposition 5 — A simple majority voting process will increase board accountability.

Most of the contemporary corporate governance recommendations properly suggest that the content of board meetings should include all key matters brought forward by the CEO and other senior executives to the board, such as the results of operations, the status and outlook of financial and strategic plans, proposed or rejected business deals, the current financial forecast and early signals on changing trends, the economic and competitive environment, public policy issues, shareholder matters, and any proposals for board approval (Moore, 2002). However, when it comes to the process by which boards make decisions, many current recommendations falter when they suggest that board members can make decisions by consensus rather than employing a formal vote. However, decision-making by consensus allows the most powerful board member or coalition on the board to sway other members into complacency by stature rather than the merits of a decision. In such situations, consensus tends to yield a decision which most can live with rather than which the majority supports. Therefore, having a formal voting process by simple majority (except in cases where a larger majority is called for) allows for a process where issues are discussed at length and members' opinions are polled, thereby creating greater accountability.

Additionally, in order to capture all relevant information on key matters brought forward by management, as well as decisions made by the board, every board should have a board manual. The board manual can also include pertinent corporate information (values, vision, plan, director and officer biographies, organization bylaws, job descriptions, committee assignments, contact numbers, important dates, and so on), as well as the goals of the board and system of evaluation. Not only will the board manual aid board members in organizing the content and process of its decision-making, it will help in the evaluation process as well.

Board member selection

Market-model bias — A nominating committee should be formed to help in the board selection process.

Propositions for control-model accountability

Model family firms — A nominating committee should be held accountable for eliciting all board members' input in the board selection process.

Proposition 6 — Use of a nominating committee will result in the selection of directors who can provide greater accountability.

Most corporate governance experts properly advocate the creation of a nominating committee to help aid in the selection of board members. Duties of the nominating committee include setting board and committee performance goals, nominating directors and committee members with the qualifications and time to meet these goals, and monitoring board composition and operations (Moore, 2002).

However, as a clarification of these duties, we believe the nominating committee's role in family businesses is to build unity among the board, as its job is also to elicit opinions from all board members and share ideas on board needs and criteria, as well as build agreement on proposed nominees based on all the board members' input. We believe the nominating committee should be formed with board and non-board members (for example, significant shareholders not represented on the board), and its role, in turn, is to be accountable for running the board selection process.

Board commitment

Market-model bias — Nominal participation is adequate for most boards.

Proposition for control-model accountability

Model family firms — Active participation is necessary for family firms.

Proposition 7 — Active board participation will increase management accountability.

Board members should devote sufficient time to their responsibilities and actively contribute to the company's performance. With regard to family businesses, however, the

Table 17.2 Board of directors degree of involvement

Phantom	Never knows what to do, if anything; no degree of involvement
Rubber stamp	Permits officers to make all decisions; it votes as the officers recommend on action issues
Minimal review	Formally reviews selected issues that officers bring to its attention
Nominal participation	Involved to a limited degree in the performance or review of selected key decisions, indicators, or programmes of management
Active participation	Approves, questions, and makes final decisions on mission, strategy, policies, and objectives. Has active board committees. Performs fiscal and management audits
Catalyst	Takes the leading role in establishing and modifying the mission, objectives, strategy, and policies. It has a very active strategy committee.

2002 Mass Mutual/Raymond Institute: American Family Business Survey indicates that a substantial percentage of family business respondents reported weak board performance. The results indicate that 15 per cent of respondents regarded their board's contribution as 'fair,' 2 per cent rated the board's contribution as 'poor,' and 25 per cent of respondents cited 'no contribution' by the board. An explanation as to why boards do not make a more positive contribution is likely due in large part to inappropriate board participation.

Boards can be involved in the performance of a business in varying degrees. Table 17.2 delineates Wheelen and Hunger's (1994) six degrees of board involvement, ranging from low (phantom) to high (catalyst).

A study conducted by Judge and Zeithaml (1992) indicates that 70 per cent of boards participate at levels from phantom to nominal participation in the US. Despite many corporate governance experts observing that a board's role is reactive, thereby indicating that this level of participation is adequate, the boards of typical family businesses should actively participate in their respective businesses. Accountability requires more than nominal participation, but a catalyst role may undermine the authority of management and appropriate checks and balances.

A related issue is that service on too many boards can interfere with the performance of board members (OECD, 1999). Because boards have such difficulty evaluating the performance of companies and managers, many directors believe they should spend more time on a single directorship (Felton and Watson, 2002). Some companies have limited the number of board positions that can be held to make directors more committed to their work (Felton and Watson, 2002). This may help ensure that members of the board enjoy legitimacy and confidence, and in turn create greater accountability (OECD, 1999). We believe a prudent rule of thumb is that active participation on three boards may be a limit of effective board service.

Board term and turnover

Market-model bias Directors have predetermined terms or no term limits.

Propositions for control-model accountability

Model family firms Directors are reviewed regularly and kept if performing well; directors are thanked and let go if no longer capable of ensuring accountability.

Proposition 8 Regular annual reviews of the board increases board effectiveness and accountability.

Studies conducted by Shen and Cannella (2002) have proven that a board member's effectiveness (and therefore accountability) decreases after 14 years, thereby indicating that in order to keep a director for a long period of time, he or she must be making significant contributions to the business. This is why many market model proponents have advocated predetermined terms for directors. Some experts have recommended that the term for directors, although reviewed annually, should be for two to three years with a mandatory retirement age set between 62 and 65 years old (Ward, 1991).

However, the reason why boards of most companies, particularly publicly held companies, have term limits is largely a political one, and at best loosely related to the accountability of their members. Therefore, we believe that a board member's term should be for a limited, yet not necessarily predetermined, time period. Although many boards of the typical family firm are filled with individuals who are close friends of members of the family, or are family members themselves, this close relationship does not preclude the fact that board members should serve for a limited period of time. Board members who serve indefinitely become entrenched in the business and generally lose their ability to render objectivity and promote accountability. As a result, a length of time that promotes accountability would be a minimum term limit of approximately three years (in order to get the director acclimated to the business) and an extensive review process after that three-year period. There must also be a process for evaluating the contribution of the director, as well as criteria for 'keep/let go' decision-making.

Board evaluation process

Market-model bias Aggregate board evaluations are sufficient, otherwise collegiality suffers.

Propositions for control-model accountability

Model family firms Aggregate, committee level, and individual board performance evaluations are necessary to promote accountability.

Proposition 9 Board evaluations will increase board accountability.

Board evaluations can be powerful tools to develop and support high-performing board members. Boards should regularly and formally evaluate the performance of the CEO, other senior managers, and the board against their goals and performance standards. Evaluations should include annual assessments of the functioning of board processes and board committees. The board should conduct reviews of its own processes annually and make changes where necessary. While not absolutely necessary, the use of independent third parties for these reviews is recommended.

However, although evaluating aggregate board performance is necessary, it is not sufficient. In order to promote the accountability of the board, board members of all firms, in particular family firms, must also be evaluated individually. Many directors do not want to conduct individual evaluations because they feel they must share in the responsibility, or they feel that they will be disturbing the collegiality present in board rooms. However, in order for directors to be held accountable for creating and maintaining a high performance board, they must be able to distinguish good contributions from poor (Felton et al., 1995), and, above all, ensure that all directors act to hold themselves and the company accountable.

Leadership: role of chairman and CEO

Market-model bias   CEO/chairman duality is a matter of debate.

Propositions for control-model accountability

Model family firms   Both CEO/chairman roles can be combined if, and only if, one person is the best option for handling two jobs.

The chairman, as the conductor of the board, can play a central role in ensuring the effective governance of the enterprise (OECD, 1999). The chairman acts as the parliamentarian for the meeting and is responsible for agenda-setting and controlling discussion on agenda items, while allowing appropriate discussion of essential items (Nash, 1995). Whether the role of the chairman should be separate from the role of the CEO has become a matter of debate. The results of the McKinsey *US Directors Survey* (2002) indicate that currently, 75 per cent of S&P 500 companies have a single person serving as both chairman and CEO. Many corporate governance experts believe that the business culture of the US lends itself to having the CEO be the board chairman in order to focus a company's leadership (Moore, 2002). In addition, other advocates for the CEO/chairman role claim that a company risks added divisiveness by splitting the role of the chairman and CEO and that it reduces the CEO's freedom of action (Felton and Watson, 2002). However, many others, particularly in the UK, believe that the separation of the CEO and chairman roles ensures an appropriate balance of power. Many commentators have viewed a lack of separation as impeding the supervisory ability of the board; others claim that if the leader of the supervisory body is also the leader of the managerial body under supervision, he or she faces a significant conflict of interest (Gregory and Simmelkjaer, 2002).

We believe that the roles of chairman and CEO should only be combined when the single person can do the two jobs effectively. Since the role of the chairman is to guide board processes, moderate meetings, ensure that the board completes all tasks in a timely and effective manner, and counsel the CEO (not direct the CEO), an increase in the role of the CEO with the additional responsibilities of the chairman functions presents a limited number of conflict of interest issues. As long as the CEO is the best person to handle this additional moderator job effectively and be held accountable, the CEO may be able to carry on both roles without causing an imbalance of power. The implementation of the single CEO/chairman role begs the question, however, of 'who evaluates the CEO?' We advise that the board institute evaluations where the opinions of individual directors and managers are sought and then synthesized in a report presented to the board and discussed privately with the CEO, and that whoever chairs the committee responsible for this task is not the CEO nor reports directly or indirectly to the CEO.

An additional area of controversy concerning board leadership is whether the retired CEO should stay on the board in the capacity of the chairman as part of the succession transition. Many times the prior CEO is not able to surrender all of his or her authority when staying on as chairman and often, as soon as there is a crisis, the departed CEO feels compelled to return to 'rescue' the organization. However, as long as the successor is able to initiate change and assert his or her leadership, and as long as sufficient knowledge transfer occurred from the old leadership to the new, then we believe the prior CEO can stay on as chairman without undermining the authority of the new leader. However, we recommend that separate committees monitor the situation and act to remedy any problems if the former CEO undermines the new CEO.

Board compensation

Market-model bias Board members should be compensated according to what the market dictates.

Propositions for control-model accountability

Model family firms Board members not employed in the company should be compensated at a rate equivalent to the CEO.

Proposition 10 Equating board compensation to a CEO pay rate will increase company performance.

Even though the *American Family Business Survey* (2002) found that fewer than 61 per cent of family businesses compensate their board members, compensation is one way to attract and retain the most qualified and accountable board. How board members should be compensated has increasingly become a controversial topic in corporate governance, owing to the current collapse of investor confidence in companies' performance. Generally, companies operating under the market model compensate board members according to market norms. Many experts claim that creating investor confidence requires performance-based compensation rather than retainers or salary (Felton and Watson, 2002). For instance, they claim that issuing restricted stock rather than offering stock options may help meet the investors' demand that directors be accountable for the actions that they implement. Even for non-family companies we believe this approach is short-sighted in that it may stimulate rather than eliminate the manipulation of financial statements.

Compensating directors with stock may hold other disadvantages for the family-controlled company. Linking directors' judgment to share price presumes that intermediate term share value is the goal, which is not always the case as family businesses often provide owning family members with non-financial returns. It also rewards directors for industry effects rather than objective corporate performance. Furthermore, family members often have a bias against stock in non-family hands, for which such an approach cannot account. To send a message that the board is as important as the CEO, we recommend that directors should be paid for their time commensurate with that of the company's CEO. A simple heuristic is to divide the CEO's annual pay by 250 working days and then each board member should be paid the resulting amount per day for each day spent on board matters (Ward, 1991). In addition to financial compensation, opportunities to learn, grow, and make a contribution are among many of the other benefits of a position on a strong and vital board.

- **The right to influence and affect control** the corporation through shareholder participation in general meetings and shareholder votes
- **The right to information** about the corporation
- **The right to participate** in the profits and positive cash flow of the corporation

Figure 17.2 Inalienable rights of shareholders

Family-owned boards must be accountable to shareholders

Regardless of whether the market or control model is utilized, in corporate governance codes around the world shareholders are viewed as embodying the following rights: the right to participate in the profits and cash flow of the corporation; the right to influence the corporation through shareholder participation in general meetings and voting; the right to affect control over the corporation; and the right to information about the corporation (OECD, 1999) (see Figure 17.2). It is within these rights that shareholders affect the performance of the board, and, in turn, force the board to be held accountable for its actions.

Even though shareholders are encouraged to participate and positively influence the corporation, particularly in the family firm, shareholders must not inappropriately seek to influence and control the board. Unlike firms that operate in the market model of corporate governance, generally shareholders of family firms are not averse to participating in the governance of the firm. While in market-model jurisdictions the shareholding body is made up of individuals and institutions whose interests, goals, investment horizons, and capabilities vary, in control-model jurisdictions, the concentrated base of shareholders often exercise a degree of control over the corporation disproportionate to the shareholders' equity ownership in the company (OECD, 1999). This disproportionate degree of control occurs because shares have more significance for family shareholders (LaPorta et al., 1999). Family firms engender more participation from shareholders because they have more than a financial stake and thus the governance processes that promote accountability will differ.

However, unlike shareholders operating under the market model, shareholders of family firms are often unable to separate their ownership duties from managerial duties or parental obligations. Family-member shareholders often become too involved. This level of activism may weaken the decision-making process, decrease accountability as confusion increases, increase the risk of passive management, or cause the sale of the company (Aronoff, 1996). This section highlights the mechanisms that shareholders of family businesses can utilize to influence the corporation, encourage accountability, protect their interests, and promote family unity, while not becoming overly involved or intrusive. Once again, it is the appropriate balance between monitoring mechanisms and collaborating mechanisms that creates the best form of accountability for family firms.

Influence the composition of the board

Market-model bias Shareholders utilize cumulative voting and market liquidity to protect minority interests.

Propositions for control-model accountability

Model family firms Shareholders of family firms *should* utilize cumulative voting and shareholders agreements to protect their interests and participate in the accountability process.

Proposition 11 Cumulative voting will increase accountability.

Shareholders operating under both market and control models can help promote their interest in the company by influencing the composition of the board. After all, the shareholder's right to influence the corporation is essentially their right to influence the decision-makers, that is, the board, by being able to select its members and approve extraordinary transactions, which, in turn, leads to greater accountability (OECD, 1999). While companies operating under the market model have been encouraging their shareholders to adopt mechanisms to ensure adequate board composition and representation of shareholders, companies operating under the control model, including family firms, have been known to use unorthodox methods, such as favoritism of family members, for determining board composition, which can be detrimental to minority shareholders.

One significant way of influencing board composition, while not alienating minority shareholders in the process, is through cumulative voting. Cumulative voting is a procedure used in the election of directors whereby the minority shareholders have a greater opportunity to secure representation of their interests on the board because they can accumulate their vote for a single candidate. In the absence of cumulative voting, a shareholder, or group of shareholders, with *50 per cent of the shares plus one* can elect the entire board of directors, assuming each share has only one vote. However, under cumulative voting, the shareholder multiplies the number of shares he or she holds by the number of directors to be elected in order to obtain the aggregate number of votes he or she is able to use. The Business Corporation Act in the US allows shareholders under cumulative voting to spread these votes evenly among an entire slate of directors, or give them all to one director – possible assuring the election of that director, depending upon the percentage of shares held and the number of directors to be elected (Murdock, 1996). While we believe cumulative voting is an important mechanism to ensure accountability, it should be a last resort mechanism that is only invoked after a family breakdown or extremely poor corporate performance. We believe an effective nominating committee that works in concert with the board review process is the best and least disruptive way to insure the protection and promotion of interests.

Another way to influence board composition is through shareholder agreements. Shareholder agreements can codify issues such as how the board or the chairman is selected, succession planning issues, estate planning issues, and dispute resolution mechanisms. Some experts recommend that in order to ensure shareholders' interests are represented on the board, shareholder agreements should include provisions stating that a family member shareholder can seek a contractual right to select a specified number of directors (Friedman and Friedman, 1994). Even though family firms do not always advocate these mechanisms for fear of divisiveness, we believe they must encourage shareholders to protect their interests so that they can participate in the accountability process and perpetuate the family's goals.

Create communication channels/forums

Market-model bias	Minimal communication between minority and majority shareholders is required including quarterly and annual financial statements.

Propositions for control-model accountability

Model family firms	Communication between minority and majority shareholders is necessary. Communicating with shareholders is necessary in order to perpetuate the family's vision, for more information is needed in addition to financial reports. Basic strategic plans, values, and industry, supplier, and customer information are required. Regular surveys, questionnaires, and meetings with shareholders to gain their views are also needed.
Proposition 12a	Increased communication between minority and majority shareholder groups will increase company performance.
Proposition 12b	Increased communication between shareholders and the board will increase company performance.

Corporations must develop better channels of communication with shareholders in order for them to affect control over the company's decision-making. Many market model proponents advocate communication in order to remove the artificial barriers to participation that are prevalent by controlling shareholders against minority shareholders, such as charging fees for voting, having prohibitions on proxy voting, or having requirements of personal attendance at general shareholder meetings to vote (Gregory and Simmelkjaer, 2002). These proponents advocate voting by proxy, as well as the enlarged use of technology to increase access to and simplify voting (OECD, 1999).

However, market-model proponents have not addressed the specific role that the board must play in furthering communication. Particularly in family businesses, communication is critically important between the board and family shareholders. The board must understand the needs of the family shareholders whom it represents. Additionally, within the typical family business, the board is dealing with a family that has a culture with some collective needs, such as family unity, that the board must understand. Shareholders need to be reassured that the board is attending to those needs. Therefore, in order to facilitate this communication, mechanisms must be in place to ensure that the voice of these shareholders is heard. Regular letters to the board, family meetings and/or dinners with the board, and family councils are all mechanisms that can be utilized to enhance this communication. This consistent, timely, meaningful two-way communication becomes the key to obtaining accountability from board members with respect to owners.

Involvement in strategic decision-making

Market-model bias	Shareholders leave decision-making up to the board and management.

Propositions for control-model accountability

Model family firms — Shareholders of family firms should be informed of all strategy conversations and decisions, and the board should take family goals and desires into consideration.

Proposition 13 — Family shareholder 'involvement' will increase board performance.

Shareholders operating under the market model tend to leave strategic decision-making entirely up to the board and management. Conversely, shareholders of family firms operating under the control model tend to direct management, make decisions on their own, or lobby individual board members. This form of decision-making weakens the power of both the board and management, and may results in factionalizing the shareholding body. We believe appropriate involvement by shareholders would entail establishing the values, vision, and goals of the business as well as being a 'partner' in strategy (Aronoff and Ward, 2002a). This means helping management and the board to understand owner goals as a basis for developing business strategy, and then embracing and supporting the strategy that is proposed by management and endorsed by the board (Aronoff and Ward, 2002a). This form of involvement in strategic decision-making allows shareholders, the board, and management to become united in their decision-making in order to add richness and strength to their business culture.

Family-owned boards must hold management accountable for their actions

The role of management in family-owned businesses is to develop and execute the business's strategy and to meet the expectations of the board and the owners through their leadership of the company (Aronoff and Ward, 2002a). The board's role, as a result, is to monitor management and hold them accountable for their actions. The McKinsey *US Directors Survey* (2002) indicates that directors have been dissatisfied with the way in which they have been monitoring management (Felton and Watson, 2002). The survey indicates that many directors do not have enough time to get to know top managers other than the CEO and tend to shy away from conflict (Felton and Watson, 2002). Directors of control-model family businesses, conversely, know the owners and managers, or are themselves managers, of the business. In addition, conflict may be prevalent, particularly among family members. Therefore, boards of family businesses need to have the ability and mechanisms in place to effectively monitor management's performance and manage conflict.

Monitoring strategic execution

Market-model bias — Boards often leave strategic decision-making to management.

Propositions for control-model accountability

Model family firms — Boards must approve the strategic plans of management and monitor them regularly.

Proposition 14 — Increased board approval of strategic plans correlates to an increase in management accountability.

One way that boards can effectively monitor management is by approving the strategic plans of management. While boards of companies operating under the market model often 'rubber stamp' proposals from management or leave the strategic decision-making process up to management, boards of companies operating under the control model tend to involve themselves too much in the decision-making process. However, even though board members maybe experts in their respective fields, they should not be formulating company strategy; rather they should be critically evaluating the strategy that management creates. Therefore, if management's strategic plan is not appropriate, we believe the board should make management amend the strategic plan – that being the most effective form of accountability.

Additionally, another aspect to monitoring management is ensuring that management continues to be held accountable for fully executing its plans. The board must periodically check in with executives in the short-term to ensure that plans are executed, as well as benchmark results against long-term indicators, such as market share, budgets and margins. By continuing this monitoring function, board members can help identify obstacles and figure ways to overcome them when performance falls short.

Executive compensation

Market-model bias	Executive compensation should be transparent. Boards should tie compensation to stock performance.

Propositions for control-model accountability

Model family firms	Compensation should be transparent among family members. Boards should tie compensation to organization's mission, annual business performance (financial and non-financial), and long-term financial results.
Proposition 15	Correlating management compensation to business performance will increase management accountability.

Owing to continued reports of excessive executive compensation, particularly across EU member states, there is an increasing call for disclosure of executive remuneration among market-model boards, such as those in the UK (Gregory and Simmelkjaer, 2002). Even though privately held family businesses are not obligated to announce executive remuneration, family businesses must also create an atmosphere of trust and transparency among family members. Many family business owners are too comfortable keeping pay undisclosed. However, nondisclosure does not prevent family members from forming strong emotions, suspicions, and beliefs about one another's compensation (Aronoff and Ward, 1993). Therefore, we believe that information concerning compensation should be easily available to family member shareholders and reviewed annually by the board in order to preserve an atmosphere of accountability.

Additionally there have been debates over how best to compensate executives. Many institutions operating under the market model primarily define performance simply as stock performance. This emphasis on stock performance, however, tends to stimulate extreme emotions – greed when things are going well, demoralization when the market falls (Elson et al., 2003). Further, business performance is only one ingredient in stock performance, which also includes economic, industry, acquisition prospects, and other factors. Therefore, for market-model companies it is now being recommended that businesses rebalance elements of executive compensation and tie compensation more closely to the organization's mission, annual business performance, and long-term financial results, which, in turn, that will create real shareholder value over time (Elson et al., 2003).

Shareholders of family businesses have a vested interest in long-term performance, and this is quite consistent with tying compensation to mission and long-term financial results. We also recommend that compensation be tied to the performance of non-financial measures, which are based on what the owning family values. These can include opportunities for the family, company reputation, employee satisfaction, information availability, family educational opportunities, and community and philanthropic activities. Some advisors recommend a compensation philosophy that can help family businesses make difficult decisions, such as whether to pay family members equally or according to market value (Aronoff and Ward, 1993). A written compensation policy, according to the family's philosophy of compensation, assures that the family's value system and vision are parallel to the way that they are operating the business.

Evaluation of corporate officers

Market-model bias — Evaluations should be conducted and emotions cast aside.

Propositions for control-model accountability

Model family firms — Expectations for executives must be clarified in order to be able to objectively evaluate and promote accountability.

Proposition 16 — Annual evaluations of corporate officers will increase accountability.

Another way for control-model boards to promote accountability of management is for boards to ensure that the evaluations of top managers are in line with performance standards. Most companies operating under the market model conduct evaluations assuming emotions and prejudice are cast aside. However, for family businesses emotions are unavoidable. Evaluating performance and making difficult decisions about promotions can be more difficult when managers are also family members. Many families use the same values to operate the business as the family, such as promoting family members based on birth order or gender, rather than on skills (Loeb, 2001). The difficulties associated with evaluating family members often causes performance management systems in family businesses to be overlooked entirely or implemented only for non-family employees (Driscoll and Korman, 2001). As the family and business grow, however, these decisions can have a negative impact on business growth as well as on the emotional health of family members (Loeb, 2001).

As a result of its potential for conflict and role confusion, we believe that performance management for family businesses is a system and not a once-a-year event (Driscoll and Korman, 2001). As such, an effective performance management system, implemented by the board, should have clear performance criteria and consequences for performance, position descriptions for family and non-family employees, and documentation of performance successes, issues, and results (Driscoll and Korman, 2001). Performance standards need not neglect the family dimension of the family business, yet they need to be set and monitored. In addition, boards need to ensure that management has a plan for building a reserve of new leaders and that the plan is followed.

Summary

The thesis of this chapter can be summarized as 'accountability, accountability, accountability'. Because of the unique characteristics of concentrated ownership and power in

SHAREHOLDERS

Should …

- Influence composition of board
- Create communication channels
- Be involved in strategic decision-making

BOARD

Should …

- Be mid to large-sized
- Not focus on independent status
- Enhance communication and conflict resolution
- Actively participate
- Have limited/not predetermined terms
- Conduct individual board member evaluations
- Allow CEO/chairman role
- Compensate equivalently to CEO
- Monitor company and hold it accountable

MANAGEMENT

Should …

- Be monitored by the board
- Tie compensation to mission, vision, long-term indicators
- Clarify expectations for evaluations

Figure 17.3 Accountability's three tiers

control model firms, the mechanisms of accountability through board governance will not necessarily be the same as those of market-model companies. All the propositions herein are directed toward better understanding these differences.

The board is the governor of the family business and as such sits squarely between the owners and their leadership. While the owners are ultimately responsible for strategic direction and this can clearly be seen in their investment, the board must insure that strategy as detailed by management is in keeping with the family owners' desires and that company leaders execute the strategy in a timely and complete fashion. To achieve accountability as described here, a competency-based board with a focus on balancing the appropriate

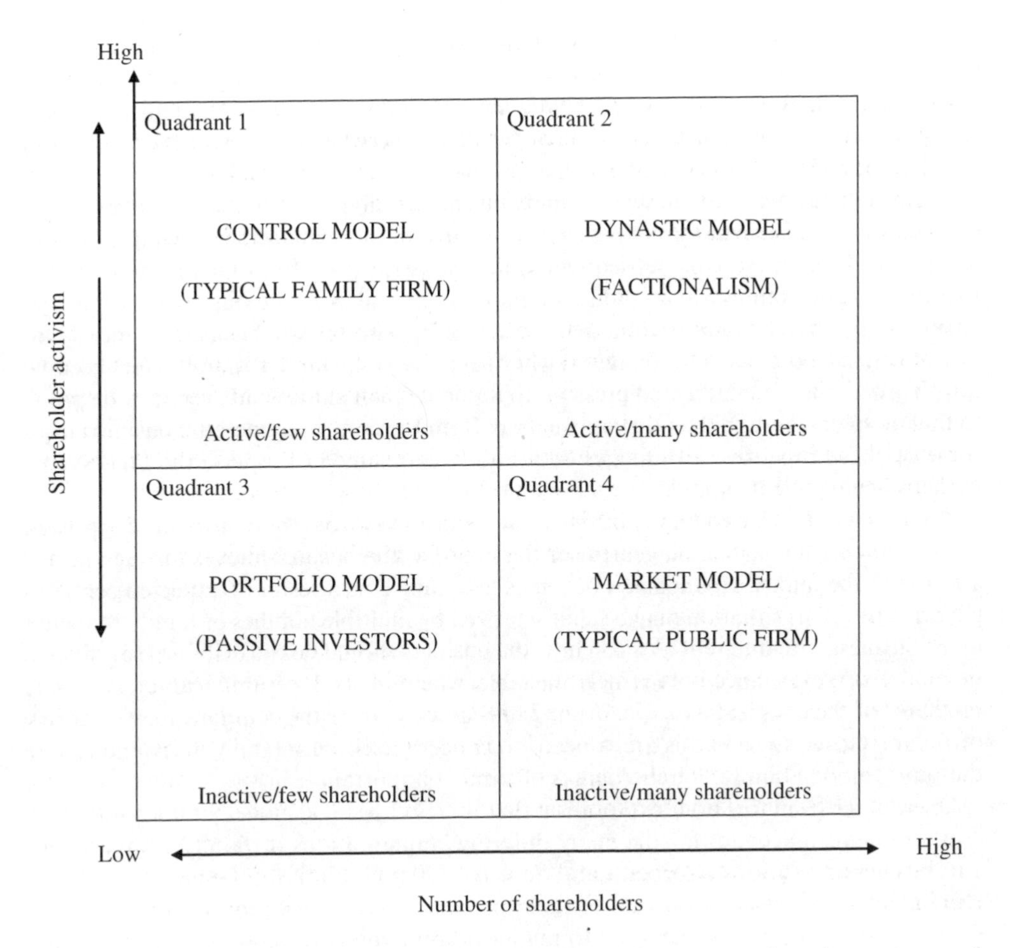

Figure 17.4 Four quadrants of family firm governance: dimensions of market and control

amounts of monitoring and collaboration is needed for the typical family-owned business operating under the control model (see Quadrant 1 in Figure 17.4).

The emotionality, overlapping roles, and oft-perceived unbusiness-like behavior have caused market model corporate governance specialists great skepticism about family-owned businesses' ability to be held accountable. However, as long as the family is psychologically mature and at least moderately unified, and the board executes the task of holding family business members accountable to clear standards and well-understood expectations that are mutually set, accountability is likely, and the probability of underperforming lessened (Gimeno et al., 2004).

However, as family businesses move away from the typical control model, which occurs when the number of active family members decrease and ownership becomes more widespread, family businesses begin to move closer towards the market model (see Quadrant 4 in Figure 17.4), and assuring appropriate accountability may need to shift. Family firms operating closer to the market model need control mechanisms, such as independence, discipline and objectivity, to sustain the business. Such controls are needed to insure that independent management acts in the best interests of the owners. Family firms operating under the market model may not have some of the strategic advantages of family companies as they lack the ability to efficiently align the interests of management with ownership in order to create a shared sense of purpose, since the identities of the family and the identities of the business can become distanced.

There are two other ways in which family businesses migrate from the control model of governance. One is when family become less active in the management owing to retirement or lack of interest by subsequent generations (moving from Quadrant 1 to 3 in Figure 17.4), but family shareholders maintain their stakes in the business. In this scenario, family shareholders become passive investors who see the business as merely an investment in a portfolio. One danger is when such 'portfolio model' shareholders become disinterested, they feel increased pressure to liquidate their shares and even may be prone to litigate to do so. Again, as they move away from the control model, the business loses some of the competitive advantages of a family company, in this case the biggest loss perhaps being patient capital.

Another way in which a family business may move away from the control model is when family remains active in management or the board as the business moves through generations, but the family expands and becomes less unified as a result of expected centrifugal pressures. This situation may be characterized by multiple families or family branches in the business with different ideas on how the business should be run. This type of 'dynastic model' of governance is extremely unstable, where the lack of unity can create a sale of shares of the company (see Quadrant 2 in Figure 17.4), or the company may be at risk of failure. Governance in this arena needs to concern itself with family unity and ensure that unity among family shareholders is of utmost importance.

One can see from this brief exploration that there is no single model for corporate governance that can account for the many differing configurations of family, shareholders, and business conditions (Corbetta and Salvato, 2004; Gubitta and Gianecchini, 2002). Our intent herein was to provide testable propositions for how control-model family firms' boards of directors may need to pursue accountability in ways different than their market-model counterparts.

It is our belief that, based on the literature, each propositions will contribute to a foundation of accountability for the particular configuration of family, shareholder, and family involvement in the business. It is our hope that each of these propositions will be further explored using empirical research models. Such in-depth research will help us better understand the effects of a concentrated ownership base in relation to corporate accountability. For example, future research can examine the relationships provided in the propositions as they relate to company performance (profits, ROE, and other financial measures) and family variables, such as those in the F-PEC scale (Astrachan et al., 2002). Using these models we might see that the levels of 'familiness' in the F-PEC scale may be a more important variable than ownership dispersion when comparing market- and control-model companies. Challenges to investigating these propositions include operationalizing and measuring the concept of accountability, as well as developing clearer measures of concepts, such as board performance.

It is clear that the present study is limited by it's exploratory nature. While the lack of research studies devoted to understanding the uniqueness of family firms and their governance necessitated the nature of this study, we feel that through using and measuring the concept of accountability in relation to the propositions herein, we can deepen our understanding of how the board governance of control-model family firms should differ from non-family market-model firms.

References

2002 Mass Mutual Financial Group/Raymond Institute: American Family Business Survey (No. DMD9500R), Fort Meyers, FL: Raymond Family Business Institute.

Aguilera, R.V. and G. Jackson (2003), 'The cross-national diversity of corporate governance: dimensions and determinants', *Academy of Management Review*, **28**(3), 447–65.

Aronoff, C.E. (ed.) (1996), 'New thinking on corporate governance: a critical analysis for family firms', *Family Business Advisor*, **5**(1), 8.

Aronoff, C.E. and J.L. Ward (1993), *Family Business Compensation*, Marietta, GA: Business Owner Resources.

Aronoff, C.E. and J.L. Ward (2002a), *Family Business Ownership: How to be an Effective Shareholder*, Marietta, GA: Family Enterprise Publishers.

Aronoff, C.E. and J.L. Ward (2002b), 'Outside directors: how they can help you', in C.E. Aronoff, J.H. Astrachan and J.L. Ward (eds), *Family Business Sourcebook*, 3rd edition, Marietta, GA: Family Enterprise Publishers, pp. 254–5.

Astrachan, J.H. and M.C. Shanker (2003), 'Family businesses contributions to the U.S. economy: a closer look', *Family Business Review*, **15**(3), 211–19.

Astrachan, J.H., S.B. Klein and K.X. Smyrnios (2002), 'The F-PEC scale of family influence: a proposal for solving the family definition problem', *Family Business Review*, **15**(1), 45–58.

Atkinson, A. and S. Salterio (2002), 'Shaping good conduct', *Management*, **49**(1), 18.

Berle, A.A. and Gardiner C. Means (1932), *The Modern Corporation and the Private Property*, New York: Harcourt Brace.

Coombes, P. and M. Watson (2001), 'Corporate reform in the developing world', *The McKinsey Quarterly*, **4**, 89–92.

Corbetta, G. and C.A. Salvato (2004), 'The board of directors in family firms: one size fits all?', *Family Business Review*, **17**(2), 119–34.

Daily, C.M., D.R. Dalton and A.A. Canella (2003), 'Corporate governance: decades of dialogue and data', *Academy of Management Review*, **28**(3), 371–82.

Dalton, D.R., C.M. Daily, A.E. Ellstrand and J.L. Johnson (1998), 'Meta-analytic reviews of board composition, leadership structure, and financial performance', *Strategic Management Journal*, **19**, 269–90.

Davis, J.H., F.D. Schoorman and L. Donaldson (1997), 'Toward a stewardship theory of management', *Academy of Management Review*, **22**, 20–47.

De Visscher, F.M., C.E. Aronoff and J.L. Ward (1995), *Financing Transitions: Managing Capital and Liquidity in the Family Business*, Marietta, GA: Business Owner Resources.

Demb, A. and F.F. Neubauer (1992), *The Corporate Board: Confronting the Paradoxes*, New York: Oxford University Press.

Donaldson, W.H. (2003), 'Corporate governance: what has happened to where we need to go', *Business Economics*, **38**(3), 16–20.

Driscoll, M. and M. Korman (2001), 'Achieving competitive advantage through performance management', in B. Spector (ed.), *Family Business Compensation Handbook*, Philadelphia, PA: Family Business Publishing, pp. 61–3.

Elson, C., E. Roiter, P. Clapman, J. Bachelder, J. England, G. Lau, E.S. Woolard Jr, P. Meyer, B. Hill, H. Barnette, W. Batts and E.N. Veasey (2003), 'What's wrong with executive compensation?', *Harvard Business Review*, **81**(1), 68–78.

Felton, R., A. Hudnut and V. Witt (1995), 'Building a stronger board', *The McKinsey Quarterly*, **2**, 162–75.

Felton, R.F. and M. Watson (2002), 'Change across the board', *The McKinsey Quarterly*, **4**, 31–46.

Friedman, M. and S. Friedman (1994), *How to Run a Family Business*, Cincinnati, OH: Betterway Books.

Gimeno, A., Labadie, G. and Saris, W. (2004), 'The effects of management and corporate governance on performance in family businesses: a model and test in Spanish firms', in S. Tomaselli and L. Melin (eds), *Research Forum Proceedings, FBN 2004*, 147–69.

Gregory, H. and R.T. Simmelkjaer (2002), *Comparative Study of Corporate Governance Codes Relevant to the European Union and its Members States on Behalf of the European Commission*, New York: Weil, Gotshal, & Manges, LLP.

Gubitta, P. and M. Gianecchini (2002), 'Governance and flexibility in family-owned SMEs', *Family Business Review*, **15**(4), 277–97.

Habbershon, T.G. and M.L. Williams (1999), 'A resource-based framework for assessing the strategic advantages of family firms', *Family Business Review*, **12**(1), 1–26.

Hillman, A.J. and T. Dalziel (2003), 'Boards of directors and firm performance: integrating agency and resource dependence perspectives', *Academy of Management*, **28**(3), 383–96.
Jensen, M.C. and W.H. Meckling (1976), 'Theory of the firm: managerial behavior, agency costs, and ownership structure', *Journal of Financial Economics*, **3**, 305–60.
Judge Jr, W.Q. and C.P. Zeithaml (1992), 'Institutional and strategic choice perspectives on board involvement in the strategic choice process', *Academy of Management Journal*, **35**(4), 766–94.
Kets de Vries, M. (1993), 'The dynamics of family controlled firms: the good and the bad news', *Organizational Dynamics*, **21**, 59–71.
La Porta, R.F., F. Lopex-de-Silanes and A. Shleifer (1999), 'Corporate ownership around the world', *Journal of Finance*, **54**(2), 471–518.
Loeb, M.E. (2001), 'Employee evaluations and promotions in the family firm', in B. Spector (ed.), *Family Business Compensation Handbook*, Philadelphia: Family Business Publishing, pp. 59–60.
Lynall, M.D., B.R. Golden and A.J. Hillman (2003), 'Board composition from adolescence to maturity: a multitheoretic view', *Academy of Management Review*, **28**(3), 416–31.
McKinsey & Company (2002), *US Directors Survey*, April/May.
Mobius, M. (2001), 'Good governance is a global challenge', *Corporate Board*, **22**(131), 1–4.
Moore, M.T. (2002), 'Corporate governance: an experienced model', *Director's Monthly*, **26**(3), 1–9.
Murdock, C.W. (1996), *Illinois Practice: Business Organizations*, vol. 7, St Paul, MN: West Publishing.
Nash, J.M. (1995), *Boards of Privately Held Companies: A Guide*, National Association of Corporate Directors Governance Series: Private Company Boards, Washington, DC.
Neubauer, F. and A.G. Lank (1998), *The Family Business: Its Governance for Sustainability*, New York: Routledge.
Newell, R. and G. Wilson (2002), 'A premium for good governance', *The McKinsey Quarterly*, **3**, 20–23.
Organisation for Economic Co-operation and Development (OECD) (1999), *Principles of Corporate Governance*, SG/CG(99)5, Paris: OECD.
Plender, J. (2003), 'Restoring trust after the bubble: lawmakers are addressing only one side of the problem', *Business Economics*, **38**(3), 21–4.
Robinson, A. (2002), 'Is corporate governance the solution or the problem?', *Corporate Board*, **23**(133), 12–16.
Rosenstein, S. and J.G. Wyatt (1990), 'Outside directors, board independence, and shareholder wealth', *Journal of Financial Economics*, **26**(2), 175–92.
Shen, W. and C. Cannella, (2002), 'Revisiting: the performance consequences of CEO succession: the impacts of successor type, postsuccession senior executive turnover, and departing CEO tenure', *Academy of Management Journal*, **45**(4), 717–33.
Shleifer, A. and R. Vishny (1997), 'A survey of corporate governance', *Journal of Finance*, **52**(2), 737–84.
Sonnenfeld, J.A. (2002), 'What makes great boards great', *Harvard Business Review*, **80**(9), 106–13.
Sundaramurthy, C. and M. Lewis (2003), 'Control and collaboration: paradoxes of governance', *Academy of Management Review*, **28**(3), 397–415.
Taguri, R. and J. Davis (1996), 'Bivalent attributes of the family firm', *Family Business Review*, **9**, 199–209.
Ward, J.L. (1991), *Creating Effective Boards for Private Enterprises*, San Francisco, CA: Jossey-Bass.
Wheelen, T.I. and J.D. Hunger (1994), *Strategic Management and Business Policy*, 8th edition, Upper Saddle River, NJ: Prentice Hall.
Yermack, D. (1996), 'Higher market valuation of companies with a small board of directors', *Journal of Financial Economics*, **40**(2), 185–212.

18 Effective knowledge transfer in family firms

Rosa Nelly Trevinyo-Rodríguez and Josep Tàpies

One of the most critical organizational changes family businesses deal with at some stage in their lives is the succession process. When evaluating it, two main targets are sought: quality and effectiveness. To meet these quality-effectiveness standards three elements should be transferred from the predecessor to the next generation member(s): (1) ownership control/power, (2) management responsibility and (3) competence/knowledge. This chapter focuses on the third element, knowledge, since most of the times, it is 'the taken-for-granted' factor. How effective intergenerational knowledge transfer in family firms takes place – under which conditions and through which variables – is the heart of this writing. We have developed the Knowledge Transfer Model in Family Firms (KTFF) which sets out several internal and external relationships in the family–enterprise–next generation system. And, although this is a conceptual text, it may drive future empirical research projects in order to provide support for the proposed interactions (relationships).

Introduction

One of the most common ways organizational changes are brought about in any business is through the replacement of key personnel. This process is generally called administrative succession. Indeed, organizationally, succession is important for two basic reasons: (1) it always leads to organizational instability, and (2) it is a phenomenon that all organizations must cope with (Grusky, 1960). One reason why all organizations must cope with the succession process is more than obvious: we are all mortal beings. However, it is important to note that not all businesses face the same problems when entering the succession stage. Family firms are a particular case where succession is extremely delicate, since some additional, special dimensions have to be taken into account; for instance, the family and its dynamic relationships.

Without doubt, the succession processes in family firms represent the most critical period confronting family businesses, since this is precisely when the business is transferred from one generation to the next. And of course, one of the main dreams a founder typically has is to hand down his or her legacy to his or her offspring, given that he or she has accumulated capital, commitment of potential members, entrepreneurial skills, and legitimacy (Stinchcombe, 1965). In addition, the founder has already generated learning curves (diminishing costs), as well as built on a reputation that can serve as a springboard for the family to jump on and promote. In truth, the founder has created a highly reliable and accountable business. Nevertheless, the very factors that make a system reliable and accountable (reproducible – it has the same structure today that it had yesterday) make it also resistant to change. That is precisely why succession is so crucial.

Owing to the importance and inevitability of succession processes, much attention has been drawn to highlight the needs of family firms to develop formal succession plans, and to start planning the let-go stage as early as possible. In fact, assuming all else is equal, the extent of organizational instability following succession tends to be inversely related to the amount of control over the process (Grusky, 1960). Thus, progress in explaining an organizational change, such as succession, requires understanding both its nature and the degree to which it can be planned and controlled (Hannan and Freeman, 1984).

Focusing on the organizational change planning and control, Handler (1990) affirms that when evaluating the succession process two targets are sought: quality and effectiveness. Quality is a reflection of how the involved family members personally experience the process, while effectiveness is more related to how others judge the outcome of the succession. To meet these quality-effectiveness standards, the transfer of three elements between the generations should take place: (1) ownership/power, (2) management responsibility, and (3) competence/knowledge (Varamäki et al., 2003). It is worth stressing that besides these three elements, there are various psychological factors that may also influence the effects of the succession process, such as the successor's (or potential candidate's) personal skills and administrative experience outside the business, his or her commitment to the goals of the family and the organization, and the legitimacy he or she can generate for the position in the eye's of the employees of the company.

In spite of the fact that little attention has been devoted to the study of how knowledge is transferred in family businesses, usually most of the studies regarding the succession topic take this transmission for granted. On the other hand, there are signs that perspectives related to transfer of knowledge and learning processes are important because one sees them appearing constantly across disciplines. Yet, in spite of the considerable advance on bridging disciplines, there is almost no organizational learning and/or family business literature focusing on 'How effective intergenerational knowledge transfer in family firms takes place'. Consequently, we embark in this study intending to find out variables and conditions that promote effective knowledge transfer in family businesses, thus bridging the organizational learning and family business areas.

How is knowledge transferred within the same firm across generations is an issue that may explain to some extent (in addition to the leadership-keeping process) why most family businesses do not survive until the third generation, the mortality/death rate being much higher during the owner–second generation transition. Although we keep our approach completely theoretical through this conceptual writing, the final idea is to set a knowledge transfer framework (with variables and conditions) that can be used later in empirical research to provide support for the proposed relationships.

This chapter is organized as follows. First, we analyse whether or not knowledge transfer in family firms is different from knowledge transfer in other firms, and if so, we will proceed to examine the concept of individual learning and how is it related to knowledge acquisition and transmission. Second, we develop a knowledge transfer model applied to family firms exposing the variables and conditions that interact during knowledge transmission processes.

Is the knowledge transfer in family firms different from the knowledge transfer in other firms?

The primary roots of knowledge are the philosophy and sociology disciplines. Writers such as Popper, Lakatos, Feyerabend, Berger, Luckmann, Polanyi, Kuhn, Wieck and Giddens influenced the thinking on knowledge in organizational and economic life. Yet, despite the fact that there is a lot written regarding this subject, there's no single standard definition of knowledge. In this writing, we will use Tsoukas and Vladimirou's (2001) interpretation of knowledge, from the organizational learning literature: knowledge is the individual capability to draw distinctions, within a domain of action, based on an appreciation of context or theory, or both; it presupposes values and beliefs, and is closely connected with action.

Knowledge is mainly 'personal': it is mostly a property of the people who acquired and use it. Actually, it has been detected that people in organizational environments are much less likely to share what they know with their colleagues, since they consider them 'proximate competitors' in the labor market. Employees usually regard certain areas of their knowledge as part of their power base within the company; thus their willingness to share is limited (Probst et al., 2000). Moreover, the human being is generally reluctant to communicate his own knowledge stock because of self-interested, opportunistic reasons (agency theory). In fact, generally individuals would tend more to share their knowledge with people outside their organization (and even with those in other organizations) than with their own colleagues. As the motto attributed to Machiavelli says, 'People make wars with neighborhoods and alliances with far located partners'.

However, the main point here is that 'unless the implications of experience can be transferred from those who experience it to those who did not, the lessons of history are likely to be lost' (Levitt and March, 1988, p. 328). So, what do organizations do in order to preserve experience and transmit knowledge? Well, most of them encapsulate it in computers via databases and systems. Others implement such norms and procedures that dictate how people should behave in almost every case. At the other extreme, there are many family-owned organizations that, instead of writing each single rule in an organizational memory, pass on their knowledge from generation to generation by means of tradition and value systems.

However, family firms have an advantage on the 'competitive-non sharing issue', since all owning members (usually from the family) make a living out of the same source: their business. In addition, family owners and next generation members (NGMs) tend to share the same objective or dream (assuming no family conflicts). Moreover, an element of trust exists that holds the family and the business together. And, even when some authors may argue that there is no difference between a family business and a non-family firm, the fact is that the strength of personal relations within a family is greater than in any business relation. As Granovetter (1985) pointed out:

> Departing from pure economic motives, continuing economic relations often become overlaid with social content that carries strong expectations of trust and abstention from opportunism . . . [However] it would never occur to us to doubt this last point in more intimate relations, which make behavior more predictable . . . In the family, there is no Prisoner's Dilemma because each is confident that the other can be counted on. In business relations the degree of confidence must be more variable. (Granovetter, 1985, pp. 490–91)

It is well known that one important factor that affects transferability of knowledge is the perceived trustworthiness of the source of knowledge (Szulanski et al., 2004; Wathne et al., 1996). In fact, experiments in the communication field have demonstrated that a trustworthy source could substantially influence a recipient's behavior (Allen and Stiff, 1989; Hovland et al., 1949; Perry, 1996; Szulanski et al., 2004). Likewise, it has to be said that trust develops over time as a consequence of individual interaction; in the end, trust is placed in a person, not in that person's specific actions (Rempel et al., 1985). When dealing with such 'trust' factors, family firms tend to be ahead of non-family businesses, since owners and NGMs in the former tend to be more closely related, and therefore know each other. The latter is much more difficult in big non-family firms where the shareholders' power is diluted and represented by board members. Even when these board members (top executives) may be connected by networks of personal relations and may be able to generate trust among their own employees and shareholders, their ties will surely not be as strong as those developed among family members.

The kind of strength of the relationships between parents and children (NGMs) that we are underlining is what Coleman (1988) called social capital. In fact, there is a lack of social capital in the family if there are no strong relations between NGMs and parents (owners). The development of social capital within the family depends on two factors: the physical presence of parents in the family (owners) and the attention given by the parents to the NGMs. In this text, we are assuming the presence of both.

Similarly, we have to consider not only the trustworthiness of the source, but also its availability and desire to transfer knowledge. In general, family business owners are inclined to teach NGMs *everything they know about the business*, since their main dream is that their children continue and build on it. The founder wants to teach his or her siblings (and/or in-laws when the case applies) how to learn faster than their competitors, in order to improve the chances of survival of the family business. However, in other kinds of organizations this may not happen, first because of the high personnel turnover and, second, because of agency problems. Besides, it has to be highlighted that for a family firm, business reputation is closely linked to family reputation. Why? Because it is the owner(s) and his or her family who represent the values and ideas promoted there. More often than not the brand name of the enterprise is precisely the surname of the family. Thus, a special consideration regarding their own reputation as individuals is attached to the family firm, making their commitment much higher and their continuation dream more vivid.

Consequently, trustworthiness, time availability and desire of the source (among others) enhance knowledge transfer, influencing the likelihood of behavioral change by the recipient. In addition, knowledge transfer is not mechanical, but interactive and embedded in the existing capabilities on both sides and in the social relationships between them. Thus, if there is not goodwill and commitment on both sides knowledge transfer is more complex and complicated. One of the obvious principles of human communication is that the transfer of ideas occurs most frequently between a source and a receiver who are alike or homophilous (Rogers and Shoemaker, 1971). If we consider that 'homophily' is the degree to which pairs of individuals who interact are similar in certain attributes, such as beliefs, values, education, from the same family, and so on, we suggest that there may be better understanding, goodwill, and commitment between family members than between non-family members, since communication is easier (assuming no conflict). As the old adage goes: 'Birds of a feather flock together.'

Having asserted that knowledge transfer in family firms is different from knowledge transfer in non-family firms, we proceed to analyse the individual learning process, in order to understand how knowledge is acquired, assimilated and used by NGMs. To do so, we use sociological and psychological literature, as well as organizational theory writings.

Learning and knowledge

> One must learn by doing the thing, for though you think you know it – you have no certainty, until you try. (Sophocles, 400 BC)

To understand how knowledge is transferred, we first have to study how individual learning takes place, since knowledge is the outcome of learning. As stated by Piaget (1968) 'learning' can be defined as a continuous genesis, a process of creation and re-creation where gestalts and logical structures are added or deleted from memory over time. If this is so, learning has an implicit assumption of change. Therefore, we may say that individual learning is dynamic and cyclic; it is a process by which relatively permanent changes occur in a person's behavior as a result of some experience or knowledge the person has acquired (Bass and Vaughn, 1966). However, the success of learning consists in the acquisition of knowledge (Polanyi, 1975) not in the change of behavior, since it may rarely happen that knowledge be gained without any accompanying change in explicit behavior (Fiol and Lyles, 1985), but only in the individual private knowledge.

When analysing the individual learning literature, we found there was a tacit agreement among the reviewed authors on the fact that individual learning usually happens through experimentation, evaluation and assessment, it being a process rather than an outcome. Moreover interesting insights, such as that of March and Olsen (1975), pointed out that roles, duties and obligations were behaviorally important to involvement in the learning process, affirming that without doubt 'People attend to decisions not only because they have an interest at stake, but because they are expected to or obliged to' (p. 151). Indeed, numerous research studies (Bower, 1981; Gadanho and Custódio, 2002; LeDoux, 1998; Picard, 1997; Underwood, 1983; Wickens and Clark, 1968) have proposed that emotional responses (affect) are a critical component of memory, representing an associative network connected to certain events, such as family actions, rituals, expectations, decisions, and so on. In fact, Damasio (1994) suggested that humans associate high-level cognitive decisions (for instance, the decision to make a conscious effort in order to actively remember/learn what's being experienced) with special feelings (emotions) which have good or bad connotations dependent on whether choices have been emotionally associated with positive or negative long-term outcomes. Furthermore, Bahrick (1984), Salasoo et al. (1985) and Johnson and Hasher (1987) suggest that as level of learning increases and new connections of events are made, some portion of this type of knowledge becomes permanent, and will be indefinitely maintained, even in the absence of further rehearsals, and regardless of potential interference encountered during the retention interval. Therefore, expectations, ideals and shared dreams, connected with emotions and values are important, though not easily detected, intangible factors that influence the nature of learning.

In addition, learning (as well as adaptation) must consider the complexity of the environment relative to time, energy, uncertainty, relative importance of attainment of success versus avoidance of failure, experience or non-experience of the individual, preferences, goals and aspiration levels, incentives, and so on (Cangelosi and Dill, 1965). If the environment is too complex and dynamic for the individual to handle, overload may occur, inhibiting learning and knowledge acquisition (Lawrence and Dyer, 1983). So, it is better to learn step by step, so that assimilation is long lasting and errors (misleading signals produced by noise) are reduced. This view perfectly matches the findings from some well-known psychology studies regarding individual learning, which note among other things:

1. Distributed learning activities seem more effective than massed learning ('step by step').
2. Learning which is rewarded (positively reinforced) will tend to be repeated, remembered, and used in other situations. Time is a critical factor to foster repetition.

3. Knowledge of results (feedback) enhances learning.
4. The more the learner is ready and motivated to learn, the more learning will occur.
5. Learning must be transferable to the job and future job situations to be of practical value.
6. For learning to be most effective, the new behaviors must be practiced.

Applying the latter to family firms, we realize that for NGMs previous attitudes and beliefs also form the structure for future cognition, affecting learning commitment, attitude and behavior. Undeniably, learning is influenced and determined by attitude, values, trust and commitment. The behavior of the NGMs will depend on the relationship between the outcomes they observe and the aspirations they have for those outcomes (Levitt and March, 1988) as well as on the rewards and punishments administered. Likewise, we can observe that since learning is a dynamic process that takes time (for assimilation of knowledge purposes, repetition and incentives need to be 'a given'), predecessors should start transmitting their knowledge to NGMs at early stages in life in order to achieve a more effective knowledge transfer. On the other hand, in order for learning to occur, a certain amount of stress is necessary (Cangelosi and Dill, 1965; Fiol and Lyles, 1985; Hedberg, 1981; Hedberg et al., 1976). Thus, predecessors must create situations where NGMs can act, reflect and name their findings. 'In directing their explorations and naming their observations, they begin to understand their environment and become more able to manipulate and change their situations' (Freire, 1970b, p. 462). Although financial gain is sometimes the cause of searches for opportunities (Lewin and Wolf, 1975), it is usually scarcity, conflict, and substandard performances that lead to actions, whereas wealth, harmony and goal accomplishment breed complacency and reinforce current behaviors. Learning is typically triggered by problems (Hedberg, 1981). Thus, when thinking about transferring knowledge to NGMs, predecessors must take into account that the learning and assimilation of concepts and ideas NGMs may acquire will be based on the challenges they face, the constant reward and punishment they receive (Lieberman, 1972), as well as their own actions and findings, in joint conjunction with the expectations they perceive and share, the values they have internalized and the commitment they show (related to their own interests and personality).

Finally, the level of stress and the degree of uncertainty about past successes will also determine the effectiveness of the conditions of learning, as well as how the environment is perceived and interpreted (Daft and Weick, 1984; Fiol and Lyles, 1985; Starbuck et al., 1978; Weick, 1979a). And, for learning to be effective, an interaction between NGMs, the predecessor and the situation must take place. Acting, or learning by doing is the means to acquire knowledge (Freire, 1973).

'Acknowledging that all knowledge contains a personal element, or to put it differently, "[recognizing] personal participation as the universal principle of knowing" (Polanyi, 1975, p. 44), implies that knowing always is, to a greater or lesser extent, a skillful accomplishment, an art' (Tsoukas and Vladimirou, 2001, p. 982). Consequently, the learning process, or acquisition of knowledge, will be subjected to the individual's attitude. In effect, to know something, the individual must act and integrate a set of details or 'particulars' of which he or she is subsidiarily aware (unconscious). Basically, what happens is that to make sense of his or her experience, the individual relies on some parts of it subsidiarily in order to attend to the main objectivefocally. In the real world, we comprehend something as a whole (focally) by tacitly integrating certain particulars, which are known by the actor subsidiarily (Tsoukas and Vladimirou, 2001).

The intriguing question of whether unconscious stimuli influence thought and behavior has been constantly studied in psychology (Dixon, 1971; Erdelyi, 1984, 1985; Johnson and Hasher, 1987; Kihlstrom, 1984), and important demonstrations of meaningful processing of unconscious stimuli have been reported by Marcel (1983), Fowler et al. (1981) and McCauley et al. (1980). Moreover, indirect tests do seem to reveal long-lasting consequences of unconscious processing (Johnson and Hasher, 1987). Consequently, we suggest that NGMs unconsciously acquire certain 'tacit knowledge' (Polanyi, 1975) – intangibles such as family values and conceptions – just by interacting with the family, the business and the predecessor (in some cases, the founder). This is in line with what Hasher and Zacks (1979, 1984) observed concerning automatic encoding.

Thus, learning or knowledge acquisition consists of three elements: subsidiary particulars, a focal target and, crucially, a person who links the two (Tsoukas and Vladimirou, 2001; in accordance with Polanyi, 1975). This idea is pretty much related to and in fact reinforces, cultural-historical activity theory (CHAT), which provides an orientation toward learning by stating that human action (in this case, knowledge acquisition) has a tripartite structure: subject, object and mediating artifact (Figure 18.1).

Cultural-historical activity theory was initially developed during the 1920s and 1930s, by Lev Vygotsky, based on Feuerbach theses (by Marx, 1886), which presented the agenda of overcoming the opposition between idealism and mechanistic materialism by means of the concept of activity. The key concept of CHAT was mediation: artifact-mediated and object-oriented action. A person never reacts directly to the environment, but through the use or mediation of tools, cultural means, signs, and so on. Vygotsky characterized the development of the intellect as an ongoing dialectic of instrumental and abstract processes, through which the mind took shape by means of the reactive and transformative relations in which people engage with the world (Keller and Dixon Keller, 1996).

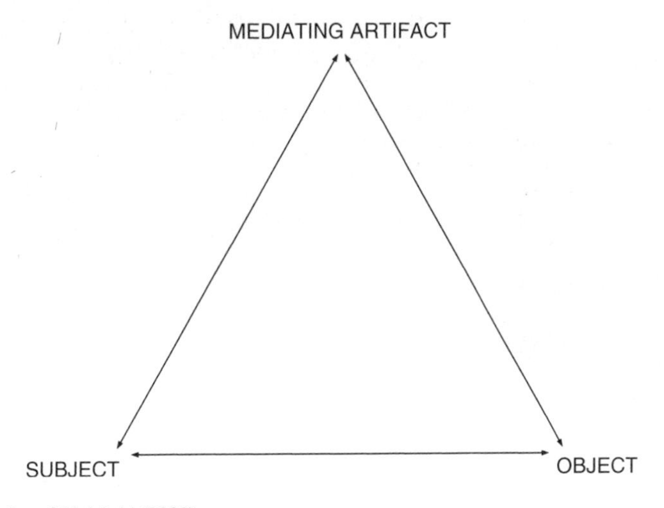

Source: University of Helsinki (2002).

Figure 18.1 Cultural-historical activity theory (CHAT)

Therefore, based on Tsoukas and Vladimirou's model and on cultural-historical activity theory – Vygotsky's model of mediated action – about how learning or knowledge acquisition (action) takes place, we apply the same tripartite structure of knowledge acquisition, which tacitly implies 'transmission' to our main issue: knowledge transfer in family firms. We have thus developed a transmission model of knowledge in family businesses, which takes into account not only the three main elements in the learning/knowledge acquisition process, but also the most relevant groups that interact in the family firm when knowledge transfer occurs: family, enterprise, and NGMs. Each of the components is treated as an independent, yet related and influential element, being:

1. Enterprise = object/focal target.
2. Family = mediated artifact/subsidiary particulars.
3. NGM = subject/person that integrated the focal target and particulars.

The model relates the subject/person, the object/focal target and the mediated artifact/subsidiary particulars (unconscious) in the following way. First, the subject/person, who is represented by the NGMs must learn/ acquire knowledge (action) about how to run the family firm, which is represented as the object/focal target. In order to do so, the subject/person needs an artifact or tool that helps him or her carry on the human activity (knowledge acquisition). This mediating artifact is represented by the family. Thus, knowledge transfer (acquisition) takes place in and through a social context.

One of the central ideas in Polanyi's concept of knowledge is tradition. Tradition describes how knowledge is transferred precisely in social contexts. The tradition is a system of values outside the individual. Indeed, tradition is a social system which takes up, stores and conveys the knowledge of society – or family (Sveiby, 1997b). This kind of knowledge-transfer is in fact represented by the master–apprentice relationship (craftsmanship) which has been in use for centuries. 'To learn by example is to submit to authority. By watching the master and emulating his efforts in the presence of his example the apprentice unconsciously picks up the rules of the art, including those which are not explicitly known to the master himself' (Polanyi, 1962, p. 53). Thus,

> The individual lets the . . . cultural patterns of the tradition form his own idiosyncrasies into an image of reality, irrespective of whether his tools are patterns of thought, patterns of action or social institutions. As time passes, some of the values are validated and transformed cognitively into beliefs about how things are. They are therefore no longer in need of being tested so they become a taken-for-granted tacit knowledge shared by the members of the group. (Sveiby, 1997b)

As a result, the family acts as a mediating artifact in conjunction with tradition (an intertwined tool embedded in the family context). Undeniably, tools are created and transformed during the development of the activity itself and carry with them a particular culture (historical remains from their development). So, the use of tools is an accumulation and transmission of social knowledge by itself. Consequently, tool use (tradition) influences the nature of external behavior and the mental functioning of individuals (Kaplelinin and Nardi, 1997).

Accordingly, the family acts as a reference system which contains a well-defined structure of values, ideas, behaviors, norms and general knowledge, which is transmitted unconsciously to the individual by means of traditions (NGM) from the time he or she is born. The next generation capitalizes on those traditions, acquiring the knowledge, integrating it, applying it and increasing it. Of course, we have to take into account that next generation level of commitment to the family business will definitely affect the knowledge assimilation process and patterns, it being a determinant factor (variable) in the context of effective and qualitative knowledge transfer. In fact, this transmission effort only makes complete sense when the three actors are completely aligned. Based on the work done by Fiol and Lyles (1985), alignment implies a potential to learn, unlearn or re-learn based on past behaviors. Essentially, the work of Chakravarthy (1982), Chandler (1962), Cyert and March (1963), Hambrick (1983), Miles and Snow (1978) and Miller and Friesen (1980) regarding firms and their strategies recognizes the widespread acceptance of this premise. Here, we are not applying the term only to firms, but also to the whole family–firm–next generation system. Consequently, if the three corresponding components do not show the same direction – potential to learn, unlearn and re-learn – the final output may vary.

In Figure 18.2 each square represents 'a system balance – depending on the faced environment', each big dashed circle stands for 'one different dynamic environment', while the small double-lined circle symbolizes the dynamistic-cause effect interrelationship among actors.

As we can see, learning and adaptation occurs along several simultaneous dimensions (Herriot et al., 2001). From a population ecology perspective and a strategic management view, the ultimate criterion of the system performance is its long-term survival and growth. To achieve this, the system must find different balances depending on the environment it moves/coexists in. In addition, we have to take into account that those environments are also dynamic; changing continuously and being affected by the time, uncertainty, energy imposed, past events, and so on. Furthermore, the system needs to find an alignment among its elements (actors), in order to adapt to the general ambience. However, this alignment is also cyclical and dynamic, preserved by a sustained strong cause–effect relationship among actors. System adaptation, alignment and actors' assimilation of knowledge (learning) regarding the environments, as well as the force of the interrelationship system balance–environment–actors will determine the family firm's survival/death and fit/misfit in the community.

The successor (or candidates) – NGMs – is embedded in the family system from the time he or she is born, and as the individual develops his or her personality, habits and beliefs, the family transmits to him or her certain views and ideologies as well as memories and values that preserve certain behaviors, mental maps, norms and ideas. The same happens with the other members embedded in the family system, who experience this isomorphic adaptation up to a point where cognitive systems, associations and memories are developed and shared among them. All these systems, associations and memories are kept alive and passed from generation to generation as family traditions. However, even when traditions are transmitted to us from the past, they are our own interpretations, which we have arrived at within the context of our own immediate problems – environments (Polanyi, 1983). Our perceptions and received stimuli from the environments we face affect our reinterpretation of traditions, and as a byproduct, impact our actions (behavior) and learning processes. Therefore, the NGM consequently affects the 'environments' back, since what came in as perceptions and stimuli, went out as redefined, re-created and changing actions that modify the surroundings. This is a never-ending cyclic process

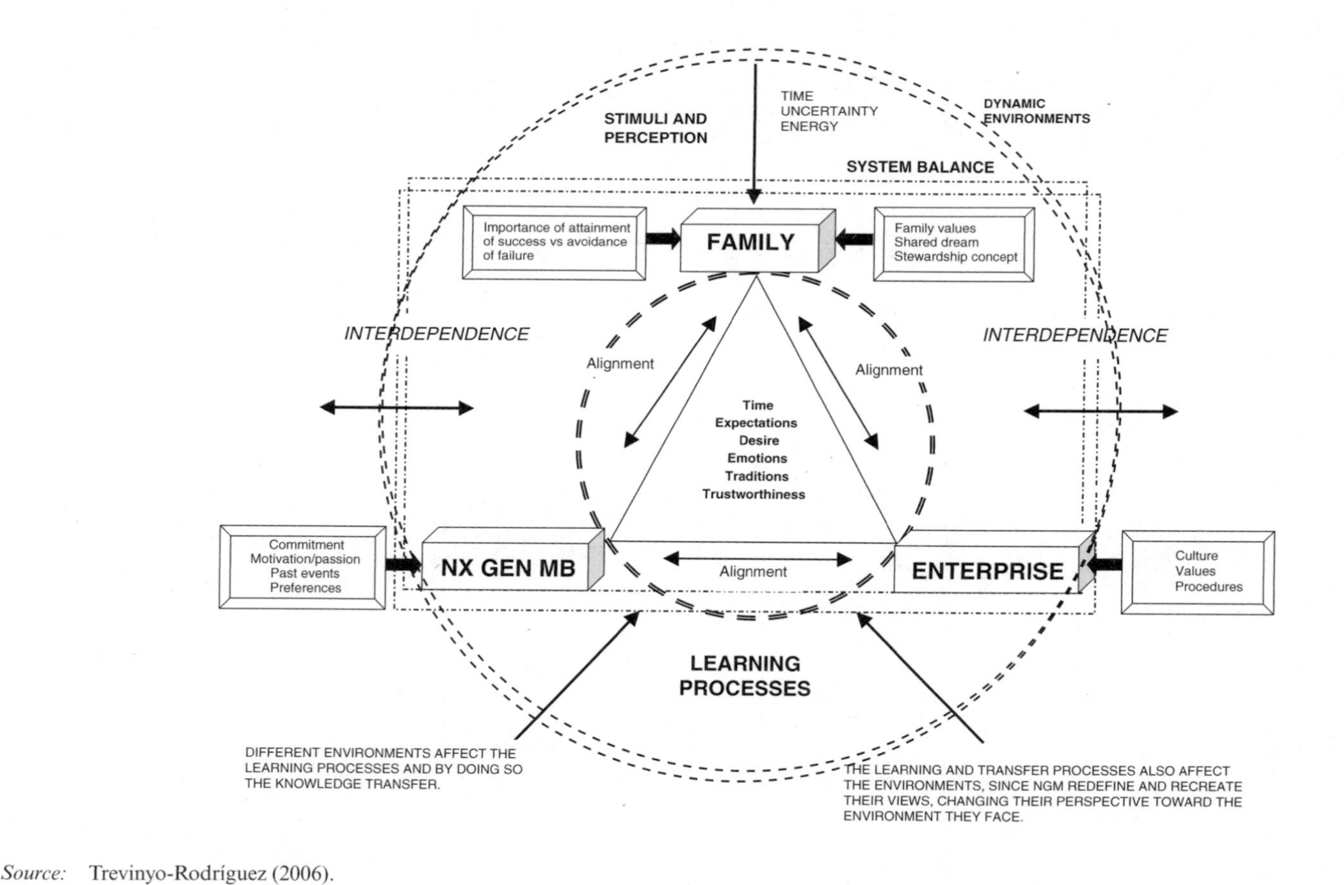

Source: Trevinyo-Rodríguez (2006).

Figure 18.2 Knowledge transfer model in family firms (KTFF) through which continuous feedback is acquired and by which traditions are adapted and redefined, though not completely changed (unless in special and extreme circumstances).

Conclusion

When evaluating succession processes one of the most important variables related to the quality-effective standards of the procedure is knowledge transfer. Nonetheless, as we pointed out before, knowledge transmission in family firms takes place differently than knowledge transmission in non-family firms, owing to intangibles related to family value systems and traditions. Some of the most relevant variables and conditions that, according to our conceptual model, affect knowledge transmission in family firms are synthesized below.

Regarding the variables, we identified two classes: (1) those that are common to/affect the whole system (essential variables of the whole system) and (2) those that are particularly focused on a specific actor (family and next generation variables). For practical reasons and easy remembering, we grouped the common variables (class 1) together and constructed the educate standard (education of NGMs – Figure 18.3), while the punctual/specific variables (class 2) are listed accordingly to the actor they have an effect upon (Table 18.1).

On the subject of the conditions needed in order for effective knowledge transfer to take place across generations in the context of family firms, we detected that some desirable circumstances/settings:

- Next generation members' learning process must start in early stages (from childhood), since 'learning' is enhanced through time with such practices as repetition, rewards and punishments, feedback and practice.
- For effective learning to take place commitment, expectations, values and perceptions have to be shared between the predecessor and NGM (explicit and implicit communication is needed).
- Learning processes must be 'constructed' in such a way that they show challenging and solvable situations highly related to real life and to the future work.

Yet, although learning/knowledge acquisition is a process intended to be helpful during critical periods, for instance, succession in family firms, it is essential to note that

Educate standard (education of next generation members)

Effective time – Rewards, punishments, repetition, feedback
Desire – Shared dream, proud of being family and enterprise member
Uncertainty – Degree
Communication – Trustworthiness
Attitudes – Emotions – Previous beliefs, successes, failures
Traditions – Value systems — Unconscious stimuli
Expectations – Motivations

Figure 18.3 Essential variables of the whole system

Table 18.1 Other specific variables affecting knowledge transmission

Family variables (Source of knowledge)	Next generation variables (Knowledge acquirer/learner)
Values, expectations, ideas	Commitment and disposition
Stewardship concept	Motivation
Social capital (how strong relationships between NGM and parents are?)	Passion and expectations
Availability of the source of knowledge	Past events
Desire to transfer knowledge	Preferences
Trust	Time dedicated to learn

learning does not always lead to intelligent behavior. Experience or wisdom can not be completely communicated, consequently the next generation has to 'live' their own knowledge processes and apply their judgment and reasoning depending on the problem facing them, making their own mistakes and learning from them. What predecessors do is just to set a context where NGMs can develop their potentials in an orderly, not so costly, and guided way. But in the end, 'Passing the torch involves doing on the part of the next generation'. Knowledge is personal; it's an ART.

Further research

Since the overall KTFF model we have proposed here includes many different variables and/or constructs which are difficult not only to measure, but also to examine jointly, we propose certain specific ideas in order to test the model 'by parts', breaking it down into more manageable pieces of research.

In order to understand how to shape and/or transform the learning processes accomplished by NGMs during their life, we need first to map out those processes/relationships. To facilitate this task, we propose to base that mapping procedure on the proposed relationships (interactions) presented in the KTFF model. Thus, a research testing knowledge acquisition (learning) during different stages of the NGMs life – for instance, childhood, adolescence, early maturity and maturity – may be insightful, showing whether the interactions among the family, the enterprise and the NGMs are different or not at separate points in time. Basically, the idea is to dynamically map how the KTFF model changes over time (along these different stages) in order to detect if there are some points where learning patterns have changed. If so, a consequent question would be why those patterns changed from one stage to another?

Another focus of research could be to assess the antecedents of such constructs as next generation commitment and/or trust (among others) and then to see how these antecedents are related either to the enterprise, to the family (context), to NGMs' individual personality, or to a pair/all of the basic variables, as well as to the learning patterns that the NGMs exhibit over time (stages). Factor analysis and structural equation modeling could help considerably in order to group antecedents and to detect causal relationships.

On the other hand, it would be equally attractive to analyse the literature related to the 'psychology of child development' and to observe how the learning processes children carry out during childhood can be effectively fostered or inhibited – and if so, to what extent – by the contexts/conditions they have grown up in (relationship family–enterprise), as well as by the use of such practices as repetition, rewards and punishments, feedback and practice.

Similarly, an attention-grabbing line of investigation would be to test how NGMs develop special skills for family businesses (tacit knowledge) in the early stages of their life, and to examine how early and up to when (time) NGMs are more receptive to learning those abilities.

Finally, we also suggest testing how/which values overlap between predecessor and NGMs, and their relationship to NGMs' commitment to the business and to the family. The latter may be useful to foresee as well as to understand several conflicts that arise not only during the succession phase, but also during the day-to-day relationships between predecessors and NGMs.

Although recommended here, these future research ideas are not intended to be exhaustive; many more lines of investigation may be – and we expect are – opened/developed in order to contribute to our knowledge and understanding on processes in family businesses.

References

Allen, M. and J. Stiff. (1989), 'Testing three models for the sleeper effect', *Western Journal of Speech Communication*, **53**, 411–26.

Bahrick, H.P. (1984), 'Semantic memory content in permastore: fifty years of memory for Spanish learned in school', *Journal of Experimental Psychology: General*, **113**, 1–29.

Bass, B.M. and J.A. Vaughn (1966), *Training in Industry: the Management of Learning*, Belmont, CA: Wadsworth.

Bower, G.H. (1981), 'Mood and memory', *Am. Psychol*, **36**, 129–48.

Cangelosi, V. and W. Dill (1965), 'Organizational learning: observations toward a theory', *Administrative Science Quarterly*, **10**, 175–203.

Chakravarthy, B.S. (1982), 'Adaptation: a promising metaphor for strategic management'. *Academy of Management Review*, **7**, 735–44.

Chandler, A. (1962), *Strategy and Structure*, Cambridge, MA: MIT Press.

Coleman, J. (1988), 'Social capital in the creation of human capital', *American Journal of Sociology*, **94**, 95–120.

Cyert, R. and J. March. (1963), *A Behavioral Theory of the Firm*, Englewood Cliffs, NJ: Prentice Hall.

Daft, R.L. and K.E. Weick (1984), 'Toward a model of organizations as interpretation systems', *Academy of Management Review*, **9**, 284–95.

Damasio, A.R. (1994), *Descartes' Error: Emotion, Reason and Human Brain*, New York: Grosset/Putnam.

Dixon, N.F. (1971), *Subliminal Perception: The Nature of a Controversy*, London: McGraw-Hill.

Erdelyi, M.H. (1984), 'The recovery of unconscious (inaccessible) memories: laboratory studies of hypermnesia', In G.H. Bower (ed.), *Psychology of Learning and Motivation*, vol. 18, New York: Academic Press, pp. 95–127.

Erdelyi, M.H. (1985), *Psychoanalysis: Freud's Cognitive Psychology*, New York: Freeman.

Fiol, C.M and M.A. Lyles (1985), 'Organizational learning', *Academy of Management Review*, **10**(4), 803–13.

Fowler, C.A., G. Wolford, R. Slade and L. Tassinary (1981), 'Lexical access with and without awareness', *Journal of Experimental Psychology: General*, **110**, 341–62.

Freire, P. (1970a), 'The adult literacy process as cultural action for freedom', *Harvard Educational Review*, **40**, 205–25.

Freire, P. (1970b), 'Cultural action and conscientization', *Harvard Educational Review*, **40**, 452–77.

Freire, P. (1973), *Education for Critical Consciousness*, New York: Seabury Press.

Gadanho, S.C. and L. Custódio (2002), 'Asynchronous learning by emotions and cognition', proceedings of the seventh international conference on simulation of adaptive behavior on 'From animals to animats', 5–9 August, Portugal.

Granovetter, M. (1985), 'Economic action and social structure: the problem of embeddedness', *The American Journal of Sociology*, **91**(3), 481–510.

Grusky, O. (1960), 'Administrative succession in formal organizations', *Social Forces*, **39**(2), 105–15.

Handler, W. (1990), 'Succession in family firms: a mutual role adjustment between entrepreneur and next-generation family members', *Entrepreneurship Theory and Practice*, **15**, 37–51.

Hannan, M.T. and J. Freeman (1984), 'Structural inertia and organizational change', *American Sociological Review*, **49**(2), 149–64.

Hasher, L. and R.T. Zacks (1979), 'Automatic and effortful processes in memory', *Journal of Experimental Psychology: General*, **108**, 356–88.
Hasher, L and R.T. Zacks. (1984), 'Automatic processing of fundamental information: the case of frequency of occurrence', *The American Psychologist*, **39**, 1372–88.
Hedberg, B. (1981), 'How organizations learn and unlearn', in P. Nystrom and W. Starbucks (eds), *Handbook of Organizational Design*, vol. 1, Oxford: Oxford University Press, pp.. 3–27.
Hedberg, B., P. Nystrom and W.H. Starbuck (1976), 'Camping on seesaws: prescriptions for a self designing organization', *Administrative Science Quarterly*, **21**, 41–65.
Herriott, S.D. Levinthal and J.G. March (2001), 'Learning from experience in organizations', *American Economic Review*, **75**, 298–302.
Hovland, C., A. Lumsdaine and F. Sheffield (1949), *Experiments in Mass Communication*, Princeton, NJ: Princeton University Press.
Johnson, M.K. and L. Hasher (1987), 'Human learning and memory', *Annual Review of Pyschology*, **38**, 631–68.
Kaptelinin, V. and B.A. Nardi (1997), 'Activity theory: basic concepts and applications', CHI 97 Electronic Publications: Tutorials, www.acm.org/sigchi/chi97/proceedings/tutorial/bn.htm
Keller, C. and J. Dixon Keller (1996), *Cognition and Tool Use: The Blacksmith at Work*, Cambridge: Cambridge University Press.
Kihlstrom, J.F. (1984), 'Conscious, subconscious, unconscious: a cognitive perspective', in G.H. Bower (ed.), *Psychology of Learning and Motivation*, vol. 18, New York: Academic Press, pp. 149–211.
LeDoux J.E. (1996), *The Emotional Brain*, New York: Simon and Schuster.
Levitt, B. and J.G. March (1988), 'Organizational learning', *Annual Review of Sociology*, **14**, 319–40.
Lewin, A.Y. and C. Wolf (1975), 'The theory of organizational slack: a critical review', *Proceedings, Twentieth International Meeting of the Institute of Management Sciences*, Tel Aviv: North Holland.
Lieberman, A. (1972) 'Organizational learning environments: effects and perceptions of learning constraints', doctoral dissertation, Stanford University.
Marcel, A. (1983), 'Conscious and unconscious perception: experiments on visual masking and word recognition', *Cognitive Psychology*, **15**, 197–237.
March, J. and J.P. Olsen. (1975), 'The uncertainty of the past: organizational learning under ambiguity', *European Journal of Political Research*, **3**, 147–71.
Marx, K. (1886), 'Theses on Feuerbach', as an appendix to Engels', *Ludwig Feuerbach and the End of Classical German Philosophy*, from *Marx/Engels Selected Works*, vol. 1, 1969, Moscow: Progress Publishers, pp. 13–15. www.marxists.org/archive/marx/works/1845/theses/theses.htm.
McCauley, C., C.M. Parmelee, R.D. Sperber and T.H. Carr (1980), 'Early extraction of meaning from pictures and its relation to conscious identification', *Journal of Experimental Psychology. Human Perception and Performance*, **6**, 265–76.
Miles, R.E. and C.C. Snow (1978), *Organizational Strategy, Structure and Process*, New York: McGraw-Hill.
Miller, D. and P.H. Friesen (1980), 'Momentum and revolution in organization adaptation', *Academy of Management Journal*, **23**, 591–614.
Perry, D.K. (1996), *Theory and Research in Mass Communication: Contexts and Consequences*, Hillsdale, NJ: Lawrence Erlbaum Associates.
Picard, R.W. (1997), *Affective Computing*, Cambridge, MA: MIT Press.
Polanyi, M. (1962), *Personal Knowledge*, Chicago, IL: University of Chicago Press.
Polanyi, M. (1983), *Personal Knowledge: Towards a Post-critical Philosophy*, London: Routledge and Kegan Paul.
Probst, G., S. Raub and K. Romhardt (2000), *Managing Knowledge: Building Blocks for Success*, Chichester: Wiley.
Rempel, J.K., Holmes, J.G. and M.P. Zanna (1985), 'Trust in close relationships', *Journal of Personality and Social Psychology*, **49**(1), 95–112.
Rogers, E.M. and F.F. Shoemaker (1971), *Communication of Innovations: A Cross-Cultural Approach*, New York: Free Press.
Salasoo, A., Shiffrin, R.M. and T.C. Feustel (1985), 'Building permanent memory codes: codification and repetition effects in word identification', *Journal of Experimental Psychology: General*, **114**, 50–77.
Sharma, P. and P.G. Irving (2005), 'Four bases of family business successor commitment: antecedents and consequences', *Entrepreneurship Theory and Practice*, **29**(1), January, 13–33.
Starbuck, W.H., Greeve, A. and B. Hedberg (1978), 'Responding to crisis', *Journal of Business Administration*, **9**(2), 112–37.
Stinchcombe, A. (1965), 'Social structure and organizations', in J.G. March (ed.), *Handbook of Organizations*, Chicago, IL: Rand McNally, pp. 153–93.
Sveiby, K.E. (1997a), *Capital Intelectual: La nueva riqueza de las empresas: Cómo medir y gestionar los activos intangibles para crear valor*, Barcelona: Gestión 2000.
Sveiby, K.E. (1997b), 'Tacit knowledge', www.sveiby.com/portals/o/articles/polanyi.html#Tradition.

Szulanski, G., R. Cappetta and R.J. Jensen (2004), 'When and how trustworthiness matters: knowledge transfer and the moderating effect of causal ambiguity', *Organizational Science*, **15**(5), 600–613.
Trevinyo-Rodriguez, Rosa Nelly (2006), 'Knowledge transfer model in family firms' (revised), January, Barcelona, Spain.
Tsoukas, H. and E. Vladimirou (2001), 'What is organizational knowledge?', *Journal of Management Studies*, **38**(7), 973–93
Underwood, B.J. (1983), *Attributes of Memory*, Glenview, IL: Scott Foresman.
University of Helsinki (2003), 'Cultural-historical activity theory', Center for Activity Theory and Developmental Work Research, www.edu.helsinki.fi/activity/pages/chatanddwr/chat/.
Varamäki, E.T. Pihkala and V. Routamaa (2003), 'Knowledge transfer in small family businesses', paper presented during the 2003 FBN Conference.
Wathne, K., J. Roos and G. von Krogh (1996), 'Towards a theory of knowledge transfer in a cooperative context', in G. von Krogh and J. Roos (eds), *Managing Knowledge: Perspectives on Cooperation and Competition*, London: Sage publications, pp. 55–81.
Weick, K.E. (1979a), *The Social Psychology of Organizing*, New York: McGraw-Hill.
Weick, K.E. (1979b), 'Cognitive Processes in Organizations', in B.M. Straw (ed.), *Research in Organizational Behavior*, vol. 1, Greenwich, CT: JAI Press, pp. 41–74.
Wickens, D.D. and S. Clark (1968), 'Osgood dimensions as an encoding class in short-term memory', *Journal of Experimental Psychology*, **78**, 580–84.

19 Feuding families: the management of conflict in family firms

Franz W. Kellermanns and Kimberly A. Eddleston

Introduction

Conflict has often been portrayed as a recurring characteristic that diminishes the performance and survival of family firms (for example, Levinson, 1971). Research on conflict, however, notes that not all conflict is inherently bad. Studies have focused on a beneficial type of conflict, task conflict, which has been found to be positively associated with performance when it occurs in moderate levels (for example, Jehn, 1995, 1997a). Task conflict involves the open discussion of the merits of ideas, thereby improving the range of options provided to decision-makers. In contrast, relationship conflict has been identified as a detrimental type of conflict (Jehn, 1995, 1997a). Relationship conflict is associated with resentment, animosity, anger, frustration and hostile behaviors. This dysfunctional nature of relationship conflict has a devastating effect on a firm's performance.

The above-mentioned distinction between task and relationship conflict is critical in understanding the consequences of conflict, however it falls short in taking into account the unique psychodynamic effects of the family firm. That is, family involvement must be considered to fully understand how conflict occurs and can be managed in family firms (Dyer Jr, 1986, 1994; Ling et al., 2002; Schulze et al., 2001, 2003b; Schwenk, 1990). While progress has been made in understanding the antecedents of both types of conflict in family firms (for example, Kellermanns and Eddleston, 2004), our understanding on how these types of conflict can be successfully managed is still lacking.

Our chapter thus focuses on five popular conflict management strategies, namely, avoiding, contending, compromising, collaboration, and third-party intervention (for example, De Dreu, 1997; De Dreu and Van Vianen, 2001; Sorenson, 1999; Wall Jr and Callister, 1995), as they relate to the occurrence of task and relationship conflict in family firms. This chapter contributes to the literature in at least two ways. First, we add to the family firm literature by showing that both task and relationship conflict need to be properly managed in family firms. Giving particular attention to the complexity perspective of family firms, we argue that while relationship conflict needs to be minimized, an optimum level of task conflict needs to be achieved in order for family firms to succeed. Second, taking into account the unique psychodynamic effects of family firms, we add to the conflict literature by discussing integrated conflict management strategies that address both task and relationship conflict. To our knowledge, this is the first theoretical paper that matches conflict management strategies with both task and relationship conflict while explicitly taking into account the unique influences of the family firm. Our chapter concludes with a discussion of the implications for theory and practice, and a presentation of future research opportunities on family firm conflict management.

A family firm perspective

Conflict in family firms is more complex than conflict in other types of organizations owing to the overlapping of the business and the family subsystem (Kellermanns and Eddleston, 2004). Gersick et al.'s (1997) work, which describes the transitions through the different life-stages of the family firm from controlling-owner, to sibling partnership to multi-generational family firm, and the popular three-circle model of family firms that describes the interplay between ownership, management and the family (for an extended model see Hoy and Verser, 1994) clearly portray the added level of complexity that characterizes family firms.

Family firms must often grapple with conflicts not present in non-family businesses, such as sibling rivalry, children's desire to differentiate themselves from their parents, marital discord, and ownership dispersion conflicts among family members (Dyer Jr, 1986, 1994; Schulze et al., 2001, 2003b). Furthermore, family members are often not able to readily leave their organizations owing to the inability to sell their shares in the family firm and the risk of losing status, inheritance rights, and other privileges the family firm provides (Gersick et al., 1997; Schulze et al., 2003b, 2003a). This often results in family members being locked into the business and family conflicts (Lee and Rogoff, 1996). Accordingly, conflict in family firms tends to be more persistent than in non-family firms.

Owing to the interconnection and frequent contact among family members working in the business with those who are not but may still have an ownership stake, recurring conflict is highly probably in family firms (Gersick et al., 1997; Harvey and Evans, 1994). The dominant presence of the family and the overlapping of control and management activities complicate the understanding of family firm conflict. In addition, the combination of conflict coming from within the business and conflict initiated within the family compounds the effects of conflict in family businesses (Harvey and Evans, 1994). As such, the causes and implications of conflict in family firms appears to be more complex than in non-family businesses, and therefore, how conflict can be effectively managed in family firms needs to be understood.

Task and relationship conflict

While most of the family firm literature tends to assume that conflict is unhealthy and disruptive, it should be noted that conflict can have a positive effect on a family firm's performance (Kellermanns and Eddleston, 2004). Indeed, conflict can have both a positive and a negative impact on the financial and non-financial performance of family firms. Positive forms of cognitive conflict can enhance decision quality because attention is focused on the discussion of goals and strategies (Kellermanns and Eddleston, 2004) and the subsequent implementation of decisions is often smooth owing to the heightened involvement in the decision-making process and the increased acceptance of the final decision (Amason, 1996; Jehn, 1995, 1997b). Furthermore, positive conflict can strengthen the cohesion of the family unit and facilitate choices for both the business and the family.

The most common, beneficial, cognitive conflict examined in the conflict literature is task conflict (for example, Amason, 1996; Jehn, 1997a). Task conflict is free of negative emotions, encourages open debate, increases understanding and facilitates task accomplishment (for example, Amason, 1996; Jehn, 1997a, 1997b). Indeed, research has shown that moderate levels of task conflict improve the performance of top-management teams, work groups and non-family business settings (Amason, 1996; Jehn, 1997a, 1997b; Jehn and Mannix, 2001). Moderate levels of task conflict appear to be the most beneficial to performance given that particularly high levels of task conflict can prohibit task completion and very low levels diminish the development of new ideas and strategic options and can encourage group-think (Janis, 1972; Kellermanns and Eddleston, 2004). However, this view of task conflict does not consider the psychodynamic effects of the family, particularly when family member interactions are dysfunctional and hostile.

In particular, negative conflict, otherwise known as relationship conflict, leads to low levels of decision quality, resentment, anger and worry (Amason, 1996; Jehn, 1995, 1997b) and threatens the stability and viability of the business as well as the family unit. Relationship conflict often interferes with work efforts by redirecting efforts related to work toward the reduction of threats, politics and coalitions (Jehn, 1997a). Squabbles and infighting within a family firm can result in insufficient attention being focused upon business needs and performance (Kets de Vries, 1993). Relationship conflict is laden with negative emotions and personal animosity, and is thus considered extremely dysfunctional (Amason, 1996; Amason and Schweiger, 1994; Brehmer, 1976). Relationship conflict revolves around disagreements and frustrations that are not related to work or a business task (De Dreu and Van Vianen, 2001). Since relationship conflict is associated with frustration, irritation and annoyance (Jehn and Mannix, 2001), it has been linked with negative performance effects such as ineffectiveness and dissatisfaction (Amason, 1996; Jehn and Mannix, 2001). Such negative conflict within the family firm may contribute to the high mortality rate of family businesses (Beckhard and Dyer Jr, 1983). Therefore, when investigating conflict, both task and relationship conflict should be considered simultaneously, given that one relates directly to the task at hand and can increase performance, while the other conflict relates to personality clashes and can diminish performance.

The complexity perspective of family firm conflict, which was proposed by Kellermanns and Eddleston (2004), emphasizes the importance of understanding the relationship between task and relationship conflict. They argue that the benefits of task conflict can be lost when relationship conflict is present in family firms. Relationship conflict distracts family members from their work and deflects attention away from tasks, preoccupying them with feelings of anger, frustration and resentment (Filbeck and Smith, 1997). Relationship conflict is particularly devastating for family firms since family members have access to key processes and information in the organization, and thus tend to yield significantly more power than normal employees in non-family businesses (Dyer Jr, 1986; Sorenson, 1999). Thus, optimal performance effects in family firms can be expected when low levels of relationship conflict and moderate levels of task conflict are experienced in family firms (see also Kellermanns and Eddleston, 2004). Formally stated:

> *Proposition 1* Low levels of relationship conflict and moderated levels of task conflict will lead to superior performance in family firms.

However, this balance between task and relationship conflict is not easy to achieve in family firms, and conflict management interventions might become necessary. Indeed, conflict management has been considered extremely important for family firms (for example, Dyer Jr, 1986; Sorenson, 1999; Ward, 1987). While the need for future research has been identified (for example, Kellermanns and Eddleston, 2004) and current research suggests integrated management approaches for non-family firms (for example, Van de Vliert et al.,

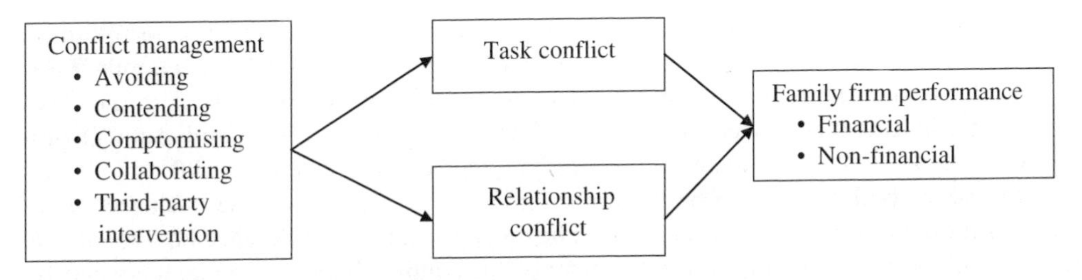

Figure 19.1 Conflict management in family firms

1999), research has yet to investigate conflict management approaches for family firms in relation to creating optimal levels of task conflict while minimizing relationship conflict. Figure 19.1 summarizes the suggested relationships between conflict management, the occurrence of conflict and family firm performance proposed in this chapter. Next, we review the five conflict management approaches suggested in Figure 19.1 under the consideration of the complexity perspective of conflict in family firms.

Conflict management approaches

Managing conflict is important to the success of family firms (Dyer Jr, 1986; Sorenson, 1999). Recent research has identified five major conflict management approaches, namely, avoidance, contending, compromising, collaborating (De Dreu and Van Vianen, 2001) and third-party intervention (Wall Jr and Callister, 1995). There is also another strategy sometimes discussed in the conflict management literature, accommodation, but because it is infrequently used in family firms (Dean, 1992; Sorenson, 1999) it will not be included in our discussion. While most studies focus exclusively on managing task conflict (for example, De Dreu et al., 1999; Jehn, 1997b) or relationship conflict (for example, De Dreu and Van Vianen, 2001) studies have yet to take into account a complexity perspective in family firms that considers both conflicts simultaneously. Family firms tend to use conflict management strategies that cannot resolve multiple concerns; that is, they use strategies that lessen input and accommodate only limited interests (Sorenson, 1999). Therefore, effective conflict management in family firms will need to resolve relationship conflict while also encouraging optimum levels of task conflict (Kellermanns and Eddleston, 2004).

Avoiding

The avoiding conflict management strategy involves a lack of response to conflict, whereby the problem is ignored. Here, individuals recognize the disruptive potential of certain issues and thus avoid and forestall dealing with them (Murninghan and Conlon, 1991). Some research has found that an avoiding conflict management strategy can be quite effective (De Dreu and Van Vianen, 2001; Jehn, 1997b; Murninghan and Conlon, 1991). For example, a study by Murninghan and Conlon (1991) found that the avoiding conflict management strategy was characteristic of highly successful string quartettes. Similarly, De Dreu and Van Viannen (2001) showed that teams that avoided conflict perceived higher levels of performance. In contrast, when conflict is not ignored and occurs frequently and openly, the intensity of relationship conflict increases significantly (Jehn, 1997a). However, these studies did not take place in family firm settings, but rather in a team context. Because of the co-mingling of the family and the business, and the consistent interactions of family members, the avoiding conflict management strategy may not be as effective in family firms.

While avoidance might allow family members to focus on the tasks at hand, the overarching problems that created the negative affect will continue to linger and fester. Owing to the interrelationship between the business and the family (Whiteside and Brown, 1991), relationship conflict needs to be minimized in both the family and the business, which makes it even more difficult to suppress. The most pronounced difference between teams and family firms is the temporal component of the conflict. While team members may be able to 'sit out' conflicts and go their own way after a project is completed, family firm members are locked into the organization and are unlikely to leave (Gersick et al., 1997; Schulze et al., 2003a, 2003b). Managing conflict in family firms requires special handling because family members must continue to work together, they tend to share the same long-term goals and they hold simultaneous memberships in several intersecting systems, not just the one in which the dispute has occurred (Kaye, 1991). Thus, it is much more difficult for family members to escape relationship conflict.

When family firms ignore conflicts, they simply 'forestall coming to terms with those undisclosed issues' (Kaye, 1991, p. 22). A high use of avoidance in family firms has been associated with low family satisfaction, high sibling rivalry and low mutual trust (Kaye and McCarthy, 1996). Indeed, effective relationships in family firms thrive on honesty, explicit communication where feelings are shared, and issues negotiated (Whiteside et al., 1993). Healthy family firms do not view arguments as bad, but instead focus on how conflicts are resolved (Whiteside et al., 1993). Accordingly, we suspect that the avoidance of relationship conflict cannot be sustained and will be detrimental in the long-run to family firms.

In addition, an avoiding conflict management strategy is not expected to encourage optimal levels of task conflict. An avoiding conflict management strategy is likely to leave a conflict as it stands, without solving or escalating it (Van de Vliert and Euwema, 1994). By avoiding task conflict, family firms become stagnant and lack the development of new strategies (Kellermanns and Eddleston, 2004). Without conflict, a family firm does not re-evaluate its tasks or create new objectives (Kaye, 1991). Opinions are not integrated into decision-making and thus the quality of decisions may suffer. Family members are likely to feel discounted when their opinions are ignored, which may even lead to relationship conflict. For example, avoiding conflict can escalate frustrations and lead to negative spillover that hurts relationships between family members and, thus, avoidance can heighten tensions and limit productive action (Sorenson, 1999). Indeed, the use of the avoiding conflict management strategy in family firms has been found to have a negative effect on business outcomes and to hurt family relationships (Sorenson, 1999). Therefore:

> *Proposition 2* The avoidance conflict management strategy is negatively related to family firm performance. Specifically, relationship conflict will remain, while the optimal level of task conflict is not realized.

Contending

The contending conflict management strategy, sometimes referred to as competing, is when individuals attempt to impose their will, wishes and perspectives upon others (De Dreu and Van Vianen, 2001). Individuals do not take into account others' concerns and, thus, contending is seen as an individualistic strategy that creates competition among family members (Sorenson, 1999). In regard to relationship conflict, because contending keeps other individuals from achieving their goals, contending is often associated with negative affect such as anger, stress and distrust (Jehn, 1997b). Contending hurts family relationships (Sorenson, 1999) because it leads to misunderstandings, negative attitudes and harmful interactions (Jehn, 1997b; Pruitt and Rubin, 1986). Therefore, a contending conflict management strategy is unlikely to resolve relationship conflict and will most likely exacerbate relationship conflict in family firms.

In terms of task conflict, research has shown that contending is particularly negative for team compliance and team effectiveness (Alper et al., 2000; De Dreu and Van Vianen, 2001). Contending can create competing coalitions within a family firm that may help to promote personal agendas in the short run but in the long run will be harmful to a business (Sorenson, 1999). In other words, a conflict management approach that puts pressure on others to conform and uses threats to accomplish the tasks at hand might lead to short-term action but will have devastating long-term consequences. Contending is unlikely to encourage optimal levels of task conflict, given that the underlying causes of the conflict are not resolved (Sorenson, 1999) and lower quality decisions result owing to diverse opinions being suppressed. For example, the competing coalitions that result from a contending strategy are likely to heighten task conflict beyond the optimum moderate level prescribed, thereby creating a tug-of-war that can lead to inactivity or an inconsistent strategy. Furthermore, when contending is found in family firms with a high degree of control concentration and owners with a strong desire for leadership, the participation in the discussion of goals and strategies is often lacking, thus leading to poor decision-making (Daily and Dollinger, 1993; Harvey and Evans, 1994). Such competitive behavior will likely suppress the expression of diverse opinions, interfere with the emancipation of younger generations and destroy the intimacy of the firm (Hilburt-Davis and Gibb Dyer, 2003). Thus, in these scenarios, task conflict is likely to be deficient and below optimum levels. Accordingly, we hypothesize:

Proposition 3 The contending conflict management strategy is negatively related to family firm performance. Specifically, relationship conflict is likely to increase, while task conflict is likely to be either deficient or too excessive to be beneficial to performance.

Compromising

Compromising refers to a conflict management strategy that tries to partially satisfy all parties in a dispute through compromises. Compromising is often seen as a distributive conflict management strategy since neither party's needs are fully met, rather, their needs and goals are partially achieved (Sorenson, 1999). However, because each party's concerns are partially addressed in the conflict resolution, this strategy also has an integrative component (Sorenson, 1999). Research has shown that this is not the most effective conflict management strategy, since the root causes of the conflict tend to persist (Murninghan and Conlon, 1991). When conflicts are rooted in differences regarding social attitudes and ideologies, comprising is unlikely to resolve disputes (De Dreu and Van Vianen, 2001). However, while compromising may not be the most effective strategy, it has been found to be associated with positive family relationships (Sorenson, 1999). In fact, a family firm that rarely compromises is likely to have harmful and negative family interactions; sometimes family members have to give and take simply to keep the peace

(Sorenson, 1999). Therefore, while compromising is unlikely to completely resolve relationship conflict, it may help lessen relationship conflict in family firms.

In contrast, a heavy reliance on a compromising conflict management strategy has been found to be negatively related to family firm business performance (Sorenson, 1999). When compromising and consensus-style approaches are used, decision quality tends to be inferior to other forms of decision-making (for example, Priem and Price, 1991; Schweiger et al., 1986). In order for task conflict to be effective, issues need to be debated and ideas integrated so as to devise the best possible solution to business problems. Compromising does not promote the satisfaction of the parties involved and does not foster the search for the best solution, but rather focuses efforts on give and take. Indeed, family firms that are too compromising may overlook important business issues and make less effective business decisions (Sorenson, 1999). Therefore, compromising is not a beneficial conflict management style in regards to task conflict. Accordingly:

> *Proposition 4* The compromising conflict management strategy has both a positive and negative effect on family firm performance. Specifically, relationship conflict is minimized, while the benefits of task conflict are not realized.

Collaborating

The collaborating conflict management style involves individuals trying to work out a mutually acceptable solution to their problem that fully satisfies the concerns of all parties (De Dreu and Van Vianen, 2001; Sorenson, 1999). Collaborating has been found to improve team effectiveness and to lead to mutually beneficial solutions, better goal achievement and less likelihood of future conflicts (Pruitt and Rubin, 1986; Tjosvold, 1997). Although some research on teams has shown that collaborating is not effective in dealing with relationship conflict (De Dreu and Van Vianen, 2001; Murninghan and Conlon, 1991), research on family firms has shown that this is the most effective conflict management strategy in terms of both family and business outcomes (Sorenson, 1999). Collaborating facilitates the understanding of others and promotes positive relationships (Sorenson et al., 1998). Because it requires open communication, trust and mutual support (Seymour, 1993), collaborating is well suited to the unique needs of family firms and may be the most important conflict management strategy for family businesses (Sorenson, 1999). As such, since collaboration is a relationship-enhancing conflict management strategy that promotes cooperation and commitment, it may be the most effective strategy for reducing relationship conflict and minimizing its negative effects.

Similarly, collaboration is also a business-enhancing strategy because it promotes participation, teamwork and learning (Sorenson, 1999). Collaboration is effective in encouraging individuals to focus on the work problem as opposed to positions (Fisher and Ury, 1981). Because it requires mutual sharing and openness, collaboration is associated with organizational learning and adaptation that should enhance the performance of a family business (Dyer Jr, 1986). Indeed, task conflict has been argued to require a collaborating response to be effective (De Dreu and Van Vianen, 2001).

> *Proposition 5* The collaborating conflict management style has a positive effect on family firm performance. Specifically, relationship conflict is more likely to be resolved, while an optimal level of task conflict is created.

Third-party intervention

While avoiding, contending, compromising and collaborating involve the disputing parties resolving conflicts themselves, third parties must often become involved when individuals are unable or unwilling to resolve the conflict themselves (Wall Jr and Callister, 1995). Third-party intervention is sometimes necessary because conflicts in family firms often 'require an outsider to discern their unspoken, hidden, barely-hinted at apprehensions' (Kaye, 1991, p. 22). There are a multitude of third-party mediation and arbitration strategies. In the case of mediation, the success rates in reaching an agreement or resolution vary substantially; in the case of arbitration, an agreement is reached by definition (Wall Jr and Callister, 1995). Mediators are helpful to family firms because they improve communication between parties and point out to each party how he or she has been contributing to the conflict (Haynes and Usdin, 1997; Kaye, 1991). Arbitrators often employ the same techniques as mediators, yet they also have the option of dictating a solution to the conflict (Wall Jr and Callister, 1995). However, the most common form of third-party intervention is conciliation and consultation. These interventions are voluntary and do not require the family firm to relinquish control to the third party. Thus, while amiable solutions are sought, they are not binding to the parties involved.

Simply getting family members to agree to consult a professional for guidance improves the chances of resolving conflicts (Whiteside et al., 1993). Furthermore, the actual step to seek help from a counselor plays an important role in conflict resolution among family members because it signals their willingness to work together and shows that they admit to the need for change (Whiteside et al., 1993). Regardless of the type of third-party intervention, it is particularly important that the intervention tries to resolve the root of the conflict and bring family members closer together. Third-party intervention often reduces or resolves the sustained relationship conflict in family firms leading to a healthy climate within the firm and family. The third party might also be able to function as an information broker between the family members and thus allow the positive effects of task conflict to develop. As such, third-party intervention is often one of the most effective conflict management strategies in terms of changing a family's system of interactions and reframing family members' goals and contributions as well as previous conflictual behaviors (Kaye, 1991). Indeed, third-party intervention has long been a popular tool of conflict resolution in family firms (Haynes and Usdin, 1997; Hilburt-Davis and Gibb Dyer, 2003). Accordingly, we propose:

> *Proposition 6* Third-party invention conflict management strategies are positively related to family firm performance. Specifically, relationship conflict is more likely to be resolved, while an optimal level of task conflict is created.

Discussion

In order to succeed, family firms need to learn how to foster task conflict, while avoiding the negative consequences of relationship conflict (Kellermanns and Eddleston, 2004). We argued that avoiding and contending conflict management approaches are negatively related to family firm performance, that compromising has a mixed effect, and that collaborating and third-party intervention are positively related to family firm performance.

The unique psychodynamic effects of the family firm are most prevalent in terms of the negative performance effects associated with the use of an avoidance conflict management style. Although avoidance is often effective in resolving conflicts in non-family firms or team settings, we argued that because of the complex and sustained interactions among family members, the avoidance conflict management style is often counterproductive in family firms (see also Sorenson, 1999). We further highlighted the effects of compromising, contending and collaborating on the occurrence of task and relationship conflict and subsequent family firm performance. Finally, we argued in favor of third-party intervention in family firms by highlighting their positive effects on performance. This is not to say that we believe that when family firms are faced with conflicts that they should relinquish control and hire professional management as soon as possible (Levinson, 1971), nor are we advocating an increase in the use of consultants in family firms (Hilburt-Davis and Gibb Dyer, 2003). We are, however, advocating that when family firms are entangled in destructive conflicts that may be spiraling out of control, they can certainly benefit from an outside view.

Implications and future research

Our arguments suggest that conflict management approaches that apply to teams or non-family business settings cannot readily be transferred to family firms because of the unique psychodynamic effects of the family. Unlike other conflicting parties, family members continue to work with each other for indefinite periods of time and they often must interact in both business and social settings. Accordingly, a temporal aspect needs to be considered when investigating conflict management in family firms. Family firms persist for generations and negative conflict can escalate for years if not through generations and thus, can threaten the viability of the family firm. While avoidance might be a short-term solution and might actually create positive short-term effects in team settings, such effects are not likely to be realized in the long run in family firms. This insight has profound implications for the empirical study of family firm conflict management; that is, a cross-sectional approach may not capture temporal effects. Therefore, longitudinal designs that track the development of conflict and the usefulness of conflict management approaches over time might be very beneficial. Alternatively, qualitative studies could be utilized to better understand the phenomena of conflict in family firms. Indeed, qualitative studies could provide valuable insights on how the types of conflict and its management might not only affect family firm performance, but the dynamics of successful successions as well.

We outlined the direct effect of the conflict management approaches on family firm performance; however, we did not investigate potential moderators that might influence the effectiveness of the conflict management strategies. For example, research has shown that cultural attributes can affect the preferred conflict management style (Morris, 1998). Indeed, it is possible that the cultural background of a family firm may strongly influence how conflict is handled. This is further complicated by cultural influences from the external environment the family firm is exposed to. Thus, not only might the ethnic background of the family influence processes (Gersick, 1992), but also the national culture the firm operates in (Hofstede, 2001) can affect the occurrence and management of conflict. Furthermore, other family specific variables, like altruism, generational involvement, or control concentration should be investigated as family-firm specific moderators in the relationship between conflict management and performance.

Kellermanns and Eddleston (2004) argued for a complexity perspective in studying family firm conflict, indicating that task and relationship conflict need to be considered simultaneously to understand conflict's affect on family firm performance. A similar logic applies to conflict management approaches. Accordingly, future research should investigate a complexity perspective of conflict management (for an investigation of dual conflict management approach see Van de Vliert et al., 1999) and examine how conflict management approaches interact with each other, since they are often not mutually exclusive.

In conclusion, we hope that this chapter serves as a starting point for family firm conflict management. Only if family firms are able to encourage optimum levels of task conflict while minimizing relationship conflict, will family firms be able to prosper and succeed.

References

Alper, S., Tjosvold, D. and Law, K.S. (2000), 'Conflict management, efficacy, and performance in organizational teams', *Personnel Psychology*, **53**(3), 625–42.

Amason, A.C. (1996), 'Distinguishing the effects of functional and dysfunctional conflict on strategic decision making: resolving a paradox for top management teams', *Academy of Management Journal*, **39**(1), 123–48.

Amason, A.C. and Schweiger, D.M. (1994), 'Resolving the paradox of conflict, strategic decision making and organizational performance', *International Journal of Conflict Management*, **5**, 239–53.

Beckhard, R. and Dyer Jr, W.G. (1983), 'SMR Forum: managing change in the family firm – issues and strategies', *Sloan Management Review*, **24**, 59–65.

Brehmer, B. (1976), 'Social judgment theory and the analysis of interpersonal conflict', *Psychological Bulletin*, **83**(6), 985–1003.

Daily, C.M. and Dollinger, M.J. (1993), 'Alternative methodologies for identifying family- versus nonfamily-managed businesses', *Journal of Small Business Management*, **31**(2), 79–90.

De Dreu, C.K.W. (1997), 'Productive conflict: the importance of conflict management and conflict issue', in C.K.W. De Dreu and E. Van de Vliert (eds), *Using Conflict in Organizations*, London: Sage, pp. 9–22.

De Dreu, C.K.W. and Van Vianen, A.E.M. (2001), 'Managing relationship conflict and the effectiveness of organizational teams', *Journal of Organizational Behavior*, **22**, 309–28.

De Dreu, C.K.W., Harinck, F. and Van Vianen, A.E.M. (1999), 'Conflict and performance in groups and organizations', in C.L. Cooper and I.T. Rovertson (eds), *International Review of Industrial and Organizational Psychology*, vol. 14, Chichester: John Wiley and Sons, pp. 369–414.

Dean, S.M. (1992), 'Characteristics of African American family-owned businesses in Los Angeles', *Family Business Review*, **5**(4), 372–95.

Dyer Jr, W.G. (1986), *Cultural Change in Family Firms: Anticipating and Managing Business and Family Transition*, San Francisco, CA: Jossey-Bass.

Dyer Jr, W.G. (1994), 'Potential contributions of organizational behavior to the study of family-owned businesses', *Family Business Review*, **7**(2), 109–31.

Filbeck, G. and Smith, L.S. (1997), 'Team building and conflict management: strategies for family businesses', *Family Business Review*, **10**(4), 339–52.

Fisher, R. and Ury, W. (1981), *Getting to YES: Negotiating Agreement without Giving In*. Boston, MA: Houghton Mifflin.

Gersick, K.E. (1992), 'Ethnicity and family enterprise', *Special Issue of Family Business Review*, **5**(4).

Gersick, K.E., Davis, J.A., Hampton, M.M. and Lansberg, I. (1997), *Generation to Generation: Life Cycles of the Family Business*, Boston, MA: Harvard Business School Press.

Harvey, M. and Evans, R.E. (1994), 'Family business and multiple levels of conflict', *Family Business Review*, **7**(4), 331–48.

Haynes, J.M. and Usdin, T.M. (1997), 'Resolving family business disputes through mediation', *Family Business Review*, **10**(2), 115–34.

Hilburt-Davis, J. and W. Gibb Dyer, J. (2003), *Consulting to Family Businesses: A Practical Guide to Contracting, Assessment, and Implementation*, San Francisco, CA: Jossey-Bass/Pfeiffer.

Hofstede, G. (2001), *Cultures Consequences: Comparing Values, Behavior, Institutions, and Organizations across Nations*, 2nd edn, Thousand Oaks, CA: Sage Publications.

Hoy, F. and Verser, T.G. (1994), 'Emerging business, emerging field: entrepreneurship and the family firm', *Entrepreneurship Theory and Practice*, **19**(1), 9–23.

Janis, I.L. (1972), *Victims of Groupthink: A Psychological Study of Foreign-Policy Decisions and Fiascoes*, Boston, MA: Houghton Mifflin.

Jehn, K.A. (1995), 'A multimethod examination of the benefits and detriments of intragroup conflict', *Administrative Science Quarterly*, **40**, 256–82.

Jehn, K.A. (1997a), 'Affective and cognitive conflict in work groups: increasing performance through value-based intragroup conflict', in D. Dreu and E. Van de Vliert (eds), *Using Conflict in Organizations*, London: Sage, pp. 87–100.

Jehn, K.A. (1997b), 'A quantitative analysis of conflict types and dimensions in organizational groups', *Administrative Science Quarterly*, **42**(3), 530–58.

Jehn, K.A. and Mannix, E.A. (2001), 'The dynamic nature of conflict: a longitudinal study of intragroup conflict and group performance', *Academy of Management Journal*, **44**(2), 238–51.

Kaye, K. (1991), 'Penetrating the cycle of sustained conflict', *Family Business Review*, **4**(1), 21–44.

Kaye, K. and McCarthy, C. (1996), 'Healthy disagreements', *Family Business*, autumn, 71–2.

Kellermanns, F.W. and Eddleston, K. (2004), 'Feuding families: when conflict does a family firm good', *Entrepreneurship Theory and Practice*, **28**(3), 209–28.

Kets de Vries, M.F.R. (1993), 'The dynamics of family controlled firms: the good and the bad news', *Organizational Dynamics*, **21**(3), 59–71.

Lee, M.-S. and Rogoff, E.G. (1996), 'Research note: Comparison of small businesses with family participation versus small businesses without family participation: an investigation of differences in goals, attitudes, and family/business conflict', *Family Business Review*, **9**(4), 423–37.

Levinson, H. (1971), 'Conflicts that plague family businesses', *Harvard Business Review*, **49**, 90–98.

Ling, Y., Lubatkin, M. and Schulze, B. (2002), 'Altruism, utility functions and agency problems at family firms', in C.S. Galbraith (ed.), *Volume 3: Strategies and Organizations in Transition*, Kidlington: Elsevier Science, pp. 171–88.

Morris, M.H. (1998), *Entrepreneurial Intensity*, Westport, CT: Quorum Books.

Murninghan, J.K. and Conlon, D.E. (1991), 'The dynamics of intense work groups: a study of British string quartets', *Administrative Science Quarterly*, **36**, 165–86.

Priem, R. and Price, K. (1991), 'Process and outcome expectation for the dialectic inquiry, devil's advocacy, and consensus techniques of strategic decision making', *Group and Organization Studies*, **16**, 206–25.

Pruitt, D.G. and Rubin, J.Z. (1986), *Social Conflict: Escalation, Stalement, Settlement*, New York: Random House.

Schulze, W.S., Lubatkin, M.H. and Dino, R.N. (2003a), 'Toward a theory of agency and altruism in family firms', *Journal of Business Venturing*, **18**(4), 473–90.

Schulze, W.S., Lubatkin, M.H. and Dino, R.N. (2003b), 'Exploring the agency consequences of ownership dispersion among inside directors at family firms', *Academy of Management Journal*, **46**(2), 179–94.

Schulze, W.S., Lubatkin, M.H., Dino, R.N. and Buchholtz, A.K. (2001), 'Agency relationship in family firms: theory and evidence', *Organization Science*, **12**(9), 99–116.

Schweiger, D.M., Sandberg, W.R. and Ragan, J.R. (1986), 'Group approaches for improving strategic decision making: a comparative analysis of dialectical inquiry, devil's advocacy, and consensus', *Academy of Management Journal*, **29**(1), 51–71.

Schwenk, C.R. (1990), 'Conflict in organizational decision making: an exploratory study of its effect in for-profit and not-for-profit organizations', *Management Science*, **36**(4), 436–48.

Seymour, K.C. (1993), 'Intergenerational relationships in family firms: the effect of leadership on succession', *Family Business Review*, **6**(3), 263–81.

Sorenson, R.L. (1999), 'Conflict management strategies used in successful family businesses', *Family Business Review*, **12**(4), 325–39.

Sorenson, R.L., Morse, E.A. and Savage, G.T. (1998), 'What motivates choice of conflict strategies?', *International Journal of Conflict Management*, **10**, 25–44.

Tjosvold, D. (1997), 'Conflict within interdependence: its value for productivity and individuality', in C.K.W. De Dreu and E. Van de Vliert (eds), *Using Conflict in Organizations*, Thousand Oaks, CA: Sage, pp. 23–37.

Van de Vliert, E. and Euwema, M. (1994), 'Agreeableness and activeness as components of conflict behaviors', *Journal of Personality and Social Psychology*, **66**, 674–87.

Van de Vliert, E., Nauta, A., Giebels, E. and Janssen, O. (1999), 'Constructive conflict at work', *Journal of Organizational Behavior*, **20**, 475–91.

Wall Jr, J.A. and Callister, R.R. (1995), 'Conflict and its management', *Journal of Management*, **21**(3), 515–58.

Ward, J.L. (1987), *Keeping the Family Business Healthy: How to Plan for Continuing Growth*, San Francisco, CA: Jossey-Bass.

Whiteside, M., Aronoff, C.E. and Ward, J.L. (1993), *How Families Work Together*. Marietta, GA: Family Enterprise Publishers.

Whiteside, M.F. and Brown, F.H. (1991), 'Drawbacks of a dual systems approach to family firms: can we expand our thinking?', *Family Business Review*, **4**(4), 383–95.

PART V

FAMILY BUSINESS SUCCESSION

20 Lost in time: intergenerational succession, change and failure in family business

Danny Miller, Lloyd Steier and Isabelle Le Breton-Miller

Executive summary

Much of the time, successions in family owned businesses (FOBs) simply do not work out. In part, this is because personal and emotional factors determine who the next leader will be. This is especially true in the case of father to son successions where the desires of a family for their children dictate that a son take over the business regardless of suitability. Here, remedial paths consist not in finding the right person for the job but getting the successor the help he/she needs. That can only come from knowing the early warning signs of problematic successions and combating them quickly and directly. This research sought to identify those warning signs.

It found that at the core of problematic succession lies an inappropriate relationship between an organization's past and its present. Either there is too strong an attachment to the past on the part of the successor, too wholesale a rejection of it, or an incongruous blending of past and present. We call these very common patterns conservative, rebellious, and wavering. Each is characterized by distinctive tendencies in strategy, organization, and governance, and each has its typical symptoms. FOB board members must be alert to these symptoms and consider them as focal points for change. It is encouraging that each of the three succession patterns is common and thematic; but each is also distinctive and requires its own characteristic changes in strategy, organization, and governance. These changes might be made with the help of a family council or board of directors while the successor remains in place. But our examples show that action may have to be taken early so that there are enough resources left to effect a turnaround.

The chapter also draws conjectures about the potential causes and performance implications of these patterns in different environments. Although tentative, these may help managers recognize even earlier potential trouble spots in an intergenerational succession.

1 Introduction

Family owned and controlled businesses account for an enormous percentage of employment, revenues, and GDP in most capitalist countries (Morck et al., 2000; Sharma et al., 1996; Shepherd and Zacharakis, 2000). Although many are small, FOBs in aggregate represent one-third of the Fortune 500 and about half of the US gross domestic product (Aronoff et al., 1996). They also employ over 80 per cent of the work force (Neuberg and Lank, 1998).

In many of these businesses, founders try to perpetuate their legacy and ensure continued family control via intergenerational succession, as when they hand over leadership to their children. Unfortunately, recent evidence indicates that 'a mere 30% of family businesses survive past the first generation' and that many intergenerational successions fail soon after the second generation takes control (Davis and Harveston,

1998, p. 32; Handler, 1990, 1992; Sonnenfeld, 1988, p. 238; Ward, 1997, p. xvi). This represents a serious blow not only to family businesses and their employees, but also to the health of an economy. Intergenerational succession failure, then, is a challenge that merits investigation.

Past research suggests that there are many reasons such successions fail. They include unclear succession plans, incompetent or unprepared successors, and family rivalries (Dyer, 1986; Handler, 1990, 1992, 1994; Hugron, 1993; Lansberg, 1999; Morris et al., 1997; Pitts, 2000). Often, however, the choice of a successor is predetermined by blood. Then it becomes a matter not of finding the right successor, but quickly identifying the problematic outcomes of a succession and helping the incumbent deal with them. The first places to look for these outcomes are in the aspects of strategy, organization, and governance that are universally believed to underlie performance (Fuchs et al., 2000). Discovery of these concrete and 'proximate' manifestations of troubled successions might help managers address them directly and provide clues about their fundamental causes.

Our approach was to track poorly performing or failed family enterprises for several years after succession to determine what happened. Of particular interest were characteristic post-succession themes in competitive tactics and capabilities, goals and values, organizational designs and processes, and top management boards and teams. We sought inductively to discover common patterns in the data – to identify early warning signs so that these can be directly addressed. We also speculate on the sources of the patterns and their performance implications in different contexts. But because we studied only failing enterprises, we cannot speak to the determinants of successful succession (see Handler, 1994; Morris et al., 1997).

1.1 The literature on CEO succession

A vast body of literature suggests that top management succession is an especially challenging event for all kinds of firms (Helmich, 1975; Miller, 1993; Reingenum, 1985). That fact is reflected in part by the significant stock price fluctuations that accompany many succession announcements (Freidman and Singh, 1989). Where hand-appointed insiders take over and where there is little change in the top management team, insider successions were found to presage periods of continuity, especially if the previous CEO remains as a board member. This was shown to delay necessary adaptation in competitive environments (Miller, 1993). Outsider successions, by contrast, could elicit too much change, as someone with a new perspective tries to leave their mark (Hambrick and Fukutomi, 1991; Helmich and Brown, 1972; Miller and Shamsie, 2001; Wiersema, 1995). Outsiders also may promote indecision and vacillation, as rookie CEOs grope to find their way (Brady and Helmich, 1984; Miller, 1993).

We suspected that intergenerational successions in FOBs might well mirror these reactions of too much or too little change. But the unusually strong influence exercised by leaders of family businesses and the intimate relationship and vast experience gaps between the old and new leaders might render these reactions more extreme (Lansberg, 1999). Reactions from an immature successor may include overdependence and conservatism, rebellion and excessive change, or ambivalence and confusion (Kets de Vries and Miller, 1984, 1987). Our working hypothesis was that such behaviors could manifest in many aspects of strategy and structure: stagnation or abandon, compulsive consensus or conflict, and stifling bureaucracy or chaos.

One reason for such extremes is that an intergenerational succession creates an unusually large age and experience gap between old and new CEOs – often 25–30 years (Handler, 1994). This gap, the immaturity of the successor, and emotion-fraught parental relationships make dysfunctional reactions of submission and rebellion all the more likely (Kets de Vries and Miller, 1984; Kimhi, 1997). Another reason is that many FOBs are centralized in power and ownership. Whether the tendency is toward conservatism or action, few can stop it. In fact, founders often see their businesses as extensions of themselves that they want to control completely (Dyer, 1986; Lansberg, 1999). As these reactions stem more from personality and emotional relationships than competitive demands, they can give rise to inappropriate strategies, unsuitable organization cultures, and flawed governance. We wished to study these post-succession problems as they developed to determine their concrete manifestations and symptom patterns.

2 The pilot study and its findings

We studied only intergenerational successions followed by poor performance that ended either in successor dismissal or bankruptcy.[1] These are very common among FOBs for the reasons we discussed. Our database consisted of case and historical book accounts of 13 major firms, as well as a series of newspaper and journal articles we compiled on what happened to the strategy and organization of these firms during the 5–10 years after succession. Sample selection was driven by the availability of ample public information, and so our firms tended to be large and well known.[2] To offset this size bias, we added an anonymous sample of three smaller family firms with which we have worked personally over the years. It should be noted that our sample includes mostly father to son successions as these are by far the most prevalent. Our findings from these 16 successions, therefore, may not generalize beyond this common situation – for example, to where mothers, daughters, or cousins are involved.

Extremes of change or conservatism expected in problematic FOB successions may manifest in numerous aspects of strategy, organization, and governance. To describe these categories, we were guided by the general model of competitive capability outlined by Miller (1990) and Fuchs et al. (2000). The model proposes that strategies must create distinctive capabilities and match them to the market via adaptive behavior. It also argues that robust organization culture, design, and governance are needed to make this happen. Following Fuchs et al. (2000), we characterized strategy according to the degree of goal consensus and consistency, capability development, innovation, market renewal, proactiveness, and risk taking (see also Miller, 1993; Miller and Friesen, 1984). Following the same authors, organization was described by values, goal clarity, stability and traditions, bureaucracy, centralization of power, and level of conflict. Finally, governance was assessed according to the power still held by the old guard and by board and managerial turnover (see Kets de Vries and Miller, 1987; Lansberg, 1999). These variables engendered the symptom lists of Appendix B.

Having read materials on each firm, two raters performed double-blind scoring of the variables and symptoms of Appendices A and B. Scores ranged from *low* = 1 to *medium* = 2 to *high* = 3 for variables, *yes* = 1 or *no* = 0 for symptoms. Rater's scores agreed over 90 per cent of the time. Any disagreements were discussed by the raters after all the scoring had been done and were resolved or averaged (a yes vs. no conflict was resolved by assigning no).

Table 20.1 The organizational implications of three succession patterns

	Conservative	Wavering	Rebellious
Strategy	Stagnation, risk aversion, insularity	Indecisive, inconsistent, start–stop	Revolutionary change – often for its own sake
Organization and culture	Tradition-bound, bureaucratic, centralized	Confused culture, conflict-ridden units	New units, new values, chaotic organization
Governance	Old guard still powerful	Mix of old and new managers	Significant turnover; new sheriff in town
Performance	Loss of market share, dying markets	Abortive projects, shrinking margins	Cost and expenditure overruns

Given our expectation of extremes along the change dimension and to discover common patterns in the data, raters were asked, again double blind, to classify the firms based only on the overall level of change (again H, M, or L). Raters agreed on all 16 classifications. As change was the classificatory variable, it is tautological that groups differed along this criterion. More striking from our accounts were the extremes that were reached of obsessive conservatism, dramatic rebelliousness, or a third wavering category that was an odd combination of these extremes rather than an effective middle ground. Remarkable too was the tremendous quantitative score similarity of firms within each category (see Appendix A). Most important were the widespread differences in virtually all variables across the three succession categories – differences that were by no means predetermined by the classification according to change. Indeed, a few traits of any of our succession events would be enough to classify patterns, and thereby predict many of their other strategic, organizational, and governance tendencies.

The modal tendencies we discovered for each group are summarized in Table 20.1 and Appendices A and B. In the descriptions that follow, we discuss these patterns and present a sample case study for each one that tries to make more concrete the nature of each type of succession. It is important to remember that this is an exploratory study. Our patterns do not constitute a validated taxonomy but are simply very common patterns that emerged in the research.

3 Conservative successions

In conservative successions, the new CEO remains in many ways dependent on the old – even after the latter has quit or died. So the shadow of the parent lingers. As a result, a period of strong leadership may be followed by one of conservatism in which strategies and organizations are locked in the past.

3.1 Strategy

Time stands still. Relative to their competitors, firms led by conservative successors undergo little change in their goals, business scope, product lines, or markets. This strategic stability is reflected by a relatively fixed functional and value chain emphasis: firms keep stressing or differentiating themselves via the same activities and policies. There is more emphasis on addressing problems than seizing opportunities. In fact, even in the face of external turbulence or competition, there is great resistance to adaptation. The focus of managers is mostly on internal matters – efficiency, operations, and quality – rather than evolving market needs.

3.2 Organization and governance
Values and patterns of interaction remain frozen. The same kinds of people are hired and promoted. In addition, the same hierarchy, rituals, compensation schemes, and modes of communication remain. Many members of the top management team, moreover, stay in place – old advisors of the founder, longstanding board members, and so on. Often, in fact, the founder himself stays on the board, second guessing and checking up on his successor.

Conservative successions can have dire consequences, especially in the changing environments. They punish people who try to get things done differently, reject freethinking employees, and alienate customers wanting new products. Before long, market share erodes, sales stagnate, and margins shrink. Such successions seemed more likely in firms that had been strong and steady performers, those having ample market power, or those not facing major competition. They also appeared more common where the old CEO had been powerful and well respected, where much of the old management team and board remained in place, and where an old or new CEO retained financial control.

3.3 Bata Shoe Company
The global giant Bata Shoe, which once had almost 100 000 employees, provides an archetypal example of a conservative succession. The 1990s at Bata was considered a lost decade. That was when Thomas Bata Jr took over as CEO from his domineering father. In the biography of Tom Sr, he describes the challenge he faced when becoming CEO – the same predicament that befell his own son.

> Being the only son of a legendary father can be a mixed blessing. On the minus side, there is an enormous weight of expectations, along with a sometimes frustrating struggle to acquire one's own identity, away from a parent's shadow. When the son is successful, he is perceived as a chip off the old block; when he makes a mistake, he is perceived not to measure up to the father's standards. (Pitts, 2000, pp. 45–6)

Tom Jr took over as CEO in 1984. Tom had been groomed to assume command for much of his life. He had been sent to the same English public school as the father and had apprenticed both at a competing firm and at Bata. But Tom Jr had had little independent experience as a manager before taking the helm at Bata. While the father was gregarious and confident, the son was an introvert. In addition, although Jr saw business opportunities, he did not pursue these with enough confidence or resources to update the Bata strategy. According to one manager who had worked at Bata, 'If [Jr] had been a strong person, he wouldn't even have been in the company. If he told me to do something, the next day the old man was sure to contradict him' (Pitts, 2000, p. 53).

3.3.1 Strategy and organization By the 1980s and 1990s, competition for Bata was increasing from firms such as Nike and Adidas, which were building market share with their global brands. Tom Jr did little to meet these challenges. Product lines remained staid, utilitarian, and anonymous in a market that would only pay a premium for fashion and branding. Consequently, Bata's offerings grew stale while rivals surged ahead and stole market share.

Given the weak leadership at Bata, different country managers were left to their own devices. Each had their own designs, manufacturing facilities, and types of outlets. While this was necessary when there were barriers to trade, it got in the way of rationalizing manufacturing operations and building and leveraging global brands. Tom Jr did make a few tentative efforts to consolidate manufacturing and invest in a more integrated global marketing effort. But he had not amassed the confidence or resources to make these changes work. To aggravate matters, both his parents still interfered with Tom's decisions. Things came to a head when Jr wanted to take advantage of high Asian stock prices in the early 1990s to raise money to improve North American operations. His parents blocked the move. In 1994, the board replaced Tom Jr with an outside manager. It is estimated that Bata employment fell from about 85 000 15 years ago to about 50 000 worldwide.

4 Wavering successions

Wavering successors are characterized by indecision. They want to make their mark on their firms but are uncertain as to how. On the one hand, they respect the policies and traditions of the founders. On the other, they wish to exert influence and show their independence. In addition, they vacillate between these attitudes, manifesting doubt and reversing their own initiatives.

4.1 Strategy

Strategies take the form of lesser, often unsuitable initiatives, which are grafted onto older strategies and traditions. For example, a leader may make corporate acquisitions, introduce products, or enter new markets. But these initiatives are too incongruent with the established strategy, or market focus to be effective. There is also a tendency to terminate initiatives in midstream – before they have had a chance to be tested. This start–stop aspect of strategy wastes resources and creates confusion.

4.2 Organization and governance

The old organization is normally left standing but new units may be added – new divisions for example or project groups often made up of recently hired employees. These groups, however, are not well integrated with the rest of the firm and frequently conflict with it. In addition, because they operate on the periphery, they tend to be ineffective. This split into 'old guard' and 'young Turk' cliques may eventually erode organization culture, values, and traditions. Yet there is no consensus behind any new vision that could consolidate the firm or unify managerial efforts. Governance too mixes the old and the new with many veteran directors and managers staying, but a number of recruits coming on board to take charge of the new initiatives.

A major consequence of wavering successions is that the firm is never able to converge on a sound new strategy. The uncomfortable admixture of old and new upsets employees and clients alike and squanders resources on changes that do not amount to anything. This blurring of the organization's strategy and image allows more distinctive competitors to steal market share. Profits also tumble because of the unproductive experimentation.

4.3 The T. Eaton Company

The T. Eaton Company was one of Canada's oldest, largest, and most successful dry goods chains. It was odd therefore that George Eaton could take over the business essentially by default – as his brothers became involved in other pursuits. George, the iconoclast of the family, spent years as a racing car driver. He was neither deeply interested in his family's business nor seen by those in the know as CEO material. But by 1988, he found himself president of Eaton's (McQueen, 1998).

Certainly, George respected many Eaton's traditions – quality merchandising, guaranteed customer satisfaction, and good service. But Eaton's, gradually and subtly, was becoming stuck in an ever narrower slice of the market – between specialty and luxury merchandisers such as Holt Renfrew and Banana Republic, and stores such as The Bay and Wal-Mart that were offering merchandise at more attractive prices.

Although George preserved much of the Eaton strategy, he made a few disastrous changes. None fit Eaton's traditional markets, its image or capabilities, or its overall strategy. In the fall of 1990, for example, George launched a program called 'everyday value pricing' (EVP). He decided that there would be no more off-price promotions or sales. Consumers were now told that the prices they would be getting would always be the lowest possible. Unfortunately, Eaton's sold fashions whose popularity was tough to predict and EVP made it impossible to get any return from less popular merchandise. Also, EVP was inconsistently carried out among the stores and it rendered useless Eaton's outstanding promotional machine – one of its core assets. Moreover, Eaton's did not have the operating efficiency to make EVP viable. By the end of 1993, after much damage to image, margins, market share, and sales, the policy was terminated.

To set things right, George spent great sums on consultants, all the while ignoring his own staff and making little attempt to bring more competent people into the firm. Some of the consultants created expensive computer systems that were rarely used. Others spawned new merchandising ideas. McKinsey & Company counseled setting up 'destination' stores focussing on high margin, high fashion items, in the hope of generating the cash needed for updating the other stores. But George was reluctant to invest the necessary resources to make the destination stores work. The store revamps that did occur were half baked, and the units no longer carried items many Eaton's customers had come to expect. Eaton's also began moving upscale in locations that did not have enough wealthy clients, and it did so in a half-hearted way that could not attract these jaded shoppers. This upgrading strategy focussed only on a few stores while the rest were virtually abandoned – leaving many employees feeling hopeless.

In spite of these changes and threats from new entrants such as Gap and Wal-Mart, George Eaton kept his firm too close to where it always had been. 'Everything to everybody, stuck in the middle in terms of price, merchandise, quality, assortment and service . . . Eaton's came to stand for nothing – not the lowest price, not the best depth in merchandise assortment and not the best service' (Thompson quoted in McQueen, 1998, p. 238). By 1995, Eaton's was in serious trouble. But even with sales and profits plummeting (a loss of US$80 million by 1995) and banks cutting off credit, George did not revisit his strategy. Ultimately, an executive management committee was established to deal with the crisis. However, disagreement among its members led to indecision and still more stop–go initiatives. By 1997 the firm was insolvent.

4.3.1 Organization and governance Eaton's organization was a fractured and confusing one, with the finance people at odds with the merchandisers and the old guard fighting the new. Some of George's key blunders came in recruiting new executives who, like himself, knew too little about Eaton's business. George and the rest of the Eaton's family began more and more to depend on advice from finance people rather than the merchants and marketers who had in the past successfully managed the firm, and these factions battled constantly (McQueen, 1998, p. 231).

George directed Eaton's from the top down, yet he and his team lacked both creative and managerial talent. Those recruited, while having just enough power to alienate the old guard with many of their decisions, usually did not have enough influence to implement their more important initiatives.

5 Rebellious successions

In rebellious successions, a new CEO rejects the legacy of the prior generation. There is wholesale eradication of the past and its practices. Rebellious successions are rarer than those of conservatism or wavering because normally, renegade offspring decide not to go into the business or are discouraged from doing so by the older generation. But if rebels do ascend to power, they want to do things very differently from the 'old man'.

5.1 Strategy

Strategy is characterized by far-reaching changes: in the product–market scope of the firm and in functional business strategies. There are, for example, significant acquisitions, divestments, expansions, product or market changes, and shifts in functional emphasis (for example, from an operations to a marketing focus). Unfortunately, these changes are often off the mark. They are spawned more by the new CEO's desire to leave an imprint and escape the past than by a judicious perception of new opportunities.

5.2 Organization and governance

Chaos reigns. Changes in values and goals are common, as are those in roles and reporting relationships, patterns of communication, information systems, compensation and hiring policies. New units and divisions may be set up to house the new businesses. But again, the changes do not relate to the needs of the business. There also tend to be radical upheavals in the top management team and among board members.

Rebellious successions deplete resources. They incite managerial turnover, load a firm with debt, alienate traditional clients, and pull firms into businesses that surpass core competencies. They also are more apt to occur when there has been conflict 'between father and son'; when circumstances or the board keep the previous CEO out of the picture; and when performance has been poor and the need for change is apparent.

5.3 Barneys, New York

Barneys, New York, was a high-end men's haberdashery with a reputation for taste. The store had long been successful, earning far higher margins than its competitors with its judicious buyers, European styles, and compelling promotions. During the 1980s, the third generation of the Pressman family – brothers Bob and Gene – began to take over responsibilities from their father Fred, and started to play more substantial roles in the business. While Bob contented himself with finances, Gene became the visionary fashion guru and prime strategist at Barneys. Ambitious and strong willed, Gene was intent on transforming Barneys as soon as he entered the business. According to an associate, 'Fred was the merchant and he watched over the business, while Gene was the guy throwing shit on the wall and seeing what stuck' (Levine, 1999, p. 95). The only way for Fred to keep Gene in the business was to leave him alone. According to Fred, 'Gene's got his own idea of where he's going, and I don't want to upset him' (Levine, 1999, p. 96).

From the outset, Gene began to push for an expanded women's department. He sought out radical new designers and embraced the most costly and provocative fashions. Gene wanted Barneys to become the New York trendsetter in women's fashion, and his tastes could be outlandish. To showcase his growing women's collection, Gene undertook to build a second, far more grandiose uptown store. He recruited the most extravagant architects to design magnificent New York premises and spent enormous amounts of money, doubling, then quintupling the budget, and severely taxing Barneys' finances.

The women's store was never a success: volume was always disappointing, and from 1985 to 1992, the business was bleeding money. For this, Gene always blamed 'the people in the store,' never admitting that his daring fashions and exorbitant tastes might be the culprits (Levine, 1999, p. 145). Resources continued to erode as Gene tapped the business to feed his lavish lifestyle and renovate his Larchmont mansion. To fund these and other extravagances, he entered into a partnership with Isetan, the Japanese retailing giant. The aim was to take Barneys into cities such as Los Angeles and Chicago. But these markets had decidedly different tastes than New York, and Gene's grandiose, impulsive expansion strategy failed. To disguise the resulting losses and personal withdrawals from Isetan, Bob and Gene initiated shady accounting practices and kept multiple sets of books. Ultimately, the changes Gene wrought in business scope, product lines, and staff, coupled with the huge expenditures required to implement them, put Barneys on the ropes. By 1996, the Pressmans had lost control of a Barneys that was in essence bankrupt.

5.3.1 Organization Under Gene's tenure, Barneys hemorrhaged talent. It lost superb marketing and sales people, alienated excellent suppliers, and chased away potential leaders. Some people were fired simply because Gene did not like the way they dressed. The administrative structure was as chaotic as the strategy. Factions grew up around Bob and Gene as each blamed the other for the poor results (Levine, 1999, pp. 206–207). This rendered impossible collaboration and even coordination between marketing and finance. Controls were almost nonexistent as neither executive 'took the slightest interest in such banalities as computer systems or organization charts' (Levine, 1999, p. 207). Inventories mushroomed and buyers did not know which items were hot and which were not. But Gene just kept spending.

6 Hypotheses about the causes of problematic successions

Although the focus of this study was on the nature of problematic successions rather than their causes, the qualitative historical data we gathered for each of our enterprises did provide hints as to the roots of many of these successions. These are summarized as conjectures on Table 20.2 and include intergenerational interaction patterns, CEO personalities and experiences, and the organizational and market contexts of a firm.

Table 20.2 Hypothesized drivers of the succession patterns

	Conservative	Wavering	Rebellious
Intergenerational family dynamics	Idealization, subservience	Conflicted, unresolved	Rejection, independence
CEO personality and managerial style	Conservative, risk averse, obsessive	Indecisive, suspicious, reactive	Dramatic, proactive, action oriented
Organizational context	Steady performance, strong culture and traditions	Politicized, divided, factionalized, conflictual	Unsettled, in crisis, deteriorating performance
Market context	Stable, protected, tradition bound	New challenges or market discontinuities	Turbulent, dynamic, competitive

6.1 Family dynamics

Our three succession patterns parallel three classic dysfunctional parent–child interaction patterns described in the family therapy literature. These dynamics occur long before the transfer of power – and reside in the family history of a new leader and his or her parents. Research suggests that such parent–child interactions may influence how a new leader runs an organization (Kets de Vries and Miller, 1984, 1987).

Interactions with parents in early life can have an enduring impact on the personality, values, and behavior of the child. During the normal process of development, children alternately identify with and individuate themselves from their parents. These reactions can be pronounced, with offspring either adoring or rebelling against especially forceful parents. By the time children are in late adolescence, these relationships stabilize into enduring patterns of interaction (Boszormenyi-Nagy and Spark, 1973; Minuchin, 1974; Stierlin, 1974; Winnicott, 1971).

The founders of many successful businesses are resolute and imposing figures. They are wealthy, respected, perhaps even feared; and in the case of entrepreneurs, they may be narcissistic (Kets de Vries, 1996; Kohut, 1971). The children of these busy parents have to compete with a powerful ego and a thriving enterprise to get attention. The literature on family psychodynamics and developmental psychology suggests that children may fall under the spell of such parents and have a hard time establishing an independent identity. In fact, they idealize their parents (Kernberg, 1975). Where this reaction continues for too long, as often occurs with a domineering parent, it can weaken a child's independence and engender timidity (Mahler et al., 1975). To make matters worse, some forceful business leaders view their firms and their children as extensions of themselves and try to control both. The result is a dependent and conservative child lacking the courage to pursue his own course – especially in matters concerning the family business – the sacred bastion of the parent (Kets de Vries, 1996, pp. 40–41; Minuchin, 1974).

Other children go through periods of parental idealization and subjugation only to realize its costs to their egos and freedom. Distancing or disappointing behavior by a parent also may drive children – often through negation of parental behavior and beliefs – to mark out their own paths. Idealization turns to opposition in an attempt to gain independence. This individuation process can engender rebelliousness – an attraction to opposites (Kets de Vries and Miller, 1984; Stierlin, 1974). In addition, because the family business represents such an important symbol of the powerful father, a rebellious child's succession can lead to the wholesale dismantling and rebuilding of an enterprise. At last, the child can demonstrate competence and originality in the very sphere that will confirm his or her independence.

A third dysfunctional pattern emerges from an unresolved combination of idealization and opposition, and it is largely one of vacillation. The child at once admires the parent but also wants to achieve some level of independence. In normal circumstances, respect and independence coexist; but where the parent has been narcissistic, inattentive, or overly controlling, the child may waver back and forth between extremes of passivity and action (Minuchin, 1974). When children take over a family business, these patterns of conservatism, rebelliousness, and vacillation may contribute to our succession patterns.[3]

Family dynamics also come into play in the interactions between brothers and sisters as they fight to control or manage a business (Lansberg, 1999; Pitts, 2000). The stagnation that characterizes conservative successions may be caused by rivalry between brothers and sisters as they block one another's actions. Wavering successions in which opposing camps pursue conflicting initiatives are another possible result.

> *Hypothesis 1* Our problematic succession patterns of conservatism, wavering, and rebellion are more likely to occur in the presence of parent–child relationships characterized by dependency and idealization, vacillation, and conflict and opposition, respectively.

6.2 *Personality and experience*

Clearly, personality is a product of far more than upbringing. Other life experiences may be crucial, as are genetic influences that help determine intelligence, stamina, optimism, anxiety, and even confidence (Kramer, 1995). These aspects of personality can drive a leader's behavior. The effects are especially marked in centralized and tightly controlled firms where CEOs have a major impact on organizational strategy and processes (Miller and Droge, 1986; Miller and Toulouse, 1986; Miller et al., 1982).

Kets de Vries and Miller (1984), for example, have shown how obsessive and depressive personality styles can give rise to many of the traits we associate with our conservative successions. These include strategic stagnation, preoccupation with irrelevant detail, and an internal rather than a market focus. The authors also found 'dramatic' leaders who implemented rebellious transitions. Changes by these executives were often too major and too irrelevant to allow their firms to survive. Kets de Vries and Miller (1984, 1987) even identified suspicious types of leaders, who made myriad stop–start changes in response to environmental challenges, but never developed any coherent strategies of their own. Here, we find many of the traits of our wavering successions.

Professional and educational experience can also influence how a successor runs the business. Successors without significant formal or practical business education are more apt to be subject to the extremes we described: they lack a compass or standard for comparison. They have not seen how other organizations within their industry operate and so have difficulty judging the appropriateness of their behavior. This was true of Eaton, Bata, and the Pressmans. A dearth of varied experience within the family firm itself can contribute to this problem.

Hypothesis 2 Our problematic succession patterns are more apt to occur where past and current leaders exhibit leadership styles characterized by dramatic, suspicious, obsessive, and depressive behavior, and where these styles are not mitigated by broad business and educative experiences.

6.3 Firm context

A third set of influences on our succession patterns is the context of the firm. Some new leaders walk into firms that are mired in the past because of their strong cultures and values, their long ensconced board members and top management team, and their rigid policies and routines. As at Bata, these companies are notoriously resistant to anything but conservatism. At the other rarer extreme, we find companies that are plagued with problems whose product lines have foundered or are in flux, say, or that are experiencing turnover in the executive ranks. These conditions promote 'rebellious' change (recall, however, that at Barney's, these factors were more a consequence than a cause of the rebellious succession). Finally, we have factional and politicized organizations with warring fiefdoms in the different departments or divisions. This is an environment in which wavering – sequentially addressing the needs of the different camps – is hard to avoid.

Hypothesis 3 The culture and governance structure of an organization may influence the succession patterns that occur, with conservative patterns more likely in stable and traditional contexts, rebellious patterns more likely in problem-ridden or turbulent situations, and wavering patterns more common in politicized and fragmented organizations.

6.4 Market context

A final group of factors influencing our succession patterns are market forces. Stagnant or highly stable industries, such as agricultural or commodity products or distilling, lend themselves to conservative successions. Here, there is little felt or actual need to change. By contrast, turbulent, highly competitive environments, and those in which technologies, fashions, or customers are rapidly changing, are more conducive to rebelliousness. In addition, industries undergoing deregulation or experiencing the shock waves of a new entrant may evoke wavering – a questioning of the past coupled with a reluctance to stray too far from it. That was the case at Eaton's as it found itself squeezed between rivals such as the Gap and Wal-Mart.

Note, however, that what characterizes most of our problematic successions is inattention to the marketplace and the larger external environment. The problem is not simply that firms are changing too much or too little, but that they are changing inappropriately given their competencies and the challenges and opportunities in their environments.

Hypothesis 4 The level of competitiveness, change, and uncertainty in the environment may influence which of our succession patterns is more likely. This relationship is apt to be stronger in successful organizations that try to match a successors with the problems and issues he or she are more likely to face.

One suspects, in the end, that our three succession patterns will be products of all of these categories of influences, and that they will be especially extreme where there is confluence among these influences.

7 The performance implications of the three succession patterns

Because we have studied only failures, we cannot establish the relationship between our succession patterns and performance. We do believe, however, that the performance impact of the patterns will be moderated by contextual factors such as the market environment, the history of the firm, and the way the patterns are implemented.

7.1 Market context

Conservative successions are apt to be especially dangerous in rapidly changing ('uncertain') environments where firms must change their products and capabilities to stay competitive. They may be fairly innocuous, however, when there is little need to change. Indeed, if not taken to extremes, the continuity bred by conservatism can help firms deepen capabilities (Miller et al., 2002). One would make essentially the opposite predictions for rebellious successions, which may be less harmful in turbulent contexts and destructively disruptive in a stable one – where the old strategy still works. Unfortunately, rebellious successors often do things more to please themselves and shock their dad than to adapt to objective forces in the environment (Kets de Vries and Miller, 1987), and this can be damaging even in a dynamic setting. Wavering successions may evoke the worst of both worlds as they combine conflicting strategy and organizational elements. Consequently, the firm is stuck in the middle.

> *Hypothesis 5* Conservative successions will be less harmful in stable than uncertain environments; for rebellious successions, the opposite will hold; and for wavering successions, performance will be unrelated to environmental uncertainty and change.

7.2 Firm historical context

Firms that have not changed much for a long time may be less harmed by rebellious successors than those that had undergone a period of upheaval and no longer have the resources to endure change. By contrast, firms that have been enduringly stable and in need of strategic renewal may be especially harmed by a conservative succession; but they may benefit from a well-aimed rebellious one (Miller and Shamsie, 2001). Again, waverers could be in the worst place, as their actions are not apt to be sufficiently concerted to take advantage of any context. The performance history of a firm may come into play as well, with successful firms being less harmed by conservative successions than unsuccessful ones; the opposite being true for rebellious successions.

> *Hypothesis 6* Conservative successions will be less harmful in stable and high-performing historical contexts; for rebellious successions, the opposite will hold; and for wavering successions, performance will be unrelated to historical context.

7.3 Execution

Not all successions within our types are created equal. They vary in their execution and in how extreme they are. Insular hidebound conservatism across many of our strategy and organizational variables will be more harmful than moderate conservative practices, or great conservatism in a few practices. The same logic holds for rebellious successions where chaos is apt to be far more damaging than coherent opportunity seeking.

Hypothesis 7 The more consistent and extreme (high or low) the scores on the variables for our types, the worse the performance of the firm.

8 Conclusions and directions for further research

Our findings suggest that intergenerational successions are very much plagued by problems of passage – by an inappropriate relationship between past and future (Gersick et al., 1997; Hugron, 1993; Kets de Vries, 1996; Kets de Vries and Miller, 1984, 1987; Lansberg, 1999; Miller, 1991, 1993). This is evidenced by excessive attachment to the past by an overly dependent and conservative successor, a rejection of the past by a rebellious one, or an incongruous blending of past and present by an unsure and wavering new leader. The fact that these patterns occur with such regularity in our sample suggests that they be studied further as syndromes that will have to be combated by a great many FOBs.

A primary next step for researchers would be to study whether patterns thematically akin to ours occur for successful successions – and to determine how these patterns differ from those we found. This could establish the overall performance implications of our succession patterns and their constituent variables. Also, because this study was based mostly on father to son successions, it would be worthwhile for subsequent researchers to expand the research to include a broader sample of succession types – father to daughter, to brothers, to cousins, and so on. Such a more encompassing project could be used to test many of our hypotheses. Another worthwhile avenue might be to examine dysfunctional successions in nonfamily businesses using the variables and symptoms we have identified. Although the findings from this qualitative study are suggestive, they require significant follow-up work to establish their range, reliability, and validity.

Acknowledgements

The authors are grateful to the Social Sciences and Humanities Research Council of Canada for Grant #410-98-0405 and the Center for Entrepreneurship and Family Enterprise at the University of Alberta School of Business for funding. They would also like to thank Bill Bone of Loram, Jim Chrisman of Mississippi State University, Kenneth Craddock of Columbia University, and Jean-Marie Toulouse of HEC, Montreal, for their helpful comments.

Notes

1. We avoided successions with unclear passage of power and diluted ownership as these are both different and well explored (Gersick et al., 1997; Handler, 1994; Hugron, 1993).
2. Firms studied included Bata (Tom Jr), Ford (Edsel), Disney (Roy), Schwinn (Ed Jr), Steinway (fourth generation), and Leathercraft (disguised) (all conservative successions); Eaton's (George), Cuddy (Peter and so on), Birks (Drummond), Yamaha (Kawakami), Bingham (Barry Jr), and Sharksboro (disguised) (all wavering successions); and Barneys (Gene Pressman), Seagrams (Edgar Bronfman Jr), Gucci (Maurizio), and Wanton Electronics (disguised) (all rebellious successions).
3. Cases compiled using public sources typically contain little information about dysfunctional family dynamics. Firms we have studied up close and confidentially, however, showed similar succession patterns connected with particular family histories.

References

Aronoff, C.E., Astrachan, J.H. and Ward, J.L. (1996), *Family Business Sourcebook II*, Marietta, GA: Business Owner Resources.

Boszormenyi-Nagy, J. and Spark, G. (1973), *Invisible Loyalties: Reciprocity in Intergenerational Family Therapy*, New York: Harper and Row.

Brady, G. and Helmich, D. (1984), *Executive Succession*. Prentice-Hall, New Jersey.
Davis, P.S. and Harveston, P.D. (1998), 'The influence of family on the family business succession process: a multi-generational perspective', *Entrepreneurship Theory and Practice*, **22**(3), 31–49.
Dyer Jr, W.G. (1986), *Cultural Change in Family Firms: Anticipating and Managing Business and Family Transitions*, San Francisco,CA: Jossey-Bass.
Freidman, S. and Singh, H. (1989), 'CEO succession and stockholder reaction', *Academy of Management Journal*, **32**, 718–44.
Fuchs, P., Mifflin, K., Miller, D. and Whitney, J. (2000), 'Strategic integration: competing in the age of capabilities',*California Management Review*, **42**(3), 118–47.
Gersick, K., Davis, J., McCollom, M. and Lansberg, I. (1997), *Generation to Generation*, Boston, MA: Harvard.
Hambrick, D. and Fukutomi, G. (1991), 'The seasons of a CEO's tenure', *Academy of Management Review*, **16**, 719–42.
Handler, W. (1990), 'Succession in family firms', *Entrepreneurship Theory Practice*, **15**(1), 37–51.
Handler, W. (1992), 'Succession experience of the next generation', *Family Business Review*, **5**(3), 283–307.
Handler, W. (1994), 'Succession in family business: a review of the research', *Family Business Review*, **7**, 133–74.
Helmich, D.L. (1975), 'Leader succession: an examination', *Academy of Management Journal*, **18**, 429–41.
Helmich, D.L. Brown, W. (1972), 'Successor type and organizational change', *Administrative Science Quarterly*, **17**, 371–81.
Hugron, P. (1993), *L'Entreprise familiale*, Institute for Research on Public Policy, Montreal Presses HEC.
Kernberg, O. (1975), *Borderline Conditions and Pathological Narcissism*, New York: Aronson.
Kets de Vries, M. (1996), *Family Business*, London: International Thompson.
Kets de Vries, M. and Miller, D. (1984), *The Neurotic Organization*, San Francisco, CA: Jossey-Bass.
Kets de Vries, M. and Miller, D. (1987), *Unstable at the Top*, New York: NAL.
Kimhi, A. (1997), 'Intergenerational succession in small family businesses', *Small Business Economics*, **9**(4), 309–18.
Kohut, H. (1971), *The Analysis of the Self*, New York: International Universities Press.
Kramer, P. (1995), *Listening to Prozac*, New York: Penguin.
Lansberg, I. (1999), *Succeeding Generations*, Boston, MA: Harvard.
Levine, J. (1999), *The Rise and Fall of the House of Barneys*, New York: Morrow.
Mahler, M., Pine, F. and Bergman, A. (1975), *The Psychological Birth of the Human Infant*, New York: Basic Books.
McQueen, R. (1998), *The Eatons*, Toronto: Stoddart.
Miller, D. (1990), *The Icarus Paradox*, New York: HarperCollins.
Miller, D. (1991), 'Stale in the saddle: CEO tenure and the match between organization and environment', *Management Science*, **37**, 34–52.
Miller, D. (1993), 'Some consequences of CEO succession', *Academy of Management Journal*, **36**, 644–59.
Miller, D. and Droge, C. (1986), 'Traditional and psychological determinants of organization structure', *Administrative Science Quarterly*, **31**, 539–60.
Miller, D. and Friesen, P.H. (1984), *Organizations: A Quantum View*, Englewood Cliffs, NJ: Prentice Hall.
Miller, D. and Shamsie, J. (2001), 'Learning across the life cycle: experimentation and performance among the Hollywood studio heads', *Strategic Management Journal*, **22**, 725–45.
Miller, D. and Toulouse, J.-M. (1986), 'CEO personality and corporate strategy and structure in small firms', *Management Science*, **32**, 1389–409.
Miller, D., Eisenstat, R. and Foote, N. (2002), 'Strategy from the inside out', *California Management Review*, **44**, Spring, 37–54.
Miller, D., Kets de Vries, M. and Toulouse, J.-M. (1982), 'Top executive locus of control and its relationship to strategy making, structure, and environment', *Academy of Management Journal*, **25**, 237–53.
Minuchin, S. (1974), *Families and Family Therapy*, Cambridge, MA: Harvard.
Morck, R.K., Stangeland, D.A. and Yeung, B. (2000), 'Inherited wealth, corporate control, and economic growth: the Canadian disease?', in R.K. Morck (ed.), *Concentrated Corporate Ownership*, Chicago: University of Chicago Press, pp. 319–71.
Morris, M., Williams, R., Allen, J. and Avila, R. (1997), 'Correlates of success in family business transitions', *Journal of Business Venturing*, **12**(5), 385–401.
Neuberg, F. and Lank, A.G. (1998), *The Family Business: Its Governance for Sustainability*, London: Macmillan.
Pitts, G. (2000), *In the Blood*, Toronto: Doubleday.
Reingenum, M. (1985), 'The effect of CEO succession on stockholder wealth', *Administrative Science Quartely*, **30**, 46–60.
Sharma, P., Chrisman, J.J. and Chua, J.H. (1996), *A Review and Annotated Bibliography of Family Business Studies*, Boston, MA: Kluwer Academic.
Shepherd, D.A. and Zacharakis, A. (2000), 'Structuring family business succession: an analysis of the future leader's decision making', *Entrepreneurship Theory and Practice*, **24**(4), 25–39.

Sonnenfeld, J. (1988), *The Hero's Farewell*, New York: Oxford University Press.
Stierlin, H. (1974), *Separating Parents and Adolescents*, New York: Quandrangle.
Ward, J.L. (1997), *Keeping the Family Business Healthy: How to Plan for Continuing Growth, Profitability and Family Leadership*, Marietta, GA: Business Owner Resources.
Wiersema, M. (1995), 'Executive succession as an antecedent to corporate restructuring', *Human Resource Management*, **34**(1), 185–202.
Winnicott, D.W. (1971), *Playing and Reality*, New York: Basic Books.

Appendix A: Means along variables of interest (scale = 1 to 3)

	Conservative ($N = 6$)		Wavering ($N = 6$)		Rebellious ($N = 4$)	
	M	S.D.	M	S.D.	M	S.D.
Strategy						
Clarity and coherency	3.0	0.0	1.3	0.5	1.3	0.5
Consensus	2.7	0.5	1.0	0.0	1.8	0.5
Market scope renewal	1.0	0.0	2.2	0.4	3.0	0.0
Product line change	1.0	0.0	2.0	0.0	3.0	0.0
Change in processes	1.2	0.4	1.8	0.4	2.3	0.5
Risk taking	1.0	0.0	2.2	0.8	3.0	0.0
Market proactiveness	1.0	0.0	2.0	0.0	3.0	0.0
New capability development	1.0	0.0	1.5	0.6	2.0	0.0
Organization						
Value stability and traditions	2.5	0.6	1.8	0.4	1.3	0.5
Goal clarity and consensus	2.8	0.4	1.3	0.5	1.5	0.6
Strength in values	2.7	0.5	1.8	0.8	1.3	0.5
Bureaucracy and rigidity	2.7	0.5	2.0	0.0	1.0	0.0
Centralization of power	2.8	0.4	2.5	0.6	2.3	0.5
Conflicts and factions	1.3	0.5	3.0	0.0	2.3	0.5
Governance						
Old guard power	3.0	0.0	2.3	0.5	1.5	1.0
Board turnover	1.0	0.4	2.7	0.5	2.3	0.5
Managerial turnover	1.4	0.6	3.0	0.0	2.8	0.5

Appendix B: Symptoms within succession patterns

Conservative	Out of six	Wavering	Out of six	Rebellious	Out of four
Strategic stagnation	6	No coherent strategy	6	Unstable strategy	4
Insularity from market	6	Blurred market focus	5	Expanded market focus/ diversification	4
Stagnant market focus	6	Inconsistent decisions	6	Dramatic, risky decisions	4
No product innovation	5	Stop–go initiatives	6	Rapid, daring innovation	2.5
No process innovation	4	Disconnect from market	5.5	Too far ahead of market	3
Traditional culture	6	Conflictual culture	5	Cowboy culture	3.5
Frozen goals	6	Inconsistent goals	5.5	Aggressive goals	4
Weakening values	4	Unclear values	4	Changing values	4
Rigid organization	5.5	Fragmented organization	5.5	Chaotic organization	3.5
Top down management	5	Politicized decision-making	4	New business units	4
Influence of old guard	6	Mix of old and new guard	6	Impulsive decision-making	3.5
Flight of new managers	4	High management turnover	5	Exodus of experienced people	4
Loss of market share	6	Decline in margins and sales	6	High management turnover	4
				Cost overruns, debt, losses	3

21 Towards a business family dynasty: a lifelong, continuing process

Johan Lambrecht and Rik Donckels

Many articles on family business open with the assertion that fewer than 30 per cent of family businesses are passed on to the second generation and that only 10 per cent make it to the third generation (Lansberg, 1999). The average lifespan of a family business is 24 years, which coincides with the number of years that the founder remains at the helm of the business (Welles, 1995). After this time, the business may continue to exist, but the ownership and the leadership do not belong anymore to the family. It is often stated that a family business 'goes to the dogs' in three generations, an observation that is expressed quite baldly in some countries; for example, in Mexico the statement is 'father–entrepreneur, son–playboy, and grandson–beggar' (Davis, 1997). These observations have spurred many researchers and consultants to study succession in family business. Succession and interpersonal family dynamics appeared to be the most frequently occurring subjects in 1998, when the *Family Business Review* drew up a balance sheet of ten years of scientific study on family business (Dyer and Sánchez, 1998). Even today, the impression remains that family business and succession are like a pair of Siamese twins. For example, current seminars about family business and for family business members deal with succession. Attention is focused on a timely succession plan, in which the financial, fiscal–legal, and emotional issues are drawn out and resolved. Planning appears to be the magic formula for succession in family business.

Recently, a new approach to research on family business has emerged. Several researchers have highlighted the fact that there is no connection between planning and successful succession (Aronoff, 1998; Astrachan, 2001; Keating and Little, 1997; Lansberg, 1999; Murray, 2003). According to these researchers, succession rarely involves only an incumbent and a successor. Instead, the process requires the perspective of a multigenerational time frame and takes place in a rich stew of social, cultural, financial, legal, strategic, moral, and other dimensions that resist neat, linear thinking. Our knowledge about how family business is successfully transferred to following generations is still in its infancy (Lansberg, 1999; Sten, 2001).

Family business and succession are often approached from the perspective of ownership (Lansberg, 1999; Ward, 2004). Nevertheless, there are cases where the ownership of the business is no longer in the hands of the family, but the family conducts the day-to-day management. The family resembles then more a business family than the business to a family business. In this chapter, we refer to a family business and a business family when the family holds the ownership and/or the day-to-day management of the business.

The dominant method of research in the family business discipline is to use large-scale surveys among a representative sample of businesses (Bird et al., 2002; Davidsson et al., 2001; Dyer and Sánchez, 1998; Fillis, 2001; Grant and Perren, 2002). To gain insight and be able to offer an explanatory model, a qualitative research approach using (for example) case studies merits consideration.

In this chapter, we want to explain the transfer of the family business to the following generations by using case studies. We wish to offer an answer to the central research question: *how is it that one family succeeds in passing the business down to following generations while another family does not succeed in doing that?* Thus, we are studying the complete transitional path and not just the last stage where the physical, financial, fiscal, and legal aspects of the transfer are taken into account. Moreover, we want to arrive at an explanatory model for transfer to following generations that is not limited to transfer to the next generation. Therefore, we quite intentionally use the plural 'generations' and not the singular 'generation' in the research question.

The chapter is structured as follows: first, we explain the method and work procedures; secondly, we present the frame of reference for the interview topics; thirdly, we present our empirical results; and finally, by way of a conclusion we offer our explanatory model of business transfer.

Method and work procedures

This research arises from a self-enriching process of reading, analysis, observation, interviewing and writing. The reviewed literature includes both scientific and popular articles. The latter recount (among other things) stories of family businesses and business families from Belgium and elsewhere. In addition, we read biographies of business families that are (were) veritable dynasties. We are convinced that their life stories confirm and supplement our case studies. Furthermore, they make a substantial contribution to the understanding of changes between different generations (first to second, second to third and so on). We reviewed biographies of the Ford family (Collier and Horowitz, 2002), the Ochs-Sulzberger family (the *New York Times*) (Tifft and Jones, 1999), the Italian Gucci family (luxury products) (Forden, 2001), the American Bingham family (the *Courier-Journal* and *Louisville Times*) (Tifft and Jones, 1991) and the Vanderbilts (steamship and rail transport) (Vanderbilt II, 1989). Admittedly, these are (or were) all large companies, but they were all very small with a single founder when they began. Moreover, they all have an interesting family and business dynamic.

We drew from these sources to test our own findings. Figure 21.1 illustrates our working method.

We set out in the following our arguments for intentionally making use of case studies. First, Yin (1989) describes a case study as an empirical research method, which, with the aid of multiple sources of evidence, studies a contemporary phenomenon within its real-life context. He believes that case studies are ideal when the borders between the phenomenon and the context are not entirely clear. Thus, case studies lend themselves to answering *how* and *why* questions (Chetty, 1996; Eisenhardt, 1989).

We selected our case studies based on articles in newspapers and periodicals. They provide knowledge about sectors, business families, and family businesses. We did not limit ourselves to family businesses that successfully transferred to the next generations. We also paid attention to businesses where the family influence has disappeared, to avoid becoming fixated solely on success stories. In the composition of the case studies, we considered several elements (see Table 21.1). We looked for businesses of different sizes: all were very small when they began. We studied businesses from the industrial, commercial

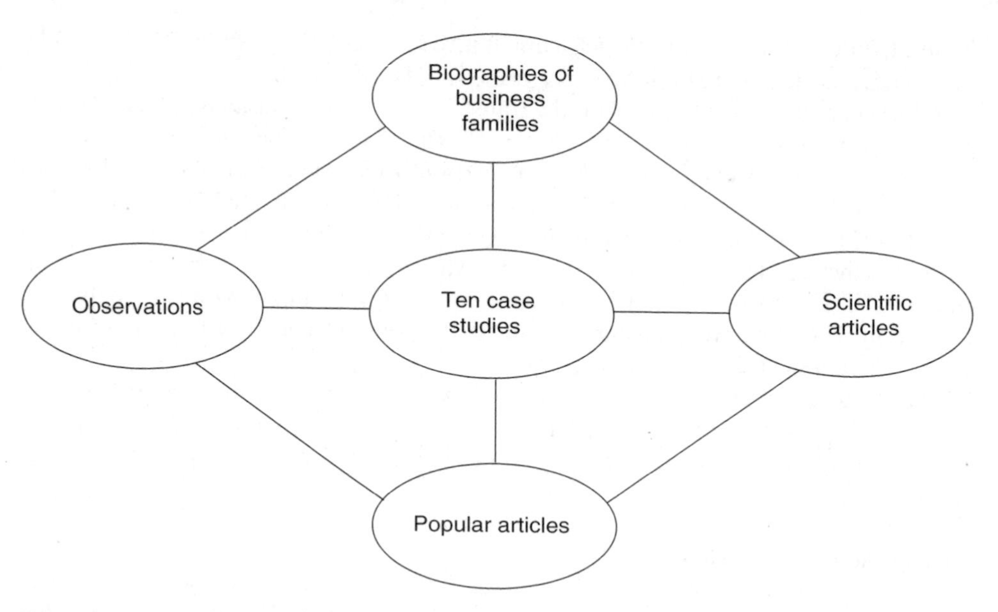

Figure 21.1 Working method

and service sectors. Our objective was to have a number of businesses with the same activity in each sector to provide good comparison. We achieved the objective for the industrial and service sectors. However, the diversity is so great in the commercial sector that we had to study businesses with different activities. We analysed businesses that were in various generations (second, third, . . .). A majority of the businesses transferred ownership and management to the following generation. Nevertheless, we also chose businesses that transferred only management to the next generations (business families) with the ownership of the business no longer belonging to the family. We looked primarily for businesses where several family members were active. Various structures recurred: owner-managed businesses, sibling partnerships and cousin consortiums. External managers were involved in three case studies (Cases 1, 5 and 9). From the selected families, which received a personal letter, two families refused cooperation owing to lack of time.

The criterion for choosing the case studies was not the number but the diversity, as we wanted to explore similarities and differences. Four to ten case studies are deemed sufficient to arrive at an explanation or theory (Eisenhardt, 1989). We set the number of case studies when we reached the point where adding more case studies did not generate any additional knowledge; that is, saturation had occurred. We attained our objective with ten case studies.

We interviewed 19 persons. The number of family members we spoke with in each case is shown in Table 21.1. We had a conversation with both the current and the previous generation where possible. We consciously chose to interview different family members of the same business separately. Our picture of the nature of the family business and the ties with the family could have been clouded if we limited ourselves to interviewing a single person in each case study (Winter et al., 1998). The cases from the literature and our ten case studies provided us with the experiences of 44 national and international business families. To illustrate our qualitative findings, the interviewees are quoted; the source of the quotation (father-transferor, son-successor, . . .) is then mentioned.

Table 21.1 Characteristics of the ten cases

Case	Size class*	Sector	Generation	Number of interviewed family members	Still family ownership	Still day-to-day family management	Generation change(s)
1	Large	Industry	Sixth	1	Yes	No	Founder→?→?→ husband of great-granddaughter of founder→ son→ nephew
2	Very small	Industry	Sixth	2	Yes	Yes	Founder→ son→ son→ son→ 2 sons→ 2 nephews
3 (Sold)	Small	Industry	Fifth	1	No	No	Founder→ son→ 2 sons→ son→ 2 sons
4	Medium-sized	Industry and trade	Fourth	5	Yes	Yes	Founder→ son→ 2 sons→ 2 nephews and niece
5	Medium-sized	Trade	Second	2	Yes	Yes	Founder→ son and daughter
6	Very small	Industry and trade	Sixth	2	Yes	Yes	Founder→ son→ son→ son→ son→ son and daughter
7	Very small	Trade	Fourth	1	Yes	Yes	Founder→ son→ 2 sons→ daughter and son-in-law
8	Small	Industry and trade	Third	2	Yes	Yes	Founder→ son→ son
9	Large	Service	Second	2	No	Yes	Founder→ son and daughter
10 (Sold)	Large	Service	Second	1	No	No	Founder→ 2 sons and daughter

Note: * Very small, less than 10 employees; small, 10–49 employees; medium, 50–249 employees; large, at least 250 employees (in accordance with the European definition).

We derived an explanatory pattern for multigenerational transition in family businesses by using different sources (literature, biographies, interviews and observations). Common characteristics emerged. The key terms are presented (later in this chapter) in a figure to illustrate an explanatory model. A limitation of our design is that the model has not been tested on a larger scale through a survey.

We ensured reliable research results by using four researchers. Except in one case, two researchers conducted the interview. The researchers continuously shared their insights with one another. Because of the absence of a theory, new elements could be identified, thus strengthening reliability. Validity was achieved by using an advisory committee (composed of a representative of the sponsor of the study, a representative of an employer organization and an academic) to give feedback on interim reports, and using different sources.

Frame of reference

The scientific literature review led to a frame of reference (see Figure 21.2). That framework furnished us with the interview topics for the case studies. We present the frame of reference and end this section with the interview questions.

The three main players in the transition of family businesses – the individual, the family and the business – are displayed in Figure 21.2. The dotted lines between the columns make

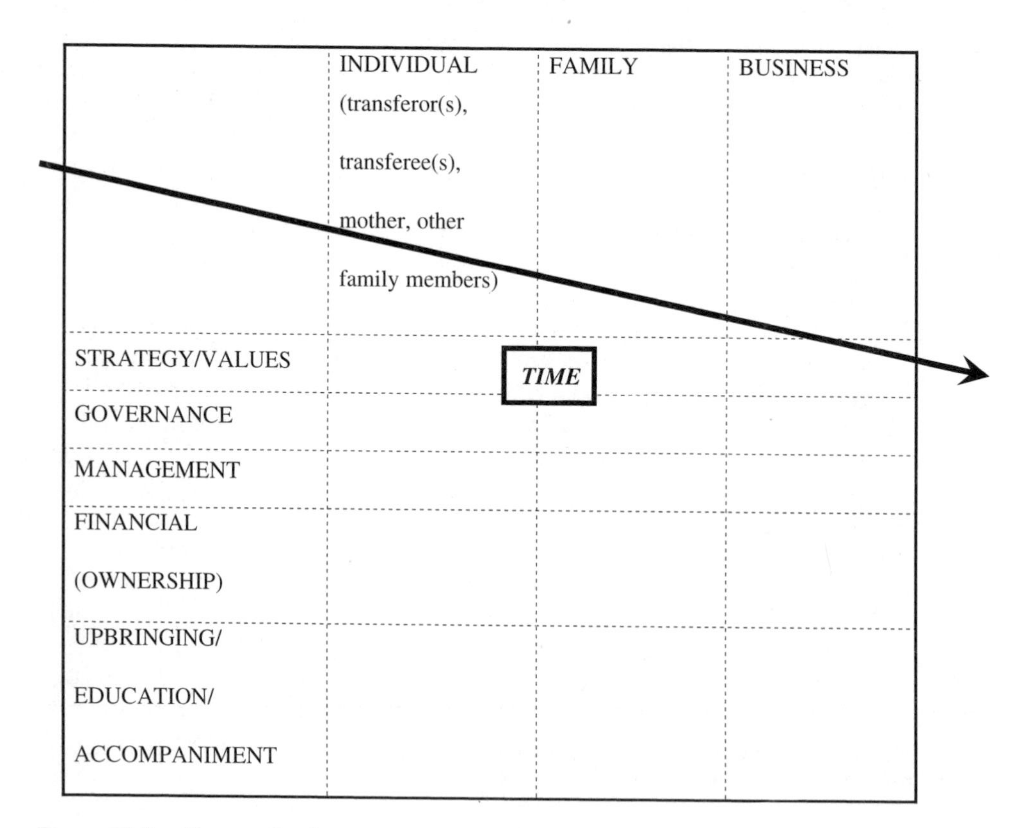

Figure 21.2 Frame of reference

clear the fact that the protagonists are connected with one another. The rows indicate possible levers. Dotted lines are also drawn to indicate that they also influence one another. The main players and the levers can evolve, and this is represented by the time axis.

There is constant interaction between the individual, the family and the business during transfer. Individuals are the transferor(s), the transferee(s) and other family members (for example, the mother, in-laws and family members who are not active in the business). The family in its entirety also plays a leading role. Research on family business must always be done through the lens of the family. Thus, the success of the family business appears to depend partly on family processes and partly on how the family deals with disruptions (Olson et al., 2003). However, the family has been relatively neglected in research. Dyer (2003) draws attention to the fact that the family is missing in organizational research. The business, as the third player, forms the object of the transfer. Some families even regard the family business as a family member (Valera, 2002).

The three main players can have a strategy or can foster values. Thus, the strategy of the mother at the individual level can consist of watching over the shared family dream (Lansberg, 1999). Behind the scenes, she frequently plays a very important role in the business (Muson, 2002). Poza and Messer (2001) distinguish six types of women in family businesses: jealous wife, CEO (chief emotional officer), business partner or co-entrepreneur, employee, guardian of the family values and free agent. At the level of the family, we can explore whether it has a common vision of the future. Lansberg (1999) talks about a shared dream, which derives from the fundamental values and wishes of the family. At the level of the business, we study whether the business strategy is decisive in the transfer process. Ward (2002) emphasizes that the strategy of the family business must be linked to the values of the family.

The board of directors governs implementation of the strategy. Owing to the diptych, family and business, it merits recommendation to govern the family as well, particularly if the family tree has several branches. In this case, a family charter can be drawn up, a sort of family constitution, in which the rights and duties of the family members are set out. Some families hold formal meetings. According to Habbershon and Astrachan (1997), such family councils ensure greater involvement of the family in the business.

Many questions arise from the issue of day-to-day management. Is the manager a family member or not? Is the manager also the owner of the family business? What is the involvement of family members in management? Is management in the hands of one person or a team? Aronoff (1998) finds that a team increasingly performs the management of a family business, especially from the second generation. In a study of 400 family businesses, Lansberg (1999) showed that a majority of owner-managers expressed the wish that several children lead the family business in the next generation. Another issue is the extent to which family values seep through into management of the business. Values are often attributed to the founder of the business (Cappuyns, 2002). These values can then place a permanent stamp on its continued existence.

We think primarily of ownership in examining the financial aspects of a family business. Ownership raises many questions. Who from the family can hold shares? Are ownership and management transferred simultaneously? Does the transferor keep a finger in the ownership pie? Is ownership a team event?

Gallo (2002) says that the upbringing children receive at home is crucial for a future career in the family business. The children learn about the values, the history and the culture of the business. They observe how the family deals with problems that develop over time. Whether the children ultimately embrace or reject the family business depends on the example given by parents, brothers and sisters.

The frame of reference leads us to the following questions for the interviews: how does the family succeed in multigenerational transition, why does the family prefer the business to be continued by the family, how is multigenerational transition prepared and how is the transfer of ownership, management and governance regulated?

Results

This section follows the order of the questions for the interviews. The results derive from our ten case studies and the study of the literature. The study of the literature illustrates and confirms the results. The number of the case study referred to in the following discussion corresponds to the number in the first column of Table 21.1.

How does the family succeed in multigenerational transition?

Five ways for successfully transferring a business to the next generation (in-laws are included as successors) were identified. Several of these occurred concurrently in some case studies.

The first way was when the successor(s) took the lead (Case 4). The succeeding generation took the initiative in continuation of the family business. Because of their interest, motivation and ambition, they viewed themselves as suitable successors: 'I did a traineeship in the family business and realized that I would feel at home in the business of my father and uncle' (daughter-successor in Case 4).

The second way was at the explicit request of the transferor (Cases 1, 3 and 5). The transferors preferred the business to be continued by the family or wanted to work with family members to grow or maintain the health of the business. They asked potential successors directly whether they wanted to join the family business or not: 'My uncle had no children, but wanted to keep the business in family hands. To assure family continuity, he asked me whether I would participate in the daily management of the family business' (nephew-successor in Case 1).

The third way arose from a moral sense of duty among the successor(s) (Cases 1, 7 and 8). Successors reported that they chose to remain involved with the family business not to disappoint the transferor. Furthermore, they would not passively stand by and watch as the family business wasted away or was abandoned. They often had other individual plans: studying, developing their own skills and starting a career. Circumstances (for example, illness of someone in the transferring generation or developments in the sector) caused them ultimately to end up in the family business: 'I chose for the family business, because I saw it was under pressure of evolutions in the market and in the sector. I did not want that the family business would collapse' (son-successor in Case 8).

The fourth way involved predestination (Cases 2, 4, 6 and 9). In about one-half of the family businesses, the successors labeled the transfer as being self-evident. They were tried and tested in the family business. The chosen successors explained that they were 'slipped' or 'sucked' into the business. Generally, in their early years they lent a hand in the business during vacations, weekends and even on school evenings: 'I grew up in the world of the family business. As a child, I walked about in the business and helped my father and uncle. I find it self-evident that I work here. Due to the family tradition in the sector, I have beer in my veins instead of blood' (son-successor in Case 2).

The fifth way was transferors giving possible successors an indirect soft push from behind (Case 6); for example, influencing the studies of possible successors. Unlike the second way, the transferors do not ask possible successors explicitly in the fifth way of transfer. They influence possible successors indirectly. Transferors do not want to coerce their children, unlike the practice in earlier years. Transfer cannot be forced, and coercion only serves to discourage the transfer.

Why does the family prefer the business to be continued by the family?

We identified three reasons why the interlocutors found it important for the family to continue the business.

The first reason was the fulfillment of values. Three values formed part of the family heritage: wanting the best for the work team (Cases 1 and 9), love for the product (Case 3) and independence (Case 5). In Case 5, 'independence' was one of the arguments of the father: 'I made it clear to my son that it is better to be a "little" boss than to be a "big" employee'.

The second reason was preservation of the family name and a reassuring feeling (Cases 4, 5, 6, 9 and 10): 'You create something and you want that the own color remains. As the children are a continuation of yourself, you want that they succeed you' (father in Case 10). The family name has great emotional and symbolic significance. It stands for the rich history of the business and the family, the achievements, heroic deeds and sacrifices of the previous generations, and the authenticity and excellence of the product. Because the family name stands on the product, the family feels responsible for it. Hence, the family name summons up a feeling of pride. This is made clear in the promotional material of family businesses (for example, websites and brochures) and in the business buildings. They direct attention to the history of the family business, its founders and the family name.

The third reason was exploitation of the advantages of a family business, such as long-term vision, versatility, life engagement and the rich family history as a sales argument, to retain the influence of the family on the business (Cases 4, 5, 8 and 9).

How is multigenerational transition prepared?

The business families made it clear that the gateway to the family business cannot be entered with seven-league boots. We distinguished six stepping-stones to the transfer of the family business: interpreneurship, studies, formal internal education, external experience, official start in the family business (beginning at the bottom of the ladder, freedom for and by the successor(s)), and written plans and agreements (see Figure 21.3).

The first stepping-stone was interpreneurship, which stands for the transfer of professional knowledge, management values, entrepreneurial characteristics and the soul of the family business to following generations. We distinguished three life stages of the child that influenced the transfer of professional knowledge (Cases 1, 6, 8 and 9). The business was a playground and a plaything up to age 11. Possible successors performed light activities in the family business (during weekends and vacations) from age 11 to 15. They performed work that was more serious in the family business from age 15 to 17. In this way, potential successors at a youthful age learned the secrets of the product and the tricks of the trade. It was important that the child not be a slave of the business and that the transferring generation not transmit stress from the business to the child. Otherwise, the young family members would experience the family business as a yoke, which they would later want to

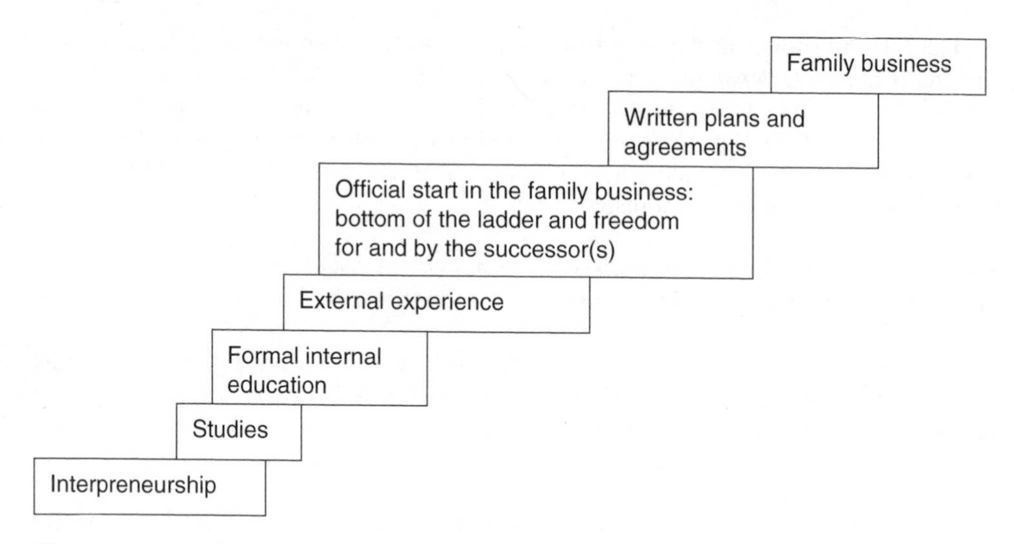

Figure 21.3 Six stepping-stones to the transfer of the family business

shake off. In other words, the transferring generation must display enthusiasm. Management values were transferred via upbringing, where the mother often played the major role. Honesty and respect were family values honored in dealing with personnel (Cases 5, 7 and 8). The family values of honesty, partnership and restraint punctuated commercial policy (Cases 7 and 9). Hard work, perseverance and networking were entrepreneurial characteristics central in childrearing (Cases 6 and 7). Symbols such as the family home, portraits and the recurrence of the first name of the founder in every generation were conduits for the soul of the family business. They forged the relationship between the family and the business and between the different generations (Cases 4 and 8).

Studies formed the second stepping-stone (Cases 1, 4, 8 and 9). Most successors earned an advanced degree *before* full-time entry into the family business. In a number of cases, the studies were oriented towards the sector of the family business. In other cases, potential successors were free to choose a discipline. Often, the diploma was a condition for being able to work in the family business: 'The successor has to obtain a diploma from university. In this way, the successor can prove that he/she deserves the family business' (son-successor in Case 4).

The literature demonstrates that larger family businesses sometimes provide formal internal education for family members at a young age (Bibko, 2003; Stern, 2003a, 2003b; Tifft and Jones, 1999). This formed the third stepping-stone. They learn about the figures of the business, its future and its values during organized training sessions and attendance at meetings of the board of directors. Along with learning about the conduct of the family and the business, this formal internal education seeks to identify the brighter ones and to allow directors to become familiar with potential future managers. The mentor can be either a family member or a non-family member. A transferor explains the importance of formal internal education: 'My grandchildren are still young: the oldest is ten. In twenty years, they can be ready for the family business. However, I believe strongly in education for them about the business. Therefore, we shall organize meetings for them with the family and with external people' (Bibko, 2003).

The fourth stepping-stone was the acquisition of outside work experience in other companies, whether or not abroad (Cases 1, 2, 3, 4 and 6). Seven reasons were mentioned for taking this fourth stepping-stone of external experience: gaining self-confidence, enhancing management and professional knowledge, gaining worldly wisdom, increasing motivation, gaining trust and admiration of the transferor, avoiding the thought that the grass is greener on the other side and contributing to the credibility of the family member as a successor.

The fifth stepping-stone occurred with the official start in the business. We distinguished between *beginning at the bottom of the ladder* and *freedom for and by the successor(s)*. Before the succeeding generation held a management position, it generally passed through the various departments in the business (Cases 5, 6, 8 and 9). In this way, the successors proved themselves, won the confidence of employees and discovered the business, the sector and the customers. Leadership by the succeeding generation meant the acquisition of credibility. Both transferors and successors underscored the importance of freedom for the new generation when they officially started in the family business (Cases 3, 5, 8 and 9). Successors must receive the necessary breathing and maneuvering room from transferors to learn from their mistakes, to give innovative impulses and to discover who best assumes the respective responsibilities. Why is freedom for the successors so crucial? First, it avoids the disappearance of the family business. Case 3 is no longer a family business, because the father could not let go: 'Letting go was not possible for father. He continued to control the family business, was worried when there was a problem in the business, etc. He wanted that his sons steered the business, but he could not let go the steering wheel' (son-successor in the fifth generation). Secondly, freedom for the successors increases the probability on good cooperation and paves the way for successful transfer to the following generations. Successors indicated that by receiving room to move, they in turn learned to give room, which already smoothes the path to a following successful transfer. In addition, it facilitates team management. Thirdly, if the successors were held tightly in the transferors' grip, they could act destructively when they suddenly gained their freedom. Some would not even recoil from dragging the business down with them (as occurred with the Gucci's (Forden, 2001)). While the acquisition of credibility is identified with leadership by the succeeding generation, leadership means the ability to let go for the transferring generation. Transferors who are masters in the art of letting go significantly increase the chance of a successful transfer. Successors in Cases 8 and 9 described their masters as 'Buddy' or 'Big Manitu'. The son in Case 9 gave evidence:

> Father was never a boss towards us. He enabled us to take responsibility. He never said 'I shall show you how I do it'. If we had questions, we could address these to him. In this way and according to our ambitions, we could grow and learn. Father has done a great job.

Freedom for the successor(s) entailed taking responsibility, respecting the previous generations, asking for advice from the transferor(s), and understanding that the past denotes the foundations and provides a lead to the future.

Stepping-stone six related to written planning and agreements, with the regulation of the legal, financial and fiscal issues. There must be an eye for measures in the event of doomsday scenarios, such as the death or resignation of a family member (Cases 4 and 8). A a successor in Case 4 made clear, it guarantees reassurance and family harmony: 'Paying attention to these doomsday scenarios was not easy and unpleasant. I still have the shakes if I think of that exercise. But the clear official regulation is a kind of "life insurance" for the family business'. Written plans were not an absolute guarantee for a successful transfer, but poor planning could prove costly for the business and the family. For the sake of family peace, a taboo might rest upon timely written agreements. This was the experience in Case 3, that has been sold:

> Aspects like ownership, management, function, objectives, etc. should be regulated in a clear and formal way. This was never the case in our family business. Agreements were only made orally, never on paper. I thought that everything would be settled after a couple of years, but the agreements that were not made returned like a boomerang (son-successor in the fifth generation).

How is the transfer of ownership, management and governance regulated?
Transition of a family business entails transfer of ownership, management and governance. Therefore, it is necessary to study how families in business have regulated these transfers.

For the sake of financial independence, most families wished to keep ownership of the business in their own hands (Cases 1, 7 and 8). Because of the many sacrifices made for the business, loss of ownership would be emotionally intolerable. However, there are families for which the possession of shares is not an absolute requirement. In Case 9, the family no longer owned the business, but the son and daughter constituted the top management. Because of market conditions (internationalization of the sector and imminent elimination by several big players), the family saw itself obliged to abandon ownership gradually. It is important to note that they did so in the interest of the business. In all cases where the family still held the ownership of the business, only the active family members were shareholders. In this way, they wished to prevent inactive family members collecting dividends and profiting from the work of the active family members. In most of the cases, the successors were co-shareholders at the time of their official entry into the business. In this way, the successors immediately became a part of the family business: 'In this way, the children realize that the family business is part of them. In addition, they are then compensated for the work they have carried out in the business' (father-transferor in Case 9). Gradual expansion of the shareholding served to motivate them.

In most of the cases, the day-to-day management was in the hands of the family. In a number of cases, the day-to-day management was shared with outside managers. Sometimes these outsiders remained until the new generation was ready to run the family business (Case 1). In Case 5, a clear distinction was made between transfers of ownership and management: 'There is a big difference between ownership and management. I want to transfer ownership to my children, but transfer of day-to-day management is dependent on the ability of the succeeding generation. The interests of the business come first' (son-successor in Case 5). Day-to-day management remained in the hands of several family members in many cases (see Table 21.1). The business families found that a clear delineation of responsibilities was necessary, often functionally (commercial, technical, and so on). Delineation of responsibilities requires open communication to avoid separate kingdoms developing.

Only Case 5 had outsiders on the board of directors. It was primarily the transferring generation that was reticent about outside directors, supposedly because they had insufficient knowledge of the sector. The younger generation was more open to the idea of outside directors, because they can counter business blindness and can promote cross-fertilization with knowledge from other sectors.

Conclusion: towards a new explanatory model

From the results of the case studies, we derived an explanatory model for transferring the family business (leadership, ownership or both) to following generations. Figure 21.4 presents the principle of sound governance on which a family dynasty is based: the understanding that the individual belongs to the family, which belongs to the business. The new explanatory model replaces the conventional representation with three overlapping circles: business, family and ownership. The conventional model misses the fact that there are business families, which no longer have ownership but still have day-to-day management. Moreover, in the conventional model management or governance can just as well replace ownership. The keywords in the three concentric circles in our explanatory model are the levers that can move the family business to further generations and are derived from the use of a variety of research sources. The time axis indicates that the individual, the family and the business, and the interaction between them, are not static. Because they evolve constantly, they form a dynamic whole.

From this study, we conclude that transfer of the family business to following generations is a lifelong, continuing process. Planning, whereby the financial and fiscal–legal issues are dealt with, is an intrinsic part of that process. It is a necessary but insufficient

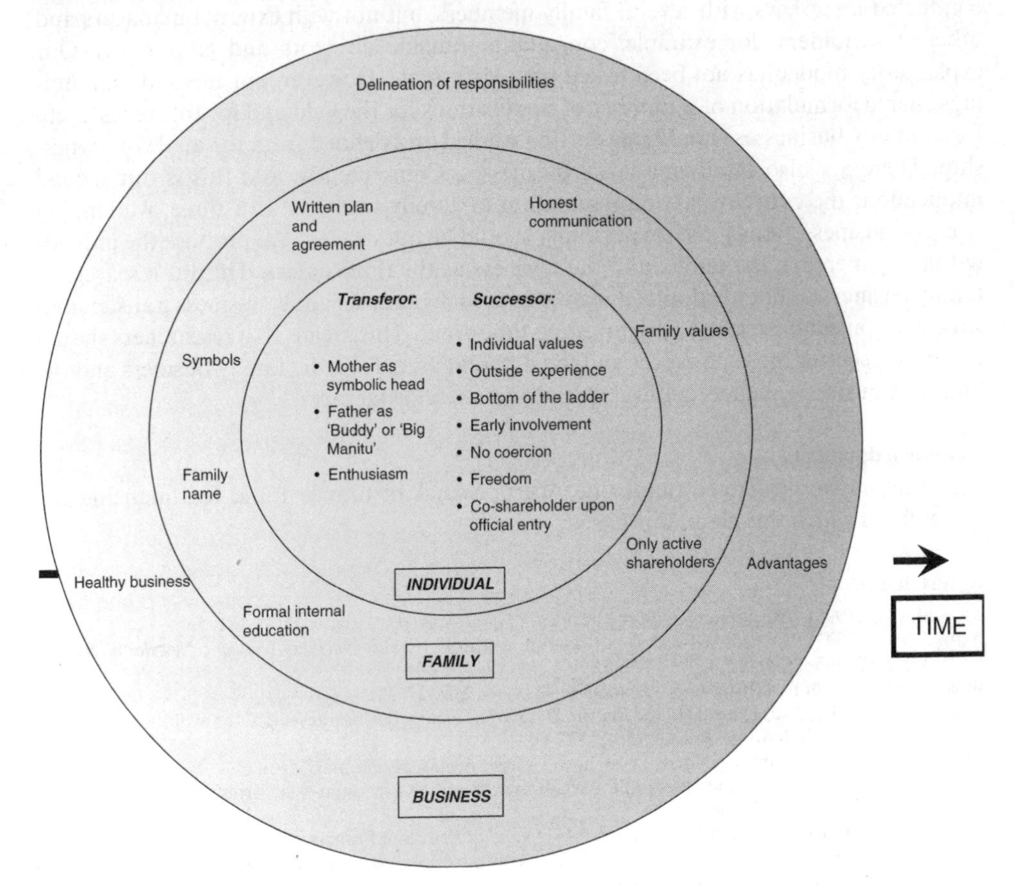

Figure 21.4 An explanatory model for transfer to further generations

condition. Moreover, specialists can be hired to work out financial and fiscal–legal regulations. The other soft elements of the transfer process – interpreneurship, freedom, values, outside experience, upbringing, education – must be addressed and fostered by the family as part of the process. Thus, succession is not about a process that can be tied up in a fixed time frame (for example, three to five years before the transferor lets go). It starts much earlier and never ends. Succession is not about transfer from one generation to the next generation repeated in discontinuous time intervals of about 25 years (the duration of one generation); it is about continuous transfer to further generations (plural). The stepping-stones – interpreneurship, studies, formal internal education, external experience, official start in the family business (beginning at the bottom of the ladder, freedom for and by the successor(s)), and written plans and agreements – pave the way for the transfer of the business to further generations and to a family dynasty. During the process, families in business should be guided by the principle of sound governance: the individual family member belongs to the family, which belongs to the business. Business advisers and business families should be more aware of this principle of sound governance and of the transfer of the family business as a lifelong, continuing process.

We are aware of a number of limitations in the research. In our case studies, we conducted interviews with several family members, but not with external managers and other stakeholders (for example, consultants, outside directors and employees). Our explanatory model has not been tested on a large scale. However, our method and findings enable formulation of a number of implications for the field and for future research. First, family businesses should not only be studied and defined from the angle of ownership. There are also family-managed businesses. Consequently, and this is our second implication, the conventional representation of family business with three overlapping circles – business, family and ownership – should be questioned. We propose the individual family member, the family and the business as the three circles. Thirdly, research on family businesses should adopt a dynamic approach because family business transfer constitutes a continuing process and not a one-time event. This means that researchers should take into account the time aspect and the different aspects of the family business and the family in business, such as cultural, social and strategic factors.

Acknowledgements
The authors would like to thank the Acerta Social Insurance Fund for financing the research on which this chapter is based.

References

Aronoff, C.E. (1998), 'Megatrends in family business', *Family Business Review*, **11**(3), 181–5.
Astrachan, J. (2001), 'Critical self-review will mature research', *Family Business Network Newsletter Special Conference Edition*, (28), 22–4.
Bibko, S. (2003), 'Off to a flying start', *Families in Business*, **2**(2), 27–31.
Bird, B., Welsch, H., Astrachan, J.H. and Pistrui, D. (2002), 'Family business research: the evolution of an academic field', *Family Business Review*, **15**(4), 337–50.
Cappuyns, K. (2002), 'Family business know-how: values', *Families in Business*, **1**(6), 65.
Chetty, S. (1996), 'The case study method for research in small- and medium-sized firms', *International Small Business Journal*, **15**(1), 73–86.
Collier, P. and Horowitz, D. (2002), *The Fords: An American Epic*, San Francisco, CA: Encounter.
Davidsson, P., Low, B. and Wright, M. (2001), 'Editor's introduction: Low and Macmillan ten years on: achievements and future directions for entrepreneurship research', *Entrepreneurship Theory and Practice*, **25**(4), 5–17.
Davis, J. (1997), 'The best practices of successful family businesses', working paper.

Dyer, W.G. Jr (2003), 'The family: the missing variable in organizational research', *Entrepreneurship Theory and Practice*, **27**(4), 401–16.
Dyer, W.G. Jr and Sánchez, M. (1998), Current state of family business theory and practice as reflected in *Family Business Review* 1988–1997', *Family Business Review*, **11**(4), 287–95.
Eisenhardt, K. (1989), 'Building theories from case study research', *Academy of Management Review*, **14**(4), 532–50.
Fillis, I. (2001), 'Small firm internationalisation: an investigative survey and future research directions', *Management Decision*, **39**(9), 767–83.
Forden, S.G. (2001), *The House of Gucci: A Sensational Story of Murder, Madness, Glamour and Greed*, New York: HarperCollins.
Gallo, M.A. (2002), 'Preparing the next generation: education', *Families in Business*, **1**(6), 62–3.
Grant, P. and Perren, L. (2002), 'Small business and entrepreneurial research: meta-theories, paradigms and prejudices', *International Small Business Journal*, **20**(2), 185–211.
Habbershon, T.G. and Astrachan, J.H. (1997), 'Research note, perceptions are reality: how family meetings lead to collective action', *Family Business Review*, **10**(1), 37–67.
Keating, N.C. and Little, H.M. (1997), 'Choosing the successor in New Zealand family farms', *Family Business Review*, **10**(2), 151–71.
Lansberg, I. (1999), *Succeeding Generations: realizing the dream of families in Business*, Boston, MA: Harvard Business School Press.
Murray, B. (2003), 'The succession transition process: a longitudinal perspective', *Family Business Review*, **16**(1), 17–33.
Muson, H. (2002), 'The Houghtons of Corning', *Families in Business*, **1**(6), 71–5.
Olson, P.D., Zuiker, V.S., Danes, S.M., Stafford, K., Heck, R.K.Z. and Duncan, K.A. (2003), 'The impact of the family and the business on family business sustainability', *Journal of Business Venturing*, **18**(5), 639–66.
Poza, E.J. and Messer, T. (2001), 'Spousal leadership and continuity in the family firm', *Family Business Review*, **14**(1), 25–36.
Sten, J. (2001), 'Examining loss and replacement in business families: toward a conceptual framework on transfers of family businesses', paper presented at RENT XV, Research in Entrepreneurship and Small Business, Turku.
Stern, M. (2003a), 'Heraeus', *Families in Business*, **2**(3), 31–3.
Stern, M. (2003b), 'Bridging the bloodline', *Families in Business*, **2**(3), 35–7.
Tifft, S.E. and Jones, A.S. (1991), *The Patriarch: The Rise and Fall of the Bingham Dynasty*, New York: Summit.
Tifft, S.E. and Jones, A.S. (1999), *The Trust: The Private and Powerful Family Behind The New York Times*, Boston: Little, Brown and Company.
Valera, F. (2002), 'The next generation', *Families in Business*, **1**(6), 46–50.
Vanderbilt II, A.T. (1989), *Fortune's Children: The Fall of the House of Vanderbilt*, New York: Perennial.
Ward, J.L. (2002), 'How family vision drives business strategy', *Families in Business*, **1**(6), 67.
Ward, J.L. (2004), *Perpetuating the Family Business*, New York: Palgrave Macmillan.
Welles, E.O. (1995), 'There are no simple businesses anymore', *Inc. Magazine*, May.
Winter, M., Fitzgerald, M.A., Heck, R.K.Z., Haynes, G.W. and Danes, S.M. (1998), 'Revisiting the study of family businesses: methodological challenges, dilemmas and alternative approaches', *Family Business Review*, **11**(3), 239–52.
Yin, R.K. (1989), *Case Study Research: Design and Methods*, London: Sage.

22 Using the strategic planning process as a next-generation training tool in family business

Pietro Mazzola, Gaia Marchisio and Joseph H. Astrachan

Introduction

The extant family business literature on the topic of generational succession has enabled researchers and practitioners to disseminate best practices to the family business community. There is consensus, for example, that generational transfer is an ongoing process and not a single event. As a part of this process, power, by way of management, ownership and control, is handed over from one generation to the next. There is also agreement that this process occurs in stages, and that critical moments can be identified, and that both can influence the behaviour of a variety of stakeholders.

In this research, we focus our attention on the next-generation family members who have entered the family company. These future leaders have overcome the decision stage, completed relevant studies and have ideally gained work experience outside the family company before taking up their, usually middle-management, positions.

Therefore, the present chapter attempts to assist families in business overcome the strains of succession. For the most part, literature on succession, concentrates on the stages that precede entry into the company and, in particular, on the conditions that influence the decisions of the younger generation regarding the most appropriate choices of university and work experience. What happens to the young members after they enter the company remains an area to be studied in greater depth.

To address that gap, we focus on the first phase after entry into the company, a period that can have differing temporal lengths, and propose a strategic planning tool. We suggest that, under certain conditions, the strategic planning process itself can be used as a training tool for the next-generation members.

Even if the real impact of strategic planning on firm performance is still inconclusive (Brews and Hunt, 1999), the strategic plan is one of the most wide spread managerial tools used within companies (Grant, 2003). Most organizations periodically engage in a process of strategic planning; although a new organizational vision might not be generated every year, contemporary budgeting norms and practices call for an annual assessment of performance goals (Ketokivi and Castañer, 2004, p. 340).

The objective of this chapter is to study whether the strategic plan presents particular benefits in the case of family business, in the particular phase of the succession process. Theoretically the role of the strategic planning as communication tool, consensus gathering and learning effects (Mintzberg, 1994b) and as an integrative device (Ketokivi and Castañer, 2004) has been acknowledged and investigated empirically (Langley, 1988). What has not yet been investigated in depth is whether and how such strategic planning process (whose outcome we assume here is the strategic plan) contribute differently to family business, and more specifically during the succession.

Family business literature is generally dominated by the issue of succession planning.

Poza (1995) affirmed that strategic planning, by helping to promote both appropriate new growth of the business and common ground in the family, can be more supportive of business continuity than traditional succession planning (Zahra, 2005)

Our findings support the idea that strategic planning process, besides producing known benefits at (family) firm level (Barringer et al., 1998; Chrisman et al., 2003; McCann et al., 2001), can also play an important role during the succession process at two levels.

First, the strategic plan can help next generation in terms of:

1. content, and in particular in helping young members in acquiring knowledge in a systematic way, thanks to the internal and external learning through collecting spread knowledge (Brews and Hunt, 1999), and to the effort of quantifying phenomena (Grant, 2003; Mintzberg, 1994);
2. interpersonal relationships, both with family and non-family members, through improving legitimization before family and non-family external stakeholders, and through allowing executives and directors to evaluate the job done, above all in terms of strategic intent (Grant, 2003; Mintzberg, 1994).

Secondly the strategic plan, under certain conditions, maybe be helpful in conflict solution. Next-generation involvement during strategic planning and its use in sharing goals, risks and needs, facilitates the overcoming of family position bias and creates collaboration and consensus among owning family members. In this respect, strategic planning works like an important integration device (Grant, 2003; Ketokivi and Castañer, 2004).

This chapter is organized in five sections, including the introductory section. The second section reviews previous studies on succession in family business and strategic planning process. The third provides information on the research methodology, the sample and the variables. The fourth section illustrates the most important findings, while the fifth offers some concluding remarks and contributions for theory and practice.

Theoretical background

Succession passage

Authors agree in considering the succession passage as the most important topic with which family businesses have to deal, and it is one of the more deeply analysed topics in the research literature. In a family business, succession is the passage of the 'leadership baton' from a generation to the following (Beckhard and Burke, 1983). Alcorn (1982) specified that succession is referred to as the changes at the top of the organization. Since the topic is so widely analysed, many perspectives have been posited, with the following being the most pertinent to the topic presented in this chapter.

There is a group of studies on succession passage whose object is the succession as a process (Farquhar, 1989; Friedman, 1987; Gabarro, 1979; Gilmore and McCann, 1983; Vancil, 1987), research that deals with the subsequent phases of this process (Gordon and Rosen, 1981), and research on the specific problems with each (McGivern, 1978).

Barach and Gantisky (1988) introduced the concept of strategy for young family members within the family company.

Ward (1987) emphasizes the characteristics of successful succession passage, introducing the concept of planning as a condition for a successful and smooth succession passage.

Many authors then underlined the importance of planning the entire succession process (Kimhi, 1997; Lansberg, 1988; Morris et al., 1997) and identified the benefits that planning can bring in terms of making the succession process easier and of facilitating relationships among those involved in the process.

Looking at this review of the available literature on succession passage, it seems evident that the focus has been mainly on the process, and in particular on the importance and the benefit of succession planning process (Davis, 1983; Handler, 1994; Upton and Heck, 1997; Ward, 1987).

Within this process a still underexplored relevant issue concerns the growing path of next-generation family members once they enter the company.

As far as this phase is concern, in the literature there are some evident open issues, for example:

- attaining legitimacy from the incumbents, family members and owners, non-family executives and other external stakeholders. Firms are embedded in a social context and next-generation entrepreneurs within family firms are inserted into an existing network structure. They assume positions by virtue of who they are. Within these networks, legitimacy partly is conferred and partly has to be earned (Steier, 2001). Again, incumbents are often reluctant to let the next generation join in the decision-making process of the business (Handler, 1989; Lansberg, 1988);
- the transfer of (tacit) knowledge from one generation to the following (Cabrera-Suárez et al., 2001; Steier, 2001).

This study aims to fill these two gaps in the literature by exploring the role of strategic planning process as a training tool itself for next-generation family members, able to solve, among the others, these issues.

Strategy and strategic planning process

The issues mentioned above can find a possible solution by taking part in a strategic planning process. Strategy is the positioning a company occupies in its environment (Porter, 1996). This positioning emerges as a consequence of decisions and behaviours of management, some of which are not planned (Mintzberg, 1994b). However, an effective strategy and a high and careful level of operating efficiency are not sufficient to ensure durable success. For success to be durable, the company must also make an effort to systematically manage its strategy realizing a strategy process (Foster and Kaplan, 2001). That is, it must observe, reflect, conceptualize, and experiment to find answers to crucial questions concerning the actual, or desired, identity of the company. And, importantly, then adopt the most appropriate actions to close the gap between the actual and the desired situation (Coda and Mollona, 2001).

It is a widely shared assumption in literature that the strategy process involves a consensus-building process (Dess and Oringer, 1987; Lyles, 1981; Nielson, 1981) during which organizational members develop a general level of agreement on the content (Markóczy, 2001, p. 1013) and on the 'fundamental priorities of the organization' (Floyd and Wooldridge, 1992, p. 28). The importance of consensus formation has been suggested both in the strategic decision-making process as well as in the implementation process (Markóczy, 2001, p. 1013).

Regarding the necessity of creating a consensus during strategic decision-making, Whyte (1989, p. 41) emphasized that the 'task, after all, of all decision making groups is to produce consensus from the initial preference of its members', while on the latter issue, Floyd and Wooldridge (1992, p. 27) assess that 'successful execution [of strategy] means managers acting on a common set of strategic priorities'. This is achieved through the development of some shared understanding and common commitment, namely by the formation of 'strategic consensus' (Markóczy, 2001, p. 1014).

Several decades of strategy research and practice show that two themes have consistently surfaced in literature: discipline and imagination. Discipline is the consistent application of rules to evaluate the full set of given alternatives (Szulanski and Amin, 2001, p. 541). In this sense the main role of a strategist is to collect information diligently, develop alternatives and choose the one that maximizes the value.

On the other hand, imagination is a central driver of evolution. According to Weick, (1989) an imaginative strategy-making process exhibits 'deliberate diversity' in the way problems are defined, alternative solutions are generated, and the rules by which they are selected (Szulanski and Amin, 2001, p. 543). It is imagination which generates a large variety of distinct options in response to each formulation of the problem (Weick, 1989). The notion of deliberate diversity determines not only a large number of alternatives but also differences among them. In fact, alternatives must be varied and distinct rather than mere variations on the same theme. Diversity affects the cognitive frames used to examine and define the problem (Schoemaker and Russo, 1996). Deliberate diversity also affects the number of rules used to select from alternatives. Hamel and Prahalad (1989) exhort explicitly for imagination in every aspect of the business, calling for the creation of new rules, and not just breaking existing ones. The strategy-making process should then be led not by those with the most experience, but by those with the greatest ability to envision the future (Hamel and Prahalad, 1989).

Discipline and imagination meet then in the strategy-making process. When dealing with strategy-making process we refer to the following four distinct moments, which result from that activity: realized strategy, strategic intents, action plans and emerging strategies. Following a brief explanation for each:

1. Realized strategy is the simultaneous and structural positioning that has taken place as the result of past decisions and actions and that consolidates over time once a structure is created and operating mechanisms and a coherent corporate culture are in place (Mintzberg, 1994b).
2. Strategic intents are the explicit directions and the intents declared by management with regard to the company's mission, its management philosophy, the field of activity it has chosen, the size it wants to achieve and the role it wants to play in the competitive arena and towards its main social stakeholder (Burgelman, 1983; Hamel and Prahalad, 1989; Mintzberg, 1990; Quinn, 1981). Because strategic intents reflect the company's long-term orientation, they generally have a certain stability over time and a relevant influence on the future results of the company.
3. Action plans: there can be a gap between realized strategy and strategic intent that can be closed only by carrying out the appropriate action plans. These are actions that, on the whole, aid the implementation of intentional strategies and thus make closing of the strategic gap possible (Brews and Hunt, 1999).

4. Emerging strategies are models of equilibrium that arise 'from the bottom up' – and are, in this sense, opposed to the models developed and achieved according to the top down logic – as the result of the learning acquired, at the individual and collective level, by people active in different positions in the operations management of a business unit or a corporate function. Day-to-day operations can, in fact, allow a person to see opportunities that are not as evident at the governance level and to suggest original solutions for unforeseen problems and threats (Mintzberg, 1994b).

Within this process of strategy-making, strategic planning takes place. Indeed, the strategic planning movement was the first broad effort to instill discipline in the strategy-making process and concurs in forming strategic intent.

In particular strategic planning provides:

- information on the realized strategy, including its clarification and articulation, and the company's past performance (Langley, 1988; Mintzberg, 1994b);
- a view of the external environment. This typically includes guidance relating to some features of markets in the planning period that are not so much forecasts as a set of assumptions relating to the environmental framework (Grant, 2003);
- information on the 'ends' which can be hierarchically ranked from the broad, higher level ends, that is, the 'grand design' (Granger, 1964), the mission or 'strategic intent' (Hamel and Prahalad, 1989) of an organization, to the lower-level, more limited and specific operational objectives or goals (Brews and Hunt, 1999);
- information on the 'means', conceived as the patterns of action through which organizational resources are allocated. This type of information typically includes action plans, programs and resource allocation activities (Brews and Hunt, 1999);
- simulations of the possible effects of the plans and forecast building (Mintzberg, 1994b). In order to foster a better understanding of the corporate performance drivers, strategic plans may include maps that show the causal relationships and the links through which specific improvements create the desired outcomes (Epstein and Westbrook, 2001; Kaplan and Norton, 2000).

However, management scholars have heavily criticized the effectiveness of strategic planning in the process of defining strategies, and observed how environmental uncertainty and the cognitive limits of planners tend to hamper the accuracy of forecasts, favoring logical incrementalism rather than formal planning, especially in unstable environments (Mintzberg, 1994b).

Besides, strategic planning can perform an important role, acting as catalyst for consensus inside and outside the company (Langley, 1988; Mintzberg, 1994a) and can:

1. facilitate internal and external learning through collecting spread knowledge (Brews and Hunt, 1999);
2. offer a context for strategic decision-making by applying methodologies and technique and facilitate communication and dialogue (Grant, 2003);

3. help legitimization before external stakeholders: impressing or influencing outsiders through the disclosure of relevant and complete set of information (Higgins and Diffenbach, 1995; Langley, 1988);
4. reduce position bias and enhance goal convergence thus acting as an integrative mechanism (Grant, 2003; Ketokivi and Castañer, 2004);
5. stimulate effort of quantifying phenomena (Grant, 2003; Mintzberg, 1994b);
6. allow executives and directors to evaluate the job done, above all in terms of strategic intent (Grant, 2003; Mintzberg, 1994b).

All these benefits seem to fit properly with the open issues highlighted above regarding the post-entrance phase for family next-generation members in the succession passage.

Research method

In order to increase our understanding of the specific benefits of the strategic planning process in family business during succession, we focused on those processes in which next-generation members took part.

Methodological approach and research process

Since there is little knowledge of the role of strategic planning in succession passage and the concept itself needs to be defined more precisely, we decided to use an interpretative paradigm using grounded theory (Glaser and Strauss, 1967; Strauss, 1987; Strauss and Corbin, 1990). Indeed, a distinctive feature of grounded theory compared with other research methods is that it is explicitly emergent (Glaser, 1994): the aim is not to test hypotheses, but to find what theory accounts for the research situation as it is. In this respect it is like action research: the aim is to understand the research situation; to discover, as Glaser in particular states, the theory implicit in the data (1994).

Following grounded theory, generating a theory is a process of research requiring data collection, coding and analysis. The three operations should blur and intertwine continually, from the beginning of the investigation to its end (Glaser and Strauss, 1967).

Grounded theory begins with a phenomenon observed which then is further explored in a research situation. In our case this is the question of whether and how strategic planning has an idiosyncratic benefits during succession in family business. Within this situation, we try to understand the underlying processes. Grounded theory suggests an approach of observation, conversation and interview. Constant comparison is at the heart of this process. First, cases have to be coded and then compared against each other. The task is to identify categories and their properties. As the categories and properties emerge, they and their links to the core category provide the theory.

Sampling

Grounded theory uses theoretical sampling. Cases are not randomly chosen from the population of interest, but are chosen on purpose so to increase the diversity of the sample in the search for different properties (Eisenhardt, 1989b; Pettigrew, 1990; Yin, 1994).

Grounded theory states that if the information gathered on these groups begins to be saturated (redundant information), one has to look for cases with different characteristics until all necessary information to build the theory is collected. Therefore the size of the sample is basically irrelevant in Grounded theory research.

Table 22.1 The sample

Company	Revenues in million €	Industry	Leading generation	New generation involved in strategic planning process
Case A	50	Shoes	II	III
Case B	25	Shoes	II	III
Case C	20	Fashion	I	II
Case D	45	Electronic	II	III
Case E	90	Distribution	I	II
Case F	25	Mechanic	II	III
Case G	200	Packaging	I	II
Case H	80	Distribution	I	II
Case I	100	Food	III	IV
Case J	120	Mechanic	I	II
Case L	160	Electronic	I	II

For all these reasons, we analysed 11 family-owned companies, which are small to medium sized, and have revenues ranging from 20 million to 200 million euros. They operate mainly in the shoes, fashion, packaging, distribution and mechanical industries.

We initially restricted the selection to the cases of strategic plans realized among the members of the Italian Association of Family Business. Within this restricted population, we selected companies where next-generation members had been involved in the process. We followed Andrew Pettigrew's principle of 'planned opportunism', choosing firms that represented extreme situations, combining highly visible and much debated cases to less scrutinized, more 'ordinary' cases (Pettigrew, 1990). Following Pettigrew's recommendations, we purposefully selected firms that, to our knowledge, seemed to disconfirm patterns from previous studies. In this theory-building phase, we considered heterogeneity as a way to as much variation as possible in the data, in order to grasp the complexity of the phenomenon and, as a consequence, to develop a richer and more refined conceptual framework. The selection was somewhat sequential as some cases were included in the study after the collection and analysis of data had already started. Following common prescriptions for multiple case studies (Eisenhardt, 1989b), we replicated the study until we had evidence that we had reached what Glaser and Strauss (1967) refer to as 'theoretical saturation'. In other words we stopped when the incremental learning coming from each additional case had become minimal, because what we observed did not seem to improve our emerging framework further.

Concerning the owning families, our choice has been concentrated on those cases of family-owners who are the first or second generation Table 22.1 shows the main characteristics of the sample.

Within selected cases, analysis has been directed towards the following critical issues (see Table 22.2): role of the next generation in the strategic planning process (leading or just team member); presence of conflict situation among the family members and the owners (we coded with 'zero' the absence of conflict, with 'one' the presence of latent conflict and with 'two' the presence of open conflicts); reason why the strategic plan was

Table 22.2 Needs for and benefits from SPP

Company	Role of next gen in SPP L = leading P = participating	Presence of conflicts 0= none 1= latent 2= open	Need for SPP	Benefits from SPP*					
				1	2	3	4	5	6
Case A	L	0	• Convincing non-family managers that next-generation members were capable through concrete elements • Getting the consensus from non-active owners	X	X	X		X	
Case B	P	0	• Inserting III generation members in the top management team • Getting a contribution from next-generation members	X	X				X
Case C	P	0	• Legitimizing next-generation members in front of banks and external stakeholder • Acquiring a deep knowledge of the business • Defining the strategic intent of the business	X	X	X		X	
Case D	P	1	• Defining the strategic intent of the business • Legitimizing next-generation members in front of owners • Quantifying business knowledge	X	X	X		X	X
Case E	P	0	• Insert and train next-generation members which were very young and had to get business knowledge	X		X	X	X	X
Case F	L	1	• Getting the consensus of non-active shareholders • Improve next-generation legitimization in order to be appointed managing director	X	X	X	X	X	X
Case G	P	2	• Train the young family member • Getting the consensus of family shareholders on a growth project	X	X	X		X	

Table 22.2 (continued)

Company	Role of next gen in SPP L = leading P = participating	Presence of conflicts 0= none 1= latent 2= open	Need for SPP	Benefits from SPP* 1	2	3	4	5	6
Case H	P	1	• Involving juniors in future business strategy • Shareholder had to decide whether to sell or to continue together and in that case to invest money to grow	X	X	X	X	X	X
Case I	P	1	• The company was not growing and one of the juniors decided to invest in a diversified business which was not performing well	X	X	X	X	X	X
Case J	L	1	• Different family branches had to decide whether to continue together or not • Family shareholder had to decide the amount of money to invest • Next generation had to enter and learn about the business	X	X	X		X	
Case L	P	0	• Inserting the new generation • Going public • Involving the new generation in the IPO process	X	X	X	X	X	X

Note:
*Benefits from SPP:
1. facilitate internal and external learning (Brews and Hunt, 1999);
2. offer a context for strategic decision making (Grant, 2003);
3. help legitimization before external stakeholders (Higgins and Diffenbach, 1995; Langley, 1988);
4. reduce position bias and enhance goal convergence (Grant, 2003; Ketokivi and Castañer, 2004);
5. stimulate effort of quantifying phenomena (Grant, 2003; Mintzberg, 1994b);
6. allow executives and directors to evaluate the job done (Grant, 2003; Mintzberg, 1994b).

realized (more family need or business related need); and, finally, the benefits obtained through the strategic planning process (SPP).

Focusing on the possible different role that the younger generation can play in drawing up the plan, we noticed that these roles varied between two extremes: complete responsibility for the process and the final result, or, participation in the work team with the role of executor. Different factors influence this role. The more responsibilities the young member has in the management of the process, the more he/she will learn not only from the contents but also from directly exercising leadership skills (or verifying the existence of such skills) and the stronger will be the legitimization he/she obtains (if the process has been completed with success) in the eyes of third parties.

The benefits of the plan in terms of learning will be different when the young member has no prior experience and therefore needs to learn everything – both content and process – 'from scratch'.

The approach chosen implied the combined use of various methods of data collection (multiple data collection methods): field observations, interviews, analysis of corporate archives, and so on. We combined qualitative and quantitative evidence collected through different methods (triangulation) so as to reach a deeper understanding of the investigated phenomena.

Data analysis

Data analysis was based on common techniques for grounded theory building and combined within-case analysis to cross-case comparison (Eisenhardt, 1989b; Glaser and Strauss, 1967; Lee, 1999). Within-case analysis was initially conducted to identify a number of core constructs. The identification of core constructs was based on a content analysis of the interviews. Therefore, we searched interviews for passages that contained references to needs and benefits obtained from strategic planning. The search was conducted independently by the researchers; later comparison of independent analysis showed a substantial agreement. This coding procedure helped us to identify, for each case, a number of key themes. Following indications from Eisenhardt (1989a), we referred to the existing literature to develop and to enrich these inductively derived insights. In this phase, we often relied on data collected from our archival research to go beyond our informant's accounts, and to extend and refine the emerging framework. Provisional interpretations and tentative propositions were refined in several iterations between theory and data until we were able, for each case, to provide a plausible explanation of the observed patterns.

In a second stage, in order to refine emerging constructs and verify how strongly each of these contributed to explain the observed phenomenon, we conducted a cross-case comparison. Cross-case comparison helped us to verify the robustness of our provisional interpretations across cases. In some cases, the comparison required a further homogenization of concepts, as some themes were grouped into a more general concept. In other cases, propositions were refined, to include the effect of intervening variables. Again, the process followed an iterative path, until the emerging conceptual framework fit the observed patterns across cases. At the end of this operation we were able to identify a number of core issues related to the benefit of strategic planning. As often happens in inductive research, these findings in part confirm and in part extend past literature, and are discussed in the next section.

Building on these results, we investigated the possible benefits that the strategic planning process has in succession passage in family business. In particular, a double set of benefits emerged, further explain in the next section.

Next-generation members' learning from the strategic planning process

In many family businesses, founders try to perpetuate their legacy and ensure continued family control via intergenerational succession, as when they hand over leadership to their children. Recent evidence suggests that 'a mere 30% of family businesses survive past the first generation' and that many intergenerational successions fail soon after the second generation takes control (Miller et al., 2003, p. 514). According to past research there are many reasons why such successions fail (Davis and Harveston, 1998; Handler, 1992; Miller et al., 2003; Sonnenfeld, 1987; Ward, 1987): in our view, some of them can be managed through the strategic plan.

Both drawing up a strategic plan and beginning strategic planning processes can help with the main difficulties related to the succession process.

The benefits which strategic planning provides concern two levels: on one hand, next generation members learn from the content of the strategic plan and, on the other hand, they learn form the process.

Learning related to strategic planning content

Taking part in the strategic planning process, next-generation members receive the double benefit of getting business-related knowledge that a family member needs to employ in order to avoid running foul of business values and norms (Astrachan and Keyt, 2003), and of learning managerial tools to develop their professional capabilities. These two benefits derives from the possible roles that strategic planning can perform, and in particular they refer to point 1 and 5 (see above, pp. 406–7).

In order to understand the learning benefits deriving from the strategic planning, it is useful to start from the typical table of content of a strategic plan. We use as a benchmark the table of content of the listing guide adopted by the Italian Stock Exchange (Table 22.3), which represents the standard used in Italy (Mazzola and Marchisio, 2003).

Leaving out the executive summary, the involvement of the next-generation members in drafting the other points, allow the following key learning.

Realized strategy Analysing realized strategy allows the young member who takes part in it to develop an in-depth knowledge of:

- the environmental context in which the family business operates;
- the resources and capabilities of the family business;
- its culture, history, traditions, and the main events and decisions that led to certain results. Available information and figures can make comparison easier because they are based on objective elements and the younger members thus learn to formulate communications based on data rather than on emotional elements that can easily lead to conflict. Moreover, understanding the history and the reasons that determined a certain positioning is valuable for the younger members, because when they attempt to renew they will have to decide what traditions to keep and how, and at what level to bring innovation into the business;

Table 22.3 Table of content of strategic plan

1. Executive summary
1.1 The proposed strategic project
1.2 Action plans
1.3 Expected results
1.4 The management team
2. The realized strategy
2.1 Competitive strategy at the corporate level
2.2 Group performance
2.3 The competitive strategy of individual strategic business areas
2.4 Performance recorded at the strategic business area level
3. The intended strategy (strategic intent)
3.1 Need for and convenience of strategic rethinking
3.2 Internal reasons: the limits of current strategy
3.3 External reasons: processes of change underway: developing threats and opportunities
3.4 The mission
3.5 Portfolio strategy
3.6 Competitive strategy of the strategic business area
3.7 Expected results
3.8 Resource requirements for the achievement of the plan
4. Action plans
4.1 Planned changes in organizational structure
4.2 Plans for increasing corporate productivity
4.3 Plans for size development
4.4 Summary framework: actions, times, organizational responsibilities, economic and financial impact, criticalities and obligations
5. Financial assumptions, key value driver and financial forecast
5.1 The model linking strategic decisions and economic variables
5.2 The assumptions on which financial forecasts are based
5.3 Plan results
5.4 Assessment of economic convenience and financial feasibility of the plan
5.5 Sensitivity analysis

Source: Strategic Plan Listing Guide – Italian Stock Exchange (Mazzola and Marchisio, 2003).

- tacit knowledge;
- using managerial tools for strategic appraisal of the family business;
- its need for renewal in a more objective way not totally influenced by typical dynamics present in pathological succession patterns which might be either too conservative, too wavering or too rebellious (Miller et al., 2003).

In case L, new generation, after some outside working experience, entered the company. In order to help his entrance and involvement, he took part to the strategic planning process, realized in order to verify the hypothesis to go public. Next generation worked full time in order to fully understand and describe the business model, to catch the main differences between them and the competitors, to grasp both competitive advantage sources, and improvement areas. At the end of process the young member described that period as a critical one for him in order to have 'a quick and proper understanding of the family company, and his fast introduction in the organization'.

Intended strategy Making the strategic intent explicit represents an important stimulus for the next generation in order to enable the process of generation, evaluation and selection of alternatives, which require two conditions: on one hand, the imagination to develop a great number of different alternatives so to avoid many variations on the same theme and, on the other, the discipline necessary to consistently evaluate the options. Szulanski and Amin (2001) synthesized these two concepts into one they called 'disciplined imagination', a fundamental skill in creating continually innovative business strategy, which in turn is a prerequisite for creating value and, very often, for the very survival of companies. It is a skill learnt by doing. Involving the younger members in this process helps them develop, right from the beginning, a way of thinking that values creativity but also the discipline thanks to which the best alternative possible can be chosen. The younger members must not necessarily possess both characteristics, although they should learn that both components are important and that they need to create conditions that will guarantee both (also in the future).

Taking part in this phase allows next-generation members have the opportunity to learn and experience tools and methodologies for scenario building, visioning and scouting possible sources of innovation.

The main benefit of this phase is the reduced risk of undergoing either a conservative or rebellious succession (Miller et al., 2003) and to provide the family business with the proper change in goals, business scope, product lines, or markets and willingness to seize opportunities.

In case A, next generation member took part to the formulation of the strategic plan to present to the family council in order to tackle with the deep industry crisis. The biggest challenge of the strategic planning process was the definition of credible and valuable vision. 'We spent a lot of time in exploring new business opportunities able to counterbalance the problems in the traditional scope.' At the end of the process the young family member was used to say 'I had a unique opportunity to discuss really relevant strategic issues: and I clearly understood how important is strategic renewal for assuring family business continuity'.

Action plan Action plans imply an in-depth knowledge of numerous elements among which are the corporate framework, its resources and obligations, the persons to involve, the responsibilities to be assigned and the time necessary to accomplish the proposed objectives. In terms of tools acquired, the definition of an action plan help young members to learn both project management and resource management. In particular, regarding the former in this phase, the new generation is educated, right from the beginning, to work in a team, by objectives and to check the effectiveness of their activity by confronting themselves with the results; while regarding the latter, in this phase young family members learn how to add, shed and allocate resources, to bundle and to leverage them, which represent a basis for reaching the desired competitive advantage (Sirmon and Hitt, 2003).

Family business I was facing a delicate problem. The company has not been growing for more than five years. One family member launched a diversified business which was both losing money and absorbing cash. This situation created a tension among owners. This situation had been solved by defining a clear action plan for that division: a clearly defined amount of resources was allocated to the initiative, a break even goal was settled and a time frame was fixed. The young family member involved remembered this phase as the most critical of the whole process. 'While there was a shared agreement among owners about the necessity of exploring growing opportunities, the main problems emerged when we had to decide the amount of resources we had to allocate for the growth project and not distribute as dividends.'

Financial assumptions, key value driver and financial forecast Together with the aforementioned qualitative analyses, financial forecast offers an additional benefit for a deeper understanding of strategic intentions, for choosing among alternative action plans and for controlling the progress of the chosen projects. Next-generation members have the opportunity to fully comprehend the quantitative implications of the strategic intent.

In this phase they learn also how organizations create value, developing strategy maps such as a balanced scorecard, which show how an organization will convert its initiative and resources into tangible outcomes (Kaplan and Norton, 2000). Thank to this activity the young member is able to understand the most important company value drivers and the key indicators through which monitoring them. This fills quickly the gap of knowledge usually acquired through long-term experience in the business.

Another benefit consists in the development of awareness of the constraints deriving from aspiration levels of owning family for growth and payout and its willingness to assume debt. The interdependence of growth and payout-aspiration levels drives any business, but its ability to facilitate goal-setting in a private business is highly significant. Profit growth and profit payout drive family wealth creation and must be disciplined by levels of profitability grounded in reality (Adams et al., 2004).

Family business E was growing very fast and had to decide whether to continue investing in the growth process also through the acquisition of a couple of minor competitors. The son strongly believed in the attractiveness of the strategic intent and supported this project in front of the other owners. 'I finally understood not only where and how to grow, but more importantly, I realized that given our profitability, the expected growth rate could be reached only if all the owners were prepared to significantly reduce their dividend expectations in the short period, and to accept to double our traditional debt/equity ratio. This latter issue was considered not acceptable in my family. So we had to change our strategic project in order to meet owners' will.'

Learning related to strategic planning process

According to Quinn (1981) 'the most important contributions of these corporate planning systems are actually in the "process" ' (p. 53). We found that a good number of benefits are related to the involvement of the next generation in family firms.

Considering the benefits offered by strategic planning (listed on pp. 406–7), we see that taking part in the process, allows next generation to experiment benefit related to points 2, 3, 4 and 6.

As far as the single phases of the process are concerned, we identified main benefits for the young family members involved in the strategic planning process. In order to

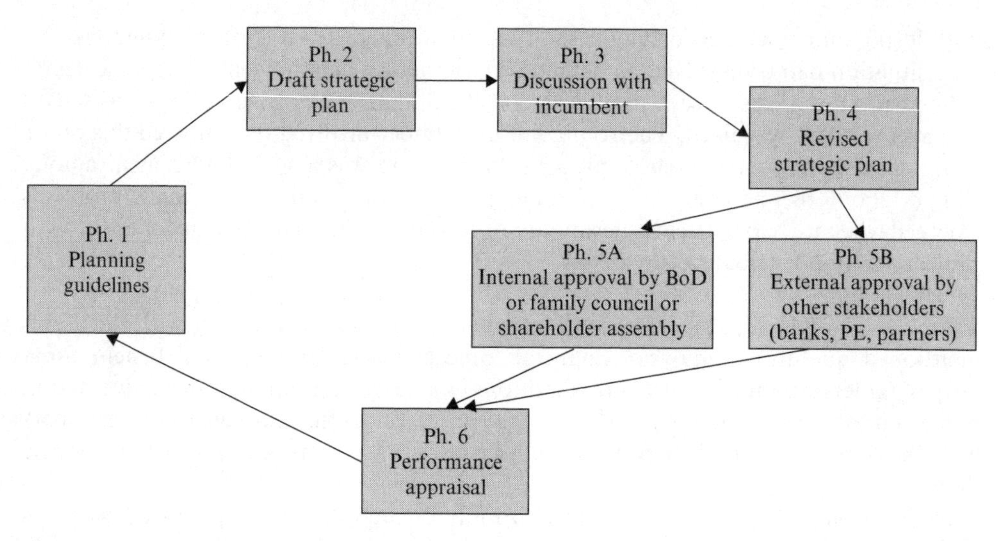

Figure 22.1 Strategic planning process: an adaptation of Grant model (2003)

represent more clearly the benefits deriving from the strategic planning process, we represent the main phases of the process itself. In Figure 22.1 we have adapted the model presented by Grant (2003). As Figure 22.1 shows, the strategic planning process includes several phases and improves the interaction with both internal and external stakeholders.

Considering the whole process, we identify two main benefits. First, young members who took part in the process had the opportunity to exercise leadership skills (Mumford et al., 2000, p. 26). The authors maintain that:

> leadership ultimately depends on one's capability to formulate and implement solutions to complex (i.e. novel, ill-defined) social problems. [. . .] it is argued that the skills needed to solve organizational leadership problems include complex creative problem-solving skills associated with identifying problems, understanding the problem and generating potential solutions; social judgment skills associated with the refinement of potential solution and the creation of implementation frameworks within a complex organizational setting; and social skills associated with motivating and directing others during solution implementation.

It is evident that the possibility of taking part in the realization of the plan is a crucial opportunity for the younger members to exercise leadership skills. This benefit is particular important if we consider that founding generations are reluctant to let the next generation join in the decision-making for the business (Handler, 1989; Lansberg, 1988).

Of course this is not the only opportunity the next generation has to exercise leadership skills. Other possibilities exist both on the business or family side. Among the former we can imagine situations where juniors have the responsibility for a specific project or team or business unit, or taking part in board of directors meetings; regarding the latter, the next generation member's leadership skills can benefit from taking part in and/or having the responsibility of organizing or managing family council or other family meetings.

Second, according to Brews and Hunt (1999), how to plan, making incremental adjustments, and the quality of planning improve with time, so the early years are particularly important for learning. Next-generation members who participate in the process have the opportunity to learn from the beginning a structured methodology and several tools.

Finally, taking part in the whole process helps young members to take a more knowledgeable decision about their staying in the company and in what role.

Considering phases 1–4, the young member has the opportunity to:

1. be exposed and collaborate both with the incumbent and with other internal and external relevant stakeholders. This exposure and collaboration helped the young to:
   (a) develop teamwork capabilities;
   (b) be introduced within the organization;
   (c) communicate and share ideas with and get feedback from acknowledged and experienced internal and external stakeholder (the entrepreneur, family and non-family executives and, eventually, customers, suppliers, bankers and consultants, encouraging mutual learning, which, according to Handler (1992), is typical in successful family firms.
2. increase their sense of belonging and 'feel to be part of the realization of a great project' (Case B), which favour:
   (a) increasing his/her commitment thanks to the fact that being involved is a recognition in itself of the ability or worthiness to contribute to the organization, instilling a sense of appreciation and a stronger identification with resulting organizational goals (Ketokivi and Castañer, 2004; Kogut and Zander, 1996);
   (b) consequently increasing commitment should reduce both position bias (Tannenbaum and Massarik, 1950), and the risk of wavering or rebellious succession (Miller et al., 2003, p. 521);

Considering phase 5a and 5b, the young member has the opportunity to:

1. earn legitimization before family owners and family and non-family executives. This helps the young member in:
   (a) exercising negotiation skills – resources have to be obtained;
   (b) honing communication skills – consensus and trust have to be gained.
2. become known by non-family external stakeholders, earning legitimization on the basis of a concrete result produced and earn their respect through the presentation of well-structured information.

Finally, phase 6 makes the performance measurement an objective one, reducing the situation where family members are measured 'by blood'.

Conclusions and implications for theory and practice

In this chapter, we have tried to shift the attention from the wide studies on succession planning and strategic planning, to the underestimated benefits that strategic planning can have as a training tool itself for family business.

Building on evidence from a series of case studies, we have argued that strategic planning is an important way to train next-generation members during a specific delicate phase of succession process. More specifically, our findings suggest that the benefits derive both from the content of the strategic plan and from the participation to the strategic process itself, and that the benefits occur simultaneously for the family company, for the next-generation members and for the owners.

Evidence from the cases suggests that the well-documented roles that strategic plans usually have in big organizations, can be used also in family businesses, offering an idiosyncratic contribution above all during the first years after a young member enters the company. In particular these benefits have to do with helping a systematic acquisition of crucial business (tacit) knowledge and skill, facilitating interpersonal working relationships between incumbent and the new generation, reinforcing the next generation's image and reputation in front of third parties, non-family members or non-family external stakeholders, and allowing family members and non-family executives and directors to evaluate the job done by the next generation.

Furthermore, evidence from our study suggests that a broader range of benefits can be obtained by drafting and presenting a strategic plan in family business. In some of the cases analysed, we saw a positive effect on latent conflict among family members, owing to the fact that the plan helped in quantifying phenomena, making them more objective and facilitating the decision process.

In order to obtain these benefits at least one condition has to be respected. This concerns the methodology and skills used in order to realize the strategic planning process and to write the plan. Considering the learning experience it represents, it is very important to do it properly. For this reason it is strongly recommended to have capable people leading the process, be they executives working in the company or external consultants, if needed. The presence of consultants is not only a condition, but also an added benefit. Collaborating with external consultants can be a further opportunity to learn not only techniques and analysis methodologies directly applied to corporate reality under the supervision of authoritative professional figures, but also to learn different approaches and ways of thinking, which are in certain cases comparable to the benefits obtained through work experiences outside the family business. Besides being potential mentors, consultants can also be precious allies when the young members are ready to introduce and sell new ideas to the company.

If the company does not use the services of consultants in the preparation of the plan, the process is an opportunity to cooperate with (family or non-family) managers from which useful reciprocal knowledge can be gained, as mentioned above.

We believe that the implications of our findings touch both theory and practice. On the theoretical side, our objective was to offer a contribution to the studies regarding the phase after the younger members enter the company, offering a concrete tool to develop the leadership and managerial skills of the new generations – a hot topic among family business researchers – and to facilitate family relations.

From a practical point of view, our findings can provide both practitioners and the families with some suggestions. As far as practitioners are concerned, and on the basis of the conclusions we have reached, we would like to remind those consultants who do not carry out a specific activity for family businesses but who find themselves in the position of drawing up a strategic plan, that they have a great responsibility towards the younger generation, since they have the potential opportunity to be important mentors for the younger members. Slightly different is the recommendation for practitioners who are instead dedicated to family businesses: here, the preparation of a strategic plan is a further learning tool for the younger members and can provide a solution for possible conflicts among the members of the family.

As far as families are concerned, we are able to offer them a concrete training tool which can help the next-generation members with the above mentioned learning, and which simultaneously offer a benefit for the development of the company and for the owners' decision process.

Of course this tool is not free from risks. The high visibility to which the next generation is exposed during the process can become a threat in case of (big) mistakes, above all in terms of wrong attitudes held by juniors. All the attention in fact is devoted to the young members and if they behave in an inappropriate way everybody will have the chance to notice it. A second possible risk is where for any reason the process does not work and the next generation does not feel involved. If this is the case, all the benefits related to the sense of belonging would be lost. Finally there is the risk that by introducing a structured activity of planning, the room left for emergent strategy reduces too much and the strategy process becomes too rigid. These dangers identify the need for further research necessary on the topic. This work is a first effort that stems from an anecdotal approach that will have to be further examined by studying further cases, establishing a control sample and then by identifying some hypothesis which has to be tested.

In terms of content we suggest further areas of investigation related to this, such as individuating conditions for success so that all potential benefits can be achieved. Among them, we think it is particularly important to focus on the characteristics and attitude of the incumbent in terms of giving freedom to act to the juniors. Very autocratic incumbents are unlikely to leave the next-generation member free to present new ideas and are even more unlikely accept them. A second possible condition to further investigate is the role of trust in the whole process.

Other areas for investigation concern better understanding the threats to the success of the process, the possible roles that the juniors, the incumbents, non-family managers and consultants can and should play in the strategic planning process, and eventually check whether there is any relationship between the role the next generation plays and the benefits they, the owning family and the company can get; verifying the possible influence of a particular choice of timing for the preparation of the plan in generational succession. Finally, a further area to be analysed is that of the plan's influence on the performance of the younger members (a measure also to be defined) and of the company in terms of growth and the identification and exploitation of new opportunities.

References

Adams, F.A., G.E. Manners Jr, J.H. Astrachan and P. Mazzola (2004), 'The importance of integrated goal setting: the application of cost-of-capital concepts to private firms', *Family Business Review*, **17**(4), 287–302.

Alcorn, P.B. (1982), *Success and Survival in the Family Owned Firms*, New York: McGraw-Hill.

Astrachan, J.H. and A.D. Keyt (2003), 'Commentary on: the transacting cognitions of non-family employees in the family business setting', *Journal of Business Venturing*, **18**(4), 553–8.

Barach, J.A., J.B. Gantisky, J.A. Carson and B.A. Doochin (1988), 'Entry of the next generation: strategic challenge for family business', *Journal of Small Business Management*, **26**(2), 49–56.

Barringer, B.R., F.F. Jones and P.S. Lewis (1998), 'A qualitative study of the management practices of rapid growth firms and how rapid growth firms mitigate the managerial capacity problem', *Journal of Developmental Entrepreneurship*, **3**(2), 97–122.

Beckhard, R. and W. Burke (1983), 'Preface', *Organizational Dynamics*, p. 12.

Brews, P. and M.R. Hunt (1999), 'Learning to plan and planning to learn: resolving the planning school/learning school debate', *Strategic Management Journal*, **20**(10), 889–913.

Burgelman, R.A. (1983), 'A model of the interaction of strategic behaviour, corporate context, and the concept of strategy', *Academy of Management Review*, **8**(1), 61–70.

Cabrera-Suárez, K.P., P. De Saa-Pérez and D. García-Almeida (2001), 'The succession process from a resource- and knowledge-based view of the family firm', *Family Business Review*, **14**(1), 36–43.
Chrisman, J.J., J.H. Chua and P. Sharma (2003), 'Current trends and future directions in family business management studies: toward a theory of the family firm', part of Coleman Foundation White Paper Series Coleman Foundation, Chicago.
Coda, V. and E. Mollona (2001), 'Managing processes of strategic change: a strategy dynamic model', working paper, Bocconi University.
Davis, P. (1983), 'Realizing the potential of family business', *Organizational Dynamics*, Summer, 47–56.
Davis, P. and P.D. Harveston (1998), 'The influence of family on the family business succession: a multi-generational perspective', *Entrepreneurship Theory and Practice*, **22**(3), 31–53.
Dess, G.G. and N.K. Oringer (1987), 'Environment, structure, and consensus in strategy formulation: a conceptual integration', *Academy of Management Review*, **12**(2), 313–30.
Eisenhardt, K.M. (1989a), 'Making fast strategic decisions in high-velocity environments', *Academy of Management Journal*, **32**(3), 543–76.
Eisenhardt, K.M. (1989b), 'Building theories from case study research', *Academy of Management Review*, **4**, 532–50.
Epstein, M.J. and R.A. Westbrook (2001), 'Linking actions to profits in strategic decision making', *MIT Sloan Management Review*, spring, 39–49.
Farquhar, K.A. (1989), 'Employee responses to external executive succession: attributions and the emergence of leadership', unpublished doctoral dissertation, Department of Psychology, Boston University.
Floyd, S.W. and B. Wooldridge (1992), 'Managing strategic consensus: the foundation of effective implementation', *Academy of Management Executive*, **6**(4), 27–39.
Foster, R. and S. Kaplan (2001), *Creative Distruction: Why Companies that Are Built to Last Underpreform the Market – and How to Successfully Transform Them*, New York: Doubleday.
Friedman, S. (1987), 'The succession process: theoretical considerations', paper presented at the annual meeting of the Academy of Management, New Orleans.
Gabarro, J. (1979), 'Socialization at the top: how CEOs and subordinates evolve interpersonal contacts', *Organizational Dynamics*, **7**(3), 3–23.
Gilmore, R.N. and J.E. McCann III (1983), 'Designing effective transitions for new correctional leaders', in J.W. Doig (ed.), *Criminal Corrections: Ideals and Realities*, Lexington, VA: Lexington Books.
Glaser, B. and A. Strauss (1967), *The Discovery of Grounded Theory: Strategies of Qualitative Research*, London: Wiedenfeld and Nicholson.
Glaser, B.G. (1994), *Basics of Grounded Theory Analysis: Emergence versus Forcing*, Mill Valley, CA: Sociology Press.
Gordon, G.E. and N. Rosen (1981), 'Critical factors in leadership succession', *Organizational Behavior and Human Performance*, **27**, 227–54.
Granger, C.H. (1964), 'The hierarchy of objectives', *Harvard Business Review*, **42**(3), 63–74.
Grant, R.M. (2003), 'Strategic planning in a turbulent environment: evidence from the oil majors', *Strategic Management Journal*, **24**, 491–517.
Hamel, G. and C.K. Prahalad (1989), 'Strategic intent', *Harvard Business Review*, **67**(3), 63–76.
Handler, W.C. (1989), 'Methodological issues and considerations in studying family businesses', *Family Business Review*, **2**(3), 257–76.
Handler, W.C. (1992), 'The succession experience of the next generation', *Family Business Review*, **5**(3), 283–307.
Handler, W.C. (1994), 'Succession in family businesses: a review of the research', *Family Business Review*, **7**(2), 133–57.
Higgins, R.B. and J. Diffenbach (1985), 'The impact of strategic planning on stock prices', *Journal of Business Strategy*, **6**(2), 64–72.
Kaplan, R.S. and D.P. Norton (2000), 'Having trouble with your strategy? Then map it', *Harvard Business Review*, September–October, 167–76.
Ketokivi, M. and X. Castañer (2004), 'Strategic planning as an integrative device', *Administrative Science Quarterly*, **49**, 337–65.
Kimhi, A. (1997), 'Intergenerational succession in small family business: borrowing constraints and optimal timing of succession', *Small Business Economy*, **9**(4).
Kogut, B. and U. Zander (1996), 'What firms do? Coordination, identity, and learning', *Organization Science*, **7**, 502–18.
Langley, A. (1988), 'The role of formal strategic planning', *Long Range Planning*, **21**(3), 40–50.
Lansberg, I. (1988), 'The succession conspiracy', *Family Business Review*, **1**(2), 119–43.
Lee, T.W. (1999), *Using Qualitative Methods in Organizational Research*, Thousand Oaks, CA: Sage.
Lyles, M.A. (1981), 'Formulating strategic problems: empirical analysis and model development', *Strategic Management Journal*, **2**(1), 61–75.

Marckóczy, L. (2001), 'Consensus formation during strategic change', *Strategic Management Journal*, **22**, 1013–31.
Mazzola, P. (2003), *Il piano industriale*, Milano: Università Bocconi Editore.
Mazzola, P. and G. Marchisio (2003), *Guida al piano industriale*, Milano: Borsa Italiana Spa.
McCann III, J.E., A.Y. Leon-Guerero and J.D. Haley Jr (2001), 'Strategic goals and practices of innovative family businesses', *Journal of Small Business Management*, **39**(1), 50–59.
McGivern, C. (1978), 'The dynamics of management succession', *Management Decision*, **16**(1), 32–42.
Miller, D., L. Steier and I. Le Breton-Milller (2003), 'Lost in time: intergenerational succession, change and failure in family business', *Journal of Business Venturing*, **18**(4), 513–51.
Mintzberg, H. (1990), 'The design school: reconsidering the basic premises of strategic management', *Strategic Management Journal*, **11**, 171–95.
Mintzberg, H. (1994a), 'Rethink strategic planning Part II: new role for planners', *Long Range planning*.
Mintzberg, H. (1994b), *The Raise and Fall if Strategic Planning. Reconciling roles for planning, plans, planners*, New York, Free Press.
Morris, M.H., R.O. Williams, J.A. Allen and R.A. Avila (1997), 'Correlates of success in family business transitions', *Journal of Business Venturing*, **12**, 385–401.
Mumford, M.D., S.J. Zaccaro, F.D. Harding, T.O. Jacobs and E.A. Fleishman (2000), 'Leadership skills for a changing world: solving complex social problems', *Leadership Quarterly*, **11**(1), 155–70.
Nielson, R.P. (1981), 'Toward a method of building consensus during strategic planning', *Sloan Management Review*, **22**(4).
Pettigrew, A. (1990), 'Longitudinal field research on change', *Organization Science*, **1**(3), 267–92.
Porter, M. (1996), 'What is strategy', *Harvard Business Review*, November–December, 61–78.
Poza, E.J. (1995), 'Global competition and the family-owned business in Latin America', *Family Business Review*, **8**(4), 301–11.
Quinn, J.B. (1981), 'Formulating strategy one step at a time', *Journal of Business Strategy*, **1**(3), 42–63.
Schoemaker, P.J.H. and J.E. Russo (1996), 'It's all in how you frame it: simple steps to make the right decision', mimeo.
Sirmon, D.G. and M.A. Hitt (2003), 'Managing resources: linking unique resources, management, and wealth creation in family firms', *Entrepreneurship Theory and Practice*, **27**(4), 339–58.
Sonnenfeld, J. (1987), 'Chief executives as the heroes or villains of the executive process', paper presented at the meetings of the Academy of Management, New Orleans, LA.
Steier, L. (2001), 'Next generation entrepreneurs and succession: modes and means of managing social capital', *Family Business Review*, **14**(3), 259–76.
Strauss, A. and J. Corbin (1990), *Basics of Qualitative Research: Grounded Theory Procedures and Techniques*, Newbury Park, CA: Sage.
Strauss, A.L. (1987), *Qualitative Analysis for Social Scientists*, New York: Cambridge University Press.
Szulanski, G. and K. Amin (2001), 'Learning to make strategy: balancing discipline and imagination', *Long Range Planning*, **34**, 537–56.
Tannenbaum, R. and F. Massarik (1950), 'Participation by subordinates on the managerial decision making process', *Canadian Journal of Economics and Political Science*, **16**, 408–18.
Upton, N.B. and R.K.Z. Heck (1997), 'The family business dimension of entrepreneurship', in D.L. Sexton and R.W. Smilor (eds), *Entrepreneurship: 2000*, Chicago, IL: Upstart Publishing Company, pp. 243–66.
Vancil R.F. (1987), *Passing the Baton: Managing the Process of CEO Succession*, Boston, MA: Harvard Business School Press.
Ward, J.L. (1987), *Keeping the Family Business Healthy*, San Francisco, CA: Jossey-Bass.
Weick, K.E. (1989), 'Theory construction as disciplined imagination' *Academy of Management Review*, **14**(4), 516–31.
Whyte, G. (1989), 'Groupthink reconsidered', *Academy of Management Review*, **14**(1), 40–56.
Yin, R.K. (1994), *Case Study Research: Design and Methods*, Thousand Oaks, CA: Sage.
Zahra, S.A. (2005), 'Entrepreneurial risk taking in family firms', *Family Business Review*, **18**(1), 23–40.

23 An integrated framework for testing the success of the family business succession process according to gender specificity

Vassilios D. Pyromalis, George S. Vozikis, Theodoros A. Kalkanteras, Michaela E. Rogdaki and George P. Sigalas

Introduction

The family firm has always been and continues to be an increasingly vital player in the economy (Duman, 1992). One of the events that may disrupt the smooth evolution of a family business is a generation transition and succession. Moreover, the enlarging role that women play in family businesses and their increasing presence during the succession process renders an investigation into the relationship between succession issues and gender specificity quite promising. This is because there have been very few studies dealing with gender issues in family firm ownership and management, even though family firms account for an estimated 80 per cent of all American businesses, and about one-third of these family firms are owned by women (Sonfield and Lussier, 2003). In addition, the global rise in female entrepreneurship and self-employment has also been noted by OECD studies (OECD, 2000) and the National Federation of Women Business Owners (National Federation of Women Business Owners, 1997). More particularly, in the USA alone, it is anticipated that women will soon own 50 per cent of all US businesses and that there will be an upsurge in female inheritance, ownership and management of companies founded immediately post-war, over the period 2000–2020 (Achua, 1997; Daniels, 1997). Finally, succession issues receive extensive attention (Dyer and Sanchez, 1998), as related articles have grown twofold in the 1990s (Wortman, 1994). It is therefore quite obvious that owing to the enlarging role that women play in family businesses, and the significance of the succession process in family firm, it seems quite worthwhile to attempt to correlate succession issues with gender specificity.

This chapter draws on existing literature and seeks to integrate gender specificity and succession in a conceptual framework that will highlight this fundamental relationship. More specifically, as the trend for women to take over leadership positions in firms is evolving, our framework is used to investigate whether the success of the succession and satisfaction from the process per se depends on the successor's gender. For this purpose, we make use of the Analytic Hierarchy Process and base our findings on evidence from the literature.

However, it is important before proceeding with the core body of our research propositions, to provide some basic definitions and clarify the basic context within which family businesses operate, as well as the most important dimensions of the issues to be examined.

Theoretical background

Family businesses are defined as 'businesses in which ownership and/or policymaking are dominated by members of an emotional kinship group' (Carsrud, 1994). They differ from other businesses in that ownership and/or control overlap with family membership (Lank, 1997), as all major operating decisions and plans for leadership succession are influenced by family members in management positions or on the board of directors (Handler, 1989a). Hoover and Hoover (1999) claim that above all else, family business is the business of relationships, and thus, one should think and talk about 'business families' rather than 'family businesses' (Le Van, 1999).

On the other hand, the succession process is 'the transfer of leadership, ownership, or managerial control from one family member to (preferably) another' (American Family Business Survey, 1997). The basic stakeholders in the transition process are similarly identified as the incumbent, the successor, and the other family members (Handler, 1989b). In the United States, 92 per cent of all firms are family firms and 18 million among them are family dominated according to Duman (1992). It is also estimated that in the next few years 47 per cent will change their top leadership. Among those, 45 per cent have not chosen a successor and lack a succession plan, while 65 per cent do not have a strategic plan. As far as ownership transition is concerned, it is interesting to note that 35 per cent of the family business owners are considering a co-CEO perspective, 34 per cent are proclaiming that a woman will be the next CEO, and 20 per cent do not have an estate plan (DiMatteo, 2004).

The difficulty of the succession process is obvious, as 30 per cent of family businesses survive past the first and 15 per cent past the second generation (Davis and Harveston, 1998). These percentages can be confirmed by the Family Business Network, a non-profit organization based in Lausanne, Switzerland, which estimates that around 70 per cent of family firms fail to survive to the second generation. One of the most important reasons for that of course, is the lack of a suitable successor. This shows the importance of succession and cooperation pools within the family itself. Two-thirds of the family business owners want their businesses to be transferred, either through sale or as a gift to the next generation of the family (American Family Business Survey, 1997), because it is believed that this way the firm would maintain its existing competitive advantage through the preservation of the 'idiosyncratic knowledge of family character' (Bjuggren and Sudd, 2001, p. 11).

Research findings on the interactive factors that affect the outcome of the leadership/ownership transition are significant and revolve around the satisfaction with the succession process as well as the effectiveness of the process per se, as Handler (1989b) asserts. These findings are summarized on Table 23.1, and are of extreme value for the development of our research propositions. Regarding the 'satisfaction' dimension, an empirical study identifies as critical success factors the 'incumbent's propensity to step aside', the 'successor's willingness to take over', the 'agreement among family members to maintain family involvement', the 'acceptance of individual roles', and 'succession planning' (Sharma et al., 2003, p. 671).

Most of the literature, however, concentrates instead on factors that affect the 'effectiveness' dimension of the succession process. Research findings here suggest that the characteristics of successful successions are the 'well prepared successors', the 'positive relationships' and the 'succession planning attempts' (Morris et al., 1997, p. 392). Additionally, Miller et al. (2003, p. 514) advocate that the 'successor's relation with the past' also affects the effectiveness and outcomes of the succession, since things to be avoided are a 'too strong attachment to the past', a 'wholesale rejection of it', and/or an 'incongruous blending of present and past'. Other factors that have a negative impact on the succession process are 'family rivalries' (Dyer, 1986) and 'incompetent or unprepared successors'

Table 23.1 Literature review of factors that affect the succession process

Researcher	Dimension	
	Satisfaction with the succession process	Effectiveness of the succession process
Sharma et al., 2003	The incumbent's propensity to step aside, the successor's willingness to take over, agreement among family members to maintain family involvement, the acceptance of individual roles and succession planning (positive influence)	
Morris et al., 1997		Well-prepared successors, the positive relationships (trust, shared values) and the succession planning and controlling attempts (positive influence)
Miller et al., 2003		Owner's relationship with the past
Dyer, 1986		Family rivalries (negative impact)
Kets De Vries and Miller, 1987		The incompetent or unprepared successors (negative impact)
Handler, 1990		Previous positive succession experience, fulfilled career, psychosocial, and life stage opportunities in the context of the family firm, the capability to exercise personal influence in the family business, mutual respect and understanding with the incumbent and high commitment to the continuation of the family business (positive influence)
Dyck et al., 2002		Sequence (appropriate skills and experience of successor), timing (effective passing), baton passing techniques (succession details) and the communication (positive impact)
Dascher and Jens, 1999		The desire to pass on a business, the ability to carry out the desire, and the willingness of heirs to accept responsibility (positive impact)
Wong et al., 1992		Size of the family business (positive influence)

exhibiting 'overdependence, conservatism, rebellion, excessive change, ambivalence, confusion, stagnation or abandon, compulsive consensus or conflict, stifling bureaucracy or plain chaos' (Kets De Vries and Miller, 1987, p. 1). Others point out that the key factors for an effective transition are the 'preparation of heirs', 'family relationships' and 'planning and control activities' (Morris et al., 1997, p. 392).

From another researcher's perspective, the 'positive succession experience', such as, 'fulfilled career, psychosocial, and life stage opportunities in the context of the family firm, the capability to exercise personal influence in the family business, the achievement mutual respect and understanding with the incumbent, and finally the high commitment to the continuation of the family business' play an important role in the succession process (Handler, 1990, p. 40). Furthermore, the 'sequence' (appropriate skills and experience of the successor), the 'timing' (effective passing), the 'baton passing technique' (succession details) and 'communication' are also crucial for the effectiveness of the transition (Dyck et al., 2002, p. 144). Others identified as additional important issues the 'desire to pass on a business', the 'ability to carry out the desire', the 'willingness of heirs to accept responsibility' (Dascher and Jens, 1999, p. 2), and the size of the family firm (Wong et al., 1992). Finally, a psychographic approach to succession is advanced by DiMatteo (2004) who views as critical tasks in the succession process the following:

- breaking through assumptions, beliefs, and the psychological barriers of succession;
- creating dialogue by managing the psychological tasks of succession;
- strategic planning by creating the future;
- aligning strategic planning with succession planning;
- tactical by training, coaching, mentoring, and estate planning;
- succession transition; and
- managing change.

Regarding the role of gender in the succession process, it is obvious from the literature that male offspring are favoured (Allen and Langowitz, 2003; Family Business Network research, 1995) regardless of suitability (Miller et al., 2003). Unfortunately, there still exist perceptual barriers to women's advancement to senior management positions in family firms, literally known as 'glass ceilings' (Crampton and Mishra, 1999). Men and women also seem to differ in many attributes and characteristics. For instance, as far as self-employment motives are concerned, men's main drive is wealth creation and economic advancement, while women's drive is to achieve a family-related lifestyle, some flexibility to balance work and family, as well as 'constructivism', and economic parity (DeMartino and Barbato, 2003). However, as the number of women in top positions is constantly increasing (American Family Business Survey, 1997; Taylor, 2002), it is important to examine whether this obvious succession bias against women is founded on real traits, on performance outcomes, or on a perception as a consequence of socio-cultural values.

Finally, there is no information in the research literature about gender-specific outcome measures of succession in family firms (Astrachan et al., 2002). Men and women have also been compared by specific traits, but there is no integrated approach and thus, no evidence as to who can handle more effectively and efficiently a succession process. The findings of the literature review on gender individualities as they relate to succession in family firms are summarized in Table 23.2.

Table 23.2 *Review of gender individualities literature*

Researcher	Incumbent's propensity to leave	Successor's will to take over	Factor positive relations – communication	Succession planning	Successor's appropriateness
American Family Business Survey, 1997	1/3 of owners still involved in the business even after retirement			More men write succession plans than women	
Axelrod-Contrada, 2004	Sons are more intimidating to father than daughters (natural-born competitiveness)				
Allen and Langowitz, 2003	Many respondents claim that the CEO will 'never' retire	Men are more committed to the family business than women	Woman-owned firms have more loyalty, agreement with goals, and company pride	More female owned firms have already chosen a successor	
Olson, 2001		Many female successors think they are ill 'groomed' for taking over			Men are more impatient to take over than women
Powell, 1990			No significant differences in management values and styles between the two genders		
Crouter, 1984			Women experience more family-business related conflicts		
Loscocco, 1997			Women are less able to manage the work–family interface		
Buttner, 2001			Women are more collaborative, relations oriented and participative		
Cole, 1997			Women are more dependent they undertake more frequently the nurturer, mediator, or peace-maker role		Men are considered to be more independent than women

Moore and Buttner, 1997	Women are more collaborative, interactive, team workers, participative and democratic		Women rely more on social networks than systematic decision-making
Brenner et al., 1998	Women have more interpersonal skills		
Astrachan et al., 2002		Women are more interactive, collaborative and can lead a smoother transition	
Miller et al., 2001		Women are less confident, use advisors, and form networks	
Johnson and Powell, 1994			Women are more cautious, less confident, aggressive, easier to persuade, provide inferior leadership and problem-solving, and are risk averse
Sonfield et al., 2001			Men are more strategic and entrepreneurial
Eagly et al., 1995			There is no gender difference in leadership effectiveness
Kaplan, 1988			Women emphasize more non-financial personal goals

Table 23.2 (continued)

Researcher	Incumbent's propensity to leave	Successor's will to take over	Factor positive relations – communication	Succession planning	Successor's appropriateness
Lee-Gosselin and Grise, 1990					Women prefer small and stable business models
Fischer et al., 1993					Men are more entrepreneurial due to social/ liberal feminism
Bailyn, 1993					Women are less entrepreneurial and run the business around their personal life

It appears from the literature review that there is a lack of an integrated conceptual framework, which deals with both dimensions and critical success factors of the succession process, namely satisfaction and effectiveness as well as gender issues. What is missing is a synthesis of the various perspectives in order to come up with a coherent and concise two-dimensional succession framework. The model developed by Morris et al. (1997) deals with satisfaction with the succession process but concentrates mostly on the effectiveness of succession. Even the ambitious conceptual framework by Sharma et al. (2001), which attempts to create such a two-dimensional integrated approach, focuses only on factors that affect the initial satisfaction with the succession process, and does not provide weights for ranking the importance of the success factors.

Conceptual framework

The conceptual framework described in this section is a result of a synthesis of the existing literature on succession, and will provide the basis for the test of our hypotheses through three stages, namely: the identification of the basic dimensions of a succession's success; the identification of the critical success factors that influence a succession's success, and finally the development of an integrated conceptual framework.

Identification of the basic dimensions of a succession's success

First, the term 'successful succession' needs to be defined. Two dimensions characterize the success of the transition: the satisfaction with the process of all parties involved and the effectiveness of the process per se (Handler, 1989b). As both dimensions are amplified and improved, so is the possibility of a successful succession. The satisfaction dimension represents the subjective assessment of individuals about the process, while the effectiveness dimension represents the objective determination of the process's impact on a family firm's performance (Sharma et al., 2001).

With this critical insight to the succession issue, we can establish a two fold causal relation between the two dimensions. If everybody is satisfied with the transition and the succession process, then it follows that they will be more committed to it, more participative, more flexible during negotiations, and therefore more effective in accomplishing an effective 'baton passing'. Furthermore, if the transition process is performed on time, as planned, and in an efficient manner, it is more than likely that everyone, or at least almost everyone, will be satisfied with it. Sharma et al. (2001) confirm this interaction by establishing a sequential cause–effect model of the relationship between initial satisfaction, effectiveness, retrospective satisfaction and succession's success, as depicted in Figure 23.1.

Identification of critical success factors that influence a succession's success

Along with the above distinction, we grouped different issues identified in the literature into five critical success factors (CSFs) that affect either the satisfaction with the succession process or the effectiveness of the process per se. In turn these five CSFs are affected by a number of other criteria that are also analysed below.

The incumbent's propensity to leave This factor has been cited as one of the most important in the business literature (Brady and Helmich, 1984; Christensen, 1953; Lansberg, 1988; Malone, 1989; Pitcher et al., 2000; Sharma et al., 2001; Vancil, 1987). This factor is affected by multiple criteria, such as the owner's fear of losing power both within the

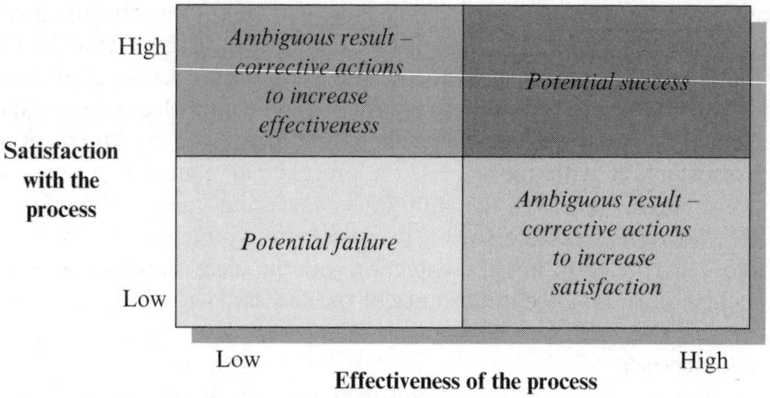

Satisfaction with the process	Low effectiveness	High effectiveness
High	*Ambiguous result – corrective actions to increase effectiveness*	*Potential success*
Low	*Potential failure*	*Ambiguous result – corrective actions to increase satisfaction*

Effectiveness of the process

Figure 23.1 Satisfaction, effectiveness and succession success

business and within the family, since withdrawal from the leadership position in the business could also mean an automatic withdrawal as head of the family. This may explain why over one-third of former family firm owners are still involved in the business even after retirement (American Family Business Survey, 1997). Moreover, owners usually link their personality and way of life with the family business and fear that by leaving the business they will lose their identity and status. Allen and Langowitz (2003) showed that 13.4 per cent of family business members claim that the CEO will 'never' retire. A safe conclusion from this discussion is that a low propensity of the incumbent to leave will affect satisfaction with the succession process in a negative way.

Successor's willingness to take over This factor has also been cited by the literature as very important (Barry, 1975; Bowen, 1978; Goldberg and Woolridge, 1993; Morris et al., 1997). It seems that the successor's willingness to take over depends on three main variables: commitment to the family; the maturity of the successor; and the degree of responsibility of the successor. The higher these three variables are, the higher the successor's willingness to take over, and consequently the higher the overall satisfaction with the succession process.

Successor's appropriateness and preparation The successor's appropriateness and preparation depends on a number of variables that are easily measurable and refer to the knowledge, skills and overall grounding of the successor (Kets De Vries and Miller, 1987; Morris et al., 1997). This critical success factor ensures that the successor is chosen not by gender but rather according to his/her abilities, namely, leadership, managerial and entrepreneurial skills, and preferably a degree of formal education. In addition, it is important for the owner to involve the successor in the business as early as possible in order to gain experience and commitment to the business through on-the-job training. On the one hand, there is a constant need for valuing everything that connects the business to its tradition, but on the other hand, as mentioned earlier, over-dependence on the past should be avoided. It is safe to assume that there is a positive relationship between the successor's appropriateness and preparation and the effectiveness of the succession process.

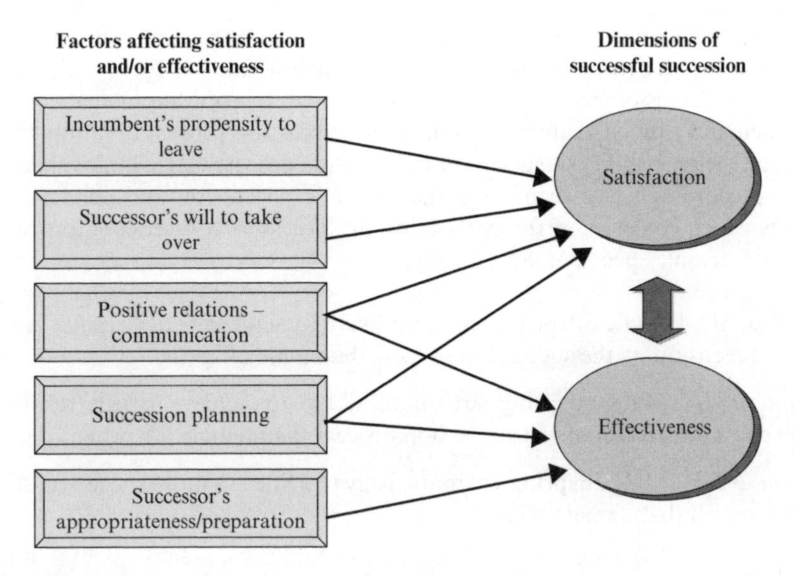

Figure 23.2 *Critical success factors affecting a successful succession*

Positive relations and communication It is obvious that if family members share the same values and show mutual respect satisfaction will be higher, and the transition will be handled more effectively (Dyer, 1986; Morris et al., 1997). Trust must be built among family members and everyone should clearly identify, acknowledge and accept their roles in the business as well as in the succession process through positive communication, and unmistakably know 'what's in there for them' in terms of personal gains in exchange for their support, so conflicts and rivalries that may affect the succession effort negatively are avoided. It is apparent that a positive relationship exists between good relations and communication and satisfaction with and overall effectiveness of the succession process.

Succession planning There is a great deal of evidence in the literature about the positive effects of good succession planning on the success of the succession transition (American Family Business Survey, 1997; Hayes and Adams, 1990; Lansberg, 1988; Morris et al., 1997). Business advisors strongly suggest incorporating a succession planning process and an exit strategy into the business plan very early, because the longer the family business succession planning, the smoother the transition process is likely to be (Ward, 1999), especially when the whole family is involved in the business succession planning discussions. Making a succession plan and then announcing it, however, is the surest way to sow family discord. Therefore, outside help with succession planning such as professional family-firm advisors, lawyers, or accountants injects credibility and objectivity into the process.

These contradictory and somewhat unrelated findings throughout the literature regarding the role of gender in the successful outcome of a succession process (Astrachan et al., 2002) and the integration of the discussion above produces the conceptual framework presented in Figure 23.2, which encompasses the critical success factors that affect a successful succession.

As a direct reflection of the aforementioned considerations about the nature of the problem and in order to cope with this issue and investigate whether gender affects the success of the succession process, we formulated research propositions that compare and contrast men and women's capability to lead the succession process in family businesses, and focus on the probability of success of a succession process with a male versus a female successor. Accordingly, we segmented the issues of succession and gender into three hypotheses which coalesce on the probability of success of a succession process with a male versus a female successor. Specifically:

Hypothesis I Female offspring are more likely to achieve better results in the satisfaction dimension of the succession process than male offspring.

Hypothesis II Male offspring are more likely to achieve better results in the effectiveness dimension of the succession process than female offspring.

Hypothesis III Male offspring are more likely to achieve a more successful succession process overall than female offspring.

Methodology

It is obvious that the issue we are dealing with is marked by:

1. Lack of concise and absolute measurements of the influence of each factor.
2. The existence of hierarchical relations, as described through the proposed conceptual framework.
3. Simultaneous manipulation of both quantitative and qualitative data.

In order to obtain an alternative-specific answer to our research hypotheses, which are based on the integrated conceptual framework of Figure 23.2, we used a multi-criteria analysis as the most appropriate analytical method to take advantage of the existing literature evidence. A novel methodological approach was chosen, namely, the Analytic Hierarchy Process (AHP) which is an Eigen value approach, and constitutes one of the most widely used multiple criteria decision-making tools. It provides a logical and scientific basis to decision-making, in which pair-wise comparisons of components are made with respect to a common goal or objective (Harker, 1988). Although the AHP tends to have problems when choosing among very close alternatives (Brugha, 2004), it enables decision-makers to exemplify the interaction of multiple factors in complex situations (Venkata Rao, 2004). Contrary to other multi-criteria decision-making methods however, the AHP is designed to incorporate tangible as well as non-tangible factors, especially when the subjective judgment of different individuals constitutes an important part of the decision process (Saaty, 1977). It has therefore been extraordinarily successful in resolving a vast array of problems (Steuer and Na, 2003), and its applications cover a wide range of fields, such as planning, resource allocation, conflict resolution, optimization, and financial decision-making, to name just a few (Harker, 1989; Steuer and Na, 2003; Vaidya and Kumar, 2003; Vargas, 1990; Zahedi, 1986).

Since the AHP is an Eigen value approach to the pair-wise comparisons (Saaty, 1980), the decision problem is structured hierarchically at different levels, each level consisting of a finite number of decision elements (Mikhailov, 2004). The top level of the hierarchy embodies the overall goal, while the lowest level is composed of all possible alternatives. One or more intermediate levels represent the decision criteria and/or sub-criteria. The relative importance of the decision elements (weights of the criteria and scores of the alternatives) is assessed indirectly from pair-wise comparison judgments (Saaty, 1977). The decision-maker provides a prioritized ranking order indicating the overall preference for each of the decision alternatives, by comparing criteria, sub-criteria and alternatives, with respect to the overall goal or objective. The process for determining both weights and scores is the same, so they are often called 'priorities'. The derived local priorities are further aggregated into 'global priorities', which are in turn used for final ranking of the alternatives and selection of the best. To accomplish this process, the existing literature was reviewed in order to weigh as accurately as possible the dimensions and the critical success factors of the succession process, as well as to make pair-wise comparisons between male and female successors.

In summary, the AHP is based on three principles (Saaty, 1990): the principle of constructing hierarchies, the principle of establishing priorities, and the principle of logical consistency. The empirical effectiveness and theoretical validity of the AHP, as well as its flaws have been discussed by many authors (Barzilai, 1998; Belton and Gear, 1983; Dyer, 1990; Harker and Vargas, 1987; Lai, 1995), but according to Zografos and Giannouli (2001), the AHP:

1. provides a structured way of judgment;
2. provides a uniform level of reliability of the results;
3. provides the ability of justification of the outcome;
4. provides a causal thinking;
5. combines qualitative and quantitative criteria;
6. takes into account the research literature expertise, but allows 'compromising solutions' when unavoidable; and
7. allows for sensitivity analysis.

Thereby, the selection of the Analytical Hierarchy Process is justified on the grounds that it is a decision-making tool that is quite suitable for selection problems (El-Wahed and Al-Hindi, 1998; Lai et al., 1999; Schniederjans and Garvin, 1997; Tam and Tummala, 2001; Tummala et al., 1997), and is capable of overcoming the difficulties of imprecise measurements while at the same time providing an alternative-specific answer to our propositions.

Research design

As mentioned above, in order to carry out the procedural steps of the AHP, a synthesis of the existing literature was undertaken to establish as accurately as possible weights for the satisfaction and effectiveness dimensions, as well as the critical success factors of the succession process so pair-wise comparisons between male and female successors can be completed. The exact weighting entailed the subjective judgment of the authors, somewhat limiting the scope of the research potential. As no significant prior research exists that correlates the outcome of the succession process with the successor's gender (Astrachan et al., 2002), the literature review that was undertaken by the authors examined over 250 studies and critically evaluated them in order to construct the conceptual framework first, and then develop comparative inter-gender weights for the critical success factors of the successful family firm succession process. To accomplish this, the screening criteria that were used were mainly the publication journal's focus and prominence and the number of references retrieved. Secondary parameters were the type of research and the date of the corresponding survey, placing more emphasis on empirical research and more recent articles. Similar criteria were used for the comparative assessment of the two genders on the derived attributes, but more importance was placed on the study's date as more recent research studies tend to equate the attributes of the two genders (Sonfield and Lussier, 2002). In addition and because of these circumstances, this part of the analysis was extended beyond family business and general management papers to more distinct areas, such as psychology, human relations, and organizational dynamics. Finally, owing to the absence of direct comparisons in the desired attributes, indirect implications were established between contiguous factors, and final classifications were drawn accordingly.

The theoretical establishment of the first setup step of the AHP thus, was developed from the conceptual framework section, where the objective (goal), the criteria (factors), and the alternatives were determined from the literature review. The transformation of the integrated conceptual framework into an AHP evaluation model is realized through a hierarchical decomposition of the evaluation of the critical success factors that influence the succession process and its successful or unsuccessful outcome (Figure 23.3).

For the purpose of our analysis, the hierarchical decomposition consists of three levels. The first level, which is the goal of the hierarchy, reflects the outcome of the succession process, that is, whether the realization of the succession process can be identified as successful or not. The second level in the hierarchy consists of the two dimensions that determine the success of the succession, that is, satisfaction with the succession process and effectiveness of the succession process. The third level consists of the factors that influence causally the dimensions of the second level. An extra level differentiates between male and female successors, which, contrary to previous studies, integrates the gender issue into the model in order to build a coherent and concise two-dimensional succession framework stressing both satisfaction with the process and the effectiveness of the overall succession process, but also the role of the successor's gender in the outcome of the succession process.

Next, the very significant second weighting step in our analysis consists of determining the relative significance of the two dimensions of the success of the succession process in order of importance. However, except for the distinction between satisfaction with and overall effectiveness of the succession process (Handler, 1989b), there is no actual research study or related findings about the relative importance of the two dimensions. Because of the interdependence between the two dimensions (Sharma et al., 2001), equal weights were assigned to the satisfaction and the effectiveness dimensions of the succession process, and with the use of the numerical scale of the pair-wise comparisons of the Expert Choice software (Expert Choice Inc., 1995), the weights for these two interdependent dimensions were both set to equal 1.00. As for as the critical success factors that are crucial to the successful outcome of an intergenerational transition process in a family firm, a lot has been written (Dyck et al., 2002; Sharma et al., 2003). However, little is known about the relative importance of each individual factor. Morris et al. (1997) suggest that the first priority of the succession process is the effort to build trust, encourage communication, and foster shared values, and that relationship issues account for

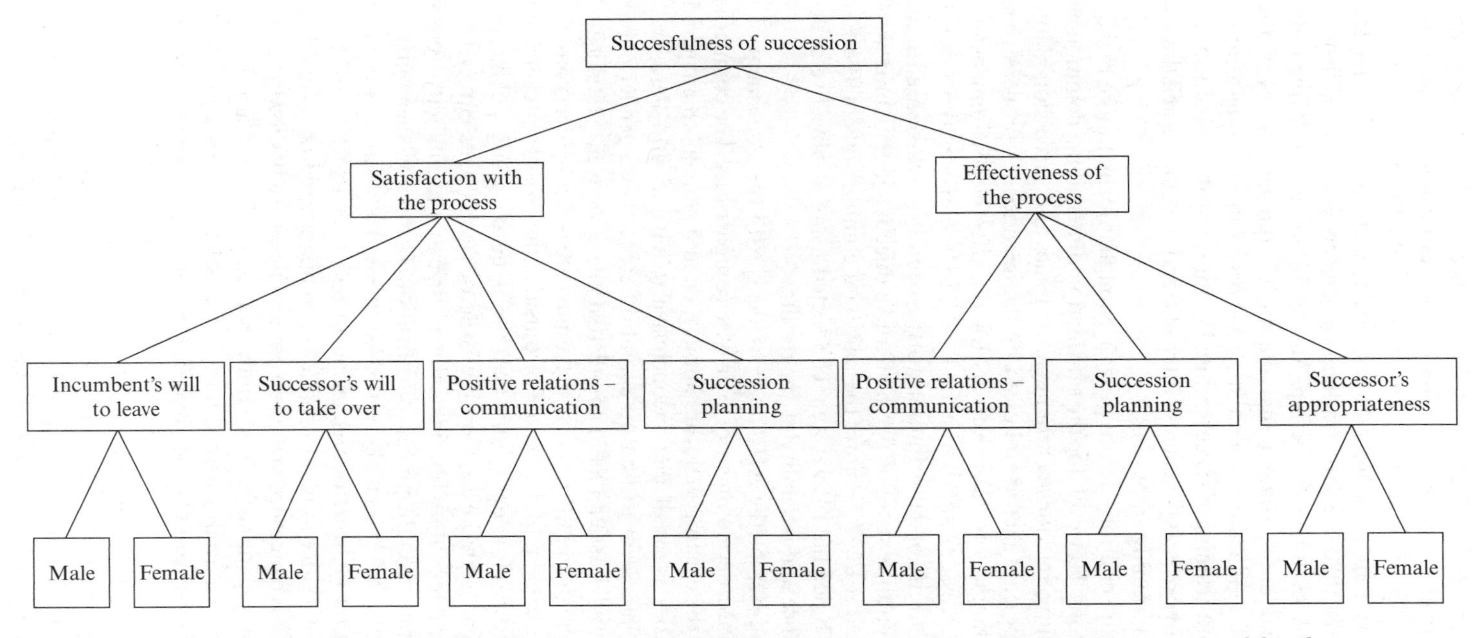

Figure 23.3 Hierarchical decomposition of the assessment of the critical success factors affecting the success of family business succession

approximately 60 per cent of the succession's success, the preparation of the heirs accounts for 25 per cent, and finally, planning and controlling activities for 10 per cent. In another study Birley and Godfrey (1999) state that the three most important factors in family business governance, in order of importance, are the heirs' freedom to choose whether they wish to take over, the training they receive, and the ability to distinguish between family and business affairs. An Australian survey by Barach and Ganistsky (1995), on the other hand, suggest that the most important issues in the succession process, in order of significance, are balancing short-term and long-term business decisions, preparing and training a successor, maintaining loyalty to non-family managers, balancing family concerns and business interests, and developing relationships between successor(s) and non-family employees.

These few literature citations do not provide enough information to make quantitative pair-wise comparisons. However, for the satisfaction dimension a common element in all of them is that positive relations/communication issues and succession planning are probably the most important factors, followed by the incumbent's propensity to leave, and the successor's willingness to take over. As for the effectiveness dimension, succession planning and the successor's appropriateness seem to be considered stronger factors than communication in terms of contribution to the overall succession effectiveness. Since these final weights entail a sense of personal and subjective judgment it was decided not to use a quantitative scale. Therefore, the pair-wise comparisons of the Expert Choice software (Expert Choice Inc., 1995) were conducted using a subjective scale in order to increase the coherence and reliability of the results.

The third ranking step of the AHP analysis, deals with the assessment of the alternatives' potential and the development of pair-wise comparisons. The comparison between the two genders in the existing literature about men and women's managerial and entrepreneurial capabilities is small and inconclusive (Chaganti and Parasuraman, 1996). Studies after the 1980s seem to find no actual differences in a gender comparison. For instance, Sonfield and Lussier (2003) found that there were no statistically significant differences between men and women in all examined areas (group decision-making, family member conflict, succession planning, use of outside advisors, long-term planning, financial management tools, founder's influence, the issue of going public, formal versus informal management style), except on the issue of debt versus equity funding, which shows that women prefer equity funding. A firm's size and profitability were also cited as significant factors that affect the process of transition. In this area men outperform women (Allen and Langowitz, 2003), as male-owned businesses have a greater profitability versus women at a ratio of 6 to 5, with average means of $30.4 versus $26.9 respectively. Women on the other hand own twice as many productive companies. As far as number of employees as a size indicator is concerned, a women entrepreneurs study shows that men have four times as many employees as women in family businesses (Santa Clara University, 2000). Another exploratory study in Spain (Shim and Eastlick, 1998) agrees and found that women-owned businesses are younger, have fewer employees, and lower revenues. Finally, with regards to counselling, some studies have found that women are less confident in their potential, and tend to form outside social networks on which to base their decision-making. This suggests that they make greater use of outside advisors than do men (Miller et al., 2001). Owing to the limited research literature, and to the ambiguity and inconclusiveness of the available studies as presented earlier in Table 23.2, in order to be able to perform as reliable comparisons between men and women on each of the aforementioned critical success factors as possible, the final weights again did not include the use of a quantitative scale but rather a reasonable application of personal judgment. Therefore, the pairwise comparisons were also conducted using a subjective scale in order to increase the rationality and reliability of the results. The conclusion was that in terms of successor managerial and entrepreneurial skills men are less conservative. Men also seem to be twice as likely to seek venture capital indicating a stronger growth orientation. From the discussion above it seems that the 'successor's appropriateness' factor favours male over female successors, and thus the male gender was given a 'moderate' superiority.

Finally, as mentioned earlier, the numerical calculations for the final fourth evaluating step were performed by the Expert Choice software package (Expert Choice Inc., 1995). The multi-criteria analysis conducted by the Expert Choice came up with a weighing of the two alternatives (male and female successors) as to their capability to produce a smooth generation transition. The sum of the two weights equals 1 and can be perceived of as a percentage, or a comparative indicator. After the evaluation was realized, sensitivity analysis was performed using the corresponding option of the Expert Choice software, with the objective of understanding the underlying basis of the overall conclusion.

Research findings

Men scored an overall rating of 0.501 while women were rated with 0.499 in their potential to undertake a successful succession process. The results suggest that in an 'ideal' literature synthesis, men and women are almost equal as far as their potential to realize a successful succession in a family business. With an overall inconsistency of the model equal to 0.00, it is quite surprising that we see the capacity of the two genders on the succession issue to be as close as it is. If we consider instead a 'distributive' model of the literature synthesis rather than the ideal one, we find very similar results with men at 0.495 and women at 0.505. The sum of the two weights equals 1 and can be perceived as a percentage, or a comparative indicator. No matter what aspect we examine, the results are very close to the absolute equation of one. The results suggest that men and women are almost equal as far as their potential to realize a successful succession in a family business. It is evident that our research leads to a solid conclusion that the success of the succession process does not depend on the gender of the successor. Therefore, Hypothesis III which asserted that male offspring are more likely to realize a more successful succession process overall than female offspring, is rejected.

Finally, another interesting aspect of the issue is to examine the outcomes of the sensitivity analysis tables, with the objective of understanding the underlying basis of the overall conclusions. The results show that women seem to outperform men on the 'satisfaction with the succession' dimension, while men seem to outperform women on the 'effectiveness of the succession' dimension. Therefore, both Hypotheses I and II which stated respectively that: female offspring are more likely to achieve better results in the satisfaction dimension of the succession process than male offspring, and male offspring are more likely to achieve better results in the effectiveness dimension of the succession process than female offspring, are both accepted.

If validated through further research, these findings can justify those who claim that men and women have complementary skills. Despite contradicting research studies

(Songfield and Lussier, 2003), men and women do not seem to perform comparatively and with equal degrees of effectiveness and efficiency on every managerially specific challenge or capability. On the contrary, women seem to possess different competencies, abilities, and modi operandi which are not only useful to the management of family businesses but also beneficial because they bring different perspectives to the various managerial, financial and emotional challenges of a family firm and, to a greater degree, any business organization trying to overcome internal and external challenges in general.

Conclusions

This chapter essentially attempts to make an important theoretical contribution to the family business literature. The conceptual framework presented here tries to fill the gap that exists in the understanding of the succession process by integrating efforts as far as critical success factors leading to a successful succession process, satisfaction with the process and effectiveness of the succession process specifically as they all relate to gender. On the other hand, the managerial implications of this framework could be of value to family business owners in their attempt to identify the factors that are critical for a smooth generation transition for their firm.

The hypotheses that were tested through the conceptual framework provided a pioneering insight into the gender-specific succession literature. If the results are validated through empirical field research, they will determine that as far as the issue of successful succession is concerned, the bias against women and the fact that in most cases they are being passed over for the ownership/control of the business is not reasonably founded. More importantly, our research suggests that men and women possess different skills that lead to different strategic competencies, all of which are very valuable to any business organization. Specifically, it seems that women outperform men on the relational issues, while men outperform women on performance issues.

One limitation of the research effort is the imprecise and contradictory state of affairs of the family business literature, especially the lack of a field research and the lack of non-English research studies and settings in general. More information and more non-US studies and papers could have influenced the results by bringing more data and an alternative or complementary view to the issue. Finally, as in every survey that depends on literature to provide new insights, the issue of accessibility of all important research studies is crucial and is thereby referred to in the limitations of the research.

The most fundamental limitation of the study, however, is that owing to the fact that the body of related existing research studies is small and somewhat inconclusive (Chaganti and Parasuraman, 1996; Sharma et al., 2001; Sonfield et al., 2001), this posed an exogenous constraint on the literature review and, consequently, the classification of the dimensions and factors of succession, as well as the pair-wise comparisons between men and women were approximated using a subjective pair-wise comparison through personal judgment by the authors. In addition, methodologically speaking, the Analytic Hierarchy Process did not include the interaction between the two dimensions of the succession process. Therefore, until the findings of this study are tested in practice the validity and reliability of the conceptual framework constructed here should be considered exploratory in nature, because of the assumptions and the subjectivity involved.

Finally, and as a result of the limitations cited above, the direction that future research could and should take is toward empirical field research studies focusing on the critical success factors in order to obtain a more accurate evaluation of their 'weight' toward a successful succession, and demonstrate and measure their importance as they relate to gender. The essence of this pioneering research study has been to provide an overall research direction, and not delve into the matter in a generalizable depth of analysis. Hopefully, the basis of sound future research effort direction has been set, and one can expect more accurate and in-depth findings about this so important but so thinly investigated issue in the future.

References

Achua, C. (1997), 'Changing the guard for family business as boomers retire and busters take over', 11th Annual Conference of the US Association for Small Business and Entrepreneurship, San Francisco, CA, June.

Allen, I.E. and N.S. Langowitz (2003), 'Women in family-owned businesses', report, Center for Woman's Leadership, Babson College-Mass Mutual Financial Group.

American Family Business Survey (1997), 'The Arthur Andersen/Mass Mutual American family business survey', www.arthurandersen.com/CFB.97surv.asp.

Astrachan, J.H., E. Allen, S. Spinelli, C. Wittmeyer and S. Glucksman (2002), *American Family Business Survey 2002*, Alfred, NY: The George and Raymond Family Business Institute.

Axelrod-Contrada, J. (2004), 'More daughters take lead role in family businesses', *Boston Globe 'Boston Works' section*, The UMass Family Business Center, www.umass.edu/fambiz/media_quotes.htm.

Bailyn, L. (1993), 'SMR forum: patented chaos in human resource management', *Sloan Management Review*, **34**(2), 77–83.

Barach, J.A. and J.B. Ganistsky (1995), 'Successful succession in family business', *Family Business Review*, **8**(2), 131–55.

Barry, B. (1975), 'The development of organization structure in the family firm', *Journal of General Management*, autumn, 42–60.

Barzilai, J. (1998), 'On the decomposition of value functions', *Operations Research Letters*, **22**, 159–70.

Belton, V. and T. Gear (1983), 'On a short-coming of Saaty's method of analytic hierarchies', *Omega*, **11**, 228–30.

Birley, S., D. Ng and A. Godfrey (1999), 'The family and the business: the governance of smaller businesses', *Long Range Planning*, **32**(6), 598–608.

Bjuggren, P.O. and L.G. Sudd (2001), 'Strategic decision making in intergenerational succession of small and medium sized family owned businesses', *Family Business Review*, **14**, 11–23.

Bowen, M. (1978), *Family Therapy in Clinical Practice*, New York: Aronson.

Brady, G. and D. Helmich (1984), *Executive Succession*, Englewood Cliffs, NJ: Prentice Hall.

Brenner, O., J. Tomkiewicz and V. Schein (1989), 'The relationship between gender role stereotypes and requisite management characteristics revisited', *Academy of Management Journal*, **32**, 662–9.

Brugha, C.M. (2004), 'Phased multicriteria preference finding', *European Journal of Operational Research*, **158**, 308–16.

Buttner, E.H. (2001), 'Examining female entrepreneurs' management style: an application of a relational frame', *Journal of Business Ethics*, **29**(3), 253–69.

Carsrud, A., C. Gaglio and K. Olm (1986), 'Entrepreneurs, mentors, networks, and successful new venture development: an exploratory study', in R. Ronstadt, J. Hornaday, R. Peterson and K. Vesper (eds), *Frontiers of Entrepreneurial Research*, Wellesley, MA: Babson College, pp. 199–235.

Carsrud, A.L. (1994), 'Meanderings of a resurrected psychologist, or lessons learned in creating a family business program', *Entrepreneurship Theory and Practice*, **19**(1), 39–40.

Chaganti, R. and S. Parasuraman (1996), 'A study of the impacts of gender on business performance and management patterns in small business', *Entrepreneurship Theory and Practice*, **21**(2), 73–5.

Christensen, C. (1953), *Management Succession in Small and Growing Enterprises*, Boston, MA: Division of Research, Harvard Business School.

Cole, P. (1997), 'Women in family business', *Family Business Review*, **10**(4), 353–71.

Crampton, S.M. and J.M. Mishra (1999), 'Women in management', *Public Personnel Management*, **28**(1), 87.

Crouter, A.C. (1984), 'Spillover from family to work: the neglected side of the work–family interface', *Human Relations*, **37**, 425–42.

Daniels, D. (1997), 'Women on top survey finds shift at family firms', *Atlanta Business Chronicle*, March, **24**, p. 4.

Dascher, P. and W. Jens (1999), 'Family business succession planning: executive briefing', *Business Horizons*, September–October, 2–4.

Davis, P.S. and P.D. Harveston (1998), 'The influence of family on the family business succession process: a multi-generational perspective', *Entrepreneurship Theory and Practice*, **22**(3), 31–53.

DeMartino, R. and R. Barbato (2003), 'Differences between women and men MBA entrepreneurs: exploring family flexibility and wealth creation as career motivations', *Journal of Business Venturing*, **18**, 815–32.
DiMatteo, B.C. (2004), 'Succession in family firms: opportunities and challenges', *Family Business Corner*, CMC, www.atlanticconsultants.com/ss_jan04.html.
Duman, R. (1992), 'Families are different', *Entrepreneurship Theory and Practice*, **17**, 13–21.
Dyck, B., M. Mauws, F. Starke and G.A. Mischke (2002), 'Passing the baton: the importance of sequence, timing, technique, and communication in executive succession', *Journal of Business Venturing*, **17**, 143–62.
Dyer, J. (1990), 'Remarks on the Analytic Hierarchy Process', *Management Science*, **36**, 274–5.
Dyer, W.G. Jr (1986), *Cultural Change in Family Firms: Anticipating and Managing Business and Family Transitions*, San Francisco, CA: Jossey Bass.
Dyer, W.G. Jr and M. Sanchez (1998), 'Current state of family business theory and practice as reflected in the *Family Business Review* 1988–1997', *Family Business Review*, **11**(4), 287–95.
Eagly, A., S. Karau and M. Makhajani (1995), 'The science and politics of comparing women and men', *American Psychologist*, **50**(3), 145–58.
El-Wahed, W.A. and H. Al-Hindi (1998), 'Applying the Analytic Hierarchy Process on selection of an expert system shell', *Advances in Modelling and Analysis*, **40**(2), 45–57.
Expert Choice Inc. (1995), *Expert Choice: User Support Manual*, Expert Choice software, Expert Choice Inc.
Family Business Network research (1995), *Massachusetts Mutual Life Insurance Company*, www.businessforum.com/family01.html.
FDU Family Business Forum (1995), www.fdu.edu/academic/rothman/article15.htm.
Filbeck G. and S. Lee (2000), 'Financial management techniques in family businesses', *Family Business Review*, **13**(3), 201–16.
Fischer, E., A. Reuber and L. Dyke (1993), 'A theoretical overview and extension of research on sex, gender, and entrepreneurship', *Journal of Business Venturing*, **8**(2), 151–68.
Goldberg, S.D. and B. Woolridge (1993), 'Self-confidence and managerial autonomy: successor characteristics critical to succession in family firms', *Family Business Review*, **6**(1), 55–73.
Handler, W.C. (1989a), 'Methodological issues and considerations in studying family businesses', *Family Business Review*, **2**(3), 257–76.
Handler W.C. (1989b), 'Managing the family firm succession process: the next generation family member's experience', doctoral dissertation, School of Management, Boston University.
Handler, W.C. (1990), 'Succession in family firms: a mutual role adjustment between the entrepreneur and next-generation family members', *Entrepreneurship: Theory and Practice*, **15**(1), 37–51.
Harker P.T. (1988), *The Art and Science of Decision Making: The Analytic Hierarchy Process*, Working Paper 88-06-03, Decision Science Department, The Wharton School, University of Pennsylvania, Philadelphia, PA.
Harker, P.T. (1989), 'The art and science of decision making: the Analytic Hierarchy Process', in B.L. Golden, E.A. Wasil and P.T. Harker (eds), *The Analytic Hierarchy Process: Applications and Studies*, Berlin: Springer, pp. 3–36.
Harker, P. and L. Vargas (1987), 'The theory of ratio scale estimation: Saaty's Analytic Hierarchy Process', *Management Science*, **33**, 1383–403.
Hayes, J.T. and R.M. Adams (1990), 'Taxation and statutory considerations in the formation of family foundations', *Family Business Review*, **3**(4), 383–94.
Hoover, E.A. and C.L. Hoover (1999), *Getting along in Family Business*, New York: Routledge.
Johnson, J. and P. Powell (1994), 'Decision making, risk and gender: are managers different?', *British Journal of Management*, **5**, 123–38.
Kaplan, E. (1988), 'Women entrepreneurs: constructing a framework to examine venture success and business failures', in B.A. Kirchoff, W.A. Long, W.E. McMullan, K.H. Vesper and W.E. Netzel, Jr (eds), *Frontiers of Entrepreneurial Research*, Wellesley, MA: Babson College, pp. 625–37.
Kets De Vries, M.F.R. and D. Miller (1987), *Unstable at the Top*, New York: NAL.
Lank, A. (1997), 'Making sure the dynasty does not become a Dallas', in S. Birley and D. Muzyka (eds), *Mastering Enterprise*, London: Pitman.
Lansberg, I. (1988), 'The succession conspiracy', *Family Business Review*, **1**(2), 119–43.
Lai, V.S. (1995), 'A preference-based interpretation of AHP', *Omega*, **23**, 453–62.
Lai, V.S., R.P. Trueblood and B.K. Wong (1999), 'Software selection: a case study of the application of the Analytical Hierarchical Process to the selection of a multimedia authoring system', *Information and Management*, **36**(4), 221–32.
Le Van, G. (1999), *The Survival Guide for Business Families*, New York: Routledge.
Lee-Gosselin, H. and J. Grise (1990), 'Are women owner-managers challenging our definitions of entrepreneurship? An in-depth study', *Journal of Business Ethics*, **9**(4, 5), 11–22.
Loscocco, K. (1997), 'Work–family linkages among self-employed women and men', *Journal of Vocational Behavior*, **50**, 204–26.

Malone, S.C. (1989), 'Selected correlates of business continuity planning in the family business', *Family Business Review*, **2**(4), 341–53.
Mikhailov, L. (2004), 'A fuzzy approach to deriving priorities from interval pair-wise comparison judgements', *European Journal of Operational Research*, **159**, 687–704.
Miller, D., L. Steier and I. Le Bretton Miller (2003), 'Lost in time: intergenerational succession, change, and failure in family business', *Journal of Business Venturing*, **18**, 513–31.
Miller, N., H. McLeod and K. Oh (2001), 'Managing family businesses in small communities', *Journal of Small Business Management*, **39**(1), 73–87.
Moore, D. and E.H. Buttner (1997), *Women Entrepreneurs: Moving Beyond the Glass Ceiling*, Thousand Oaks, CA: Sage.
Morris, M.H., R.O. Williams, J.A. Allen and R.A. Avila (1997), 'Correlates of success in family business transitions', *Journal of Business Venturing*, **12**(5), 385–401.
National Federation of Women Business Owners (1997), 'Women entrepreneurs are a growing international trend', www.nfwbo/org.
Olson, H. (2001), 'Dads, daughters and the family business', *Star Tribune*, 6 May, http://www.cebcglobal.org/Newsroom/News?News_050601.htm.
Organisation for Economic Co-operation and Development (OECD) (2000), 'Realising the benefits of globalisation and the knowledge-based economy', *The 2nd OECD Conference on Women Entrepreneurs in SMEs*, November, Paris: OECD, pp. 29–30.
Pitcher, P., S. Chreim and V. Kisfalvi (2000), 'CEO succession research: bridge over troubled waters', *Strategic Management Journal*, **21**, 625–48.
Powell, G. (1990), 'One more time: do male and female managers differ?', *Academy of Management Executive*, **4**(3), 68–75.
Saaty, T.L. (1977), 'A scaling method for priorities in hierarchical structures', *Journal of Mathematical Psychology*, **15**, 234–81.
Saaty T.L. (1980), *The Analytical Hierarchy Process: Planning, Priority Setting, Resource Allocation*, New York: McGraw-Hill.
Saaty T.L. (1990), Decision-making for leaders: the Analytic Hierarchy Process for decisions in a complex world, Pittsburgh, PA: RWS Publications.
Santa Clara University for Innovation and Entrepreneurship (2000), *Women Entrepreneurs Study*, Santa Clara, CA: Santa Clara University.
Schniederjans, M.J. and T. Garvin (1997), 'Using the Analytic Hierarchy Process and multi-objective programming for the selection of cost drivers in activity-based costing', *European Journal of Operational Research*, **100**(1), 72–80.
Sharma, P., J.J. Chrisman and J.H. Chua (2003), 'Predictors of satisfaction with the succession process in family firms', *Journal of Business Venturing*, **18**, 667–87.
Sharma, P., J.J. Chrisman, A.L. Pablo and J.H. Chua (2001), 'Determinants of initial satisfaction with the succession process in family firms: a conceptual model', *Entrepreneurship Theory and Practice*, **25**(3), 17–35.
Shim, S., and M.A. Eastlick (1998), 'Characteristics of Hispanic female business owners: an exploratory study', *Journal of Small Business*, **36**(3), July, 18–34.
Sonfield, C.M. and N.R. Lussier (2003), 'Family firm ownership and management: a gender comparison', paper presented at the Proceedings of the National Conference of the Small Business Institute Directors Association, University of Central Arkansas, Hofstra University, Springfield College, February.
Sonfield, M., R. Lussier, J. Corman and M. McKinney (2001), 'Gender comparisons in strategic decision-making: an empirical analysis of the entrepreneurial strategy matrix', *Journal of Small Business Management*, **39**(2), 55–63.
Steuer, R.E. and P. Na (2003), 'Multiple criteria decision making combined with finance: a categorized bibliographic study', *European Journal of Operational Research*, **150**, 496–515.
Tam, M.C.Y. and V.M.R. Tummala (2001), 'An application of the AHP in vendor selection of a telecommunications system', *Omega*, **29**, 171–82.
Taylor, E. (2002), 'Women reach the top at most family firms', *Startup Journal – The Wall Street Journal Center for Entrepreneurs*, 14 March, www.startupjournal.com/.
Tummala, V.M.R., K.S. Chin and S.H. Ho (1997), 'Assessing success factors for implementing CE: a case study in Hong Kong electronics industry by AHP', *International Journal of Production Economics*, **49**, 265–83.
Vaidya, O.S. and S. Kumar (2003), 'Analytic Hierarchy Process: an overview of applications', *European Journal of Operational Research*, **169**, 1–29.
Vancil, R.F. (1987), *Passing the Baton: Managing the Process of CEO Succession*, Boston, MA: Harvard Business School Press.
Vargas, L. (1990), 'An overview of Analytic Hierarchy Process: its applications', *European Journal of Operational Research*, **48**(1), 2–8.

Venkata Rao, R. (2004), 'Evaluation of metal stamping layouts using an Analytic Hierarchy Process method', *Journal of Materials Processing Technology*, **152**, 71–6.
Ward, S. (1999), 'Family business succession planning: succession planning issues for family-run businesses. Your guide to small business', sbinfocanada.about.com/cs/buysellabiz/a/succession1.htm.
Wong, B., W. McReynolds and W. Wong (1992), 'Chinese family firms in the San Francisco Bay areas', *Family Business Review*, **5**, 355–72.
Wortman, Jr, M.S. (1994), 'Theoretical foundations for family-owned businesses: a conceptual and research based paradigm', *Family Business Review*, **7**(1), 3–27.
Zahedi, F. (1986), 'The Analytic Hierarchy Process: a survey of methods and its applications', *Interfaces*, **16**(4), 96–108.
Zografos, K. and M. Giannouli (2001), 'Development and application of a methodological framework for assessing supply chain management trends', *International Journal of Logistics: Research and Applications*, **4**(2), 153–90.

PART VI

FAMILY BUSINESS PERFORMANCE: GLOBAL AND TRANS-CULTURAL ISSUES

24 Internationalization of family businesses through strategic alliances: an exploratory study

Kristin Cappuyns

Introduction

There are a number of phenomena, aside from competitive pressures, such as globalization and rapid product life cycles, that encourage companies to embark on internationalization strategies by exporting their products, establishing wholly owned foreign subsidiaries or entering into strategic alliances. Strategic alliances offer numerous potential benefits to small firms, including the ability to tap into new markets, access economies of scale, obtain complementary resources in under-developed value chain activities, respond to environmental uncertainties, and receive endorsements from reputable incumbents (Ariño and Reuer, 2003; D'Souza and McDougall, 1989; Deeds and Hill, 1996; Dickson and Weaver, 1997; Eisenhardt and Schoonhoven, 1996; Gomes-Casseres, 1997; Hara and Kanai, 1994; Larson, 1991; Shan, 1990; Stuart et al., 1999).

Hence, at some stage of their business life cycles, both family businesses (FBs) and non-family businesses (NFBs) need to trust outside business partners, despite the fact that these businesses may have conflicting interests or may even be competitors. In other words, entrepreneurial risk-taking is important for the survival and successful performance of family firms (Rogoff and Heck, 2003), even though these activities are time consuming and their payoffs are uncertain (Zahra, 2005, p. 35). Researchers have expressed their concerns on several occasions about the fact that family businesses become more risk averse as they grow older, and prefer conservative strategies that limit their growth and profitability (Sheperd and Zahra, 2003).

In fact, a number of recent publications have demonstrated how this risk aversion is manifested, in comparisons of the international activities of family businesses with those of non-family businesses. Family businesses are seen to produce a lower level of exports and direct foreign investments, and when a family business enters the international marketplace, the pace of the process tends to be slower than in non-family businesses (Fernández and Nieto, 2005; Flören, 2001; Gallo and García Pont, 1993; Gallo and Sveen, 1991).

These publications, moreover, agree that there are certain internal characteristics of family businesses, such as concern about loss of control, risk aversion, the overlap of the family, the business and the ownership system, the delayed succession process, and the prolonged presence of the same people at the head of the organization, which can be identified as the causes of the slowness of family businesses compared to non-family businesses (Flören, 2001; Gallo et al., 2004; Ward, 1987).

At the same time, we must bear in mind that growing advances in the development of strategic alliances among companies in different countries, along with an increase in the knowledge of how to make these alliances successful, may provide opportunities for driving the internationalization of FBs at greater speed.

Nevertheless, the literature on strategic alliances does not consider the influence which the characteristics of the individual partners (for example, whether or not they are FBs) can have on a decision to form strategic alliances, as well as on the alliance's development process. In other words, even if being a family business does not appear, a priori, to influence the strategic goals pursued, entering into a strategic alliance does affect these characteristics of the company: its values, management process, and use of control mechanisms. Therefore, it is appropriate to expect that some of unique characteristics of family businesses will have a bearing on the formation and development of the strategic alliance. In fact, results have indicated that when both partners are family businesses, this helps them to bridge cultures in international contexts better than other organizations, as they share the universality of family (Swinth and Vinton, 1993).

At the same time, in order to determine the impact of these characteristics, this study analyses strategic alliances with partners in emerging markets, as this will add another dimension, resulting from the significant cultural distance between partners, which considerably challenges the family businesses' risk aversion as well as their dislike of sharing control with partners with substantial cultural differences.

This chapter proceeds as follows: in the next section we give an extended review of literature in both the family business and strategic alliance fields. This is followed by a section about the sample and research methodology. After that, the main findings are presented in two different sections: the first part of the findings illustrates that when the family business embarks on internationalization projects, the firm's product, financial resources and organizational capacity are key. The second part of the results goes a bit further; we have found taxonomies which confirm that family businesses, in order to succeed in forming and developing strategic alliances, must in the first place improve their ability to manage in contexts where objectives are not shared; they must build personal preferences for the use of alliances; and they must develop trust towards their partners.

The status of the field: conceptual framework

The field of family business internationalization has not yet been very closely studied, as the general idea exists that family business simply do not grow. Contrary to this popular perception, family businesses can grow, but they need to take into account their specific characteristics when doing so, by following some simple but critical steps, as mentioned by Ward (1987). In the first place, several authors have illustrated the general phenomenon that family businesses, compared with non-family businesses, are slower, start later, and are more prudent in making international commitments. Among the small number of current publications with a clear connection to family businesses, the following are worth mentioning.

Gallo and Estapé (1992), working with a rather small and partly opportunistic sample of businesses, found that family businesses tended to internationalize later and much more slowly than non-family businesses. Another study, also in Spain, led by Gallo and García Pont (1993), using factorial analysis and analysis of declines in a sample of 57 companies, affirmed that a focus on products aimed principally at the local market and an inadequate level of technology appear to be the main causes of the 'rigidity' of family businesses with respect to internationalization strategies.

The results of a study using a representative sample of Spanish small and medium-sized enterprises (SMEs) family businesses from 1991 to 1999 show that there is a clear negative relationship between family ownership and international involvement, measured both by export propensity and export intensity. In other words, there are few family businesses that export, and those that do so export to a lesser extent than do other SMEs (Fernández and Nieto, 2005, p. 86). One of the main reasons is the difficulties that family firms face in accessing the resources and capabilities essential to building competitive advantages at an international level (Fernández and Nieto, 2005, p. 86).

In fact, the literature seems to identify the factors that might justify this delay or slower progress. Gallo and Sveen (1991) discuss the important change that internationalization signifies for a family business, and point to features of the company's culture, strategy and organization that may hinder the internationalization process. They also discuss the characteristics of the different stages of the company's life cycle and the qualities of the owning family that may assist the process. But what are some of the most common characteristics to be considered?

One basic characteristic is the fact of having one family with a controlling interest in the business, whose members are unwilling, or even afraid, to lose the control and power that comes with ownership, when undertaking an international venture (Donckels and Fröhlich, 1991; Gallo et al., 2001).

The results of a study of US manufacturing companies highlighted that it is easier for the owner family to appreciate the financial and strategic benefits of market expansion than it is to form an alliance whereby they have to share their knowledge and capabilities with other companies. Alliances can leak information about a company's operations to other firms; they also require a great deal of integration and coordination among partners and take time to contribute to the family firm's profitability (Zahra, 2005, p. 36).

Another study, of 109 Dutch family businesses, shows that a very strong barrier to the internationalization of these businesses comes in the form of resistance from family and shareholders, as they do not want shares to be held by third parties (Flören, 2001, p. 22).

Apart from these findings, a recent study based on a sample of 222 firms from the south of Spain (Andalusia), representing a total of seven industrial sectors with a significant level of exporting companies, illustrates, through a multidimensional model including variables at both the individual and firm level, to what extent family involvement affects the internationalization process. The results show that the influence of family involvement on a company is a mediating value within the model, together with other variables such as the size and age of the firm, the individual's demographic characteristics or the perception of risk (Casillas and Acedo, 2004, p. 22).

Still, not all studies defend the same point of view, as shown by the outcome of a study by Swinth and Vinton (1993), which questions to what extent family businesses have strategic advantages in undertaking international joint ventures, and argues that although family businesses desire control, they can tolerate a joint venture with its inherent lack of control because each firm controls part of the venture (Swinth and Vinton, 1993, p. 24).

A second relevant feature of family businesses is the prolonged presence of family members in management and governance roles. On the one hand, this offers flexibility and can speed up strategic decision-taking, but it has also some negative consequences as shown below. On the other hand, recent studies clearly have shown that the length of a CEO's tenure is negatively associated with entrepreneurial risk-taking, especially a family firm's emphasis on innovation and venturing into domestic and international markets (Zahra, 2005, p. 36).

Another very recent study (Graves and Thomas, 2004), which is based on the results of a longitudinal database of Australian family and non-family businesses, highlights that, compared with non-family businesses, family businesses lag behind in building their managerial capabilities as they progress towards advanced stages of internationalization. Besides, the results suggest that family businesses are less likely to appoint outside managers to take the lead in their internationalization processes, counting instead on suitable family members to become the leading figures.

Nevertheless this is not a priori harmful to the family businesses' continuity; on the contrary, recent findings clearly underline the way the firm can benefit from the incorporation of qualified family members.

Members of the owner family have a special incentive to encourage a firm's focus on innovation, because the success of their company increases their wealth. The results echo the call for greater participation by the family in the life of the firm as a way of achieving strategic renewal (Fernández and Nieto, 2005, p. 86; Gallo and García Pont, 1993; Gersick et al., 1997; Miller et al., 2003; Ward, 1987; Zahra, 2005, p. 38).

A third important characteristic is the impact of a strong business culture. There are many contradictions in the abundant literature on this subject, but we will limit ourselves to a few works that have related culture to the internationalization process.

Family businesses in general are often characterized as having a culture that is inward-looking and resistant to change, and where decision-makers are constrained by the firm's history and tradition (Dyer and Handler, 1994; Gersick et al., 1997; Kets de Vries, 1993). On the contrary, many of these businesses have learned that it is possible to gain strategic advantages by building management systems based on trust and loyalty, a possibility not, however, inherent in all families (Swinth and Vinton, 1993). At the same time, many of the cultural characteristics of family businesses are the same all over the world, making it easier to form strategic alliances, as noted by Gallo and Sveen (1991, p. 187).

Similar results are provided by a study about strategic alliances among family businesses, in which Swinth and Vinton state that there are always some very powerful cultural differences between countries, especially between partners in emerging markets, and the fact that businesses are family owned does not eradicate these differences. But as they share the unique characteristics of being family-owned businesses, this might at some level ease the partnership (Swinth and Vinton, 1993).

Furthermore, the literature on strategic alliances could maybe help us to learn more about the role of family business partners. Unfortunately, so far this field (Gulati, 1995; Parkhe, 1993; Ring and Van de Ven, 1994) has not paid sufficient attention to the influence that certain of the partners' characteristics, such as the fact of being a family business, can have on the decision to form a strategic alliance, as well as on the evolution of the alliance. Research carried out in this area shows that the decision to enter into a strategic alliance is influenced by the partners' perceptions of compatibility between their strategic goals and corporate cultures, in the widest sense of the word. Although the fact of being a family business may not, a priori, influence a company's strategic goals, it will influence its cultural characteristics; its values, management processes and use of control mechanisms (Kogut and Singh, 1988, p. 4).

It is reasonable, therefore, to expect that the intrinsic characteristics of family businesses definitely have a bearing on the formation of strategic alliances. In spite of the unique capabilities and challenges associated with family businesses, there has been very little research on the ability of these businesses to create, reconfigure and exploit their resources and capabilities to undertake strategic alliances in the global marketplace. Consequently, the precise objective of this study is to detect which characteristics do have an influence and to what extent they can positively or negatively influence the formation and development of the alliances.

Research methodology

The prompt to carrying out this research in an exploratory way was the scarcity of research into the internationalization of family businesses in general, and particularly their involvement in strategic alliances. The methodology used was the analysis of qualitative data collected in semi-structured interviews.

The sample includes Spanish family businesses that had formed strategic alliances in order to enter emerging markets, or had unsuccessfully tried to engage in strategic alliances, or that were seriously thinking about forming them. Latin American alliances were left out of the study because the cultural distance between these countries and Spain is smaller than the cultural distance between Spain and other countries with emerging economies. Hence, the formation and development of strategic alliances with countries outside Latin America more clearly reflects the challenges posed by such initiatives.

The decision to initiate strategic international alliances is usually rooted in a firm's desire to enter emerging markets and expand geographically. The family businesses examined belonged to different stages in the formation of strategic alliances in order to identify the characteristics of family businesses that influence the different phases of the process.

A total of 13 family businesses participated in the study, representing three industrial sectors: the food and beverage sector (4), textiles (4) and mechanical equipment (5). We considered a business to be an FB when the following three conditions are met: (1) the proportion of equity owned by one family is large enough for the family to control the company; (2) family members are involved in the management and/or governance of the company; and (3) at least the second generation of the family is involved in the company.

Interviews were conducted between December 1999 and March 2000 with managers of Spanish family businesses that had formed strategic alliances in order to penetrate emerging markets. By strategic alliance (SA) we understand an explicit agreement among firms to collaborate in a limited aspect of their activity for a relatively long term, and this may or may not result in a separate organizational entity. In fact, some, but not all, have managed to achieve their objectives via these alliances.

One person was interviewed per company. This person belonged to the top management team that had been involved in the strategic alliances. All the family businesses had at least some members of the second generation working in the business, but three of them were in the fourth generation already. Tables 24.1 to 24.5 provide more numeric data about the 13 family businesses of the sample, such as; generation involved at the moment the strategic alliance was formed, current generation incorporated (at the moment of the interviews), ownership, alliance and the nationality of the partners.

Data were collected from semi-structured interviews with a person from the top-level management position in the company. The information from the interviews was complemented by archive data obtained during visits to the company, and from press cuttings. The

Table 24.1 Generation at the time of making the strategic alliance

Generation	Number	%
1	1	8
1–2	2	17
2	5	42
3	3	25
4 or more	1	8
Total	12*	100

Note: * In one case the alliance was not achieved.

Table 24.2 Current generation

Generation	Number	%
1 and 2	3	23
2	4	31
2 and 3	2	15
3	2	15
4 or more	2	15
Total	13	100

Table 24.3 Percentage of ownership controlled by the family

Family ownership	Number	%
100%	7	54
99–50%	4	31
<50%	2	15
Total	13	100

Table 24.4 Type of alliance

Type	Characteristics	Number	%
Capital alliance	Majority	6	50
Capital alliance	Equality	3	25
Capital alliance	Minority	1	8
Contractual alliance	Contractual	2	17
Total	*	12	100

Note: * In one of the cases the alliance was never formally constituted.

Table 24.5 Nationality of the partner

Emerging markets	Number	%
Asia	5	42
W. Europe	4	33
E. Europe	3	25
Total	12	100

data were analysed using the data analysis tools and techniques suggested by Miles and Huberman (1984). The procedures suggested by these authors include using descriptive and analytic matrices designed ad hoc by researchers to facilitate data reduction, deduction of conclusions and verification. The research team that worked on this project held various meetings at this stage to compare partial conclusions and draw further conclusions.

Results: positive synergies and behavioural patterns in the formation and development process

The results of this study were twofold: on the one hand, all family businesses in the sample had a very similar set of variables that seemed to be fundamental when a family business was seriously considering the possibility of initiating and developing strategic alliances; on the other hand, the sample presented different forms of behaviour with respect to the formation and development of these strategic alliances. In the following paragraphs we illustrate these two set of results.

Variables of synthesis

The global economy has obliged both family-owned and non-family-owned firms to enter international markets to compete. It is important for family businesses to realize that they possess distinctive resources related to the influence of the family and ownership developmental cycles that may assist them in their international expansion. It is a question of being aware of, and taking maximum advantage from, the right allocation (Gersick et al., 1997; Habbershon and Williams, 1999).

The presence of family businesses in the global marketplace illustrates that not all family businesses are averse to growth and risk-taking. On the contrary, some of them undertake domestic and international strategic alliances to upgrade their existing capabilities or to acquire and develop new skills that expand their growth options. These risky moves require a significant resource allocation and demand major changes in these companies' internal decision-making process, without guarantees of financial success (Zahra, 2005, p. 24).

In the following paragraphs we illustrate how the family businesses in our sample proceed when initiating alliances, as well as during their development process. The findings allowed us to think in terms of three 'variables of synthesis'. By the expression 'variables of synthesis' we mean that each variable gives rise to a 'virtuous spiral', resulting, on the one hand, from positive synergies between different dimensions and, on the other, from the learning ability of the many players (owners, managing family members from different generations, and so on) who have so much influence on the strategic management of the FB.

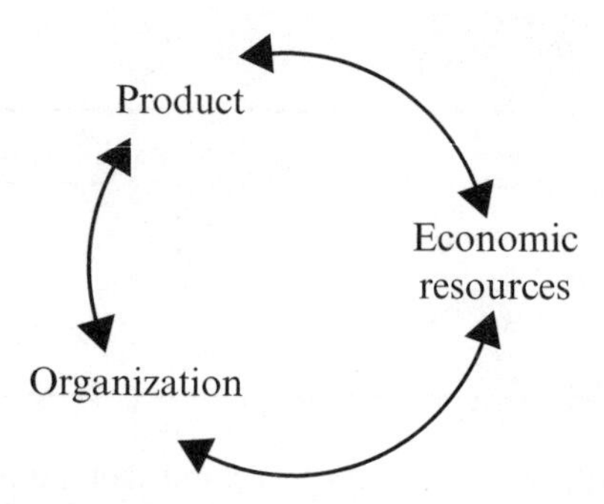

Figure 24.1 Variables of synthesis

These variables turn out to be critical to the success of the internationalization process (Figure 24.1), both because of what each means individually, and because of the relationships between the variables, which tend to strengthen them.

Product The first step in the learning process comes when the product or service becomes a leader in the niche market of the business activity. This position encourages an examination of opportunities in other competitive markets where this leadership position will be put to the test.

Economic resources As a consequence, family businesses are forced to invest financially in those markets as much as they can. At the same time, family businesses observed how they learnt at this stage that for efficient cooperation with a partner, the political power resulting from ownership of the majority of the share capital is not the right element to be taken into account. However, high levels of both professional competence and integrity in both partners will be determining factors in an alliance's success. This has been confirmed in a study that states that those family businesses that open their equity to other shareholders are more active in the international marketplace (Fernández and Nieto, 2002). It is not only the entrance of new capital that is an advantage; the shareholder can also help the family firm acquire knowledge of foreign markets, so reducing the uncertainty of the process (Fernández and Nieto, 2005). In addition, the entrance of a new shareholder involves the professionalization of management, as the family business will be accountable to third parties for its actions on a regular basis (Fernández and Nieto, 2005). This brings us to the third variable: organization.

Organization All the family businesses interviewed have very strong organizational structures, most of them headed by the founder or a successor, who gives an example of the ELISA values in his way of living and directing the business. The ELISA values are 'Excellence, Labor ethic, Initiative, Simplicity of life style, Austerity'. This set of values, which has been found to be present in very successful Spanish family businesses (Gallo and Cappuyns, 1999, p. 11), is implemented in the business culture, and because the family businesses share the universality of family, this may enable them to better bridge cultures in international contexts than other organizations (Swinth and Vinton, 1993, p. 19).

The 'virtuous spiral' resulting from positive synergies between the three dimensions of product, economic resources and organization, and the fact that it happens to be a

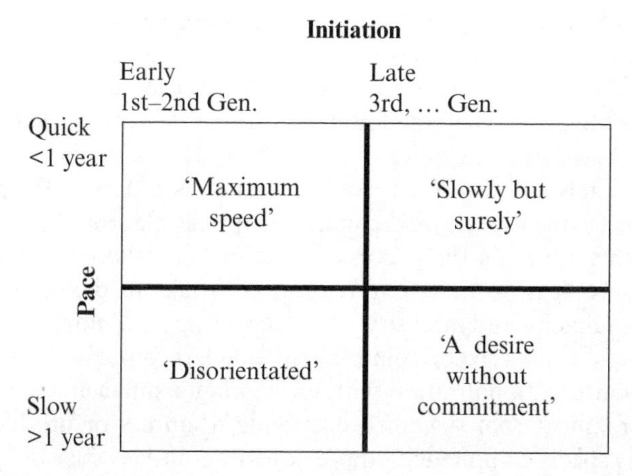

Figure 24.2 The start-up process

learning process, confirms findings from previous studies in the field of entrepreneurship, where Eriksson found that when the firm decides to go abroad, it must develop routines compatible with its internal resources and competence, and that can guide the search for experiential knowledge about foreign markets and institutions (Eriksson et al., 1997, p. 353).

As the company advances along this virtual spiral, it acquires three strengths that are key to success in strategic alliances: a very solid organizational structure which makes these businesses attractive for capable managers by fostering trust, especially for shareholders, and focussing the personal preferences of owner-managers towards a viable international strategy.

In the following paragraphs we illustrate how the family businesses in our sample benefit from these positive synergies during the start-up and formation process of the strategic alliance.

Taxonomies in the processes

The internationalization of companies is not, except in very exceptional circumstances, a change from 'nothing to everything', that is to say, an immediate step from 'today I am local and tomorrow I will be international'. Internationalization is a process in which different stages must be reached and in which advances and setbacks occur, depending on knowledge and on how the commitment made by the company in order to reach external markets is materializing. The figures provide more details about the importance of this commitment and the different variables that are important in order to measure it.

The start-up process An analysis of the companies in our sample shows how there are broad differences between them in two frequently quoted aspects of internationalization processes. Figure 24.2 illustrates the aspects of 'when this process will really get started in the FB life cycle', and 'the pace of the initiation of the strategic alliance'. The combination of these two aspects can be termed the 'level of commitment to internationalization'.

The four different patterns of behaviour illustrate a decreasing level of commitment to the internationalization process:

- 'Maximum speed' is the type of commitment observed in family businesses with a founder convinced of the opportunity and need for internationalization. Strategic alliances in this group are very successful. The decision is taken at an early stage in the business life cycle, when the founder is still at the top of the firm, and the pace of the process is extremely quick.
- 'Slowly but surely' is a positive commitment, not as certain as the previous one, but one which is frequently the only commitment possible from family businesses that have problems initiating the process. However, once these problems are resolved, once they have 'learnt', these family businesses take to the road to internationalization with tenacity and intensity, even recovering 'lost' time.
- 'Disorientated' is the type of commitment which is found in family businesses that launch into internationalization without gaining a sufficient level of development of either organizational structure, economic resources or product leadership (as regards the aspects commented on previously about variables of synthesis).
- 'A desire without commitment', that is to say, lacking true commitment but justifying its position using arguments. Reasons are found such as 'none of the potential partners in the target country realizes the importance of "our" contribution to the alliance', and so on. This attitude finally leads to failure in setting up a strategic alliance, and all possible excuses are used to justify it.

Moreover, in the interviews it has been commented that the businesses in the first two categories mentioned above, 'Maximum speed' and 'Slowly but surely', started the internationalization process at a very early stage. Their attitude confirms the results of a study about effects of the age at entry into the global marketplace, in which the authors proposed that as firms get older, they develop learning impediments that hamper their ability to successfully grow in new environments and that the relative flexibility of newer firms allows them to rapidly learn the competencies necessary to pursue continued growth in foreign markets (Autio et al., 2000, p. 921). For family businesses that do not start the internationalization process in the first or second generation it is very unlikely that they will do so in the future (Okoroafo, 1999, p. 154).

Furthermore, the same study states that early venturing across borders establishes a self-reinforcing pattern (Amburgey et al., 1993; Cohen and Levinthal, 1990) that implants a proactive culture in the firm, enhancing not only the ability to see and realize foreign opportunities, but also the willingness to do so (Penrose, 1959). This interpretation is consistent with the work of Brush (1992), who found that early internationalizers held more positive attitudes towards non-domestic markets than did late internationalizers; and this has been confirmed with the family businesses from this Spanish sample, where the age of the business (in function of generation) and the pace of time for the start-up reflect the level of commitment to internationalization.

Nevertheless, for those family businesses in our sample that took the decision to go abroad in the first generation, this was not a coincidence, but a well considered decision, taken by leaders who have tremendous personal energy and passion for growth and have been defined as 'masters of growth' (Ward, 1987).

The development process Our analysis also indicates a second important difference between the family businesses in the sample when dealing with the execution of these

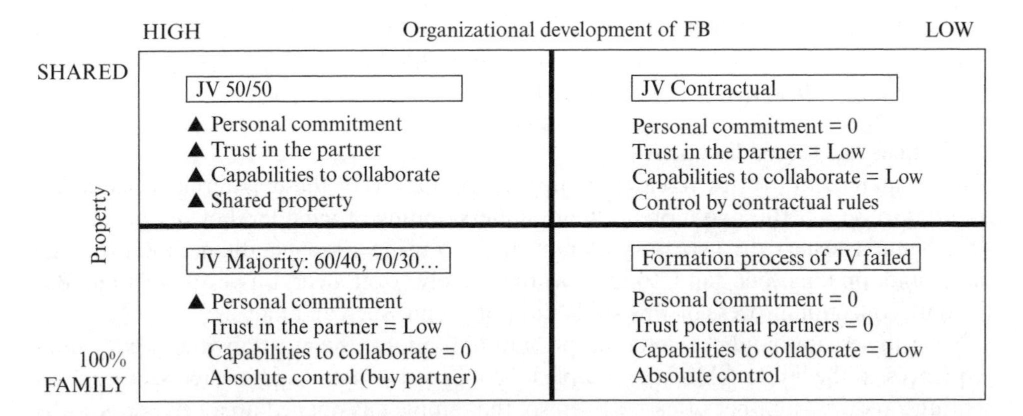

Figure 24.3 Organizational development of a family business

alliances. There are two determining factors at the execution level: the first is the level of family ownership, that is to say, whether the family is the 100 per cent owner of the company's share capital, or if there are other partners from outside the family. The second factor relates to the level of organizational development of the FB, that is, whether at the governing level there is an active and useful board of directors (who are not there merely to cover legal requirements) and whether at the management level there is a professional and well-prepared organizational structure.

Figure 24.3 illustrates the far-reaching impact that these two factors – ownership and organizational development – can have on the content and performance of strategic alliances in which family businesses are engaged. The results presented in Figure 24.3 suggest that the use of strategic alliances as a vehicle for internationalization, apart from a strong personal commitment, requires other factors such as trust in the partner and managerial capabilities.

At a higher level, there is a need to trust that allies will maintain a relationship with the ownership structure and the developments of the family business. An ownership structure shared with third parties outside the family facilitates the development of the ability to manage in the context of unshared aims as a strategic alliance, as well as the capacity to trust the partners. 'Commitment to trust, reliability and mutual benefit is exactly the kind of old-fashioned thinking that ideally positions family firms for the future' (Ward and Aronoff, 1991, p. 44). Family firms that retain these fundamental values as guides to decision-making and operations find themselves at a strategic advantage (Lansberg and Perrow, 1991).

The very successful family businesses in our sample, as can be appreciated in Figure 24.3, had a strong commitment as well as trust in their partners, and that was especially reflected in the fact that they did not strive for a majority percentage of the partnership. As the level of trust decreases, the strategic alliances are bound by a set of contractual rules evolving to absolute control by keeping a comfortable majority of property.

Another aspect of our findings is related to managerial capacity. This finding is completely in line with the results from another Spanish study that concluded that the managers' backgrounds are one of the determining factors to be taken into account, apart from the firm's size and age, for understanding the large set of causal relationships with regard to the company's level of internationalization (Casillas and Acedo, 2004).

Finally, Figure 24.3 illustrates how different combinations of these influential characteristics such as commitment, trust, property and organizational development, result in the use of different types of alliances, or in some cases their non-use.

Limitations and future directions

There is material for future research in the further analysis of these patterns of behaviour in order to get a better and more in-depth understanding of what level of organizational development fosters the necessary commitment to using strategic alliances, on the one hand, and, on the other hand, to see how these evolve over a certain period of time. But the data collection process of this study did not permit such an analysis.

Nevertheless this study has other important limitations; the most obvious are the consequences of the limitation of the sample by country, by sector and in each sector representing a reduced number of cases. Even so, the sample has opened up a new set of ideas that could be a good starting point for future studies.

Hence there is first an important need to get these results tested to a more heterogeneous sample of family businesses, by sector, country of origin, size, and so on, as this will offer a more specific and much more precise frame of the interfering factors than has been presented in this study. At the same time, as researchers are considering multiple dimensions, such as trust, commitment, property, and so on, it is important to learn about the impact of each of these and it would be useful to present the results not only in a matrix, but in a figure that reflects the weight of each of these factors in the start-up process, as well as during the further development process.

Conclusion

In contrast with the majority of research findings, the data analysed in this study have clearly initiated a breakthrough in the understanding of the failure of the internationalization processes in family businesses; above all, it calls into question the tendency to blame the intrinsic characteristics of these businesses for the delayed or the reduced activity in the international arena.

The main results of this study are threefold: in the first place, it indicates that family businesses as well as non-family businesses are able to grow through internationalization processes, and more especially through strategic alliances. At the same time, it highlights the fact that there are three critical variables that together drive the internationalization process: possession of a product whose potential exceeds that of the local market; desire to grow the company, even if this entails greater indebtedness or accepting new equity partners; adoption of an organizational structure that enables growth, without stagnating in traditional forms. When all three conditions are met, the company will develop strengths to enable it to successfully engage in strategic alliances, simultaneously allowing the development of the necessary managerial capabilities for internationalization, which in turn favours the formation of personal preferences and a strong conviction of the desirability of internationalization. An Australian survey, based on a longitudinal database of Australian family and non-family businesses could not find significant differences with respect to financial performance in the internationalization process in both groups (Graves and Thomas, 2004, p. 210). So, even if family businesses follow a specific pattern, they can reach equal outcomes as non-family businesses.

Secondly, this study shows that although internationalization is often discussed, not many companies tackle it effectively. To internationalize quickly, a company needs a high level of personal commitment to the process, combined with self-confidence and the necessary managerial capabilities. If a company lacks these capabilities, it may take steps to develop them and may eventually succeed. If it lacks trust in the process, all talk of internationalization is so much hot air – worthy intentions with no substance.

Thirdly, the results presented in this study suggest that the use of strategic alliances as a vehicle for internationalization requires commitment, managerial capabilities and trust that the partners will maintain a relationship consistent with the ownership structure and the development of the family business. Sharing ownership with non-family partners will help to develop the ability to manage in a context of unshared goals, as is the case in a strategic alliance, and the capacity to trust in the partners.

Nevertheless the principal contribution of this study lies in the fact that it provides some important challenges for family business owner-managers who are strongly committed to the FB's prosperity. For years the family business literature has been emphasizing the role of the founder in shaping the family firm's culture (Gersick et al., 1997; Schein, 1995). It also indicates how the concentration of power in the founder's hands, might intensify conservatism and stifle entrepreneurial innovation, especially when the CEO's tenure is very long; long tenure gives CEOs time to institutionalize their personal systems and processes, with the possibility of limiting adaptability to change and growth if they so want (Zahra, 2005, p. 36).

Contrary to this general perception, the 13 family businesses of this sample show that it is possible for owner-managers to grow and, especially in these cases, undertake strategic alliances with partners in emerging markets. Apart from some general best practices that indicate that they are a powerful tool facilitating growth, there are two more conditions important for long-term growth: the motivation of leadership to follow these practices and the commitment of the owning family to support the sacrifices necessary for growth (Ward, 1987, p. 334).

This is what finally led us to our main conclusion: that cultural distance is not the real factor that slows down the process, nor is it intrinsic characteristics, but rather it is the owner-manager's personal commitment to the continuity of the business. An important deal will depend on how founders or owner-managers use their formal and informal powers vis-à-vis their companies' entrepreneurial activities as well as in controlling the influence of family involvement on a company which has been defined as a mediating value within the internationalization model.

References

Amburgey, T.L., D. Kelly and W.P. Barnett (1993), 'Resetting the clock: the dynamics of organizational change and failure', *Administrative Science Quarterly*, **38**, 51–73.

Ariño, A. and J.J. Reuer (2003), *Alliance Dynamics for Entrepreneurial Firms*, Research paper No. 526, IESE Business School, Barcelona: IESE.

Autio, E., H.J. Sapienza and J.G. Almeida (2000), 'Effects of age at entry, knowledge intensity, and imitability on international growth', *Academy of Management Journal*, **43**(5), 909–25.

Brush, C. (1992), 'Factors motivating small firms to internationalize: the effect of firm age', doctoral dissertation, Boston University.

Casillas, J.C. and F.J. Acedo (2004), 'Internationalization of family SMEs: how family involvement affects?', *Research Forum Proceedings of the 15th FBN Annual World Conference*, Copenhagen: IFERA Publications, pp. 22–38.

Cohen, W. and D. Levinthal (1990), 'Absorptive capacity: a new perspective on learning and innovation', *Administrative Science Quarterly*, **35**, 128–52.

D'Souza, D.E. and P.P. Mc Dougall (1989), 'Third world joint venturing: a strategic option for the smaller firm', *Entrepreneurship Theory and Practice*, **13**, 19–33.
Deeds, D.L. and C.W.L. Hill (1996), 'Strategic alliances, complementary assets and new product development: an empirical study of entrepreneurial biotechnology firms', *Journal of Business Venturing*, **11**, 41–55.
Dickson, P.H. and K.M. Weaver (1997), 'Environmental determinants and individual-level moderators of alliance use', *Academy of Management Journal*, **40**, 404–25.
Donckels, R. and E. Fröhlich (1991), 'Are family businesses really different? European experiences from STRATOS', *Family Business Review*, **4**(2), 149–60.
Dyer, G.W. and W. Handler (1994), 'Entrepreneurship and family business: exploring the connections', *Entrepreneurship Theory and Practice*, Fall, 71–83.
Eisenhardt, K.M. and C.B. Schoonhoven (1996), 'Resource-based view of strategic alliance formation: strategic and social effects in entrepreneurial firms', *Organization Science*, **7**, 136–50.
Eriksson, K., J. Johanson, A. Majgard and D. Sharma (1997), 'Experiential knowledge and cost in the internationalization process', *Journal of International Business Studies*, **28**, 337–60.
Fernández, Z. and M.J. Nieto (2002), *International Involvement of SMEs: The Impact of Ownership*, Working Paper No. 02-58 (21) Business Economic Series, Universidad Carlos III de Madrid, Spain.
Fernández, Z. and M.J. Nieto (2005), 'Internationalization strategy of small and medium-sized family businesses: some influential factors', *Family Business Review*, **18**(1), 77–89.
Flören, R. (2001), *Internationalization of Family Business in the Netherlands: Research Results about Implementation of New International Strategies and Barriers to Growth on the International Markets*, BDO Accountants and Adviseurs, University Nyenrode, Breukelen, Netherlands.
Gallo, M.A. and K. Cappuyns (1998), *Internationalization of Family-Owned Businesses: Strategic Alliances*, Research paper No. 540, IESE Business School, Barcelona: IESE.
Gallo, M.A. and K. Cappuyns (1999), 'Characteristics of Successful Family Businesses', in M.ªA. Gallo, *Family Business No. 6*, Barcelona: Estudios y Ediciones IESE S.L., pp. 165–82.
Gallo, M.A. and M.J. Estapé (1992), *Internationalization of the Family Business*, Research paper No. 230, IESE Business School, Barcelona: IESE.
Gallo, M.A. and C. García-Pont (1993), *Important Factors in the Internationalization of Family-Owned Businesses*, Research paper No. 256, IESE Business School, Barcelona: IESE.
Gallo, M.A. and J. Sveen (1991), 'Internationalizing the family business: facilitating and restraining factors', *Family Business Review*, **4**(2), 181–90.
Gallo, M.A., G. Corbetta, G.W. Dyer, S. Tomaselli, D. Montemerlo and K. Cappuyns (2001), *Success as a Function of Love, Trust and Freedom in Family Businesses*, IESE, Family Business Chair, Monographic, No. 4, Barcelona: IESE.
Gallo, M.A., J. Tàpies and K. Cappuyns (2004), 'Comparison of family and non-family business: financial logic and personal preferences', *Family Business Review*, **17**(4), 303–18.
Gersick, K.E., J.A. Davis, M.M. Hampton and I. Lansberg (1997), *Generation to Generation: Life Cycles of the Family Business*, Boston, MA: Harvard Business School Press.
Gomes-Casseres, B. (1997), 'Alliance strategies of small firms', *Small Business Economics*, **9**, 33–44.
Graves, C. and J. Thomas (2004), 'Internationalisation of the family business: a longitudinal perspective', *Research Forum Proceedings of the 15th FBN Annual World Conference*, Copenhagen: IFERA Publications, pp. 196–214.
Gulati, R. (1995), 'Social structure and alliance formation patterns: a longitudinal study', *Administrative Science Quarterly*, **40**, 619–52.
Habbershon, T.G. and M.L. Williams (1999), 'A resource-based framework for assessing the strategic advantages of family firms', *Family Business Review*, **12**(1), 1–25.
Hara, G. and T. Kanai (1994), 'Entrepreneurial networks across oceans to promote international strategic alliances for small businesses', *Journal of Business Venturing*, **9**, 489–507.
Kets de Vries, M.F.R. (1993), 'The dynamics of family controlled firms: the good and the bad news', *Organizational Dynamics*, **21**(3), 59–71.
Kogut, B. and H. Singh (1988), 'The effect of national culture on the choice of entry mode', *Journal of International Business Studies*, **19**(3), 411–32.
Lansberg, I. and E. Perrow (1991), 'Understanding and working with leading family businesses in Latin America', *Family Business Review*, **4**(2), 127–47.
Larson, A. (1991), 'Partner networks: leveraging external ties to improve entrepreneurial performance', *Journal of Business Venturing*, **6**, 173–88.
Miles, M.B. and M.A. Huberman (1984), *Qualitative Data Analysis*, Newbury Park, CA: Sage.
Miller, D., L. Steier and I. Le Breton-Miller (2003), 'Lost in time: intergenerational succession, change, and failure in family business', *Journal of Business Venturing*, **18**(4), 513–31.
Okoroafo, S.C. (1999), 'Internationalization of family businesses: evidence from Northwest Ohio, USA', *Family Business Review*, **12**(2), 147–58.

Parkhe, A. (1993), 'The structuring of strategic alliances: a game-theoretic and transaction-cost examination of interfirm cooperation', *Academy of Management Journal*, **36**(4), 794–829.
Penrose, E. (1959), *The Theory of the Growth of the Firm*, New York: Wiley.
Ring, P.S. and A.H. Van de Ven (1994), 'Developmental processes of cooperative interorganizational relationships', *Academy of Management Review*, **19**, 90–118.
Rogoff, E.G. and R.K.Z. Heck (2003), 'Evolving research in entrepreneurship and family business: recognizing family as the oxygen that feeds the fire of entrepreneurship', *Journal of Business Venturing*, **18**(5), 559–66.
Schein, E. (1995), 'The role of the founder in creating organizational culture', *Family Business Review*, **8**(3), 221–38.
Shan, W. (1990), 'An empirical analysis of organizational strategies by entrepreneurial high-technology firms', *Strategic Management Journal*, **11**, 129–39.
Shepherd, D. and S. Zahra (2003), 'From conservatism to entrepreneurialism: the case of Swedish family firms', unpublished paper, University of Colorado.
Stuart, T.E., H. Hoang and R.C. Hybels (1999), 'Interorganizational endorsements and the performance of entrepreneurial ventures', *Administrative Science Quarterly*, **44**, 315–49.
Swinth, R.L. and K.L. Vinton (1993), 'Do family-owned businesses have a strategic advantage in international joint ventures?', *Family Business Review*, **6**(1), 19–30.
Ward, J.L. (1987), 'Growing the family business: special challenges and best practices', *Family Business Review*, **10**(4), 323–37.
Ward, J.L. and C.E. Aronoff (1991), 'Trust gives you the advantage', *Nation's Business*, **79**(8), August, 42–4.
Zahra, S.A. (2005), 'Entrepreneurial risk taking in family firms', *Family Business Review*, **18**(1), 23–40.

25 Family and cultural forces: shaping entrepreneurship and SME development in China

David Pistrui, Wilfred V. Huang, Harold P. Welsch and Zhao Jing

This study profiles the characteristics, attributes and growth orientations of Mainland Chinese entrepreneurs including the relationships, roles, and contributions family and culture play in the development of private small and medium-size enterprises (SMEs). Drawing on a sample of 222 entrepreneurs' psychographic motives, demographic attributes and business activities are revealed. Family and enterprise relationships related to employment, investment, and active family participation and growth orientations are empirically tested. The findings suggest that entrepreneurs are motivated by the need for independent-based achievement and continuous learning around a family focus. Entrepreneurs were found to rely heavily on family member participation to establish, develop and grow their enterprises. The majority of the entrepreneurs surveyed employed at least one family member on a full-time basis. Entrepreneurs were also found to use family finances as the primary source of start-up capital. Family participation in the form of employment and investment was found to have a positive impact on entrepreneurial growth intentions and expansion plans.

Introduction and overview

Family businesses are the engine that drives socio-economic development and wealth creation around the world, and entrepreneurship is a key driver of family businesses. Entrepreneurial thinking and leadership are fundamental factors in the creation of new enterprises and the sustained competitive advantages of both large and small businesses. The ability to create and foster an entrepreneurial mindset across generations is a major element of family business continuity and longevity and is instrumental in effective strategic execution, innovation and growth.

Cantillon (1755) recognized the entrepreneur as an individual who accepts an element of uncertainty in the course of profit-seeking business activity. Timmons and Spinelli (2004) defines entrepreneurship as creative human action that builds something of value from almost nothing through the pursuit of opportunity beyond the resources one actually controls. Entrepreneurship links vision, commitment, passion, and people to a common cause. Entrepreneurs and families create family-based business networks in response to economic and social needs. Family businesses can be defined as owner-managed enterprises with family members exercising considerable financial and/or managerial control (Ward and Aronoff, 1990). An estimated 85 per cent of businesses in the European Union and 90 per cent of US businesses are family controlled (Burns and Whitehouse, 1996; Shanker and Astrachan, 1996).

In market economies (where resources are allocated via supply and demand), entrepreneurial family businesses are a primary source of job creation. In transition economies (in countries moving from a state-planned to a mixed market-based system), the family unit is often the only intact socio-economic institution capable of supporting entrepreneurial activities. The family also plays a leading role in new venture formation, often serving as the primary source of start-up capital, not to mention in expensive labor and know-how.

Research question and direction of study

The People's Republic of China (PRC) provides a unique living laboratory in which to explore entrepreneurship, family business, and SME development. Although there is an emerging body of knowledge about entrepreneurship and private-enterprise development, there are few in-depth empirical investigations. Siu and Kirby (1999) point out that the opening of the Chinese economy provides an opportunity for extended research into Mainland China, where small firms are beginning to play an increasing and important role in the development of the economy. Consequently, researchers have a unique opportunity to identify, probe, and analyse the characteristics of new Chinese entrepreneurs, the enterprises they are developing, and family network involvement.

The general research question posed is: 'What are the characteristics, attributes and growth orientations of new Mainland Chinese entrepreneurs; and what relationships, roles and contributions do family and culture play in the development of private SMEs?'

This study explores four dimensions shaping entrepreneurial characteristics and orientations: (1) the psychographic motives and demographic attributes of the entrepreneur, (2) the types of businesses being started, as well as their ownership structure and method of establishment, (3) family and enterprise relationships related to participation and influence including employment, investment, and advice, and (4) how family participation shapes and influences the growth intentions and expansion plans of Mainland Chinese entrepreneurs.

By probing the individual and social forces shaping entrepreneurial motives, we hope to gain a sense of how culture influences the decision to start a private enterprise. This study explores demographic attributes associated with education, age, and experience. Such exploration provides insights into the types of personal resources entrepreneurs are drawing from as they start new businesses. The study also probes entrepreneurial spirit as it relates to business idea conception and actual business start-up.

In addition, the study explores the types of business activities that Chinese entrepreneurs are pursuing. Looking at this pursuit provides insight into the sectors of the economy on which entrepreneurs are focusing. Next the study investigates ownership structure, including type of business organization and method of establishment. Such investigation reveals which strategies entrepreneurs rely on to start private enterprises in a socialist economy. This is followed by an empirical investigation related to family participation and its influences on growth orientation and expansion plans.

By exploring family and enterprise relationships, we can gain further insights into how Chinese culture affects entrepreneurship and private-enterprise development.The study profiles full- and part-time family employment characteristics and how family involvement shapes entrepreneurship and SME development.

Entrepreneurship in a socialist market economy

Entrepreneurship and the development of small and medium-size enterprises continue to be at the forefront of economic development in virtually all economies today.

Entrepreneur-led SMEs provide social stability and serve as the engine of economic growth. One such example can be found in the post-communist transition of Eastern Europe where entrepreneur-led family businesses have been found to be the engine driving economic growth while fostering social stability (Hisrich and Fulop, 1997; Pistrui et al., 1997b; Poutziouris et al., 1997).

As far back as 1978 China began to realize some of the advantages of mixing state and private enterprise. China has adopted a different approach toward entrepreneurship and private-enterprise development, opting for the development of a mixed 'socialist market economy.'

In an effort to stimulate economic growth and development Chinese leaders encouraged the formation of rural enterprises and private businesses, liberalized foreign trade and investment, relaxed state control over some prices, and invested in industrial production and the education of its work force (Hu and Khan, 1997).

By the end of 1990s it was estimated that more than 12 million private enterprises were operating in China (Quanyu et al., 1997). Many of the newly emerging entrepreneur-led private sector businesses are micro-enterprises. Starr (1998) pointed out that most Chinese SMEs employ, on average, fewer than 15 workers and hold less than \$40 000 in registered capital. Both Dana (1999) and Davis (2000) cited examples of a series of emerging areas such as financial services, real estate, subcontracting, restaurants and the entertainment sectors where entrepreneurial activities have begun to flourish.

The family and the entrepreneurial socio-economic process

The family plays an important role in social and economic value creation and transgenerational wealth perpetuation processes (Habbershon and Pistrui, 2002). Families serve a number of important roles providing seed capital, employees, managers, and advisors during start-up and business development. As an area of research, family business offers unique research questions outside the boundary conditions of standard entrepreneurship research (Bird et al., 2002). This research probes such opportunities by exploring two general areas: (1) the direct role of family in the new venture development process (for example, participation and involvement in the business), and (2) the direct role and influence of family participation on growth intentions and expansion plans.

The family has been linked to the foundation of virtually all cultures. Organized around a host of functions the family can be characterized as the clustering of individual roles around common needs (Benedict, 1991; Parsons, 1955). Traditionally the family serves three primary functions within its social system. First, the family plays an economic role. Steier (2003) pointed out the substantial role familial ties play in the entrepreneurial process as the family represents a valuable repository of socio-economic resources that entrepreneurs rely on and draw from as they create new businesses outside the household or family unit. The family represents the unit of learning economic activity, teaching and passing on skills which encourage economic development.

Secondly, the family establishes a moral system which helps guide the conduct of the unit. The moral system is primarily engaged in providing a balance between ideas and realism. Finally, the family unit creates its own culture. Families adapt, and build networks based on needs and skills. Within this cultural setting the family creates a motivating force which is central to private enterprise formation as well as sustaining the enterprise across successive generations.

Family dynamics, entrepreneurship, and SME development

Developing and transforming countries have adopted entrepreneurship as a growing, visible, and vibrant economic activity (Honing, 1998). Entrepreneurs and families create family-based business networks in response to economic and social needs. Benedict (1991) argues that family-based kinship networks serve as foundational resources in private-enterprise development, especially in developing and less stable economies.

The liberation of Central and Eastern Europe, including the reunification of Germany, created the opportunity for the rebirth and accelerated development of entrepreneur-led family firms. In their study of Central and Eastern European countries, Donckles and Lambrecht (1999) report that the immediate family was the most important source of labor and capital for the establishment and development of private businesses during the post-communist transition. Pistrui et al. (2000) discovered that established West German SMEs were found to have a higher level of family investment than those in the post-communist East, while East German SMEs had more family members working full time than did West German SMEs.

According to researcher Cobianu-Bacanu (1994), in Romania, the family was the driving force behind entrepreneurship and enterprise development during the post-communist transition. Poutziouris et al. (1997) describe how family businesses in the Balkans develop adaptive family business systems in order to prosper and grow during the turbulence and uncertainty associated with transition to a mixed socialist market economy.

Business systems relying on family and other personal ties for security are also common throughout most of East and Southeast Asia (Perkins, 2000). Masurel and Smit (2000) found that family members often serve as employees and provide capital to new, private Vietnamese SMEs. In Indonesia, small enterprises play a critical role in creating employment and generating income, especially in the manufacturing sector and in rural areas (Tambunan, 2000).

Growth orientation, entrepreneurship, and family participation

Dunkelberg and Cooper (1982) argued that growth orientation in and of itself represents an important entrepreneurial characteristic. Carland et al. (1984) suggested that planned growth is an important method of differentiating entrepreneurs from small business owners. In his study of 410 Romanian entrepreneurs (of which 94 per cent were identified as family businesses) Pistrui (2002) reports the following:

- Growth intentions and expansion plans are a fundamental component of entrepreneurship and small business development.
- Entrepreneurial values play a special role in supporting entrepreneurial growth intentions and expansion plans.
- Cultural dynamics, including family, age, and education, impact entrepreneurial growth intentions and expansion plans.
- The macroeconomic environment may not support entrepreneurship, yet entrepreneurs may strive for growth nevertheless.

Culture has also been found to impact growth orientation. For example, Koiranen (2002) reported that established Finnish family businesses scored lower than expected on values related to economic return such as growth and social recognition, and higher on values and ethical behavior.

Martin and Lumpkin (2003) discovered that as US family businesses mature, their entrepreneurial orientation is replaced with a family orientation where stability and inheritance concerns become the businesses' primary drivers. In a study of 305 Spanish firms Gallo et al. (2004) concluded that family businesses have less growth in domestic and international sales and equity. Virtually all the cited researchers identify the need for additional research to advance the understanding of growth orientation in relationship to entrepreneurship and family business.

This research intends to build on the work of Pistrui et al. (1997a and b), Gundry and Welsch (2001), and Pistrui (2002) by advancing the understanding of how family participation and involvement affects growth orientations and expansion plans of Chinese entrepreneurially led enterprises.

Chinese cultural contexts and the role of the family

Chinese culture is centered on Confucian values, which are built around family, social ethics, education, centralized authority, and conformity. Confucianism places great importance on family, promotion of collective values, deep respect for age, hierarchy, authority, and an importance on reputation achieved through hard work and successful enterprise (Zapalska and Edwards, 2001).

The Chinese view the family as the fundamental social unit that promotes collectivism and harmony. The family works together to build and sustain a network of social relationships. Family norms are adopted as formal codes of conduct and members are bound to these principles (Zapalska and Edwards, 2001).

In China, family-centered, extended networks support a cultural orientation of relationships or connections called *Guanxi*. Given the high marginal cost of cultivating new relationships, it makes sense to do business first with close family, then with extended family, then neighbors, then former classmates, and only then, reluctantly, with strangers ('A survey of China', 2000).

***Guanxi* and the Chinese business network**

Chinese business mentality is grounded in the concepts associated with *Guanxi* which literally means 'relationships' (www.chinese-school.com, 2005). The Chinese prefer to do business with people they know and trust. In the Chinese business context *Guanxi* represents a network of relationships among various parties that cooperate together and support one another. Wellman (2001) pointed out that *Guanxi* forms a multidimensional continua of interpersonal behavior between both companies and people.

At the very foundation of the *Guanxi* relationship networks are the family and kinships groups. *Guanxi*-like relationships have been documented in other communist and mixed socialist economies such as Romania by Pistrui (2002; Pisturi et al., 1997a and b), Hungary by Sik and Wellman (1999), and across the Balkans by Poutziouris et al. (1997).

Most often the Chinese feel obligated to do business with friends and family first. *Guanxi* relationships grounded in the family can reduce uncertainty, lower transactions costs and provide usable resources (Wellman, 2001). *Guanxi* networks provide ways to reduce environmental uncertainty and exploit opportunistic behavior such as entrepreneurship and private-enterprise development. Based upon these cultural traditions it seems natural for family members to be a critical component of, and to actively participate in, the development of new and emerging entrepreneur-led Chinese private enterprises.

Chinese culture, family, and small business development

Although Confucianism does not promote the concept of small business it does encourage hard work, thoroughness and thriftiness. Wu (1983) suggested and Dana (1999) supported the idea that Chinese cultural principles include the following traits:

- a high propensity to save and reinvest business earnings;
- a universal drive for the education of children who are expected to carry on the business; and
- a strong sense of loyalty and mutual obligation within the extended family.

Chinese entrepreneurs view work as more important than leisure and as contributing to family welfare rather than competing with it (Zapalska and Edwards, 2001). Tsang (2001) pointed out that Chinese businesses are perceived as essentially a family possession where the head of the family has the final say in decision-making.

The nature of Chinese familism and the strong inclination to trust only people related to them has been found to create difficulty in Chinese businesses' abilities to move from small owner-managed firms to larger professionally managed enterprises (Fukuyama, 1995). Nonetheless entrepreneur-led small and medium-size family businesses will continue to be at the forefront of the emerging PRC socialist market economy.

Although small entrepreneur-led family-run enterprises can proliferate in China, their growth into more complex organizations may be held back by the disastrous legal inheritance of socialism (A Survey of China, 2000). Accordingly, the aspects associated with family business development represents an area for more in-depth analysis.

Research methodology and setting of study – central China

Researchers usually face a number of challenges when it comes to collecting data in developing countries. These well-documented challenges include low response rates and low percentages of usable questionnaires, to name just a few. To improve response rates, we adopted a focused method of investigation by choosing Wuhan – a major urban area and the provincial capital of Hubei, China.

Located on both the Yangtze and Hanshui rivers, Wuhan serves as a major transportation hub in central China. Wuhan has two international harbors, two airports and a major railway network. With a population of 7.3 million and an area of 8467 square kilometers, Wuhan serves as the largest financial and commercial center in central China (www.chinapages.com/hubei/wuhan, 2000) and was one of the earliest cities to be industrialized. Metallurgy, automobiles, machinery and high technology are the key economic sectors of Wuhan.

In recent years, the Wuhan region has established a number of major joint ventures with foreign multinationals including Citröen (France), Budweiser and Coca-Cola (US), NEC (Japan) and Philips (Holland).

China is a huge country with diverse local culture and uneven economic development across different regions. Even though conducting research in one location would not be sufficient to grasp the complexity of Chinese entrepreneurship and family business development in general, Wuhan represents a unique vantage point. On the one hand, Chinese coastal cities such as Shanghai, Qingdao, and Guangzhou are among the first few cities opened to outside world during the early stage of economic reform. Historically, people in those cities are more entrepreneurial and family business orientated than those in inner cities such as Wuhan.

On the other hand, compared with those in the western areas of China such as GanSu, and Chongqing, people in Wuhan are more entrepreneurial and have some tradition of family participation in business development. Therefore, by sampling Wuhan entrepreneurs, we expect to have a nominal view of Chinese entrepreneurs and family business development in terms of their demographics, family and enterprise relationships, entrepreneurial spirit, type of business activities, financing, personal attributes, growth aspirations and motivations, among other characteristics.

Survey instrument and data collection – the Entrepreneurial Profile Questionnaire

The Entrepreneurial Profile Questionnaire (EPQ) was utilized as a data collection instrument. The EPQ was designed to survey the effect of individual, societal and environmental factors on entrepreneurship and family business development by collecting a combination of demographic information and extensive detail related to characteristics and orientations.

The EPQ collects information on two fronts. First it asks a series of questions related to the demographic profile of the entrepreneur and business characteristics of the enterprise. Using a nominal scaling technique this data provides measures of the social, educational, and enterprise dimensions. This approach has been supported by the work of Kerlinger (1964), Bauer (1966), and Campbell (1976) as an accepted method of behavioral measurement.

Second the EPQ contains 10 sections which ask a series of questions placed on an agreement continuum. Using a summed rating scale, specifically a five-point Likert-type scale, the study identifies and measures the properties and characteristics associated with entrepreneurial growth intentions. The determinants of small business growth are likewise measured. The approach of assigning numerals to behavioral incidents and properties provides the framework to measure the relationships between different types of growth intentions and expansion plans.

Using the Likert (1932) model of item analysis, a number of statements related to the topic of interest were asked. From this point an item analysis technique can be used to sort through the data and select the best items for the final scale. This measurement technique has been widely used in behavioral research and thus represents a valid measurement methodology (Kerlinger, 1964; Mahler, 1953; Rosenthal and Rosnow, 1984).

The EPQ was successfully piloted and validated through a series of studies in Romania (Pistrui, 2002; Pistrui et al., 1997a), Turkey, Russia, Poland, the Czech Republic, Hungary, Lithuania, Estonia, and Germany (Pistrui et al., 2000; Wintermantel, 1999), Venezuela (Pistrui et al., 2000) as well as South Africa (Welsch and Pistrui, 1996), Mexico and the United States (Gundry and Welsch, 2001). The EPQ has been independently validated as a valuable data collection tool in transition economies such as post-communist Eastern and Central Europe, North and South America, Africa, as well as China (Liao et al., 2003; Pistrui et al., 2001). The EPQ was professionally translated and edited into Mandarin, pre-tested and then translated back to clear up ambiguities or idiosyncratic terminology.

Personal interviews were determined to be the optimum data collection technique. The EPQ was administered through personal interviews conducted by the authors and a team of professors and graduate students. The decision was made to develop a set of procedures to guide the EPQ interview process. This included training sessions for interviewers where a series of policies and procedures were established and reinforced. Furthermore, a system of random checking back both interviewees and interviewers was established and implemented to ensure quality data points. All written correspondence including the EPQ and accompanying covering letter would carry the logo and endorsement of the Chinese institution along with the US partner who was a co-sponsor of the research project.

With the assistance of the local chamber of commerce, we randomly selected 500 operating businesses from a firm registration database. In China, the Chamber of Commerce (COC) is a government agency with tremendous political influence. The introduction by the COC provided us access to local entrepreneur-led SMEs in a way that we would not otherwise have had. This ensured a reasonable response rate. All interviewees were assured anonymity. Many questionnaires were disqualified owing to incomplete data and missing information. Out of 500, we received usable EPQ data from 222 (a 44 per cent response rate) operating entrepreneur-led SMEs.

We use the results in three ways: to explore the types of entrepreneurial activity taking place in central China; to uncover the roles and the relationships of how cultural and family forces are related to entrepreneur-led SME development in a socialist market economy; and to provide direction for further study and analysis related to entrepreneurship and family business development in transforming economies.

Chinese entrepreneurial motives

The 'socialist market economy,' which strives to blend state and private enterprise, seems to be encouraging entrepreneurial thinking in China. In Table 25.1, the mean ratings of the top 10 motive-based attributes were arranged in descending order, including the standard deviations. Given the fact that the attributal items are not independent, a standard t-test of means was used to determine whether the overall mean ratings were different. Results verify that the attributes differ significantly in importance.

Table 25.1 Top 10 reasons and motives for entrepreneurship

Item	Mean (SD)
1. Desire to have high earnings	3.89 (1.10)***
2. Desire to have fun	3.63 (1.18)***
3. Desire to be challenged by the start up and growth of a business	3.63 (1.18)***
4. Desire to make a better use of my training and skills	3.61 (1.07)***
5. Desire to be my own boss, to work for myself	3.57 (1.16)***
6. Desire to have freedom to adopt my own approach to work	3.52 (0.94)***
7. Desire to keep learning	3.54 (1.02)***
8. Desire to give myself, my husband/wife, and children security	3.47 (1.12)***
9. Desire to make a direct contribution to the success of a company	3.44 (1.27)***
10. Desire to have a greater flexibility in my personal and family life	3.44 (1.06)***

Notes:
Range 1–5; N = 222.
*** $\alpha = 0.001$, ** $\alpha = 0.01$, * $\alpha = 0.05$.

The data suggest that Chinese entrepreneurs are motivated by the need for personal achievement and the desire to make a direct contribution to the success of an enterprise. The desire for higher earnings was the primary motive. Family security also appeared as a central motivating force.

Chinese entrepreneurs indicated an eagerness to make better use of their training and skills and to keep on learning. There is a strong drive to achieve a personal sense of accomplishment, foster family well-being, and develop new skills. These findings appear to be consistent with important components of the Chinese value system, such as Confucianism, which encourages continuous self-improvement, hard work, and diligence. The data further validate the work of Wu (1983), Dana (1999), Pye (2000), and (Zapalska and Edwards, 2001) who set forth the theory that Chinese cultural orientations rooted in Confucianism were fundamental in shaping entrepreneurial orientation and family business development.

Although this analysis is descriptive in nature, these basic findings provide an empirical foundation from which to develop further analyses. These findings (see Table 25.1) suggest that Chinese entrepreneurs from the Wuhan region appear to be motivated by the need for independent-based achievement and continuous personal development around a family focus.

Chinese entrepreneurial motives were found to be quite different from their counterparts in post-communist Romania and East Germany. For example, Pistrui (2002), and Pistrui et al. (1997b) documented the fact that Romanian entrepreneurs were driven by the motivations of increasing security, social status, and prestige of the family unit, whereas Chinese entrepreneurs are driven by lifestyle, financial motives and personal challenges associated with business start up and development.

Wintermantel (1999), and Pistrui et al. (2000, 2003) found that new entrepreneurs in the former communist East Germany were motivated by a blend of individual and social aspects including lifestyle, the desire for independence and family well-being. Furthermore, significant differences were found in the motivations between East and West German entrepreneurs, verifying that culture and the political economy can shape entrepreneurial motivations. Chinese entrepreneurs clearly have a unique set of motivational attributes when compared with their post-communist Central and Eastern European counterparts. Attention will now be given to some general demographics and entrepreneurial spirit.

Demographic profile, entrepreneurial spirit, and enterprise formation

Table 25.2 indicates the average age of Chinese entrepreneurs was approximately 37 years. In comparison with other communist transition economies Pistrui (2002) reported Romanian entrepreneurs' average age to be just over 39. Wintermantel (1999) found new East German entrepreneurs to be 45 years of age, a full eight years older than Chinese entrepreneurs. Chinese male entrepreneurs tended to be both slightly older than the average and approximately two years older than Chinese females.

Survey results suggest that entrepreneurship is not exclusively a male activity in central China. Over 20 per cent (23% of the entrepreneurs surveyed) were women, which is similar to findings in Romania (Pistrui et al., 1997b), and in Hungary (Hisrich and Fulop, 1997). Under communism, women were encouraged to participate in work activities and are, thus, fairly well integrated into the Chinese economic system. For example, Quanyu et al.

Table 25.2 Demographic profile, entrepreneurial spirit and enterprise formation

Category	Total	Male Mean (SD)	Female Mean (SD)
Years of education	13.09	13.19 (2.77)	12.75 (2.71)
Years of business experience	7.83	8.21 (6.23)	6.62 (5.95)
Years of work experience	10.08	10.37 (8.58)	9.12 (7.67)
Age	36.66	37.15 (8.12)	35.08 (8.69)
Total	222	170	52

Question 1: When did you first know you wanted to go into your own business?
Question 2: What year was your business started?

Question	Before 1987	1988–90	1991–93	1994–96	1997–99
Question 1	23.68%	23.00%	18.42%	30.00%	4.74%
Question 2	11.98%	11.06%	15.21%	29.49%	32.26%

	Mean (SD)
Question 1 − Question 2	2.90 (3.86)***

Notes:
N = 222.
*** α = 0.001, ** α = 0.01, *α = 0.05.

(1997) point out that state-owned enterprises actually maintain special Women's Commissions, which share offices and activities with enterprise unions.

These findings suggest that entrepreneurial development is taking place across gender lines. As Table 25.2 illustrates, no statistically significant differences were found based on gender. This may reflect the impact of socialism, which encourages the active participation of woman in the economy. Women entrepreneurs may benefit from the support of their extended families that provide assistance with children and household chores. The fact that female entrepreneurs tended to be plentiful and a bit younger may indicate the emergence of a new enterprising class of young Chinese women.

With an average of 13.09 years of education, it does not appear that Wuhanian entrepreneurs have much formal training beyond high school. This is in contrast to what Pistrui (2002) found in Romania where entrepreneurs had just over 15 years of education, and by Wintermantel (1999) who documented that East German entrepreneurs had a similar level of slightly over 15 years.

The average number of years of business experience, 7.83 years, suggests that these entrepreneurs have been successfully self-employed over an extended period of time. This is significantly more experience than Romanian entrepreneurs (3.4 years of business experience, Pistrui, 2002; Pistrui et al., 1997a). While Wintermantel (1999) and Pistrui et al. (1997a, 2003) reported that East German entrepreneurs had 5.8 years of business experience. Men were found to have slightly (1.59 years) more business experience than women (see Table 25.2). By the same token, men were found to have only slightly more work experience – a difference of only 1.25 years. Both Romanian and East German entrepreneurs were found to have approximately 14 years of work experience, almost four years more on average than their Chinese counterparts.

Although male and female entrepreneurs appeared to have different combinations of education and experience, statistically the findings were not significant. These finding suggest the need for further focused study of how education and experience impact entrepreneurship and SME development in different socio-economic contexts.

Idea conception and the act of enterprise creation

There appears to be a significant time lag of approximately three years between the idea of starting a business and the actual creation (see Table 25.2). For example, in the period between 1988 and 1990, 23 per cent of the respondents indicated that they aspired to start a business, but only approximately 11 per cent did so. Although the 1982 constitution established the private sector as a legitimate economic complement to the socialist economy, the results of this survey suggests that from 1988 onward, entrepreneurship began to be viewed as a favorable occupational alternative. In comparison to Romania and East Germany researchers Pistrui et al. (1997b, 2002) and Wintermantel (1999) discovered that over 90 per cent of the entrepreneurs surveyed started their businesses after 1990 and the fall of the Soviet Bloc. Thus it appears that Chinese officials recognized the value of entrepreneurship a full 12 years prior to the Central and Eastern European politicians.

Business start-ups seemed to increase during the mid to late 1990s, when approximately 77 per cent of the entrepreneurs surveyed started their business. The decade of the 1990s, with the consumer revolution and advent of the Internet, seemed to foster the desire to start new private enterprises.

This view seems to concur with Davis's (2000) – that the rapid commercialization of consumption during the 1990s broke the monopolies that had cast urban consumers in the role of supplicants to the state. These findings suggest that the desire to establish a new business appears to be accelerating in central China, while the time lag between conception and launch is becoming shorter. This may very well be a positive result of the socialist market economic policies in China.

Family and enterprise relationships

The foundation of Chinese society continues to be the family, often consisting of several generations united under a single roof (Starr, 1998). As Table 25.3 indicates, the family plays an important role in enterprise formation and development. These findings support the studies of Wellman (2001), Sik and Wellman (1999), Poutziouris et al. (1997), and Pistrui et al. (1997a, 2002) that family plays a central role in venture development in post-communist and socialist economies. Furthermore, based upon the role of *Guanxi* relationships it seems logical that the family and kinship group represent the cornerstone and thus will be actively involved in entrepreneur-led enterprise development.

Of the total sample, slightly over 67 per cent of the enterprises had at least one family investor; while just a little over 32 per cent had none. Further indication of the importance of family involvement in funding start-up is the fact that over 40 per cent of those surveyed had more then one family investor. This is in contrast to the research of Pistrui (2002) who reported that 95 per cent of Romanian entrepreneurs relied on family investors. One explanation might be that Chinese culture views financial resources as familial property. Wintermantel (1999) discovered that just less than half (49 per cent) of East German entrepreneurs relied on family investors.

Table 25.3 Family and enterprise relationships – investment and employment

Question 1: How many family members are investors in your enterprise?
Question 2: How many family members are full-time employees in your firm?
Question 3: How many family members are part-time employees in your firm?

Question	0	1	2	3	4	Mean (SD)
Q1: Investors	32.44%	23.42%	31.53%	10.81%	1.80%	0.75 (1.05)
Q2: Full-time employees	49.10%	26.13%	17.57%	5.41%	0.45%	0.88 (1.16)
Q3: Part-time employees	78.38%	16.22%	3.60%	1.80%	0%	0.29 (0.62)

Note: N = 222.

Family employees are also active in enterprise creation, development and operation. At least half (50.9 per cent) of all businesses in the sample had at least one family member employed full time in the enterprise (see Table 25.3). Approximately 40 per cent were found to have two or more family members employed on a full-time basis. Given the government restrictions on family size, this finding suggests that multiple generations including cousins might be working together to establish and grow entrepreneur-led family-managed SMEs.

Family members were not found to be engaged as much in part-time employment. More then 78 per cent indicated that they had no family members employed part-time in their firm. With just over 20 per cent (21.62 per cent) reporting a family member employed part time, 16 per cent had only one family member serving as a part-time employee.

Most likely, entrepreneurs perceive employment as part of the family's obligation, reflecting the traditional Chinese cultural terms associated with Confucianism and *Guanxi* relationships. Pistrui et al. (2001) reported that there was not a strong relationship between family investment and family employment as Chinese entrepreneurs had a difficult time distinguishing between investment and active participation in the business, and see the two as one in the same. These findings point to the need for further study of the relationships between entrepreneurs and family members in different cultural settings.

Sources of financing and start-up capital

Family savings are the primary resources used to establish private enterprises. In fact, extended family networks represented two of the top five sources of start-up capital (see Table 25.4). Close friends, partners, and trusted colleagues also provided financial resources to help new Chinese entrepreneurs get their ventures started. These data suggest that entrepreneurs strongly agree that seed capital is likely to be obtained from family networks rather than public institutions. These *Guanxi* relationships reduce uncertainty and lower transactions costs and depict the multidimensional network described by Wellman (2001). These findings support the notion that the Chinese prefer to deal with people they know and trust (www.chinese-school.com, 2005).

The reliance on family investment for startup also supports the work of Davis (2000), and Liao et al. (2003), who theorize that closely knit Chinese families accumulate substantial savings that they tended to invest in business start-ups and home-buying. As

Table 25.4 Sources of financing and start-up capital

Category	Mean importance (SD)
1. Family savings	3.36 (1.19)***
2. Close friends	3.18 (1.19)***
3. Suppliers	2.85 (1.28)***
4. Extended family	2.76 (1.06)***
5. Someone you work with	2.63 (1.21)***
6. Investor	2.62 (1.34)***
7. Banks	2.53 (1.24)***
8. Employer	2.52 (1.21)***
9. Vendor credits	2.31 (1.06)***
10. Credit card	2.11 (1.19)***
11. Government	2.00 (1.11)***
12. Former owner	1.98 (1.18)***
13. Partners	1.63 (0.96)***

Notes:
*** $\alpha = 0.001$, ** $\alpha = 0.01$, * $\alpha = 0.05$.
Range 1–5; N = 222.

expected, new Chinese entrepreneurs were found to rely much less on formal institutions, such as banks and government assistance, to get started. This is similar to the research of Pistrui et al. (1997b) who found that Romanian entrepreneurs relied primarily on family and personal networks for start-up financing during the transition from communism to a mixed socialist economy. Hisrich and Fulop (1997) found a similar phenomenon in Hungary where over 70 per cent of women entrepreneurs relied on personal saving and the support of family and friends. Attention will now turn to the types of businesses being started.

Type of business organization

Over half (53.85 per cent) of those businesses surveyed were sole proprietorships. Another 32 per cent were found to be limited liability companies (LLCs) (see Table 25.5). These findings indicate that most entrepreneur-led SMEs were closely held private enterprises built on family and extended family financial support. Perkins (2000) points out that even the LLCs that sold their shares on local stock exchanges tended to be closely held and family controlled, where even non-family majority shareholders had little say in business operations. This is different than what has been reported in Romania (Pistrui, 2002) where LLCs and partnerships were more prevalent. By the same token Wintermantel (1999) discovered similar patterns to Romania in post-communist East Germany, where LLCs and partnerships prevailed. These distinctions may be a reflection of cultural differences and represent a worthy area for further comparative studies.

Method of establishment

The vast majority (86.48 per cent) of the entrepreneurs surveyed originated their enterprises. As Table 25.5 illustrates, another 8.1 per cent of the entrepreneurs purchased their businesses from someone else. Only 5.4 per cent inherited the business, which further

Table 25.5 Business organization and activities of Chinese entrepreneurs

	Percentage
Type of business organization	
Sole proprietorship	53.85
Limited liability company (LLC)	32.13
Partnership	10.41
Corporation	4.07
Total	100.00
Method of establishment	
Originate	86.48
Purchase	8.11
Inherit	5.41
Total	100.00
Business activities	
1. Retail	27.94
2. Service organization	20.27
3. Distribution	16.67
4. Manufacturer	12.61
5. Computer/technology	10.81
6. Construction	4.95
7. Financial/insurance	3.15
8. Transportation	2.25
9. Professional services	1.35
Total	100.00

Note: N = 222.

suggests that this sample of Wuhanian entrepreneurs represents a new urban entrepreneurial class. These finding are similar to those reported by Wintermantel (1999) who found new East German led family businesses were originated by new entrepreneurs in the post-communist period following the fall of the Berlin Wall. However, Wintermantel (1999), and Pistrui et al. (2002) also reported that over 90 per cent of entrepreneurs surveyed originated their enterprise.

The findings of this study suggest that Chinese entrepreneurs search for localized market opportunities and create closely held inward-looking sole proprietorships to serve them. Given the youth of these enterprises, it appears highly likely that these newly emerging SMEs will face increased competition as the private sector of the Chinese economy grows. There will also be an emerging need for succession and continuity assistance as these businesses mature and integrate additional family members into their operations.

Business activities of Chinese entrepreneurs

Entrepreneurs are beginning to identify profitable niches where they can earn more income and have greater independence. Table 25.5 provides a ranking and percentage breakdown of the various enterprise types. Three types of business activity, retail (27.94 per cent), service organizations (20.27 per cent), and distributors (16.67 per cent) were found to dominate the sample. These three indexes represented well over half (64.88 per cent) of new start-ups in the Wuhan area.

The domination of retail activity is further evidence of the emerging entrepreneur-led, private-enterprise sector focused on consumer markets. The findings in this study support the work of Davis (2000), who set out examples of private restaurateurs or subcontractors (*chengbao*) that have replaced state canteens as well as the development of leisure and entertainment businesses. Similar patterns of retail and service start-ups have been documented in Romania (Pistrui et al., 1997b), and East Germany (Wintermantel, 1999) during the transition from communism, suggesting that entrepreneur-led family firms have seized a moment of opportunity to fill a void in the emerging marketplace.

Manufacturing and computer/technology activities represented over 20 per cent of the sample. Most likely, these sectors are examples of the tertiary markets developing to serve the retail and technology sectors. Given the abundance of cheap labor it should not be surprising that manufacturing represents almost 13 per cent of the enterprises. Most likely entrepreneurs with state connections are leveraging *Guanxi* relationships in combination with family resources to launch and develop new private enterprises.

The data also suggest that the advent of the Internet, mobile telephones, and personal computers seems to be creating a flurry of entrepreneurial activity around computer hardware and software markets. Chinese officials opened the Internet to the public in 1996 (a time period which shows a significant increase in the number of new business start-ups; see Table 25.2). The China Network Information Center (CNNIC) estimated 16.9 million Internet users as of June 2000, up from a mere 600 000 in 1997 (Hartford, 2000). Another example of entrepreneur-based activity related to the Internet is found in Internet connections. Hartford (2000) also points out that by late 1999, 500 private companies held licenses as Internet service providers. Consequently, part of the sample data may encapsulate this type of activity.

Whereas some markets, such as banking, telecommunications, and general distribution, are off limits to entrepreneurs, other consumer service sectors are being developed ('A survey of China', 2000). Examples from this study include a small, yet important, level of professional services, transportation, insurance and real estate.

Family participation and growth intentions and expansion plans

Attention will now turn to the impact family participation has on entrepreneurial growth intentions and expansion plans. The study will focus on the question: 'How does family participation in the form of active employment and financial investment shape and influence entrepreneurial growth intentions and expansion plans of mainland Chinese entrepreneurs?'

Pistrui (2002) and Pistrui et al. (1997a) developed, tested and confirmed the validity of a growth model based on entrepreneurs' intentions to implement specific attributes associate with market expansion, technological upgrades, and operation/production expansion. They found that those entrepreneurs surveyed seemed to have definite, well-defined expansion plans for the future.

Building on the work of Pistrui (2002) and Pistrui et al. (1997a) Gundry and Welsch (2001) used the same 'implementable attributes of planned growth' model to study high-growth entrepreneurs. Results of their empirical investigation showed that a group labeled 'ambitious entrepreneurs' had distinctive strategic intentions that emphasized market growth, technological change, and a greater willingness to sacrifice on behalf of the business.

To expand on this research we hypothesize: 'family participation in the business has a positive impact and encourages entrepreneurial growth intentions and expansion plans.'

Attention will now turn to the operationalization of variables and empirical testing of the research question and hypothesis.

Implementable attributes of planned growth

Inspired by the five characteristics Schumpeter (1934) associated with entrepreneurial behavior, a series of implementable attributes of planned growth (IAPG) can be established. Figure 25.1 provides an overview of 18 specific entrepreneurial behaviors reflecting strategic characteristics associated with the IAPGs. The IAPGs serve to identify and operationalize the specific types of new combinations entrepreneurs intend to pursue.

The Chinese socio-economic transition process is creating the need for technological upgrading, developing specialized human resources, and acquiring new business skills.

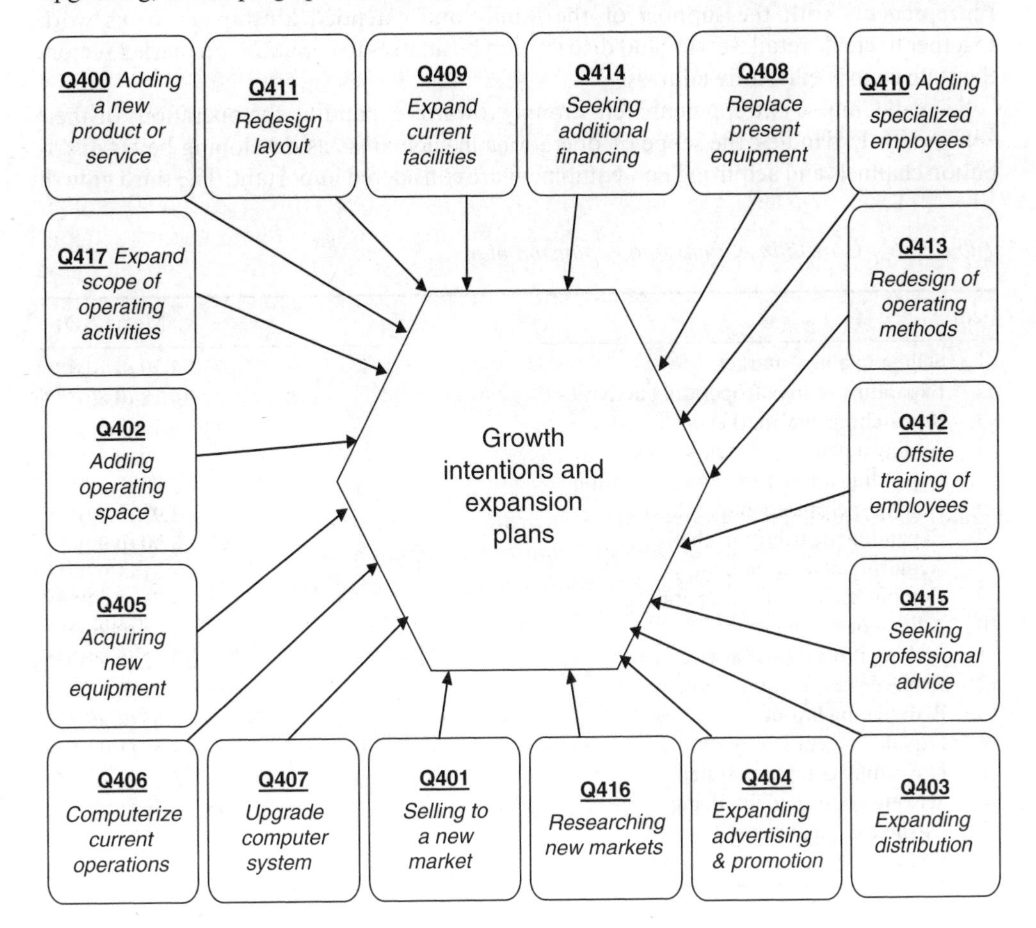

Figure 25.1 *Implementable attributes of planned growth: strategic behavioral characteristics – growth intentions and expansion plans*

Perhaps more importantly, transition is creating vast new markets and fostering the introduction of new goods and services. Methods of production are also radically changing. In general Chinese entrepreneurs are redesigning how enterprise is conducted.

Mainland China provides an excellent field site in which to investigate entrepreneurial growth intentions and expansion plans. The transition from communism toward a socialist market based economic system offers a rich environment to investigate entrepreneurship and small business growth.

Chinese entrepreneurial growth intentions and expansion plans

Chinese entrepreneurs appear to be strongly committed to entrepreneurial expansion and business growth (see Table 25.6). Three growth themes emerged among these entrepreneurs. First, entrepreneurs desired to expand into new markets by offering new products and services. This includes researching new markets and expanding advertising and promotion. Pistrui (2002) and Pistrui et al. (1997b) reported a similar pattern they called 'Market Expansion' in Romania's post-communist socialist market economy. Entrepreneurs with the support of the family and extended kinship networks work together to create retail, service and distribution businesses to capitalize on market sectors the state cannot efficiently address.

Second, Chinese entrepreneurs felt strongly towards expanding the operations of their enterprises. Expanding the scope of operations in such areas as developing better distribution channels and acquiring new equipment are considered important. The third growth

Table 25.6 Growth intentions and expansion plans

Item	Mean (SD)
1. Selling to a new market.	4.20 (0.74)***
2. Expanding scope of operating activities.	4.18 (0.85)***
3. Researching new markets.	4.14 (0.85)***
4. Adding a new product or service.	3.99 (0.95)***
5. Expanding advertising and promotion.	3.98 (0.85)***
6. Adding specialized employees.	3.93 (1.09)***
7. Expanding distribution channels.	3.90 (0.90)***
8. Acquiring new equipment.	3.81 (1.07)***
9. Computerizing current operations.	3.78 (1.13)***
10. Adding operating space.	3.74 (0.86)***
11. Seeking professional advice.	3.72 (0.99)***
12. Replace present equipment.	3.71 (0.96)***
13. Redesigning layout.	3.63 (0.98)***
14. Expand current facilities.	3.59 (1.04)***
15. Upgrading computer systems.	3.58 (1.20)***
16. Seeking additional financing.	3.46 (1.07)***
17. Offsite training for employees.	3.36 (1.09)***
18. Redesigning operating methods.	3.28 (1.04)***

Notes:
*** $\alpha = 0.001$, ** $\alpha = 0.01$, * $\alpha = 0.05$.
N = 222; Range 1–5.

theme was the intention to expand internal resources and capabilities. This includes adding specialized employees, developing information technology, and seeking special advice.

Chinese entrepreneurs seem to have definite, well-defined plans for the future. Although vestiges of communism remain, entrepreneurs and their families are becoming masters of their own destiny by focusing on the emergent private sector business opportunities in China's socialist market economy.

In a study that employed the EPQ to sample 1427 entrepreneurs across five countries (China, US, Romania, Germany, and Venezuela), Pistrui (2005) discovered that Chinese entrepreneurs were found to have modest levels of growth intentions and expansion plans as compared to the other nations. They led only three of the 18 categories including:

- Selling to a new market 4.20 (mean 3.80)
- Adding a new product or service 3.99 (mean 3.95)
- Seeking professional advice 3.72 (mean 3.51)

From this point attention will turn to exploring the proposition and testing the hypothesis with multivariate statistical tools.

Structural equation modeling

The research model was tested using a covariance-based structural equation modeling (SEM). This is a powerful multivariate technique that facilitates the testing of psychometric properties of the scales used to measure unobserved variables (constructs) as well as estimate the parameters of a structural model, which is the magnitude and direction of the relationships among the model variables (Bollen, 1989; Gefen et al., 2000; Hair et al., 1998). Structural equation modeling embodies two inter-related models. The measurement model represents the relationships between the observed items and their constructs measured by these items, while the structural model represents the paths among a set of dependent and independent variables.

In order to assess the model, we shall perform evaluations on the measurement model, the structural model and the overall model. The evaluation on the measurement model includes an exploratory factor analysis to identify the constructs, and to examine the convergent and discriminant validity of the research instrument. The evaluation on the structural model consists of estimation of path coefficients and their associated significance p-value. Squared multiple correlations (SMC) are calculated to know the proportion of explained variance in the each construct. Finally, the evaluation of the overall model is on the overall goodness-of-fit for SEM.

Exploratory factor analysis

From the 18 questions (see Tables 25.7 and 25.8) in the survey, an explanatory factor analysis was performed using Principal Component Analysis (PCA) and Varimax rotation methods. Principal Component Analysis was used since it detects the existing latent structure in data and serves as a useful data reduction technique (Pedhazur and Schmelkin, 1991). Six constructs were identified as shown in Table 25.9.

As Tables 25.9 and 25.10 illustrate six growth constructs were identified. The constructs range from equipment and information technology upgrades to business development activities. These findings suggest that Chinese entrepreneurs have some focused areas of growth orientation defined around specific subset themes.

Table 25.7 Rotated component matrix

	Component					
	1	2	3	4	5	6
Q405	*0.696997*	0.229397	0.090072	0.075331	0.055791	0.164158
Q408	*0.677611*	0.289496	0.129154	−0.00874	−0.01052	0.233306
Q410	*0.662826*	0.13584	0.179559	0.326973	0.218882	0.033297
Q409	*0.658423*	0.014342	0.135884	0.09164	0.340662	−0.12031
Q407	*0.262109*	*0.790139*	0.041563	0.110632	0.190906	0.03194
Q406	0.286511	*0.748468*	0.011189	0.086158	0.196459	0.055385
Q413	0.139998	0.171646	*0.80743*	−0.00108	0.064003	0.110485
Q414	0.027254	−0.14912	*0.639916*	0.351181	0.12406	−0.02033
Q411	0.289439	−0.13936	*0.577899*	−0.07161	−0.00741	0.42918
Q412	0.210592	0.359911	*0.549978*	0.09934	0.250277	−0.21704
Q416	0.063031	0.204153	0.059255	*0.795849*	0.200018	0.21935
Q417	0.165275	−0.13706	0.029659	*0.635582*	0.390886	0.229306
Q415	0.247958	0.376833	0.330661	*0.629607*	−0.20545	−0.00307
Q401	0.135413	0.222546	0.128909	0.144989	*0.668172*	0.122955
Q400	0.228402	0.30601	0.079432	0.086151	*0.648681*	−0.04097
Q402	0.18084	0.119731	0.00555	0.244248	−0.02105	*0.690047*
Q403	0.039917	−0.11647	0.112877	0.164493	0.500906	*0.583126*
Q404	−0.03609	0.544743	0.061192	0.057353	0.083684	*0.557024*

Notes:
Extraction method: Principal Component Analysis.
Rotation method: Varimax with Kaiser Normalization.
Rotation converged in 10 iterations.

Table 25.8 Relationship between constructs and observed variables

Construct	Observed variables	Abbreviation
Equipment/facility upgrade	Q405, Q408, Q409, Q410	EquipUpg
IT upgrade	Q406, Q407	ITUpg
Operational expansion	Q411, Q412, Q413, Q414	OpExp
Business development	Q415, Q416, Q417	BusDev
Market expansion	Q400, Q401	MktExp
Infrastructure development	Q402, Q403, Q404	InfraDev

Chinese entrepreneurs are keen to expand operations and develop new markets for their products and services. There is also the desire to improve infrastructure to support enterprise growth and development. Chinese entrepreneurs were found to have a clearly defined subset of growth intentions and expansion plans. In a similar study of Romanian entrepreneurs Pistrui et al. (1997b) and Pistrui (2002) documented growth intentions and expansion plans around market expansion, technological upgrades, and operations/production constructs. The finding of this research suggests that Chinese entrepreneurs have an even more defined pattern of growth intentions and expansion plans.

Table 25.9 Validity and reliability analysis

Construct	Observed variable	Factor loading	Cronbach α	Eigenvalue	Composite reliability	Variance explained
EquipUpg			0.733	5.307	0.745	29.48
	Q405	0.697				
	Q408	0.678				
	Q409	0.658				
	Q410	0.663				
ITUpg			0.836	1.643	0.761	9.13
	Q406	0.748				
	Q407	0.790				
OpExp			0.642	1.469	0.700	8.16
	Q411	0.578				
	Q412	0.550				
	Q413	0.807				
	Q414	0.640				
BusDev			0.679	1.150	0.762	6.39
	Q415	0.630				
	Q416	0.796				
	Q417	0.636				
MktExp			0.581	1.052	0.672	5.84
	Q400	0.649				
	Q401	0.668				
InfraDev			0.557	0.940	0.682	5.22
	Q402	0.690				
	Q403	0.583				
	Q404	0.557				

Table 25.10 Squared correlations and AEV square-roots

	EquipUpg	ITUpg	OpExpan	BusDev	MarketExp	InfraExp
EquipUpg	*0.674*					
ITUpg	0.252	*0.770*				
OpExpan	0.219	0.168	*0.652*			
BusDev	0.305	0.234	0.203	*0.691*		
MarketExp	0.423	0.325	0.282	0.393	*0.658*	
InfraExp	0.303	0.233	0.203	0.282	0.391	*0.613*

Assessment of construct validity

All scales were tested for various validity and reliability properties. Construct validity was assessed by both convergent and discriminant validity. Convergent validity was evaluated by examining if the questions loaded on the theorized constructs. Unidimensionality is an underlying assumption for reliability calculation; therefore it should be verified for all multiple-indicator constructs (Hair et al., 1998). Unidimensional measures must load on only one construct. Convergence implies that all within-construct correlations are high with similar magnitude. Discriminant validity was assessed by examining the rotated component matrix to ensure that items did not cross load on multiple factors.

Results of the analysis indicated that a priori assumptions were substantiated with a six-factor solution: EquipUpg, ITUpg, OpExp, BusDev, MktExp and InfraDev. Table 25.9 presents the test of convergent validity including standardized Cronbach α's, eigenvalues, composite reliabilities and variances explained by each construct. All variables have high factor loadings on their respective construct (> 0.55), the Cronbach α's for each construct are above generally accepted guidelines (Nunnally, 1978), and the composite reliabilities of the measures exceed the minimum value of 0.60 (Bagozzi and Yi, 1988). This supports the reliability of the measures integrated in the hypothesized model. Also, the convergent validity is supported because all loadings are highly statistically significant ($p < 0.001$).

Convergent validity can be assessed by the degree of association among the items measuring the constructs (Table 25.7). To evaluate discriminant validity, Fornell and Larcker (1981) suggest a comparison between the average extracted variance (AEV) of each factor and the variance shared between the constructs (the squared correlations between the constructs). In Table 25.10, squared correlations are reported on the off-diagonal and AEV square-roots are on the diagonal. An AEV value should be higher than 50 per cent (Rivard and Huff, 1988) and should be greater than its squared correlations.

Family participation construct: defining the family business

The construct of family participation (FP) consists of three observed variables: number of family members employed full time (FT), number of family members employed part time (PT) and number of family members invested in the firm (FamInv). The FP construct corresponds to Ward and Aronoff's (1990) definition that includes owner-managed enterprises where the family exerts considerable financial and managerial control.

Assessment of the structural model

The structural model shown in Figure 25.2 provides the hypothesized relationships between FP and the growth constructs. As previously discussed we hypothesized that Chinese cultural traditions encourage family participation, which in turn has a positive impact on entrepreneurial growth intentions and expansion plans Thus we will further define and dissect the specific types of growth constructs and how FP impacts these subsets. The hypotheses were tested by SEM using the input model in AMOS (Analysis for MOments Structures) as shown in Figure 25.3. The Maximum Likelihood function was used to estimate the model parameters.

Using AMOS 5.0.1, we obtained the results presented in the Table 25.11. For example, the squared multiple correlation (SMC) of 0.572 in H1 reveals that FP explains 57.2 per cent of variance in EU. The path coefficient in H1 is 0.756. All paths are statistically significant at the 0.001 level.

Assessment of the overall model

The chi-square test (X^2) reports as 405.912, degrees of freedom (df) is 184. The ratio of chi-square to degrees of freedom (X^2/df) is 2.206 (see Table 25.12). Although the ratio is above the recommendation to be between 1 and 2 (Hair et al., 1998), the behavior of X^2 is very much a function of sample size and the model complexity. The root mean square error of approximation (RMSEA) index is 0.074, which is less than 0.10 recommended by

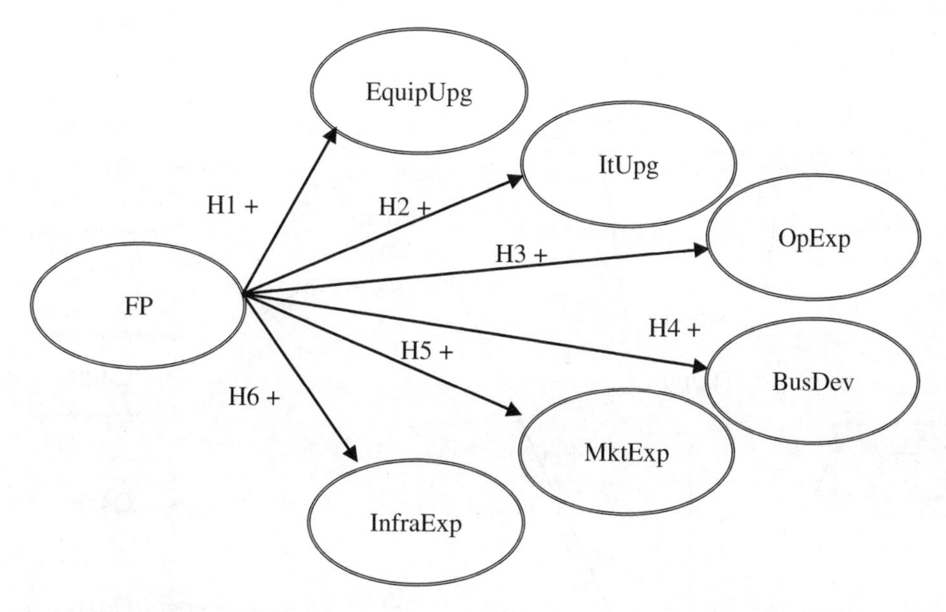

Figure 25.2 Research model

Kelloway (1998). By these criteria, we have verified that our model has a good fit with the empirical data.

Carmines and McIver (1981) state that the ratio of chi-square to degrees of freedom should be in the 2:1 or 3:1 range for an acceptable model. Kline (1998) says 3 or less is acceptable. Even though some indices in our model fall short of the recommendation, the existing goodness-of-fit measures are related to the ability of the model to account for the sample covariances and therefore assume that all measures are reflective.

The SEM procedures that have different objective functions and/or allow for formative measures would, by definition, not be able to provide such fit measures. In actuality, models with good-fit indices may still be considered poor based on other measures such as the R-square and factor loadings. The fit measures only relate to how well the parameter estimates are able to match the sample covariances. They do not relate to how well the latent variables or item measures are predicted.

The SEM algorithm takes the specified model as true and attempts to find the best-fitting parameter estimates. If, for example, error terms for measures need to be increased in order to match the data variances and covariances, this will occur. Thus, models with low R-square and/or low factor loadings can still yield excellent goodness of fit (Chin, 1998a and b). Therefore, our model fit is adequate because of the strong loadings, significant weights, high R-squares and significant structural paths.

The impact of family participation and growth intentions and expansion plans

The hypothesis that that family participation in the business has a positive impact and encourages entrepreneurial growth intentions and expansion plans was confirmed.

Family participation in the form of employment and investment was found to have a positive impact on entrepreneurial growth intentions and expansion plans. These findings

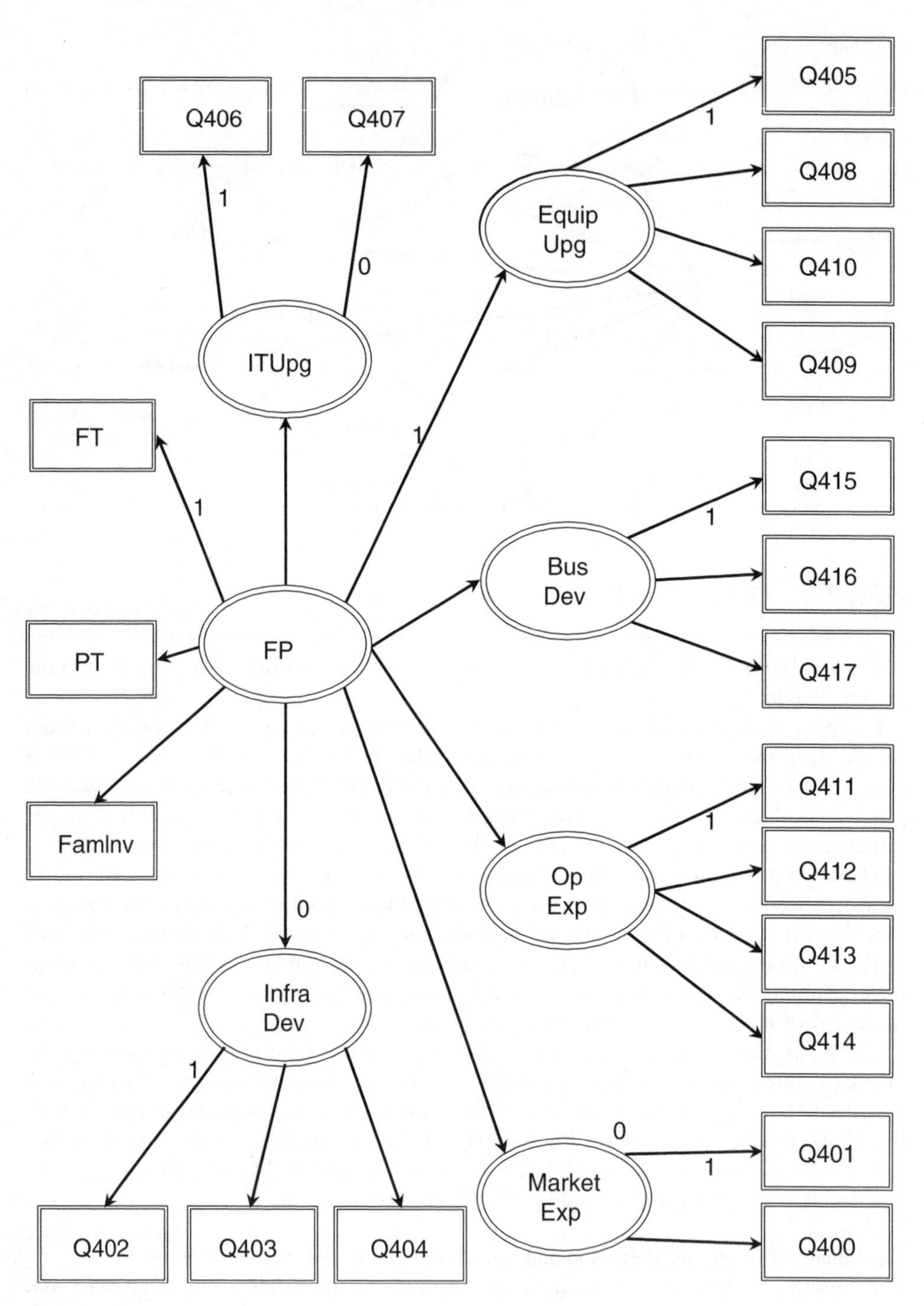

Figure 25.3 Input model

Table 25.11 Summary of the parameters for the research model

Hypothesis	Testing the relationship between	Squared multiple correlation	Standardized regression coefficient
H1	FP has a positive impact on EquipUpg	0.572	0.756***
H2	FP has a positive impact on ItUpg	0.441	0.664***
H3	FP has a positive impact on OpExpa	0.382	0.618***
H4	FP has a positive impact on BusDev	0.532	0.729***
H5	FP has a positive impact on MktExp	0.739	0.859***
H6	FP has a positive impact on InfraExp	0.529	0.728***

Note: *** denotes $p<0.001$.

Table 25.12 The goodness-of-fit indices

Chi-square (χ^2) = 405.912	df = 184	χ^2/df = 2.206
NFI = 0.695	RFI = 0.652	IFI = 0.806
CFI = 0.802	NCP = 221.912	FMIN = 1.853
RMSEA = 0.074	Hoelter .05 = 117	Hoelter .01 = 125

suggest that newly emerging Chinese family businesses intend to expand and grow. These enterprising families are at the forefront of the transition from communism to a socialist market economy.

Six specific growth constructs have been identified and validated in these entrepreneur-led family enterprises. Family participation was found to shape these growth and expansion subsets. As outlined in the previous sections the model has been statistically validated suggesting that family participation creates growth personalities that reflect a clustering of implementable attributes of planned growth around specific categorical orientations such as market expansion, equipment upgrades, and business development.

These findings are also empirical validation that Chinese cultures including Confucianism and *Guanxi* networks connect the family and entrepreneur to their enterprise. Family participation reflects the Confucian values and *Guanxi* ties associated with family, collective values, centralized authority, conformity and the importance of reputation achieved through hard work and successful enterprise as described by Zapalska and Edwards (2001). The family is the social unit that often helps blend the myriad business tasks together and helps smooth out the rough edges when conflicts arise with respect to growth and policy questions. The obligation to share wealth with family members and to provide support and security may deter the entrepreneur from investing in the growth aspects of the business. Thus the research question arise as to whether the family was a deterrent to or facilitator of entrepreneurial growth.

The results indicate that the family may not only be instrumental in helping to establish the business but also is a significant factor in the growth of the firm. Various forms of family support could include growth capital, room and board, building space, tools, encouragement and moral support to growth initiatives. Kuratko and Welsch (2004) suggest that the family influence helps obtain access to suppliers, merchants, creditors, market authorities, local officials and persons with economic power and influence. Many of these factors provide the medium and vehicles for the growth of the firm. Thus families have proven to be a valuable mechanism for entrepreneurial growth and governments should encourage families in this direction.

Families have been found to play a central role in the growth and development of entrepreneurial led SMEs in Mainland China. Although the findings of this study are significant there is a clear need for further research.

Limitations and directions for further research

There are several limitations in this study. First, the study is based only on survey data conducted in Wuhan, China. Even though it is believed that Wuhan is a good representation of China, the study should be broadened to other parts of China. Second, the research model consists of only the growth attribute. It is believed that other psychographic attributes, such as intensity and sacrifice, should enter the model to make this research more robust. Third, there is a need to compare entrepreneurs across cultures. One immediate opportunity is to compare Romanian and Chinese entrepreneurs to identify, compare and contrast cultural and family forces in post-communist socialist market economies.

Conclusion

The findings of this study suggest that Mainland Chinese entrepreneurs located in the Wuhan area are motivated by the need for personal achievement, the desire to make a direct contribution to the success of an enterprise, and the desire for family security.

New-venture creation grew during the mid to late 1990s, when approximately 77 per cent of the entrepreneurs surveyed actually started new enterprises. The results of this study found that the overwhelming majority (85.98 per cent) of new private firms were closely held sole proprietorships or LLCs. Over 86 per cent of the entrepreneurs surveyed indicated that they had originated their enterprises.

The family and Chinese culture played a central role in supporting entrepreneur-led SME growth and development. The household and extended family tended to provide primary sources of start-up capital. Over 67 per cent of the businesses had at least one family investor; while 31.5 per cent had two. Fifty per cent of these new enterprises had a family member employed full time, while 23.4 per cent had more than one family member working full time.

Family participation was found to have a positive impact on entrepreneurial growth intentions and expansion plans. This represents one of the first empirical validations of the impact of Confucian values and *Guanix* ties on entrepreneurial led business development in a post-communist transition socio-economic system.

One thing that seems certain is that entrepreneur-led, family-centered private SMEs will continue to play an increasing role in the development of China's socialist market economy. Although small, entrepreneur-led, family-run private enterprises can proliferate in China, their growth into more complex organizations may be held back by the legal inheritance of socialism ('A survey of China', 2000).

From a policy standpoint, this study indicates a low degree of support from the Chinese government. It therefore highlights the urgency for the Chinese government to channel financial resources and services, and to create innovative programs to support entrepreneurial family centered business development. Historically, the Chinese government has devoted much of its attention to state-owned enterprise and scant attention has been paid to the development of private-owned enterprise. Not until after 1992, did the Chinese government start to accept and realize that entrepreneurs are the engine of economic development.

One notable caveat in this study is related to our sample, which is based on entrepreneurs in Wuhan. We note that Chinese entrepreneurs in more entrepreneurial areas such as Shanghai and Guangzhou may exhibit different socio-economic characteristics and demographics. We expect there may be some differences across various regions both in terms of family dynamics and the socio-economic environment. Nevertheless, no attempt is made here to generalize findings from this study to other areas of China.

A call for more research

This study represents an initial effort in identifying the family and cultural forces shaping entrepreneurship and SME development in China. Accordingly, the aspects associated with entrepreneurship and family business development need more in-depth analysis and further study.

In terms of future research topics, there is a need for additional research on the relationships between entrepreneurs and family-business development. Traditionally, researchers have expected private Chinese enterprises to remain small; accordingly, another important area to explore is how private Chinese SMEs approach growth over time. If the private sector is to live up to the expectations of the Chinese government's plan of a socialist market economy, it will be vital for entrepreneur-led family firms to grow and prosper.

Future research might focus on the following four areas. First, researchers can further explore entrepreneurial orientation including how intensity and sacrifice shape SME growth intentions and expansion plans. Second, studies can also be devoted to comparative analysis between different regions in China of how family participation shapes the development of entrepreneurial SMEs. By the same token, cross-country studies comparing Chinese entrepreneurial led family centered SMEs to those in other nations can yield some valuable insights and learning.

Thirdly, the findings of this study suggest that women-led enterprises represent a unique socio-economic microcosm worthy of additional study and analysis. Finally, family size and business continuity are another area that needs study. Modern Chinese households tend to be small, often with only a single child per family. How will this fact impact family business development? Will small Chinese family firms create extended networks to build larger, more diversified SMEs or stay small and succumb to short life cycles?

References

'A survey of China: Now comes the hard part (special section on China)' (2000), *The Economist*, 8 April, p. 116.

Bagozzi, R. and Yi, Y. (1988), 'On the evaluation of structural equation models', *Journal of the Academy of Marketing Science*, **16**(1), 74–94.

Bauer, R. (ed.) (1966), *Social Indicators*, Cambridge, MA: MIT Press.

Benedict, B. (1991), 'Family firms and economic development', in C. Aronoff and J. Ward (eds), *Family Business Source Book*, Detroit, MI: Omingraphics, Inc.

Bird, B., Astrachan, J., Welsch, H. and Pistrui, D. (2002), 'Family business research: the evolution of an academic field', *Family Business Review*, **15**(4), 337–50.

Burns, P. and Whitehouse, O. (1996), 'Family ties', Special Report of the 3I European Enterprise Center, August.

Campbell, A. (1976), 'Subjective measures of well-being', *American Psychologist*, **31**, 117–24.

Cantillon, R. (1755), *Essay on the Nature of Commerce in General*, ed. H. Higgs, London: Macmillan.

Carland, J., Hoy, F., Boulton, W. and Carland, J.C. (1984), 'Differentiating entrepreneurs small business owners: a conceptualization', *Academy of Management Review*, **9**(2), 73–84.
Carmines, E. and McIver, J. (1981), 'Analyzing models with unobserved variables: analysis of covariance structures', G.W. Bohmstedt and E.F. Borgatta (eds), *Social Measurement*, Thousand Oaks, CA: Sage Publications, pp. 65–115.
Chin, W. (1998a), 'Issues and opinion on structural equation modeling', *MIS Quarterly*, **22**(1), vii–xvi.
Chin, W. (1998b), 'The partial least squares approach to structural equation modeling', in G.A. Marcoulides (ed.), *Modern Methods for Business Research*, Mahwah, NJ: Lawrence Erlbaum Associates, pp. 295–336.
Cobianu-Bacanu, M. (1994), 'Transition social norm and family values', *Sociologie Romaneasea, Serie noua*, **5**(5), 519–31.
Dana, L. (1999), 'Small business as a supplement in the People's Republic of China (RPC)', *Journal of Small Business Management*, **37**(3), 76–80.
Davis, D. (2000), 'China's consumer revolution', *Current History*, September, 248–54.
Donckles, R. and Lambrecht, J. (1999), 'The re-emergence of family-based enterprises in East Central Europe: what can be learned from family business research in the western world?', *Family Business Review*, **12**(2), 171–88.
Dunkelberg, W. and Cooper, A. (1982), 'Entrepreneurial typologies', in K. Vesper (ed.), *Frontiers Of Entrepreneurship Research 1982*, Wellesley, MA: Babson College, pp. 1–15.
Fornell, C. and Larcker, D. (1998), 'Structural equation models with unobserved variables and measurement error', *Journal of Marketing Research*, **18**, 39–50.
Fukuyama, F. (1995), *Trust: The Social Virtues and the Creation of Prosperity*, New York: Simon and Schuster.
Gallo, M., Tapies, J. and Cappuyns, K. (2004), 'Comparison of family and nonfamily businesses: financial logic and personal preferences', *Family Business Review*, **17**(4), 303–18.
Gundry, L. and Welsch, H. (2001), 'The ambitious entrepreneur: attributes of firms exhibiting high growth strategies of women-owned enterprises', *Journal of Business Venturing*, **16**(5), 453–70.
Habbershon, T. and Pistrui, J. (2002), 'Enterprising families domain: family-influenced ownership groups in pursuit of transgenerational wealth', *Family Business Review*, **15**(3), 223–37.
Hair, J., Anderson, R., Tatham, R. and Black, W. (1998), *Multivariate Data Analysis*, Upper Saddle River, NJ: Prentice Hall.
Hartford, K. (2000), 'Cyberspace with Chinese characteristics', *Current History*, September, 255–62.
Hisrich, R. and Fulop, G. (1997), 'Women entrepreneurs in family business: the Hungarian case', *Family Business Review*, **10**(3), 281–302.
Honing, B. (1998), 'What determines success? Examining the human, financial, and social capital of Jamaican microentrepreneurs', *Journal of Business Venturing*, **13**, 371–94.
Hu, Z. Khan, M. (1997), 'Why is China growing so fast?', *International Monetary Fund, Economic Issues 8*, Washington, DC: IMF.
Kelloway, E. (1998), *Using LISREL for Structural Equation Modeling: A Researcher's Guide*, Thousand Oaks, CA: Sage Publications.
Kerlinger, F. (1964), *Foundations of Behavioral Research*, 3rd edn, Orlando, FL: Harcourt Brace College Publishers.
Kline, R. (1998), *Principles and practice of structural equation modeling*, New York: Guilford Press.
Koiranen, M. (2002), 'Over 100 years of age but still entrepreneurially active in business: exploring the values and family characteristics of old Finnish family firms', *Family Business Review*, **15**(3), 175–87.
Kuratko, D. and Welsch, H. (2004), *Strategic Entrepreneurial Growth*, 2nd edn, Cincinnati OH: Thomson South-Western.
Liao, J., Welsch, H. and Pistrui, D. (2003), 'Patterns of venture financing: the case of Chinese entrepreneurs', *Journal of Entrepreneurial Finance & Business Ventures*, **8**(2), 55–69.
Likert, R. (1932), 'A technique for the measurement of attitudes', *Archives of Psychology*, (140).
Mahler, I. (1953), 'Attitudes towards socialized medicine', *Journal of Social Psychology*, (38), 273–82.
Martin, W. and Lumpkin, T. (2003), 'From Entrepreneurial Orientation to "Family Orientation": Generational Differences in the Management of Family Businesses', in W. Bygrave, C. Brush, M. Lernes, P. Davidsson, D. Meyer, J. Fiet, J. Sohl, P. Greene, A. Zacharakis and R. Harrison (eds), *Frontiers of Entrepreneurship Research*, Wellesley, MA: Babson College, pp. 309–21.
Masurel, E. and Smit, H. (2000), 'Planning behavior of small firms in Central Vietnam', *Journal of Small Business Management*, **38**(2), 95–102.
Nunnally, J. (1978), *Psychometric Theory*, New York: McGraw-Hill.
Parsons, T. (1955), *Family, Socialisation and Interaction Process*, Glencoe, IL: The Free Press (with Robert Bales).
Pedhazur, E. and Schmelkin L. (1991), *Measurement, Design and Analysis: An Integrated Approach*, Hillsdale, NJ: Erlbaum.
Perkins, D. (2000), 'Law, family ties, and the East Asian way of business', in L. Harrison and S. Huntington (eds), *Culture Matters: How Values Shape Human Progress*, New York: Basic Books, pp. 232–43.

Pistrui, D. (2002), 'Growth intentions and expansion plans of new entrepreneurs in transforming economies: an investigation into family dynamics, entrepreneurship and enterprise development', doctoral dissertation, Universitat Autonoma de Barcelona, Barcelona.

Pistrui, D. (2005), 'Acumen Growth Index', Acumen Dynamics, LLC, Chicago, IL, www.acumendynamics.com.

Pistrui, D., Huang, W., Oksoy, D., Jing, Z. and Welsch, H. (2001), 'Entrepreneurship in China: characteristics, attributes and family forces shaping the emerging private sector', *Family Business Review*, **14**(2), 141–52.

Pistrui, D., Welsch, H. and Roberts, J. (1997a), 'Growth intentions and expansion plans of new entrepreneurs in the former Soviet Bloc', in R. Donckels and A. Miettinen (eds), *Entrepreneurship and SME Research: On Its Way to The Next Millennium*, Aldershot: Ashgate, pp. 93–111.

Pistrui, D., Welsch, H. and Roberts, J. (1997b), 'The [re]-emergence of family business in the transforming Soviet Bloc', *Family Business Review*, **10**(3), 221–37.

Pistrui, D., Welsch, H., Wintermantel, O., Liao, J. and Pohl, H. (2000), 'Entrepreneurial orientation and family forces in the new Germany: similarities and differences between East and West German entrepreneurs', *Family Business Review*, **13**(3), 251–63.

Pistrui, D., Welsch, H., Wintermantel, O., Liao, J. and Pohl, H. (2003), 'Entrepreneurship in the New Germany', in D. Kirby and A. Watson (eds), *Small Firms and Economic Development in Developed and Transition Economies: A Reader*, Aldershot: Ashgate, pp. 116–30.

Poutziouris, P., O'Sullivan, K. and Nicolescu, L. (1997), 'The [re]-generation of family entrepreneurship in the Balkans', *Family Business Review*, **10**(3), 239–62.

Pye, L. (2000), 'Asian values: from dynamos to dominoes?', in L. Harrison and S. Huntington (eds), *Culture Matters How Values Shape Human Progress*, New York: Basic Books, pp. 244–55.

Quanyu, H., Leonard, J. and Tong, C. (1997), *Business Decision Making in China*, New York: International Business Press.

Rivard, S. and Huff, S. (1988), 'Factors of success for end-user computing', *Communications of the ACM*, **31**(5), 309–29.

Rosenthal, R. and Rosnow, R. (1984), *Essentials of Behavioral Research: Methods and Data Analysis*, 2nd edn, 1991, New York: McGraw-Hill.

Schumpeter, J. (1934), *The Theory Of Economic Development*, trans. Redvers Opie, New York: Oxford University Press.

Shanker, M. and Astrachan, J. (1996), 'Myths and realities: family businesses' contributions to the US economy', *Family Business Review*, **9**(2), 107–23.

Sik, E. and Wellman, B. (1999), 'Network capital in capitalist, communist and post-communist countries', in B. Wellman (ed.), *Networks in the Global Village*, Boulder, CO: Westview Press, pp. 225–54.

Siu, W. and Kirby, D. (1999), 'Small firm marketing: a comparative study of Chinese and British marketing practices', *Research into Entrepreneurship(RENT)*, Proceedings of the 13th RENT Conference, London, 25–26 November.

Starr, J.B. (1998), *Understanding China: A Guide to China's Economy, History, and Political Structure*, 2nd edn, New York: Hill and Wang.

Steier, L. (2003), 'Unraveling the familial sub-narrative in entrepreneurship research', in W. Bygrave, C. Brush, M. Lerner, P. Davidsson, D. Meyer, J. Fiet, J. Sohl, P.G. Greene, A. Zacharakis and R. Harrison (eds), *Frontiers of Entrepreneurship Research*, Wellesley, MA: Babson College, pp. 258–72.

Tambunan, T. (2000), 'The performance of small enterprises during economic crisis: evidence from Indonesia', *Journal of Small Business Management*, **38**(4), 93–101.

Timmons, J. and Spinelli, S. (2004), *New Venture Creation Entrerpreneurship for the 21st Century*, New York: McGraw Hill Irwin.

Tsang, E. (2001), 'Internationalizing the family firm: a case study of a Chinese family business', *Journal of Small Business Management*, **39**(1), 88–94.

Ward, J. and Aronoff, C. (1990), 'To sell or not sell', *Nations Business*, p. 73.

Wellman, B. (2001), 'Networking Guanxi', in T. Gold, D. Guthrie and D. Wank (eds), *Social Networks in China: Institutions, Culture, and the Changing Nature of Guanxi*, pp. 200–223.

Welsch H. and Pistrui, D. (1996), 'Essential Elements in Entrepreneurship Development in South Africa: An Analysis of Affecting Factors', *International Council for Small Business, 41st World Conference Proceedings*, Stockholm, Sweden, 16–19 June, pp. 105–27.

Wintermantel, O. (1999), 'East Meets West: entrepreneurship in the "New" Germany', unpublished graduate thesis, Univeritaten Mannheim and Heidelberg.

Wu, Y. (1983), 'The role of alien entrepreneurs in economic development', *American Economic Review*, **73**(2), 112–17.

www.chinapages.com/hubei/wuhan(2000).

www.chinese-school.com (2005).

Zapalska, A. and Edwards, W. (2001), 'Chinese entrepreneurship in a cultural and economic perspective', *Journal of Small Business Management*, **39**(3), 286–92.

26 Board of directors in Italian public family-controlled companies

Guido Corbetta and Alessandro Minichilli

This chapter aims to explore the features of the boards of directors in the publicly listed companies of the Italian stock exchange. It analyses the board along its classical three dimensions, that is, the composition, the structure, and the processes. The purpose is to understand if the family companies present specific characteristics with respect to the functioning of corporate boards. The analysis has been conducted over the 114 family-controlled Italian listed companies, and focused mostly on the independence of the board of directors from the controlling family. Results indicated that in public family-controlled companies the compliance with the code prevails over their family nature in order to explain their board characteristics. Results showed some correlations between the board characteristics and the company stock performance. Nonetheless, no evidence was found that public family-controlled companies have strong specific characteristics descending from the family representation within the board of directors.

Research background

Few topics like corporate governance and boards of directors found such considerable attention in the past decade (Daily et al., 2003). Some would say that we are experiencing the 'golden age' of boards and governance (Huse et al., 2005). A superficial glance suggests that the corporate scandals and bankruptcies worldwide compelled academics and managers to became aware of the relevance of corporate governance. We maintain that this interest comes from other reasons.

First, several theoretical developments and fresh ideas widened the meaning of corporate governance. Corporate governance traditionally has been defined as the ways suppliers of finance to corporations assure themselves of getting a return on their investment (Shleifer and Vishny, 1997). Therefore, the governance research has so far been dominated by the agency theory paradigm in explaining the moderating role the board of directors plays between managers and shareholders (Eisenhardt, 1989; Fama and Jensen, 1983; Jensen and Meckling, 1976; Shleifer and Vishny, 1997). This theoretical perspective has its roots in Berle and Mean's seminal work (1932) on the separation between ownership and control, which emphasized the contribution of the board of directors in lowering the risk of managerial opportunistic behaviours. Hence, it reduced the role of the board to a mere monitor against funds' expropriations inspired by a managerial self-serving attitude. The ongoing research on governance mechanisms shows that people could have different behaviours. It also shows how the board has to interface with a broader bundle of actors than the sole shareholders. The stakeholder theory arguments (Blair, 1995; Freeman and Reed, 1983; Huse, 1998, 2001), for instance, proposed a more comprehensive definition of corporate governance as a set of interactions that occur among the internal stakeholders (including the management), the external stakeholders (including the shareholders), and the board in directing the corporation to create value (Huse, 2002). This new theoretical perspective created room for alternative interpretations of the governance matter. Stewardship theory, for instance, emphasized a diverse view of human behaviour. In this view, managers are motivated to act in the best interest of their principals, founded on the conviction that pro-organizational behaviours have higher utility than individualistic and self-serving behaviours (Davis et al., 1997; Stiles and Taylor, 2001).

The family business literature followed the same path, and benefited from these developments. Scholars in this field traditionally faced the governance problem in terms of agency relationships between the board and the management (Schulze et al., 2000). They also focused on the comparison of agency costs in family and non-family companies (Chrisman et al., 2004). Theoretical advancements showed how the incidence of agency costs in family companies is likely to depend on the type of person prevailing within the firm (Corbetta and Salvato, 2004). In other words, the family contexts characterized by an individualistic person would have experienced higher agency relationships than the family contexts featured by a self-actualizing person. In those contexts, the prevalence of stewardship behaviours would have determined a stronger emphasis on the maximization of potential performance (Corbetta and Salvato, 2004).

Another reason that determined the emphasis on boards and governance is the changing role of the board of directors. We maintain that boards of directors are substantially changing from ineffectual pawns to actual leaders (Lorsch, 1989). Their behaviour is different now than only few years ago. Directors on boards are starting to hire, fire and compensate top executives. They are also starting to perceive themselves as value-creating corporate actors. These behaviours are taken for granted in the literature on corporate governance and boards of directors. According to various scholars, the two basic roles each board should perform are the control role and the service role (for example, Huse and Rindova, 2001; Johnson et al., 1996; Monks and Minow, 2004; Stiles and Taylor, 2001; Zahra and Pearce, 1989). What has happened in the recent past is a rather different story. Boards have long been considered 'ornaments on the Christmas three' whose relevance was confined within a pure legalistic perspective. This belief was even stronger in the context of family companies, although there was a complete lack of evidence with respect to board characteristics in those situations.

This study aims to combine the traditional corporate governance research on large public corporations with the specific characteristics of family-controlled companies. We focused on the board of directors as the most relevant governance mechanism, whose primary purpose is to protect stakeholders against managerial misconduct, and to provide management with counselling and advice. We tried to understand which are the typical board characteristics in a public family-controlled company, and how much does the family matter in the design of the board structure. We therefore described the board characteristics and the functioning of the Italian family-controlled companies under the perspectives highlighted. Our purpose has been to understand the board design in large public family companies. The qualitative description used specific indicators to appreciate both the board characteristics and the family representation. The results have been used to make comments on the board functioning in the companies we examined, and explore the potential link between board characteristic, family representation, and company stock performance.

Board characteristics

Board composition, structure and processes

Board characteristics refer to what the literature commonly addressed as board composition, board structure, and board processes. To describe the board composition we used the classical indicators like the board size and the proportion of independent directors on the board. The board size is widely considered a relevant feature that can have much to do with a board's ability to control and to serve the management. Nonetheless, it is impossible to identify an optimal size, even for boards operating in listed company (Bowen, 1994; Charan, 1998). It depends upon contingency measures such as the company's size, its ownership structure, its organizational complexity, and the number and features of the markets served. Moreover, before being a governance body, the board of directors is first of all a team where persons with different backgrounds, experiences, competencies and cultures meet together to deal with complex problems. This outlines the classic trade-off between the effectiveness of decision-making by small boards, and the breadth of ideas and expertise large boards have. Board composition deals also with the proportion of independent directors compared with the total number of directors on the board. Independent directors are supposed to provide greater protection to stakeholders, broaden the strategic view of the board, resolve the potential internal conflicts of interests, and properly appoint, monitor, replace and evaluate the CEO and his or her executive team (Clarke, 1998). Independent directors are non-executive directors without business ties to the company, or personal ties to anyone with ties to the company itself. They should therefore guarantee the required externality, objectivity and independence of judgement to control the insiders, but also enhance the board's value creation. The matter of directors' real independency has always been subject to debate. Each board establishes which directors can be considered independent, and each national code sets different requisites for independency (Cuervo, 2002). In this study we relied on the information collected in the companies' corporate governance reports, which followed the provisions of the Italian national code (Committee for the Corporate Governance of Listed Companies, 2002).

The board structure deals mainly with the CEO duality and the presence of board committees. As to the *CEO duality*, both the codes of best practice across the world and the literature agreed to split the CEO from the chairman of the board (Cadbury, 2002; Conger et al., 2001; Daily and Dalton, 1997; Donaldson and Davis, 1991; Lorsch, 1989). The arguments which favour such separation relate basically to three aspects. First, the CEO duality concentrates a great deal of power in the hands of one person, and makes it difficult for the board to supervise the management and the CEO. Secondly, the two positions require different mixes of abilities and experiences. The chairman should be an organizational figure with a leadership profile inside the team, and a strong ability to externally represent the company on behalf of its stakeholders. The CEO should have mostly the ability to properly run the business. Third, since the organizational activity the chairman is required to perform is extremely time-consuming, it is advisable to seek out a person to specifically devote themselves to this task.

The other element of the board structure is the presence of committees. Literature on the topic is rather wide-ranging, but the basic assumptions of the scholars are essentially: (1) the most important board decisions originate at the committee level (Kesner, 1988); (2) despite the number of possible board committees, the greater influence derives from the audit, compensation and nomination committees (Vance, 1983); (3) while the overall board composition is substantially unrelated to firm performance, the structure and composition of its committees does impact on it (Klein, 1988). The results we obtained from our analyses confirmed the last hypothesis.

The emphasis on board processes finally stemmed from the need to go beyond the demographic approach to board of directors. This tendency followed the need to open the board black box (Pettigrew, 1992). Notwithstanding, the complexity of exploring mechanisms and interactions inside the board, limited the research on this topic. This study used the number of meetings as a proxy of the board processes and activities. Even though it only partially addresses the topic, it represents a first step towards this emerging research trend.

Family representation

The other element we considered in our study is family representation. We considered three possible indicators of family representation, that is, the presence of a family CEO, the presence of a family chairman, and the number of family members among the total board members. In the context of family companies, the appointment of a family versus non-family CEO (that is, external) is an important indicator of the wish to separate ownership and management. Along this line, the appointment of an external CEO is a clear signal of the will to select the key actors based on their competences, rather than on family ties. This position is yet not prescriptive. Several Italian companies with a family CEO showed positive performances and were managed with competence and technical ability. The second indicator is the family versus non-family chairman. As we anticipated earlier, the chairman has different characteristics than the CEO, since the chairman should have an internal leadership and an external ability to represent the company. We thus hypothesized that for a family member the chairman position could be more suitable, at least in principle. The third indicator refers to the number of family members among the total number of directors. Coupled with the family or non-family nature of the CEO and the chairman, it completed the picture on the family overlap with the company. Even though we did not have hypotheses on the family directors, we argue that an overrepresentation of family members on the board should be avoided for at least two reasons. First, a high number of family members on the board limits the potential contribution in terms of advice and counsel that outside directors can give. This is not a criticism of family members' ability to effectively counsel management. Nonetheless, their excessive representation in the boardroom clearly narrows the variety of perspectives external directors can express. Secondly, the presence of too many family members could limit the will of independent directors to control management, especially if the management itself is strongly family based.

Sample and data collection

The sample we used for the study is made up of all the non-financial public family-controlled companies listed at the Borsa Italiana (Italian Stock Exchange). The family connection has been defined through an analysis of the ownership structure of each of the listed companies considered. Data on the ownership structures were collected through the Consob (Italian Commission for the Stock Exchange). We used a classification criterion according to which a family company is a company where one or few families have the control over the board of directors, both directly and through financial holdings. We identified 114 family companies out of 168 non-financial listed companies, ordered by turnover. We later identified three subsamples: (1) the majority controlled family companies, where the family/families held the absolute majority of shares; (2) the minority controlled family companies, where the family/families controlled the board of directors without a strict majority of the shares; (3) the first 30 companies for turnover. The information about the turnover were collected through the AIDA – Italian Digital Database of Companies. The majority-controlled public family companies were 96 out of the 114, with an average turnover of 1.820 million euros. The minority-controlled public family companies were 18, with an average turnover of 4.842 million euros. The highest mean was clearly that of the first 30 companies for turnover, with an average of 8.126 million euros.

To analyse the board characteristics we codified and collected the information contained in the various companies' corporate governance reports for the year 2003. We collected information about: (1) the size of the board; (2) the separation between the CEO and the chairman; (3) the percentage of independent directors over the total number of directors; (4) the presence of the three main committees (audit, remuneration and nomination); (5) the number of meetings held in the past year by the board of directors; and (6) the number of meetings for each of the board committees. The analysis of family representation was realized through the three indicators earlier defined, that is: (1) the family CEO; (2) the family chairman; (3) the number of family members within the board. To define the family connection of each board member across the 114 companies selected we considered different sources. Among them, we used the companies' websites, direct phone interviews with stakeholder relations departments, and a database of newspaper articles from a major Italian financial newspaper.

Descriptive results

The descriptive analyses of the indicators we considered gave a whole picture of the board characteristics in Italian family-controlled companies, with a focus on the major differences across the subsamples we defined. This picture is presented in the appendix to this chapter. Our purpose is to sketch out and to comment some of the most interesting results.

The size of the board, although correlated to the size of the companies (the average is 9.1 directors in the majority-controlled companies against 12.5 in the first 30 companies), still showed a remarkable variance across the samples. A considerable number of companies (30.7 per cent) had up to 7 directors in the board, and the 9.6 per cent had more than 15 directors. We would have wished a more homogeneous distribution around the mean of the whole sample, which is 9.6.

The results also clearly demonstrated how larger companies presented the highest compliance with the code provisions. The first 30 companies showed a split between the CEO and the chairman in 90 per cent of the cases, against the 62.5 per cent for the majority-controlled companies. The same occurred for the presence of committees of the board. Of the first 30 companies 86.7 per cent had an audit committee, against the 73.9 per cent of the majority-controlled companies and 77.7 per cent of the minority-controlled companies. The difference increased with respect to the remuneration committee. This committee was established in the 76.7 per cent of the first 30 companies, against the 60.4 per cent of the majority-controlled companies. Finally, there were fewer nomination

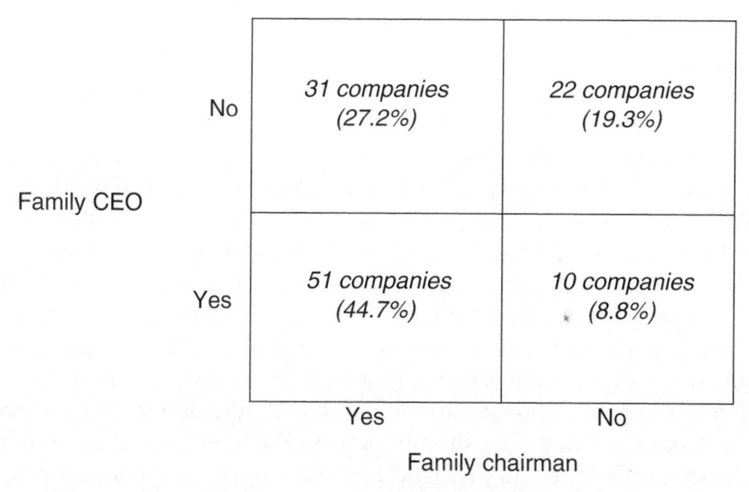

Figure 26.1 *The overlap between the family and the company*

committees in each of the subsamples we considered. It existed in only 12 per cent of cases. This is a remarkable problem, since the nomination committee represents the vehicle through which new outside directors are brought in for selection, in order to limit the discretional power of the CEO to nominate new directors from friendship or family ties (Westphal and Zajac, 1995).

Family representation, then, varied a lot across companies. The first 30 family companies by turnover nominated an external CEO in 70 per cent of cases, while the majority-controlled family companies did so only in 45.8 per cent of cases. The data on the percentage of family members on the board confirmed this tendency. Of the first 30 companies 53.3 per cent had an incidence of family directors lower than one-fifth of the entire board, against only 36.4 per cent of the majority-controlled companies. A similar result came from the minority-controlled companies (38.8 per cent), and in the whole sample (36.8 per cent). The same trend emerged for the chairman, who was external to the family in 36.6 per cent of the first 30 companies, and in 27.1 per cent of the majority-controlled companies. Anyhow, all the subsamples confirmed the strong tendency to have a family chairman on the board of directors. The matrix (Figure 26.1) synthesizes the family representation with respect to the family or non-family nature of the CEO and the chairman. It confirms what we argued earlier, and also implicitly gives a measure of CEO duality for the whole sample (44.7 per cent of cases). It is interesting to observe how in 22 companies (19.3 per cent) both CEO and chairman are external to the family.

The proportion of independents was quite similar in all the subsamples. The average measure over all 114 companies showed how the 64.9 per cent had less than one-third of the board made up of independents, while 25.4 per cent had a proportion of independents between one-third and half of the board. It was not an encouraging result. Given the tendency to appoint outside directors who are generally independent, the percentages above are a proxy for the proportion of outsiders on the board. It hides the risk that boards are generally insider dominated, lowering the potential of service to management, and the potential of control.

Board of directors and stock performance

The most popular criticism of the stream of research on boards of directors has been to have no implications in terms of the company's performance. The adoption of the governance mechanisms and provisions on boards of directors seems to be almost exclusively oriented to the formal respect of equity criteria towards the company's stakeholders. This is the reason why the board of directors has so far been studied by focusing mostly on the linkage with corporate financial results (Baysinger and Butler, 1995; Baysinger and Hoskisson, 1990; Dalton et al., 1998, 1999; Hermalin and Weisbach, 1991). Nonetheless, the research of any relationship between the characteristics of the board of directors and the company's performance often gave equivocal results (Huse, 2000). The theoretical arguments for these results are immediate. The design of fair, transparent and effective governance mechanisms should per se reduce expropriation, the extraction of private benefits, and corruption. Moreover, the design of balanced, competent and critic boards of directors should increase the likelihood the company they serve will successfully compete and perform. In a short time, good corporate governance should create value. Nonetheless there are contingency variables, both internal to the company (quality of management, organizational structure and so on), and external (macroeconomic trends, industry trends, political scenarios), which moderate the relationship between the quality of boards and the financial performance of firms. In other words, corporate governance is a necessary but not a sufficient condition to gain superior performances. We hypothesized an intermediate position. From one side, we recognized the moderate nature of the relationship between the board characteristics and the company performance; from the other side, we maintained the existence of some form of direct relationship with a particular type of performance, that is, stock performance. This choice has clear motivations. Although stock performance is strongly influenced by the financial and the competitive results of a company, it reflects also the stability of these results, the reliability of the company, and its future potential for growth. In this sense, the composition of a quality board of directors can increase the trustworthiness of the company, its reputation in the financial stock market, and ultimately its stock value. The analysis we realized considered the percentage spread of capitalization that each of the 114 family-controlled companies realized in the period from January 2003 to October 2004. It consisted of a bivariate correlation analysis, which included all the variables previously discussed in the descriptive section. Synthetic results are shown in Table 26.1.

The results of the correlation analysis are as follows. First, the spread of capitalization in the considered period is significantly related to the size of the board (0.38*), and to the number of independent directors in the boardroom (0.29*). This evidence goes in the direction of what a large part of the literature hypothesized, although never definitely demonstrated. Other interesting results came from the impact the size of the company had on the variables analysed. Particularly, when the size increased there was a reduction in the tendency of companies to appoint both a family CEO (−0.19*) and a family chairman (−0.24*). This is a natural result, since the large and complex realities often need the intervention of external professionals at the top. An increase in size was then positively related to the establishment of a nomination committee (0.24*), for the same reason we explained before. An obvious result is the relationship between the number of family members in the board, and the appointment of a family CEO (0.32**), and a family

Table 26.1 Bivariate correlation analysis

	Turnover	Family chairman	Chairman = CEO	Family CEO	No. of directors	No. of family members	No. of independents	Audit C.	Nominat. C.	Remuner. C.	Board meetings	Δ capitaliz.
Turnover	1	–	–	–	–	–	–	–	–	–	–	–
Family chairman	−0.24*	1	–	–	–	–	–	–	–	–	–	–
Chairman = CEO	−0.14	0.31*	1	–	–	–	–	–	–	–	–	–
Family CEO	−0.19*	0.28*	0.63**	1	–	–	–	–	–	–	–	–
No. of directors	0.18	−0.03	−0.19*	−0.16	1	–	–	–	–	–	–	–
No. of family members	−0.12	0.31**	0.21*	0.32**	0.21*	1	–	–	–	–	–	–
No. of independents	0.09	−0.06	−0.12	−0.05	0.58**	0.05	1	–	–	–	–	–
Audit Committee	−0.04	0.08	−0.11	0.04	0.27*	0.08	0.42**	1	–	–	–	–
Nomination Committee	0.24*	−0.02	−0.15	−0.11	0.03	−0.08	0.17	0.19*	1	–	–	–
Remuneration Committee	0.01	0.07	−0.10	−0.02	0.36*	0.14	0.42**	0.61**	0.26**	1	–	–
Board meetings	0.18	−0.13	−0.15	−0.24*	−0.02	−0.06	−0.11	0.02	0.01	−0.03	1	–
Δ capitalization	0.22*	0.04	−0.07	−0.04	0.38*	0.09	0.29**	0.18	0.09	0.18	0.00	1

Note: Pearson correlation coefficient, 1-tailed: *< 0.05; **< 0.01; N = 114.

chairman (0.31**). The companies with a high number of family directors showed a tendency to identify the family with the company itself. As a consequence, they preferred to grant the guidance of the company to professionals selected within the family. Finally, the presence of a family CEO was negatively related to the number of meetings the board annually had (−0.24*). When the CEO is a family member, the board of directors acted as a 'rubber stamp' for its proposals. Thus, the number of meetings is likely to be fewer than in companies where the board effectively acts as a decision-maker. Nonetheless, none of the three measures of family representation we built showed any significant correlation with the company stock performance.

Conclusions

The evidence from the previous analyses suggested three main conclusions for the study. First, the public family-controlled firms presented a tendency to select the CEO, the chairman, and the other board members from among the family group. This is not necessarily a negative stance. On the contrary, there is a plethora of examples of successful CEOs selected from within the family. Secondly, we found correlations between the company stock performance and some board characteristics, such as the board size and the presence of independent directors. As we argued before, independents on the board have basically a signalling power, which increases the company reputation and its stock value. Third, the company stock performance had no significant relationship with the measures of family representation we adopted. This is the most unexpected result, since we hypothesized the stock market would have appreciated the introduction external directors and top executives in the context of a family-run businesses. This conclusion suggested that the board composition in family companies is basically different between non-listed and listed ones. Among the listed companies the prevailing guidance for the composition the board of directors is the code of best practice, and the impact of the family is less important than in privately held family businesses.

The study suffered some limitations. The first is the context of where it took place. Italian companies are traditionally dominated by large controlling families, whose interference in governance choices is often relevant. In other words, the Italian managerial culture is used to a strong family presence in governance bodies, and that could have biased the results of our analyses. Even though the conclusion showed no significant relationships between family representation on the board and the stock performance, the study would benefit from proper comparisons with those of other countries. Secondly, the study represented only a first and mostly descriptive attempt to explore the effects of board characteristics on firm performance in the setting of family-run businesses. A further theoretical development towards a model of causal relationships which link board characteristics, family representation in the board, and firm performance is thus required.

Nonetheless, the study provided some preliminary insights on the board design that need to be highlighted. It opened interesting perspectives in understanding which companies' features favour a greater presence of professional managers on the board, and which managerial profile they present. Regardless of the immediate impact on stock performance, the family versus non-family members on the board are thought to give different contributions to the firm success. Family executives should have greater motivation to create value for the company, often owing to a significant shareholding.

Non-family executives should, conversely, provide an external point of view to internals, a diverse background, and sometimes greater professionalism. These further extensions of the study could be particularly beneficial for the nomination process, both internally and externally. They could represent guidance for the nomination committees and for the external governance consultancy agencies in the appointment of non-family members. But mostly, they could represent guidance in the definition of the proper mix of family and non-family directors accordingly to each company's features.

References

Baysinger, B. and R.E. Hoskisson (1990), 'The composition of boards of directors and strategic control: effects on corporate strategy', *Academy of Management Review*, **15**, 72–87.

Baysinger, B.D. and H.N. Butler (1995), 'Corporate governance and the board of directors: performance effects of changes in board composition', *Journal of Law, Economics and Organization*, **15**, 72–87.

Berle, A.A. and G.C. Means (1932), *The Modern Corporation and Private Property*, New York: Macmillan.

Blair, M.M. (1995), *Ownership and Control: Rethinking Corporate Governance for the Twenty-First Century*, Washington, DC: The Brookings Institution.

Bowen, G. (1994), *Inside the Boardroom*, New York: John Wiley and Sons.

Cadbury, Sir A. (2002), *Corporate Governance and Chairmanship: A Personal View*, Oxford: Oxford University Press.

Charan, R. (1998), *Boards at Work: How Corporate Boards Create Competitive Advantage*, San Francisco, CA: Jossey Bass.

Chrisman, J.J., J.H. Chua and R.A. Litz (2004), 'Comparing the agency costs of family and non-family firms: conceptual issues and exploratory evidence', *Entrepreneurship Theory and Practice*, Summer, 335–54.

Clarke, T. (1998), 'The contribution of non-executive directors to the effectiveness of corporate governance', *Career Development International*, **3**, 118–24.

Committee for the Corporate Governance of Listed Companies (2002), *Corporate Governance Code ('PredaReport')*, Milan: Italian Stock Exchange.

Conger, J.A., E.E. Lawler and D.L. Finegold (2001), *Corporate Boards. Strategies for Adding Value at The Top*, San Francisco, CA: Jossey-Bass.

Corbetta, G. and C.A. Salvato (2004), 'Self-serving or self-actualizing? Models of man and agency costs in different types of family firms: a commentary on "Comparing agency costs of family and non-family firms: conceptual issues and exploratory evidence"', *Entrepreneurship Theory and Practice*, Summer, 355–62.

Cuervo, A. (2002), 'Corporate governance mechanisms: a plea for less code of good governance and more market control', *Corporate Governance: An International Review*, **10**, 84–93.

Daily, C.M. and D.R. Dalton (1997), 'CEO and board chair roles held jointly or separately: much ado about nothing', *Academy of Management Executive*, **11**(3), 11–20.

Daily, C.M., D.R. Dalton and A.A. Cannella (2003), 'Corporate governance: decades of dialog and data', *Academy of Management Review*, **28**, 371–82.

Dalton, D.R., C.M. Daily, A.E. Ellstrand and J.L. Johnson (1998), 'Meta-analytic reviews of board composition, leadership structure, and financial performance', *Strategic Management Journal*, **19**, 269–90.

Dalton, D.R., C.M. Daily, J.L. Johnson and A.E. Ellstrand (1999), 'Number of directors and financial performance: a meta-analysis', *Academy of Management Journal*, **42**, 674–86.

Davis, J.H., D.F. Schoorman and L. Donaldson (1997), 'Toward a stewardship theory of management', *Academy of Management Review*, **22**, 20–47.

Donaldson, L. and J.H. Davis (1991), 'Stewardship theory or agency theory: CEO governance and shareholder returns', *Australian Journal of Management*, **6**(1), 49–64.

Eisenhardt, K. (1989), 'Agency theory: an assessment and review', *Academy of Management Review*, **14**, 57–74.

Fama, E. and M. Jensen (1983), 'Seperation of ownership and control', *Journal of Law and Economics*, **26**, 301–25.

Freeman, R.E. and D.L. Reed (1983), 'Stockholders and stakeholders: a new perspective on corporate governance', *California Management Review*, **25**, 88–106.

Hermalin, B.E. and M.S. Weisbach (1991), 'The effects of board composition and direct incentives on firm performance', *Financial Management*, Winter, 101–12.

Huse, M. (1998), 'Researching the dynamics of board–stakeholders relations', *Long Range Planning*, **31**, 218–26.

Huse, M. (2000), 'Boards of directors in SMEs: a review and research agenda', *Entrepreneurship & Regional Development*, **12**, 271–90.

Huse, M. (2002), 'Corporate governance and corporate entrepreneurship: revisiting the governance-performance links', paper presented at the Corporate Governance Conference 'Corporate Governance and Firm's Organizations: Nexus and Frontiers', Bocconi University.

Huse, M. and V. Rindova (2001), 'Stakeholders' expectations of boards of directors: the case of subsidiary boards', *Journal of Management and Governance*, **5**, 153–78.
Huse, M., A. Minichilli and M. Schoning (2005), 'Corporate boards as assets for operating in the new Europe: the value of process-oriented boardroom dynamics', *Organizational Dynamics*, **34**(3), 285–97.
Jensen, M. and W. Meckling (1976), 'Theory of the firm: managerial behaviour, agency costs and ownership structure', *Journal of Financial Economics*, **3**, 305–60.
Johnson, J.L., C.M. Daily and A.E. Ellstrand (1996), 'Boards of directors: a review and research agenda', *Journal of Management*, **22**, 409–38.
Kesner, I. (1988), Directors' characteristics and committee membership: an investigation of type, occupation, tenure, and gender, *Academy of Management Journal*, **31**(1), 66–84.
Klein, A. (1998), 'Firm performance and board committee structure', *Journal of Law and Economics*, **41**, 275–303.
Lorsch, J.W. (with E. MacIver) (1989), *Pawn or Potentates: The Reality of America's Corporate Boards*, Boston, MA: Harvard Business School Press.
Monks, R. and N. Minow (2004), *Corporate Governance*, Cambridge: Blackwell Business.
Pettigrew, A. (1992), 'On studying managerial elites', *Strategic Management Journal*, **13**, 163–82.
Schulze, W.S., M.H. Lubatkin, R.N. Dino and A.K. Buchholtz (2001), 'Agency relationships in family firms: theory and evidence', *Organization Science*, **12**, 99–116.
Shleifer, A. and R.W. Vishny (1997), 'A survey of corporate governance', *Journal of Finance*, **52**, 737–83.
Stiles, P. and B. Taylor (2001), *Boards at Work*, Oxford: Oxford University Press.
Vance, S.C. (1983), *Corporate Leadership: Boards, Directors, and Strategy*, New York: McGraw-Hill.
Westphal, J.D. and E.J. Zajac (1995), 'Who shall govern? CEO/board power, demographic similarity, and new director selection', *Administrative Science Quarterly*, **40**, 60–83.
Zahra, S.A. and J.A. Pearce (1989), 'Boards of directors and corporate financial performance: a review and integrative model', *Journal of Management*, **15**(2), 291–334.

Appendix: Descriptive analyses

Size of the board

Number and type of companies		Minimum no. of directors	Maximum no. of directors	Average
Majority controlled	96	4	16	9.1
Minority controlled	18	4	22	11.3
First 30 for turnover	30	6	22	12.5
All the sample	114	4	22	9.6

Separation between chairman and CEO

Number and type of companies		Chairman ≠ CEO (number)	Chairman ≠ CEO (percentage)
Majority controlled	96	60	62.5
Minority controlled	18	13	72.2
First 30 for turnover	30	27	90.0
All the sample	114	73	64.0

Family/non-family CEO

Number and type of companies		Non-family CEO (no.)	Non-family CEO (%)
Majority controlled	96	44	45.8
Minority controlled	18	9	50.0
First 30 for turnover	30	21	70.0
All the sample	114	53	46.5

Family/non-family chairman

Number and type of companies		Non-family chairman (no.)	Non-family chairman (%)
Majority controlled	96	26	27.1
Minority controlled	18	6	33.3
First 30 for turnover	30	11	36.6
All the sample	114	32	28.1

Family members on the board

Number and type of companies		Less than 1/5		Between 1/5 and 1/3		Between 1/3 and half		More than half	
		No.	%	No.	%	No.	%	No.	%
Majority controlled	96	35	36.5	34	35.5	23	23.9	4	4.1
Minority controlled	18	7	38.8	5	27.7	5	27.7	1	5.8
First 30 for turnover	30	16	53.3	11	36.7	3	10.0	0	0.0
All the sample	114	42	36.8	39	34.2	28	24.6	5	4.4

Independent directors

Number and type of companies		Less than 1/3		Between 1/3 and half		Between half and 2/3		More than 2/3	
		No.	%	No.	%	No.	%	No.	%
Majority controlled	96	62	64.5	24	25.0	9	9.5	1	1.0
Minority controlled	18	12	66.7	5	27.8	0	0.0	1	5.5
First 30 for turnover	30	21	70.0	7	23.4	1	3.3	1	3.3
All the sample	114	74	64.9	29	25.4	9	7.9	2	1.8

Presence of board committees

Number and type of companies		Audit		Remuneration		Nomination	
		No.	%	No.	%	No.	%
Majority controlled	96	71	73.9	58	60.4	10	10.4
Minority controlled	18	14	77.8	13	72.2	2	11.1
First 30 for turnover	30	26	86.7	23	76.7	3	10.0
All the sample	114	87	76.3	74	64.9	13	11.4

*Average number of meetings for the whole board, the audit committee and the remuneration committee**

Number and type of companies		Board	Audit	Remuneration
Majority controlled	96	7.5 (on 91 comp.)	3.5 (on 62 comp.)	1.7 (on 51 comp.)
Minority controlled	18	8.1 (on 18 comp.)	3.8 (on 15 comp.)	2.1 (on 13 comp.)
First 30 for turnover	30	8 (on 30 comp.)	3.9 (on 25 comp.)	2.3 (on 23 comp.)
All the sample	114	7.6 (on 109 comp.)	3.5 (on 77 comp.)	1.8 (on 64 comp.)

Note: * The average number of meetings has been calculated on the number of companies that indicated it (among parenteses).

27 Family-firm relationships in Italian SMEs: ownership and governance issues in a double-fold theoretical perspective

Luca Gnan and Daniela Montemerlo

This chapter offers a comparative study of family versus non-family small and medium-sized enterprises (SMEs) in terms of ownership and governance issues, and is based on a survey on 620 incorporated companies of small and medium size.

The survey was started in 2000 and was aimed at making an in-depth exploration of Italian small and medium-sized enterprises, with a special focus on family firms. Previous surveys had already analysed family small and medium-sized enterprises in Italy (Corbetta and Montemerlo, 1999); this survey intended to update some information and to go deeper into some issues, particularly into ownership and governance in terms of both structure and evolution. Another goal of the survey was to make some preliminary tests of different theoretical perspectives. On one side, we wanted to verify whether agency theory assumptions may apply to small and medium-sized enterprises and not only to large corporations; on the other side, we maintain that agency is a relevant perspective by which to interpret family small and medium-sized enterprises, but that it is not sufficient and should be integrated with a relational perspective.

The main findings we obtained confirmed, even at an exploratory level, that both perspectives are relevant. Family small and medium-sized enterprises are becoming more complex, especially in terms of ownership, which is increasingly fragmented. Fragmentation typically brings about differentiation between managing and non-managing owners, which raises a number of critical topics to be coped with to prevent agency problems, such as shares' transfers, dividends, appointment criteria for future leaders, company control through governance bodies. Family small and medium-sized enterprises do appear to cope with such topics by sharing rules for ownership and leadership, and by articulating governance systems. But this is not enough to retain ownership unity, which explains why, in a relational perspective, unofficial governance bodies such as family councils are used to nurture trust and shared vision.

First, we present the main theoretical references and the propositions that have been derived from both the agency and the relational perspectives. Then, we illustrate the methods, sample and data collection. The next section reports main findings. Finally, we discuss such findings, offering some concluding remarks to researchers and owning families.

Theoretical references

We maintain that, in order to read the structure and evolution of small and medium-sized family firms from a theoretical point of view, a double perspective is necessary. Specifically, we think we need two approaches that are traditionally used alternatively, but that should be complementary to enable a more in-depth understanding of family businesses.

The first perspective is the contractual one of agency theory that is not often applied to small and medium-sized enterprises, on the assumption that these firms do not have to cope with substantial agency problems; our point of view is that, on the contrary, small and medium-sized enterprises can be complex enough to feature such problems.

The second perspective is the relational one centred on the special links that exist between family owners and that may make the contractual devices that are typical of ownership insufficient. According to this perspective, a 'social capital' of trust, shared vision and networks exists, which has to be nurtured by means of relational devices that integrate with the contractual devices.

The contractual perspective

Agency theory is one of the literature mainstreams that look at companies from a contractual perspective, that is, as nexuses of contracts where the main counterparts are owner-principals and manager-agents; the key issue is how to align their interests, and particularly how to guarantee that agents behave in the interest of owners and not in their own. To overcome this threat (which is reinforced by contextual and behavioural conditions such as self-utility maximization, information asymmetry, bounded rationality, prevalence of economic goals and moral hazard) it is necessary to afford various agency costs in order to perform activities and operating systems that either monitor or bond agents, such as pay incentives, strategic planning, boards of directors, formal control systems, and so on (Jensen and Meckling, 1976; Morck et al., 1988).

For agency theory, small and medium-sized firms are considered to be the companies where agency problems are minimized, as owners and managers' roles are often played by the same people, which reduces costs related to conflicts (Fama and Jensen, 1983). But when the company evolves from the 'owner-manager' model towards the classical archetype of the corporation featuring complete separation between ownership and control, the typical agency threats re-emerge. In fact, different actors and corresponding interests may be involved even when the company is still small and medium sized: this happens, for instance, when managers not involved in ownership are hired, or when ownership becomes more numerous and differentiated and, by this means, only some owners play management roles.

The family nature that characterizes most small and medium-sized enterprises is also traditionally assumed to reduce agency costs for a number of reasons, particularly: in family firms, relations are based on kinship and blood; as such, these relations are made of emotions, sentiments, trust and altruism that are supposed to counter-balance opportunistic behaviours; family firms' long-term horizon reduces moral hazard problems (Daily and Dollinger, 1992; Gomez-Mejia et al., 2001; Harvey, 1999; Kang, 2000). But some studies show that the family nature can actually bring about special agency costs owing to problems of incongruity between executives' and family goals, lacking market discipline, self-control, adverse selection, managerial entrenchment and moral hazard (Buchanan, 1975; Gomez-Mejia et al., 2001; Jensen, 1998; Morck et al., 1988). On top of that, many of these problems can be originated by altruism itself (Schulze et al., 2001).

All this leads us to assume that agency threats do have to be coped with in both family and non-family small and medium-sized enterprises. Particularly, we assume that the more companies become complex in size and ownership, the more it is necessary to delegate tasks to agents at various levels and, consequently, to monitor them. To this purpose, a number of tools can be used, such as:

- policies to handle ownership issues like shares' transfers and dividend distribution. These are typical tools that are put in place to cope with agency problems, especially between managing and non-managing owners. In fact, both transfers and dividends may be regulated to guarantee that the former will not cheat the latter by putting up obstacles to exit and by 'hiding' company performances. Differentiation between managing and non-managing owners generally occurs when ownership becomes fragmented, that is when the number of owners increases (Corbetta and Montemerlo, 2003);
- policies to handle another key ownership issue such as the criteria for appointing future company leaders. Again, fragmented ownership may tend to agree on 'teams at the top' as leadership models for generations to come, as this model better guarantees full ownership representation in company governance. Later on, we will see that the agency perspective is not sufficient to understand teams at the top, and another interpretation will be offered;
- articulation of governance systems, typically by appointing bodies that appoint and monitor other bodies, and so on. By governance systems, we mean the combination of bodies that can be involved in governance at ownership, board and top management level (Montemerlo et al., 2004; Rediker and Seth, 1995).

This basic assumption can be translated into the following propositions:

Proposition 1 In family firms, extent of ownership fragmentation is related positively to policies adopted to regulate ownership issues such as shares' transfers and dividend distribution;

Proposition 2 In family firms, extent of ownership fragmentation is related positively to policies adopted to create a 'team at the top' leadership model;

Proposition 3 In family firms, size of the company and extent of ownership fragmentation are related positively to the articulation of their governance systems.

The relational perspective

In family firms, for the reasons mentioned above, agency costs should be lower; but in fact, the debate on their level in this type of company has not yet led to definite results. What emerges, nevertheless, is that agency theory represents a fundamental perspective, but at the same time an insufficient one to understand family firms in depth (Corbetta and Salvato, 2004; Mustakallio, 2002).

The relational perspective integrates the contractual perspective; within the former perspective, the social capital stream of theories seems to be particularly useful. Social capital can be defined as an asset that is rooted in social relations and networks (Leana and Van Buren, 1999; Nahatapiet and Goshal, 1998); such an asset appears to be critical in family firms, given the strong relational component that contracts feature in these companies. Nevertheless, application of this conceptual category in studies about governance, and particularly about governance of small and medium-sized enterprises, is relatively recent.

As to family firms, Mustakallio has offered a comprehensive framework of 'social capital including structural, relational and cognitive dimensions' (2002, p. 107). Within the structural dimension, family institutions such as family meetings and family councils can play a role in both family and company governance. As regards company governance, family councils represent the 'unofficial' part of governance structures, as they do not exist either in law or in management practice. In literature, they have been analysed both as complements and as substitutes to 'official' bodies and, especially, to the shareholders' meeting (Gersick et al., 1997; Lank and Ward, 2000; Lansberg, 1999; Ward, 1987, 1991). It has to be noted that these studies refer to large family firms and that relatively little attention is given to family councils in family small and medium-sized enterprises (Moores and Mula, 2000). A few studies have examined the role of family institutions such as family councils in the creation of trust and shared vision (that represent, respectively, the relational and cognitive dimensions of social capital – Gilding, 2000; Habbershon and Astrachan, 1997; Neubauer and Lank, 1998; Tsai and Goshal, 1998). Trust and shared vision determine unity and commitment of family and non-family actors; and together with quality of decisions, unity and commitment are acknowledged by several authors as the key conditions for family firms' success (Davis and Harveston, 1998; Gallo et al., 2001; LaChapelle and Barnes, 1998).

The presence of family councils in family business highlights the importance of the relational perspective to interpret the structures and dynamics of family firms. Our study researched such presence in Italian family small and medium-sized companies, based on the following propositions:

Proposition 4 In family firms, 'official' governance bodies coexist with 'unofficial' bodies.

Proposition 5 In family firms, 'official' governance bodies are utilized less than 'unofficial' bodies.

Method

Sample and data collection

As mentioned above, the sample comprised 620 incorporated companies of small and medium size, representative of the Italian population in terms of size, industries and geographical location. By small companies we mean firms with less than 250 employees and turnover of 50 million euros; medium-sized firms are considered to be those employing 251 to 500 employees and turnover totalling 50 to 250 million euros.

To build up the sample, 15 157 companies were randomly extracted in such a way as to be also representative of the reference population by region, range of employees and industries. Then a questionnaire was mailed to the extracted companies in October 2000; it was a complex questionnaire, comprising six sections investigating some anagraphical data on companies and respondents, ownership and governance structure, strategy, performance, and succession issues. All 620 responses were collected through January 2001; respondents held a leading position in 95 per cent of cases. Collected data have been elaborated by using descriptive statistics, t-tests and cluster analysis.

Of the 620 companies, 513 (83 per cent) were identified as family businesses. We define family businesses as those companies that meet at least one of the following requirements: (1) 51 per cent of equity or more is owned by the family; (2) the family owns less than 51 per cent but controls the company in partnership with friends, other entrepreneurs,

Table 27.1 Mailing list and sample by size and macro-industry

		Mailing list		Sample		Response rate
		Number	%	Number	%	%
Manufacturing	Small	6048	40	266	43	4.4
	Medium	887	6	45	7	5.1
Non-manufacturing	Small	6181	41	242	39	3.9
	Medium	2041	13	67	11	3.3
		15157	100	620	100	4.1

employees; (3) respondents perceive the company to be a family business, whatever the family share (which actually happened in 14 cases, see Greenwald and Associates, 1995).

Focusing on the sub-sample of family businesses, the redemption rate is indicated in Table 27.1 and is in line with the rates which are normally obtained in Italy. We compared the industries represented in our sample with those of the database used and found no differences in the industries represented (see Table 27.1). We also compared early respondents (first half) with late respondents (second half), following the Armstrong and Overton procedure (1977); differences here were also not significant. The same happened with other variables, for example, company age, size (employees and turnover), market and industry characteristics. All this suggested that non-response bias might not be a problem and that control variables were not necessary.

The incidence of family businesses on the whole sample confirms the worldwide acknowledged relevance of family firms, as these firms represent the predominant model.

What is more, Italian family firms perform the same in terms of continuity and the same, or even better, in terms of profitability; besides, they feature similar strategies. However, family companies are structurally smaller than non-family ones.

The ages of both types of firms resulted to be analogous, that is, the mean of year of foundation for family businesses is 1966 (median: 1974) and 1968 (median: 1976) for non-family firms.

Profitability indicators show that, from 1994 to 1999, family and non-family firms feature similar ratios in terms of ROS (return on sales), ROI (return on investment) and ROE (return on equity), as shown in Table 27.2. It has to be noted that the difference between family and non-family firms is significant only for return on investment in 1995 and 1996 and for return on equity in 1995 (sig. $<=0.05$).

Strategic behaviours are also similar in family and non-family firms, comprising a common trend towards increasing complexity. In fact, main strategic changes in the past decades and those envisioned for the next appear similar. For example, for the 10 years preceding the year of this survey, owners highlighted growth in sales (79 per cent) and employees (48 per cent), entry into new segments of the same industry (38 per cent), internationalization (34.6 per cent), diversification into new industries (25 per cent), and strategic alliances with other companies (21 per cent), as key changes.

For the following decade, entrepreneurs figure out they will establish many more alliances (47 per cent), they will increase the degree of internationalization (41.4 per cent), they will diversify more into new industries (37.3 per cent) than new segments (35 per cent). As to growth, 65 per cent of companies expect to increase their size in term of sales;

Table 27.2 Profitability indicators in family and non-family firms

		Mean		Std. deviation	
		Non-family firms	Family firms	Non-family firms	Family firms
ROS	1995	3.55	4.79	4.62	7.73
	1996	6.12	5.98	4.50	6.89
	1997	4.18	5.45	5.81	4.74
	1998	4.22	4.58	6.64	4.49
	1999	5.25	5.36	5.88	5.25
ROI	1995	4.10	6.89	6.58	5.72
	1996	7.39	8.37	5.68	6.53
	1997	5.00	7.29	6.14	5.72
	1998	5.51	6.26	6.31	5.35
	1999	6.78	7.14	6.66	6.37
ROE	1995	2.08	12.08	18.09	17.51
	1996	17.31	14.37	23.41	19.70
	1997	7.34	9.25	12.55	16.52
	1998	5.63	7.19	23.53	17.83
	1999	9.62	7.44	20.89	20.21

in general, the expected growth is more in terms of sales than of employees (24 per cent, $t = 17.614$, sig. $<= 0.001$).

As to size, family firms are smaller, on average, than non-family firms; 84.5 per cent of family firms are small whilst the corresponding percentage of non-family is 69.1 per cent ($t = 3.246$, sig. $<= 0.001$).

Moreover, most family firms are concentrated in the lower-size ranges: for example, 73.4 per cent of family firms and 58.3 per cent of non-family ones have less than 50 employees ($t = 2.932$, sig. $<= 0.001$). Presence of family firms, and particularly of small ones, is higher in more consolidated, manufacturing industries: 43.1 per cent of family firms are manufacturing, while the incidence of manufacturing non-family firms is 30.8 per cent ($t = 2.475$, sig. $<= 0.001$, the difference seems to be especially due to the stronger presence of non-family firms in service industries: 25 per cent versus 15.5 per cent, $t = 2.120$, sig. $<= 0.001$); small family firms account for 87.4 per cent of manufacturing and 81.6 per cent of non-manufacturing companies.

Results

Family ownership between tradition and change

Family ownership is still quite 'traditional' in terms of both prevalence of family and scarce openness to outsiders. We found a number of similarities with previous studies in terms of family stake, concentration of control power, average number of family shareholders, relevance of managing ones and scarce presence of non-family shareholders (Corbetta and Montemerlo, 1999); particularly:

- family owns 100 per cent of equity in 72 per cent of cases, and more than 51 per cent in another 18.5 per cent;

- average share owned by the family as a whole is 89 per cent today (std. dev. 22.5) and has not changed over the past 10 years;
- present average share held by single-family owners is 37.9 per cent (std. dev. 19.4), while the average share of the most important family shareholder is 51.3 per cent (std. dev. 23.4). Again, no change occurred with respect to 10 years before;
- the average number of family shareholders is 3 (std. dev. 2.2); with respect to 10 years ago, the increase (+0.35 per cent) is not significant; neither is this number expected to change in the future. Besides, it turned out to be positively correlated (sig. <= 0.001) with size, both for turnover (r = 0.215) and number of employees (r = 0.191) (Gnan and Montemerlo, 2001);
- most family business shareholders (on average, 69 per cent of total shareholders) work in the company;
- non-family partners exist in about 28 of cases and in almost two-thirds of these cases they are friends of the controlling family, with no changes with respect to 10 years before. This actually represents an increase with respect to previous surveys (Corbetta and Montemerlo, 1999);
- 53 per cent of total family assets, on average, are invested in the company.

This does not mean that ownership structure is not going to change. On one hand, in the past 10 years, some family owners exited from 20 per cent of companies (and in 81 per cent of these cases this occurred during the succession process). As a whole, 42 per cent of shares was transferred (28 per cent per shareholder on average). Exits were correlated (sig. <= 0.001) with generation (r = 0.210), age (r = 0.141) and fragmentation, that is, number of family owners (r = 0.205), confirming that family business' evolution naturally brings about transfers of shares, either within the same generation or from one generation to the next. Some companies are going to significantly change the structure owing to family exit; in 6.6 per cent of cases, respondents declared there will not be any family shareholders in 10 years' time.

Fragmentation was already being experienced in the past decade, and it is increasing; family ownership groups comprising four to six shareholders increased from 20.4 per cent to 25.7 per cent in the past 10 years (t = 2.019, sig. <= 0.001), and are expected to increase to 31 per cent in the next 10 years (t = 1.887, sig. <= 0.001).

Test of proposition 1: In family firms, extent of ownership fragmentation is related positively to policies adopted to regulate ownership issues such as shares' transfers and dividend distribution Thirty-six per cent of cases featured policies for shares' transfer and, particularly, pre-emption and option rights. Such rules are mainly formalized in articles of associations, more rarely in owners' agreements. They appear to be more frequent the more numerous the family owners (r = 0.167, sig. <= 0.001), and especially managing ones (r = 0.220, sig. <= 0.001).

Besides, in 38 per cent of cases explicit dividend policies were declared. Again, a positive correlation has been found between such policies and the number of owners (r = 0.163, sig. <= 0.001), plus generation (r = 0.122, sig. <= 0.050) and age (r = 0.134, sig. <= 0.001).

In sum, the more family ownership becomes fragmented and generations advance, the more owning families feel the need to regulate critical issues through appropriate rules. Proposition 1 is thus verified.

Test of propositions 2: In family firms, extent of ownership fragmentation is related positively to policies adopted to create a 'team at the top' leadership model The survey has measured for the first time the presence of 'teams at the top', that is, groups of peers co-leading the company: respondents declared that such teams characterize 55.4 per cent of companies today. Such teams resulted to be composed by 1.69 people on average (0.34 of which are women) and they include non-family members in 34 per cent of cases and a majority of them in 12 per cent.

Presence of teams at the top was positively correlated with number of owners ($r = 0.225$, sig. $<= 0.001$), and particularly of managing ones ($r = 0.310$, sig. $<= 0.001$), thereby confirming proposition 2.

Looking into the future, multiple leaders are expected to increase further up to 64 per cent of companies ($t = 2.819$, sig. $<= 0.001$). As ownership fragmentation is also expected to increase, the two trends might be correlated as well, thereby further supporting proposition 2 also for the future.

Governance structures' composition and functioning: five archetypes

Governance bodies can be divided into three groups: (1) at ownership level, the shareholders' meeting; (2) at board level, the board of directors of the holding company in case of groups, the board of directors and executive committee of operating companies, the operating companies' chairman and chief executive officer (CEO), a 'sole CEO' who is the alternative to the operating company board as, according to the Italian law, neither the board of directors nor the chairman exist when he/she is nominated; and (3) at top management level – the general manager of operating companies, the managing committee of operating companies. It was not possible to locate at which level the 'team at the top' operates (see above).

Each body, in general, may appoint members of the bodies it delegates tasks to, define their functioning mechanisms, approve their proposals, advise and monitor them, formulate some decisions itself (Huse, 2000). Bodies at levels (1) and (2) are more often devoted to decision control (ratification and monitoring), while at level (3) they are also delegated decision management functions (that is, initiation and control; see Fama and Jensen, 1983, Huse, 2000; Rediker and Seth, 1995).

These bodies can be regarded as 'official', as they are acknowledged as 'corporate organs' by law or by practice. We decided to consider also some 'unofficial' bodies, that is, bodies that normally do not appear in company organization chart, namely: (4) the family council, that is a collegial body composed by adult family members, whether they are owners of the company or not, either formal (that is, structured and organized with its own regulation) or informal (family members just meet when they need to); and (5) third parties (consultants, chartered accountants, lawyers) – empirical evidence highlights they may have critical influence, especially in advising and monitoring governance decisions but, sometimes, also taking part in their formulation.

The analysis on governance was conducted on a more restricted sample of companies, namely, 450 family small and medium-sized enterprises. A cluster analysis was made on the restricted sample and it led to identification of five 'archetypes', that is, five macro-structures encompassing similar governance structures.

The archetypes featured different complexity in terms of both number and relative frequency of their composing bodies. That is, complexity is higher when the number of

*Table 27.3 Archetypes of governance structures in family small and medium-sized enterprises**

Archetype	1	2	3	4	5
Number of firms	123	79	143	76	29
Percentage	27.3	17.6	31.8	16.9	6.4
Family council	26.8%	**50.6%**	0.7%	**100.0%**	24.1%
	23.6%	*45.6%*	*0.0%*	***100.0%***	*17.2%*
Shareholders' meeting	**100.0%**	**100.0%**	**100.0%**	**100.0%**	**100.0%**
	41.5%	*21.5%*	***69.9%***	***81.6%***	***65.5%***
Holding board of directors	0.0%	2.5%	7.0%	6.6%	31.0%
	0.0%	*1.3%*	*4.9%*	*6.6%*	*31.0%*
Operating board of directors	0.0%	**100.0%**	**100.0%**	**100.0%**	**100.0%**
	0.0%	*13.9%*	*43.4%*	*40.8%*	***62.1%***
Chairman	0.0%	49.4%	**97.9%**	**93.4%**	**82.8%**
	0.0%	*3.8%*	***91.6%***	***80.3%***	***65.5%***
CEO	0.0%	**83.5%**	**90.2%**	**96.1%**	**72.4%**
	0.0%	*22.8%*	***64.3%***	***76.3%***	***65.5%***
Sole CEO	**100.0%**	0.0%	0.0%	0.0%	0.0%
	77.2%	*0.0%*	*0.0%*	*0.0%*	*0.0%*
Executive committee	2.4%	3.8%	4.2%	13.2%	13.8%
	1.6%	*3.8%*	*4.2%*	*13.2%*	*13.8%*
General manager	14.6%	11.4%	22.4%	19.7%	**96.6%**
	5.7%	*5.1%*	*15.4%*	*15.8%*	***93.1%***
Managing committee	4.9%	1.3%	5.6%	10.5%	**51.7%**
	4.1%	*1.3%*	*4.2%*	*9.2%*	***51.7%***
Third parties	17.1%	2.5%	23.1%	44.7%	**93.1%**
	14.6%	*2.5%*	*23.1%*	*42.1%*	***93.1%***

Note: * For each governance body, the two percentages indicate existence and functioning (italic). Percentages of 50% and more are indicated in bold.

bodies that make up the archetype is higher, or when the number is the same but at least most of the bodies that make up the archetype feature a higher frequency in terms of existence and/or functioning. Two dichotomized indicators were used to measure these two phenomena, based on the answers given to a question that, for each of the 11 official and unofficial bodies listed above, asked respondents to indicate: (1) if the body did exist in the company or not; and (2) whether it was functioning, that is, actually utilized or not.

A synthesis of archetypes' characteristics is sketched out in Table 27.3. Details are offered in the remaining part of this section.

Archetype I: Single leader This is family firms' simplest archetype, present in 27.3 per cent of companies, where the shareholders' meeting is always there but is actually used in less than half the cases (41.5 per cent). In fact, the family council is existing in 26.8 per cent and utilized in 23.6 per cent of companies. The stronger body in this archetype is the 'sole' CEO, always present, and almost always functioning (77.2 per cent). Presence of other bodies is very small.

Archetype 2: Family council over official collective bodies This archetype accounts for 17.6 per cent of cases. The stronger body appears to be the family council, present in half the companies (50.6 per cent) and almost always utilized (45.6 per cent). The family council might partially act as a substitute for the shareholders' meeting (actually used in 21.5 per cent of cases) but also of other bodies such as the board of directors (which is also there in all cases, but is functioning only in 13.9 per cent). The chairman and CEO are quite frequent (49.4 per cent and 83.5 per cent respectively) but not much used (3.8 per cent and 22.8 per cent, respectively). All other bodies feature a very low presence.

Archetype 3: Active ownership and board in a single company Archetype 3 is the most common within its sub-set of companies as well, as it accounts for 31.8 per cent of cases. With respect to archetypes 1 and 2 of family firms, archetype 3 features greater prevalence and a higher utilization of official bodies. The family council is almost absent; the shareholders' meeting is used in 69.9 per cent of cases; the board of the operating company plays an actual role in less than half the companies (43.4 per cent); the chairman and CEO are the most prevalent (97.9 per cent and 90.2 per cent respectively) and functioning (91.6 per cent and 64.3 per cent); again, overlaps are possible. Other bodies feature a very small presence, with the partial exception of the general manager and third parties.

Archetype 4: Active ownership, board and unofficial bodies in a single company This archetype accounts for 16.9 per cent of cases. It is quite similar to archetype 3 but for a greater articulation owing to a higher utilization of the family council and third parties. In particular, the family council is always present and used, and might partially substitute for the board of directors, which is always there, but is functioning in only 40.8 per cent of cases.

Archetype 5: All active, both official and unofficial bodies, in single companies and groups This is the most articulated archetype, including governance bodies at all levels; in particular, holding boards, top management bodies and third parties are much more utilized than in previous archetypes. This archetype accounts for 6.4 per cent of cases.

Test of other propositions – Proposition 3: In family firms, size of the company and extent of ownership fragmentation are related positively to the articulation of their governance systems; Proposition 4 In family firms, 'official' governance bodies coexist with 'unofficial' bodies; Proposition 5: In family firms, 'official' governance bodies are utilized less than 'unofficial' bodies Results of this part of the study show (see Tables 27.4 and 27.5), first, that governance archetypes may be quite complex even in small and medium-sized enterprises, while they are generally supposed to feature very simple structures in governance literature. It also emerged that there is always a gap between existence of and actual utilization of, which confirms that governance bodies may be present, but not functioning.

Secondly, matching archetypes with company size showed that family small and medium-sized enterprises increase the existence and actual use of governance bodies as long as they grow larger and the number of family owners increases, which supports proposition 3. Thus, a further complexity might be expected in the future as companies expect to grow more and to experience more fragmentation (see above).

Thirdly, our study offers insights on the presence and importance of the family council. In particular it shows that family councils do exist in a number of small and medium-sized

Table 27.4 Test of proposition 3

Archetype		1	2	3	4	5
Number of firms		123	79	143	76	29
Percentage		27.3	17.6	31.8	16.9	6.4
Sales ('000 euros)	Mean	5879.2	8544.2	15 029.8	17 760.8	35 482.6
	Std. dev.	14 352.7	9917.3	17 578.4	22 853.8	43 702.0
Employees	Mean	45.9	39.7	79.9	77.0	143.2
	Std. dev.	68.3	69.1	87.4	79.4	132.2
No. of shareholders	Mean	2.9	3.0	5.4	4.4	4.3
	Std. dev.	0.9	1.3	5.3	2.3	4.6
Family shareholders	Mean	2.4	2.8	3.1	4.0	4.2
	Std. dev.	0.9	1.2	1.9	1.9	4.3

*Student t values for means differences**

Sales	Archetype 1	Archetype 2	Archetype 3	Archetype 4	Archetype 5
Archetype 1	–				
Archetype 2	**4.736**	–			
Archetype 3	**8.162**	**8.162**	–		
Archetype 4	**6.733**	**6.733**	**6.732**	–	
Archetype 5	**9.172**	**9.172**	**9.172**	**9.172**	–
Employees					
Archetype 1	–				
Archetype 2	**6.240**	–			
Archetype 3	**6.176**	**6.105**	–		
Archetype 4	**7.474**	**7.417**	**7.418**	–	
Archetype 5	**9.097**	**9.035**	**9.043**	**8.945**	–
No. of shareholders					
Archetype 1	–				
Archetype 2	0.310	–			
Archetype 3	**3.736**	1.852	–		
Archetype 4	0.368	0.356	1.668	–	
Archetype 5	1.314	0.679	0.630	0.430	–
Family shareholders					
Archetype 1	–				
Archetype 2	0.467	–			
Archetype 3	**4.213**	**2.058**	–		
Archetype 4	**2.822**	1.550	**4.811**	–	
Archetype 5	**3.085**	1.617	**5.382**	**4.524**	–

Note: * All t values in bold feature a $p < 0.005$.

enterprises, thus partially supporting proposition 4, and it highlights that they may sometimes substitute the shareholders' meeting and the board of directors, thereby partially supporting proposition 5. This raises some issues for the future that will be discussed in the next section.

Table 27.5 *Test of propositions 4 and 5*

Functioning % in each archetype	Number of firms	Family Council	Shareholders' Meeting	Holding Board of Directors	Operating Board of Directors	Chairman	CEO	Sole CEO	Executive committee	General Manager	Managing committee	Third parties
Family firms	450	32.4%	55.3%	4.9%	27.1%	47.6%	41.6%	21.1%	5.6%	16.0%	7.6%	24.9%
Archetype 1	123	23.6%	41.5%	0.0%	0.0%	0.0%	0.0%	77.2%	1.6%	5.7%	4.1%	14.6%
Archetype 2	79	45.6%	21.5%	1.3%	13.9%	3.8%	22.8%	0.0%	3.8%	5.1%	1.3%	2.5%
Archetype 3	143	0.0%	69.9%	4.9%	43.4%	91.6%	64.3%	0.0%	4.2%	15.4%	4.2%	23.1%
Archetype 4	76	100.0%	81.6%	6.6%	40.8%	80.3%	76.3%	0.0%	13.2%	15.8%	9.2%	42.1%
Archetype 5	29	17.2%	65.5%	31.0%	62.1%	65.5%	65.5%	0.0%	13.8%	93.1%	51.7%	93.1%
*t values associated with % difference between presence of a family council and presence of other bodies**												
Family firms	450	0.000	**7.110**	**11.341**	1.752	**4.683**	**2.843**	**3.871**	**10.944**	**5.867**	**9.821**	**2.515**
Archetype 1	123	0.000	**3.050**	**6.160**	**6.160**	**6.160**	**6.160**	**9.974**	**5.496**	**4.102**	**4.622**	1.796
Archetype 2	79	0.000	**3.311**	**7.715**	**4.637**	**6.960**	**3.110**	**8.133**	**6.960**	**6.616**	**7.715**	**7.325**
Archetype 3	143	0.000	**18.236**	**2.713**	**10.462**	**39.511**	**16.061**	0.000	**2.503**	**5.099**	**2.503**	**6.550**
Archetype 4	76	0.000	**4.143**	**32.851**	**10.503**	**4.323**	**4.857**	0.000	**22.396**	**20.133**	**27.370**	**10.223**
Archetype 5	29	0.000	**4.282**	1.244	**3.926**	**4.282**	**4.282**	**2.458**	0.363	**8.981**	**2.964**	**8.981**
*t value of the difference of % between the presence of third parties and the presence of other bodies**												
Family firms	450	**2.515**	**9.802**	**8.781**	0.760	**7.279**	**5.393**	1.348	**8.382**	**3.326**	**7.256**	0.000
Archetype 1	123	1.796	**4.907**	**4.592**	**4.592**	**4.592**	**4.592**	**12.660**	**3.843**	**2.347**	**2.895**	0.000
Archetype 2	79	**7.325**	**3.836**	0.584	**2.664**	0.455	**4.019**	1.432	0.455	0.834	0.584	0.000
Archetype 3	143	**6.550**	**8.997**	**4.593**	**3.728**	**16.248**	**7.734**	**6.550**	**4.839**	1.658	**4.839**	0.000
Archetype 4	76	**10.223**	**5.482**	**5.606**	0.165	**5.245**	**4.577**	**7.435**	**4.217**	**3.738**	**5.012**	0.000
Archetype 5	29	**8.981**	**2.758**	**6.337**	**3.053**	**2.758**	**2.758**	**19.786**	**9.981**	0.000	**3.977**	0.000

Note: *All t values in bold feature a $p < 0.005$.

Concluding remarks

Synthesis of main findings
Both theoretical perspectives, contractual and relational, seem to be relevant to understanding family small and medium-sized enterprises. The contractual perspective is pushing owning families to establish rules and to make their governance systems more complex to cope with the agency problems brought about by company growth and ownership fragmentation. The relational perspective encourages owning families to keep unofficial governance bodies such as family councils, and sometimes to substitute official bodies with these, to keep trust and shared vision between family members. Interestingly, teams at the top may result from a mix of agency needs (such as full ownership representation) and relational needs (for example, keeping cohesion between family executives).

Some hints for researchers
We envision a few possible directions for future research.

First, our propositions might be turned into hypotheses to be extensively tested.

Second, studies on other countries might test the influence of national culture on the results we obtained. In particular, the very close relationship between family and firm that is typical of the Italian context might give unofficial governance bodies such as the family council a superior importance with respect to other countries.

Third, analysis of governance structures might go deeper from many points of view. For instance, to prevent the questionnaire being too onerous for respondents, we could not ask whether the same or different people played governance roles at different levels (ownership, governance and top management) and also in different positions (chairman, CEO, co-CEO, general manager), which is typical of family firms. The presence of the same people in various bodies might cause us to question our findings about governance archetypes. Another area that is worth exploring further is the substitution effect that was assumed for family councils over other bodies; to go deeper and verify it, it would be useful to analyse what types of decisions are taken by various official and unofficial governance bodies. Another relevant topic is agency costs and their measures, which would be worth working on to go deeper into the open issue of whether these costs are lower or not in family firms. Another challenging issue is the relationship between governance structures and company performances, which has not been definitely identified so far.

Fourth, given their present and perspective relevance, study of origination, composition and functioning of teams at the top represents another key topic to be further analysed in order to help families best design and manage them. Also, it was not possible to investigate where such teams are located in the governance system (At board level? At top management level?); to go deeper into this issue might be useful for a better understanding of both leadership and governance in family firms.

Finally, as we stated above, teams at the top could be a consequence of fragmented ownership's diffusion, that is, they could be used as an agency-based tool to settle possible conflicts of interest within family ownership. But this is not sufficient: the success of teams at the top requires a relational perspective as well, as it depends on personal, family and company variables such as team members' complementarity, mutual esteem, commitment to working together, listening and mediation aptitudes, clear division of roles but shared responsibility on key decisions, and family education centred on cooperation rather than competition (Ward, 1997). In sum, teams at the top may result from a mix of agency needs (such as full ownership representation) and relational needs (such as retaining cohesion between family executives); this might also be further explored to better understand and support leadership and governance structures.

Some hints for families

Our survey results show that, at a 'macro' level, the family model has been 'keeping its position' so far with respect to relevance, profitability and strategies.

The 'size issue', instead, has not been solved, which might be quite a weakness in a more and more difficult environment, to the extent that family business prominence and performance might be threatened. Some 'good news' is that family small and medium-sized enterprises are making progress: their strategic priorities are in line with the evolution of competitive systems, governance systems are developed in complexity consistently with size, and the importance of preparing for succession seems to be increasingly acknowledged. But it is critical not to 'lower one's guard' from all the mentioned points of view.

As far as ownership is concerned, important dynamics will have to be managed. In particular, it will become more important to facilitate relations between more numerous (and perhaps more diversified) shareholders, to govern exit processes and at the same time to activate all structures and processes that may be necessary to preserve unity and commitment and their most direct antecedents such as love for the company and trust between various actors.

Family small and medium-sized enterprises have already been doing some work in this direction by setting up shared rules on topics such as shares' transfers and dividends, which are critical when ownership gets extended. But this work has taken place especially in articles of association, whose formulating process is generally top-down and formal: perhaps more involving family agreements might be necessary to share principles and rules and to foster communication and cohesion (Montemerlo and Ward, 2005).

Some other work has been done on governance structures. Our study shows that Italian small and medium-sized enterprises do make efforts to keep their governance structures consistent with their ownership structure as well as their size, especially activating shareholders' meetings and boards of directors. But this 'upper part' of the official structure is still not utilized in many cases. This is likely owing to the fact that, in this kind of firms, ownership is generally concentrated and very much involved within the company. But in the future, with further fragmentation, the need to govern the agency relationships that will be created through a more active role of shareholders' meetings and boards could increase as well.

Another important part of the 'governance work' has been that on family councils. We have seen that, in many companies, the family council 'cuts across' the official structure, replacing the shareholders' meeting and the board and, by this mean, likely mixing up company and family issues. In these cases, the family council may represent a strength from a relational point of view, creating trust and shared vision in the owning family. But it might also be a weakness as it could lead family owners to neglect agency problems: for instance, mixing family and company through the family council might not be good for governing increasingly numerous family ownership, and particularly ownership groups composed of both managing and non-managing shareholders; also, it might make the company less attractive to qualified contributions of external actors such as managers and partners, who might feel excluded from company governance.

So, a big challenge for owning families and their leaders could be to use unofficial bodies such as the family council to complement rather than replace official corporate ones (Corbetta and Montemerlo, 2003). Given their influence, this is a big challenge for third parties as well.

Last, but not least, many family small and medium-sized enterprises are envisioning team leadership for the next generation. As mentioned above, teams at the top should be looked at in a 'mixed' perspective, that is, with the lenses of both agency and relational approaches. So, it is necessary to guarantee teams at the top's sense of responsibility and accountability to family owners and to preserve organizational clarity and transparency in order to attract the most qualified resources from both family and outside. But it is also necessary to share values that preserve team members' unity and commitment, and to translate values into rules that enable teams at the top successfully to complement other governance bodies, making the most of all stakeholders' contributions.

Acknowledgement
This work benefited from the financial support of the Italian Ministry of University and Scientific Research, Bocconi University and Catholic University to a co-financed project: 'Generational transitions in medium-size Italian family firms: successful experiences and best practices', which is hereby gratefully acknowledged.

References

Armstrong, J.S. and T.S. Overton (1977), 'Estimating non-response bias in mail surveys', *Journal of Marketing Research*, **14**, 396–402.
Buchanan, J.M. (1975), 'The Samaritan's dilemma', in E.S. Phelps (ed.), *Altruism, Morality and Economic Theory*, New York: Russell Sage Foundations.
Corbetta, G. and D. Montemerlo (1999), 'Ownership, governance and management issues in small and medium sized family businesses: a comparison of Italy and the United States', *Family Business Review*, **12**(4), 361–74.
Corbetta, G. and D. Montemerlo (2003), 'Leading family firms: a double challenge', in P. Poutziouris and L. Steier (eds), *New Frontiers in Family Business Research: the Leadership Challenge*, Family Business Network Research Forum Proceedings, Manchester Business School and Alberta School of Business.
Corbetta, G. and C. Salvato (2004), 'Self-serving or self-actualizing? Models of man and agency costs in different types of family firms', *Entrepreneurship Theory and Practice*, **28**(4), 355–62.
Daily, M. and M.J. Dollinger, (1992), 'An empirical examination of ownership structure in family and professionally managed firms', *Family Business Review*, **5**(2), 117–36.
Davis, P.S. and P.D. Harveston (1998), 'The influence of the family on the family business succession process: a multi-generational perspective', *Entrepreneurship Theory and Practice*, **22**, 31–53.
Fama, E. and M. Jensen M (1983), 'Separation of ownership and control', *Journal of Law and Economics*, **26**, 301–25.
Gallo, M.A., G. Corbetta, K. Cappuyins, G. Dyer Jr, D. Montemerlo and S. Tomaselli (2001), *Love as a Function of Love, Trust and Freedom in Family Businesses*, ed. IESE Monographics, Spain.
Gersick, K.E., J.A. Davis, I. Lansberg and M. McCollon Hampton (1997), *Generation to Generation: Life Cycles of the Family Business*, Cambridge, MA: Harvard Business School Press.
Gilding, M. (2000), 'Family business and family change: individual autonomy, democratization and the new family institution', *Family Business Review*, **13**(3), 239–49.
Gnan, L. and D. Montemerlo (2001), 'Structure and dynamics of ownership, governance and strategy: role of family and impact on performance in Italian SMEs', in G. Corbetta and D. Montemerlo (eds), *The Role of Family in Family Business*, Milano: EGEA.
Montemerlo, D., L. Gnan, W. Shultze and G. Corbetta (2004), 'Governance structures in Italian family SMEs', in S. Tomaselli and L. Melin (eds), *Family Firms in the Wind of Change*, Research Forum Proceedings, FBN 2004.
Gomez-Mejia, L., M. Nuñez-Nickel and I. Gutierrez (2001), 'The role of family ties in agency contracts', *Academy of Management Journal*, **44**(1), 81–95.
Greenwald and Associates (1995), *1995 Research Findings*, Massachussetts Mutual Life Insurance Company, Washington, DC.

Habbershon, T.G. and J.H. Astrachan (1997), 'Perceptions are reality: how family meetings lead to collective action', *Family Business Review*, **10**(1), 37–52.
Harvey, J.S. Jr (1999), 'What can the family contribute to the business? Examining contractual relationships', *Family Business Review*, **12**(1), 61–71.
Huse, M. (2000), 'Boards of directors in small firms: a review and research agenda', *Entrepreneurship and Regional Development*, **12**(4), 271–90.
Jensen, M.C. (1998), 'Self-interest, altruism, incentives, and agency', *Foundations of Organizational Strategy*, Cambridge, MA: Harvard Business School Press.
Jensen, M.C. and W.H. Meckling (1976), 'Theory of the firm: managerial behavior, agency costs and capital structure', *Journal of Financial Economics*, **3**, 305–60.
Kang, D. (2000), 'The impact of family ownership on performance in public organization: a study of the U.S. Fortune 500, 1982–1994', 2000 Academy of Management Meetings, Toronto, Canada.
LaChapelle, K. and L.B. Barnes (1998), 'The trust catalyst in family-owned businesses', *Family Business Review*, **11**(1), 1–17.
Lank, A.G. and J.L. Ward (2000), 'Governing the business-owning family', *FBN Newsletter*, **26**(May), 1–6.
Lansberg, I. (1999), *Succeeding Generations*, Boston, MA: HBS Press.
Leana, C.R. and H.J. Van Buren, III (1999), 'Organizational social capital and employment practices', *Academy of Management Review*, **24**(3), 538–55.
Montemerlo, D. and J.L. Ward (2005), *The Family Constitution: Agreements to Perpetuate Your Family and Your Business*, Marietta, GA: Family Enterprise Publishers.
Moores, K. and J. Mula (2000), 'The salience of market, bureaucratic and clan controls in the management of family firm transitions: some tentative Australian evidence', *Family Business Review*, **13**(2), 91–106.
Morck, R., A. Schleifer and R.W. Vishny (1988), 'Management ownership and market valuation: an empirical analysis', *Journal of Financial Economic*, **20**, 293–315.
Mustakallio, M. (2002), *Contractual and Relational Governance in Family Firms: Effects on Strategic Decision-making Quality and Firm Performance*, Helsinki: Helsinki University of Technology.
Nahatapiet, J. and S. Goshal (1998), 'Social capital, intellectual capital, and the organizational advantage', *Academy of Management Review*, **23**(2), 242–66.
Neubauer, F. and A.G. Lank (1998), *The Family Business: Its Governance for Sustainability*, London: Macmillan.
Rediker, K.J. and A. Seth (1995), 'Boards of directors and substitution effects of alternative governance mechanisms', *Strategic Management Journal*, **16**, 85–99.
Schulze, W.S., M.H. Lubatkin, R.M. Dino and A.K. Bucholtz (2001), 'Agency relationships in family firms: theory and evidence', *Organization Science*, **12**, 99–116.
Tsai, W. and S. Goshal (1998), 'Social capital and value creation: the role of intrafirm networks', *Academy of Management Journal*, **41**(4), 464–76.
Ward, J.L. (1987), *Keeping the Family Business Healthy: How to Plan for Continuing Growth, Profitability, and Family Leadership*, San Francisco, CA: Jossey-Bass.
Ward J.L. (1991), *Creating Effective Boards for Private Enterprises*, San Francisco, CA: Jossey-Bass.
Ward, J.L. (1997), 'Growing the family business: special challenges and best practices', *Family Business Review*, **10**(4), 323–37.

28 Longevity of Japanese family firms

Toshio Goto

Introduction

This chapter presents an overview of Japanese family firms, highlighting their longevity based on empirical data to provide a broad view of their uniqueness as well as their commonality when compared with their counterparts in other countries. The study aims to fill some gaps in the research on family firms from a Japanese point of view by proposing that, although there are important differences in the factors contributing to the longevity of Japanese family firms compared with their counterparts in other parts of the world, there are also some factors which are held in common with family firms in the rest of the world. These factors will be examined to test the validity of this proposition.

The first part of the chapter focuses on aspects of the longevity of Japanese family firms. As a part of this research, a total of 1157 family firms that have been in operation for more than 200 years were identified. The major attributes, including age, industry segment and geographical distribution of these long-lived firms are given, along with a brief description of the background of selected industries. In order to better understand the magnitude of Japanese family firms' longevity, a preliminary survey was also conducted on the longevity of family firms worldwide.

The second part is devoted to the analysis of three major factors contributing to the longevity of Japanese firms, with a special focus on the Tokugawa period of 1603 to 1867 (also known as the Edo period). These factors are economic development, which provided the capacity for family firms to sustain growth, the existence of relatively advanced management systems, which contributed significantly to the longevity and prosperity of family firms, and the philosophical background, namely Confucianism, which required the giving of precedence to the family business over the family itself.

While Confucianism itself as an ethical philosophy and a prescription for family and business relationships is recognized as having made a unique contribution to the longevity of Japanese family firms, it can also be said that the specifics of its teachings and the business ethics it incorporated are factors that have similarly contributed to the longevity of family firms worldwide. These are family unity, a commitment to continuing the family legacy as the bedrock of survival, a product catering to basic human needs, allowing the business, rather than the family, to come first, an obligation to community and customer service, conflict arrangement and a system of governance. The nature of these factors, their roots in Confucian philosophy and the way they manifested themselves in the practices of long-lived family firms will be examined to emphasize the uniqueness and commonality of Japanese firm's longevity.

Literature review

Until relatively recently, the subject of Japanese family businesses has been given insufficient attention both in academic circles as well as in the world of business. Little research had been done on the relevance of family firms in Japan until a study by Goto (2005) indicated that family firms account for 96.5 per cent of total firms and 77.4 per cent of total employment in Japan. These research findings are based on a random sampling of all types of business entities headquartered in Shizuoka Prefecture, which, located as it is in the centre of the main island of Japan, is often viewed as representing Japan overall both quantitatively as well as qualitatively.

Family firms are defined in Goto (2005) as those where members of the founder's family are involved in both management and ownership to the extent of having two or more family members either as top executives and/or shareholders.[1] This definition, following Neubauer and Lank (1998) and other preceding works, is kept rigorous enough to describe the voting power of the family, while giving a practical consideration to the data availability.

The magnitude of Japanese family businesses in Japan's domestic economy, as suggested by this study, is comparable to or even greater than that in other developed countries, as shown in previous research such as that on Australia by Owens (1994), Germany by Reidel (1994) and Klein (2000), Italy by Corbetta (1995), Spain by Gallo (1995), Sweden by Morck and Yeung (2003), the United States by Shanker and Astrachan (1996), as well as in Astrachan and Shanker (2003), and in an overall comparison made by Reynolds et al. (2002).

While the above studies serve as important background for this research, there are various limitations to it. Most of these limitations were caused by the lack of accessible public records, together with the secretive attitude of family firms (Neubauer and Lank, 1998), and these same problems have hitherto also constituted difficulties for researchers in Japan.

In one of a limited number of studies carried out in Japan, Kurashina (2003) indicated there are 1074 family firms out of 2515 firms listed on the stock exchange, (excluding JASDAQ and OTCs) or 42.7 per cent, but did not give further details nor disclose the source of his data. For example, the resulting data resembles counterparts in the United States as reported by Anderson and Reeb (2003). This therefore makes this research a significant contribution to the field because it is based on a more comprehensive and better sourced database of the writer's own compiling.

In his research into the longevity of Japanese firms, Shimizu (2002) analysed the public companies listed in the first section of the Tokyo Stock Exchange, and indicated that the mean duration of listings is 228.28 years and the median is 145.22 years, using the listed duration as the surrogate for firm longevity. The rate of delisting was nearly constant; this finding is inconsistent with the 'liability of newness' theory in organizational ecology (Hannan and Freeman, 1989).

Shimizu's (2002) analysis used event history analysis and the Cox regression model. Since the average listed duration is more than a century, Japanese big businesses would seem to have considerably longer life spans than small firms in Japan as well as in the United States. Reporting on the ability of Japanese small firms to survive, the Small and Medium Enterprise Agency (1999) showed that the median survival time of business establishments (place of business) is approximately five to ten years. Since the majority of Japanese small firms are family firms, the survival time of family firms is also considered to be short.

Goto (2005), on the other hand, suggests that the average age of family firms is 52 years, which is significantly higher than in the United States, where the average life span of a family firm is 24 years (Lansberg, 1983). It has been said that only 30 per cent of family firms reach the second generation (Beckhard and Dyer, 1983), while less than 16 per cent survive to a third generation (Applegate, 1994). These observations lead to the hypothesis that non-family firms will be older than family firms principally because the vast majority of family firms fail to survive the first generation (Ward, 1987).

Family Business (2003) released a preliminary list of the world's hundred oldest firms and included six Japanese firms that have been in business since 1800 or earlier. Yokozawa et al. (2000), however, had already shown that there are at least 746 long-lived firms in Japan that have been in operation for more than two centuries, including 25 firms founded in the ancient era (in or before 1191), 65 in the middle era (1192 to 1573), while Goto (2005) identified 1157 firms[2] that have been in business for two centuries or more. Expanding on Yokozawa et al. (2000), Yokozawa and Goto (2004) claimed that the vast majority of the long-lived firms are family firms.

These discrepancies in findings naturally give rise to questions about the validity of comparisons of the longevity of family firms beyond national boundaries. In the case of long-lived family firms in operation for centuries, O'Hara (2004), in one of the pioneering works on this subject, explored 12 family firms from around the world, each of which provided lessons on the key factors for a family firms' ongoing success. Pointing to family unity and a commitment to continue the legacy as the bedrock of survival, O'Hara (2004) also identified as recurring principles and practices the following: a product catering to basic human needs; primogeniture; which brings about a role for women; the use of adoption as a means to perpetuate family ownership; allowing the business, rather than the family, to come first; an obligation to community and customer service; conflict arrangement; plans in writing; and a system of governance. In an interview with Karofsky (2003), O'Hara also remarked on the importance of an ability to change without forsaking basic family values.

Prior to this work, O'Hara and Mandel (2002) had already published 'The World's Oldest Family Companies', which gives useful data, even though – as the authors admit – it is quite limited in its coverage. This list is composed of 88 family firms that have sustained their longevity for at least 200 years and is based upon various public sources including lists of members of Les Hénokiens and Tercentenarians.[3]

A number of studies have been done outside Japan which take up the subject of the longevity of Japanese firms, including those by Bellah (1957) on the value system that came into being during Tokugawa period, Fruin (1983) on Kikkoman Company, Hirschmeiner and Yui (1975) on the development of the merchant philosophy and Roberts (1973) on the Mitsui family business. The existence of these studies, which mainly examine historical aspects, indicates the importance placed by researchers into family businesses worldwide on various factors considered to be in some way peculiar to Japan.

To gain an understanding of reasons for the longevity of Japanese family firms, an issue of definite importance is the contribution philosophical and religious factors have made, and whether these differ significantly from similar factors affecting family firms in other parts of the world. It is interesting to note that Neubauer and Lank (1998), when citing the Mogi family's constitution, remarked that values here were influenced by religious manifestos and strivings derived from moral psychology. The Mogi family's constitution[4] was written in line with the so-called 'Seventeen Articles Constitution' documented by Prince Shotoku (574–622) in ancient times, and was thus strongly subject to the influence of Buddhism.

These factors will be taken up in the second part of this chapter in which the key issues are the way in which religious philosophy, education, management skills, separation of ownership and control were influential factors in the longevity of long-lived family firms in Japan.

Research methodology

For the purposes of this research, a database, the Long Lived Family Firms in Japan (LFFJ) database,[5] which had already been accumulated by Yokozawa and Goto (2004), was further expanded by adding in additional data from relevant trade publications and homepages, focusing on those industry segments which were thought by Yokozawa et al. (2000) to include a high proportion of long-lived firms. Because of these additions, LFFJ2, or the expanded version of LLFJ, is skewed towards the sake brewing, hotel and confectionery industry segments. In this chapter, family firms are defined, to follow Goto (2005), as those where members of the founder's family are involved in both management and ownership to the extent of having two or more family members either as top executives and/or shareholders.

Verification of data, and of the year of foundation in particular, was one of the most challenging aspects of this research. Points which needed to be verified were threefold: existence of the firm, self-statement by the subject firm and, finally, verification based upon some other form of evidence. For this research, the first verification was made by checking the existence of either the subject firms' homepage[6] or telephone number. The second verification was made by accessing the firm's homepage to find some statement about its history and foundation. For those firms without a homepage, this process is omitted. The third verification could not be made available for the purposes of writing this chapter, primarily because of time and resource limitations.[7]

As part of this research a preliminary comparative survey was conducted on the longevity of family firms worldwide in order to better understand the magnitude of Japanese family firms' longevity. Kompass Online and Amadeus (Online) were chosen as the primary sources of data[8] to compile an extensive and reliable database for this purpose.

Findings and discussion

In total 1157 family firms in business since 1804 or earlier were identified in the LLFJ2 database. More comprehensive than any other published data of this kind, LLFJ2 reveals that there are a surprisingly large number of long-lived family firms in Japan for one country alone, although it needs to be kept in mind that this figure is still preliminary due to the inevitably restricted coverage of the database.

The LLFJ2 database is used here to identify attributes of long-lived firms, including age, industry segment, geographical distribution and the number of generations involved. Distribution of the age of the firms is summarized in Figure 28.1.

There are only 32 firms which have been in operation since the fifteenth century or earlier. From the sixteenth century on, the number of family firms increased exponentially. The oldest identifiable firm is Kongo-gumi, the origins of which can be traced back to 578 and a group of temple construction craftsmen (O'Hara, 2004). This firm, named as the oldest firm in the world, is still actively expanding business throughout Japan and has its headquarters in Osaka Prefecture. The second in the list is Hoshi, which is, according to the *Guinness Book of Records*, the world's oldest hotel, and still has a high reputation. The next three firms are located in Kyoto City, the ancient capital of Japan

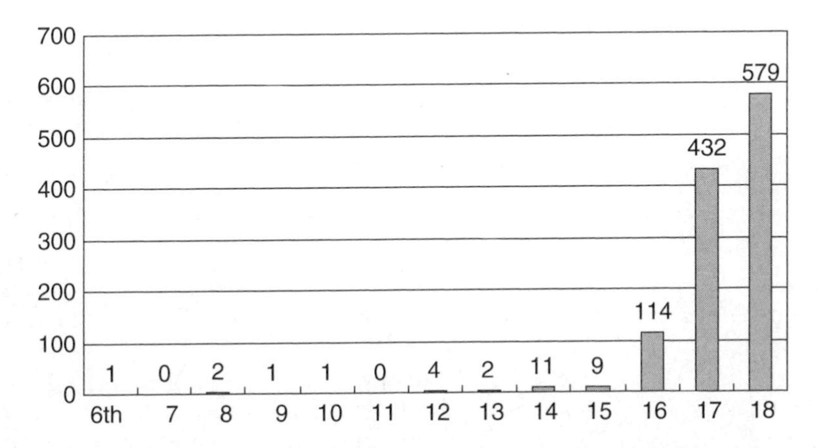

Notes: The horizontal axis shows the firms' ages as according to century of foundation.

Figure 28.1 Number of long-lived firms broken down by firms' age

during the late eighth to twelfth centuries. These firms are manufacturers and suppliers of ritual accessories, religious items and confectioneries.

By way of comparison, Table 28.1 shows the total number of long-lived firm in other countries. As is clearly apparent, Japan has by far the largest proportion at 32.7 per cent, followed by Germany (856 firms), Netherlands (240), France (177), Austria (167) and Russia (155). Table 28.2 shows the oldest firm in each country. Japan is listed at the top as having the oldest, followed by Germany (founded in 768), Austria (1074), Italy (1074), the United Kingdom (1189), Latvia (1211), France (1239) and so forth. This is only a preliminary comparison, but further research may well prove that Japan is exceptional both for the proportion and longevity of long-lived family firms.

Japanese longest-lived family firms, that is, those founded during or before the fifteenth century are shown in Table 28.3 in descending order of foundation. As shown in Table 28.3, industry segments are concentrated in hotel management and confectionary-making partially due to the way the LFFJ2 database was compiled, as already explained. Table 28.4 shows the distribution of all long-lived firms in operation for at least two centuries. Such a concentration is considered to reflect the extent of the development of these industries in pre-modern Japan.

Distribution by geographical area shows a relatively high concentration of firms in Kyoto. This is to be expected because of Kyoto's long history as Japan's ancient capital. When viewed historically, industries can generally be seen to have started either in a marketplace where there was a relatively high density of population, especially one with a high consumption rate, or in an area with competitive superiority such as access to innovation or raw materials.

Geographical distribution of all firms contained in LFFJ2, however, shows changes over time. Kyoto Prefecture alone had 37.5 per cent of all the long-lived family firms nationwide during the twelfth century or earlier, but the proportion decreased to the level of 23.0 per cent in the eighteenth century. While the proportion in the area of the ancient capital decreased, the proportion in the Kanto area (Tokyo and its adjacent six prefectures) steadily increased. This trend reflects the diffusion of industry from Kyoto, the

Table 28.1 Total number of long-lived firms by country

Country	Number of firms	Distribution (%)
Oceania		
Australia	5	0.1
China	9	0.3
India	3	0.1
Japan	1146	32.7
New Zealand	2	0.1
Subtotal	1165	33.3
Europe		
Austria	167	4.8
Belgium	29	0.8
Croatia	4	0.1
Czech	53	1.5
Denmark	32	0.9
Estonia	1	0.0
Finland	10	0.3
France	177	5.0
Germany	856	24.4
Hungary	13	0.4
Ireland	21	0.6
Italy	104	3.0
Latvia	6	0.2
Lithuania	2	0.1
Luxembourg	2	0.1
Malta	1	0.0
Netherlands	240	6.8
Norway	32	0.9
Poland	51	1.5
Portugal	11	0.3
Romania	5	0.1
Russia	155	4.4
Serbia and Montenegro	11	0.3
Slovakia	2	0.1
Spain	43	1.2
Sweden	69	2.0
Switzerland	73	2.1
UK	108	3.1
Ukraine	35	1.0
Subtotal	2313	66.0
America		
Mexico	1	0.0
Canada	2	0.1
Chile	5	0.1
USA	14	0.4
Virgin Islands	1	0.0
Subtotal	23	0.6

Table 28.1 (continued)

Country	Number of firms	Distribution (%)
Africa		
South Africa	2	0.1
Syria	2	0.1
Subtotal	4	0.2
Total	3505	100.0

ancient national political and commercial center, to its adjacent areas first and then to remoter areas.

A quick glimpse of three industry segments, religion-related businesses, inns, confectioners and sake breweries, will provide a good understanding of the development of industry from the ancient to the pre-modern era and underline the point that one of the factors that long-lived businesses have in common is that they deal in a product catering to basic human needs.

The religion-related industries listed in Table 28.1 are Kongo-gumi (currently Kongo Gumi Co. Ltd[9]) and Tanaka Iga Butsuguten. Members of the Kongo family were brought from Korea to Japan by Prince Shotoku more than 1400 years ago to build the Buddhist Shitennoji Temple, which still stands. Buddhism, first introduced from China in 538, became the state religion by the end of the century, that is, soon after, the patronage of the Prince. Tanaka Iga Butsuguten grew along with temples in Kyoto, providing religious items and accessories to the faithful. Buddhism was a strong part of the establishment during the ancient and medieval era.

The inn system was created as a part of the station system set up during the Taika Reform between 645 to 649. Stations established every 30 li (nearly 20 kilometers), provided both horses and accommodation for officials delivering urgent correspondence. This station system was subsequently reorganized and expanded to become the Tokaido (literally, the 'Eastern Sea Road') and other official main roads of feudal Japan. Besides the official inns, there were also privately built inns for the use of commoners. Saint Gyoki (668–749), one of the pioneers in this business, built nine *fuseya*, charitable institutions, to house and care for sick travelers. His endeavors were mostly centered in the Kinki area. Major temples such as Todaiji built hostels to ease the travel of the faithful, and these resemble the hospitals and hostels connected to churches in medieval Europe.

The establishment of the confectionery industry can also be traced back to the ancient capital, Kyoto. Shosoin or the Imperial Repository, located in Nara, the capital of Japan from 710 to 784, preserves a document that includes a sugarless candy in a list of foods of the eighth century. Sugar was first imported from China in 754 as a precious flavoring at that time, it was mostly consumed for medicinal purposes, while leftovers were used to make candies and sweets. Only nobles and high ranked officers had access to these confectioneries.

Toraya[10] was already supplying confectionery to the Imperial Family in the sixteenth century as it still does today. Legend tells that Toraya was already in this trade as long ago as the Nara period (710–794) and by the 1600s Toraya's proprietor, Enchu Kurokawa, considered to be the founding father of the present-day Toraya, had established a prosperous confectionery business in Kyoto.

Table 28.2 List of the oldest long-lived family firms listed by the year of foundation

Name	Founded in	Country	Location	Industry	SIC
Kongo-gumi	578	Japan	Osaka	Contractor	15
Schloss Johannisberg	768	Germany		Wine	20
Benediktierstift Admont	1074	Austria	Admont	Woodwork	24
Frescobaldi	1106	Italy		Wine	20
Faversham Oyster Fishery	1189	UK	Faversham	Fishery	9
Rigas 1. gimnazija	1211	Latvia	Riga	School	72
Eyguebelle	1239	France	Grignan	Alcoholic spirits	20
Beijing Glazed Products Factory	1267	China	Beijing	Porcelain	32
Browar Namyslow	1321	Poland	Namyslow	Brewery	20
Pivovar Broumov	1348	Czech	Broumov-Olivt ín	Brewery	20
Stella Artois	1366	Belgium	Leuven		
Neemrana Fort-Palace	1464	India	Rajasthan	Hotel	70
Pivovar Steiger	1473	Slovak	Vyhne	Brewery	20
Zanabili Beg Khan	1510	Syria	Allepo	Soap	28
Christoffel BV	1516	Netherlands	Plasmolen	Brewery	20
Orell Füssli Holding AG	1519	Switzerland	Zürich	Holding	74
Treschow-Fritzøe AS	1540	Norway	Larvik	Holding	74
Codorniu	1551	Spain		Wine	20
Alföldi Nyomda Rt.	1561	Hungary	Debrecen	Printing	27
Berte Qvarn AB	1569	Sweden	Slöinge	Cereals	20
Hotel Dagmar	1581	Denmark	Ribe	Hotel	70
Zildjian Cymbal Co.	1623	USA	Norwell	Musical instruments	39
Reusner	1633	Estonia	Tallinn	Printing	27
Tysmenytsia Fur Company, PubJSC	1638	Russia	Tysmenytsia	Fabric	22
Frenckellin Kirjapaino Oy	1642	Finland	Espoo	Printing	27
Pusser's	1655	Virgin Islands		Alcoholic spirits	20
Rustenberg Wines (Pty) Ltd	1682	South Africa	Cape Town	Wine	20
Belje D.D.	1697	Croatia	Osijek	Food and beverage	20
Quinta do Noval	1715	Portgal	Vila Real	Wine	20
Lviv Brewery, PubJSC	1715	Ukraine	Lviv	Brewery	20
Villeroy & Boch Sàrl	1748	Luxembourg	Luxembourg	Porcelain	32
Vi a Los Vascos S.A.	1750	Chile	Santiago	Wine	20
Becej-Pivara A.D.	1754	Serbia and Montenegro	Becej	Brewery	20
Tequila Cuervo, S.A.	1758	Mexico		Alcoholic spirits	20
Waterford Wedgwood	1759	Ireland		Glassware	32
U.C.M.Resita	1771	Romania	Resita	Steel	34
Svyturio Alaus Darykla Uzdaroji akcine bendrove 'Svyturys-Utenos alus'	1784	Lithuania	Klaipeda	Brewery	20
Molson	1786	Canada		Brewery	20
Corrective Services Industries – CSI	1788	Australia	New South Wales	Textile	22
Vincent Cuschierei Ltd	1795	Malta	Paola	Food	20
Benjamin Bowring Printing	1803	New Zealand	Auckland	Printing	27

Table 28.3 List of the oldest long-lived Japanese family firms by the year of foundation

Name of firm	Year	Location (city/town, prefecture)	Industry	*
Kongogumi	578	Shitenoji-shi, Osaka	Contractor	39
Hoshi	718	Komatsu-shi, Ishikawa	Hotel	46
Mizuhiki Motsuyui Genda	771	Kyoto-shi, Kyoto	Ritual accessories	
Tanaka Iga Butsuguten	889	Kyoto-shi, Kyoto	Religious items	
Ichiwa	1000	Kyoto-shi, Kyoto	Confectionery	24
Sudo Honke	1141	Tomobe-cho, Ibaragi	Sake brewery	55
Tsuen	1160	Kyoto-shi, Kyoto	Tea manufacturing	23
Gorobei Ame Sohonpo	1181	Aizu Wakamatsu-shi, Fukushima	Confectionery	
Goshobo	1190	Kobe-shi, Hyogo	Hotel	
Masamoto Gama	1207	Imada-cho, Hyogo	Ceramic pottery	15
Sankomaru Honten	1319	Gosho-shi, Nara	Pharmacy	
Gurenya	1327	Matsushima-cho, Miyagi	Confectionery	
Kanbukuro	1329	Sakai-shi, Osaka	Confectionery	
Keiunkaku	1332	Hayakawa-cho, Yamanashi	Hotel	
Kuroda Sennedo	1333	Yasuki-shi, Shimane	Confectionery	
Maruya Haccho Miso	1337	Okazaki-shi, Aichi	Fermented bean paste	
Shiose Sohonke	1341	Chuo-ku, Tokyo	Confectionery	34
Takada Shozoku	1346	Tokyo	Apparel	
Matsumaeya	1392	Kyoto-shi, Kyoto	Seaweed products	
Naruoka Construction	(2)	Shizuoka-shi, Shizuoka	Contractor	
Uiro	(3)	Odawara-shi, Kanagawa	Confectionery	24
Kameya Mutsu	1421	Kyoto-shi, Kyoto	Confectionery	
Sohonke Surugawy	1461	Wakayama-shi, Wakayama	Confectionery	
Sohonke Surugawy	1461	Kyoto-shi, Kyoto	Confectionery	
Honke Owariya	1465	Kyoto-shi, Kyoto	Confectionery	14
Hishigo Nakabayashi Chikuzaiten	1467	Kyoto-shi, Kyoto	Bamboo products	
Mizuta Gyokuundo	1477	Kyoto-shi, Kyoto	Confectionery	
Yamamura Juhodo	1487	Kyoto-shi, Kyoto	Pharmacy	
Hiraizumi Honpo	1487	Nigaho-cho, Akita	Hotel	
Sasaya Hotel	(4)	Numata-shi, Gunma	Hotel	

Source: compiled by the author based upon LFFJ2 database.

Notes:
* shows the generation currently in charge of management.
Entries (2) and (3) have been in operation for 600 years.
Entry (4) started business in the 1500s.

Sake was often associated with ritual ceremonies in the ancient era (the first written record of sake dates from the third century), therefore it is not surprising that it was primarily made by the imperial court or by large temples and shrines for ceremonial purposes. At first only the surplus was preserved for private consumption, but then sake

Table 28.4 The distribution of all long-lived firms by industry

SIC		No.	Min.	Max.	Median	Mean	SD
1	Agricultural crops	4	1603	1804	1604	1682	
9	Fishing, hunting	1	1705	1705	1705	1705	
15	Construction contractors	21	578	1798	1716	1653	
20	Food and kindred products	564	1000	1804	1704	1169	
22	Textile mill products	11	1597	1799	1717	1724	
23	Apparel and other textile products	60	771	1804	1716	1619	
24	Lumber and wood products	15	1596	1799	1704	1703	
25	Furniture and fixtures	16	1599	1784	1695	1608	
26	Paper and allied products	18	1579	1804	1704	1713	
27	Printing and publishing	7	1579	1787	1602	1656	
28	Chemical and allied products	64	1319	1800	1680	1673	87.7
31	Leather and leather products	1	1582	1582	1582	1582	
32	Stone, clay and glass products	44	1207	1804	1695	1687	103.7
34	Fabricated metal products	34	1293	1800	1687	1682	99.7
35	Industrial machinery and equipments	5	1560	1746	1642	1668	90.0
36	Electronic and electrical products	1	1746	1746	1746	1746	
37	Transportation equipment	1	1535	1535	1535	1535	
38	Instruments and related products	6	1613	1760	1746	1677	53.9
39	Miscellaneous manufacturing products	126	889	1804	1704	1687	110.3
42	Trucking and warehousing	1	1795	1795	1795	1795	
47	Transportation services	2	1615	1800	1615	1708	130.8
50	Wholesale trade – durable goods	15	1560	1798	1669	1676	70.3
51	Wholesale trade – nondurable goods	7	1604	1717	1688	1679	39.4
53	General merchandise stores	5	1615	1788	1751	1725	66.4
54	Food stores	72	1465	1804	1716	1706	79.4
56	Apparel and accessory stores	1	1751	1751	1751	1751	
58	Eating and drinking places	15	1532	1804	1716	1702	85.4
59	Miscellaneous retail	34	1568	1804	1751	1725	67.5
61	Nondepository institutions	1	1689	1689	1689	1689	
65	Real estate	11	1585	1765	1697	1685	62.7
70	Hotels and other lodging places	127	717	1804	1675	1647	169.3
72	Personal services	6	1624	1787	1751	1736	60.2
73	Business services	11	1584	1764	1689	1687	61.4
76	Miscellaneous repair services	2	1648	1688	1648	1668	28.3
80	Health services	1	1773	1773	1773	1773	

brewing was gradually expanded into a commercial business. This resembles the origins of wine and liquors associated with religious institutions in Europe. In Japan the government controlled the brewing industry by granting special permission to the makers in order to secure important tax revenue.

Factors contributing to longevity

How were so many family firms in Japan not only able to survive but also to continue to prosper for more than two centuries? Is there anything special about these firms? As a first step to answer these questions, the rest of the chapter aims to identify several factors contributing to the longevity of family firms in Japan. Factors examined here include economic development, management systems and philosophical background.

When reading the analysis it is important to take note of the particular period of Japanese economic development to which the points taken up apply in particular and keep in mind the question of whether these dimensions might nowadays harm the ability of Japanese businesses to further sustain growth.

The analysis essentially focuses on the Tokugawa period (1603 to 1868), the pre-modern era, which formed a strong foundation for Japan's rapid growth in the modern era. The Tokugawa period, although providing the strength of one of the most rigid feudalistic societies in the world, was not without its weaknesses and limitations. Deeply rooted philosophical beliefs held back the healthy growth of the Japanese economy instead of allowing it to free itself from a feudalistic legacy. In this sense, it would be appropriate to say that Tokugawa period served both as strong basis for and a threat to the longevity of Japanese businesses.

Economic development

Economic development is indispensable to provide the capacity for new businesses to start up and maintain growth. Although limited in both land area and natural resources, Japan showed steady economic growth during the Tokugawa period. Between 1600 and 1872, the population grew from 12 million to 31.3 million at an annual growth rate of 0.4 per cent. Edo, with a population of well over 1 million, was the largest city in the world.

The area of land under cultivation increased from 2065 thousand *chobu* (approximately one hectare or 2.45 acres) to 3234 *chobu*, an annual growth rate of 0.2 per cent, while agricultural production grew 0.3 per cent annually during the same period. The first half of the Tokugawa shogunate (1603–1868) experienced a somewhat higher population growth rate of around 1 per cent, while the latter half experienced stagnation.

This is not to say, however, that Japan was more developed than Western countries. In fact, Japan lagged far behind. The country's gross national product (GNP) per capita was only 25 per cent of that of England and 36 per cent of that of the United States in 1870. When the level of industrialization is judged by the percentage of the total labor force employed in agriculture, Japan is seen to have had 72 per cent of its total labor force in agriculture, compared with England (19 per cent) and the United States (51 per cent) and was thus still in an under-industrialized state.

This not withstanding, it is important for our discussion to note that Japan, despite its relative backwardness with regard to economic development, had ample capacity for business to start up and develop.

Management skills

Here, it is important to be aware that economic development aside, seventeenth-century Japan was advanced in many other ways, comparing favorably with Elizabethan England economically and with the Celestial Empire culturally (Roberts, 1973). Family firms in Japan had developed fairly advanced management systems by the eighteenth century, including those related to business organization, separation of ownership and management, personnel management, accounting systems and risk management. These systems were necessitated by the aggressive economic and commercial activity of the seventeenth century and strong efforts on the part of firms to protect themselves from possible future calamities of one kind or another.

Organization A unique feature of the House of Mitsui, founded in 1673 and the wealthiest merchant house of the Edo period, was the creation of a kind of joint-stock corporation. Hachirobei, the founder of Mitsui, which is also one of the biggest family firms, combined the six houses of his sons into a more comprehensive economic unit named Mitsui-gumi. By forming this super-house composed not only of individuals but of houses whose unifying element was a common ancestry, he made possible the formation of a corporate body capable of engaging in a capitalistic enterprise within a feudal economy. Mitsui-gumi was, to some extent, a joint-stock corporation of limited liability, and this house of houses was structurally a partnership built upon mutual trust and corporation.

In 1710, Mitsui-gumi established a main headquarters, the Omotokata in Kyoto to coordinate central administrative functions. A meeting was held twice a month, formally attended by three family members and two deputy heads from among the employees. Instead of accumulating capital separately, the six brothers pooled it in the Omotokata, which they owned jointly and from which they could borrow operating funds. They also turned in fixed dividends in accordance with the annual net income of all the businesses comprising Mitsui-gumi. Financial progress was examined not only semi-annually but also every three years.

One of the most important mechanisms contributing to the longevity of Mitsui-gumi was its system of reserves, which actually saved it several times when other companies went bankrupt. There were four forms of reserves to be resorted to in descending order of the level of the emergency being faced, starting with 'the shop's foundation' reserve for working capital and ending with 'the cellar silver' reserve, a hoard of coins and bullion to be dug up only when every other resource had been exhausted.

This level of organization was to be seen not only in Mitusi but also in many other firms. Ohmi merchants, well known for their long tradition of making 'business trips', created a long-term partnership to avert risk, an arrangement which is comparable to the commenda[11] in medieval Italy. This arrangement, and other similar ones that were also in place at this time, was created to assure the durability of houses and their business.

Separation of ownership and management As early as the late seventeenth century, family firms in Japan were characterized by the nearly complete separation of ownership and management, together with a high level of specialization of functions, which is usually considered to be a unique feature of modern managerial capitalism (Fruin, 1983).

In Noda soy sauce breweries, including the forerunner of Kikkoman, Japan's largest soy sauce producer, owners did not directly manage or even oversee their production facilities. They did little more than finance and 'supervise' the front office. Overall management was delegated to managers and was divided into separate spheres: general management, production management, and labor management.

Decision-making by consensus was stressed as a means to protect family firms from risk as well as any evil-doing by the family heads. House rules encouraged subordinates to voice their opinion, good or bad, about their heads. In cases where the heads ignored these opinions, another meeting was required to be called to officially request the head to improve. If no improvement was made, the head was asked to retire as a last resort. Such retirements actually occurred in many houses.

Bookkeeping Larger houses such as Mitsui and Konoike developed rather elaborate systems of bookkeeping, which had some similarities to the double-entry bookkeeping developed by Italians. The most advanced system was established by the Nakai family of mid-eighteenth century, where the capital account and the income-expenditure account both gave the same final profit and loss figures.

The merchant houses had rules for bookkeeping; transfers from the journal to the main ledger had to be done under supervision and had to be verified by official seals. A variety of books were kept: for sales, for purchases, for cash in- and out-flow, for capital account, for shipments, and so on, all written with a brush and ink and verified with seals.

However, no single prevailing method evolved during the Tokugawa period; each house had its own way of bookkeeping and thus the practice of double-entry bookkeeping, even when it existed in some rudimentary form, did not spread (Hirschmeier, 1975). In Italy, Luca Pacioli (1445–1517), the 'father of double-entry bookkeeping', published a study of the books of the treasury of the Republic of Genoa of 1340, which presumably accelerated the introduction of double-entry bookkeeping and resulted in a creation of a highly sophisticated standard accounting system.

In contrast to this, in Japan, each house kept its bookkeeping confidential and this prevented the fast and unanimous introduction of the method. The existence of sophisticated double-entry bookkeeping in some form or other, however, indicates a high level of commercial activity, demanding elaborate methods.

Personnel management After delegating much authority to subordinate managers, the major function of house management was the selection and training of staff who would dedicate themselves to guarding the business of the House. Employees in a merchant house consisted of *detchi* (apprentice), *tedai* (shop assistant), and *banto* (a clerical manager).

A *detchi*, recruited at age 12 to 13, would in time be elevated to *tedai* at a specified salary and a senior position at age 17 or 18, but promotion to the next rank was based on an evaluation of performance, and did not occur automatically. A *tedai* became eligible for the position of *banto* at the age of about 30 and was then granted the right to establish his own business, a *bekke* (independent house), remaining loyal to his former house.

In this manner on-the-job-training was integrated with promotion and the family management system in a systematic manner. There was also off-the-job training: *detchi*, for example, were inducted into the teachings of Shingaku (literally 'Heart Learning'), the merchant philosophy, which will be described separately, as part of their human development. There were also other types of schools for merchants such as the Kaitokujuku school in which as many as several thousands students were enrolled.

Risk management It is not an exaggeration to say that family firms in the Tokugawa period were fully equipped with risk management policies. One such policy was that houses avoided dividing their business, but rather gave shares to their sons who would own, but not manage, the growing family concern.

Also, family constitutions were drafted codifying the principles that had maintained the house safely for generations. After the clampdown on extravagance, a feature of the

Genroku era (1688–1704), which took place under the rule of shogun Yoshimune, the typical merchant became rather conservative, and showed a willingness to comply with the rulings of the Tokugawa government and the clans. It was during this period of retrenchment on the part of the merchants that more successful merchants compiled sets of rules for civic and business behavior, which are a good reflection of their overall mentality and approach to business.

Education and philosophical background

An important factor contributing to Japan's economic development was education. The most important places of education were *terakoya*, private schools for the children of commoners, more than 10 000 of which had been established not only in major cities but also in rural areas by the time of the Meiji Restoration starting in 1868. Forty-three per cent of the young boys and 10 per cent of the young girls in Japan were enrolled in these schools, and the level of education was the highest in the world at that time; in major industrial cities in England one in every four or five boys were attending school in 1837.

The Tokugawa shoguns adopted Jukyo, a version of Confucianism as the official ethical philosophy of Japan and this contributed significantly to the longevity of family firms. Less a religion than an ethical philosophy and a prescription for social relationships, Confucianism, after its introduction from China, and together with Buddhism, has traditionally been the major philosophical backbone of Japan and it received its greatest support during the Edo period.

It is important to understand, when examining the reasons for longevity of Japanese family firms, that under the tenets of Jukyo, the family was recognized as the basic social unit, and loyalty to one's superiors was considered the highest virtue. In the merchant community, *ie* (house as in household) means both family and family business entity in an integrated manner. In the mindset of merchants, the top priority was put on the preservation of family business enterprises by no matter what means and, as such, the living family was subservient to the continuous economic entity called the house. Hirschmeier (1975) concluded that only this priority of the house as a going concern over blood considerations can fully explain, given the high mortality rates, the long-term continuity of the business houses.

Because of the prevalence of primogeniture, to become head of a house meant taking up the holy responsibility of exerting all efforts toward the prosperity and continuity of the house, and never defiling the house's name. Desperate efforts were exerted to preserve family business enterprises, including taking an adopted son as successor when necessary, delegating management authority to non-family *banto*, and even forcing the family head to retire as a last resort to save the enterprise from major crises.

The most influential ideologist at this line was Ishida Baigan (1685–1754), who emphasized persistence, frugality and, most importantly, the commitment to the basic calling of the merchant to render service to society. Denying any intrinsic differences between human beings, Baigan urged merchants to remain true to their vocation as a merchant.

Bellah (1957) places great value on Baigan, discovering in his philosophy something akin to Western this-worldly mysticism. The essence of Baigan's ideas is summarized in the following citations from his first book, *Toimondo* (*Dialogues with his Disciples*), and his last book, *Seikaron* (*An Essay on Household Management*):[12]

> To sell rice is nothing more or less than a commercial transaction. It may therefore be said that all the people, from the feudal lords of the great provinces down, are in a sense engaged in commerce . . . Commerce is absolutely indispensable in daily life, hence it is wrong to despise money or hold commerce in contempt . . . There is nothing shameful about selling things. What is shameful is the conduct of men who fail to pay their debts to merchants.

> Obtaining profit from sale is the way of the merchant . . . The merchant's profit from sale is like the samurai's stipend. No profit from sale would be like the samurai serving without a stipend.

> The family is handed down by ancestors and passed to descendants. Money does not belong to just one individual. If money belongs to society as a whole it is not to be spent by one person for his own sake. If small it must be spent for the whole family.

> In looking over conditions in the world in general, there is nothing, which decays so easily as a merchant family. If you seek the cause of this, it is an illness called foolishness. This foolishness quickly becomes extravagance. Though foolishness and extravagance are two, we must say that they are hard to distinguish.

Whilst being cautious not to criticize the fundamentals of the feudalistic class system, Baigan strongly emphasized the equality of the merchant's work with that of any social class, even including the feudal lord and samurai. Contrary to the common perception of commercial activities as being shameful, Baigan claimed that they were in fact as noble as any other social function since they were of value to society. In this manner, honest profit was justified as the proper reward given by the empire for such value added services. By remaining true to his vocation of unselfishly selling at a just profit, a merchant could fulfill his social duty as expected; this was the only way for a merchant to carry out his religious duties.

After Baigan's death, the philosophy of Shingaku built upon his theories and became influential nationwide partly because the Tokugawa government made use of it to discipline the behavior of the merchants. And the merchants, since they were positioned as the very bottom of the feudalistic hierarchy, not unexpectedly supported Baigan's idea of equality fervently.

Shingaku had an important influence on the merchants' ideology in various ways. Many Shingaku schools were established throughout Japan and merchants solicited the teachers to draft their family percepts and constitutions as a part of their determined efforts to preserve the family and family business for future generations. Therefore, it is natural that family percepts and constitutions in Japan have strong philosophical as well as religious overtones, reflecting the background of Shingaku.

Shingaku, built up and strengthened under the Tokugawa feudal regime, remained significant into the modern era even after the Meiji Restoration in 1868, and although its overt influence gradually decreased it still exerts a covert influence in present-day Japan. It is necessary to recognize both the positive as well as negative aspects of the influence of this traditional philosophical background.

In Japan the teachings of Shingaku strongly recommended merchants to remain frugal, honest and diligent and were thus instrumental in preserving family businesses for generations. The hardworking and ascetic spirit imbued in the teachings of Confucius is comparable to the Protestant ethic, which is viewed as important in the modern transformation of the West as analysed by Tsuchiya (1989). It is in this comparison that we begin to see the way in which differing philosophical backgrounds from different parts of the world contributed a system of governance and a belief in the firm's obligation to community and customer as well as educational and organizational systems of various kinds which contributed each in their specific ways to achieve similar ends: the longevity of long-lived family firms around the world.

Long-lived firms and the future

Long-lived family firms in Japan not only survived the stagnant business environment of the nineteenth and first half of the twentieth centuries, but they also experienced the rapid post-war growth in the economy as well as the current business slowdown after the burst of the bubble economy in 1991.

Greiner (1972) describes the development of business as a series of evolutionary stages of growth and revolutionary stages of crisis. The appropriate organizational skills, management attitudes and control systems are required in order to overcome the problems specific to each stage. Often the solutions best fitted to a revolutionary stage sow the seeds of decay and lead to another period of evolution; management needs to be preparing for the next change while adapting itself to the current stage. Thus, business sustainability is a function of both permanence and flexibility and this is what is codified in the house precept.

Long-lived family firms consciously or unconsciously have continuously adapted themselves to their business environment. There are, however, family firms, which, after enjoying three centuries of prosperity, have gone bankrupt in more recent times because of excessive speculative investment in real estate, while others have ceased operation because of their inadaptability to ever-changing technology. Unless they are prepared for the next change, whether it is revolution or evolution, there is no guarantee that long-lived family firms will continue to prosper only on the basis of their remarkable longevity even if their business is currently sound.

Conclusion

The databases used in this research, even in their preliminary forms, served to provide valuable clues as to the characteristics of Japan's long-lived family firms and identified factors contributing to the longevity of family firms. The analyses of the data in the databases gave a clear indication that there are more family firms in Japan than in other countries in the world. It has already been noted, however, and must be emphasized again here, that the confirmation of this conclusion awaits the compilation and analysis of similar databases for other countries. Such research in other countries is also necessary to validate the second provisional conclusion of this research, which is that Japanese family firms are longer lived than other countries.

On the assumption that it is valid to argue that long-lived family firms are a notable feature of Japan's economic environment, it is also shown that it is possible to identify three important reasons for this: the steady economic growth of Japan since the Tokugawa (Edo) period which provided the capacity for the setting up of new businesses which prospered and eventually established themselves as the long-lived family firms discussed here; the management skills of these family firms which had already attained a marked level of sophistication by the eighteenth century; and the philosophical background based on the teaching of Shingaku which was such an integral part of the spirit of the merchant houses in the Edo period and contributed significantly to the longevity of family firms.

This philosophical background, while characterizing the uniqueness of Japanese long-lived family firms, substantiate the common key factors for long-lived family firms' success which include: family unity, sense of commitment and willingness to sacrifice self for business and out of a sense of obligation to society which supported systems of governance and contingency arrangements which kept firms alive for centuries.

Longevity is without doubt a subject of keen interest for existing family firms worldwide and this research implies the possibility of identifying a large body of long-lived family firms if similar data were compiled so that comparisons and in-depth analyses of long-lived family firms can be made on a global scale. A global comparative study of longevity of family firms, the possible findings of which can be previewed in this chapter, would enhance understanding of the key factors for success of family firms. It is hoped that this research will contribute to future comparative studies on a wider scale.

Notes

1. Firms are recognized as family firms, if more than two individuals with the same family name appear either as owner or management. In Japan, it is obligatory for women to change their family names on marriage and the maiden names are never kept or added as is the practice in some other countries, which makes it difficult to identify founding family members under different family names. Daughters, after marriage, may play an important role either as an owner or in some cases in management in a family firm. Parents and siblings of the married daughter also need to be surveyed as potential participants in a family firm. Since these possibilities are of necessity disregarded with the databases used in this research, data presented in this chapter should be interpreted as being skewed downward.
2. These firms are among the 748 firms in Yokozawa et al. (2000).
3. Les Hénokiens is an international association of family-owned businesses with a history of 200 or more years, which was established in 1981 with the purpose of preserving and sharing the traditions learned through years of experience. The Tercentenarians Club is an international association of business that have been trading continuously for 300 years or more and retain links with the founding family.
4. The first article starts with 'All family members desire peace. Never fight, and always respect each other. Ensure progress in business and the perpetuality of family prosperity.'
5. LLFJ was prepared based on 103 Chambers of Commerce directories listing the year of foundation. It should be noted that the coverage of LLFJ is significantly limited since: (1) more than half of the total number of Chambers of Commerce are not covered, including the Tokyo Chamber of Commerce and (2) rural areas where there are no Chambers of Commerce are not covered.
6. In Japan, the majority of long-lived firms have homepages accessible to the public.
7. This work involves complicated procedures as well as various issues related to the definition of foundation and longevity, such as how to deal with intermission of operation, transfer of location and so on.
8. Kompass Online covers 76 countries, and includes 1 886 792 firms, of which the European region comprises 43 countries and 1 376 753 firms, while Amadeus covers 38 countries and 7 376 186 firms in Europe.
9. The current president is Toshitaka Kongo, the thirty-ninth generation.
10. www.toraya-group.co.jp.
11. A very early form of partnership, in which one merchant provided capital for another who actually undertook the voyage.
12. As translated by Bellah (1957).

References

Anderson, R. and Reeb, D.M. (2003), 'Founding-family ownership and firm performance: evidence from the S&P500', *Journal of Finance*, June.

Applegate, J. (1994), 'Keep your firm in the family', *Money*, **23**, 1301–28.

Astrachan, J. and Shanker, M. (2003), 'Family businesses' contribution to the U.S. economy: a closer look', *Family Business Review*, **16**(3), 211–19.

Beckerd, R. and Dyer, G. (1983), 'Managing continuity in the family-owned business', *Organizational Dynamics*, **12**(1), 5–12.

Bellah, R. (1957), '*Tokugawa Religion: The Values of Pre-Industrial Japan*', Glencoe, IL: Free Press.
Corbetta, G. (1995), 'Patterns of development of family businesses in Italy', *Family Business* Review, **8**(4), 255–65.
Davis, J.A., Pitts, E. and Cormier, K. (2000), 'Challenges facing family companies in the Gulf region', *Family Business* Review, **13**(3), 217–38.
Family Business (2003).
Fruin, M. (1983), *Kikkoman Company, Clan and Community*, Cambridge, MA: Harvard University Press.
Gallo, A. (1995), 'Family businesses in Spain', *Family Business Review*, **8**(4).
Goto, T. (2005), 'Family business no genjo to kadai' ('Perspectives and issues of family business'), *Bulletin of Shizuoka Sangyo University*, **7**, 225–337.
Greiner, L. (1972), 'Evolution and revolution as organizations grow', *Harvard Business Review*, **50**(4), 37–46.
Hannan, M. and Freeman, J. (1989), *Organizational Ecology*, Cambridge, MA: Harvard University Press.
Hirschmeiner, J. and Yui, T. (1975), *The Development of Japanese Business, 1600–1973*, Cambridge, MA: Harvard University Press. (Second edition, 1981).
Karofsky, P. (2003), 'Interview with Dr. William O'Hara', *Family Business Review*, **16**(3), 221–3.
Klein, S. (2000), 'Family businesses in Germany: significance and structure', *Family Business Review*, **13**(3), 157–81.
Kurashina, T. (2003), *Famiri Kigyo no Keieigaku* (*Management of Family Business*), Tokyo: Toyo Keizai Shinposha.
Lansberg, I. (1983), 'Managing human resources in family firms: the problem of institutional overlap', *Organizational Dynamics*, **12**(1), 39–46.
Morck R. and Yeung B. (2003), 'Agency problems in large family business groups', *Entrepreneurship Theory and Practice*, **27**(4), 367–82.
Neubauer, F. and Lank, A. (1998), *The Family Business: It's Governance for Sustainability*, New York: Routledge.
O'Hara, W. (2004), *Centuries of Success*, Avon: Adams Media.
O'Hara, W. and Mandel, P. (2002), 'The world's oldest family companies', *Family Business Review*, Spring, available at http://www.familybusinessmagazine.com/oldworld.html, accessed 15 January 2004.
Owens, R. (1994), 'Australian family business, ethics, energy and long-term commitment: the hallmarks of success', *Family Business Network Newsletter*, No. 9, p. 4.
Reynolds, P., Bygrave, W., Autio, T., Cox, L. and Hay, M. (2002), *Global Entrepreneurship Monitor, 2002*, www.emkf.org/GEM2002.
Reidel, H. (1994), 'Family business in Germany', *Family Business Network Newsletter*, No. 9, p. 6.
Roberts, J. (1973), *Mitsui: Three Centuries of Japanese Business*, New York: and Tokyo: Weatherhill.
Shanker, M. and Astrachan, J. (1996), 'Myths and realities: family businesses' contribution to the US economy – a framework for assessing family business statistics', *Family Business Review*, **9**(2), 107–23.
Shimizu, T. (2002), 'The Longevity of the Japanese Big Businesses', *Annals of Business Administrative Science*, **1**(3), 39–46.
Small and Medium Enterprise Agency (1999), *The White Paper on Small and Medium Enterprises in Japan*, Tokyo: Small and Medium Enterprise Agency.
Tsuchiya, T. (1989), Nihon Keiei Rinenshi (History of Japanese Management Philosophy), Tokyo: Reitaku Shuppankai.
Ward, J.L. (1987), *Keeping the Family Business Healthy: How to Plan for Continued Growth, Profitability and Family Leadership*, San Francisco, CA: Jossey-Bass.
Yokozawa, T. and Goto, T. (2004), 'Some characteristics of Japanese long-lived firms and their financial performance', *Proceedings of the 15th FBN-IFERA Academic Research Conference*, IFERA Publications.
Yokozawa, T., Mori, M., Taomoto, K., Takeda, S., Goto, T., Sonehara, K., Hiroi, T. and Arata, K. (2000), *Shinie Kigyo no Kenkyuu* (*Study on Long-lived Firms*) Tokyo: Seisansei Shuppan.

PART VII

FAMILY BUSINESS FINANCE

29 Family firms and financial behavior: how family shareholder preferences influence firms' financing

Myriam Lyagoubi

This chapter investigates the relationship of family ownership and debt financing behavior in privately held and public companies. Several studies (Agrawal and Mandelker, 1987; Demsetz, 1983; McConaughy et al., 1996; McConnell and Servaes, 1990; Shleifer and Vishny, 1986; and, for empirical research in France, Charreaux, 1991) have been carried out to determine how a firm's ownership structure, especially the shareholding stake, can affect financial performance by mitigating agency conflicts between management and shareholders. Empirical results suggest that the organization of ownership structure can have an impact on corporate performance. However, few studies have looked at the relationship between ownership structure and capital structure outside the USA and UK. In particular, there are few academic works on the influence of family shareholders on financing decisions (Gallo and Vilaseca, 1996; Poutziouris and Sihar, 2001; Romano et al., 2000). The present research belongs to this analytical framework. It focuses on the financing choices of French family businesses. The objective of the chapter is to investigate if and how family control of shareholding induces a firm's specific financial behavior. In particular, our aim is to identify how these shareholders' preferences can help to explain the differences observed between family firms and other firms in their choice of capital structure and debt maturity.

We consider a company to be a family business when a family or an individual is the leading shareholder and is involved in the ownership and control of the company, whatever the manager's identity. The chapter uses a theoretical framework based on agency theory (Fama and Jensen, 1983; Jensen and Meckling, 1976) and the financial contracting theory (Hart, 1995) to develop several hypotheses. These theories offer a multifaceted analytical framework which make it possible to study the relations among several actors who have an influence on the allocation of the firm's resources and its investment decisions. Moreover, this framework takes into account certain psychological factors, such as the risk of a transfer of control from the shareholder to creditors and the preference for autonomy, that are important for the family shareholder. The nature of these relationships influences the strategic and financial decisions taken in the firm, decisions that will have an impact on debt and stock issuance, and therefore on the firm's value.

An empirical study is conducted for the period 1995–2000 on a sample of French firms composed of 477 private firms and 236 public companies. The main results show that family firms exhibit specific financial behavior influenced by the objectives and characteristics of their leading shareholder. Moreover, it emerges that going public reduces the problems of information asymmetry that characterize investment family business. Furthermore, the professionalization of management via delegation to outsiders induces a higher level of free cash flow risk. As a result, managerial family firms run by a hired CEO are more indebted than owner-managed family firms. Finally, significant differences are not found between the debt maturity structure of family and non-family firms. No evidence is found for a non-monotonic relationship between debt level and family ownership structure when managerial family firms are studied.

The chapter continues with a review of the literature and investigations into the factors governing family shareholder influence on financing decisions, leading to the hypothesis-building. Next is a detailed discussion of sample data and research methodology, followed by empirical results and conclusions.

Family shareholders' behavior and firms' financing choices: a risk attitude insight

Agency theory (Jensen and Meckling, 1976) offers a theoretical framework for analysing the relations between the various interested parties in a company, namely, the shareholders, the creditors and the managers. It makes it possible to analyse more precisely the underlying factors governing such conflict, which arise when one of the interested parties (the principal) delegates the management of his or her interests – generally financial – to one of the other actors (the agent) in the firm (Jensen and Meckling, 1976). However, agency theory takes into account only the conflicts that arise among the firm's actors owing to the way financial resources are allocated; whereas conflicts in family firms can also arise from the manner of allocating control and specific private benefits (Hart, 1995). Therefore, the study of the financial behavior of family firms requires an analysis of the financing relationship through both agency theory and the financial contracting theory.

Three types of agency relationship can be defined (Jensen and Meckling, 1976). In the first, the shareholders can delegate the ongoing business decisions to a manager whose mission is to maximize their wealth. The performance of the business is therefore heavily influenced by the strategic, financial and commercial choices made by the manager. The effectiveness of these decisions depends not only on the professional skills of managers, but also on their specific incentive to make the optimal choices to maximize shareholders' wealth.

In the second type, the company's financial creditors delegate to the shareholder the management of the funds that are made available to the business. The creditors' interests are satisfied since the company's aggregate risk is not modified between the date the funds are made available and the date they are reimbursed. Some choices made by shareholders (risky strategic projects, changes in the dividend policy, or an increase in the level of indebtedness) may modify the company's level of aggregate risk without the creditors being in a position to intervene (Jensen and Meckling, 1976; Myers, 1977).

Finally, the third relationship exists where there is opposition between shareholders. The origin of shareholder related conflict is twofold. The first is common to all public firms and is the result of differences between the interests of family shareholders and those of external shareholders. This conflict is over and above those which may emerge between family owners. This second type of conflict is specific to family firms and can be dangerous for the perpetuation of the business in the family.

In this context, the intensity of agency conflicts between the agent and the principal depends on their preference set, in which the degree of risk aversion is an important determinant (Jensen and Meckling, 1976).

Aversion to risk of the interested parties: the case of family firms

The family firm has several stakeholders among which four – the family shareholder, external shareholders, the manager and lenders – influence the company's financing process. The degree of risk tolerance of each interested party depends on the extent to which the risk attached to their investment can be diversified. Financial literature makes it possible to establish a 'typology' of the firm's interested parties as a function of their degree of risk aversion.

Shareholders limit their financial commitment to the value of the shares they own. According to financial theory, most investors prefer to invest in a fully diversified portfolio to minimize their risk. Thus, as long as shareholders are able ideally to diversify their risks, they would be risk-neutral. Shareholders therefore have the possibility of increasing the risk they take in each company through this diversification mechanism. Their preference would then tend towards strategies that result in maximizing their monetary wealth; namely, towards risky strategies and especially those using firms' leverage.

On the other hand, managers cannot take the same level of risk, since a large part of their wealth derives from their significant investment in human capital specific to the firm. Managers' employment, employability and reputation depend on this human capital investment (Diamond, 1989, 1991). Unlike with financial capital, the risks associated with human capital are hard to diversify (Amihud and Lev, 1981). The risk in this case is strongly linked to the viability of the company that employs the manager. For this reason, when decisions are taken about the firm, managers will choose solutions that ensure the survival of the business, even if those choices do not coincide with the interests of their principal, the shareholder. Managers can protect their investment by reducing the firm's default risk, and especially by lowering the company's level of debt (Friend and Lang, 1988).

Finally, from the loan that they grant, creditors hope to gain future revenues determined by the loan contract. The more risky the project undertaken by the business the greater is the risk borne by creditors of not being reimbursed. They are therefore in opposition to the interests of the shareholder, whose wealth (in terms of dividends and capital gain) increases in proportion to profit growth. Any risky investment taken after the lending operation is unacceptable to the creditors because, if it succeeds, they will earn the same revenue whatever type of project the company undertakes, but they will bear a higher risk. The financing terms they offer will reflect their perception of the risk incurred and the uncertainty of their future reward, particularly the high costs of the supervision that they exercise. These elements of risk and cost increase with the burden of debt, and are generally included *ex ante* in the interest they charge.

Differences highlighted in how a firm's actors assume risk affect the firm's recourse to debt, because the latter represents a significant element in the company's aggregate risk (Harris and Raviv, 1991). The shareholder is thus the actor ready to assume the highest level of risk, the manager and the creditor having a greater aversion to risk. Still, we cannot generalize the behavior described above to all types of shareholder. Actually the principal hypothesis in this behavior is shareholders' ability to diversify their portfolio of investments and thus to minimize their risk. Accordingly, we should question whether family shareholders can easily diversify the risk associated with their financial and personal investment, and, if not, the impact this might have on the financing choice of family firms.

The family shareholder is characterized by a close relationship between his personal wealth and the value of the company. Therefore, his or her risk tolerance is one of the key determinants of the firm's risk-taking strategy and financing. According to Xiao et al. (2001) the decision process depends on two factors: the attitude towards risk and the risk-taking behavior. If family owners seem to have a more open attitude towards risk than other individual shareholders, their risk-taking behavior in the firm is more cautious, owing to their low wealth diversification. Indeed, unlike the shareholders described in the financial literature, family shareholders invest an important part of their financial net worth and wealth in the business. The cost of their risk diversification is greater if they have little cash left at their disposal to invest in other assets. This cost is greater for family owners than for other kind of blockholders.

Moreover, the association of family shareholders to the firm is not only monetary driven. Various studies carried out in Europe and in the United States show that family owners favor their decision-making autonomy and their firm's continued viability rather than achieving a growth target or realizing liquidity for the shareholders. The determinants of their risk behavior depend, first, on their preference for maintaining the firm's viability, secondly on their desire for managerial and strategic autonomy and, finally, on the protection of their control and their reputation. These factors represent a set of private benefits (Hart, 2001; Hart and Moore, 1989) that are dependent on the control power of the family owner. As a result, family owners do not have the same attitude to the company's aggregate risk as that of classical diversified shareholders. To the extent that the risk of their investment is closely linked to their firm's aggregate risk, any increase in the risk of bankruptcy tied to an increase in debt augments their potential loss of wealth. Because their attitude is influenced by this closeness to their patrimony, family owners' risk aversion is closer to that of managers, and differs from the risk neutrality of the diversified shareholder (Nagar et al., 2000). Financing decisions therefore depend on the desire and the ability of family owners to protect their monetary and private benefits (Dietsch and Godbillon, 1998; Zingales, 2000).

Actors' relationships and family firms' capital structure

The family owner's aversion to risk appears to be similar to that of managers. For this reason there is a similarity in the interests of managers (whether or not they belong to the owner's family) and family shareholders, as distinct from the hypothesis of 'alignment of interests' described by Jensen and Meckling (1976), since it is no longer the managers who align their interests with those of the shareholders but the reverse. Consequently, we assume that this similarity of interests implies that, like managers, family shareholders would be reluctant to contract debt and to increase the firm's gearing. This attitude towards debt leverage complies with the interests of creditors. Moreover, several studies (McConnell and Servaes, 1990, 1995; Shleifer and Vishny, 1996) demonstrate that blockholders exert better control over management compared with other owners. Therefore, family owners' direct control over management stands for indirect debt control.

> *Hypothesis 1* Family firms are less indebted, than their mainstream peers, whatever the manager's identity (that is, insider or outsider)

Nevertheless, the delegation of a firm's operations to a hired manager results in distinctive financial behavior among family businesses. These managers agree to run the firms only if their wages and private benefits are higher than they could obtain outside the firm (Burkhart et al., 2002). They bear the risk of *ex post* over-control by family owners, which can reduce their ability to extract monetary and non-monetary personal profits. According to Gómez-Mejia et al. (2001), this risk represents an additional agency threat for the manager. The separation of control and management decisions in family businesses generates, despite the similarity of their risk aversion, a free cash flow risk (Jensen and Meckling, 1976) owing to the manager's opportunistic behavior; it also induces agency costs for family owners (Ang et al., 2000). However, this free cash flow risk is assumed to be lower than in mainstream professionally run counterparts, owing to the similarity of risk aversion in family companies. Debt represents an indirect control mechanism to reduce the opportunistic behavior of the external manager.

Hypothesis 2 Family firms managed by a member of the owner family are less indebted than others family firms.

In contrast, debt is costly for family owners because it increases the risk of default and the risk of transfer of shareholders' control to creditors in case of default. That is why the use of debt as a control mechanism depends on a compromise for family owners between debt costs and direct control costs.

Nagar et al. (2000) show that there is a negative relation between the family shareholding stake and its control costs. Therefore, at low shareholding levels, family shareholders' direct control costs could be higher than debt costs. Consequently, family owners would prefer the use of debt to reduce the free cash flow risk: at low shareholding levels, debt would be a positive function of family shareholdings. In contrast, at high shareholding levels, the cost of contracting debt is higher for family owners than the cost of direct control. As a result, they would prefer to effect direct control of managers rather than increasing the level of debt: at high shareholding levels, debt leverage would be a negative function of family shareholding when an external manager runs the firm. Therefore, family firms' debt leverage, when an external manager is in charge, is a non-monotonic function of the family shareholding stake.

Hypothesis 3 The relation between debt level and family shareholding stake is non-monotonic for family firms run by external non-family managers.

Recent developments in financial theory suggest that the choice of debt maturity can mitigate the agency conflict between the family owner and lenders, and alleviate the problem of information asymmetries. Bolton and Scharfstein (1990) suggest that lenders have enough power to exclude the firm from the capital market, and hence to stop future financing altogether, if family owners default. This is called the explicit lender threat. Under the assumption that family owners value highly their firms' continued viability, the authors show that repeated debt refunding reduces information asymmetry and related agency conflicts. In that context, borrowers have to come back at regular, short intervals for more funds. Myers (1977) and Jensen (1986) propose to reduce debt maturity so that frequent renewals require the family owner to justify the firm's results to creditors. These regular contractual renegotiation requirements make it possible to reduce information asymmetry and to mitigate agency costs. Overall, significant short-term indebtedness forces the family owner to repay the creditors and to choose the most profitable investment project. As a result of having a whole range of controls, creditors combine cash flow rights with the ability to regularly interfere in the major decisions of firms using short-term debt. Therefore, when creditors anticipate a high information asymmetry risk, they will prefer short-term lending. Although the main costs of short-term debt are such that family owners may be prevented from undertaking good projects because debt covenants keep them from raising additional funds, they may also be forced by creditors to liquidate when it is not efficient to do so (Diamond, 1991; Hart, 1995; Stulz, 1990). In these conditions short-term debt can accelerate the transfer of control from the family owner to creditors. These debt agency costs, together with the possibility that lenders might investigate the company books, reduce the family's decision-making autonomy, and deprive it of the benefits of control.

The family owner would hence be reluctant to rely on short-term debt. In order to take advantage of future growth opportunities and to preserve their decision-making autonomy, family owners would rather rely on long-term debt contracts. The future payment flows of the latter are the slowest and the time horizon the furthest. However, it is clear that this is not in the creditors' interests (Hart and Moore, 1989). For this reason, the family firm's debt maturity would depend on the creditor's perception of information asymmetry and the negotiating power of both family owner and creditor. This last factor would determine whose interest would predominate.

In summary, since family firms appear to bear a higher information asymmetry (Catry and Buff, 1996; Dietsch and Godbillon, 1998), associated with a strong preference for decision-making autonomy, short-term debt would be too restricting for family owners. If they were free to choose the debt maturity, they would rely on long-term financing. However, this debt maturity arbitrage is influenced by the related information asymmetry costs. Thus when the cost of contracting long-term debt cannot be assumed, the family would rather rely on short-term debt.

Hypothesis 4 Family firms would rather contract more long-term debt, *ceteris paribus*.

Empirical methodology and data collection

Samples and data collection

In order to investigate the relationship between ownership structure and the firm's financial behavior, an array of data is needed. Data was obtained from two French databases: on the one hand, the SCRL-Diane database enables us to collect firms' financial data from 1995 to 2000; on the other hand, the Dafsa liens database is used to extract information about firms' ownership and organizational structure.

The SCRL-Diane database provides financial information on more than 840 000 French firms. We set several criteria in order to build our sample. Banks, financial institutions and public utilities are excluded owing to their specific financing structure. Moreover, defaulted companies are eliminated together with those under receivership during the study period. Furthermore, only firms are selected for which there are financial data for each year of the study period. Finally, we extract limited liability companies achieving a minimum turnover of 8 million euros in 1995. The resulting sample is composed of 12 539 private firms and 262 public firms. In order to reduce the number of private firms in our sample, in view of the difficulties of collecting data and calculating for a five-year period, 500 firms are selected at random from the initial sample of private firms. The random selection process suggests that this sample might well offer a good representation of French private medium-sized firms. The final sample is made up of 477 private firms and 236 public firms.[1]

Description of variables and the model

Prior research (Shanker and Astrachan, 1996) offers a broad set of definitions of family firms. Discussions are still going on about the identification of family firms. However, we choose a broad definition that identifies family companies as firms in which a family or an individual is the leading shareholder, and held at least 10 per cent[2] of the company's shares in the year 2000. The main reason for choosing this definition is that the ownership level may not represent the main influence that family members exert on the firm (Anderson and Reeb, 2003). In order to investigate the influence of family ownership, we create a variable measuring the level of the family ownership stake. We create a *family firm* dummy variable that equals one when the firm is family owned.

A second dummy variable (*family managed* variable), that equals 1 when a family member runs the family firm, enables us to study the influence of management delegation. The manager is identified as a family member if his surname overlaps with that of the controlling family. There is no claim that this approach is free of any caveats.

Since the underlying hypothesis is that capital structure depends on the nature of the shareholders, the ratio of the firm's financial debt to equity (gearing) and the ratio of the firm's short-term debt to equity (debt maturity) are regressed on measures of ownership structure and other control variables. A firm's capital structure is likely to be affected by many factors other than the allocation of equity ownership, although several variables are included to attempt to control these other effects. The control variables are close to those used in earlier studies of capital structure (Stohs and Mauer, 1996; Titman and Wessels, 1988) and research conducted on ownership structure (Agrawal and Knoeber, 1996; Short and Keasey, 1999). We introduce organizational variables such as the number of known shareholders and the identity of the firm's CEO because these factors may influence the firm's financial behavior. We also incorporate a dummy variable in order to measure the effect of being listed on the financial behavior of family firms. Other variables are integrated to control the effects of risk, agency costs, asset specificity, tax effect and performance. Finally, 22 control variables are incorporated in our empirical analysis and are presented in Table 29.1.

Our objective is to study the influence of the family shareholder on the firm's financial behavior. First, we conduct a descriptive analysis of our sample and, secondly, in order to identify the determinants of family firms' financial behavior, we use two ordinary least squares (OLS) regressions for each dependent variable (gearing and debt maturity).

Family ownership and firm financial behavior: results and discussion

Family firms' organizational features: a descriptive analysis

We notice that family firms represent the main organizational form in France: 72 per cent of listed French firms and 60 per cent of private firms are family-owned businesses (Table 29.2).

We find that family firms are much older than other firms. The oldest was created in 1820, 60 years before the oldest non-family firms. Furthermore, family owners show entrepreneurial dynamics because the youngest family firm was created in the mid-1990s. Age does not appear as a discriminating factor among firms in relation to accessing financial markets. Listed firms have on average the same age, which is 60 years.

Family firms tend to be open structure in term of their capital base and management team. Table 29.2 shows that the average family shareholding stake is 55 per cent for public firms and 73 per cent for private firms, whereas, on average, other firms have respectively 54 per cent and 79 per cent. Family firms have a significant preference for majority control mainly for public family firms for which the difference with non-family firms is significant at 1 per cent. This suggests that family shareholders impose strong control over the firm. The analysis *free float* – the proportion of shares offered to the public – reveals insignificant differences between family-controlled (27 per cent of shares) and non-family firms (28 per cent of shares). Thus , there is a tendency for public family firms to assimilate and adhere to market mechanisms. However, they prefer widespread distribution of the shares put on the market compared with concentration in the hands of financial investors. Table 29.3 shows that on average only 13 per cent of family firms' shares belong to financial investors compared to 43 per cent on average for non-family businesses. These results highlight the importance given by family owners to seeking a close relationship with their financial stakeholders and to the preservation of family control. Dilution of the floating

Table 29.1 Definition of variables and measurements

Description	Variables	Calculation
Financial structure (dependent variables)	Gearing Debt maturity	Financial debt divided by equity Ratio of short-term debt to equity
Ownership structure	Family firm Ownership stake	1 if the firm is a family firm 0 elsewhere % of capital hold by the leading shareholder
Organizational structure	Number of owners Family managed	Ln (1+ Number of known shareholders) 1 if the manager belongs to the family 0 elsewhere
	Public firms	1 if the firm is listed, 0 elsewhere
Operating risk	Ebitda growth	Standard deviation of EBITDA annual change
	Employee expenses Size Age	Employees expenses/operating profit Size = Ln (sales) Difference between 2000 and the year of incorporation
	Industry	Two-digit NAF code (French SIC code)
Agency costs	*Growth variables*	
	Capital employed growth	Mean of capital employed annual change
	Free cash flow risk	
	Free cash flow	Free cash flow/capital employed
	Management expenses in excess	[Expenses – cost of external goods – employees expenses – financial expenses]/ sales
	Information asymmetry risk	
	Proportion of operating debt	Operating debt/capital employed
	Financial expenses on value added	Financial expenses/added value
	Debt cost	Financial expenses/financial debt
	Interest cover	Financial expenses/EBITDA
	Cash flow cover	Firm's net cash flow/financial debt

Table 29.1 (continued)

Description	Variables	Calculation
Assets specificity	Intangible assets	Fixed assets/capital employed
Tax effect	Non-debt tax shield	Depreciation expenses/capital employed
Performance	Operating return	EBIT/capital employed
	ROE	Net profit/shareholders equity book value

Notes for variables:
Family firm is a binary variable that equals one when the firm is a family firm.
Ownership stake is the percentage of a firm's shares in the hands of the main shareholder.
Number of owners is the natural log of (1 + number of the firm's known shareholders).
Family managed is a binary variable that equals one if the CEO is a member of the family shareholder.
Public firms is a binary variable that equals one when the firm is listed.
Ebitda growth is the standard deviation of Ebitda annual growth for the study period.
Employee expenses is the average for the study period of the ratio of employee expenses to firm operating profit.
Age is the number of years since the firm's inception.
Size is the natural log of the firm's turnover.
Industry is a dummy variable for two-digit NAF codes (which is the French code equivalent to SIC code).
Capital employed growth is the average of annual capital employed growth rate. Free cash flow is the average of annual free cash flow on capital employed ratio.
Management expenses in excess is the average ratio of total firm expenses less the cost of external goods less employee expenses less financial expenses divided by sales.
Debt costs is the average of interest paid on financial debt.
Interest cover is the average of annual financial expenses on Ebitda.
Cash flow cover is the ratio of firm net cash flow to financial debt.
Proportion of operating debt is the ratio of operating debt to capital employed.
Financial expenses on value added is the ratio of financial expenses divided by value added.
Intangible assets is the ratio of fixed assets to capital employed.
Non-debt tax shield is the ratio of depreciation expenses to capital employed.
Operating return is the ratio of EBIT to capital employed.
ROE or return on equity is the ratio of net income divided by equity.

Table 29.2 Number of family firms in selected samples

	Firm samples			
	Listed		Private	
Leading shareholder's identity	Number	Proportion (%)	Number	Proportion (%)
Family owner (EP)	171	72.5	285	59.7
Financial owner	22	9.3	9	1.9
Holding entity	14	6	23	4.8
Industrial firm	29	12.3	152	31.9
Miscellaneous			8	1.7
Total	236	100	477	100

shares enables the power of the other investors to be diluted to the advantage of family shareholders.

Moreover, family firms are not closed managerial entities. As demonstrated in Table 29.3: some family firms have a non-family CEO who is not hired from the owner's family group. Nevertheless, these observations highlight the importance given by family

Table 29.3 Samples' organizational characteristics

	Samples			
	Listed		Private	
Firms characteristics (on average)	Family (n = 171)	Non-family (n = 65)	Family (n = 285)	Non-family (n = 192)
Ownership stake	54.94%[a]	53.76%	73.04%	79.6%
Free float	27.97%	27.07%		
Institutional shareholding	13.35%[a]	42.78%		
Number of identified owners[1]	2.99	3.51	1.42	
Managerial identity[2]	1.43		1.6	

Notes:
(a) $p < 1\%$.
1. Family shareholders taken as a unique group.
2. If the manager is a family member the number is 1, 0 elsewhere.

shareholders to the control of their firm and their management. This especially enables them to preserve their decision-making autonomy.

Family firms are less indebted than other firms
Table 29.4 presents results using accounting measures of gearing (first column) and debt maturity (second column). The results of the OLS regression show that ownership structure and the identity of the leading shareholder are significant factors in determining firms' gearing: the family firm variable is significant and negatively linked to gearing. This finding contradicts the theory of neutrality of ownership structure proposed by Demsetz (1983) and validated for French firms by Charreaux (1991). As family firms are less indebted than non-family firms, our first hypothesis is validated. Thus the preferences of family owners constitute a major factor in determining financial behavior. Their propensity to preserve control over the firm and to ensure the firm's viability, coupled with their risk aversion – given the risk of transfer of control linked to contracting debt – induce a higher cost of debt financing for family firms. Moreover, in public family firms debt does not represent a way for minority shareholders to achieve control over family owners. This suggests two interpretations. First, the family shareholder acts in the interests of all shareholders, and minority shareholders in particular. In these conditions, the latter do not need to use any means of control (Jensen and Meckling, 1976). This argument is not so straightforward when we take into account the fact that the interests of financial markets generally differ from those of family shareholders, and when we recognize that the private benefits that family owners extract from the exercise of control are not transferable to others. Secondly, the family firm's reduced level of debt might be a consequence of an entrenchment phenomenon on the part of family owners (Shleifer and Vishny, 1986). Family owners want to preserve their private benefits and as a result they are not open to enhancing the level of gearing. Nevertheless, family-owner entrenchment does not occur at the expense of minority shareholders (Charreaux, 1991).

However, the ownership stake is not a significant determinant of the level of a firm's financial debt. This result is surprising when compared with previous studies, among them

Charreaux (1991). We conduct two separate regressions for, on the one hand, listed family firms and, on the other, private firms. The results do not alter the significance and the sign of the relationship between gearing and family shareholder ownership. We find that the ownership stake has a significant (at 5 per cent significance level) and negative effect on gearing. Nevertheless, this result is only validated for public family firms, not for private ones. Therefore, this finding confirms that the cost of the family owner's direct control is higher as capital is diluted: debt is used as an additional control mechanism for the family owner (Table 29.4).

Being listed does not induce a specific financial behavior. The *public firms* variable is positive but insignificant. This result suggests that being listed does not enable family firms to diversify their financial resources. Therefore, the financial behavior between listed and private family firms are not hugely different.

Management delegation induces specific financial behavior on the part of family firms. Our results underline that, when management is delegated to an external CEO, family firms are more indebted than those run by a family member. In Table 29.3 the coefficient of the *active family management* variable is significant and negative. Thus delegation of decisions induces a free cash flow risk and managerial opportunistic behavior. Our Hypothesis 2 is confirmed.

In order to study the possibility of a non-monotonic relationship between gearing and family ownership when the CEO is an external manager, we add the square of family ownership as a continuous variable to our OLS regression specification model.[3] The results show that the relationship between family ownership and gearing is linear, invalidating our third hypothesis. It seems that for public firms, the family owner has the same interest as other shareholders and the main agency conflict is that between managers and all shareholders, mainly because going public induces an increased risk of managerial opportunism. Therefore, the relation between debt and family shareholding is negative: at low stakes, debt is an additional control mechanism; at high stakes, the family owner exercises direct control over managers.

Regarding control variables, we find that agency cost variables are significant determinants of the firm's gearing. The debt level is negatively related to debt cost and cash flow cover and positively linked to capital employed growth and interest cover. Moreover, tangible assets are a means of enhancing the level of debt. Finally, we find that the gearing is negatively related to the operating profit in accordance with the Pecking Order theory (Myers and Majluf, 1984). The results of our analysis of the control variable are generally consistent with the findings of previous research.

Family firms and debt maturity

We examine whether debt maturity is a function of family ownership control. The second column of Table 29.4 presents the results of our regression. The findings reveal that there is no significant relationship between family ownership and identity and the firm's debt maturity. Moreover, it appears that family firms and non-family firms have no differences in their debt maturity structures. These results invalidate our fourth hypothesis.

However, only two variables are significant at the 1 per cent level. The *public firms* variable is significant and negatively related to debt maturity. The financial market seems to reduce the information asymmetry risk faced by creditors and enables public family firms to contract more long-term debt. The financial stock exchange allows family firms to take

Table 29.4 Financial structure and family ownership

	Gearing	Debt maturity
Intercept	−4.278	−3.111
(t statistic)	(−0.790)	(−0.464)
Family firm	−0.388[a]	−0.018
	(−2.539)	(−0.096)
Ownership stake	0.001	−0.001
	(0.230)	(−0.203)
Organizational structure		
Number of owners	0.152	−0.237
	(0.724)	(−0.914)
Family managed	−0.244[b]	−0.033
	(−1.646)	(−0.180)
Public firms	0.187	−0.642[a]
	(0.872)	(−2.417)
Operating Risk		
Ebitda growth	−0.017	0.004
	(−0.573)	(0.131)
Employees remuneration	0.002	0.014
	(0.293)	(1.194)
Age	0.001	0.001
	(0.405)	(0.041)
Size	0.026	0.063
	(0.703)	(0.750)
Industry	0.001	−0.004
	(0.217)	(−0.928)
Agency costs		
Capital employed growth	0.781[b]	−0.234
	(1.869)	(−0.451)
Free cash flow	1.253	0.764
	(1.522)	(0.749)
Management expenses in excess	−0.867	0.524
	(−1.132)	(0.552)
Debt cost	−4.961[a]	3.664
	(−2.423)	(1.445)
Interest cover	1.267[a]	0.074
	(4.320)	(0.206)
Cash flow cover	−0.195[a]	0.045
	(−7.641)	(1.451)
Proportion of operating debt	0.504[a]	0.273
	(2.793)	(1.224)
Financial expenses on value added	2.725[a]	0.907
	(2.657)	(0.714)
Intangible assets	0.981[a]	−1.486[a]
	(2.949)	(−3.605)
Non-debt tax shield	1.728[a]	−0.954
	(2.080)	(−0.927)

Table 29.4 (continued)

	Gearing	Debt maturity
Profitability		
Operating return	-1.232^{a}	-0.904
	(-2.505)	(-1.483)
ROE	1.358^{b}	0.220
	(1.675)	(0.219)
Number of observations	733	733
R square	0.368	0.144
F test	8.466^{a}	2.244^{a}

Notes:
(a) $p < 1\%$, (b) $p < 10\%$.
This table reports the results of regressing on the one hand firm gearing on family ownership and on the other hand firm debt maturity on family ownership. All ratios were calculated for each year from 1995 to 2000. The average ratio of each variable was integrated in the regression. t-values are in parentheses.

benefits from the availability of other kinds of long-term financing. Bankers are therefore in competition with other financial investors, and lose their monopolistic position as resource providers for family firms. Debt costs would therefore decrease. Furthermore, debt maturity is negatively linked to tangible assets. This result highlights the guarantee role played by fixed assets when family firms want to borrow in the long term. However, despite the use of variables defined by several previous studies of firms' debt maturity structure (Barclay and Smith, 1995), our model do not reveal any significant relationship. This calls for further research.

Conclusion

The aim of this research is to determine whether the identity of the leading shareholder influences the company's financing decisions. More specifically, we intend to contrast the debt-financing behavior of public and private family companies with that of non-family businesses (that is, managerial firms). Comparing the financing decisions of French family firms with those of French managerial firms enables us to do this. Attention is also given to investigating the determinants of the debt-financing decisions of family and managerial firms. Using the agency theory and the financial contracting theory frameworks, we suggest that the family shareholder is more risk averse than the shareholder described by the classical theory.

Our results provide evidence that the type of the leading shareholder is not neutral to the decision to contract debt. The empirical results suggest that family companies are less indebted. This emphasizes several factors. First, family owners' preference for decision-making autonomy and their fear of the risk of control being transferred to creditors reduce their willingness to contract debt. Secondly, their direct control over managers is less costly than for other blockholders. Finally, the family owner exercises control over managers more efficiently; this reduces the latter's opportunistic behavior. We do not find evidence of a non-monotonic relation between debt level and family owner shareholding for managerial family firms. This indicates that when a family firm's equity is diluted, the cost of direct control is high for family owners: they therefore behave like other shareholders and rely on debt control over management. On the contrary, the debt costs such as the risk of transfer of control to the creditor – the default risk – are higher than the costs of the direct exercise of control of management. The debt level would decrease with family owners' shareholding. Furthermore, we point out that there is a similarity of debt maturity structure in family and non-family firms. Therefore, creditors do not differentiate significantly between the two categories when we consider short-term debt. Finally, the study of public family firms suggests that they have fully integrated financial market rules. Nevertheless, the identity of their leading shareholder still constitutes a differentiating factor.

The observations borne out in the course of this study have enabled us to define several axes of development. More precisely, it would be interesting to focus on an aspect of agency conflict that we have ignored, namely, the conflict among shareholders. These conflicts arise when one shareholder has enough control of the firm to be able to take actions that benefit him at the expense of the non-controlling shareholders (Shleifer and Vishny, 1996). The conflicts which arise between a blockholder and minority shareholders are added in family firms to those generated by family members' opposing points of view. Family firm shareholders' agency conflicts might be more complex than those of non-family businesses.

Notes

1. After selecting firms, we recheck the data for each firm and each year. We exclude several firms for which financial data are still missing.
2. This ownership stake is consistent with the definition given by several works of research relating to large shareholders' influence and patrimonial firms. In this regard see Short (1994), Shleifer and Vishny (1996) and Shanker and Astrachan (1996).
3. The results of this OLS regression are not presented in the chapter owing to the insignificance of *the square family ownership* variable and the non-modification of the previous regression main results.

References

Agrawal, A. and C. Knoeber (1996), 'Firm performance and mechanisms to control agency problems between managers and shareholders', *Journal of Financial and Quantitative Analysis*, **31**(3), 377–97.

Agrawal, A. and G. Mandelker (1987), 'Managerial incentives and corporate investment and financing decisions', *Journal of Finance*, **42**(September), 823–37.

Amihud, Y. and B. Lev (1981), 'Risk reduction as a managerial motive for conglomerate mergers', *Bell Journal of Economics*, **12**(2), 605–17.

Anderson, R. and D. Reeb (2003), 'Founding-family ownership and firm performance: evidence from the S&P 500', *Journal of Finance*, June, (3), 1301–28.

Ang, J., R. Cole and J. Lin (2000), 'Agency costs and ownership structure', *Journal of Finance*, February (1), 81–106.

Barclay, M. and C. Smith (1995), 'The maturity structure of corporate debt', *Journal of Finance*, **50**(June), 609–31.

Bolton, P. and D. Scharfstein (1990), 'A theory of predation based on agency problems in financial contracting', *American Economic Review*, **80**(March), 93–106.

Burkhart, M., F. Panunzi and A. Shleifer (2002), *Family Firms*, US National Bureau of Economic Research, Working Paper No. 8776.

Catry, B. and A. Buff (eds) (1996), *Le Gouvernement de l'Entreprise Familiale*, Paris: Publi-Union éditions.

Charreaux, G. (1991), 'Structure de propriété, relation d'agence et performance financière', *Revue Economique*, (3), 521–52.

Demsetz, H. (1983), 'The structure of ownership and the theory of the firm', *Journal of Law and Economics*, **26**(June), 375–90.

Diamond, D. (1989), 'Reputation acquisition in debt markets', *Journal of Political Economy*, **97**, 828–62.

Diamond, D. (1991), 'Debt maturity structure and liquidity risk', *Quarterly Journal of Economics*, **43**, 1027–54.

Dietsch, M. and B. Godbillon (1998), 'Choix de structure financière et performances: le cas des PMI', *Working Paper, Centre d'Etudes des Politiques Financières*, May, Strasbourg: Institut d'Etudes Politiques de Strasbourg.

Fama, E. and M. Jensen (1983), 'Separation of ownership and control', *Journal of Law and Economics*, **26**(2), 301–25.

Friend, I. and L. Lang (1988), 'An empirical test of the impact of managerial self-interest on corporate capital structure', *Journal of Finance*, **43**(June), 271–81.

Gallo, M. and A. Vilaseca (1996), 'Finance in family business', *Family Business Review*, **9**, 387–401.

Gómez-Mejia, L., M. Núñez-Nickel and I. Gutiérrez (2001), 'The role of family ties in agency contracts', *Academy of Management Review*, **44**, 81–95.

Harris, M. and A. Raviv (1991), 'The theory of capital structure', *Journal of Finance*, March, 297–355.

Hart, O. (ed.) (1995), *Firms, Contracts and Financial Structure*, Oxford: Clarendon Press.

Hart, O. (2001), *Financial Contracting*, US National Bureau of Economic Research, Working Paper No. 8285.

Hart, O. and J. Moore (1989), 'Debt and seniority: an analysis of the role of hard claims in constraining management', *American Economic Review*, **85**(3), 567–85.

Jensen, M. (1986), 'Agency cost of free cash-flow, corporate finance and takeovers', *American Economic Review*, **76**(2), 323–29.

Jensen, M. and W. Meckling (1976), 'Theory of the firm: managerial behavior, agency costs and ownership structure', *Journal of Financial Economics*, **3**(4), 305–60.

McConaughy, D., D. Mendoza and C. Mishra (1996), 'Loyola University Chicago family firm stock index', *Family Business Review*, **9**, 125–37.

McConnell, J. and H. Servaes (1990), 'Additional evidence on equity ownership and corporate value', *Journal of Financial Economics*, **27**, 595–612.

McConnell, J. and H. Servaes (1995), 'Equity ownership and the two faces of debt', *Journal of Financial Economics*, **39**, 131–57.

Myers, S. (1977), 'Determinants of corporate borrowing', *Journal of Financial Economics*, **5**(2), 147–75.

Myers, S. and N. Majluf (1984), 'Corporate financing and investment decisions when firms have information that investors do not have', *Journal of Financial Economics*, **13**, 187–221.

Nagar, V., K. Petroni and D. Wolfenzon (2000), *Ownership Structure and Firm Performance in Closely-held Corporations*, Ann Arbor, MI: University of Michigan Business School.

Poutziouris, P. and S. Sihar (2001), 'The financial structure and performance of family and non-family companies revisited: evidence from the UK private economy', in G. Corbetta and D. Montemerlo (eds), *The Role of Family in Family Business*, Book Proceedings: 12th Annual Family Business Network World Conference, Rome, pp. 416–35.

Romano, C.A., G.A. Tanewski and K.X. Smyrnios (2000), 'Capital structure decision-making: a model for family business', *Journal of Business Venturing*, **16**, 285–310.

Shanker, M. and J. Astrachan (1996), 'Myths and realities: family businesses' contribution to the US economy: a framework for assessing family business statistics', *Family Business Review*, **9**, 107–23.

Shleifer, A. and R. Vishny (1986), 'Large shareholders and corporate control', *Journal of Political Economy*, **94**(3), 461–88.

Shleifer, A. and R. Vishny (1996), *A Survey of Corporate Governance*, US National Bureau of Economic Research, Working Paper No. 5554.

Short, H. and K. Keasey (1999), 'Managerial ownership and the performance of firms: evidence from UK', *Journal of Corporate Finance*, **5**, 79–101.

Short, H. (1994), 'Ownership, control, financial structure and the performance of firms', *Journal of Economics Surveys*, **8**, 203–49.

Stohs, M. and D. Mauer (1996), 'The determinants of corporate debt maturity structure', *Journal of Business*, **69**(3), 279–312.

Stulz, R. (1990), 'Managerial discretion and optimal financing policies', *Journal of Financial Economics*, **26**, 3–27.

Titman, S. and R. Wessels (1988), 'The determinants of capital structure choice', *Journal of Finance*, **40**(March), 1–19.

Xiao, J., M. Alhabeeb, G. Hong and G. Haynes (2001), 'Attitude toward risk and risk-taking behavior of business-owning families', *Journal of Consumer Affairs*, **35**(Winter), 307–25.

Zingales, L. (2000), 'In search of new foundations', *Journal of Finance*, **55**, 1623–53.

30 The structure and performance of the UK family business PLC economy*

Panikkos Zata Poutziouris

This chapter reports on the profile of UK family-controlled quoted companies and, via the Family Business Index, reports on their performance vis-à-vis that of their mainstream counterparts. It comes in response to recent empirical investigations on the role and performance of family-controlled firms in stock markets across leading economies, notably those of the US, France, Germany and Spain.

The investigation focuses on UK quoted family-controlled companies that are constituents of the Financial Times Stock Exchange (FTSE) All-Share Index and draws comparative evidence on their structure, growth, profitability and share price performance over a five-year period, 2000–04.

Interestingly, the research reveals that the index of family business public limited companies (FB-PLCs) capitalization performs better than that of mainstream FTSE indices, despite the continued diminishing role of family shareholding (Franks et al., 2003). The outperformance of UK FB-PLCs mirrors US findings where quoted family firms (with founding families playing a active role in ownership and management control) financially outperform their Standard and Poors (S&P) counterparts (Anderson and Reeb, 2003).

In the light of recent concerns about the effectiveness of corporate governance mechanisms to master agency costs, it emerges that the quoted family-controlled PLC model is not a 'deficient' organization structure. Quoted family firms have their own approach in mastering their long-term growth and development, and, moreover, in managing principal versus agency conflicts, in building relations with financial agents, in mitigating risks and in charting effective strategic decision-making – all in all resulting in *valued added familiness.*

Introduction

Worldwide, the family firm, is the most prevalent form of business organization. For most developed market economies, the family business sector is estimated to represent from 60 per cent to 75 per cent of all enterprises – including, in certain cases, up to a third of quoted public limited companies, accounting for about half of GDP economic activity and private employment (Astrachan and Shanker, 2003).

As industrial statistics indicate, the influential role of the family in the business activity is more central at the early stages of the corporate life cycle, where the founders and owners managers – and their family ties – are the main source of entrepreneurial drive and capital. Notwithstanding this, with the emergence of managerial capitalism, which is fuelled by the separation of ownership and control owing to the growth and financial development of large private companies via flotations, the role of families in terms of ownership and control remains important.

A series of investigations indicate that the proportion of family controlled quoted companies in main equity capital markets across OECD economies is very substantial ranging from 10 per cent to over 50 per cent (Faccio and Lang, 2002; La Porta et al., 1999). It is argued that family shareholding in quoted family companies is variant depending on the development of the capital market and its legal-regulative model, and, of course, the business cultural paradigm conditioning the appetite for business families to sustain control via voting ordinary shareholding, often reinforced by pyramidal cross-shareholding, and multi-classification of shares with enhanced blockholding voting power.

In contrast to the persistence of family capitalism in other continental capital markets (Becht et al., 2001; Faccio and Lang, 2002; La Porta et al., 1999) the UK family shareholding according to Franks et al. (2003) has been in decline during the past century. Despite the diminishing role of insiders and families in the UK PLC economy, the *Investors Chronicle* (2003), has cited evidence revealing that family-controlled quoted companies were outperforming their counterparts. This was in line with the findings of the Stoy Hayward and BBC 1992 business survey and more recent investigations which focused on the financial performance of family quoted firms in the US (*Business Week*, 2003), and in Europe (Miller, 2004).

More specifically, from striking new analysis by Thomson Financial (Miller, 12 April 2004), it emerges that family-controlled quoted companies are outperforming their rivals on all six major stock indexes in Europe, from London's FTSE to Madrid's IBEX, and often dramatically so. The Thomson research team created an index for both family and non-family PLCs and tracked their performance over a 10-year period to December 2003. In addition, they also produced a list of the top 10 fastest growing family-company shares. Hereby, in summary, are the results:

- In Germany the family index soared 206 per cent, led by BMW, while the non-family stocks climbed just 47 per cent.
- In France, the family index surged an equally breathtaking 203 per cent, led by the likes of Sanofi-Synthelabo, L'Oreal and LVMH, while its counterpart rose only 76 per cent.
- Family controlled PLCs also outperformed their peers in Switzerland, Spain, Britain and, even, Italy.

Indeed, this business analysis comes in contrast to recent scandals in North America and Europe (for example, US based Adelphia, Swiss Erb Group and Italian Parmalat, to name a few) that portray family-controlled businesses as scandalous, in search of mechanisms to 'shine the family silver' and of course to sustain family control . However, the evidence about the outperformance of family companies could offer comfort to investors who are worrying about the practice of expropriation of special benefits for the controlling family owners and, of course, for the despotism and altruistic nepotism characterizing business family dynasty entrepreneurs.

In the light of renewed interest in the financial affairs of quoted family firms, the aim of this investigation is to direct the microscope to the UK family business PLC in order to monitor its structure and performance over time. This research report continues with a brief literature review relating to the topic of quoted family firms and their financial affairs. This is followed by an outline of the research methodology and data. Then there is a comparative analysis of the UK Family Business Index vis-à-vis London main stock market indices. Finally, in conclusion, a set of implications are briefly discussed emanating from the explorative Phase A of the empirical study.

Finance theories and family firms

According to the pecking order hypothesis (Myers, 1984), privately owned companies finance their capital needs in a hierarchical fashion, first using internally available funds, followed by debt and, finally, external equity. Arguably, the pecking order hypothesis is particularly relevant to closely held family firms, characterized by an aversion to outside capital infusions (Gallo and Vilaseca, 1996; Poutziouris, 2001; Poutziouris et al., 1998; Romano et al., 2000) as they experience relatively more restrictive transactional and behavioural costs in raising external equity (Pettit and Singer, 1985). In the case of the growing family firm, heavy investments in organic and/or acquisitive expansion, innovation-enabling technologies and global marketing (niche) strategies, could result in the exhaustion of debt facilities and so compel the owner family to seek external (private and public) equity.

A stock market flotation, would widen the share ownership of the firm, and ultimately could lead to the dilution of control by the original founding and/or descendant family owner-managers, or even bring about an ultimate loss of family control owing to a hostile takeover. Zingales (1995) and Pagano et al. (1996, 1998) established that the decision of the owners of a firm to go public depends on whether they are likely to succeed in simultaneously raising capital and retaining power (shareholding and managerial control). This objective can be achieved through the *free float* of a limited portion of total shares combined with dilution of outside shareholdings. Alternatively, the issue of preference shares, the use of different classes of shares (which can multiply their voting power – the Ford approach) can allow controlling family shareholders to externally issue additional capital without diluting their blockholding/majority position (see Schürmann and Körfgen, 1997, p. 104).

There is plethora of academic studies into the impact of family ownership and managerial control on performance. The theoretical base of such inquiries has been agency theory which expound problems of monitoring costs involved in restraining the opportunistic behavior of managers, expropriation by controlling owner-managers of private benefits at the expense of minority shareholders and entrenchment and tunnel vision of controlling families and so on.

Agency costs theorem

Agency theory argues that the separation of ownership and management creates conflicting goals between principals (that is, shareholders) and agents (that is, managers), which could arise from divergent utility functions and information asymmetries about their views on growth, variant investment horizons, different attitude to risk diversification and takeovers, and so on (Jensen and Meckling, 1976). A number of scholars argued that the owner-managed family firm model, characterized by intra-familial altruistic elements and clan control, tends to develop more goal congruence among stakeholders, and thus could be exempted from serious problems of traditional agency conflict (Ang et al., 2000; Chrisman et al., 2004; Daily and Dollinger, 1992; Jensen and Meckling, 1976.). Arguably, a large family stakeholding could reduce agency problems owing to the incentives and higher capabilities of highly committed (especially founding) shareholders to monitor management.

On the other hand, another group of scholars demonstrated that agency problems exist in the closely held owner-managed family firm and family-controlled business groups, often resulting from internal dysfunctions owing to the autonomy of controlling shareholders (La Porta et al., 1999), nepotistic altruism (Schulze et al., 2001, 2003b) and tolerance of honest incompetence (Hendry, 2002). Morck et al. (1988) argued that agency costs arise for minority shareholders from having an entrenched dominant shareholder. More specifically, Bebchuck et al. (2000) and Morck (2000) and Morck and Yeung (2003) argue that conflicts in family firms are associated with the fact that managers may act solely for one single stakeholder, the family, and neglect the interests of the rest shareholders.

Ownership control is the central issue in the agency cost theoretical framework, since ownership influences the magnitude of the separation or enmeshment of owners and managers. Increasing managerial ownership would encourage goal alignment, however, the extent to which managers can invest in the equity-residual claims of the firm is constrained by their personal wealth and risk diversification considerations (Jensen and Meckling, 1976). Moreover, the emergence of a closed ownership regime could exacerbate the problem of information asymmetries.

The underlying assumptions of asymmetric information are that owner-managers know more about the company's current earning and investment opportunities than do outside investors, and they act in the best interests of the firm's existing shareholders. This condition obscures the ability of external investors to distinguish between good and bad projects, and creates undervaluation and underinvestment problems. On the one hand, undervaluation occurs when investors or shareholders assign a low average value to the shares of all firms and will buy new equity issues only at a large discount from their equilibrium values without informational asymmetries. On the other hand, underinvestment occurs when the management refuse to accept positive net present value (NPV) investment opportunities if this would entail issuing new equity (which could further dilute founding family shareholding!) since this would give away too much of the project's value to the new shareholders at the expense of the existing shareholders (Megginson, 1997) and, thus, relinquish control.

Consequently, the cost of financing good projects with external funds exceeds that of financing such projects with internal funds. In order to minimize these problems, companies have to retain sufficient financial slack (that is, cash and marketable securities plus unused debt capacity) to be able to internally fund investments with positive returns. Firms with sufficient financial slack can thus issue risky debt or equity securities in order to fund their investment projects, and they have the ability to alleviate asymmetric information problems between management and shareholders (Megginson, 1997). Therefore, owner-managers tend to favour internally generated funds as a source of additional capital, followed by external debt and, finally, by external equity (Myers, 1984).

Family ownership in public equity markets

La Porta et al. (1999) investigated the top 20 publicly traded firms in 27 countries and found that the role of families as main shareholders, based on the cut-off of 10 per cent of shares in the hands of the family (and its units), ranged from 5 per cent (for the UK) to 70 per cent (for Hong Kong). The percentages for other countries varied as follows: 20 per cent in the US, 30 per cent in Canada, 10 per cent in Germany, 20 per cent in Italy, 55 per cent in Sweden and 20 per cent in France (data referring to the end of 1995). The same study also analysed, for each of the 27 countries, the ownership of 10 medium-sized listed companies (that is, 10 smaller companies with capitalisation of at least $500 million, in order to have comparable sizes). Families were present in from 10 per cent (Japan) to 100 per cent (Greece) of the cases. In the UK, 60 per cent of those companies had family owners, 30 per cent in the US, 40 per cent in Germany, 80 per cent in Italy, 60 per cent in Sweden and 50 per cent in France.

Faccio and Lang (2002) conducted a comprehensive study of ultimate ownership and control in 13 Western European economies and established that families (again based on the ownership cut off of 20 per cent) were the most pronounced type of controlling shareholders. They found that 44.3 per cent of Western European corporations were family controlled. However, family control was the lowest in the UK (23.68 per cent) and Ireland (24.63 per cent); in continental Europe, the lowest percentages were recorded in Norway (38.55 per cent), Sweden (46.9 per cent), Switzerland (48.1 per cent) and Finland (48.8 per cent). In the remaining countries, such as Austria, Belgium, France, Germany, Italy, Portugal and Spain, family-controlled firms were in the majority.

Blondel et al. (2002) investigated the 250 largest publicly traded companies in France, the so-called SBF 250 and reported on the prevalence, evolution, and degree of control of patrimonial firms. Patrimonial firms were defined as companies where individuals or families were identified as major ultimate shareholders with at least 10 per cent of equity at each level of the ownership chain. The study established that, even in this group of quoted companies, where spread ownership would be expected to be the norm, patrimonial firms are the majority – representing 57 per cent of all companies in the SBF 250. Patrimonial firms were present in most sectors of the economy, and their presence increased from 1993 to 1998. Their share of capitalization was lower than their importance in terms of business numbers, reflecting their concentration within the 'smaller range'. Stakes owned by families and individuals were quite high, the use of cross-holdings and voting rights further increased corporate control by patrimonial family businesses and business families.

Klein and Blondel (2002) in a comparative study of the role of family controlled companies in the Paris and Frankfurt stock markets established that over 50 per cent of the 250 largest quoted companies were patrimonial – family-controlled companies – concentrated in the lower range of capitalizations. They found that family shareholding was relatively lower and on the decrease (during 1993–98) for the German case.

Navarro and Ansón (2004), building on previous investigations into the degree of concentrated ownership in Spanish quoted companies (Crespí-Cladera and García-Cestona, 2001; Faccio and Lang, 2002; La Porta et al., 1999), reinforce the importance of individuals and families as owners of large Spanish quoted companies. They found, using 20 per cent of ownership as the threshold, that for 56 per cent of the sample firms the largest shareholder is an individual or family. Families in control of business groups tend to use pyramids and indirect ownership and other complicated chains and cascades of intermediate firms in order to *defend* their investments and ensure control rights exceed cash flow rights; moreover, in the majority of cases, members of the controlling family take the CEO role. Finally, the study offers conclusive evidence that Spanish quoted family firms are facing a serious survival crisis, with more and more families having to relinquish control and ownership power.

The performance of family-controlled PLCs

Family Business Index – approaches

Initially, practitioners directed the microscope to the role and performance of family-controlled public limited companies. In the UK, the Stoy Hayward/BBC Family Business Index in 1992 analysed a sample of 71 public family companies and compared their share price against the FTSE All Share Index – representing the top 500 UK public companies – during the 1970 to 1991 period. The study established that family companies with at least 25 per cent of shares in the hands of family shareholders, outperformed the FTSE All Share Index.

In a similar study, the Pitcairn Financial Management Group investigated the financial performance of 165 US companies, with a minimum of 10 per cent family shareholding, which had been in existence from 1969 to 1989. The Pitcairn basket of family quoted companies outperformed the Standard & Poors Index. More specifically, the cumulative returns for family companies over the period were more than double that of the S&P 500.

Again, with a US focus two indexes were developed to track the share performance of publicly held family firms – the Family Business Stock Index (FBSI), which follows more than 200 of the largest family-controlled companies nationwide and the Loyola University Chicago Family Firm Stock Index (LUCFFSI), which tracks 38 publicly traded, family-controlled firms headquartered in the Chicago area.

The FBSI was developed by NetMarquee Online Services Inc., of Needham, Massachusetts, and Robert Kleiman, an associate professor of finance at Oakland University, in Rochester, Michigan. A study of the 20-year performance of FBSI companies showed average annual returns of 16.6 per cent, compared with 14 per cent for the Standard & Poor's 500-stock index. (During some shorter time periods, however, the FBSI lagged the S&P – for example, for the year that ended with the first quarter of 1996, the FBSI showed a total return of 18.20 per cent, compared with 31.98 per cent for the S&P.)

McConaughy et al. (1996) developed the Loyola University Chicago Family Firm Stock Index (LUCFFSI) to track the performance of publicly traded, family-controlled firms headquartered in the Chicago area. The LUCFFSI, over the period from 28 September 1990, to 28 July 1995 outperformed local and national indices. The second study compared the performance of the LUCFFSI with the Dow Jones industrial average and Crain's Chicago Stock Index from 28 September 1990 to 28 July 1995. The LUCFFSI increased 94 per cent during that period, compared with 92 per cent for the Dow and 65 per cent for Crain's.

According to family-business enthusiastic practitioners, the main source of family business competitiveness was attributed to their consistent management objectives and long-term strategic view of family owners which restrain *short-termist* opportunistic managerialism, the influence of the family network and culture, the reduced vulnerability to takeovers, and a conservative approach to risk. Subsequently, the aforementioned pioneering studies stimulated academic interests in the affairs of family firms.

The family ownership effect on performance under the microscope
In the widely publicised investigation into US quoted family firms Anderson et al. (2003) examined the relationship between founding-family ownership and firm performance. They established that family shareholding is both prevalent and substantial, as family control was identified in one-third of the S&P 500 and on average it accounted for 18 per cent of outstanding equity. In their models, to resolve the contestable issue of what is family ownership control in the quoted companies, they employed a dummy variable that equated to 1 when founding families hold shares in the firm *or* when founding family members were present on the board of directors.

Contrary to their conjecture, they found family firms to outperform their non-family counterparts, in terms of financial performance metrics. Moreover, their research revealed that the relationship between family shareholding and financial performance is nonlinear and that when family members serve as CEOs, performance is better than with outsiders as CEOs. Such findings come in antithesis to the line that minority shareholders are adversely affected by family ownership. They concluded that family ownership is an effective organizational structure.

In a similar investigation into the financial affairs of publicly quoted companies Anderson and Reeb (2003) established that in general the debt costs of family businesses are lower compared with the debt costs of non-family businesses. They argued that families represent a special class of large shareholders that potentially operate unique incentive structures, have a strong voice in the firm and, of course, powerful motives to steward one particular firm. The unique incentives of founding family owner-managers suggest that they may alleviate agency conflicts between the firm's debt and equity claimants and consequently benefit from lower debt costs.

More recently, Villalonga and Amit (2004) utilizing Fortune 500 data (and using as a threshold of 5 per cent or more of the firm's equity in the hands of blockholders – founders and descendants) have found that family ownership creates value only when the founder is active in the business either serving as the CEO of the family firm or as its chairman with a hired CEO. Their study also revealed that family voting enhancement tactics, such as dual share classes, pyramids, and voting agreements reduce the founder's premium. Paradoxically, they found that family descendants serving as CEOs, tend to destroy firm value. Moreover, family firms run by descendant-CEOs were characterized by more costly agency costs between family and non-family shareholders compared with the agency costs in the owner-manager conflict characteristic of non-family firms.

In the light of the aforementioned review, this explorative investigation aims to address the following questions:

1. What is the role of family control in the UK equity capital markets?
2. Is there a diminishing role of family shareholding in the UK PLC economy?
3. How do UK family-controlled PLCs perform against their peers?

This chapter will only offer the preliminary findings of the MBS-IFB-UBS investigation into the structure and performance of UK family business PLCs. It is planned that the next phase of the research will attempt to examine the generic hypothesis about the positive effect that family ownership has on business performance, and the key parameters governing such family business superiority.

Data and methodology

The review of family business PLC focused literature has enlightened us on research methodologies employed in similar studies. It emerges that any definition used to categorize family-controlled quoted firms will be contestable as the family business economy is very heterogeneous. The role and performance of quoted family business PLCs companies is governed by a number of internal (that is, family role in terms of cash and voting ownership rights, management control and so on) and external environmental-regulative and cultural factors which evolve across countries, capital markets, taxonomies of securities and sectoral distributions.

The definition of family-controlled quoted PLCs

For the purpose of this study, our definition of what is a family-controlled public limited company is based on the key criteria of minimum 10 per cent family ownership control, but the active role of the family at the board level has also been considered.

Family firms These are quoted companies with at least 10 per cent family ownership, that have experienced generational transition and where there is at least one family member on the board. In a couple of cases a partnership involving siblings (founders) has been considered a family business, for example, Antofagasta, CLS Holdings and Goldshield. Also, a more subjective judgement is made that family owner-managers are compelled to sustain the level of family control.

One classic example of a family firm, according to our definition, is Associated British Foods (ABF), where the Weston family has held a major interest since its foundation, and generation after generation is geared to preserve the control of the company. A succession process was recently completed at ABF when George Weston was announced as the replacement for the CEO, Peter Jackson, who was expected to retire in April 2005. Peter Jackson was appointed CEO in 1999 after the ill health of George's father, the late Garry Weston, forced him to step down as CEO.

Patrimonial firms These are defined as companies where the family controls at least 10 per cent of the voting shares, but do not conform fully to the other aforementioned criteria (that is, generational succession and family board involvement). However, there is the intention to keep the business in the hands of the owner family.

One example of patrimonial firm is J Sainsbury plc, whose executive control is no longer with the Sainsbury family. However, the Sainsbury family is still the key shareowner, either directly or as major beneficiaries of family trusts, and there is no evidence that the family might exit the business.

Entrepreneurial firms These are defined as firms where the founders or other individuals (for example, directors after a management buyout – MBO) hold a substantial proportion of shares but there is no clear evidence that the company will be passed on to the next generation of family owner-managers. One example of an entrepreneurial firm is Matalan where the founder and group chairman, John Hargreaves, and his family are the major shareholders but there is uncertainty as to future plans of family perpetuation.

For the purpose of this report the sample of family business PLCs is restricted to family-controlled and patrimonial firms, which can provide the solid base for further investigation involving the longitudinal monitoring of an index of family business capital. There are two the reasons to do so. Such established family companies are less prone to big market capitalization variations or to family exiting the business.

Sampling criteria

The first guideline employed is to restrict the sample to companies that were constituents of the FTSE All-Share Index and have been listed on the London Stock Exchange for at least five years, representing the period 1999–2004. This includes FTSE 100, FTSE 250 and FTSE SmallCap companies but FTSE Fledgling companies were excluded.

The decision to exclude FTSE Fledgling constituents, despite the fact that smaller quoted family firms relatively proliferate in this category, was based on two reasons. First, Fledgling constituents have less than 0.2 per cent of market capitalization compared with the full market capitalization of the FTSE SmallCap and, more importantly, they do not meet the liquidity criteria – turnover of at least 0.5 per cent of shares in issue per month – for their participation in the FTSE All-Share index (see FTSE, 2004, and visit www.ftse.com). The second guideline was to select FTSE-quoted companies where more than 10 per cent of the issued ordinary shares were ultimately owned by a family or by an individual (directly or by means of a family trust or another investment vehicle).

Sources of information

All information regarding share prices and market capitalization for the FTSE indices positions was retrieved from Thomson's Datastream database. The financial information for the construction of the sample of family firms and for the FTSE All-Share constituents was retrieved from the Bureau van Dijk's FAME database. FAME was also used for retrieving the financial data. The FTSE webpage was the main source for information about the UK series of the FTSE Actuaries Share indices. The London Stock Exchange webpage was used to retrieve communications to the market that could help towards the verification of family shareholding. More historical information regarding the role of families in quoted companies was also sourced from Hemscot, the *Sunday Times* Rich List, the *Investors Chronicle* studies and the Stoy Centre for Family Enterprise. Finally, the Institute for Family Business (UK), with its growing network of family business stakeholders, corroborated our classification process.

The profile of the UK family business PLC economy

In an examination of the 687 companies that are constituents of the FTSE All-Share Index, only 48 met the criteria for the classification of a quoted company either as a family-controlled PLC or as a patrimonial business. Thus, the sample of family companies represent only 7 per cent of the number of FTSE-quoted companies, and are categorized in Table 30.1.

It emerges that relatively a higher concentration (10 per cent) of sample family business PLCs, that is, both familial and patrimonial, are in the FTSE 100 category (Table 30.2).

On the other hand, the distribution of FTSE All-Share sample quoted family business PLCs indicates that 50 per cent are categorized in the SmallCap; more specifically 54.8 per cent of family-controlled companies are in the FTSE SmallCap, while 47.1 per cent of patrimonial quoted firms are in the FTSE 250 (Table 30.3).

Table 30.1 Distribution of family business PLCs by category (absolute numbers)

	FTSE 100	FTSE 250	FTSE SmallCap	FTSE All-Share
Family-controlled firms	8	6	17	31
Patrimonial firms	2	8	7	17
Familial and patrimonial	10	14	24	48
FTSE constituents	100	250	337	687

Table 30.2 Distribution of family business PLCs across FTSE indices (percentages)

	FTSE 100	FTSE 250	FTSE SmallCap	FTSE All-Share
Family-controlled firms	8.0	2.4	5.0	5.0
Patrimonial firms	2.0	3.2	2.1	2.0
Familial and patrimonial	10.0	5.6	7.1	7.0
FTSE constituents	100	100	100	100

Table 30.3 Distribution of family business PLCs by category (percentage of FTSE All-Share)

	FTSE 100	FTSE 250	FTSE SmallCap	FTSE All-Share
Family-controlled firms	25.8	19.4	54.8	100
Patrimonial firms	11.8	47.1	41.2	100
Familial and patrimonial	20.8	29.2	50.0	100
FTSE constituents	14.6	36.4	49.1	100

Market capitalization of family business PLCs

The total adjusted market capitalization of family business PLCs in the selected sample is approximately £45 billion (Table 30.4). The market capitalization for family-controlled and patrimonial companies and all FTSE constituents are adjusted by free float factor, as follows: market capitalization = share price × number of ordinary shares issued × free float factor (see ftse.com for methodology).

In terms of capitalization, family-controlled and patrimonial companies represent 3.5 per cent of the market capitalization of the FTSE All-Share constituents (Table 30.5).

Distribution of the capitalization of quoted family business PLCs across the FTSE indices, reveals that FTSE 100 index accounts of 81.2 per cent of the market capitalization (Table 30.6). The discrepancy between the role of quoted family business PLCs, as measured by the percentage of quoted companies (7 per cent) and percentage of market capitalization (3.5 per cent) in relation to the FTSE All-Share, is mainly because they have a higher incidence in the lower ranks (in terms of market capitalization) of each of the FTSE indexations. Taking into account that we are working with adjusted market

Table 30.4 Market capitalization of family-controlled companies per category (£ million)

	FTSE 100	FTSE 250	FTSE SmallCap	FTSE All-Share
Family-controlled firms	22 848.30	2774.20	1870.20	27 492.70
Patrimonial firms	13 624.50	3233.20	586.10	17 443.80
Familial and patrimonial	36 472.80	6007.40	2456.30	44 936.50
FTSE constituents	1 083 169.40	171 782.20	41 635.30	1 296 586.90

Table 30.5 Market capitalization across FTSE indices by category (percentages)

	FTSE 100	FTSE 250	FTSE SmallCap	FTSE All-Share
Family-controlled firms	2.1	1.6	4.5	2.2
Patrimonial firms	1.3	1.9	1.4	1.3
Familial and patrimonial	3.4	3.5	5.9	3.5
FTSE constituents	100	100	100	100

Table 30.6 Market capitalization by category and in relation to the FTSE All-Share (percentages)

	FTSE 100	FTSE 250	FTSE SmallCap	FTSE All-Share
Family-controlled firms	83.1	10.1	6.8	100
Patrimonial firms	78.1	18.5	3.4	100
Familial and patrimonial	81.2	13.4	5.5	100
FTSE constituents	83.5	13.2	3.2	100

capitalizations, it is quite natural that family business PLCs have relatively, on average, a *lower free float multiplier* compared with non-family FTSE constituents, thus contributing lower to adjusted market capitalizations.

The trend, for quoted family/patrimonial firms, to account disproportionately for a lower proportion of market capitalization in relation to the number of quoted family firms in the stock market was also observed in France (Blondel et al., 2001) and Germany (Klein and Blondel, 2002).

Sectoral distribution of quoted family business PLCs

When compared with the distribution of FTSE All-Share constituents in the different business sectors, family and patrimonial businesses present some interesting divergences. The presence of family business PLCs in the highly conglomerated Resources and Utilities sectors is considerably low. Blondel et al. (2001) and Klein and Blondel (2002) found the same trend in France and Germany. Normally, companies in these sectors are

Table 30.7 Sectoral distribution of family business PLCs by SIC (percentages)

	FTSE All-Share	Sample of FB-PLCs
Agriculture, hunting and forestry; fishing	–	0.00
Mining and quarrying	3.41	2.08
Manufacturing	24.78	39.58
Electricity, gas and water supply	2.23	0.00
Construction	4.60	4.17
Wholesale and retail trade; repairs	10.39	18.75
Hotels and restaurants	2.23	4.17
Transport, storage and communication	6.08	6.25
Financial intermediation	26.56	8.33
Real estate, renting and business activities	15.88	14.58
Education	0.15	0.00
Health and social work	0.45	0.00
Other community, social and personal services	3.26	2.08

Table 30.8 Age distribution of family business PLCs (percentages)

Year of incorporation	FTSE All-Share	Sample of FB-PLCs
Prior 1899	7.41	12.50
1900–24	8.44	12.50
1925–49	9.78	18.75
1950–74	13.93	29.17
1975–99	51.11	27.08
2000 +	9.33	*

Note: * this is owing to sampling criteria.

in general former state-owned monopolies characterized with high capital intensity and regulation, which constitute barriers of entry for family business PLCs. It is evident that the family-controlled and patrimonial companies are relatively more active in cyclical consumer goods services (Table 30.7).

Further analysis of the distribution of family business PLCs by SICs demonstrates the high concentration of family and patrimonial companies in manufacturing (about 40 per cent of the sample). The sector is hospitable to more traditional and mature companies.

Age distribution in family business PLCs

There is a significant difference between the age distribution of FTSE All-Share constituents and that of the sample of family and patrimonial companies (Table 30.8).

According to the age distribution analysis – based on year of incorporation – it is evident that family-controlled/patrimonial companies tend to be older that mainstream FTSE All-Share companies. It can also be argued that family capitalism in the UK is relatively more active in traditional industries, with the new wave of family firms exhibiting less enthusiasm for flotation on the main market.

Concentration of family share ownership

The share ownership of families in the sample of family-controlled quoted companies (Table 30.9) tends to be low, with more than 70 per cent of the companies having less than 40 per cent of ownership concentrated in the hands of a family and more than 50 per cent of the companies having less than 25 per cent of family ownership in the hands of a family.

This might be symptomatic of the factors (offering additional rounds issued shares, coupled with strict regulation governing minority protection, takeovers, and so on) conditioning the diminishing family ownership as suggested by Franks et al. (2003).

Dividend yield and P/E ratio

Overall, there are no statistically significant differences between the dividend yields and P/E ratios of family business PLCs compared to those of FTSE constituents (Tables 30.10 and 30.11).

Table 30.9 Distribution of family business PLCs by shareholding (percentages)

Family control – % of issued ordinary shares	Sample of family business PLCs
10–25	51
25–40	19
40–55	17
55–70	11
70–100	2

Table 30.10 Dividend yields of family controlled (FC) and patrimonial companies (PC)

	FTSE 100		FTSE 250		FTSE SmallCap		FTSE All-Share	
	FCs	PCs	FCs	PCs	FCs	PCs	FCs	PCs
% average – FB-PLCs	2.36	4.04	2.59	3.34	2.89	2.32	2.68	3.00
% average – FTSE All	3.32		2.65		2.46		2.66	
t-statistic	−1.16	0.36	−0.17	1.38	1.77	−0.22	0.16	−0.99
Significance	0.31	0.78	0.87	0.21	0.08*	0.83	0.88	0.33

Table 30.11 P/E ratios of Family Controlled (FC) and Patrimonial Companies (PC)

	FTSE 100		FTSE 250		FTSE SmallCap		FTSE All-Share	
	FCs	PCs	FCs	PCs	FCs	PCs	FCs	PCs
Average – FB-PLCs	16.27	13.90	17.66	11.91	11.84	15.66	14.18	13.69
Average – FTSE All	15.58		16.06		16.72		16.30	
t-statistic	0.28	−0.29	0.19	−1.79	−2.48	−0.17	0.84	−0.95
Significance	0.81	0.82	0.86	0.10	0.02*	0.87	0.41	0.35

The only exception relates to lower dividend yield and P/E ratios for family-controlled companies categorized in the FTSE SmallCap, which are statistically different at 10 per cent and 5 per cent level of significance, respectively.

Although not statistically confirmed, seemingly there is a tendency for the P/E ratios of family/patrimonial companies to be lower than the P/E ratios for FTSE All-Share constituents. This is again symptomatic of the poor market confidence in the long-term performance for family-controlled and patrimonial companies, perhaps owing to their renowned family control concerns and culture of restrained growth.

Comparative analysis of balance sheet structures

Table 30.12 is the balance sheet structure comparative analysis of our sample family-controlled and patrimonial companies compared with the average of FTSE All-Share companies and with that of a filtered list of FTSE All-Share companies.

The filtered list eliminates financial institutions which tend to have different asset and capital structures. All variables calculated represent the average for the 1999–2003 period

Table 30.12 Balance sheets of family business PLCs versus FTSE companies

	Sample of FB-PLCs	FTSE All-Share	t stats	Sign. levels	FTSE All-Share*	t stats	Sign. levels
Fixed assets							
Tangible assets	42.25	25.98	7.546	0.000	35.67	3.016	0.003
Intangible assets	10.22	11.80	−1.272	0.205	14.97	−3.745	0.000
Total fixed asset	*59.87*	*59.37*	*0.264*	*0.792*	*53.77*	*3.192*	*0.002*
Current assets							
Stock and WIP	10.42	8.60	1.821	0.070	12.47	−1.989	0.048
Trade debtors	12.78	10.75	1.794	0.074	14.11	−1.169	0.243
Bank and deposit	7.59	8.60	−1.482	0.139	9.80	−3.139	0.002
Total current assets	*40.13*	*40.63*	*−0.264*	*0.792*	*46.23*	*−3.192*	*0.002*
Current liabilities							
Trade creditors	11.43	8.06	3.702	0.000	10.64	0.862	0.389
Short-term loan	5.30	5.41	−0.226	0.822	5.99	−1.401	0.162
Total current liabilities	*31.51*	*29.31*	*1.523*	*0.129*	*33.77*	*−1.567*	*0.118*
LR liabilities							
LR loan	18.35	14.68	2.929	0.004	17.21	0.895	0.372
Other LR liabilities	5.43	6.92	−2.084	0.038	5.26	2.625	0.811
Total LR liabilities	*24.84*	*22.08*	*1.883*	*0.061*	*23.10*	*0.239*	*0.239*
Capital and reserves							
Issued capital	5.90	7.32	−2.659	0.008	7.34	−1.179	0.009
Share premium	10.62	21.90	−8.576	0.000	26.24	−10.171	0.000
Revaluation reserves	5.93	2.08	5.675	0.000	2.59	4.879	0.000
Retained profits	17.07	1.90	5.472	0.000	1.59	5.199	0.000
Shareholder funds	*43.65*	*48.61*	*−2.635*	*0.009*	*43.12*	*0.280*	*0.780*

Notes:
* Excluding financial institutions.
Rounding of figures explains discrepancies.

and are expressed as a percentage of total assets. In summary, the comparative analysis of the financial structures of family-controlled companies (familial and patrimonial) versus that of their FTSE All-Share Index PLCs demonstrate the following key statistically significant differences:

- Family business PLCs have the tendency to invest more in tangible assets – they are often regarded as symbol of financial autonomy – as they can be used as collateral for external debt.
- Family business PLCs tend to use more long-term loans – perhaps because they can secure better deals and, more importantly, they can refrain from issuing more external equity at the cost of family control.
- Evidently, and in line with the pecking order hypothesis, family business PLCs issue relatively lower share capital, including additional rounds, and of course are more prudent with profits – as they enthusiastically reinvest it.
- The higher 'Revaluation reserves' could be due to keeping tangibles up to market values in order to comfortably command external debt.

Performance of family business PLCs versus FTSE companies

The performance of family business PLCs have been compared to that of the list of FTSE All-Share constituents, which again has been filtered to exclude financial institutions (Table 30.13). The analysis is based on performance parameters covering the period 1999–2003.

In summary, the following key statistically significant results emerge:

- Family business PLCs exhibit a lower growth rate, in terms of sales and assets, compared with their mainstream FTSE All-Share companies (although not statistically confirmed).

Table 30.13 Comparative analysis of growth and profitability

	Sample of FB-PLCs	FTSE All-Share	t stats	Sign. levels	FTSE All-Share*	t stats	Sign. levels
% growth rates							
Sales	12.40	19.11	−2.41	0.02	17.95	−2.01	0.05
Employment	8.92	11.54	−1.27	0.21	11.36	−1.17	0.24
Assets	11.24	17.44	−2.82	0.01	18.84	−3.38	0.00
Profitability ratios							
Return on total asset	7.71	2.76	6.92	0.00	4.87	3.97	0.00
Return on total equity	19.30	9.81	5.03	0.00	13.35	2.91	0.00
Return on capital employed	11.28	5.64	2.34	0.02	8.34	1.19	0.23
Profit margin	10.59	8.01	2.76	0.01	7.69	3.08	0.00
Gearing ratio	79.61	147.02	−6.85	0.00	115.26	−3.55	0.00

Note: * Excluding financial institutions.

- Interestingly, they are more profitable in each of the performance parameter (although not statistically confirmed for ROCE when compared with the filtered list of companies).
- Family business PLCs are less leveraged, as their gearing ratio (defined as short-term loans and overdraft + long-term liabilities over shareholder's funds) is considerably lower. These results indicate that, in general, sample family controlled and patrimonial companies are characterized by 'cash cow' features as they tend to be more mature, slow in growth and yield superior profitability margins.

The research team envisaged computing multivariate econometric models to examine the impact of family control on performance; and on the level of agency costs. It emerges that the level of 'director's remuneration' as a percentage of 'gross profit' in family-controlled and patrimonial companies is much lower than that of FTSE All-Share constituents. This is an indication that agency costs in family-controlled quoted companies is lower – the results are in line with postulations of some scholars such as (Chrisman et al., 2004; Jensen and Meckling, 1976). More recently, Villalonga and Amit (2004) suggest that the the conflict between family and non-family shareholders in descendant-CEO firms is more costly than the owner-manager conflict in non-family firms.

The UK family business index

The final part of Phase A of the investigation deals with the computation of a family business index. Using a basket of ordinary shares representing family business PLCs, the aim is to compare the growth in capitalization of family business PLCs against the FTSE All-Share Index during 1999–2004.

Previous studies have indicated that family-controlled quoted companies consistently outperformed the market in terms of share price performance. The Stoy Hayward/BBC Family Business Index has beaten the FT All-Share Index by nearly 30 per cent over a period of 20 years (1970–91). In 1990, the Pitcairn Financial Management Group of Philadelphia, in a comparative study of US quoted family-controlled companies, again concluded that the cumulative returns for family companies over a period of 20 years were more than double that of mainstream Standard and Poors 500.

Since early 1990s, the FTSE All-Share Index replaced the FT All-Share Index and the methodology for the index computation has been revised. Besides that, the landscape for quoted family firms has changed dramatically, given the increased mergers and acquisition activity, tightening of capital market regulations, and the economy enduring different cycles that certainly has put to test the market attractiveness of certain family business PLCs. In the light of these changes, a family business index is developed and its performance is benchmarked against FTSE UK indices covering (at this phase of the project) the years 1999–2004. *The aim is to test whether the superior share performance of UK family quoted companies still holds.*

Indexation of the Capitalization of family business PLCs – the methodology

In fact, two versions of the Index for Family Business were formulated: one with the 30 securities representing higher market capitalization weights and the other with all 50 securities (from a sample of 48 family-controlled companies), labelled FB 30 and FB All-Share, respectively.

The selection of quoted family companies follows the methodology outlined before in the discussion of criteria for the categorization of family-controlled and patrimonial companies. For the selection of the 30 securities that are constituents of the FB 30 index, the securities were ranked at each period (on a weekly period) according to their absolute market capitalization in relation to the sum of absolute market capitalizations of all securities in the sample. This ranking was calculated for each week in the five-year period from 1 October 1999 to 1 October 2004.

The composition of the FB 30 index was reassessed at the end of each interval of six months (reassessment dates: 31 March 2000, 29 September 2000, 30 March 2001, 28 September 2001, 29 March 2002, 27 September 2002, 28 March 2003, 26 September 2003, 26 March 2004, 21 September 2004). The securities chosen for each of the periods were those that featured in the top 30 positions during the previous six-month period (understandably the selection for the first period, was based on that period itself).

Market capitalization adjusted for free float For the calculation of adjusted market capitalization, the free float factor was taken into account. The free float factor represents the proportion of shares for a specific security that is readily available for trading, that is, is not held by a major shareholder (for example, a family).

The case of Associated British Foods is used by the FTSE (2004) to exemplify the use of free float adjustments:

> Free float is not purely restricted to which listed companies own what proportion of other listed companies but also take into consideration interests held by other parties. An example of this case could be Gary Wesland,[1] who owns 53% of Associated British Foods. This would lead to a possible free-float of 50%

To be consistent in the benchmarking with the FTSE UK indices, the time series of free float factor for each of the securities was derived from information available in the FTSE website. For the latest free float factors, the 'FTSE UK Index Constituent Rankings' document was used. For each company under consideration, the information regarding number of ordinary shares issued and fully paid was extracted from the notes to the latest accounts available. The number of shares of each company was then multiplied by the share price available in the FTSE document, resulting in the absolute market capitalization of the company. The free float factor was then determined dividing the market capitalization informed in the FTSE document (adjusted for free float) by the market capitalization resulting from the calculation.

The changes in the free float factor throughout the period under analysis were determined from the annual notes released by the FTSE with the modifications made to the UK Actuarial Series during each year. The dates (in the weekly series) in which there were changes to the free float factor for the securities under consideration were: 15 June 2001, 17 May 2002, 21 June 2002, 21 March 2003, 19 September 2003, 19 December 2003, and 17 September 2004. With the time series of free float factor in hand and the matching time series of absolute market capitalization for each of the securities obtained from Datastream, the adjusted market capitalizations were determined.

Computation of the FB 30 and FB All-Shares The value of each of the indices was determined weekly by the division of the sum of market capitalizations (adjusted for the free float factor) by a divisor. The first divisors were calculated so that indices could start at 100. Thereafter, every date with changes in a free float factor or with changes to the basket of companies in the index (in the case of FB 30) called for a new divisor to be calculated for the next week. The new divisor was calculated by simply dividing the week's total market capitalization by the value of the index in the previous week.

The UK Index for Family Business versus FTSE indices

Figure 30.1 illustrates the comparative performance of the growth in the capitalization of family-controlled and patrimonial quoted companies as represented by FB 30 and FB All-Shares during the 1999–2004 period. The graph shows that family-controlled and patrimonial companies performed quite well in terms of share price/market capitalization during the five-year period, with an increase of around 30 base points for the FB 30 index and around 40 for the FB All-Shares index.

Finally, Figure 30.1 reveals that family-controlled and patrimonial quoted companies outperform mainstream FTSE counterparts by up to 80 per cent. More specifically, FB All-Shares does better than the FTSE All-Shares by about 50 per cent.

Further investigation is warranted in order to establish the factors that govern the outperformance by family business PLCs:

- To what extent do smaller family-controlled quoted companies, perhaps with more concentrated family shareholding and active founding family business entrepreneurs, have a tendency to perform better, that is, in terms of share price movement or share capital issuance?
- Does the methodology used to select companies for the FB 30 Index, which is based on the ranking in the preceding six months, tend to influence the comparatives?

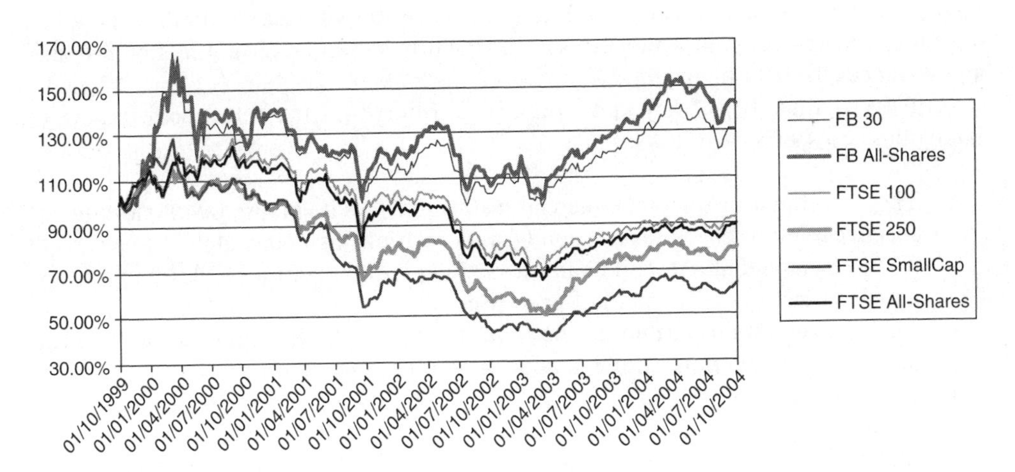

Figure 30.1 Performance of FB 30 and FB All-Shares versus FTSE indices

Conclusion

This investigation provided evidence about the demographic profile, financial structure and performance of UK family-controlled quoted companies (familial and patrimonial PLCs). The share price/capitalization performance of constituents of the UK family business PLC economy, was established via the computation of the FB 30 and the FB All-Shares, which were benchmarked against the FTSE UK indices. Evidence suggests that family business PLCs, overall outperform their FTSE counterparts.

In a previous explorative investigation, the research team, established that controlling for sectoral effect the share performance of family-controlled companies can be a mixed blessing: generally a third of family-controlled companies outperforms, another third underperforms and the remaining companies registering similar performance. The message is that, although evidence about the outperformance of family business PLCs can offer ammunition to the defenders of family capitalism, more research is needed to corroborate some of the findings.

What are the factors driving family-controlled quoted firms to outperform their counterparts?

Scholars have been arguing the family firms possess dynamic capabilities, value-adding familiness and access to idiosyncratic social capital advantages which fuel their competitive advantage. The competitiveness of family firms, has been accounted for in the context of social capital (Steier, 2001) and resource based view frameworks (Habbershon et al., 2003) and the dynamic capabilities perspectives (Salvato and Melin, 2003).

This edge is more evident when the economy and capital markets underperform and relatively suppressed when market conditions are buoyant. Stein (1989) demonstrated that quoted companies with shareholders characterized with longer investment horizons suffer less from managerial myopia and opportunism, as they are less likely to forgo good investments for the sake of boosting short-term profits. James (1999) argued that family firms, because of family commitment to perpetuating ownership onto succeeding generations, provides inherent incentives to invest more efficiently and prudently according to the market rules. Anderson et al. (2003) found that one implication of families maintaining a long-term presence in family firms is that the firm will enjoy certain economies, such as a lower cost of debt financing.

Tentatively, a number of key factors can be attributed for the better performance of family business PLCs, as follows:

- They develop as a socio-commercial institution – well endowed with elements of social capital such as trust, bonds and relationships; the ability to marshal resources/capital and to build alliances; strategic and operational flexibility; transferable knowledge and a pool of expertise.
- They concentrate in certain business sectors; this sectoral congregation is evident mainly in construction, retailing, and food-processing.
- They are tuned to the long-term view as epitomized by the Morrison's philosophy – avoidance of too much innovative and growth adventures, such as going online, expanding abroad, and so on. Instead, they cherish traditional practices, such as investing in customer service, and building up an enviable land bank.

- They are into niche-orientated market penetration and reap the economies of specialization (the Clinton Cards phenomenon).
- They can tap into family wealth and easily use family assets to secure competitively priced financial solutions.
- They can manage to limit the negative implications of agency problems.

While this five-year retrospective benchmark provides valuable insight into the outperformance of the FTSE indices by quoted family business PLCs, there is scope to broaden the study on many parameters:

- Extend retrospectively the study, in order to allow the comparative analysis to incorporate more business cycles, as it is believed that family companies do not do well relatively – because of restrained growth culture during booming economic conditions; and do not suffer heavily during recessionary periods – because of financial prudence and familiness-based capabilities.
- Extend the study in the future in order to update the basket of family business securities with new initial public offerings (IPOs) as the economics of the new wave of quoted family firms should be carefully examined across equity markets. Evidence from Germany and Spain, reveal that family business IPOs, underperform when compared with the performance of mainstream IPOs (Jaskiewicz et al., 2005).
- Consider the building of other family business indices to represent smaller quoted family firms active in the Fledgling market and other secondary equity markets, such as AIM and OFEX. Broadening the database of family business PLCs will allow more rigorous statistical analysis, which can allow the control of the interplay of certain covariates such as ownership regime, sector, governance mechanisms and so on.

This investigation, in order to overcome some inherent dilemmas (for example, definition of what constitute a family controlled firm), had to experiment with certain parameters; therefore certain dimensions and methodological caveats need revisiting:

- The survivors bias – the performance of the family business index does include the fate of de-listings and companies that were taken private.
- There is a need to further explore the ownership regime of sample companies. How do family owners gain ownership control rights in addition to cash rights?
- Do they operate pyramidal structures and cross-shareholdings via multiple control chains? Family owners as a large shareholder may exercise control over a firm through a chain of intermediate firms.
- Do they use classes of shares which can allow the multiplying of voting power?
- Will the adoption of the more sophisticated F-PEC measurements for the selection and classification of family firms prove more effective?
- What evidence can we get to justify family business outperformance? Other areas which deserve further investigation include transaction costs, agency costs, corporate governance issues and aspects of familiness.
- The documentation of representative studies for the paradigmatic and problematic family business PLCs, will contribute towards the demystification of certain trends, such as the chronic fall in the number of family business PLCs.

The role of family companies in capital markets – a longitudinal picture
Despite the outperformance of family-controlled quoted companies, evidence suggests that there is a diminishing role for family business PLCs in certain countries (for Germany, Klein and Blondel, 2002; and for Spain, Navarro and Ansón (2004) . For the case of the UK family capitalism, Franks et al. (2003) explored the dynamics of the UK capital markets and provided historical evidence and convincing arguments about the demise of the British family business PLC economy as epitomized by the dilution of family ownership in the twentieth century. The main driver for this trend has been the issuance of equity shares by family-controlled companies in order to finance acquisitive growth. This financial practice occurred in the absence of minority investor protection and, normally, families were able to retain control by occupying a disproportionate number of seats on the boards of firms. Over time, with smaller stakeholdings, rising hostile takeovers, demanding institutional shareholders and increased capital market regulation and takeover reforms, families found it very challenging to sustain control. Thus, while acquisitions facilitated the growth of family-controlled firms (in the first half of the twentieth century), it resulted in dilution of their ownership and, ultimately, loss of control (in the second half of the twentieth century) – this period was marked by more capital market regulations geared to offer protection to minority shareholders.

It has been argued that the prevalence of quoted family business is diminishing. In recent years, we read about public-to-private deals involving family companies, for example William Jackson & Son Limited of Hull and Silentnights; the sale of family-controlled firms, for example Weetabix, Brake Bros, and so on, and the struggling state of affairs of Courts (just gone into receivership) and Moss Bros, which undermines the leading role of the controlling families. In order to establish some longitudinal trends charactering the UK family business PLC, Poutziouris (2006) tracked the 1990 BDO Stoy list of family-controlled companies active in primary and secondary equity markets. The key finding of the research into the fate of the 'family controlled, patrimonial and entrepreneurial' quoted companies is that the primary factor governing the shrinking role of families in PLCs is takeovers (42 per cent of family-controlled companies have been taken over). This longitudinal trend epitomizes the recent demise of the UK family business economy, and constitutes the main factor explaining why the great majority of family business owner-managers are sceptical about the flotation route.

Nonetheless , there is scope to launch an investigation into the Fledgling and secondary equity markets in order to round up the inquiry into their financial structure and performance of UK family-controlled companies.

Notes

* The MBS research team benefited from the financial support of the Institute for Family Business (UK), and UBS-Wealth Management which are hereby gratefully acknowledged.

1. Gary Wesland is a reference to Garfield Howard Garry Weston who died in 2002. The share ownership of ABF is kept under the family investment company called Wittington Investments.

References

Anderson, R. and Reeb, D. (2003), 'Founding-family ownership and firm performance: evidence from the S&P 500', *Journal of Finance*, **58**(3), 1301–28.

Anderson, R.C., Mansi, S.A. and Reeb, D.M. (2003), 'Founding family ownership and the agency cost of debt', *Journal of Financial Economics*, **68**, 263–355.

Ang, J., Cole, R. and Lin, J. (2000), 'Agency costs and ownership structure', *Journal of Finance*, February (1), 81–106.
Astrachan, J. and M.C. Shanker (2003), 'Family businesses' contribution to the US economy – a closer look', *Family Business Review*, **16**(3), 211–19.
Bebchuck, K., Kraakman, R. and Triantis, G. (2000), 'Stock pyramids, cross-ownership and dual class equity: the mechanisms and agency costs of separating control from cash flow rights', in R. Morck (ed.), *Concentrated Corporate Ownership*, Chicago, IL: University of Chicago Press, ch.10.
Becht, M., Chapelle, A. and Renneboog, L. (2001), 'Shareholding cascades: the separation of ownership and control in Belgium', in F. Barca and M. Becth (eds), *The Control of Corporate Europe*, Oxford: Oxford University Press.
Blondel, C., Rowell, N. and Van der Heyden, L. (2001), 'Prevalence of patrimonial firms in the SBF 250: evolution from 1993 to 1998', INSEAD, France.
Business Week (2003), 'Defining family – how did BW come up with its list? Sometimes it wasn't easy', *Business Week*, 10 November, 111–14.
Chrisman, J.J., Chua, J.H. and Litz, R.A. (2004), 'Comparing the agency costs of family and non-family firms: conceptual issues and explorative evidence', *Entrepreneurship Theory and Practice*', Summer, 355–54.
Crespí-Cladera, R. and García-Cestona, M. (2001), 'Ownership and control of Spanish listed firms', in F. Barca and M. Becht (eds), *The Control of Corporate Europe*, Oxford: Oxford University Press.
Daily, M. and Dollinger, M.J. (1992), 'An empirical examination of ownership structure in family and professionally managed firms', *Family Business Review*, **5**(2), 117–36.
Ehrhardt, O. and Nowak, E. (2003), 'The Effect of IPOs on German Family-Owned Firms: Governance Changes, Ownership Structure, and Performance', *Journal of Small Business Management*, **41**(2), 222–32.
Faccio, M. and Lang, L. (2002), 'The ultimate ownership of western European corporations', *Journal of Financial Economics*, **65**, 365–95.
Franks, J., Mayer C. and Rossi, S. (2003), 'Spending less time with the family: the decline of family ownership in the UK', Working Paper, London Business School.
FTSE (2004), 'Guide to calculation methods for the UK series of the FTSE Actuaries Share Indices', *FTSE – the Independent Global Index Company*, February.
Gallo, M.A. and Vilaseca, A. (1996), 'Finance in family business', *Family Business Review*, **9**, 387–402.
Gallo, M.A. and Villaseca, A. (1998), 'A financial perspective on structure, conduct, and performance in the family firm: an empirical study', *Family Business Review*, **11**(1), 35–47.
Galve, C. and Salas, V. (1996), 'Ownership structure and firm performance: some empirical evidence from Spain', *Managerial and Decision Economics*, **17**(6), 575–86.
Gomez-Mejia, L., Nuñez-Nickel, M. and Gutierrez, I. (2001), 'The role of family ties in agency contracts', *Academy of Management Journal*, **44**, 81–95.
Habbershon, T.G., Williams, M.L. and MacMillan, I. (2003), 'A unified systems perspective of family firm performance', *Journal of Business Venturing*, **18**(4), 451–65.
Hendry, J. (2002), 'The principal's other problems: honest incompetence and management contracts', *Academy of Management Review*, **27**, 98–113.
Investors Chronicle (2003), 'Shut up and get rich', *Investors Chronicle*, August.
James, H.S. (1999), 'Owner as manager, extended horizons and the family firm', *International Journal of the Economics of Business*, **6**(1), 41–55.
Jaskiewicz, P., González, V.M., Menéndez, S. and Schiereck, D. (2005), 'Long-run IPO performance analysis of German and Spanish family-owned businesses', *Family Business Review*, **18**(3), 179–202.
Jensen, M. and Meckling, W. (1976), 'Theory of the firm: managerial behaviour, agency costs and ownership structure', *Journal of Financial Economics*, October, 305–60.
Klein, S.B. and Blondel, C. (2002), 'Ownership structure of the 250 largest listed companies in Germany', Working Paper, INSEAD.
La Porta, R., Lopez-de-Sinales, F. and Shleifer, A. (1999), 'Corporate ownership around the world', *Journal of Finance*, **54**(2), 471–517.
McConaughy, D., Mendoza, D. and Mishra, C. (1996), 'Loyola university Chicago family firm stock index', *Family Business Review*, **9**, 125–37.
McConaughy, D.L., Walker, M.C., Henderson Jr, G.V. and Mishra, C.S. (1998), 'Founding family controlled firms: efficiency and value', *Review of Financial Economics*, **7**(1), 1–19.
Megginson, W.L. (1997), *Corporate Finance Theory*, Reading, MA: Addison-Wesley.
Miller, KarenLowry (2004), 'Thomson financial: Europe's best companies', *Newsweek*, 12 April.
Morck, R. (ed.) (2000), *Concentrated Corporate Ownership*, Chicago, IL: University of Chicago Press.
Morck, R. and Yeung, B. (2003), 'Agency problems in large family business groups', *Entrepreneurship Theory and Practice*, **27**(4), 367–82.
Morck, R., Shleifer, A., Vishny, R.W. (1988), 'Management ownership and market valuation: an empirical analysis', *Journal of Financial Economics*, **20** (January–March), 293–323.

Myers, S.C. (1984), 'The capital structure puzzle', *Journal of Finance*, **39**(3), 575–92.
Navarro, M. and Ansón, S. (2004), 'Ultimate owners of the Spanish listed firms: the role of families', in S. Tomaselli and L. Melin (eds) (2004), *Family Firms in the Wind of Change*, Research Forum Proceedings, Book of Research Forum Proceedings, 15th Annual Family Business Network World Conference, FBN-IFERA Copenhagen Publication.
Pagano, M., Panetta, F. and Zingales, L. (1996), 'The stock market as a source of capital: some lessons from initial public offerings in Italy', *European Economic Review*, **40**, 1057–69.
Pagano, M., Panetta, F. and Zingales, L. (1998), 'Why do companies go public? An empirical analysis', *Journal of Finance*, **53** (1), 27–64.
Pettit, R. and R. Singer (1985), 'Small business finance: a research agenda', *Financial Management*, Autumn, 47–60.
Poutziouris, P. (2001), 'The views of family companies on venture capital: empricial evidence from the UK small to medium-size enterprising economy', *Family Business Review*, **14**(3), 277–91.
Poutziouris, P. (2006), *The UK Family Business Economy*, London: Institute of family Business (UK-UBS Wealth Management Publication).
Poutziouris, P., Chittenden, F. and Michaelas, N. (1998), *The Financial Affairs of Private Companies*, Tilney Fund Management Publication, June.
Poutziouris, P., Chittenden, F. and Michaelas, N. (1999), *The Financial Development of Smaller Private and Public Ltd Companies*, Liverpool: Tilney Fund Management Publication, September.
Poutziouris, P., Sitorus, S. and Chittenden, F. (2002), *The Financial Affairs of Family Companies*, London: Sand Aire Private Equity and Grant Thornton.
Romano, C.A., Tanewski, G. and Smyrnios, K.X. (2000), 'Capital structure decision making: a model for family business', *Journal of Business Venturing*, **16**, 285–310.
Salvato, C. and Melin, L. (2003), *Competitive Advantages of Family Businesses: A Dynamic Capabilities Perspective*, Research Forum Proceedings of the 14th FBN World Conference, 319–36.
Schulze, W.S., Lubatkin, M.H. and Dino, R.N. (2003a), 'Exploring the agency consequences of ownership dispersion among the directors of private family firms', *Academy of Management Journal*, **46**(2), 174–94.
Schulze, W.S., Lubatkin, M.H. and Dino, R.N. (2003b), 'Toward a theory of agency and altruism in family firms', *Journal of Business Venturing*, **18**(4), 473–90.
Schulze, W.S., Lubatkin, M.H., Dino, R.N. and Buchholtz, A.K. (2001), 'Agency relationships in family firms', *Organization Science*, **12**(2) March/April, 99–116.
Schürmann, W. and Körfgen, K. (1997), *Familienunternehmen auf dem Weg zur Börse*, 3rd edition, München.
Steier, L. (2001), 'Next-generation entrepreneurs and succession: an exploratory study of modes and means of managing social capital', *Family Business Review*, **14**(3), 259–76.
Stein, J. (1989), 'Efficient Capital Markets, Inefficient Firms: A Model of Myopic Corporate Behaviour', *Quarterly Journal of Economics*, **103**, 655–69.
Stoy Hayward/BBC (1992), *Family Business Index*, London: Stoy Hayward.
Thompson Financial (2001), *Major UK Companies – Handbook*, October, University Press.
Villalonga, B. and Amit, R.H. (2004), 'How do family ownership, control, and management affect firm value?', EFA 2004 Maastricht Meeting, Paper No. 3620.
Zingales, L. (1994), 'The value of a voting right: a study of the Milan stock exchange', *Review of Financial Studies*, **7**, 125–48.

31 Ownership structure and firm performance: evidence from Spanish family firms*

Susana Menéndez-Requejo

This chapter explores the relationship between founding-family ownership and firm performance. Starting from the agency theory approach, the potential benefits and costs of family ownership, and their influence on firm performance, are considered. The database under investigation constitutes a 8000 large and medium size Spanish firms. Several univariate analyses are conducted, as well as a cross-sectional analysis and a data envelopment analysis (DEA) in order to compare family and non-family firms performance. The empirical analysis shows that Spanish family firms perform better, in terms of return on equity, than non-family firms of the same size and in the same industry. Family involvement in management of the firm does not prove to have a positive impact on firm performance.

Introduction

Founding-families are a special class of shareholder, because often they hold poorly diversified portfolios but, at the same time, they are long-term investors, as their aim is to bequeath the firm to the next generations (Anderson and Reeb, 2003). In this context, the aim of this chapter is to evaluate the effects of family governance and ownership on firm performance, starting from the agency theory approach:

1. On the one hand, having large shareholders such as founding families, could negatively influence firm performance. Founding families have concerns and interests of their own, such as stability and capital preservation, compensation, related-party transactions, nepotism in manager selection or special dividends, that may not align with the interests of other investors in the firm. Large shareholders have the incentives and the power to pursue objectives like firm growth, firm survival or in general taking decisions that benefit themselves at the expense of firm value. According to these arguments, family-controlled firms should have poorer performances than firms with a dispersed ownership.
2. On the other hand, concentrated shareholders can mitigate managerial expropriation. For instance, large non-diversified equity positions and control of management favours monitoring managers in family firms and diminishes agency conflicts between owners and managers while facilitating firm value maximization. Founding families that hold management positions can better align the manager's interest with those of the owners. Moreover, the family owners interest in maintaining family reputation and firm control also reduces agency conflicts with suppliers and creditors, given that the same governance supports longer relationships. Taking into account these arguments, large investors and especially family firms have incentives to maximize firm performance.

As there is potential benefits and costs of family ownership, the family's influence on firm performance is an empirical issue, which is explored in this study. Starting from these approaches, several univariate analyses are conducted, as well as a cross-sectional analysis and a DEA, in order to compare family and non-family firms performance. The impact of large equity blockholders on firm performance is examined, in particular, whether founding-family presence hinders or facilitates it.

The database includes all the large and medium size Spanish firms in 2002 (the last year available) (SABI database), which is around 8000 firms. First, the ownership and managerial structure of each firm are examined, in order to classify firms according to their ownership. The use of this database is a differential point in this chapter, since, the empirical analysis in relation to Spanish family firms usually takes as its database the firms quoted on the Spanish stock market or in the Spanish Business Strategy Survey. In the first case, we should take into account that Spanish public firms represent a very small group of firms, since there were just 151 non-financial firms quoted on the Spanish stock market in 2002, versus 8018 large and medium size non-financial firms that there were in Spain in 2002, which is the database that we analyse. The Spanish Business Strategy Survey database, which considers a sample of 1600 firms instead of the whole population, classifies firms according to the answer to the question in relation to if the firm has family owners as managers, directors or in other positions. However, in this study, the firms are classified into family and non-family, beginning with the ownership and management structures analysis, one firm after another.

Starting with the firm's financial statements, different methodologies are followed in order to test the influence on firm performance of the ownership structure and control characteristics of the firms, distinguishing between family and non-family firms. This is another point that differentiates this chapter, that is, the comparison of the results of univariate analyses, matched-pairs analysis, regression analysis and DEA. Active and passive family control of the firm are pointed out (in relation to the presence of family members as CEOs) and the family generations involved in the firm.

The remainder of this chapter is organized as follows. First, the arguments on the impact of family ownership on firm performance are presented. Next, the methodology is shown, in relation to the data sample and the statistical procedures, presenting the results of univariate analysis, regression analysis and DEA, and finally, the conclusions are presented.

Family business performance

Potential negative influence of family ownership on firm performance

Founding families have concerns and interests of their own, such as stability and capital preservation, excessive compensation, related-party transactions or special dividends, that may not align with the interests of other investors of the firm (Anderson and Reeb, 2003; DeAngelo and DeAngelo, 2000). Large owners in firms may extract private benefits from the firm, owing to their superior control rights, pursuing their own interests but not the firm value maximization for all the shareholders (Demsetz, 1983; Fama and Jensen, 1983; Shleifer and Vishny, 1997). In this case, the investment decisions can be influenced by their personal utility instead of by market value maximization (Fama and Jensen, 1985).

Large shareholders, such as founding families, have the incentives and the power to pursue objectives such as firm growth, firm survival or in general to take decisions that benefit themselves at the expense of firm value.

Furthermore, nepotism and lack of professionalization can characterize family businesses, when they prefer family members as CEOs instead of more capable, qualified and talented outside professional managers. The family firms bias towards family members entering the business as managers can result in suboptimal investments and lower profitability, that is, negatively influencing firm performance. Besides, nepotism in family firms negatively influences senior non-family executives, if they feel that merit and talent are not considered to reach top management positions.

Following these proposals, family-controlled firms are expected to take suboptimal decisions, resulting in poorer performance than firms with a dispersed ownership. In this case, family ownership would be a less effective organizational form.

Some empirical studies show poorer performance for family firms. Morck et al. (1988) obtain that continued founding-family ownership in US corporations is an organizational form that leads to poor firm performance. Faccio et al. (2001) report that family control leads to wealth expropriation in the presence of less than transparent financial markets, in their case, the East Asian firms (corporate governance and the political-regulatory environment). Pagano and Roell (1988) note how the presence of other large blockholders can reduce the concerns of controlling shareholder wealth expropriation.

Potential positive influence of family ownership on firm performance

Large concentrated shareholders can mitigate managerial expropriation (Demsetz and Lehn, 1985). For instance, large undiversified equity position and long-term relationships between the family and the managers, diminish agency conflicts between owners and managers, and facilitate firm value maximization (Anderson and Reeb, 2003; DeAngelo and DeAngelo, 1985; Fama and Jensen, 1983; Jensen and Meckling, 1976; McConaughy et al., 1998). In addition, founding families sometimes hold management positions, aligning the manager's interest with those of the owners (family). In these cases, the need for costly monitoring by outside shareholders is reduced, increasing firm value.

Moreover, the family owners' interest in maintaining control of the firm in the long term also reduces agency conflicts with suppliers and creditors, given that the same governance supports longer relationships. The stability in company control and family reputation can even determine lower cost of debt financing compared with non-family firms (Anderson et al., 2003).

In addition, it must be taken into account that firm survival is a principal concern for families, given their aim to pass the firm on to succeeding generations. Consequently, family businesses have longer investment horizons, leading to greater investment efficiency (James, 1999). Dispersed ownership favours shorter managerial horizons.

Family influence can also provide competitive advantages to the firm (Burkart et al., 2002), in particular, family CEOs can bring special skills and attributes to the firms that outside managers do not possess (Anderson and Reeb, 2003), owing to their identification with firm values, reputation and performance. McConaughy et al. (1998) find that founders and their descendants run their firms more efficiently than managers without founding-family ties, with descendants being the most efficient. They also find that corporate efficiency and value are unrelated to the managerial ownership level. Consistent with the ideas of Fama and Jensen (1983) and DeAngelo and DeAngelo (1985), in relation to agency explanation, family ties apparently serve to better align management and company interests than does managerial ownership in the absence of such ties, resulting in more efficient operations and superior results (McConaughy et al., 1998). Nevertheless, Yermack (1996) shows that the presence of a founding family CEO is negatively related to company value.

According to the previous arguments, large concentrated investors, and especially family firms, can have incentives that positively influence firm performance, motivating them to perform better than non-family firms.

The empirical research also provides favourable evidence of a better performance for family firms. McConaughy et al. (1998) and Anderson and Reeb (2003) conclude that family firms perform better than non-family firms, controlling for industry and firm characteristics, since firms with continued founding-family presence exhibit significantly better accounting and market performance than non-family firms. These results are relatively unaffected by the consideration of other blockholders or by the discrepancy between the family's ownership and control rights (Anderson and Reeb, 2003).

Starting from the previous theoretical proposals, family's influence on firm performance is analysed empirically in this study of Spanish firms, in order to examine the prevalence of the opposing arguments.

Methodology

Database

The database taken to analyse the performance of family firms includes all the large and medium Spanish non-financial firms for the year 2002. For this sample, corporate-level accounting and performance information has been collected from the SABI tapes, an electronic database that provides the information that firms are required to deposit in the Spanish Mercantile Registry for their accounts. Information on the ownership and management structures have been also collected from this database.

Following the European Union Commission Recommendation, firms are classified as a large firm when they generate annual sales up 40 million euros and have more than 250 employees. Medium firms are considered those with annual sales between 7 million and 40 million euros and between 50 and 250 employees. In 2002 there were 7775 large and medium size anonymous and limited companies in Spain. It is preferable to take this large sample of firms, instead of only those listed firms on the stock market, since just 5 per cent of the Spanish large firms are public firms and none of the medium firms are listed. In this way, a more representative sample of Spanish family firms is going to be considered.

The fractional equity ownership of the founding family and the presence of family members on the board of directors[1] are examined to identify family firms. In Spain people have two surnames (the first one of the father and of the mother), so it is easier than in other countries to identify family members after the founder, even to include distant relatives such as second or third cousins. In the case of shareholders being firms, it has been necessary to search into their ownership data to find the last owner. The firms are classified into family ones when a family group has the control of the firm, that is, they are the major blockholder and/or they have family members on the board of directors.

One thousand six hundred and thirty-four firms could not be classified with the available information, being cases of firms with just one shareholder, or with two shareholders who may or may not be married. It is preferable to develop the analysis taking the 6141 firms that can be clearly classified into family and non-family firms.

Table 31.1 *Distribution of family and non-family firms in Spain by sales (2002)*

(Annual sales)	Family firms Number of firms (Percentage)	Non-family firms Number of firms (Percentage)	Without classification	Total
Large	532	1046		
(More than 40 million €)	(34%)	(66%)		(100%)
Medium	2870	1693		
(7 million to 40 million €)	(63%)	(37%)		(100%)
Total	3402	2739	1634	7775

Note: The firms are classified into family ones when a family group has control of the firm, in other words, they are the major blockholder and/or they have family members on the Board of Directors. All the large and medium sized firms in Spain in 2002 were analysed.

Table 31.1 shows the data sample distribution. Family firms constitute 34 per cent of Spanish large firms, but 63 per cent of the medium-sized companies, in the year 2002.

In order to compare the characteristics of family firms with non-family firms on the basis of size, it has been tested if family firms are significantly smaller than non-family firms in terms of sales, total assets and number of employees, between both groups, differentiating between large and medium companies. Table 31.2 shows the descriptive statistics and the results of the mean differences tests for two independent samples (t-test). Family firms are significantly smaller than non-family firms in each size group (large firms and medium-sized firms), according to sales, total assets, and number of employees. In relation to the firm's age, significant differences were found in favour of family firms, which are older than their homologous non-family firms.

Table 31.3 summarizes the distribution of firms in each group according to their main activity sector. The firms have been grouped according to two-digit industry codes (the group named 'other sectors' is the total of the numerous remaining).

With the database being described, the differences in performance between family and non-family firms are now analysed, for large and medium-sized companies in Spain. Theoretical proposals have established positive and negative factors that can determine the performance of a family firm. The prevalence of these opposing arguments is an empirical question.

Statistical procedures

Univariate analysis

As a first step, a univariate analysis of the differences between performance of family and non-family firms is developed. Accounting measures of firm performance are considered, given that our sample includes all the large and medium size Spanish non-financial firms, instead of just firms quoted on the Spanish Capital Market. After the classification of the firms into family and non-family ones, 6141 firms constitute the sample of this study (only 150 non-financial firms are listed on the Spanish stock market; they are also included in the sample, as they are large firms). The typical lower investor protection that characterizes a 'continental' stock market, like the Spanish one, with a civil law tradition, makes a broader database more representative than only listed firms.

Table 31.2 Differences by size between family and non-family firms in Spain (2002)

Variable	Family business			Non-family business			Differences test
	Mean	Median	Stand. Dev.	Mean	Median	Stand. Dev.	t-test
Panel A: Large firms (more than 40 million €)							
Sales (million €)	257	95	727	502	124	1710	−3.967***
Total assets (million €)	238	88	608	630	108	2582	−4.645***
No. employees	1441	526	4505	2035	610	5528	−2.280**
Age	26	23	18	24	17	21	1.740*
Panel B: Medium firms (sales between 7 million and 40 million €)							
Sales (million €)	17	15	8	18	15	9	−3.417***
Total assets (million €)	16	11	15	24	12	66	−4.927***
No. employees	104	89	49	113	98	52	−5.536***
Age	23	21	14	19	15	15	8.092***

Notes:
Significant at 1%, 5% and 10% (***, **, *) level.
Summary statistics for the size of the firms in the analysis, comprised of 532 large family firms, 1046 large non-family firms, 2870 medium family firms and 1693 medium non-family firms, in Spain for the year 2002. Family firms are those firms with family groups being the major blockholders or/and family presence on the board of directors (non-family firms in other case).

Table 31.3 Percentage of family and non-family firms in each industry, by two-digit CNAE code

Activity sectors	Large family firms	Medium family firms	Large non-family firms	Medium non-family firms
Food and beverage	11	8	5	5
Whole saling	10	14	8	13
Construction	10	11	6	10
Retailing	8	3	4	3
Chemical and paper	8	10	13	13
Metal	6	9	7	9
Miscellaneous manufacturing	5	6	12	8
Transportation	5	6	7	6
Real estate	4	2	1	2
Mineral	4	6	3	4
Hotel and restaurant	4	3	1	2
Textile	3	4	1	2
Auto sales and repair	2	7	3	4
Electricity, gas and water	0	0	3	2
Miscellaneous	13	2	12	6
Other sectors	8	9	12	11

Notes:
The group 'other sectors' is the total of the numerous remaining.
The total number of firms in each group are: 532 large family firms, 1046 large non-family firms, 2870 medium family firms and 1693 medium non-family firms, in Spain for the year 2002.

The measures analysed, in order to evaluate the performance of family firms versus non-family firms, are the following one (calculated for the year 2002 in all cases):

- *Return on assets* (ROA) is computed using earnings before interest, tax and amortization divided by the book value of total assets.
- *Return on equity* (ROE) is measured as net earnings divided by the book value of shareholder equity. It measures shareholders return. This is the main variable for evaluating firm performance of non-listed firms.
- *Sales growth rate* (gSales) is calculated as the increase in sales in 2002 divided by sales at the beginning of the period. This variable approximates operating efficiency.
- *Sales per employee* (SalEmp) is the ratio of sales to the number of employees, at the end of the year 2002.
- *Cash flow per employee* (CashEm) is the ratio of cash-flow to the number of employees at the end of the period.
- *Leverage* (Lev) is measured as the ratio of total debt to total liabilities, giving information about financial risk and leverage taken, and as a critical factor of the differences between economic and financial returns (ROA vs ROE).

Better performance should mainly be reflected in higher returns on equity, and in higher returns on assets, sales growth rate, sales per employee and cash flow per employee. Table 31.4 illustrates the univariate analysis, showing the descriptive statistics of the

Table 31.4 Differences in performance between family and non-family firms in Spain by size groups (2002)

	Family business			Non-family business			Differences test
Variable	Mean	Median	Stand. Dev.	Mean	Median	Stand. Dev.	t-test
Panel A: Large firms (more than 40 million €)							
ROE	12.3%	10.8%	317.4%	14.5%	11.8%	870.4%	−0.731
ROA	7.0%	5.9%	8.0%	5.9%	5.4%	10.9%	2.111**
Sales growth	13.4%	8%	30.8%	19.3%	7.1%	58.1%	−2.622***
Sales per employee	236	178	209	336	191	855	−3.553***
Cash flow per employee	17	12	20	23	13	85	−2.202**
Leverage	63.2%	65.4%	19.7%	66.0%	68.6%	21.3%	−2.514**
Panel B: Medium firms (sales between 7 million and 40 million €)							
ROE	12.1%	10.0%	48.4%	10.8%	11.3%	69.2%	0.688
ROA	7.1%	5.7%	8.8%	5.8%	5.5%	13.7%	3.534***
Sales growth	16.4%	7.5%	54.9%	19.1%	7.5%	62.1%	−1.493
Sales per employee	187	157	112	185	156	115	0.676
Cash flow per employee	11	9	50	13	9	30	−1.244
Leverage	61.3%	64.2%	21.6%	65.0%	67.7%	23.2%	−5.102***

Notes:
Significant at 1% (***) and 5% (**) levels.
Summary statistics for the size of the firms in the analysis, comprised of 532 large family firms, 1046 large non-family firms, 2870 medium family firms and 1693 medium non-family firms, in Spain for the year 2002. Family firms are those firms with family groups being the major blockholders or/and family presence on the board of directors (non-family firms in other case).

Table 31.5 2-way ANOVA: family versus non-family firms, large versus medium firms (Spain 2002)

Variable	Family	Size	Family* size
ROE	0.06	1.04	0.87
ROA	12.69	0.00	0.21
Sales growth	3.73*	0.40	0.02
Sales per employee	19.46***	74.15***	20.83***
Cash flow per employee	5.64**	23.83***	1.96
Leverage	20.97***	2.93*	0.00

Notes:
Significant at 1 per cent (***), 5 per cent (**) and 10 per cent (*) levels.
F values are included. Data sample is comprised of 532 large family firms, 1046 large non-family firms, 2870 medium family firms and 1693 medium non-family firms, in Spain for the year 2002.

accounting performance measures for the firms in the study, comparing family to non-family firms, and separately for large and medium-sized firms. Statistics for differences in means test (t-test) is also included.

There are no differences in the univariate analysis between family and non-family firms in relation to their return on equity. Large family firms have a 12.3 per cent median ROE versus 14.5 per cent for large non-family firms, and medium size family firms have a 12.1 per cent median ROE versus 10.8 per cent for medium non-family firms, but these differences are not statistically significant.

On the other hand, statistically significant differences were found, between family and non-family firms, in relation to higher return on assets for family firms, both for large and medium size companies. Nevertheless, as Table 31.4 shows, this higher economic performance for family firms, coincides with significantly smaller leverage for family firms. That is to say, although family firms show a better economic performance, in the end they do not obtain better financial performances for their owners, because of their financial structure and/or financial conditions.

In relation to efficiency in terms of sales growth, sales per employee level and cash flow per employee, Table 31.4 illustrates significantly worse levels for large family firms in comparison to large non-family firms, while no significant differences were observed for medium size companies.

The information of t-test is completed with a 2-way ANOVA, comparing family versus non-family and large versus medium firms, as well as their interaction effects. Table 31.5 shows, again, that there are no statistically significant differences between family and non-family firms, in relation to their return on equity, neither between large and medium companies, nor is the interaction effect of family ownership and size significant. Nevertheless, a deeper multivariate analysis is needed, as we show in Tables 31.8 and 31.9.

Table 31.6 shows the univariate analysis for different age groups of firms, distinguishing young firms (first generation in the case of family firms), medium-age firms (second generation) and old firms (third and more generations in the case of family firms). There are no statistical differences between family and non-family firms in relation to their return on equity.

Table 31.6 Differences in performance between family and non-family firms in Spain by age groups (2002)

Variable	Family business			Non-family business			Differences test
	Mean	Median	Stand. Dev.	Mean	Median	Stand. Dev.	t-test
Panel A: Young firms (less than 30 years)							
ROE	13.6%	11.1%	43.7%	11.1%	11.9%	77.4%	1.303
ROA	7.5%	6%	9%	5.8%	5.4%	13.2%	4.977***
Sales growth	19.1%	8%	67.4%	23.5%	8.1%	77.9%	−1.984**
Sales per employee	194.2	158.7	124.2	239.8	164.8	596.3	−3.342***
Cash flow per employee	11.8	8.8	54.0	15.5	9.5	62.2	−2.078**
Leverage	64.6%	67.5%	20.9%	68%	69.8%	25.7%	−4.734***
Panel B: Medium-age firms (age between 30 and 60 years)							
ROE	8.3%	8.6%	43.8%	14.9%	10.5%	72.1%	−1.843*
ROA	6.3%	5.4%	7.5%	6.4%	6.0%	11.0%	−0.145
Sales growth	10.2%	6.6%	33.6%	10.1%	6.5%	29.4%	0.050
Sales per employee	195.0	160.7	139.1	245.1	181.9	282.5	−3.883***
Cash flow per employee	13.3	9.7	16.5	18.8	11.5	32.9	−3.590***
Leverage	57.4%	58.3%	20.5%	60%	61.1%	23.1%	−2.182**
Panel C: Old firms (more than 60 years)							
ROA	5.6%	5.3%	8.3%	5.8%	5.4%	12%	0.173
ROE	10.0%	7.8%	93.8%	20.5%	9.4%	79.9%	−0.923
Sales growth	9.7%	6.7%	20.0%	12.0%	5.3%	27.9%	−0.742
Sales per employee	199.2	171.1	117.3	277.5	189.3	343.9	−2.488**
Cash flow per employee	12.6	10.4	21.2	26.5	14.3	65.5	−2.398**
Leverage	53.1%	50.2%	21.5%	59.8%	60.4%	23.5%	−2.184**

Notes:
Significant at 1% (***), 5% (**) and 10% (*) levels.
Summary statistics for the age of the firms in the analysis, comprised of 2458 young family firms, 2053 young non-family firms, 824 medium-age family firms, 515 medium-age non-family firms, 107 old family firms and 135 old non-family firms, in Spain for the year 2002. Family firms are those firms with family groups being the major blockholders or/and family presence on the board of directors (non-family firms in other case).

Matched-pairs methodology
In order to consider the sensitiveness of the results in the methodology employed, a matched-pair methodology is now developed. Each family firm is compared with the closest non-family firm, in terms of industry and size, in order to control for these effects (McConaughy et al., 1998). First, the firms are organized according to their three-digit CNAE code (sector code, equivalent to SIC numbers), and then, in each sub-sector, the firms are ranked by sales. Inside each three-digit sector, the closest non-family firm to each family firm is looked for, taking into consideration that sales were, as a maximum, 25 per cent above or below the sales level of each family firm, and that the total assets and the number of employees were also similar. Nevertheless, the average of the absolute value of the differences between the sales of each family firm considered and their non-family firm pair found is 7 per cent for large firms and 4 per cent for medium firms. However, there is not always another firm in the same sector with a similar size. This is why, starting from the 532 large and 2870 medium size family firms identified in Spain in 2002, 1535 matched pairs of family and non-family firms can be defined, distributed as 285 pairs of large firms and 1250 of medium size companies.

Table 31.7 shows the results of the differences means t-test, for two related samples, as in this case we have matched pairs. The matching process appears to work well; in no case is there a statistically significant difference between family firms sample and control sample sales level nor in the number of employees, according to the t-test. In relation to the total assets level, the t-test still shows a significantly smaller total asset size for family firms, although they have a similar size in terms of sales and employees.

In relation to the performance comparisons, no significant differences are observed in terms of return on equity, again, for this sample that exhaustively controls industry and size. That is to say, family firms perform like non-family firms. The differences again appear in a higher return on assets and a lower leverage for family firms, in comparison with their homologous non-family firms, in terms of activity sector and size.

Relationship between family ownership and firm performance: multivariate analysis
The univariate analyses inform about the characteristics of family firms, but they do not control for additional variables influencing performance. In order to consider the influence of different variables in the firm performance, controlling for firm-specific characteristics, a regression analysis is developed, conducting a time-series cross-sectional comparison of family and non-family firms. Again, the database is composed of the large and medium size companies in Spain for the year 2002. All the variables relate to the end of 2002.

The regression equation for the multivariable analysis takes the form:

$$Performance = \alpha_0 + \beta_1\, Family + \beta_2\, Age + \beta_3\, Fam{*}Age + \beta_4\, FamMan + \gamma_5\, (Control\ Variables) + \varepsilon$$

where,

- Firm *Performance* is return on equity, calculated as net earnings divided by the book value of shareholder equity.
- *Family* firm is a binary variable that equals one when the firm is a family firm and zero otherwise.

Table 31.7 Pair comparisons for family firms and non-family firms matched on activity sector and size

Variable	Family business			Non-family business			Differences test
	Mean	Median	Stand. Dev.	Mean	Median	Stand. Dev.	t-test
Panel A: Large firms (more than 40 million €)							
Sales (million €)	216 636	95 735	498 817	212 363	94 327	462 228	1.035
Total assets (million €)	199 703	89 466	389 243	246 110	79 326	514 050	−2.128**
No. employees	1050	540	1749	1028	503	1731	0.512
Age	26	23	18	25	18	21	0.644
ROE	13.6%	11.9%	33%	19.6%	12.3%	99%	−0.962
ROA	7.9%	6.6%	7.7%	6.5%	5.5%	10.3%	1.978**
Sales growth	13.5%	7.7%	33%	27.5%	6.2%	134%	−1.717*
Sales per employee	236.86	175.29	206.4	251.65	187	250.5	−1.006
Cash flow per employee	18.55	12.19	23.9	22.72	12.1	54.3	−1.243
Leverage	63%	65.8%	19.7%	66.9%	69.4%	26.2%	−1.991**
Panel B: Medium firms (sales between 7 million and 40 million €)							
Sales (million €)	17 359	14 829	8497	17 374	14 702	8677	−0.387
Total assets (million €)	16 078	11 893	14 677	20 695	11 723	55 040	−3.086***
No. employees	106	92	49	109	95	54	−2.204**
Age	23	21	14	20	16	15	5.894***
ROE	13.2%	10%	57.7%	11.4%	11.5%	66%	0.748
ROA	7.1%	5.7%	9.4%	6.3%	5.6%	11.3%	1.905*
Sales growth	41%	7.7%	6.3%	33%	8%	2.1%	0.420
Sales per employee	185.98	155.5	107.3	185.91	155.4	115.9	0.031
Cash flow per employee	9.56	8.9	73.6	12.02	8.6	23.8	−1.138
Leverage	62%	63.9%	21.7%	65.4%	68.1%	24.7%	−3.821***

Notes:
Significant at 1% (***) and 5% (**) levels.
The control group is matched on activity sector three-digit code classification and by size (considering sales first and also number of employees and total assets).
285 pairs of large firms and 1250 pairs of medium size companies were analysed for the database of Spanish firms in 2002.

- Firm *Age* is measured as the Neperian logarithm of the number of years since the firm was founded until 2002. McConaughy et al. (1998) observe that the value of founder-controlled firms is negatively related to the firm's age. Alternative definitions of this variable are considered as dummy variables that classify the firms by whether the firm was founded less than 30 years ago, between 30 and 60 years ago, or more than 60 years ago (other cut-off points taken are, 20 and 40 years ago, and 25 and 50 years ago).
- *Fam**Age* is a multiplicative variable, that multiples the family firm binary variable by the firm age variable. This interaction term considers the possible differential impact of age on family and non-family firms.
- *FamMan* (family management), approximates active family control of the firm, and is defined as a dummy variable that takes the value of 1 if founding family members are present on the board of directors, that is, family members acting as CEOs (active family involvement in firm management). Eighty-six per cent of the CEOs are family members and 14 per cent are outsiders, among the family firm sample. Anderson and Reeb (2003) find that accounting profitability is better with outside CEOs when family members (founders or founder descendants) serve as CEOs (market performance appears to be better only in the presence of founder CEOs and outside CEOs, while founder descendants serving as CEOs have no effect on market performance).

Controlling for industry and firm characteristics, *Control Variables* include the following:

- Firm *Size* is the Neperian logarithm of the book value of total assets. This variable is included, as the univariate analysis shows that on average family firms are smaller than non-family firms.
- *Leverage* is measured as the relation between total debt and total liabilities.
- *Listed* is a dummy variable that equals one if the firm is listed on the Spanish stock market.
- *Activity sector* is identified by dummy variables to denote each two-digit SIC code (excluding one of them).

Table 31.8 shows the results of the regression analysis (the second column includes only the statistically significant variables). This multivariate analysis is developed in order to examine if being a family firm influences firm performance, but not a global model for firm performance, so the F-score is a better indicator than R square.

The coefficient estimate on family firms is positive and statistically significant, in accordance with a positive impact of family ownership on firm performance. Firm age has also influence on performance, but positive for non-family firms and negative for family firms. That is, younger family firms show better performance than older firms. This results is in accordance with the previous ones obtained in the univariate analysis (Table 31.6). Nevertheless, family involvement in the management of the firm does not impact on firm performance.

The control variables size and leverage, which the univariate analysis showed significantly different for family firms, are statistically significant also. Smaller size and more

Table 31.8 Performance and family firms: regression analysis (Spain, 2002)

Variables	Coefficient	t-value	Coefficient	t-value
Intercept	0.057	0.704	0.040	0.512
Family	0.164	2.634***	0.130	2.256**
Age	0.036	2.628***	0.036	2.713***
*Fam*Age*	−0.046	−2.347**	−0.048	−2.466**
FamMan	−0.048	−1.540		
Size	−0.014	−2.184**	−0.012	−2.042**
Leverage	0.169	4.714***	0.169	4.712***
Listed	0.031	0.516		
Sector		No significant		
F	5.899***		6.604***	
Adjusted R square	0.006		0.006	

Notes:
Significant at 1% (***) and 5% (**) levels.
Family is binary variable that equals one when founding family is present in the firm. *Age* is the Neperian logarithm of firm's age. *Fam*Age* is a multiplicative variable of *Fam* and *Age*. *FamMan* is a dummy variable that takes the value of one when any family member act as CEOs. *Size* is the Neperian logarithm of total asset book value. *Leverage* is the ratio of total debt divided by total liabilities. *Listed* is a dummy variable that equals 1 if the firm is listed on the Spanish stock market. *Sector* is defined as eight dummy variables of the nine two-digit codes sector activity.
Number of observations is 6094.
The independent variable is *Return on equity*.

debt are variables that positively influence performance. Activity sector is unrelated to firm performance.

Being a family firm proved to be a differential factor determining firm performance. Theoretical arguments in relation to the positive influence of family ownership on firm performance are shown to be the more relevant ones. Nevertheless, family involvement in management does not influences firm performance.

In any case, we should take into account that different goals to firm value maximization can coexist with financial objectives in family firms (Gersick et al., 1997). Firm growth, reputation, good name, financial security for the family in the future, firm survival, social and business esteem are some of these possible additional aims for family firms (Chaganti et al., 1995; Mathews et al., 1994; Romano et al., 2000).

Data envelopment analysis

Finally, family and non-family firm performance is compared through a non-parametric analysis, so a functional form for the relation between performance and its determinants does not need to be assumed. A DEA is developed, in order to analyse the efficiency of each group of firms.

The DEA methodology consists in the construction of the efficient frontier with all the firms that, starting with the same inputs combination, generate the optimum output ('objective firm'). The efficiency for each firm is calculated as the ratio of its output in relation to the objective firm output. When this ratio takes the value of 1, the firm has an optimum efficiency level, that is to say, it is in the efficient frontier (Seiford and Thrall, 1990). The statistical program gives a value between 0 and 100 to each firm, being 100 the value of the firms in the efficient frontier.

Another advantage of DEA is that several performance indicators can be considered simultaneously. The firm outputs included are the following, relating to the year 2002:

- *Return on assets* (ROA). Computed as the earnings before interest, tax and amortization divided by the book value of total assets.
- *Return on equity* (ROE). Measured as net earnings divided by the book value of shareholder equity.
- *Sales growth rate* (gSales). Variation of total sales for the year 2002 in relation to the sales level at the end of the year 2001.
- *Employees growth rate* (gEmpl). Variation of total number of employees for the year 2002 in relation to the sales level at the end of the year 2001.

In order to group the firms, according to a similar level of inputs used, and then compare outputs that they obtain (their performance), the following inputs are considered for the year 2002:

- total assets;
- number of employees;
- firm age;
- operating costs;
- ownership structure, as the number or people who own at least 51 per cent of the firm shares;
- book-equity to total assets ratio;
- interest coverage, as the operating profit divided by the interest expenditures.

Table 31.9 shows the average efficiency of family firms versus non-family firms, separately for large and medium size companies. The Kruskal-Wallis test is the non-parametric test for the differences in efficiency between the family and non-family firms. The efficiency of family firms is better than for non-family firms, for large and medium size companies and for all the measurements employed. Taking into account that the maximum value given to efficiency in the DEA is 100, while the average ROA and ROE for large family firms is 75.26 and 61.65, these values are 69.44 and 51.43 respectively for large non-family firms, with the difference being statistically significant. The same conclusions can be established in relation to sales and employees growth and for the medium size firms. That is to say, the DEA show that family firms are more efficient than non-family firms.

The better performance of Spanish family firms in comparison with non-family firms, is in agreement with the results of other empirical studies such as Leach and Leahy (1991) for the United Kingdom and Anderson and Reeb (2003) for listed firms on the New York Stock Exchange (S&P 500).

Table 31.9 Efficiency of family firms versus non-family firms (data envelopment analysis), Spain 2002

		Large firms				Medium firms			
		ROA	ROE	gSales	gEmpl	ROA	ROE	gSales	gEmpl
Family firms	Average efficiency*	75.26	61.65	13.07	2.72	78.17	71.75	23.95	16.74
	Standard Deviation	4.83	7.21	12.24	7.70	8.24	12.91	26.05	24.03
	Minimum efficiency	63.69	45.59	0.05	0.84	55.02	11.44	0.08	1.37
	Number of firms	332	332	332	332	1752	1752	1752	1752
	No. firms in the frontier	2	2	2	2	103	109	104	81
Non-family firms	Average efficiency*	69.44	51.43	1.88	2.21	75.06	63.49	9.42	11.12
	Standard deviation	6.8	9.19	10.95	10.84	10.18	15.15	22.41	22.85
	Minimum efficiency	41.14	2.67	0.17	0.09	0.46	0.11	0.27	1.07
	Number of firms	830	830	830	830	1213	1213	1213	1213
	No. firms in the frontier	5	4	10	9	80	49	63	67
Test	Kruskal-Wallis	291.126	432.675	44.277	495.241	138.740	306.775	124.236	239.223
	Significance	0.000	0.000	0.000	0.000	0.000	0.000	0.000	0.000

Notes:
* Maximum efficiency = 100.
ROA = return on assets, ROE = return on equity, gSales = sales growth rate, gEmpl = employees growth rate.
Variable returns are considered.

Summary and conclusion

Family firm ownership and control have potential benefits and costs on firm performance. On the one hand, founding families can have concerns and interests of their own, such as stability, compensation, nepotism or special dividends, that benefit these families at the expense of firm value, influencing firm performance negatively. On the other hand, family firm control can mitigate managerial expropriation, supporting longer relationships with firm stakeholders, and contributing positively to firm performance.

In order to test empirically the positive and negative arguments to family firm performance, a large sample of 7775 large and medium size Spanish firms was analysed for the year 2002. While 34 per cent of the large firms can be classified as family firms, this percentage rises to 63 per cent for the medium size companies. A significantly smaller size is observed for family firms.

In terms of performance, family firms have a higher return on assets and lower leverage, but, they do not have a significantly different return on equity as do non-family firms of the same size and in the same industry. Univariate analysis does not observe differences on performance for family firms, in terms of return on equity. Nevertheless, the regression analysis and the data envelopment analysis, show that family firms perform better and are more efficient than non-family firms.

The main implication of the chapter is to highlight that Spanish family firms perform better, in terms of return on equity, than non-family firms of the same size and in the same industry. Family involvement in management does not prove to have a positive impact on firm performance. In accordance with Leach and Leahy (1991) for the United Kingdom and Anderson and Reeb (2003) for listed firms on the New York Stock Exchange (S&P 500), Spanish family firms show that the theoretical potential benefits of family ownership have a higher impact on firm performance, in comparison with the theoretical potential costs. Nevertheless, further theoretical analyses is needed in relation to the differences between family and non-family firms in their capital structure and the impact of this decision on firm performance. Future research could also explore the relationship between firm performance, family ownership and firm growth decisions, given the smaller size that characterizes family firms.

From a practical point of view, the main implication for practitioners is the worse performance observed as family firms get older, in such a way that it is crucial to solve the typical succession problems.

Some limitations of this study are geographical, since the sample is drawn exclusively from Spain, and the unavailability of more detailed ownership and management data. In this sense, future research could consider a broader data sample of family firms from different countries. Another possible extension is to consider additional theoretical approaches to the analysis of family-firm performance.

Notes

* The author acknowledges financial support from the Spanish Institute of Family Businesses, the Asturian Association of Family Businesses (AAEF) and the Asturian Federation of Entrepreneurs (FADE) in Spain.

1. Astrachan et al. (2002) propose the F-PEC power subscale for the definition of family-owned businesses, as a continuous scale with its three subscales: power, experience and culture. Nevertheless, there is not enough information to calculate F-PEC with this database.

References

Anderson, R.C., S.A. Mansi and D.M. Reeb (2003), 'Founding family ownership and the agency cost of debt', *Journal of Financial Economics*, **68**, 263–85.

Anderson, R.C. and D.M. Reeb (2003), 'Family founding ownership and firm performance: evidence from S&P 500', *Journal of Finance*, **58**(3), 1301–27.

Astrachan, J.H., S.B. Klein and K.X. Smyrnios (2002), 'The F-PEC scale of family influence: a proposal for solving the family business definition problem', *Family Business Review*, **15**(1), 45–58.

Burkart, M., F. Panunzi and A. Shleifer (2002), 'Family firms', Working paper, Harvard University.

Chaganti, R., D. Decarolis and D. Deeds (1995), 'Predictors of capital structure in small ventures', *Entrepreneurship Theory and Practice*, **20**, 7–18.

DeAngelo, H. and L. DeAngelo (1985), 'Managerial ownership of voting rights: a study of public corporations with dual classes of common stock', *Journal of Financial Economics*, **14**(1), 33–69.

DeAngelo, H. and L. DeAngelo (2000), 'Controlling stockholders and the disciplinary role of corporate payout policy: a study of the Times Mirror Company', *Journal of Financial Economics*, **56**, 153–207.

Demsetz, H. (1983), 'The structure of ownership and the theory of the firm', *Journal of Law and Economics*, **25**, 375–90.

Demsetz, H. and K. Lehn (1985), 'The structure of corporate ownership: causes and consequence's', *Journal of Political Economy*, **93**, 1155–77.

Faccio, M., L. Lang and L. Young (2001), 'Dividends and expropriation', *American Economic Review*, **91**, 54–78.

Fama, E. and M. Jensen (1983), 'Separation of ownership and control', *Journal of Law and Economics*, **26**, 301–25.

Fama, E. and M. Jensen (1985), 'Organizational forms and investment decisions', *Journal of Financial Economics*, **14**, 101–19.

Gersick, K.E., J.A. Davis, M.M. Hampton and I. Landsberg (1997), *Generation to Generation: Life Cycles of the Family Business*, Boston, MA: Harvard Business School Press.

James, H.S. (1999), 'Owner as manager, extended horizons and the family firm', *International Journal of the Economics of Business*, **6**(1), 41–55.

Jensen, M. and W. Meckling (1976), 'Theory of the firm: managerial behavior, agency costs and ownership structure', *Journal of Financial Economics*, **4**, 305–60.

Leach, D. and J. Leahy (1991), 'Ownership structures, control and the performance of large British companies', *Economic Journal*, **101**, 1418–37.

Matthews, C.H., D.P. Vasudevan, S.L. Barton and R. Apana (1994), 'Capital structure decision making in privately held firms: beyond the finance paradigm', *Family Business Review*, **7**(4), 349–67.
McConaughy, D., M. Walker, G. Henderson and C. Mishra (1998), 'Founding family controlled firms: efficiency and value', *Review of Financial Economics*, **7**(1), 1–19.
Morck, R., A. Shleifer and R. Vishny (1988), 'Management ownership and market valuation: an empirical analysis', *Journal of Financial Economics*, **20**, 293–315.
Pagano, M. and A. Roell (1988), 'The choice of stock ownership structure: agency costs, monitoring and the decision to go public', *Quarterly Journal of Economics*, **113**, 187–225.
Romano, C.A., G.A. Tanewski and K.X. Smyrnios (2000), 'Capital structure decision making: a model for family business', *Journal of Business Venturing*, **16**, 285–310.
Seiford, L. and R. Thrall (1990), 'Recent developments in DEA: the mathematical programming approach to frontier analysis', *Journal of Econometrics*, **46**, 7–38.
Shleifer, A. and R. Vishny (1997), 'A survey of corporate governance', *Journal of Finance*, **52**, 737–83.
Yermack, D. (1996), 'Higher market valuations of companies with a small board of directors', *Journal of Financial Economics*, **40**, 185–211.

32 Family ownership, corporate governance and firm value: evidence from the Spanish market

María Sacristán Navarro and Silvia Gómez Ansón

Introduction

This chapter analyses how ultimate family ownership, pyramids, and corporate governance affect a company's value. We measure how the presence of families within company management structures, that is, a family's participation in company management and control, influence company value and how the management and control by founders affects company performance. Using a sample of 86 Spanish non-financial companies, our results show that family ownership does not affect company value per se, whereas a company's corporate management structure does influence company value. When the managers or the chairpersons of the board belong to a family, the firms' performance appears to be affected. The presence of descendants in the firm's management and/or in the firm's control seems to affect negatively firm value.

Family-run businesses are a very common ownership structure in continental markets (Faccio and Lang, 2002; La Porta et al., 1999). In this respect, Spain is no exception. Family-run businesses represent 86 per cent of the business panorama in Spain and feature serious management problems. Only 5 per cent of Spanish family companies use family boards (*Gaceta de los Negocios*, 29 April 2005) and only some have effective boards of directors (Gallo, 1998). Recent studies (IEF, 2005) highlight the importance of the professionalization of decision-making mechanisms within family companies (with the creation of family boards and board of directors).

In this respect, the question of how family ownership and a company's corporate management structure affect its value is currently a topic of interest and, although having received increased attention from scholars, stills remains a very open issue. The aim of this chapter is to relate family ownership and the effect of pyramids (measured by the excess of voting rights over cash flow rights) to company value, and to assess the implication of the ultimate family owner group on its management, corporate management structure and value. In addition, we explore how the presence of founders or descendants within the company's management and control mechanisms affect its performance.

This study contributes to the existing literature on family companies and corporate management in a number of important ways. First, as family business research is still overshadowed by definition issues, we use the term family-controlled company, whose ultimate owning group is a family (which is the largest shareholder in the company) that owns more than 10 per cent of the voting rights at the end of the company control chain (Sacristán Navarro and Gómez Ansón, 2004). Although the influence of family ownership control on company value has been the subject of numerous research investigation, scholars sporadically address the ultimate owner perspective (Claessens et al., 2000; Du and Dai, 2005; Faccio et Lang, 2002; La Porta et al., 1999). This chapter aims to contribute in this direction. Secondly, regarding corporate management structure, we look at whether companies are managed or controlled by the ultimate owner group and how this may affect company value. Family involvement in company management or control (Anderson and Reeb, 2003, 2004; McConaughy et al., 2001; Villalonga and Amit, 2005) has not previously been measured in such a way (as far as we know). Thirdly, our analysis of the founder/descendant's influence on company value has not been studied for the Spanish market.

The results suggest that the observed differences in the companies' performance are more due to the origin of management and its relation to the founding family than to its ownership structure, that is, whether it is a family or a non-family company.

The chapter is structured as follows. In the next section we describe the theoretical framework regarding the possible influence of family ownership, the role of the family in active management and/or control on company value. In the third Section we present the methodology (data, sample and variables) employed in the study. In the fourth section we analyse the main results and, finally, we present the main conclusions and implications of the chapter.

Theoretical framework

Family ownership, pyramids and company value

The relationship between family ownership and company value has traditionally received quite a lot of theoretical and empirical attention from scholars, since the very beginning of research on family-run businesses (Chaganti and Damampour, 1991; Daily and Dollinger, 1992). This is owing to the fact that families are in an uncommon position to exert influence and control over a company, which may potentially lead to differences in performance, when comparing family and non-family companies (Anderson and Reeb, 2003, p. 1304).

According to agency theory, Berle and Means (1932) and Jensen and Meckling (1976) suggest that a company's ownership concentration should have a positive effect on its value, as it reduces the conflicts of interest between owners and managers resulting from a greater shareholder incentive to monitor managers. Nevertheless, when large shareholders are families or individuals, they may also have an incentive to extract private benefits at the expense of minority shareholders. The possibility of managerial entrenchment may be caused by the pursuit of family interest, thus neglecting those of other shareholders. Finally, the relationship between family members could also be worse than between non-family members, triggering agency problems and enhancing agency costs (Schulze et al., 2001).

Consequently, family ownership could have positive or negative effects on company value. In the economic literature, these effects have been referred to as the monitoring (positive effect) and the expropriation (negative effect) hypotheses. We will first refer to the monitoring hypothesis.

From a theoretical point of view, the economic literature has claimed that concentrated family ownership may positively influence company performance. Concentrated ownership would give the owners, for example families and individuals, an incentive to monitor managers and to assume the task of monitoring (Shleifer and Vishny, 1997). Theoretically, Pollak (1985) and Coleman (1990) argue that in family-run businesses, the relationships within such families, characterized by loyalty and trust, may promote operational flexibility, ease decision-making and reduce shirking. Besides, when monitoring requires knowledge of the company's technology, families may provide greater oversight when their lengthy tenure enables them to move further along the company's learning curve (Anderson and Reeb, 2003, p. 1305). The long-term horizon that characterizes family companies may also affect their efficiency. Family ownership may provide incentives for a company to invest according to market rules (Anderson and Reeb, 2003) and would lead to lower debt financing costs, as compared with non-family companies (Anderson et al., 2003).

Nevertheless, there are also arguments that suggest that family-owned companies may be less efficient than non-family companies (expropriation hypothesis). Concentration of ownership reduces the possible diversification of financial risk, increases the risk premium and consequently the cost of capital (Demsetz and Lehn, 1985). In addition, agency problems may also occur between family members (Schulze et al., 2001), and the presence of family groups, when a company obtains outside equity financing, may also trigger agency issues owing to the use of pyramidal groups to separate ownership from control, the entrenchment of controlling families and non-arm's length transactions, as well as 'channelling' between related companies (Morck and Yeung, 2003). Moreover, founding families may have the incentives and power to take action and adopt investment decisions that benefit themselves, to the detriment of other shareholders (Demsetz and Lehn, 1985).

An important question when analysing the influence of families on company value is their use of pyramids to channel ownership. In this sense, family ownership can also be considered by measuring voting and cash flow rights instead of property rights (Claessens et al., 2002; La Porta et al., 1999). Empirical evidence shows that the separation between voting and cash flow rights is more pronounced in family-controlled companies and small companies, and may have an influence on company performance (Claessens et al., 2002).

Empirical evidence

The empirical relationship between family ownership and company value is not conclusive and therefore requires further attention. Certain papers report a positive relationship, for example, for the Spanish market (Camisón, 2001; Santana Martín and Cabrera Suárez, 2001). For the US market, Anderson and Reeb (2003a) find that family companies outperform non-family companies; McConaughy et al. (1998) report that founding family-controlled companies are more efficient and valuable than others and that descendant-controlled family companies are more efficient than founder-controlled companies, while Villalonga and Amit (2005) report that family ownership creates value when founders work as chief executive officers (CEO). Other papers report no differences in performance between family and non-family companies (Alcalde et al., 2001; Galve Górriz and Salas Fumás, 1993, 1994) or find that family companies perform worse than non-family companies. For example, Morck et al. (1988) find that Tobin's q is lower for older companies managed by a member of the founding family and Morck et al. (1998) report that companies controlled by heirs of the founder are less profitable than others in the same industry. Amongst large US corporations, Holderness and Sheehan (1988) document that family companies have a lower Tobin's q than non-family companies. Wall (1998) and Barth et al. (2005) find that family companies are less productive than non-family companies.

Most of these papers use the return on assets (ROA), return on equity (ROE) ratios and Tobin's q as a measure of company performance and one has to consider that the way they measure performance may affect the results of these studies. For instance, Galve Górriz and Salas Fumás (1996), document higher productivity for family-owned companies in Spain, although no difference in profitability is found. While the ROA and ROE ratios are based on accounting data, Tobin's *q* is based on market values (see Demsetz and Villalonga, 2002, p. 213, for a critical review), and therefore different results may be obtained. Differences in the results may also be the result of sample selection biases. Probably owing to the difficulty in obtaining data on family companies, most of the papers employ samples of large quoted companies, and only a minority use samples of small and medium-sized companies or even non-listed family companies.

Another problem refers to the endogenous nature of the relationship between family ownership and company value. In this sense, the studies by Demsetz (1983), Demsetz and Lehn (1985) and Demsetz and Villalonga (2001) suggest that ownership concentration does not affect company value. They claim that the company ownership structure is the endogenous outcome of decisions that reflect both the influence of shareholders and of market trends. Thus, as a company's ownership structure reflects decisions made by those who own the shares, the ownership that emerges should be influenced by the profit-maximizing interest of shareholders, and consequently there should be no systematic relationship between variations in the family ownership structure and variations in company performance.

Family management, family control and company value

The involvement of families in the management and/or control of the companies of which they hold large shares has also been substantially researched but results are contestable. A family's involvement in the company management structure has been theoretically explained as having both a positive and negative influence on its value. The positive line of argument claims that owner-managers should minimize the agency costs arising from the separation of ownership and control. For instance, the need to monitor a family's agent should be reduced as 'family members have many dimensions of exchange with one another over a long horizon that leads to advantages in monitoring and disciplining family-related decision agents' (Fama and Jensen, 1983, p. 306). Moreover, kinship and altruism amongst family members will temper manager's self-interest and provide an incentive to family directors to adopt a long-term perspective and undertake investments that will benefit coming generations of owners (Schulze et al., 2003). Besides, Morck et al. (1988) point out that founder chief executive officers bring innovative and value-enhancing expertise to the company.

However, owner-managed companies are also particularly vulnerable to managerial entrenchment, and there may also be agency threats under family contracting as affective ties between the parties may reduce the presence of formal safeguards designed to mitigate threats to company performance (Gómez-Mejía et al., 2001). For instance, families' chief executive officers have an average tenure of 24 years according to Beckhard and Dyer (1983), which is twice that observed in widely owned companies (Hambrick and Fukutomi, 1991). In addition, the choosing of top managers from a more restricted pool of talent may lead to lower quality of owner-managers, as compared to professional managers, and Bukart et al. (1997) observe that families acting on their own behalf can adversely affect employees' efforts and productivity. Summing up, as Anderson and Reeb (2003) state, previous research suggests that large shareholders, that is, founding families, will ensure that management serves their own interests, the family's interests.

Thus, the effect of a company's professionalization on its value is ambiguous. Although professionalization allows the company to choose its managers from a broad pool of talent, the reduction of the agency costs that may arise owing to the coinciding owners and managers may not balance the costs of professionalizing the company, and a professionalized family company's agency costs may exceed those of non-professionalized companies.

Family control of the board of directors could also lack the use of good management practice recommendations (such as the presence of independent directors, or of an inoperative board of directors – Gallo, 1998), thus negatively affecting company value. How the board is comprised is important, since the influence of independent directors may represent an important line of defence that minority shareholders can employ in protecting themselves against the opportunism of large shareholders. Therefore, an effective board structure in companies in which families hold a significant stake would require a balance between family directors' interests and independent director objectivity (Anderson and Reeb, 2004).

The effect of family presence on company value may also depend on the involvement of the family in company management and control, as suggested by empirical studies that show the differential effect on performance of family founders or descendants.

Empirical evidence

The results of the studies that try to link a family company's professionalization to company value are non-conclusive. Family involvement in the company's management tends to be measured by whether the chief executive officer belongs to the family group or not, and family involvement in company control tends to be analysed by determining whether or not the chairperson of the board belongs to the family group. While some studies report a positive impact of family chief executive officers on company performance, others do not. In this sense, examples of studies that show a positive influence on company value by family involvement in its management (family chief executive officers) are those of Anderson and Reeb (2003), Camisón (2001), Sraer and Thersmar (2004) as well as Villalonga and Amit (2005). Other authors (Anderson and Reeb, 2004; Villalonga and Amit, 2005) report a positive influence on company value by family management and family control – family chief executive officer plus family chairman – or by founding managers (Adams et al., 2005; McConaughy et al., 1998; Sraer and Thersmar, 2004), or young founding managers (McConaughy et al., 1998).

On the contrary, other authors report a negative influence on company value by active family management (Barth et al., 2005), old founding managers (Morck et al., 1988), second or third generation managers (Pérez-González, 2001) and family control (Cabrera-Suárez et al., 2001; Schulze et al., 2001), whereas, for example, Jayaraman et al. (2000) finds no significant influence of the involvement of founders as managers on company share returns. Finally, other authors report that family companies managed by professional management hired from outside the owner family are equally productive as non-family owned companies (Barth et al., 2005).

Methodology

Data

The initial database used for the analysis comprised all companies quoted on the electronic markets of Spain's four stock exchanges: Madrid, Barcelona, Bilbao and Valencia, at the end of 2002 (31 December). The information was provided by the Spanish Supervisory Agency (CNMV) and the four stock exchanges provided the platform upon which the sample was constructed. The following data were employed in the study: company accounting data supplied by the Spanish stock exchanges, stock quotes published by the daily stock bulletins of the Spanish stock exchanges, data on major shareholders and board composition published by the Spanish Supervisory Agency, the Official Company Register and the SABI database.

The following filters were applied to this initial database:

1. Investment companies were not included in the sample.
2. Foreign companies were excluded (four cases) as we were unable to follow the chains of control of these companies.
3. As we employ adjusted market values to measure company performance, we exclude companies which are unique in their industry (according to the SIC code at a two-digit level).
4. All financial corporations (SIC codes 60–60 except Real Estate companies – SIC code 65) were excluded due to their special corporate management characteristics.

These filters reduced the final sample down to 86 companies, all of which are listed on the electronic markets, with a mean market capitalization of 2026 million euros. The number of companies included in the sample is similar to other empirical studies that also employ samples of non-financial Spanish companies (Galve Górriz and Salas Fumás, 1996). Among the sample, 37 companies (43 per cent) are family companies from an ultimate owner perspective (mean market capitalization of 990 million euros), and 49 (57 per cent) are non-family companies (mean capitalization of 2825 million euros) see Table 32.1.

Procedure

In order to define a family company, we employ part of the F-PEC scale (Astrachan et al., 2002) by exclusively considering the ownership variable. We therefore consider a broad definition of family companies, defining a family company as one whose main ultimate owner group is a family or an individual who holds voting rights above the 10 per cent threshold. A cut-off point of 10 per cent is conventionally used in literature (Du and Dai,

Table 32.1 Sample description

	Total sample	Family companies	Non-family companies	t-test (sig.)
No. of companies	86	37	49	
Percentage	100	43	57	
Mean capitalization (million euros)	2026	990.11	2825.11	1.618 (0.109)
St. Dev. capitalization (million euros)	5234	2358	6573	

Note: The sample is comprised of 86 non-financial companies quoted on the electronic market of Spain's four stock exchanges: Madrid, Barcelona, Bilbao and Valencia, at the end of 2002.

2005) because: (1) it provides a significant threshold of votes; and (2) most countries mandate disclosure of 10 per cent ownership stakes, and usually even lower (in Spain the percentage goes down to 5 per cent). Whenever possible, each company's chain of control was analysed in order to identify the ultimate owners.

Regarding the methodology, we conducted mean difference analyses and cross-sectional analyses in order to test whether company performance was affected by family ownership, the existence of pyramidal groups, the level of professionalism of management and control of the family companies, and the founder effect. In this sense we conducted multiple regressions where the dependant variable was the firm performance and the independent variables were those related to family ownership, pyramids, corporate governance variables and founder-descendants variables, as well as several control variables (size, leverage, financial risk and age).

Variables definition

With respect to the definition of variables (see Table 32.2), company performance – the dependant variable – is measured as its market performance, the market to book value of common equity ratio (MB) and as a its accounting performance, the ratio of operating income to total assets (ROA). Given that previous studies have shown that industry factors affect company performance (King, 1966; Livingston, 1977), and in order to avoid multiple co-alignment problems that could arise when running the regressions with dummy variables representing the companies' industry, we use the industry-adjusted market to book value of common equity (AMB) and the industry adjusted ratio of operating income to total assets (AROA) as dependant variables. Both measures are computed by subtracting the industry median ratio from each company's ratio.

We have included different variables in regressions as explanatory variables. The ownership variable that measures if the firm is a family or a non-family one under the ultimate owner perspective is a dummy variable (UOFAM) that adopts a value of 1 when the ultimate owner group, using the 10 per cent threshold, is a family or individual, and otherwise a value of zero. A continuous variable (DIVVRCF) measures the excess of voting rights over cash flow rights of the ultimate owner group, meaning the existence of pyramids.

Family's presence in the control of the firm is measured through the variable CONTROLUO which adopts a value of 1 when any member of the ultimate owner family group is the chairperson of the board of directors, and zero otherwise. The variable that measures family management is MANUO, a variable that takes value 1 when the ultimate owner manages the company, that is, when the company's chief executive officer belongs to the ultimate owner's group, and otherwise zero. The total presence of ultimate family owners is measured through the variable, FAMANCON, a dummy variable that takes value of 1 when the company is managed and controlled by the ultimate family owners, and zero otherwise.

Founders/descendants presence on corporate governance is measured through variable FOUNDESC, a dummy variable that identifies whether or not the company is managed or controlled by its founders or by their descendants. This variable takes value 1 when the company is managed or controlled by the family's descendants, and zero when it is managed or controlled by the founders. In order to define this variable correctly, we first considered founders as individuals that (solely) own significant shares in a company, and descendant-managed companies as those in which several individuals from the same family are present. Nevertheless, as we are aware that companies can also be founded by

Table 32.2 Definition of variables

Variable	Description
1 Tobins q (MB)	MB denotes the market to book value of common equity ratio
2 Return on assets (ROA)	ROA denotes the ratio of EBITDA to total assets
3 Industry adjusted q (AMB)	The MB ratio adjusted by the industry median (SIC codes at a two-digit level)
4 Adjusted ROA (AROA)	AROA the industry median adjusted ROA ratio adjusted by the industry median
5 Family company (UOFAM)	Companies whose ultimate owners are families or individuals who own more than the 10% threshold
6 DIVVRCF	Excess of control rights over cash flow rights for all companies
7 Family controlled (CONTROLUO)	A variable that adopts a value of 1 when any member of the ultimate owner family holds the position of chairperson of the board and zero in all other cases
8 Family managed (MANUO)	A variable that takes on a value of 1 when any member of the ultimate owner family holds the position of chief executive officer and zero in all other cases
9 FAMANCON	Dummy variable that adopts a value of 1 when the chief executive officer and the president of the board belong to the ultimate owner's family, and zero in all other cases
10 FOUNDESC	Dummy variable that adopts a value of 1 when the company is managed or controlled by descendants, and zero when it is managed or controlled by its founders
11 FOUNDMAN	A variable that adopts a value of 1 when the manager is the founder and zero in all other cases
12 DESCDMAN	A variable that adopts a value of 1 when the manager is the descendant and zero in all other cases
13 FOUNDCON	A variable that adopts a value of 1 when the president of the board of directors is the founder, and zero in all other cases
14 DESCDCONTROL	A variable that adopts a value of 1 when the president of the board of directors is a descendant, and zero in all other cases
15 SIZE (Ln TA)	TA represents total company assets (in millions of euros)
16 AGE (Ln company age)	The number of years since the foundation of the company
17 Leverage (LEV)	LEV the ratio of leverage defined as the ratio of total debt to equity
18 RISK	Standard deviation of the daily stock return, multiplied by the square root of the number days of the year

brothers, or by members of the same family, we have crossed this data with the age of the company. For all companies classified as managed by a descendant, but less than 25 years old, the company's website was checked and its founder identified, in order to correctly define the variable. The variables that captures the presence of founders/descendants within the firm's management are FOUNDMAN and DESCDMAN. Both variables are dummy variables that adopt a value of 1 when the company is managed by its founder (FOUNDMAN), or a descendant (DESCDMAN), and zero otherwise. The variables that capture the presence of founders/descendants within the firm's control are FOUND-

CON and DESCDCONTROL. They are dummy variables that adopt a value of 1 when a company is controlled by its founders (FOUNDCON) or by a descendant (DESCD-CONTROL), and zero otherwise.

Finally, as control variables we included several variables in the regression models: company size measured by the logarithm of its assets (SIZE), company age measured as the logarithm of the number of years since its foundation (AGE) and company leverage measured as the ratio of total debt to equity (LEV). We also included the company's financial risk (RISK) measured as the standard deviation of its daily stock return. This data was obtained from the SABI database and the daily stock bulletins published by the Spanish stock exchanges. Management-related variables were taken from the companies' corporate management reports submitted to the Spanish Supervisory Agency.

The definitions of all variables are summarized in Table 32.2.

Results

Descriptive analysis

The results of the descriptive analysis (Table 32.3) show that family companies underperform non-family companies when the industry adjusted ratio of return on assets is used to represent company performance (statistically significant at a 5 per cent level), this

Table 32.3 Summary statistics

Panel A: Summary statistics on quantitative variables

	All companies		Family companies		Non-family companies		U-Mann Whitney test
Variable	Mean	Std. Dev.	Mean	Std. Dev.	Mean	Std. Dev.	Z statistic (Sig.)
MB	1.70	1.29	1.67	1.57	1.72	1.047	−0.624 (0.533)
AMB	0.34	1.25	0.36	1.56	0.32	0.96	−0.536 (0.592)
ROA	0.10	0.07	0.10	0.07	0.11	0.07	−0.859 (0.390)
AROA	0.01	0.06	−0.002	0.06	0.020	0.06	−1.984** (0.047)
TA	4.853876	13.408.817	3.848.650	14.157.836	5.612.923	12.910.813	−1.688 (0.091)
LEV	1.84	2.90	1.97	4.18	1.74	1.35	−0.109 (0.913)
AGE	44	25	43	26	45	25	−0.637 (0.524)
DIVRCF	2.38	6.21	3.64	7.37	1.42	5.04	−3.009 (0.003)***
RISK	37.95	12.12	37.52	11.95	38.28	12.36	−0.226 (0.821)
Number of companies	86		37		49		

Table 32.3 (continued)

Panel B: Summary statistics on qualitative variables (percentages)

Variable	All companies		Family companies
	N	Mean	Mean
UOFAM	86	0.43	
MANUO	85		0.129
CONTROLUO	85		0.329
FOUNDESC	38		0.632
FAMANCON	85		0.116
FOUNDMAN	86		0.036
DESCMAN	86		0.081
FOUNDCON	86		0.128
DESCONTROL	86		0.186
Number of companies	86		37

Note: P-values in brackets. *** Statistically significant at a 1% level. ** Statistically significant at a 5% level. * Statistically significant at a 10% level. The number of companies is 86 except for the risk variable, where it is 76. MB denotes the market to book value of common equity ratio. AMB denotes the MB ratio adjusted by the industry median. ROA denotes the ratio of EBITDA to total assets. AROA denotes the ROA ratio adjusted by the industry median. TA represents total company assets. LEV is defined as the ratio of total debt to equity. AGE represents the number of years since foundation. RISK represents the Standard Deviation of the daily stock return, multiplied by the square root of the number days of the year. DIVRCF represents the excess of control rights over cash flow rights for all companies. UOFAM is a dummy variable which reflects those companies whose largest ultimate owners are families or individuals who own more than the 10% threshold. MANUO is a dummy variable that takes on a value of 1 when any member of the ultimate owner family holds the position of chief executive officer and zero in all other cases. CONTROLUO is a dummy variable that adopts a value of 1 when any member of the ultimate owner family holds the position of chairperson of the board and zero in all other cases. FOUNDESC is a dummy variable that adopts a value of 1 when the company is either managed or controlled by descendants, and zero when it is managed or controlled by its founders. FAMANCON is a dummy variable that adopts a value of 1 when the chief executive officer and the president of the board belong to the ultimate owner's family, and zero in all other cases. FOUNDMAN is a dummy variable that adopts a value of 1 when the manager is the founder and zero in all other cases. DESCMAN is a dummy variable that adopts a value of 1 when the manager is the descendant and zero in all other cases. FOUNDCON is a dummy variable that adopts a value of 1 when the president of the board of directors is the founder, and zero in all other cases. DESCONTROL is a dummy variable that adopts a value of 1 when the president of the board of directors is a descendant, and zero in all other cases.

not being the case when the industry adjusted market to book value of the common equity ratio or the market to book ratio or return on assets is used to represent company performance. Although there are differences in the other measures of company performance, they are not statistically significant. These results suggest, as we previously mentioned, that the way of measuring company performance may affect results. Caution is therefore necessary when comparing and discussing the results of empirical studies that employ different measures of company performance.

Family companies are also smaller than non-family companies (but this difference is not statistically significant). Previous empirical Spanish studies report a smaller size for family firms (Galve Górriz and Salas Fumás, 1996). Family companies also have a greater excess of voting rights over cash flow rights (the difference being statistically significant at a 1 per cent level). This evidence is consistent with the results reported by Claessens et al. (2002) for Asian markets. In addition, family companies have similar ages, a similar leverage and similar risk, as opposed to non-family companies.

Interestingly 43 per cent of sample companies are family companies, according to our definition. In summary, the profile of the database can be characterized as follows:

- 13 per cent of companies are managed by the main family ultimate owner (MANUO) and in 33 per cent, the chairman of the board is a member of the main family ultimate owner group (CONTROLUO);
- 63.2 per cent of family companies are managed or controlled by descendants, whereas 36.8 per cent are managed or controlled by founders or descendants (FOUNDESC);
- in 11.6 per cent of the total sample, the family held the position of chief executive officer or chairman of the board, that is, families took part in the firms' corporate governance structures;
- 3.5 per cent of all sample companies are managed by their founders, while 8.1 per cent of the sample's companies are managed by descendants (DESCMAN);
- 12.8 per cent of the total sample's firms are controlled by the founder (FOUND-CON) and 18.6 per cent are controlled by descendants (DESCONTROL).

The analysis suggests that ultimate family presence is larger within the firm's control (as chairpersons of the board) than within the firm's management (as chief ececutive officers). For sample's firms, firm's management and control is based mostly on descendants rather than on founders.

The bivariate correlations amongst variables included in the study are shown in Table 32.4. As it is expected significant correlations (at a 1 per cent level) are observed between the performance indicators. Adjusted market value is also positively correlated with firm's leverage. At a 5% level, there are negative correlations between the adjusted return on assets and corporate governance variables (MANUO, CONTROLUO) and the founder/descendants effect (FOUNDESCD, DESCDMAN, DESCDCONTROL). As observed, the presence of a family as ultimate owner (UOFAM) is highly correlated (at a 1 per cent level) with the variables MANUO, CONTROLUO or FAMANCON. Therefore, due to possible multiple co-alignment problems, we opted not to include the variable UOFAM jointly with these other variables in the analyses. We also ran step-wise regressions in order to avoid such problems.

Analyses

Analysis of mean differences

By using mean difference analyses, we study the possible differences in company performance and in the presence of pyramids between family and non-family companies, companies controlled by the ultimate owners and the rest, companies managed by the ultimate owners and the rest, and companies that are managed and/or controlled by families and the rest (see Table 32.5).

Results do not show significant differences in company performance (when measured by the industry adjusted market to book value of common equity ratio) between family and non-family companies, between companies controlled by the ultimate owner and the rest of sampled companies, between companies managed by family groups or non-family groups

Table 32.4 Pearson correlations

	(1)	(2)	(3)	(4)	(5)	(6)	(7)
1. AMB	1						
2. AROA	0.545***	1					
3. TA	0.022	0.215**	1				
4. LEV	0.367***	0.037	0.152	1			
5. AGE	−0.051	−0.076	0.017	0.142	1		
6. DIVVRCF	−0.134	−0.054	0.128	−0.036	0.033	1	
7. RISK	0.040	0.167	0.061	−0.002	−0.188	0.039	1
8. UOFAM	0.018	−0.181	−0.165	0.040	−0.041	0.178	−0.031
9. FOUNDESD	−0.122	−0.381**	−0.254	0.124	0.332**	−0.452**	−0.185
10. MANUO	−0.114	−0.236**	−0.166	−0.022	−0.051	0.114	0.055
11. CONTROLUO	0.070	−0.158	−0.089	0.161	−0.052	0.069	−0.013
12. FAMANCON	−0.111	−0.214**	−0.114	−0.016	−0.085	0.117	0.296***
13. FOUNDMAN	0.015	−0.031	−0.028	−0.009	−0.039	0.229**	0.173
14. DESCMAN	−0.135	−0.249**	−0.212**	−0.017	0.006	−0.094	−0.085
15. FOUNDCON	0.127	0.102	0.061	−0.082	−0.204***	0.196	0.126
16. DESDCON	0.011	−0.269**	−0.186	0.239**	0.071	−0.074	−0.093

Note: *** Statistically significant at a 1% level. ** Statistically significant at a 5% level. AMB denotes the MB ratio adjusted by the industry median. AROA denotes the return on assets ratio adjusted by the industry median. TA represents total company assets. LEV is defined as the ratio of total debt to equity. AGE represents the number of years since foundation. DIVRCF represents the excess of control rights over cash flow rights for all companies. RISK represents the Standard deviation of the daily stock return, multiplied by the square root of the number days of the year. UOFAM is a dummy variable which reflects those companies whose largest ultimate owners are families or individuals who own more than the 10% threshold. FOUNDESC is a dummy variable that adopts a value of 1 when the company is either managed or controlled by descendants, and zero when it is managed or controlled by its founders. MANUO is a dummy variable that takes on a value of 1 when any member of the ultimate owner family holds the position of chief executive officer and zero in all other cases. CONTROLUO is a dummy variable that adopts a value of 1 when any member of the ultimate owner family holds the position of chairperson of the board and zero in all other cases. FAMANCON is a dummy variable that adopts a value of 1 when the chief executive officer and the president of the board belong to the ultimate owner's family, and zero in all other cases. FOUNDMAN is a dummy variable that adopts a value of 1 when the manager is the founder and zero in all other cases. DESCMAN is a dummy variable that adopts a value of 1 when the manager is the descendant and zero in all other cases. FOUNDCON is a dummy variable that adopts a value of 1 when the president of the board of directors is the founder, and zero in all other cases. DESCONTROL is a dummy variable that adopts a value of 1 when the president of the board of directors is a descendant, and zero in all other cases.

and between ultimate owner controlled and managed companies and the rest of the sample. Consistent with the results shown in Table 32.4, when the industry adjusted ratio of return on assets is used to measure company performance, we do find significant differences between family and non-family companies (UOFAM) (statistically significant at a 5 per cent level), between companies that are controlled and managed by their ultimate owners (FAMANCON) and the rest of the sample (the difference being almost statistically significant at a 5 per cent level), between the companies managed by families (MANUO) and the rest of sampled companies (the difference being statistically significant at a 5 per cent level), or between companies that are controlled by families (CONTROLUO) and the rest of the sample (nearly statistically significant difference at a 5 per cent level). These results tend to suggest that family companies, family-controlled companies and family-managed companies perform worse than non-family, professionally controlled and professionally managed companies.

(8)	(9)	(10)	(11)	(12)	(13)	(14)	(15)	(16)
1								
0.215	1							
0.439***	0.006	1						
0.798***	−0.085	−0.085	1					
0.417***	−0.39	0.947**	0.443**	1				
0.219**	−0.383**	0.496***	0.273**	0.524**	1			
0.343***	0.363**	0.777**	0.245**	0.688**	−0.057	1		
0.441***	−0.836***	0.269**	0.55***	0.296***	0.496***	−0.114	1	
0.55***	0.651***	0.263**	0.687***	0.293***	−0.091	0.404***	−0.183	1

Table 32.5 Mean difference analyses

	N	AMB	AROA	DIVVRCF
UOFAM	37	0.36	−0.002	3.64
REST	49	0.32	0.020	1.42
Z (Sig.)		−0.536	−1.984**	−3.009***
		(0.592)	(0.047)	(0.003)
CONTROLUO	28	0.47169	−0.003220	3.01337
REST	57	0.28517	0.017001	2.10575
Z (Sig.)		−1.225	−1.898**	−2.456**
		(0.221)	(0.058)	(0.014)
MANUO	11	−0.0208	−0.027282	4.24173
REST	74	0.40125	0.015932	2.13167
Z (Sig.)		−1.322	−2.042**	−2.338**
		(0.186)	(0.041)	(0.019)
FAMANCON	55	0.29246	0.017474	2.14591
REST	30	0.44590	−0.002740	2.87925
Z (Sig.)		−1.177	−1.922**	−2.959***
		(0.239)	(0.055)	(0.003)

Note: *** Statistically significant at a 1% level. ** Statistically significant at a 5% level. UOFAM is a dummy variable which reflects those companies whose largest ultimate owners are families or individuals who own more than the 10% threshold. CONTROLUO is a dummy variable that adopts a value of 1 when any member of the ultimate owner family holds the position of chairperson of the board and zero in all other cases. MANUO is a dummy variable that takes on a value of 1 when any member of the ultimate owner family holds the position of chief executive officer and zero in all other cases. FAMANCON is a dummy variable that adopts a value of 1 when the chief executive officer and the president of the board belong to the ultimate owner's family, and zero in all other cases.

We also find significant statistical differences in the divergence between cash flow rights and voting rights between the sub-samples of companies. The excess of voting rights over cash flow rights is larger for family versus the non-family companies (UOFAM and the rest) (statistically significant at a 1 per cent level), in companies controlled or managed by the ultimate owner group as opposed to the rest of the sample (statistically significant at a 5 per cent level) and in companies which are controlled (CONTROLUO) and managed (MANUO) by the ultimate owner group (FAMANCON) (statistically significant at a 1 per cent level). These results suggest the need to explore the possible influence of pyramids on company value and the need to study the reasons behind the reported statistical differences. We therefore ran multiple regression models.

Regression analysis

We next test whether the presence of families and individuals as ultimate owners (UOFAM), the excess of voting rights over cash flow rights, that is pyramids (DIVVRCF) held by ultimate owners, corporate management variables (FAMANCON), the presence of founders or descendants in management and control (FOUNDESCD) and several control variables (SIZE, AGE, RISK, LEV) may influence company value (estimated using the measure of company market value as a dependant variable: the industry adjusted market to book value of common equity ratio – in Reg.1 to Reg. 3, Table 32.6 and the industry adjusted ratio of operating income to total assets as a measure of the company accounting value – in Reg. 4 to Reg. 6, Table 32.6.

Regressions were estimated step-wise. For both dependant variables, the first block of regression models (Reg. 1 and Reg. 4) are estimated without including the corporate governance variable-FAMANCON and the founder/descendant variables, FOUNDESC, but considering the ownership variable (UOFAM); the second block of regression models (Reg. 2 and Reg. 5) include the variable FAMANCON, which is the effect of corporate governance variables without considering the ownership variable (UOFAM) and the third block of regression models (Reg. 3 and Reg. 6) consider the possible influence on company value of founders or descendants (FOUNDESC) that manage or control such companies, also without considering the ownership variable (UOFAM). These regressions are estimated for the total sample of companies (86 companies). All regression models turn out to be statistically significant.

When the adjusted market value is used as dependant variable, leverage turns out to affect positively a firm value (at a 1 per cent level). This means that the market values positively firm's leverage, possibly owing to the less free cash flow left for owner-managers.

When the industry adjusted return on assets ratio is used as dependant variable, the results suggest a negative influence on company performance resulting from family involvement in the company's management and control (FAMANCON) (statistically significant at a 5 per cent level) – in Reg. 5. These results suggest the need for further research into the interactions between variables for a more detailed analysis of the underlying reasons for the observed negative relation. We endeavour to explore these interactions in the regression models shown in Table 32.7.

The results of these regressions suggest for both dependant variables no significant influence of family ownership and of pyramids on company performance. This finding is in line with the results of Galve Górriz and Salas Fumás (1996), which obtain the same profitability for Spanish family and non-family companies.

Table 32.6 *Company ownership, pyramids and company value*

	Dependant variable: adjusted market value			Dependant variable: adjusted return on assets		
	Reg. 1	Reg. 2	Reg. 3	Reg. 4	Reg. 5	Reg. 6
(Constant)	0.807	1.141	2.255	−0.1	−0.083	0.109**
	(0.560)	(0.397)	(0.087)	(0.080)	(0.219)	(0.046)
DIVVRCF	−0.024	−0.019				
	(0.261)	(0.360)				
UOFAM	0.098			−0.019		
	(0.720)			(0.155)		
SIZE	−0.047	−0.086		0.015	0.015	
	(0.788)	(0.622)		(0.089)	(0.090)	
AGE	−0.166	−0.185	−0.570		−0.006	−0.024
	(0.416)	(0.361)	(0.136)		(0.536)	(0.122)
RISK	0.003	0.004		0.001	0.001	
	(0.762)	(0.688)		(0.161)	(0.124)	
LEV	0.162***	0.164***	0.162***			
	(0.001)	(0.000)	(0.007)			
FAMANCON		−0.428			−0.42**	
		(0.297)			(0.045)	
FOUNDESC			−0.272			−0.040
			(0.594)			(0.065)
N	86	86	86	86	86	86
F	2.516	2.709	3.699	2.782	2.7	4.509
Sig.	0.028**	0.019**	0.021**	0.046**	0.036**	0.018**

Note: P-values in brackets *** Statistically significant at a 1% level. ** Statistically significant at a 5% level. * Statistically significant at a 10% level. As we have run step-wise regressions we report the first step in which the model is statistucally significant at a 5% level or below. Adjusted market value denotes the market to book ratio adjusted by the industry median. Adjusted return on assets denotes the return on assets ratio adjusted by the industry median. DIVRCF represents the excess of control rights over cash flow rights for all companies. UOFAM is a dummy variable which reflects those companies whose largest ultimate owners are families or individuals who own more than the 10 % threshold. SIZE represents total company assets. AGE represents the number of years since foundation. RISK represents the Standard Deviation of the daily stock return, multiplied by the square root of the number days of the year. LEV is defined as the ratio of total debt to equity. FAMANCON is a dummy variable that adopts a value of 1 when the chief executive officer and the president of the board belong to the ultimate owner's family, and zero in all other cases. FOUNDESC is a dummy variable that adopts a value of 1 when the company is either managed or controlled by descendants, and zero when it is managed or controlled by its founders.

Table 32.7 shows the results of the regression models when we consider alternatively the influence on company performance of pyramids, corporate management variables, founder effect variables and the control variables. Reg. 1 to 5 use the market representative measure for company performance as dependant variable, adjusted market value, while Reg. 6 to 9 use the accounting measure adjusted return on assets. All the regressions models are statistically significant at a 5 per cent level or below (1 per cent).

First, we considered the possible influence of corporate governance variables (the effect of the ultimate family owner as manager – MANUO – or as chairman of the board of directors – CONTROLUO) on company value (Reg.1 and Reg. 6). Secondly, we separated

Table 32.7 Company management and control and company performance

	Dependant variable: adjusted market value					Dependant variable: adjusted return on assets			
	Reg. 1	Reg. 2	Reg. 3	Reg. 4	Reg. 5	Reg. 6	Reg. 7	Reg.8	Reg. 9
(Constant)	1.060	0.882	0.777	1.357	0.751	−0.099	0.015	−0.090	−0.094
	(0.442)	(0.516)	(0.450)	(0.232)	(0.571)	(0.078)	(0.029)	(0.111)	(0.076)
DIVVRCF	−0.021	−0.029	−0.032	−0.025	−0.028			−0.001	
	(0.331)	(0.167)	(0.129)	(0.240)	(0.199)			(0.363)	
SIZE	−0.085	−0.085	−0.117	−0.112	−0.080	0.013		0.014	0.018**
	(0.634)	(0.627)	(0.494)	(0.528)	(0.643)	(0.125)		(0.126)	(0.047)
AGE	−0.167	−0.109		−0.170	−0.1				
	(0.413)	(0.592)		(0.399)	(0.626)				
RISK	0.004	0.001		0.002	0.002	0.001		0.001	
	(0.736)	(0.901)		(0.853)	(0.824)	(0.110)		(0.202)	
LEV	0.156***	0.172***	0.165***	0.165***	0.168			0.002	
	(0.001)	(0.000)	(0.000)	(0.001)	(0.000)***			(0.512)	
CONTROLUO	0.250								
	(0.211)								
MANUO	−0.545					−0.040**			
	(0.211)					(0.046)			
FOUNDMAN					−0.398				
					(0.628)				
DESCMAN			−0.623	−0.672			−0.056**		
			(0.187)	(0.204)			(0.022)		
FOUNDCON		0.635	0.662		0.763				
		(0.118)	(0.087)		(0.095)				
DESCONTROL		−0.148		−0.026				−0.043**	
		(0.678)		(0.946)				(0.020)	
N	86	86	86	86	86	86			86
F	2.355	2.605	4.049	2.458	2.617	3.506	5.464	2.569	4.082
Sig.	0.031**	0.018**	0.003***	0.025**	0.018**	0.019**	0.022**	0.033**	0.047**

Note: P-values in brackets *** Statistically significant at a 1% level. ** Statistically significant at a 5% level. As we have run step-wise regressions we report the first step in which the model is statistically significant at a 5% level or below or once the model is significant, the best iteration – after excluding correlated variables – to obtain statistically significant variables. Adjusted market value denotes the market to book ratio adjusted by the industry median. Adjusted return on assets denotes the return on assets ratio adjusted by the industry median. DIVVRCF represents the excess of control rights over cash flow rights for all companies. SIZE represents total company assets. AGE represents the number of years since foundation. RISK represents the Standard Deviation of the daily stock return, multiplied by the square root of the number days of the year. LEV is defined as the ratio of total debt to equity. MANUO is a dummy variable that takes on a value of 1 when any member of the ultimate owner family holds the position of chief executive officer and zero in all other cases. CONTROLUO is a dummy variable that adopts a value of 1 when any member of the ultimate owner family holds the position of chairperson of the board and zero in all other cases. FOUNDMAN is a dummy variable that adopts a value of 1 when the manager is the founder and zero in all other cases. DESCMAN is a dummy variable that adopts a value of 1 when the manager is the descendant and zero in all other cases. FOUNDCON is a dummy variable that adopts a value of 1 when the president of the board of directors is the founder, and zero in all other cases. DESCONTROL is a dummy variable that adopts a value of 1 when the president of the board of directors is a descendant, and zero in all other cases.

the influence of founder control (FOUNDCON) and descendant control (DESCDCON) and management effects (FOUNDMAN and DESCMAN) on company value (Reg. 2 to Reg. 5 for the adjusted market value and Reg. 7 to Reg. 9 for the adjusted return on assets).

For this purpose, we ran several regressions combining the identity of the persons who controls and manages the firms, although in Table 32.7 we only refer to those that were significant. Reg. 2 and Reg. 8 take into account the presence of founders or descendants in company control. Reg. 3 and Reg. 7 include the combination of companies run by descendant managers that retain founder control. Reg. 4 reflect the descendant effect when taking into account the combination of descendant manager and descendant control, and Reg. 5 and 9 include the founder effect by measuring the presence of founders as managers and as chairpersons of the board of directors.

These results again suggest that the factors influencing company performance are different when company value is represented by the variables adjusted market value or adjusted return on assets.

Starting with adjusted market value, the results show that what affects company value the most is leverage (a positive effect), suggesting a possible monitoring effect of debt. Corporate governance variables and the effect of founders/descendants do not affect firm value, neither do pyramids. These results suggest that the market does not value who manages or controls the firm. The external market control mechanism debt, seems to be preferred by investors as monitoring device.

When using the AROA ratio as dependant variable, the results of the regression models change. In this case, the control variable that significantly influences company value is company size (statistically significant at a 5 per cent level). Ultimate family management negatively affects company value (statistically significant at a 5 per cent level), and this negative influence is mainly due to the negative effect of descendant management on company value (statistically significant at a 5 per cent level). These results are in line with those provided by Morck et al. (1998) who show that Canadian heir-controlled companies have poor financial performance in their industries, as compared to other companies.

All the reported results suggest that what really affects company value is not the presence of family groups as large shareholders, but the way companies are controlled or managed, as well as the identity of the managers and persons who control such companies (founders versus descendants). In particular, the descendants' effect seems to be negative for a firm's performance. This evidence supports the theoretical arguments that suggest that agency problems may also occur between family members (Schultze et al., 2001) and that the presence of family groups may trigger agency issues that are detrimental to minority shareholders (Morck and Yeung, 2003). Our findings contradict the results from other markets, that is, McConaughy et al. (1998) or Anderson and Reeb (2003) that suggest family companies are more efficient, valuable and perform better than non-family companies.

Our results point towards the arguments that suggest that owner-managed companies are particularly vulnerable to entrenchment and that, by choosing the senior executive from a more restricted pool of talent, families may have lower-quality owner-managers. These arguments are supported by empirical evidence provided by Barth et al. (2005), which report that family-owned companies managed by members of the owning family are less productive, Morck et al. (1988) who claim that Tobin's q is lower for older companies managed by members of the founding family and by Morck et al. (1988) that report that companies controlled by heirs of the founder are less profitable.

Conclusions

This chapter explores how the fact of being a family company and the use of pyramids by families affect company value. It also explores how the presence of ultimate family owners in the management of companies (as chief executive officers) and in the control of such companies (as chairpersons), either as a founder or descendant, also affect company value.

The results of the analyses carried out suggest that being a family company or not, from an ultimate owner perspective, does not seem to affect company value. Neither does the use of pyramids. What really influences company value is the identity of its top management – board executives and the chairperson. The presence of descendants seems to reduce firm value. Companies that are chaired and/or managed by heirs show lower levels of accounting performance than others. Thus, our results suggest that the descendants of family groups, when in charge of both the management and control of a company, may use their power to extract income from minority shareholders. Thus, what seems to be crucial for a family company is how its management and control is organized.

These findings contradict the hypotheses that propose a positive or negative influence of family ownership on company value. In this sense our results differ from those reported by Anderson and Reeb (2003) or McConaughy et al. (1998), but are similar to the findings reported by Alcalde et al. (2001) or Galve Górriz and Salas Fumás (1993, 1996).

We must say that our study suffers from several limitations, resulting from the measurement of certain variables, that must be taken into account when interpreting the results. Just to name a few, we are aware that we may be missing something by measuring the effect of founder management or founder control by only considering the positions of chief executive officer and chairman of the board. In addition, we are also aware that the results vary when we use different performance measures. The study would also benefit from the consideration of other corporate management related variables, that is, type of boards of directors or the existence of anti-takeover devices.

Further research needs to be undertaken to understand the underlying reasons for the observed negative effect of descendants on firm value. For example, one open question is, what factors determine that descendants turn out, on average, to be bad managers? Case studies may help us to understand such issues.

We believe that our study offers interesting implications for managers and family companies. In order to prevent future problems, families should stay as qualified shareholders delegating to professional managers the day-to-day running of the company. The positions held by family members as well as the identity of such family members must be subject to special consideration. Family companies should try to overcome potential personal gains of descendant managers by hiring external managers or establishing adequate monitoring devices that impede heirs from extracting income, to the detriment of other shareholders. This will allow the company to survive over time. According to a Spanish proverb: 'From hardworking grandfathers, rich fathers and poor grandchildren'. Family companies should try to turn the proverb around.

References

Adams, R.B., H. Almeida and D. Ferreira (2005), 'Understanding the relationship between founder-CEOs and firm performance', available at SSRN, http://ssrn.com/abstract = 470145 or 001, 10.2139/ssrn.470145.

Alcalde, N., C. Galve and V. Salas (2001), 'Análisis económico y financiero de la gran empresa familiar', *I Congreso Nacional de Investigación sobre la Empresa Familiar*, OPVI, Valencia, 171–88.

Anderson, R.C. and D.M. Reeb (2003), 'Founding-family ownership and firm performance: evidence from the S&P 500', *Journal of Finance*, **58**(3), 1301–28.

Anderson, R.C. and D.M. Reeb (2004), 'Board composition: balancing family influence in S&P 500 firms', *Administrative Science Quaterly*, **49**, 209–37.

Anderson, R.C., S.A. Mansi and D.M. Reeb (2003), 'Founding family ownership and the agency cost of debt', *Journal of Financial Economics*, **68**, 263–355.

Astrachan, J., S. Klein and K. Smyrnios (2002), 'The F-PEC scale of family influence: a proposal for solving the family business definition problem', *Family Business Review*, **15**(1), 45–58.

Barth, E., T. Gulbrandsen and P. Schone (2005), 'Family ownership and productivity: the role of owner management', *Journal of Corporate Finance*, **11**, 107–27.

Beckhard, R. and G. Dyer (1983), 'Managing continuity in the F-owned business', *Organizational Dynamics*, Summer, 5–12.

Berle, A. and G. Means (1932), *The Modern Corporation and Private Property*, New York: Macmillan.

Bukart, M., D. Gromb and F. Panunzi (1997), 'Large shareholders, monitoring and the value of the firm', *Quaterly Journal of Economics*, **112**, 693–728.

Cabrera-Suárez, R., P. De Sau-Pérez and D. Garcia-Almeida (2001), 'The succession process from a resource based view of the family firm', *Family Business Review*, **XIV**(1), 37–47.

Camisón, C. (2001), 'Estructura de propiedad. Control de la EF y desempeño organizativo: un análisis dentro de la población de empresas industriales valencianas', *I Congreso Nacional de Investigación sobre la Empresa Familiar*, OPVI, Valencia, Septiembre.

Chaganti, R. and F. Damanpour (1991), 'Institutional ownership, capital structure, and firm performance', *Strategic Management Journal*, **12**, 479–92.

Claessens, S., S. Djankov and L. Lang (2000), 'The separation of ownership and control in East Asian corporations', *Journal of Financial Economics*, **58**, 81–112.

Claessens, S., S. Djankov, J. Fan and L. Lang (2002), 'Disentangling the incentive and entrenchment effects of large shareholdings', *Journal of Finance*, **57**(6), 2741–71.

Coleman, J.S. (1990), *Foundations of Social Theory*, Cambridge, MA: The Belknap Press of Harvard University Press.

Daily, C. and M. Dollinger (1992), 'An empirical examination of ownership structure in family and professionally managed firms', *Family Business Review*, **5**(2), 237–50.

Demsetz, H. (1983), 'The structure of ownership and the theory of the firm', *Journal of Law and Economics*, **26**, 375–90.

Demsetz, H. and K. Lehn (1985), 'The structure of corporate ownership: causes and consequences', *Journal of Political Economy*, **93**(6), 1155–77.

Demsetz, H. and B. Villalonga (2001), 'Ownership structure and corporate performance', *Journal of Corporate Finance*, **7**, 209–33.

Du, J. and Y. Dai (2005), 'Ultimate corporate ownership structures and capital structures: evidence from East Asian economies', *Corporate Governance*, **13**(1), 60–71.

Faccio, M. and L. Lang (2002), 'The ultimate ownership of western European corporations', *Journal of Financial Economics*, **65**, 365–95.

Fama, E. F. and M.C. Jensen (1983), 'Separation of ownership and control', *Journal of Law and Economics*, **26**, 301–26.

Gaceta de los Negocios (2005), El Buen Gobierno alcanza a las Empresas, familiares, p. 1, 29th April.

Gallo, M.A. (1998), 'Los consejos de administración en las empresas familiares: presencia, actividad y utilidad', *Iniciativa emprendedora y empresa familiar*, **9**, March/April, 12–15.

Galve Górriz, C. and V. Salas Fumás (1993), 'Propiedad y Resultados de la gran empresa española', *Investigaciones Económicas*, **27**(2), 207–38.

Galve Górriz, C. and V. Salas Fumás (1994), 'Propiedad y Resultados de la empresa: una revisión de la literatura teórica y empírica', *Economía Industrial*, **300**, 171–98.

Galve Górriz, C. and V. Salas Fumás (1996), 'Ownership structure and firm performance: some empirical evidence from Spain', *Managerial and Decision Economics*, **17**(6), 575–86.

Gómez-Mejía, L., M. Nuñez Nickel and I. Gutierrez (2001), 'The role of family ties in agency contracts', *Academy of Management Journal*, **44**, 81–95.

Hambrick, D.C. and G. Fukutomi (1991), 'The seasons of a CEO's tenure', *Academy of Management Review*, **16**(4), October, 719–42.

Holderness, C. and D. Sheehan (1988), 'The role of majority shareholders in publicly held corporations', *Journal of Financial Economics*, **20**, 317–46.

Instituto De La Empresa Familiar (IEF) (2005), 'Report of good governance in family firms', Paper no. 128, Barcelona.

Jayaraman, N., A. Khorana, E. Nelling and J. Covin (2000), 'CEO founder status and firm financial performance', *Strategic Management Journal*, **21**, 1215–24.

Jensen, M. and W. Meckling (1976), 'Theory of the firm: managerial behaviour, agency costs and ownership structure', *Journal of Financial Economics*, **3**, 305–60.
King, B.F. (1996), 'Market and industry factors in stock price behaviour', *Journal of Business*, January, 139–90.
La Porta, R., F. López De Silanes and A. Shleifer (1999), 'Corporate ownership around the world', *Journal of Finance*, **54**(2), 471–517.
Livingston, M. (1977), 'Industry movements of common stock', *Journal of Finance*, June, 861–74.
McConaughy, D., M. Walker, G. Henderson and C. Mishra (1998), 'Founding family controlled firms: efficiency and value', *Review of Financial Economics*, **7**(1), 1–19.
McConaughy, D.C. Matthews and A. Fialko (2001), 'Founding family controlled firms: performance, risk and value', *Journal of Small Business Management*, **39**(1), 31–49.
Morck, R. and B. Yeung (2003), 'Agency problems in large family groups', *Entrepreneurship Theory and Practice*, Summer, 367–82.
Morck, R., A. Shleifer and R.W. Vishny (1988), 'Management ownership and market valuation: an empirical analysis', *Journal of Financial Economics*, **20**(January–March), 293–323.
Morck, R., D. Stangeland and B. Young (1998), 'Inherited wealth, corporate control, and economic growth, the Canadian disease?', NBER Working paper 6814, National Bureau of Economic Research, Cambridge (at http://ideas.repec.org/p/wdi/papers/1998-209.html).
Pérez-Gonzalez, F. (2001), 'Inherited control and firm performance', working paper, University of Colombia.
Pollak, R.A. (1985), 'A transaction cost approach to families and households', *Journal of Economic Literature*, **23**, 581–608.
Sacristán Navarro, M. and S. Gómez Ansón (2004), 'Ultimate owners of the Spanish listed firms: the role of families', Research Forum Proceedings 15 World Conference FBN-IFERA, Copenhagen, 366–85.
Santana Martín, D. and K. Cabrera Suárez (2001), 'Comportamiento y resultados de las empresas cotizadas familiares versus las no familiares', XI Congreso Nacional ACEDE.
Schulze, W., M. Lubatkin and R. Dino (2003), 'Toward a theory of agency and altruism in family firms', *Journal of Business Venturing*, **18**(4), 473–90.
Schulze, W., M. Lubatkin, R. Dino and A. Buchhaltz (2001), 'Agency relationships in family firms: theory and evidence', *Organization Science*, **12**(2), 99–116.
Shleifer, A. and R.W. Vishny (1997), 'A survey of corporate governance', *Journal of Finance*, **52**, 737–83.
Sraer, D. and D. Thersmar (2004), 'Performance and behaviour of family firms: evidence from the french stock market', working paper, Centre de Reserche en Economie et Statistique, Tolouse, France.
Villalonga, B. and R. Amit (2005), 'How do family ownership, control and management affect firm value?', *Journal of Financial Economics*, in press.
Wall, R.A. (1998), 'An empirical investigation of the production function of the family firm', *Journal of Small Business Management*, **6**, 24–32.

Epilogue: theory building and the survival of family firms – three promising research directions

Shaker A. Zahra, Sabine B. Klein and Joseph H. Astrachan

These are times of profound change around the globe. Family businesses, the most enduring and venerable organizational form, are under immense pressure to adapt in order to exploit emerging opportunities in their domestic and international markets. Without successful adaptation, these firms can lose their competitive standing and sources of competitive advantage, which could lead to their demise. Regrettably, family business owners looking at the voluminous literature for guidance are apt to be disappointed. Research on family firms is fragmented and non-cumulative, and lacks good theoretical grounding (Zahra and Sharma, 2004), suggesting a need for more creative theory building by capitalizing upon the unique qualities of family firms, especially their cultures and histories that determine family dynamics and decision-making. A 'systems approach' is clearly needed to build and test such a theory. Furthermore, the field has to develop a common language to ensure that communication is taking place between the disciplines involved such as organizational behavior, strategy and finance.

In this chapter, we sketch out some ways that could expedite our progress in developing theory that guides futures research into family firms and helps their owners (managers) achieve superior performance. We believe that one theory, or meta-theory, cannot fully capture the richness and diversity of family firms, their cultures, systems and managerial styles. Instead, in-depth research of the different variables is needed before a theory of the family firm can emerge and gain wide acceptance. Understanding these variables can be the foundation to determine the emergence, evolution and survival of family firms. Chrisman et al., propose that 'the ultimate aim of research about family business is to develop a theory of the family firm' (2005, p. 566). Such theory would be valuable if it helps families and their businesses do better, create wealth and add value to society.

How can we develop effective theories that will guide future research on family firms and their decision-making processes? We believe that the answer lies in encouraging both descriptive *and* prescriptive research, selectively importing theories from sister disciplines, and developing and testing our original theories. As family firms are more complex than their non-family counterparts, we also have to be careful to observe where imported theories fall short and how we can further develop them or whether a new approach is needed in order to grasp the whole complexity. Our success in achieving these objectives lies in our ability to better contextualize our research in the life and nature of the family firms, with their complexities and ambitions. *Contextualizing* our research means thoroughly understanding the phenomena and issues at hand, factoring these issues into our theory building and testing, and considering the particulars of the family firm situation in our interpretations of our findings (Zahra, 2006). These processes will make theory building and testing in family firms more challenging. Family firms are more complicated in many respects compared with their non-family counterparts, which might explain why they have been overlooked by mainstream researchers. Next, we discuss three ways to contextualize future family research and make it more relevant and useful.

1 Developing first-hand familiarity with family firm challenges

Family firm research in its early years was driven by both consultants and practitioners dealing with and/or coming from family businesses (Bird et al., 2002). As a result, there was no obvious need to talk about how to develop familiarity with family businesses and their challenges. With the development of the field and rising interest from outsiders, familiarity with family businesses cannot be taken for granted. So, how do we help researchers to develop a deep appreciation of the uniqueness of family firms and their cultures?

There are three ways to develop familiarity with family businesses. The first and seemingly easiest approach is to be born into a family business. Yet, this may not develop a profound familiarity within academia. The tendency to go into academia as well as the ability to do so among family business children is low, and there is a danger of being biased by one's own history. To come from a business-owning family and get involved in high-level family business research requires a solid theoretical foundation and maturity to deal with one's own story.

A second way to develop familiarity is to study family businesses in depth through theory-based education *and* case study research. Few places around the globe offer this combination of skills. Rather, it is more common to find individual coaching of incoming academics in the family-firm field by experienced scholars. A third approach to gaining first-hand experience with family firms, one that should be combined with the second, is to spend time in family businesses and with business family members. This could take place through work shadowing family-firm managers and employees, and observing them as they make decisions, or through work experience. Case writing under the supervision of experienced scholars could enhances these first-hand observations and work experience. Consulting and/or teaching assignments could also enrich one's knowledge of family firms.

2 Selective theory importation

As with other young fields of research, family-business scholars have borrowed key theories from other disciplines. Examples include the growing use of agency, transaction cost, stewardship, resource-based and knowledge-based views of the firm. Researchers have shown versatility in using, extending and integrating these views into the analysis of diverse phenomena (Zahra and Sharma, 2004). This versatility, however, has its own costs insofar as some scholars have not given attention to the boundary conditions and assumptions of original theories.

Researchers have tended to assume that because original theories have proven to be robust in other settings, they would apply equally well to family firms. As a result, researchers have undermined the quality of their own findings. As family firms account for the majority of companies in most countries, it can be – and should be – assumed that they are a heterogeneous group in itself. Testing original theories in family-firm contexts requires that these contexts be described and understood clearly. What might hold true in one type of family business does not necessarily hold true in other types of family firms. It is unfortunate that some researchers have not challenged their own assumptions about the nature of family firms or the influence of the owner family, contending that family firms behave differently from other types of firms. This might be true, but many family firms share important similarities with other firms. This highlights the importance of what is really distinctive about family firms. Schulze et al.'s (2001) analyses of the various manifestations of agency problems in family firms is a notable example of the theoretical richness that could be gained from probing and questioning our beliefs about family firms. Turning our assumptions upside-down can open the door for greater depth and rigor in future family firm research.

3 Family-firm theory construction: uniting the individual, institution and context

A shortcoming of theories imported from other fields is ignoring the 'human agent'. To be sure, theories borrowed from other disciplines speak to the motives and behaviors of individuals and groups. Transaction cost economics highlights the potential for opportunism. Agency theory stresses self-motivated and self-centered behavior. Yet, many of the studies that have used these theories refrain from considering human agents or their motivations. They also ignore the learning that occurs from success and failure, and from repeated experiences. Strategy researchers have fallen into this trap, ignoring the effect of managers' aspirations, personalities, and skills. As a consequence, companies' strategic moves are often depicted as an almost reflexive response to external changes. Rarely are managerial motives factored into strategic analyses. Yet, these motives are often highlighted as an area that is worthy of future inquiry.

Constructive theory building should creatively unite the individual (the decision-maker), the institution (the firm) and the context in a coherent fashion. It should also take into account that there is a multitude of motivations, aspirations and value sets to be found in the real world. The model of man depicted in economic analyses has proven to be too narrow to explain the behavior in family companies that have survived and thrived over several generations. A psychological perspective (for example, Argyris, 1964) might be more appropriate to grasp the distinctiveness of business family firms and their owner/managers.

Integrating the individual, the institution and the context is a major and challenging task that requires a 'system' view. This view looks into the interrelatedness of these dimensions and pays close attention to their interactions and the various combinations that might characterize these interactions. This is important as variations in these interactions (or combinations thereof) are likely to yield very different results and shape the evolution of the family firm and its decisions. These interactions are a key source of variability in family firm research and therefore should be thoroughly analysed. Systems thinking is inherently complicated and complex. But using systems' logic can increase the realism of theory-building, the rigor of our ideas, and the relevance of the findings.

Adopting systems thinking to future theory construction has implications for how future knowledge in the family business field is developed and disseminated. There is a need for scholars who bring forward bold and innovative ideas and theories about these firms and their operations. These theorists have the important task of envisioning where the field should be and telling us how to get there. There is also a need for a second group of scholars who test these ideas and theories, to establish their boundaries and determine their efficacy. These researchers are the builders of effective frameworks *and* the developers of empirically grounded observations about the ways family firms work. A third group of scholars would focus more on translating the research findings for family-firm owners and managers, clarifying for them what we know and why it matters.

The good news is that the three groups of researchers we have just identified already exist in the field, but there is a clear shortage of 'great idea' people and theorists. This shortage is understandable. Theory-building skills are difficult to master and take years to develop. 'Systems' thinking complicates the accumulation of these rare and complicated skills. Our focus on borrowing from other disciplines also stands in the way of having to face the fact that we need to innovate while theorizing about family firms and their operations. The survival of these firms is important not only for their owners but for society at large.

Conclusion

Research on the factors that determine the success and failure of family firms is in transition. It is becoming more theoretically grounded while making use of rigorous research methods and analytical tools. Building theories that could enhance the success and survival of family firms can be demanding, as we have stated throughout this chapter. Our challenge then is to better develop a more intimate understanding of the dynamic interrelationships that connect individuals, institutions and the context as we explore new issues of importance to family firms or tackle persistent debates using fresh insights and newly developed theories. We need, therefore, to move from borrowing and imitation to effective theory construction and innovation in our research. In this way, we can more competently address relevant issues with rigor and thus enrich our findings. By building our theories, we can give back to the disciplines from which we have borrowed heavily. Knowing more about what makes family firms unique and special can provide an important platform from which we can theorize more accurately and richly about the differences that might exist between these companies and non-family firms. Such a platform could give us an opportunity to enrich our dialog with sister disciplines while speaking to the owners of family firms about issues of great importance to them, especially the survival of their companies.

References

Argyris, (1964), *Integrating the Individual and the Organization*, New York: Wiley.

Bird, B., Welsch, H., Astrachan, J.H. and Pistrui, D. (2002), 'Family business research: the evolution of an academic field', *Family Business Review*, **15**(4), 337–50.

Chrisman, J.J., Chua, J.H. and Sharma, P. (2005), 'Trends and directions in the development of a strategic management theory of the family firm', *Entrepreneurship Theory and Practice*, September, pp. 555–75.

Schulze, W.G., Lubatkin, M.H., Dino, R.N. and Buchholtz, A.K. (2001), 'Agency relationships in family firms: theory and evidence', *Organization Science*, **12**, 99–116.

Zahra, S. and Sharma, P. (2004), 'Family business research: a strategic reflection', *Family Business Review*, **17**(4), 331–46.

Zahra, S.A. (2006), 'Contextualizing theory building in entrepreneurship research', *Journal of Business Venturing*, in press.

Index

Titles of publications appear in *italics*.

Abbreviations used in the index:
CLR – critical leader relationships
SFB – Sustainable Family Business